F. M. Faranda · L. Guglielmo · A. Ianora (Eds.)

*Ross Sea Ecology*

**Springer**

*Berlin*
*Heidelberg*
*New York*
*Barcelona*
*Hong Kong*
*London*
*Milan*
*Paris*
*Singapore*
*Tokyo*

F. M. Faranda · L. Guglielmo · A. Ianora (Eds.)

# Ross Sea Ecology

## Italiantartide Expeditions (1987–1995)

With 281 Figures and 114 Tables

Springer

Professor Dr. Francesco Maria Faranda
Università di Genova
Instituto di Scienze Ambientali Marine
Viale Benedetto XV 5
I-16132 Genova
Italia

Professor Dr. Letterio Guglielmo
Università di Messina
Dipartimento di Biologia ed Ecologia Marina
Salita Sperone 31
I-98166 Messina-S. Agata
Italia

Dr. Adrianna Ianora
Stazione Zoologica "Anton Dohrn"
Villa Comunale
I-80121 Napoli
Italia

ISBN-13: 978-3-642-64048-3 Springer-Verlag Berlin Heidelberg New York

Library of Congress Cataloging-in-Publication Data

Ross Sea ecology : Italiantartide Expeditions (1987–1995) / Francesco Faranda, Letterio Guglielmo, Adrianna Ianora (eds.).
p. cm.
Includes bibliographical references.
ISBN-13: 978-3-642-64048-3 e-ISBN-13: 978-3-642-59607-0
DOI: 10.1007/978-3-642-59607-0
1. Marine ecology–Antarctica–Ross Sea. I. Faranda, Francesco, 1933– . II. Guglielmo, L. (Letterio), 1945– . III. Ianora, A. (Adrianna), 1953–
QH84.2.R68 1999
577.7'09167'4–dc21 99–33291


Softcover reprint of the hardcover 1st edition 2000



Cover design: Design & Production, Heidelberg
Cover photo: Adumussium colbecki, photo by M. Chiantore et al.
Typesetting: Best-set Typesetter Ltd., Hong Kong
SPIN 10567973 31/3136 – 5 4 3 2 1 0 – Printed on acid-free paper

*In memory of Professor Felice Ippolito,*
*mentor and founder*
*of the Italian Antarctic Program*

# Foreword

In 1987, Italy started a series of oceanographic cruises to the Southern Ocean. These Italiantartide Expeditions, from 1987–1995, have concentrated mainly on the Ross Sea which has been crossed by countless supply vessels but rather little explored in a systematic way. F.M. Faranda, L. Guglielmo and A. Ianora have jointly compiled many of the results of these expeditions in a single volume consisting of 43 contributions from about 120 different authors, mostly Italians. Guglielmo and Ianora are well known for their editing of the *Atlas of Marine Zooplankton in the Straits of Magellan*. In this new book, many aspects of modern ecology and biological oceanography, including trophic interactions and relations to the biological and physical environment, are treated and an extensive descriptive biogeography of Antarctic plankton and benthos is provided. Four papers deal with water mass characteristics and large-scale circulation in the Ross Sea. Nutrient uptake, particulate matter, phytoplankton, microzooplankton and physiological aspects of primary production are the subject of 13 other contributions. Macrozooplankton and its spatial structure including krill are dealt with in eight papers. The work on zoo- and phytobenthos concentrates on Terra Nova Bay (nine papers). The remaining contributions are devoted to a multitude of subjects like biogeochemical processes, pollution, climate, marine fish, birds and mammals and palaeoecology.

Scientists from many institutes participated in these expeditions and in the analysis of their rich collections. *Ross Sea Ecology* demonstrates the scientific wealth of Italy's marine research with emphasis on marine ecology and zoogeography. The book is meant as a pendant to *Weddell Sea Ecology* (C. Hempel, ed., Springer 1993) on the European Polarstern Study (EPOS) 1988/1989. In EPOS we enjoyed the participation of Italian physiologists and population geneticists as part of the large international team of scientists from all over Western Europe. While EPOS covered all seasons, from early spring to late autumn of one year, the Italiantartide Expeditions provide information on late spring and summer over several years. Together with a number of original papers by Italian authors, mostly published in *Polar Biology*, *Ross Sea Ecology* contributes greatly to our knowledge of the Ross Sea and will be appreciated by everyone working in those waters. The book, however, goes beyond these specific regional interests as it provides comparisons with the Weddell Sea and general insights into the ecology of the seasonal sea ice zone and the Marginal Ice Zone (MIZ) of the Southern Ocean in general. At the same time, this book is a fine demonstration of the vivid entry of Italian marine science into international Antarctic research. The three editors should be congratulated and thanked for their great efforts in bringing together these 43 papers.

Center for Tropical Marine Ecology (ZMT)
Bremen, Germany

COTTHILF HEMPEL

# Preface

The Antarctic represents the last of the world's still unexplored continents. Since 1985, Italy has sent 14 expeditions to this region, making it one of the most scientifically active nations in the Antarctic. Three of these expeditions have been exclusively devoted to research on the marine ecology of the Ross Sea region, from studies of the physical environment and its flora and fauna, to understanding the physiological processes that govern its functioning. This volume attempts to present a global picture of this research, based on the late spring-summer cruises made between 1987 and 1995. The main objectives of these cruises were the characterization of the physical, chemical and biological environment (POLAR QUEEN Cruise I), the structure and functioning of the Antarctic ecosystem (CARIBOO Cruise II) and the ecology of the Ross Sea marginal ice zone (ROSSMIZE Cruise III). This series of collected reprints compiles the results of these cruises, covering the major areas of research by Italian scientists. The contents include papers on water mass characteristics, particulate matter and nutrient utilization, physiological aspects of primary production, zooplankton spatial structure, and krill and top predator interactions in relation to physical and biological parameters. Other papers deal with ecological features of coastal fish communities, spatial and temporal variability of benthic biocoenoses and biogeochemical processes in the Ross Sea. The 43 papers which comprise the book have been subjected to an extensive peer review process and scientific editing. The aim of the book is to serve as a sound platform for future research in this area and to draw attention to the research being promoted by Italian scientists in the Antarctic.

F.M. Faranda
L. Guglielmo
A. Ianora

# Referees

The following scientists provided reviews of one or more of the papers considered for publication:

| | |
|---|---|
| Alcaraz M | Institut de Ciences del Mar (C.S.I.C.), Barcelona, Spain |
| Angel MV | Southampton Oceanography Centre, Southampton, UK |
| Ansell A | The Scottish Assoc. for Marine Sciences, Oban Argyll, UK |
| Archambeau AS | Université Pierre et Marie Curie, Paris, France |
| Arnaud PM | Station Marine d'Endoume, Marseille, France |
| Arntz W | Alfred Wegener Institute, Bremerhaven, Germany |
| Azam F | Scripps Istitution of Oceanography, San Diego, California, USA |
| Bathmann UV | Alfred Wegener Institute, Bremerhaven, Germany |
| Berkman P | Byrd Polar Research Center, Columbus, Ohio, USA |
| Bianchi M | Laboratoire de Microbiologie Marine, Marseille, France |
| Buffle J | Université de Geneve, Geneve, Switzerland |
| Cauwet G | Laboratoire Arago, Banyuls-sur-mer, France |
| Clarke A | British Antarctic Survey, Cambridge, UK |
| Coccioni R | Istituto di Geologia, Università Urbino, Urbino, Italy |
| Di Prisco G | Institute of Protein Biochemistry and Enzymology, Naples, Italy |
| Dunbar R | Rice University, Texas, Houston, USA |
| El-Sayed S | Dept. of Oceanography, Texas A&M University, College Station, Texas, USA |
| Estrada M | Institut de Ciences del Mar, Barcelona, Spain |
| Everson I | British Antarctic Survey, Cambridge, UK |
| Falkowski P | Oceanographic and Atmospheric Sciences Division, Upton, USA |
| Fanucci F | Dip. Geodinamica e Sedimentologia, Università Urbino, Urbino, Italy |
| Froneman PW | Dept. of Zoology, Rhodes University, Grahamston, South Africa |
| Garcia MA | Universitat Politecnica de Catalunya, Barcelona, Spain |
| Goyens L | Vrije Universiteit Brussel, Brussels, Belgium |
| Hutchings P | The Australian Museum, Sydney, Australia |
| Iosune Uriz MJ | Centro de Estudios Avanzados de Blanes (C.S.I.C.), Blanes, Spain |
| Joiris R C | Free University of Brussels, Brussels, Belgium |
| Kaminski MA | University College London, London, UK |
| Kannan K | Pesticide Research Center, Michigan State University, East Lansing, Michigan, USA |
| King J | British Antarctic Survey, Cambridge, UK |

| | |
|---|---|
| Locarnini R | Center for Coastal Physical Oceanography, Norfolk, Virginia, USA |
| Moreno J | C.S.I.C., Madrid, Spain |
| Moyano GHI | Depto. de Zoologia, Concepcion, Chile |
| Naganobu M | National Research Institute of Far Sea Fisheries, Shizuoka, Japan |
| Nelson D | College of Oceanographic & Atmospheric Sciences, Corvallis, Oregon, USA |
| Platt T | Bedford Institute of Oceanography, Dartmouth, Canada |
| Poulet S | Station Biologique de Roscoff, Roscoff, France |
| Priddle J | British Antarctic Survey, Cambridge, UK |
| Pudsey C | British Antarctic Survey, Cambridge, UK |
| Rintoul S | Australian Antarctic Division of Oceanography, Tasmania, Australia |
| Sameoto D | Bedford Institute of Oceanography, Dartmouth, Canada |
| Schnack-Schiel SB | Alfred Wegener Institute, Bremerhaven, Germany |
| Sicinski J | University of Lodz, Lodz, Poland |
| Siegel V | Institut für Seefisherei, Hamburg, Germany |
| Smetacek V | Alfred Wegener Institute, Bremerhaven, Germany |
| Smith R | University of California, Santa Barbara, California, USA |
| Stan Van den Berg CMG | Oceanography Laboratories, University of Liverpool, Liverpool, UK |
| Stoecker D | Horn Point Environmental Labs, Cambridge, Massachusetts, USA |
| Tanabe S | Ehime University, Matsuyama, Japan |
| Tomas CR | Florida Marine Research Institute, Florida, USA |
| Treguer P | Université de Bretagne Occidentale, Brest, France |
| Tripodi G | Istituto di Botanica, Università di Messina, Messina, Italy |
| Turner J | University of Massachussetts, Dartmouth, Dart mouth, Massachusetts, USA |
| Vacelet J | Station Marine d'Endoume, Marseille, France |
| White M | British Antarctic Survey, Cambridge, UK |
| Wiencke C | Alfred Wegener Institute, Bremerhaven, Germany |
| Wyatt T | Istituto de Investigaciones Marinas, Vigo, Spain |

## Contributors

# Acknowledgements

The papers accepted for publication in this volume were reviewed by an international board of eminent scientists. The editors wish to thank these reviewers for their assistance and cooperation which have greatly facilitated the relatively prompt publication of this volume. The authors also wish to thank Prof. Nunzio Crescenti and Giuseppe Arena, Vincenzo Bonazinga, Nicola Donato, and Francesco Soraci for their assistance in preparing the data analyses, graphics and text elaborations. This volume was financially supported by the Italian National Antarctic Program PNRA.

# Contents

## Introduction



## General Characteristics and Biogeochemical Processes



## Water Masses and Circulation Patterns



## Nutrients, Particulate Organic Matter, and Microbial Biomass

## Phytoplankton and Primary Production



## Zooplankton and Micronekton Distribution

## Ecotoxicity Studies

CHAPTER 1

# The Italian Oceanographic Cruises in the Ross Sea (1987–95): Strategy, General Considerations and Description of the Sampling Sites

**F.M. Faranda[1], L. Guglielmo[2], and A. Ianora[3]**

## Introduction

The Antarctic has often been considered a paradox in terms of primary productivity. That is, the rich water masses in this region, with their elevated nutrient concentrations, are rarely matched by correspondingly high phytoplankton standing stocks and productivity. Exceptions to this general condition are the productive coastal areas and frontal systems that are considered as productive as most regions in temperate latitudes. The Ross Sea may be another exception to this general *scenario* of low plankton productivity. As of the early 1980s, there has been increasing evidence that this area may represent one of the most productive regions in the Antarctic. Elevated phytoplankton biomass and productivity have, in fact, been measured both in situ (Smith and Nelson 1985; Wilson et al. 1986; Innamorati et al. 1990, 1992; Smith et al. 1996; Saggiomo et al. 1998; Lazzara et al., this Vol.) and via satellite images (Arrigo and McClain 1994). Further, there is also evidence of elevated flux of biogenic material accumulated over the continental shelf (Dunbar et al. 1985; DeMaster et al. 1992; Fabiano et al. 1996; Nelson et al. 1996; Langone et al. 1997; Ceccaroni et al. 1998; Frignani et al., this Vol.). The elevated productivity of this region has also recently been confirmed for its top predators such as krill, birds, seals and whales (Saino and Guglielmo, this Vol.; Azzali et al., this Vol.).

This special volume is devoted to the recent studies in this area conducted by Italian scientists from 1987–1995, presenting a global picture of the results of three of the oceanographic cruises conducted in the Ross Sea during the Antarctic late spring-summer periods. The work was part of a series of expeditions funded by the Italian government devoted to research on the marine ecology of the Ross Sea region, from studies of the physical environment and its flora and fauna, to understanding the physical and biological processes that govern its functioning. This introductory chapter provides information on the study site and describes the logistic features common to the three cruises to avoid unnecessary repetition of information in the accompanying chapters of this volume. It also describes the main objectives of these cruises and attempts to give a synopsis of the main results obtained to date and the perspectives for research in this region in the future.

### Polar Queen Cruise

The first Italian oceanographic expedition from 20 December 1987 to 29 February 1988, on board the R/V Polar Queen, was conducted in Terra Nova Bay between Cape Washington and Drygalsky Ice Tongue, Western Ross Sea (70°30′ and 75°30′S; 164°00′ and 172°00′E). Since this was the first expedition to the region and since there was practically no previous information on the study site, the objective of this expedition was to define the spatial and temporal variability of physical, chemical and biological characteristics in this region. It was our hope that the program would extend our knowledge and provide a better understanding of the processes through which this coastal ecosystem functions.

Terra Nova Bay is an 80 × 30-km-wide inlet in the southwestern part of the Ross Sea, delimited to the north by the narrow peninsula of Cape Washington and to the south by the Drygalski Ice

[1] Istituto Scienze Ambientali Marine, Università di Genova, Viale Benedetto XV 5, 16132 Genova, Italy
[2] Dipartimento Biologia Animale ed Ecologia Marina, Università di Messina, Salita Sperone 31, 98166 Messina, Italy
[3] Stazione Zoologica "A. Dohrn", Villa Comunale, 80121 Napoli, Italy

Tongue arising from David Glacier. The area is characterized by polynyas persisting during wintertime, formed and preserved by strong katabalic winds that keep the area free of ice. According to Patterson and Whitworth (1990), the highest salinities seem to occur in coastal areas of persistent wintertime leads and polynyas that prolong the exposure of the sea to the cold Antarctic atmosphere.

The bay is a tortuous continental shelf with numerous banks and trenches. The mean depth of the shelf is about 450 m, with greater depths close to the coast, and up to 1000 m depths in the Drygalski basin. The Bay is entirely covered by sea ice for at least 9 months of the year. To study this area, two phases were delineated. A first phase included a pelagic study whereby a spatial grid was designed large enough to include representative habitats with synoptic sampling of both physical and biological parameters. Physical variables were sampled in 40 stations by CTD casts to the bottom with the aim of defining the principal water masses present and identifying the influence of melted water and surface heating on the thermohaline structure of these water masses (Fig. 1). Station locations, depths, dates and times visited are given in Table 1. Simultaneously to the CTD casts, continuous vertical profiles were made of in vivo chlorophylls and downward quantum PAR radiation. Downwelling, upwelling and scalar underwater radiation at depths of 0, 5, 10, 25 and 50 m, and an upwelling spectral radiance at a depth of 5 m were also carried out (LI-COR).

Water samples were taken with Nisken bottles at depths of 0, 10, 25, 50, 100, 200 and 300 m for measurements of salinity, dissolved oxygen, nutrients ($NO_3$, $NO_2$, $NH_3$, $PO_4$ and $SiO_2$), photosynthetic pigments (chl a, b, c and phaeopigments), phytoplankton and microzooplankton taxonomy. Spectral particle analyses were made on board by means of a Coulter Counter Multisizer. Zooplankton samples were taken at all stations at 10 discrete depth intervals from the surface to 800 m by a BIONESS multinet (0.25 $m^2$ mouth area). A total of 314 samples were collected using 250 and 500 μm mesh sizes. Simultaneously to the tows, continuous profiles were made of temperature and salinity. Current measurements were conducted using CTD and ADCP current meters.

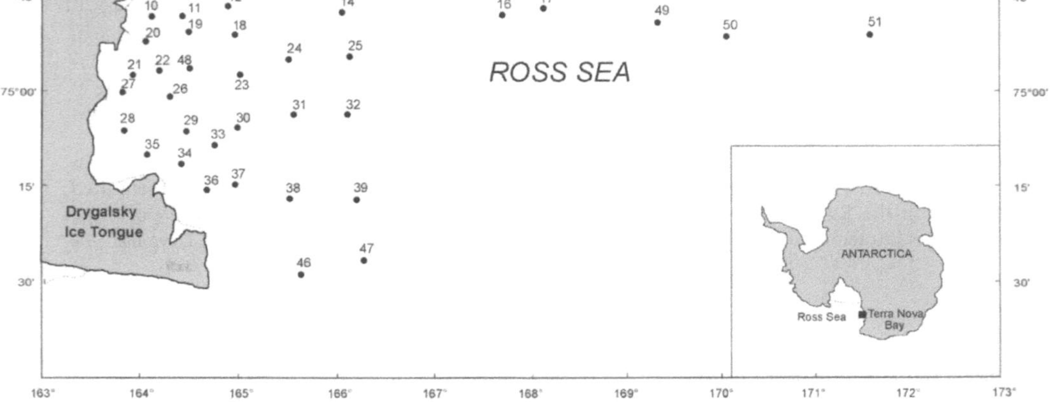


**Fig. 1.** Sampling area and location of the hydrological stations during the 1987–88 Polar Queen cruise

A second phase of this expedition was conducted near the Italian base in late summer when the bay was ice-free. The aim of this research was to compile a first checklist of the benthic species of Terra Nova Bay and to provide a preliminary map of species distributions together with a sedimentological map of the region. Benthic organisms analyzed during this first cruise, from the quantitative and qualitative point of view, included molluscs, polychaetes, bryozoans, foraminifers, ostracods, echinoderms, sponges, hydroids and pycnogonids. Also, research was conducted on the distribution of the phytobenthic communities along the coast, between Cape Washington and Inexpressible Island. The coastal fish community was studied from samples collected off the Italian base with the objective to acquire new data on the biology and population dynamics of some species suitable for commercial exploitation. This was accomplished also in the light of the recommendations of the Commission for the Conservation of the Antarctic Marine Living Resources (CCAMLR) asking the governments of countries with scientific programs in the Southern Ocean to include studies on the biology of species of potential interest to commercial fisheries. Further

**Table 1.** Location by latitude (°S) and longitude (°E), date, time (NZ) and bottom depth (m) at all hydrological stations sampled during the Polar Queen cruise

| Station | Date | Time | Bottom Depth | Latitude | Longitude |
|---|---|---|---|---|---|
| 08 | 01/05/88 | 17:30 | 550 | 74°43′.10 | 164°26′.00 |
| 09 | 01/06/88 | 09:15 | 370 | 74°43′.10 | 164°13′.80 |
| 10 | 01/06/88 | 15:50 | 370 | 74°48′.10 | 164°11′.10 |
| 11 | 01/07/88 | 10:30 | 550 | 74°48′.40 | 164°29′.80 |
| 20 | 01/09/88 | 17:45 | 270 | 74°53′.20 | 164°12′.60 |
| 19 | 01/10/88 | 20:20 | 742 | 74°50′.00 | 164°30′.00 |
| 21 | 01/12/88 | 10:45 | 570 | 74°58′.10 | 164°05′.60 |
| 22 | 01/12/88 | 16:15 | 794 | 74°58′.20 | 164°13′.70 |
| 26 | 01/13/88 | 10:30 | 940 | 75°02′.40 | 164°15′.50 |
| 27 | 01/13/88 | 17:04 | 850 | 75°01′.90 | 163°52′.10 |
| 29 | 01/05/88 | 09:30 | >1000 | 75°08′.00 | 164°24′.00 |
| 28 | 01/14/88 | 14:00 | 985 | 75°07′.20 | 163°52′.20 |
| 35 | 01/14/88 | 24:00 | 1010 | 75°11′.50 | 163°59′.80 |
| 34 | 01/15/88 | 08:30 | 1100 | 75°12′.30 | 164°16′.70 |
| 33 | 01/15/88 | 12:30 | 1100 | 75°11′.00 | 164°37′.00 |
| 36 | 01/15/88 | 17:30 | 812 | 75°18′.00 | 164°33′.00 |
| 12 | 01/24/88 | 19:30 | 820 | 74°47′.90 | 164°49′.80 |
| 07 | 01/25/88 | 01:30 | 578 | 74°44′.50 | 164°48′.00 |
| 03 | 01/25/88 | 10:00 | 160 | 74°38′.90 | 165°36′.70 |
| 13 | 01/25/88 | 15:30 | 670 | 74°45′.80 | 165°30′.90 |
| 24 | 01/26/88 | 01:30 | 1100 | 74°58′.00 | 165°38′.00 |
| 05 | 01/26/88 | 10:15 | 710 | 74°42′.00 | 166°04′.50 |
| 14 | 01/26/88 | 14:15 | 1010 | 74°47′.90 | 166°08′.50 |
| 25 | 01/26/88 | 21:30 | >1000 | 74°57′.20 | 166°17′.20 |
| 17 | 01/27/88 | 09:00 | 525 | 74°48′.00 | 167°57′.00 |
| 18 | 01/28/88 | 20:45 | 723 | 74°50′.90 | 164°55′.30 |
| 23 | 01/29/88 | 0030 | 909 | 74°58′.80 | 164°59′.80 |
| 48 | 01/29/88 | 10:00 | 953 | 74°57′.20 | 164°37′.20 |
| 30 | 01/29/88 | 16:30 | 850 | 75°09′.00 | 165°00′.30 |
| 37 | 01/29/88 | 21:30 | 770 | 75°16′.70 | 164°51′.80 |
| 46 | 01/30/88 | 14:00 | 849 | 75°29′.00 | 165°31′.80 |
| 47 | 01/30/88 | 18:00 | 680 | 75°30′.00 | 166°13′.20 |
| 31 | 01/31/88 | 08:15 | >1000 | 75°06′.40 | 165°29′.10 |
| 38 | 01/31/88 | 13:10 | 716 | 75°17′.80 | 165°20′.60 |
| 39 | 02/02/88 | 12:30 | 850 | 75°17′.60 | 166°08′.80 |
| 32 | 02/02/88 | 15:45 | 835 | 75°07′.00 | 166°08′.40 |
| 16 | 02/02/88 | 22:30 | 681 | 74°48′.20 | 167°17′.30 |
| 51 | 02/10/88 | 03:00 | 516 | 74°51′.20 | 171°29′.20 |
| 49 | 02/12/88 | 04:30 | 325 | 74°50′.40 | 169°12′.70 |
| 50 | 02/14/88 | 05:30 | 309 | 74°51′.70 | 170°02′.50 |

details on data collections during this cruise are available in the data reports published by the Italian Antarctic Commission in 1990 (National Scientific Commission for Antarctica, Data Reports Part I and Part II, see references).

## Cariboo Cruise

The second oceanographic expedition 1 year later, from 25 November 1989 to 20 February 1990, investigated a larger geographic area in the Pacific sector of the Southern Ocean, from 50°S to the Balleny Islands and from these to the Italian base at Terra Nova 75°S in the Ross Sea on board the R/V Cariboo. The general objectives of this expedition were to better understand the functioning of the Antarctic pelagic ecosystem through the study of hydrodynamic features such as macroscale water mass distribution associated with frontal systems, ice coverage and the study of the vertical structure of the water column during the spring. The Polar Queen cruise the preceding year had furnished a series of preliminary data on the structure of the pelagic and benthic communities in Terra Nova Bay. However, since Terra Nova Bay is marginal to large-scale circulation patterns in the Southern Ocean, it was necessary to amplify the sampling area. Also, the absence of krill (*Euphausia superba*) in plankton samples during the preceding cruise indicated the necessity to expand the sampling area in order to define krill distribution patterns in this region. The following aspects of the pelagic ecosystem were considered: distribution of water masses and determination of thermohaline structures involved in biological processes, distribution of nutrients, dissolved oxygen and inorganic carbon resulting from photosynthetic activity and heterotrophic metabolism, particulate organic carbon, nitrogen and nucleic acids (DNA, RNA), biochemical composition of particulate organic matter, distribution of phaeopigments (HPLC) and biochemical liposoluble components in phytoplankton and krill, activity and potential productivity of pico- and phytoplankton, biomass and composition of phytoplankton, structure and composition of micro- and zooplankton communities, and estimation of krill stocks.

CTD profiles were collected with a Neil Brown MK III system coupled with a 10-l, 12-bottle GO Rosette. Twenty-seven stations were sampled to the bottom to a maximum depth of 3509 m (Fig. 2, Table 2). The CTD was equipped with a dissolved oxygen sensor and fluorometer. On board, continuous measurements with a submerged pump at 2 m depth were taken of salinity and in vivo fluorescence by using a multiparametric KMS (ME) CTD probe and an underwater fluorimeter. Quantum PAR irradiance downwelling in air and downwelling, upwelling and scalar underwater irradiance were measured by means of a cosine quantum sensor and light probe equipped with two cosine and one spherical quantum sensor (LI-COR), connected to a computer through a monitor (mod. 801, Idronaut). Water samples were taken over more restricted depths from the surface to 100 m, at 8-m intervals, to better support productivity studies. Thereafter, samples were more spaced out downwards to 3000 m. These were used for nutrients, particulate organic matter and biochemical analyses as well as for phytoplankton and microzooplankton taxonomy. Both prokaryotic and eukaryotic picophytoplanktonic cells and autofluorescent organisms >2 μm were counted. HPLC analysis was used to characterize phytopigments.

Zooplankton samples were taken at all stations to a maximum of 1000 m with a BIONESS multinet (0.25 $m^2$ mouth area) equipped with a Back-Scat fluorometer and a multiparametric KMS II probe which continuously measured salinity, oxygen, temperature and light attenuation (PAR). Flow through the nets was monitored by external and internal acoustic Doppler flowmeters mod. ME-SM 11H. The depth intervals were chosen as a function of the main water masses present in the column (10–100 m). Zooplankton organisms examined were copepods, euphausiids, ostracods, mysids, salps, pelagic molluscs and larval fishes. A total of 572 samples were collected using 10 nets with 500 μm mesh size.

An echosurvey, covering about 1700 nautical miles, was conducted in the following periods: 4–6 December 1989 in the vicinity of the Balleny Islands; 26 December 1989 to 5 January 1990 from the Ross Sea to the Italian Antarctic Base "Terra Nova" and 12–26 January 1990 from "Terra Nova" up to 65°S. Investigations were carried out using a Bio-Sonics Echosounder mod. 102 operating at a frequency of 38 and 120 kHz in a multiplexing mode equipped with an ESP mod. 221 echo Integrator and a 3-ft towed body transducer system.

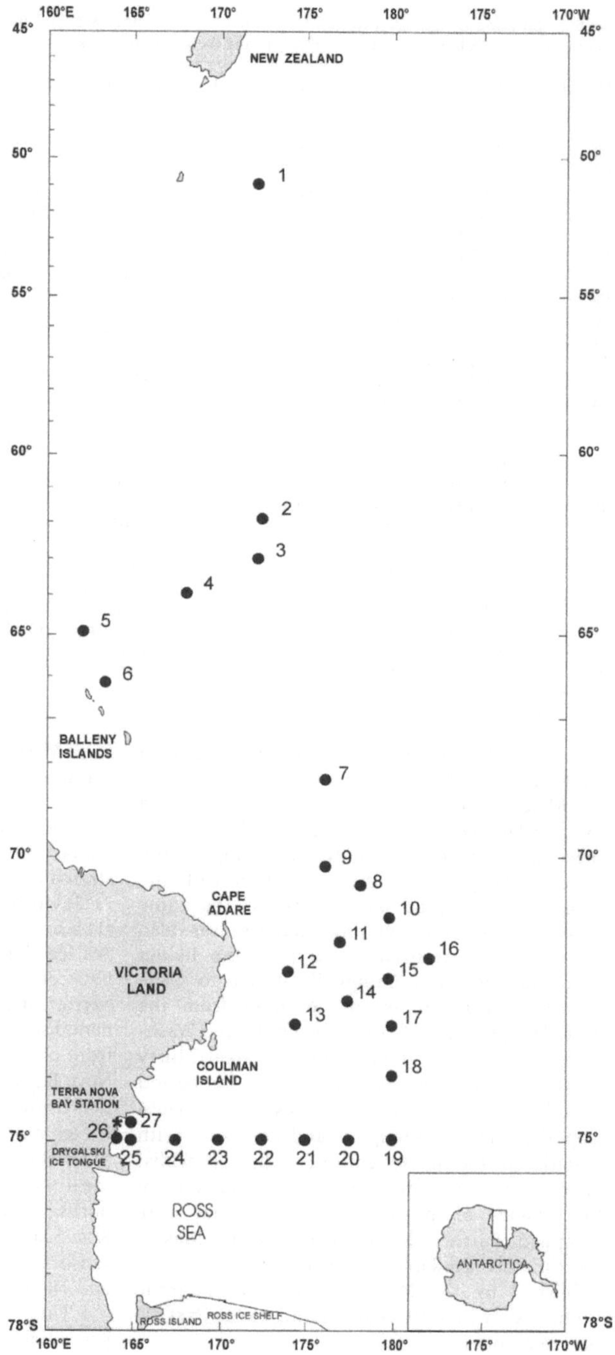


**Fig. 2.** Sampling area and location of the hydrological stations during the 1989–90 Carbioo cruise

**Table 2.** Location by latitude (°S) and longitude (°E), date, time (NZ) and bottom depth (m) of hydrological stations sampled during the Cariboo cruise

| Station | Date | Time | Bottom Depth | Latitude | Longitude |
|---|---|---|---|---|---|
| 01 | 11/25/89 | 17:52 | 518 | 50°58′.00 | 171°57′.00 |
| 02 | 11/29/89 | 13:45 | 3800 | 61°55′.00 | 172°28′.00 |
| 03 | 12/02/89 | 20:33 | 2700 | 63°01′.00 | 172°12′.00 |
| 04 | 12/03/89 | 00:42 | 2980 | 63°58′.00 | 168°03′.00 |
| 05 | 12/05/89 | 12:57 | 3190 | 64°54′.00 | 162°00′.00 |
| 06 | 12/06/89 | 11:24 | 2670 | 66°03′.00 | 163°25′.00 |
| 07 | 12/07/89 | 10:35 | 2800 | 66°32′.00 | 165°40′.00 |
| 08 | 12/14/89 | 18:29 | 3446 | 66°16′.00 | 174°05′.00 |
| 09 | 12/21/89 | 09:28 | 3280 | 70°12′.00 | 176°17′.00 |
| 10 | 12/23/89 | 09:03 | 1550 | 71°12′.00 | 179°57′.00 |
| 11 | 12/24/89 | 14:13 | 870 | 71°39′.00 | 176°59′.00 |
| 12 | 12/25/89 | 17:38 | 585 | 72°14′.00 | 173°59′.00 |
| 13 | 12/26/89 | 12:19 | 314 | 73°10′.00 | 174°17′.00 |
| 14 | 12/27/89 | 12:51 | 1567 | 72°44′.00 | 177°19′.00 |
| 15 | 12/28/89 | 11:42 | 2138 | 72°22′.00 | 179°43′.00 |
| 16 | 12/29/89 | 13:47 | 746 | 71°55′.00 | 177°52′.00 |
| 17 | 12/30/89 | 13:26 | 535 | 73°13′.00 | 179°53′.00 |
| 18 | 12/31/89 | 12:46 | 273 | 73°59′.00 | 179°56′.00 |
| 19 | 01/01/90 | 14:41 | 460 | 74°59′.00 | 179°56′.00 |
| 20 | 01/02/90 | 10:57 | 380 | 75°01′.00 | 177°31′.00 |
| 21 | 01/03/90 | 12:12 | 295 | 75°00′.00 | 174°57′.00 |
| 22 | 01/04/90 | 00:03 | 535 | 75°01′.00 | 172°38′.00 |
| 23 | 01/05/90 | 12:29 | 335 | 75°00′.001 | 170°00′.00 |
| 25 | 01/05/90 | 20:03 | 935 | 75°01′.00 | 165°02′.00 |
| 24 | 01/06/90 | 11:05 | 560 | 74°58′.00 | 167°25′.00 |
| 26 | 01/11/90 | 13:32 | 420 | 74°56′.00 | 164°01′.00 |
| 27 | 01/12/90 | 10:04 | 610 | 74°46′.00 | 164°59′.00 |

The BIONESS net was used for sampling krill swarms.

Meteorological observations were conducted to correlate certain parameters such as wind speed and direction, humidity, air temperature, atmospheric pressure, and irradiation with the main oceanographic parameters. At the same time, intense oceanographic activity was also observed in a restricted area near the Italian base, starting from Campbell Glacier to Cape Russel, at a maximum of 16 miles from the base. The presence of sea ice around Cape Washington prevented sampling in this area. Thirty benthic communities were studied in this period including demospones, molluscs, polychaetes, isopods, tanaids, amphipods and decapods. With regards to the phytobenthos, a first checklist of the diatom genera and an evaluation of the biomass of the dominant macroalgae were compiled. Further details on data collections at the time are available in two other data reports published by the National Scientific Antarctic Commission in 1991 (Part I) and 1992 (Part II) (see references).

## ROSSMIZE (Ross Sea Marginal Ice Zone Ecology) Cruise

The third expedition, from 2 November 1994 to 3 January 1995, covered, from south to north, a wide stretch of the western sector of the Ross Sea (lat. 71°56′ to 76°30′S). The area investigated included: (1) a first group of stations located along the Ross Sea Ice Shelf; (2) a second large group of stations between longitudes 180° and 170°W and from the barrier of the Ross Sea Ice Shelf up to the continental slope (around lat. 72°S); (3) a third group in front of Cape Adare; and (4) a last group in Terra Nova Bay (Fig. 3, Table 3). The cruise was made on board the R/V Italica, jointly organized with the US team of the R/V Palmer, and represents one of the first oceanographic cruises ever carried out in the Ross Sea during the austral spring. In addition, during navigation from New Zealand (47°S) to the Ross Sea (64°S), XBT sippicam T7 and sippicam T5 sensors were dropped to a maximum depth of 760 and 1840 m, respectively.

The scientific results obtained during the preceding 1989–90 cruise had emphasized that

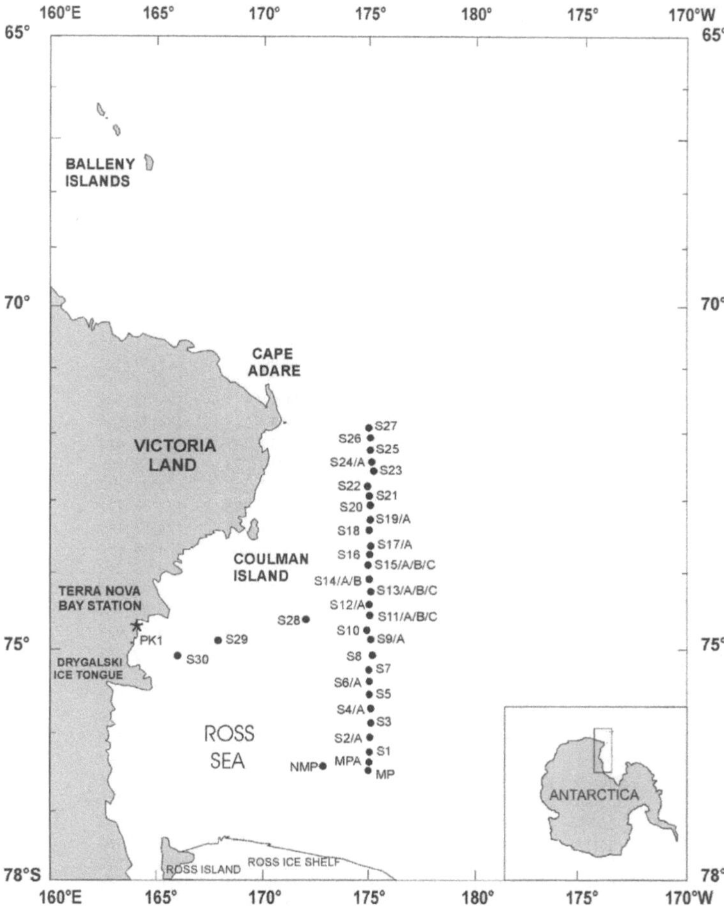


**Fig. 3.** Sampling area and location of the hydrological stations during the 1994–95 ROSSMIZE cruise

phytoplankton productivity in this region was greatly influenced by ecohydrodynamic processes occurring in the ice-edge area. It was shown that ice melting and retreat in the Ross Sea, propagating from the south to the north (polynya), strongly increased the extension of the productive marginal ice zone (Fig. 4). This greatly influences the global productivity in the area, favoring intense phytoplankton blooms in the water column. The algae contained in the ice are released by melting, mixed in the upper layers of the water column and consumed by zooplankton grazing. With the specific aim of better understanding this aspect of the ecology of the Ross Sea, a mutidisciplinary study was developed taking into account the main physical, chemical and biological features of the area, integrating experimental, field and prediction model studies.

The program focused on the following topics: nutrient distribution and dynamics, new and regenerated production, role of mesozooplankton, krill and top predators in controlling primary production, biochemical composition and fate of particulate organic matter, biogenic flux in the water column, biogenic sedimentation and seafloor bioturbation. A main transect was delineated including 52 hydrological stations from south to north to examine the evolution of the water column from a

**Table 3.** Location by latitude (°S) and longitude (°E), date, time (NZ) and bottom depth (m) at all hydrological stations sampled during the ROSSMIZE cruise

| Station | Date | Time | Bottom Depth | Latitude | Longitude |
|---|---|---|---|---|---|
| MP | 11/14/94 | 09:21 | 464 | 76°30′.103 | 174°59′.886 |
| 01 | 11/15/94 | 09:38 | 496 | 76°19′.997 | 174°57′.763 |
| 02 | 11/15/94 | 15:30 | 579 | 76°09′.824 | 174°59′.734 |
| 03 | 11/16/94 | 08:42 | 561 | 75°59′.724 | 175°01′.968 |
| 04 | 11/16/94 | 14:18 | 457 | 75°49′.607 | 175°02′.125 |
| 05 | 11/16/94 | 20:03 | 317 | 75°39′.633 | 174°59′.303 |
| 06 | 11/17/94 | 08:55 | 304 | 75°29′.405 | 174°58′.760 |
| 07 | 11/17/94 | 16:30 | 309 | 75°20′.468 | 174°58′.437 |
| NMP | 11/18/94 | 16:48 | 618 | 76°30′.538 | 172°47′.654 |
| MPA | 11/20/94 | 10:07 | 469 | 76°26′.900 | 175°02′.200 |
| 2A | 11/20/94 | 22:09 | 585 | 76°09′.591 | 174°56′.949 |
| 4A | 11/21/94 | 10:31 | 468 | 75°50′.394 | 175°01′.144 |
| 6A | 11/21/94 | 16:57 | 305 | 75°30′.487 | 174°58′.965 |
| 08 | 11/22/94 | 09:12 | 313 | 75°09′.540 | 175°10′.322 |
| 09 | 11/22/94 | 14:18 | 314 | 74°57′.673 | 175°02′.343 |
| 10 | 11/22/94 | 17:45 | 388 | 74°50′.543 | 174°52′.655 |
| 11 | 11/23/94 | 09:24 | 438 | 74°39′.245 | 175°01′.414 |
| 12 | 11/23/94 | 19:05 | 480 | 74°30′.758 | 174°59′.969 |
| 13 | 11/25/94 | 08:49 | 539 | 74°19′.927 | 175°05′.162 |
| 14 | 11/25/94 | 14:27 | 552 | 74°10′.380 | 174°58′.791 |
| 15 | 11/26/94 | 21:02 | 576 | 73°59′.070 | 174°57′.145 |
| 11A | 11/28/94 | 16:19 | 438 | 74°40′.466 | 174°56′.480 |
| 9A | 11/28/94 | 22:29 | 281 | 75°00′.399 | 174°58′.293 |
| 12A | 11/29/94 | 12:59 | 508 | 74°29′.681 | 175°01′.653 |
| 13A | 11/29/94 | 18:41 | 540 | 74°20′.145 | 175°00′.764 |
| 15A | 11/30/94 | 09:31 | 575 | 73°58′.956 | 174°58′.943 |
| 14A | 11/30/94 | 15:50 | 557 | 74°09′.141 | 174°59′.643 |
| 17 | 12/03/94 | 17:56 | 487 | 73°42′.067 | 175°01′.318 |
| 16 | 12/03/94 | 22:06 | 563 | 73°49′.590 | 175°02′.625 |
| 11B | 12/04/94 | 08:28 | 437 | 74°39′.210 | 174°58′.199 |
| 13B | 12/04/94 | 14:34 | 546 | 74°19′.103 | 174°58′.280 |
| 15B | 12/04/94 | 20:44 | 576 | 73°59′.643 | 175°01′.124 |
| 18 | 12/05/94 | 18:40 | 389 | 73°29′.229 | 174°59′.036 |
| 19 | 12/05/94 | 21:21 | 380 | 73°19′.979 | 175°00′.233 |
| 20 | 12/06/94 | 13:28 | 368 | 73°07′.733 | 174°59′.198 |
| 21 | 12/06/94 | 16:43 | 351 | 72°59′.969 | 175°00′.162 |
| 22 | 12/06/94 | 21:08 | 324 | 72°50′.881 | 174°53′.528 |
| 23 | 12/07/94 | 08:59 | 428 | 72°37′.692 | 175°09′.267 |
| 24 | 12/07/94 | 11:14 | 446 | 72°29′.006 | 175°02′.643 |
| 25 | 12/07/94 | 14:19 | 673 | 72°17′.802 | 174°58′.780 |
| 26 | 12/07/94 | 19:04 | 1396 | 72°06′.984 | 175°01′.436 |
| 27 | 12/08/94 | 08:45 | 1810 | 71°57′.122 | 174°58′.622 |
| 24A | 12/10/94 | 17:00 | 439 | 72°29′.264 | 174°59′.092 |
| 19A | 12/11/94 | 08:44 | 384 | 73°18′.924 | 175°01′.848 |
| 17A | 12/11/94 | 20:42 | 437 | 73°39′.633 | 174°58′.736 |
| 15C | 12/12/94 | 11:55 | 576 | 73°59′.802 | 175°00′.400 |
| 14B | 12/12/94 | 16:45 | 552 | 74°09′.798 | 175°01′.474 |
| 13C | 12/13/94 | 13:44 | 539 | 74°20′.123 | 175°00′.904 |
| 11C | 12/13/94 | 19:24 | 439 | 74°40′.360 | 174°59′.559 |
| 28 | 12/14/94 | 16:41 | 522 | 74°42′.136 | 172°00′.528 |
| 29 | 12/15/94 | 09:25 | 382 | 74°59′.004 | 167°59′.326 |
| 30 | 12/15/94 | 14:57 | 825 | 75°08′.740 | 166°07′.589 |
| PK1 | 12/16/94 | 17:00 | 350 | 74°41′.300 | 164°10′.500 |

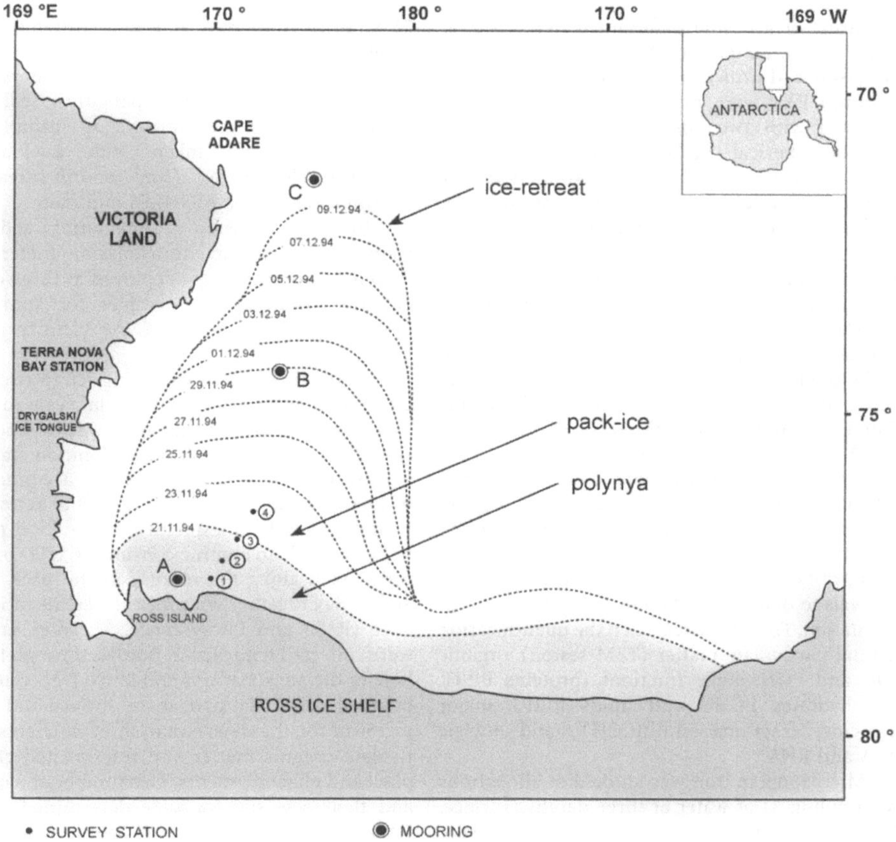


**Fig. 4.** Map of the expected pack-ice retreat during the 1994–95 ROSSMIZE cruise

situation of sea-ice coverage to one of open water. A Sea-Bird Electronics SBE 9/11+ was employed to perform CTD profiles coupled with a 12-l 24-Niskin bottle SBE 32 Corousel for water sampling during the upcasts. The CTD was equipped with double temperature-conductivity sensors flushed at a constant rate by a pump, a parascientific pressure sensor, a Chelsea fluorometer, a Sea-Tech 25-cm transmissometer, a Yellowspring dissolved oxygen sensor and a Datasonics altimeter.

Seawater was sampled at hydrological stations to determine dissolved oxygen and nutrients (reactive silicate, reactive phosphorus, reactive nitrate and nitrite). Total chlorophyll a and phaeopigment concentrations of total and fractionated phytoplankton (pico <2 μm, nano 2–10 μm, and micro >10 μm) were determined. At three stations, ice samples were also collected to determine sympagic algal pigments. In addition, total surface chlorophyll a and size-fractionated phytoplankton were monitored along a contour track off Cape Adare where a total of 579 samples were collected. Continuous fluorometry was performed connecting a Turner Design (10-AU-005) fluorometer to a continuous water circulation system fed through an intake located in the ship hull at about the 3 m depth.

Underwater PAR irradiance was measured along a transect covering the polynya and the drifting, broken and closed pack-ice zones. Light measurements were recorded at 45 stations. Surface downwelling quantum PAR was measured by means of a cosine quantum meter (LI-COR, LI-192SA) mounted on the ship's upper deck. Underwater downwelling, upwelling and scalar quantum PAR were measured by a photoprobe equipped with two cosine (LI-COR, LI-192SA) and one spherical (LI-COR, LI-193SA) quantum meter. Measurements of primary production were carried out at almost all of the stations after identification of seven optical depths (50, 20, 10, 5, 1, 0.1% of incident PAR and surface). Samples were incubated on deck in a series of seven seawater circulating tanks simulating in situ conditions.

Thirty-two surface samples were collected to determine abundance and species composition of the phytoplankton community. Other samples were collected at five optical depths to determine abundance and carbon biomass of nanoplankton (2–10 μm) and picophytoplankton (0.2–2 μm). Moreover, 135 samples were collected at five optical depths to evaluate microbial (ATP) and bacterial (LPS) biomass, and respiratory activity (ETSa).

Fifty-two stations were sampled at 12 depth intervals to determine POM distribution and sedimentation. This study regarded the quantification of total particulate matter (TSM seston), organic POC and PON, labile fractions (proteins PPRT, carbohydrates PCHO and lipids PLIP), major refractory fractions (chitin CHT) and nucleic DNA and RNA.

Microzooplankton was studied at all stations by sampling 5 l of water at three depths (surface, 30–40 m and 150–300 m). Twelve depths were sampled in the vicinity of mooring B and five depths near mooring C. Mesozooplankton was sampled at 36 stations using a BIONESS (1 $m^2$ mouth area) equipped with sensors for the measurement of temperature, conductibility, dissolved oxygen, PAR, fluorescence, Mie Back Scattering and an Optical Plankton Counter (OPC). A total of 372 samples were taken at depth intervals from 20–350 m with 12 nets (250 μm mesh). An additional 31 samples were taken from 40 m to the surface with a WP2 (200 μm mesh) net to obtain experimental animals for grazing and excretion studies.

Biomass and density of krill were estimated in ice-free waters using a Bio-Sonic Ecosounder at two frequencies (38 and 120 kHz) and under the pack ice with a hull-mounted transducer (120 kHz). The area covered during this survey was about 65 000 square nautical miles between 170°E (coastal area) and 179°W (annual pack) and from 75°S (Terra Nova Bay) and 63°S (Antarctic Convergence). A total of 3934 miles were covered acoustically during 520 sessions to survey top predators including birds, penguins, seals and whales. During the survey, 36 samples of micronekton were taken with a Plankton Hamburg Net (PHN) (5 $m^2$ mouth area and 500 μm mesh) and a BIONESS multinet.

To study biogenic sedimentation, sediment traps, current meters, transmission meters and CTD autorecorders were deployed at three different sites. Bottom core samples for studies of bioturbation were taken in the vicinity of the moorings.

Production processes during ice retreat were also examined on a smaller spatial scale in Terra Nova Bay. The aim of this work was to study the strong coupling between phytoplankton standing stocks, particulate matter flux and accumulation of sinking material in the sediments to better estimate the quality and quantity of the food potentially available to benthic consumers. CTD profiles were taken along the margin of the pack ice in Tethys Bay together with measurements of irradiance (PAR) and fluorescence. Seawater samples were collected with Niskin bottles, through the ice hole in the pack stations and in ice-free waters on board the R/V Malippo, at the surface and down to 150 m for the determination of nutrients, particulate organic matter and fractionated chlorophyll and phaeopigments. The number of particles and their size spectra were determined with a Coulter Counter Multisizer (140 μm orifice tube). Phytoplankton samples were taken for microscopic analysis of sympagic and pelagic algae.

A total of 48 zooplankton samples were taken vertically at the 150–100, 100–50, 50–25 and 25–0 m depths and 4 samples were taken horizontally using a WP2 (200 μm) net. Another 24 samples were taken to determine heterotrophic bacterial and total picoplankton numbers and lipopolysaccharide and ATP content in the 0.2–2 and 2–250 μm size fractions. Biogenic material flux was estimated using submerged traps at 40 m depths, preferably in the vicinity of areas dominated by the bivalve *Adamussium colbecki*. To determine TSM, POC, PON, PPRT, PCHO, PLIP, CHT, Chl, DNA and RNA 2300 samples were taken.

The zonation of littoral benthic populations was studied using a 65-l Van Veen grab and a triangular grab. To quantify the role of *Adamussium colbecki* in the transfer of energy along the water column, several laboratory experiments were conducted on clearance, filtering and biodeposition rates. Finally, 24 SCUBA immersions to the 26 m depth were conducted to collect hard bottom samples. Further details on data collections during this cruise are provided in the data report published by the Italian Antarctic Commisssion in 1997 (Faranda et al. 1997).

## A Synopsis of the Results

The Ross Sea represents a mosaic of functionally different marine subsystems. For example, the coastal areas of this sea, including Terra Nova Bay, are characterized by a diminution in nutrient concentrations, especially in the euphotic zone, that are more marked than in other regions (Catalano et al., this Vol.). In summer, due to melting of the ice pack, the stability of the water column becomes crucial for the assimilation of nutrients in the mixed upper layers. Under such conditions, silica may be a limiting factor, especially in coastal areas (Catalano et al. 1997). High levels of primary productivity during the austral spring are not a consequence of the stability of the water column (Smith and Nelson 1985) but the outcome of release of sympagic algae from the pack ice (Carrada et al., this Vol.; Saggiomo et al., this Vol.).

In terms of spatial distribution, the Ross Sea is characterized by different levels of primary production: the area west of 175′ is the most productive, probably due to the presence of micronutrients (e.g. Fe). Primary production values are very high and optimum irradiance levels are low (Lazzara et al., this Vol.). The west coast of the Ross Sea is characterized by two blooms, in the spring-summer and late summer periods (Innamorati et al., this Vol.) and the phytoplankton community shows a well-defined biogeographic distribution (Nuccio et al., this Vol.; Saggiomo et al., this Vol.). The spatial and temporal distribution of particulate organic matter is strongly linked to ice-melting processes. In fact, biochemical studies have shown that POM in pelagic waters is rapidly degraded by higher trophic levels (Fabiano et al. 1993, 1995, 1996, 1997). By contrast, the greater part of this detrital material more or less settles unaltered to the seafloor in neritic areas ((Fabiano et al. 1996; Fabiano and Pusceddu 1998; Fabiano et al., this Vol.). Studies on the abundance, biomass and microbial activity of the meiobenthic community have shown that there is a strong relation between the input of organic material from the water column to the sediments and the response of benthic organisms (Fabiano and Danovaro 1999; Fabiano et al., this Vol.).

When the pack ice is compact, zooplankton biomass is scarce and the population is poorly structured. Nonetheless, large numbers of juvenile krill (*Euphausia superba*) have been recorded in the pack ice. As summer advances, copepods reach high biomass values from 40–60 m, below the thermocline (Hecq et al., this Vol.). The major species include *Calanoides acutus*, *Calanus propinquus*, *Metridia gerlachei* and *Rhincalanus nasutus* (Zunini Sertorio et al., this Vol.). In Terra Nova Bay, the summer zooplankton community is characterized by the pteropod mollusc *Limacina helicina*, post-larval stages of the teleost *Pleurogramma antarcticum* and calyptopis and furcilia stages of the euphausiid *Euphausia cristallorophia* (Guglielmo et al. 1998; Granata et al., this Vol.). With regards to the mesozooplankton and micronekton, the most important constituents include copepods, hyperiid and gammarid amphipods, clionid molluscs, chaetognaths, ostracods and the mesopelagic mysid *Antarctomysis ohlini* (Carli et al., this Vol.; Crescenti et al. 1994, this Vol.).

The results of the ecosurvey conducted during the ROSSMIZE cruise indicate that krill biomass can reach high values (>100 t/nm), particularly in areas of partial ice cover (72–73′S, 177–178′W; 73–74′S, 174–175′E), at the entrance between Coulmann Island and Cape Adare, and in proximity to Scott Island (66′S, 179′W). In coastal areas, as well as in open waters and in the area of polynas, the number of individuals recorded is much lower and the population is more dispersed. On the other hand, dense krill swarms have been recorded below the pack ice (Azzali et al., this Vol.; Azzali and Kalinowski, this Vol.). Highest krill predator populations of Imperial and Adelie penguins and carnivorous seals were located between 70–71′S, progressively diminishing eastwards (Saino and Guglielmo, this Vol.).

In general, the hard bottom benthic community of Terra Nova Bay is relatively rich in terms of both biomass and species numbers. Of particular interest is the biocenosis of Porifera and Anthezoa. This community is dominated by filter-feeders and can be considered one of the most unique benthic communities of the Antarctic (Gaino et al. 1994; Cattaneo-Vietti et al. 1996; Cattaneo-Vietti et al.

this Vol.; Chiantore et al., this Vol.). The most striking difference compared to the benthic populations of Adelie Land and Cape Hallet is the absence of large perennial brown algae probably due to the persistence of pack ice in this area and, consequently, the lower intensity of sunlight and greater mechanical disturbance. Under such conditions, the red algae Phyllophora has been observed even at very shallow depths. The presence of a highly simplified macroalgal community in terms of biomass and structure, the high disturbance derived from ice melting and scouring favor the establishment of poorly structured communities with a high biomass but low number of species. Several of these species have peculiar life cycles with ample interannual fluctuations in biomass (Gambi et al. 1994) Notwithstanding such fluctuations, the composition and basic structure of the communities are stable over time (Gambi et al., this Vol.).

### Future Perspectives

Based on past experience and on the results that have emerged from the ROSSMIZE cruise, we now understand that the functioning of the Ross Sea largely depends on ice formation and retreat, and how these processes affect species succession patterns. Future programs in this area will therefore focus on the flux of energy through the food chain and how this relates to the presence of ice. For example, what processes control the spatial distribution of diatoms as opposed to *Phaeocystis*? What relationship exists between zooplankton (micro and meso) and *Phaeocystis* blooms? Also, the discovery of large krill swarms at such low latitudes suggests that this organism is not strictly confined to areas of divergence, as often considered in the past. The presence of large numbers of fertilized female *Eupahusia superba* places in doubt the need of great depths for the sedimentation of their eggs. What factors promote the development of krill so early in the season? And what role do ice algae have on krill development? Future programs must address such problems by focusing on feeding, excretion, respiration and fecal pellet production rates of key species, as well as studying egg production, viability and development of juveniles. Much remains to be answered also with regards to planktonic and bacterial production within the pack ice and how this production is transferred to the pelagic domain. Many of these "unanswered" questions are currently being investigated within the framework of several international projects such as EASIZ, SO-JGOFS and SO-GLOBEC. We hope to be able to contribute to the understanding of these problems in the future within the framework of the Italian Antarctic program PNRA.

## References


Arrigo KR, McClain CR (1994) Spring phytoplankton production in the western Ross Sea. Science 266: 261–263

Catalano G, Povero P, Fabiano M, Benedetti F, Goffart A (1997) Nutrient utilisation and particulate organic matter changes during summer in the upper mixed layer (Ross Sea, Antarctica). Deep Sea Res 44: 97–112

Cattaneo-Vietti R, Bavestrello G, Cerrano C, Sará M, Benatti U, Giovine M, Gaino E (1996) Optical fibres in an Antarctic sponge. Nature 383: 397–398

Ceccaroni M, Frignani M, Langone L, Ravaioli M, Frank M, Mangini A (1998) Late Quaternary fluctuations of biogenic component flux on the continental slope of the Ross Sea, Antarctica. J Mar Syst 17: 516–526

Crescenti N, Costanzo G, Guglielmo L (1994) Developmental stages of *Antarctomysis ohlini Hansen*, 1908 (Mysidacea) in Terra Nova Bay, Ross Sea, Antarctica. J Crustacean Biol 14: 383–395

DeMaster DJ, Dunbar RB, Gordon LI, Leventer AR, Morrison JM, Nelson DM, Nittrouer CA, Smith WO Jr (1992) Cycling and accumulation of biogenic silica and organic matter in high latitude environments: the Ross Sea. Oceanography 5(3): 146–153

Dunbar RB, Anderson JB, Domack EW, Jacobs SS (1985) Oceanographic influences on sedimentation along the Continental Shelf. In: Jacobs SS (ed) Oceanology of the Antarctic Continental Shelf. Antarct Res Ser, AGU, Washington, DC, 43: 291–313

Fabiano M, Danovaro R (1999) Enzymatic activity, bacterial distribution and organic matter composition in sediments of the Ross Sea (Antarctica). Appl Environ Microbiol 64: 3838–3845

Fabiano M, Pusceddu A (1998) Total and hydrolizable particulate organic matter (carbohydrates, proteins and lipids) at a coastal station in Terra Nova Bay (Ross Sea, Antarctica). Polar Biol 19: 125–132

Fabiano M, Povero P, Danovaro R (1993) Distribution and composition of particulate organic matter in the Ross Sea (Antarctica). Polar Biol 13: 525–533

Fabiano M, Danovaro R, Crisafi E, La Ferla R, Povero P, Acosta Pomar L (1995) Particulate matter composition and bacterial distribution in Terra Nova Bay (Antarctica) during Summer 1989–1990. Polar Biol 15: 393–400

Fabiano M, Povero P, Danovaro R (1996) Particulate organic matter composition in Terra Nova Bay (Ross Sea, Antartica) during summer 1990. Antarct Sci 8: 7–13

Fabiano M, Chiantore M, Povero P, Cattaneo-Vietti R, Pusceddu A, Misic C, Albertelli G (1997) Short-term variations in particulate matter flux in Terra Nova Bay, Ross Sea. Antarct Sci 9: 143–149

Faranda FM, Guglielmo L, Povero P (1997) ROSSMIZE (Ross Sea Marginal Ice Zone Ecology) 1993–95. National

Scientific Commission for Antarctica, Data Report I, pp 1–430

Gaino E, Bavestrello G, Cattaneo-Vietti R, Sará M (1994) Scanning electron microscope evidence for diatom uptake by two Antarctic sponges. Polar Biol 14: 55–58

Gambi MC, Lorenti M, Russo GF, Scipione MB (1994) Benthic associations of the shallow hard bottoms off Terra Nova Bay, Ross Sea: zonation, biomass and population structure. Antarct Sci 6: 449–462

Guglielmo L, Granata A, Greco S (1998) Distribution and abundance of postlarval and juvenile *Pleuragramma antarcticum* (Pisces, Nototheniidae) off Terra Nova Bay (Ross Sea, Antarctica). Polar Biol 19: 37–51

Innamorati M, Mori G, Lazzara L, Vanucci S (1990) Eutrofia ed Oligotrofia nell'Oceano Antartico. Oebalia (Suppl) XVI-1: 153–162

Innamorati M, Lazzara L, Massi L, Mori G, Nuccio C, Saggiomo V (1992) Indagine sulla Biomassa Fitoplanctonica nel Mare di Ross in Relazione ai Fattori Ambientali. In: Gallardo VA, Ferretti O, Moyano HI (eds) Oceanografia in Antartide. ENEA-PNRA-EULA Concepcion, Chile, pp 235–252

Langone L, Frignani M, Cochran JK, Ravaioli M (1997) Scavenging processes and export fluxes close to a retreating seasonal ice margin (Ross Sea, Antarctica). Water Air Soil Pollut 99: 705–715

National Scientific Commission for Antarctica (1990) Oceanographic Campaign 1987–88. Data Report I, pp 1–503

National Scientific Commission for Antarctica (1990) Oceanographic Campaign 1987–88. Data Report II, pp 1–277

National Scientific Commission for Antarctica (1991) Oceanographic Campaign 1989–90. Data Report I, pp 1–409

National Scientific Commission for Antarctica (1992) Oceanographic Campaign 1989–90. Data Report II, pp 1–507

Nelson DM, DeMaster DJ, Dunbar RB, Smith WO Jr (1996) Cycling of organic carbon and biogenic silica in the Southern Ocean: estimates of water-column and sedimentary fluxes on the Ross Sea Continental Shelf. J Geophys Res 101: 18519–18532

Patterson SL, Whitworth T (1990) Physical oceanography. In: Glasby GP (ed) Antarctic Sector of the Pacific. Elsevier, Amsterdam, pp 55–93

Saggiomo V, Carrada GC, Mangoni O, Ribera d'Alcala M, Russo A (1998) Spatial and temporal variability of size fractionated biomass and primary production in the Ross Sea (Antarctica) during austral spring and summer. J Mar Sys 17: 115–128

Smith WO Jr, Nelson DM (1985) Phytoplankton bloom produced by a receding ice edge in the Ross Sea: spatial coherence with the density field. Science 227: 163–166

Smith WO Jr, Nelson DM, Di Tullio GR, Leventer AR (1996) Temporal and spatial patterns in the Ross Sea: phytoplankton biomass, elemental composition, productivity and growth rates. J Geophys Res 101: 18455–18465

Wilson DL, Smith WO Jr, Nelson DM (1986) Phytoplankton bloom dynamics of the western Ross Sea ice edge-I. Primary productivity and species-specific production. Deep Sea Res 33: 1375–1387

CHAPTER 2

# Climatic Characterization of the Terra Nova Bay Region

**M. Colacino[1], E. Piervitali[1], and P. Grigioni[2]**

**ABSTRACT**

Italian research on Antarctic atmosphere started in a systematic way in 1986 when the Programma Nazionale di Ricerche in Antartide began its activity. Studies on the atmosphere have been developed in five different projects: planetary boundary layer physics; meteorology and climatology; aerosols radiation and clouds; minor components and greenhouse gases; physics and chemistry of the stratosphere. The monitoring activity has been developed considering two different aspects: (1) the need to have continuous and reliable measurements of the meteorological parameters in the area of the Italian base and (2) the possibility to carry out some specific studies concerning the physics, the dynamics and chemistry of the atmosphere. A climatological observatory has been arranged with a network of fixed automatic weather stations, that record meteorological data throughout the year. During the summer upper level soundings are also conducted. In the present chapter the climatological data, collected in about a decade, are analyzed to give a climatic picture of Terra Nova Bay. The parameters considered are radiation, pressure, temperature, humidity and wind. The obtained results have been compared, when possible, with the data collected at foreign stations, to verify their reliability. A short discussion about Antarctica and climate change closes the work.

## Introduction

The Programma Nazionale Ricerche Antartiche (PNRA hereafter) started its activity in 1986 when the Italian Parliament voted a law to promote the program and to insure the necessary financial support. A project on the physics and chemistry of the atmosphere was also included among the research fields fostered by the PNRA and work in this field already started in the first expedition.

The aim of the research was twofold: (1) to collect data and information useful to define the climate of the Terra Nova Bay region, where the Italian base is located, and (2) to carry out experiments, observational surveys and develop models to achieve a better knowledge of the complex atmospheric phenomenology and of the role of Antarctica in the global climate system.

Having established the main objectives and taking also into account the already existing international programs, different research topics were identified and diversified into the following five subprojects:

- Physics of the planetary boundary layer.
- Meteorology and climatology.
- Radiation and clouds.
- Greenhouse gases and minor components.
- Physics and chemistry of the stratosphere.

The data collection was organized by installing a network of automatic weather stations (AWS) which worked all year-around and measured surface classical meteorological parameters: solar radiation, pressure, temperature, humidity and wind (dd, vv). These data, that were inserted in the GTS and have contributed to the synoptic observations system, were integrated during the summer period when the base was open, either by upper-level soundings or by specific measurements taken yearly. In this way a data set has been achieved, which, even if relative to a short period (~10 years), allows for a first description of climatology of Terra Nova Bay.

[1] Istituto di Fisica dell'Atmosfera (CNR), Area di Ricerca di Tor Vergata, via del Fosso del Cavaliere, 100-00133 Rome, Italy
[2] ENEA Progetto Antartide C.R. Casaccia, S.P. Anguillarese, 301-00060 Santa Maria di Galeria Rome, Italy

In the present chapter, after a summary of the network of Terra Nova Bay stations, the patterns for main meteorological parameters are illustrated. The analysis regards solar radiation, pressure, temperature, humidity and wind, and gives a first overview of the climatic conditions that are experienced by the people working in and around the Italian base. The results are compared, when possible, with data collected at the nearby stations of McMurdo and Scott, to check for the reliability of the obtained results. Finally, a short discussion about trends in air temperature, which seem to have increased in recent years, closes the chapter.

## Description of the Terra Nova Bay Network

Terra Nova Bay is a large inlet lying in front of the Ross Sea in the region known as Victoria Land in the eastern Antarctica. Its length, between Cape Washington (74°39′5) and the tongue of the Drygalski glacier (75°15′5), is around 80 km, while the other boundaries are represented to the east by the meridian 165°30′E and to the west by the coastline. The main feature of the region is the orography: on the northern side Mount Melbourne is dominant, while to the east the Transantarctic Range is present, which represents the boundary of the polar ice cap and is crossed by outlet glaciers reaching the sea (glaciers David, Priestley, Reeves, Campbell). Orographic factors have a noticeable impact on the climate of the region since they drive the atmospheric flows, giving rise to the dynamic systems prevailing in the area such as the barrier and katabatic winds. Figure 1 shows a map of the region and the location of the different AWS, the data of which have been used in this work. Table 1 summarizes for the different stations the geographical coordinates and the measured parameters. The periods of availability are not always the same for all of the observations, ranging from a minimum of 2 years to a maximum of 8 years.

## Data Analysis

According to the rules of the World Meteorological Organisation (WMO), the climatology of a given place can be defined only where a data set lasts for about 30 years. Since the files we have collected are relative to shorter periods, the obtained results must be considered with great caution to avoid misleading conclusions. When possible, the obtained patterns have been checked with those recorded at other sites near Terra Nova (McMurdo and Scott bases), where the data collections cover a longer period of time. The parameters considered are those usually examined in climatic studies: solar radiation, atmospheric pressure, temperature, humidity, precipitation and wind.

### Solar Radiation

Figure 2 shows the pattern of solar radiation in $W\,m^{-2}$, recorded at station 7351 located near Cape King. The radiation appears in August when a value of $10\,W\,m^{-2}$ was measured and increases in time reaching maximum values in December with $340\,W\,m^{-2}$. The radiation then decreases and is absent during May, June and July, which correspond to the polar night. Maximum values were strongly different from the mean with variations from 25 to 50%. The amount of solar radiation reaching the ground was variable from one year to the next, ranging from a minimum of $90\,W\,m^{-2}$ in 1989 to a maximum of $130\,W\,m^{-2}$, the mean value of around $110\,W\,m^{-2}$.

**Table 1.** List of automatic weather stations

| ARGOS ID (Name) | Geographical location | Latitude | Longitude | Height (m) | Sensors[a] |
|---|---|---|---|---|---|
| 7350 (Sofia) | Nansen Ice Sheet | 74°48′S | 163°19′E | 40 | P T RH W |
| 7351 (Alessandra) | Cape King | 73°35′S | 166°37′E | 160 | P T RH W RAD |
| 7352 (Zoraida) | Priestley Glacier | 74°15′S | 163°10′E | 640 | P T RH W |
| 7353 (Eneide) | Terra Nova Bay | 74°42′S | 164°06′E | 90 | P T RH W RAD |
| 7354 (Aeroporto) | Enigma Lake | 74°43′S | 164°02′E | 210 | P T RH W |
| 7355 (Modesta) | Priestley Source | 73°38′S | 160°39′E | 1900 | P T RH W |
| 7356 (Lola) | Tourmaline Plateau | 74°08′S | 163°26′E | 1700 | P T RH W |
| 7357 (Arelis) | Cape Ross | 76°43′S | 162°58′E | 150 | P T RH W |
| 7359 (Silvia) | Cape Philips | 73°03′S | 169°36′E | 550 | P T RH W |

[a] P, Atmospheric pressure; T, air temperature; RH, relative humidity; W, wind; Rad, solar radiation.

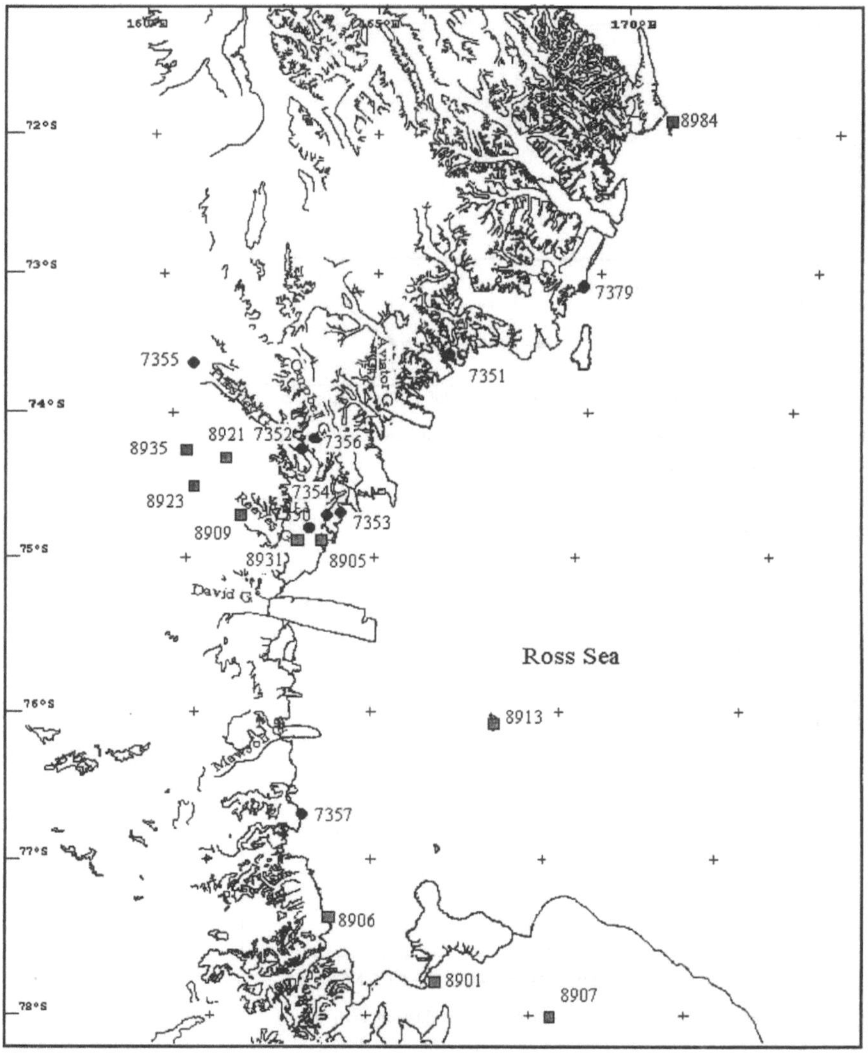


■ Italian stations
● USA stations

**Fig. 1.** Map of the Terra Nova Bay region and location of the AWS

## Atmospheric Pressure

Figure 3 illustrates the data for atmospheric pressure, recorded at the same station in the years from 1987 to 1995, together with those collected at the McMurdo and Scott bases, reduced to msl (Jones and Limbert 1989) to obtain a more complete description of the pressure field. The comparison was not always homogenous, since the data at McMurdo are relative to the period 1956–1986 and

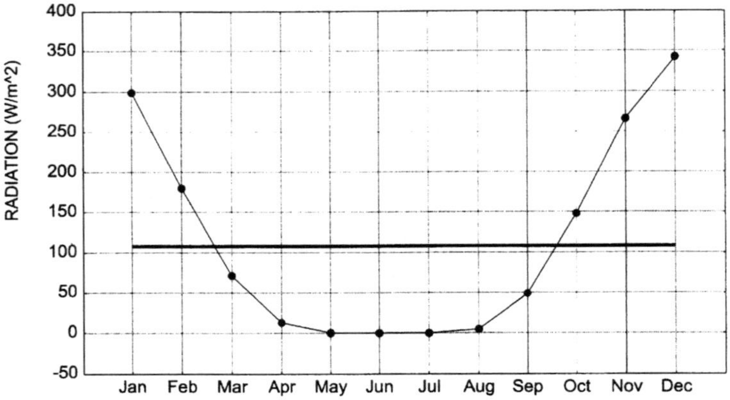


**Fig. 2.** Mean monthly values of solar radiation recorded at Cape King station

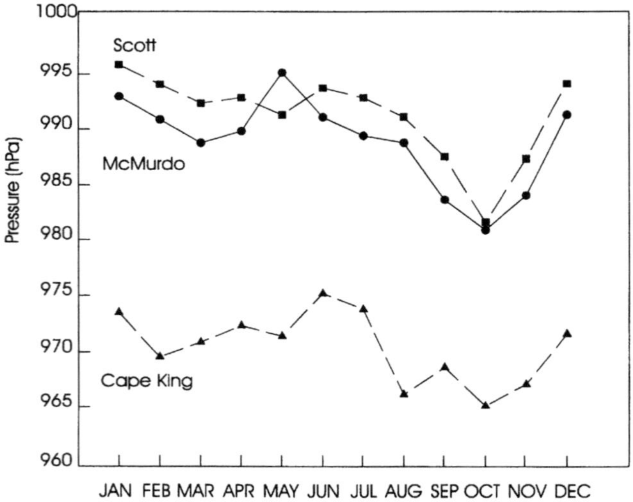


**Fig. 3.** Mean monthly values of atmospheric pressure at different stations

those at Scott to the period 1957–1984. Mean yearly values at the Cape King station were around 971 hPa with a very small oscillation between the minimum 967.9 hPa recorded in 1984 and maximum 973 hPa recorded in 1991. The monthly trend showed strong variations only during the transition from the cold to warm season when a decrease in pressure was recorded.

In general, pressure values at McMurdo and Scott (mean yearly values around 988 and 990 hPa, respectively) were higher than at the Italian station. The difference was probably due to the different altitude of the barometers used. In the surface layer, atmospheric pressure decreases with height at a rate of about 1 hPa/8 m. Since the difference in altitude between the Cape King and McMurdo and Scott stations is about 140 m, the variation in pressure was about 18 hPa, which more or less corresponds to the difference between the analyzed data sets. Very interesting results were also obtained by monthly values. Also in McMurdo and Scott, a sharp jump was observed during the transition from winter to austral summer, which was probably related to variations in the dynamics of the polar vortex that occurs during the change in seasons.

## Air Temperature

Figure 4 presents the yearly mean values for air temperature recorded at different stations surrounding the Italian base. At some stations, records date back to 1987, whereas at other stations the files are shorter. Therefore, data analyses have been carried out only for those having a longer temporal series.

The yearly values seem to be more or less the same from one year to another. Higher temperatures at all stations occurred only in 1988 when an increase from a minimum of 1 °C (station 7353) to a maximum of 3.5 °C (station 7352) was recorded. Another easily foreseeable feature was the reduction in temperature with height. Table 2 reports the number of stations, height above sea level, and yearly mean and extreme values of temperature.

Air temperature in the lower layer (0–200 m) did not follow the trend in temperature with height given by the law

$$T = T_o - \gamma z,$$

but it presented a random distribution probably related to other forcing factors such as distance from the sea, prevailing wind regime and so on. Evaluation of the $\gamma$ parameter was therefore calculated with reference to the higher altitude stations. The observed value was around 10–11 °C/1000, which indicates that the atmosphere was usually slightly unstable.

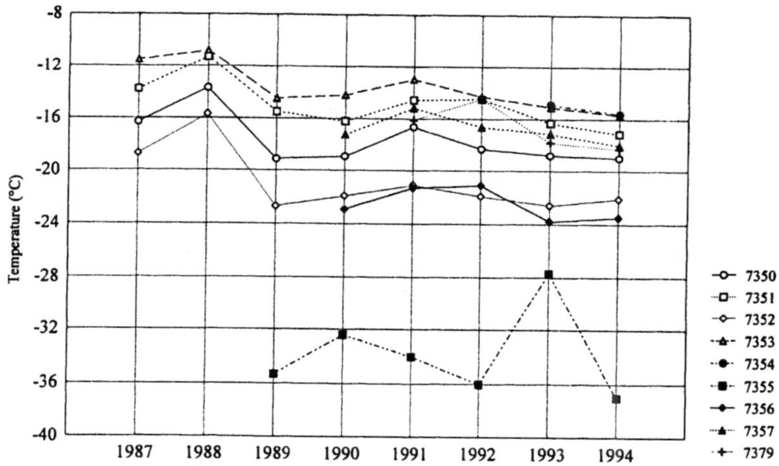


**Fig. 4.** Mean yearly values of the air temperature at the different AWS

**Table 2.** Mean yearly temperature and height above sea level

| Station | Height (m) | T (°C) | $T_{min}$ (°C) | $T_{max}$ (°C) | ΔT (°C) $T_{max} - T_{min}$ |
|---|---|---|---|---|---|
| 7350 | 40 | −17.6 | −26.5 | −4.5 | 22.0 |
| 7351 | 160 | −14.9 | −25.0 | −3.0 | 22.0 |
| 7352 | 640 | −20.9 | −29.5 | −8.0 | 21.5 |
| 7353 | 90 | −13.6 | −23.5 | −2.0 | 21.5 |
| 7354 | 210 | −15.1 | −24.5 | −4.0 | 20.5 |
| 7355 | 1900 | −33.8 | −42.5 | −20.5 | 22.0 |
| 7356 | 1700 | −28.1 | −29.0 | −14.0 | 15.0 |
| 7357 | 150 | −16.7 | −27.0 | −4.0 | 23.0 |
| 7379 | 550 | −16.6 | −27.0 | −5.5 | 21.5 |

The mean monthly pattern for the different stations shown in Fig. 5 indicates that minimum values were reached in May while the maximum occurred in December. The annual wave of temperature was not symmetrical. The cold season spanned from March to September, while the warm period was reduced to only December and January. The warming period lasted longer than the period of cooling. The first started in September and ended in December while the second was sharp and occurred in only 2 months from January to March. During the cold period, temperature oscillations were reduced. The same also occurred during the warm months when temperature did not change at all.

The yearly excursion in air temperature increased from the coastline towards the continent. Maximum values were measured at station 7355, lying on the rim of the Priestley glacier, where the value recorded was 22 °C.

Analysis of the seasonal trend indicates that winters have features typical of Antarctic climates (Schwerdtfeger 1970, 1984) such as the absence of a well neat minimum (coreless winter). On the synoptic scale, this may depend on an increase in warm air addiction due to strong cyclonic activity and, on the local scale, to the breaking of the inversion by strong winds. The asymmetry between warming and cooling may be due to the radiative balance at the air-ground interface. The high albedo of the ice reflects incoming solar radiation and the longwave radiation emitted by the surface cools the first layers of the atmosphere producing strong inversions. In autumn, these push the atmosphere to rapidly assume the temperature values typical of winter, while in spring they slow down the process of air warming.

Using the data also obtained by the US meteorological stations, we plotted maps of air-surface temperature in the region around the Italian base in cold and warm seasons. Figure 6a,b shows that the patterns obtained were similar in both seasons, with the obvious exception of the different temperatures. The main feature was the strong decrease in temperature from the coast inland where values ranged from 0 to −20 °C during the austral summer and from −20 to −42 °C in winter. The warmest area was that of Cape Philips, while the coldest was at the top of the mountains above the Reeves and Priestley glaciers. The yearly thermal excursion, for the Italian stations only, was around 20–22 °C, increasing inland.

## Humidity

Values for relative humidity (RH) were usually around 50% throughout the year, with small differences among the various AWS. The winter–summer excursion was low, in the range

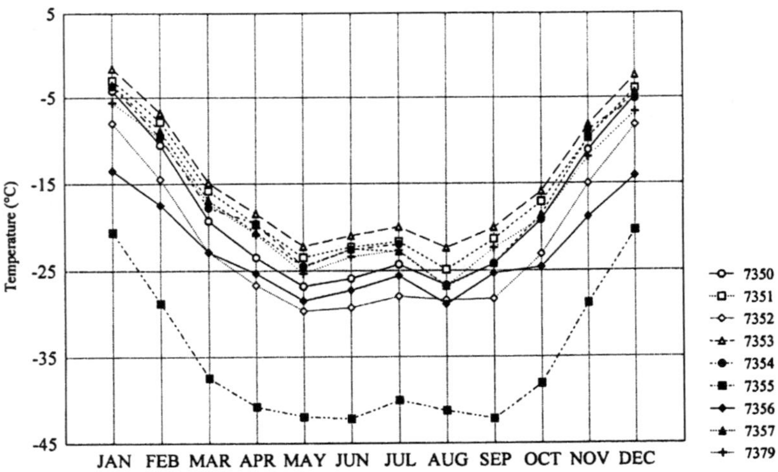


**Fig. 5.** Mean monthly trend of the air temperature at the different AWS

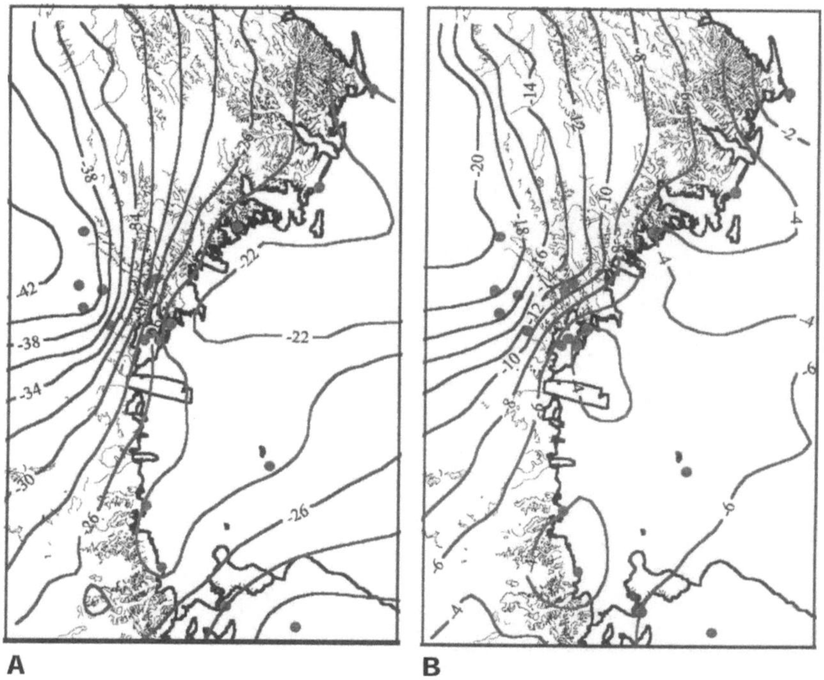


**Fig. 6a,b.** Map of the air surface temperature. **a** Cold season. **b** Warm season

5–10%. The station with the highest values of RH was located near the coast at Cape Philips, while lower values were recorded in Terra Nova Bay and surrounding areas. Depending on the air temperature, the water content in the atmosphere was not the same for a given value of RH. For example, an RH of 50% with a temperature of −10 °C corresponded to a water content of $0.88\,g\,kg^{-1}$, while the same RH with a temperature of −20 °C gave a specific humidity of $0.32\,g\,kg^{-1}$. In any case, values for water vapour content in the atmosphere were very low making Antarctica an ideal site for astronomical observations. Figure 7a,b illustrates the maps of seasonal patterns, obtained using also the data collected at the US stations. Values of RH presented a small variation and higher values were found during the warm season. This corresponded to a higher content of water vapour in the atmosphere, in agreement with the more intense evaporation that occurs during the austral summer due to the melting of the ice shield.

## Winds

The collection of the anemological data indicate that the Terra Nova Bay region has three wind categories:

- katabatic winds,
- barrier winds,
- eastern flows.

The katabatic winds originate from gravity currents generated by strong cooling occurring at the surface of the continent. They originate over the plateau and blow down toward the coast following the valleys which intersect the trans-Antarctic range. In Terra Nova Bay, these winds are particularly intense (Parish and Bromwich 1991) representing a forcing factor for the formation and maintenance of the polynya (Bromwich and Kurtz 1982, 1984). Such winds usually blow toward the Ross Sea either from SW, if they come from the Priestley glacier, or from NW if from Reeves Glacier. The wind speed is around $15–20\,m\,s^{-1}$ also

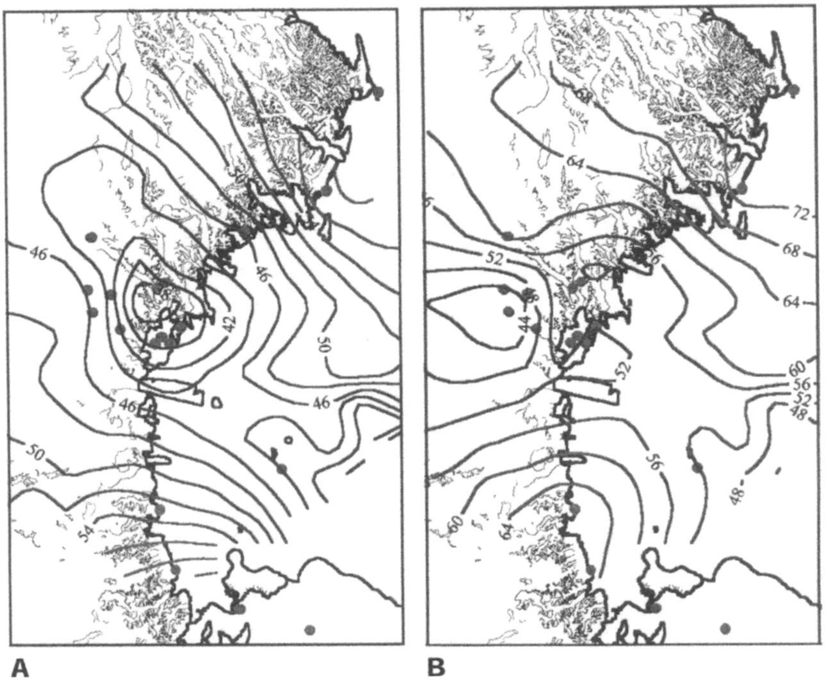


**Fig. 7a,b.** Map of the relative humidity. **a** Cold season. **b** Warm season

**Table 3.** Resultant wind and constancy for each AWS

| AWS | Resultant wind | | Constancy |
|---|---|---|---|
| | Speed (kts) | Direction (°) | |
| 7350 | 8.8 | 284 | 0.65 |
| 7351 | 2.1 | 268 | 0.51 |
| 7352 | 26.3 | 329 | 0.94 |
| 7353 | 8.7 | 273 | 0.74 |
| 7354 | 13.5 | 273 | 0.83 |
| 7355 | 9.9 | 305 | 0.83 |
| 7356 | 6.6 | 226 | 0.65 |
| 7357 | 8.6 | 203 | 0.72 |
| 7379 | 13.0 | 234 | 0.81 |

in summer, with strong gusts that bring clear skies and lee waves as confirmed by the presence of orographical clouds. In summer, their frequency is around 30 and 50% in winter. Stations lying along the line plateau-Priestley-Nansen are subject to katabatic winds. This is supported by the data relative to wind constancy that are reported in Table 3 for the different AWS.

Stations over the plateau and Priestley glacier show a higher constancy in wind behaviour, because of the stability in wind direction. Over the Nansen, a greater variability was found since the station was exposed to the winds coming from Priestley and Reeves glaciers. Whereas the Priestley station records winds at a speed of about 27 knots coming more or less always from NW-NNW (Fig. 8), corresponding to the valley axis, the Nansen station shows about 30% of the winds coming from Reeves (wind speed greater than 41 knots) and 18% coming from Priestley (wind speed about 27 knots) (Fig. 9). This was the pattern of katabatic winds at the surface. A detailed analysis of the wind behaviour during the austral summer in the boundary layer was carried out in 1988–1989 by using a triaxial Sodar (Argentini et al. 1995) that was installed during the 1988 survey in the Italian base at a height of 80 m (asl) and 500 m inland from the coast. In the following year, the same equipment was deployed on the Nansen Ice Sheet at the confluence of the Priestley and Reeves glaciers, 15 km inland. The surrounding

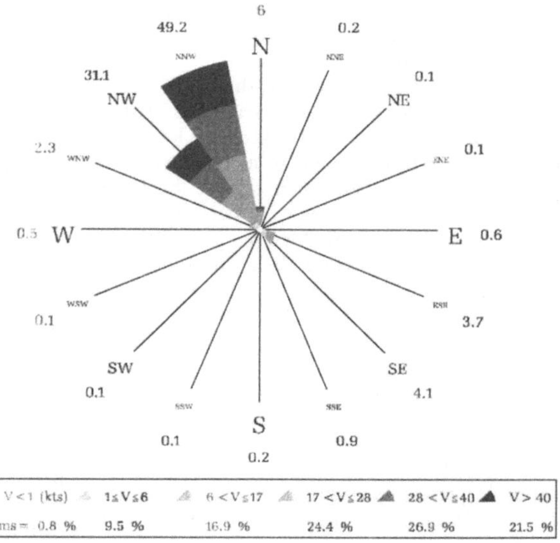


**Fig. 8.** Wind rose at station 7352 (Priestley glacier)

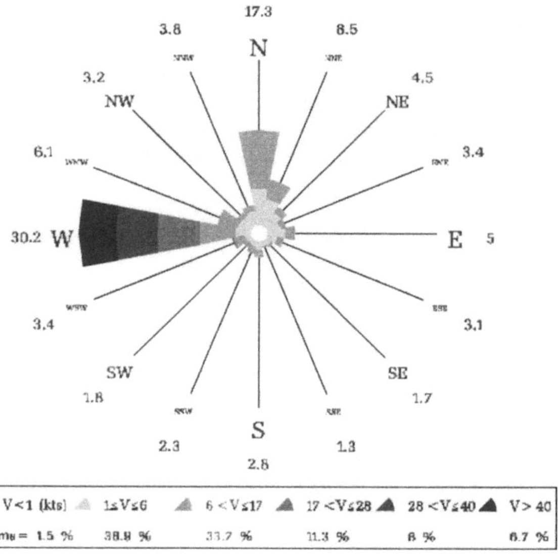


**Fig. 9.** Wind rose at station 7350 (Nansen glacier)

area was almost flat for a distance of about 10 km. The results obtained confirm the strong dependence of surface layer winds on the orography and indicate that a noticeable variation in wind intensity and direction may also occur in upper levels. In particular, the local circulation connected to the katabatic winds from Reeves or Priestley glaciers presents some peculiarities during the warm season. The wind coming from the Priestley glacier was deviated by northern foothills and flowed into the Nansen Ice Sheet partially losing its intensity. In addition, the northern foothills prevented barrier winds from reaching the coastal site where the Italian base is located, where low wind velocities are usually recorded.

The second type of wind is the barrier wind, that is present along a mountain range as a consequence of the advection of cold air at low levels perpendicular to the range, in a stratified stable atmosphere. In the Ross Sea region, the trans-Antarctic mountains create a barrier against which the cyclone of the Ross Sea pushes the air travelling over the Ross Ice Shelf. The wind, deviated northward, sometimes reaches Terra Nova Bay with low currents from the south. The humidity is high, and the sky coverage and precipitation are variable but frequent. During the summer, barrier winds represent 13% of the wind events. These winds and the atmospheric circulation that drives them were recorded at stations 7354, 7367, 7356, 7379 (Fig. 10) that lie along the coastline. The synoptic pattern was characterized by frequent cyclones going from south to north, giving rise to the barrier winds.

Finally, variable winds were also connected to local effects which were driven by katabatic flows. At times, such winds also reached the sea. Other wind systems were due to particular synoptic situations or to thermally induced circulation. During the summer, a high number of wind calm events were recorded due either to the shelter of northern foothills or to the anticyclone that often occurred during the season in this area (Grigioni 1996; G. Frustaci and A. Pellegrini, pers. comm.).

## Antarctica and Climate Change

Recent studies concerning the evolution in climate indicate that over the past century the air temperature at the global scale has increased at a rate of 0.5–0.6 °C 100 year$^{-1}$. According to some climatologists, this increase is a consequence of the enhancement of the greenhouse effect due to the anthropogenic activity. Since Antarctica is a region very sensitive to climate change, it is interesting to analyze the trend in air temperature using the data series that are continuous and reliable since 1958. In the area delimited by the 60°S parallel that includes the Antarctic continent, the trend in temperature anomalies is increasing from −0.3 to 0.9 °C at a rate of about 4 °C 100 year$^{-1}$, in the period 1958–1992 (Fig. 11). In particular, it is interesting to note that the greatest increase has been recorded during the cold season, as shown in Fig. 12, where summer and winter anomalies are reported. The analyzed temperature data in this area are reported in Angell (1994).

The increase is not the same throughout the continent. In the Terra Nova Bay region in the period 1957–1990, the mean temperature has risen by about 1.5 °C, or at a rate of 5 °C 100 year$^{-1}$ at the

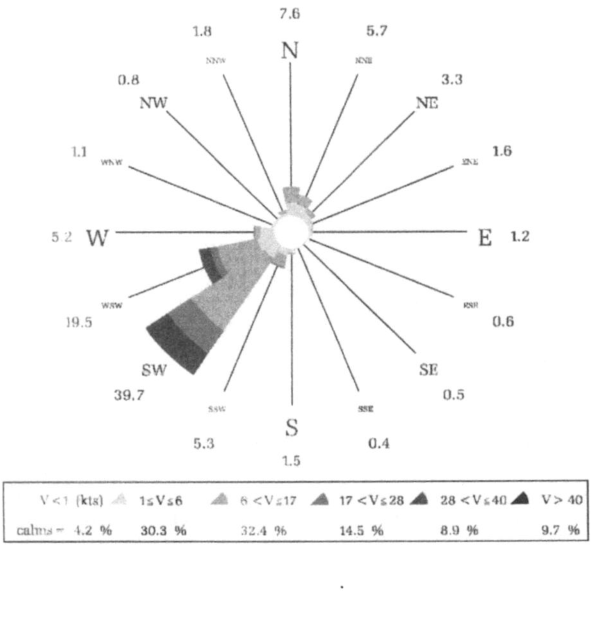


**Fig. 10.** Wind rose at station 7379 (Cape Philips)

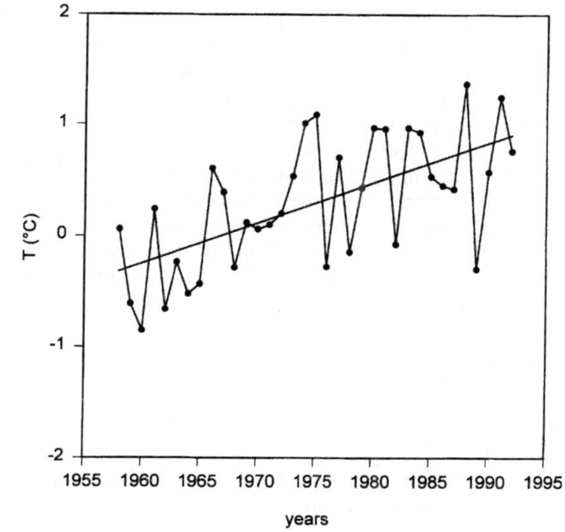


**Fig. 11.** South Polar (60°–90°S) annual temperature anomalies in the period 1958–1992

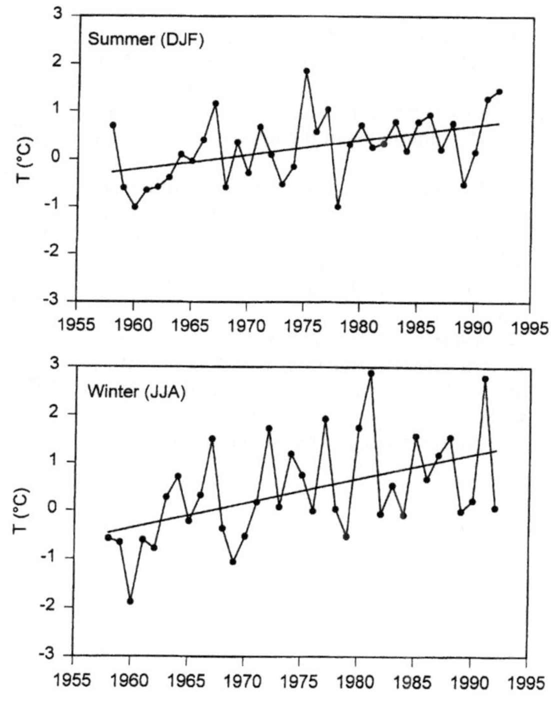


**Fig. 12.** South Polar (60°–90°S) seasonal temperature anomalies (summer *top*; winter *bottom*) in the period 1958–1992

McMurdo station. Braaten and Dreschhoff (1992) analyze the trend of both the annual maximum and minimum temperature at this station and find that the pattern is strongly different between the two data series. Minimum temperature trend is null (0.01 °C $year^{-1}$ in the examined period), while the maximum one is about 1.51 °C $year^{-1}$. The greatest increase has been recorded over the Antarctic Peninsula where, according to Stark (1994), in the period 1947–1990, the temperature has increased by about 2.7 °C at a rate of 6.1 °C 100 $year^{-1}$.

## Concluding Remarks

From the above results it is possible to draw the following conclusions:

1. The climate of Terra Nova Bay is typical of the coastal zones of Antarctica.
2. The climate is strongly influenced by the presence of the sea and of the mountains that surround the region.
3. The sea influences, above all, the air temperature. The yearly thermal excursion is, in fact, more reduced than that found in other areas and this is valid also for the stations located over the mountains in the Reeves and Priestley glaciers.
4. On the other hand, the orography influences instead the atmospheric circulation. The katabatic winds are generated by air masses flowing from the glaciers towards the sea, while the barrier winds are driven by the trans-Antarctic range.
5. The Antarctic is sensitive to climatic variations as indicated by the increase in temperature recorded over the continent. This increase seems to be in agreement with the projections of *scenario* models. This suggests the opportunity to intensify climate observations and meteorological measurements over Antarctica, as suggested also by international agencies and programs.

## References

Angell JK (1994) Global, hemispheric and zonal temperature anomalies derived from radiosonde records. In: Boden TA, Kaiser DP, Sepanski RJ, Stoss FW (eds) Trends '93: a compendium of data of global change. Environmental Science Division, Oak Ridge National Laboratory, Publ no ORNL/CDIAC-65, ESD publ no 4195, pp 636–672

Argentini S, Del Buono P, Della Vedova AM, Mastrantonio G (1995) A statistical analysis of wind in Terra Nova Bay for austral summers 1988 and 1989. Atmosph Res 39: 145–156

Braaten DA, Dreschhoff GAM (1992) Maximum and minimum temperature trends at McMurdo Sound Station. Antarct J US XXVII (5): 282–283

Bromwich DH, Kurtz DD (1982) Experiences of Scott's northern party: evidence for a relationship between winter katabatic wind and the Terra Nova Bay polynya. Polar Rec 21: 137–146

Bromwich DH, Kurtz DD (1984) Katabatic wind forcing of the Terra Nova Bay polynya. J Geophys Res 89: 3561–3572

Grigioni P (1996) The Italian Antarctic meteorological observatory: technical activity and some climatological results. In: Meloni A, Morelli A (eds) Italian geophysical observatories in Antarctica. Editrice Compositori, Bologna, pp 135–149

Jones PD, Limbert DWS (1989) Antarctic surface temperature and pressure data. ORNL/CDIAC-27 NDP-032 Environmental Science Div Publ no 3215, pp 1–31 (Appendix, pp A1–A60)

Parish TR, Bromwich DH (1991) Continental-scale simulation of the Antarctic katabatic wind regime. J Climate 4: 135–146

Schwerdtfeger W (1970) The climate of the Antarctica. In: Landsberg HE (ed) Climate of the polar regions. World survey of climatology, vol 4. Elsevier, Amsterdam, pp 253–355

Schwerdtfeger W (1984) Weather and climate of the Antarctic. Elsevier, Amsterdam, 261 pp

Stark P (1994) Climatic warming in the central Antarctic Peninsula area. Weather 49 (6): 215–220

CHAPTER 3

# Oxygen Isotope Characterization of Terra Nova Bay Seawater

M. Dini and B. Stenni[1]

**ABSTRACT**

During the 1987–88 and 1989–90 Italian expeditions to Antarctica, seawater samples were collected in Terra Nova Bay (Western Ross Sea) in order to distinguish different water masses. The aim of this work was to identify, by means of stable isotope measurements, the presence of the High Salinity Shelf Water (HSSW), which seems to form in the polynya present in this region.

The presence of two main water types was confirmed by the results obtained: Antarctic Surface Water (AASW) and High Salinity Shelf Water (HSSW); a new water type was identified in the hydrological station close to the edge of the Hell's Gate Ice Shelf. On the basis of its isotopic and physical characteristics it is ascribed to Ice Shelf Water (ISW). It is worth pointing out the presence of ISW even beneath a small ice shelf such as Hell's Gate because it implies a Deep Thermohaline Circulation.

The most probable origin and evolution of the water masses is discussed by comparing $\delta^{18}O$ values with salinity data on $\delta^{18}O$/S diagrams. The mean $\delta^{18}O$ values ranged from –0.33 to –0.23‰ for AASW and were close to –0.36‰ for HSSW and –0.45‰ for ISW.

## Introduction

The global mean $\delta^{18}O$ value of seawater is fairly steady around –0.08‰ V-SMOW (Craig 1965). The most significant variations of $\delta^{18}O$ in seawater reflect primarily the water mass balance between evaporation and dilution processes and mixing processes between different water masses. The evaporation and exchange at the sea surface represent the most important part of the isotopic cycle. The vapour forming at the sea surface is depleted in $^{18}O$ and consequently leaves the seawater enriched in $^{18}O$ and salinity. On the contrary, dilution processes such as rainfall and continental input and, at high latitudes, glacial meltwater, will deplete the isotopic content and salinity at the sea surface. This is due to the fact that meteoric waters are characterized by negative values of $\delta^{18}O$. These values are highly variable, with the most negative at higher latitudes and the least negative in equatorial regions (Dansgaard 1964). At high latitudes, minor isotopic fractionation may occur in continental shelf areas where freezing of surface seawater takes place seasonally. The freezing processes result in a marked increase in the salinity of the residual water and only a slight decrease of its $\delta^{18}O$ due to the low fractionation factor between ice and water (O'Neil 1968). As a consequence of all the processes mentioned above, a linear relationship has been observed between oxygen (or hydrogen) isotope composition and salinity (Craig and Gordon 1965). By plotting the $\delta^{18}O$ and salinity values on a bidimensional diagram, with a method similar to that used for a T/S diagram, it is possible to outline mixing processes among water masses with different characteristics.

One of the advantages of using the isotope method at high latitudes is related to the high $^{18}O$ depletion of glacial meltwater relative to seawater: the $\delta^{18}O$ is thus a useful variable to trace glacial melting and related mixing with seawater in continental shelf areas.

We report here a set of isotopic measurements carried out on samples collected during the 1987–88 and 1989–90 Italian expeditions to Antarctica. These measurements were carried out in order to characterize the $\delta^{18}O$ of the different

[1] Department of Geological, Environmental and Marine Sciences, University of Trieste, Via E. Weiss 2, 34127 Trieste, Italy

water masses and to obtain information about their origin and evolution in the area of Terra Nova Bay, particularly by comparing the isotopic data with salinity data. As already pointed out by several authors (Craig and Gordon 1965; Redfield and Friedman 1969; Jacobs et al. 1985), in this way it is possible to recognize the origin and the evolution of each water mass.

Terra Nova Bay is a continental shelf section of the Western Ross Sea, extending from the Drygalski Ice Tongue to Cape Washington along the Northern Victoria Land coast. Even though the Weddel Sea is commonly assumed to be the primary source of Antarctic Bottom Water (AABW), the Ross Sea has been recognized as another important source area (Jacobs et al. 1970). In particular, High Salinity Shelf Water (HSSW) forms in Terra Nova Bay by sea surface freezing in an extended polynya located at about 74°50′S, 164°30′E and, flowing down the continental slope, contributes to AABW formation.

## Materials and Methods

The locations of the sampling sites studied are reported in Figs. 1 and 2, while the geographical positions of the stations are reported in Table 1.

The oxygen isotopic composition of the water samples was double-measured according to the $CO_2$-water equilibration technique (Epstein and Mayeda 1953). The analyses were conducted by means of a Finnigan Delta S mass spectrometer and the values obtained are reported in the δ terminology versus V-SMOW isotopic standard (Gonfiantini 1978) after normalization with GISP and SLAP IAEA standards. The average standard deviation of our measurements is of the order of ±0.05‰ (1σ).

## Results and Discussion

The physical and chemical characteristics of Ross Sea water have already been reported and discussed by Jacobs et al. (1970, 1985); Stocchino and Lusetti (1990) and Stocchino and Manzella (1991) reported and discussed these parameters in detail in the case of Terra Nova Bay water.

The surface circulation in the Ross Sea consists of a cyclonic gyre (Klepikov and Grigor'yev 1966) originating a northward flow along the Victoria Land coast. The deep circulation is controlled by a relatively warm Circumpolar Deep Water (CDW) which flows southward on the continental shelf interacting with the Ross Ice Shelf

**Table 1.** Geographical coordinates of the 1987–88 and 1989–90 sampling sites

| Station (east) | Date | Bottom (m) | Latitude (south) | Longitude |
|---|---|---|---|---|
| 87-3 | 25/01/88 | 160 | 74°38.90 | 165°36.70 |
| 87-9 | 06/01/88 | 370 | 74°43.10 | 164°13.80 |
| 87-21 | 12/01/88 | 570 | 74°58.10 | 164°05.60 |
| 87-22 | 12/01/88 | 794 | 74°58.20 | 164°13.70 |
| 87-25 | 26/01/88 | 1100 | 74°57.20 | 166°17.20 |
| 87-32 | 02/02/88 | 885 | 75°07.00 | 166°08.40 |
| 87-36 | 15/01/88 | 812 | 75°18.00 | 164°33.00 |
| 89-2 | 15/01/90 | 540 | 74°41.88 | 164°24.24 |
| 89-3 | 26/01/90 | 180 | 74°39.06 | 164°14.66 |
| 89-4 | 26/01/90 | 233 | 74°41.34 | 164°04.36 |
| 89-5 | 13/01/90 | 343 | 74°40.61 | 164°08.59 |
| 89-6 | 23/01/90 | 75 | 74°46.35 | 163°57.70 |
| 89-7 | 23/01/90 | 185 | 74°46.59 | 164°03.77 |
| 89-8 | 28/01/90 | 160 | 74°42.70 | 164°08.84 |
| 89-9 | 20/01/90 | 112 | 74°50.51 | 164°01.45 |
| 89-10 | 18/01/90 | 750 | 74°52.56 | 163°48.36 |
| 89-11 | 18/01/90 | 250 | 74°53.26 | 163°57.60 |
| 89-12 | 29/01/90 | 100 | 74°45.26 | 164°06.26 |
| 89-14 | 29/01/90 | 200 | 74°49.04 | 164°07.34 |
| 89-16 | 31/01/90 | 232 | 74°38.21 | 164°04.75 |
| 89-17 | 02/02/90 | 516 | 74°40.13 | 164°11.75 |
| 89-18 | 02/02/90 | 183 | 74°39.72 | 164°24.00 |

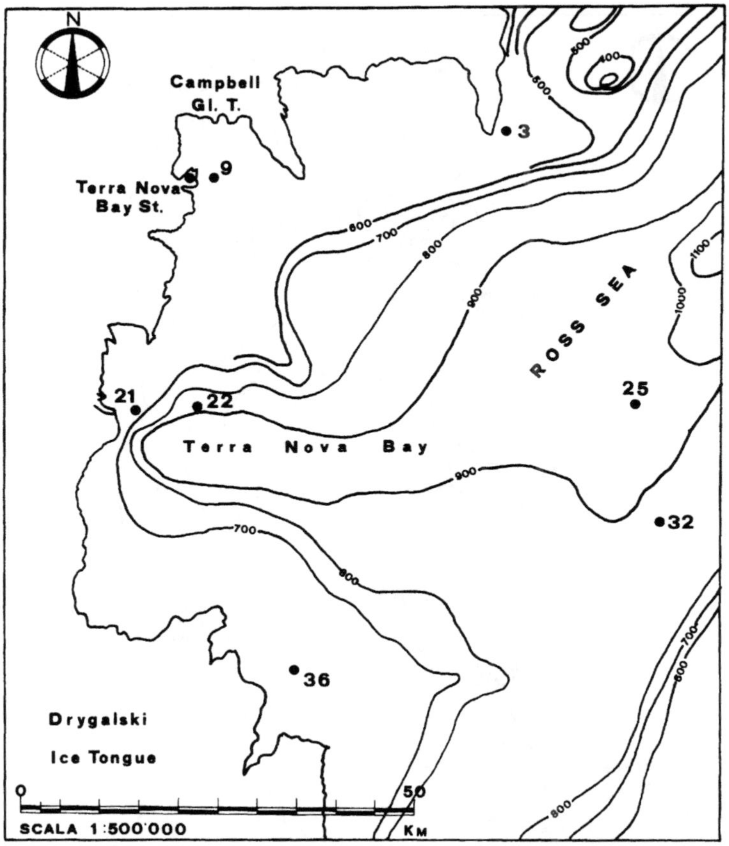


Fig. 1. Location of the 1987–88 hydrological stations

barrier. This movement helps in confining the HSSW, formed in the western Ross Sea in the depressions of the shelf and, particularly, in those of Terra Nova Bay. Stocchino and Lusetti (1990) pointed out the existence, in the northernmost coastal region of Terra Nova Bay, of a local anticyclonic gyre originating from a current entering from the east and flowing toward Campbell Ice Tongue and Gerlache Inlet, and of a small cyclonic gyre seaward of Cape Russell.

The most important feature of Terra Nova Bay is the persistent presence of polynyas during winter. This anomaly is probably due to the combined actions of katabatic winds which shift the forming sea ice eastward, and the Drygalski Ice Tongue, which prevents the northward drifting of pack ice (Bromwich and Kurtz 1984). The presence of open water areas is very important because it increases heat and moisture flows into the atmosphere and enhances higher biological activity (Jacobs and Comiso 1989). This process could also be responsible for the upwelling of denser water whose presence has been detected during summer especially in the coastal area (Stocchino and Manzella 1991). The surface layer (about 50 m thick) is ascribed to Antarctic Surface Water (AASW), which may have two origins: an evolution from modified CDW or the mixing of shelf water and sea-ice meltwater (Jacobs et al. 1985). The physical and chemical characteristics of AASW are highly variable owing to the different processes they undergo

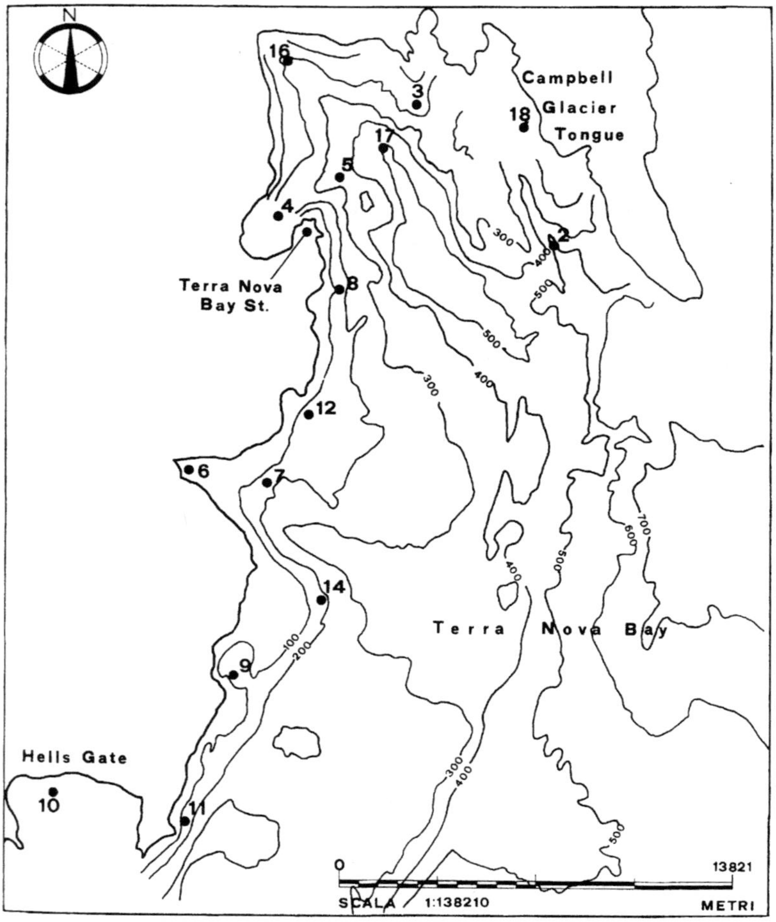


Fig. 2. Locations of the 1989–90 hydrological stations

on the continental shelf: principally surface freezing and interactions with meltwater of continental origin and sea ice. This water mass generally has relatively high temperature values (around 0 °C with a maximum of 2 °C) (Boldrin and Stocchino 1990) and is delimited downward by a temperature minimum (Stocchino and Manzella 1991), which seems to be of hydrodynamic importance since it reduces the vertical movements of the water column during summer. Below the surface layer, High Salinity Shelf Water, characterized by a higher salinity and lower temperature, occupies the remaining depth of the water column. HSSW is the densest water type present in the Antarctic Ocean (Jacobs et al. 1985) and is formed by sea-surface freezing, as mentioned above, along the Victoria Land coast. HSSW is present on the continental shelf of Terra Nova Bay throughout the year. During winter it forms in the polynya after sea-ice freezing; during summer its presence is preserved for three different reasons: (1) its high density (Jacobs et al. 1970, 1985), (2) the southward movement of CDW that acts as barrier for HSSW outflow (Stocchino and Lusetti 1990) and (3) upwelling processes forced by katabatic winds that explain the presence of HSSW at relatively shallow depths. The upwelling process is evidenced by relatively low temperature values and high nutrient concentrations (typical of HSSW) recorded at shallower depths by Boldrin and Stocchino (1990) Catalano and Benedetti (1990) and Catalano et al. (1991) in Terra Nova Bay.

The $\delta^{18}O$ values obtained are reported in Table 2. Table 3 shows $\delta^{18}O$ as the mean values calculated for each identified water mass along with the mean temperature and salinity values reported by Boldrin and Stocchino (1990) and by Fabiano et al. (1991) for the 1987–88 and 1989–90 expeditions, respectively. Temperature and salinity data were collected along with water samples from which the oxygen isotope measurements were performed. The identification of the different water masses and their depth range was made on the basis of temperature and salinity values. In order to characterize the isotopic composition of the different water masses, their mean $\delta^{18}O$ values were calculated for each identified depth interval and are reported in Table 3.

By plotting on a $\delta^{18}O$/S diagram (Fig. 3) the overall mean $\delta^{18}O$ and salinity values of the different water bodies detected in the area studied (AASW and HSSW), it can be seen that they occupy well-defined positions.

Considerable differences were apparent between the $\delta^{18}O$ and the salinity values obtained during the 1987–88 and 1989–90 oceanographic cruises. These differences were particularly marked in the case of AASW, since higher salinity and temperature values and lower $\delta^{18}O$ values were recorded during the first cruise, even though lower salinity values were measured at stations 87-25 and 87-32, probably affected by a slightly larger addition of sea-ice meltwater. In fact, during the

**Table 2.** $\delta^{18}O$ values (vs. V-SMOW) reported for each station

| Depth (m) | 87-3 | 87-9 | 87-21 | 87-22 | 87-25 | 87-32 | 87-36 | |
|---|---|---|---|---|---|---|---|---|
| 0 | −0.34 | −0.40 | −0.42 | −0.36 | −0.23 | −0.31 | −0.30 | |
| 10 | −0.34 | −0.37 | −0.38 | −0.36 | −0.26 | −0.27 | −0.25 | |
| 25 | −0.36 | −0.37 | −0.42 | −0.33 | −0.33 | −0.32 | −0.28 | |
| 50 | −0.37 | −0.38 | −0.38 | −0.36 | −0.33 | −0.37 | −0.30 | |
| 100 | −0.41 | −0.44 | −0.38 | −0.44 | −0.26 | −0.36 | −0.32 | |
| 150 | −0.44 | | | | | | | |
| 200 | | −0.40 | −0.36 | −0.47 | −0.34 | −0.36 | −0.36 | |
| | 89-2 | 89-3 | 89-4 | 89-5 | 89-6 | 89-7 | 89-8 | |
| 0 | −0.11 | −0.26 | −0.22 | −0.13 | −0.25 | −0.22 | −0.15 | |
| 10 | −0.11 | −0.30 | −0.30 | −0.21 | −0.21 | −0.28 | −0.22 | |
| 25 | −0.29 | −0.28 | −0.24 | −0.31 | −0.19 | −0.28 | −0.24 | |
| 50 | −0.30 | −0.26 | −0.26 | −0.27 | −0.15 | −0.21 | −0.27 | |
| 75 | | −0.30 | −0.29 | −0.20 | −0.25 | −0.37 | −0.37 | |
| 100 | −0.35 | −0.38 | −0.30 | −0.29 | | −0.36 | −0.21 | |
| 150 | | −0.32 | | | | | −0.24 | |
| 170 | | | | | | −0.38 | | |
| 200 | −0.35 | | −0.43 | | | | | |
| 320 | | | | −0.39 | | | | |
| 450 | −0.38 | | | | | | | |
| | 89-9 | 89-10 | 89-11 | 89-12 | 89-14 | 89-16 | 89-17 | 89-18 |
| 0 | −0.14 | −0.20 | −0.33 | −0.26 | −0.23 | −0.28 | −0.27 | −0.15 |
| 10 | −0.01 | −0.20 | −0.31 | −0.30 | −0.20 | −0.31 | −0.05 | −0.35 |
| 25 | −0.21 | −0.27 | −0.28 | −0.23 | −0.31 | −0.29 | −0.06 | −0.35 |
| 50 | −0.26 | −0.24 | −0.36 | −0.29 | −0.27 | −0.30 | −0.04 | −0.21 |
| 75 | −0.25 | | −0.26 | −0.36 | −0.37 | −0.30 | | −0.26 |
| 90 | −0.23 | | | −0.29 | −0.36 | −0.37 | | |
| 100 | | −0.34 | | | | | −0.13 | −0.24 |
| 160 | | | | | −0.36 | | | −0.32 |
| 200 | | −0.40 | | | | −0.41 | −0.22 | |
| 500 | | −0.45 | | | | | −0.24 | |

**Table 3.** Mean values of the different water masses and their standard deviations reported for the 1987–88 and 1989–90 stations

| Station | Water mass | Delta-18O (permil) | Salinity (psu) | Temperature (°C) |
|---|---|---|---|---|
| 87-3 | AASW | −0.35 ± 0.01 | 34.83 ± 0.02 | 0.52 ± 0.36 |
| 87-9 | AASW | −0.38 ± 0.02 | 34.86 ± 0.02 | 1.83 ± 0.39 |
| | HSSW | −0.42 ± 0.03 | 34.92 ± 0.01 | −1.85 ± 0.07 |
| 87-21 | AASW | −0.41 ± 0.02 | 34.79 ± 0.002 | −0.52 ± 0.03 |
| | HSSW | −0.37 ± 0.01 | 34.90 ± 0.01 | −1.92 ± 0.02 |
| 87-22 | AASW | −0.36 | 34.81 ± 0.01 | 0.52 ± 0.11 |
| | HSSW | −0.45 ± 0.02 | 34.92 ± 0.02 | −1.89 ± 0.02 |
| 87-25 | AASW | −0.24 ± 0.02 | 34.07 ± 0.25 | 0.52 ± 0.32 |
| | HSSW | −0.30 ± 0.06 | 34.92 ± 0.05 | −1.90 ± 0.01 |
| 87-32 | AASW | −0.30 ± 0.03 | 34.75 ± 0.06 | 0.58 ± 0.38 |
| | HSSW | −0.36 | 34.93 ± 0.04 | −1.89 ± 0.01 |
| 87-36 | AASW | −0.27 ± 0.03 | 34.62 ± 0.11 | 1.40 ± 0.45 |
| | HSSW | −0.36 | 34.92 ± 0.04 | −1.88 ± 0.03 |
| 89-2 | AASW | −0.17 ± 0.10 | 33.49 ± 0.63 | 0.53 ± 0.27 |
| | HSSW | −0.37 ± 0.02 | 34.79 ± 0.02 | −1.93 ± 0.02 |
| 89-3 | AASW | −0.28 ± 0.02 | 33.82 ± 0.17 | 0.73 ± 0.06 |
| | HSSW | −0.32 | 34.77 ± 0.01 | −1.83 ± 0.03 |
| 89-4 | AASW | −0.25 ± 0.04 | 33.90 ± 0.16 | 0.55 ± 0.24 |
| | HSSW | −0.43 | 34.75 ± 0.01 | −1.86 ± 0.06 |
| 89-5 | AASW | −0.23 ± 0.08 | 33.98 ± 0.48 | 0.79 ± 0.53 |
| | HSSW | −0.39 | 34.78 ± 0.01 | −1.90 ± 0.03 |
| 89-6 | AASW | −0.23 ± 0.03 | 34.00 ± 0.16 | 0.61 ± 0.34 |
| 89-7 | AASW | −0.25 ± 0.04 | 33.88 ± 0.32 | 0.75 ± 0.76 |
| | HSSW | −0.37 ± 0.01 | 34.73 ± 0.01 | −1.85 ± 0.02 |
| 89-8 | AASW | −0.2 ± 0.05 | 34.00 ± 0.21 | 0.07 ± 0.38 |
| 89-9 | AASW | −0.12 ± 0.10 | 34.18 ± 0.28 | 0.15 ± 0.67 |
| 89-10 | AASW | −0.20 | 34.55 ± 0.06 | −1.29 ± 0.31 |
| | HGW1 | −0.25 ± 0.02 | 34.69 ± 0.01 | −1.77 ± 0.01 |
| | HGW2 | −0.37 ± 0.04 | 34.77 ± 0.02 | −1.92 ± 0.04 |
| | ISW | −0.45 | 34.81 ± 0.01 | −2.02 ± 0.01 |
| | HSSW | | 34.84 ± 0.01 | −1.98 ± 0.01 |
| 89-11 | AASW | −0.32 ± 0.01 | 34.39 ± 0.14 | −0.38 ± 0.79 |
| 89-12 | AASW | −0.26 ± 0.03 | 34.03 ± 0.16 | −0.10 ± 0.34 |
| 89-14 | AASW | −0.25 ± 0.06 | 33.98 ± 0.16 | 0.09 ± 0.36 |
| | HSSW | −0.36 | 34.74 ± 0.01 | −1.88 ± 0.02 |
| 89-16 | AASW | −0.29 ± 0.01 | 34.07 ± 0.11 | −0.09 ± 0.43 |
| | HSSW | −0.41 | 34.76 ± 0.01 | −1.84 ± 0.05 |
| 89-17 | AASW | −0.16 ± 0.15 | 33.92 ± 0.31 | −0.26 ± 0.22 |
| | HSSW | −0.23 ± 0.01 | 34.78 ± 0.01 | −1.94 ± 0.04 |
| 89-18 | AASW | −0.25 ± 0.14 | 33.76 ± 0.25 | −0.06 ± 0.14 |
| | HSSW | −0.32 | 34.74 ± 0.01 | −1.87 ± 0.01 |

1987–88 oceanographic cruise pack ice was confined to the more seaward area, while the more coastal area was already ice-free (G. Catalano, pers. comm.). Hence, the discrepancy between the two years must have been related to different processes affecting the AASW during the two sampling periods or to processes affecting AASW at different rates (e.g. different melting or freezing rates or different amounts of upwelling of warmer and saltier water). A higher sea ice melting rate in the coastal region during the 1989–90 expedition (G. Catalano, pers. comm.) may thus explain the slightly lower salinity and slightly higher $\delta^{18}O$ values observed.

By means of a $\delta^{18}O/S$ diagram (Fig. 3) it was possible to outline the origin and the evolution of the reported water masses. During the freezing process the $^{18}O$ content of sea ice is higher than that of liquid seawater. The $^{18}O$ enrichment factor between liquid water and ice under isotopic equilibrium conditions is of the order of 3‰, as calculated by O'Neil (1968). Winter freezing of surface seawater can thus significantly increase the salinity of the residual liquid water and slightly

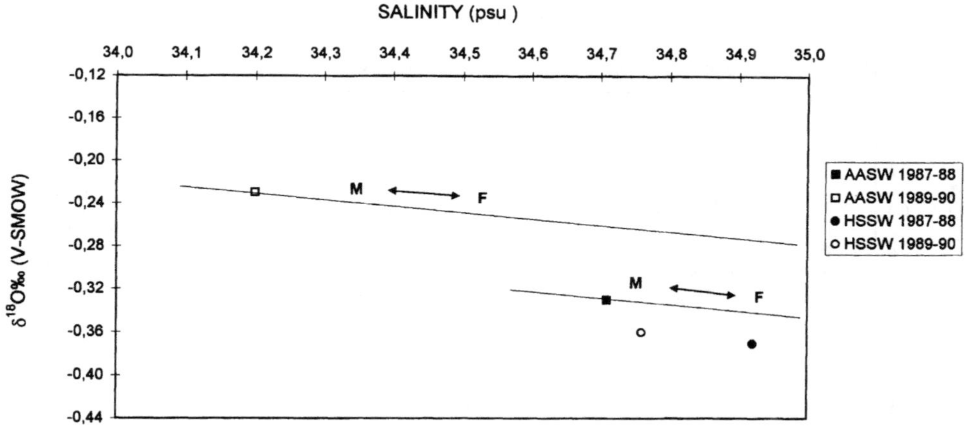


**Fig. 3.** $\delta^{18}O$/Salinity diagram showing the overall mean data for AASW and HSSW during the 1987–88 and 1989–90 cruises. The *solid lines* represent freezing-melting lines with a slope of 0.08‰ $\delta^{18}O$/‰S. A water mass undergoing sea-ice freezing (*F*) or melting (*M*) moves to the *right* or to the *left*, respectively

decrease its $\delta^{18}O$, while summer melting should reverse such an effect. Jacobs et al. (1985) reported a decrease of 0.08‰ in $\delta^{18}O$ for an increase in salinity of one part per thousand. In Fig. 3 two freezing-melting lines with a slope of 0.08‰ $\delta^{18}O$/‰ S were drawn through the AASW data. The variables which characterize a water body should then move along the freezing-melting line, to the right for sea-ice freezing (F) or to the left for sea-ice melting (M). These processes could be responsible for the differences in the mean values of $\delta^{18}O$ and salinity measured in AASW between the different sampling periods. We can see that HSSW exhibits slightly more negative $\delta^{18}O$ values than those of surficial waters, suggesting a direct formation from the AASW freezing process. In the case of the 1989–90 values, HSSW lies considerably below its freezing-melting line, suggesting a greater contribution of continental water with highly negative $\delta^{18}O$ values. According to the measurements carried out in our laboratory over a period of several years, the mean $\delta^{18}O$ values of summer and winter precipitations in the area of Terra Nova Bay range from −14 to −30‰ at sea level (Caprioli et al. 1997), with progressively lighter values moving inland.

The mean $\delta^{18}O$ and salinity data obtained from our samples are reported in Fig. 4 along with similar data for the Southern Ocean obtained over time by several authors. The isotopic composition of AASW agrees well with the other $\delta^{18}O$ values, even though the 1987–88 samples are shifted toward higher salinity values, suggesting that peculiar environmental conditions heavily affected the distribution of pack ice and perhaps there was also an extremely low contribution of continental meltwater. In this case, the distinction between 1987–88 AASW and HSSW in the field can be made on the basis of temperature data, the AASW temperature being highly different (see Table 3). On the other hand, HSSW data are comparable to the values reported by previous authors for "oceanic samples".

Special attention should be drawn to those stations affected by particular local conditions; for example, if we consider the surface pattern of the values observed at station 89-2, it probably represents a significative input of sea-ice meltwater since it shows higher $\delta^{18}O$ and lower salinity values. At the same station, Catalano et al. (1991) measured extremely low $SiO_2$, $PO_4$ and total inorganic carbon and high dissolved $O_2$ contents that are typical of a situation affected by pack ice melting. A similar feature is also represented by the isotopic values of the surface layer measured at station 89-5.

We should also mention the relatively high $\delta^{18}O$ values (about −0.05‰) obtained in the subsurface layer (10–50 m) of station 89-17, situated above the northern sector of a 550-m-deep trench just outside the Gerlache Inlet. The influence of seafloor topography on water mass circulation,

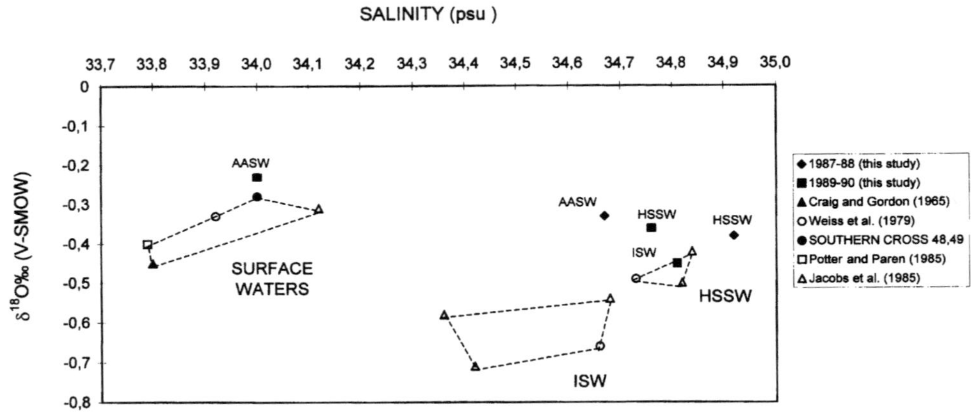


**Fig. 4.** $\delta^{18}O$/Salinity diagram reporting our mean data in comparison with those reported over time by several authors for the Southern Ocean. Data from Southern Cross stations 48 (68°00′S, 170°07′W) and 49 (69°29′S, 169°57′W) are converted from deuterium measurements (Horibe and Ogura 1968) using the relationship $\delta D = 8.3\delta^{18}O$-0.8 (Weiss et al. 1979). The *dotted lines* encompass the data field of ASSW, ISW and HSSW

with the formation of ascending currents, seems to be an important feature of this station, as reported by Stocchino (1990), but we cannot exclude the possibility of a local advection of sea-ice meltwater (even though the salinity values do not support this hypothesis). Indeed, higher $\delta^{18}O$ values, not found elsewhere in the Terra Nova Bay region, may be only partially explained in this way.

Figure 5 reports the $\delta^{18}O$, temperature and salinity profiles obtained at station 89-10, situated in front of the edge of the Hell's Gate Ice Shelf (HGIS); the station shows very distinctive hydrological and isotopic characteristics. Hell's Gate is a small ice shelf located in Terra Nova Bay and is influenced by seawater-ice interaction processes, as pointed out by many authors (Baroni et al. 1991; Souchez et al. 1991; Tison et al. 1993). The surface water (AASW) was characterized by lower temperatures and higher salinities than the values measured at all other stations. Such waters were also limited to the uppermost part of the water column to a depth of 40 m. Below the surface layer, four different water types were distinguished.

Following the models proposed by Jacobs et al. (1979) for the Ross Ice Shelf and by Nicholls et al. (1991) for the Ronne Ice Shelf, a deep circulation is driven by HSSW which passes under the ice shelf causing the melting of ice. The HSSW provides heat for melting due to the reduction of the seawater freezing point with increasing pressure (Doake 1976). The mixture of HSSW and ice-meltwater ascends, becoming supercooled and, when the in-situ freezing point is reached, it begins to freeze at shallower depth. As a consequence, this mechanism implies a transfer of ice along the ice shelf from higher to lower depths. Ultimately, this plume emerges at mid-depths as Ice Shelf Water (ISW). As pointed out by Jacobs et al. (1992), the melting potential provided by this thermohaline circulation could be an important mechanism to accelerate the retreat of the grounding line or, on the contrary, the accumulation of dense water in a sub-ice shelf depression could limit the penetration of warmer water close to the grounding line.

If we apply this model to our station, ISW was detected from 250 to 525 m, the most typical feature being a temperature close to −2.01 °C and a $\delta^{18}O$ of about −0.45‰ (according to the mean values reported by Jacobs et al. 1985 for the Ross Ice Shelf Water). The $\delta^{18}O$/S diagram in Fig. 6 represents the data relative to the different water masses identified in station 89-10. The freezing-melting line (Fig. 6, solid line) and the envelopes encompassing probable $\delta^{18}O$ values for glacial meltwater (dashed lines) are reported in order to clarify the origin and evolution of the different water masses. According to this

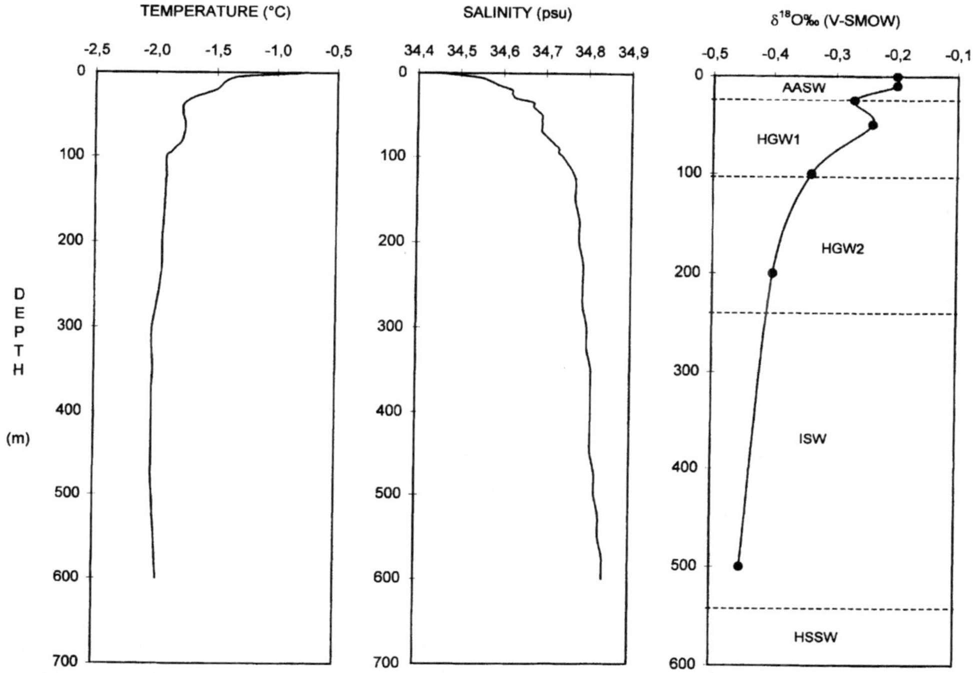


**Fig. 5.** $\delta^{18}O$, salinity and temperature vertical profiles for station 10. The $\delta^{18}O$ profile reports the boundary between different water masses as proposed by Tison et al. (1997); see text for abbreviations

figure, ISW seems to form from HSSW after continental basal ice melting and subsequent marine ice freezing.

Below the ISW the presence of HSSW can be identified on the basis of its temperature and salinity values, since no $\delta^{18}O$ values are available.

The upper 100 m of the water column is characterized by a step-like configuration of both temperature and salinity profiles. This structure is generally ascribed to a process of basal melting near the ice front (Jacobs et al. 1981, 1992). Indeed, the surficial water is above the in-situ freezing point and by tidal action can induce a circulation driven by melting of the ice shelf (van Heijst 1987). Whether this circulation is also capable of producing frazil ice at the edge of the ice shelf or below it by a double-diffusion process is still under discussion (Tison et al. 1993).

In this way the unusual high salinity and low temperature measured at the surface could be explained by successive melting and refreezing processes, even though the isotopic values seem to be ambiguous. Below the AASW, two water layers may be detected on the basis of their salinity and $\delta^{18}O$ values (labelled in Fig. 6 as HGW1 and HGW2, where HGW = Hell's Gate Water). The origin of HGW1 and HGW2 (Fig. 6) can be explained in two different ways:

1. The salinity and isotopic values of HGW1 and HGW2 could be attributed to a mixing process involving ISW ($S = 34.81‰$, $\delta^{18}O = -0.45‰$) and AASW ($S = 34.55‰$, $\delta^{18}O = -0.22‰$), their values being intermediate.
2. Melting of continental basal ice and successive freezing underneath the shelf, even though this possibility should be excluded by the continuous presence of marine ice below Hell's Gate found by Souchez et al. (1991) and Tison et al. (1993).

Further investigations which are being carried out on seawater profiles and ice cores at the front

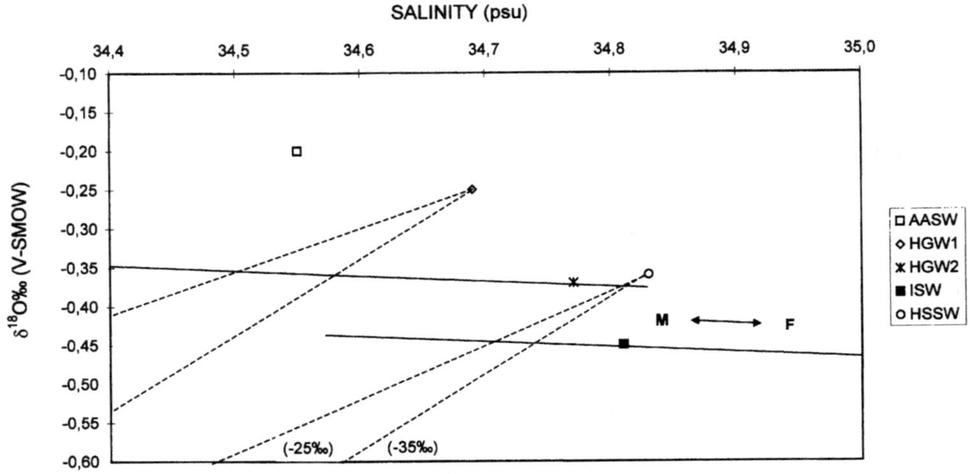


**Fig. 6.** $\delta^{18}O$/Salinity diagram showing the positions of the different water masses at station 10. The *dashed lines* encompass the $\delta^{18}O$/S data field representing the evolution of a water mass undergoing dilution with glacial meltwater with $\delta^{18}O$ between −25 and 35‰ and with salinity = 0 psu. A mean value of HSSW is used here. The *solid line* represents freezing-melting line (see Fig. 3 for explanations)

of Hell's Gate Ice Shelf (Tison et al., pers. comm.) could help to achieve a better understanding of the seawater-ice interactions beneath the HGIS.

## Conclusions

The set of $\delta^{18}O$ data reported clearly suggests that the isotopic method can be a useful tool to distinguish different water types. By a comparison of $\delta^{18}O$ and salinity data we were also able to identify the origin and outline the evolution of water masses that undergo freezing processes, dilution processes with ice-meltwater and interactions with the atmosphere. In particular, in the area of Terra Nova Bay a substantial presence of HSSW, apparently originating in this region, has been pointed out. According to this model, the persistent polynya acts as a "pump" in renewing shelf water mass circulation. The presence of Ice Shelf Water even below a small ice shelf such as Hell's Gate has been highlighted.

The importance of studying the hydrological characteristics, both physicochemical and isotopic, of polar seawater is clear, taking into account that the Southern Ocean plays an important role in determining the features of deep and bottom waters, and that it also represents a net heat source for the atmosphere. For this reason, the most recent studies have focused on a better understanding of the role played by the abovementioned mechanisms in bringing about climate changes.

*Acknowledgements.* The authors are indebted to Prof. Antonio Longinelli for his continuous assistance during the development of this study and to Dr. Giulio Catalano for his helpful comments and suggestions. This work was carried out in the framework of the Italian Antarctic Research Program financially supported by ENEA (CAP. 2/12/05 COD.040026). The authors are grateful to the members of the 1987–88 and 1989–90 Italian expeditions to Antarctica, who carried out the fieldwork and collected the samples studied.

## References

Baroni C, Stenni B, Iacumin P (1991) Oxygen isotopic composition of ice samples from the Hell's Gate and Backstairs Passage Ice Shelves (Victoria Land, Antarctica): evidence of bottom freezing. Mem Soc Geol Ital 46: 45–48

Boldrin A, Stocchino C (1990) On the hydrological characteristic of Terra Nova Bay. National Scientific Commission for Antarctica, Oceanographic Campaign 1987–88, Data Rep Part 1, Genova, pp 11–57

Bromwich DH, Kurtz DD (1984) Katabatic wind forcing of the Terra Nova Bay polynya. J Geophys Res 89 (C3): 3561–3572

Caprioli R, Gragnani R, Grigioni P, Maggi V, Serra F, Stenni B, Torcini S (1997) Stratigraphic, isotopic and chemical profiles of a firn core from Drygalski Ice Tongue and of

a snow pit from Aviator Glacier (Northern Victoria Land, Antarctica). Terra Antart Rep 1: 71–76
Catalano G, Benedetti F (1990) Distribution of nutrients in the Terra Nova Bay and in the Ross Sea. National Scientific Commission for Antarctica, Oceanographic Campaign 1987–89, Data Rep Part 1, Genova, pp 61–83
Catalano G, Benedetti F, Iorio M (1991) Coastal oceanography from Cape Russel to Campbell Ice Tongue (Terra Nova Bay). Dissolved oxygen, nutrients, pH, total alkalinity and total inorganic carbon distribution. National Scientific Commission for Antarctica, Oceanographic Campaign 1989–90, Data Rep Part 1, Genova, pp 25–32
Craig H (1965) The measurements of oxygen isotope palaeotemperatures. In: Tongiorgi E (ed) Stable isotopes in oceanographic studies and paleotemperatures. CNR, Pisa, pp 161–182
Craig H, Gordon LI (1965) Deuterium and oxygen 18 variations in the ocean and marine atmosphere. In: Schink D, Corliss J (eds) Proceedings of a symposium on marine chemistry, Occ publ 3. University of Rhode Island, Narragansett, pp 277–374
Dansgaard W (1964) Stable isotope in precipitation. Tellus 16: 436–468
Doake CSM (1976) Thermodynamics of the interaction between ice and the sea. Polar Rec 18 (112): 37–41
Epstein S, Mayeda T (1953) Variation of $^{18}O$ content of waters from natural sources. Geochim Cosmochim Acta 4: 213–224
Fabiano M, Povero P, Catalano G, Benedetti F (1991) Hydrological data collected during the biological, chemical and geological sampling in Terra Nova Bay. National Scientific Commission for Antarctica, Oceanographic Campaign 1989–90, Data Rep Part 1, Genova, pp 35–71
Gonfiantini R (1978) Standards for stable isotope measurements in natural compounds. Nature 271: 534
Horibe Y, Ogura N (1968) Deuterium content as a parameter of water mass in the ocean. J Geophys Res 73 (4): 1239–1249
Jacobs SS, Comiso JC (1989) Sea ice and oceanic processes on the Ross Sea continental shelf. J Geophys Res 94 (C12): 18195–18211
Jacobs SS, Amos AF, Bruchhausen PM (1970) Ross Sea oceanography and Antarctic Bottom Water formation. Deep Sea Res 17: 935–962
Jacobs SS, Gordon AL, Ardal JL Jr (1979) Circulation and melting beneath the Ross Ice Shelf. Science 203: 441–443
Jacobs SS, Huppert HE, Holdsworth G, Drewry DJ (1981) Thermohaline steps induced by melting of the Erebus Glacier Tongue. J Geophys Res 86 (C7): 6547–6555
Jacobs SS, Fairbanks RG, Horibe Y (1985) Origin and evolution of water masses near the Antarctic continental margin: evidence from $H_2{}^{18}O/H_2{}^{16}O$ ratios in seawater. Antarct Res Ser 43: 59–85
Jacobs SS, Helmer HH, Doake CSM, Jenkins A, Frolich RM (1992) Melting of ice shelves and the mass balance of Antarctica. J Glaciol 38 (130): 375–387
Klepikov VV, Grigor'yev YA (1966) Water circulation in the Ross Sea. Info Bull Soviet Antarctic Expedition 6: 52–54
Nicholls KW, Makinson K, Robinson AV (1991) Ocean circulation beneath the Ronne Ice Shelf. Nature 354: 221–223
Potter JR, Paren JG (1985) Interaction between ice shelf and ocean in George VI Sound, Antarctica. Antarct Res Ser 43: 35–58
O'Neil JR (1968) Hydrogen and oxygen isotope fractionation between ice and water. J Phys Chem 72: 3683–3684
Redfield AC, Friedman I (1969) The effect of meteoric water, meltwater and brine on the composition of polar sea water and of the deep waters of the ocean. Deep Sea Res 16: 197–214
Souchez R, Meneghel M, Tison JL, Lorrain R, Ronveaux D, Baroni C, Lozej A, Tabacco I, Jouzel J (1991) Ice composition evidence of marine ice transfer along the bottom of a small antarctic ice shelf. Geophys Res Lett 18(5): 849–852
Stocchino C (1990) Oceanografia fisica e idrografia. In: PNRA (ed) Rapporto sulla campagna antartica, Estate Australe 1989–90. Programma Nazionale Ricerche in Antartide, Genova, pp 15–22
Stocchino C, Lusetti C (1990) Prime osservazioni sulle caratteristiche idrologiche e dinamiche di Baia Terra Nova (Mare di Ross-Antartide). Ist Idr della Marina, Genova
Stocchino C, Manzella G (1991) Analisi delle correnti marine di Baia Terra Nova (Mare di Ross-Antartide). CNR, Area della Ricerca di Genova
Tison JL, Ronveaux D, Lorrain RD (1993) Low-salinity frazil ice generation at the base of a small Antarctic ice shelf. Antarct Sci 5(3): 309–322
Tison JL, Barbante C, Bondesan A, Lorrain R, Capra A (1997) Ice shelf/ocean interactions at the front of Hell's Gate Ice Shelf (Terra Nova Bay-Antarctica). Terra Antart Rep 1: 29–32
van Heijst GJF (1987) On the oceanic circulation near a shelf edge. In: van der Veen CJ, Oerlemans J (eds) Dynamics of the West Antarctic Ice Sheet. Reidel, Dordrecht, pp 37–56
Weiss RF, Ostlund HG, Craig H (1979) Geochemical studies of the Weddell Sea. Deep Sea Res 26 (10A): 1093–1120

CHAPTER 4

# Biogeochemical Processes in the Ross Sea (Antarctica): Present Knowledge and Perspectives

**M. Frignani, L. Langone, L. Labbrozzi, and M. Ravaioli**[1]

**ABSTRACT**

Fluxes of particles, biogenic silica and organic carbon through the water column were obtained at three sites in the northwestern Ross Sea using $^{238}U/^{234}Th$ disequilibrium, drifting traps, and moored sediment traps. Burial rates of biogenic components were calculated by determining sediment mass accumulation rates, while inferences on the past evolution of the system were obtained through the record stored in sediments. Particle flux was characterized by a high seasonality and was strictly dependent on the ice coverage. Furthermore, the data suggested a high interannual variability and a rapid sinking of particles, which implies a minimal degradation of the material, mainly aggregates and faecal pellets, below the euphotic zone. High fluxes, measured occasionally, were linked to particular post-bloom conditions or to the presence of marginal ice. Time lags between production and trap collection can be explained by post-bloom mechanisms. Sediments at mooring sites, A B and C showed different lithologies, with C representing conditions of no present sediment accumulation. Composition, density and structure of benthic communities also differed significantly, mostly as a function of water depth. Fluxes from the water column at both mooring sites, A and B, did not account for sediment accumulation and burial rates of the biogenic components, hence lateral advection must have provided important contributions. Depth profiles of magnetic susceptibility and biogenic components in cores showed a dramatic change corresponding to the onset of present seasonally open marine sedimentation. Present-day fluxes of biogenic silica and organic carbon were higher in the Joides basin than in the Drygalski basin and other areas of the northwestern Ross Sea. The results of various research approaches are discussed in the framework of the present knowledge of processes in the Ross Sea, and views on the perspectives of the research are suggested.

## Introduction

In the Southern Ocean sea-ice dynamics, thermohaline convection with consequent formation of cold, salty water masses, and biological productivity all influence the exchange of $CO_2$ between the ocean and atmosphere, and are key processes relating climate and the Antarctic environment. As a consequence, biogeochemical changes affecting the carbon balance in the upper water column of this region can potentially modify the level of atmospheric $CO_2$ and influence global climate. In this respect, the study of particle fluxes in the oceans is important for assessing the transfer of organic carbon as well as biogenic carbonate and opal, through the system. This information can be used to better understand the chemistry of the ocean, the biogeochemical cycles of key elements, the food supply to the benthos, and the character of the sedimentary record. The information stored in sediments provides insight into the processes at work in the past and the history of environmental changes.

The Southern Ocean appears to be an area where great amounts of biogenic silica are accumulating in modern sediments. According to DeMaster (1981), over one half of the silica (about $2.5 \times 10^{14}\,g\,SiO_2\,year^{-1}$) supplied annually to the ocean by rivers and hydrothermal emanations is removed in this area. This means that Southern Ocean processes dominate the global oceanic silica budget and that the Ross Sea, in particular, is a preferential site of biogenic silica production and accumulation (Ledford-Hoffman et al. 1986).

[1] Istituto di Geologia Marina (CNR), Via Gobetti 101–40129 Bologna, Italy

These large fluxes to the sediments are due to the unusually high proportion of biogenic silica which passes through the water column without dissolving. A similar efficiency is not observed for particulate organic matter, and this implies a decoupling of the cycles of biogenic silica and organic matter, arising from marked differences in the rates of key processes: dissolution for silica and oxidative degradation for carbon. Despite previous studies focused on the cycling of biogenic silica and organic carbon in the water column and uppermost sediments, many questions remain (DeMaster et al. 1992, 1996; Nelson et al. 1996). In this chapter we discuss our contributions to the understanding of the biogeochemical cycles of opal and organic carbon in the Ross Sea and the response of the system to natural climate changes during the late Quaternary.

## Background

The Ross Sea is representative of the Antarctic continental shelf, which is a dominant centre of productivity in the Southern Ocean (Comiso et al. 1993). It is characterized by a large, deep (250–1200 m) shelf, the presence of basins which act as sediment traps, and the recurrent development of the largest coastal polynya in the Antarctic. Results of a major research effort to study this area have been discussed by Nelson et al. (1996). These authors report estimates of annual production, near-surface recycling, sinking particle flux, delivery to the sea bed, benthic regeneration, and long-term burial of organic carbon and biogenic silica integrated over most (75–80%) of the Ross Sea shelf. Their estimated spatially-averaged annual primary productivity is $142\,g\,m^{-2}\,year^{-1}$, about four times the mean productivity of the whole Southern Ocean but considerably less than that on continental shelves at low latitudes. According to these authors, the integrated annual sinking flux of biogenic opal to 250 m is 50–100% of the annual net production of silica in the upper 50 m, while the sinking flux of organic carbon is only 5–10%. These results imply a relatively intense regeneration of organic matter, but not silica, between the 50- and 250-m depths, and an annual biogenic silica export that is quite high, despite the modest annual productivity. Comparison of fluxes calculated from core data and integrated annual fluxes measured by sediment traps at 250 m and near the sea floor suggests that biogenic particulate matter that reaches the 250-m depth is delivered to the sea floor (500–850 m) with only minor loss of either organic or siliceous material. In the sections that follow we review the biogeochemical processes operating in the Ross Sea.

### Primary Production

An important source for particles in the Ross Sea is primary production in the upper 100 m of the water column. The Ross Sea is an area of large phytoplankton blooms that begin in early summer in the polynya close to the Ross Ice Shelf and expand northward, following the sea-ice retreat. Early studies had suggested that the blooms resulted from vertical stability imparted by melting ice (Smith and Nelson 1985). Phytoplankton growth and biomass may also be controlled by irradiance (Smith and Sakshaug 1990; Nelson and Smith 1991), dissolved iron (Martin et al. 1990), or grazing by microzooplankton (Mitchell and Holm-Hansen 1991). It is unlikely that macronutrients play a significant role given the high levels found around the Antarctic continent (Smith et al. 1996).

A distinct S-N transition has been observed both in productivity and phytoplankton biomass, with the lowest values occurring in the northern Ross Sea. East-west gradients in phytoplankton biomass and composition occurred within the southern Ross Sea (Smith et al. 1996). Nelson et al. (1996) estimated primary production in three subareas of the Ross Sea. The highest values occurred in November and early December 1994 in the southeastern region (up to $6\,g\,C\,m^{-2}\,year^{-1}$). In the southwest the seasonal cycle was dominated by the development and breakdown of an ice-edge diatom bloom. Primary productivity in the north, mainly attributable to diatoms, started later, due to the larger duration of the ice cover. Nelson et al. (1996) estimated that, on an annual basis, primary productivity ranged from $91\,g\,C\,m^{-2}\,year^{-1}$ in the north to $216\,g\,C\,m^{-2}\,year^{-1}$ in the southeast.

### Particle Degradation and Removal

A large portion of the particulate silica formed in the surface layer dissolves and is recycled (DeMaster et al. 1992). The remainder leaves the surface layer primarily via sinking faecal pellets in the west and as large, low density aggregates in the east (Dunbar et al. 1991). Diatoms are probably removed largely by herbivorous grazing, which is decoupled in time with respect to production, whereas *Phaeocystis* cells are removed by aggregation and passive sinking, thereby decoupling the removal of carbon and silica from the surface (Smith et al. 1996). Faecal pellets and tabular aggregates that are probably produced by zooplankton are important contributors to the vertical particle flux (Jaeger et al. 1996). Faecal pellet and aggregate fluxes are greatest in the southwestern Ross Sea and decrease with depth (Nelson et al. 1996). Decreasing pellet fluxes are indicative of degradation and breakup of faecal pellets and/or grazing loss within the deep water column. Distribution patterns of $\delta^{13}C$ of organic matter in surface waters are faithfully recorded at the seafloor (Rogers 1995). This is compatible with rapid particle settling rates since preservation of site-specific surface water signals at the seafloor implies minimal horizontal transport distances.

### Particle Sinking and Lateral Transport

Long-term measurements of particle fluxes and marine currents at three separate mooring sites indicate significant spatial differences in the character of biogenic particle production and current dynamics (Jaeger et al. 1996). These differences lead to distinct trajectories for biogenic particles: in the southwestern Ross Sea, slow currents and large particles allow local biogenic material to reach the seabed in close proximity (<20 km) to its origin. In this case, a one-dimensional model is applicable to explain the fate of biogenic material, and a good coupling exists between high rates of surface productivity and the high silica accumulation rates at the seabed (Jaeger et al. 1996). In the northwestern Ross Sea, local biogenic material is transported many tens of kilometres (perhaps off the shelf) from its source location, seabed sediments do not reflect inputs from the overlying water column, and the moderate particle production in surface waters is not apparent as appreciable sediment accumulation rates. The south-central region of the Ross Sea is intermediate in character between these.

### Input of Lithogenic Material and Particle Distribution

Despite the high accumulation of biogenic particles in the Ross Sea, the largest component of sediment is terrigenous detritus. At present, this material is supplied principally by glaciers, which play the greatest role in delivering sediments to the continental shelf of this area (Anderson et al. 1991). Large ice shelves deposit the bulk of their debris near the grounding lines, and most debris laden icebergs melt over the continental shelf. Sediment of the inner shelf consists mainly of terrigenous silt, with variable concentrations of siliceous biogenic material (mostly diatom frustules). On shallow banks and at the shelf edge and upper slope, marine currents winnow fine-grained sediments to produce coarse-grained deposits. Sediment gravity flow processes are active sedimentary agents on the continental slope and on those portions of the shelf where glacial erosion has resulted in steep gradients (Anderson et al. 1979).

### Particle Deposition and Accumulation

The seabed is the ultimate sediment trap and may record the long-term history of particle production and transport in the overlying water column. However, this record can be easily biased by other processes that influence sediment deposition and accumulation at any given site. In fact, particle size and composition and mass accumulation rates are often a function of water-column dynamics because particles can be resuspended on banks and deposited on the bottom of basins, gravitative mass transport can affect sediment accumulation in basins, and lateral advection can cause the biogenic material produced in the euphotic layer to be exported from the region. Because of these possibilities, sediment mass accumulation rates often are different from rain rates of material produced by biogeochemical processes in the water column. Furthermore, the effects of bioturbation make it difficult to obtain accurate chronologies and to separate the different influences on the profiles when studying the stratigraphic record for paleoenvironmental reconstruction. On the basis of $^{14}C$ measurements in Kasten cores, sediment accumulation rates and burial fluxes of biogenic material were established by DeMaster et al. (1996) for a variety of depositional environments in the Ross Sea. They obtained mass accumulation rates between 0 and $70\,g\,cm^{-2}\,ka^{-1}$, whereas biogenic silica accumulation ranged from 0 to $31\,g\,cm^{-2}\,ka^{-1}$. Seabed preservation efficiencies, calculated by comparing pore water fluxes and burial fluxes, correlated very well with sediment accumulation rates. The authors estimated that biogenic silica and organic carbon accumulation rates on the Ross Sea Shelf (west of 175°W), during the Holocene, were $2.3 \times 10^{12}$ g $year^{-1}$ and $8.4 \times 10^{10}$ g $year^{-1}$, respectively. Biogenic silica accumulation rates increased during the mid and late Holocene, reaching their maximum values during the past 500–1000 years (DeMaster et al. 1996).

### Quaternary Sediments and Paleoenvironmental Reconstruction

Piston cores penetrate through surface sediments into relict deposits that are mainly of glacial origin. These include glacial marine sediments as well as till (Anderson et al. 1984). Radionuclide proxies such as the activity ratios $^{231}Pa_{xs}/^{230}Th$ and $^{10}Be/^{230}Th_{xs}$ in sediments from the Atlantic sector of the Southern Ocean over the last 140 ka indicate that glacial periods were characterized by greatly increased fluxes of biogenic detritus out of surface waters. This increase in export production, which may have contributed to lower concentrations of carbon dioxide in the glacial atmosphere, was accompanied by more than a fivefold increase in accumulation of lithogenic iron (Kumar et al. 1995). These observations support the hypothesis of the iron limitation of primary production in Antarctic high-nutrient waters. Analyses of marine sediment cores have provided important constraints on the extent of the West Antarctic Ice Sheet during the Last Glacial Maximum (LGM) and also the chronology of deglaciation in the Western Ross Sea. Published $^{14}C$ data (Licht et al. 1996) suggest that by 11 ka B.P. the grounding line has already reached the latitude of Terra Nova Bay and the glacial withdrawal was accomplished by about 7 ka B.P. Further information by Baroni and Orombelli (1994) dates an abandoned penguin rookery south of the Drygalski Ice Tongue to 13–11 ka B.P. (uncorrected). Brambati et al. (1997) confirm the onset of open marine conditions in Drygalski Basin at about 11 ka B.P. Regarding the position and age of the grounding line of the LGM in the Western Ross Sea, the problem is unresolved because of the difficulty in dating barren basal tills and overlying low-carbon sediments. Reid (1989) places the grounding line about 30–40 km north of Coulman Island at 18 ka B.P. (uncorrected). Licht et al. (1996) demonstrate that the grounded ice did not cover the outer shelf between 20 and 29 ka B.P., and estimate the position of the grounding zone south of Coulman Island (about 74°S). Brambati et al. (1997) demonstrate that the decoupling of the ice sheet from the bottom of Drygalski Basin began before 18.1 ka B.P.

## Materials and Methods

Sample collection was carried out using large volume filtration of seawater, drifting and moored traps, gravity coring, and box coring. In November–December 1994, large volumes of water (599–772 l) were filtered in the upper water column (50, 100, 150 and 200 m depth) at three stations (Fig. 1A) using in-situ pumps. $^{234}Th$ was retained by a series of three cartridges: the first for particles and the other two, $MnO_2$ impregnated, for dissolved Th (Buesseler et al. 1992). $^{238}U$ concentrations were calculated as a function of salinity, and $^{234}Th$ ($t_{1/2}$ = 24.1 days) was measured by gamma spectrometry 30–48 days after sampling. Drifting traps were deployed, at the same depths, to measure export fluxes from the photic zone. Two stations (15c and 11c) were sampled (Fig. 1A). These samplings were nearly synchronous with the study of the $^{238}U/^{234}Th$ disequilibrium and lasted for 12–13 h.

During the ROSSMIZE cruise in November–December 1994, we deployed two moorings in the centre of the northern part of the Joides Basin (site B, 586 m water depth) and on the northern flank of the Mawson Bank, very close to the shelf break (site C, 456 m water depth), respectively. Their locations are shown in Fig. 1A. Site B was characterised by accumulation of biosiliceous sediment, whereas at site C the reworked sediment was representative of a high energy environment. Site A, located in the SW Ross Sea (810 m water depth), had been studied since February 1994 via a collaborative effort between the IGM-CNR and the team of R.B. Dunbar (Stanford University, Stanford, CA). Each mooring was configured with two time-series sediment traps, current meters, transmissometers, and recorders of temperature and conductivity. Instruments were put at about 200 m depth below the sea surface and 30–40 m depth above the sea floor. Oceanographic parameters were measured every 30 minutes. We obtained a 1-year time series for site B and C, and a 2-year time series for site A. In January–February 1996, moorings A and B were serviced and redeployed, whereas mooring C was definitively recovered.

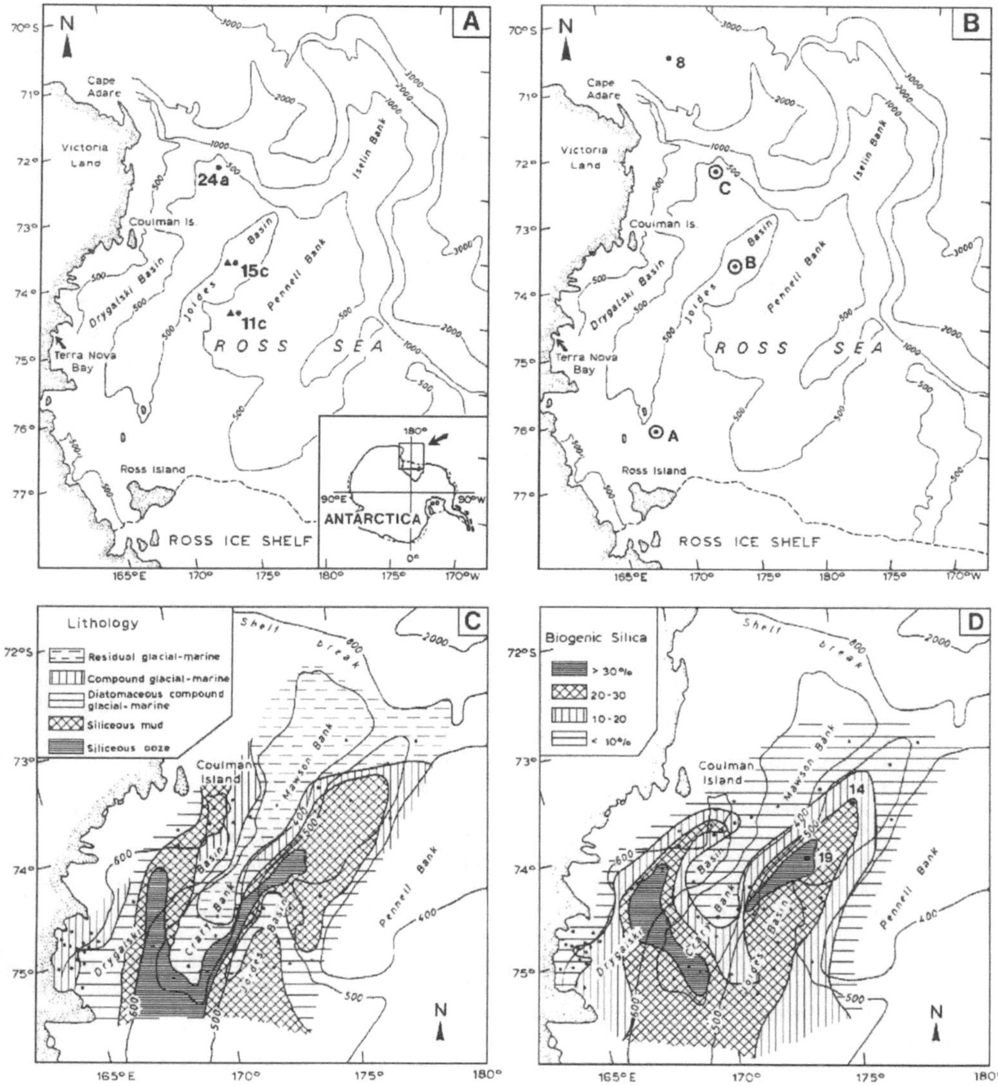

**Fig. 1A–D.** Study area and sampling locations. **A** Sites for the study of particle behaviour and fluxes using $^{238}U/^{234}Th$ disequilibrium (*filled circles*) and drifting traps (*triangles*). **B** Mooring sites and position of core 8. **C** Locations of surface sediment samples and lithology. **D** Locations of cores for paleoenvironmental reconstruction (14 and 19) and biogenic silica distribution in surface sediments

Trap material was treated and measured with a method similar to that suggested by Heussner et al. (1990).

Gravity cores and box cores were collected during several cruises between 1987 and 1996. Magnetic susceptibility and other mineral magnetic parameters were measured for core logging and correlation. $^{14}C$ dates and $^{210}Pb$ activity-depth profiles were used to evaluate the rates of sediment accumulation and bioturbation. Mass accumulation rates were used to calculate burial fluxes of biogenic materials. Biogenic silica and organic carbon were determined through the progressive alkaline dissolution suggested by DeMaster (1981) and using a CHN Analyzer (Frignani et al. 1992), respectively. $^{230}Th$ analyses followed the procedure of Frank et al. (1994), with sample dissolution, radiochemical purification and separation, and counting of alpha-emitting isotopes of uranium and thorium. $^{14}C$ activities were obtained from the Physics Division of ANSTO (Australia) which measured the organic matter dispersed in sediment layers by AMS.

## Results and Discussion

### Fluxes Through the Water Column

The results of experiments based on $^{238}U/^{234}Th$ disequilibrium provide insight into $^{234}Th$ removal and particle flux during the opening of the polynya in the Ross Sea (Langone et al. 1997). The depth profiles at stations 11c and 15c were scattered, due to particular post-bloom conditions, whereas site 24a showed deficiencies of $^{234}Th$ due to the uptake onto particles and particle sinking. Using a steady-state, one-dimensional model (Coale and Bruland 1985), we calculated residence times of $^{234}Th$ due to scavenging ($\tau = 130$–$247$ days) and sinking ($\tau =$ 6.6–8.1 days). Sinking fluxes of particles were obtained from the inventories of particles and residence times of $^{234}Th$, assuming that these are similar to the bulk particle residence times. We obtained very high values, ranging from 1.2 to 5 g $m^{-2}\,day^{-1}$ as a function of depth. At the other two stations, drifting traps revealed low values of particle fluxes, probably in relation to post-bloom conditions. Very high fluxes, such as those estimated for station 24a, were probably due to the particular environmental conditions caused by the presence of marginal ice, and were limited in time.

The material collected by moored sediment traps at station A (Fig. 1B) from 31/01/94 to 19/01/95 was largely made of faecal pellets and highly diatomaceous (Langone et al. 1996). Opal and carbon fluxes followed similar trends, suggesting a diatom source also for organic carbon. The maximum C and biogenic silica fluxes varied from 126.2 $mg\,C\,m^{-2}\,day^{-1}$ and 486.0 $mg\,Si\,m^{-2}\,day^{-1}$ in the surface trap to 22.8 $mg\,C\,m^{-2}\,day^{-1}$ and 387.2 $mg\,Si\,m^{-2}\,day^{-1}$ in the near-bottom trap. The fluxes of biogenic phases peaked in late summer, between March 14 and April 30. These high vertical fluxes were delayed with respect to the productivity maximum, and the time interval was long compared to previous observations. The lag was probably due to processes following summer productivity (decomposition of phytoplankton cells, aggregate formation, grazing and faecal pellets). It is worth noting the large difference of C fluxes between the surface and near-bottom traps. The decrease of the C flux was fourfold greater than the opal flux and was most likely related to the different rates of destruction of the two biogenic phases.

Bianchi (1997) analyzed samples collected from 15/12/94 to 12/01/96 by sediment traps moored at site B in the Joides Basin (Fig. 1B). The shallow trap was placed at 211 m depth, and the bottom one at 546 m depth, 42 m above the seafloor. Biogenic fluxes showed a marked seasonal variability (Fig. 2), and there was a large difference between fluxes recorded in the two traps and accumulation rates of the bottom sediments. The shallow trap measured a maximum flux in early January 1995 (110 $mg\,m^{-2}\,day^{-1}$), whereas the flux was nearly zero (0.1 $mg\,m^{-2}\,day^{-1}$) between April and December, due to the ice cover. The flux integrated over the entire 1995 was 3.9 $g\,m^{-2}$ $year^{-1}$. The patterns of Fig. 2 show that the record of the bottom trap was quite different: (1) the integrated flux was 30 $g\,m^{-2}\,year^{-1}$, an order of magnitude higher than the flux registered below the photic zone; (2) fluxes decreased after the summer maximum, were low but discrete in the intermediate period, and increased again to a maximum in December 1995. On the other hand, the mass accumulation rate estimated for site B (381 $mg\,m^{-2}$ $year^{-1}$) solving the advection diffusion equation both for $^{14}C$ and $^{210}Pb$ (Labbrozzi et al. 1998), was more than an order of magnitude higher than the flux measured by the bottom trap. The high flux of particles at depth was probably due to lateral advection of sediment resuspended from banks. Therefore, it was linked more to the hydrodynamics of the area than to the production in surface

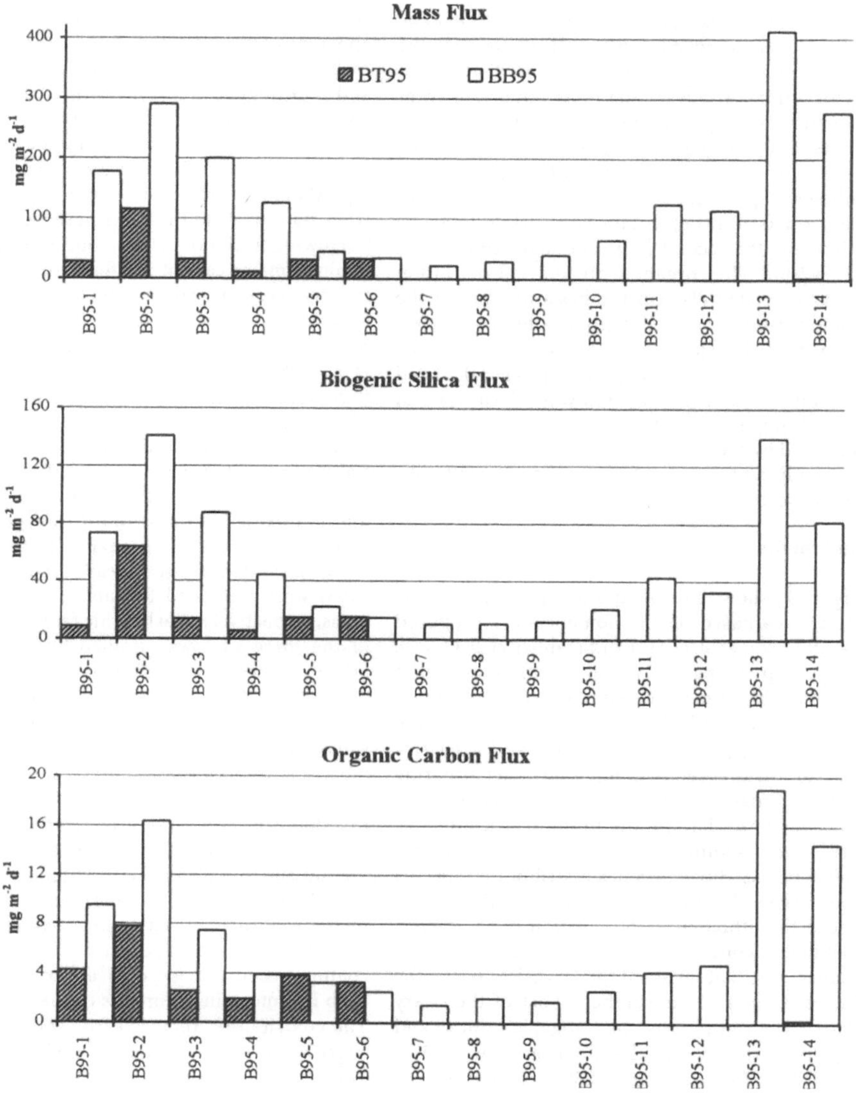


**Fig. 2.** Fluxes of material, OC and biogenic silica at mooring B (211 and 546 m depth) during the period December 1994–January 1996. *BT95* and *BB95* refer to the top and bottom trap, respectively

waters. This was also confirmed by $^{210}Pb$ data: while traps recorded low fluxes, the delivery to the sediment was much higher than predicted. Biogenic silica was always the most important component (34–53%) in the shallow trap, and the content of lithogenic particles accounted for only 6–24% of the total dry weight. On the contrary, lithogenic particles represented the most abundant fraction at depth (40–62%), while biogenic silica was the second most abundant component (29–47%). The organic carbon concentration never exceeded 8%. On the basis of data on the composition of materials from different sources, Bianchi (1997) suggested a model to account for the flux measured by the deep trap. She calculated that only 13% of the material was a local vertical flux through the water column, the rest being advected by focusing of biogenic particles raining from adjacent areas and resuspended from banks and advected to the site.

## Sediments

Figure 1c shows a detailed lithological map of the study area based on the nomenclature proposed by Anderson et al. (1984) and Dunbar et al. (1985). This map (Langone et al. 1998) shows that residual glacial-marine sediments cover the top of banks, compound glacial marine sediments are confined along the western coast, and a narrow belt of diatomaceous compound glacial-marine sediment marks the passage to siliceous mud and siliceous ooze that are the prominent product of present-day sedimentation in the Joides and Drygalski basins. Biogenic silica distribution (Fig. 1d) was driven by the same processes that influenced the lithological composition of sediments: maximum concentrations were found in an U-shaped area that included the Drygalski Basin, the Joides Basin and the southern part of the Crary Bank. Concentrations of biogenic silica and organic carbon were 0.5–32.7% (average, 13.5%) and 0.21–2.78 % (average, 0.88%), respectively.

Surface sediments at sites A and B consisted of brown or olive-grey diatomaceous mud, respectively. Dropstones were present below 10 cm in the box core of site A. At the shelf break station (site C), sediment is sandy with a high content of biogenic carbonate debris. At the surface, a pebble layer suggested a lag deposit. Average organic carbon contents were 1.09, 0.95 and 0.13%, for sites A, B and C, respectively. For the same sites the average contents of biogenic silica were 26.0, 26.0 and 0.3. Based on $^{210}Pb$ profiles, we calculate apparent accumulation rates of sediment, organic carbon and opal, that are similar for stations A and B (97 mg cm$^{-2}$ year$^{-1}$, 0.98 mg C cm$^{-2}$ year$^{-1}$ and 25 mg bioSi cm$^{-2}$ year$^{-1}$). However, sediment, organic C and Si accumulation rates could be overestimated, if the $^{210}Pb$ profiles were affected by biological mixing. From X-radiographs, the sediment from stations A and B looks massive. Mottling or burrowing were not evident, but in the absence of lamination, we cannot say for sure that bioturbation was very low. Solving the advection diffusion equation both for $^{14}C$ and $^{210}Pb$, we estimated a mixing rate of 0.53 cm$^{-2}$ year$^{-1}$ for site B (Labbrozzi et al. 1998). Although significant inventories of $^{210}Pb$ were present in sediments taken from the top of the banks and at station C, sediment accumulation rates were practically zero because of strong bottom currents (Picco et al. 1999). Distributions of $^{210}Pb$ at such sites were due to physical or biological mixing with downward transport of $^{210}Pb$. Mixing rates of 0.06 cm$^2$ year$^{-1}$ (Labbrozzi et al. 1998) and 0.10–0.46 cm$^2$ year$^{-1}$ (Langone et al. 1998) were estimated for site C and for shallow areas, respectively. The benthic fauna in sediments of the three sites was studied by Ravaioli et al. (1999). There were differences in composition, density and structure of benthic communities: the presence of infaunal forms, mainly surface and subsurface deposit feeders with reworking capability, characterize sites A and B, whereas at site C the community was mainly composed of epibenthic filter feeders and was strongly influenced by sediment instability. These differences were mainly due to bottom-water dynamics.

We have used $^{210}Pb$ distribution in the sediments as a tracer of sediment dynamics (Langone et al. 1998). The production of $^{210}Pb$ in the water column accounted for only a part of the excess $^{210}Pb$ inventory in sediments of the NW Ross Sea. The contribution required from outside the area suggests a water residence time of 50 years. Furthermore, the comparison between expected and measured $^{210}Pb$ inventories and surface activities in the study area enables us to distinguish several sedimentary settings: (1) over shallow banks, physical resuspension causes an enhanced scavenging of $^{210}Pb$ from the bottom water and penetration of $^{210}Pb$ in the sediment; (2) a fraction of this resuspended sediment is winnowed from banks and deposited on flanks and basins; (3) in the northern part of the Drygalski Basin, low inventories suggest an area of modern erosion.

$^{210}Pb$ chronologies give apparent accumulation rates ranging between 14 and 80 mg cm$^{-2}$ year$^{-1}$ (0.02–0.12 cm year$^{-1}$) in the shelf basins. Values of $^{210}Pb$-derived accumulation rates were higher than those obtained using $^{14}C$ (DeMaster et al. 1996). The discrepancies can reflect both an overestimate of $^{210}Pb$-derived accumulation rates due to mixing and an increase in sediment accumulation rates in modern time (Langone et al. 1998).

## Palaeoenvironmental Reconstruction

The paleoenvironmental record of eight cores collected in the northwestern Ross Sea was discussed by Frignani et al. (1998). Lithologies of cores 14 and 19, retrieved from the Joides Basin (Fig. 1d), and depth profiles of magnetic susceptibilities, TOC and biogenic silica are shown in Fig. 3. Uncalibrated $^{14}C$ ages also are reported for core 14. Correlation of the lithological compositions, magnetic susceptibility depth profiles and porosities, between cores 14 and 19 provides a chronostratigraphy for core 19 (Fig. 3). The sediment sequence consists of three principal lithologic units. The deeper one, massive and overconsolidated, was a diamicton (Cooper and Davey 1987), which can represent basal till or glacial-marine sediment. This facies was characterized by high magnetic susceptibility and nil or low content of biogenic components and was probably formed as basal till during the LGM by the grounding ice sheet (Anderson et al. 1984; Dunbar et al. 1985). Above the diamicton we found glacial-marine sediments, composed of muddy sand interbedded with mud and characterized by an upward increase in the biogenic components and a corresponding decrease in magnetic susceptibility. The uppermost unit was a biosiliceous mud, with relatively high contents of biogenic silica and organic carbon. Pebbles were found throughout the sediment columns and represented the coarsest fraction of the continental input. Magnetic susceptibility was high in sediments with an important detrital component, and low where a significant fraction of the sediment was biogenic, thus marking the shift in sediment input during the retreat of the ice sheet. Biogenic silica, organic carbon and total nitrogen contents followed the same behaviour. These cores showed organic carbon and biogenic silica concentrations ranging from 0.23 to 1.41% and 0.5 to 33.9%, respectively.

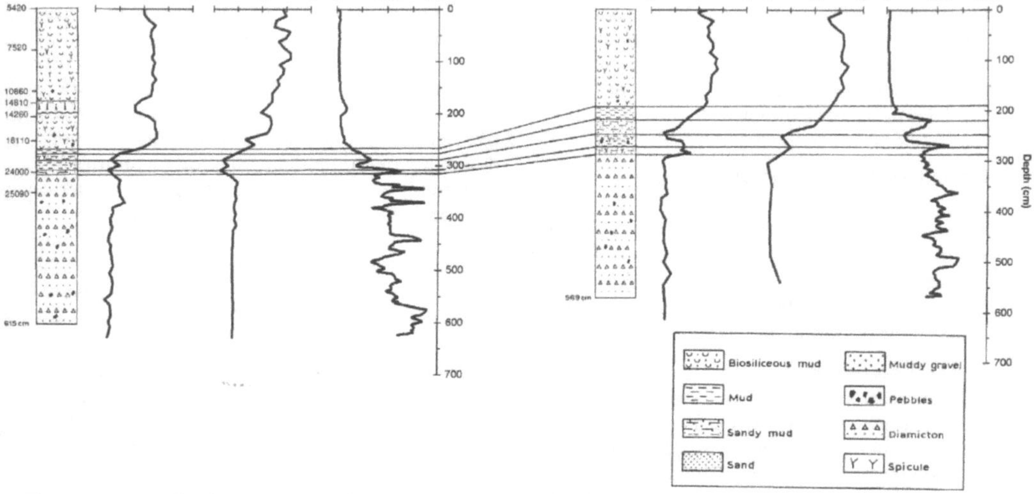


**Fig. 3.** Cores 14 and 19 (Joides Basin): sediment lithologies and depth profiles of magnetic susceptibility, organic carbon and biogenic silica concentrations. (From Frignani et al. 1998)

Sediment accumulation rates were calculated assuming constant sedimentation between dated levels. In the Joides Basin, sediment accumulation rates vary between less than 1 cm $ka^{-1}$ and 38 cm $ka^{-1}$. The highest values characterized the modern sediment in the deepest part of the basin and decreased downcore, with the exception of the interval between 270 cm and 320 cm at site 14. The fluxes of biogenic components through time were mainly related to the patterns of accumulation rates and changed dramatically with the establishment of open marine sedimentation. Both organic carbon and biogenic silica fluxes increased during the Holocene, though slightly differently. For example, in the uppermost 80 cm of core 14 (7520 ± 50 uncorrected $^{14}C$ year B.P.) opal concentration and flux increased more than organic carbon ones (Fig. 3). Present-day fluxes of organic carbon and biogenic silica were higher in the Joides Basin (up to 0.2 and 5.0 g $cm^{-2}$ $ka^{-1}$, respectively) than in the Drygalski Basin (up to 0.06 and 1.4 g $cm^{-2}$ $ka^{-1}$, respectively) and other areas of the northwestern Ross Sea.

Core ANTA91-8 (Fig. 1b), collected from the western part of the continental slope of the Ross Sea at 2380 m water depth, was studied by Ceccaroni et al. (1998). In this case an excess $^{230}Th$ based chronology was established for the last two climatic cycles (250 ka), and the effects of sediment redistribution were characterised by the comparison between measured and expected fluxes of excess $^{230}Th$. Sediment focusing prevailed during the most isotope stages and paleoproductivity, estimated by focusing-corrected accumulation rates of biogenic components, was low (1–10 gC $m^{-2}$ $year^{-1}$), relative to the Antarctic Polar Front in the Atlantic sector (Kumar et al. 1995).

## Conclusions and Perspectives

These results from the ROSSMIZE cruise and related studies must be considered as a first step toward a better understanding of the organic carbon and silica cycles in the Ross Sea. The main conclusions of our studies are:

1. The mass flux of particles was characterized by high seasonality and was strictly dependent on the ice coverage. High fluxes measured in surface waters appeared to be linked to the presence of marginal ice. The data collected by sediment traps suggested a high interannual variability and a rapid sinking of particles below 250 m water depth, which implies a minimal degradation of the material, mainly aggregates and faecal pellets. The bottom trap collected the vertical flux but also laterally advected material. At site A, high fluxes were measured after the summer bloom. Time lags between production and sinking fluxes can be explained by delays in particle flux related to kinetics of aggregation, grazing by zooplankton, etc.
2. Bottom sediments at the mooring sites (A, B and C) showed different lithologies, with C representing conditions of no present sediment accumulation. Composition, density and structure of benthic communities also differed significantly, mostly as a function of depth of the site. Fluxes from the water column at mooring sites A and B did not completely account for sediment accumulation and burial rates of the biogenic components, hence lateral advection provided important contributions to these sites.
3. Depth profiles of magnetic susceptibility and biogenic components in cores showed dramatic changes corresponding to the onset of present conditions of seasonally open marine sedimentation. Present fluxes of organic carbon and biogenic silica were higher in the Joides Basin than in the Drygalski Basin and in other areas of the northwestern Ross Sea.

Though the controlling processes were basically known, estimates and inferences of how these processes interrelate are still based on limited data, and several times the authors had to modify their own former conclusions. Therefore, future research should focus especially on the quantitative reconstruction of the spatial and temporal dynamics of the system. New opportunities are presented by the recovery of long time-series data and the study of sediment cores. This will provide a better understanding of the coupling between the processes in the water column and the corresponding sediment record. Both perspectives: "present as key to the past", and "past as key to the present" can be combined for a better description of the system, its role in the context of the Southern Ocean and its variability in response to climate changes. Accuracy in producing mass balances, a key step in this strategy, can be improved using analyses of sediment cores in conjunction with high resolution seismic profiles. The role of seasonal ice must be addressed in order to quantify the contribution of production in the sea ice to the overall fluxes. Future work should focus on the following objectives: (1) determining the timing and the pattern of the onset of present conditions of seasonally open sea; (2) assessing the role of water dynamics in driving production and export (vertical and lateral) of biogenic particles; (3) understanding how post-bloom conditions determine magnitude and timing of biogenic particle fluxes; (4) develop a three-dimensional model describing processes of production-degradation and accumulation, and taking into account lateral advection, when effective; (5) using radionuclide tracers to calculate bioturbation rates, and correlate these with the abundance and composition of benthic fauna. Additional questions to be addressed include: (a) Does the type of food web determine the relative importance and the composition of particles? (b) Are there links between the interannual variability in the beginning and intensity of production and meteorological and climate factors, especially with the cyclic perturbation, called the Antarctic Circumpolar Wave by White and Peterson (1996), which affect the Southern Ocean with a period of 4–5 years?

*Acknowledgements.* The research was carried out within the framework of the Projects Ecology and Biogeochemistry of the Southern Ocean and Glaciology and Paleoclimatology of the Italian Programma Nazionale di Ricerche in Antartide, and financially supported by ENEA. The authors wish to thank F. Giglio, E. Lipparini, G. Rovatti, G. Marozzi, and V. Palumbo who helped in the sampling operations and/or carried out analytical work. Drawings were provided G. Zini. J.K. Cochran read the paper. This is contribution no. 1163 from the Istituto di Geologia Marina, CNR, Bologna.

## References

Anderson JB, Kurtz DD, Weaver FM (1979) Sedimentation on the Antarctic continental slope. SEPM Spec Publ 27: 265–283

Anderson JB, Brake CF, Myers NC (1984) Sedimentation on the Ross Sea continental shelf, Antarctica. Mar Geol 57: 295–333

Anderson JB, Kennedy DS, Smith MJ, Domack EW (1991) Sedimentary facies associated with Antarctica's floating ice masses. Geol Soc Am Spec Paper 261: 1–25

Baroni C, Orombelli G (1994) The retreat of the Antarctic ice sheet from the Ross Sea continental shelf and the Holocene diffusion of the Adélie penguins in Victoria Land. Terra Antarct 1: 151–152

Bianchi C (1997) Flussi di particelle e processi biogeochimici in un'area interessata dall'arretramento stagionale del margine di ghiaccio (Mare di Ross Settentrionale). Thesis, University of Bologna, 105 pp

Brambati A, Fanzutti GP, Finocchiaro F, Melis R, Frignani M, Ravaioli M, Setti M (1997) Paleoenvironmental record in core ANTA91-30 (Drygalski Basin, Ross Sea, Antarctica). Antarct Res Series 71: 137–151

Buesseler KO, Cochran JK, Bacon MB, Livingston HD, Casso SA, Hirschberg D, Hartman MC, Fleer AP (1992) Determination of thorium isotopes in sea water by non-destructive and radiochemical procedures. Deep Sea Res 39: 1103–1114

Ceccaroni LM, Frank M, Frignani M, Langone L, Ravaioli M, Mangini A (1998) Late Quaternary fluctuations of biogenic component flux on the continental slope of the Ross Sea, Antarctica. J Mar Syst 17: 515–525

Coale KH, Bruland KW (1985) $^{234}Th/^{238}U$ disequilibria within the California current. Limnol Oceanogr 30: 22–33

Comiso JC, McClain CR, Sullivan CW, Ryan JP, Leonard C (1993) Coastal zone color scanner pigment concentrations in the Southern Ocean and relationship to geophysical surface features. J Geophys Res 98: 2419–2451

Cooper AK, Davey FJ (1987) The Antarctic continental margin: geology and geophysics of the Western Ross Sea. CPCEMR Earth Sci Ser 5B: 1–15

DeMaster DJ (1981) The supply and accumulation of silica in the marine environment. Geochim Cosmochim Acta 45: 1715–1732

DeMaster DJ, Dunbar RB, Gordon LI, Leventer AR, Morrison JM, Nelson DM, Nittrouer CA, Smith WO (1992) Cycling and accumulation of biogenic silica and organic matter in high latitude environments: the Ross Sea. Oceanography 5: 146–153

DeMaster DJ, Ragueneau O, Nittouer CA (1996) Preservation efficiencies and accumulation rates for biogenic silica and organic C, N, and P in high-latitude sediments: the Ross Sea. J Geophys Res 101:18501–18518

Dunbar RB, Anderson JB, Domack EW (1985) Oceanographic influences on sedimentation along the continental shelf. Antarct Res Ser 43: 291–312

Dunbar RB, Mucciarone DA, Leventer AR (1991) Sediment trap experiments in the central and western Ross Sea, January and February 1990. Antarctic J US 26: 115–117

Frank M, Eckhardt JD, Eisenhauer A, Kubik PW, Dittrich-Hannen B, Mangini A (1994) $^{10}Be$, $^{230}Th$ and $^{231}Pa$ in Galapagos microplate sediment: implications for hydrothermal activity and paleoproductivity changes during the last 100 000 years. Paleoceanography 9: 559–578

Frignani M, Langone L, Ragaglia L, Ravaioli M (1992) Organic matter and biogenic silica in bottom sediments of the western Ross Sea (Antarctica). In: Albertelli G, Ambrosetti W, Piccazzo M, Ruffani Riva T (eds) Atti IX Congr Ass Ital Oceanogr Limnol. Lang, Genova, pp 573–582

Frignani M, Giglio F, Langone L, Ravaioli M, Mangini A (1998) Late pleistocene-holocene sedimentary fluxes of organic carbon and biogenic silica in the northwestern Ross Sea (Antarctica). Ann Glaciol 27: 697–703

Heussner S, Ratti C, Carbonne J (1990) The PPS 3 time-series sediment trap and the trap sample processing techniques used during the ECOMARGE experiment. Cont Shelf Res 10: 943–958

Jaeger JM, Nittrouer CA, DeMaster DJ, Kelchner C, Dunbar RB (1996) Lateral transport of settling particles in the Ross Sea and implications for the fate of biogenic material. J Geophys Res 101: 18479–18488

Kumar N, Anderson RF, Mortlock RA, Froelich PN, Kubik P, Dittrich-Hannen D, Suter M (1995) Increased biological productivity and export production in the glacial Southern Ocean. Nature 378: 675–680

Labbrozzi L, Langone L, Frignani M, Ravaioli M (1998) Burial rates for biogenic silica, organic C and N at three sites of the Ross Sea (Antarctica). In: Albertelli G, Ambrosetti W, Piccazzo M, Ruffani Riva T (eds) Atti XII Congr Ass Ital Oceanogr Limnol, vol 2. Lang, Genova, pp 121–130

Langone L, Dunbar RB, Labbrozzi L, Ravaioli M, Frignani M (1996) Preliminary results from the ROSSMIZE cruise on biosiliceous sediment accumulation. Int Workshop Ross Sea Ecology, Taormina, May 14–16, pp 98–100 (Abstr)

Langone L, Frignani M, Cochran JK, Ravaioli M (1997) Scavenging processes and export fluxes close to a retreating ice margin (Ross Sea, Antarctica). Water Soil Air Pollut 99: 705–715

Langone L, Frignani M, Labbrozzi L, Ravaioli M (1998) Present-day biosiliceous sedimentation in the northwestern Ross Sea, Antarctica. J Mar Syst 17: 459–470

Ledford-Hoffman PA, DeMaster DJ, Nittrouer CA (1986) Biogenic silica in the Ross Sea and the importance of Antarctic continental-shelf deposits in the marine silica budget. Geochim Cosmochim Acta 50: 2099–2110

Licht KJ, Jennings AE, Andrews JT, Williams KM (1996) Chronology of late Wisconsin ice retreat from the western Ross Sea, Antarctica. Geology 24: 223–226

Martin JH, Fitzwater SE, Gordon RM (1990) Iron-deficiency limits phytoplankton growth in Antarctic waters. Global Biogeochem Cycles 4: 5–12

Mitchell BG, Holm-Hansen O (1991) Observations and modeling of the Antarctic phytoplankton crop in relation to mixing depth. Deep Sea Res 38: 981–1007

Nelson DM, Smith WO Jr (1991) Sverdrup revisited: critical depth, maximum chlorophyll levels, and the control of the Southern Ocean productivity by the irradiance-mixing regime. Limnol Oceanogr 36: 1650–1661

Nelson DM, DeMaster DJ, Dunbar RB, Smith WO (1996) Cycling of organic carbon and biogenic silica in the Southern Ocean: large-scale estimates of water-column sedimentary fluxes in the Ross Sea. J Geophys Res 101: 18519–18532

Picco P, Amici L, Langone L, Meloni R, Ravaioli M (1999) Temporal variability of currents in the Ross Sea (Antarctica). In: Manzella GMR, Spezie G (eds) Oceanography of the Ross Sea – Antarctica, Springer, Milano, Italy, 103–117

Ravaioli M, Frignani M, Gambi MC, Labbrozzi L, Langone L (1998) Particle fluxes and sediment characteristics at three selected sites in the Ross Sea (Antarctica). In: Manzella GMR, Spezie G (eds) Oceanography of the Ross Sea – Antarctica. Springer, Milano, Italy, 209–222

Reid DE (1989) Late cenozonic glacial-marine, carbonate, and turbidite sedimentation in the northwestern Ross Sea, Antarctica. MS Thesis, Rice University, Houston, TX, 179 pp

Rogers JC (1995) $^{13}C/^{12}C$ ratios in suspended and sea ice organic matter: variability and contributions to the sediments of the Ross Sea, Antarctica. MS Thesis, Rice University, Houston, TX, 108 pp

Smith WO Jr, Nelson DM (1985) Phytoplancton bloom produced by a receding ice edge in the Ross Sea: spatial coherence with the density field. Science 227: 163–166

Smith WO Jr, Sakshaug E (1990) Polar phytoplankton. In: Smith WO Jr (ed) Polar oceanography. Academic Press, San Diego, CA, pp 477–525

Smith WO Jr, Nelson DM, DiTullio GR, Leventer AR (1996) Temporal and spatial patterns in the Ross Sea: phytoplankton biomass, elemental composition, productivity and growth rates. J Geophys Res 101: 18455–18465

White WB, Peterson RG (1996) An Antarctic circumpolar wave in surface pressure, wind, temperature and sea-ice extent. Nature 380: 699–702

# CHAPTER 5

# Some Palaeoecological Remarks on the Ross Sea Shelf, Antarctica

A. Brambati, G.P. Fanzutti, F. Finocchiaro, R. Melis, N. Pugliese, G. Salvi, and C. Faranda[1]

**ABSTRACT**

Three gravity cores collected during the austral summer 1990–91 in different areas of the Ross Sea continental shelf have been studied. Sedimentological, micropalaeontological and chronological data show evidence for the different evolution of the Drygalski Basin, the Joides Basin and the Cape Adare area, during the late Quaternary. The advance of glaciers during LGM is well documented in the Drygalski Basin, where a basal till testifies the presence of an ice sheet anchored on the seafloor. In contrast, a glacial marine diamicton with calcareous foraminifers in the basal part of the core collected in the Joides Basin indicates that the West Antarctic Ice Sheet was not completely grounded on the seafloor during LGM. The successive deglacial phase is more or less synchronous in both basins (about 17 ka B.P.) and is documented in the overlaying units by the occurrence of well-sorted sands and foraminifer colonization. A very different sediment accumulation rate between the Drygalski and the Joides Basins is probably related to higher bioproductivity and mud accumulation in the latter during the Holocene. The Cape Adare area is a different example of environmental evolution, as shown by the presence of biogenic calcareous deposits. These deposits occur throughout the entire core, but they decrease in phase with increasing terrigenous inputs, probably during interglacial times. Within this evolution, foraminiferal and ostracod assemblages are sensitive to glacial/interglacial conditions. Their higher diversity is mainly recorded during glacial time.

## Introduction

Marine geology research in the Antarctic area has already highlighted the influence of glacial features (glaciers, ice sheets and shelves) on continental shelf evolution. In particular, the Ross Sea and Weddell Sea shelves have been the object of numerous studies in recent years, since the Quaternary sediments are much better represented here and more widespread than in other Antarctic areas. Therefore, numerous works have addressed the study of these sediments to define the environmental evolution of these areas during the late Cenozoic (e.g. Anderson et al. 1991; Licht et al. 1995, 1996; Brambati et al. 1997). Furthermore, international programs have focused their attention on the past 20000 years, since palaeoclimate reconstruction in this time interval may be useful to obtain forecasts for the near future.

The Ross Sea continental shelf is characterized by a mean depth of about 500 m and a rather irregular bathymetry and morphology which are related to the Cenozoic glacial erosion and deposition. In the western part of the Ross Sea shelf, the position and morphology of basins and banks are also controlled by the main structural features, oriented NW-SE near Cape Adare and NNW-SSE in the southern part of western Ross Sea (Fig. 1; Marchetti et al. 1993).

The present-day sedimentation in the Ross Sea shelf is mainly represented by both the terrigenous inputs of glaciers (ice shelves, sheets and tongues) and the accumulation of biogenic (siliceous and/or calcareous) remains (Anderson et al. 1984). The sediment distribution of a wide sector of the western Ross Sea continental shelf has already been studied by many authors (see Anderson et al. 1984 for a review) and also by Brambati et al. (1989). Their work, based on a multivariate analysis of textural characteristics, permitted distinguishing the zoning of the surface sediments: banks (top and flanks) and deep basins

[1] Dipartimento di Scienze Geologiche, Ambientali e Marine, v. Edoardo Weiss, 2-34127 Trieste, Italy

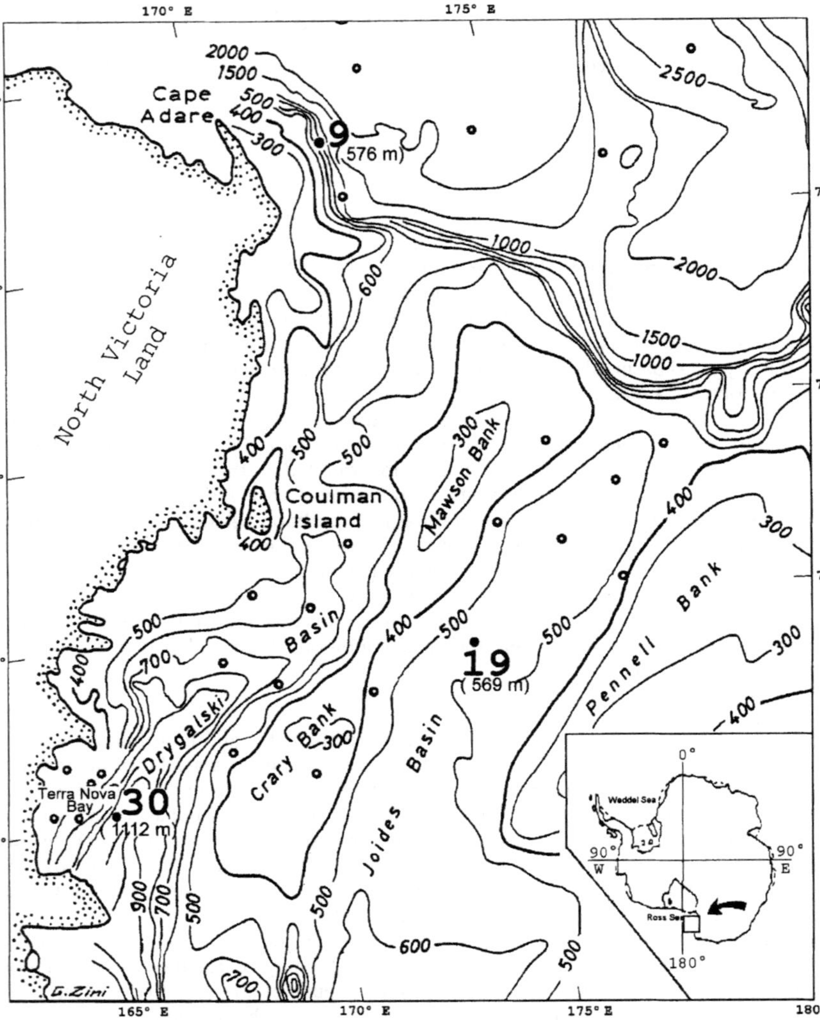


Fig. 1. Location of studied cores and bathymetric sketch map of western Ross Sea. *Open circles* show locations of other cores collected during Cariboo 1990–91 cruise

are characterized by poorly sorted coarse sediment (till) and by fine grained sediments, respectively. Till sediment is thought to be the primary sediment. The detrital component of deep basin sediment originates by the reworking of the till: on the top of the banks bottom currents winnow fine detrital mud, leaving pockets of well-sorted residual sand. Another important component of deep basin sediments is biogenic silica remains.

The coarse deposits characterize the inner shelf of Terra Nova Bay where the terrigenous inputs are due to the David and Campbell glaciers (Brambati et al. 1995). Preliminary geophysical data in the Ross Sea shelf indicate that the recent sediments of the Joides Basin area are thicker than those of the Drygalski Basin area (Stefanon 1989; Pescatore and Senatore 1989).

Past climatic variations have certainly influenced the extent of the main Antarctic glaciers, consequently conditioning the oceanographic circulation, the bioproductivity and the sedimentation processes. Since Quaternary sedimentation is the result of several complex depositional processes, the interdisciplinary study of the relative proportion of these deposits can provide useful palaeoenvironmental information (Grobe and Mackensen 1992; Domack et al. 1995).

During the late Quaternary, the Ross Sea area experienced a complex evolution, since its different sectors were conditioned by the fluctuation of the ice sheets (West Antarctic Ice Sheet, WAIS, and East Antarctic Ice Sheet, EAIS). During the Last Glacial Maximum (LGM), the ice sheets reached a more advanced (seaward) position with respect to the present one (Anderson et al. 1992; Baroni and Orombelli 1994). Although knowledge of the environmental evolution of the western sector of the Ross Sea shelf is quite advanced, several problems still remain. In particular, the exact position of the grounding line during the LGM and its successive evolution is still under discussion (Licht et al. 1995; 1996; Shipp and Anderson 1995).

Several sediment cores collected in different areas of the central and western continental shelf, where the ice sheet reached the seafloor during its maximum expansion, indicate a basal till overlain by glacial marine sediments (Anderson et al. 1991; Licht et al. 1996; Brambati et al. 1997). Other cores located north of 74°S are characterized by a glacial marine diamicton instead of basal till as the basal unit (Licht et al. 1996). Thus, these authors suggest a grounding line located about 250 km south of the shelf break (about 74° lat. S), even if the relationships between grounding line position and morphological features of the seabed (sequence of basins and banks) at this latitude are not discussed. However, preliminary geotechnical data on the northern Joides Basin suggest a grounded ice sheet along the northern edge of the basin which becomes floating ice above its deepest part (Corradi et al. 1997). In the central part of this basin, fossiliferous diamictons remain still difficult to date and interpret (Anderson et al. 1991; Dini and Salvi 1995; Licht et al. 1995; Melis et al. 1997a,b).

The complexity outlined is probably due to the fact that the WAIS, draining into the Ross Sea, consists of several ice streams which have moved independently during the climatic pulsations (Anderson et al. 1992; Jahns 1994; Shipp and Anderson 1995). In this context, several studies have already demonstrated that the combination of sedimentological and micropalaeontological data is very useful to reconstruct these fluctuations (Kellogg et al. 1979; Grobe and Mackensen 1992; Taviani et al. 1993; Domack et al. 1995).

This study concerns three gravity cores selected among those collected during the 6th PNRA oceanographic cruise (1990–1991) from three areas of the western Ross Sea shelf, characterized by very different morphological features: the Drygalski Basin (core ANTA91-30, water depth: 1112 m), the Joides Basin (core ANTA91-19, water depth: 569 m) and the Cape Adare area (core ANTA91-9, water depth: 576 m; Fig. 1). Among those already studied within this project, these cores are considered the most representative of the late Quaternary evolution for the previously cited areas. Besides, in these cores we would expect to note the different sedimentological processes already identified in the recent settings.

The aim of this work was to define three different reference sequences combining sedimentological, micropalaeontological (foraminifers and, where present, ostracods) and chronological data.

## Materials and Methods

The cores were submitted to X-ray radiographs; cores were then split, described and photographed. Subsamples of about 10–15 g were taken for grain-size analysis. Granules and small pebbles (2–8 mm) were separated from the sand and mud fractions by sieving at 2 mm. Since the small amount of subsamples was not representative of the gravel class, the % gravel was considered only an estimate. For this reason we prefer, in the text, to discuss grain-size data recalculated only on the <2 mm fraction. Each sample was wet-sieved at 50 µm, using procedures of Holland laboratories in order to separate sand and mud. The sandy fraction (2000–50 µm) was analyzed in a settling tube (tube length: 182.5 cm; internal diameter 8 cm) and the mud fraction on a Micromeritics ET 5000 Sedigraph, using a 0.5% sodium hexametaphosphate dispersing solution. Grain-size data were represented at 1/2 phi intervals. Carbonate content of the sand fraction was determined by weight loss after 18% HCl treatment, while that of the mud fraction was calculated using a Dietrich Fruling calcimeter.

The foraminifera and ostracod analyses were carried out on the >50 µm fraction on samples of about 15 g. In the samples where the number of specimens were <300, they were all counted and identified. In the case of very abundant assemblages, the sandy fraction was split in order to count a minimum of 300 specimens. Where present, all ostracods were separated. Adults were counted while the presence or absence of juveniles was noted.

Radiocarbon ages were determined on decalcified organic carbon and carbonate material using the AMS facility at the University of Groningen and one radiocarbon age was determined on a large mollusc (core ANTA91-9), at the University of Trieste. The conventional ages were corrected using a conventional reservoir correction of 1400 years and also using the calibration program CALIB 3.0 (Stuiver and Reimer 1993; Table 1).

## Results

### Drygalski Basin

The Drygalski Basin is located in the westernmost part of the Ross Sea and presents a maximum depth of 1500 m. Its morphology results from the extensive erosion of the David Glacier during its maximum expansion. The recent stratigraphy in this area has been studied by Anderson et al. (1991), Brambati et al. (1997), and Cassidy et al. (1988). These authors recognize the complete sequence from the LGM up to the present only in the deepest part of the basin.

Core ANTA91-30 (lat.: 75°57.05′S, long.: 165°46,96′E, length: 315 cm, water depth: 1112 m) is located in the central and deepest part of the basin along its longitudinal axis.

### Sedimentology

The complete stratigraphic sequence of the ANTA91-30 core was characterized by the following units, from the bottom to the top: basal till, glacial marine sediments and Holocene siliceous mud (Brambati et al. 1994, 1997).

The lower unit (315–103 cm) was characterized by the lack of stratification, low water content (22.4 ± 3.8%), very low organic carbon and biogenic silica content, homogeneity of textural and compositional characteristics (Fig. 2), low sorting, and the fair presence of well-elaborated quartz grains (Protopsalti 1997). These features have led us to classify the basal till, in agreement with Anderson and Molnia (1989). The glacial marine sequence (103–30 cm) consisted of a well-sorted sand (103–88 cm) in its lower part. In the upper levels (88–30 cm), the changing of textural characterstics and in particular the presence of laminations (40–30 cm), as well as the fluctuations of biological parameters, suggest a first period of climatic warming, followed by a period of cooling. The upper unit (30–0 cm) consisted of siliceous mud, with an increase of organic carbon and biogenic silica. This was related to the establishment of open-marine conditions.

### Micropalaeontology

The micropalaeontological record of this core included benthic foraminifers, radiolarians, diatoms and sponge spicules. The biogenic remains were generally scarce. In the basal till

**Table 1.** $^{14}C$ dating of cores. Corrected ages are calculated subtracting from the uncorrected values the reservoir correction proposed by Gordon and Harkness (1992). Calibrated ages are calculated using the software Calib 3.0 (Stuiver and Reimer 1993)

| Core | Level (cm) | Uncorrected age | Corrected age | Calibrated age |
|---|---|---|---|---|
| ANTA91-30 | 0–2 | 4935 | 3535 | 3396 |
| ANTA91-30 | 22–24 | 11365 | 9965 | 10881 |
| ANTA91-30 | 71.5–74 | 16950 | 15550 | 18071 |
| ANTA91-19 | 24–26 | 4470 | 3070 | 2815 |
| ANTA91-19 | 179–185 | 9070 | 7670 | 8077 |
| ANTA91-19 | 206–209 | 12440 | 11040 | 12570 |
| ANTA91-19 | 376–382 | 19020 | 17620 | 20428 |
| ANTA91-09 | 109–119 | 21500 | 20100 | – |

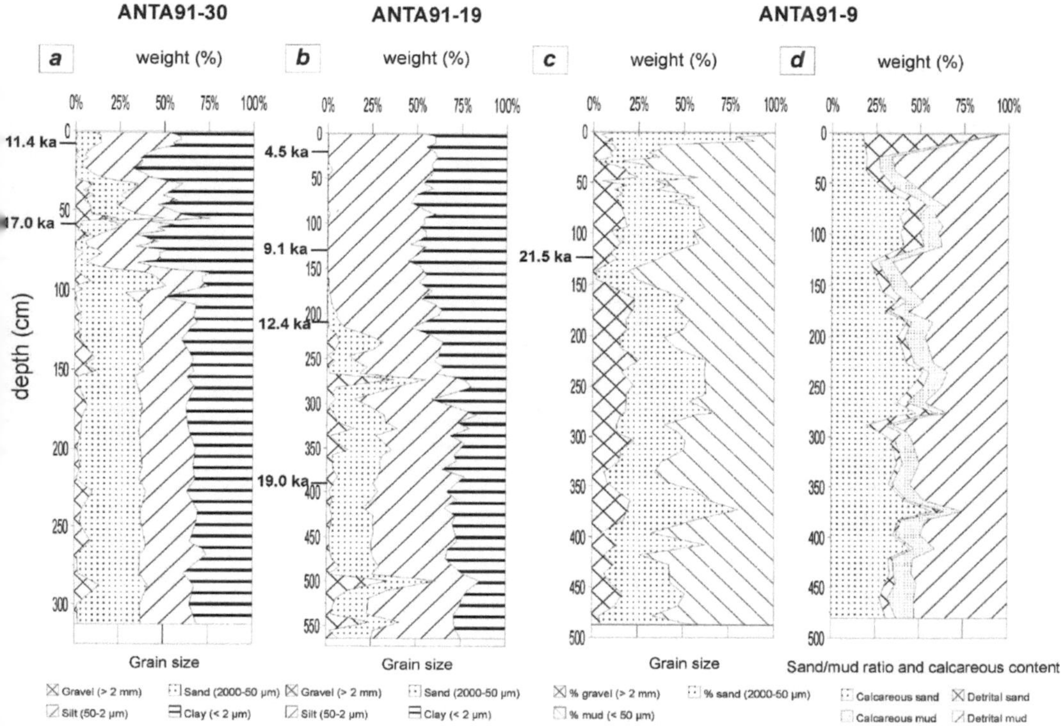


**Fig. 2.a,b** Sand/silt/clay in cores ANTA91-30 and ANTA91-19. **c** Gravel/sand/mud in core ANTA91-09. **d** Carbonate content in core ANTA91-09. Uncorrected ages are reported

unit only very rare and abraded spicules were present.

Foraminifers were present in the upper two units, where 24 species belonging to 16 genera were recognized, all belonging to the suborder Textulariina. On the basis of their good test preservation, they are considered autochthonous. *Miliammina* spp. and *Trochammina multiloculata* appeared sparsely immediately above the basal till (at 100 cm). They may be considered the first colonizers of the seafloor. From 100 cm to 30 cm, foraminifers were present in very low numbers of specimens and species, being absent from several intervals (Fig. 3). In the upper, Holocene unit, they became more diversified, with the occurrence of species of *Cyclammina*, *Hyperammina*, *Miliammina*, *Reophax* and *Trochammina*. This specific diversity trend may be related to the increase of organic carbon and biogenic silica, which testify to a higher bioproductivity. In this unit the foraminiferal concentration was still very low, ranging from 0.44 to 11.0 specimens $g^{-1}$ of sand with some peaks at 64, 20, 14 and 0 cm. These low values were comparable with those of other cores collected in the central Ross Sea shelf (Kellogg et al. 1979) and in the Antarctic Peninsula (Domack et al. 1995).

The foraminiferal assemblage of ANTA91-30 core was similar to the Deep Basin one reported by Milam and Anderson (1981) in the recent deep sediments of the George V shelf. In particular, the prevalence of *Miliammina earlandi* (also as *Miliammina arenacea*) characterized the recent Drygalski Basin sediments, where the deposition occurred below the CCD depth (D'Onofrio and Pugliese 1989) and other areas where the waters were very corrosive (Finger and Lipps 1981).

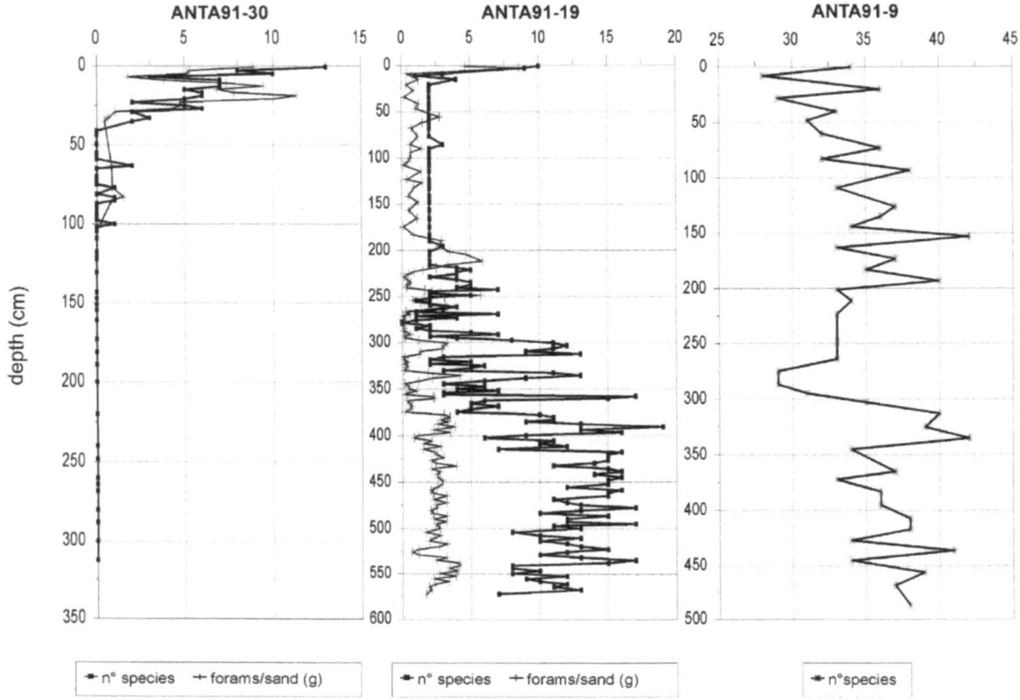


**Fig. 3.** Foraminifer distribution: number of species (no. species) and ratio between number of specimens on weight of the sand fraction [*forams/sand* (g)] in studied cores

## Joides Basin

The Joides Basin presents an elongated shape with a sill near the 74°30′ parallel (Fig. 1). The northern area of the basin, with a depth of 600 m, extends nearly to the continental shelf break. Southwards the area is deeper and, south to Franklin Island, it opens and connects with the basin area north of Ross Island. Bottom sediments are mainly composed of mud (Anderson et al. 1984; Brambati et al. 1989), with more than 30% of biogenic silica.

Core ANTA91-19 is located in the deepest sector of the basin (lat.: 74°26.04′S, long.: 173°46,14′E, length: 575 cm, water depth: 569 m).

## Sedimentology

Through textural analysis (Fig. 2) core ANTA91-19 has been subdivided into three units. The lower unit (from 575 to 300 cm) has a fine grained matrix with the inclusion of gravel sized clasts. The sand content averages 20% with some levels (around 500 cm) reaching values of 40%. This unit was characterized by rough and faint stratification, an average water content of 35.4 ± 4.4%, and homogeneity of textural and compositional characteristics with a constant ratio sand/silt/clay. Only one coarse-grained level (gravel and sand) at 506–498 cm was evidenced. The morphometric features of sand grains appeared to be less elaborated, if compared to those of the core ANTA91-30 basal till (Protopsalti 1997).

The central unit (from 300 to 180 cm) showed variable percentages of sand and water content. In particular an increase of the sand fraction (values up to 34%) occurred from 250 to 300 cm. The upper unit (from 180 to top) was characterized by a fine grain-size sediment in which a very low sand content was noted.

## Micropalaeontology

The ANTA91-19 core yielded 72 species of benthic foraminifers belonging to 35 genera. Among these, 57 calcareous, including the only planktonic species *Neogloboquadrina pachyderma*, and 15 arenaceous species were found.

Calcareous foraminifers were present only in the lower interval of this core (575–217 cm, lower unit and lower part of the central unit), whereas they totally disappeared above 217 cm and were substituted by arenaceous forms associated with siliceous microfossils. In decreasing order, *Globocassidulina subglobosa, Trifarina angulosa, Globocassidulina biora* and *Globocassidulina crassa* were the most common calcareous foraminifers. These were often associated with other species, such as *Cassidulina laevigata, Cassidulinoides porrectus, Ehrenbergina glabra* and, very subordinately, *Astrononion antarticus, Astrononion echolsi, Pullenia subcarinata* together with *Fissurina* spp., *Lagena* spp. and *Oolina* spp. *Miliammina earlandi* was the only arenaceous species present in this basal unit, in low abundance.

This assemblage showed a relatively constant specific diversity from 517 to about 400 cm, and successively tended to decrease in diversity even if some evident peaks occurred (Fig. 3); above 218 cm no calcareous forms were found. This trend seems to correspond to the period ranging from the end of LGM to the beginning of the Holocene. In particular, the occurrence of very abundant and well-sorted specimens of *Trifarina angulosa* and/or *Globocassidulina subglobosa* in correspondence to some sandy levels (300–280 cm) can be related to the occurrence of strong bottom currents during the ice-shelf retreat. At the base of the upper interval (217 cm), only *Miliammina earlandi* and successively *Miliammina lata* evidently survived the changed oceanographic conditions. In the uppermost part of the core (24 cm to top), corresponding to 4470 years B.P., the specific diversity increased with the exclusive presence of other arenaceous foraminifers, such as *Cribrostomoides wiesneri, Lagenammina laguncola, Reophax* spp. and *Trochammina* spp.

## Cape Adare Area

Towards the external margin of the Ross Sea shelf, the sediments showed a progressive increase in their calcareous biogenic component characterizing them as true bioclastic deposits on the shelf margin and in some basinal areas. These deposits, which are contemporary with the maximum expansion of the WAIS, were several metres thick (Taviani et al. 1993; Taviani and Anderson 1995; Brambati et al. 1996; Melis et al. 1997a). The presence of these sediments of predominantly calcareous composition is rather singular in the Antarctic environment, which is normally characterized by a scarcity, if not complete absence, of such remains, even at shallow depths. In fact, the low temperature of the water exerts a determining effect on the solution of calcareous remains, forcing the CCD limit generally to about 350–550 m on the shelf (Kennett 1966; Fillon 1974).

The presence of such a richness of calcareous taxa in relation to cold climatic cycles, and the reason for their particular concentration in this sector of the Ross Sea, are object of ongoing studies (Taviani et al. 1993; Taviani and Anderson 1995).

At present, knowledge on the polar carbonate is increasing; the main factors that control their origin are considered water masses circulation, nutrient supply, sea-ice cover, presence/abundance of biota, and rate of $CaCO_3$ dissolution (Rao 1996). In the outer Ross Sea shelf, the Circumpolar Deep Water (CDW) flowing into the shelf contributes significantly to the nutrient supply and the productivity increase of this area. On the contrary, the High Salinity Shelf Water (HSSW) flowing northwards creates a well developed frontal system near the shelf break (Jacobs et al. 1985). For this reason, strong currents can occur in the area, producing coarse bioclastic carbonate deposits (Taviani et al. 1993).

Core ANTA91-9 (lat.: 71°30.77′S, long.: 171°54,22′E, length: 496 cm, water depth: 576 m) is located in the northwestern area of the mentioned biogenic sector, near Cape Adare. Here the shelf is narrow and deepens rapidly basinward (Fig. 1).

## Sedimentology

The core was characterized by an abundance of coarse calcareous bioclastic material (Fig. 2). The calcareous fragments were more than 2 mm in size and were constantly present throughout the core, showing a mean percentage of about 10%. The middle part of the core (270–170 cm) presented a gravel mean content of about 18%, showing a maximum content of 29% at 173–178 cm. In the

0–40 cm interval, gravels were also composed of detrital clasts.

The maximum percentage of sand (up to 91.8%) was at the top (0–10 cm). In the remaining part of the core the sand values ranged from 30 to 50% (average: 40.7 ± 11.3), with a decreasing trend downward. The sand fraction was mainly composed of calcareous organic remains (average percentage 83.8 ± 13.5). Two levels showed only a low organic content, i.e. the core top and 289–287 cm. In general, the upper part of the core (160–0 cm) had the lowest values (mean value 79.6 ± 17.6) if compared to those of the lower part (from bottom to 160 cm: 90.0 ± 8.2). At 156–120 cm and 60–10 cm the mean grain-size became finer, with an increase in mud content (65–80%).

The calcareous content in mud fraction was low (average:18.4 ± 13.5%). Its downcore trend was very similar to that of the sand fraction, with the exception of the top level.

## Micropalaeontology

In the sandy fraction of this core, a rich calcareous faunal assemblage was found, including abundant molluscs, bryozoans, ostracods, corals, barnacles, brachiopods and foraminifers (both benthic and planktonic). Among these, foraminifers and ostracods were normally very diverse and well preserved. Biogenic silica was also present with common sponge spicules, diatoms and very subordinate radiolarians. Conventional $^{14}C$ analysis, carried out on a mollusc valve, indicates a date of 21 500 years B.P. at a depth of 110 cm from the top.

Forty-four genera and 59 species of foraminifera were found in the core; of these genera 31 belong to rotalids, 8 to miliolids and only 5 to arenaceous taxa. Different species of small *Discorbis*, *Fissurina*, *Lagena* and *Entosolenia* were provisionally grouped as genera without species identification, according to several authors (i.e. Milam and Anderson 1981; Mead 1985). *Neogloboquadrina pachyderma* was very common in all the levels. In general, planktonic foraminifers ranged between 13 to 40% of the total population with higher values in the bottom and top of the core.

Benthic foraminifers were widely diffused in the entire core and at least 30 taxa were generally present; among these, *Astrononion antarcticus*, *Fissurina* spp., *Globocassidulina rossensis*, *G. subglobosa*, *Lagena* spp., *Miliolinella subrotundata* and *Patellina corrugata* were always present. Other species, such as *Cibicides* spp., *Epistominella exigua*, *Rosalina globularis*, *Triloculina subrotunda*, *Trifarina angulosa* and *Tubinella funalis*, were very common. Other species (i.e. *Lenticulina* spp., *Pyrgo* spp., *Globocassidulina biora* and some arenaceous forams) appeared haphazardly, but in general they were subordinate.

This assemblage was composed of a similar occurrence of both epifaunal and infaunal species which were generally found in the upper shelf-bathyal settings. In particular, *G. subglobosa* and *T.earlandi* were related to well-oxygenated water masses of the shelf, while the *E. exigua* assemblage characterized the deepest environment of Antarctica (Milam and Anderson 1981; Mackensen et al. 1993). Most of these species belong to the Neogene polar assemblage which has exhibited a notable stability over the past 30 Ma (Webb 1988).

Although the foraminifer assemblage composition noted in this core seems to be very constant, in correspondence to the levels where the carbonate content was high, the species diversity increased. On the contrary, in correspondence to the levels where the carbonate content was low, a slight decrease in specific diversity and a reorganization of some species was noted (Fig. 3). In these latter levels, *Globocassidulina* spp., *M. subrotunda* and other miliolids were the most abundant species. During the LGM (21.5 ka at about 110 cm), the assemblage reached a great species diversity. On the basis of a certain sorting of the foraminiferal tests of the uppermost level only, the taxa can be considered displaced.

The ostracode fauna which occurred in the whole core was represented by 64 species belonging to 38 genera. Most of these (54 species) are believed to be autochthonous, since they are simultaneously represented in the core samples by adult and juvenile specimens. On the contrary, ten species were considered displaced since they were represented by adults without juveniles, or juveniles only.

Among the autochthonous species, two groups were identified. The first group included the species almost constantly present throughout the core: *Patagonicythere devexa*, *Pseudocythere similis*, *Loxoreticulatum fallax*, *Xestoleberis rigusa* and, subordinately, *Bairdoppilata* cf. *B. simplex*, *Hemicytherura irregularis*, *Kangarina* sp. In addition, some of these were at times the dominant species in the assemblage (*P. devexa*, *P. similis*, *L. fallax*).

The second group consists of species which were absent in the intervals characterized by high terrigenous inputs (289–287, 144–137, 22–10 cm). Such species were *Nealocythere antarctica*, *Bairdia* sp., *Macrocypris* cfr. *M. similis*, *Paracytherois* sp., *Paradoxostoma* spp., *Polycope* spp., *Sclerochilus reniformis*, *Xestoleberis setigera*.

On the basis of these preliminary data, a provisional hierarchy of opportunism can be realized (sensu Dodd and Stanton 1990). The most opportunistic species belong to the first group, since they were always present, also in levels with increased terrigenous inputs. *Patagonicythere devexa*, *Pseudocythere similis* and *Loxoreticulatum fallax*, which were sometimes dominant, may be considered the best examples of opportunistic species. Species of the second group included forms which were less opportunistic, since they did not appear in correspondence to the levels of strong terrigenous inputs.

On the basis of the recent literature (Benson 1964; Neale 1967; Briggs 1978; Whatley et al. 1988; Hartmann 1990, 1991; D'Onofrio and Pugliese 1989; Melis et al. 1997a,b) and taking into account the occurrence of the ostracode species in the uppermost levels of the core at a water depth of 596 m, most of the above-mentioned autochthonous species normally showed a wide depth distribution. On the contrary, the very rare autochthonous species (*Antarctiloxoconcha frigida*, *Cativella bensoni*, *Convexochilus meridionalis* and *Sclerochilus antarcticum*) up to now have been found at shallower depths. Thus, their exclusive occurrence in some intervals (bottom-303 cm; 119–97 cm, 75–40 cm and 30–20 cm) might indicate settings of decreased depth.

## Conclusions

The study of selected cores from three sectors of the Ross Sea shelf highlights the different pattern of environmental evolution of these areas during the late Quaternary. The Drygalski Basin normally shows a type evolution characterized by a sequence of basal till, glacial marine and marine muds. Within such an evolution, the lack of biogenic remains, together with other geochemical and sedimentological evidence, indicates that during the LGM the marine life conditions were not favourable, due to the presence of a grounded ice sheet (basal unit). The appearance of arenaceous foraminifers in the glacial marine unit clearly indicates the beginning of the conditions for the seafloor colonization after the retreat of this ice sheet. Nevertheless, during the Holocene, sedimentological processes and the depth of the seafloor constantly below the CCD level in this area do not allow a great foraminiferal diversification.

The radiometric data on three levels of core ANTA91-30 confirm the retreat of the grounding line from the latitude of Drygalski Basin before 18.1 ka (basal unit), while the open sea conditions began at about 10.9 ka B.P., showing that the glacial marine unit was deposited in a time span of about 7 ka, with a sedimentation rate of 6.9 cm/ka. Holocene sedimentation rates decrease to 2.3–2.4 cm/ka (Brambati et al. 1997).

The Joides Basin evolution is different from that of the Drygalski Basin, because the basal unit of core ANTA91-19 was characterized by a glacial marine diamicton (sensu Licht 1996) with a high content of relatively well-preserved foraminifers. The dating of 19.0 ka B.P. (uncorrected) of the upper part of glacial diamicton suggests that, in the deepest part of this basin, the WAIS was not completely grounded during the LGM, in contrast with the situation on the flanks of the basin, where evidence of overconsolidation of the sediments has been observed (Corradi et al. 1997).

The successive deglaciation phase was marked by the occurrence of some very sandy levels where the finding of particularly abundant and well-sorted specimens of *Trifarina angulosa* and/or *Globocassidulina subglobosa* suggests the occurrence of strong bottom currents during the ice-shelf retreat. Dates ranging from 12.4 and 9.1 ka B.P. (uncorrected) indicate the accomplishment of deglaciation in the Joides Basin area and the successive onset of open sea conditions.

The difference in the evolution between the Drygalski and Joides Basins was also recorded by their sedimentation rates. In the Joides Basin, the sedimentation rate and the thickness of marine muds was rather high. The first 25 cm of marine muds showed a sedimentation rate of 8.1 cm/ka, while the remaining muds indicated an increase to 34.1 cm/ka. These sedimentation rates can be related both to the increase of silica biogenic remains and to the contribution of detrital muds. The former are probably related to an increasing productivity of siliceous taxa strongly influenced by oceanic circulation; the mud is winnowed by bottom currents from the top of the banks and accumulates in the lower-energy shelf basin.

During the deposition of marine muds, the CCD rose, reaching shallower depths. In this upper unit only arenaceous foraminifers were observed. Their species diversity increased in recent sediments, where the environmental conditions were similar to the present-day ones.

The Cape Adare area is a quite different example of environmental evolution linked to the presence of mostly biogenic calcareous deposits. This results from its position north of a grounding line, as pointed out by the recent literature. The grounding line acts as a division between two environments: the northern one characterized by calcareous sedimentation, the southern one by glacial sedimentation (diamicton or basal till). In core ANTA91-09, calcareous skeletal deposits occurred throughout the core except for some levels where terrigenous (both tephra and ice rafted debris) inputs increased. Thus, in general, the glacial phases were characterized by calcareous deposits; the interglacial ones showed an increase in terrigenous input. Within this evolution, a stock of opportunistic foraminifers and ostracods were always present. These tolerated environmental conditions linked to glacial and interglacial phases. Other ostracode species were present in correspondence to glacial episodes only. Among the latter, scattered ostracods may indicate settings of decreasing depth. This faunal distribution suggests that at least in this sector of the Ross Sea the biogenic calcareous deposits may be autochthonous.

The date of 21.5 ka B.P. (at 114 cm) reveals a mean sedimentation rate of 5.3–5.4 cm/ka; this may indicate that the sequence has a late Quaternary thickness greater than that reported by Taviani et al. (1993), who reported a thick calcareous sequence truncated by the LGM event.

*Acknowledgements.* We thank two anonymous reviewers for their useful suggestions to improve the manuscript. This work was financed by the Italian Progetto Nazionale di Ricerche in Antartide, research project Glaciologia e Paleoclima.

## References

Anderson JB, Molnia B (1989) Glacial-marine sedimentation. Short course in geology, 28th Int Geol Congr 9. American Geophysical Union, Washington, DC, 127 pp

Anderson JB, Brake CF, Myers NC (1984) Sedimentation on the Ross Sea continental shelf, Antarctica. Mar Geol 57: 295–333

Anderson JB, Kennedy DS, Smith MJ, Domack EW (1991) Sedimentary facies associated with Antarctica's floating ice masses. Geol Soc Am Spec Paper 261: 1–25

Anderson JB, Shipp SS, Bartek LR, Reid DE (1992) Evidence for a grounded ice sheet on the Ross sea continental shelf during the late Pleistocene and preliminary paleodrainage reconstruction. Antarct Res Ser 57: 39–62

Baroni C, Orombelli G (1994) The retreat of the Antarctic ice sheet from the Ross Sea continental shelf and the Holocene diffusion of Adélie penguins in Victoria Land. Terra Antarct 1: 151–152

Benson RH (1964) Recent cytheracean ostracodes from McMurdo Sound and the Ross Sea, Antarctica. Paleontol Contrib Univ Kans 6: 1–36

Brambati A, Fanzutti GP, Finocchiaro F, Simeoni U (1989) Sediments and sedimentological processes in the Ross Sea continental shelf (Antarctica): results and preliminary conclusions. Boll Oceanol Teor Appl 7(1–1): 159–188

Brambati A, Ceccaroni L, D'Onofrio S, Fanzutti GP, Finocchiaro F, Frignani M, Langone L, Melis R, Ravaioli M (1994) Paleoenvironmental inferences from the core ANTA91-30 (Drygalski Basin—Ross Sea, Antarctica). Terra Antarct 1: 335–337

Brambati A, Fanzutti GP, Finocchiaro F, Simeoni U, Tsakiridou E (1995) A textural seafloor map in Terra Nova Bay (Ross Sea, Antarctica): an approach. In: Ricci CA (ed) The Antarctic region: geological evolution and processes. Terra Antarctica Publication, Siena, pp 1167–1169

Brambati A, Fanzutti GP, Finocchiaro F, Melis R, Frignani M, Ravaioli M, Setti M (1997) Paleoenvironmental record in Core ANTA91-30 (Drygalski Basin, Ross Sea, Antarctica). In: Barker PF, Cooper AC, Brancolini G (eds) Geology and seismic stratigraphy of the Antarctic Margin, part 2. Antarct Res Ser 71: 137–151

Briggs WM (1978) Ostracoda from the Pleistocene Taylor Formation, Ross Island, and the Recent of the Ross Sea and McMurdo Sound region, Antarctica. Antarct J 13(4): 27–29

Cassidy DS et al. (eds) (1988) The United States Antarctic Research Program in the Western Ross Sea, 1979–1980: the sediment descriptions. Sedim Res Lab Contrib 53. Tallahasee, Fl

Corradi N, Fierro G, Mirabile L, Ferrari M, Ivaldi R (1997) Geotechnical, sedimentological characteristics and seismic stratigraphy of Northern Joides Basin (Ross Sea—Antarctica): preliminary results. In: Ricci CA (ed) The Antarctic region: geological evolution and processes. Terra Antarctica Publication, Siena, pp 885–887

Dini M, Salvi G (1995) A paleontological and geochemical study of the core Anta 19—Joides Basin (Antarctica). 7th Int Symp on Antarctic Earth Sciences, p 113 (Abstr)

D'Onofrio S, Pugliese N (1989) Foraminiferal and ostracod fauna from the Ross Sea, Antarctica: preliminary results. Boll Oceanol Teor Appl 7(1–2): 129–138

Dodd RJ, Stanton RJ Jr (1990) Paleoecology. Concepts and application. Wiley, New York, 502 pp

Domack EW, Ishman SE, Stein AB, McClennen CE, Jull AJ (1995) Late holocene advance of the Muller Ice Shelf, Antarctic Peninsula: sedimentological, geochemical and palaeontological evidence. Antarct Sci 7(2): 159–170

Fillon RH (1974) Late Cenozoic foraminiferal paleoecology on the Ross Sea, Antarctica. Micropaleontology 20(2): 129–151

Finger KL, Lipps JH (1981) Foraminiferal decimation and repopulation in an active volcanic caldera, Deception Island, Antarctica. Micropaleontology 27(2): 111–139

Grobe H, Mackensen A (1992) Late Quaternary climatic cycles as recorded in sediments from the Antarctic continental margin. Antarct Res Ser 56: 349–376

Hartmann G (1990) Antarktische benthische Ostracoden. VI. Auswertung der Reise der Polar Stern Ant. VI-2 (1. Teil, Meiofauna und Zehnerserien) sowie Versuch einer vorläufigen Auswertung aller bislang vorliegenden Daten. Mitt Hamb Zool Mus Inst 87: 191–245

Hartmann G (1991) Antarktische benthische Ostracoden. VII. Ostracoden aus dem Oberen Litoral von King George Island. Mitt Hamb Zool Mus Inst 88: 211–220

Jacobs SS, Fairbanks RG, Horibe J (1985) Origin and evolution of water masses near the Antarctic continental margin: evidence from $H_2\,^{18}O/H_2\,^{16}O$ ratio in seawater. Antarct Res Ser 43: 59–85

Jahns E (1994) Evidence for a fluidized till deposit on the Ross Sea continental shelf. Antarct J 29(5): 139–141

Kellogg TB, Osterman LE, Stuiver M (1979) Late quaternary sedimentology and benthic foraminiferal paleoecology of the Ross Sea, Antarctica. J Foraminiferal Res 9: 322–335

Kennett JP (1966) Foraminiferal evidence of a shallow calcium carbonate solution boundary, Ross Sea, Antarctica. Science 153: 191–193

Licht KJ, Hilfinger MF, Franceschini MJ, Domack EW, Jennings AE, Andrews JT (1995) New marine stratigraphic evidence for glaciation across the Ross Sea continental shelf: ice sheet or ice shelf? 7th Int Symp on Antarctic Earth Sciences, p 244 (Abstr)

Licht KJ, Jennings AE, Andrews JT, Williams MK (1996) Chronology of late Wisconsin ice retreat from the western Ross Sea, Antarctica. Geology 24(3): 223–226

Mackensen A, Futterer DK, Grobe H, Schmield G (1993) Benthic foraminiferal assemblage from the eastern South Atlantic Polar Front region between 68 and 73°: distribution, ecology and fossilization potential. Mar Micropaleontol 22: 33–69

Marchetti A, Baroni C, Brancolini G, Lodolo E, Orombelli G, Stefanon A (1993) Sea bottom morphologies of the Ross Sea area. Boll Geofis Teor Appl 35: 203–218

Mead GA (1985) Recent benthic foraminifera in the Polar Front region of the southwest Atlantic. Micropaleontology 31(3): 221–248

Melis R, Pugliese N, Salvi G (1997a) Foraminifera and ostracoda of Antarctic Late Quaternary shelf sediments: their utilization in paleoenvironmental variations. 1st Int Conf on Application of Micropalaeontology in Environmental Sciences, Tel Aviv, pp 84–85 (Abstr)

Melis R, Salvi G, Dini M, D'Onofrio S, Pugliese N (1997b) Micropaleontological aspects of some cores of the Western Ross Sea (Antarctica). Terra Antarct 1: 97–102

Milam RW, Anderson JB (1981) Distribution and ecology of recent benthonic foraminifera of the Adelie—George V continental shelf and slope, Antarctica. Mar Micropaleontol 6: 297–325

Neale JW (1967) An ostracod fauna from Halley Bay, Coat Land, British Antarctic Territory. Br Antarct Surv Sci Rep 58: 1–50

Pescatore T, Senatore MR (1989) Analysis of high-resolution seismic profiles in the western Ross Sea, Antarctica. Boll Oceanol Teor Appl 7(1–2): 77–82

Protopsalti I (1997) Applicazione di un metodo automatico per l'estrazione di parametri morfometrici da clasti per una caratterizzazione di sedimenti marini e costieri antartici in ottica paleoambientale. PhD Thesis, University of Trieste, 103 pp

Rao CP (1996) Modern carbonates: tropical, temperate, polar—introduction to sedimentology and geochemistry. Carbonates, Hobart, Australia, 206 pp

Shipp S, Anderson JB (1995) Quaternary deglacial history of Ross Sea, Antarctica; results from recent seismic investigation. 7th Int Symp on Antarctic Earth Sciences, p 347 (Abstr)

Stefanon A (1989) First Italian oceanographic expedition in the Ross Sea, Antarctica: the SBP survey. Boll Oceanol Teor Appl 7(1–2): 55–70

Stuiver M, Reimer PJ (1993) Extended $^{14}C$ data base and revised CALIB 3.0 $^{14}C$ age calibration program. Radiocarbon 35(1): 215–230

Taviani M, Anderson JB (1995) Response of biogenic carbonate factories to ice sheet and ice shelf dynamics: a largely deglaciated Ross Sea since isotopic stage 3? 7th Int Symp on Antarctic Earth Sciences, p 367 (Abstr)

Taviani M, Reid D, Anderson J (1993) Skeletal and isotopic composition and paleoclimatic significance of late Pleistocene carbonates, Ross Sea, Antarctica. J Sedim Petrol 63 (1): 84–90

Webb PN (1988) Upper Oligocene-Holocene benthic foraminifera of the Ross Sea region. Rev Paleobiol 2: 589–603

Whatley R, Chadwick J, Coxill D, Toy N (1988) The ostracod family Cytheruridae from the Antarctic and south-west Atlantic. Rev Esp Micropaleontol 20(2): 171–203

CHAPTER 6

# Water Masses and Suspended Matter Characteristics of the Western Ross Sea

G. Budillon[1], S. Tucci[2], A. Artegiani[3], and G. Spezie[1]

**ABSTRACT**

As part of an interdisciplinary field project performed during the austral summer 1994/95 devoted to examine the physical and bio-geochemical characteristics of the Ross Sea, hydrological stations were operated in the western sector off Victoria Land and water samples were collected at different depths of the water column to determine the particulate matter content. These investigations were carried out in the framework of the CLIMA (Climatic Long-Term Interaction for the Mass Balance in Antarctica) project of the Italian PNRA (National Program of Research in Antarctica).

The three-dimensional thermohaline field was analyzed to define the most relevant physical features of this region and it has been linked with the geological data.

Hydrological data reveal that the Antarctic Slope Front was altered by a 200-km southward penetration of Circumpolar Deep Water at 174°E. This warm intrusion onto the shelf appears to be continuous and may be a typical feature, as inferred by our geological data and also by the recent long-term current meter series collected by other Italian groups.

The thermohaline field shows two distinct circulations in the Drygalsky Basin associated with the southward patterns of the modified Circumpolar Deep Water and the northward flow of the High Salinity Shelf Water.

Water masses have been characterized also on the basis of their suspended matter contents. The geological information is in good agreement with the hydrological data as shown by means of a comparative analysis.

## Introduction

The Southern Ocean is the only place where ocean currents flow unconstrained around the globe. This ocean has the land mass of Antarctica as its southern boundary, but has no land boundary to the north. Traditionally, the Subtropical Front is considered the northern boundary of the Southern Ocean (Gordon 1988).

One of the most important dynamic features of the Southern Ocean is the Antarctic Circumpolar Current (ACC), which flows eastward around Antarctica and mixes with different water masses along its path. Below the surface and extending to a depth of around 4000 m is the Circumpolar Deep Water (CDW), the most voluminous water mass of the Southern Ocean; this water mass is carried round Antarctica by the ACC (Whitworth and Nowlin 1987).

South of the ACC are vast areas of sluggish flow, many of which are organized into cyclonic gyres. One of them is located north of the Ross Sea, in the Pacific sector of the Southern Ocean. In this region the dynamic height anomaly of the sea surface relative to 2500 dbar shows an elongated cyclonic circulation extending northeast from the Ross Sea, which is referred to as the Ross Gyre (Gordon and Molinelli 1982; Locarnini 1994). This cyclonic feature develops between 160°E and 140°W. CDW flows in the Ross Gyre and, reaching the continental slope in the southern limb, mixes isopycnally with the shelf waters of the continental shelf (Locarnini 1994).

[1] Istituto di Meteorologia e Oceanografia, IUN Naples, Italy
[2] Dipartimento di Scienze della Terra, University of Genoa, Italy
[3] Istituto di Ricerche sulla Pesca Marittima, CNR Ancona, Italy

Reaching the Antarctic continental shelf, the CDW moves upward in the water column and, at the continental shelf break, mixes with the shelf waters, forming deep and bottom waters generically named Antarctic Bottom Waters (AABW). AABW entering in the Southern Ocean mix with deep waters to form new CDW.

The mixing of the CDW with the surface and shelf waters in the Ross Sea forms a distinct water mass, the Modified Circumpolar Deep Water (MCDW) or Warm Core (WMCO), characterized by a subsurface potential temperature maximum and dissolved oxygen minimum (Jacobs et al. 1985). MCDW actively interacts with the cold atmosphere, sea and glacial ice to form shelf waters in the Ross Sea. Water masses of the Ross Sea have been accurately described by different authors (Jacobs et al. 1985; Trumbore et al. 1991; Locarnini 1994).

The shelf waters in the Ross Sea are formed during the austral winter when the upper layers cool down and freeze, thus delivering part of their saline content which increases the salinity of the subsurface waters (Jacobs et al. 1985). Shelf waters generally have temperatures close to the surface freezing point (between −1.95 and −1.75 °C) and show salinity values higher in the western sector than in the eastern one (Locarnini 1994). The high salinity in the western sector could be explained by taking into account the large extent of ice-free areas (polynya) even during the winter period (Kurtz and Bromwich 1983, 1985; Bromwich and Kurtz 1984). These areas, despite being involved in a large ice-formation process, are kept open by the wind action, which takes away the ice as soon as it is formed. The rejected brine increases the salinity of the subsurface waters, forming the densest waters in the Southern Ocean. These waters, with salinity values which increase with depth from 34.75 to 35.00, are named High Salinity Shelf Waters (HSSW, Jacobs et al. 1985). The presence of the MCDW in the subsurface layer of the western sector of the Ross Sea play an important role in the HSSW formation. Thus, when the surface water freezes during the winter, the released brine is added to subsurface waters that have relatively high salinity values due to the direct influence of the MCDW (Locarnini 1994). HSSW is supposed to be formed during the winter particularly in Terra Nova Bay (Jacobs et al. 1985).

Part of the HSSW moves north, reaching the continental shelf break, and takes part in the formation of the AABW; another branch goes southward, flows under the Ross Ice Shelf and takes part in the formation of a different water mass named Ice Shelf Water (ISW, Jacobs et al. 1970) characterized by temperatures lower than the freezing point at the surface. These waters are primarily located in the central side of the Ross Sea; they also move northward, bringing a further contribution to the formation of the Antarctic Bottom Waters (Jacobs et al. 1985).

CDW is generally separated from the shelf waters by a front – the Antarctic Slope Front (ASF) – a common oceanographic feature near the Antarctic continental shelf break (Jacobs 1991). It is supposed to be topographically controlled and related to high biological productivity and is a deep ocean source or sink region of heat, salt, nutrients, particulate, sediments and atmospheric gases (Trumbore et al. 1991). Temperature is the best indicator to characterize the presence of the ASF, showing horizontal gradients of 2–3 °C in 20–30 km (Jacobs 1991).

Traditionally, the circulation of the Ross Sea was supposed to be formed by a single cyclonic gyre, but more recent studies do not support this idea. The subsurface circulation seems to be formed by two different anticyclonic gyres at each end of the Ross Sea connected by a U-like cyclonic feature in the vicinity of the Ross Ice Shelf front (Pillsbury and Jacobs 1985; Locarnini 1994).

More recent current meter measurements collected in the Ross Sea have shown the barotropic characteristic of the general circulation with an important mesoscale activity and a weak seasonal variability (Pillsbury and Jacobs 1985; Jaeger et al. 1996; Picco et al. 1998). An application of barotropic model forced by the action of the East Wind Drift and by the local wind revealed the importance of the wind drag on the circulation in the Ross Sea (Commodari and Pierini 1999).

Previous studies of the Ross Sea have described water masses in terms of geological parameters, using content in silicates or estimating the vertical fluxes by means of sediment traps (Carter et al. 1981; Gardner et al. 1985; Ledford-Hoffman et al. 1986). In this work, for the first time, water masses of the Ross Sea have been characterized also on the basis of their suspended matter concentration using water samples collected during a hydrological cruise.

The Antarctic environment offers very interesting characteristics for the study of the natural phenomena of accumulation and spreading of the suspended load in waters for two reasons: the absence of polluting elements and the presence of unsorted terrigenous contributions essentially glacial in origin (Kurtz and Anderson 1979).

Nearly all the coastline and a large part of the continental shelf waters are under the ice cover for a period varying from 8 to 12 months per year, with the result that terrigenous contributions are exclusively linked to glacier transport: due to these conditions, sediments are influenced by offshore currents and by tides and wave motion only episodically; but they are redistributed almost exclusively by deep currents (Carter et al. 1981; Anderson et al. 1984; Singer and Anderson 1984).

It must be remembered that, in general, the suspended load in waters reflect terrigenous contributions, biogenic productivity and the transport by superficial currents, together with the influence of vertical and horizontal patterns of water masses during the sedimentation phase; the importance of each of these factors is variable and is due to the different environmental conditions.

In this chapter an effort is made to describe thermohaline and suspended matter characteristics in the western Ross Sea as derived by a modern dataset collected in the framework of the CLIMA expedition during the austral summer 1994/95. In particular HSSW, CDW and MCDW were identified by their hydrological parameters and characterized on the basis of their suspended matter contents. In this context, their general pathways and frontal areas, separating these water masses, were identified on the continental shelf.

## The Study Region

The Ross Sea is the body of water on the Antarctic continental shelf between Cape Colbeck at 158°W and Cape Adare at 170°E. On the southern side it is bound by the front of the Ross Ice Shelf (RIS) at around 78.5°S. This broad ice cover, which extends over nearly half the continental shelf, is about 250 m thick on its northernmost side (Jacobs 1989). The Ross Ice Shelf therefore limits only the uppermost waters, while the deeper waters can circulate freely under the ice shelf.

The western sector of the Ross Sea is characterized by alternating banks and depressions, the latter deeper than the continental shelf edge. Indeed, the bottom topography of the studied zone (Fig. 1) is rather irregular, with the shelf of about 500 m deep isolated from the shore by a canyon, named Drygalski Basin, deeper than 1000 m. This basin has a northeastward axis and is separated from the Joides Basin (that has the same orientation) by the Mawson Bank and Crary Bank (sill depth 250 m).

The region studied here is located on the northwestern sector of the Ross Sea. A bathymetric map of the area which superimposed the position of the stations collected during X expedition (austral summer 1994/95) of the Italian PNRA is shown in Fig. 1.

The expedition (II Leg-CLIMA project) was focused on the study of the hydrodynamics of the northeastern Ross Sea. The 61 hydrological stations were operated during the 1995 austral summer, from February 6 to 20, on board the R/V Italica. Stations close to Victoria Land were operated, from the continental shelf break to the formation area of the HSSW close to Terra Nova Bay (Jacobs et al. 1985), in order to identify and describe the HSSW spreading.

## Data and Methods

The hydrological stations were operated using a Sea Bird Electronics SBE 9/11 plus with double temperature and conductivity sensors, coupled with a Carousel water sampler SBE 32 carrying 24 bottles of 12 l each.

Calibration of temperature and conductivity sensors was performed at the SACLANT CENTRE of La Spezia (Italy) before and after the cruise. During the cruise, CTD temperature was controlled by means of two SIS RTM4200 digital reversing platinum thermometers. At every station several samples of water at different depths were collected and analyzed on board by means of an Autosal Guildline Salinometer. According to the pre- and after-cruise calibrations, an offset of −0.0028 °C was applied to the raw data. Concerning the conductivity calibration, only the bottom samples were used for the salinometer analysis. According to these data, an offset of 0.00095 $S^{-1}$ m was applied to the raw data. Hydrological data were then corrected and processed according to international procedures (UNESCO 1988), and standard algorithms (UNESCO 1983) were used to compute derived quantities.

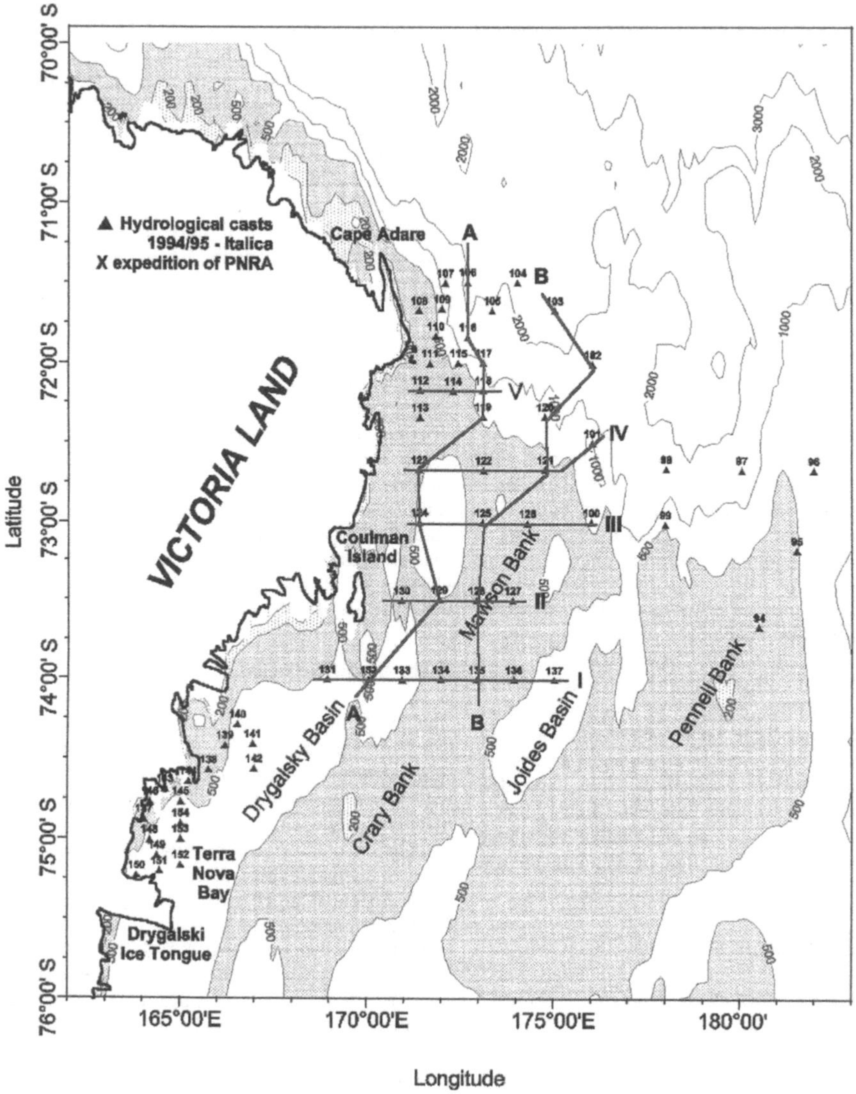


**Fig. 1.** Bathymetry of the northwestern Ross Sea with station locations of the 1994/95 CLIMA cruise. Depths are in m and the *shading* indicates depths below 500 m

Water samples collected at different depths in 40 stations were subdivided and underwent different filtrations in order to prepare them for the subsequent laboratory investigations to be performed on the suspended load. Determination of total particulate matter (TPM) concentration was carried out according to the methods reported in Strickland and Parsons (1968). In this phase, the total concentration data were used, and the organic fraction was differentiated from the inorganic (inorganic particulate matter - IPM). Suspended sediments were also filtered onto a

0.45-μm Millipore filter, then plated with 10–15 μm of gold coating. The filters were scanned and photographs of all grains and aggregates were taken; magnification ranged from 200× to 7000×.

## Results

Both horizontal and vertical sections of potential temperature, salinity and potential density are utilized to describe the three-dimensional structure and pathways of the HSSW and MCDW in the northwestern sector of the Ross Sea. In particular, we focused on the intrusion of the CDW onto the shelf and on the spreading of the HSSW from the formation area to the continental slope eastward of Cape Adare. According to Jacobs et al. (1985), CDW temperature is greater than 1.0 °C and salinity is greater than 34.68, whereas HSSW shows temperature ranging between −1.93 and −1.88 °C with salinity greater than 34.75.

Figure 2 shows the θ/S diagram of selected stations at different latitudes of the western Ross Sea, some of which are also in the vertical section of Fig. 8 discussed later. The potential temperature in this region varies between the in situ freezing point and the characteristic values of the CDW found in the southern limb of the Ross Gyre (θ > 1.0 °C close to the shelf break); while the salinity of these shelf waters varies between 33.2 (surface waters, not shown) and 34.87 (HSSW). The thermohaline modifications of the CDW entering the shelf area can be traced along the isopycnal of about 27.78 kg m$^{-3}$. Figure 2 reveals the progressive warming of the surface layer spanning from north to south and a cooling and freshening of a water mass derived from the CDW (stations 125 and 135) that we call Western Modified Circumpolar Deep Water (WMCDW). This water mass is found in ten stations in the western continental shelf of the Ross Sea; average core values show a potential temperature of −0.037 °C with a standard deviation of 0.5 °C and salinity of 34.607 with a standard deviation of 0.025. WMCDW salinity values are greater than the MCDW (or WMCO) because the former interacts with the western shelf waters that are saltier than in the eastern sector of the Ross Sea (Locarnini 1994).

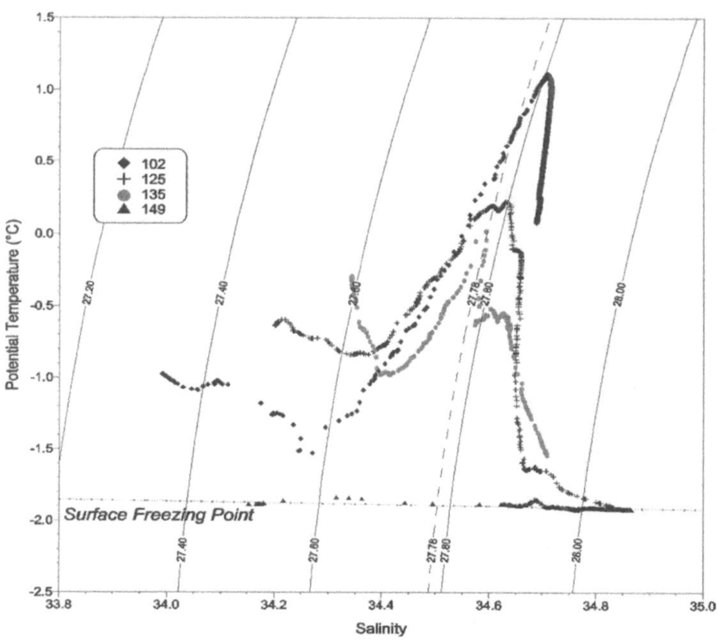


**Fig. 2.** Potential temperature (°C)-salinity diagram for selected stations. Station locations are shown in Fig. 1

Intermediate layers of station 125 are also occupied by the WMCDW and only near the bottom is HSSW found. However, WMCDW is not detected in the southern stations, close to Terra Nova Bay, where the water column is nearly isothermal with stability preserved by salinity that increases with depth (Fig. 2, station 145).

During the observation period, the sea surface temperature (SST, 5–10 m averaged, Fig. 3a) varied between the surface freezing point and just below −0.4 °C. SST above −0.6 °C were found in the central region of the sampled area, and SST below −1.8 °C were measured close to Terra Nova Bay. The sea surface salinity (SSS, 5–10 m averaged) distribution (Fig. 3b) seems very much related, with the SST field showing a high salinity nucleus eastward of Coulman Island.

The vertical structure is analyzed by means of potential temperature, salinity and potential density sections. Five sections are west-east, perpendicular to the expected flows of HSSW and WMCDW (Fig. 1, sections I, II, III, IV and V); two are north-south-oriented (Fig. 1, sections A and B).

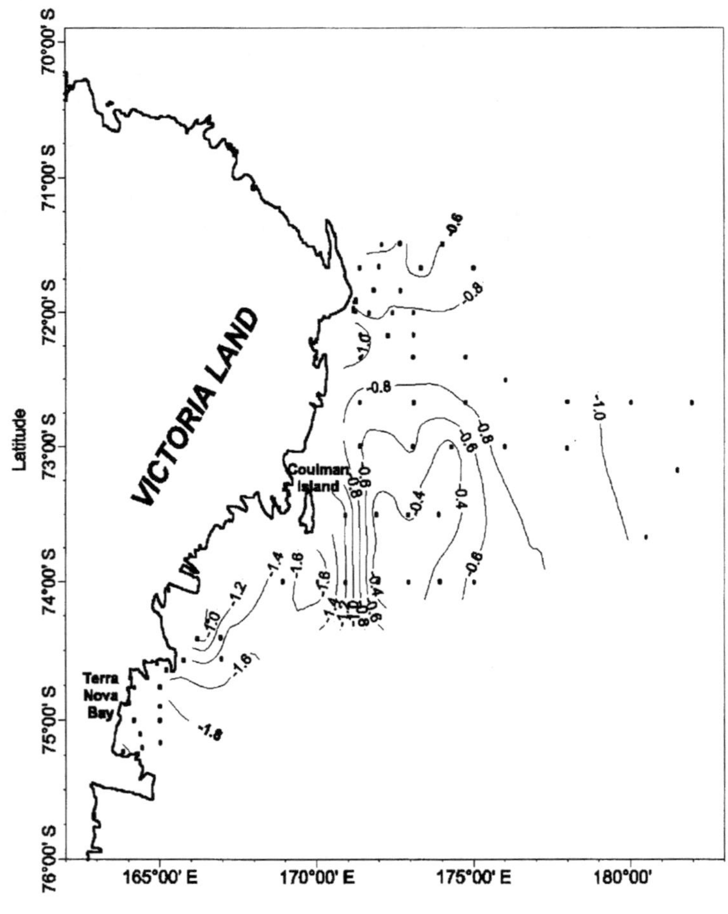


a

Fig. 3. a Potential temperature (°C). b Salinity (psu) for the surface layer (5–10 m). *Dots* indicate the position of the stations used to contour the data

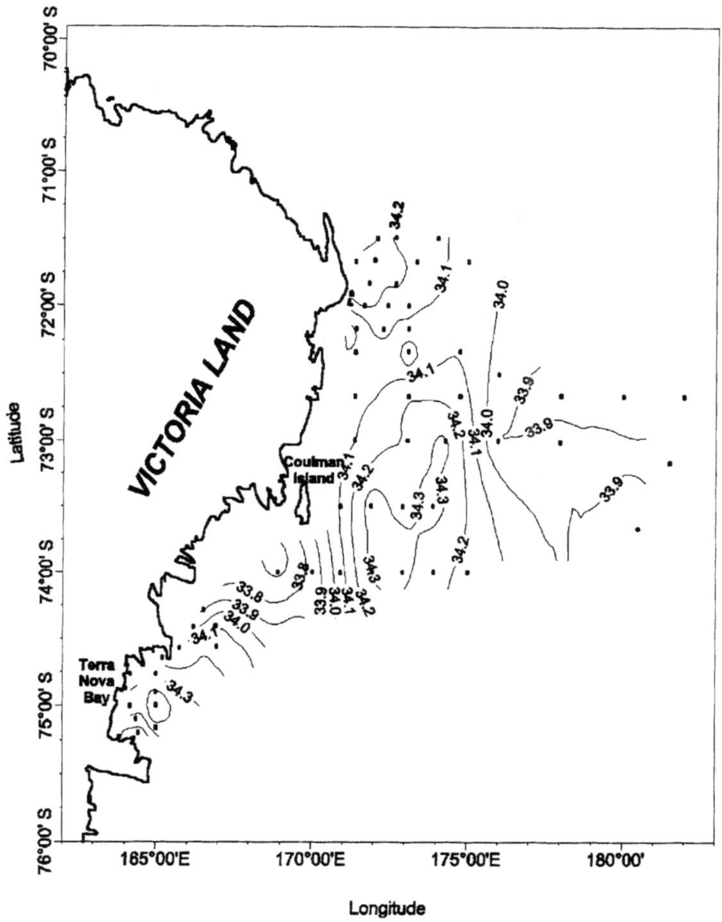


b

**Fig. 3.** (*Continued*)

The northward path of HSSW can be tracked up to the shelf break in the different zonal sections off Victoria Land (Figs. 4, 5 and 6). The southern most section (section I) shows, in the western part, a water column about 400 m thick at the same potential temperature ($\theta < -1.88\,°C$), or close to the surface freezing point. This nearly isothermal layer shows, in the deepest part, salinity values greater than 34.70 increasing with depth (Fig. 5, section I). The bottom layer, with potential density greater than 28.00 kg/m$^3$ and salinity greater than 34.80 (Figs. 5 and 6, section I) is filled with HSSW formed during the winter in Terra Nova Bay. The progressive freshening of HSSW is clear in Fig. 5, where this dense water occupies more than 200 m of the water column close to the bottom in section I and only a thin layer in the most northern sections (IV and V).

In our dataset, the evidence of the WMCDW is given with the temperature maximum detected in the sections of Fig. 4. The most southward influence of WMCDW is observed at stations 135 and 137, with the potential temperature maximum core at a depth of 150 dbar, occupying most of the water column over the Mawson Bank (Fig. 4, section I). In section III (Fig. 4) the strong

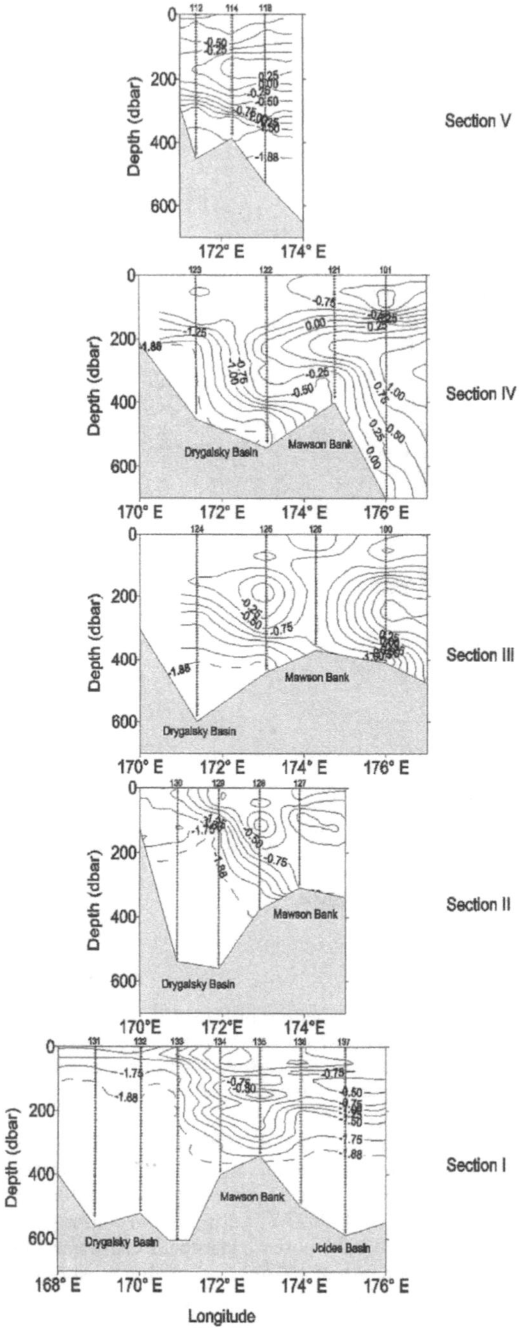


**Fig. 4.** Zonal vertical sections of potential temperature (°C) off Victoria Land across the continental shelf and slope of the western Ross Sea from south (*Section I*) to north (*Section V*). See Fig. 1 for transect locations

**Fig. 5.** Zonal vertical sections of salinity off Victoria Land across the continental shelf and slope of the western Ross Sea from south (*Section I*) to north (*Section V*). See Fig. 1 for transect locations

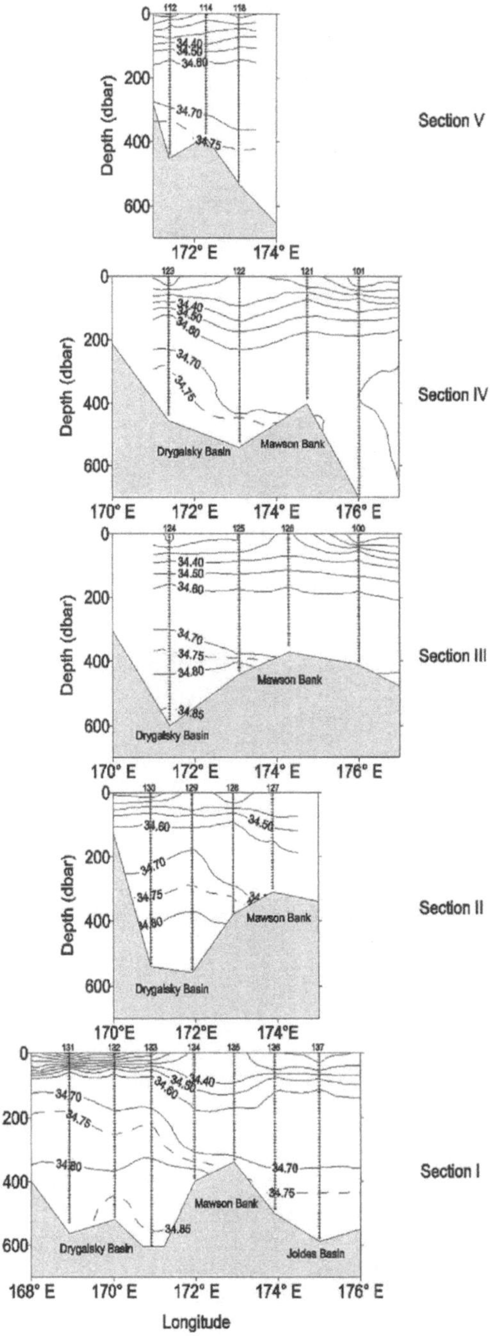

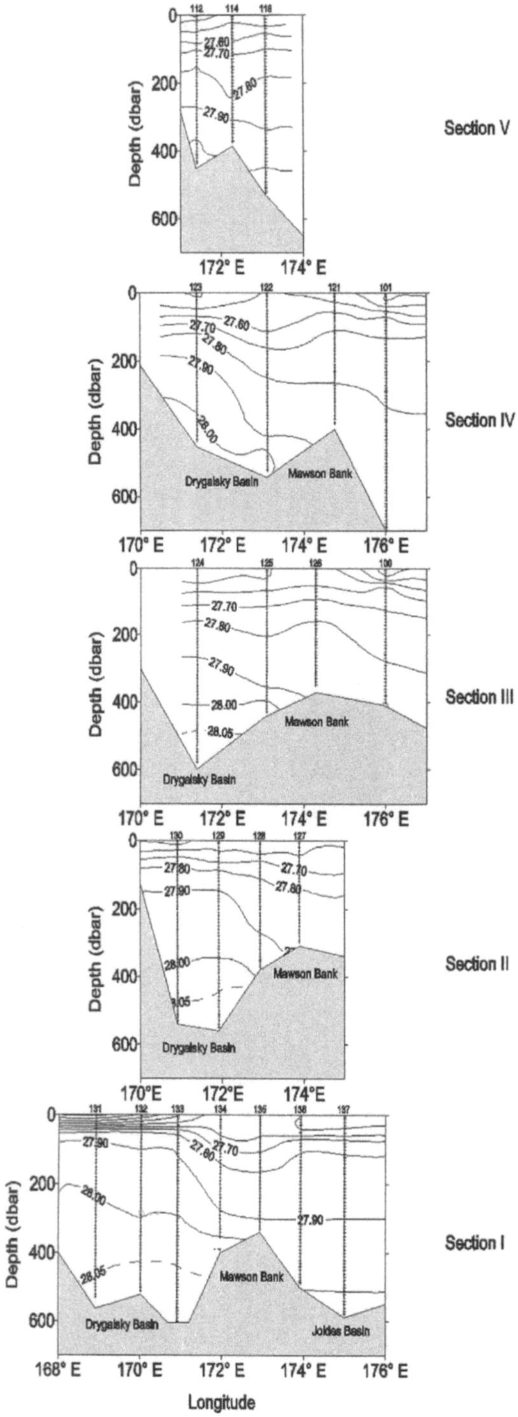


**Fig. 6.** Zonal vertical sections of potential density (kg m$^{-3}$) off Victoria Land across the continental shelf and slope of the western Ross Sea from south (*Section I*) to north (*Section V*). See Fig. 1 for transect locations

horizontal gradient in temperature reveals the presence of the ASF between stations 126 and 100, where a nucleus of CDW is detected.

The progressive modification of the HSSW on its northward path, shown above by means of the five sections of Fig. 5, is also clear on section A (Fig. 7). Section A was carried out connecting the stations where the higher concentration of HSSW was found. In this section the HSSW occupies overall the water column in the southern part and only the bottom layer close to the shelf break and there is no evidence of this dense water along the slope (Fig. 7). The ASF, which is located between stations 118 and 117, is characterized by a very sharp gradient of potential temperature from about 300 m until 1200 m along the continental slope with a maximum gradient of 2.8 °C over 20 km at 400 m. At station 106, under a subsurface cold water layer, we can see the CDW which, moving upward in the water column, slightly intrudes onto the continental shelf, modifying its properties (stations 117, 118 and 119) and spreading over the HSSW. A nucleus of WMCDW is isolated in station 124 ($\theta > -0.5$ °C). At station 116, close to the bottom, a vein of cold water is present; this layer (about 70 m thick) is not very salty (S = 34.67) and shows potential temperature ($\theta < -0.6$ °C) colder than those observed in the neighbouring stations ($\theta > -0.3$ °C, S = 34.68).

The intrusion of the WMCDW onto the continental shelf is more clear in section B (Fig. 8). This section was identified detecting the potential temperature core of the CDW and WMCDW for each station of our data set and connecting those where the maximum values were found. The potential temperature reveals a subsurface maximum characteristic of CDW approaching the shelf. CDW primarily follows the shelf edge and entering onto the shelf modify its thermohaline properties forming WMCDW. Core depth varies considerably from 300 dbar and more at the deepest stations and up to 200–120 dbar inside the shelf.

We calculated the average concentrations and variances for the TPM and their percentage content in IPM for all samples collected in the water column. Results of samples related to CDW, WMCDW and HSSW are reported in Table 1. The content in suspended matter ranges between 0.4 mg $l^{-1}$ and 2.4 mg $l^{-1}$ with an average value of 1.94 mg $l^{-1}$. The results obtained for CDW and HSSW distinguish the different particulate matter concentrations of these water masses, but the small number of samples and the high standard deviation do not allow us to characterize the WMCDW in the particulate matter concentration.

**Table 1.** Mean and variance of the TPM (mg $l^{-1}$) and inorganic particulate matter (%)

| | CDW | HSSW | WMCDW |
|---|---|---|---|
| TPM mean | 1.938 | 0.553 | 0.733 |
| TPM variance | 0.445 | 0.035 | 0.354 |
| IPM (%) | 69.4 | 59.8 | 27.3 |
| Samples | 10 | 12 | 3 |

In order to correlate the hydrological with the geological data, two vertical sections of TPM are compared with similar hydrological sections.

The first section (Fig. 9) is west-east-oriented and is made using the same stations of section III (Figs. 4, 5 and 6). The highest concentrations were detected close to the shelf break (station 100), whereas the water column inside the Drygalski Basin reveals a low concentration, less than 1.0 mg $l^{-1}$. The second vertical section, north-south-oriented, is shown in Fig. 10; this section is the same as the hydrological section B already discussed and reported in Fig. 8. Also clear in this section is the different particulate matter concentration in the subsurface layers between the shelf waters (low concentration, less than 1.0 mg $l^{-1}$) and CDW (high concentration, greater than 1.0 mg $l^{-1}$).

We analyzed the particulate matter composition in order to correlate it with the different concentrations measured. Our analysis shows the presence of both terrigenous input and biogenic load, with a presence of aggregates with a strong organic component. By investigation with electron microscopy and microprobes, different types of particles were found:

1. Biogenic debris (40–80% of the IPM).
2. Particles of terrigenous origin (10–60% of the IPM).
3. Organic fragments (40–50%).
4. Volcanic ash (5%).

The biogenic suspended matter has a very abundant siliceous component and, contrary to bibliographical data (Carter et al. 1981; Gardner et al. 1985), is not constituted of fragments of diatoms, but mostly of fragments of sponge spicules and dinoflagellates; this important cluster is composed of particles about 1–3 μm in size.

The amount of terrigenous matter is limited and is mainly constituted of phyllosilicates. Quartz

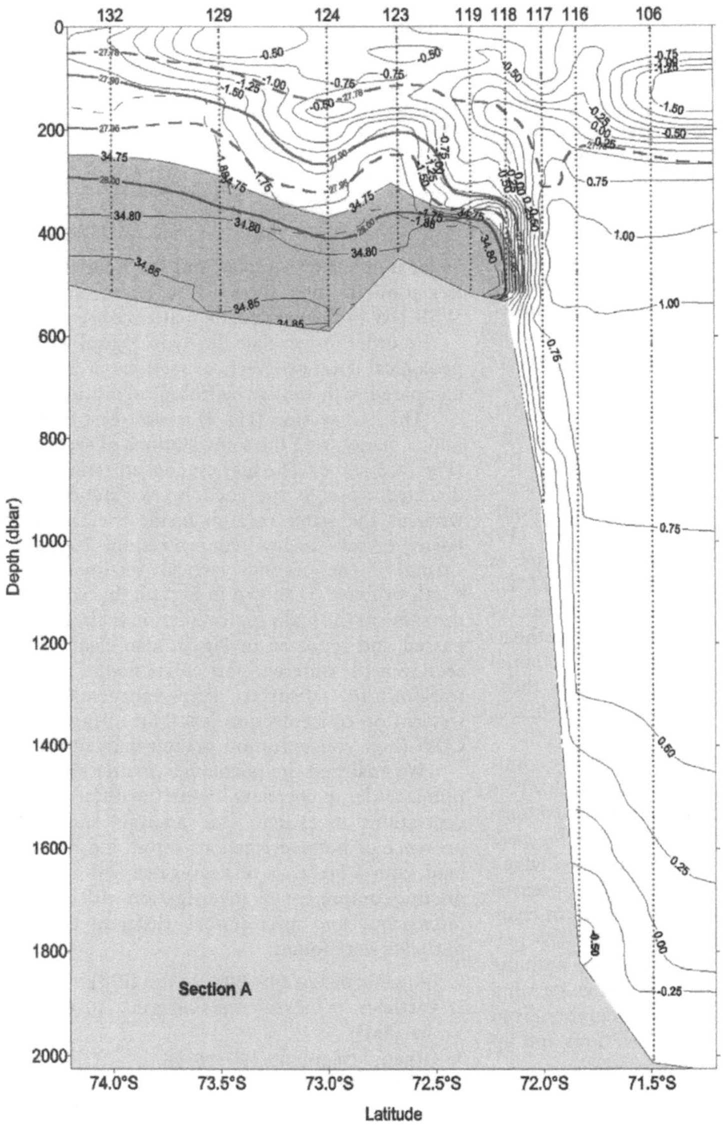


**Fig. 7.** Section A. North-south vertical sections of potential temperature (°C) and potential density across the continental shelf to the slope of the western Ross Sea. See Fig. 1 for transect locations

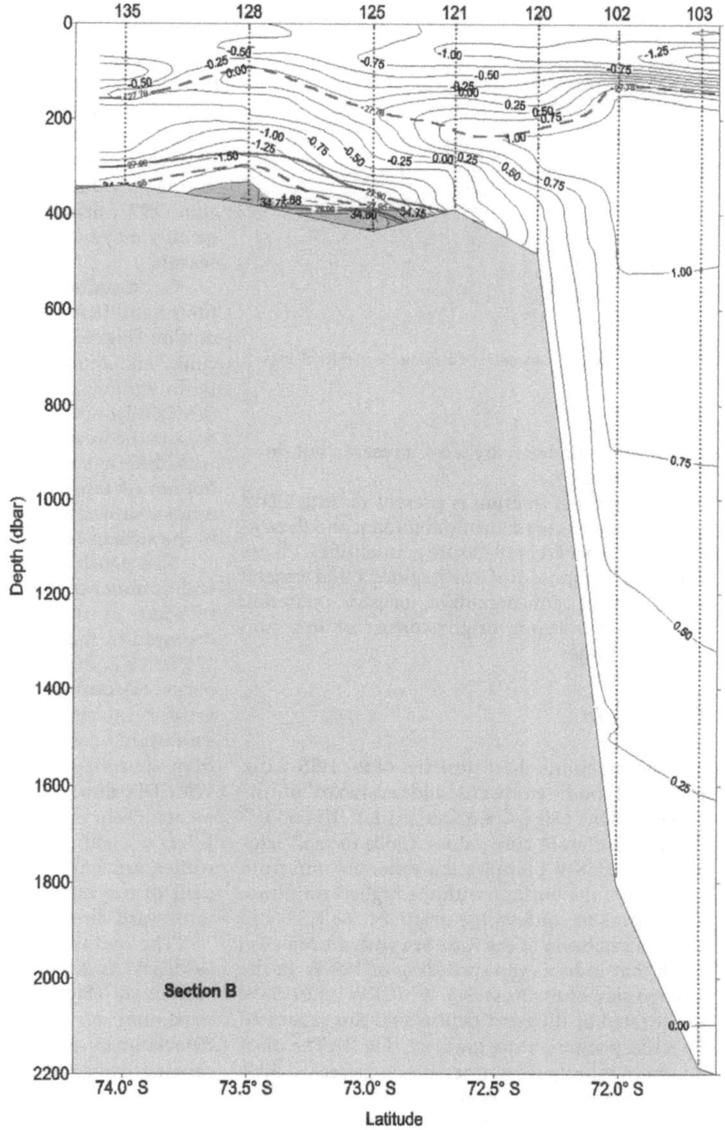


**Fig. 8.** Section B. North-south vertical sections of potential temperature (°C) and potential density across the continental shelf to the slope of the western Ross Sea. See Fig. 1 for transect locations

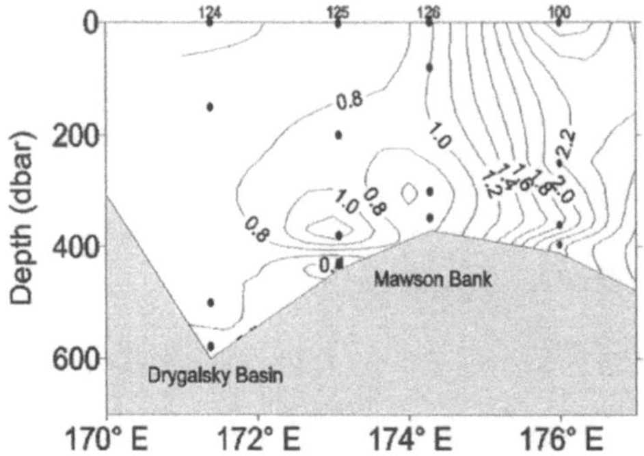


**Fig. 9.** TPM ($mg\,l^{-1}$) concentration along section III of Figs. 4, 5 and 6

and femic minerals are also present, but in a smaller quantity.

The organic fraction is present in both CDW and shelf waters in relatively abundant and dimensionally elevated (10–20 μm) quantifies. It is generally composed of cartilaginous and vegetal remains and non-organized organic material. Particles of volcanic origin consist of iron and titanium oxides.

## Discussion

The thermohaline characteristics of the HSSW can be found both westward and eastward of the Mawson Bank (Figs. 4, 5, 6 sections I, II, III and IV) but with different core values. Close to the Victoria Land, HSSW occupies the water column from 200 dbar to the bottom, with the highest salinities. This seems to confirm the origin of the HSSW in the western basin of the Ross Sea with the Mawson Bank that reduces the spreading of HSSW in the eastern side of the Ross Sea. WMCDW and HSSW are located at different depths and are separated by a sharp temperature gradient (Fig. 4). The most southern salinity section reveals an increased dilution in the upper layers close to the coast, probably caused by the ice melting in stations 131 and 132 (Fig. 4, section I).

The dome-shaped formation of isopycnal in sections I ($\sigma_\theta = 28.05\,kg\,m^{-3}$) and in section II ($\sigma_\theta = 28.00\,kg\,m^{-3}$) on the west side of these sections across the Drygalski Basin (Fig. 6), together with a similar dome-shaped formation present in the north-south sections (Fig. 8, section B), centered at station 128, suggest an elongated cyclonic circulation interesting the deep layers inside the Drygalski Basin. A cyclonic flow in this region would bring cold and salty water to the surface layer, but the presence of a warm and salty nucleus at the surface in our dataset (Fig. 3a,b) does not support this idea. A very similar warm and salty region southeast of Coulman Island was also observed during previous Italian cruises (Artegiani and Paschini 1992; Spezie et al. 1993, 1994) and consequently may be a common feature of the summer season.

Geostrophic velocities were calculated for sections I and II, taking the bottom as the level of no motion (Fig. 11). Results reveal an opposite baroclinic circulation pattern inside the Drygalski Basin with a southward flow associated with the WMCDW on the eastern side and a northward flow on the western side. This pattern is preserved, with different velocities, from the surface to the bottom of the water column in section II while, more southward, a poleward coastal flow appears in the surface layer (Fig. 11, section I).

The geostrophic field is in good agreement with summer current meter data recently reported by Picco et al. (1998) measured by a mooring deployed in the Joides Basin ($\phi = 74°01.36'S$, $\lambda = 175°03.19'E$, depth = 586 m). During the summer period (December 1994–March 1995), the current at the intermediate depth (220 m) shows a southward component (average values range from $2\,cm\,s^{-1}$ to $6\,cm\,s^{-1}$) representative of the WMCDW flow. During the hydrological measurements (February 1995), the mean speed was $2.7\,cm\,s^{-1}$ with a direction of 142°. During the winter, current measurements reveal a different path of the intermediate layer with a prevailing northward direction.

The southward intrusion of the CDW and the WMCDW inside the shelf is identified along the section B (Fig. 8) following the temperature maximum on isopycnal 27.78. This tongue is detectable for more than 200 km with a thickness ranging from 200 m close to the shelf break and less than 30 m at the southern station (135). The warm intrusion onto the shelf appears to be continuous and may be a semipermanent feature in this region. It seems unlikely that this 200-km-long tongue could have been set up only during the brief sampling period.

The presence of this warm water is crucial for the heat and salt budgets of the shelf waters of this area. WMCDW evidence was found down to 74 °S, i.e. throughout the southernmost transect, but we

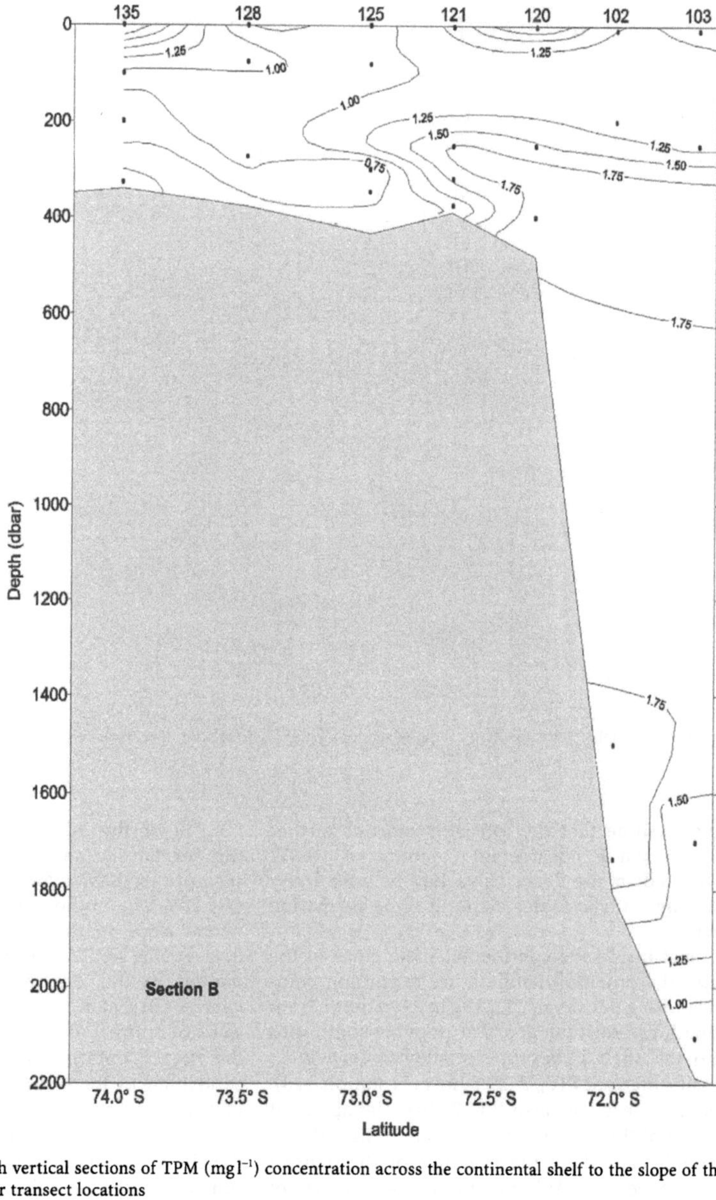


**Fig. 10.** Section B. North-south vertical sections of TPM ($mg\,l^{-1}$) concentration across the continental shelf to the slope of the western Ross Sea. See Fig. 1 for transect locations

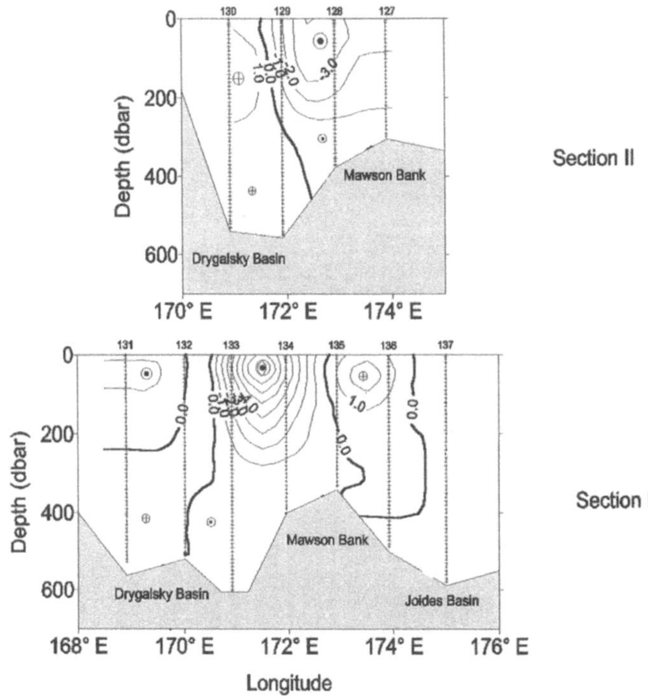


**Fig. 11.** Geostrophic velocities ($cms^{-1}$) along sections I and II. No motion level is referred to the bottom

may presume that the intrusion extends farther south when a considerable amount of HSSW is formed in the Terra Nova Bay polynya likely reaching with a colder signature, the formation area.

Just inside the continental shelf, close to the slope, the potential temperature maximum coincides with a density of 27.78 kg m$^{-3}$ (station 121) at a depth greater than 250 dbar; moving southward onto the shelf it becomes shallower even if its density increases ($\sigma_\theta \cong 27.80$ kg m$^{-3}$, station 128). It seems clear (see also Fig. 2) that the apparent conflict of the increase in density and the reduced depth is caused by the presence of the densest water type (HSSW) in the deeper layers of this section. HSSW can be detected in this section by the potential temperature values less than −1.88 °C and salinity greater than 34.75 (respectively, the thin dashed line and the shaded area in Fig. 9).

The northward flow of HSSW identified along the coastal stations forming section A (Fig. 7) seems to establish a sort of barrier to the intrusion of CDW. The outflow of HSSW along the continental slope in this region is not clear in our dataset. The small internal Rossby radius of deformation in this region (about 5 km) implies a narrow current, hence the sampling of the survey was not optimal for resolving these features. Also, the rugged bottom topography of this area can contribute to force the outflow of the HSSW in very narrow currents not resolved by our sampling. It can also be supposed that the outflow of the HSSW is not continuous but, on the contrary, can be released from the shelf by episodic outflow down to the continental slope.

The cold vein close to the bottom of station 116 (Fig. 7, depth = 1800 m) could be a signature of the cold HSSW coming from the shelf. However, the low salinity values of this layer (S = 34.67), very similar to those observed in the neighbouring stations (not shown), do not support this idea, and consequently may have a different origin.

Suspended matter data always identified higher concentration in the CDW layer and minimum values in the intermediate and bottom waters inside the shelf.

Vertical sections of Figs. 9 and 10 show the boundaries between the shelf waters poor in TPM and CDW with high concentration in TPM. These boundaries are very similar and located in the same position as the ASF that separates the shelf waters from the CDW along the continental shelf break; consequently, the presence of the ASF is crucial for the lateral transport of suspended matter.

The poleward section B (Fig. 10) confirms the strong relationship between suspended matter and thermohaline fields also where the CDW penetrates onto the continental shelf, modifying the ASF. On the contrary, the section of Fig. 9, which is perpendicular to the most significant flows, clearly shows, in the same location of the ASF, a sharp gradient in TPM (stations 100 and 126).

The difference in particulate matter between HSSW and CDW is also defined by the distinct composition of these water masses. Even if the percentages of IPM of CDW and HSSW are quite similar (70 and 60%, respectively), their compositions obtained with SEM (scanning electronic microscope) photographs are significantly different. Typically, shelf waters contain mostly particles of terrigenous origin and also organic fragments and volcanic ash; CDW, essentially, carries biogenic debris (fragments of sponge spicules and diatoms) and organic matter.

Therefore, organic matter components do not distinguish these water masses; on the contrary, the concentration of total particulate matter and inorganic composition are two tracers that confirm the patterns of HSSW and CDW. Biogenic debris are abundant in CDW and are carried poleward (stations 103 to 131) and terrigenous particles are only carried offshore only in the HSSW (stations 135 to 125). Both these transports have been estimated to be longer than 100 km at least (Fig. 10). It follows that for the suspended matter parameters the horizontal transport by current is more important than the vertical one. Current meter data measured in January and February at a mooring positioned on the Joides Basin show values in the range 3–5 cm $s^{-1}$ in the intermediate water (220 m), while close to the bottom values are 2–3 cm $s^{-1}$ (Picco et al. 1998). Using the former velocities to estimate the transport by the CDW, it follows that it was carried in 23–38 days from stations 103 to 121. Similarly, suspended matter in HSSW moved from station 135 to 125 in 38–58 days. These results, that can be described only as indicative due to the position of the mooring, reveal the persistence of the described hydrological field. Furthermore, as also inferred before, the WMCDW southward flow was most likely present also some weeks before the cruise.

## Conclusion

The analysis of hydrological and geological data collected in the northwestern Ross Sea during the 1994/95 CLIMA cruise reveals the patterns of the most important water masses in this region. HSSW is mostly present in the Drygalski Basin and flows northward to the continental slope, occupying most of the water column in the western side of the basin. The CDW flows close to the continental shelf break, preserving the ASF that separates it from the colder shelf waters. The CDW penetrates onto the shelf at about 174 °E and, modifying its properties, follows a southward path for about 200 km. This water mass is a cooled form of CDW and is designated Western Modified Circumpolar Deep Water (WMCDW) to distinguish it from the MCDW (or WMCO) that has a lower salinity. This warm intrusion onto the shelf appears to be continuous and may be a typical feature, as inferred by our geological data. The poleward evidence of WMCDW has been identified at 74 °S, but it can be supposed that most southern intrusion occurs when considerable amounts of HSSW are formed in the Terra Nova Bay polynya.

The thermohaline field produces a weak baroclinic circulation inside the Drygalski Basin associated with the HSSW and WMCDW opposite flows. The geostrophic velocities inferred from the thermohaline field are in good agreement with the long-term current meter measurements collected in the same period in a neighbouring area by other Italian groups.

The outflow of HSSW along the continental slope in this region is not clear in our data set due to the sampling scale being too coarse in relation to the small Rossby radius of deformation in this area, and also the rugged bottom topography.

This dense water can also be released from the shelf by episodic outflow down to the continental slope.

Geological parameters collected during the 1994/95 CLIMA cruise helped to characterize the water masses as HSSW and CDW but, due to the scarce number of samples, it is not possible to define the WMCDW.

Biogenic debris in CDW are carried southward and terrigenous particle are carried offshore in the HSSW; both these transports were estimated to be about 100 km long. Using the neighbouring current meter data, the characteristic diffusion time was estimated. All data are consistent with a structure having about a monthly recurrent pattern.

Generally, hydrological and suspended matter data are in good agreement even if the horizontal and vertical sampling strategies were dissimilar.

It is well known that the upwelled CDW and their modified versions (as WMCDW) provides the heat source that maintains the ice melting and hence the circulation (Hofmann and Klinck 1998). Therefore, a careful mapping of the distribution of the WMCDW is crucial for understanding the oceanographic and biological features in this region of the Ross Sea.

Further understanding of the dynamics and mixing processes in this region will require sampling programs designed to capture the small but crucial scales associated with these features.

*Acknowledgements.* This study was performed as part of the Italian National Program for Research in Antarctica and was financially supported by ENEA through a joint research program.

## References

Anderson JB, Brake CF, Myers NC (1984) Sedimentation on the Ross Sea continental shelf, Antarctica. Mar Geol 57: 295–333

Artegiani A, Paschini E (1992) Condizioni oceanografiche nel settore Neo-Zelandese dell'Oceano Antartico durante l'estate australe 1989–90. In: Gallardo VA, Ferretti O, Moyano HI (eds) Oceanografia in Antartide, ENEA – PNRA-EULA, Concepcion, Chile, pp 33–38

Bromwich DH, Kurtz DD (1984) Katabatic wind forcing of the Terra Nova Bay polynya. J Geophys Res 89: 3561–3572

Carter L, Mitchel JS, Day NJ (1981) Suspended sediment beneath permanent and seasonal ice, Ross Ice Shelf, Antarctica. N Z J Geol Geophys 24: 249–262

Commodari V, Pierini S (1999) A wind- and boundary-driven circulation model of the Ross Sea. In: Spezie G, Manzella GMR (eds) Oceanography of the Ross Sea – Antarctica. Springer, Milano, Italia, 135–144

Gardner WD, Southard JB, Hollister CD (1985) Sedimentation, resuspension and chemistry of particles in the northwest Atlantic. Mar Geol 65: 199–242

Gordon AL (1988) Spatial and temporal variability within the Southern Ocean. Springer, Berlin Heildelberg New York, pp 41–56

Gordon AL, Molinelli E (1982) Southern Ocean atlas: thermohaline and chemical distributions and the atlas data set. Columbia University Press, New York, 11pp, 233 plates

Hofmann EE, Klinck JM (1998) Hydrography and circulation of the Antarctic continental shelf: 150 °E to the Greenwich meridian. In: Robinson AR, Brink KH (eds) The sea, vol 2, pp 997–1042

Jacobs SS (1989) Marine controls on modern sedimentation on the Antarctic continental shelf. Mar Geol 85: 121–153

Jacobs SS (1991) On the nature and significance of the Antarctic slope front. Mar Chem 35: 9–24

Jacobs SS, Amos AF, Bruchhausen PM (1970) Ross Sea oceanography and Antarctic bottom water formation. Deep Sea Res 17: 935–962

Jacobs SS, Fairbanks RC, Horibe Y (1985) Origin and evolution of water masses near the antarctic continental margin: evidence from $H_2^{18}O/H_2^{16}O$ ratios in seawater. Antarct Res Ser 43: 59–85

Jaeger JM, Nittrouer CA, DeMaster DJ, Kelchner C, Dunbar RB (1996) Lateral transport of settling particles in the Ross Sea and implications for the fate of biogenic material. J Geophys Res 101(C8): 18479–18488

Kurtz DD, Anderson JB (1979) Recognition and sedimentologic description of recent debris flow deposits from the Ross and Weddel Seas, Antarctica. J Sedim Petrol 49(4): 1159–1170

Kurtz DD, Bromwich DH (1983) Satellite observed behaviour of the Terra Nova Bay polynya. J Geophys Res 88: 9717–9722

Kurtz DD, Bromwich DH (1985) A recurring atmospherically forced polynya in Terra Nova Bay. Antarct Res Ser 43: 177–201

Ledford-Hoffman PA, DeMaster DJ, Nittrouer CA (1986) Biogenic-silica accumulation in the Ross Sea and the importance of the Antarctic continental-shelf deposits in the marine silica budget. Geochim Cosmochim Acta 50: 2099–2110

Locarnini RA (1994) Water masses and circulation in the Ross Gyre and environs. PhD Thesis, Texas A&M, 87 pp

Picco P, Bergamasco A, De Micheli L, Langone L, Manzella GMR, Meloni R (1998) Current, temperature and salinity measurements in the Ross Sea shelf (Antarctica) during 1995–96: data presentation. Assoc Ital Oceanol Limnol XII 2: 109–120

Pillsbury RD, Jacobs SJ (1985) Preliminary observations from long-term current meter moorings near the Ross Ice Shelf, Antarctica. Antarct Res Ser 43: 87–107

Singer JK, Anderson JB (1984) Use of grain-size distributions to define bed erosion and transport for poorly sorted sediment undergoing simulated bioturbation. Mar Geol 57: 335–359

Spezie G, Tucci S, Budillon G, Corradi N, Piccazzo M, Fanucci F, Firpo M, Mirabile L, Ferrari M (1993) Water mass characteristics, sedimental dispersal and seismic stratigraphy in the continental margin of the Ross Sea (Antarctica). Ann Ist Univ Navale 60: Poster

Spezie G, Tucci S, Budillon G, Corradi N, Piccazzo M (1994) Caratteristiche fisiche delle acque del Mare di Ross e dispersione del materiale sospeso. Soc Geogr Ital 51: 291–305

Strickland JDH, Parsons TR (1968) A practical handbook of seawater analysis. Bull Fish Res Board Can 167: 1–311

Trumbore SE, Jacobs SS, Smethie WM Jr (1991) Chlorofluorocarbon evidence for rapid ventilation of the Ross Sea. Deep Sea Res 38(7): 845–870

UNESCO (1983) The acquisition, calibration and analysis of CTD data. A report of SCOR WG 51. UNESCO Tech Pap Mar Sci vol 54, 59 pp

UNESCO (1988) Algorithms for computation of fundamental properties of seawater. UNESCO Tech Pap Mar Sci vol 44, 53 pp

Whithworth T, Nowlin WD (1987) Water masses and currents of the Southern Ocean at the Greenwich meridian. J Geophys Res 92: 6462–6476

CHAPTER 7

# Water Mass Characteristics During the ROSSMIZE Cruise (Western Sector of the Ross Sea, November–December 1994)

A. Russo[1]

**ABSTRACT**

From 14 November to 27 December 1994, the multidisciplinary ROSSMIZE (Ross Sea Marginal Ice Zone Ecology) cruise was carried out in the western sector of the Ross Sea (Antarctica) on board the M/N Italica. This chapter describes the hydrological conditions encountered on the basis of 78 CTD vertical profiles acquired during the cruise. Most of the CTD casts were localized along a transect centred on 175°E, spanning about 520 km from 76°30'S to 71°50'S; casts were generally distanced by 10 nautical miles (18.5 km). At some locations, CTD casts were repeated up to five times during the sampling period. In this area, the period of the cruise corresponded to late austral spring-beginning of summer. Ross Sea polynya was expanding, and a marginal ice zone was encountered on the Pennell Bank. Interestingly, an intense Warm Core (WMCO) was present below it. WMCO occupied also the northern bank crossed by the transect. Lower Circumpolar Deep Water (LCDW) intruding on the continental shelf and its mixing with surface waters in correspondence with the slope front originating WMCO were also noted. Below the LCDW, a bottom boundary layer of Low Salinity Bottom Water (LSBW) was detected; it could have originated from Low Salinity Shelf Water (LSSW) outflowing from the northern end of the Joides Basin. On the two depressions crossed by the transect, both belonging to the Joides Basin, High Salinity Shelf Water (HSSW) formed a dome. HSSW occupied almost the whole water column, with an evident upwelling, in the southern depression, while in the northern one HSSW was present mainly as a bottom layer, with LSSW over it.

## Introduction

The Ross Sea is the southernmost of the Antarctic Seas, having a line ideally joining Cape Adare to Cape Colbeck as a northern limit and the Ross Ice Shelf (RIS) as the southern end (Fig. 1), but water can freely circulate below the RIS (which is about 250 m thick on its northern edge) and reach much higher latitudes. During the austral winter, this sea is completely covered by pack ice, with the important exception of polynya areas such as Terra Nova Bay (Bromwich and Kurtz 1984). The Ross Sea has a relevance from a physical oceanographic point of view because it is one of the main sources of Antarctic Bottom Water. To describe other than the physical oceanography of this area would be beyond the objectives of the present work, so it will be sufficient to say that the main aim of this multidisciplinary cruise was to investigate the marginal ice zone (for more details on the cruise strategy see Faranda et al. this Vol.). The marginal ice zone is the area where the pack ice is melting, releasing matter possibly trapped during the icing process. The melting occurs on a wide scale during the austral spring, when the Ross Sea polynya begins to form close to the Ross Ice Shelf and becomes larger and larger (Fig. 2 shows the ice cover during the cruise). The period November–December corresponds to the end of the spring and the beginning of the austral summer, when most of the Ross Sea is still covered by pack ice, so that few measurements were collected previously in this period. After some days of navigation between the pack ice (following a Russian icebreaker), the M/N Italica reached the polynya area, sampling the first station (MP) on 14 November 1994.

[1] CNR Istituto di Ricerche sulla Pesca Marittima (IRPeM). L.go Fiera della Pesca 1, 60125 Ancona, Italy

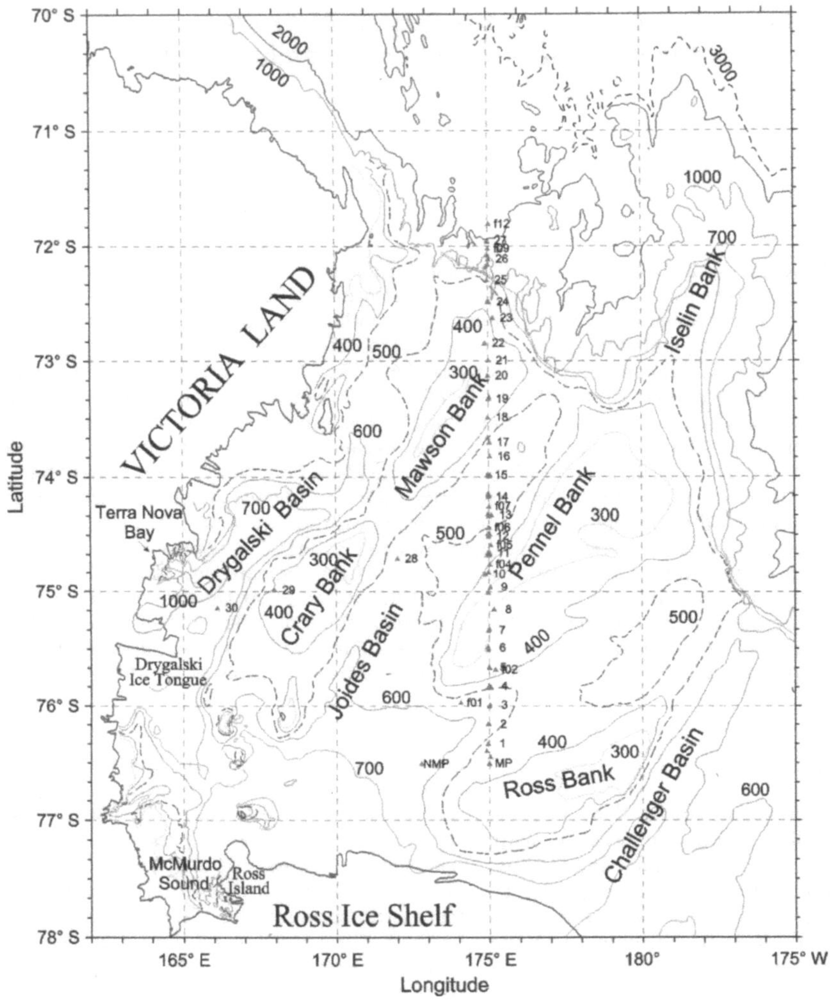


**Fig. 1.** Western sector of the Ross Sea with CTD stations collected during the ROSSMIZE cruise

## Materials and Methods

A CTD Sea Bird SBE 9/11plus, equipped with redundant temperature-conductivity sensors, paroscientific pressure sensor, Chelsea fluorometer, Sea Tech transmissometer, Yellowspring dissolved oxygen sensor and Datasonics altimeter (main characteristics of these sensors are given in Table 1), was used. The CTD was coupled to a SBE 32 Carousel water sampler, equipped with 24 hydrological bottles with a capacity of 12 l, to collect water samples for biogeochemical analysis made by other research groups involved in the cruise. Before and after the cruise, temperature and conductivity sensors were calibrated, and during the cruise were checked by means of digital reversing thermometer, on-board salinometer analysis and comparison between the primary sensors and the secondary ones. CTD data were acquired with the maximum fre-

**Fig. 2.** Daily sea-ice concentration (reported at 5-day intervals) during the ROSSMIZE cruise from Nimbus-7 SMMR and DMSP SSM/I Passive Microwave Data. Data and images provided by the EOS Distributed Active Archive Center (DAAC) at the National Snow and Ice Data Center, University of Colorado, Boulder, CO

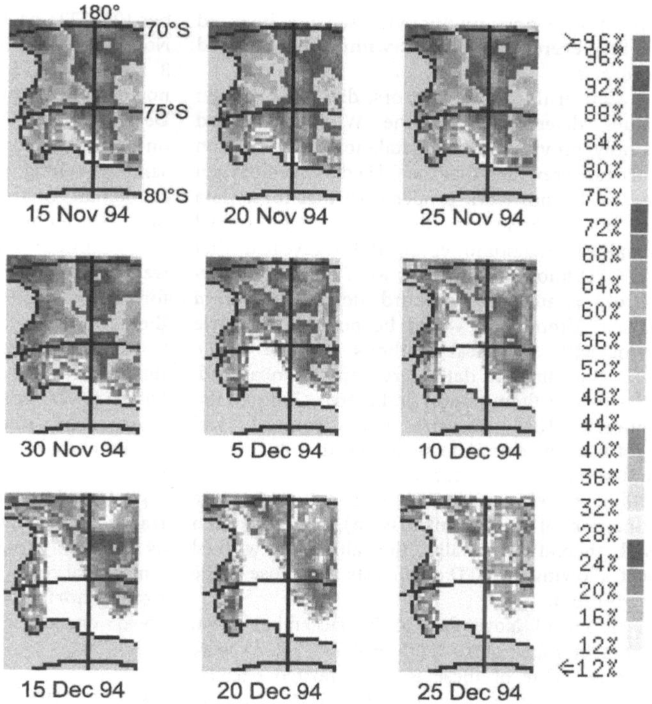


**Table 1.** Accuracy of the CTD sensors utilized

| Sensor | Constructor | Model | Accuracy |
|---|---|---|---|
| Pressure | Digiquartz | Series 4000 | 0.015% f.s. (10 000 psia) |
| Temperature | Sea Bird Electronics | SBE 3 | 0.002 °C |
| Conductivity | Sea Bird Electronics | SBE 4 | 0.0003 S $m^{-1}$ (0.003 mmho $cm^{-1}$) |
| Dissolved Oxygen | Sea Bird Electronics | SBE 13 Y | 0.1 ml $l^{-1}$ |
| Fluorometer | Chelsea | Acquatracka III | 0.01 |
| Transmissometer | Sea Tech | 25 cm Transm | 0.5% |
| Altimeter | Datasonics | PSA-900 | 0.1 m |

quency allowed by the CTD probe (24 Hz) from the sea surface (1–2 m below it) until the sea bottom (generally 1–2 m above it). They were subsequently processed following standard procedures (SBE 1995). A slight correction was applied (a few millesimals for both temperature and conductivity data) to the calibration files computed on the basis of calibration data and salinometer checks. Data were then corrected for cell thermal mass effects and averaged at 1-dbar bins; finally, potential temperature, salinity and potential density anomaly were computed (UNESCO 1981; Fofonoff and Millard 1983). On 18 November 1994, an intercalibration station (NMP) was carried out jointly by the Italian research groups onboard the M/N Italica and the USA research groups on board the R/V N. B. Palmer. Differences between the three pairs of temperature and conductivity sensors involved in this exercise remained within the certified accuracies.

Concerning other sensors, dissolved oxygen values determined by the Winkler method (kindly provided by G. Catalano for calibration purposes) revealed that the CTD dissolved oxygen sensor did not work properly, so that these data are not discussed in this chapter. Chlorophyll *a* values (Saggiomo et al. 1998) revealed that the CTD fluorometer values were not straightforwardly related to on-board determined chl-*a* values. Although it would be possible to relate them with statistical methods (Galassi 1997), CTD fluorometric data were kept as measured, considering them simply indicative. Comparison between TSM (Povero, pers. comm.) and light transmission data suggested that during the first seven casts (from MP to 5) there were some problems with the transmissometer (undere stimation of light transmission), which worked well thereafter. Finally, the altimeter worked well, allowing for CTD downcasts very close to the sea bottom.

From 14 November to 22 December 1994, a total number of 79 CTD profiles (Fig. 1) were acquired. One of these is not reported because the pump that ensured water flow inside the SBE temperature and conductivity sensors was blocked by ice formed inside it. Another six profiles were immediate repetitions of previous casts for water-sampling needs and one was a repetition for instrumentation test. Finally, two profiles (f12 and 26R) did not reach the sea bottom because these casts were interrupted by pack ice approaching the CTD sea cable. With the exception of six stations (NMP, f01, f02, 28, 29 and 30), all CTD casts were performed along the 175°E meridian, starting from 76°30′S until 71°50′S. Spacing between standard stations (i.e. CTD cast with biogeochemical sampling) was about 10 nautical miles. Two segments were sampled with an increased spatial resolution; in the first, a CTD cast (f04, f05, f06 and f07) was added between standard stations in the central area where marginal ice zone conditions were encountered; in the second, a high resolution CTD investigation on the continental slope was performed at the northern end of the transect (stations f08, f09, f10 and f11; this investigation was unfortunately interrupted due to strong winds). The main transect was sampled in five periods: from 14 to 18 November (15 CTD profiles from MP to 7), 20 to 26 November (20 CTD profiles from MPA to 15), 28 to 30 November (8 CTD profiles from 11A to 15A), 3 to 8 December (23 CTD profiles from 16 to f12, northernmost station of the transect) and 10 to 13 December (8 CTD profiles from 24A to 11C). Then on 14 and 15 December, stations 28, 29 and 30 were sampled along the ship route track towards the Italian Antarctic base of Terra Nova Bay, and on 22 December 1994 the last CTD profile (26R) was acquired. Some stations along the transect were sampled several times (28 profiles were performed on previously sampled locations) during the period of measurements.

This sampling strategy, with several repetitions of single stations and of entire parts of the transect, was chosen to pursue the main aim of the cruise, i.e. to find and follow the evolution of a marginal ice zone. In this context, various kinds of pack ice conditions were encountered. Simplifying, it could be considered that the first part of the transect, from station MP to station 8, was sampled as a polynya area, stations 9 to 11 were considered a marginal zone (Saggiomo et al. 1998), while continuing northwards, more compact pack ice was always encountered.

## Results and Discussion

Various water masses were identified during this cruise (Fig. 3) and classified according to Jacobs et al. (1970, 1985) and Locarnini (1994). In Fig. 3a, together with seven representative stations, the water types of interest are reported on the basis of the definitions given by Jacobs et al. (1985, modified); in Fig. 3b the cumulative θ-S diagram for all the stations is shown. Some of the water masses sampled (WMCO, LSSW and also LSBW) were generally saltier than the water types of Jacobs et al. (1985). This should be mainly due to the fact that the data used for the definitions of Jacobs and coworkers were collected in the eastern sector of the Ross Sea, while the present cruise was entirely carried out in the western sector, and it is well known (Locarnini 1994) that salinity progressively increases going from the eastern to the western edge of the Ross Sea.

Surface Waters on the continental shelf, indicated as RSSW (Ross Sea Surface Water), were generally characterized by relatively low salinity values (around 34.3–34.4 psu). Almost no trace of a summer temperature increase in the surface layer (due to solar heating) was evident. A slightly

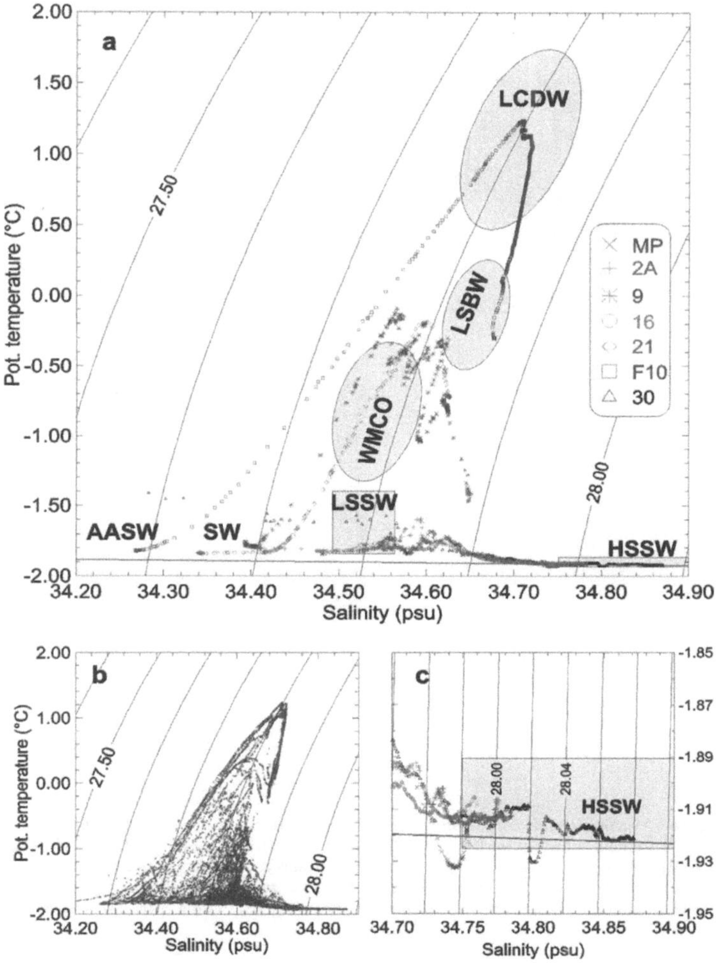


**Fig. 3.** **a** Diagram of seven representative stations: *MP* southernmost station, located on the northern flank of the Ross Bank; *2A* southeastern corner of the Joides Basin; *9* off the Pennell Bank; *16* northeastern corner of the Joides Basin; *21* on the Mawson Bank; *F10* out of the Ross Sea continental shelf; *30* Drygalski Basin. **b** Cumulative θ-S diagram. **c** Zoomed area of **a**

different type of surface water was found on the northern edge of the transect, defined as Antarctic Surface Water (AASW) to distinguish its location (outside the continental shelf). In any case, RSSW also belongs to AASW.

Outside the Ross Sea continental shelf, below the surface water, LCDW was observed. LCDW flows around Antarctica; it is observed on the northern end of the transect, after the shelf break, and appears (Fig. 4) to intrude on the shelf. The mixing occurring on the slope front (Jacobs 1991), identified by the V-shaped deformation of isolines, transforms the LCDW in WMCO.

WMCO is the only water mass (except surface waters) that enters in the Ross Sea shelf, thereby playing a crucial role in the renewal of shelf waters and in the global heat budget. Moreover, WMCO may have a major role in the opening of the Ross Sea open-sea polynya. In our case, WMCO was observed mostly on the

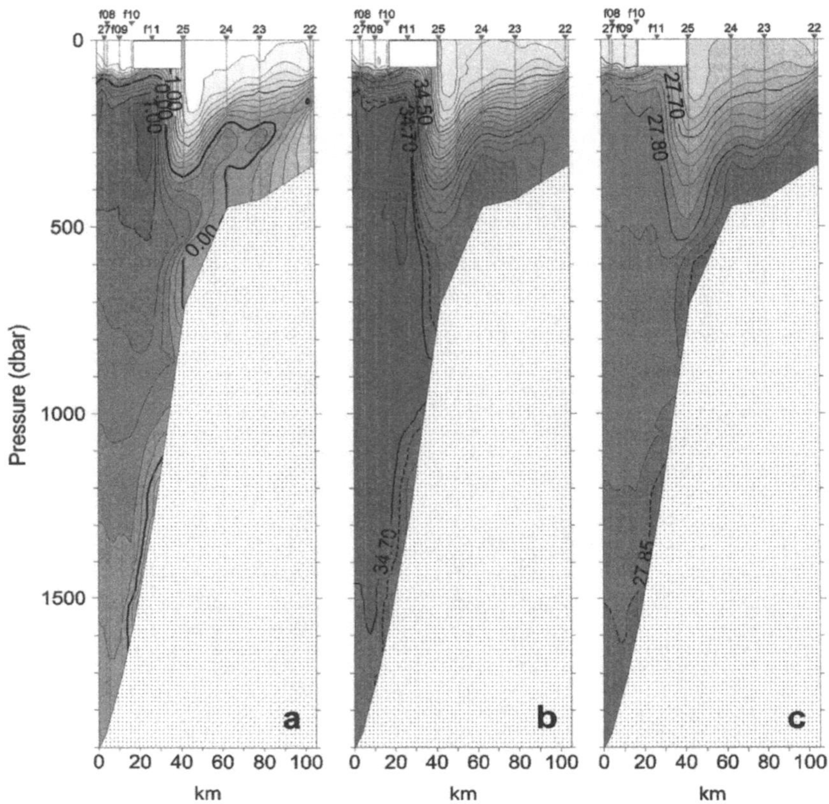


**Fig. 4a–c.** Vertical sections on the shelf slope of **a** potential temperature (°C), **b** salinity (psu) and **c** potential density anomaly (kg m$^{-3}$). Contour intervals: for $\theta$ 0.2 units; for S 0.02 units (with the 34.69 dotted isohaline added); for $\sigma_\theta$ 0.02 units (with the 27.85 dotted isopicna added). The upper part of the transect between stations f10 and 25 is not shown because interpolation was not corrected due to lack of data in the upper part of station 11 and to the particular interpolation parameters chosen to take in account the pronounced slope of the sea bottom (see Fig. 5 for a correct representation of the *blank area*)

two banks (Pennell and Mawson) crossed by the main transect.

HSSW is the water mass most characteristic of the Ross Sea western sector and is the densest of the entire Antarctic. It is produced during the austral winter on the wind-generated coastal polynyas of Victoria Land (Terra Nova Bay polynya in particular), where catabatic winds continuously displace recently formed pack ice offshore. In this way, salt is added, by means of brine release and evaporation, to a water that has a temperature equal to the surface freezing point. Therefore, the main quantity of HSSW was found close to Victoria Land, in the Drygalski Basin (station 30 in our case), with high salinity values (34.8 psu and more) and a thick layer (500 m on station 30) occupied by it. On the transect, HSSW was observed in the Joides basin only, consisting in a thick layer with slightly lower salinity values (34.7–34.8 psu) in the southern part and a thin bottom layer in the northern part. HSSW outflowing from the shelf, along the Drygalski Basin, gives rise to High Salinity Bottom Water (not observed during this cruise due to the lack of sampling in the HSSW outflow area).

Low Salinity Shelf Water (LSSW) is generated mostly in the Ross Sea eastern sector, where the absence of wind-generated polynyas does not give rise to the high salinities observed in the western sector. Different mechanisms of LSSW formation have been proposed, and at present there is no widely accepted theory. LSSW outflowing from the shelf mixes with LCDW and gives rise to the Low Salinity Bottom Water. In our case, LSSW was observed as a layer overlying the HSSW, mainly on the northern part of the Joides basin.

At a few stations (MP, MPA, 30) a thin layer (20 to 50 m) of ISW (Ice Shelf Water) intruded within the HSSW layer (see zoomed θ-S diagram in Fig. 3c). These layers were characterized by a potential temperature slightly less (about 0.01°C) than the surface freezing point. This may suggest that the layer of ISW (about 50 m) observed at stations MP and MPA (same location of MP, but performed 6 days after) at a depth around 400 m originated from beneath the Ross Ice Shelf, while the thinner layer (20 m) observed at station 30 at 250 m depth was likely formed beneath the Victoria Land ice shelves (Drygalski ice tongue in particular).

In a few other stations, a bottom boundary layer of Low Salinity Bottom Water (LSBW) was observed on the shelf slope (Fig. 4). With the usual station spacing adopted during the cruise (10 nautical miles), this layer would have been badly observed (in fact, CTD stations f08, f09, f10 and f11 were performed to investigate the slope area which is supposedly highly dynamic). Considering its depth and θ-S characteristics, the observed LSBW could be derived by LSSW exiting from the northern end of the Joides Basin (the presence of a sill with a depth between 450 and 500 m partially traps the HSSW in this basin). The relatively low depth of the core (between 1200 and 1500 m) is not in good agreement with the alternative hypothesis (Locarnini 1994) that this water is derived by LSSW outflowing from the middle-eastern area of the Ross Shelf. In that area, the LSBW was observed at depths greater than 1500 m.

In Fig. 5, the distribution along the main transect of potential temperature, salinity, potential density anomaly and a scheme of the water masses is shown. The stations used for this figure are the first sampled for each location. The data cannot be considered as synoptic, because they span 24 days from 14 November (MP) to 8 December (f12). Notwithstanding, this was the best way to obtain a general overview of the entire transect. Some characteristics already reported can be better described. On the shelf break, the slope front where RSSW and AASW appeared to converge and to mix with LCDW, forming WMCO, was again evident. In the middle of the transect there was a surface area with low salinity, and the relative minimum was between stations f04 and 11. This very likely indicates a recent melting of the pack ice, as indirectly confirmed by the echo-survey group that recoded a super-swarm of krill in the area between f04 and 11 (Azzali and Kalinowski, this Vol.) and by the multitude of minke whales seen from the ship. It is evident that the WMCO was present only on the two banks. On the Pennell Bank, at station 7, a thin layer of WMCO detached from the main one, arriving near the surface, where the temperature increased to values greater than −1.7°C. A similar behaviour was observed at station 18. Such layers could have a role in the melting of pack ice, and it is interesting to note that a nucleus of WMCO was located just below the marginal ice zone of the Pennell Bank. HSSW almost entirely occupied the southeastern corner of the Joides Basin, forming a great dome, and the bottom layer of its northeastern part. The cyclonic domes on these two depressions may indicate an upwelling area, in particular the southernmost one, where the 27.85 isopicnal (Fig. 5c) reached the surface and the 27.95 isopicnal was very close to it.

In Fig. 6a, the five profiles for potential temperature, salinity, potential density anomaly and fluorescence collected at station 11 (some 100 m around it) are shown. There was an evident variability also in a time interval of just 18 h (casts 11a and 11R). This was mainly because station 11 was at the northern limit of the WMCO on the Pennell Bank (Fig. 5). Therefore, even very slight variations of the cast position could have caused these different profiles. An alternative or concurrent reason could be a tide-induced horizontal motion, because tidal dynamics on the Ross Sea banks may not be as negligible as in deeper areas. The first (11) and last (11c) casts show almost no WMCO, while it was present in a very large portion of the three intermediate profiles. Fluorescence shows a subsurface maximum located just above the pycnocline, a characteristic observed also at the other stations; the fluorescence maximum decreased with time. The last temperature profile shows the beginning of the seasonal thermocline.

Figure 6b is as Fig. 6a, but for station 14, 30 miles north of station 11, in the middle of the

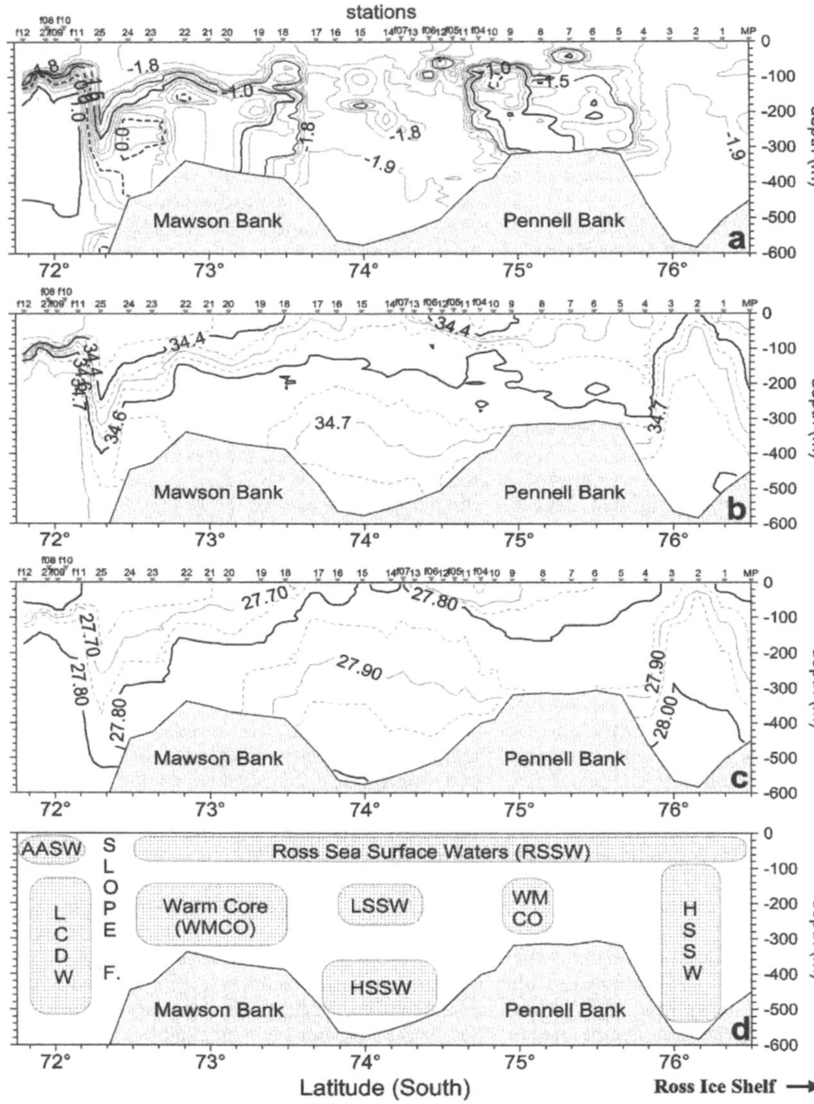


**Fig. 5a–d.** Vertical sections on the main transect of **a** potential temperature (°C), **b** salinity (psu), **c** potential density anomaly ($kg\,m^{-3}$) and **d** scheme of the water masses. Contour intervals: for θ 0.1 units below −1.5 and 0.25 units above −1.5; for S 0.05 units; for $\sigma_\theta$ 0.05 units

**Fig. 6a, b.** Vertical profiles of potential temperature (°C), salinity (psu), potential density anomaly (kg m$^{-3}$) and fluorescence (a.u.) for **a** five CTD casts performed at different times (see legend) at station 11 and **b** four CTD casts at station 14

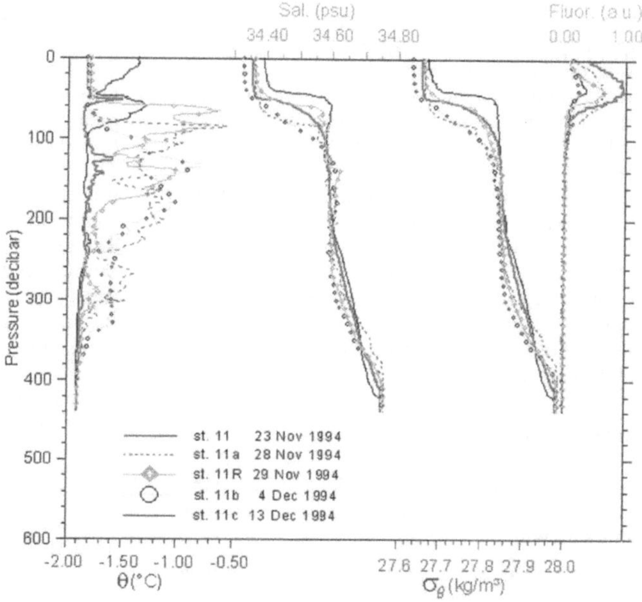


(a)

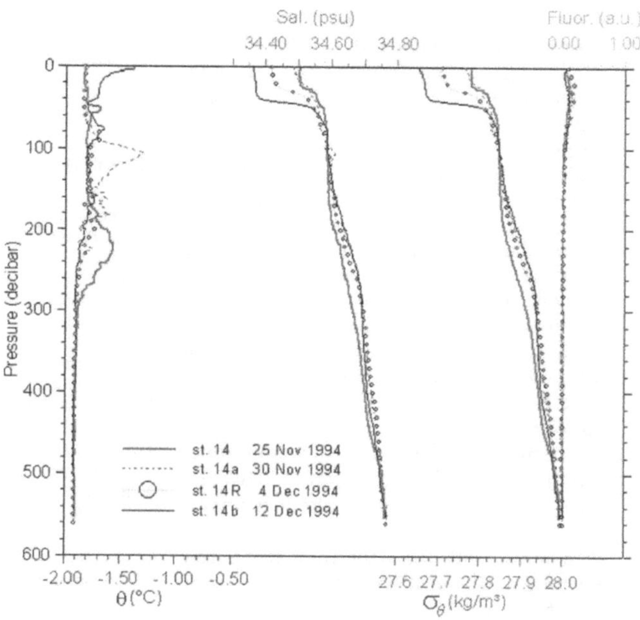


(b)

northern portion of the Joides Basin. Here, temperature variations were very limited with respect to station 11; surface salinity and density progressively decreased, the mixed layer deepened and stratification became stronger. There was an appreciable variation of the salinity of the intermediate layer, where the LSSW layer initially present appeared to be replaced by a saltier and colder water, while the bottom layer of HSSW did not exhibit variations. Again, the last temperature profile showed a thermocline that had just formed. At this location, as in all the area (mostly covered by pack ice) sampled northward of station 11, fluorescence values were very low; maximum values were measured at stations 28, 30 and in the polynya area.

## Concluding Remarks

From the analysis of CTD data collected during the ROSSMIZE cruise, the following main conclusions can be drawn:

- Water masses observed during the cruise seem to be well related (except RSSW) with the topography of the area. Surface waters consisted in a layer of depth variable from 250 m on the slope front to almost 0 on the southeastern corner of the Joides Basin. Absolute minimum salinity values (<34.3 psu) were outside the Ross Sea continental shelf, indicating the presence of Antarctic Surface Waters (AASW) formed outside the Ross Sea. A relative salinity minimum (<34.35) was observed on the northern flank of the Pennell Bank, indicating the presence of the marginal ice zone. HSSW occupied almost the entire southeastern corner of the Joides Basin and the bottom layer of its northeastern one; the maximum presence of HSSW was obviously observed on the Drygalski Basin (station 30). LSSW was in a layer above the HSSW of the northern part of the Joides Basin, from which it could exit and, by mixing with LCDW, generate the observed LSBW bottom boundary layer. LCDW was located just outside the shelf break, with the slope front that mixed it with AASW and RSSW when intruding on the shelf, thereby forming the WMCO core observed on the Mawson Bank. The other WMCO core, observed on the Pennell Bank, was most likely advected by a similar intrusion of LCDW on the northeast flank of the Pennell Bank.
- Salinity differences between water types defined by Jacobs et al. (1985) and θ-S characteristics observed during the cruise may be mainly due to the different geographical positions of the samplings.
- The cyclonic doming on the two depressions indicates an upwelling area, particularly in the southernmost one.
- The density structure was determined by the salinity, and the stratification was generally weak. On the northern part of the Joides Basin, the mixing layer progressively deepened and gained buoyancy by the freshening of the surface layer (evidently due to pack ice melting). The seasonal thermocline in November was not yet formed, and the beginning of the summer surface water heating was only observed starting from December 12. The fluorescence subsurface maximum was detected just above the pycnocline, usually around 30 m.

*Acknowledgements.* Thanks are due to Prof. Tom S. Hopkins for helpful discussions, and to two anonymous referees for their useful remarks. This work was carried out in the framework of the Italian National Research Programme in Antarctica (PNRA), whose support is gratefully acknowledged.

## References

Bromwich DH, Kurtz DD (1984) Katabatic wind forcing of the Terra Nova Bay polynya. J Geophys Res 89: 3561–3572

Fofonoff NP, Millard RC Jr (1983) Algorithms for computation of fundamental properties of seawater. UNESCO Tech Pap Mar Sci 44, 53 pp

Galassi M (1997) Influenza dei fattori ambientali nella determinazione in vivo della clorofilla-a con profilatori in continuo. Il caso di un ambiente polare: il Mare di Ross (Antartide). Dissertation for the Laurea degree, University of Ancona

Jacobs SS (1991) On the nature and significance of the Antarctic slope front. Mar Chem 35: 9–24

Jacobs SS, Comiso JC (1989) Sea ice and oceanic processes on the Ross Sea continental shelf. J Geophys Res 94 (C12): 18195–19211

Jacobs SS, Amos AF, Bruchhausen PM (1970) Ross Sea oceanography and antarctic bottom water formation. Deep Sea Res 17: 935–962

Jacobs SS, Fairbanks RC, Horibe Y (1985) Origin and evolution of water masses near the antarctic continental margin: evidence from H218O/H216O ratios in seawater. Antarct Res Ser 43: 59–85

Locarnini RA (1994) Water masses and circulation in the Ross Gyre and environs. PhD Thesis, Texas A&M University, College Station, Texas

National Snow and Ice Data Center (1989) DMSP SSM/I brightness temperatures and sea ice concentration grids for the polar regions, 1987. Digital data available from nsidc@kryos.colorado.edu. NSIDC Distributed Active Archive Center, University of Colorado, Boulder

Saggiomo V, Carrada GC, Mangoni O, Ribera d'Alcalà M, Russo A (1998) Spatial and temporal variability of size fractioned biomass and primary production in the Ross Sea (Antarctica) during austral spring and summer. J Mar Syst 17: 115–127

SBE (1995) Application note No. 31. Sea bird electronics, 1808 136th Place Northeast, Bellevue, WA 98005, 5 pp

UNESCO (1981) Background papers and supporting data on the Practical Salinity Scale 1978. UNESCO Tech Pap Mar Sci 37, 144 pp

CHAPTER 8

# Large-Scale Circulation Features in the Central and Western Ross Sea (Antarctica)

P. Picco[1], A. Bergamasco[2], L. Demicheli[3], G. Manzella[1], R. Meloni[4], and E. Paschini[5]

**ABSTRACT**

Circulation patterns in the central-western part of the Ross Sea were inferred from CTD casts, ADCP profiles and from a 1-year recording of currents, temperatures and salinities from moorings deployed from February 1995 to January 1996 in the framework of the Italian National Program of Antarctic Research (PNRA). The spreading paths of two important water masses produced in the Ross Sea were also studied. Ice Shelf Water (ISW) was found outflowing from beneath the Ross Ice Shelf at 177 °W, with two preferential directions, westward and northward. The northward flow of ISW in the Challenger Basin was not observed beyond 76°S. The High Salinity Shelf Water (HSSW) found in Terra Nova Bay was mainly flowing in the northeast direction along the coast, but a significant southward flow was detected by currentmeters during November. Bathymetric constraints were significant for spreading of the continental shelf water masses. The seasonal temperature and salinity cycle due to ice formation and melting processes was detected in all the recorded time series at depths around 250 m, but does not seem to affect in a significant way the local circulation, as indicated from currentmeter records.

## Introduction

The Ross Sea is a wide continental shelf, with an area of about 500 000 km$^2$, bounded west by the Victoria Land Coast and east by Edward VII Land; the northern boundary can be considered the line from Cape Adare to Cape Colbek and the southern boundary is the Ross Ice Shelf. Due to the isostatic reaction to ice load during the last Great Ice Age, the mean depth of the continental shelf is about 450 m, with deeper values close to the coastal boundary, up to 1000 m in the Drygalski basin. The topography of the continental shelf is locally modified by canyons and banks, reaching a minimum depth of 106 m in the Pennell Bank. The bottom topography plays an important role in determining the main general circulation features; the deepest part acts as a reservoir of salty and denser waters, determining the west-east gradient of the salinity field, while the morphology of marine mountains near 180°E is an important constraint for the interaction of Modified Circumpolar Deep Water and newly formed Ice Shelf Water outflowing below the Ross Ice Shelf.

The Ross Ice Shelf, with an area of about 530 000 km$^2$, is the largest ice shelf in the world; between its base and the sea bottom there is a cavity of variable depth from a few tenths to more than 300 m where seawater can freely circulate (Hellmer and Jacobs 1995). Processes occurring in this cavity strongly influence the circulation of this area, modifying the characteristics of the incoming waters. Temperature and salinity measurements under the Ross Ice Shelf (Jacobs et al. 1979) indicate the presence, at the ice-seawater interface, of water with a 34.39 PSU salinity and an in-situ freezing point of −2.16 °C, and with −1.86 °C and 34.83 near the sea floor.

Ice melting during the short austral summer, fast ice formation and a long period of total ice coverage mainly determine the seasonal variability of this area, influencing the thermohaline characteristics of surface and subsurface waters. The seasonal evolution of the ice coverage in the Ross

[1] ENEA CRAM, p.o. Box 316, 19100 La Spezia, Italy
[2] CNR ISDGM, S. Polo 1364, 30125 Venice, Italy
[3] DISTER, Univ.di Genova, V.le Benedetto XV, 16100 Genova, Italy
[4] CNR ISOF, Pozzuolo di Lerici, La Spezia, Italy
[5] CNR IRPEM, Largo Fiera della Pesca, 60125 Ancona, Italy

Sea has been studied on the base of microwave radiometer and ocean colour (Comiso et al. 1990), but the total amount of ice produced each year is still debated. The Ross Sea seems to be free from ice for a longer period than other Antarctic areas. The rapid increase in ice concentration takes place at the end of February. Interannual variability has been detected at the onset of ice formation (Jacobs and Comiso 1989), while winter sea-ice concentrations seem to have a less significant long-term variability. Ice growth starts early in the eastern part of the Ross Sea, advancing successively from all directions toward the northwestern area, which can be considered the one with lowest ice concentration.

Warm and relatively salty waters stemming from the Antarctic Circumpolar Current are transported toward the Ross Sea by the Ross Gyre. These are modified during their permanence on the continental shelf by interaction with sea ice and mixing with the resident waters, thus producing a modified circumpolar water characterized by a Warm Core with a temperature near −1 °C and salinity of 34.54. The presence of a recurrent coastal polynya in Terra Nova Bay (Kurtz and Bromwich 1985) and of other open leads and polynyas (Jacobs and Comiso 1989) is crucial for the formation of High Salinity Shelf Water (HSSW), the densest water mass of the Southern Ocean, characterized by a temperature of −1.9 °C and salinity higher than 34.8. It is mainly found in the deep layers of the western part of the Ross Sea, while, in the eastern sector, a second type of shelf water is formed, the so-called Low Salinity Shelf Water (LSSW) (Jacobs et al. 1985) characterized by a temperature of −1.7 °C and salinity of 34.5. Waters circulating in the Ross Ice Shelf cavity are cooled by the interaction with its base and transformed into Ice Shelf Water (ISW), which is characterized by an in-situ temperature below the surface freezing temperature and salinity around 34.67. HSSW and LSSW, together with warm core water inflowing from the deep ocean, account for 75% of the water circulating in the Ross Sea (Carmack 1977). Dense waters leaving the Ross Sea also represent an important contribution to the total amount of Antarctic Bottom Waters.

The aim of this chapter is to describe water mass distribution and circulation in particular regions of the Ross Sea as inferred by CTD and ADCP data collected during the austral summer 1995 cruise and from 1-year-long current, salinity and temperature measurements in selected locations.

## Data and Methods

An oceanographic cruise was performed in the Ross Sea from December 1994 to February 1995 within the framework of the Italian National Program of Antarctic Research CLIMA (Climatic Long-Term Interaction for the Mass balance in Antarctic) Project (PNRA 1995) by the R/V Italica. During the expedition, 154 CTD casts were collected in different areas of the Ross Sea (Fig. 1) by means of a Seabird 9/11 Plus multiparameter probe, calibrated before and after the cruise at the SACLANTCEN Institute in La Spezia, Italy. The spatial coverage allowed the description of the water masses in a wide region of the continental shelf and slope. However, a significant spatial gap in the spatial coverage of sampling locations exists in the central and south-western portion of the continental shelf. A transect along the Ross Ice Shelf, from longitude 168°E to 166°W, at about 2 miles from the front, was carried out to study the main water masses circulating below the Ross Ice Shelf and other stations were distributed along the Challenger Basin to follow the path of the emerging ISW. Some stations were located along the shelf break to identify upwelling areas where Circumpolar Deep Water intrudes on the continental shelf and where water masses modified during their permanence in the Ross Sea leave it to join Antarctic Bottom Waters. Another area investigated was Terra Nova Bay and the westernmost boundary of the Ross Sea, up to Cape Adare, to support the study of Terra Nova Bay polynya processes and to follow the northern spread of the salty water mass there formed.

Vertical profiles of currents in the upper 250 m were obtained by a ADCP-RDI 150 kHz narrow-band current meter hung outboard the ship during hydrological stations every time weather conditions allowed. The instrument was set up to measure the three components of current speed (referred to instrument case) every second with a vertical resolution of 10 m; current profiles, averaged every minute, were stored in an ensemble together with instrument pitch, roll and fluxgate compass heading. Current data were corrected of the ship drift computed by GPS integrated with TRANSIT and OMEGA position and transformed from instrumental to absolute refer-

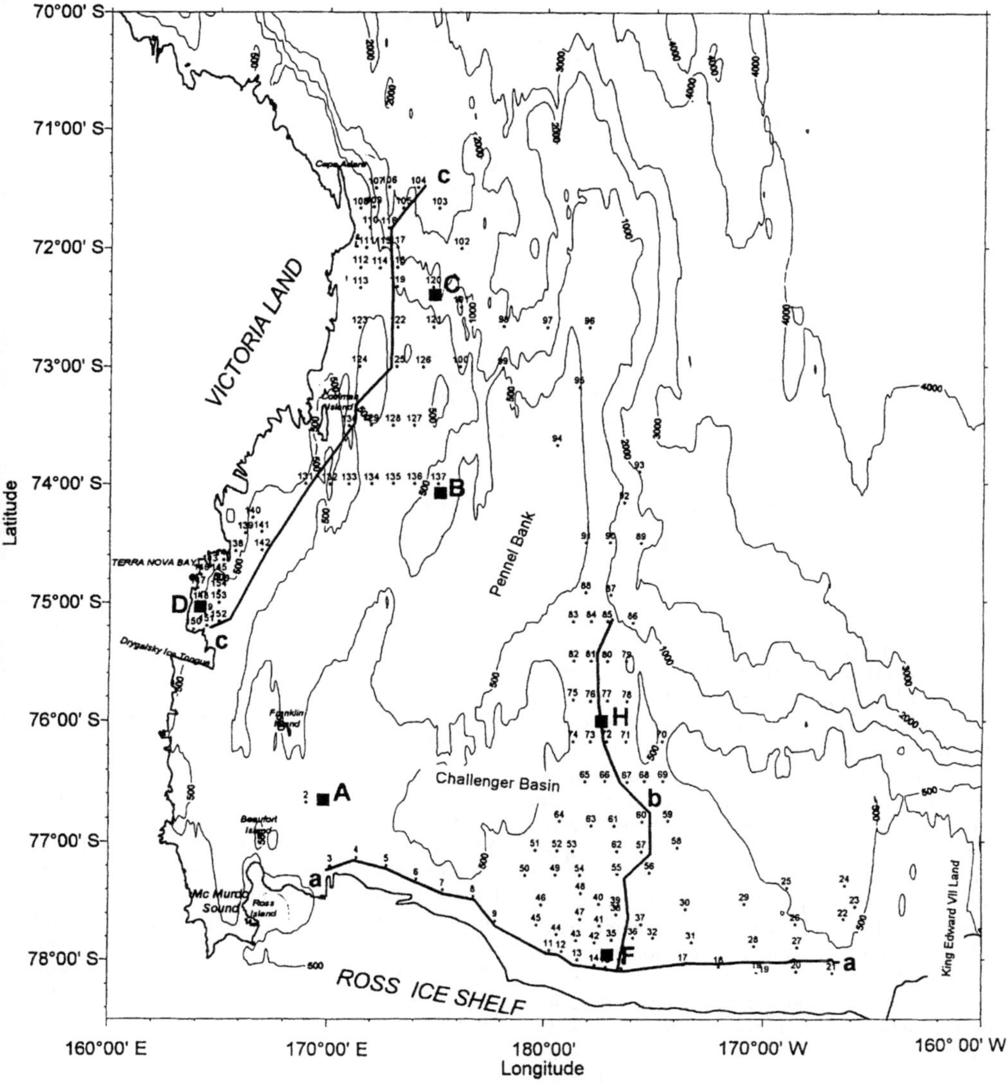


**Fig. 1.** Ross Sea with CTD stations and mooring locations

ence by pitch, roll and heading correction. The flux-gate heading of the instrument, that was checked to be parallel with the ship, was corrected for the magnetic declination obtained by a comparison between the compass and the gyro compass of the ship. After check and spike elimination, data for each deployment were averaged for the whole period of measurements, namely about 1 h. Further corrections were applied by eliminating the tidal components evaluated from current meters moored in close proximity.

Three moorings composed by currentmeters, temperature and conductivity sensors, sediment traps and turbidimeters were deployed (Table 1).

Table 1. PNRA moorings in the Ross Sea during 1995. Position, ocean depth, depth of sensors and record lengths. The sampling time of current meters was 30 min

| | A | B | C |
|---|---|---|---|
| Latitude | 75°06′.10S | 77°59′.99S | 75°56′.20S |
| Longitude | 164°13′.04E | 177°01′.62W | 177°35′.98W |
| Sea depth | 815 m | 588 m | 456 m |
| Deployment | 22/01/95 | 11/12/94 | 04/02/95 |
| Recovering | 19/01/96 | 12/01/96 | 12/01/96 |
| sensors depth (m) | | | |
| Temperature | 231, 780 | 225, 547 | 225, 415 |
| Conductivity | 231, 780 | 225 | |
| Currentmetrer | 231, 780 | 221, 552 | 230, 425 |
| | **D** | **F** | **H** |
| Latitude | 75°06′.10S | 77°59′.99S | 75°56′.20S |
| Longitude | 164°13′.04E | 177°01′.62W | 177°35′.98W |
| Sea depth | 912 m | 602 m | 625 m |
| Deployment | 17/02/95 | 27/01/95 | 04/02/95 |
| Recovering | 14/01/96 | 21/01/96 | 23/01/96 |
| Sensors depth (m) | | | |
| Temperature | 55, 140, 402, 748 | 244, 390, 579 | 282, 450, 476, 503, 597 |
| Conductivity | 140, 404, 753 | 244, 390 | 282, 476 |
| Current meter | 55, 140, 402, 748, 882 | 244, 390, 579 | 282, 450, 503, 597 |

One of these (F) was located at about 2 miles from the edge of the Ross Ice Shelf at a longitude of 177 °W, another one (H) further north, at about the same longitude near the shelf break; the third (D) was located in the Terra Nova Bay polynya area for the study of dense water formation processes. Some information about temperature, salinity and currents from three moorings deployed during the same period in the framework of the ROSSMIZE Project were added to obtain a more complete description of the general circulation (Picco et al. 1999). One was located near Ross Island (A), another between Mawson and Pennell Banks (B), and a third in front of Cape Adare near the shelf break.

In order to avoid damage from drifting icebergs, instrumentation was left well below the surface. Current meters were Aanderaa RCM7, also equipped with temperature and conductivity sensors. Seacat sensors, which assure better accuracys and stability (respectively, 0.002 °C year$^{-1}$ and 0.001 S/m month) were also used for the temperature and conductivity measurements. All the moorings were recovered, maintenanced and deployed again the following year, during the 1995/96 expedition (PNRA 1996). Time series of about 1 year of current, temperature and salinity were analyzed. After a preliminary check for gross errors and lack of data, each series was filtered and resampled to have a data set containing daily mean values.

## Summer Distribution and Circulation of Water Masses

The summer distribution and circulation of water masses were inferred by the analysis of CTD profiles and ADCP measurements. Particular attention was devoted to the analysis of water masses in intermediate and deep layers, for which also temporal series of current, temperature and salinity were available. The main focus of the project, was, in fact, to investigate the spreading path of the Ice Shelf Water flowing from below the Ross Ice Shelf, and High Salinity Shelf Water from Terra Nova Bay.

### Ross Ice Shelf Barrier and Challenger Basin

A transect in front of the Ross Ice Shelf barrier (a, Fig. 1) clearly shows two main water masses circulating beneath it: the MCDW (Modified Circumpolar Deep Water) and ISW (Fig. 2). The MCDW warm core with temperatures higher than −1.6 °C and salinities between 34.4 and 34.54 psu (maximum T = −1.14 °C at 360 m with corresponding salinity of 34.48) was found extending

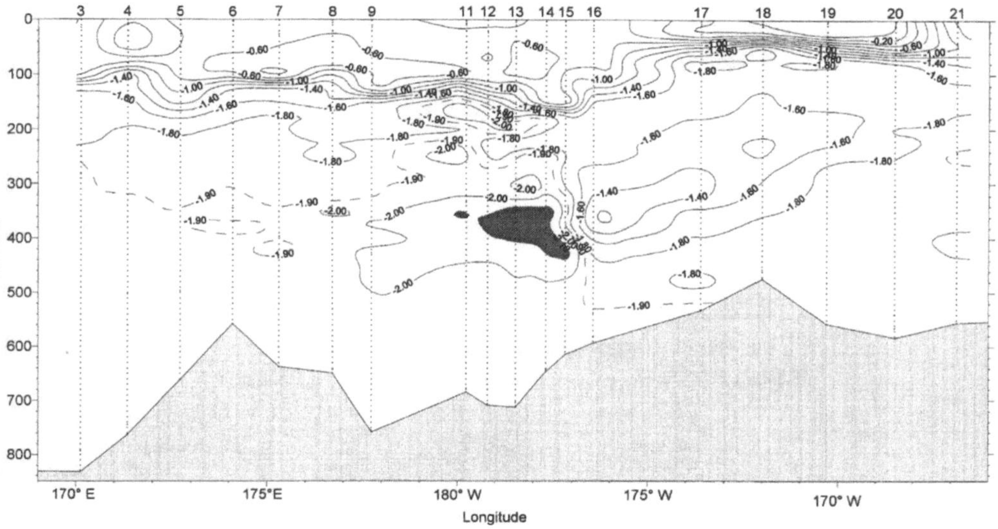


**Fig. 2.** Temperature distribution along transect a in front of the Ross Ice Shelf

from 171° to 177°W in a layer between 180 m in the eastern side and 420 m in the western. It was located in a position similar to that in previous cruises (e.g. Jacobs et al. 1970; Pillsbury and Jacobs 1985), but about 0.3 °C warmer. Separated by an intense thermal front, at 177°W, the presence of ISW was detected by a cold area with temperatures below −2 °C (minimum −2.15 °C at 380 m) extending eastward around 178°E and in a thick layer between 320–500 m. Salinities increased with depth reaching 34.79 below 700 m. This water, cooled by the interaction with the base of the Ross Ice Shelf, was found spreading northward in the Challenger Basin.

A transect along meridian 177°W, from 78 °S to 75°S (b, Fig. 1) shows the northward path of ISW. Its signature, a tongue of water colder than −2 °C, can be observed between 400 and 500 m (Fig. 3). This water deepens and moves northward up to about 75°48′S, (T = −2.0 °C, S = 34.68 at 440 m) probably without reaching the shelf, as indicated from temperature and salinity time series at mooring H which did not detect its passage. The presence of two distinct nuclei of cold water (with minimum temperatures, respectively, of −2.19 and −2.18 °C and salinity of −34.66) indicate an intermittent outflow. Near the bottom, a thin layer of water with a salinity of 34.78 and temperatures of −1.9 °C indicate possible mixing with HSSW. The warm core of MCDW extends north up to about 77°30′S (T = −1.32 °C, S = 34.46 at 300 m) in a layer between 190 and 380 m. An intrusion of warm water (T = −0.45 °C, S = 34.53 at 190 m) stemming from the Antarctic Circumpolar Current can also be observed in the northernmost edge near the shelf break.

In the western stations along the Ross Ice Shelf barrier, where the bottom reaches more than 800 m, salinities tend to increase; waters having the characteristics near those of HSSW (T = −1.91 °C and salinity higher than 34.78) were observed below 450 m; further north (station 2), in the direction of Terra Nova Bay, at about the same depth, salinity was 0.01 psu higher. Waters saltier than 34.85 were observed in the past (Jacobs and Comiso 1989) along the transect from Terra Nova Bay to Franklin Island below 600 m, thus supporting the hypothesis of a southward flow of HSSW.

## Terra Nova Bay and Victoria Land Coasts

Some CTD casts in Terra Nova Bay were taken during the first week in February. Vertical profiles indicated a shallow layer at the surface of about 40 m with a mean salinity of 34.3 remaining after the summer melting. The layer above the pycnocline was characterized by a strong salinity gradi-

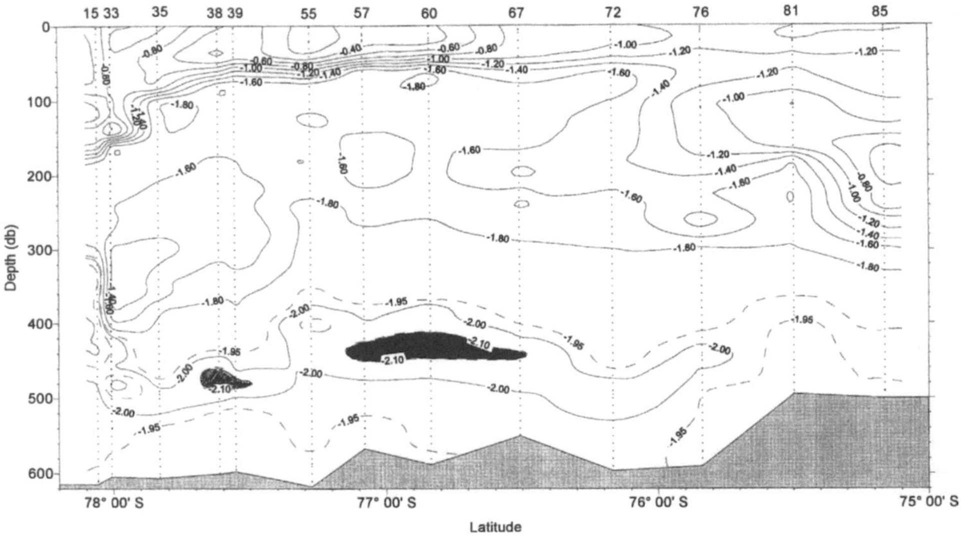


**Fig. 3.** Temperature distribution along transect b in Challenger Basin

ent (34.7 at depths between 100 and 150 m). Below this level, salinity slowly increased with depth reaching 34.78 at 500 m in near coastal waters and 34.86 moving east. Surface temperatures in this period were between −1.84 and −1.88 °C, indicating well-developed cooling processes. Relative maxima between −1.7 and −1.6 °C were found in some stations at depths around 120 m. Stations located inside the Drygalsky Basin, more than 1100 m deep, showed an almost uniform layer, from 700 m to the bottom, filled with HSSW having temperatures of −1.92 °C and salinities of 34.87, among the higest values observed in Antarctica. The northward spreading of this water mass moving along the coast of Victorian Land is shown in the transect from Terra Nova Bay to Cape Adare (c, Fig. 1). The rising of the sea floor from 1100 m to less than 500 m represents a constraint for the spreading of waters with salinities higher than 34.85 which reach their northernmost location at about 74°12′S in a thin layer below 600 m (Fig. 4). A layer with a salinity higher than 34.83 is found below 400 m; its signature vanishes at 72°24′S before reaching the shelf slope. CTD profiles taken along the shelf break did not indicate a deepening of this water since the salinity of this water was not higher than 34.7, but a mixing with the lower edge of MDCW may have been possible. This water mass was found between 320 and 500 m with its core at 400 m (maximum temperature T = 1.06 °C and S = 34.71) enlarging to 72°30′S. A strong thermal front separated open seawaters from those of the continental shelf, probably confining them to recirculate southward. Only limited information on the eastward spreading of HSSW can be obtained from some CDT casts along the west-east transects, since a wide central area of the basin was not investigated. Also in this western region of the Ross Sea the role of the bathymetry was crucial for the spreading of this water mass due to the presence of Crary and Mawson banks which separate coastal deep regions from the central basin. The easternmost location where a tongue with a salinity of 34.83 still survives near the bottom at 172°E.

## Summer Circulation Scheme

The summer circulation scheme in the upper thermocline was inferred by analyzing both ADCP and CTD data. The spatial coverage allowed the description of the dynamics in the area close to the shelf break. The resulting current fields from the surface to 250 m presented the same dynamical features, decreasing in strength with increasing depth. For the calculation of the dynamic heights, a reference level of 250 m was assumed, since this

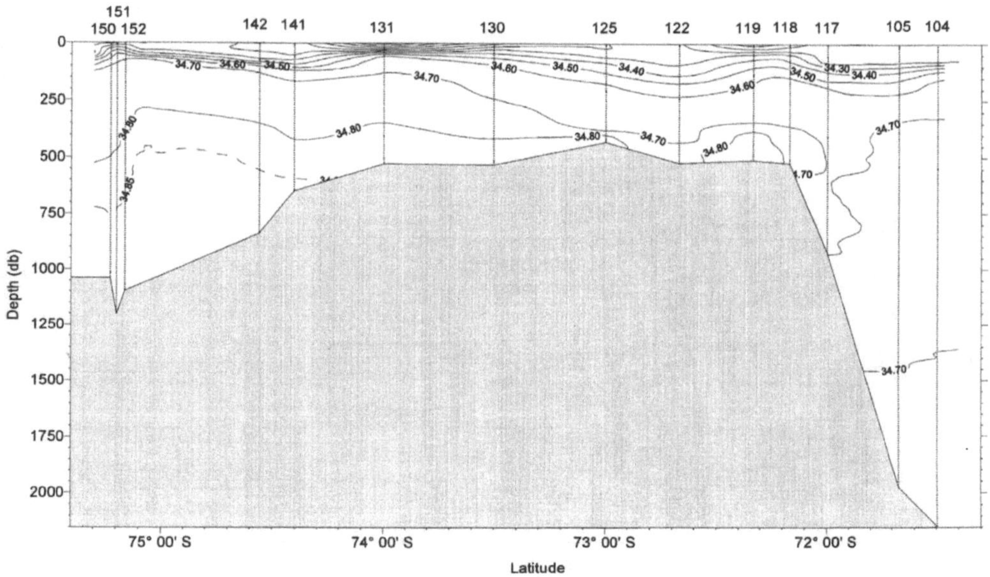


**Fig. 4.** Salinity distribution along transect c from Terra Nova Bay to Cape Adare

was the maximum common depth to all CTD stations. In any case, the use of higher reference levels did not show significant differences. Geostrophic currents show a general northward movement of water masses in the upper thermocline, deviating toward the open ocean in front of Victoria Land.

ADCP profiles near the Ross Ice Shelf barrier indicated that currents were mainly flowing in a western direction; further north the water movements were along the isobaths, indicating a MCDW inflow from the deeper basin (Fig. 5). In the outer northern stations, in front of Victoria Land, there was an outflow of water masses, with an intensification of the currents due to the narrowing of the continental shelf; this was also observed from currentmeter time series (mooring C). A slight modification of the general circulation scheme may be attributed to mesoscale activity, local bathymetry or local wind effects.

## Temporal Time Series

The temporal variability of currents and water mass characteristics in the intermediate and deep layers have been inferred from the analysis of time series of currents, temperature and salinity from fixed moorings. The mean annual currents are reported in Table 2.

Currents in Terra Nova Bay from mooring D show a prevailing northward direction in the upper layer with a clockwise rotation with depth (Fig. 6). Strongest currents with daily mean velocities up to 50 $cm\,s^{-1}$ were observed at 55 m, but high velocities, up to 30 $cm\,s^{-1}$, were also found in the deeper layers. High variability and rotations were the main features from August to October, while a clear southward flow was observed at all levels in March and November (Manzella et al. 1999). Highest stratification in Terra Nova Bay was reached during March and April, when mean salinity at 140 m was 34.64 with a minimum of 34.58 (Fig. 7). Dense water formed at the surface during winter gave origin to deep convective movements; from late August to early December the water column up to about 600 m was well mixed, with a salinity around 34.82. Although there are no direct observations of the Terra Nova polynya during winter 1995, this deep mixing was consistent with the intense heat exchanges at the air-sea interface which allow a high rate of ice formation, such as those occurring in polynyas. Nevertheless, other than a peak of 34.86 at 140 m

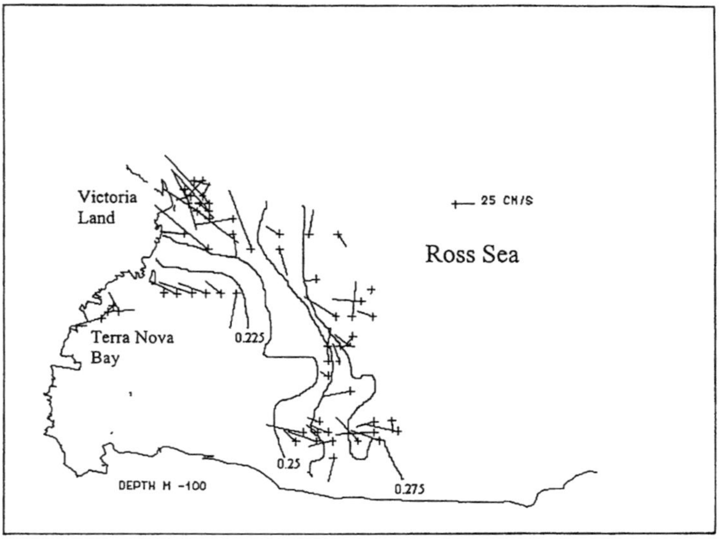


**Fig. 5.** Summer circulation from ADCP measurements and dynamic heights

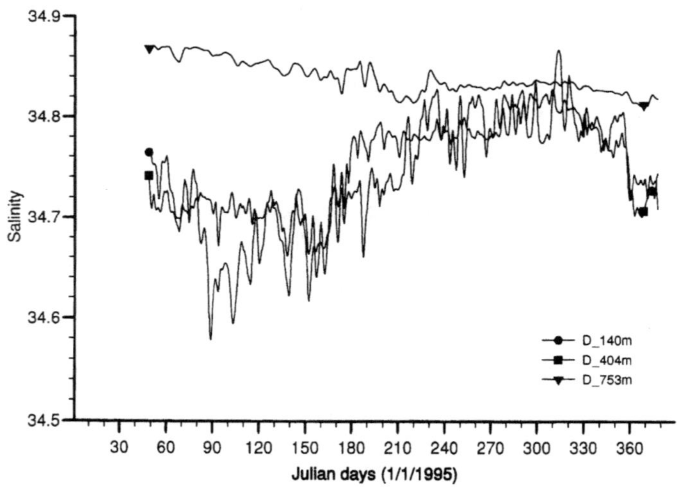


**Fig. 6.** Temporal series of salinity in Terra Nova Bay

in mid-November, salinity values such as those of resident deep water masses were never observed in the time series data, suggesting that this salty water had been produced during past years or was coming from other regions such as the Ross Sea polynya. Summer heating started in mid-November, with a maximum temperature near 0 °C at 55 m in December; a winter minimum was reached in early April.

A northwestward movement of the water masses outflowing from below the Ross Ice Shelf was detected by mooring F (Fig. 8). Starting from

**Table 2.** Mean current intensity and direction (obtained from averaged vx and vy components) over the entire investigation period

| Mooring | A | | B | | C | | F | | |
|---|---|---|---|---|---|---|---|---|---|
| Depth (m) | 231 | 780 | 221 | 555 | 230 | 425 | 244 | 390 | 579 |
| Intensity (cm s$^{-1}$) | 0.7 | 1.3 | 0.6 | 2.3 | 8.0 | 8.4 | 4.3 | 3.0 | 5.0 |
| Direction (°N) | 129 | 127 | 147 | 89 | 343 | 316 | 306 | 312 | 337 |
| **Mooring** | **D** | | | | | **H** | | | |
| Depth (m) | 55 | 140 | 402 | 748 | 882 | 282 | 450 | 503 | 597 |
| Intensity (cm s$^{-1}$) | 7.8 | 3.2 | 2.1 | 4.3 | 4.9 | 3.6 | 4.1 | 3.1 | 2.9 |
| direction (°N) | 354 | 340 | 15 | 71 | 86 | 62 | 69 | 61 | 72 |

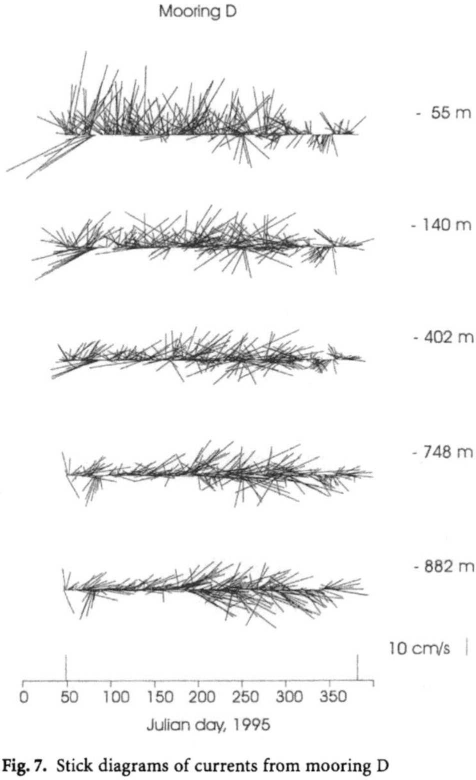


**Fig. 7.** Stick diagrams of currents from mooring D

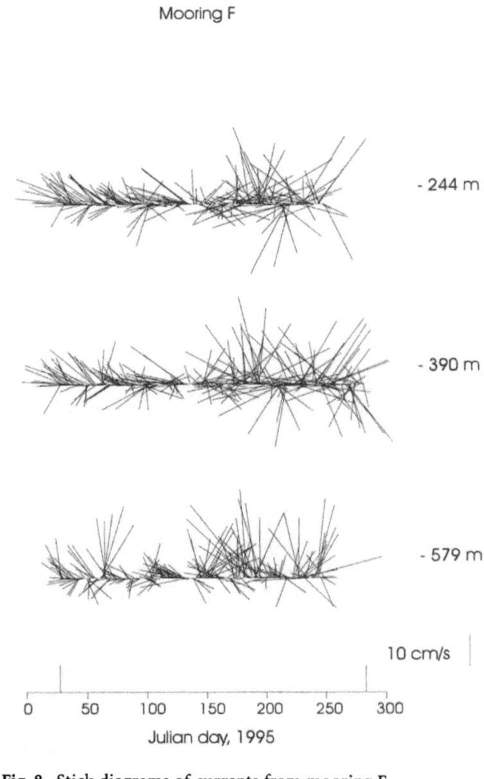


**Fig. 8.** Stick diagrams of currents from mooring F

June, currents were more intense with daily mean peaks of more than 25 cm s$^{-1}$ and were mainly directed north. Summer CTD profiles along the Ross Ice Shelf demonstrated that mooring F was located between two water masses: the ISW to the west, and the Warm Core to the East. The cores of these water masses were probably oscillating in an east-west direction, leaving their signatures in the sensors. Episodes lasting about 1 week, during which salinity increased to 34.64 with corre-

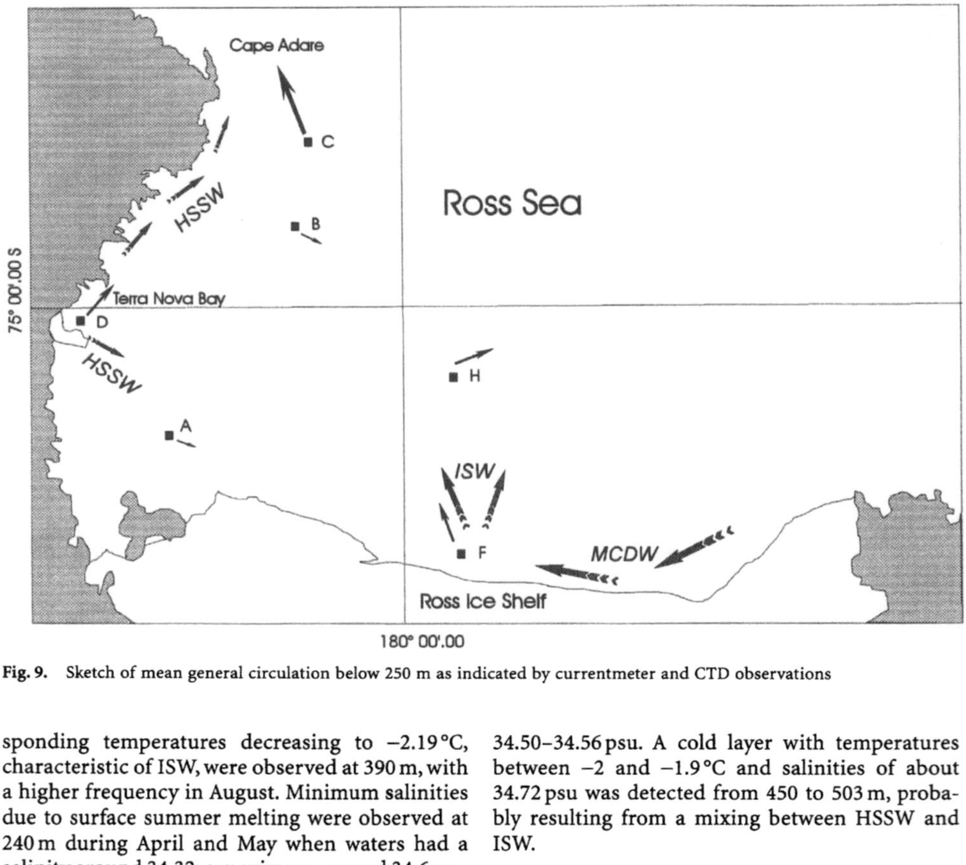


**Fig. 9.** Sketch of mean general circulation below 250 m as indicated by currentmeter and CTD observations

sponding temperatures decreasing to −2.19 °C, characteristic of ISW, were observed at 390 m, with a higher frequency in August. Minimum salinities due to surface summer melting were observed at 240 m during April and May when waters had a salinity around 34.32; a maximum around 34.6 was recorded from September to early December. The temperature seasonal signal at the same depth was from −1.4 °C in March to −1.85 °C during winter, with a high variability probably due to the mixing of MCDW and ISW below.

Currents in position H showed a well-defined barotropic northeast flow with some mesoscale activity denoted by complete rotations and meanderings. Mean velocities were around 6.5–8 cm $s^{-1}$ with daily mean peaks of more than 25 cm $s^{-1}$. Although the system was layered, as detected by the salinity time series, movements were strongly barotropic. The water at 282 m had a limited temperature seasonal cycle, with maximum values in April-May (−1.68 °C) and minimum values in August - October (−1.9 °C). However, many peaks up to −1.5°C were observed for most of the year. Salinity was quite constant, with values of about 34.50–34.56 psu. A cold layer with temperatures between −2 and −1.9 °C and salinities of about 34.72 psu was detected from 450 to 503 m, probably resulting from a mixing between HSSW and ISW.

## Discussion and Conclusions

Current, temperature and salinity data collected during 1995 in the central and western parts of the Ross Sea have allowed us to propose a general scheme for the large-scale circulation in this area (Fig. 9). In the western sector, the resulting mean circulation was mainly westnorthwest along the coast and mainly southeastward along the border of the shelf break, describing a cyclonic gyre. Intense northwestward currents were observed all year long off Cape Adare, while a southward flow was observed during some periods of the year in the southwest region.

Waters stemming from the Circumpolar Current, forming a branch of the Ross Gyre, intruded on the continental shelf, where they were subjected to important modifications. In the western Ross Sea, areas of penetration of this water were identified by some CDT measurements on the shelf break. One was located around 75°30′S, another at 73°30′S. Two important water masses, MCDW and ISW, circulating inside the Ross Ice Shelf cavity, were observed emerging from below, separated by a strong thermal front at 177°W. WCDW, which probably comes from the eastern side of the Ross Sea, moved west along the barrier.

ISW was mainly directed northwest spreading in the Challenger Basin where waters with temperatures below −2 °C and with nuclei colder than −2.1 °C were detected up to 76°15′S. HSSW filled the bottom layer, below 500 m, and was mainly concentrated in the western side of the Ross Sea, where the sea reached depths up to 1100 m, thus acting as a reservoir of salty dense water. The main flow of HSSW from Terra Nova Bay was directed northeast along the coast of Victoria Land up to the shelf break in front of Cape Adare. However, a significant southward transport was also observed in March and November. The examined time series of salinity in the Terra Nova Bay polynya did not indicate production of waters with the same characteristics as those of resident deep ones during winter 1995, suggesting that important contributions may originate from other areas such as the Ross Sea polynya, and that there was an interannual variability in the characteristics and amount of HSSW formed.

The coarse spatial resolution of CTD casts along the shelf break did not allow us to identify areas where this salty and dense water mass leaves the Ross Sea to deepen into the open ocean. Bathymetric constraints are crucial for limiting the eastern spreading; the general westward increase of salinity seems consistent with both the observed and computed circulation (Bergamasco et al. 1999).

## References


Bergamasco A, Carniel S, Caldesi Valeri L (1999) Reconstructing the general circulation of the Ross Sea (Antarctica) using a robust diagnostic model. In: Manzella GMR, Spezie G (eds) Oceanography of the Ross Sea (Antarctica). Springer, Milano, Italia, pp 119–134

Carmack EC (1977) Water characteristics of the southern ocean south of the Polar Front. In: Angel M (ed) A voyage of discovery (1977) George Deacon 70th anniversary volume. Deep Sea Res 24 (Suppl): 15–41

Comiso JC, Maynard NG, Smith WO Jr, Sullivan CW (1990) Satellite ocean colour studies of Antartic ice edges in summer and autumn. J Geophys Res 95(C6): 9481–9496

Hellmer HH, Jacobs SS (1995) Seasonal circulation under the eastern Ross Ice Shelf, Antartica (1995). J Geophys Res 100(C6): 10873–10855

Jacobs SS, Comiso JC (1989) Sea ice and oceanic processes on the Ross Sea continental shelf. J Geophys Res 94(C12): 18195–19211

Jacobs SS, Amos FA, Bruchhausen PM (1970) Ross Sea oceanography and Antartic bottom water formation. Deep Sea Res 17: 935–962

Jacobs SS, Gordon AL, Arday JL Jr (1979) Circulation and melting beneath the Ross Ice Shelf. Science 203: 439–443

Jacobs SS, Fairbanks GR, Horibe Y (1985) Origin and evolution of water masses near the Antarctic continental margin: evidence from H218O/H216O ratio in sea water. Antarct Res Ser 43: 59–86

Kurtz DD, Bromwich DH (1985) A recurring atmospherically forced polynya in Terra Nova Bay. Antarct Res Ser 43: 177–202

Manzella GMR, Meloni R, Picco P (1999) Current, temperature and salinity observations in Terra Nova Bay polynya area. In: Manzella GMR, Spezie G (eds) Oceanography of the Ross Sea (Antarctica). Springer, Milano, Italia, pp 165–173

Picco P, Amici L, Demicheli L, Langone L, Meloni R, Ravaioli M (1999) Temporal variability of currents in the Ross Sea (Antarctica). In: Manzella GMR, Spezie G (eds) Oceanography of the Ross Sea (Antarctica). Springer, Milano, Italia, pp 103–117

Pillsbury RD, Jacobs SS (1985) Preliminary observation from long-term current meter moorings near the Ross ice shelf, Antartic. Antarct Res Ser 43: 87–107

Stocchino C, Manzella GMR (1991) Analisi delle correnti marine di Baia Terra Nova (Mare di Ross, Antartide). Consiglio Nazionale delle Ricerche, Area della Ricerca di Genova, 56 pp

CHAPTER 9

# Spatial and Temporal Patterns of Nutrient Distributions in the Ross Sea

G. Catalano[1], F. Benedetti[2], S. Predonzani[3], A. Goffart[4], S. Ruffini[5], P. Rivaro[6], and C. Falconi[1]

**ABSTRACT**

The Ross Sea is one of the most peculiar zones in the World Ocean because of its physical, chemical and biological characteristics. On its continental shelf, it is influenced by only one water mass of external origin, the Circumpolar Deep Water (CDW), which, in turn, generates all the other water types by cooling or mixing with water coming from ice melting. From the biological and chemical point of view, the Ross Sea is considered as a typical Continental Shelf Coastal Zone (CSCZ) where, during summer, occasional high nutrient consumption occurs particularly in the Marginal Ice Zone (MIZ) and in some protected coastal areas as Terra Nova Bay.

As a part of the ITALIANTARTIDE Polar Research Project, four oceanographic cruises were carried out in the late austral spring or summer in different years (1987–88, 1989–90, 1994–95 and 1995–96) with the aim of extending our knowledge about the variation in spatial and temporal nutrient patterns in the Ross Sea, in relation with physical and biological processes. Due to a slower recycling rate of Si compared to the N and P cycles, high Si concentrations and different evolution of the N/P and Si/N ratios are typical of CDW and of that part of the Ross Sea which is affected by this water mass. Advection of melt water can have a very important role in the nutrient concentration in the upper mixed layer because the decrease in salinity forces a strong vertical stratification, thereby preventing phytoplankton cells from leaving the euphotic layer. Indeed, nutrient plots suggest that the decrease of N and Si in the surface layer cannot be explained by simple dilution due to sea-ice melting water advection, but phytoplankton uptake must also be taken into account. The High Salinity Shelf Water was also present in some stations. In the Terra Nova Bay, however, this surface water can be modified by summer heating and reach a temperature of 2 °C under particular meteorological conditions. A thermal stratification which is able to support an increased phytoplankton production and a nutrient decrease is thus originated.

## Introduction

The Ross Sea can be considered the largest Continental Shelf and Coastal Zone (CSCZ) of Antarctica where seawater and continental ice can interact directly by ice melting, and thereby modify the characteristics of the water masses involved. In addition, the katabatic winds in the southern and western parts of the Ross Sea make these zones particularly active in producing sea ice and brine.

Although the importance of these processes is mainly reflected on physical features of the Ross Sea (water-mass modification, sea-ice formation, polynya maintenance, bottom-water formation and ventilation of the deep ocean) some effects involve chemical and biological processes, such as nutrient distribution and uptake, phytoplankton blooms, grazing by krill and other zooplankton and, through the trophic net, the ecology of marine birds and mammals.

The Ross Sea is influenced by only one water mass of external origin: the Circumpolar Deep Water (CDW), characterized by relatively high temperature and salinity, high silicic acid and low oxygen concentration. All other water types are

[1] CNR, Istituto Talassografico, Trieste, Italy
[2] Dipartimento di Scienze Chimiche, Università di Trieste, Trieste, Italy
[3] Laboratorio di Biologia Marina, Trieste, Italy
[4] Institut de Physique, Université de Liege, Liege, Belgium
[5] Dipartimento di Chimica, Università di Genova, Genova, Italy

derived from modifications of the CDW by mixing, cooling and by inputs from precipitation, melting and brine formation (Jacobs et al. 1985; Russo, this Vol.). Therefore, a certain variation of nutrient concentrations in the Ross Sea must be expected, due to these different modifications of CDW.

The processes responsible for polynya maintenance are also responsible for the modification which leads to the formation of the High Salinity Shelf Water (HSSW), located in the southwestern Ross Sea and at Terra Nova Bay (Jacobs et al. 1985). The HSSW is the most salted and dense water found in the Ross Sea.

Supercooled Deep Ice Shelf Water (DISW) originates from glacial ice melting, operated by the HSSW at the base of the Ross Ice Shelf. Its domain is mainly located in the southern zone of the Ross Sea, near the Ross Ice Shelf edge. It can be detected by a temperature minimum below freezing point, the lowest temperature reached in the Ross Sea. Its density is lower than that of HSSW but higher than that of any other water type in the Ross Sea.

The Low Salinity Shelf Water (LSSW), which is slightly higher in temperature but considerably lower in salinity, is found between the surface Antarctic waters AASW and the HSSW, mainly on the eastern part of the Ross Sea shelf (Jacobs et al. 1985; Russo, this Vol.).

Another type of bottom water, the Low Salinity Bottom Water (LSBW), is characterized by intermediate properties between those of the CDW portion which intrudes on the continental shelf and of the LSSW. It has been found in some stations carried out during Italian expeditions along the continental slope (Russo, this Vol.).

An important modification of the CDW, resulting from the mixing which occurs on the slope front, is its transformation in the Warm Core (WMCO). The WMCO is an intermediate water mass with a temperature above freezing point, and is considered a major factor responsible for the opening of the Ross Sea (Russo, this Vol.).

Coming now to the surface waters, the Antarctic Surface Water (AASW) evolves from CDW which, on upwelling, has been modified by interactions with atmosphere, ice and Winter Water. Alternatively, AASW can also be produced in summer, on the continental shelf, from the mixing of shelf water and sea-ice meltwater. All these factors make AASW the most variable water type in the Ross Sea.

From a chemical and biological point of view, the Southern Ocean is considered a typical High-Nutrient Low-Chlorophyll (HNLC) zone (Dugdale and Wilkerson 1991). Abundant primary nutrients are accompanied by a low primary productivity, and usually the primary productivity stops before the nutrient bulk has been completely depleted. As a HNLC zone, however, the Ross Sea is peculiar in that it can exhibit nutrient exhaustion in certain inner parts during the summer. Especially at the ice edge, $NO_3$ and $PO_4$ concentrations can drop to undetectable values, while $Si(OH)_4$ concentrations can decrease to near 5 $\mu$M due to biological uptake (Nelson and Tréguer 1992; Tréguer and Jacques 1992).

More generally, estimates of primary productivity indicate that about 40% of annual primary production of the Southern Ocean may derive from ice-edge blooms (Smith and Nelson 1986). The Ross Sea is also peculiar in this aspect with its marginal ice that recedes from south (the Ross Sea polynya) to north (Fig. 1).

The depth and vertical stability of the upper mixed layer, together with the irradiance regime, are among the parameters to be considered as typifying the variation in the CSCZ characteristics with time. Especially the depth and the vertical stability of the upper mixed layer are able to control the growth and loss rate of phytoplankton in the euphotic layer (Smith and Nelson 1985; Sambrotto et al. 1986; Smetacek and Passow 1990; Mitchell and Holm-Hansen 1991; Nelson and Smith 1991; Cota et al. 1992; Catalano et al. 1997a). In turn, depth and stability are strongly influenced by the input of low-density melt water from the receding pack ice. Thus, the abundance of nutrients in the upper layer is mainly controlled by phytoplankton dynamics which depends on hydrological structures and thereby on the pack-ice dynamics.

Production/regeneration processes also influence the nutrient concentrations. Tréguer and Jacques (1992) found that the C, N, Si cycles are uncoupled in the Ross Sea. According to Smith and Harrison (1991), C/N-specific uptake rates approach the C/N ratio for standing stock (C/N in particulate matter stands between 5 and 8, i.e. in the classical range) only in the upper part of the euphotic zone. At low irradiance, the C/N ratio drops to values as low as 1–2. Reasons for such imbalance are not clear at present, but biochemical processes, like the release of N-rich dissolved organic matter from the particulate pool after production and bacterial uptake of inorganic N and removal of organic C, might be involved.

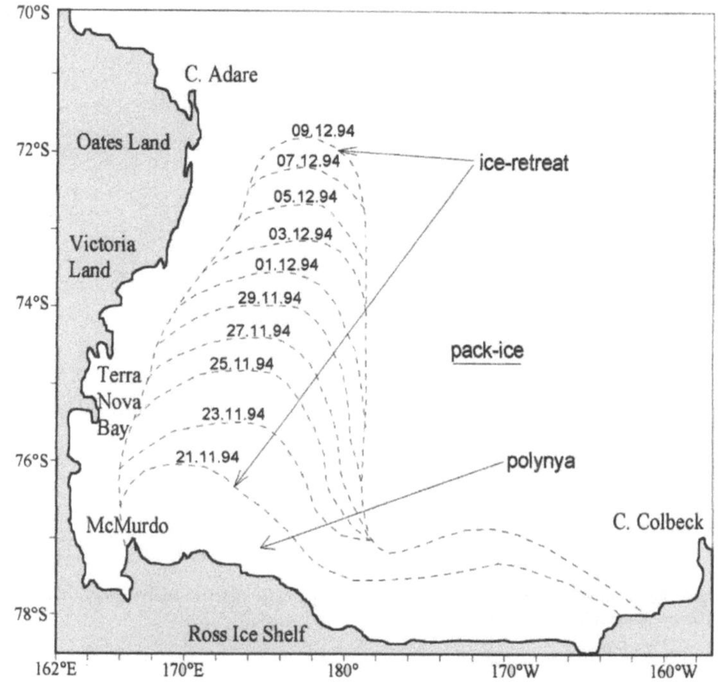


**Fig. 1.** The pack ice retreating from south to north in 1994, reconstructed from satellite images received on board the R/V Italica

The uncoupling of the Si and N cycles is, on the one hand, due to a higher removal rate of silicic acid with respect to nitrate (Zentara and Kamykowski 1981). On the other hand, it can also be due to a faster recycling rate of $NH_4$ and $NO_3$ in the surface layer (Lancelot et al. 1992). According to Nelson et al. (1991), the vertically integrated silica dissolution rate in the euphotic layer accounts for 64% of the gross biogenic silica flux. This means that the dissolution rates of biogenic Si are high enough to support an intense recycling of Si within the upper 50-m layer, except in the immediate vicinity of the ice edge, where Si can drop to limiting concentrations (Nelson and Tréguer 1992). The effects of such imbalance between C and Si and between C and N will thus be reflected in the deeper layers, influencing the concentration of nutrients and especially their ratios.

The aim of this chapter is to present some typical patterns of nutrient distributions found in the Ross Sea, and identify physical and biological factors which control these patterns, depending on seasonal or local events. Other aspects concerning a more comprehensive description of the water masses in the Ross Sea, are dealt with elsewhere in this Volume (Artegiani et al.; Picco et al.; Russo, this Vol.).

## Methods

### Cruises

In the framework of the ITALIANTARTIDE Polar Research Project, four oceanographic cruises were carried out during the late austral spring or summer in the years 1987–88, 1989–90, 1994–95 and 1995–96. The 1987–88 cruise covered mainly the area of Terra Nova Bay (Fig. 2) and a transect extending 120 NM eastwards in the Ross Sea at a latitude of about 75°S. It was carried out on board the R/V Polar Queen from 5 January to 14 February 1988.

In 1989–90, the cruise was divided into two legs. The first leg was carried out on board

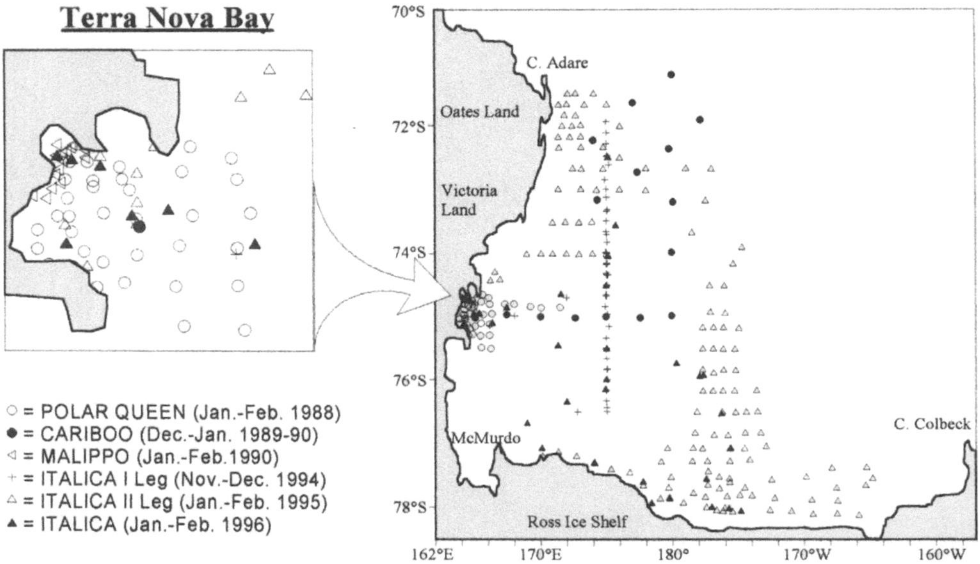


**Fig. 2.** Representation of all stations where water sampling for nutrient analysis was carried out during the four ITALIANTARTIDE oceanographic expeditions

the R/V Cariboo and lasted 1 month, from 5 December 1989 till 6 January 1990. It covered the area of the continental Ross Shelf margin (Fig. 2), eastwards from Cape Adare, and a transect in the Ross Sea at the same latitude of the one carried out during the 1987–88 cruise. The second leg, from 13 January till 2 February 1990 with the R/V Malippo, concentrated on a set of stations along the coast (ca. 100–200 m offshore) from Evans Cove to Campbell Ice Tongue at Terra Nova Bay (Fig. 2).

In 1994–95, the cruise consisted of two legs, both with the R/V Italica. The first leg in late spring, from 14 November till 15 December 1994, was entirely devoted to following the seasonal ice retreat from the Ross Sea polynya northwards along the longitude of 175°E (Fig. 2). The second leg, from 20 January till 20 February 1995, comprised a rather wide area of the Ross Sea, from the Ross Ice Shelf barrier to the Ross Sea Continental Shelf and slope, and again in Terra Nova Bay (Fig. 2).

Finally, in 1995–96 from 12 January till 9 February 1996, areas already studied in the previous cruises were again covered, each one with a limited set of stations, to examine the interannual variability (Fig. 2).

According to the classification for the dynamics of nutrient and phytoplankton (Tréguer and Jacques 1992), the areas investigated in the different cruises mainly belonged to the Continental Shelf and Coastal Zone (CSCZ) with some stations typical of a Marginal Ice Zone (MIZ) in a phase of massive melting.

## Sampling

Seawater samples were collected by means of a General Oceanics rosette sampler or a Seabird carousel water sampler fitted with 12 or 24 12-l Niskin bottles. The vertical profiles of temperature, salinity, fluorescence and light transmission were logged during downcasting.

In the upper 200 m, standard sampling depths were: surface, 10, 25, 50, 100 and 200 m. Other sampling depths were added according to the water column stratification monitored from the CTD profil, the fluorescence maximum and the depths corresponding to 50, 20, 10, 5 and 1% attenuation of incident PAR. Below 200 m, the sampling depths were chosen on the basis of the CTD profile only.

## Methodologies and Instruments

Different CTD probes were used during the cruises: EG&G Mk3, ME mod. KMS and SBE 9/11 Plus. Whenever possible, salinity was checked with an Autosal 8400 Guildline or Minisal 2001 AGE salinometer. Reversing digital (SIS) or analogical thermometers were mounted on the Niskin bottles as temperature control.

The subsamples for dissolved oxygen and nutrient (reactive silicic acid, reactive phosphorus, reactive nitrate and nitrite) determinations were collected directly from the Niskin bottles. Samples collected in the layer of high fluorescence were filtered on Whatman GF/C filters prior to nutrient analysis. All chemical analyses were carried out on board.

Dissolved oxygen was determined according to the Winkler method (Strickland and Parsons 1972) adopting a potentiometric detection of the end point with a Metrohm 682 Titroprocessor. Until 1990, nutrients were determined by means of a Technicon II Autoanalyzer (Hansen and Grasshoff 1983), then an Alpkem segmented flow analyzer was used according to Alpkem Flow Solution Methodologies (ALPKEM 1992a–c). Throughout this chapter, depths are expressed in dbar of pressure, temperature in degree Celsius, salinity is a dimensionless parameter according to the practical salinity scale (PSS-78) and all concentrations are in $\mu mol\,dm^{-3}$, shortened to $\mu M$ (UNESCO 1985).

# Results and Discussion

The following discussion is based on original data reported in Boldrin and Stocchino (1990), Catalano and Benedetti (1990), Catalano et al. (1991a,b), Fabiano et al. (1991), Artegiani et al. (1992) and Catalano et al. (1997a–c).

A typical profile for $Si(OH)_4$, relatively to the N-S transect, carried out in spring 1994 (first leg R/V Italica), is shown in Fig. 3. This profile was obtained before the summer advection of melt water from the pack ice and shows essentially the intrusion of CDW [higher $Si(OH)_4$ concentration], or its modifications, at the edge of the the Ross Sea Continental Shelf (northwards from Mawson Bank) and on the Pennell Bank. No significant vertical gradient of $Si(OH)_4$ concentration is evident, and the same is also true for the other nutrients.

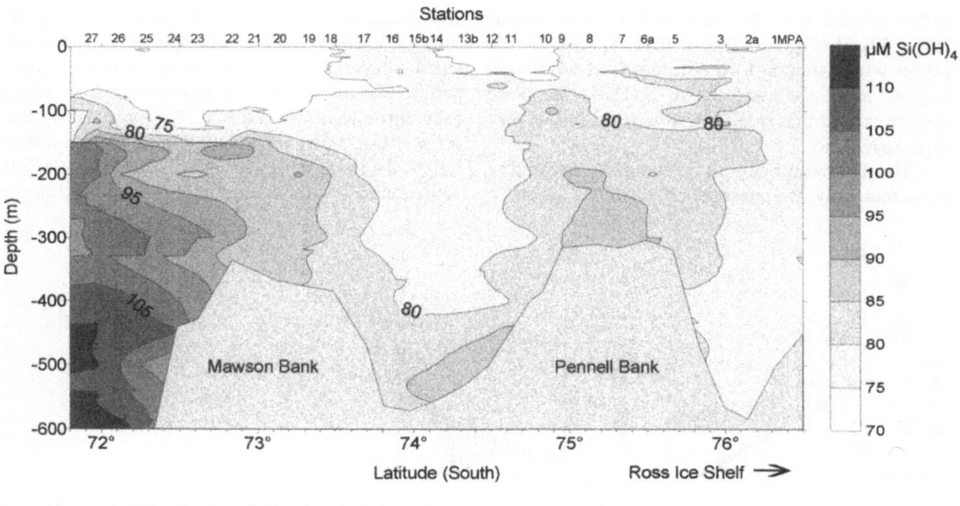


**Fig. 3.** The vertical distribution of $Si(OH)_4$, relatively to the N-S transect carried out in spring 1994 (first leg R/V Italica). This profile was obtained before the summer advection of meltwater from the pack ice and shows essentially the intrusion of CDW, or its modifications, on the Ross Sea shelf

## Nutrient Patterns Linked to the Circumpolar Deep Water

Temperature and dissolved oxygen are good tracers for CDW. The core of this water mass has a temperature of 1.17 °C (Jacobs et al. 1985). As a long segregated water body, its $O_2$ concentration is very low for the Ross Sea. A diagram of the $O_2$/T data, for the stations where an intermediate layer with a T > 0 °C was found (Fig. 4a), shows a very clear trend suitable for characterizing the CDW (see also Fig. 4b) in the temperature range from 0.5 to 1.5 °C and in the oxygen range from 250 to 180 µM. The location of stations where the temperature was above 0 °C and the oxygen less than 250 µM is reported in Fig. 5, showing the CDW dominance in the NW Ross Sea.

The identification of CDW, based on temperature and oxygen concentrations, is also useful for understanding the different variation in N/P and Si/N ratios, from the surface to the deeper layers. Figure 6 shows the plots of $NO_3 + NO_2$ vs $PO_4$ and $Si(OH)_4$ vs. $NO_3 + NO_2$, as well as the theoretical lines representing the phytoplankton utilization ratios according to values estimated by Catalano et al. (1997a), for the R/V "Cariboo" cruise. While the line corresponding to an N/P ratio of 11.3 fits well with the experimental distribution of $NO_3 + NO_2$ and $PO_4$ (Fig. 6a), an Si/N ratio of 2.7 gives an acceptable fit only for silicic acid concentrations below 80 µM and for nitrogen concentrations below 32 µM (Fig. 6b). Above these values, for the points which belong to the domains of CDW and bottom layer, the increase of $Si(OH)_4$ seems to greatly exceed that of $NO_3 + NO_2$, resulting in very high Si/N ratio.

This behaviour cannot be explained by mixing processes only. On average, only $Si(OH)_4$ shows a monotonic increase (up to more than 120 µM) in the entire temperature range used to trace the CDW, as shown in the plots of $Si(OH)_4$, $NO_3 + NO_2$ and $PO_4$ vs. temperature (Fig. 7). On the contrary, N and P reach a concentration maximum (33 and 2.2 µM, respectively) at a temperature around 0 °C, which is outside the CDW temperature range. This could be an indirect confirmation of the different recycling rates of nitrogen, phosphorus and silica. In other words, assuming temperature as a conservative tracer of the history of this water mass, the different behaviour of nitrogen and phosphorus can be explained if we consider that a complete remineralisation occurred in the upper layers, while silicic acid, because of its slower recycling rate, reaches its maximum in the "older" deep water masses. In the CDW this process is enhanced because the dissolution rate of diatom frustules increases with warmer temperature (Tréguer et al. 1989; Tréguer and van Bennekom 1991). On the contrary, a constant N/P ratio close to 11.3 in the entire concentration range (Fig. 6a) indicates that the nitrogen and phosphorus cycles are coupled in the whole water column.

## Surface Nutrients and Sea-Ice Melting Water Advection

According to Jacobs et al. (1985), the Antarctic Surface Water (AASW) corresponds to the mixed surface layer characterized in summer by a lower salinity (34.1 < S < 34.5) and a temperature which may sometimes exceed 0 °C. Unless massive sea-ice melting is present, salinity should not decrease below 34.1. In the first leg of the 1994–95 R/V Italica cruise, carried out along the longitude of

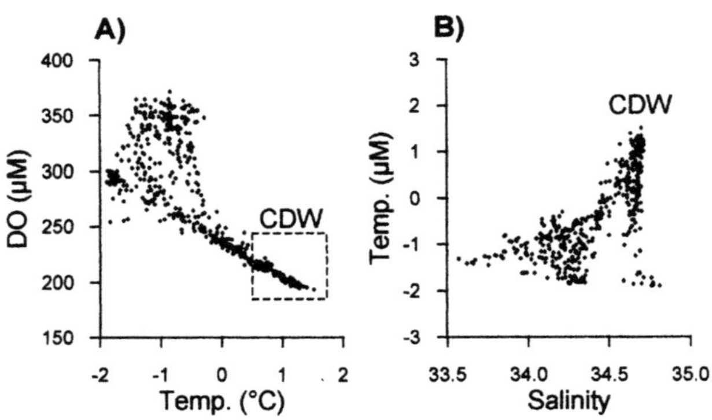


Fig. 4. Dissolved oxygen (DO) vs. temperature (A) and T/S diagram (B) of those stations where a temperature >0 °C and DO < 250 µM were found

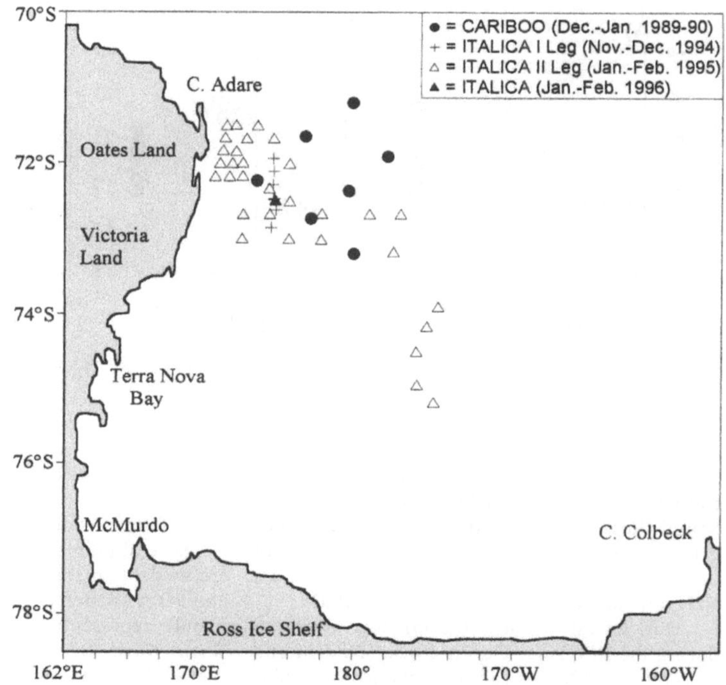


**Fig. 5.** Location of the stations where CDW (T > 0 °C and DO < 250 μM) was detected and sampled during the Italian expeditions in the Ross Sea

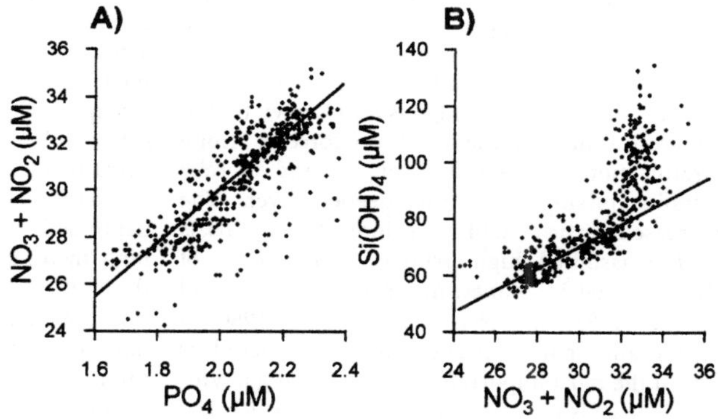


**Fig. 6.** Plots of and $NO_3 + NO_2$ vs. $PO_4$ (**A**) and $Si(OH)_4$ vs $NO_3 + NO_2$ (**B**) for all the stations where CDW was detected. The *lines* stating the estimated ratios of 11.3 and 2.7, respectively, for N/P and Si/N (Catalano et al. 1997a) are also shown

175°E in late spring, from 14 November to 15 December 1994, the salinity never dropped below 34.3 (Russo, this Vol.). In the other cruises, a lower surface salinity was found in correspondence with summer sea-ice melting. We can therefore presume that all surface salinity values lower than 34.1 are due to local mixing of ice melt water with the preexisting AASW.

The advection of melt water can have a very important influence on the nutrient concentration of the upper mixed layer. In the Weddell Sea, Ackley et al. (1979) attributed to sea ice an average

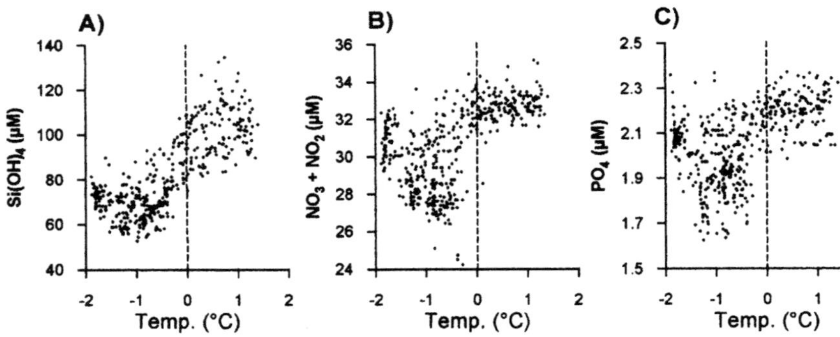


**Fig. 7.** Plots of $Si(OH)_4$ (**A**), $NO_3$ + $NO_2$ (**B**) and $PO_4$ (**C**) vs. temperature for all the stations where CDW was detected. Above 0 °C, only $Si(OH)_4$ continues to increase

salinity of 6. With regard to the nutrient concentrations, Arrigo et al. (1995) in the McMurdo Sound (Ross Sea) have attributed to sea ice (congelation ice) an average nutrient content of 4.4 µM of $NO_3$, 0.43 µM of $PO_4$, 5.66 µM of $Si(OH)_4$ and a $NH_4$ content varying between 0.6 and 4.1 µM.

Our lowest salinity (32.45) was observed during the R/V Polar Queen cruise (Boldrin and Stocchino 1990). This can be explained by a dilution of the surface layer, caused by the advection of 5.9% of melting water to an original water with a salinity of 34.1. Hence, any significant depletion of nutrients (greater than 5.9%), whenever observed in the upper mixed layer, cannot be explained by dilution and must be attributed to phytoplankton assimilation (Catalano et al. 1997a). On the other hand, it is also known that in Antarctica the decrease in salinity forces a strong stratification of the water column and thereby inhibits mixing with the deeper layer. The increased vertical stability is, indeed, at the basis of the assumed high primary productivity of the Marginal Ice Zones. According to Smith and Nelson (1985) and Nelson and Smith (1986), a strong and shallow pycnocline is considered as one of the most important factors allowing phytoplankton cells to remain in the euphotic layer. The combination of inhibited vertical mixing and increased phytoplankton assimilation can therefore, in summer, often lead to marked decrease of nutrients in the upper mixed layer.

Even if situations of increased productivity in surface layers with salinity greater than 34.1 cannot be excluded, we will focus on the effects of the advection of melting water on nutrient concentrations. For this reason, the value of 34.1 has been adopted as a threshold to identify stations affected by melt water; only these are considered in the following discussion.

Several points with a salinity lower than 34.1 are evident in the plots of $Si(OH)_4$, $NO_3$ + $NO_2$ and $PO_4$ vs. salinity (Fig. 8). Many of these points, mainly representing surface water collected in Terra Nova Bay during the Malippo cruise (Jan.–Feb. 1990), correspond to nutrient concentrations as low as 2.10 µM of $Si(OH)_4$, 2.56 µM of $NO_3$ + $NO_2$ and 0.12 µM of $PO_4$, thus confirming a high biological assimilation.

Actually, all areas in the Ross Sea are, to some extent, influenced by sea-ice melting. What characterizes each zone is the thickness of the surface-melt water mixed layer thus originated and the strength of the resulting pycnocline. These properties will be enhanced in some coastal zones, like Terra Nova Bay, by the synergic effect of reduced circulation and solar heating or, on the contrary, depressed in other zones characterized by a high hydrodynamism. To evaluate and compare these effects on nutrient and oxygen concentrations, the depth coresponding to a salinity of 34.1 has been chosen as the boundary of the layer influenced by melt-water advection, and the means of these depths were calculated for selected areas of the Ross Sea (Table 1). To obtain a threshold value for nutrients, which can be matched to a salinity of 34.1, the mean concentrations of $Si(OH)_4$, $PO_4$ and $NO_3 + NO_2$ have been calculated from all the samples having a salinity between 34.095 and 34.104. These thresholds have been evaluated to be 62.0 µM for $Si(OH)_4$, 1.62 µM $PO_4$ and 22.4 µM for $NO_3 + NO_2$. In Table 1 are reported the mean values of temperature, salinity, oxygen,

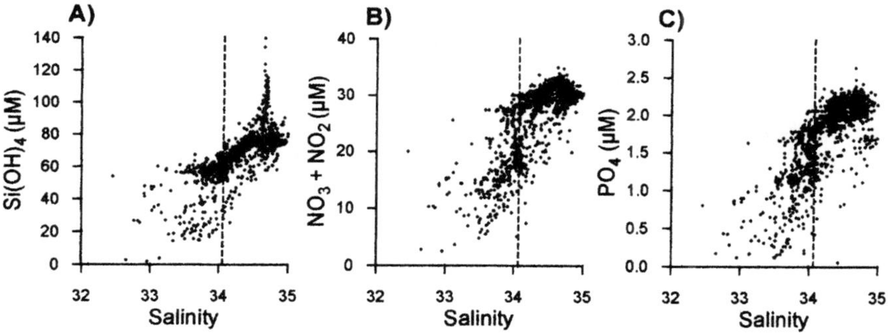


**Fig. 8.** Plots of $Si(OH)_4$ (**A**), $NO_3 + NO_2$ (**B**) and $PO_4$ (**C**) vs. salinity for the stations where a surface salinity <34.1 was observed. The *dotted lines* show the threshold of 34.1

**Table 1.** Mean depths, values and depletion of nutrients calculated in the surface layer, affected by meltwater (S < 34.1). The depletions have been estimated by difference from the mean nutrient concentration at salinity of 34.1. Only positive depletions have been taken into account for calculating the respective averages

| Zone | Depth[a] | Temp.[b] | Salinity[b] | Oxygen[b] | Silica[b] | Phosph.[b] | $NO_3 + NO_2$[b] | ΔSilica[c] | ΔPhosph.[c] | $\Delta(NO_3 + NO_2)$[c] |
|---|---|---|---|---|---|---|---|---|---|---|
| Ross Ice | 48 | −0.5 | 33.9 | 374.3 | 56.9 | 1.3 | 17.3 | 5.5 | 0.4 | 5.3 |
| Shelf edge | | 0.3 | 0.2 | 11.3 | 3.7 | 0.2 | 2.7 | 3.2 | 0.1 | 2.5 |
| Eastern Ross | 24 | −0.8 | 33.9 | 356.7 | 59.3 | 1.6 | 23.5 | 3.8 | 0.1 | 3.1 |
| Sea | | 0.4 | 0.1 | 15.2 | 4.0 | 0.2 | 3.6 | 3.0 | 0.1 | 1.8 |
| Northern Ross Sea | 32 | −1.1 | 34.0 | 348.0 | 61.0 | 1.8 | 27.6 | 4.3 | – | – |
| Continental Break | | 0.3 | 0.1 | 13.5 | 5.3 | 0.1 | 1.2 | 1.9 | – | – |
| Victoria Land | 12 | 0.1 | 33.7 | 377.0 | 50.6 | 0.9 | 18.8 | 14.6 | 0.8 | 6.3 |
| Coast | | 1.1 | 0.4 | 16.1 | 12.0 | 0.5 | 5.0 | 11.0 | 0.5 | 3.7 |
| Terra Nova | 16 | 0.1 | 33.6 | 385.4 | 29.0 | 0.7 | 11.3 | 33.0 | 1.0 | 1.1 |
| Bay | | 0.7 | 0.3 | 19.9 | 11.5 | 0.3 | 3.8 | 11.5 | 0.3 | 3.8 |

[a] Mean depth (dbar) of the surface layer affected by meltwater (S < 34.1).
[b] Mean values (µM) and standard deviations in the upper mixed layer.
[c] Average (µM) and standard deviations of all positive depletions of nutrients.

reactive silicic acid, phosphate and nitrate plus nitrite, calculated in the layer influenced by melt water, as well as the mean nutrient decrease (compared to each threshold value) due to dilution plus phytoplankton uptake.

Among the zones listed in Table 1, those corresponding to the part of the Ross Sea along Victoria Land Coast and Terra Nova Bay show the shallowest mixed layer (12 and 16 dbar, respectively) and the highest average temperature (0.1 °C). Terra Nova Bay is further characterized by the lowest average salinity of the upper mixed layer (33.6) and the highest dissolved oxygen concentration (385.4 µM) and nutrient decrease [33.0, 1.0 and 11.1 µM for $Si(OH)_4$, $PO_4$ and $NO_3 + NO_2$ respectively].

In the stations approximately located on the continental break in the northern Ross Sea (Fig. 9), the upper mixed layer is, on the contrary, characterised by the highest salinity (34.0), close to the threshold value of 34.1, and the lowest oxygen content (348.0 µM) and nutrient utilization [4.3 µM for $Si(OH)_4$ and no significant depletion for the other nutrients]. It is possible that in this particular location, which can be considered as a slope front (Jacobs 1991), a greater hydrological dynamism favours water mixing and thereby the vertical instability of the water column. Iron limited phytoplankton growth near the continental shelf break could be another explanation for the lower depletion of nutrients. Bottle enclosure experiments carried out by Martin et al. (1990) revealed a very severe iron limitation in phytoplankton growth near the shelf break compared to stations near Victoria Land, where little or no iron limitation was found. However, a reanalysis

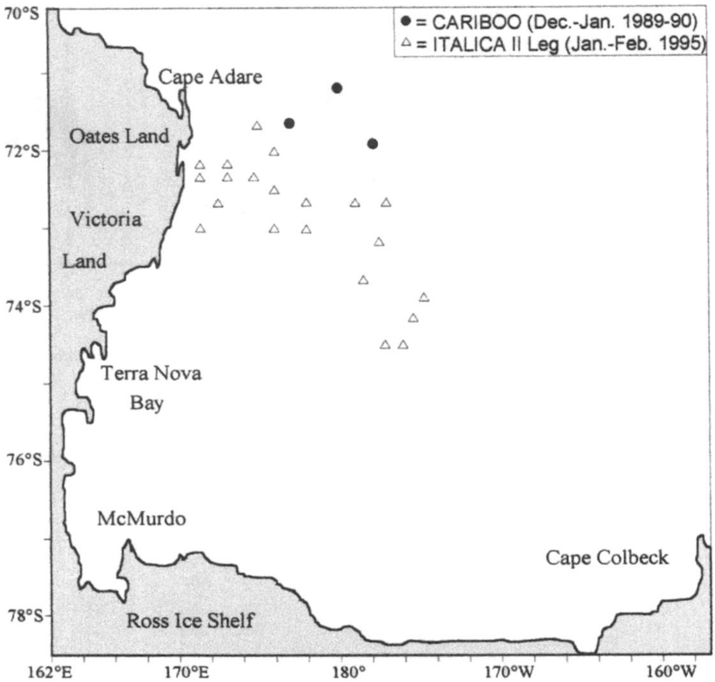


**Fig. 9.** In the stations approximately located on the continental break, in the northern Ross Sea, a greater instability of the water column characterizes the upper mixed layer for its higher salinity (34.0), its lower oxygen content (348.0 μM) and nutrient utilization

of their iron addition experiments by Dugdale and Wilkerson (1990) suggested that other experimental designs are needed in order to avoid a possible misinterpretation of the results due to bottle effects when working with natural seawater samples containing microzooplankton and bacteria.

It has been pointed out already that the melting water has an important role in determining nutrient utilization by phytoplankton. Further considerations can be drawn on N, P and Si cycling, by comparing the concentrations of nutrients (Fig. 10) in the upper (assimilation prevailing) and deeper layers (regeneration prevailing). While the estimated N/P utilization ratio of 12.8 (Catalano et al. 1997a) fits the whole set of the $NO_3 + NO_2$ vs. $PO_4$ points (Fig. 10a), the calculated Si/N ratios of 2.7 (Cariboo cruise), or 3.4 (Malippo cruise), are consistent only with the lower portion of the $Si(OH)_4$ vs. $NO_3 + NO_2$ plot (Fig. 10b). This confirms that the surface decrease in N, P and Si was due to phytoplankton assimilation and shows that Si and N are not linearly correlated for $Si(OH)_4$ values higher than 50 μM, namely in the section of the plot corresponding to the bottom water domain. The progressive increase of Si/N ratio in the deeper layers can be attributed to the slower rate of Si recycling which uncouples this element from the N and P. In the case of the Ross Sea, a considerable amount of biogenic silica not yet recycled can thus reach the bottom. Estimates of water-column and sedimentary fluxes on the Ross Sea Continental Shelf (Nelson et al. 1996) indicate that the integrated annual sinking flux of biogenic opal to 250 m is 50–100% of the annual net production of silica in the upper 50 m, and that about 5.8% of the opal produced annually by diatoms is long-term buried in the seabed. The burial/production ratio of biogenic silica on the Ross Sea Continental Shelf is thus about twice the apparent global average estimated by Tréguer et al.

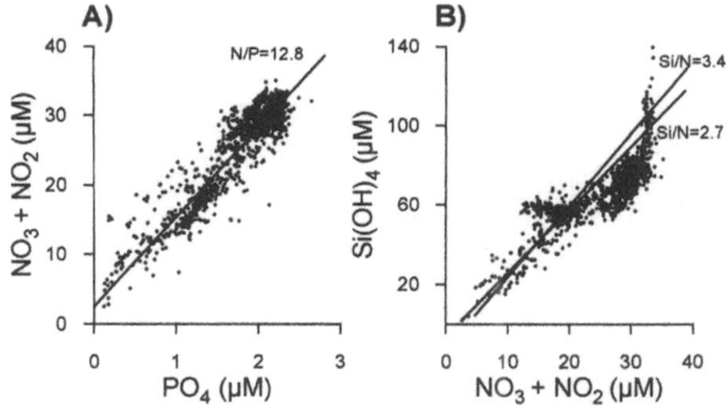


**Fig. 10.** Plots of $NO_3 + NO_2$ vs. $PO_4$ (**A**) and $Si(OH)_4$ vs $NO_3 + NO_2$ (**B**) in all the stations affected by meltwater advection ($S < 34.1$). The *lines* corresponding to the estimated N/P and Si/N ratios (Catalano et al. 1997a) are also shown

(1995), but less than 1/3 of the estimated burial/production ratio for the Southern Ocean as a whole (Nelson et al. 1996).

## Nutrient Concentrations Related to High Salinity Shelf Water (HSSW)

According to Jacobs et al. (1985), the main feature of HSSW is its high salinity (34.75–35.00) caused by the combined effects of surface freezing and brine formation. In Fig. 11 are reported the locations with the salinity higher than 34.75 and temperature around the freezing point ($-1.95 < T < -1.80$). It can be seen that HSSW fills the bottom layer in a wide area including the western and southern parts of the Ross Sea.

Compared to the upper and intermediate layers, a nutrient concentration increase and a dissolved oxygen decrease are typical features expected in a bottom water. For HSSW, these features are quantitatively less pronounced than for the intermediate, less dense CDW and its intrusions on the Ross Sea continental shelf. In HSSW, the mean concentrations of nutrients reach values of 78.0, 2.05 and 30.5 µM, respectively, for $Si(OH)_4$, $PO_4$ and $NO_3 + NO_2$ vs. 104.5, 2.22 and 33.0 µM in CDW; the average $O_2$ concentration is 299.4 µM in HSSW vs. 199.7 µM in CDW. This indicates that HSSW, although more dense, was formed more recently than the CDW.

According to Kurtz and Bromwich (1985), a large amount of HSSW forms along the Victoria Land coast, especially in Terra Nova Bay, where a recurring polynya can be maintained because of the blocking effect of the Drygalski Ice Tongue. Therefore, only in the western Ross Sea, can a bottom layer of high salinity water, thicker than 500 metres, be found depending upon the bottom morphology.

Terra Nova Bay is even more particular, because water with a salinity higher than 34.8 has been found, even up to the sea surface, in a wide part of the bay. This water cannot always be strictly considered as HSSW, because summer warming sometimes raises its temperature far above 0 °C, and consequently reduces its density significatively. In the summer of 1987–88, strong solar heating allowed this water to reach a temperature of 2 °C in the upper layer at some stations, thus originating a thermal stratification (Fig. 12) able to support an increased phytoplankton production. This phenomenon was observed during that cruise only and resulted in a characteristic drop in nutrient concentrations in the region of the highest salinity of the plot (Fig. 13). This drop was more evident for $PO_4$ and $NO_3 + NO_2$ (down to 0.6 and 15 µM, respectively; Fig. 13bc) than for $Si(OH)_4$ (Fig. 13a). It is also noteworthy that several values of dissolved $O_2$ in this warm and salted surface water were higher than 390 µM. Oxygen concentrations between 390 and 417 µM are normal for surface water during the phase of a phytoplankton bloom, but unusual in a water as salted as bottom water. This indicates that Terra Nova Bay can be strongly affected by the presence of HSSW which, in turn, can be strongly modified by the effects of unusual summer conditions combined with the particular morphology of the bay. Terra Nova Bay was again investigated in January–February 1990 and February 1996, but this phenomenon did not occur because ice-covering and meteorological conditions were different, and an advection of less salted ice-melting water was by far predominant in the upper layer.

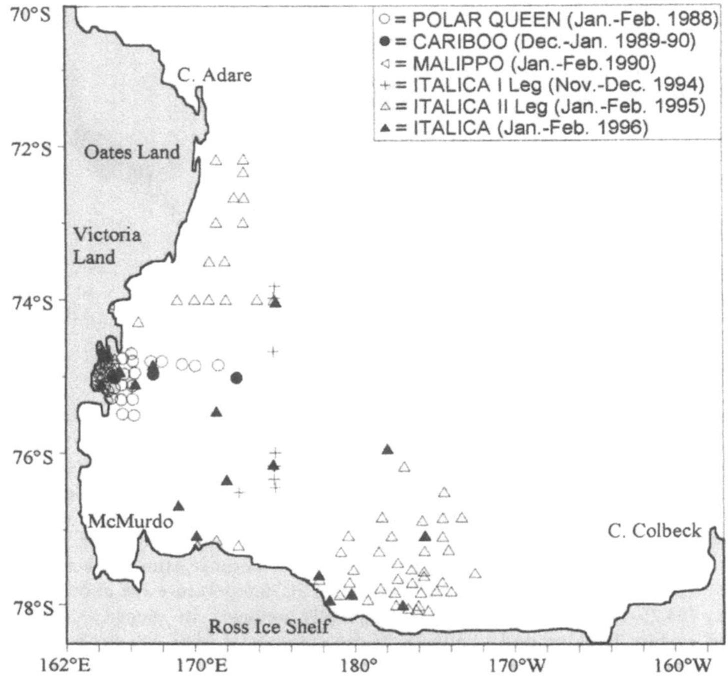


**Fig. 11.** Location of the stations where HSSW (S > 34.75) was detected and sampled for nutrient analysis

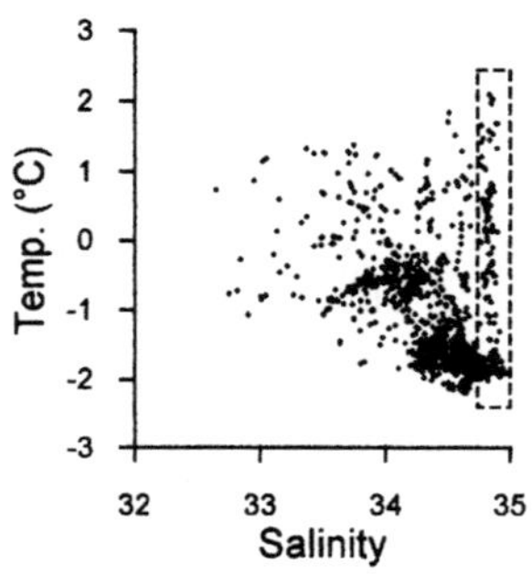


**Fig. 12.** T/S diagram of the stations at Terra Nova Bay. The case of a strong summer heating (T > 2 °C) of water with a salinity >34.8 is highlighted in the *dotted area*

## Conclusions

1. Very low nutrient concentrations were found in the Ross Sea, corresponding to low-salinity melt-water zones. The data show, however, that dilution by melting ice alone cannot account for nutrient concentrations as low as those observed, which must therefore be attributed to enhanced biological assimilation. Thus, the low nutrient content, a consequence of biological uptake, characterizes well the Antarctic Surface Water (AASW) when the upper water column stability and irradiance regime favour phytoplankton blooms. Nutrient assimilation is increased by summer heating, sea-ice melting and low hydrodynamism, which is rather common in the Marginal Ice Zone (MIZ) and in coastal zones. In Terra Nova Bay the nutrient depletion is very pronounced for the concurrence of all these factors, with nutrients sometimes reaching concentration levels which might be limiting for phytoplankton growth. These features are recurrent in the Ross Sea in summer, with a high variability in nutrient depletion and in spatial and temporal extent.
2. The dissolved N and P concentrations are linearly related. This simple relationship is not observed with Si, and both Si/N and Si/P ratios vary non-linearly in the direction of a substan-

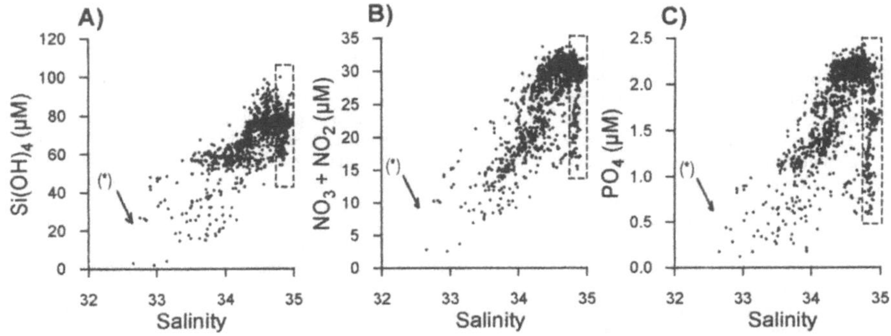


**Fig. 13A–C.** The nutrient concentration decrease observed in surface water having a high salinity is evidenced *inside the dotted rectangles*. The decrease of nutrients (*) in low salted surface water lies instead in the *lower left side* of each diagram

tial excess of Si in the nutrient-rich, deep waters. We attribute this to net dissolution of silica in deeper layers, resulting from its lower recycling rate, and, in some cases, from higher water temperatures.

3. Some features linked with hydrological circulation, such as the presence of Circumpolar Deep Water (CDW), and its intrusions on the continental shelf are well reflected by nutrient, oxygen and temperature distributions. These features were especially evident during the Italian cruises in the NW part of Ross Sea, near the continental shelf break.
4. A particular case of thermal stratification, characterized by surface waters of high salinity, was observed in Terra Nova Bay in 1988. In that year, strong summer heating induced the stratification of surface water as salted as HSSW, in which substantial nutrient assimilation by phytoplankton was clearly detectable.
5. No significant patterns of nutrient distribution were found related to the spreading of HSSW or Low Salinity Shelf Water (LSSW) and Low Salinity Bottom Water (LSBW) on the Ross Continental Shelf; these water masses were defined on the basis of T/S data.
6. Iron limitation of phytoplankton growth has been proposed as a factor controlling nutrient depletion. While this problem is beyond the aim of this work, further studies are needed in order to evaluate its relevance for nutrient concentrations in Ross Sea surface waters.

*Acknowledgements.* The authors are grateful to the physical oceanography teams directed by Giancarlo Spezie, Antonio Artegiani and Aniello Russo for their data on temperature and salinity from the 1994–95 and 1995–96 cruises. This study was supported by the National Scientific Commission for Antarctica of the Italian Government in the framework of the Italian National Programme for Antarctic Research (PNRA).

## References

Ackley SF, Buck KR, Taguchi S (1979) Standing crop of algae in the sea ice of Weddel Sea region. Deep Sea Res 26: 269–281

ALPKEM (1992a) Nitrate plus nitrite in seawater. In: ALPKEM (ed) The flow solution methodology. Document no 000630, 8/92 Rev. B. Wilsonville, Oregon, USA 19 pp

ALPKEM (1992b) Orthophosphate in seawater. In: ALPKEM (ed) The flow solution methodology. Document no 000626, 4/92 Rev. A. Wilsonville, Oregon, USA 16 pp

ALPKEM (1992c) Silica in seawater. In: ALPKEM (ed) The flow solution methodology. Document no 000671, 5/92 Rev. A. Wilsonville, Oregon, USA 16 pp

Arrigo KR, Dieckmann G, Gosselin M, Robinson DH, Fritsen CH, Sullivan CW (1995) High-resolution study of the platelet ice ecosystem in McMurdo Sound, Antarctica: biomass, nutrient, and production profiles within a dense microalgal bloom. Mar Ecol Prog Ser 127: 255–268

Artegiani A, Azzolini R, Paschini E, Creazzo S (1992) Physical oceanographic conditions in the Southern Pacific Ocean and in the Western Ross Sea. In: National Scientific Commission for Antarctica (ed), Oceanographic Campaign 1989–90. Data report Part II: 5–62

Boldrin A, Stocchino C (1990) On the hydrological characteristics of Terra Nova Bay (Ross Sea-Antarctica). In: National Scientific Commission for Antarctica (ed), Oceanographic Campaign 1987–88. Data report Part I: 11–57

Catalano G, Benedetti F (1990) Distributions of nutrients in the Terra Nova Bay and in the Ross Sea. In: National Scientific Commission for Antarctica (ed), Oceanographic Campaign 1987–88. Data report Part I: 61–83

Catalano G, Benedetti F, Goffart A, Iorio M (1991a) Distribution of dissolved oxygen, pH, total alkalinity and nutrients in Southern Ocean and Ross Sea (R/V Cariboo 1989–90 cruise). In: National Scientific Commission for

Antarctica (ed), Oceanographic Campaign 1989–90. Data report Part I: 11–23

Catalano G, Benedetti F, Iorio M (1991b) Coastal oceanography from Cape Russel to Campbell Ice Tongue (Terra Nova Bay). Dissolved oxygen, nutrients, pH, total alkalinity and total inorganic carbon distribution. In: National Scientific Commission for Antarctica (ed), Oceanographic Campaign 1989–90. Data report part I: 25–32

Catalano G, Povero P, Fabiano M, Benedetti F, Goffart A (1997a) Nutrient utilisation and particulate organic matter changes during summer in the upper mixed layer (Ross Sea, Antarctica). Deep Sea Res 44: 97–112

Catalano G, Predonzani S, Rivaro P, Ruffini S (1997b) Dissolved oxygen and nutrient distribution in the Ross Sea during the 1994 Austral Spring (R/V Italica cruise, leg 1). In: National Scientific Commission for Antarctica (ed), Oceanographic Campaign 1994–95. Data report (in press)

Catalano G, Predonzani S, Ruffini S (1997c) Dissolved oxygen and nutrient distribution in the Ross Sea during the 1995 Austral Summer (R/V Italica cruise, leg 2). In: National Scientific Commission for Antarctica (ed), Oceanographic Campaign 1994–95. Data report (in press)

Cota GF, Smith WO Jr, Nelson DM, Muench RD, Gordon LI (1992) Nutrient and biogenic particulate distributions, primary productivity and nitrogen uptake in the Weddell-Scotia Sea marginal ice zone during winter. J Mar Res 50: 155–181

Dugdale RC, Wilkerson FP (1990) Iron addition experiments in the Antarctic: a reanalysis. Global Biogeochem Cycles 4: 13–19

Dugdale RC, Wilkerson FP (1991) Low specific nitrate uptake rate: a common feature of high-nutrient, low-chlorophyll marine ecosystems. Limnol Oceanogr 36: 1678–1688

Fabiano M, Povero P, Catalano G, Benedetti F (1991) Hydrological data collected during the biological, chemical and geological sampling in Terra Nova Bay. In: National Scientific Commission for Antarctica (ed), Oceanographic Campaign 1989–90. Data report Part I: 35–71

Hansen HP, Grasshoff K (1983) Automated chemical analysis. In: Grasshoff K, Ehrhardt M, Kremling K (eds) Methods of seawater analysis, 2nd eds. VCh, Weinheim, pp 347–379

Jacobs SS (1991) On the nature and significance of the Antarctic slope front. Mar Chem 35: 9–24

Jacobs SS, Fairbanks RG, Horibe Y (1985) Origin and evolution of water masses near the Antarctic continental margin: evidence from $H_2{}^{18}O/H_2{}^{16}O$ ratios in seawater. Antarct Res Ser 44: 59–85

Kurtz DD, Bromwich DH (1985) A recurring, atmospherically forced polynya in Terra Nova Bay. Antarct Res Ser 44: 177–201

Lancelot C, Mathot S, Becquevort S, Dandois JM, Billen G (1992) Carbon and nitrogen cycling through the microbial network of the MIZ of the Southern Ocean with particular emphasis to the NW Weddell Sea. Rep Contract ANTAR/05, Dec 91, 106 pp

Martin JH, Fitzwater SE, Gordon RM (1990) Iron deficiency limits phytoplankton growth in Antarctic waters. Global Biogeochem Cycles 4: 5–12

Mitchell BG, Holm-Hansen O (1991) Observation and modelling of the Antarctic phytoplankton crop in relation to mixing depth. Deep Sea Res 38: 981–1007

Nelson DM, Smith WO Jr (1986) Phytoplankton bloom dynamics of the western Ross Sea ice edge. II. Mesoscale cycling of nitrogen and silicon. Deep Sea Res 33: 1389–1412

Nelson DM, Smith WO Jr (1991) Sverdrup revisited: critical depths, maximum chlorophyll levels and the control of Southern Ocean productivity by the irradiance-mixing regime. Limnol Oceanogr 36: 1650–1661

Nelson DM, Tréguer P (1992) Role of silicon as a limiting nutrient to Antarctic diatoms: evidence from kinetic studies in the Ross Sea ice-edge zone. Mar Ecol Prog Ser 80: 255–264

Nelson DM, Ahern JA, Herlihy LJ (1991) Cycling of biogenic silica within the upper water column of the Ross Sea. Mar Chem 35: 461–476

Nelson DM, DeMaster DJ, Dunbar RB, Smith WO Jr (1996) Cycling of carbon and biogenic silica in the Southern Ocean: estimates of water-column and sedimentary fluxes on the Ross Sea continental shelf. J Geophys Res 101: 18519–18532

Sambrotto RN, Niebauer HJ, Goering JJ, Iverson RL (1986) Relationship among vertical mixing, nitrate uptake, and phytoplankton growth during the spring bloom in the Southeast Bering Sea middle shelf. Cont Shelf Res 5: 161–198

Smetacek V, Passow U (1990) Spring bloom initiation and Sverdrup's critical depth model. Limnol Oceanogr 35: 228–233

Smith WO Jr, Harrison WG (1991) New production in polar regions: the role of environmental control. Deep Sea Res 38: 1463–1479

Smith WO Jr, Nelson DM (1985) Phytoplankton bloom produced by a receding ice edge in the Ross Sea: spatial coherence with the density field. Science 227: 163–166

Smith WO Jr, Nelson DM (1986) The importance of ice-edge phytoplankton production in the Southern Ocean. BioScience 36: 251–257

Strickland JDH, Parsons TR (1972) A practical handbook of sea water analysis. Bull Fish Res Board Can 167: 310 pp

Tréguer P, Jacques G (1992) Dynamics of nutrients and phytoplankton, and fluxes of carbon, nitrogen and silicon in the Antarctic Ocean. Polar Biol 12: 149–162

Tréguer P, van Bennekom AJ (1991) The annual production of biogenic silica in the Antarctic Ocean. Mar Chem 35: 477–488

Tréguer P, Gueneley S, Kamatani A, Quéguiner B (1989) Kinetics of dissolution of Antarctic diatoms and the biogeochemical cycle of silicon in the Southern Ocean. Polar Biol 9: 397–403

Tréguer P, Nelson DM, van Bennekom AJ, DeMaster DJ, Leynaert A, Quéguiner B (1995) The silica balance in the world ocean: a reestimate. Science 268: 375–379

UNESCO (1985) The international system of units (SI) in oceanography. UNESCO Tech Pap Mar Sci 45

Zentara SJ, Kamykowski D (1981) Geographic variations in the relationship between silicic acid and nitrate in the South Pacific Ocean. Deep Sea Res 28: 455–465

# Particulate Organic Matter and Nutrient Utilization in the Mixed Layer of the Ross Sea

P. Povero[1], M. Fabiano[1], and G. Catalano[2]

**ABSTRACT**

The relationship between particulate organic matter, nutrient depletion and vertical stability were studied in different areas of the Ross Sea during the Antarctic spring and summer. In the upper mixed layer, identified in relation to the vertical stability index E[Z(UML)], a significant relationship between the vertical stability index and nutrient depletion was found. A similar relationship was also observed between particulate matter distribution and composition vs. nutrient utilization. Data analysis showed that stations could be divided into different groups with differing characteristics. Summer stations in the western and central areas of the Ross Sea and in the coastal zone of Terra Nova Bay were affected by ice melting and showed strong stratification (E[Z(UML)] > 25), higher biomass and nutrient depletion values (average values of POC = 236 $\mu g\ l^{-1}$; PON = 40 $\mu g\ l^{-1}$; ΔNW, ΔPW, ΔSiW = 7, 0.7, 23 mmol $m^{-3}$, respectively). Another group of stations in the northern-central area of the Ross Sea was characterized by low nutrient utilization (ΔNW, ΔPW, ΔSiW < 4, 0.4, 10 mmol $m^{-3}$, respectively), low E[Z(UML)] values (<25) and low concentrations of POM (average values of POC = 85 $\mu g\ l^{-1}$ and PON = 11 $\mu g\ l^{-1}$). In spring, the southern and central MIZ or polynya stations were more productive, with evident POM accumulation (average values of POC > 180 $\mu g\ l^{-1}$ and PON > 29 $\mu g\ l^{-1}$) and with phytoplankton assemblages dominated by diatoms and/or *Phaeocystis antarctica*. The degree of stabilization, but also sedimentation rates and grazing pressure played an important role in the development, accumulation and spatial variability of particulate organic matter in the Ross Sea areas.

## Introduction

The Southern Ocean is a variable region in terms of productivity, with areas of high productivity often associated with retreating of sea ice and areas of very low productivity comparable with the oligotrophic areas of other oceans (El-Sayed 1984; Mitchell and Holm-Hansen 1991; Treguer and Jacques 1992; Priddlle et al. 1994; Smith and Gordon 1997). The Ross Sea is considered a site of high primary production and large surface accumulations of phytoplankton biomass (Smith and Nelson 1985; DeMaster et al. 1992; Comiso et al. 1993; Sullivan et al. 1993; Smith and Gordon 1997). Typical of the Ross Sea, classified as a Coastal and Continental Shelf Zone (Treguer and Jacques 1992), are the marginal ice zones (MIZ) that recent estimates suggest are responsible for up to 40% of the total primary production of the Antarctic Ocean (Sakshaug 1994).

During the melting of pack ice, some consistency in enhanced vertical stability has been observed, following the input of less saline and less dense meltwater, as well as an inoculum of epontic algae, plankton which initially develop within the pack ice, nutrient depletion and biogenic matter production (El-Sayed and Taguchi 1981; Sakshaung and Holm Hansen 1984; Smith and Nelson 1985; Nelson et al. 1987, 1989; Cota et al. 1992). The stability generated during ice melting favours the increase in photosynthetic activity, but this effect is not expected to be uniform. Sullivan et al. (1988) show an important spatial variability in pigment concentration in the MIZ, obtained through ocean colours (CZSC).

Other factors such as light limitation, nutrient depletion, micronutrient availability, sedimentation and grazing pressure, together with increased vertical stability, can favour or inhibit phytoplankton growth, primary production, accumula-

[1] Dipartimento per lo Studio del Territorio e delle sue Risorse, Università di Genova, Viale Benedetto XV 5, Genova, Italy
[2] Istituto Talassografico V. Vercelli, CNR, Viale Gessi 2, Trieste, Italy

tion of organic matter, and changes in its biochemical composition (Niebauer and Alexander 1985; Smetacek 1985; Smith and Nelson 1985; Nelson et al. 1989; Martin et al. 1990; Nelson and Treguer 1992; Fabiano et al. 1993).

The aim of this study was to investigate the relationship between the strength of the pycnocline in the Ross Sea and nutrient removal due to primary production, and amount and biochemical composition of the particulate organic matter.

## Materials and Methods

Water samples were taken during two cruises (Cariboo and ROSSMIZE cruises) in the austral springs and summers of 1989–90 and 1994–95 using Niskin bottles assembled on a rosette or a SBE Carousel 32 sampler or mounted on a hydrological wire. Samples were taken from the surface to the bottom; sampling depths were selected according to both hydrographical depths and the attenuation of incident radiation. Hydrological casts were supported by CTD vertical profiles recorded with either an EG&G Mk3, an ME mod. KMS or a SBE 9/11+ plus probe (Fabiano et al. 1991; Artegiani et al. 1992; Russo et al. 1997).

Chemical analyses for nutrient concentrations (reactive silicate, reactive phosphorus, reactive nitrate and nitrite) were carried out on board after filtration on Whatman GF/C filters (Hansen and Grasshoff 1983) and using a Technicon II AutoAnalyzer during the cruise in 1989–90. During the 1994–95 cruise, subsamples for nutrients were also collected directly from Niskin bottles. Nutrients were determined according to the Alpkem-Perstorp Flow Solution Methodologies (Alpkem 1992 a–c).

For particulate organic matter analysis, water samples were taken between 500 and 3000 ml and were then prefiltered through 200-µm mesh net and filtered onto (0.4-µm pore size) Nuclepore filters or Whatman GF/F for particulate protein (PRT) and particulate carbohydrate (CHO) analyses, or onto GF/C and GF/F Whatman filters (precombusted at 450 °C) for organic carbon (POC) and nitrogen (PON). After filtration, filters were immediately stored at −20 °C.

Particulate organic carbon and nitrogen (POC and PON) were analyzed by a CHNS-O EA1108 Elemental Analyzer-Carlo Erba after removal of carbonates by exposure to HCl fumes (Hedges and Stern 1984). Cyclohexanone 2,4-dinitrophenylhydrazone was used as standard. Particulate carbohydrates (CHO) were detected according to Dubois et al. (1956). Absorbance was measured at 490 nm using a Varian spectrophotometer. Solutions of D(+)glucose were used as standard. Particulate proteins (PRT) determination was conducted according to Hartree (1972). Absorbance was measured at 650 nm. Bovine serum albumine was used as standard.

Water samples were collected on a single replicate. However, a series of replicate samples were collected at different intervals at the surface water and along the water column to define the coefficient of variation for all parameters: 2.3% for POC, 2.7% for PON, 11.0% for CHO, 3.2% for PRT. Data on station locations, dates, sampling depths and values for measured parameters are reported in AA.VV. (1991, 1992), Faranda and Guglielmo (1994), Faranda et al. (1997). The depth, corresponding to the maximum density gradient (Z(UML)), the strength [E(Z(UML)], and the base of the pycnocline [Z(L)] were estimated at every station according to Steinhorn (1985), Mitchell and Holm-Hansen (1991) and Catalano et al. (1997: Fig. 2). Assuming that Z(L) corresponds to the top of winter water, salinity and nutrient concentrations determined at this depth, immediately below the UML, are taken as reference values for the winter period, prior to dilution by advection of ice-melt water or utilization by phytoplankton (Catalano and Benedetti 1990; Catalano et al. 1991a,b; Fabiano et al. 1993)

To estimate nutrient depletion (nitrate ΔNW, phosphate ΔPW, silicate ΔSiW in mmol $m^{-3}$) in the upper layers due to phytoplankton assimilation, nutrient concentrations were first integrated between the surface and Z(L) and the result was then divided by Z(L), thus giving a weighted average concentration of each nutrient in the UML. Using an average sea-ice salinity of 6‰ (Ackley et al. 1979), and an average nutrient content in the sea ice according to Dieckmann et al. (1991), nutrient concentration reduction due only to dilution by meltwater was calculated from the average salinity measured in the UML. Nutrient utilization was then determined for each station by using the difference between the calculated "diluted" nutrient value and the weighted mean calculated on the basis of the integrated value in the UML. Statistical analysis was carried out using the STATISTICA package (StatSoft 1995). Characteristic Ross Sea areas were identified on the basis of discontinuities in the

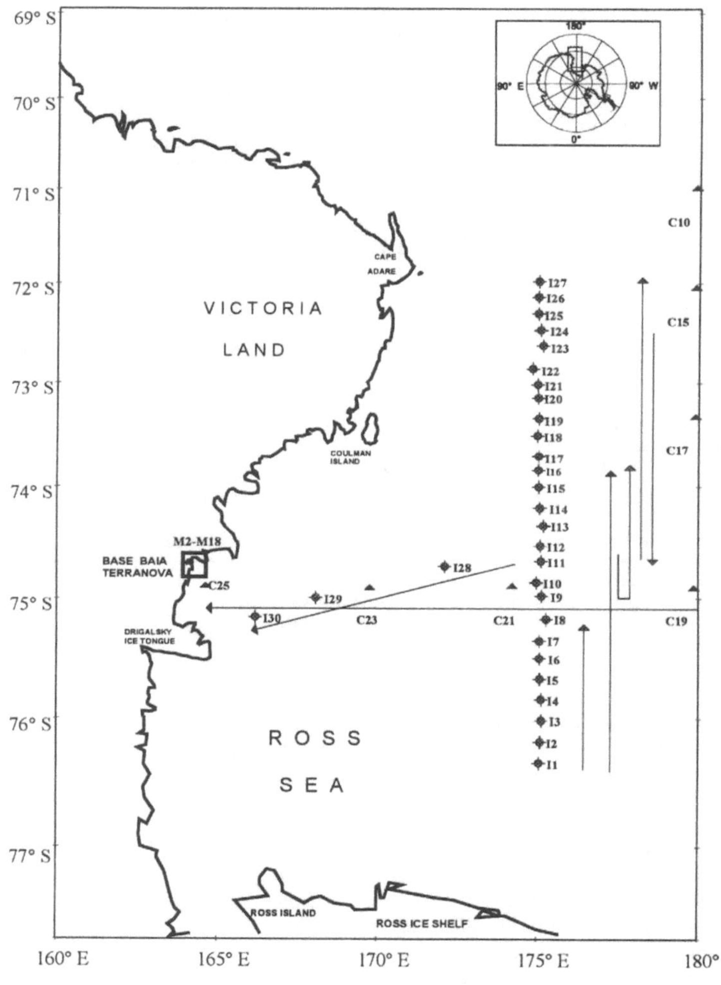


**Fig. 1.** Study area and sampling stations. ▲ Cariboo cruise (5 December 1989 to 6 January 1990); □ Cariboo cruise (coastal stations from 13 January to 2 February 1990); ✦ ROSSMIZE cruise (14 November to 15 December 1994)

data series by means of an hierarchic clustering algorithm (complete linkage, Eucledian distance).

## Results and Discussion

In Antarctica, marginal ice zones are believed to be the sites of enhanced phytoplankton activity, whereas ice-free areas with strongly mixed waters often have a low biomass (Smith and Nelson 1985; Mitchell and Holm-Hansen 1991). In the Ross Sea, some stations were located in a typical zone of MIZ, with a massive melting phase. Others were in an area with well-structured pack ice, while others had features typical of areas that are sometimes free of sea ice (Fig. 1).

Table 1 shows the mean values for different parameters measured during different cruises. The results of the first cruise, Table 1, carried out at the end of spring and the following summer, have already partly been reported in Catalano et al.

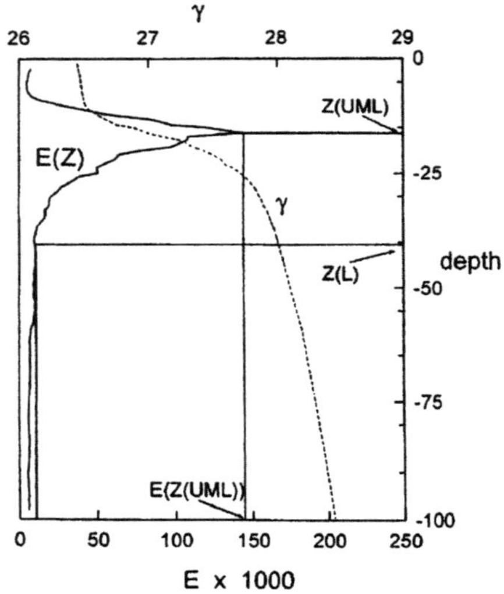


**Fig. 2.** Vertical profile of the stability index E. [$E(Z) = \rho^{-1} \cdot [\gamma T(Z + \Delta Z/2) - \gamma T(Z - \Delta Z/2)]/\Delta Z$, where ρ and γ are respectively the seawater density, expressed in $kg\,dm^{-3}$, and the density anomaly (UNESCO 1985), T is the temperature (°C), Z is the depth in dbar and ΔZ is a selected depth interval (4 dbar in our case)]

(1997). Stations with a strong vertical stratification induced by ice-melting (E[Z(UML)] values between 25 and 130) were characterized by greater nutrient utilization (average value = 7.1, 0.7, and $23.6\,mmol\,m^{-3}$ for ΔNW, ΔPW and ΔSiW, respectively), a high biomass (max POC concentrations of $360\,\mu g\,l^{-1}$) and a high PRT:CHO ratio up to values of 3, typical of "fresh" particulate organic matter of autotrophic origin with C:N 5.5–6 (Fabiano et al. 1993). Another series of stations, located in the east and north, were characterized by low E[Z(UML)] values (average value = 10), low nutrient utilization (average value of 2.0, 0.17 and $3.4\,mmol\,m^{-3}$ for ΔNW, ΔPW and ΔSiW, respectively) and lower biomass (average value POC = $100\,\mu g\,l^{-1}$), with mostly detrital particulate matter with PRT:CHO = 1.2, C:N 7–7.5. The data collected in the spring period (Table 1b), during the ROSSMIZE cruise, refer to a period in which the melting ice had not yet occurred or had just begun. The E[Z(UML)] stability index was low (<25) and tended to increase towards the end of the sampling period and near the coast (from 4 to 20). Nutrient depletion was low with values of less than 4, 0.4 and $10\,mmol\,m^{-3}$ for ΔNW, ΔPW and ΔSiW, with the only exception of station 28 with regard to ΔNW ($5.89\,mmol\,m^{-3}$). Only ΔSiW showed a pronounced increase with time and towards the coast. Average POC values were low (<$100\,\mu g\,l^{-1}$), with the exception of more coastal and central stations of MIZ ($180–250\,\mu g\,l^{-1}$); minimum values were observed at northern stations (average value = $54\,\mu g\,l^{-1}$). The PRT:CHO ratios were high (av. = 3.1) and tended to decrease moving northwards and increased or remained constant with time. The C:N ratios were higher in the north (>10), lower closer to the coast (<5.5) and showed a tendency to decrease with time.

**Table 1.** Descriptive statistics of physical, chemical and particulate matter parameters in Ross Sea: index of vertical stability, depth of the base of the mixed layer [Z(L) in dbar], salinity weighted average values SW(Z(L)) of the UML, nutrient utilization (ΔNW, ΔPW and ΔSiW in $mmol\,m^{-3}$), particulate organic matter (POC in $\mu g\,l^{-1}$), ΔSiW:ΔNW, ΔNW:ΔPW, PPRT:PCHO, C:N and POC:Chl a ratio (Chl a data from Innamorati et al. 1991; Carrada et al. 1997). A Cariboo cruise (end spring and summer 1989–90); B Italica cruise (spring 1994)

| A | Mean | Minimum | Maximum | SD |
|---|---|---|---|---|
| Z(UML) | 17.8 | 4.0 | 43.0 | 12.2 |
| Z(L) | 47.2 | 28.0 | 67.0 | 10.4 |
| E[Z(UML)] | 53.0 | 6.0 | 130.0 | 40.2 |
| SW[Z(L)] | 34.146 | 33.834 | 34.278 | 0.132 |
| ΔNW | 5.63 | 1.19 | 13.26 | 2.99 |
| ΔPW | 0.53 | 0.09 | 1.13 | 0.28 |
| ΔSiW | 17.91 | 0.35 | 40.34 | 10.78 |
| ΔSiW:ΔNW | 2.9 | 0.3 | 5.0 | 1.2 |
| ΔNW:ΔPW | 11.2 | 7.2 | 18.5 | 2.4 |
| POC | 212.2 | 93.2 | 362.9 | 83.1 |
| PON | 35.2 | 13.5 | 66.1 | 15.7 |
| PRT:CHO | 2.1 | 0.6 | 3.0 | 0.6 |
| C:N | 6.2 | 5.5 | 7.5 | 0.6 |
| POC:Chl a | 146 | 96 | 233 | 48 |

| B | Mean | Minimum | Maximum | SD |
|---|---|---|---|---|
| Z(UML) | 53.1 | 9.0 | 140.0 | 29.8 |
| Z(L) | 60.4 | 18.0 | 144.0 | 30.2 |
| E[Z(UML)] | 9.1 | 2.6 | 20.2 | 4.9 |
| SW[Z(L)] | 34.428 | 34.302 | 34.653 | 0.075 |
| ΔNW | 1.03 | 0.05 | 5.89 | 1.03 |
| ΔPW | 0.07 | 0.01 | 0.37 | 0.07 |
| ΔSiW | 2.38 | 0.21 | 7.99 | 1.84 |
| ΔSiW:ΔNW | 3.5 | 0.2 | 12.6 | 2.9 |
| ΔNW:ΔPW | 17.6 | 4.7 | 22.7 | 12.7 |
| POC | 96.4 | 33.4 | 268.2 | 54.0 |
| PON | 14.1 | 3.3 | 40.6 | 10.5 |
| PRT:CHO | 3.1 | 1.4 | 5.1 | 1.0 |
| C:N | 8.1 | 4.3 | 11.9 | 2.4 |
| POC:Chl a | 159 | 56 | 337 | 74 |

We have examined correlations of E[Z(UML)] with nutrient consumption (ΔNW, ΔPW and ΔSiW), and POM quantity and quality, to describe: (1) the impact of water-column stabilization on nutrient depletion and particulate matter accumulation; (2) the influence of the mixed-layer depth on productivity; (3) the increase in particulate organic matter and the change in biochemical composition during algal blooms (Smetacek and Hendrikson 1979; Parsons et al. 1984; Jennings et al. 1984; Tanoue 1985; Sambrotto et al. 1986; Fabiano et al. 1993, 1997; Smith and Gordon 1997).

In the Ross Sea surface waters, estimated nutrient depletion and particulate organic carbon and nitrogen (POC and PON) depended on pycnocline strength, despite the differences in sampling periods and sites (Fig. 3, Table 2). Figure 3 shows a logarithmic distribution of the data related to pycnocline strength; for E[Z(UML)] ≤ 25, an increase in utilized nitrate, phosphate and silicate was higher than those for values of E[Z(UML)] > 25. Furthermore, for E[Z(UML)] > 25, nutrient utilization ratios were more independent of pycnocline strength (Table 2). Significant correlations between nutrient consumption (ΔNW, ΔPW and ΔSiW) and both POC and PON concentrations were found (Fig. 4). Regression analysis between nutrient depletion (ΔNW and ΔPW) and POM was most significant for values of E[Z(UML)] ≤ 25 and for ΔSiW values of E[Z(UML)] > 25 (Table 2). It is interesting to note that some stations located at the central Ross Sea area with low pycnocline strength (E[Z(UML)] < 10) had high POM values in spring (Fig. 3).

Different areas of the Ross Sea were characterized by analyzing the spatial and temporal distribution of pycnocline strength, nutrient depletion and particulate matter concentrations, as well as their relationships and gradients. This classification is in agreement with other authors (DeMaster et al. 1992; Comiso et al. 1993; Leventer and Dumbar 1996; Nelson et al. 1996; Smith et al. 1996). The areas had different physical, chemical, and biological characteristics, often associated to the different evolution of pack ice and its role in production processes in surface waters. To better define these sectors, a cluster analysis was performed on the data set of pycnocline strength, nutrient depletion and particulate matter concentrations. This analysis indicated two main groups of Ross Sea stations (Fig. 5).

The first (1) were characterized by a strong stratification (E[Z(UML)] > 25), higher biomass and nutrient depletion (average values of POC = 237 $\mu g\,l^{-1}$; PON = 40 $\mu g\,l^{-1}$; ΔNW, ΔPW, ΔSiW = 7, 0.6, 23 $mmol\,m^{-3}$, respectively) (Table 3). These stations were located in the western and central area of the Ross Sea and in the coastal zone of Terra Nova Bay. This was the richest Ross Sea area, especially from the beginning of January and close to the coast, as also observed by other authors (Fabiano et al. 1993, 1996, 1997; Smith et al. 1996; Catalano et al. 1997; Smith and Gordon 1997). Low C:N, POC:Chl a and high PRT:CHO suggested that phytoplankton accounted for a significant fraction of POM (Table 3). In this group of stations, a significant correlation between stability index, particulate matter and nutrient depletion was found particularly for silicate (Table 2); moreover, the lowest ΔNW:ΔPW ratio (10.6) of the entire sampling area was observed as well as a high ΔSiW:ΔNW ratio (3.6). From December to January, in correspondence to the increased stability index, an increase in silicate depletion was

**Table 2.** Relationship between the index of vertical stability E[Z(UML)], nutrient depletion (ΔNW, ΔPW and ΔSiW in $mmol\,m^{-3}$), POC and PON($\mu g\,l^{-1}$): Pearson correlation coefficient (r) and probability level ($p$)

| | | All | | E[Z(UML)] > 25 | | E[Z(UML)] < 25 | |
|---|---|---|---|---|---|---|---|
| E[Z(UML)] | ΔNW | 0.870 | $p < 0.001$ | 0.474 | $p < 0.1$ | 0.553 | $p < 0.001$ |
| E[Z(UML)] | ΔPW | 0.887 | $p < 0.001$ | 0.676 | $p < 0.01$ | 0.540 | $p < 0.001$ |
| E[Z(UML)] | ΔSiW | 0.927 | $p < 0.001$ | 0.673 | $p < 0.01$ | 0.380 | $p < 0.1$ |
| E[Z(UML)] | POC | 0.721 | $p < 0.001$ | 0.555 | $p < 0.05$ | 0.171 | $p < 0.5$ |
| E[Z(UML)] | PON | 0.744 | $p < 0.001$ | 0.441 | $p < 0.1$ | 0.164 | $p < 0.5$ |
| POC | ΔNW | 0.799 | $p < 0.001$ | 0.368 | $p < 0.1$ | 0.777 | $p < 0.001$ |
| POC | ΔPW | 0.800 | $p < 0.001$ | 0.425 | $p < 0.1$ | 0.812 | $p < 0.001$ |
| POC | ΔSiW | 0.674 | $p < 0.001$ | 0.420 | $p < 0.1$ | −0.154 | $p < 0.5$ |
| PON | ΔNW | 0.774 | $p < 0.001$ | 0.392 | $p < 0.1$ | 0.731 | $p < 0.001$ |
| PON | ΔPW | 0.773 | $p < 0.001$ | 0.385 | $p < 0.1$ | 0.743 | $p < 0.001$ |
| PON | ΔSiW | 0.661 | $p < 0.001$ | 0.398 | $p < 0.1$ | −0.172 | $p < 0.5$ |

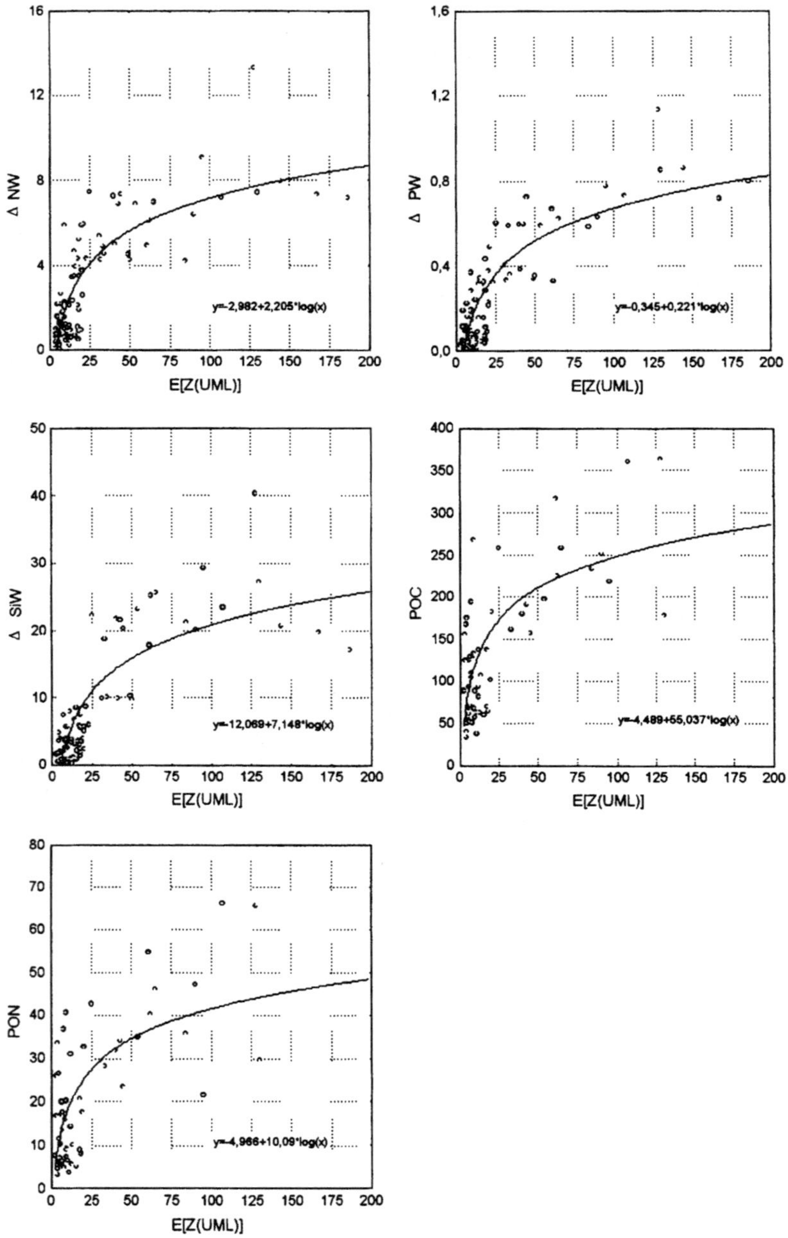


**Fig. 3.** Nutrient utilization ($mmol\,m^{-3}$): nitrate (ΔNW), phosphate (ΔPW), silicate (ΔSiW) and POM (POC $\mu g\,l^{-1}$ and PON $\mu g\,l^{-1}$) vs. the vertical stability E[Z(UML)]

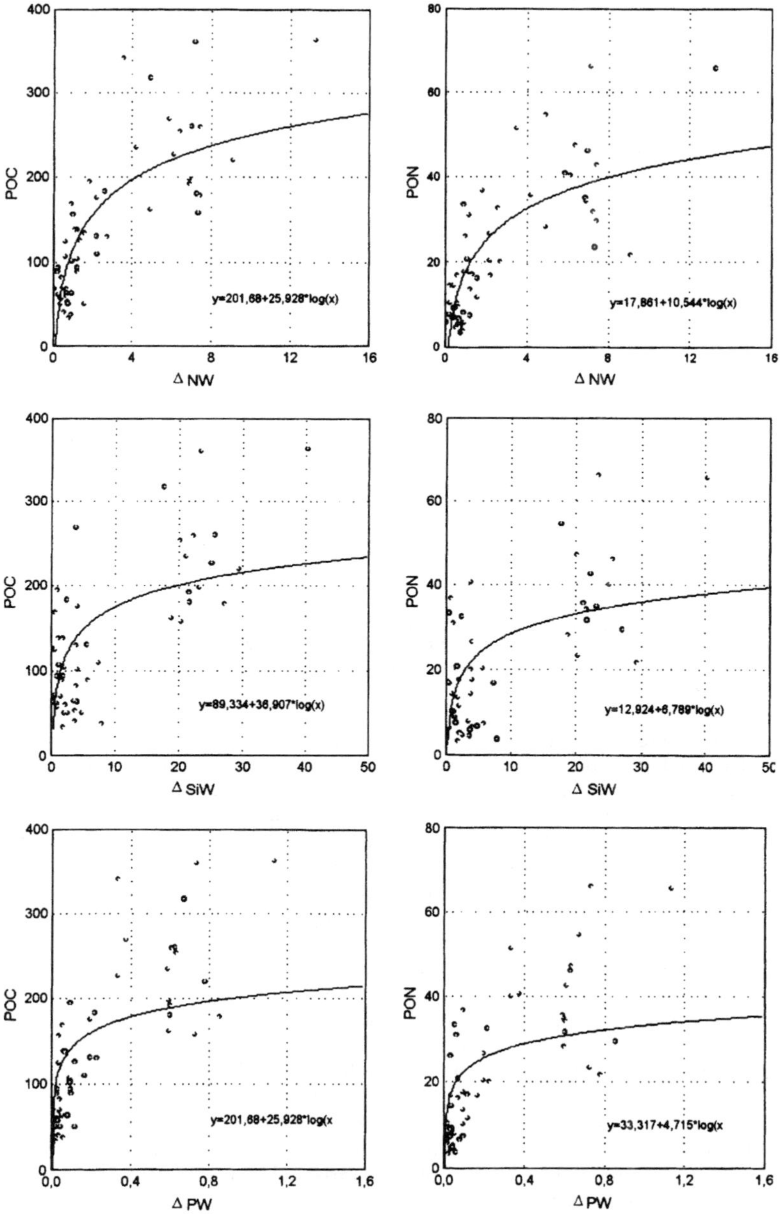


**Fig. 4.** Comparison of POC ($\mu g\,l^{-1}$) and PON ($\mu g\,l^{-1}$) with nutrient consumption in mmol $m^{-3}$: nitrate (ΔNW), phosphate (ΔPW), silicate (ΔSiW)

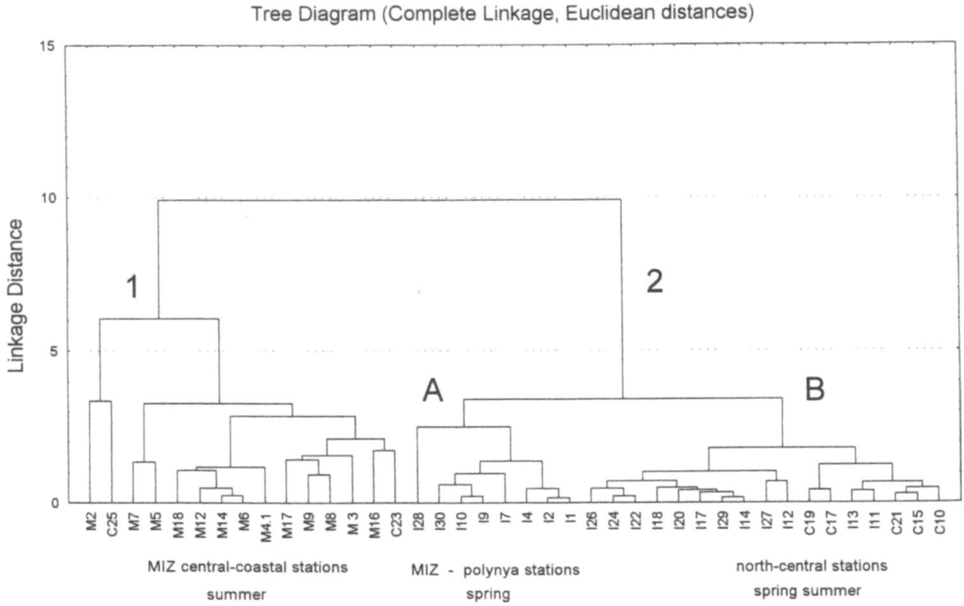


**Fig. 5.** Results of the cluster analysis on the set of data for vertical stability index E[Z(UML)], particulate organic matter (POC, PON $\mu g l^{-1}$), nutrient utilization ($\Delta NW$ and $\Delta SiW$ in mmol $m^{-3}$). *1* MIZ central-coastal stations, summer; *2A* MIZ polynya stations, spring; *2B* north-central stations spring-summer

found compared to nitrate, due to the dominance of diatoms (Innamorati et al. 1994; Marino and Cabrini 1997). Diatoms reached their maximum evolution and accumulation when water column stability was greater (E[Z(UML] > 25) (Karl et al. 1991; Leventer and Dumbar, 1996) and when regenerated production with preferential ammonium utilization began (Goeyens et al. 1995; Hecq et al. 1996; Nelson et al. 1996; Lipizer et al. 1997).

The second group (Fig. 5) included stations located in the central and northern part of the Ross Sea, mainly along a south-north transect (long. 175° E-180°) (Fig. 1). These stations were characterized prevalently by low nutrient utilization and low E[Z(UML)] values (<25). Here, we found stations without any relevant stratification, either because the melting process had begun or ended, or because they were located in areas unlikely to show strong stratification. This area included north-central stations with low concentrations of POM (POC average value = $85 \mu g l^{-1}$) and southern and central MIZ (POC average value = $189 \mu g l^{-1}$) or polynya stations, which were more productive (POC average value = $150 \mu g l^{-1}$).

North-central stations (Fig. 5) had low productivity and biomass which persisted much longer into the spring (Nelson et al. 1996; Smith et al. 1996; Carrada et al. 1997; Marino and Cabrini 1997; Saggiomo et al. 1997) (Table 3). This was attributed to the typical pattern of annual ice breakdown in the Ross Sea, which results in an ice-free season over a month shorter in the north than south (Leventer et al. 1993). Our spring data showed very low E[Z(UML)] values (av. = 6.7). In this period, we found the highest $\Delta SiW:\Delta NW$ ratio(12.5) and also a very high $\Delta NW:\Delta PW$ ratio(18.9). Also POC values were very low (av. $54 \mu g l^{-1}$). The high values of POC:Chl a (often over 200) and C:N (av. = 10) and the decrease in PRT:CHO values compared to productive stations (from 3.5 to 2.6) can be explained by the type of phytoplankton community (nano-picoplankton) (Carrada et al. 1997; Marino and Cabrini 1997) and also by the type of detritus remains from the ice (Fabiano et al. 1996). Grazing pressure may also have been involved, depending on the zooplankton community present in the area with pack ice (Smetacek et al. 1990; Azzali and Kalinowski 1996).

**Table 3.** Descriptive statistics of physical, chemical and particulate matter parameters in different areas of Ross Sea: index of vertical stability E[Z(UML)], depth corresponding to the maximum density gradient [Z(UML)], depth of the base of the mixed layer [Z(L) in dbar], salinity weighted average values SW[Z(L)] of the UML, nutrient utilization (ΔNW, ΔPW and ΔSiW in mmol $m^{-3}$), particulate organic matter (POC in $\mu g l^{-1}$), ΔSiW:ΔNW, ΔNW:ΔPW, PPRT:PCHO, C:N and POC:Chl a ratio (Chl a data from Innamorati et al. 1991; Carrada et al. 1997). A MIZ central-coastal area (summer); B north-central area (spring-summer); C polynya area (spring); D MIZ (spring)

| A | Mean | Minimum | Maximum | SD | C | Mean | Minimum | Maximum | SD |
|---|---|---|---|---|---|---|---|---|---|
| Z(UML) | 14.9 | 4.0 | 43.0 | 10.8 | Z(UML) | 27.0 | 14.0 | 41.0 | 11.5 |
| Z(L) | 46.0 | 28.0 | 67.0 | 0.4 | Z(L) | 33.0 | 18.0 | 48.0 | 12.8 |
| E[Z(UML)] | 70.8 | 25.0 | 186.0 | 33.4 | E[Z(UML)] | 7.1 | 3.6 | 11.9 | 3.5 |
| SW[Z(L)] | 34.085 | 33.834 | 34.249 | 0.171 | SW[Z(L)] | 34.505 | 3.431 | 34.653 | 0.086 |
| ΔNW | 7.10 | 4.21 | 13.26 | 2.09 | ΔNW | 1.21 | 0.60 | 1.80 | 0.48 |
| ΔPW | 0.67 | 0.33 | 1.13 | 0.17 | ΔPW | 0.06 | 0.04 | 0.09 | 0.02 |
| ΔSiW | 23.87 | 17.83 | 40.34 | 5.51 | ΔSiW | 0.71 | 0.50 | 1.05 | 0.26 |
| ΔSiW:ΔNW | 3.4 | 2.8 | 5.0 | 0.6 | ΔSiW:ΔNW | 0.7 | 0.4 | 0.9 | 0.2 |
| ΔNW:ΔPW | 10.8 | 7.2 | 18.5 | 2.7 | ΔNW:ΔPW | 18.9 | 16.7 | 22.7 | 2.4 |
| POC | 236.8 | 157.1 | 362.9 | 66.4 | POC | 151.3 | 124.6 | 193.9 | 28.8 |
| PON | 40.1 | 21.6 | 66.1 | 13.8 | PON | 26.8 | 16.1 | 36.7 | 9.7 |
| PRT:CHO | 2.3 | 1.7 | 3.0 | 0.4 | PRT:CHO | 3.4 | 2.5 | 4.1 | 0.6 |
| C:N | 5.9 | 5.5 | 6.8 | 0.4 | C:N | 6.3 | 4.5 | 8.3 | 1.9 |
| POC:Chl a | 111 | 96 | 124 | 11 | POC:Chl a | 78 | 56 | 108 | 22 |
| **B** | **Mean** | **Minimum** | **Maximum** | **SD** | **D** | **Mean** | **Minimum** | **Maximum** | **SD** |
| Z(UML) | 53.5 | 10.0 | 140.0 | 36.1 | Z(UML) | 44.8 | 9.0 | 67.0 | 25.3 |
| Z(L) | 65.8 | 26.0 | 14.0 | 33.5 | Z(L) | 5.3 | 30.0 | 73.0 | 18.1 |
| E[Z(UML)] | 7.3 | 2.6 | 13.2 | 3.1 | E[Z(UML)] | 10.1 | 4.5 | 20.2 | 6.9 |
| SW[Z(L)] | 34.352 | 34.124 | 34.510 | 0.111 | SW[Z(L)] | 34.446 | 34.379 | 34.528 | 0.065 |
| ΔNW | 0.96 | 0.05 | 2.69 | 0.71 | ΔNW | 3.34 | 2.18 | 5.89 | 1.72 |
| ΔPW | 0.07 | 0.01 | 0.23 | 0.06 | ΔPW | 0.25 | 0.19 | 0.37 | 0.08 |
| ΔSiW | 2.76 | 0.21 | 7.99 | 2.46 | ΔSiW | 3.49 | 2.38 | 3.99 | 0.75 |
| ΔSiW:ΔNW | 3.8 | 0.2 | 12.6 | 3.2 | ΔSiW:ΔNW | 1.2 | 0.6 | 1.8 | 0.5 |
| ΔNW:ΔPW | 16.6 | 6.5 | 18.9 | 11.8 | ΔNW:ΔPW | 12.7 | 11.3 | 15.8 | 2.1 |
| POC | 85.1 | 33.4 | 133.1 | 33.5 | POC | 188.7 | 129.3 | 268.2 | 57.9 |
| PON | 11.1 | 3.3 | 20.4 | 5.7 | PON | 29.9 | 19.9 | 40.6 | 8.8 |
| PRT:CHO | 2.5 | 0.6 | 5.1 | 1.1 | PRT:CHO | 3.5 | 2.9 | 4.2 | 0.5 |
| C:N | 8.6 | 5.9 | 11.9 | 2.1 | C:N | 6.4 | 5.8 | 6.8 | 0.5 |
| POC:Chl a | 169 | 56 | 284 | 64 | POC:Chl a | 96 | 64 | 132 | 28 |

Summer samples during a period free of ice pack, despite the extreme poverty of this area, showed a decreased ΔSiW:ΔNW ratio. This was due to an increase in nitrate depletion and particulate matter, which increased in biomass and changed in composition (lower PRT:CHO and C:N), showing a prevalence of detrital or heterotrophic components which are typically associated with primary production processes in late summer and autumn. (Holm-Hansen et al. 1989; Nelson et al. 1989; Treguer et al. 1990).

The productive area included stations where the open waters first appear in spring. These were characterized by a mixed upper water column due to strong winds blowing off the ice shelf, and with low E[Z(UML)] values (av. = 7.1) (Fig. 5, Table 3). Phytoplankton population at these stations were dominated by *Phaeocysts antarctica*, (Leventer and Dumbar 1996; Smith et al. 1996; Marino and Cabrini 1997). Data analysis showed that nutrient depletion was not high, even though medium-high POC values (max values 193 $\mu g l^{-1}$ at station 7) with the highest ΔNW:ΔPW ratio (av. = 19) and the lowest ΔSiW:ΔNW value (av. = 0.71) were observed. POM autotrophic component was high (POC:Chl a av. = 84), in agreement with Di Tullio and Smith (1996) who observed a POC:Chl a ratio of *P.antarctica* populations from the SE Ross Sea that was substantially low (92). A low relationship between ΔSiW and POM and the stability index was detected (Pearson correlation coefficient 0.032 $p > 0.1$). This result would suggest that the observed bloom of *Phaeocystis antarctica* was probably unrelated to the E stability index. *Phaeo-*

*cystis antarctica,* as opposed to diatoms, may be better adapted to a less stable water column owing to the positive buoyancy of its mucilaginous stage and/or its greater adaptability to a wide range of irradiances (Leventer and Dumbar 1996).

Other productive stations in spring 1994–95 were located in the central Ross Sea and more towards the coast, in an area where the beginning of ice melting was observed (Fig. 5). Here, E[Z(UML)] values were still low (av. = 10), with a tendency to increase in time (from 4.5 to 20); depletion values were not yet high but production was. At these stations, phytoplankton populations varied from those dominated by *Phaeocystis* – with low vertical stability index values and high nitrate depletion (station I 28) – to a mixture of diatoms, prymnesiophytes and dinoflagellates (station 9), to diatoms dominated prevalently by *Fragilariopsis curta* (stations 10, 30). Phytoplankton biomass (Carrada et al. 1997) and POC values, ranging between 130 and 268 $\mu g l^{-1}$, failed to reach the concentration levels recorded by other authors during an apparent bloom caused by ice melting (Smith and Nelson 1985; Innamorati et al. 1991; Fabiano et al. 1993, 1996, 1997; Smith et al. 1996). Also at these stations, C:N ratio was low (<6.5), PRT:CHO ratio was relatively high, (av. 3.5) and POC:Chl a had an average value of 95 $\mu g l^{-1}$ indicating that physiological conditions were favourable to phytoplankton, the principal component of particulate matter (Fabiano et al. 1993; Perez et al. 1994; Catalano et al. 1997).

We also examined whether nitrate depletion at the start of the season was reflected in the particulate organic matter accumulation. The comparison between POC concentrations and the nitrate depletion converted into carbon units by multiplying by a ratio of 5.6 (w/w) (Jennings et al. 1984; Fabiano et al. 1993), and between nitrate depletion and PON stocks, indicated clear differences between the groups of Ross Sea stations (Fig. 6a,b). The concentration of POM stocks (POC and PON) was not statistically different from POM concentration based on ΔNW in the north-central stations (spring and summer sampling) (Student's t-test $p > 0.05$), but showed significant differences in the MIZ (summer samples) and polynya areas (Student's t-test $p < 0.01$ and $p < 0.05$, respectively). It appears that nitrate uptake reflected the biomass present (POC and PON) only at central-north stations. This suggests that these stations reflect early or final stages of bloom development. Since the biomass at these stations was mainly characterized by nano- and picoplankton fractions (Innamorati et al. 1994; Carrada et al. 1997), the statement that nitrate uptake can be fueled into a retention system is well confirmed here (Smetacek et al. 1990).

In summer, nitrate uptake in the MIZ area was not reflected in the biomass present (Fig. 6a,b). Here, biomass content was affected by sedimentation and lateral advention (Leventer and Dumbar 1996; Jaeger et al. 1996) and also by grazing activity with the active removal of phytoplankton characterized by diatoms (Innamorati et al. 1994; Carrada et al. 1997), as recorded by Goffart and Hecq (1996) with pigment and ammonium analyses, and by Azzali and Kalinowski (1996), who observed a high standing stock of krill. Large krill stocks can rapidly reduce particulate matter concentrations (Wefer et al. 1989), as well as determine biochemical variations in particulate matter (Fabiano et al. 1993; Robins et al. 1995).

At polynya stations dominated by *Phaeocystis antarctica,* POM stocks were in excess when nitrate uptake values were considered (Fig. 6a,b). This can be related to an increase in bacterioplankton which increase by an order of magnitude during a *Phaeocystis* bloom biomass (Putt et al. 1990) and/or to low sedimentation rates and grazing. It is interesting to note that grazing occurred when the population was dominated by diatoms, confirming that motile forms of *P. antarctica* are not well consumed by macrozooplankton (Di Tullio and Smith 1996). Sediment trap data suggest that grazing on *P. antarctica* colonies is either minimal or results in easily degraded faecal material (Leventer and Dumbar 1996).

In conclusion, the analysis of vertical stability, nutrient depletion, particulate organic matter distribution and composition in the Ross Sea showed that the system responds differently depending not only on the season but also on the sampling area. The good correlation among the parameters has demonstrated that different environmental situations can lead to the same response. Different zones in the Ross Sea (MIZ, polynya, north zones) have different productivities not only related to stratification but also to other factors that surely play an important role in the regulation of primary productivity in the Ross Sea ecosystem such as grazing pressure, lateral advection, sedimentation, light limitation and micronutrient availability (Smetacek 1985; Martin et al. 1990; Di Tullio and Smith 1996).

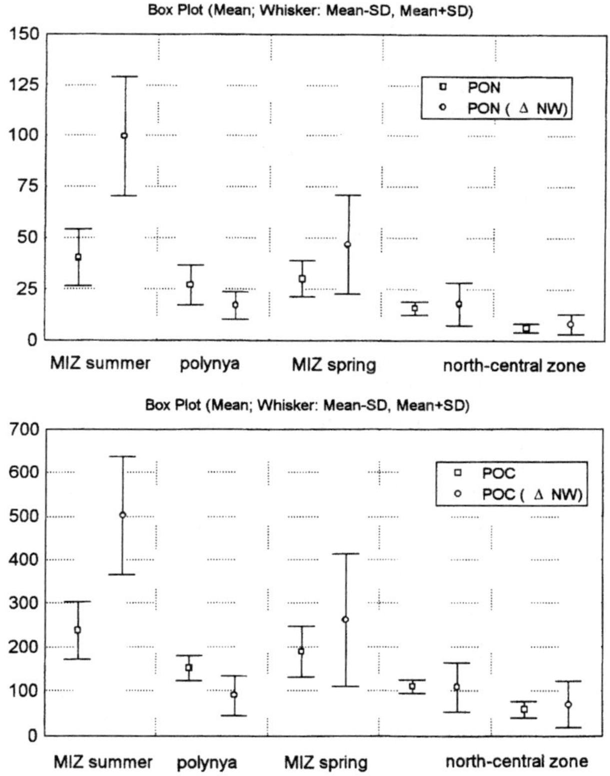


**Fig. 6.** Comparison of particulate organic matter (POC and PON in $\mu g\,l^{-1}$) with nitrate uptake in different areas of the Ross Sea

*Ackowledgements.* This work was supported by Progetto Nazionale Ricerche in Antartide (PNRA) and was funded by the Italian Government. The authors also wish to thank the scientific staff of the project for their helpful suggestions and comments.

## References


AA. VV. (1991) Oceanographic campaign 1989-90. National Scientific Commission for Antarctica (ed), Oceanographic Campaign 1989-90, Data Rep Part I, Genova, 410 pp

AA. VV. (1992) Oceanographic campaign 1989-90. National Scientific Commission for Antarctica (ed), Oceanographic Campaign 1989-90, Data Rep Part II, Genova, 508 pp

Ackley SF, Buck KA, Taguchi S (1979) Standing crop of algae in the sea ice of Weddel Sea region. Deep Sea Res 26: 269-281

Alpkem (1992a) Nitrate plus nitrite in seawater. In: Alpkem (ed) The flow solution methodology. Doc No 000630, 8/92 Rev B, Wilsonville, Oregon, 19 pp

Alpkem (1992b) Orthophosphate in seawater. In: Alpkem (ed) The flow solution methodology. Doc No 000626, 4/92 Rev A, Wilsonville, Oregon, 16 pp

Alpkem (1992c) Silica in seawater. In: Alpkem (ed) The flow solution methodology. Doc No 000671, 5/92 Rev A, Wilsonville, Oregon, 16 pp

Artegiani A, Azzolini R, Paschini E, Creazzo S (1992) Physical oceanographic conditions in the Southern Pacific Ocean and in the Western Ross Sea. In: National Scientific Commission for Antarctica (ed), Oceanographic Campaign 1989-90, Data Rep Part II, Genova, pp 5-62

Azzali M, Kalinowski J (1996) Spatial and temporal distribution of krill *Euphausia superba* biomass in the Ross Sea (1989-90 1994). In: Workshop on Ross Sea ecology, CoNISMa, Taormina, Italy, 96-97

Carrada GC, Mangoni O, Russo GF, Saggiomo V, Sgrosso S (1997) Spatial and temporal distribution of phytoplankton pigments in the Ross Sea during Austral spring 1994. In: Faranda F, Guglielmo L, Povero P (eds) National Programme for Antarctic Research, ROSSMIZE 1993-95, Data Rep Part I, Genova, pp 197-228

Catalano G, Benedetti F (1990) Distributions of nutrients in the Terra Nova Bay and in the Ross Sea. In: National

Scientific Commission for Antarctica (ed), Oceanographic Campaign 1987–88, Data Rep Part I, Genova, pp 61–83

Catalano G, Benedetti F, Goffart A, Iorio M (1991a) Distribution of dissolved oxygen, pH, total alkalinity and nutrients in Southern Ocean and Ross Sea (R/V "Cariboo" 1989–90 Cruise). In: National Scientific Commission for Antarctica (ed) Oceanographic Campaign 1989–90, Data Rep Part I, Genova, pp 11–23

Catalano G, Benedetti F, Iorio M (1991b) Coastal oceanography from Cape Russel to Campbell Ice Tongue (Terra Nova Bay). Dissolved oxygen, nutrients, pH, total alkalinity and total inorganic carbon distribution. In: National Scientific Commission for Antarctica (ed) Oceanographic Campaign 1989–90, Data Rep Part I, Genova, pp 25–32

Catalano G, Povero P, Fabiano M, Benedetti F, Goffart A (1997) Nutrient utilisation and particulate organic matter changes during summer in the upper mixed layer (Ross Sea, Antarctica) Deep Sea Res 44: 97–112

Comiso JC, McClain CR, Sullivan W, Ryan JP, Leonard CL (1993) Coastal zone color scanner pigment concentrations in the Southern Ocean relationships to geophysical surface features. J Geophys Res 98: 1816–1833

Cota GF, Smith WO Jr, Nelson DM, Muench RD, Gordon LI (1992) Nutrient and biogenic particulate distributions, primary productivity and nitrogen uptake in the Weddell-Scotia Sea marginal ice zone during winter. J Mar Res 50: 155–181

DeMaster DJ, DunbarL I, Gordon LI, Leventer AR, Morrison J, Nelson DM, Nittrouer CA, Smith WO Jr (1992) The cycling and accumulation of organic matter and biogenic silica in high latitudine environments: the Ross Sea. Oceanography 5: 146–153

Dieckmann GS, Lange MA, Ackley SF, Jennings JC Jr (1991) The nutrient status in the sea ice of the Weddell Sea during winter: effects of sea ice texture and algae. Polar Biol 11: 449–456

Di Tullio GR, Smith WO Jr (1996) Spatial patterns in phytoplankton biomass and pigment distributions in the Ross Sea. J Geophys Res 101 (C8): 18467–18477

Dubois M, Gilles K, Hamilton JK, Rebers PA, Smith F (1956) Colorimetric method for determination of sugars and related substances. Anal Chem 28: 350–356

El-Sayed SZ (1984) Productivity of the Antarctic water. In: Holm-Hansen O, Bolis L, Giles R (eds) Marine phytoplankton and productivity. Springer, Berlin Heidelberg New York, pp 19–34

El Sayed SZ, Taguchi S (1981) Primary production and the standing crop of phytoplankton along the ice-edge in the Weddel Sea. Deep Sea Res 28: 1017–1032

Fabiano M, Povero P, Catalano G, Benedetti F (1991) Hydrological data collected during the biological, chemical and geological sampling in Terra Nova Bay. In: National Scientific Commission for Antarctica (ed) Oceanographic Campaign 1989–90, Data Rep Part I, Genova, pp 35–71

Fabiano M, Povero P, Danovaro R (1993) Distribution and composition of the particulate organic matter of the Ross Sea (Antarctica). Polar Biol 13: 525–533

Fabiano M, Povero P, Danovaro R (1996) Particulate organic matter composition in Terra Nova Bay (Ross Sea, Antarctica) duing summer. Antarct Sci 8(1): 7–13

Fabiano M, Chiantore M, Povero P, Cattaneo-Vietti R, Pusceddu A, Misic C, Albertelli G (1997) Short-term variations in particulate matter flux in terra Nova Bay, Ross Sea. Antarct Sci 9(2): 143–149

Faranda F, Guglielmo L (1994) Ross Sea expeditions 1987–88 and 1989–90: Straits of Magellan expedition 1991. In: Faranda F, Guglielmo L (eds) National Scientific Commission for Antarctica, Data Rep Part II, Genova, 224 pp

Faranda F, Guglielmo L, Povero P (1997) ROSSMIZE (Ross Sea Marginal Ice Zone Ecology), oceanographic expedition 1993–95. In: Faranda F, Guglielmo L, Povero P (eds) National Programme for Antarctic Research, Data Rep Part I, Genova, 431 pp

Goeyens L, Treguer P, Bauman MEM, Baeyens W, Dehairs F (1995) The leading role of ammonium in the nitrogen uptake regime of the Southern Ocean marginal ice zones, J Mar Syst 6: 345–361

Goffart A, Hecq JH (1996) Distribution of phytopigments and degradation products detected by HPLC as tracers of planktonic fluxes in the Ross Sea. In: Workshop on Ross Sea Ecology. CoNISMa, Taormina, Italy, 90–91

Hansen HP, Grasshoff K (1983) Automated chemical analysis. In: Grasshoff K, Ehrhardt M, Kremling K (eds) Methods of seawater analysis, 2nd edn. V Ch Weinheim, pp 347–379

Hartree EF (1972) Determination of proteins: a modification of the Lowry method that gives a linear photometric response. Anal Biochem 48: 422–427

Hecq JH, Goffart A, Catalano G, Guglielmo L (1996) Modelling and simulation of the Ross Sea upper water column and planktonic ecosystem. In: Workshop on Ross Sea Ecology, CoNISMa, Taormina, Italy, 92–93

Hedges JI, Stern JH (1984) Carbon and nitrogen determination of carbonate-containing solids. Limnol Oceanogr 29: 657–663

Holm-Hansen O, Mithell BG, Hewes CD, Karl DM (1989) Phytoplankton blooms in the vicinity of Palmer Station, Antarctica. Polar Biol 10: 49–57

Innamorati M, Lazzara L, Mori G, Nuccio G, Saggiomo V (1991) Phytoplankton ecology. In: National Scientific Commission for Antarctica (ed) Oceanographic Campaign 1989–90, Data Rep Part I, Genova, pp 141–252

Innamorati M, Lazzara L, Massi L , Mori G, Nuccio G, Saggiomo V, Cartei P, Franchini L, Gamba C (1994) Phytoplankton ecology: particles, gilvin, fluorescence spectra, irradiance and phytoplankton species density in the Ross Sea. Oceanography Campaign 1989–90. In: Faranda FM, Guglielmo L (eds) National Scientific Commission for Antarctica, Ross Sea expeditions 1987–88 and 1989–90 and Straits of Magellan expedition 1991, Data Rep Part III, Genova, pp 101–178

Jaeger JM, Nittrouer CA, DeMaster DJ, Kelchner C, Dumbar RB (1996) Lateral transport of settling particles in the Ross Sea and implications for the fate of biogenic material. J Geophys Res 101 (C8): 18479–18488

Jennings JC Jr, Gordon LI, Nelson DM (1984) Nutrient depletion indicates high primary productivity in the Weddell Sea. Nature 309: 51–54

Karl DM, Tillbrook BT, Tien G (1991) Seasonal coupling of organic matter production and particle flux in the western Bransfield Strait Antarctica. Deep Sea Res A 38: 1097–1126

Leventer A, Dumbar RB (1996) Factors influencing the distribution of diatoms and other algae in the Ross Sea. J Geophys Res 101 (C8): 18489–18500

Leventer A, Dumbar RB, DeMaster DJ (1993) Diatom evidence for late Holocene climatic events in Granite Harbor, Antarctica. Paleoceanography 8: 373–386

Lipizer M, Ruffini S, Pedonzani S, Cozzi S, Goffart A, Catalano G (1997) Antarctic spring: the importance of "new production'. In: Picazzo M (ed) Atti XII Congr Ass Italian Oceanogr Limnol, Lang, Genova, pp 153–164

Marino D, Cabrini M (1997) Distribution and succession of phytoplankton population in the Ross Sea (Antarctica) during the Austral spring 1994. In: Faranda F, Guglielmo L, Povero P (eds) Nat Progr Ant Res ROSSMIZE 1993–95. Data Rep I, pp 307–341

Martin JH, Gordon RM, Fitzwater S (1990) Iron in Antarctic waters. Nature 345: 156–158

Mitchell BG, Holm-Hansen O (1991) Observation and modelling of the Antarctic phytoplankton crop in relation to mixing depth. Deep Sea Res 38: 981–1007

Nelson DM, Treguer P (1992) On the role of silicon as a limiting nutrient to Antarctic diatoms: evidence from kinetics studies in the Ross Sea ice-edge zone. Mar Ecol Prog Ser 80: 255–264

Nelson DM, Smith WO Jr, Gordon LI, Huber BA (1987) Spring distributions of density, nutrients, and phytoplankton biomass in the ice-edge zone of the Weddell-Scotia Sea. J Geophys Res 92: 7181–7190

Nelson DM, Smith WO Jr, Muench RD, Gordon LI, Sullivan CW, Husby DM (1989) Particulate matter and nutrient distributions in the ice-edge zone of the Weddel Sea: relationship to hydrography during late summer. Deep Sea Res 36(2): 191–209

Nelson DM, DeMaster DJ, Dunbar RB, Smith WO Jr (1996) Cycling of organic carbon and biogenic silica in the Southern Ocean: estimates of water column and sedimentary fluxes on the Ross Sea continental shelf. J Geophys Res 101 (C8): 18519–18532

Niebauer HJ, Alexander V (1985) Oceanographic frontal structure and biological production at an ice-edge. Cont Shelf Res 4: 367–388

Parsons TR, Takahasky M, Hargrave B (1984) Biological oceanographic processes. Pergamon Press, New York, 332 pp

Perez FF, Tokarczyk R, Figueiras FG, Rios AF (1994) Water masses and phytoplankton biomass distribution during summer in the Weddel Sea marginal ice zone. Oceanol Acta 17(2): 191–199

Priddle J, Brandini F, Lipski M, Thorley MR (1994) Pattern variability of phytoplankton biomass in the Antarctic Peninsula Region: an assessment of BIOMASS cruises. In: El-Sayed SZ (ed) Southern Ocean ecology: the BIOMASS perspectives. Cambridge University Press, Cambridge, pp 49–61

Putt M, Börseim KY , Miceli G, Stoecker DK (1990) Seasonal changes in cell size and abundance of bacterioplankton during Phaeocystis sp. bloom in McMurdo Sound. Antarct J US 25(5): 199–201

Robins DB, Harris RP, Bedo AW, Fernandez E, Fileman TW, Harbour DS, Head RN (1995) The relationship between suspended particulate material, phytoplankton and zooplankton during the retreat of the marginal ice zone in the Bellingshausen Sea. Deep Sea Res II 42: 1137–1158

Russo A, Gallarato A, Testa G, Corbo C, Pariante R (1997) Physical data collected during ROSSMIZE cruise (Ross Sea, November–December 1994). In: Faranda F, Guglielmo L, Povero P (eds) National Promgramme for Antarctica, ROSSMIZE 1993–95, Data Rep Part I, Genova, pp 25–110

Saggiomo V, Carrada GC, Mangoni O, Ribera d'Alcalà M, Russo A (1997) Spatial and temporal variability of size-fractionated biomass and primary production in the Ross Sea (Antarctica) during austral spring and summer. In: Faranda F, Guglielmo L, Povero P (eds) National Promgramme for Antarctica, ROSSMIZE 1993–95, Data Rep Part I, Genova, pp 291–305

Sakshaug E (1994) Discussant's report: primary production in the Antarctic pelagial – a view from the North. In: El-Sayed SZ (ed) Southern ocean ecolgy: the BIOMASS perspective. Cambridge University Press, Cambridge, pp 25–126

Sakshaug E, Holm-Hansen O (1984) Factors governing pelagic production in polar oceans. In: Holm-Hansen O, Bolis L, Gilles R (eds) Marine phytoplankton and productivity. Springer Berlin Heidelberg New York, pp 1–18

Sambrotto RN, Niebauer HJ, Goering JJ, Iverson RL (1986) Relationship among vertical mixing, nitrate uptake, and phytoplankton growth during the spring bloom in the Southeast Bering Sea middle shelf. Cont Shelf Res 5: 161–198

Smetacek V (1985) Role of sinking in diatom life-history cycles: ecological, evolutionary and geological significance. Mar Biol 84: 239–251

Smetacek V, Hendrikson P (1979) Composition of particulate organic matter in Kiel Bight in relation to phytoplankton succession. Oceanol Acta 2: 287–298

Smetacek V, Scharek R, Nothig EM (1990) Seasonal and regional variation in the pelagial polar ocean: relationship to the life history cycle of krill. In: Antarctic ecosystem. Ecological change and conservation. Kerry KR, Hempel G (eds) Springer Berlin Heidelberg New York, pp 103–114

Smith WO Jr, Gordon LI (1997) Hyperproductivity of the Ross Sea (Antarctica) polynya during austral spring. Geophys Res Lett 24: 233–236

Smith WO Jr, Nelson DM (1985) Phytoplankton bloom produced by a receding ice edge in the Ross Sea: spatial coherence with the density field. Science 227: 163–166

Smith WO Jr, Nelson DM, Di Tullio GR, Leventer R (1996) Temporal and spatial patterns in the Ross Sea: phytoplankton biomass, elemental composition, productivity and growth rates. Geophys Res 101 (C8): 18455–18465

StatSoft (1995) STATISTICA for Windows. Computer program manual. StatSoft, Inc. Tulsa, OK, 420 pp

Steinhorn I (1985) The disappearance of the long-term meromictic stratification of the Dead Sea. Limnol Oceanogr 30: 451–472

Sullivan CW, McClain CR, Comiso JC, Smith WO Jr (1988) Phytoplankton standing crops within an Antarctic ice edge assesed by satellite remote sensing. J Geophys Res 93: 12487–12498

Sullivan WC, Arrigo KR, McClain CR, Comiso JC, Firestone J (1993) Distribution of phytoplankton blooms in the Southern Ocean. Science 262: 1832–1837

Tanoue E (1985) Distribution and chemical composition of particulate organic matter in the pacific sector of the Antarctic Ocean. Oceanogr Trans Tokyo Univ Fish 6: 43–57

Treguer P, Jacques G (1992) Dynamics of nutrients and phytoplankton, and fluxes of carbon, nitrogen and silicon in the Antarctic Ocean. Polar Biol 12: 149–162

Treguer P, Nelson DM, Gueneley S, Zeyons C, Morvan J, Buma A (1990) The distribution of biogenic and lithogenic silica and the composition of particulate organic matter in the Scotia Sea and Drake Passage during autumn 1987. Deep Sea Res 35: 833–851

UNESCO (1985) The international system of units (SI) in oceanography. UNESCO Tech Pap Mar Sci 45. (Also: IAPSO Publ Sci 32)

Wefer A, Fisher G, Fuettterer D, Gersonde R (1989) Seasonal particle flux in the Bransfield Strait, Antarctica. Deep Sea Res A 35: 891–898

CHAPTER 11

# Spatial and Temporal Distribution of Particulate Organic Matter in the Ross Sea

**M. Fabiano, P. Povero, and C. Misic**[1]

**ABSTRACT**

Spatial distribution and biochemical composition of particulate organic matter (POM) were studied in relation to ice melting of the Ross Sea (Antarctica) during the austral spring and summer 1994 (from 14 Nov to 15 Dec 1994). POM distribution, characterized by areas with different amounts and biochemical composition of organic particles, was strongly influenced by ice melting. POM, mostly of phytoplankton origin, showed a large detrital component. On average, proteins accounted for the largest fraction of POM (40–80%), followed by carbohydrates (10–40%) and lipids (5–35%). POM vertical distribution was characterised by a concentration decrease between 20- and 60-m depth. Particulate organic matter in the surface layer and in the deep layer differed in quantitative and qualitative biochemical composition. POC and PON concentrations ranged, in the surface layer, between 36–256 and 3–41 $\mu g\,l^{-1}$, respectively, and in the deep layer between 16–50 and 1–7 $\mu g\,l^{-1}$, respectively. Changes in chemical composition were related to microbial ectoenzymatic activity. Degradation processes lead to a strong decrease in the protein fraction, as a result of the activity of protein degrading ectoenzymes, mainly leucine aminopeptidase, which showed the highest activities (>30 $nmol\,l^{-1}\,h^{-1}$) in the surface layer. The Ross Sea system acts as a retentive system since a large part of the organic matter is remineralized within the water column. POM spatial variation is considered within the frame of biochemical composition changes induced by ice-melting processes and compared with previous observations in the Ross Sea.

## Introduction

The seasonal ice margin constitutes an ecological interface between different systems, one associated with the open ocean, the other with pack ice. There is thus a growing body of evidence that ice-edge phenomena may be of great importance in the overall biomass, production and nutrient and sedimentation cycles of the Southern Ocean (Smith and Nelson 1985, 1986; Nelson et al. 1987; Treguer and Jacques 1992; DeMaster et al. 1992).

Intense phytoplankton blooms near the edge of the seasonally retreating pack ice are quite common in the Ross Sea (Smith and Nelson 1985; Nelson et al. 1987, 1996). A large amount of particulate organic matter, deriving from primary production, is available to consumers in the water column and in the sediments (Dayton 1990; Karl et al. 1991; Cadèe 1992). The high food value of suspended particulate matter accounts for higher levels of tertiary productivity despite a low primary productivity of Antarctic waters (El-Sayed 1988; Azam et al. 1991; Gonzales 1992). However, part of the organic material is not utilized and reaches the sea floor (Matsuda et al. 1990; Nedwell et al. 1993; Fabiano et al., this Vol.). Nevertheless, these blooms, due to their restricted space and time extent, can easily be missed during oceanographic surveys (El Sayed and Taguchi 1981; Smith et al. 1987; Jacques 1989).

Despite the large amount of data available on the elemental composition of suspended matter in the Ross Sea (Dunbar et al. 1985; Nelson and Smith 1986; Smith and Nelson 1985), much less is known about its biochemical composition (Dhargalkar and Bhosle 1989; Handa et al. 1992; Fabiano et al. 1993, 1996, 1997; Berdalet et al. 1997). Biochemical composition of particulate matter changes with depth by degradation processes (Liebezeit and Bölter 1986; Matsuda et al. 1990; Fabiano et al. 1993), mediated by heterotrophic microorganisms (Gillespie et al. 1976; Bölter and Dawson 1982;

[1] Dipartimento per lo Studio del Territorio e delle sue Risorse, Università di Genova, Corso Europa 26, 16132 Genova, Italy

Knox 1994). The study of hydrolytic enzymatic activity is a valuable and largely applied protocol for evaluating assimilative and productive microbial capacities (Chróst 1990; Christian and Karl 1992; Smith et al. 1992; Unanue et al. 1993). This method provides information on degradation activity efficiency and capacity (Christian and Karl 1995).

The aim of this chapter is:

1. to estimate the quantitative importance and distribution of particulate organic matter in the Ross Sea during sea-ice melting;
2. to study the vertical distribution of particulate organic matter in relation to degradation processes during sinking;
3. to investigate biochemical composition changes of particulate organic matter in relation to the bloom and the decay of phytoplanktonic populations along the marginal ice zone.

## Materials and Methods

### Sampling

Sampling was carried out by the R/V Italica between November 14 and December 15, 1994. The Ross Sea sampling schedule included six transects of stations (Fig. 1, Table 1). Five of these transects had a south-north direction and followed the retreat of the ice pack from the Ross Sea polynya to the offshore area. The sixth transect covered the area from the Ross Sea polynya to MIZ area of Terra Nova Bay polynya, passing through a closed pack-ice area (station 29).

Data on temperature and salinity were recorded using a multiparametric SBE 9/11+ plus CTD system. Water samples were collected throughout the water column from the surface to the bottom (which ranged between 281 and 1828 m) using 24 Niskin (10-l) bottles on a SBE Carousel 32 sampler. The Niskin bottles were previously washed with a solution of 0.1 N HCl according to Burney et al. (1979). Samples were collected at 10–12 depths according to bottom depths and light extinction profiles, detected by a photoprobe equipped with LI-COR quantum meters (Innamorati et al. 1991).

### Biochemical Analysis

The volumes of filtered water for each biochemical parameter ranged between 500 and 2000 ml. Water samples (prefiltered on 200-μm mesh net) were filtered on Whatman GF/F filters for total suspended matter, elemental organic carbon and nitrogen particulate carbohydrates, particulate proteins, and particulate lipids.

Total suspended matter (TSM) was assessed by the gravimetric method (Strickland and Parsons 1972), using a Mettler M3 balance (accuracy ± 1.0 μg). Particulate organic carbon and nitrogen (POC and PON) were analysed by a Carlo Erba CHNS-O EA1108 Elemental Analyzer after removal of carbonates by exposure to HCl fumes (Hedges and Stern 1984). Cyclohexanone 2,4-dinitrophenylhydrazone was used as standard. Particulate carbohydrates (CHO) were detected according to Dubois et al. (1956). Absorbance was measured at 490 nm using a Varian spectrophotometer. Solutions of D(+)glucose were used as standard. Particulate protein (PRT) determination was conducted according to Hartree (1972). Absorbance was measured at 650 nm. Bovine serum albumin was used as standard. Particulate lipids (LIP) extraction was carried out according to the Bligh and Dyer (1959) method. Analyses were performed by carbonization (Marsh and Weinstein 1966). Absorbance was measured at 375 nm. Tripalmitine was used as standard. CHO, PRT and LIP were converted into carbon equivalents

**Table 1.** Sampling schedule of the six transects, location and depth of first and last station of each transect

| Transect | Period | From station | Latitude S | Longitude E | Depth | To station | Latitude S | Longitude E | Depth |
|---|---|---|---|---|---|---|---|---|---|
| 1 | 14.11–17.11 | MP | 76°30′.097 | 175°00′.000 | 464 | 7 | 75°20′.416 | 174°58′.150 | 310 |
| 2 | 20.11–26.11 | MPA | 76°26′.624 | 175°01′.452 | 471 | 15 | 73°58′.879 | 174°56′.956 | 577 |
| 3 | 28.11–03.12 | 11A | 74°40′.424 | 174°56′.524 | 438 | 17 | 73°42′.023 | 175°01′.018 | 484 |
| 4 | 04.12–08.12 | 11B | 74°39′.124 | 175°58′.187 | 436 | 27 | 71°56′.662 | 174°58′.565 | 1828 |
| 5 | 10.12–13.12 | 24A | 72°29′.094 | 174°58′.922 | 442 | 11C | 74°40′.554 | 174°59′.833 | 439 |
| 6 | 14.12–16.12 | 11C | 74°40′.554 | 174°59′.833 | 439 | 30 | 74°41′3 | 164°10′5 | 827 |

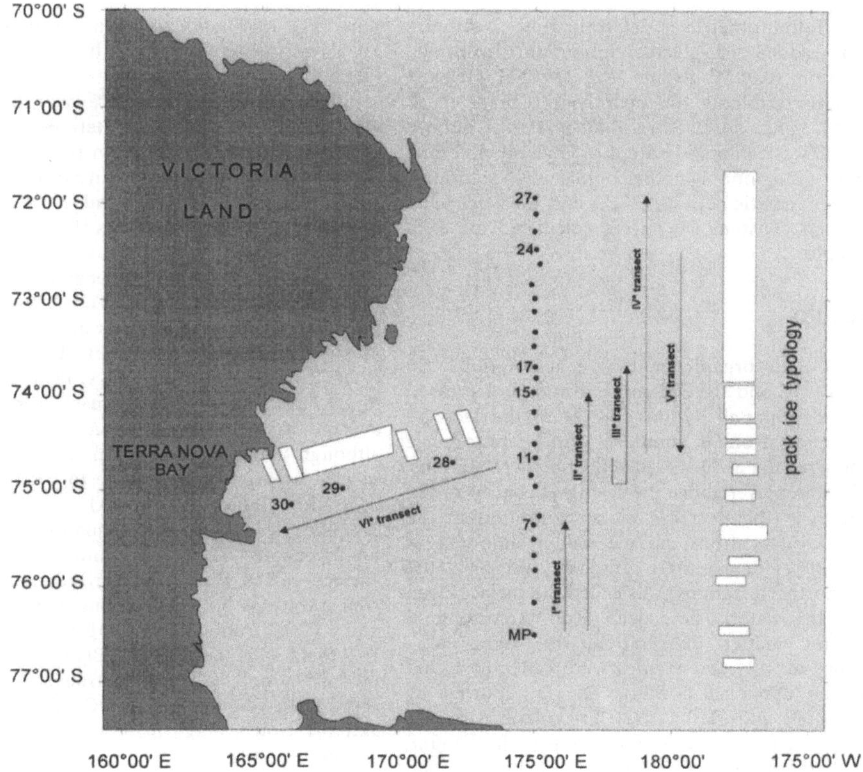


**Fig. 1.** Study area and transect location. *Arrows* show sampling direction

using 0.40, 0.49 and 0.75 μgCμg$^{-1}$ conversion factors. The labile fraction of the particulate organic matter (C-POM) was expressed as the sum of carbohydrates (C-CHO), lipids (C-LIP) and proteins (C-PRT). Water samples were collected on a single replicate. However, a series of replicate samples were collected at different intervals in surface water in order to define the coefficient of variation for all parameters: 12.8% for TSM, 2.3% for POC, 2.7% for PON, 11.0% for CHO, 3.2% for PRT and 10.7% for LIP.

## Enzymatic Analysis

Water samples, prefiltered on a 200-μm mesh net, were filtered on Nuclepore filters (0.2-μm pore diameter) to retain the enzymatic fraction linked to particulate matter (ectoenzymes, Chróst 1991). Ectoenzymatic activity evaluation was carried out immediately after sample collection, according to the methods proposed by Chróst and Velimirov (1991) and Vrba and Machácek (1994). The artificial substrates, from Sigma Co., were 4-methylumbelliferyl β-D -glucoside, 4-methylumbelliferyl N-acetyl-β-D -glucosaminide and L-leucine 7-amido-4-methyl-coumarin hydrochloride, respectively, for β-glucosidase (BG), ß-N-acetylglucosaminidase (GA) and leucine aminopeptidase (LA). The five subsamples employed for each ectoenzymatic activity determination [one filter in 5 ml of prefiltered (0.2 μm) and autoclaved seawater] and five blanks [5 ml of prefiltered (0.2 μm) and autoclaved seawater each] were supplemented with 0.5 ml of substrate (from 1 to 100 μmol l$^{-1}$ final concentration) and then incubated for 3 to 5 h at in situ temperature in the dark. Before spectrofluorometric determination, samples were added of glycine/ammonium hydroxide solution (0.05 M glycine + 0.2 M $NH_4OH$) pH 10.5 to terminate the hydrolysis (Chróst et al. 1989). The spectrofluorometer Perkin Elmer LS50B was calibrated with 4-methylumbelliferone and 7-amino-4-methylcoumarin solutions. Ectoenzymatic activity ($V_{max}$) and the Michaelis-Menten constant ($K_m$) were calculated for each sample.

## Results

Particulate organic matter concentrations (in both POC and C-POM concentrations) decreased with depth at all stations (Figs. 2, 3); the decrease was quite strong at about 30–50 m, or quite feeble and irregular at 30–120 m at the northern stations. The water layer above the strongest concentration decrease (the decrease in concentration in the deep waters from surface values amounts to >70–80% for southern stations and >40–50% for northern stations) was called the surface layer, and the water below deep layer. According to organic matter distribution, the higher $V_{max}$ values for all ectoenzymes were generally found in the upper 30 m (Table 2), as reported by Christian and Karl (1992). Enzymatic activities for particulate LA were comparable to those found in Antarctic waters for total LA (Christian and Karl 1992, 1995) and, according to these authors, LA $K_m$ was generally higher than BG and GA $K_m$ (Table 2).

With regard to the surface layer, the stations were divided in two groups. Stations belonging to transects 1, 2 and to the main part of transect 6 (except for station 29, located in a heavy pack area) were characterized by high values of particulate organic matter concentrations (POC from 86.9 to 265.7 $\mu g l^{-1}$ and C-POM from 97.2 to 196.5 $\mu g l^{-1}$) and, generally, by high ectoenzymatic activities (on average BG 0.33, GA 0.35, LA 8.39 $nmol l^{-1} h^{-1}$) (Tables 2, 3, Fig. 4A). Stations belonging to transects 3, 4 and 5 showed lower POM concentrations (POC from 36.1 to 89.7 $\mu g l^{-1}$, C-POM from 28.15 to 84.7 $\mu g l^{-1}$) and ectoenzymatic activities (on average BGV 0.13, GAV 0.19, LAV 2.74 $nmol l^{-1} h^{-1}$), although the $K_m$ of ectoenzymes, similar to those in rich areas, was favourable to organic matter utilisation (Tables 2, 3; Fig. 4A).

The deep layer was quantitatively more homogeneous (Fig. 4B) and, compared to surface values, was characterized by low mean concentrations both in stations of transects 1, 2 and 6 (POC from 18.4 to 48.3 $\mu g l^{-1}$, C-POM from 16.7 to 42.3 $\mu g l^{-1}$; on average BGV 0.04, GAV 0.28, LAV 0.57 $nmol l^{-1} h^{-1}$) and in transects 3, 4 and 5 (POC from 12.4 to 43.2 $\mu g l^{-1}$, C-POM from 9.5 to 32.1 $\mu g l^{-1}$; on average BGV 0.06, GAV 0.08, LAV 0.25 $nmol l^{-1} h^{-1}$).

**Table 2.** Ectoenzymatic activity ($V_{max}$, $nmol l^{-1} h^{-1}$) and Michaelis-Menten constant ($K_m$, $\mu M$) values in different areas of Ross Sea surface layer and deep layer. BG: β-glucosidase, GA: β-N-acetylglucosaminidase, LA: leucine aminopeptidase

| | Station | BG $V_{sat}$ | BG $K_m$ | GA $V_{sat}$ | Ga $K_m$ | LA $V_{sat}$ | LA $K_m$ |
|---|---|---|---|---|---|---|---|
| Surface layer | | | | | | | |
| Polynya | MP | 0.21 | 0.16 | 0.13 | 0.05 | 8.75 | 1.93 |
| MIZ | 6 | 0.31 | 0.26 | 0.35 | 0.07 | 4.25 | 0.44 |
| MIZ | 9 | 0.31 | 0.17 | 0.87 | 0.15 | 11.53 | 0.33 |
| Polynya | 11b | 0.08 | 0.26 | 0.27 | 0.15 | 2.11 | 1.35 |
| Pack | 24a | 0.04 | 0.12 | 0.05 | 0.06 | 0.50 | 0.62 |
| Polynya MIZ | 28 | 0.90 | 0.67 | 0.41 | 0.12 | 11.95 | 0.46 |
| MIZ | 30 | 0.38 | 0.40 | 0.94 | 0.07 | 5.56 | 0.07 |
| Deep layer | | | | | | | |
| Polynya | MP | 0.05 | 1.12 | 0.40 | 1.09 | 0.40 | 0.39 |
| MIZ | 6 | 0.04 | 0.38 | 0.11 | 0.13 | 0.40 | 0.24 |
| MIZ | 9 | 0.01 | 0.11 | 0.16 | 0.63 | 0.60 | 1.83 |
| Polynya | 11b | 0.12 | 0.24 | 0.13 | 0.18 | 0.39 | 0.22 |
| Pack | 24a | 0.00 | 0.02 | 0.04 | 0.11 | 0.03 | 0.30 |
| Polynya MIZ | 28 | 0.03 | 0.08 | 0.10 | 0.26 | 1.55 | 1.58 |
| MIZ | 30 | 0.03 | 0.29 | 0.42 | 0.83 | 0.34 | 0.03 |

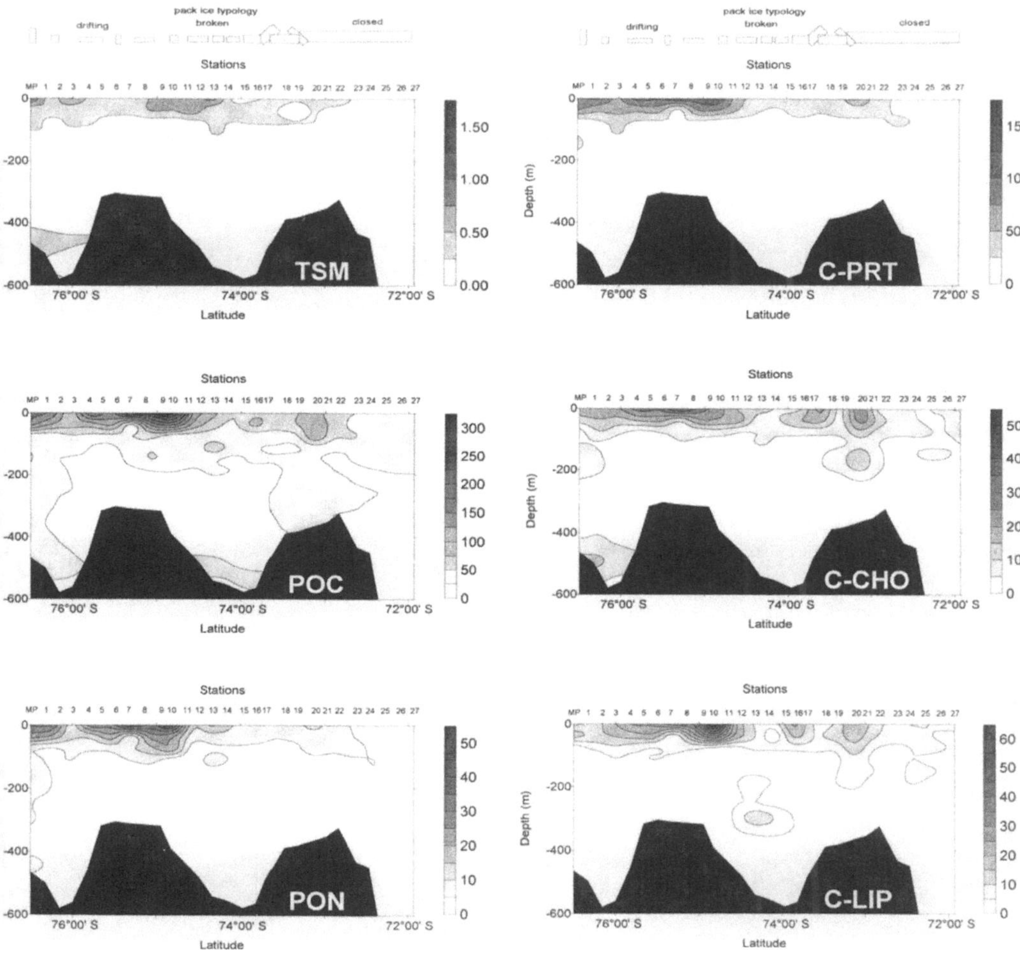


**Fig. 2.** Total suspended matter (*TSM*, mg l$^{-1}$), particulate organic nitrogen (*PON*, μg l$^{-1}$), particulate organic carbon (*POC*, μg l$^{-1}$), particulate proteins (*C-PRT*, μg l$^{-1}$), particulate carbohydrates (*C-CHO*, μg l$^{-1}$) and particulate lipids (*C-LIP*, μg l$^{-1}$) distributions from stations MP to 27

Figure 3 compares the organic matter concentrations of the surface layer and pack ice typology. High organic matter concentrations and ectoenzymatic activities values were observed at the most Southern station (from station MP to station 2, Ross Sea polynya transect 1), while lower concentrations were observed in the area of station 3 (ice drifting area). POM concentrations then increased to station 7, at the beginning of the broken ice area. Station 10 (transect 2) showed the highest values, related to the pack melting and formation of broken ice (Fig. 3), while the northern section (stations 12 to 15) showed lower concentrations of areas with heavy pack-ice coverage.

The northern area (transects 3, 4 and 5), characterized by wide areas of heavy ice cover, shows low values of all biochemical and ectoenzymatic parameters (Fig. 3, Tables 2, 3). Although BG was assessed to be quite constant throughout the

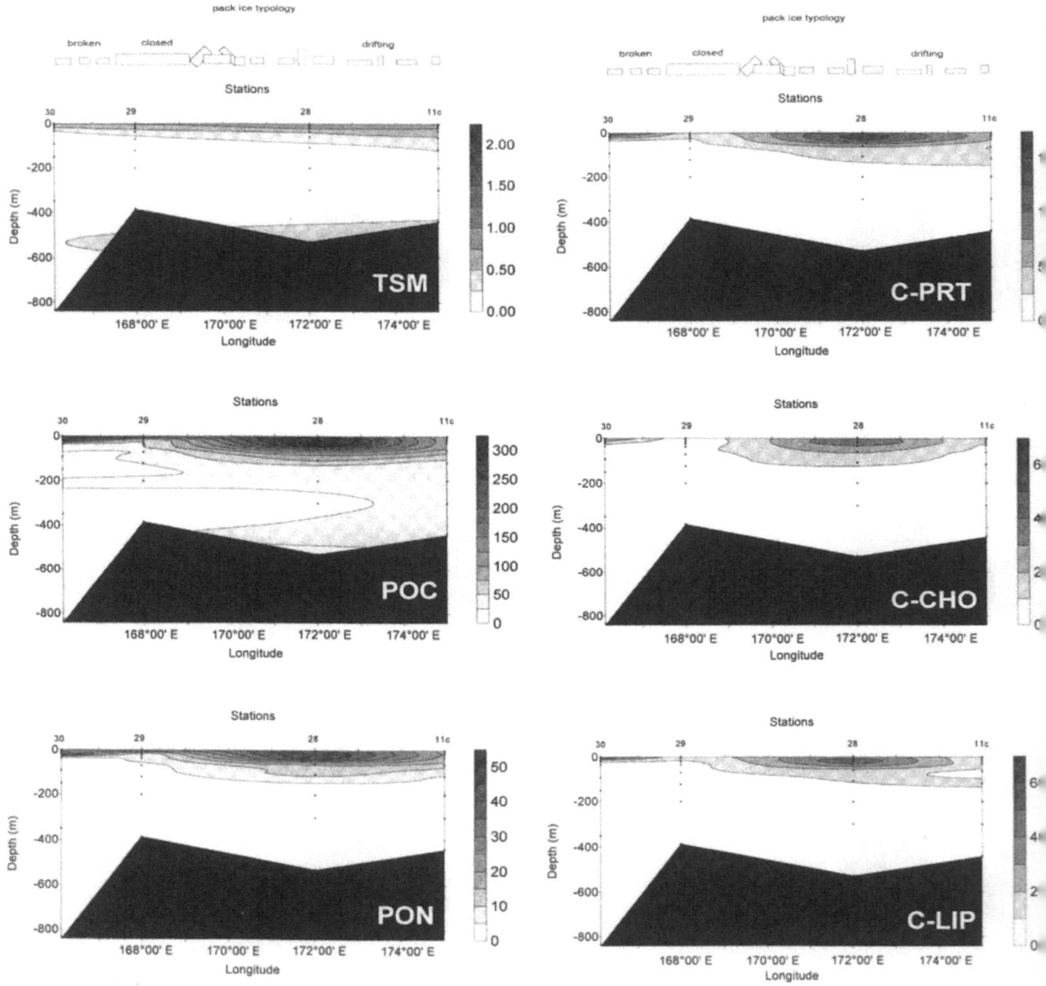


**Fig. 3.** Total suspended matter (*TSM*, $mg\,l^{-1}$), particulate organic nitrogen (*PON*, $\mu g\,l^{-1}$), particulate organic carbon (*POC*, $\mu g\,l^{-1}$), particulate proteins (*C-PRT*, $\mu g\,l^{-1}$), particulate carbohydrates (*C-CHO*, $\mu g\,l^{-1}$) and particulate lipids (*C-LIP*, $\mu g\,l^{-1}$) distributions from stations 11c to 30

transects, it showed a lower efficiency in transect 3. From stations 23 to 27, in the northern part of transect 4 (Fig. 3), organic matter concentrations within the water column were clearly lower than mean POM-poor area values.

In transect 6 (Fig. 4), chemical parameters were characterized by high values at station 30 (the innermost station of Terra Nova Bay), a decrease of concentration at station 29 (placed in a closed pack-ice area) and then an increase at station 28. Station 11c, the outer station of the transect placed in an ice-free area, showed low concentrations of POM. In stations characterized by high POM concentrations, ectoenzymatic activity, on average, was the highest found during the survey for BG and GA, while LA values were similar to transects 1 and 2.

**Table 3.** Biochemical characterization of POM in different areas of Ross Sea surface layer and deep layer. Particulate organic carbon (POC, $\mu g l^{-1}$), particulate labile fraction (C-POM, $\mu g l^{-1}$), particulate refractory fraction (Ref, $\mu g l^{-1}$), percentage of particulate proteins, carbohydrates, lipids and refractory fraction vs. POC (% C-PRT; % C-CHO; % C-LIP, % Ref), particulate proteins and carbohydrates ratio (PRT:CHO)

| Station | POC | C-POM | Ref | % Ref | % C-PRT | % C-CHO | % C-LIP | PRT:CHO |
|---|---|---|---|---|---|---|---|---|
| Surface layer | | | | | | | | |
| Polynya Ross Sea (MP) | 162.9 | 122.6 | 40.3 | 24.7 | 57.1 | 10.7 | 7.4 | 5.3 |
| MIZ Ross Sea (8) | 223.6 | 186.2 | 37.4 | 16.7 | 56.0 | 14.8 | 13.8 | 3.8 |
| MIZ Ross Sea (10) | 236.9 | 191.9 | 45.1 | 19.0 | 48.0 | 8.6 | 24.4 | 5.6 |
| Ice-free area (11b) | 65.5 | 62.4 | 3.1 | 5.0 | 61.2 | 17.7 | 22.2 | 3.5 |
| Closed pack (27) | 40.8 | 29.0 | 11.8 | 29.0 | 41.6 | 16.2 | 12.0 | 2.6 |
| MIZ Ross Sea (28) | 265.7 | 196.5 | 69.2 | 26.1 | 45.2 | 13.4 | 15.4 | 3.4 |
| Closed pack (29) | 66.0 | 39.3 | 26.7 | 40.5 | 40.6 | 6.4 | 12.5 | 2.9 |
| MIZ BTN (30) | 221.9 | 164.5 | 57.4 | 25.9 | 48.5 | 10.5 | 15.1 | 4.6 |
| Deep layer | | | | | | | | |
| Polynya Ross Sea (MP) | 48.3 | 28.2 | 20.1 | 41.6 | 40.1 | 10.0 | 8.3 | 4.0 |
| MIZ Ross Sea (8) | 45.8 | 31.8 | 14.0 | 30.6 | 44.0 | 13.4 | 11.9 | 3.3 |
| MIZ Ross Sea (10) | 34.8 | 22.6 | 12.1 | 34.9 | 37.3 | 15.5 | 12.3 | 2.4 |
| Ice-free area (11b) | 26.2 | 24.2 | 2.0 | 9.1 | 56.7 | 29.1 | 24.5 | 1.9 |
| Closed pack (27) | 18.6 | 10.9 | 7.7 | 41.3 | 29.9 | 19.3 | 9.4 | 1.5 |
| MIZ Ross Sea (28) | 35.3 | 26.9 | 8.4 | 23.7 | 42.7 | 20.6 | 12.9 | 2.1 |
| Closed pack (29) | 24.5 | 18.2 | 6.3 | 25.7 | 37.1 | 24.6 | 12.6 | 1.5 |
| MIZ BTN (30) | 27.5 | 20.6 | 6.9 | 25.1 | 45.6 | 14.9 | 14.4 | 3.1 |

## Discussion

### Space Distribution of POM and Retreating Ice Pack

In the Ross Sea, the distributions of particulate organic matter and algal biomass showed similar patterns, with higher POC and PON concentrations in stations where photosynthetic processes were more intense (Saggiomo et al., this Vol.), confirming previous observations in the Southern Ocean (Nelson et al. 1989; Lancelot et al. 1991). In the surface layer, the C:N ratio generally ranged between 5 and 6.5, except for the stations with the lowest concentrations of particulate organic matter (C:N >7); these values were similar to those measured by other authors in the euphotic layer of the Ross Sea and in other areas of the Antarctic Ocean (Bodungen et al. 1986; Nelson et al. 1989; Fabiano et al. 1993; Smith et al. 1996). The intercept of regression analyses between POC and PON concentrations (Fig. 5) is thought to be an estimate of the detrital contribution (Smith et al. 1996). The intercept values, related to POC average concentrations, suggest a larger detrital fraction of organic carbon in the pack ice area (>30%), where algal biomass was relatively low. In the MIZ areas, where phytoplankton activity was higher, the intercept value suggests a detrital fraction amount less than 15%. Moreover, C-PRT was the prevalent fraction (from 40 to 61%) of POC in the surface layer. Considering that proteins commonly account for 35–68% of organic carbon in phytoplankton (Parsons et al. 1961), these values suggest that the particulate organic matter in the Ross Sea was mainly autocthonous and of phytoplankton origin (Bodungen et al. 1986; Nelson and Smith 1986; Nelson et al. 1987; Treguer and Jacques 1992; Fabiano et al. 1993). Consequently, the biochemical composition of particulate organic matter was expected to depend largely upon the phytoplankton community.

Phytoplankton blooms are limited in space and time and strongly associated with the receding ice-edge process (Smith and Nelson 1985, 1990; Nelson and Smith 1991) and consequently to the increase in water stability and to the release of epontic algae and organic matter (Fukuchi et al. 1984; Catalano et al. 1997). In Antarctic waters, the phytoplankton is generally dominated by diatoms (Knox 1994), but relevant blooms of Prymnesiophyceae and Cryptophyceae are observed in the offshore marginal ice-edge zone and in shelf waters (Jacques and Panouse 1991; Wassman 1994; Di Tullio and Smith 1996). Even though our biochemical approach was too coarse to resolve

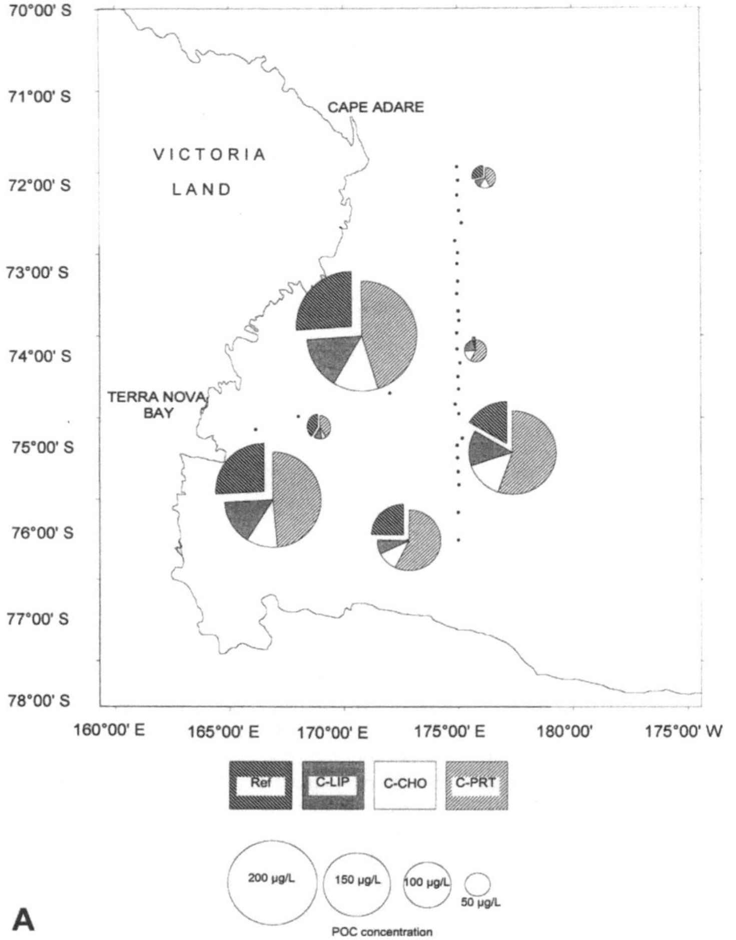


**Fig. 4A,B.** Biochemical characterization of POM in different areas of Ross Sea surface layer (**A**) and deep layer (**B**). Particulate organic carbon concentrations (*POC*, $\mu g l^{-1}$), percentage of particulate proteins (*C-PRT*) carbohydrates (*C-CHO*), lipids (*C-LIP*) and refractory fraction (*Ref*) vs. POC

differences in phytoplankton community structure, the biochemical composition of particulate organic matter indicates that stations showing high POM concentrations (in polynya and MIZ areas, for example stations MP, 8 and 28, Table 3), where Prymnesiophyceae (*Phaeocystis* sp.) were prevalent (>80%, Marino and Cabrini 1997), showed a large protein fraction on POC (>55%) while the percentage of proteins decreased (for example stations 10 and 30, Table 3) to about 50% where high POM concentrations were coupled with high diatoms abundances (>60% Marino and Cabrini 1997). It is well known, in fact, that carbohydrates linked to *Phaeocystis* rapidly decrease during the degradation process, while carbohydrates linked to diatoms are better retained within their cellular matrix (Hitchcock 1982, Moal et al. 1987). Moreover, although diatoms typically display relatively low POC: Chl a, polar diatoms have been observed as having large accumulation

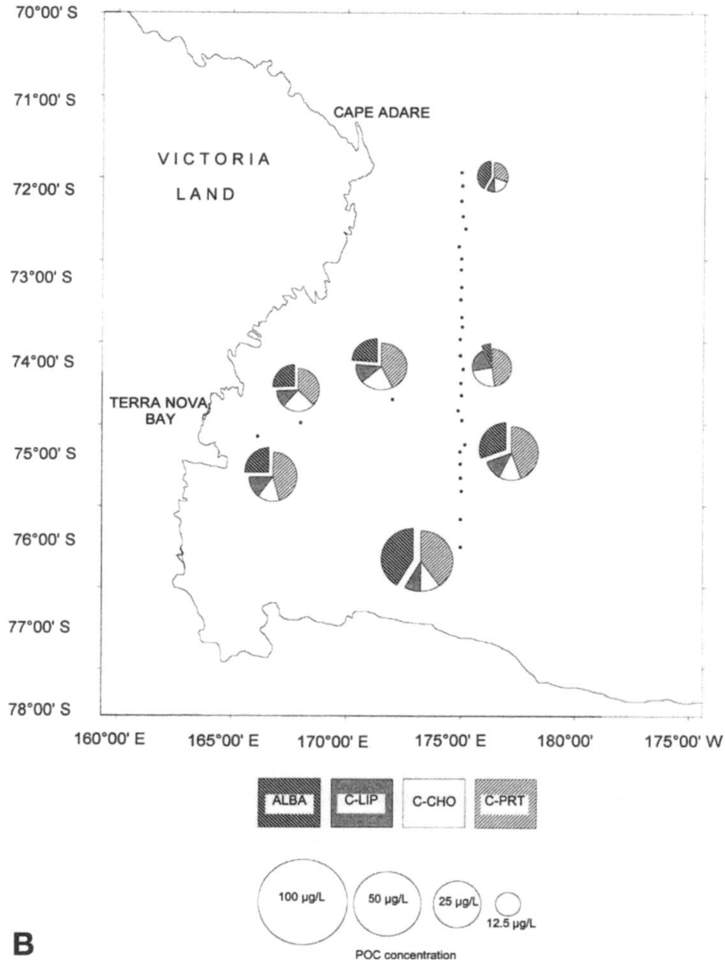


**Fig. 4A,B.** *Continued*

of lipids, partially related to temperature effects by altering the allocation of carbon into storage products, such as cytoplasmic lipid droplets and carbohydrate reserves, thus increasing the POC:Chl a ratio (Di Tullio and Smith 1996). In contrast, *P. antarctica* has no storage lipids, although it does have polysaccaride accumulation (Lancelot and Mathot 1985).

In the stations where algal biomass was low (transects 3, 4 and 5, Carrada et al. 1997), leading to POM concentrations above $100\,\mu g\,C\,l^{-1}$, the protein content decreased to percentages above 45% (for example stations 27 and 29 placed in closed pack ice areas, Table 3). Moreover, in areas where melting had removed the whole pack coverage and POM concentrations were low, like station 11b zone (Table 3), heterotrophic activities may have led to an increase of protein contribution to 60%.

In conclusion, a spatial distribution of particulate organic matter in Antarctic water, strongly influenced by melting ice pack, was identified. POM biochemical composition was related to phytoplankton biomass. Near the pack ice the

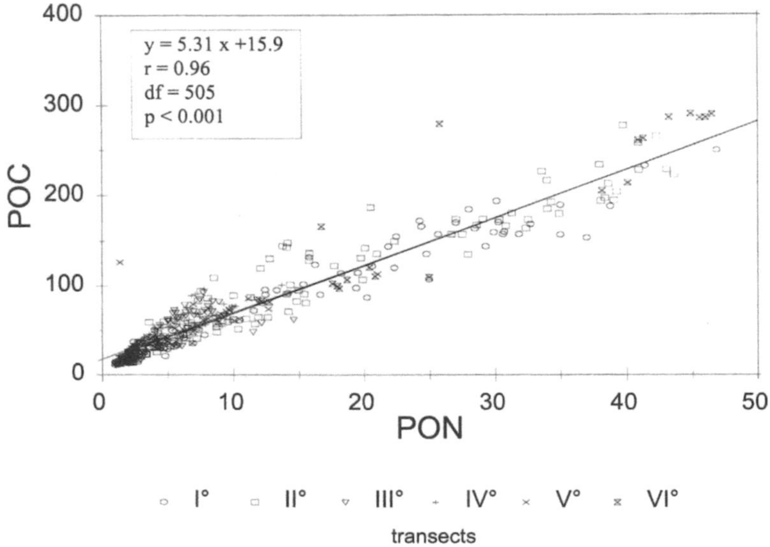


**Fig. 5.** Relationship between particulate organic nitrogen (*PON*, $\mu g l^{-1}$) and particulate organic carbon (*POC*, $\mu g l^{-1}$)

concentrations were quite high and POM was mainly detrital, characterized by higher C:N ratio and low PRT:CHO ratio. At 1 or 2° of latitude in distance (stations 8 and 10) and inside Terra Nova Bay (station 30), concentrations of POM and of its different components strongly increased. Organic matter derived from *Phaeocystis* blooms was enriched with proteins, leading to the PRT:CHO ratio increase.

## Vertical Distribution and POM Degradation

A typical feature of the particulate organic matter in Antarctic waters is the large living algal component within the photic layer. With increasing depth, POM becomes mostly composed of dead phytoplankton and heterotrophic organisms (Dhargalkar and Bhosle 1989; Fabiano et al. 1993; Knox 1994). The vertical profiles of TSM, POC, PON, C-CHO, C-PRT and C-LIP concentrations decreased with depth, showing maxima in the surface layer and low and constant values below (Table 3, Figs. 2, 3 and 4).

Within the water column, particulate organic matter was subject to hydrolysis by extracellular enzymes (Delille and Bouvy 1989; Vetter and Deming 1994). Higher values of ectoenzymatic activity were found in the surface layer above the POM concentration decrease, according to Chróst (1990), who pointed out that the highest enzymatic activity occurred in relation to products generated by phytoplankton bloom breakdown, when the water contained a large amount of particulate and dissolved organic matter. Leucine aminopeptidase, supposed to be one of the most useful enzymes for cellular energy balance (Rosso and Azam 1987), showed a clear quantitative predominance on β-glucosidase and β-N-acetylglucosaminidase, although in some areas the lowering of glycolytic $K_m$ may cause a more efficient exploitation of glucidic organic matter (Table 2). As previously reported (Christian and Karl 1992; Vetter and Deming 1994), microbial populations responded quickly to variable organic input, generating an extremely patchy distribution of enzymatic activity in space and time. In our study, the relationship between ectoenzymes and related substrates was generally characterized by a significantly positive correlation (Fig. 6). The quick response of the microbial community to POM availability, in the surface layer of the Ross Sea, finally resulted in depletion of POM amounts (the decrease from

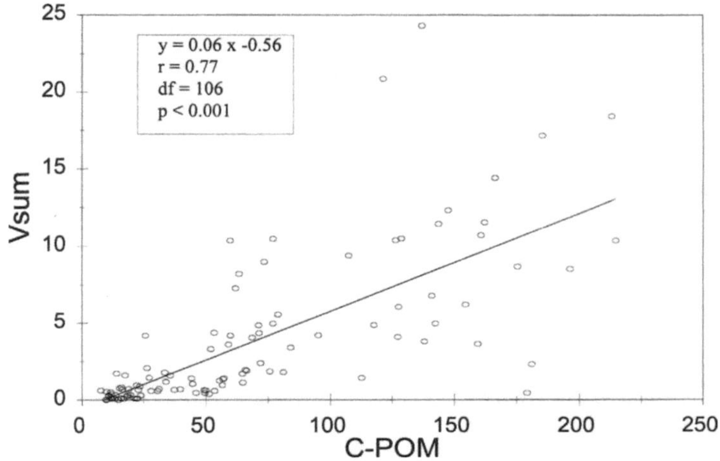


**Fig. 6.** Relationship between ectoenzymatic activity (*Vsum* sum of BG, GA and LA $V_{max}$, nmol $l^{-1}$ $h^{-1}$) and substrate availability (*C-POM*, µg C $l^{-1}$)

surface layer values was higher than 80%) in deep water.

Quantitative organic matter depletion was high mainly in the stations with higher C-POM concentrations (Fig. 4A,B). As regards the biochemical composition of POM, the lipid fraction contribution to POC remained quite constant in the whole water column, while a decrease in the percentage of protein fraction was observed, together with a slight increase of carbohydrate contribution. This trend is consistent with the results of Fabiano et al. (1993) in the Ross Sea, showing that the average value of the PRT:CHO ratio decreased with depth, and with the observations of Dawson and Liebenzeit (1982), who found that extractable sugars were lost less rapidly than cellular amino acids. Moreover, these results are consistent with previous observations by Bodungen et al. (1986) and Müller et al. (1986), indicating a preferential nitrogen loss during particle sedimentation. These observations confirm that, in the Antarctic Ocean, nitrogen recycling from sinking organic material is more rapid than carbon recycling (Treguer et al. 1990).

In conclusion we agree with other authors (Gillespie et al. 1976; Cota et al. 1990; Sullivan et al. 1990; Knox 1994), that the microbial populations of Ross Sea waters are able to respond to phytoplankton blooms utilizing and mineralizing a large part of available organic matter. Although some information on the amount of organic matter that reaches the sea bottom is available (e.g. DeMaster et al. 1992; Fabiano et al., this Vol.), the organic matter produced in the photic layer is likely to be mainly recycled within the water column (Nelson et al. 1996). Moreover, benthic microbial communities are known to decompose about 95% of the organic matter that reaches the sea floor in the upper sediments (Nelson et al. 1996). Nevertheless, the variation of PRT:CHO ratio from the surface layer (PRT:CHO ratio about 4.4 on average; Table 3) to the bottom layer (PRT:CHO ratio about 2.5 on average; Table 3) indicates a decrease in organic matter quality, thus contributing to increase the preservation efficiency of carbon in deep waters. Thus, unlike the shallow waters of Terra Nova Bay (Fabiano et al. 1997), the open waters of the Ross Sea in spring can be considered a retentive system sensu Peinert et al. (1989), where most of the organic matter is recycled and modified biochemically within the photic layer of the water column and therefore only a small portion reaches the bottom water layer (5–10% of surface values; Fig. 7) and becomes available for benthic organisms, although, owing to its refractory features, it is a limited source of food for Ross Sea benthic consumers.

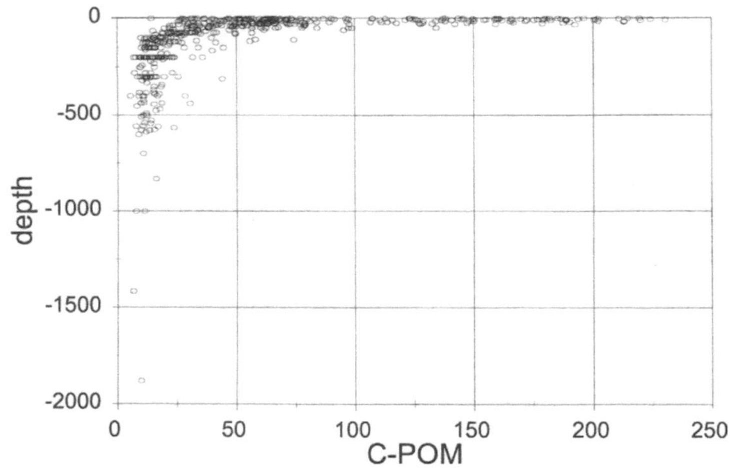


**Fig. 7.** *C-POM* distribution ($\mu g l^{-1}$) vs. depth (m) in the Ross Sea

## Time Changes of POM in Relation to Melting Ice

Our sampling strategy related to ice-edge receding and to-ice cover transformations (closed ice areas, MIZ, drifting ice and polynya areas) allowed the identification of both temporal and spatial development of biological processes. It is well known, in fact, that ice-melting processes are the starting point of particulate organic matter production (Nelson and Smith 1986) and of its biochemical variations (Fabiano et al. 1993).

Based upon the qualitative and quantitative composition of suspended matter, Fabiano et al. (1993) tried to define different time sequences occurring in the photic layer during the austral summer in relation to ice-melting processes (Table 4)

At the beginning of the melting process, high organic matter concentrations (C-POM and POC) due to epontic algae release from pack was observed (phase 1), followed by organic material accumulation in the surface layer, related to strong stratification of the water column, and by the increase of the protein fraction (phase 2). The following phase 3, when heterotrophy became the main feature of the system, showed a decrease of organic matter concentrations and the exploitation of the more labile resources of POM.

These situations were compared with the open waters of the Southern Pacific Ocean. C-POM and POC concentrations were relatively low, mainly detrital and characterized by a large carbohydrate fraction, as suggested by the low PRT:CHO ratio and high POC:Chl a ratio.

**Table 4.** Quantitative and qualitative composition of particulate organic matter related to ice-melting evolution in Ross Sea (after Fabiano et al. 1993). Particulate organic carbon (POC, $\mu g l^{-1}$), particulate labile fraction (C-POM, $\mu g l^{-1}$) particulate organic carbon and chlorophyll *a* ratio (POC:Chl a), particulate organic carbon and nitrogen ratio (C:N), particulate proteins and carbohydrates ratio (PRT:CHO)

| | POC | C-POM | POC:Chl a | C:N | PRT:CHO |
|---|---|---|---|---|---|
| Phase 1 | 428.9 | 220.3 | 108 | 6.7 | 1.5 |
| Phase 2 | 471.6 | 330.9 | 108 | 5.6 | 2.4 |
| Phase 3 | 108.1 | 47.3 | 197 | 6.8 | 1.7 |
| Phase 0 | 102.5 | 67.0 | 450 | 8.1 | 0.5 |

Our data may be evaluated within the frame of this scenario, although the different sampling periods must be considered. The 1989–90 PNRA cruise reported by Fabiano et al. (1993) was carried out later in the season (from 23 November 1989 to 07 January 1990), thus the data are only partly overlapping. Our data pointed out three quite homogeneous groups of stations, depending on ice-melting evolution (Table 3).

The first group (northernmost stations, 24 to 27, and 29 of transect 6) was time-located before the beginning of the ice melting. The second (southernmost stations, MP to 11) and third groups (mid Ross Sea, stations 12 to 23) represent two moments of the same phase related to relevant melting processes. The former, located in the polynya area, was characterized by an organic matter accumulation due to a developed bloom; the latter, less homogeneous, represented an early bloom next to the pack ice. Consequently, the melting phase, according to Fabiano et al. (1993), was a more complex period, when diatom and Prymnesiophyceae bloom succession affected POM biochemical composition.

A more synoptic investigation and longer sampling periods should allow for the identification of the entire system of phases. Moreover, complex POM spatial distribution and time changes make the system difficult to understand. However, this description, confirmed in different years, gives a reasonable picture of POM distribution and evolution in Antarctic waters.

*Acknowledgements.* This work was supported by the Progetto Nazionale Ricerche in Antartide (PNRA) funded by the Italian Government. The authors are indebted to the scientific staff of the project, who contributed suggestions and comments.

## References

Azam F, Smith DC, Hollibaugh JT (1991) The role of microbial loop in Antarctic pelagic ecosystem. Polar Res 10: 239–243

Berdalet E, Vaqué D, Arin L, Estrada M, Alcaraz M, Fernandez J (1997) Hydrography and biochemical indicators of microplankton biomass in the Bransfield Strait (Antarctica) during January 1994. Polar Biol 17: 31–38

Bligh EG, Dyer W (1959) A rapid method for total lipid extraction and purification. Can J Biochem Physiol 37: 911–917

Bodungen BV, Smetacek VS, Tilzer MM, Zeitschel B (1986) Primary production and sedimentation during spring in the Antarctic Peninsula region. Deep Sea Res 33 (2): 177–194

Bölter M, Dawson R (1982) Heterotrophic utilization of biochemical compounds in antarctic waters. Neth J Sea Res 16: 315–332

Burney CM, Davis PG, Johnson KM, Sieburth J (1979) Dependence of dissolved carbohydrate concentrations upon small-scale nanoplankton and bacterioplankton distributions in the western Sargasso Sea. Mar Biol 65: 289–296

Cadèe GC (1992) Organic carbon in the upper layer and its sedimentation during the ice-retreat period in the Scotia-Weddel Sea, 1988. Polar Biol 12: 253–259

Catalano G, Povero P, Fabiano M, Benedetti F, Goffart A (1997) Nutrient utilisation and particulate organic matter changes during summer in the upper mixed layer (Ross Sea, Antarctica) Deep Sea Res 44: 97–112

Carrada GC, Mangoni O, Russo GF, Saggiomo V, Sgrosso S (1997) Spatial and temporal distribution of phytoplankton pigments in the Ross Sea during austral spring 1994. In: Faranda F, Guglielmo L, Povero P (eds) National Programme for Antarctic Research, ROSSMIZE 1993–95, Data Rep Part I, pp 197–228

Christian JR, Karl DM (1992) Exocellular enzyme activities in Gerlach Strait, Antarctica Antarct J US 27: 170–171

Christian JR, Karl DM (1995) Bacterial ectoenzymes in marine waters: activity ratios and temperature responses in three oceanographic provinces Limnol Ocean 40: 1042–1049

Chróst RJ (1990) Microbial ectoenzymes in aquatic environments. In: Overbeck J, Chróst RJ (eds) Aquatic microbial ecology. Biochemical and molecular approaches. Springer, Berlin Heidelberg New York, pp 47–78

Chróst RJ (1991) Environmental control of the synthesis and activity of aquatic microbial ectoenzymes. In: Chróst RJ (ed) Microbial enzymes in aquatic environments. Springer, Berlin Heidelberg New York, pp 29–59

Chróst RJ, Velimirov B (1991) Measurement of enzyme kinetics in water samples: effect of freezing and soluble stabilizer. Mar Ecol Prog Ser 70: 93–100

Chróst RJ, Münster U, Rai H, Albrecht D, Witzel PK, Overbeck J (1989) Photosynthetic production and exoenzymatic degradation of organic matter in the euphotic zone of a eutrophic lake. J Plankton Res 11: 223–242

Cota GF, Kottmeier ST, Robinson DH, Smith WO Jr, Sullivan CW (1990) Bacterioplankton in the marginal ice zone of the Weddell Sea: biomass, production and metabolic activities during the austral autumn. Deep Sea Res 37: 1145–1167

Dawson R, Liebenzeit G (1982) The determination of amino acids and carbohydrates. In: Grasshoff K (ed) Methods of seawater analysis, 2nd edn. VCh, Weinheim, pp 319–345

Dayton PK (1990) Polar Benthos. In: Smith WO Jr (ed) Polar oceanography, part B. Academic Press, London, pp 637–676

Delille D, Bouvy M (1989) Bacterial responses to natural organic inputs in a marine sub-Antarctic area. Hydrobiologia 182: 225–238

DeMaster DJ, Dunbar LI, Gordon LI, Leventer AR, Morrison J, Nelson DM, Nittrouer CA, Smith WO Jr (1992) The cycling and accumulation of biogenic silica and organic matter in high latitude environments: the Ross Sea. Oceanography 5: 146–153

Dhargalkar VK, Bhosle NB (1989) Particulate organic matter in shelf waters of Princesse Astrid Kyst, Antarctica. Polar Rec 25: 229–234

Di Tullio GR, Smith WO Jr (1996) Spatial patterns in phytoplankton biomass and pigment distributions in the Ross Sea. J Geophys Res 101 (C8): 467–477

Dubois M, Gilles K, Hamilton JK, Rebers PA, Smith F (1956) Colorimetric method for determination of sugars and related substances. Anal Chem 28: 350–356

Dunbar RB, MacPerson AJ, Wefer G (1985) Water-column particulate flux and sea floor deposit in the Bransfield Strait and southern Ross Sea, Antarctica. Antarct J US 20 (5): 98–102

El-Sayed SZ (1988) Seasonal and interannual variabilities in Antarctic phytoplankton with reference to krill distribution. In: Sahrhage D (ed) Antarctic ocean and resources variability. Springer, Berlin Heidelberg New York, pp 101–118

El-Sayed SZ, Taguchi S (1981) Primary production and the standing crop of phytoplankton along the ice edge in the Weddel Sea. Deep Sea Res 28: 1017–1032

Fabiano M, Povero P, Danovaro R (1993) Distribution and composition of the particulate organic matter of the Ross Sea (Antarctica). Polar Biol 13: 525–533

Fabiano M, Povero P, Danovaro R (1996) Particulate organic matter composition in Terra Nova Bay (Ross Sea, Antarctica) during summer. Antarct Sci 8 (1): 7–13

Fabiano M, Chiantore M, Povero P, Cattaneo-Vietti R, Pusceddu A, Misic C, Albertelli G (1997) Short-term variations in particulate matter flux in Terra Nova Bay, Ross Sea. Antarct Sci 9 (2): 143–149

Fukuchi M, Tanimura A, Ohtsuka H (1984) Seasonal change of chlorophyll a under fast ice in Lutzow-Holm Bay, Antarctica. Mem Natl Inst Polar Res Special Issue 32: 51–59

Gillespie PA, Morita RY, Jones LP (1976) The heterotrophic activity for amino acids, glucose and acetate in Antarctic waters. J Oceanogr Soc Jpn 32: 74–82

Gonzales HE (1992) The distribution and the abundance of krill faecal material and oval pellets in the Scotia and Weddel Seas (Antarctica) and their role in particle flux. Polar Biol 12: 81–91

Handa N, Nakatsuka T, Fukuchi M, Hattori H, Hoshiai T (1992) Vertical fluxes and ecologial significance of organic materials durng the phytoplankton bloom during austral summer in Breid Bay, Antarctica. Mar Biol 112 (3): 469–478

Hartree EF (1972) Determination of proteins: a modification of the Lowry method that gives a linear photometric response. Anal Biochem 48: 422–427

Hedges JI, Stern JH (1984) Carbon and nitrogen determination of carbonate-containing solids. Limnol Oceanogr 29: 657–663

Hitchcock GL (1982) A comparative study of the size-dependent organic composition of marine diatoms and dinoflagellates. J Plankton Res 4 (2): 363–377

Innamorati M, Lazzara L, Mori G, Nuccio G, Saggiomo V (1991) Phytoplankton ecology. National Scientific Commission for Antarctica (ed) Oceanographic Campaign, 1989–90, Data Rep Part I, pp 141–252

Jacques G (1989) Primary production in the open Antarctic Ocean during the austral summer. A review. Vie Milieu 39: 1–17

Jacques G, Panouse M (1991) Biomass and composition of size-fractionated phytoplankton in the Weddel Sea Confluence area. Polar Biol 11: 315–328

Karl DM, Tilbrook BD, Tien G (1991) Seasonal coupling of organic matter production and particle flux in the western Bransfield Strait, Antarctica. Deep Sea Res 38 (879): 1097–1126

Knox GA (1994) The biology of the Southern Ocean. Cambridge University Press, Cambridge

Lancelot C, Mathot S (1985) Biochemical fractionation of primary production by phytoplankton in Belgian coastal waters during short- and long-term incubations with $^{14}C$-bicarbonate, II, *Phaeocystis pouchetii* in colonial populations. Mar Biol 86: 227–232

Lancelot C, Veth C, Mathot S (1991) Modelling ice-edge phytoplankton bloom in the Scotia-Weddell sea sector of the Southern Ocean during spring (1988) J Mar Syst 2: 333–346

Liebezeit G, Bölter M (1986) Distribution of particulate amino acids in the Bransfield Strait. Polar Biol 5: 199–206

Marino D, Cabrini M (1997) Distribution and succession of phytoplankton population in the Ross Sea (Antarctica) during the austral spring (1994) In: Faranda F, Guglielmo L, Povero P (eds) National Programme Antarctic Research, ROSSMIZE 93–95, Data Rep Part I, pp 307–342

Marsh JB, Weinstein WJ (1966) A simple charring method for determination of lipids. J Lipids Res 7: 574–576

Matsuda O, Ishikawa S, Kawaguchi K (1990) Seasonal variation of particulate organic matter under the Antarctic fast ice and its importance to benthic life. In: Kerry KR, Hempel G (eds) Antarctic ecosystems, ecological change and conservation. Springer Berlin Heidelberg New York, pp 143–148

Moal J, Martin Jezequel V, Harris RP, Samain J-F, Poulet S (1987) Interspecific and intraspecific variability of the chemical composition of marine phytoplankton Oceanol Acta 10 (3): 339–346

Müller PJ, Suess E, Ungerer A (1986) Amino acids and amino sugars of surface particulate and sediment trap material from waters of the Scotia Sea. Deep Sea Res 33: 819–838

Nedwell DB, Walker TR, Ellis-Evans JC, Clarke A (1993) Measurement of seasonal rates and annual budget of organic carbon fluxes in an Antarctic coastal environment at Signy Island, South Orkney Islands, suggest a broad balance between production and decomposition. Appl Environ Microbiol 59: 3898–3995

Nelson DM, Smith WO Jr (1986) Phytoplankton bloom dynamics of the western Ross Sea ice edge. II. Mesoscale cycling of nitrogen and silicon. Deep Sea Res 33 (10): 1389–1412

Nelson DM, Smith WO Jr (1991) Sverdrup revisited: critical depths, maximum chlorophyll levels and the control of Southern Ocean productivity by irradiance-mixing regime. Limnol Oceanogr 36: 1650–1661

Nelson DM, Smith WO Jr, Gordon LI, Huber BA (1987) Spring distributions of density, nutrients, and phytoplankton biomass in the ice-edge zone of the Weddell-Scotia Sea. J Geophys Res 92: 7181–7190

Nelson DM, Smith WO Jr, Muench RD, Gordon LI, Sullivan CW, Husby DM (1989) Particulate matter and nutrient distributions in the ice-edge zone of the Weddel Sea: relationship to hydrography during late summer. Deep Sea Res 36 (2): 191–209

Nelson DM, DeMaster DJ, Dunbar RB, Smith WO Jr (1996) Cycling of organic carbon and biogenic silica in the Southern Ocean: estimates of water column and sedimentary fluxes on the Ross Sea continental shelf. J Geophys Res 101 (C8): 519–532

Parsons TR, Stephens K, Strickland JDH (1961) On the chemical composition of eleven species of marine phytoplankters. J Fish Res Board Can 18: 1001–1016

Peinert R, von Bodungen B, Smetacek V (1989) Food web structure and loss rates. In: Begrer WH, Smetacek V, Weffer G (eds) Productivity of the oceans: present and past. Dahlem Konferenzen. Wiley, New York, pp 33–48

Rosso AL, Azam F (1987) Proteolytic activity in coastal oceanic waters: depth distribution and relationship to bacterial populations. Mar Ecol Prog Ser 41: 231–240

Smith DC, Simon M, Alldredge AL, Azam F (1992) Intense hydrolytic enzyme activity on marine aggregates and implications for rapid particle dissolution. Nature 359: 139–142

Smith WO Jr, Nelson DM (1985) Phytoplankton bloom produced by a receding ice edge in the Ross Sea: spatial coherence with the density field. Science 227: 163–166
Smith WO Jr, Nelson DM (1986) The importance of ice-edge phytoplankton productivity in the Southern Ocean. Bioscience 36: 251–257
Smith WO Jr, Nelson DM (1990) Primary productivity and new production in the Weddel Sea marginal ice zone during austral spring and autumn. Limnol Oceanogr 35: 809–821
Smith WO Jr, Baumann MEM, Wilson DL, Aletsee L (1987) Phytoplankton biomass and productivity in the marginal ice zone of the Fram Strait during summer 1984. J Geophys Res 92: 6777–6786
Smith WO Jr, Nelson DM, Di Tullio GR, Leventer R (1996) Temporal and spatial patterns in the Ross Sea: phytoplankton biomass, elemental composition, productivity and growth rates. J Geophys Res 101 (C8): 455–465
Strickland JDH, Parsons TR (1972) A practical handbook of seawater analysis. Bull Fish Res Board Can 167: 1–158
Sullivan WC, Cota GF, Krempin DW, Smith WO Jr (1990) Distribution and activity of bacterioplankton in the marginal ice zone of the Weddell-Scotia Sea during austral spring. Mar Ecol Prog Ser 63: 239–252
Treguer P, Jacques G (1992) Dynamics of nutrients and phytoplankton, and fluxes of carbon, nitrogen and silicon in the Antarctic Ocean. Polar Biol 12: 149–162
Treguer P, Nelson DM, Gueneley S, Zeyons C, Morvan J, Buma A (1990) The distribution of biogenic and lithogenic silica and the composition of particulate organic matter in the Scotia Sea and the Drake Passage during autumn 1987. Deep Sea Res 37 (5): 833–851
Unanue M, Aua L, Barcina I, Egea L, Iriberri J (1993) Size distribution of aminopeptidase activity and bacterial incorporation of dissolved substrates in three aquatic ecosystems. FEMS Microbiol Ecol 102: 175–183
Vetter YA, Deming JW (1994) Extracellular enzyme activity in the Arctic northeast water polynya. Mar Ecol Prog Ser 114 (23–24): 23–34
Vrba J, Machácek J (1994) Release of extracellular β-N-acetylglucosaminidase during crustacean moulting. Limnol Oceanogr 39: 712–716
Wassman P (1994) Significance of sedimentation for termination of Phaeocystis blooms. J Mar Syst 5: 81–100

CHAPTER 12

# Coupling of Surface Layer Biological Processes and Vertical Flux in the Ross Sea

**W.O. Smith, Jr.[1], H.J. Niebauer[2], and V.L. Asper[3]**

**ABSTRACT**

Primary production in the euphotic zone results in the generation of organic particles, and these particles are exported to deeper waters via the ocean's biological pump. In the Ross Sea large blooms of phytoplankton occur, and it has been hypothesized that aggregate formation plays an important role in generating rapidly sinking particles under these conditions. To assess the coupling of surface layer biological processes and the vertical flux patterns in the Ross Sea, we used a coupled biophysical model of phytoplankton production linked to a submodel of aggregate formation. The model suggests that the process of aggregate formation can be the major component of vertical flux of blooms in the Ross Sea. Furthermore, large aggregates are responsible for much of the vertical flux, and their production lags the production of chlorophyll in the surface layer. Aggregate formation is highly dependent on the parameterization of "stickiness". The role of the aggregates in the surface layer of the Ross Sea and the coupling to vertical flux of organic matter may provide a means of understanding the patterns of vertical flux for other productive areas of the Antarctic.

## Introduction

The Ross Sea is the site of the largest and most spatially extensive phytoplankton bloom in the Southern Ocean (Sullivan et al. 1993). In November and December of each year, the Ross Sea polynya expands rapidly and the haptophyte *Phaeocystis antarctica* reaches massive standing stocks (Smith and Gordon 1997). Chlorophyll concentrations in this bloom range from 10–20 $\mu g\,l^{-1}$ (Comiso et al. 1993; Arrigo and McClain 1994; Smith and Gordon 1997), but generally decline to less than 2 $\mu g\,l^{-1}$ by late January (Smith et al. 1996). The causes of the bloom's demise are uncertain, but it has been suggested that aggregate formation (one component of the oceanic biological pump or transfer of organic matter to depth) and accelerated vertical flux of larger particles is a primary cause of the particulate matter decrease (Wassmann 1994). Aggregates have been detected in deep (250 and 550 m) sediment traps in the Ross Sea (DeMaster et al. 1992); furthermore, the appearance of these particles in the traps relative to their production at the surface is consistent with the expected particle settling velocities (Smith and Dunbar 1998).

*Phaeocystis antarctica* is unusual among phytoplankton because it forms large (>1-mm) colonies. These colonies have a mucoid sheath that largely consists of polysaccharides, and it has been estimated that under some conditions the sheath comprises up to 90% of the total organic matter in the colonies (Thingstad and Billen 1994). Although the sheath might be expected to be sticky (i.e. promote the adherence of two particles upon collision), experimental evidence suggests the contrary: that *Phaeocystis* colonies are not extraordinarily sticky and do not promote the formation of aggregates any more than diatoms do (Passow and Wassmann 1994). However, because the production of aggregates depends not only on stickiness but also on particle abundance and physical conditions (such as differential shear), aggregate formation can lead to enhanced settling velocities and increased vertical flux rates during *P. antarctica* blooms. As such, the process of

[1] Virginia Institute of Marine Science, Greate Road, Box 1346, Gloucester Pt., Virginia 23062, USA
[2] Department of Atmospheric and Oceanic Sciences, University of Wisconsin, Madison, Wisconsin 53706, USA
[3] Institute of Marine Sciences University of Southern Mississippi, Stennis Space Center, Mississippi 39529, USA

aggregate genesis can be an important component of the marine carbon cycle of some systems.

Phytoplankton blooms can be modeled adequately if the initial conditions and transfer coefficients of the important processes are known. In polar oceans models of phytoplankton growth are complex, in that they must include air-sea-ice interactions and changes in solar radiation induced by ice and extreme solar angles. Regardless, a number of models have successfully simulated the growth of phytoplankton and the transformation of the organic matter produced in polar systems (e.g. Slagstad 1985; Niebauer and Smith 1989; Walsh and Dieterle 1994). Similarly, the process of coagulation and the formation of aggregates from smaller particles has been treated theoretically (e.g. Jackson 1990; Riebesell and Wolf-Gladrow 1992; Logan et al. 1995). The generation of aggregates is known to be dependent on a number of biological and physical processes such as shear, turbulence, interparticle collision probabilities, and particle stickiness. However, incorporation of the aggregation process into models depicting mesoscale oceanographic processes has not been attempted.

## Materials and Methods

We took a previously developed coupled biophysical model (which included air-sea-ice interactions; Niebauer and Smith 1989) in which phytoplankton growth was a function of temperature, irradiance, nutrient concentrations (nitrate and ammonium), and losses due to micro- and mesozooplankton grazing. The earlier model had two size classes, but we increased the number of size classes to ten in this formulation. We also included a function for the production of aggregates from the various size classes based on the equations of Jackson (1990). A particle can increase in size by sticking to another, and the production of a larger particle is a function of particle stickiness and total number of particles per unit volume. Stickiness ($\alpha$) was modeled as a function of absolute and relative growth rate (see values for Particle Stickiness, below). All maximum growth rates were calculated from the temperature-growth relationship of Eppley (1972). The equations were solved using a finite-difference method and the simulation run for 84 days.

### Initial Conditions

Initial conditions for the nutrient and hydrographic conditions were taken from early spring in the Ross Sea polynya (Smith and Gordon 1997). The model began on October 15 (and ran through January 7), and nitrate concentrations throughout the water column initially were 30 μM. Ice thickness was 30 cm, and it completely melted by November 29. Winds were initially modest ($6\,m\,s^{-1}$) and were reduced to $0.1\,m\,s^{-1}$ on November 24. The mixed layer depth was 30 m, and the change in density at that depth was 0.3 sigma-t unit. Initial water temperatures were −1.7 °C. Total chlorophyll concentrations were set at $0.1\,mg\,m^{-3}$ and distributed evenly in all size classes. Both micro- and mesozooplankton grazing were minimized, similar to what is found in the Ross Sea during this period.

We calculated the growth or losses of each of the ten size classes in terms of chlorophyll or particulate nitrogen ($mg\,m^{-3}$). After growth and losses were calculated, biomass was converted into cell numbers, and aggregation was then calculated. We assumed that $0.01\,mg\,m^{-3}$ chlorophyll equaled $10^5$ cells $m^{-3}$, and hence if there were 30 cells aggregate$^{-1}$, then $0.01\,mg\,m^{-3}$ chlorophyll = $3.3 \times 10^3$ aggregates $m^{-3}$. We chose 30 cells aggregate$^{-1}$ as the maximum number of cells per aggregate in the smallest size class, so that the largest aggregate consisted of 300 cells.

### Aggregation Submodel

To model aggregation, we started with the equations of Jackson (1990) and Riebesell and Wolf-Gladrow (1992) in modeling algal flocs through physical coagulation processes:

$$\frac{dC_i}{dt} = \frac{1}{2}\mu_{i/2}(1-b_{1/2})C_{i/2} - \mu_i C_i + 2\mu_i b_i C_i + \alpha\sum_{j-1}^{1/2}\beta_{i-j}C_j C_{i-j} - dC_i\sum_{j-1}^{i_{max}}(1+\delta_{i,j})\beta_{i,j}C_j - \frac{w_i}{H}C_i, \quad (1)$$

where i and j are the number of cells per aggregate, $C_i$ and $C_j$ are concentrations of aggregates of size i and j, μ is phytoplankton growth rate, b is the probability of aggregate disruption immediately after doubling, α is the probability of aggregates sticking upon contact (stickiness), $\delta$ is the Kronecker delta, $\beta$ the collision kernel, $w_i$ the size-specific sinking rate, and H is the depth of the upper mixed layer (Jackson 1990). The first three terms represent growth and are dependent on the phytoplankton growth rate. The summation terms represent the gain in aggregate concentration by coagulation of two smaller particles, as well as the loss by aggregation to larger particles, respectively. The last term is the loss of aggregates to vertical sinking. In our model, phytoplankton growth and sinking are handled differently than in Jackson (1990), but gains and losses of particles through aggregation are explicitly determined. The concentration changes are solved at every grid point in the model at every time step for each particle or aggregate size class.

After the numbers of particles are calculated in each size class, we then chose a cell diameter of $d_1 = 10\,\mu m$ from which we calculate the diameters of the aggregates, $d_i$:

$$d_i = d_1 * I^{0.44}, \tag{2}$$

where $d_i$ = aggregate (of size i) diameter and $d_1$= cell diameter (Jackson 1990). Once $d_i$ is calculated, the aggregate settling velocity or particle sinking rate, $w_i$, is calculated:

$$w_i = 2.29 \times 10^{-7} * d_i^{1.17}. \tag{3}$$

For 30 to 300 cells per aggregate, the aggregate diameters range from 44.7 to 123.0 µm, assuming a cell diameter of 10 µm. The calculated aggregate settling velocities then range from 1.7 to 5.5 m day$^{-1}$. This maximum sinking rate of 5.5 m day$^{-1}$ is similar to that calculated from sediment trap data in the Ross Sea (Smith and Dunbar 1998), although it is more than an order of magnitude less than the maximum rates reported for similar sized particles (ca. ~100 m day$^{-1}$; e.g. Alldredge and Gottshalk 1988). Instead of using the sinking term explicitly, we calculated the vertical chlorophyll flux at each grid point by adding the sinking rate to the vertical advection term before calculating the flux.

The collision kernel ($\beta$) consists of two parts. The first is due to shear flow ($\beta^s$) and the second due to differential settling ($\beta^d$). That due to shear flow can be written as

$$\beta^s_{i,j} = 0.136 d^3_{i,j} * EC_{i,j}, \tag{4}$$

where 0.136 is a constant that depends upon the energy dissipation rate and kinematic viscosity of water, $d_{i,j}$ is the sum of aggregate diameters, and $EC_{i,j}$ is the contact efficiency or the probability that two aggregates will come into contact.

The part of the collision kernel due to differential settling can be written

$$\beta^s_{i,j} = \pi \frac{d^{-}_{i,j}}{4} |w_i - w_j| (EC_{i,j} + ED_{i,j}), \tag{5}$$

where $w_i$ and $w_j$ are the sinking speeds of particles with i and j cells, and $ED_{i,j}$ is the diffusion efficiency (McCave 1975), which we neglected because diffusion is negligible for particles ⩾1 µm. $EC_{i,j}$ is the theoretical contact efficiency

$$EC_{i,j}(p) = EC_0\left(-\frac{0.1}{p}\right), \tag{6}$$

where $p = d_{min}/d_{max}$ and $EC_0 = 1/\exp(-0.1)$. Finally, we calculate the collision kernel as

$$\beta_{i,j} = (\beta^s_{i,j} + \beta^d_{i,j}). \tag{7}$$

The change in particle concentrations, as influenced by the aggregation terms in Eq. (1) (i.e. the fourth and fifth terms), are calculated and integrated over the time interval of interest yielding concentrations of aggregates. These aggregate concentrations ($C_i$) are proportionally adjusted so that the number of cells before and after the aggregation process are conserved. Finally, these new aggregate concentrations are converted back into chlorophyll and used in the advection, diffusion, and growth subroutines of the model at the next time step prior to further aggregation.

## Values for Particle Stickiness

To complete the calculation of aggregation terms, we multiply by a dimension-less particle stickiness, $\alpha$ (e.g. Jackson 1990; Logan et al. 1995). The range of natural phytoplankton stickiness is poorly known. In the laboratory Kiørboe et al. (1994) measured values of 0.001–0.1, which showed some dependence on nutrient concentrations. In marine snow aggregates, whose density is greater than phytoplankton, stickiness values approached 0.9 (Alldredge and McGillivary 1991). We completed a number of sensitivity studies using various values for $\alpha$, as well as modifying the functional relationship between stickiness and light, nutrients and growth. Large stickiness values (greater than 0.1) allow for a rapid generation of large aggregates and subsequent flux to depth, but surface chlorophyll values and nutrient removal rates are unrealistically low. Conversely, stickiness values less than 0.01 resulted in vertical flux rates which were unrealistically low. As a result, stickiness ($\alpha$) was modeled as a function of growth rate ($\mu$) and relative growth rate ($\mu_{rel}$; Logan et al. 1995) such that $\alpha = -0.17*\mu_{rel} + 0.06$, with the constraints such that stickiness = 0.1 for $\mu_{rel} < 0.1\,\mu$ and 0.02 for $\mu_{rel} > 0.22\,\mu$. Hence, stickiness in the model increased under conditions which reduced growth rates, such as low irradiance levels or reduced nutrient concentrations.

## Results and Discussion

The model accurately depicted the onset of the bloom in the Ross Sea. Initial conditions were similar to those observed in October and November (Smith and Gordon 1997; Smith, unpubl.), and on approximately November 15 phytoplankton biomass exceeded $1\,mg\,m^{-3}$ and reached a maximum chlorophyll concentration on December 1 ($>7\,mg\,m^{-3}$; Fig. 1). Maximum biomass has been observed via satellite imagery on December 10 in 1979 (Arrigo and McClain 1994) and by direct sampling on December 5 (Smith and Gordon 1997). The latter study also found chlorophyll concentrations similar to those predicted by the model. Modeled nitrate concentrations (Fig. 2) closely paralleled in situ observations through December 10 (Smith and Gordon 1997), but after that date continued to decrease. Nitrate concentrations in the central Ross Sea have never been observed to decrease below $10\,\mu M$, most likely due to limitation by trace elements such as iron (e.g. DiTullio and Smith 1997). Nutrient limitation by elements other than nitrogen was not included in our model, and hence the reduction in nitrate uptake could not be realistically simulated.

Vertical fluxes through 250 m were initially low ($<10\,mg\,C\,m^{-2}\,day^{-1}$), but began to increase rapidly around December 18 (day 65; Fig. 3), and continued to a maximum flux ($168\,mg\,C\,m^{-2}\,day^{-1}$) at the end of the simulation (day 84, January 7). Smith and Dunbar (1998) reported minimum fluxes of less than $2\,mg\,C\,m^{-2}\,day^{-1}$ and maximum fluxes of ca. $92\,mg\,C\,m^{-2}\,day^{-1}$, which suggests that the model overestimated the fluxes. The model does not realistically simulate remineralization below 150 m, nor does it include an explicit disaggregation term. However, the overestimate in vertical flux might be due to the excess nutrient removal and production within the bloom, and that the excess production led to the transfer of particulate matter to the larger size classes and the subsequent rapid vertical flux of large aggregates. We tested this by artificially reducing nutrient uptake when nitrate decreased to $10\,\mu M$, but even though production was greatly diminished, no effect on vertical flux was observed. This occurred because of the time delay inherent from the production of smaller particles to the generation of large, rapidly sinking aggregates, and hence the effect of reduced surface production on vertical flux was not observed during the 84 days

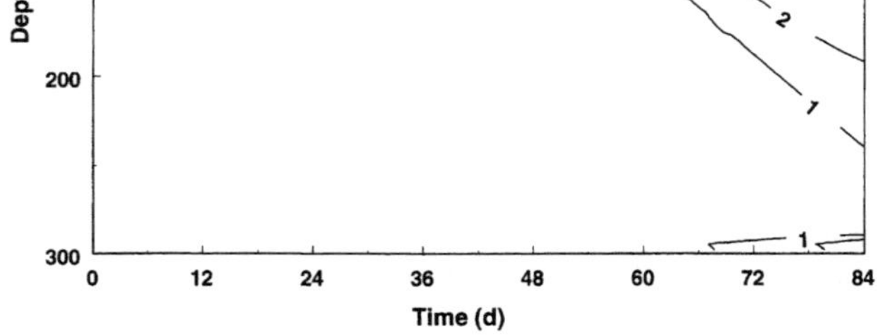


**Fig. 1.** Simulated chlorophyll concentrations in the Ross Sea polynya as a function of time and depth. *Day 0* October 15; *day 84* January 7

of the experiment. Asper and Smith (1999) also found that vertical flux of carbon at the base of the euphotic zone in January averaged 110 mg C $m^{-2}$ $day^{-1}$ (as measured by short-term deployments of floating sediment traps at 200 m), with a maximum flux of 150 mg C $m^{-2}$ $day^{-1}$. Thus, the modeled values of flux do not appear to be inconsistent with the available field data.

Larger particles contributed overwhelmingly to vertical flux as a result of their sinking rates being an exponential function of particle diameter (Table 1). This effect mirrors the results of experimental investigations of sinking rates of particles (e.g. Alldredge and Gottschalk 1988) in which larger, denser particles sank faster and represented the greatest losses of organic matter from the surface layer. Sinking of larger particles represents the greatest loss term from this size class, whereas the greatest loss term in the smaller classes is aggregation and transfer into larger particles.

The time needed to convert a small particle into a larger one was calculated from the time need for the transformation from smallest to largest particles estimated from the time differential in maxima in the two size classes. We found that the lag between the two size classes maxima was ca. 20 days, which gives an estimate of the potential rate of particle size change. We also ran a simulation in which sinking was eliminated and calculated

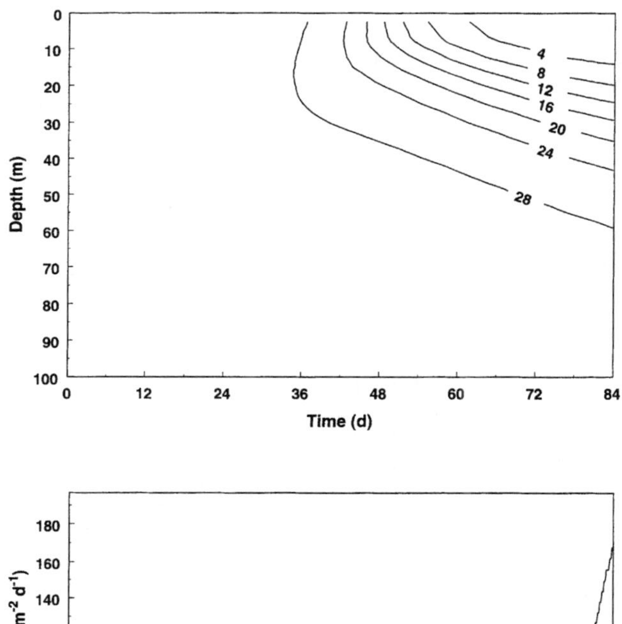


**Fig. 2.** Simulated nitrate concentrations in the Ross Sea polynya as a function of time and depth. *Day 0* October 15; *day 84* January 7

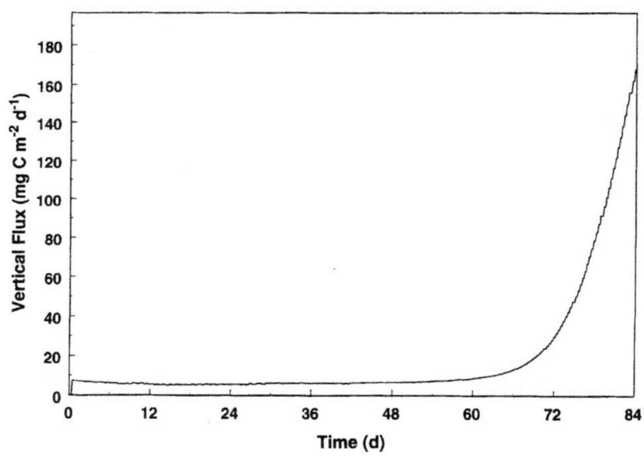


**Fig. 3.** Simulated vertical flux of organic carbon through 250 m as a function of time. *Day 0* October 15; *day 84* January 7

**Table 1.** Contribution of each size class to simulated vertical flux at 250 m on day 84

| Size class (μm) | Vertical flux ($mg\,C\,m^{-2}\,day^{-1}$) | Fraction of total flux (%) |
|---|---|---|
| 45 | 0.05 | 0.03 |
| 61 | 0.17 | 0.10 |
| 72 | 0.34 | 0.20 |
| 82 | 0.79 | 0.47 |
| 91 | 1.14 | 0.68 |
| 99 | 2.33 | 1.39 |
| 105 | 4.49 | 2.68 |
| 112 | 10.9 | 6.51 |
| 117 | 28.5 | 17.0 |
| 123 | 119 | 71.0 |

the rate of growth in each size class. Because no changes in a size class occurred other than aggregation and the generation of larger particles from smaller ones, the loss term due to aggregation can be estimated. The largest size class growth rate was 35% greater, and hence the loss rate in the smaller size classes was approximately 0.16 $day^{-1}$ when the assemblage is growing at $\mu_{max}$. Such losses could be directly compared to those of grazing to quantitatively assess the importance of the two processes to changes in the particle size spectrum.

Aggregate formation and the resultant enhanced vertical flux are clearly quantitatively important processes in the Ross Sea polynya during *Phaeocystis antarctica* blooms. Simulations show that aggregation alone can result in vertical flux rates which are similar to those measured by sediment traps, which suggests that for this region grazing and fecal pellet formation are quantitatively insignificant as components of vertical flux. Clearly the bloom's biomass is among the largest in the world's oceans, but at this extreme aggregate formation and enhanced sinking contribute significantly to the oceanic biological pump. Further studies of the system should incorporate experimental assessment of aggregate abundance and dynamics, including assessment of stickiness, to further elucidate this important aspect of the carbon dynamics in the surface layer of the Ross Sea.

*Acknowledgements.* This research was supported by grants from the National Science Foundation (OPP-9317587 to WOS, OPP-9317598 to VA) and the Office of Naval Research (to HJN). We thank C. Chu, S. Polk, and A.-M. White for their assistance.

## References

Alldredge AL, Gottschalk C (1988) In situ settling behavior of marine snow. Limnol Oceanogr 33: 339–351

Alldredge AL, McGillivary P (1991) The attachment probabilities of marine snow and their implications for particle coagulation in the ocean. Deep Sea Res 38: 431–443

Arrigo KR, McClain CR (1994) Spring phytoplankton production in the Western Ross Sea. Science 266: 261–263

Asper VL, Smith WO Jr (1999) Particle fluxes during austral spring and summer in the southern Ross Sea. J Geophys Res 104: 5345–5360

Comiso JC, McClain CR, Sullivan CW, Ryan JP, Leonard CL (1993) Coastal zone color scanner pigment concentrations in the Southern Ocean and relationships to geophysical surface features. J Geophys Res 98: 2419–2451

DeMaster DJ, Dunbar RB, Gordon LI, Leventer AR, Morrison JL, Nelson DM, Nittrouer CA, Smith WO Jr (1992) The cycling and accumulation of organic matter and biogenic silica in high-latitude environments: the Ross Sea. Oceanography 5: 146–153

DiTullio GR, Smith WO Jr (1997) Studies on dimethylsulfide in Antarctic coastal waters. In: Battaglia B, Valencia J, Walton DWH (eds) Antarctic communities: species, structure and survival. Cambridge University Press, Cambridge, pp 93–100

Eppley RW (1972) Temperature and phytoplankton growth in the sea. Fish Bull 70: 1063–1085

Jackson GA (1990) A model of the formation of marine algal flocs by physical coagulation processes. Deep Sea Res 37: 1197–1211

Kiørboe T, Lundsgaard C, Olsen M, Hansen JLS (1994) Aggregation and sedimentation processes during a spring phytoplankton bloom: a field experiment to test coagulation theory. J Mar Res 52: 297–323

Logan BE, Passow U, Alldredge AL, Grossart H, Simon M (1995) Rapid formation and sedimentation of large aggregates is predictable from coagulation rates (half-lives) of transparent exopolymer particles (TEP). Deep Sea Res 42: 203–214

McCave IN (1975) Vertical flux of particles in the ocean. Deep Sea Res 22: 491–502

Niebauer HJ, Smith WO Jr (1989) A mesoscale model of the physical-biological interactions in the Fram Strait marginal ice zone. J Geophys Res 94: 16151–16176

Passow U, Wassmann P (1994) On the trophic fate of *Phaeocystis pouchetii* (Hariot): IV. The formation of marine snow by *P. pouchetii*. Mar Ecol Prog Ser 104: 153–161

Riebesell U, Wolf-Gladrow DA (1992) The relationship between physical aggregation of phytoplankton and particle flux: a numerical model. Deep Sea Res 39: 1085–1102

Slagstad D (1985) A model of phytoplankton in the marginal sea-ice zone of the Barents Sea. In: Gray JS, Christiansen ME (eds) Marine biology of polar regions and effects of stress on marine organisms. Wiley, New York, pp 429–431

Smith WO Jr, Dunbar RB (1998) The relationship between new production and vertical flux on the Ross Sea continental shelf. J Mar Systems 17: 445–457

Smith WO Jr, Gordon LI (1997) Hyperproductivity of the Ross Sea (Antarctica) polynya during austral spring. Geophys Res Lett 24: 233–236

Smith WO Jr, Nelson DM, DiTullio GR, Leventer AR (1996) Temporal and spatial patterns in the Ross Sea: phytoplankton biomass, elemental composition, productivity and growth rates. J Geophys Res 101: 18467–18478
Sullivan CW, Arrigo KR, McClain CR, Comiso JC, Firestone J (1993) Distributions of phytoplankton blooms in the Southern Ocean. Science 262: 1832–1837
Thingstad F, Billen G (1994) Microbial degradation of *Phaeocystis* material in the water column. J Mar Syst 5: 55–65
Walsh JJ, Dieterle DA (1994) $CO_2$ cycling in the coastal ocean. I. A numerical analysis of the southeastern Bering Sea with applications to the Chukchi Sea and the northern Gulf of Mexico. Prog Oceanogr 34: 335–392
Wassmann P (1994) Significance of sedimentation for the termination of *Phaeocystis* blooms. J Mar Syst 5: 81–100

CHAPTER 13

# Bacteria, Protozoa and Organic Matter Composition in the Sediments of Terra Nova Bay (Ross Sea)

M. Fabiano[1], R. Danovaro[2,3], M. Chiantore[1], and A. Pusceddu[2]

**ABSTRACT**

Spatial distributions of bacterial and protozoan abundance and biomass were investigated in the sediments of Terra Nova Bay (Ross Sea) at depths ranging from 36 to 223 m, during January-February 1994. Microbial parameters were compared to the distribution of several food indicators (such as phytopigments, lipids, proteins, total and soluble carbohydrates) in order to provide quantitative information about the different components of the benthic microbial loop and identify which factors are responsible for microbial distribution. As suggested by the high content of chloroplastic pigment equivalents (on average 29.0 μg $g^{-1}$ dry wt), proteins (1.9 mg $g^{-1}$ dry wt) and carbohydrates (5.3 mg $g^{-1}$ dry wt), the sediments of Terra Nova Bay were characterized by large amounts of deposited primary organic material. These values are among the highest reported for coastal sediments. Sediment organic matter was mostly composed of labile compounds and 98% of the total carbohydrate content was composed of soluble carbohydrates. Benthic bacterial densities were 1 to 2 orders of magnitude lower than those previously reported in Antarctic sediments or at temperate latitudes (on average $1.7 \times 10^7$ cells $g^{-1}$ dry wt). By contrast, microprotozoa showed high densities (on average $873 \times 10^3$ cells $g^{-1}$) and accounted for 11% of the total microbial biomass. Such high protozoan densities are comparable to those reported for highly productive systems. Bacteria and protozoa were significantly correlated with phytopigment and protein concentrations indicating a response to the organic matter accumulation. Protozoan density was significantly correlated to bacterial number and biomass. Microbial communities in Terra Nova Bay sediments appeared to be mainly bottom-up controlled (i.e. dependent upon food source and substrate availability). However, the close relationship between protozoa and labile compounds and the high protozoan to bacterial biomass ratio stress the relevance of benthic protozoa that may represent, in coastal Antarctic sediments, the main vector for the direct transfer of detrital carbon and bacterial biomass to metazoan food web.

## Introduction

Abundant literature is available on the pelagic microbial loop (see Sherr and Sherr 1991), but much less is known on the structure and role of the microbial loop in benthic ecosystems (Kemp 1988; Bak and Nieuwland 1989; Hondeveld et al. 1994). To date, it has been demonstrated that benthic bacteria play an essential role in recycling sedimentary organic detritus otherwise lost to marine food webs (Deming and Barross 1993) and that the transfer of this secondary production to higher trophic levels depends, to a large extent, upon bacterial grazers such as protozoa (Fenchel 1982a,b; Deming and Baross 1993). However, quantitative information on the three main components of the benthic microbial loop (i.e. organic detritus, bacteria and protozoa) is extremely limited (Alongi 1987; Cunningham and Ustach 1992; Kemp 1994).

The scarcity of information on the benthic microbial assemblages is even more evident in the Antarctic Ocean (White et al. 1985; Lochte 1994). Coastal Antarctic sediments are generally characterized by the accumulation of large quantities of organic matter of high nutritional quality that, potentially, could support large microbial biomasses (Knox 1994; Pusceddu 1997). Despite this,

[1] Dipartimento per lo Studio del Territorio e delle sue Riserse, Università di Genova, Corso Europa 26, 16132 Genova, Italy
[2] Facoltà di Scienze, Università di Ancona, 60131 Ancona, Italy
[3] Dipartimento di Zoologia, Università di Bari, 70125 Bari, Italy

in Antarctic sediments, bacterial densities are comparable or lower than those reported in most oligotrophic sediments at temperate latitudes ($10^7$–$10^9$ cells $g^{-1}$; Delille et al. 1990; Deming and Yager 1992) and their distribution appears to be irrespective of habitat type and water depth. Moreover, benthic bacteria appear characterized by very low production rates, suggesting that a large fraction of the bacterial assemblage is inactive (White et al. 1985). These data suggest that the large accumulation of sedimentary organic detritus is also the result of an inefficient benthic microbial loop, that does not respond to the organic inputs and/or is not able to channel organic detritus to the higher trophic levels. However, the complete lack of information on benthic protozoa abundance and the few data available on benthic bacteria preclude analysis of the role of the benthic microbial loop in Antarctic sediments.

In this study we tested, through a descriptive approach, the role of different organic inputs on the abundance and relative significance of bacteria and protozoa in order to: (1) provide quantitative information on benthic microbial communities that might prove to be crucial for assessing the role of these components in the functioning of the benthic Antarctic ecosystem and (2) identify trophic interaction between bacteria, protozoa and the availability of organic detritus (measured as content of labile compouds).

## Materials and Methods

### Sampling

The study area is within Terra Nova Bay between the Drygalski Ice Tongue and Cape Washington (Fig. 1). Sediment samples were collected from ten stations (at depths ranging from 36.5 to 223 m), between 8 January and 3 February 1994, in Terra Nova Bay (Ross Sea) using a large Van Veen grab. At stations 8, 13, 19 and 26, two to three additional grabs were taken at different depths and analyzed separately (Table 1).

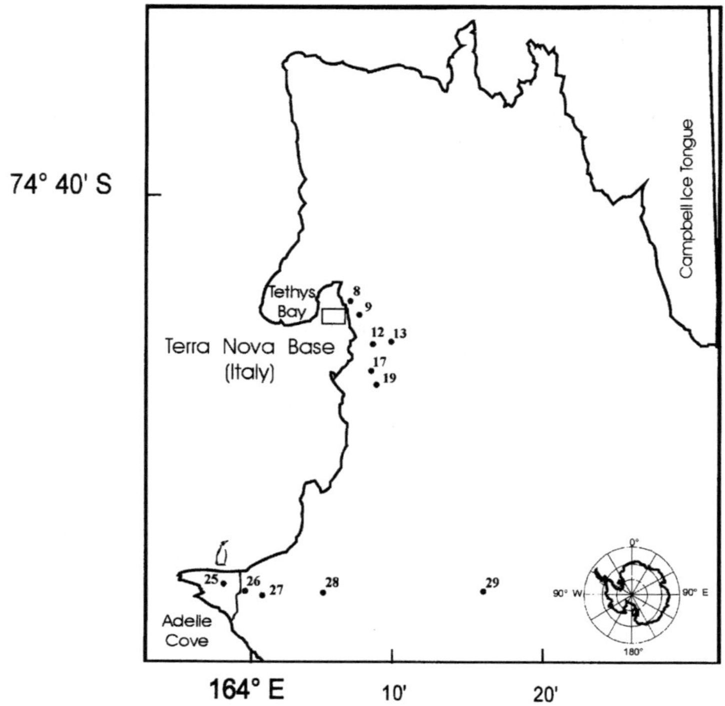


Fig. 1. Sampling area and station location

**Table 1.** Station location, depth and grain-size composition in Terra Nova Bay (Ross Sea, Antarctica)

| Station | Sample | Depth (m) | Latitude south | Longitude east | Coarse sand (>1 mm) % | Sand-mud (1–0.063 mm) % | Silt-clay (<0.063 mm) % |
|---|---|---|---|---|---|---|---|
| 8 | ADAB1 | 84 | 74°41′.99 | 164°07′.84 | nd | nd | nd |
| | ADAB3 | 55 | 74°41′.99 | 164°07′.72 | 72.9 | 27.0 | 0.1 |
| 9 | SCOB1 | 52 | 74°42′.25 | 164°07′.79 | 14.7 | 85.1 | 0.2 |
| 13 | FARB1 | 205 | 74°43′.51 | 164°09′.18 | nd | nd | nd |
| | FARB2 | 223 | 74°42′.56 | 164°09′.14 | 1.7 | 96.4 | 1.9 |
| | FARB5 | 148 | 74°42′.89 | 164°08′.78 | nd | nd | nd |
| 17 | ALB4 | 50 | 74°43′.03 | 164°07′.45 | 8.5 | 91.4 | 0.1 |
| 19 | SP1 | 132 | 74°42′.71 | 164°08′.21 | 1.2 | 97.9 | 0.9 |
| | SP2 | 155 | 74°42′.63 | 164°08′.24 | 2.3 | 97.1 | 0.7 |
| 25 | VARB1 | 81 | 74°46′.36 | 163°57′.52 | nd | nd | nd |
| 26 | PEN1A | 40 | 74°46′.32 | 163°58′.38 | 5.7 | 93.8 | 0.5 |
| | PEN1D | 36.5 | 74°46′.41 | 163°59′.06 | 0.6 | 98.3 | 1.1 |
| 27 | PEN2C | 67 | 74°46′.33 | 163°58′.03 | 0.4 | 98.4 | 1.1 |
| 28 | PEN4B | 146 | 74°47′.05 | 164°02′.64 | 0.3 | 98.3 | 1.4 |
| 29 | PEN5C | 219 | 74°46′.18 | 164°10′.78 | 0.0 | 99.1 | 0.9 |

For both bacterial and protozoan analysis, replicate samples (n = 3–5) were collected from the top 1 cm of the sediments using sterile syringes. The samples were fixed with 0.2 μm prefiltered seawater containing buffered formalin (2% final concentration) and stored at 4 °C (Sherr and Sherr 1983a,b). Two additional cores were taken to analyze the biochemical composition of sedimentary organic matter. The upper 1 cm of the sediments from two replicate cores was homogenized by simple stirring, placed in a sterile petri dish and frozen at −20 °C.

## Chloroplastic Pigments

Chlorophyll a and phaeopigment concentrations were determined according to Lorenzen and Jeffrey (1980). All analyses were carried out on three to five replicates. For each biochemical analysis, blanks were made using the same sediments previously treated in a muffle furnace (550 °C, 4 h). Chloroplastic pigment equivalents (CPE) were determined as the sum of chlorophyll a and phaeopigment content.

## Biochemical Composition of Sedimentary Organic Matter

Total carbohydrates were analyzed according to Gerchacov and Hatcher (1972) and expressed as glucose equivalents. Acid-soluble carbohydrates were extracted in 0.1 N HCl (2 h, 50 °C). Another set of replicates was extracted in NaOH 0.1 M (4 h, ambient temperature) for the determination of the NaOH-soluble carbohydrates. Insoluble carbohydrate concentrations were calculated by difference between total carbohydrates and HCl- plus NaOH-soluble carbohydrates. Data were then normalized to sediment dry weight.

Proteins were measured after extraction with NaOH (0.5 M, 4 h), according to Hartree (1972) modified by Rice (1982) to compensate for phenol interference. Concentrations are referred to as albumin equivalents.

Lipids were extracted from sediment samples by direct elution with chloroform and methanol. Analyses were carried out using the methods of Bligh and Dyer (1959) and Marsh and Weinstein (1966). Concentrations are reported as tripalmitine equivalents.

## Bacterial Parameters

For bacterial analysis, subsamples were sonicated three times (Sonifier Branson 2200, 60 W for 1 min) and diluted 100 times. Bacterial numbers and frequency of dividing cells (FDC) were estimated as described by Danovaro and Fabiano (1995) by Acridine Orange direct count using epifluorescence microscopy (Zeiss Universal Microscope, HBO 50 W, FT510, LP520). For biovolume estimations, bacteria were divided into size classes and their biovolume was converted to biomass assuming a conversion factor of 310 fg C $\mu m^{-3}$ (Fry 1990).

### Protozoan Parameters

Protozoan densities were determined with a double-staining epifluorescence technique as described by Sherr and Sherr (1983a,b). The samples were diluted in 0.2 μm prefiltered seawater, treated with Acridine Orange to stain DNA and filtered onto 2.0-μm black-stained Nuclepore polycarbonate filters. Then the filters were stained with the protein-binding fluorescein isothiocyanate (FITC) and scanned at 1000*x*. All protozoa encountered in the fields were measured for length and width with a micrometric ocular for biovolume estimates. Protozoan biovolume was converted to biomass using a conversion factor of 220 $\times 10^{-9}$ μgC μm$^{-3}$ (Borsheim and Bratbak 1987). All microbial data were normalized to dry weight after desiccation (60 °C, 24 h).

## Results and Discussion

### Quantity and Quality of the Sedimentary Organic Detritus

The photosynthetically produced organic matter reaching the sea floor represent the main food source for benthic communities. Chlorophyll a and CPE (Fig. 2a,c) are widely used as tracers for phytodetritus sedimentation (Pfannkuche 1993). The high chlorophyll a and CPE concentrations reported in Terra Nova Bay sediments indicate the presence of large inputs of primary organic matter. In particular, the high chlorophyll a concentrations (on average 10.2 μg g$^{-1}$), up to tenfold those reported at temperate latitudes (Plante et al. 1986; Fabiano et al. 1995), would also suggest that the organic input is composed of large amounts of living phytoplankton and/or fresh phytodetritus. Highest values were reported at station 25 (81 m depth) facing the Adelie Cove, where the microphytobenthic contribution is likely to be negligible. However, at shallow depths (such at stations 9, 17 and 26; 36.5–52 m depth), it is possible also that microphytobenthos contributed to the total sedimentary pigment pools. Such high chloropigment concentrations provide evidence that large amounts of fresh organic matter reach the sea floor of Terra Nova Bay during the study period, but do not quantify the fraction actually available to benthic consumers and do not discriminate between labile detritus and the refractory part.

The biochemical composition of the sedimentary organic matter was used to gather further information on the origin, quality and availability of the deposited material (Danovaro et al. 1993). In the sediments of Terra Nova Bay, lipids showed concentrations (Fig. 3a) comparable to those reported for other estuarine or highly productive environments (Fabiano and Danovaro 1994; Sargent et al. 1983). In contrast, sedimentary carbohydrates and proteins (Fig. 3b,c) showed very high concentrations, among the highest reported in the literature even for the most productive areas (Meyer-Reil 1983; Khripounoff et al. 1985; Tanoue and Handa 1987; Fabiano and Danovaro 1994; Danovaro 1996 and citations therein). The biochemical composition of the sedimentary organic detritus shows the dominance of carbohydrates (on average 5.3 mg g$^{-1}$) followed by proteins (1.9 mg g$^{-1}$) and lipids (0.2 mg g$^{-1}$). The dominance of carbohydrates over proteins is characteristic of highly oligotrophic environments (Danovaro et al. 1993), where proteins (i.e. organic nitrogen) are generally the limiting factor. The utilization of labile compounds results in highly refractory sedimentary organic material. In the highly oligotrophic Cretan Sea (N-E Mediterranean), for instance, Danovaro et al. (1996) reported that soluble carbohydrates (determined using the same extraction procedure) represented a low fraction (<6%) of the total carbohydrate content. By contrast, these coastal Antarctic sediments were characterized by the dominance of soluble compounds that accounted for 98% of the total carbohydrate concentration (Fig. 3b). The dominance of soluble carbohydrates indicates that large amounts of available (i.e. rapidly utilizable) organic matter are present in the coastal sediments of Terra Nova Bay.

The significant correlations found among the different biochemical components and, particularly, between chloroplastic pigment equivalents and lipids, proteins and insoluble carbohydrates ($r = 0.709, p < 0.01, r = 0.716, p < 0.01, r = 0.589, p < 0.05$, respectively) indicate that sedimentary organic detritus in Terra Nova Bay is mostly derived from recently deposited primary organic material. These data are in agreement with particle flux composition measured synoptically in Terra Nova Bay (Fabiano et al. 1997). Notably, only insoluble-structural carbohydrates were significantly correlated to phytopigments, whereas soluble carbohydrates showed no significant correlation with any investigated parameter. These data suggest that carbohydrates determined after

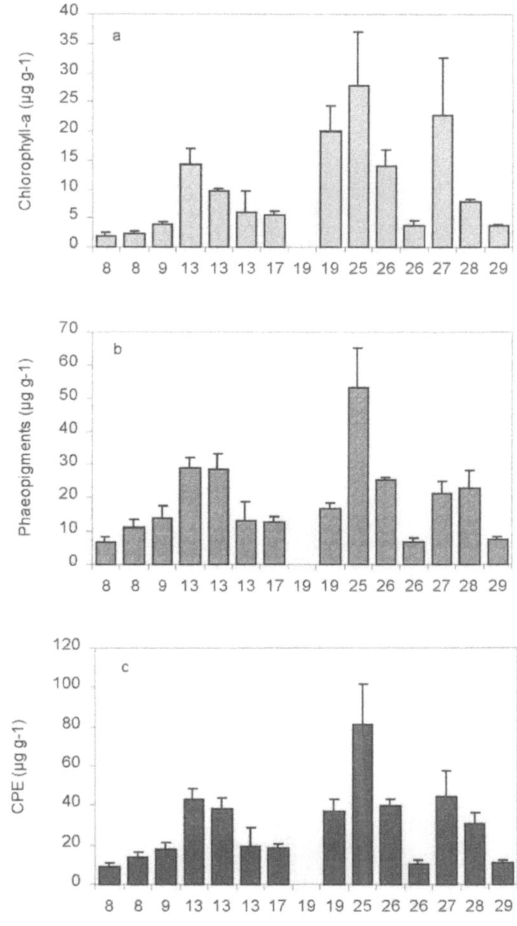


**Fig. 2a–c.** Distribution of chloroplastic pigments in Terra Nova Bay. **a** Chlorophyll a. **b** Phaeopigments. **c** Chloroplastic pigment equivalents (*CPE*). Standard deviations are reported

extraction by acids and sodium hydroxide might include an important fraction of structural carbohydrates and/or other combined sugars and therefore might not represent the actual labile fraction of the carbohydrate pools.

Organic matter distribution in Terra Nova Bay did not display clear spatial patterns (Fig. 4). Water depth had, apparently, a limited influence on organic matter distribution. The northern part of the sampling area was characterized by highly variable bottom morphology, near-bottom transport and deposition regime. All these factors may have contributed to the highly variable mesoscale distribution: station 13, for instance, displayed significant changes between replicate sampling (at depths ranging from 148 to 223 m) in most of the investigated parameters (e.g. chloropigments, carbohydrates, bacterial density, biomass, cell size and activity). However, the analysis of organic matter distribution along the stations 25–29 transect revealed a decrease of lipid and insoluble carbohydrate concentrations with increasing depth, contrasting with the opposite increase in the concentration of soluble carbohydrates. This pattern is unusual since, generally, labile compounds, such as soluble carbohydrates, are expected to decrease with depth as a result of their selective utilization. However, as previously stated, the extraction procedure applied in this study may have included also refractory carbohydrates in the so-called

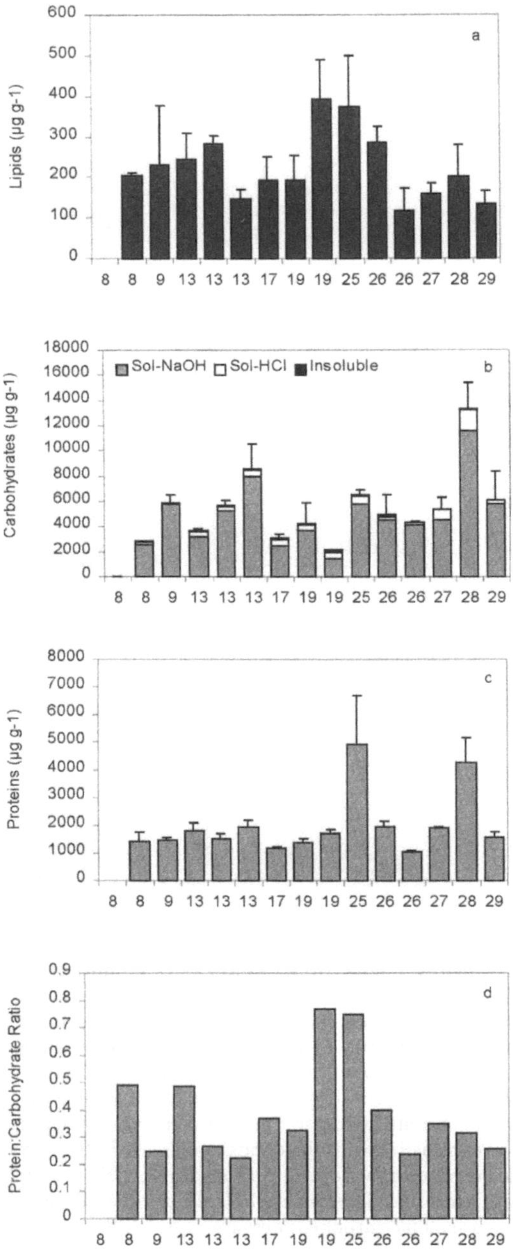


**Fig. 3a–d.** Biochemical composition of the sedimentary organic matter. Reported are: **a)** lipid concentrations; **b)** NaOH-soluble carbohydrates, HCl-soluble carbohydrates and insoluble carbohydrates; **c)** protein concentrations; **d)** protein to carbohydrate ratio. Data are expressed as $\mu g\, g^{-1}$ sed. dry wt

**Fig. 4.** Protein, lipid and carbohydrate distributions vs. depth

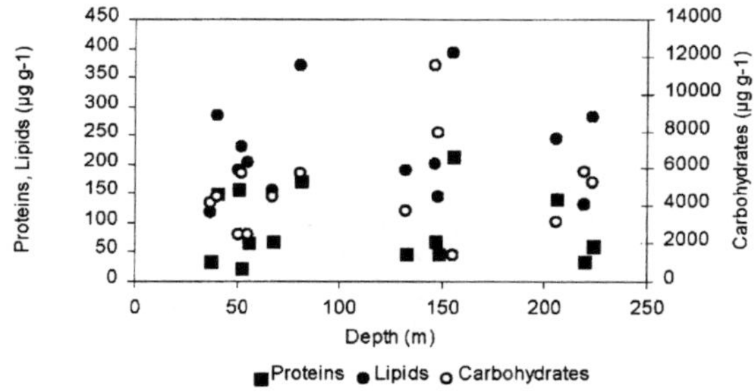


soluble fraction. At the same time, the decreasing patterns of the protein to carbohydrate ratio along the Adelie Cove transect (Fig. 3d) suggest protein depletion with increasing depth. Nonetheless, the presence of large amounts of sedimentary organic matter in the deeper stations would suggest a limited OM utilization by benthic communities.

## Detritus-Bacteria-Protozoa Trophic Interactions

High bacterial densities, ranging between $10^8$ and $10^9$ cells $g^{-1}$ of sediment dry weight, generally characterize marine sediments at temperate latitudes and in sub-Antarctic areas (Fallon et al. 1983; Delille and Bouvy 1989). Bacterial density estimates from previous studies in coastal Antarctic sediments reported values in good agreement with those reported above, ranging from 3.5 to $21 \times 10^8$ cells $g^{-1}$ of sediment dry wt (Smith et al. 1989). In contrast, the sediments of Terra Nova Bay showed bacterial densities 1 to 2 orders of magnitude lower (on average $1.73 \times 10^7$ cells $g^{-1}$ dry wt, Fig. 5a). In particular, extremely low values were reported at station 8 ($5.57 \times 10^6$ cells $g^{-1}$ dry wt), characterized by the dominance of the sandy fraction. Benthic bacterial biomass was low (on average 8.32 µgC $g^{-1}$, Fig. 5b), thus providing a negligible contribution to the sedimentary organic carbon pool (<0.3%). Bacterial biomass showed a distribution similar to the one observed for protozoan biomass (Fig. 5c). Bacterial cell size (Fig. 5d) was characterized by generally high values and was significantly related ($p < 0.05$) to grain size composition (Table 1). The frequency of dividing cells (FDC, Fig. 5e) showed values comparable to those reported in most benthic studies (Meyer-Reil 1983; Danovaro et al. 1993; Danovaro 1996) but, interestingly, appeared to respond to the organic inputs, particularly to protein content ($r = 0.501$, $p < 0.05$). Highest FDC values were observed at station 25 close to the Adelie Cove (Penguin Bay) and in areas of OM accumulation, and decreased significantly with increasing water depth ($p < 0.05$).

Microprotozoa are important links of the microbial loop that channel bacterial production into the classic food chain (Epstein et al. 1992). At the same time, bacterivorous protozoa might affect bacterial density and production (Kemp 1988; Epstein et al. 1992), thus controlling top-down (i.e. by predation) their dynamics. The role of protozoa is largely dependent upon their quantitative significance in the microbial loop. In Terra Nova Bay, microprotozoa are quite abundant with densities ranging from 18.9 to $9260 \times 10^3$ cells $g^{-1}$ (on average $873 \times 10^3$ cells $g^{-1}$). Such values are comparable to those reported in other temperate and highly productive systems such as the North Sea, where Hondeveld et al. (1994) reported 859–1100 $\times 10^3$ heterotrophic flagellates $g^{-1}$. Lower densities have been observed in oligotrophic systems, such as the Eastern Mediterranean Sea, characterized by a strong benthic microbial loop ($67.8 \times 10^3$ flagellates $g^{-1}$; Danovaro, unpubl. data).

The significance of protozoa inside the Antarctic microbial loop is highlighted by the comparison with bacteria. Cunningham and Ustach (1992) reported a bacterial to protozoan density ratio of $10^3$ to $10^8$, in environments highly dominated by bacteria. In contrast, in Antarctic sediments, the high protozoan density determined a ratio of bacterial to protozoan abundance of 195 (excluding one single anomalous peak reported at station 13). The significance of protozoa in Antarctic sediments increased in terms of biomass, as protozoa accounted for 11% of the total microbial biomass (as the sum of

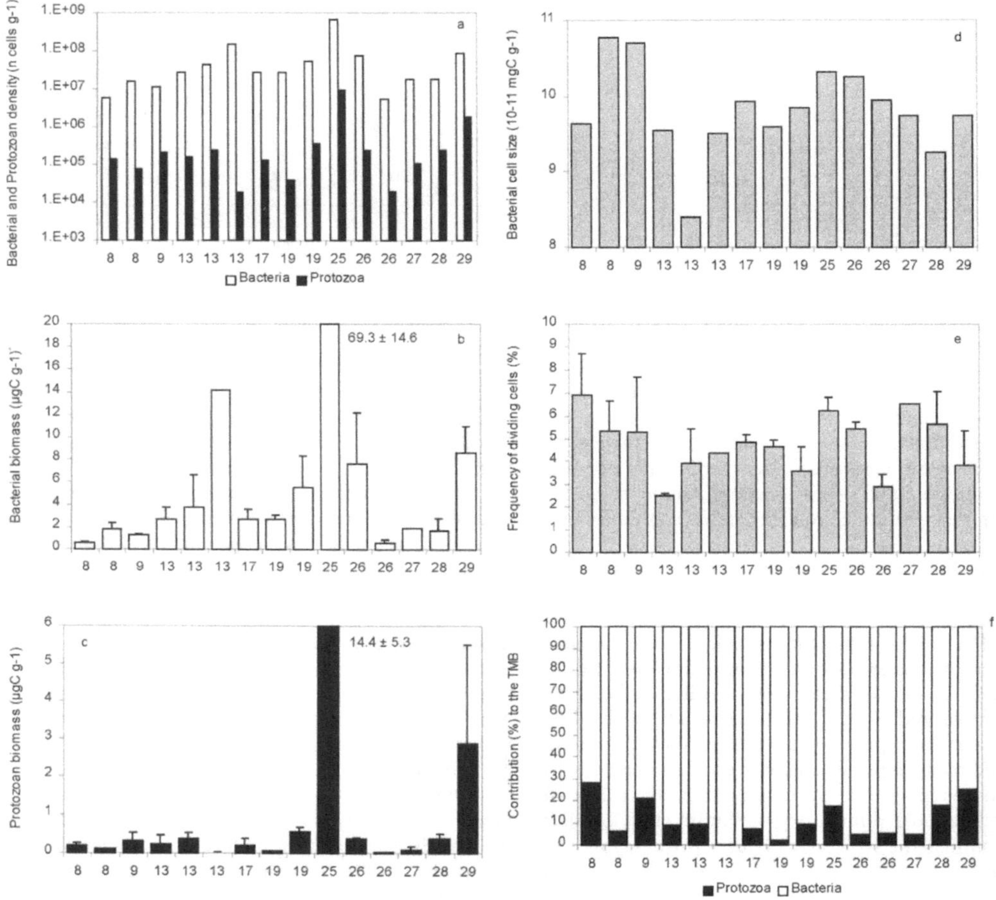


**Fig. 5a–f.** Microbial parameters in the sediments of Terra Nova Bay. **a** Bacterial and protozoan density (cells $g^{-1}$ sed. dry wt). **b** Bacterial biomass (µgC$g^{-1}$ sed. dry wt). **c** Protozoan biomass (µgC$g^{-1}$ sed. dry wt). **d** Bacterial cell size (mgC $10^{-11}$ $cell^{-1}$). **e** Frequency of dividing cells (%). **f** Bacterial and protozoan contribution to the total microbial biomass (TMB)

bacterial and protozoan biomass, Fig. 5f). These data clearly indicate that protozoa in the Antarctic microbial loop play a quantitative role by far more important than in other systems at temperate latitudes.

Recent studies have proven that benthic bacterial distribution is largely controlled by the POC flux from the water column and the concentrations of labile compounds such as soluble carbohydrates, proteins or DNA (Alongi 1987; Deming and Yager 1992; Danovaro et al. 1994). Delille and Bouvy (1989) reported a significant bacterial response to organic matter accumulation in sub-Antarctic sediments. Conversely, scant information is available on factors affecting protozoan distribution. Bak et al. (1995), with an experimental approach in controlled boxcosms, reported that heterotrophic flagellate densities were significantly related to bacterial number and production but also reacted significantly to the sedimentation of algal material. The results of this study are consistent with those previously reported, indicating a significant bacterial and protozoan response to organic matter accumulation. Both bacteria and protozoa, indeed, correlated significantly with phytopigment and protein concentrations ($r = 0.742, p < 0.01$; $r = 0.712, p < 0.01$, respectively), but not with grain size composition (Fig. 6). In addition both bacteria and protozoa showed highest densities at station 25 characterized by a large accumulation of organic matter due to the penguin guano coming from the Adelie Cove. Such OM enrichment was evident throughout all investigated parameters. At the same time, protozoan density was significantly correlated to bacterial number and biomass ($r = 0.601, p < 0.01$; Fig. 7). Hondeveld et al. (1994) reported similar positive correlations between heterotrophic flagellates and bacterial densities, concluding that either bacterial density sets limits for protozoan growth or that protozoa graze upon bacteria up to a threshold level. The coupling observed between sedimentary organic content and bacterial and protozoan distributions along the transect stations 25–29 would suggest that bacteria and protozoa are dependent upon the availability of labile compounds. These data suggest that microprotozoa in Antarctic sediments, due to low bacterial numbers, may directly utilize labile OM as a food source. However, simple correlations do not allow to distinguish between cause and effect and definitive conclusions about detritus-bacteria-protozoa trophic interactions cannot be taken on the basis of our data. On the other hand, the low bacterial to protozoan density ratio would indicate that protozoa might have a significant predatory impact over bacteria. Utilizing the protozoan grazing rate reported by Hondeveld et al. (1992) and Bak et al. (1995) for flagellates in the North Sea (2.5–67 bacteria flagellate$^{-1}$ h$^{-1}$), it can be estimated that 1–34% of bacteria might be grazed by protozoa h$^{-1}$. These data, together with the high protozoan to bacterial biomass ratio, allow us to conclude that protozoa might play a key role in the direct transfer of detrital carbon and bacterial biomass to the metazoan food web in Antarctic sediments.

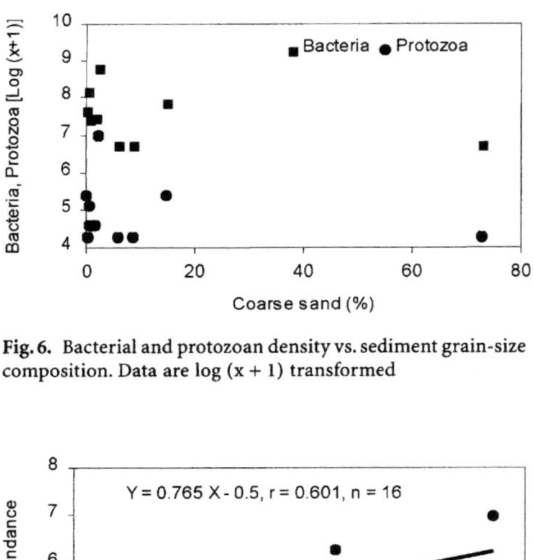


**Fig. 6.** Bacterial and protozoan density vs. sediment grain-size composition. Data are log (x + 1) transformed

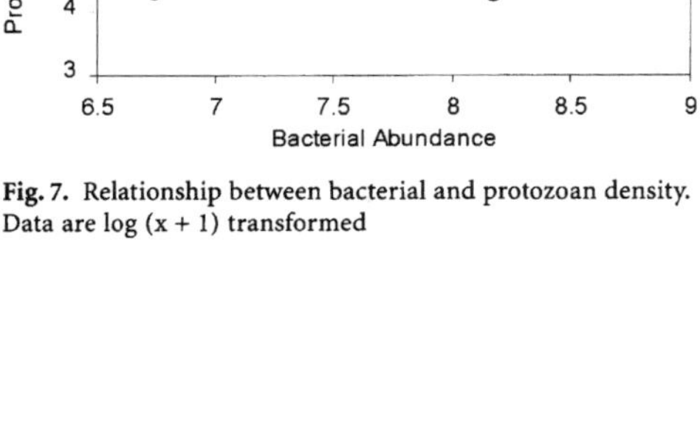


**Fig. 7.** Relationship between bacterial and protozoan density. Data are log (x + 1) transformed

## Conclusions

This study was undertaken to understand the trophic interactions between bacteria, protozoa and OM composition in the coastal sediments of Terra Nova Bay (Ross Sea). Although located at different depths and subjected to a wide range of organic inputs, all these stations were characterized by extremely high organic matter concentrations (particularly carbohydrates and proteins). The large amounts of OM and the increase of the organic matter concentrations with increasing depth suggest that most of the OM input from the water column is not efficiently utilized by benthic metabolism. Benthic bacterial densities were 1 to 2 orders of magnitude lower than in temperate and sub-Antarctic sediments while protozoan communities displayed high densities and a large contribution to the total microbial biomass. Both bacteria and protozoa appeared to be controlled by the amounts of available food but the low ratio of bacterial to protozoan density suggests that protozoa have the potential to top-down control bacterial dynamics in Antarctic sediments.

*Acknowledgements.* We thank Dr. R. Cattaneo-Vietti for help during sampling and two anonymous reviewers for their comments. We acknowledge the logistic and financial support of the Programma Nazionale di Ricerche in Antartide (PNRA).

## References


Alongi DM (1987) The distribution and composition of deep-sea microbenthos in a bathyal region of the western Coral Sea. Deep Sea Res 34: 1245–1254

Bak RPM, Nieuwland G (1989) Seasonal fluctuations in benthic protozoan populations at different depths in marine sediments. Neth J Sea Res 24: 37–44

Bak RPM, Van Duyl FC, Nieuwland G (1995) Organic matter and macrofauna as forcing factors in marine benthic nanoflagellates communities. Microb Ecol 29: 173–182

Bligh EG, Dyer W (1959) A rapid method for total lipid extraction and purification. Can J Biochem Physiol 37: 911–917

Borsheim KY, Bratbak G (1987) Cell volume to cell carbon conversion factors for a bacterivorous *Monas* sp. enriched from seawater. Mar Ecol Prog Ser 36: 171–175

Cunningham JR, Ustach JF (1992) Protozoan number and biomass in the sediments of the Blake Outer Ridge. Deep Sea Res 39: 789–794

Danovaro R (1996) Detritus-bacteria-meiofauna interactions in a seagrass bed (*Posidonia oceanica*) of the NW Mediterranean. Mar Biol 127: 1–13

Danovaro R, Fabiano M (1995) Seasonal and inter-annual variation of bacteria in a seagrass bed of the Mediterranean Sea: relationship with labile organic compounds and other environmental factors. Aquat Microb Ecol 9: 17–26

Danovaro R, Fabiano M, Della Croce N (1993) Labile organic matter and microbial biomasses in deep-sea sediments (Eastern Mediterranean Sea). Deep Sea Res 40: 953–965

Danovaro R, Fabiano M, Boyer M (1994) Seasonal changes of benthic bacteria in a seagrass bed (*Posidonia oceanica*) of the Ligurian Sea in relation to origin composition and fate of the sediment organic matter. Mar Biol 119: 489–500

Danovaro R, Della Croce N, Marrale D (1996) Microbial dynamics in continental and deep-sea sediments of the Cretan Sea (eastern Mediterranean). In: Tselepides A, Papadopoulou K-N, Polychronaki T (eds) CINCS: Pelagic-benthic coupling the oligotrophic Cretan Sea, MAST-II Mediterranean Targeted Project. IMBC, Crete, Greece, pp 135–153

Delille D, Bouvy M (1989) Bacterial response to natural organic inputs in a marine sub-Antarctic area. Hydrobiologia 182: 225–238

Delille D, Guidi LD, Cahet G (1990) Temporal variations of benthic bacterial microflora on the North Western Mediterranean continental shelf and slope. PSZNI: Mar Ecol 40: 953–965

Deming JW, Barross JA (1993) The early diagenesis of organic matter: bacterial activity. In: Engel M, Macko S (eds) Organic geochemistry. 6. Topics in geobiology. Plenum Press, New York, pp 119–144

Deming JW, Yager PL (1992) Natural bacterial assemblages in deep-sea sediments: towards a global view. In: Rowe GT, Pariente V (eds) Deep-sea food chains and the global carbon cycle. Kluwer Academic, Dordrecht, pp 11–28

Epstein SS, Burkovsky IV, Shiaris MP (1992) Ciliate grazing on bacteria, flagellates and microalgae in a temperate zone sandy tidal flat: ingestion rates and food niche partitioning. J Exp Mar Biol Ecol 165: 103–123

Fabiano M, Danovaro R (1994) Composition of organic matter in sediment facing a river estuary (Tyrrhenian Sea): relationship with bacteria and microphytobenthic biomass. Hydrobiologia 277: 71–84

Fabiano M, Danovaro R, Fraschetti S (1995) A 3-year series of elemental and biochemical composition of organic matter in subtidal sandy sediments of the Ligurian Sea (northwestern Mediterranean). Cont Shelf Res 15: 1453–1469

Fabiano M, Chiantore M, Povero P, Cattaneo-Vietti R, Pusceddu A, Misic C, Albertelli G (1997) Short-term variation in particulate matter flux in Terra Nova Bay, Ross sea. Antarctic Sci 9 (2): 143–149

Fallon RDS, Newell SY, Hopkinson CJ (1983) Bacterial production in marine sediments: will cell-specific measures agree with whole system metabolism? Mar Ecol Prog Ser 10: 265–275

Fenchel T (1982a) Ecology of heterotrophic microflagellates. I. Some important forms and their functional morphology. Mar Ecol Prog Ser 8: 211–223

Fenchel T (1982b) Ecology of heterotrophic microflagellates. II. Bioenergetics and growth. Mar Ecol Prog Ser 8: 225–231

Fry JC (1990) Direct methods and biomass estimation. Methods Microbiol 22: 41–85

Gerchacov SM, Hatcher PG (1972) Improved technique for analysis of carbohydrates in sediments. Limnol Oceanogr 17: 938–943

Hartree EF (1972) Determination of proteins: a modification of the Lowry method that gives a linear photometric response. Anal Biochem 48: 422–427

Hondeveld BJM, Bak RPM, Van Duyl FC (1992) Bacterivory by heterotrophic nanoflagellates in marine sediments measured by uptake of fluorescently labeled bacteria. Mar Ecol Prog Ser 89: 63–71

Hondeveld BJM, Nieuwland G, Van Duyl FC, Bak RPM (1994) Temporal and spatial variations in heterotrophic nanoflagellate abundance in North Sea sediments. Mar Ecol Prog Ser 109: 235–243

Kemp PF (1988) Bacterivory by benthic ciliates: significance as a carbon source and impact on sediment bacteria. Mar Ecol Prog Ser 49: 163–169

Kemp PF (1994) Microbial carbon utilization on the continental shelf and slope during the SEEP-II experiment. Deep Sea Res 41: 563–581

Khripounoff A, Crassus P, Desbruyeres D, Le Cox JR (1985) Le flux organique particulaire et ses transformations à l'interface eau-sédiment. In: Laubier L, Monniot C (eds) Peuplement profond du Golf de Gascogne. IFREMER, Brest, pp 101–108

Knox GA (1994) The biology of the Southern Ocean, Cambridge University Press, Cambridge

Lochte K (1994) Microbiological investigations in deep-sea sediments of the Weddel Sea, Antarctica. 7th Deep-Sea Biology Symp, Abstr, Heraklion, Crete, Greece

Lorenzen C, Jeffrey J (1980) Determination of chlorophyll in seawater. UNESCO Tech Papers Mar Sci 35: 1–20

Marsh JB, Weinstein WJ (1966) A simple charring method for determination of lipids. J Lipid Res 7: 574–576

Meyer-Reil LA (1983) Benthic response to sedimentation events in western Kiel Bight. II. Analysis of benthic bacterial populations. Mar Biol 77: 247–256

Pfannkuche O (1993) Benthic response to sedimentation of particulate matter at the BIOTRANS station, 47°N, 20°W. Deep Sea Res 40: 135–149

Plante R, Plante-Cuny MR, Reys JP (1986) Photosynthetic pigments of sandy sediments on the north Mediterranean coast: their spatial distribution and its effect on sampling strategies. Mar Ecol Prog Ser 34: 133–141

Pusceddu A (1997) Origine, composizione biochimica e destino della sostanza organica in ambienti marini a differente livello di trofismo. PhD Dissertation, University of Genova, 173 pp

Rice DL (1982) The detritus nitrogen problem: new observations and perspectives from organic geochemistry. Mar Ecol Prog Ser 9: 153–162

Sargent JR, Hopkins CCE, Siering JV, Youngson A (1983) Partial characterization of organic material in surface sediments from the Balsfjorden, northern Norway, in relation to its origin and nutritional value for sediment-ingesting animals. Mar Biol 76: 87–94

Sherr BF, Sherr EB (1983a) Enumeration of heterotrophic microprotozoa by epifluorescence microscopy. Estuar Coast Shelf Sci 16: 1–17

Sherr BF, Sherr EB (1983b) Double-staining epifluorescence technique to assess frequency of dividing cells and bacterivory in natural populations of heterotrophic microprotozoa. Appl Environ Microbiol 46: 1388–1393

Sherr BF, Sherr EB (1991) Planktonic microbes: tiny cells at the base of the ocean's food webs. Tree 6: 50–54

Smith GA, Davis JD, Muscat AM, Moe AM, White DC (1989) Lipid composition and metabolic activities of benthic near-shore microbial communities of Arthur Harbor, Antarctic Peninsula: comparisons with McMurdo Sound. Polar Biol 9: 517–524

Tanoue E, Handa N (1987) Monosaccharide composition of marine particles and sediments from the Bering Sea and North Pacific. Oceanol Acta 10: 91–99

White DC, Smith GA, Nichols PD, Stanton GR, Palmisano AC (1985) The lipid composition and microbial activity of selected recent Antarctic benthic marine sediments and organisms: a mechanism for monitoring and comparing microbial populations. Antarct J US 19: 130–132

CHAPTER 14

# Microbial Biomass and Respiratory Activity Related to the Ice-Melting Upper Layers in the Ross Sea (Antarctica)

E. Crisafi, F. Azzaro, R. La Ferla, and L.S. Monticelli[1]

**ABSTRACT**

In late austral spring 1994, a study was carried out in the Ross Sea, with sampling in the ice-free, ice-covered and marginal ice zones. Samples were taken along two transverse transects, to determine spatial variations in some microbiological parameters. Some stations were repeatedly sampled to study the temporal development of microbial components. Microbial biomass was measured as fractionated ATP in the following size categories: micro- (250–10 μm), nano- (10–2 μm) and picoplankton (2–0.2 μm); bacterial biomass was estimated on the basis of lipopolisaccharide concentration and microbial respiratory rate was estimated by the Electron Transport System activity of organisms <200 μm.

Microbial ATP ranged from 4.37 to 628.74 $\mu g\, m^{-3}$, with high values always associated with the upper 50-m depth; depth integrated biomass ranged from 0.283 to 9.217 $g\, C\, m^{-2}$. Biomass structure was dominated by microplankton-sized organisms (mean 59.2 and 50.3% of total biomass in ice-free and ice-covered zones, respectively). At the station with the highest standing stock, microbial biomass represented 84.4% of the total. Picoplankton biomass exceeded nanoplankton biomass in the ice-covered zone and during the evolution of the marginal ice zone. Bacterial biomass from LPS measurements ranged from 0.152 to 1.278 $mg\, C\, m^{-3}$, with higher depth-integrated values in the ice-covered zone (22% of total microbial biomass) compared to the ice-free zone (2.5% of total microbial biomass). Intermediate values were found in the marginal ice zone. Respiration ranged from 0.017 to 0.734 $ml\, O_2\, h^{-1}\, m^{-3}$ with no significant differences in the depth integrated values in the three zones. Low respiratory losses of carbon occurred at equivalents of 5.1% $day^{-1}$ and <1% $day^{-1}$ of the microbial carbon biomass in ice-covered and ice-free zones, respectively. The repeated sampling of some stations revealed that the early ice-free zones supported rapid changes of microbial biomass with variations among micro-, nano- and picoplanktonic organism ratios.

## Introduction

Recent research in the Antarctic has revealed that, more than on other continents, Antarctic biology is based on microbial processes which dominate oceanic life (Friedman and Thisle 1993). The Antarctic Ocean is characterized by a low and nearly constant temperature, but it is unclear whether this temperature regime limits the significance of microbial processes in the ecosystem dynamics of the Southern Ocean. Sorokin (1971) hypothesized that production of organic matter by phytoplankton is often uncoupled both in time and space from bacterial utilization, because low temperatures (together with low standing stock) suppress microbial processes in situ. Carbon and energy exported from the Antarctic Sea by means of horizontal transport may be exploited in temperate and subtropical ecosystems, sustaining secondary production processes. This hypothesis, though initially criticized, is consistent with subsequent studies in other oceans. Pomeroy and Deibel (1986), on the contrary, hypothesized that low temperatures suppress bacterial activity more than phytoplankton growth, thus restricting energy flow through the microbial loop and significantly affecting the fate of primary production (Pomeroy 1974; Pomeroy and Wiebe 1988). Hodson et al. (1981) and Karl et al. (1991) showed that, despite low seawater

[1] Istituto Sperimentale Talassografico, CNR. Spianata S. Raineri 86, 98100 Messina, Italy

temperature, a high-energy flux involved heterotrophic bacteria. This varied in magnitude and was not always correlated with primary production levels.

During the austral spring, sea-ice formation and ablation occur with consequent discontinuities both in biomass and biological processes of the coastal marine biota (Karl et al. 1991; Karl 1993). The marginal ice zone (MIZ) is considered a region of elevated microbial activity; however, how efficiently this energy flux is transferred to higher trophic levels is not fully understood. Much of the ice-edge biomass sinks and supports the large benthic community (Vincent 1988).

As a whole, annual production rates in the Southern Ocean are comparable to those found in oligotrophic tropical waters (Eppeley and Peterson 1979). The dramatic variability of biomass over different spatial (from kilometres to a few metres) and temporal scales (from seasons to hours) in the MIZ has been described and given different interpretations by various authors. El-Sayed (1988) suggested that three main factors were responsible for the variability in phytoplankton, namely, nutrient salts, water column stability and grazing. Karl et al. (1991) hypothesized that the possible causes responsible for the lack of coincidence between phytoplanktonic blooms and bacterial secondary production were due either to the bacteriostatic effect of massive phytoplankton exudates or to bacterial loss due to high grazing rates or, again, to direct competition for growth-limiting substrates. Bolter et al. (1988) suggested that there were two causes for the lack of interrelationships among biological, chemical and physical descriptors of the ecosystem: the complex hydrography of the region and the different scales in the activity of organisms and biochemical processes. Tupas et al. (1994) also found an apparent lack of correlation between the autotrophic and heterotrophic pathways of energy and nutrients.

Marchant and Murphy (1994) assessed that the energy flux in Antarctic microbial communities can be conveyed in two major pathways: a new production system and a regeneration system, as occurs in other aquatic environments (Jacques 1989; Hewes et al. 1985). New production, based on nitrate, is dominated by large diatoms, grazed by herbivorous zooplankton such as krill, which in turn, is preyed on by larger predators. The regeneration system is supported by microbial loop activity (Azam et al. 1983) due to microzooplankton (ciliates and flagellates) which feed on smaller autotrophic and heterotrophic organisms. Autotrophic production depends on ammonia and heterotrophic production on dissolved or particulate organic matter. In this pathway, much of the production is quickly recycled rather than exported.

Previous research on the microbial community of Terra Nova Bay (Ross Sea) during January and February 1990, showed that microbial biomass, assayed by ATP measurements, was temporally unstable. On the contrary, bacterial carbon biomass (derived from LPS measurements) was uniform both in time and space. Picoplankton biomass varied considerably compared to total microbial biomass, assuming greater importance (up to 60–65%) when larger-sized biomass decreased (La Ferla et al. 1995). The structural composition of the microbial community differed at coastal and offshore stations. Pico-sized organism abundance and biomass were the principal components at the offshore site; on the other hand, the coastal site was dominated by larger phytoplankton (Fabiano et al. 1995; Bruni et al. 1995).

In this chapter, we discuss the results of a survey carried out in a zone subject to melting ice in the Ross Sea during spring 1994, focusing on the evolution of microbial biomass and respiratory rates in both time and space. Respiratory rates (ETSa) are useful to understand the mechanisms driving the biological pump in the ocean. The distribution of respiration along the water column constitutes an index of the quantity of organic matter exported (Packard et al. 1988; Christensen et al. 1989; Naqvi et al. 1996). ATP determination quantifies microbial biomass since it represents a central compound in the energy metabolism of all living organisms, not only of primary producers. A comprehensive review on the applicability of this parameter to measure microbial standing stocks is given by Karl (1980) and is supported by more recent studies. Among the biochemical procedures to assess bacterial biomass, the LPS test is simple as well as highly specific and sensitive to Gram-negative bacteria (Watson et al. 1977; Karl and Tien 1991). ATP and LPS were converted to C biomass as was ETSa to metabolic $CO_2$ production rate. The applicability of the conversion factors for estimating the biomass of natural microbial population is still strongly criticized, but the method is usually adopted to relate different parameters to one another.

## Materials and Methods

The oceanographic survey, in the context of the ROSSMIZE interdisciplinary research project, was carried out from 16 November to 16 December 1994, on board the R/V Italica in a zone in the Ross Sea which is periodically ice-free during summer (Fig. 1). With the aim of defining the different biological conditions in the study area, we chose the sampling locations in relation to the ice cover. Along the S-N transect, three main areas were distinguished on different spatial scales: the IFZ (ice-free zone), in correspondence to the Ross Sea polynya, the MIZ (marginal-ice zone) and the ICZ (ice-covered zone). The IFZ was found from stations 2a to 11; the MIZ between Sts. 13 and 17; and the ICZ between stations 19 and 27. In addition, another transverse transect (W-E) was studied between stations 11c and 30. With the aim of better defining the evolution of microbial processes, some stations were repeatedly sampled.

Seawater samples were taken at different depths in the layer between the surface and 100–120 m. Usually, the sampling depths were defined on the basis of light penetration and the samples were taken using a rosette sampler equipped with Niskin bottles, CTD and multiparametric probe. The following parameters were investigated:

- Particulate ATP in the micro- (from 250 to 10 μm), nano- (from 10 to 2 μm) and picoplankton (from 2 to 0.2 μm) size fractions. Total ATP was obtained as the sum of the preceding size fractions, subsequently filtered from each sample. Taking into consideration possible filtration stress, sequential filtration was performed to obtain a more homogeneous water sample. The analyses were carried out on board according to Holm-Hansen and Paerl (1972). Boiling TRIS-EDTA phosphate buffer was used to extract the nucleotide, and ATP was then measured with a ATP-3000 Biospherical Luminometer. ATP values were converted to C biomass using the conversion factor of C/ATP = 250 (Karl 1980).

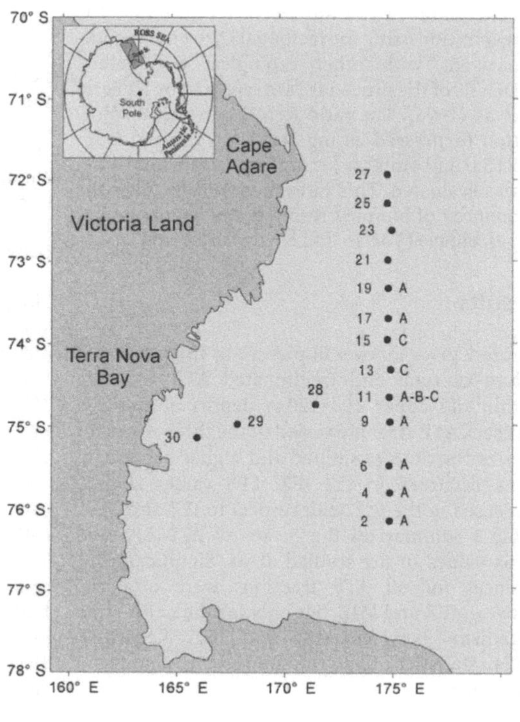


Fig. 1. Sampling stations in the Ross Sea

- LPS (lipopolysaccharide) samples were kept frozen in pyrogen-free tubes until processing. Determinations were carried out according to Watson et al. (1977), using the *Limulus* amoebocyte lysate QCL-1000 (Bio-Wittaker, inc. Walkersville, MD, USA), in a Lambda 3 Perkin Elmer Spectrophotometer. Bacterial biomass in C units was calculated from LPS data by the factor 6.35 (Watson and Hobbie 1979). As a methodological check, direct counts by epifluorescent microscopy with DAPI were carried out according to Porter and Feig (1980) on 30 seawater samples taken in representative areas along the S-N transect.
- ETSa (electron transport system activity) was evaluated on 5-l samples concentrated on GF/F Whatman glass fibre membranes and immediately frozen in liquid nitrogen, to prevent enzyme degradation. As soon as possible, at least within 3 months, samples were analyzed according to Packard (1971, 1985) as modified by Kenner and Ahmed (1975). The activity was corrected for in situ temperature with the Arrhenius equation. According to Martinez and Estrada (1992), the activation energy used was 12 kcal $mol^{-1}$. ETSa was converted to units of C respiration using the factor 0.057, to enable comparisons with other ecological parameters. Details of the procedure are reported by La Ferla et al. (1996). The ratio R:MB between respiration (expressed as mg $day^{-1}$ $m^{-2}$, derived from ETSa) and biomass (as mg C $m^{-2}$, from total ATP) was evaluated. This ratio theoretically gives the quantity of biomass respired $day^{-1}$ expressed as percentage (Vosjan and Nieuwland 1987).

## Results

Figure 2 gives an overall picture of the integrated values of total and fractionated ATP and LPS within the upper 100–120 m depths of the S-N transect. ATP strongly varied in the three zones for all size fractions examined and higher concentrations occurred in the IFZ. LPS values slightly increased in the ICZ with respect to IFZ and MIZ. Table 1 summarizes the means of ATP, LPS and ETSa values in the studied areas. Significant differences for all ATP fractions were observed between IFZ and MIZ, but only for the nano-sized organisms between MIZ and ICZ. Regarding ETSa, statistical analysis showed no significant differences.

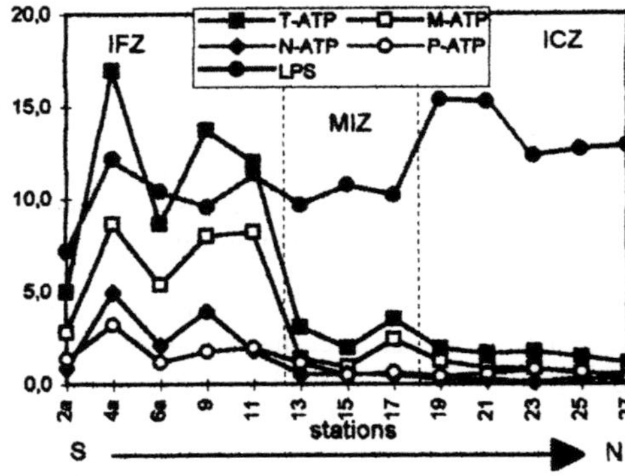


**Fig. 2.** Integrated values for total and fractionated ATP (mg $m^{-2}$) and LPS ($\mu g\, m^{-2}$) along the south to north transect: IFZ (ice-free zone) stations sampled from 20 Nov 1994 to 23 Nov 1994 (Ross Sea polynya); MIZ (marginal-ice zone) stations sampled from 25 Nov 1994 to 3 Dec 1994; ICF (ice-covered zone) stations sampled from 5 Dec 1994 to 8 Dec 1994

Figure 3 illustrates distribution values along the water column. In the IFZ, the concentrations of ATP and ETSa was stratified; highest values were found in the upper layers (<50 m). In the deeper water, ATP exceeded 50 $\mu g\, m^{-3}$ only at stations 2a and 4a, corresponding to two clear fluorescence peaks detected by a CTD-fluorometer at roughly the 100-m depth. Higher ETSa values also occurred in surface waters with the exception of stations 2a and 4a, as already observed for ATP. A slight stratification in MIZ was observed for ATP and ETSa. In the ICZ, the water column was more homogeneous with no significant difference between the first 50 m and deeper waters.

All three areas were relatively homogeneous along the water column with respect to LPS. Measurements of bacterial abundance at some stations along the S-N transept gave values ranging from $5.61 \times 10^5$ and $3.69 \times 10^4$ cells/ml (average $8.25 \times 10^4$; SD $9.19 \times 10^4$), roughly comparable to previous data for Antarctic areas (Hodson et al. 1981; Hanson et al. 1983). By comparison with LPS data, each bacterial cell appeared to contain 1.74 fg LPS (SD 0.67), corresponding to 11.06 fg C/bacterium (SD 4.27), if converted in biomass by the 6.35 ratio. The C content per cell was similar to that determined by Watson and Hobbie (1979), but about half of 20 fg C/bacterium frequently adopted in oceanographic studies (Ducklow and Carlson 1992). This questions the applicability of "universal" conversion factors in ecological studies.

**Table 1.** Means of integrated values of total and fractionated ATP, ETSa and LPS from IFZ, MIZ and ICZ

| | IFZ | | MIZ | | ICZ | | IFZ-ICZ |
|---|---|---|---|---|---|---|---|
| | Stations 2 to 11<br>n = 5 | p[a] | Stations 13 to 17<br>n = 3 | p | Stations 19 to 27<br>n = 5 | p | |
| ATP ($mg\,m^{-2}$) | | | | | | | |
| Total | 11.260 (4.595)[b] | 0.013 | 2.849 (0.801) | w.s.[c] | 1.593 (0.301) | | 0.009 |
| Micro | 6.618 (2.492) | 0.008 | 1.571 (0.764) | w.s. | 0.812 (0.276) | | 0.006 |
| Nano | 2.734 (1.674) | 0.041 | 0.521 (0.033) | 0.0005 | 0.237 (0.08) | | 0.028 |
| Pico | 1.885 (0.794) | 0.033 | 0.757 (0.346) | w.s. | 0.542 (0.199) | | 0.017 |
| ETSA ($ml\,O_2\,h^{-1}\,m^{-2}$) | 15.922 (4.023) | w.s. | 15.281 (3.541) | w.s. | 15.280 (6.046) | | w.s. |
| LPS ($mg\,m^{-2}$) | 10.139 (1.908) | w.s. | 10.209 (0.554) | 0.004 | 13.716 (1.487) | | 0.011 |

[a] Mean comparison Student's t-test.
[b] Mean (standard deviation).
[c] Without significance.

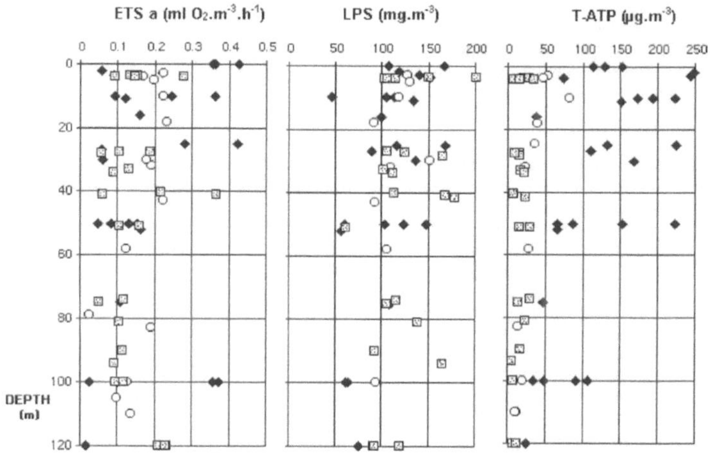


**Fig. 3.** South-north transect: spatial distribution of total-ATP, ETSa and LPS with reference to IFZ (◆), MIZ (○) and ICZ (□)

Figure 4 shows the integrated values for total and fractionated ATP, ETS activity and LPS along the E-W transect. In this transect, microplankton biomass and respiratory activity reached highest values at station 28 (M-ATP: 31.136 $mg\,m^{-2}$, ETSa: 53.96 ml $O_2\,h^{-1}\,m^{-2}$, respectively). At the ice-covered station 29, ATP values sharply decreased, reaching high values once again at station 30 located in the polynya of Terra Nova.

An attempt was made to characterize the three zones from a metabolic point of view by using the relationships between biomass and respiratory activity (Table 2). In the IFZ, bacterial biomass constituted on average 2.5% of total biomass (BB: MB) and was closely dependent on total ATP values. In the ICZ, the ratio reached a mean of 22.4% and depended on decreasing ATP and increasing LPS; in the MIZ the ratio was intermediate (mean 9.7%). The respiratory losses of carbon reached on average only <1% of the entire microbial biomass in the IFZ area (R: MB). Respiration in the ICZ consumed about 5.15% of the total biomass; it was linked to changes in C-ATP rather than to increased respiratory activity. We suppose that the quantity of available carbon, exportable to other trophic levels or sinking from the euphotic zone, is very high. This supposition is confirmed by the C availablity as determined by

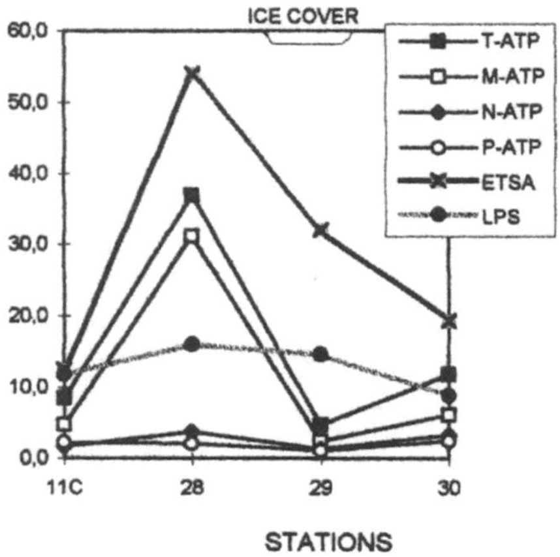


**Fig. 4.** Integrated values of total and fractionated ATP ($mg\,m^{-2}$), ETSa ($ml\ O_2\ m^{-2}h^{-1}$) and LPS ($mg\,m^{-2}$) along the east-west transect. Stations were sampled from 13 to 15 November 1994

**Table 2.** Microbial biomass, respiratory activity and bacterial biomass as carbon from total-ATP, ETSa and LPS. Stations from 2a to 27 correspond to south-north transect and stations from 11c to 30 correspond to east-west transect. For conversion factors see Materials and Methods

| | ST | Microbial biomass (MB) $g\,C\,m^{-2}$ | Respiratory activity (R) $mg\,C\,d^{-1}\,m^{-2}$ | Bacterial biomass (BB) $mg\,C\,m^{-2}$ | R:MB[a] (%) | BB:MB (%) | R:PP[a] (%) |
|---|---|---|---|---|---|---|---|
| IFZ | 2a | 1.253 | 27.647 | 45.818 | 2.21 | 3.66 | - |
| | 4a | 4.234 | 20.854 | 77.213 | 0.49 | 1.82 | - |
| | 6a | 2.167 | 13.526 | 66.143 | 0.62 | 3.05 | - |
| | 9 | 3.427 | 20.965 | 60.591 | 0.61 | 1.77 | - |
| | 11 | 2.994 | 25.909 | 72.162 | 0.87 | 2.41 | 1.82 |
| MIZ | 13 | 0.763 | 20.707 | 61.353 | 2.71 | 8.04 | 4.71 |
| | 15 | 0.491 | 25.842 | 68.382 | 5.26 | 13.92 | - |
| | 17 | 0.882 | 16.161 | 64.740 | 1.83 | 7.34 | - |
| ICZ | 19 | 0.480 | 31.082 | 97.743 | 6.47 | 20.36 | 7.03 |
| | 21 | 0.414 | 15.721 | 96.938 | 3.80 | 23.44 | - |
| | 23 | 0.441 | 28.303 | 78.320 | 6.41 | 17.75 | 30.54 |
| | 25 | 0.373 | 17.090 | 80.550 | 4.58 | 21.60 | - |
| | 27 | 0.283 | 12.312 | 81.941 | 4.35 | 28.96 | - |
| | 11c | 2.101 | 17.110 | 74.791 | 0.81 | 3.56 | 4.67 |
| | 28 | 9.217 | 73.816 | 101.355 | 0.80 | 1.10 | 1.52 |
| | 29 | 1.169 | 43.823 | 92.641 | 3.75 | 7.93 | - |
| | 30 | 2.964 | 26.459 | 56.532 | 0.89 | 1.91 | - |

[a] Ratio between respiratory activity and primary production.

primary production data (PP) (Saggiomo, pers. comm.) at some stations located in the IFZ, MIZ and ICZ areas. The R:PP ratios underline the amount of net autotrophic and net heterotrophic carbon occurring in the IFZ and ICZ, respectively.

Table 3 shows integrated data for the stations where sampling was repeated. A general increase in microbial and bacterial biomass was noted coupled with a decrease of surface layer salinity and a deepening of the surface halocline, as given by CTD profiles (not shown). Net biomass doubling rate per day was calculated from variations in biomass concentrations (Table 4). This value represents the balance between growth factors and factors which lead to a decrease in cell concentrations (e.g. grazing, sinking, cell death, lateral transport etc.). The calculated rates indicate both an increase as well as decrease in the different biomass values. In the IFZ, total biomass duplication rate was closely linked to variations in the microplankton fraction. On the contrary, at stations between the MIZ and ICZ (St. 17 and 19), pico-sized organisms showed the highest rates.

**Table 3.** Integrated values of total and fractionated ATP, ETSa, LPS and R:MB rates from resampled stations

| Station | Date | Salinity upper layer | T-ATP $mg\,m^{-2}$ | M-ATP $mg\,m^{-2}$ | N-ATP $mg\,m^{-2}$ | P-ATP $mg\,m^{-2}$ | ETSA $ml\,O_2\,h^{-1}\,m^{-2}$ | LPS $mg\,m^{-2}$ | R:MB % |
|---|---|---|---|---|---|---|---|---|---|
| 4 | 16.11.94 | 34.46 (40 m)[a] | 9.312 | 3.182 | 1.000 | 5.130 | 16.444 | 8.335 | 0.96 |
| 4a | After 5 days | 34.44 (44 m) | 16.935 | 8.669 | 4.962 | 3.192 | 15.245 | 12.159 | 0.49 |
| 6 | 17.11.94 | 34.46 (13 m) | 9.290 | 5.551 | 1.987 | 1.751 | 14.286 | 9.021 | 0.84 |
| 6a | After 4 days | 34.44 (13 m) | 8.666 | 5.396 | 2.098 | 1.172 | 9.888 | 10.416 | 0.62 |
| 11 | 23-11-94 | 34.34 (46 m) | 11.975 | 8.219 | 1.796 | 1.960 | 18.940 | 11.364 | 0.86 |
| 11a | After 5 days | 34.35 (48 m) | 2.750 | 1.304 | 0.741 | 0.705 | 15.125 | 10.429 | 3.00 |
| 11b | After 11 days | 34.31 (42 m) | 4.830 | 2.589 | 1.037 | 1.204 | 9.701 | 13.122 | 1.09 |
| 11c | After 20 days | 34.37 (40 m) | 8.404 | 4.621 | 1.673 | 2.239 | 12.508 | 11.778 | 0.81 |
| 13 | 24-11-94 | 34.43 (17 m) | 3.054 | 1.373 | 0.533 | 1.149 | 15.138 | 9.662 | 2.71 |
| 13c | After 18 days | 34.34 (39 m) | 6.574 | 3.249 | 1.721 | 1.604 | 15.707 | 12.506 | 1.29 |
| 15 | 26-11-94 | 34.52 (18 m) | 1.965 | 0.924 | 0.548 | 0.494 | 18.891 | 10.769 | 5.26 |
| 15c | After 16 days | 34.38 (41 m) | 7.692 | 4.586 | 1.573 | 1.533 | 16.807 | 16.031 | 1.19 |
| 17 | 03-12-94 | 34.48 (25 m) | 3.527 | 2.414 | 0.484 | 0.629 | 11.814 | 10.195 | 1.83 |
| 17a | After 8 days | 34.43 (51 m) | 4.744 | 2.529 | 0.884 | 1.331 | 15.956 | 13.645 | 1.83 |
| 19 | 05-12-94 | 34.35 (38 m) | 1.920 | 1.225 | 0.282 | 0.414 | 22.722 | 15.393 | 6.45 |
| 19a | After 6 days | 34.30 (48 m) | 2.708 | 0.890 | 0.310 | 1.496 | 9.417 | 14.684 | 1.90 |

[a] Superficial holocline.

**Table 4.** Biomass doubling rates per day of micro-, nano- and picoplancton and microbial assemblage (T-ATP) and bacteria (LPS) concerning resampled stations. Negative values show decrease in biomass in the period. For dates of sampling see Table 3

Biomass doubling rate ($day^{-1}$)

| Stations | Days | T-ATP | M-ATP | N-ATP | P-ATP | LPS |
|---|---|---|---|---|---|---|
| 4–4a | 5 | 0.163 | 0.344 | 0.792 | −0.075 | 0.091 |
| 6–6a | 4 | −0.016 | −0.006 | 0.014 | −0.082 | 0.038 |
| 11–11a | 5 | −0.154 | −0.168 | −0.117 | −0.128 | −0.016 |
| 11a–11b | 6 | 0.126 | 0.164 | 0.067 | 0.118 | 0.043 |
| 11b–11c | 9 | 0.082 | 0.086 | 0.068 | 0.096 | −0.011 |
| 13–13c | 18 | 0.064 | 0.076 | 0.124 | 0.022 | 0.016 |
| 15–15c | 16 | 0.182 | 0.248 | 0.117 | 0.131 | 0.030 |
| 17–17a | 8 | 0.043 | 0.006 | 0.103 | 0.140 | 0.042 |
| 19–19a | 6 | 0.068 | −0.046 | 0.017 | 0.436 | −0.007 |

## Discussion

In the Ross Sea during late austral spring 1994, high variability in plankton distribution occurred over spatial and temporal scales. The different biological systems and dynamic patterns of microbial assemblages reflected the evolution of melting processes. Obviously, the development of the most important trophic process took place in the upper 50-m layer. Differences with respect to the presence of ice were revealed mainly by ATP concentrations. In particular, in the earlier IFZ, values showed mesotrophy as classified by Karl (1980) whereas a range indicative of oligotrophy was detected in the MIZ and ICZ.

Despite the biomass variations with time and space, biomass values had a range similar to those usually observed at temperate latitudes. However, we recorded higher total ATP values with respect to those determined in the previous Antarctic survey in the Terra Nova Bay during spring 1988–89 when values ranged from 3.4 to 293.3 $\mu g\,m^{-3}$ (La Ferla et al. 1995). The extreme richness (up to 2000 $\mu g\,m^{-3}$ for ATP) recorded by Karl et al. (1991) in Gerlache and Bransfield Straits was, however, not found.

Although the microbial assemblage changed in structure between the three areas, it was always dominated by microplankton. In the ICZ, when melting processes were not very evident, microbial biomass was about 0.3 $g\,C\,m^{-2}$ in the first 100 m. This biomass was constituted mainly of microplankton and picoplankton fractions. The nanoplankton population constituted a minor fraction until the end of the melting process. When melting had already started, with an evident deepening of the surface halocline coupled with a reduction in surface salinity, the picoplankton assemblage was the first to react, reaching a net biomass doubling rate of 0.43 $day^{-1}$, followed by an increase in nanoplankton and microplankton. At the end of the melting process and during all productive pulses observed in the IFZ, the microplankton fraction developed actively, constituting the most important fraction of the microbial biomass and nanoplankton prevailed over pico-sized organisms. Variations in pico-sized populations were already reported for Terra Nova Bay where picoplankton was of greater importance, reaching 64% of the total, when larger-sized biomass decreased (La Ferla et al. 1995; Bruni et al. 1995).

Our results indicate that the IFZ is a dishomogeneous area where it is possible to find different levels of trophic development. In fact, in the southernmost station, some peculiarities of the MIZ area were revealed by the relationship between pico- and nanobiomass as well as by the calculated ratios. At the next station, a high increase of micro- and nano-sized biomass occurred as well as an availability of more than 4 $g\,C\,m^{-2}$ of microbial biomass. Further north, there was evidence of the end of a productive pulse, with an intense biomass reduction. During 23–25 November, between stations 11 and 13, a frontal zone was observed. The northern limit of the IFZ (station 11) showed a reduction in microbial biomass with a tendency to increase the R:MB ratio, while at station 13 (MIZ) microbial biomass decreased. After 20 days, both stations reached more than 1.5 $g\,C\,m^{-2}$ of microbial biomass and a less evident front was observed. Between 25 November and 13 December, at the southern edge of the MIZ (stations 13 and 15), the microbial biomass increased 2.8 times, from 0.62 to 1.78 $g\,C\,m^{-2}$ (1.75 $g\,C\,m^{-2}$, subtracting the respired biomass).

The temporal variability observed at station 11 in 20 days showed once again a diverse evolution of the microbial component, with a decrease of 0.46 $g\,C\,m^{-2}\,day^{-1}$ followed by an increase of 0.13 $g\,C\,m^{-2}\,day^{-1}$. The microbial assemblage, dominated by microplankton, showed variable ratios between the differently sized fractions.

The highest biomass and respiratory activity within the entire study area occurred over the E-W transect. Station 11c showed a moderate trophic level which rose sharply to highest values at station 28. Here, also as a consequence of a marked halocline at approx. 40 m, ATP reached the highest values for any station, 628.74 $\mu g\,m^{-3}$, indicating a high trophism (Karl 1980). There was a surprisingly enhanced value of ETSa at station 28, corresponding to the drastic increase in ATP. We speculate that this ice-free station represents the high point of a microbial biomass growth process in which higher trophic webs have not yet been established. Station 29, located under the ice, was not representative of the ICZ probably due to lateral transport from adjacent ice-free stations. At station 30, relative to the well-established polynya, biomass values increased once again.

LPS concentrations showed fewer variations along the entire transect, with a significant increase in the ICZ. LPS values were higher but similar to those previously reported for Terra Nova Bay (La Ferla et al. 1995). Bacterial and microbial biomass were correlated in the IFZ. It is logical that high production rates will cause a release in exudates and carbon which are directly or indirectly used by bacteria. However, bacteria constituted only a mean of 2.5% of total biomass in this area. In the ICZ, on the contrary, bacterial biomass constituted a mean of 87.0 $mg\,C\,m^{-2}$ and represented 22.4% of microbial biomass. Our results are in agreement with those reported by other authors that bacterial abundance is greater in the under sea-ice upper layers, and that heterotrophic activity and the release of POC is higher (Delille 1992; Gieskes et al. 1989; Palmisano and Garrison 1993). This emphasizes the importance of bacterial processes in the ice-cover zone where the ice is about to thaw.

Despite the low temperatures, respiratory activity was also consistent with observations from temperate and tropical areas. In general, both under the ice and at the ice edge, the respiratory activity was comparable to the euphotic zones of the Southern Adriatic and Ionian Seas (La Ferla et al. 1996). Our rates were not as high as those found by Martinez and Estrada (1992) in the innermost pack-ice region or the outer marginal zones. Temperature rather than substrate availability seemed to limit microbial metabolism. Novitski and Morita (1977), during a study on psycrophilic *Vibrios*, noted a reduction in respiratory activity during starvation. Such a depression due to low temperature probably allowed for the conservation of C and cellular energy for long-term survival. Pomeroy et al. (1991) also asserted that bacterial respiration accounted only for 0.8 and 8.0% of PP, respectively, during early and late bloom events.

Our research has indicated that there is notable variability in plankton biomass due to the existence of different subsystems in relation to the evolution of ice melting, from the polynya to the pack ice. Moreover, the temporal series of samplings has shown that early ice-melt zones can support rapid changes in microbial biomass so that the various microbial compartments assume changing roles in different trophic conditions.

*Acknowledgements.* The research was supported by the Italian PNRA (National Antarctic Research Programme) Project 2b.2. The authors wish to thank Dr. Antonina Allegra, Dr. Maurizio Azzaro and Dr. Giuseppina Chiodo of Istituto Talassografico of the CNR (Messina) for their valuable contribution to the research and Stephen Conway for the translation and comments of the manuscript.

## References

Azam F, Fenchel T, Field JG, Gray JS, Meyer-Reil LA, Thingstad F (1983) The ecological role of water column microbes in the sea. Mar Ecol Prog Ser 10: 157–263

Bolter M, von Bodungen B, Liebezeit G, Meyer M (1988) The pelagic ecosystem of the Bransfield Strait, Antarctica: an analysis of microbiological, planktological and chemical characteristics by multivariate analyses. In: Sahrhage D (ed) Antarctic ocean and resources variability. Springer, Berlin Heidelberg New York, pp 160–166

Bruni V, La Ferla R, Acosta Pomar MTL, Salomone L (1995) Structural differences of the microbial community in two sites of the Terra Nova Bay (Ross Sea, Antarctica): a statistical analysis. Microbiologica 18: 409–422

Christensen JP, Packard TT, Dortch FQ, Minas HJ, Gascard JC, Richez C, Garfield PC (1989) Carbon oxidation in the deep Mediterranean Sea: evidence for dissolved organic carbon source. Global Biogeochem Cycles 3: 315–335

Delille D (1992) Marine bacterioplankton at the Weddell Sea ice edge: distribution of psychrophilic and psychrotrophic populations. Polar Biol 12: 205–210

Ducklow HW, Carlson CA (1992) Oceanic bacterial production. Adv Microb Ecol 12: 113–181

El-Sayed SZ (1988) Seasonal and interannual variabilities in Antarctic phytoplankton with reference to krill distribution. In: Sahrhage D (ed) Antarctic ocean and resources variability. Springer, Berlin Heidelberg New York, pp 101–119

Eppeley RW, Peterson BJ (1979) Particulate organic matter flux and planktonic new production in the deep ocean. Nature 282 (5740): 677–680

Fabiano M, Danovaro R, Crisafi E, La Ferla R, Povero P, Acosta Pomar MTL (1995) Particulate matter composition and bacterial distribution in Terra Nova Bay (Antarctica) during summer 1989–1990. Polar Biol 15: 393–400

Friedman EI, Thisle AB (1993) Antarctic microbiology. In: Friedman EI (ed) Antarctic microbiology. Wiley, New York, pp ix–x

Gieskes WW, Tilzer MM, Heusel R, Kraay G (1989) The underwater light climate. Ber Polarforsch 62: 75–83

Hanson RB, Lowery HK, Shafer D, Sorocco R, Pope HK (1983) Microbes in antarctic waters of the Drake Passage: vertical patterns of substrate uptake, productivity and biomass in January 1980. Polar Biol 2: 179–188

Hewes CD, Holm-Hansen O, Sakshaug E (1985) Alternate carbon pathways at lower trophic levels in the Antarctic food. In: Siegfried WR, Condy PR, Laws RM (eds) Antarctic nutrient cycles and food webs. Springer, Berlin Heidelberg New York

Hodson RE, Azam F, Carlucci AF, Fuhrman JA, Karl DM, Holm-Hansen O (1981) Microbial uptake of dissolved organic matter in McMurdo Sound, Antarctica. Mar Biol 61: 89–94

Holm-Hansen O, Paerl HW (1972) The applicability of ATP determination for estimation of microbial biomass and metabolic activity. Mem Ist Ital Idrobiol 29 Suppl: 149–168

Jacques G (1989) Primary production in the open Antarctic Ocean during austral summer. A review. Vie Milieu 39: 1–17

Karl DM (1980) Cellular nucleotide measurements and applications in microbial ecology. Microbiol Rev 44: 739–796

Karl DM (1993) Microbial processes in the southern oceans. In: Friedman EI (ed) Antarctic microbiology. Wiley, New York, pp 1–63

Karl DM, Tien C (1991) Bacterial abundances during the 1989–90 austral summer phytoplankton bloom in the Gerlache Strait. Antarct JUS 26 (5): 147–149

Karl DM, Holm-Hansen O, Taylor GT, Tien G, Bird DF (1991) Microbial biomass and productivity in the western Bransfield Strait, Antarctica during the 1986–87 austral summer. Deep Sea Res 28: 1029–1055

Kenner RA, Ahmed SJ (1975) Measurement of electron transport activities in marine phytoplankton. Mar Biol 33: 119–128

La Ferla R, Allegra A, Azzaro F, Greco S, Crisafi E (1995) Observation on the microbial biomass in two stations of Terra Nova Bay (Antarctica) by ATP and LPS measurements. Mar Ecol 16: 1–9

La Ferla R, Azzaro M, Chiodo G (1996) Microbial respiratory activity in the euphotic zone of the Mediterranean Sea. Microbiologica 19: 243–250

Marchant HJ, Murphy EJ (1994) Interactions at the base of the Antarctic food web. In: El-Sayed SZ (ed) Southern Ocean ecology. Cambridge University Press, Cambridge, pp 267–285

Martinez R, Estrada M (1992) Respiratory electron transport activity of microplankton in the Weddell Sea during early spring: influence of the ice cover and the ice edge. Polar Biol 12: 275–282

Naqvi SWA, Shailaja MS, Dileep Kumar M, Sen Gupta R (1996) Respiration rates in subsurface waters of the northern Indian Ocean: evidence for low decomposition rates of organic matter within the water column in the Bay of Bengala. Deep Sea Res II 43: 73–81

Novitsky JA, Morita RY (1977) Survival of a psycrophilic marine vibrio under long-term nutrient starvation. Appl Environ Microbiol 33: 635–641

Packard TT (1971) The measurement of respiratory electron transport system in marine phytoplankton. J Mar Res 29: 235–244

Packard TT (1985) Measurement of electron transport activity of microplankton. In: Jannasch HW, Williams PJleB (eds) Advances in aquatic microbiology 3. Academic Press, London, pp 207–261

Packard TT, Denis M, Rodier M, Garfiekd P (1988) Deep-ocean metabolic CO2 production: calculations from ETS activity. Deep Sea Res 35: 371–382

Palmisano AC, Garrison DL (1993) Microorganisms in Antarctic sea ice. In: Friedman EI (ed) Antarctic microbiology. Wiley, New York, pp 167–218

Pomeroy LR (1974) The ocean's food web, a changing paradigm. BioScience 24: 499–504

Pomeroy LR, Deibel D (1986) Temperature regulation of bacterial activity during the spring bloom in Newfoundland and coastal waters. Science 233: 359–361

Pomeroy LR, Wiebe WJ (1988) Energetics of microbial food webs. Hydrobiology 159: 7–18

Pomeroy LR, Wiebe WJ, Deibel D, Thompson RJ, Rowe GT, Pakulski JD (1991) Bacterial responses to temperature and substrate concentration during the Newfoundland spring bloom. Mar Ecol Prog Ser 75: 143–159

Porter KG, Feig YS (1980) The use of DAPI for identifying and counting aquatic microflora. Limnol Oceanogr 25 (5): 948–951

Sorokin YI (1971) Bacterial populations as components of oceanic ecosystem. Mar Biol 11: 101–105

Tupas LM, Koike I, Karl DM, Holm-Hansen O (1994) Nitrogen metabolism by heterotrophic bacterial assemblages in Antarctic coastal waters. Polar Biol 14: 195–204

Vincent WF (1988) Microbial ecosystem of Antarctica. Cambridge University Press, Cambridge

Vosjan JH, Nieuwland G (1987) Microbial biomass and respiratory activity in the surface waters of the east Banda Sea and northwest Arafura Sea (Indonesia) at the time of the southeast monsoon. Limnol Oceanogr 32(3): 767–775

Watson SW, Hobbie JE (1979) Measurement of bacterial biomass as lipopolysaccharide. In: Conserton JW, Colwell RR (eds) Native aquatic bacteria: enumeration, activity and ecology. ASTM, Philadelphia

Watson SW, Novitsky TJ, Quinby HL, Valois FW (1977) Determination of bacterial number and biomass in the marine environment. Appl Environ Microbiol 33(4): 940–946

CHAPTER 15

# Ultraphytoplanktonic Community Structure in the Ross Sea During the Austral Spring 1994

**S. Vanucci and V. Bruni**[1]

**ABSTRACT**

During the cruise ROSSMIZE leg 1 (November–December 1994), the spatial and temporal variation of the vernal ultraphyplanktonic (0.2–10 µm) community structure was investigated in relation to sea-ice conditions, hydrography and light condition along a south-north transect performed from the open waters of the Ross Sea polynya through its northern marginal ice zone into pack ice. Ultraphytoplankton abundance ranged within one and one-half orders of magnitude ($10^5$–$5 \times 10^6$ cells $l^{-1}$). Four main size spectrum features of the community could be distinguished and related to irradiance-vertical-mixing regime and to the distribution of different water masses. The abundance and the relative significance of 5–10 µm cells were higher in the surface waters of the recently ice-free area. Picophytoplankton (0.2–3 µm) reached its peak and dominated at the pycnocline depth of the marginal ice zone and all along the water column of the northernmost part of the transect occupied by Open Antarctic Surface Waters. Our data suggest that 5 µm may represent a boundary size in the response of the Antarctic phytoplankton structure to physical light climate.

## Introduction

Two basic concepts developed during the past two decades have to be considered central to our present understanding of the dynamics of the Southern Ocean ecosystem (Marchant and Murphy 1994): the first is the recognition that the Antarctic marine food web is by far more complex than originally envisaged, and comparable to the food webs present in ecosystems at lower latitudes (Hewes et al. 1985; Weber and El-Sayed 1987; Garrison 1991). The second is the realization of the importance of the marginal ice-edge zones as the most productive regions of the Antarctic marine ecosystem (Smith et al. 1990; Cota et al. 1991). There is a growing body of evidence that physical processes at the ice edge are important in the overall biomass, production and nutrient and sedimentation cycles of the Southern Ocean (Marchant and Murphy 1994; Knox 1994).

In accordance with the "critical depth" concept (Sverdrup 1953; Nelson and Smith 1991), water column stability, primarily due to the addition of less saline waters derived from sea-ice melting, provides a positive balance between vertical light penetration and vertical mixing. This has been identified as the most important factor controlling the vernal ice-edge phytoplankton blooms in the Southern Ocean (Smetacek and Passow 1990; Mitchell and Holm-Hansen 1991; Nelson and Smith 1991; Gleitz et al. 1994). Recent studies carried out in various seasonal retreating pack-ice zones have emphasized the great variability of phytoplankton biomass and community structure (Jacques and Panouse 1991; Becquevort et al. 1992; Schloss and Estrada 1994; Jochem et al. 1995). However, little information (Kivi and Kuosa 1994; Mura et al. 1995; Detmer and Bathmann 1997) is available on the abundance, structure and dynamics of ultraphytoplanktonic assemblages.

The aim of this study was to investigate spatial and temporal changes of the vernal ultraphytoplanktonic community structure in relation to hydrography and light regime at the marginal ice zone of the Ross Sea polynya. Particular attention was paid to picophytoplankton assemblages in order to provide additional information on the relative importance of both classical food chain

[1] Dipartimento di Biologia Animale ed Ecologia Marina, Università di Messina, Salita Sperone 31, 98166 S. Agata, Messina, Italy[1]

and microbial loop in relation to the combined effects of mixing and irradiance during the sea-ice retreat.

## Materials and Methods

### Sampling

The first leg of the Italian interdisciplinary ROSSMIZE (Ross Sea Marginal Ice Zone Ecology) research programme was carried out by the R.V. Italica from 15 November to 15 December 1994 in the Ross Sea. Water sampling was performed every 10 nautical miles, along the 175° E meridian from the open waters of the Ross Sea polynya through its northern marginal ice zone towards the pack ice. Several stations were investigated at different dates up to four times (Fig. 1,A,B,C). Seawater samples were taken in the upper 100–150 m at optical depths using an SBE 32/24 Carousel.

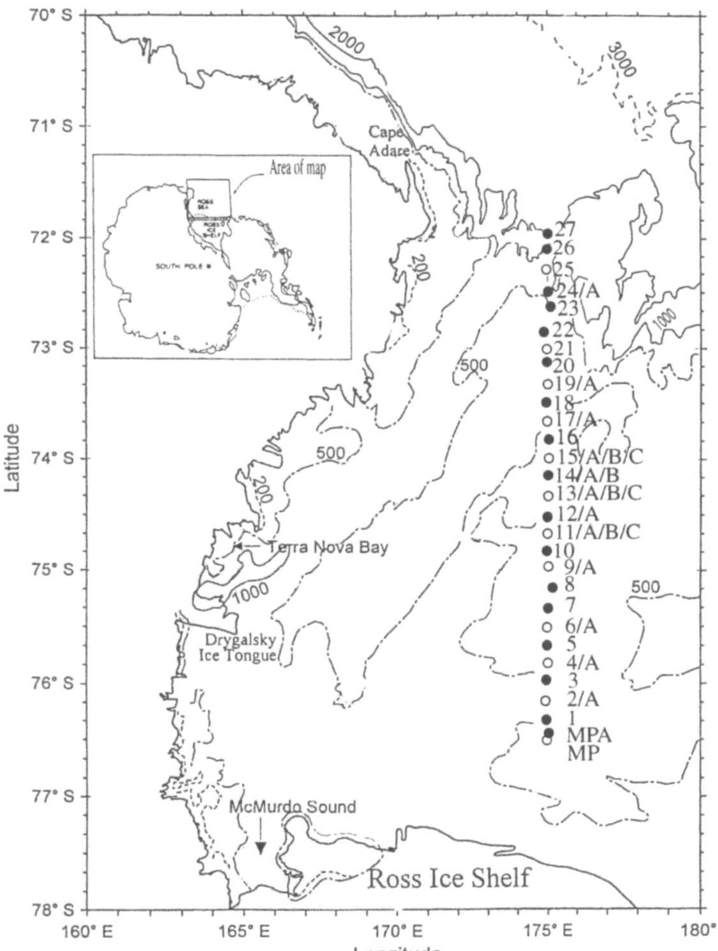


**Fig. 1.** Location of sampling stations. *Open circles* indicate stations sampled for ultraphytoplankton

### Optical and Physical Measurements

Quantum photosynthetically available radiation (PAR, 400–700 nm) was continuously recorded over the investigated period with a light sensor (LI-COR 196) mounted on the ship's bridge. Optical depths (50, 10, 5, 1, 0.1% of incident PAR) were determined by evaluation of underwater downwelling PAR profiles recorded by a photoprobe equipped with LI-COR quantum meters (Innamorati et al. 1994). Hydrographic data were recorded by means of an SBE 9/11 Plus CTD, mounted on the Carousel. The depth ($Z_{UML}$) of the upper mixed layer (UML) for every station was defined as a change in $\sigma_\theta \geq 0.05$ over a 10-m depth interval, determined by evaluation of CTD plots (Mitchell and Holm-Hansen 1991; Gleitz et al. 1994). The euphotic zone ($Z_{eu}$) was defined as the 1% light level depth and the density structure in the euphotic zone was used as a measure of stability. A stability index (SI) was calculated as proposed by Jochem et al. (1995):

$$SI = (\Delta\sigma_\theta/\Delta z) \times 1000, \text{ with } \Delta z = Z_{eu}.$$

### Microscope Observations

After collection, water samples were immediately prefiltered through 47-mm diameter 10-μm pore size Nuclepore filters to separate larger phytoplankton from the ultraplanktonic fraction <10 μm (Murphy and Haugen 1985; Tamigneaux et al. 1995). An aliquot was again filtered from each filtrate onto 47 mm diameter 2-μm pore size Nuclepore membrane to separate smaller cells (<2 μm). Both filtrates were preserved with 0.2 μm filtered paraformaldehyde 0.2% final concentration (Hall 1991) and stored in the dark at 4 °C until the slides were prepared aboard ship, within 1 h of sampling.

For larger (2–10-μm) ultraphytoplankton, subsamples were filtered at low pressure (<50 mm Hg) onto Irgalan Black-stained Nuclepore filters (2-μm pore size). Some slides were observed aboard to check cell concentration, while most of them were stored frozen until examination in the laboratory. Slides were examined for chlorophyll fluorescence using a Zeiss Axioplan epifluorescence microscope (filter set BP450–490/FT510/LP520). Cells were measured using an ocular micrometer at ×1000 magnification and classified into three size groups: <3 μm, 3–5 μm 5–10 μm. Additional SEM measurements were performed on <3-μm size cells to check for accuracy of size determinations. Although identification of autotrophic ultraplankton by epifluorescence microscopy is uncertain, most of the flagellates displayed two chloroplasts typical of the Prymnesiophyceae. Cryptophyceae, Bacillariophyceae, Dinophyceae and some large Prasinophyceae were distinguished by their typical shape.

For the determination of <2-μm size autotrophic cells, subsamples were filtered onto 0.2-μm black Nuclepore filters and stored at −20 °C in the dark until analysis. Cells were counted using the same epifluorescence microscope and filter set. In this study picophytoplankton (0.2–3 μm) was defined as cells ranging from 0.2 μm to 3 μm (i.e. the cells counted onto 0.2-μm pore size filters plus the cells <3 μm counted onto 2-μm pore size filters).

## Results

### Environmental Conditions and Physical Structure of the Water Column

Most of the investigation period was characterized by cloudy weather and modest solar irradiance flux (29 mol m$^{-2}$ day$^{-1}$, mean value). The physical structure of the water column along the study transect is shown in Fig. 2. A frontal structure off Mawson Bank, at the shelf break, between stations 24 and 26 was identified. Additional information on the hydrography of the sampling area is provided by Russo (this Vol.). On the basis of the temperature and salinity characteristics of the surface waters and ice cover, we distinguished four main areas, the boundaries of which shifted during the period of observation: (1) polynya open waters, (2) marginal-ice zone, (3) closed pack-ice zone and (4) northern area.

1. The polynya (stations 1–8) was occasionally covered by a thin (up to 20 cm) layer of ice and the salinity and the temperature of the surface waters ranged from 34.66 to 34.44 psu and from −1.80 to −1.74 °C, respectively. The upper 100–150 m of the water column was weakly stratified and a pycnocline (30 m) could only be identified at station 2. $Z_{eu}$ did not exceed 25–30 m and SI was low (Fig. 3).
2. In the marginal-ice zone (stations 9–13), ice cover ranged from 5 to 7 tenths and salinity values were lower (34.35 psu, mean value) than

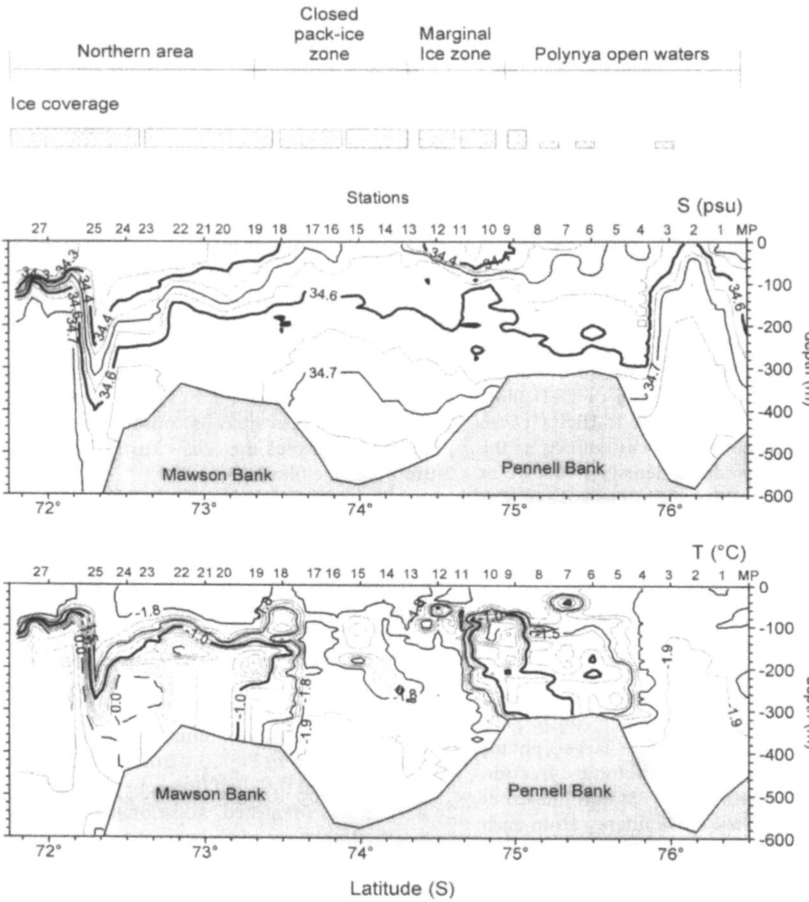


**Fig. 2.** Temperature and salinity distribution along the study transect. (modified from this Vol. Russo)

in polynya due to the ice melting. The stratification of the surface layers was generally weak and the depth of the upper mixed layer ($Z_{UML}$) ranged between 70 and 35 m at stations 9 and 12, respectively.

3. In the closed pack-ice zone (stations 15–17; ice cover was 7–8 tenths) the lowest temperature (−1.84 °C) and the highest salinity (34.52 psu) values were recorded at station 15 where the water column was deeply mixed. Stronger stabilization of the upper water layers (40–50 m) occurred 2 weeks later due to the northwards ice-edge retreat and to the atmospheric heating of ice-free surface waters (Fig. 3b).
4. In the northern area (stations 19–27), at the beginning of December, ice cover conditions were variable (6–9 tenths); heavy pack ice was interrupted by relatively broad stretches of ripple open waters due to wind stress. $Z_{eu}$ ranged between 75 and 95 m (Fig. 3a). A low density ($\sigma_\theta = 27.67$) water mass, corresponding to Antarctic Surface Water, was confined in the upper 40 m at station 19 and extended down to 200 m at station 25. At the shelf break, the intrusion of Lower Circumpolar Deep Water

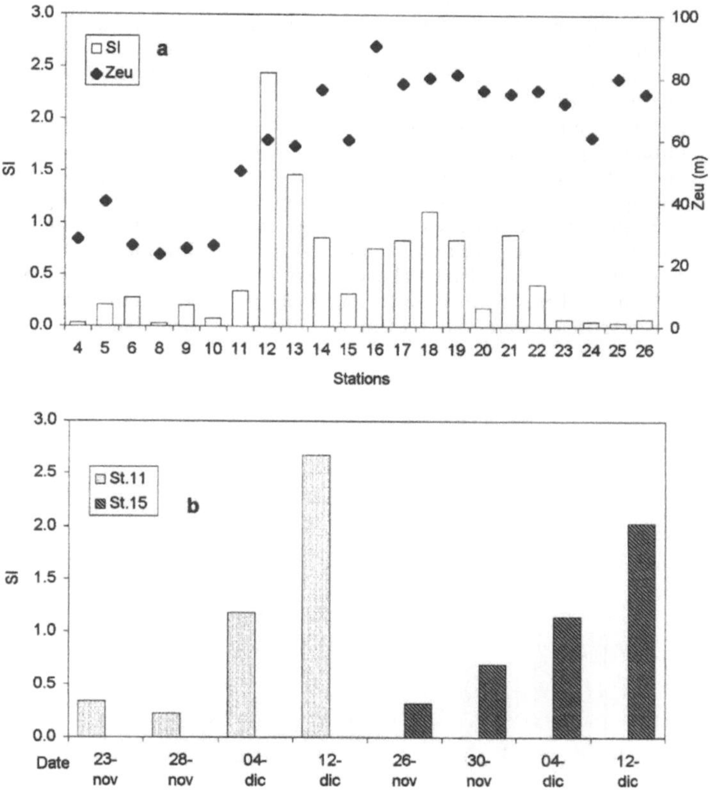


**Fig. 3a,b.** Spatial distribution of stability index and depth of the euphotic zone (**a**). Temporal variation of stability index at stations 11 and 15 (**b**)

on the Ross Sea continental shelf and its mixing with surface waters were also detected (Russo this Vol.).

## Ultraphytoplanktonic Abundance and Size Structure

During the study period, the total ultraphytoplanktonic abundance (0.2–10 μm) ranged from $1.3 \times 10^5$ to $4.9 \times 10^6$ cells $l^{-1}$ (station 6, 150 m and station 21, 4 m, respectively), with an average value of $1.4 \times 10^6$ cells $l^{-1}$. Generally, highest cell numbers were found in the upper 30 m. Cells 5–10 μm in size, mostly cryptophytes and diatoms, accounted, on average, for 10% of cell abundances while 3–5-μm size cells, mostly prymnesiophytes, contributed for 50% of the total cell density. Picophytoplankton (0.2–3 μm) showed a mean value of $5.89 \times 10^5$ cells $l^{-1}$ and accounted for 40% of the total abundance.

Ultraplanktonic communities displayed wide changes in abundance and size structure among the four hydrographic areas. Surface (4 m) and integrated values for the upper 100 m of the water column were found to be closely correlated ($p < 0.01$, $r = 0.75$) so that only the surface values are illustrated in Fig. 4.

1. In polynya open waters, in mid-November, the size structure of the ultraplanktonic community was characterized by the dominance of 3–5-μm size cells throughout the water column. A high proportion of these was represented by

colonial stage cells of *Phaeocystis* sp. which reached densities up to $1.7 \times 10^6$ cells $l^{-1}$ accounting for about 50% of the total cell density (station 2, 20 m).

2. In the marginal ice zone, at station 11, surface ultraphytoplanktonic abundance was about one order of magnitude lower than in the polynya and a sharp change in the size spectrum structure was observed. Such change was mainly due to the depletion of most 5–10 μm size cells and to the substantial decrease of the 3–5 μm flagellates. Some colonial stage cells of *Phaeocystis* sp. were still present but disappeared northwards. Changes in picophytoplanktonic density were comparatively modest and limited to the surface layer so that the picophytoplanktonic contribution to the total community increased (Fig. 4a).
3. In the upper layers close to the pack ice, relatively high contributions of larger cells (5–10 μm) were observed (25–30% of the total ultraphytoplankton density).
4. In the northern area, picophytoplankton reached highest densities ($1–4 \times 10^6$ cells $l^{-1}$) accounting for 70–90% of the total cell abundance.

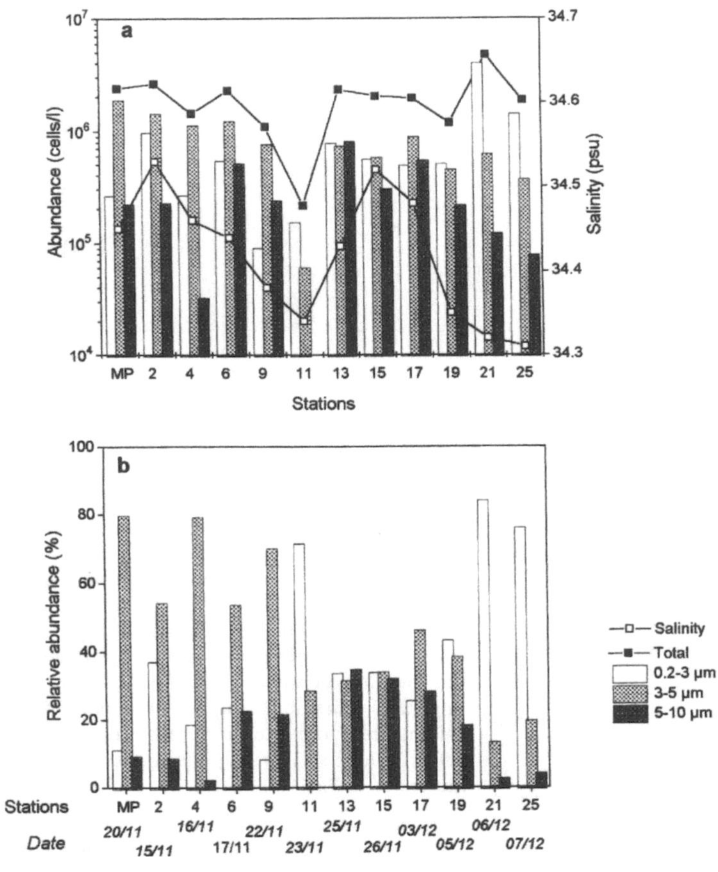


**Fig. 4a,b.** Total ultraphytoplankton abundance and densities of the three size classes at 4-m depth (**a**). Relative importance of the three size classes at 4-m depth (**b**). The superimposed continuous curves in **a** represent total ultraphytoplanktonic cell abundance and salinity at the 4-m depth

## Vertical Distribution and Temporal Variations of the Ultraphytoplanktonic Community

Vertical changes in the relative significance of the different size classes of the ultraphytoplanktonic community are reported in Fig. 5. Synoptic data indicate an increase in the significance of the picophytoplankton northwards. Temporal changes in ultraphytoplanktonic abundance and composition, at three example stations: station 11 (marginal zone), station 15 (zone close to the pack ice) and station 19 (northern area), are represented in Fig. 6. As stratification of the upper layer progressed, a clear increase of the 5–10-μm size cell abundance was observed (Fig. 6b,d). Under these conditions, vertical patterns of picophytoplankton abundance showed the highest values at the top of the pycnocline that coincided with the $Z_{eu}$ depth.

## Relationship with Environmental Factors

In order to evaluate some environmental factors driving the size structure of ultraphytoplankton, integrated cell densities were set in relation to the $Z_{UML}$, $Z_{eu}$, subsurface incident PAR, and SI. None of the biological variables showed a significant

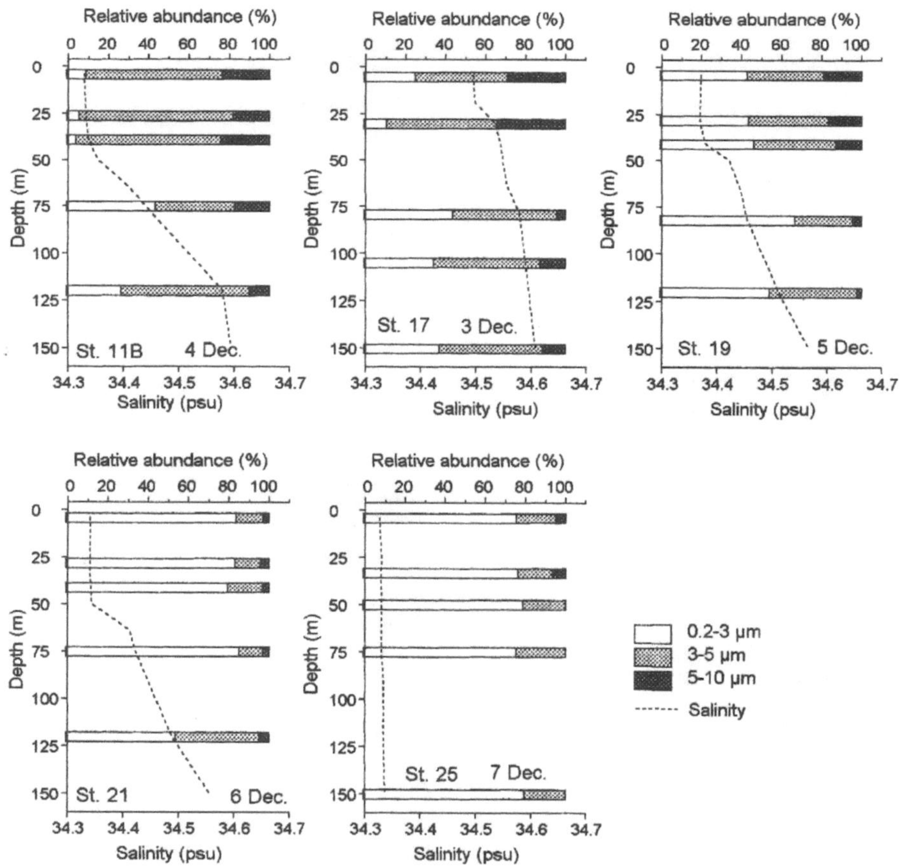


**Fig. 5.** Salinity (S) profiles and vertical size structure of ultraphytoplanktonic community – each class, 0.2–3, 3–5, 5–10 μm, is expressed as relative abundance of the total – at synoptic stations

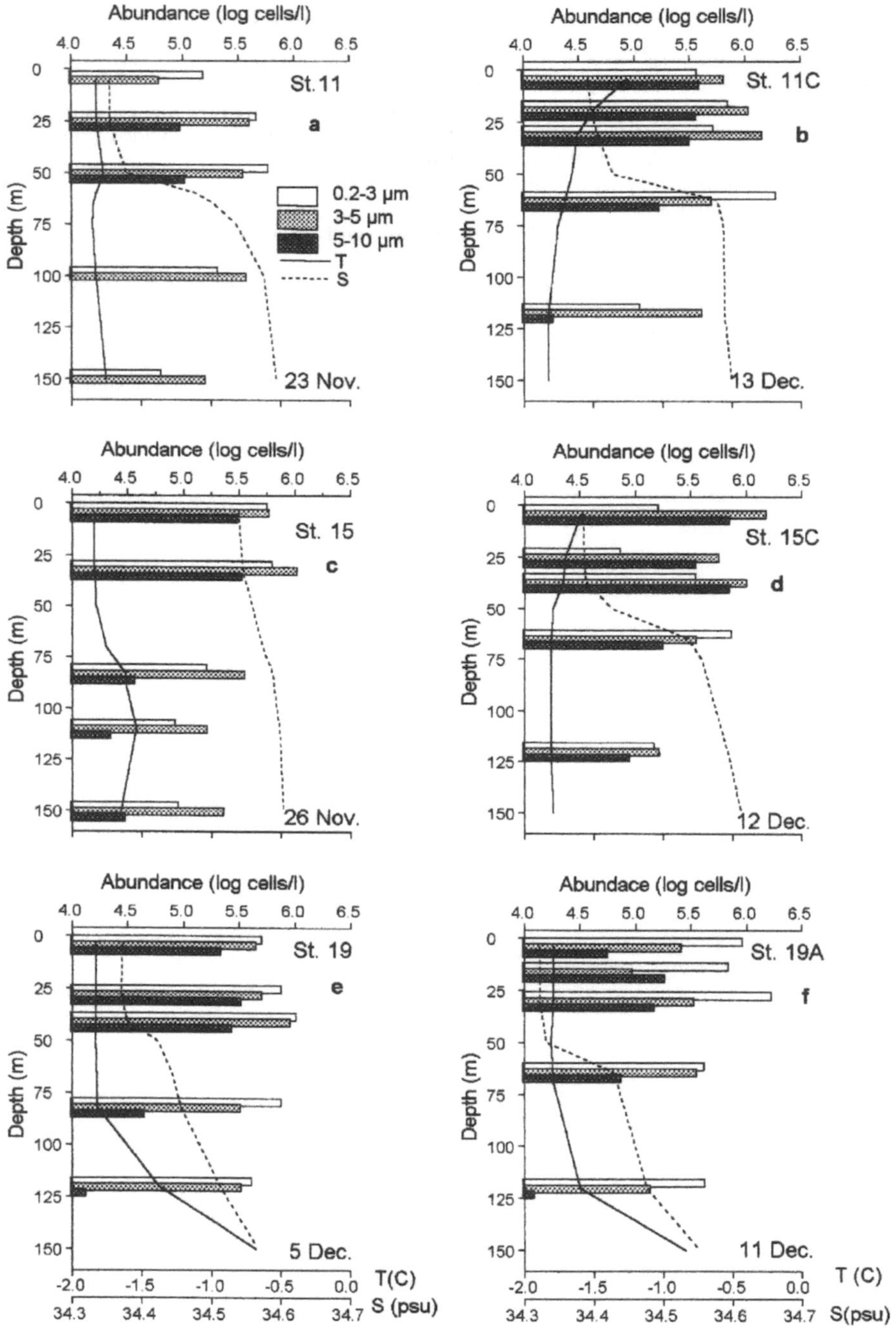


**Fig. 6a–f.** Temporal variation of the abundance of the three ultraphytoplanktonic size classes (0.2–3, 3–5, 5–10 µm) at three time-series stations. Temperature (*T*) and salinity (*S*) profiles

correlation with $Z_{UML}$ and PAR (plots not depicted).

By contrast, total ultraphytoplankton cell density (Fig. 7a,e) showed positive correlations with $Z_{eu}$ and SI (n = 19; $p < 0.05$ and $p < 0.01$, respectively). The 5–10-μm size class was positively correlated to SI (n = 19, $p < 0.01$; Fig. 7f), while 3–5-μm size cells did not show correlation with any examined variable. A positive correlation was also observed between picophytoplankton abundance and $Z_{eu}$ (n = 19, $p < 0.05$; Fig. 7d). The ratio of the 3–5-μm to 5–10-μm size cells correlated negatively with SI (n = 19, $p < 0.05$; Fig. 7k), while no significant correlations were found between SI and other cell size ratios.

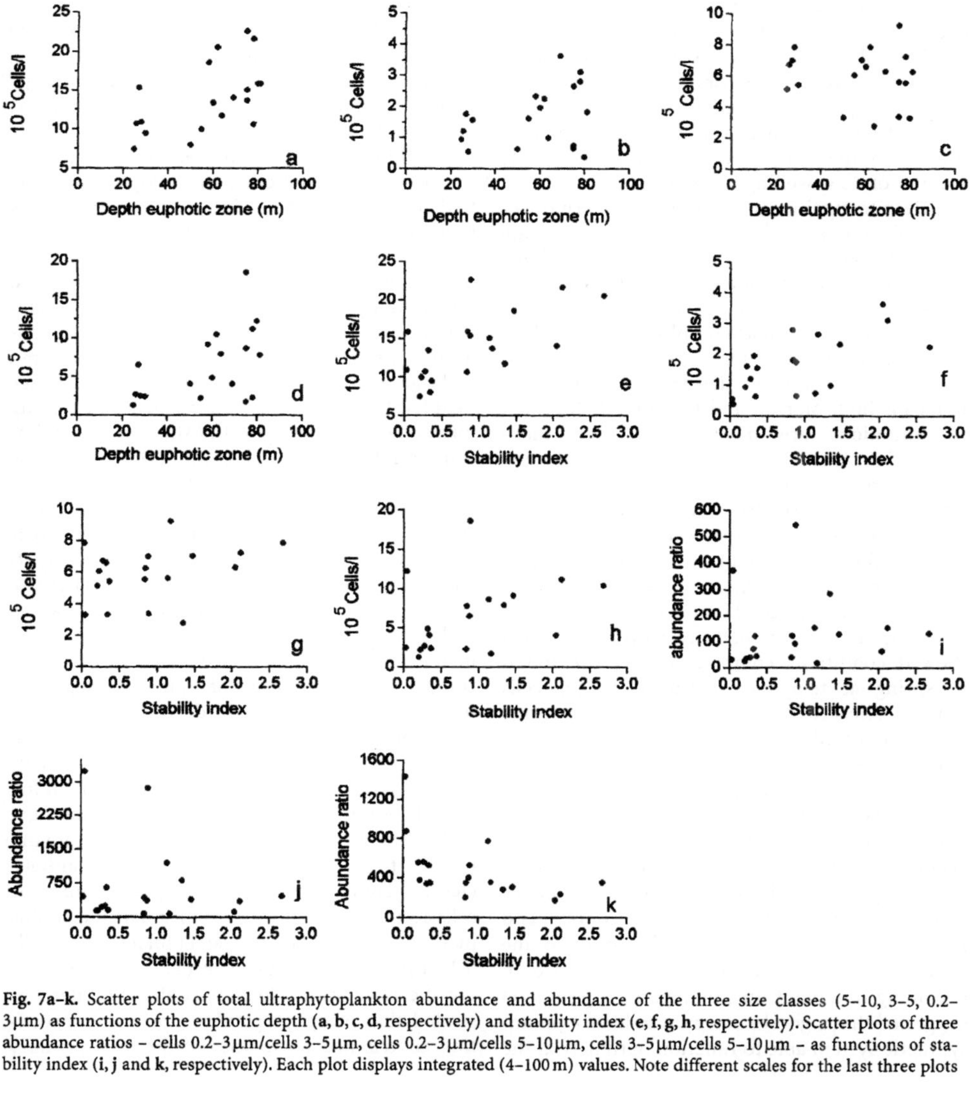


**Fig. 7a–k.** Scatter plots of total ultraphytoplankton abundance and abundance of the three size classes (5–10, 3–5, 0.2–3 μm) as functions of the euphotic depth (**a**, **b**, **c**, **d**, respectively) and stability index (**e**, **f**, **g**, **h**, respectively). Scatter plots of three abundance ratios – cells 0.2–3 μm/cells 3–5 μm, cells 0.2–3 μm/cells 5–10 μm, cells 3–5 μm/cells 5–10 μm – as functions of stability index (**i**, **j** and **k**, respectively). Each plot displays integrated (4–100 m) values. Note different scales for the last three plots

## Discussion

During the investigated period, the polynya retreating ice edge moved northward and the polynya reached its maximum extent flowing into open waters. The combined effects of the presence of different water masses into the study area and warm cores originated by the mixing of Lower Circumpolar Deep water with Antarctic Surface waters, influenced the water circulation (Russo this Vol.). This fact, together with local wind stress, could be responsible for the ice cracks and breakups observed also by NOAA satellite images. This highly variable environment, combined with the local light regime and thermodynamic processes, determined large variation, in the density profiles of the upper water column and in the amount of downwelling irradiance flux.

Despite significant ice retreat, no vernal ice-edge blooms were observed during the study period. Data on phytoplankton biomass as chlorophyll a were low (Carrada et al. 1997): highest Chl a values (1–3 $\mu g l^{-1}$) were observed in the surface waters of the polynya, where colonial stage cells of *Phaeocystis* accounted for 75–95% of the phytoplanktonic assemblages (determined by the Utermöhl method; Marino and Cabrini 1997). These concentrations are much lower than those reported in previous studies on the Ross Sea during summer (Smith et al. 1990; DeMaster et al. 1992). As can be inferred from changes in the abundance and size spectrum of the ultraphytoplanktonic community, the establishment of optimal irradiance-mixing regime for the beginning of >5-µm size cell blooms occurred in mid-December. In this period, the recently ice-free area showed conditions comparable to those found in spring at temperate latitudes. Larger cells (>5 µm) released from the melting sea ice, due to the increased SI, coupled to the lack of tight grazing pressure, would be able to "seed" the nutrient-rich water column (Knox 1994). These data suggest that our study did not last long enough for micro- and larger nanophytoplanktonic blooms to occur. By contrast, in these transiently stratified waters, picophytoplankton, having a better quantum absorption efficiency (Morel and Bricaud 1981) took advantage at the bottom of the euphotic zone, increasing their significance in becoming dominant. In this regard, Fogg (1991) suggested that a good strategy is to have a photosynthetic system able to function optimally at the depth where vertical mixing is reduced and nutrient concentrations are higher, i.e. at the bottom of the photic zone.

The evident dominance of picophytoplanktonic cells in northern stations (Fig. 5) appears to be related to surface water hydrographic constraints (Fig. 6e,f) rather than to successional stages of the ultraphytoplanktonic community associated to ice-edge processes occurring over the Ross Sea continental shelf. The dominance of picophytoplanktonic cells was also reported in terms of chlorophyll a concentration in the same area from synoptic samples (Carrada et al. 1997); in this area, indeed, the lowest total chlorophyll a concentrations (<0.3 $\mu g$ Chl a $l^{-1}$) and the highest picophytoplanktonic contribution (up to 60–80%) to the chlorophyll a content were observed.

In northern stations picophytoplankton, due to their capability to survive and grow abundantly even in deep vertical mixed waters (Throndsen and Kristiansen 1991), seem to display better competitive advantage than larger phytoplankton also under inorganic nutrient-rich conditions (Catalano et al. 1997). In fact, because of rough sea conditions, stratification of the surface water column at the ice edge is not established so that the average light levels are low within the mixed layer, limiting microphytoplankton growth (Mitchell and Holm-Hansen 1991; Nelson and Smith 1991). Under these conditions eukaryotic pico-size cells could have an advantage due to their negligable sinking rate (Fogg 1986), better quantum efficiency than larger cells (Morel and Bricaud 1981) and high light-harvesting capacity (Shapiro and Guillard 1986). In other words, they appear to be better adapted to deep vertical mixing than larger cells. Therefore in the north-ern area, the driving force for the competitive advantage of picophytoplankton over larger cells could be the deep vertical mixing, also under nutrient-rich conditions. Moreover, it cannot be excluded that inorganic nutrients would support eukaryotic picophytoplankton growth (Mousseau et al. 1996). This is consistent with data reported by Detmer and Bathmann (1997) in the Atlantic sector of the Southern Ocean. They found that picophytoplanktonic cells accounted for 60–80% of the total autotrophic pico- plus nanoplankton assemblages (i.e. cells 0.2–20 µm in size) in deep vertical mixing waters. For the same waters, Jochem et al. (1995) demonstrated that picophytoplankton primary production contributed up to 75% of the total primary production and that picophytoplankton primary production showed no relation to water column stability.

Our data suggest that picophytoplankton could play an important role in the trophic organization of the northernmost part of the system. In these waters, primary production is probably originated and recycled within the smaller components of the pelagic community. This is apparently in contrast to temperate and tropical seas where picophytoplankton dominates in strongly stratified waters and under oligotrophic conditions (Cushing 1989). On the other hand, in these waters picophytoplankton is represented mostly by prokaryotes which are more abundant than eukaryotic pico-size cells by 1 order of magnitude (Glover et al. 1985; Joint 1986). By contrast, in the Southern Ocean, cyanobacteria are much less important and picophytoplankton is mostly represented by eukaryotic cells (Kivi and Kuosa 1994; Detmer and Bathmann 1997). Such different compositions of the picophytoplanktonic assemblages imply that the two picophytoplanktonic components have different physiological characteristics (Marchant et al. 1987; Tamigneaux et al. 1995; Mousseau et al. 1996), which may result in a competitive advantage of eukaryotes over prokaryotes in polar seas.

The positive correlation found between the depth of the euphotic zone and picophytoplankton cell abundance (Fig. 7d) can be considered an indirect consequence of the picophytoplanktonic competitive advantage with respect to larger cells. In these waters, due to the dominance of small cells, both self-shading and light absorption may be considered negligible, as well as their effects on the depth of the euphotic zone.

The low cell density encountered at station 11 in the upper 50 m, within a lens of low salinity water influenced by ice melting (Fig. 6a), was accompanied by highest bacterioplankton abundance ($>1.0 \times 10^8$ cells $l^{-1}$ in the upper 100 m; Bruni et al. 1997) recorded during the cruise. Such densities followed the passage of very large krill swarms, together with the presence of remarkable populations of penguins and whales (Azzali and Kalinowsky, this Vol.). These conditions led the authors to hypothesize that changes in phytoplanktonic abundance and size structure were related to an intensive size-selective grazing pressure (Priddle et al 1986; Jacques and Panouse 1991). However, our data do not allow us to clarify whether at that time krill fed on the algae released from the melting ice or on a well-developed bloom.

## Conclusions

Recent studies in Antarctica have shown that the macro pathway is preponderant wherever total primary production is high and the micro pathway is dominant in situations where vertical mixing is deep and phytoplankton biomass low (Nelson and Smith 1991; Marchant and Murphy 1994). In this regard our data allow us to highlight that:

1. In the microbial pathway, autotrophic organisms may be represented by pico- and <5 µm size cells rather than "actual" nano-size organisms, as often reported (Holm-Hansen and Mitchell 1991; Jacques and Panouse 1991). In fact, during our study both in the polynya and in the northern area (where water stability was weak) phytoplankton >5 µm in size was negligeable. This hypothesis is also in accordance with data of Detmer and Bathmann (1997). They reported that the size class 5–20 µm of the 0.2–20 µm autotrophic cell assemblage was always of minor importance (e.g. 2–10 and 5–22% in the Antarctic Circumpolar Current and ice-covered marginal-ice zone, respectively). Our results pointed out that cells larger than 5 µm are sensitive to SI (Fig. 7f). The 5-µm size may represent a boundary size in the response of Antarctic phytoplankton to the physical/light regime; in fact, changes in abundance of cells >5 µm, rather than their relative significance, seem to be responsible for the observed ultraphytoplanktonic structure in recently stratified waters. Cushing (1989) reported that the 5-µm size was critical for discriminating between the traditional food chain and the microbial loop, identifying the predominance of the latter food web in strongly stratified oligotrophic environments. Our data allow us to implement this concept, indicating that in Antarctic waters the microbial loop may represent the dominant pathway in deeply mixed waters under non-limiting conditions.
2. The microbial pathway, considering the usual situations in the ice-free zone south of the Antarctic convergence and in the seasonal pack-ice zone as a postbloom condition (Knox 1994), may constitute the vernal-summer

autotrophic component of the open Antarctic ice edge ecosystem as suggested by other authors (Jochem et al. 1995). The northernmost area may represent an ecotone between the community associated with the Ross Sea continental shelf receding ice edge and the community with the open Antarctic receding ice edge. The two autotrophic communities differ with respect to light/mixing-history patterns and rates of organic matter transfer.

The role of picophytoplankton and, in general, of <5 μm autotrophic cells in the Antarctic phytoplankton dynamics seems to be much greater than once hypothesized especially considering the extension of the Antarctic regions where hydrographic and physical factors seem to select these size classes. Additional information is needed to evaluate the transfer efficiencies of the carbon pathways at the base of the food web in the Antarctic marine ecosystem.

*Acknowledgements.* We thank Prof. M. Fabiano, University of Genova, and Dr. R. Danovaro, University of Ancona, for useful comments and suggestions on the manuscript. This research was undertaken in the framework of the X Italian Antarctic Expedition. We acknowledge support from the Programma Nazionale di Ricerche in Antartide (PNRA).

## References

Becquevort S, Mathot A, Lancelot C (1992) Interactions in the microbial community of the marginal ice zone of the northwestern Weddell Sea through size distribution analysis. Polar Biol 12: 211–218

Bruni V, Maugeri TL, Allegra A, Gugliandolo C (1997) Spatial and temporal distribution of picoplankton and culturable heterotrophic bacteria in the Ross Sea (Antarctica). In: Faranda F, Guglielmo L, Povero P (eds), National Programme for Antarctic Research, ROSSMIZE 93–95, Data Rep Part I, Genova, pp 361–372

Carrada GC, Mangoni O, Russo GF, Saggiomo V, Sgrosso S (1997) Spatial and temporal distribution of phytoplankton pigments in the Ross Sea during the austral spring 1994. In: Faranda F, Guglielmo L, Povero P (eds) National Programme for Antarctic Research, ROSSMIZE 93–95, Data Rep Part I, Genova, pp 197–228

Catalano G, Predonzani S, Rivaro P, Ruffini S (1997) Dissolved oxygen and nutrient distribution in the Ross Sea during the 1994 Austral Spring (R/V "Italica" cruise, leg 1). In Faranda F, Guglielmo L, Povero P (eds) National Programme for Antarctic Research, ROSSMIZE 93–95, Data Rep Part I, Genova, pp 111–134

Cota GF, Kottmeier ST, Robinson DH, Smith WO Jr, Sullivan CW (1991) Bacterioplankton in the marginal ice zone of the Weddell Sea: biomass, production and metabolic activities during austral autumn. Deep Sea Res 37: 1145–1167

Cushing DH (1989) A difference in structure between ecosystems in strongly stratified waters and in those that are only weakly stratified. J Plankton Res 11: 1–13

DeMaster DJ, Dunbar RB, Gordon LI, Leventer AR, Morrison JM, Nelson DM, Nittrouer CA, Smith WO Jr (1992) Cycling and accumulation of biogenic silica and organic matter in high latitude enviroments: the Ross Sea. Oceanography 5: 146–153

Detmer AE, Bathmann UV (1997) Distribution patterns of autotrophic pico- and nanoplankton and their relative contribution to algal biomass during spring in the Atlantic sector of the Southern Ocean. Deep Sea Res 44: 299–320

Fogg GE (1986) Picoplankton. Proc R Soc Lond B Bio Sci 228: 1–30

Fogg GE (1991) The phytoplanktonic ways of life. Tansley review no 30. New Phytol 118: 191–232

Garrison DL (1991) An overview of the abundance and role of protozooplankton in Antarctic waters. J Mar Syst 2: 317–331

Gleitz M, Bathmann UV, Lochte K (1994) Build-up and decline of summer phytoplankton biomass in the eastern Weddell Sea, Antarctica. Polar Biol 14: 413–422

Glover HE, Smith AE, Shapiro L (1985) Diurnal variations in photosynthetic rates: comparisons of ultraphytoplankton with a larger phytoplankton size fraction. J Plankton Res 7: 519–535

Hall JA (1991) Long-term preservation of picophytoplankton for counting by fluorescence microscopy. Br Phycol J 26: 169–174

Hewes CD, Holm-Hansen O, Sakshaug E (1985) Alternate carbon pathways at the lower trophic levels in the Antarctic food web. In: Siegfried WR, Condy PR, Laws RM (eds) Antartic nutrient cycles and food web. Springer, Berlin Heidelberg New York, pp 277–283

Holm-Hansen O, Mitchell BG (1991) Spatial and temporal distribution of phytoplankton and primary production in the western Bransfield Strait region. Deep Sea Res 38: 961–980

Innamorati M, Massi L, Lazzara L, Cartei P (1994) Biomassa fitoplanctonica e proprietà ottiche del mare. Biol Mer Medit 11: 83–88

Jacques G, Panouse M (1991) Biomass and composition of size fractionated phytoplankton in the Weddell-Scotia Confluence area. Polar Biol 11: 315–328

Jochem F, Mathot S, Quéguiner B (1995) Size-fractionated primary production in the open Southern Ocean in austral spring. Polar Biol 15: 381–392

Joint IR (1986) Physiological ecology of picoplankton in various oceanographic provinces. Can Bull Fish Acquat Sci 214: 287–309

Kivi K, Kuosa H (1994) Late winter microbial communities in the western Weddell Sea (Antarctica). Polar Biol 14: 389–399

Knox GA (1994) The biology of the Southern Ocean. Cambridge University Press, Cambridge, 444 pp

Marchant HJ, Murphy EJ (1994) Interactions at the base of the Antarctic food web. In: El-Sayed SZ (ed) Southern Ocean ecology: the BIOMASS perspective. Cambridge University Press, Cambridge

Marchant HJ, Davidson AT, Wrigth SW (1987) The distribution and abundance of chroococcoid Cyanobacteria in the Southern Ocean. Proc NIPR Symp Polar Biol 1: 1–9

Marino D, Cabrini M (1997) Distribution and succession of phytoplankton populations in the Ross Sea (Antarctica) during the austral spring 1994. In: Faranda F, Guglielmo L, Povero P (eds) National Programme for Antarcic Research, ROSSMIZE 93-95, Data Rep Part I, pp 307–341

Mitchell BG, Holm-Hansen O (1991) Observations and modeling of the Antarctic phytoplankton crop in relation to mixing depth. Deep Sea Res 38: 981–1007

Morel A, Bricaud A (1981) Theoretical results concerning light absorption in a discrete medium and application to specific absorption of phytoplankton. Deep Sea Res 28: 1375–1393

Mousseau L, Legendre L, Fortier L (1996) Dynamics of size-fractionated phytoplankton and trophic pathways on the Scotian Shelf and at the shelf break, Northwest Atlantic. Aquat Microb Ecol 10: 149–163

Mura MP, Satta MP, Agustì S (1995) Water-mass influences on summer phytoplankton biomass and community structure. Polar Biol 15: 15–20

Murphy LS, Haugen EM (1985) The distribution and abundance of phototrophic ultraplankton in the North Atlantic. Limnol Oceanogr 30: 47–58

Nelson DM, Smith WO Jr (1991) Sverdrup rivisited: critical depths, maximum chlorophyll levels, and the control of Southern Ocean productivity by the irradiance-mixing regime. Limnol Oceanogr 36: 1650–1661

Priddle J, Hawes I, Ellis-Evans JC, Smith TJ (1986) Antarctic aquatic ecosystems as habitats for phytoplankton. Biol Rev 61: 199–238

Schloss I, Estrada M (1994) Phytoplankton composition in the Weddell-Scotia Confluence area during austral spring in relation to hydrography. Polar Biol 14: 77–90

Shapiro LP, Guillard RRL (1986) Physiology and ecology of the marine eukaryotic ultraplankton. Can Bull Fish Aquat Sci 214: 371–390

Smetacek V, Passow U (1990) Spring bloom initiation and Sverdrup's critical-depth model. Limnol Oceanogr 35: 228–234

Smith WO Jr, Kelly HP, Rich JH (1990) Chlorophyll distribution and primary production in the Ross Sea, austral summer 1990. Antarct J US 25: 179–182

Sverdrup HU (1953) On conditions for the vernal blooming of phytoplankton. J Cons Cons Perm Int Explor Mer 18: 287–295

Tamigneaux E, Vazquez E, Mingelbier M, Klein B, Legendre L (1995) Evironmental control of phytoplankton assemblages in nearshore marine waters, with special emphasis on phototrophic ultraplankton. J Plankton Res 7: 1421–1448

Throndsen J, Kristiansen S (1991) *Micromonas pusilla* (Prasinophyceae) as part of pico- and nano-plankton communities in the Barents Sea. Polar Res 10(1): 201–207

Weber LH, El-Sayed SZ (1987) Contribution of the net-, nano- and picoplankton to the standing crop and primary productivity in the Southern Ocean. J Plankton Res 9: 973–994

CHAPTER 16

# Picoplankton Abundance and Biomass at Terra Nova Bay (Ross Sea, Antarctica) During the 1989–1990 Austral Summer

**M.L.C. Acosta Pomar, T.L. Maugeri, and V. Bruni**[1]

**ABSTRACT**

Temporal, depth abundance and biomass of the picoplankton (cells ranging between 0.2 and 2 μm, which includes heterotrophic and autotrophic bacteria, coccoid Cyanobacteria and small eukaryotic algae) were investigated at two stations in Terra Nova Bay (Antarctica) during the 1989–90 austral summer. The picophytoplankton was considered separately, and prokaryotic and eukaryotic cells distinguished. Biomass was calculated from biovolumes obtained on the basis of cell counts and measurements carried out by scanning electron microscopy and laser scanning microscopy, using appropriate conversion factors. Picoplankton densities ranged from $7.2 \times 10^{7}$ to $2.9 \times 10^{10}$ cells $l^{-1}$ and biomass from 3.9 to 1584 μg C $l^{-1}$, increasing from the middle of January onwards. Picophytoplankton was not recorded below the 100-m depth and was on average only 0.4% of the whole picoplanktonic community. The Cyanobacteria represented only a small part of the phototrophic picoplankton. The neritic station (MER) had highest numbers and biomass. Total picoplankton abundance and chlorophyll a concentrations were uncoupled at both stations.

## Introduction

Studies on size-fractionated plankton have increasingly emphasized the role played by the <20-μm microbial community (Sieburth and Davis 1982; Lenz 1992; Fogg 1995). The importance of the microbial food web and of the microbial loop in the pelagic marine ecosystem is now appreciated (Azam et al. 1983). The use of differential filtration has allowed the separation of planktonic organisms into different size-classes: micro-(20–200 μm), nano-(2–20 μm) and picoplankton (0.2–2 μm), and have shown that microbial processes are dominant not only in oligotrophic but also in eutrophic waters. They have further demonstrated that the processes of larger organisms are merely superimposed upon them (Fogg 1995). Relationships among different-sized organisms and the distinction between autotrophic and heterotrophic biomasses have allowed us to evaluate the structure and trophic dynamics of the planktonic microbial community.

Picoplankton is an integral component of the microbial community, in which heterotrophic bacteria are the more abundant organisms. Its autotrophic fraction, both prokaryotic and eukaryotic, is ubiquitous and very important as primary producer in the oceans. Several studies have shown that <2-μm cells can also be an important component in the Antarctic marine ecosystem, even though their quantity and composition are highly variable (Karl 1993; Fogg 1994; Marchant and Murphy 1994).

The evaluation of bacterial biomass is of special interest as particulate carbon and nitrogen pool relative to other major particulate pools of these elements which participate in nutrient cycling and sinking flux (Cho and Azam 1990). The contribution of smaller phytoplanktonic cells to the global ocean biomass is important because picophytoplankton have been shown to contain a greater quantity of volume-specific C and N relative to larger phototrophs, since the former have a lower water content (Verity et al. 1992).

In this chapter we report the pico- and picophytoplankton biomass variations at two stations in Terra Nova Bay (Ross Sea) during the 1989–1990 austral summer and examine the degree of

[1] Dipartimento di Biologia Animale ed Ecologia Marina, Salita Sperone 31, I - 98166 S. Agata Messina, Italy

coupling between autotrophic and micro-heterotrophic pelagic assemblages. In particular, we estimated biomass by counting cells with an epifluorescent microscope, and measured their volume with a scanning electron microscope and a laser scanning microscope. Cells observed by epifluorescence and scanning electron microscopy were used directly to analyze images (Krambeck et al. 1981; Van Wambeke 1988; Sieracki et al. 1989). In order to draw inferences about algal-bacterial interactions, we also performed a simple linear regression analysis on the data.

## Materials and Methods

### Station and Sample Collection

Seawater samples were collected in Terra Nova Bay (Ross Sea) between January 2 and February 4 1990, on alternate days at two stations: Mergellina (MER: 76°41′42″S, 164°07′E), and Santa Maria Novella (SMN: 74°43′S, 164°16′E). Sampling was performed by a CTD rosette system equipped with 15-l Niskin bottles. All sampling bottles were acid-cleaned according to Burney et al. (1979). Sampling depths were 0, 10 and 25 m at the MER Station (about 250 m off the coast) and 0, 10, 25, 50, 100 and 200 m at the SMN Station, located in the centre of Terra Nova Bay (10.5 km off-shore, depth 500 m). A total of 41 samples were taken from the MER Station and 64 from the SMN Station.

### Sample Fixation and Direct Counts

Immediately after sampling, 1-l subsamples were prefiltered through a sterile 200-μm plankton net and then filtered onto 2.0-μm Nuclepore polycarbonate filters. To enumerate total picoplankton, two 20-ml subsamples were fixed with formalin (final concentration, 5%) and duplicate slides were prepared by filtering 1 ml of the fixed sample prestained with DAPI (Porter and Feig 1980) onto 0.2-μm Nuclepore black membranes.

For picophytoplankton enumeration, two 50-ml subsamples were fixed with glutaraldehyde (final concentration 2%) and duplicate slides were prepared by filtering 20–30 ml of the fixed sample onto 0.2-μm Nuclepore black membranes. The slides were kept at +4 °C in the dark until analysis after 2 months. A Zeiss Standard Microscope equipped with a 100-W mercury lamp, ×100 neofluor lens and ×10 eyepieces, and an ocular micrometer grid was used. Picoplanktonic cells (DAPI) were counted under UV excitation, (Zeiss filter - 487702), and phototrophic organisms under blue light excitation (Zeiss filter - 487709). Under blue light excitation, prokaryotic (Cyanobacteria) cells fluoresce yellow-orange while eukaryotic algae fluoresce deep red. At least 50–100 Cyanobacteria and 100–200 eukaryotes were counted on each filter.

### Biovolume and Biomass Estimation

Biovolumes were determined using the SEM and Zeiss Laser Scan Microscope (LSM 10, Universal confocal Laser Scan microscope). This later instrument was regulated at confocal reflecting light, using the excitation wavelength 488/514 nm; microscope calibration for measurements (in μm) was determined by pixels, using a stage micrometer with slide with 10 mm for the smallest divisions; the 100 magnification objective had an electronic zoom of 120×. The calibration was stored in the Scan Laser software programme and in the semiautomated VIDAS system (Kontron). The final magnification was 12 000×. Two hundred cells were counted and measured by LSM, using the same slides as for epifluorescence microscopy. Counts started from the surface samples and proceeded towards the deeper ones, in order to obtain a mean cell volume per analyzed area by a trapezoidal integration. Measurements by LSM were analyzed by an semi-automated VIDAS system (Kontron) (Sieracki et al. 1989). The criterion used to calculate the biovolume gave well-defined geometrical shapes to cells, considering their width and length (Krambeck et al. 1981).

Biovolumes using SEM were performed according to Paerl and Shimp (1973). Two hundred cells were counted and measured by SEM using the same criteria as for measurements by LSM. Picoplankton biomass was calculated using the conversion factor of Lee and Furhman (1987) ($3.8 \times 10^{-13}$ g C μm$^{-3}$). This factor was used according to Kroer (1994), who emphasized that biovolume to biomass conversion factors may be higher than previously thought. Picophytoplankton biomass was estimated according to Takahashi et al. (1985) ($4 \times 10^{-13}$ g C μm$^{-3}$) for the Cyanobacteria, and Mullin et al. (1985) and Booth (1987) ($2.2 \times 10^{-13}$ g C μm$^{-3}$) for the eukaryotes. The total biomass reported is the sum of the two values.

## Statistical Analysis

Fisher's dispersion index [$D^2 = (n - 1)\, s^2 / \bar{r}$, where $D^2$ is the dispersion index, $(n - 1)$ the degrees of freedom, $s^2$ = variance, and $\bar{r}$ = mean value of $r_1$, $r_2 \ldots r_n$; the distribution of $D^2$ is well approximated by a chi-square with $(n - 1)$ degrees of freedom] was used to establish if the picoplankton counts were fitted by a Poisson distribution and was calculated from the counts by epifluorescence, SEM and LSM. Student's t test and its distribution were also calculated from the picoplankton biovolume values obtained by our measurements and those obtained from seawater samples collected from McMurdo Sound (Fuhrman and Azam 1980) and Terra Nova Bay (Acosta Pomar et al. 1993). We used regression analysis to estimate the degrees of temporal covariance between picoplankton abundance and trophic condition, expressed as concentration of chlorophyll a measured by Innamorati et al. (1991) at both stations, and the Wilcoxon Rank test to confirm the results.

## Results

Morphometric analysis by LSM revealed that picoplanktonic cells are generally spherical (only 11.9% are not) in Terra Nova Bay. Observations by SEM gave similar results (only 7.5% have another shape). This was confirmed by binary images leading to the calculation of densitometrical and geometrical parameters. Table 1 shows the frequency distribution of spherical and ellipsoidal forms, as obtained by LSM and SEM measurements. Mean diameter and volume were 0.65 μm and 0.14 ± 0.012 $\mu m^3$, respectively, from 201 cells measured by LSM, whereas with SEM, the mean diameter and volume were 0.62 μm and 0.12 ± 0.011 $\mu m^3$, respectively, for the 200 cells measured. The calculation of Fisher's dispersion index reveals that picoplanktonic organisms follow a typical Poisson distribution (for LSM: $D^2 = 10.24$; for SEM: $D^2 = 12.36$). The Student's t calculation determined by comparing our experimental biovolumes and average volumes obtained from a theoretical diameter of 0.6 μm, (Acosta Pomar et al. 1993) shows that mean values of experimental biovolumes are not significantly different from the theoretical ones.

Table 2 reports minimum, maximum and mean values of picoplankton density and biomass for each sampling station. The <2-μm microbial population ranged between a minimum of $7.2 \times 10^7$ cells $l^{-1}$ (2.01.1990, MER station at 10 m depth) and a maximum of $2.9 \times 10^{10}$ cells $l^{-1}$ (3.02.1990, MER station at surface), which is the highest abundance reported for Antarctic seas. Densities of the order of $10^9$ cells $l^{-1}$ were frequently found at both stations not only at euphotic layers but also at depth after 12 January 1990. Picophytoplankton densities (Table 3) were 1 to 3 orders of magnitude lower, ranging between $7.5 \times 10^4$ cells $l^{-1}$ (23 January 1990, SMN at 100 m depth) and $6.6 \times 10^7$ cells $l^{-1}$ (10.01.1990, MER, surface). Acosta Pomar (1992) found that >2 μm autotrophic cells ranged between $5.9 \times 10^7$ cells $l^{-1}$ and $1.5 \times 10^8$ cells $l^{-1}$ at MER and between $3.8 \times 10^4$ cells $l^{-1}$ and $8.2 \times 10^7$

**Table 1.** Abundance of spherical and ellipsoidal cells in four volume size classes for a total of 201 observations with LSM and 200 observations with SEM

| Size class ($\mu m^3$) | LSM | | SEM | |
|---|---|---|---|---|
| | Spheres | Ellipsoids | Spheres | Ellipsoids |
| 0–0.1 | 20 | 1 | 21 | 0 |
| 0.1–0.2 | 142 | 14 | 152 | 10 |
| 0.2–0.3 | 10 | 6 | 8 | 3 |
| >0.3 | 5 | 3 | 4 | 2 |
| Total | 177 | 24 | 185 | 15 |

**Table 2.** Abundance (cells $l^{-1}$) and biomass (μg C $l^{-1}$) of picoplankton at MER and SMN

| Station | n | Density (cells $l^{-1}$) | | | Biomass (μg C $l^{-1}$)[a] | | |
|---|---|---|---|---|---|---|---|
| | | Min | Max | Average | Min | Max | Average |
| MER | 39 | $7.2 \times 10^7$ | $2.9 \times 10^{10}$ | $2.7 \times 10^9$ | 3.9 | 1584.6 | 147.1 |
| SMN | 39 | $1.0 \times 10^8$ | $7.5 \times 10^9$ | $1.6 \times 10^9$ | 5.5 | 415.3 | 85.3 |

[a] Experimental biovolume converted to biomass by multiplying $3.8 \times 10^{-13}$ g C $\mu m^{-3}$ (Lee and Fuhrman 1987).

**Table 3.** Prokaryotic and eukaryotic cellular density (cells $l^{-1}$) and biomass (ng $C l^{-1}$) of the picophytoplankton at MER and SMN

| Prokaryotic Station | Density (cells $l^{-1}$) | | | Biomass (ng $C l^{-1}$)[a] | | |
|---|---|---|---|---|---|---|
| | Min | Max | Average | Min | Max | Average |
| MER | $3.8 \times 10^4$ | $5.3 \times 10^5$ | $3.9 \times 10^4$ | 2.17 | 30.4 | 2.7 |
| SMN | $3.8 \times 10^4$ | $4.5 \times 10^5$ | $4.7 \times 10^4$ | 2.19 | 26.1 | 2.7 |
| **Eukaryotic Station** | **Density (cells $l^{-1}$)** | | | **Biomass (ng $C l^{-1}$)[b]** | | |
| | Min | Max | Average | Min | Max | Average |
| MER | $3.8 \times 10^4$ | $6.5 \times 10^7$ | $1.7 \times 10^7$ | 16.7 | 2100 | 560 |
| SMN | $4.3 \times 10^4$ | $4.3 \times 10^7$ | $9.4 \times 10^6$ | 1.4 | 1400 | 290 |
| **Total picophytoplankton Station** | **Density (cells $l^{-1}$)** | | | **Biomass (ng $C l^{-1}$)** | | |
| | Min | Max | Average | Min | Max | Average |
| MER | $7.56 \times 10^4$ | $6.5 \times 10^7$ | $1.7 \times 10^7$ | 18.8 | 2100 | 560 |
| SMN | $8.06 \times 10^4$ | $4.3 \times 10^7$ | $9.4 \times 10^6$ | 3.6 | 1400 | 290 |

[a] Experimental biovolume converted to biomass by multiplying $4.0 \times 10^{-13}$ g C $\mu m^{-3}$ (Takahashi et al. 1985).
[b] Experimental biovolume converted to biomass by multiplying $2.2 \times 10^{-13}$ g C $\mu^{-3}$ (Mullin et al. 1985; Booth 1987).

cells $l^{-1}$ at SMN, where they were mostly absent from 100 m depth down.

Picoplankton biomass averaged 147.1 and 85.3 $\mu g C l^{-1}$ at MER and SMN respectively. At MER, biomass was usually more abundant than at SMN. Figure 1a shows the temporal variation of the picoplankton biomass at the two stations. Biomass increased about by 50% through January. At MER, maximum values were found between 0 and 15 m depth. At SMN, however, from the second half of January onwards, maximum values were between 50 and 200 m. On average 0.6% of the picoplankton community consisted of phototrophic cells. Figure 1b,c shows temporal variation of the picophytoplankton expressed as biomass. These organisms were recorded down to 100 m depth. Eukaryotes were more abundant than Cyanobacteria at both stations. The latter reached 30.4 ng C $l^{-1}$ in the surface layer and were not found below 50 m. The prokaryotes, from the beginning of observation, were present at SMN, principally between 0 and 25 m, and declined at the end of the sampling period. Conversely, at MER, Cyanobacteria reached their maximum values in surface waters at the end of the sampling period. Eukaryotes were present at both stations during the entire period. Eukaryotes smaller than 2 $\mu m$ and Cyanobacteria generally showed an inverse trend (Fig. 1b,c).

The spatial-temporal distribution of picophytoplanktonic biomass (Acosta Pomar 1992) also showed a temporal cycle during which each component alternated (coccoid Cyanobacteria and eukaryotic cells); this trend has also been reported by Marchant et al. (1987), Walker and Marchant (1989) and Rivkin et al. (1990). The observed predominance of the eukaryotic component over the Cyanobacteria is in agreement with other studies and depends on latitude and temperature (Waterbury et al. 1986).

## Discussion

To differentiate picoplanktonic organisms, the use of LSM rather than SEM allowed a morphometrical analysis and a more rapid estimation of the planktonic microbial biomass. The possibility of linking the LSM to the VIDAS semiautomated system offered numerous advantages over tradi-

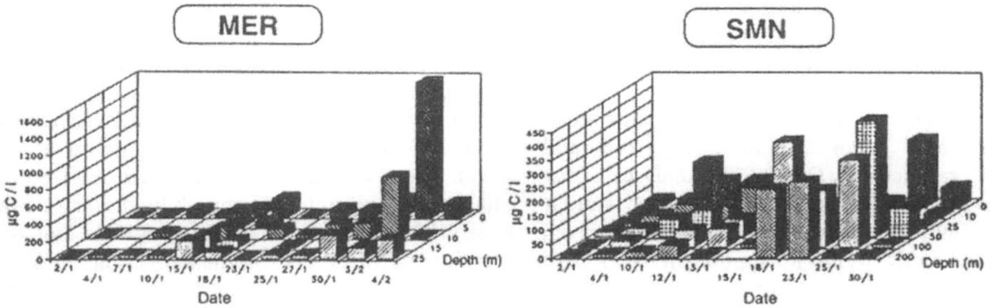


a) Variations of picoplanktonic biomass (µg C $l^{-1}$) at MER and SMN stations at different depths

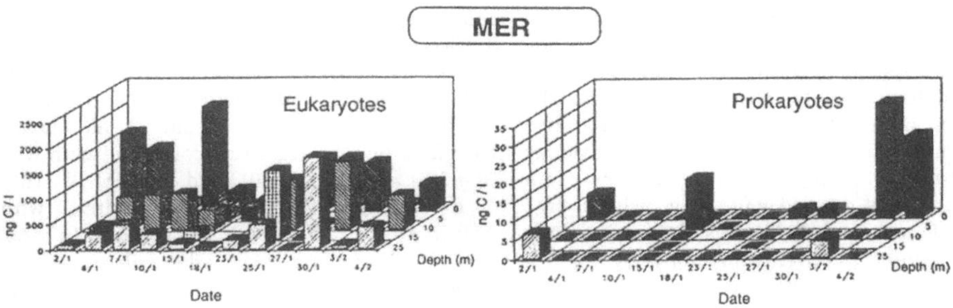


b) Variations of picophytoplanktonic biomass (ng C $l^{-1}$) at MER station at different depths

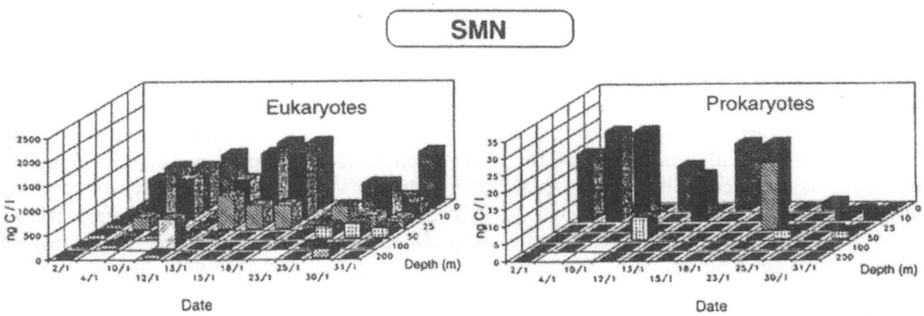


c) Variations of picophytoplanktonic biomass (ng C $l^{-1}$) at SMN station at different depths

**Fig. 1a–c.** Temporal variations of picoplanktonic ($\mu g\,C\,l^{-1}$) and picophytoplanktonic ($ng\,C^{-1}$) biomass in two stations in Terra Nova Bay tional manual methods and allowed geometrical and statistical analyses of a large number of samples (Sieracki et al. 1989; Verity et al. 1992; Kroer 1994)

Picoplankton were mostly spherical, the mean diameter and mean volume being 0.65 and 0.14 $\mu m^3$, respectively. Other SEM analyses of the same organisms in the Ross Sea yielded similar values (Fuhrman and Azam 1980). However, during the study period no significant differences were noted between the experimental and theoretical biovolumes, showing that the bacteria of ice-free Antarctic waters have cell volumes comparable with those of bacteria in warmer environments. The picoplanktonic biomass was similar to the biomass of bacterioplankton recorded in the open waters of the Ross Sea (Fuhrman and Azam 1980; El-Sayed 1987), Indian Ocean (Samyshev 1986) and Weddell Sea (Sullivan et al. 1990). The picoplanktonic values were higher than the bacterioplankton biomass of ice-covered waters of the Antarctic Ocean previously described by Gustafson et al. (1990) and Cota et al. (1990). In a previous study carried out in a pelagic area (Acosta Pomar et al. 1993), this biomass was calculated using the conversion factor ($1.21 \times 10^{-13}$ g C $\mu m^{-3}$) proposed by Watson et al. (1977), three times smaller than that used in this chapter. At that time, we obtained picoplanktonic biomass values 1 order of magnitude less.

The variations of picoplankton biomass observed during the period of observation can be related to blooms of some nanoplankton organisms (Rivkin et al. 1990; Putt et al. 1990). A bloom of *Phaeocystis* sp., which occurred at the in-shore station after the pack-ice melting at the end of December (Innamorati et al. 1991), could also explain the lower bacterial abundance owing to the bactericidal acrylic acid production of this prymnesiophyte. As reported by Innamorati et al. (1991 and also this Vol.), chlorophyll a and nutrients ($NO_3$, $NO_2$, $PO_4$, $SiO_2$) were highest at the beginning of the sampling at both stations, increasing only towards the end of observation presumably due to another bloom. The analysis of thermohaline data according to Saggiomo et al. (1992) indicated a well-defined and clearly identifiable surface layer until January 25 with an evident pycnocline at about 40 m, and from then onwards, a progressive destabilisation of the surface layer with the formation of a second pycnocline together with nitrate and phosphate. Maximum values for carbon assimilation were always observed at the 5–10 m depth interval, the highest being 6.9 mg C $m^{-3} h^{-1}$ registered at 5 m depth at SMN on January 12. Measurements carried out at 25 and 50 m, that is on the boundary to or below the euphotic layer, showed values from 0.15 to 1.98 mg C $m^{-3} h^{-1}$ (Saggiomo et al. 1992). The contribution of the <2-$\mu$m fraction to total production ranged from 17 to 28%. La Ferla et al. (1995) found by ATP measurements that, on certain occasions, the biomass of smaller-sized organisms (<2 $\mu$m) could reach about 60% of total biomass becoming the main component of the entire microbial population and that pico-sized organisms were generally the least important component of the plankton when the total biomass was highest, while they became the main component when the larger plankton biomass decreased. It is interesting, moreover, to emphasize several relationships (Table 4) between biological parameters and environmental variables, previously examined by Bruni et al. (1995). Picoplankton was negatively correlated with ATP ($n = 38$, $r = -0.50$, $P < 0.01$) and chlorophyll a ($n = 38$, $r = -0.37$, $P < 0.05$) at MER, but positively with ATP ($n = 60$, $r = 0.33$, $P < 0.05$) and LPS ($n = 60$, $r = 0.33$, $P < 0.05$) at SMN, where also the autotrophic fraction showed a significant positive correlation with Chl a ($n = 60$, $r = 0.37$, $P < 0.05$) and LPS ($n = 60$, $r = 48$, $P < 0.01$). Above all, the significant negative correlation between picoplankton abundance and Chl a indirectly suggests that heterotrophic and autotrophic processes were uncoupled in the station near the coast, while the significant positive correlation between picophytoplankton abundance and Chl a suggests that production in the off-shore station was mostly due to the pico-sized producers. Regression analysis (Fig. 2) also reveals a considerable scatter in picoplanktonic abundance versus chlorophyll a concentrations and, consequently, uncoupling between these two biological parameters, variable and different in space and time from those of warmer seas. Cota et al. (1990) and Karl et al. (1991) found similar results in other Antarctic seas. We agree with Karl and Bird (1993), who supposed that the possible causes for the uncoupling of phytoplankton and bacterioplankton population dynamics include differential effects of temperature, organic substrate limitation, chemical antagonism and direct competition between algae and bacteria. Moreover, one can note that the two relative linear regression equations (Fig. 2) are statistically distinguishable, and that the slope was

**Table 4.** Significant correlation coefficients between picoplankton, picophytoplankton and other parameters at the two station: in-shore MER and off-shore SMN

| | | | | | | | |
|---|---|---|---|---|---|---|---|
| MER station | | | | | | | |
| Picoplankton versus: | | | | Picophytoplankton versus: | | | |
| ATP | n = 38 | r = −0.50 | $P < 0.01$ | $SiO_2$ | n = 38 | r = 0.38 | $P < 0.05$ |
| Chl a | n = 38 | r = −0.37 | $P < 0.05$ | S‰ | n = 38 | r = 0.38 | $P < 0.05$ |
| | | | | T°C | n = 38 | r = −0.47 | $P < 0.05$ |
| SMN station | | | | | | | |
| Picoplankton versus: | | | | Picophytoplankton versus: | | | |
| ATP | n = 58 | r = 0.33 | $P < 0.05$ | LPS | n = 60 | r = 0.48 | $P < 0.01$ |
| LPS | n = 60 | r = 0.33 | $P < 0.05$ | $NO_2$ | n = 60 | r = 0.57 | $P < 0.01$ |
| $NO_2$ | n = 60 | r = 0.30 | $P < 0.05$ | $NO_3$ | n = 60 | r = −0.55 | $P < 0.01$ |
| $SiO_2$ | n = 60 | r = 0.28 | $P < 0.05$ | $PO_4$ | n = 60 | r = −0.49 | $P < 0.01$ |
| | | | | $SiO_2$ | n = 60 | r = −0.44 | $P < 0.01$ |
| | | | | S‰ | n = 60 | r = −0.54 | $P < 0.01$ |
| | | | | T°C | n = 60 | r = 0.42 | $P < 0.01$ |
| | | | | Chl a | n = 60 | r = 0.37 | $P < 0.01$ |

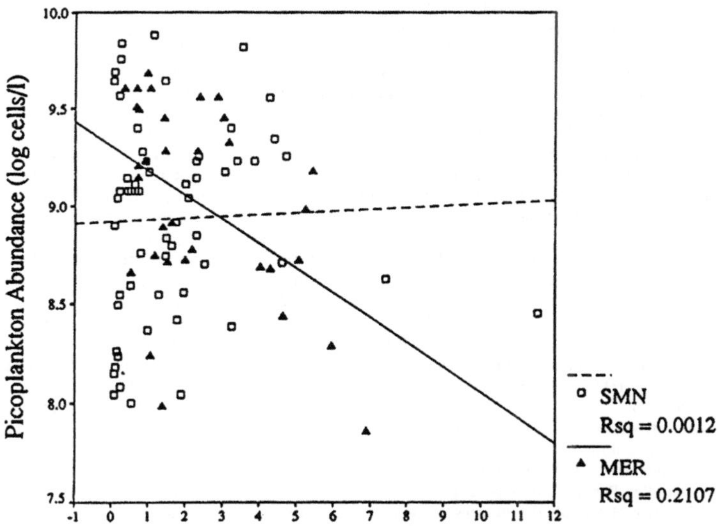


**Fig. 2.** Relationship between picoplankton abundance ($log_{10}$ cells $l^{-1}$) and chlorophyll a concentrations ($\mu g l^{-1}$) at MER (*solid line*) and SMN (*dotted line*)

negative for MER but positive for SMN. Finally, the highly significant result ($P < 0.01$) of the Wilcoxon Rank test confirms this difference between the two stations.

*Acknowledgements.* This work was supported by the Italian National Antarctic Research Programme (PNRA). The authors thank Dr. A. Feminò for his help in the use of LSM, Dr. A. Rapisarda for his skilled assistance and L. Salomone for the statistical elaboration.

## References

Acosta Pomar MLC (1992) Variazioni spazio temporali della comunità picofitoplanctonica nella Baia di Terra Nova (Mare di Ross). In: Gallardo V, Ferretti O, Moyano H (eds)

Atti Seminario di Oceanografia Antartica. Universidad de Concepcion, Chile, pp 133–142
Acosta Pomar MLC, Bruni V, Maugeri TL (1993) Picoplankton biomass in the Ross Sea (Antarctica). Polar Biol 13: 1–6
Azam F, Fenchel T, Field JG, Gray JS, Meyer-Reil LA, Thingstad F (1983) The ecological role of water-column microbes in the sea. Mar Ecol Prog Ser 10: 257–263
Booth BC (1987) The use of autofluorescence for analyzing oceanic phytoplankton communities. Bot Mar 30: 101–108
Bruni V, La Ferla R, Acosta Pomar MLC, Salomone L (1995) Structural differences of the microbial community in two sites of the Terra Nova Bay (Ross Sea, Antarctica): a statistical analysis. Microbiologica: 18: 409–422
Burney CM, Johnson KM, Lavoie DM, Sieburth JMcN (1979) Dissolved carbohydrate and microbial ATP in the North Atlantic: concentrations and interactions. Deep Sea Res 26A: 1267–1290
Cho BC, Azam F (1990) Biogeochemical significance of bacterial biomass in the ocean's euphotic zone. Mar Ecol Prog Ser 63: 253–259
Cota GF, Kottmeier ST, Smith WO Jr, Sullivan CW (1990) Bacterioplankton in the marginal ice zone of the Weddell Sea: biomass, production and metabolic activities during austral autumn. Deep Sea Res 37: 1145–1167
El-Sayed SZ (1987) Biological productivity of Antarctic waters: present paradoxes and emerging paradigms. Biomass Sci Ser Aquat Biol 7: 1–21
Fogg GE (1994) Critical appraisal of the BIOMASS Programme. In: El-Sayed SZ (ed) Southern Ocean ecology. Cambridge University Press, Cambridge, pp 383–389
Fogg GE (1995) Some comments on picoplankton and its importance in the pelagic ecosystem. Aquat Microb Ecol 9: 33–39
Fuhrman JA, Azam F (1980) Bacterioplankton secondary production estimates for coastal waters of British Columbia Antarctica and California. Appl Environ Microbiol 39: 1085–1095
Gustafson DE Jr, Anderson MR, Rivkin RB (1990) Bacterioplankton abundance and productivity at the ice-edge zone of McMurdo Sound, Antarctica. Antarct J Rev 25(5): 191–193
Innamorati M, Lazzara L, Mori G, Nuccio C, Saggiomo V (1991) Phytoplankton ecology. National Scientific Commission for Antarctica (ed), Oceanographic Campaign 1989–90. Data Rep Part I, Genova, pp 141–252
Karl DM (1993) Microbial processes in the Southern oceans. In: Friedmann EI (ed) Antarctic microbiology. Wiley-Lyss, New York, pp 1–63
Karl DM, Bird DF (1993) Bacterial-algal interactions in Antarctic coast ecosystems. In: Guerrero R, Pedrós-Alió (eds) Trends in microbial ecology. Spanish Society for Microbiology, pp 37–40
Karl DM, Holm-Hansen O, Taylor GT, Tien G, Bird DF (1991) Microbial biomass and productivity in the western Bransfield Strait, Antarctica, during the 1986–87 austral summer. Deep Sea Res 38: 1029–1055
Krambeck C, Krambeck HJ, Overbeck J (1981) Microcomputer-assisted biomass determination of plankton bacteria on scanning electron micrographs. Appl Environ Microbiol 42: 142–149
Kroer N (1994) Relationships between biovolume and carbon and nitrogen content of bacterioplankton. FEMS Microbiol Ecol 13: 217–224
La Ferla R, Allegra A, Azzaro F, Greco S, Crisafi E (1995) Observations on the microbial biomass in two stations of Terra Nova Bay (Antarctica) by ATP and LPS measurements. PSZNI: Mar Ecol 16: 307–315
Lee S, Fuhrman A (1987) Relationships between biovolume and biomass of naturally derived marine bacterioplankton. Appl Environ Microbiol 53: 1298–1303
Lenz J (1992) Microbial loop, microbial food web and classical food chain: their significance in pelagic marine ecosystems. Arch Hydrobiol Beih 37: 265–278
Marchant HJ, Murphy EJ (1994) Interactions at the base of the Antarctic food web. In: El-Sayed SZ (ed) Southern Ocean ecology. Cambridge University Press, Cambridge, pp 267–284
Marchant HJ, Davidson AT, Wright SW (1987) The distribution and abundance of chroococcoid Cyanobacteria in the Southern Ocean. Polar Biol 1: 1–9
Mullin MM, Sloan PR, Eppley RW (1985) Relationship between carbon content, cell volume and area in phytoplankton. Limnol Oceanogr 30: 307–311
Paerl HW, Shimp SL (1973) Preparation of filtered plankton and detritus for study with scanning electron microscopy. Limnol Oceanogr 18: 802–805
Porter KG, Feig YS (1980) The use of DAPI for identifying and counting aquatic microflora. Limnol Oceanogr 25: 943–948
Putt M, Borheim KY, Miceli G, Stoecker DK (1990) Seasonal changes in cell size and abundance of bacterioplankton during the *Phaeocystis* sp. bloom in McMurdo Sound. Antarctic J US 25(5): 199–201
Rivkin RB, Anderson MR, Gustafson DE Jr (1990) Seasonal patterns of bacteria and net, nano- and picophytoplankton McMurdo Sound, (Antarctica) during the austral spring. Antarctic J US 25(5): 195–196
Saggiomo V, Massi L, Moldigh M, Innamorati M (1992) Size-fractionated primary production in Terra Nova Bay (Ross Sea) during the austral summer (1989–90). In: Gallardo V, Ferretti O, Moyano H (eds) Atti Seminario di Oceanografia Antartica. ENEA-PNRA-EULA, Concepcion, Chile, pp 289–293
Samyshev EZ (1986) Bacterioplankton of Antarctic waters: concentration, production and bacterial destruction. Oceanology 26: 508–512
Sieburth J McN, Davis PG (1982) The role of heterotrophic nanoplankton in the grazing and nurturing of planktonic bacteria in the Sargasso and Carribean Seas. Ann Inst Oceanogr Paris 58: 285–296
Sieracki ME, Reichenbach SE, Webb KL (1989) Evaluation of automated threshold selection methods for accurately sizing microscopic fluorescent cells by image analysis. Appl Environ Microbiol 55: 2762–2772
Sullivan CW, Cota GF, Krempin DW, Smith WO Jr (1990) Distribution and activity of bacterioplankton in the marginal ice zone of the Weddell Sea during austral spring. Mar Ecol Ser 63: 239–252
Takahashi M, Kikuchi K, Hara Y (1985) Importance of picocyanobacteria biomass (unicellular, blue-green algae) in the phytoplankton of the coastal waters off Japan. Mar Biol 89: 63–69
Van Wambecke F (1988) Numération et taille des bactéries planctoniques au moyen de l'analyse d'images couplée à l'épifluorescence. Ann Inst Pasteur Microbiol 9: 261–272

Verity PG, Robertson CY, Tronzo CR, Andrews MG, Nelson JR, Sieracki ME (1992) Relationships between cell volume and the carbon and nitrogen content of marine photosynthetic nanoplankton. Limnol Oceanogr 37: 1435–1446

Walker TD, Marchant HJ (1989) The seasonal occurrence of chroococcoid Cyanobacteria at one Antarctic coastal site. Polar Biol 9: 193–196

Waterbury JB, Watson SW, Valois FW, Franks DG (1986) Biological and ecological characterization of the marine unicellular cyanobacterium *Synechococcus*. In: Platt T, Li WKW (eds) Photosynthetic picoplankton. Can Bull Fish Aquat Sci, Ottawa Canada, pp 71–120

Watson SW, Novitski TJ, Quimby HI, Valois FW (1977) Determination of bacterial number and biomass in the marine environment. Appl Environ Microbiol 33: 940–946

CHAPTER 17

# Phytoplankton Size-Fractionated Biomass in the Ross Sea: Spatial and Temporal Variations During the Austral Spring

G.C. Carrada[1], O. Mangoni[1], G.F. Russo[2], and V. Saggiomo[2]

**ABSTRACT**

The present work analyzes, in terms of pigment content, the spatial and temporal variations of the micro, nano- and picoplankton fractions of the Ross Sea phytoplankton biomass. Special attention is given to plankton dynamics in the area of the polynya, during austral spring 1994. In total, 50 stations were sampled in the period from November 14 to December 15. From December 18 to 27, a discrete surface sampling at 60 positions, distributed along a 1200-nautical mile track, was also performed. Three subsystems were identified in the sampling area: the polynya, the marginal-ice zone and the pack-covered waters. In the first subsystem, the microphytoplankton predominated, but its relative abundance was certainly biased by the presence of large quantities of the palmelloid form of *Phaeocystis* cfr. *antarctica*. In the marginal ice zone, the micro- and nanophytoplankton fractions were at first dominated by diatoms, which were replaced after a short time by the picophytoplankton. In the pack-covered waters, the bulk of the biomass consisted of picophytoplankton. The ratio phaeo/Chl a suggests that in these two last subsystems the dimensional structure of phytoplankton is the product of the selective removal of larger cells by grazing, probably by krill. The oscillation of Chl a concentrations between high and low levels throughout the investigated area suggests the presence of recurrent blooms. Biomass concentrations also confirm the key role of the marginal ice zone in setting the conditions for the seasonal evolution of the open water system, even when it does not induce a stabilizing effect on the water column.

## Introduction

The information gathered to date about the production of organic carbon in the Antarctic pelagic system does not yet allow drawing univocal conclusions, either as regards quantities and their spatial and temporal variations, or about their allocation to the different phytoplankton size fractions. This information is nevertheless of paramount importance to resolve the structure and functioning of the trophic web, as well as for assessing the flux of biogenic material toward the bottom. The long-debated problem of the so-called Antarctic paradox, for example, began to find an explanation when the first information about production and consumption of the different size fractions of phytoplankton became available. In fact, the importance of the size fractions of phytoplankton in determining structure and efficiency of marine food webs is a widely accepted concept that has received in the last decade an increasing amount of attention. Recently, Kiørboe (1993) has reviewed the implications of cell size on food web structure under different hydrographic contexts.

The structure of Antarctic pelagic food webs is the subject of a growing scientific literature, focused on such topics as the identification of the relative weights of the dimensional fractions of phytoplankton, their successional dynamics and the reciprocal influences with the level of grazers (Jacques and Panouse 1991; Lancelot et al. 1993). The available information on phytoplankton size-fraction composition and production stresses the importance of the nano- and picophytoplankton fractions (Hewes et al. 1985; Wilson et al. 1986; Weber and El-Sayed 1987; Jacques and

[1] Department of Zoology, University of Naples Federico II, Naples, Italy
[2] Stazione Zoologica "A. Dohrn", Biological Oceanography Laboratory, Naples 80121, Italy

Panouse 1991; Jochem et al. 1995; Savidge et al. 1995; Ning et al. 1996; Detmer and Bathmann 1997). In this context, the knowledge gained for the Ross Sea is considered by Marra (1996) to have wider implications with respect to other marine systems.

Problems such as the importance of selective grazing on the quantitative variations of the different dimensional fractions of Antarctic phytoplankton assemblages and on their accumulation and/or removal are the subject of critical debate (Di Tullio and Smith 1996). In situ measurements, mostly carried out in the last decade (Wilson et al. 1986; DeMaster et al. 1992; Innamorati et al. 1992a,b; Saggiomo et al. 1992, 1998; Smith et al. 1996; Di Tullio and Smith 1996) and satellite imagery (Comiso et al. 1993; Sullivan et al. 1993; Arrigo and McClain 1994) have both demonstrated a strong spatial and temporal heterogeneity of phytoplankton biomasses in the Ross Sea. The factors controlling such dynamics are as yet poorly known and different authors point to different factors, such as stability of the water column, irradiance regimes, micronutrient (particularly iron) concentration, grazing by micro- and macrozooplankton.

The present work was aimed at analyzing the spatial and temporal variations of the micro- (>10 μm), nano-(10–2 μm) and picophytoplankton (<2 μm) fractions of the Ross Sea phytoplankton, measured as size-fractionated pigment content, with respect to the dynamics of the marginal area of the polynya during the austral spring. It is worth noting that the information presented in this chapter was collected during a cruise that, together with a US cruise carried out at the same time (Smith and Gordon 1997), represents the earliest expeditions to the Ross Sea in the austral spring.

## Materials and Methods

During the Antarctic spring 1994, a cruise was carried out in the Ross Sea on board the R/V Italica, from 14 November to 13 December. Twenty-seven stations were sampled along a transect lying on the 175°E and stretching from 76.5°S to 72.0°S. Several stations (MP, 2, 4, 6, 9, 11, 12, 13, 14, 15, 17, 19 and 24) were repeatedly sampled at different time intervals in order to evaluate the temporal variability of pigment concentrations and of the size structure of the phytoplankton community. From December 13 to 15, three additional hydrological stations (28, 29 and 30) were situated along a transect from the Ross Sea polynya to Terra Nova Bay (Fig. 1). From December 18 to 27, continuous surface (−5 m) monitoring of in vivo fluorescence by flow fluorometry (Turner Designs, mod. 10-AU-005), as well as discrete sampling at 60 positions (indicated as CT) distributed at 20-nautical-mile intervals, were carried out along the 1200-nautical-mile track to investigate krill distribution by echo-survey techniques. The track covered an area included between 170°E to 179°W and 74°S to 68°S (Fig. 2). From the discrete samples, phytoplankton size fractions were isolated and submitted to fluorometric analysis of pigment content. At hydrographic stations, samples were collected by means of a SBE 32 Carousel sampler, equipped with 24 bottles (12-l), a SBE 9/11 Plus CTD, a SBE oxygen sensor, a Chelsea Aquatrack III fluorometer, a Sea Tech transmissometer and a Datasonics PSA 900 altimeter. At each station, profiles of down- and upwelling underwater PAR were obtained by means of a photoprobe with an LI-COR quantum meter. Sampling depths were selected according to both optical levels (100, 50, 20, 10, 5, 1 and 0.13% of incident irradiance) and hydrographical depths down to 200 m. Samples were processed as follows:

- filtration of a 500-ml sample on a Whatman GF/F filter;
- filtration of a 500-ml sample on a 10-μm Nuclepore polycarbonate membrane followed by sequential filtration on a Whatman GF/F filter;
- filtration of a 500-ml sample on a 2-μm Nuclepore polycarbonate membrane followed by sequential filtration on a Whatman GF/F filter.

This procedure allowed the isolation of the micro- (>10 μm), nano- (2–10 μm) and total picophytoplankton (<2 μm) fractions and the determination of their contribution to the total biomass.

Spectrofluorometric analyses of chlorophyll a (Chl a) and phaeopigments (phaeo) were carried out on board, according to Holm-Hansen et al. (1965), within a maximum of 12 h from collection. Filters were placed in neutralized 90% v/v acetone, sonicated with a Branson 250 sonifier and allowed to extract for 2 h. The extract was read, before and after acidification, at a Spex Fluoromax spectrofluorometer. The instrument was checked daily against a solution of Chl a from *Anacystis*

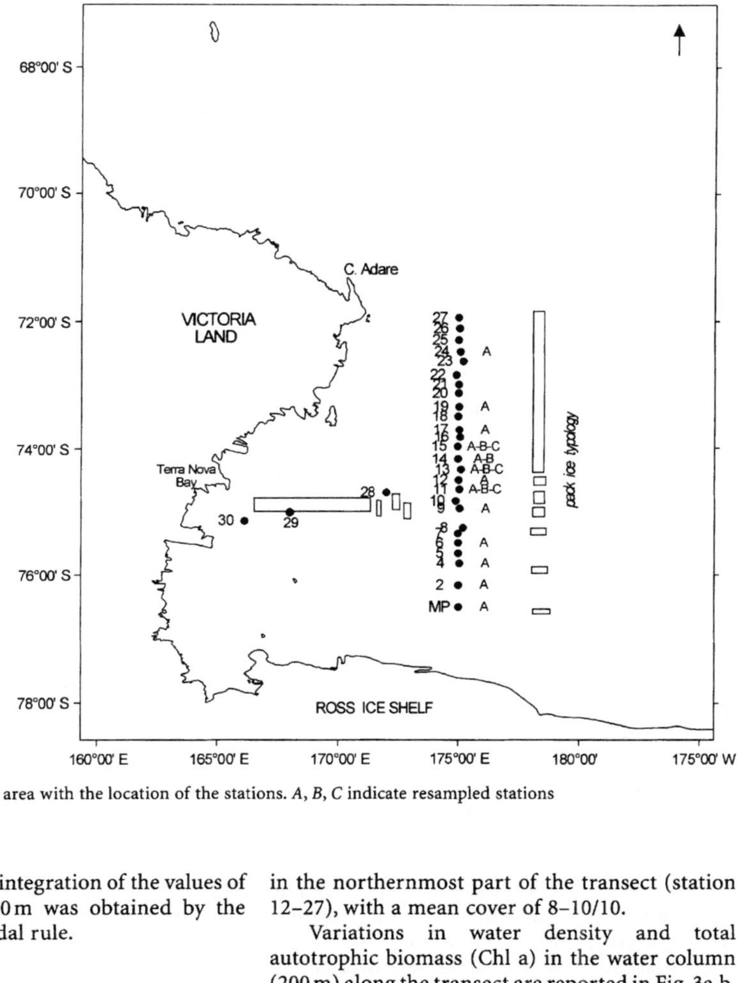


**Fig. 1.** Ross Sea. The studied area with the location of the stations. *A*, *B*, *C* indicate resampled stations

*nidulans* by Sigma. The integration of the values of Chl a in the upper 200 m was obtained by the method of the trapezoidal rule.

## Results

The entire sampling period was characterized by poor meteorological conditions, with a strong cloud cover and low irradiance levels. On the base of pack-ice cover, three main zones were identified along the transect (Fig. 1): (1) the polynya, in its southernmost part (from station MP to station 7), with drifting sea ice or a homogeneous cover of thin newly formed ice; (2) the marginal-ice zone (MIZ), in the central section (station 8–11), with a mean ice cover of 5–7/10 characterized by drifting fragmented ice, producing short-lived leads and lagoons; (3) an area with well-structured pack ice in the northernmost part of the transect (station 12–27), with a mean cover of 8–10/10.

Variations in water density and total autotrophic biomass (Chl a) in the water column (200 m) along the transect are reported in Fig. 3a,b. A well-mixed water column was present in the southern area (polynya), whereas a very slight stabilization by ice melting was observed at the MIZ (stations 8 to 11). The distribution of the integrated total biomass (Chl a), the phaeo/chl a ratio (Fig. 4a) and the relative contribution of the size fractions to total biomass (Fig. 4b), showed different patterns characterizing the three areas described above.

In the polynya and MIZ, integrated Chl a concentrations ranged from 69.9 to 164.7 and from 99.64 to 220 $mg\,m^{-2}$, respectively; maximum chl a concentrations (3.95 $\mu g\,l^{-1}$) were recorded at the

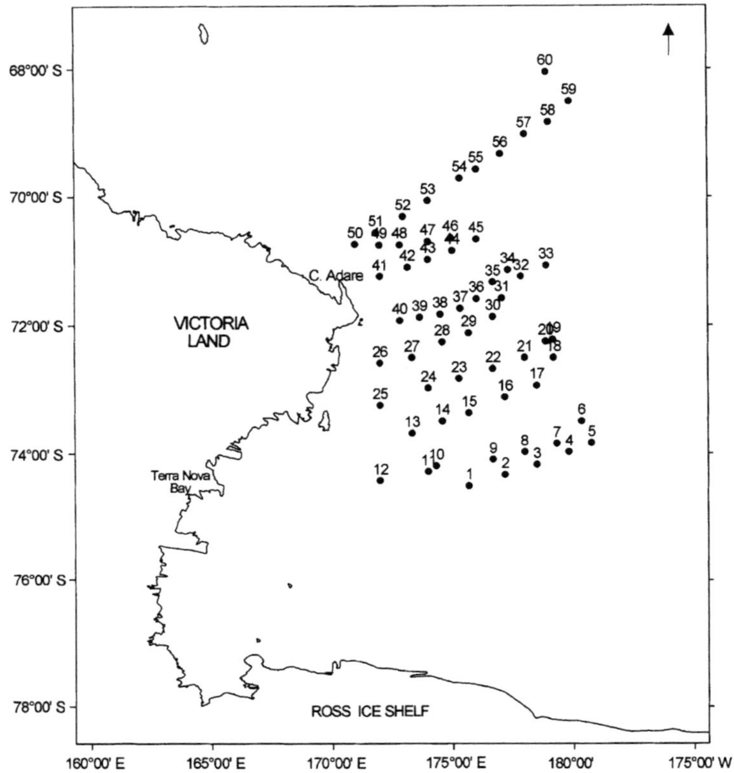


**Fig. 2.** Ross Sea. The track followed by the krill echo survey and utilized for surface Chl a monitoring by flow fluorometry. *Numbers* indicate the discrete sampling positions, referred to as CT in the test

surface at station 10. The area of compact ice was characterized by markedly lower Chl a levels, ranging from 19.4 to 37.7 $mg\,m^{-2}$, with maximum concentrations (0.59 $\mu g\,l^{-1}$) at the surface at station 24A.

As regards the quantitative importance of the different size fractions, the micro- and nanophytoplankton were strongly dominant in the polynya and in the MIZ, while the picophytoplankton fraction was by far the most abundant in the closed pack ice zone. Table 1 summarizes the dimensional structure of the phytoplankton community in the three areas by reporting the mean percentage contributions of the different fractions to total Chl a.

Stations 28, 29 and 30 were located in the western MIZ of the polynya, in a pack-ice covered area and in the eastern ice edge of the Terra Nova Bay polynya, respectively (Fig. 1). The vertical profiles of total and size fractionated Chl a and of water density are reported in Fig. 5. Integrated Chl a concentrations were 207, 29 and 58 $mg\,m^{-2}$, respectively. Maximum chlorophyll concentrations (3.38 $\mu g\,l^{-1}$) occurred at station 28 at a depth of 30 m. The microphytoplankton fraction prevailed and was particularly high at station 28, where its mean percentage contribution was

**Table 1.** Mean percentage contribution of micro-, nano- and picoplankton fractions to total biomass in the polynya, marginal-ice zone and pack-ice area

| % Fractions | Polynya | MIZ | Pack-ice area |
|---|---|---|---|
| Microphytoplankton | 44 | 61 | 26 |
| Nanophytoplankton | 26 | 22 | 22 |
| Picophytoplankton | 29 | 17 | 54 |

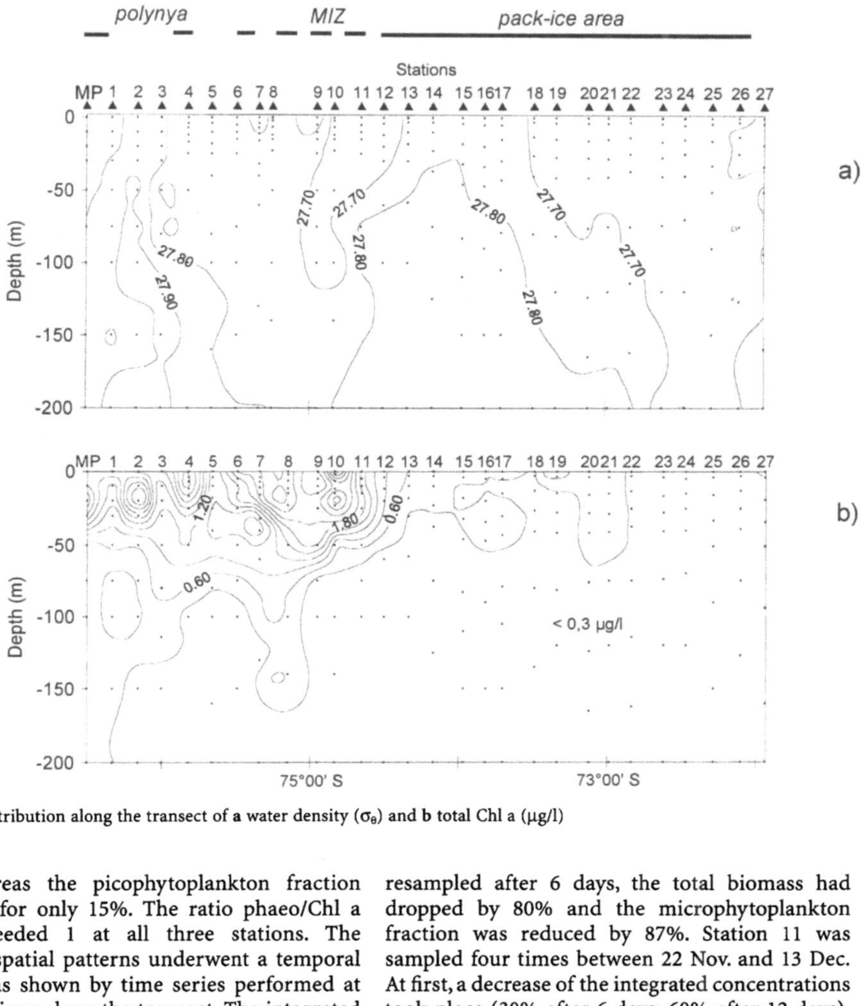


**Fig. 3a,b.** Distribution along the transect of **a** water density ($\sigma_\theta$) and **b** total Chl a (µg/l)

68%, whereas the picophytoplankton fraction accounted for only 15%. The ratio phaeo/Chl a never exceeded 1 at all three stations. The described spatial patterns underwent a temporal variation, as shown by time series performed at several stations along the transect. The integrated values of total Chl a and the percentage contribution of size fractions in three replicated sets of stations, each set characteristic of one of the three areas, are reported in Fig. 6. At the polynya stations MP and 6, resampled after 5 and 6 days, respectively, the biomass did not change either in quantitative terms or in its size spectrum, the integrated values of Chl a varying from 164.7 to 175.2 mg $m^{-2}$ at station MP and from 96.1 to 96.7 mg $m^{-2}$ at station 6.

In the marginal ice zone, on the contrary, substantial changes were observed. At station 6, resampled after 6 days, the total biomass had dropped by 80% and the microphytoplankton fraction was reduced by 87%. Station 11 was sampled four times between 22 Nov. and 13 Dec. At first, a decrease of the integrated concentrations took place (30% after 6 days; 60% after 12 days), which was followed, after 21 days, by a slight increase of the total biomass and of the micro- and nanophytoplankton fractions. Approaching the pack-covered area, at stations 11 and 12, a markedly modified dimensional structure of the community was observed, with an overwhelming presence of small cells. At station 15, resampled four times between 26 Nov. and 12 Dec., slight variations affected the integrated total biomass, while the nano- and, particularly the picophytoplankton fractions increased.

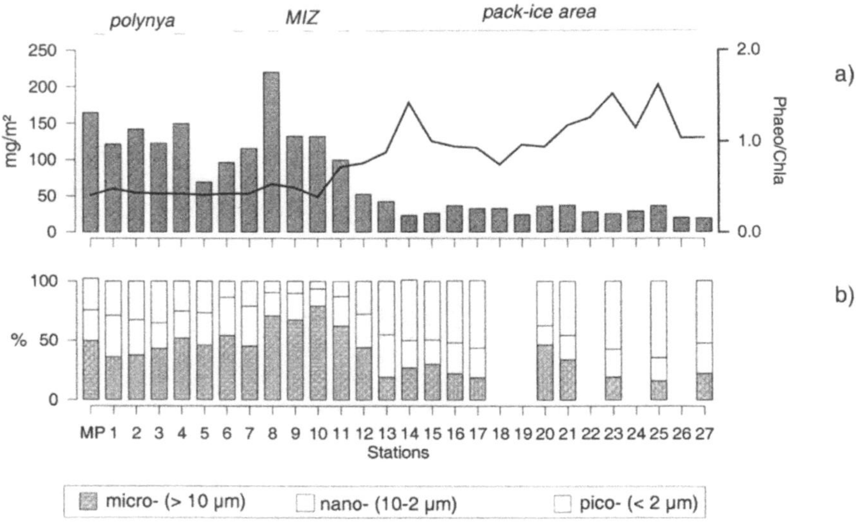


**Fig. 4a,b.** Distribution along the transect of **a** total Chl a and phaeo/Chl a ratio and **b** the relative contribution of phytoplankton size fractions to total biomass

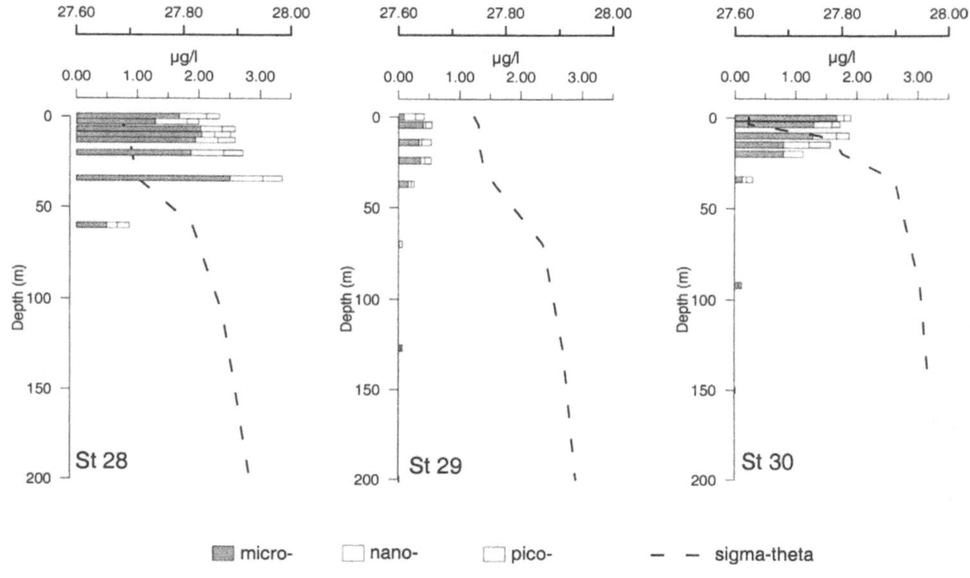


**Fig. 5.** Vertical profiles of water density ($\sigma_\theta$), total and size fractionated Chl a ($\mu g\,l^{-1}$) at stations 28, 29 and 30

**Fig. 6.** Variations in integrated values of total Chl a ($\mu g\,l^{-1}$) and percentage contribution of the different size fractions in replicated sets of stations in the three areas of the polynya, marginal ice zone and pack ice

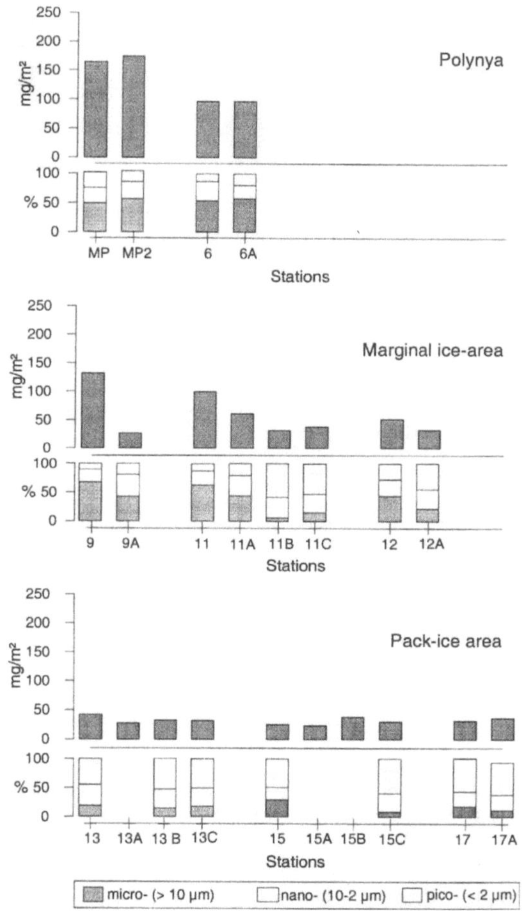


The track followed by the acoustic survey of krill stocks (Fig. 2) provided the opportunity for collecting a set of samples for further investigating the spatial patterns of phytoplankton biomass and checking temporal variations at some stations of the 175°E transect positioned within the area covered by the echo-survey track. The area facing Victoria Land was characterized by values of surface biomass ranging from 0.5 to $0.9\,\mu g\,l^{-1}$ of Chl a. Higher concentrations were found by Cape Adare, with a maximum value of $1.64\,\mu g\,l^{-1}$ at CT42. The total biomass concentrations and the contribution by the different size fractions are reported in Fig. 7, while Table 2 shows the mean values of total Chl a and percentage contribution of the size fractions. The observations at the discrete sampling points coinciding with the stations of the 175°E transect provided further information on the variation of phytoplankton biomass and its dimensional structure. As shown in Table 2, both a spatial and temporal variability was observed when spatially coinciding sampling points of the track and of the transect were compared.

## Discussion

Most of the literature on Antarctic phytoplankton reports low to moderate Chl a concentrations, with a mean of $0.5\,mg\,m^{-3}$ resulting from a large

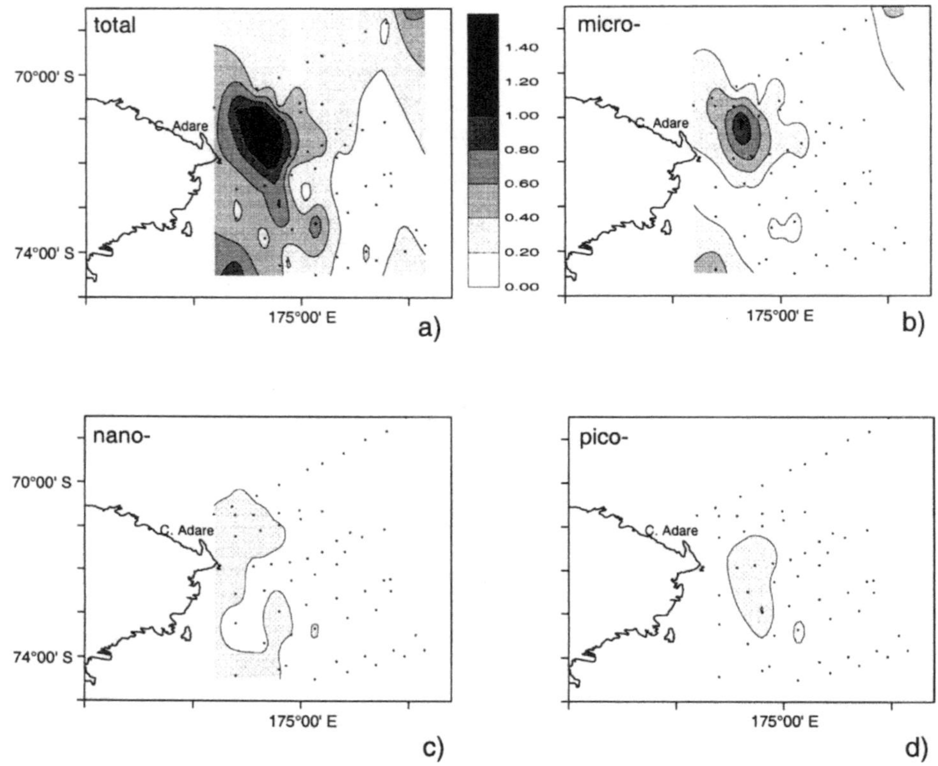


**Fig. 7a–d.** Surface distribution of concentrations (μg l$^{-1}$) of **a** total Chl a, **b** micro- **c** nano- and **d** picoplankton fractions in the area covered by the echo-survey track

**Table 2.** Temporal variation of values of total Chl a (μg$^{-1}$) and percentage contribution of size fractions in the time interval occurring between the sampling of the transect stations and the sampling of the corresponding positions (CT) along the echo-survey track

| Station | Date | Total biomass | % Micro- | % Nano- | % Pico- |
|---|---|---|---|---|---|
| 14B | Dec. 12 | 0.28 | 27 | 11 | 65 |
| CT 10 | Dec. 19 | 0.22 | 29 | 10 | 62 |
| 18 | Dec. 5 | 0.42 | 25 | 36 | 39 |
| CT 14 | Dec. 20 | 0.57 | 47 | 35 | 18 |
| 22 | Dec. 12 | 0.26 | n.d. | n.d. | n.d. |
| CT 24 | Dec. 21 | 0.13 | 19 | 47 | 34 |
| 25 | Dec. 7 | 0.16 | 7 | 23 | 69 |
| CT 28 | Dec. 22 | 0.20 | 35 | 24 | 42 |
| 27 | Dec. 8 | 0.19 | 32 | 11 | 60 |
| CT 38 | Dec. 24 | 0.80 | 50 | 17 | 32 |

geographical and temporal variability. Low biomass values in terms of pigment concentrations are commonly observed in open ocean situations, while, in waters close to the continent, higher and more variable values are reported, as for the case of areas around the Antarctic Peninsula and within the Ross Sea.

To our knowledge, for phytoplankton biomass and production during the austral spring in the Ross Sea, few satellite data (Arrigo and McClain 1994) and some field studies (Saggiomo et al. 1998) are available, but a number of sparse and fragmentary observations have been reported from different geographical and ecological situations (Nelson et al. 1987; Jacques and Panouse 1991; Lancelot et al. 1993; Savidge et al. 1995; Boyd et al. 1995; Jochem et al. 1995; Ning et al. 1996; Detmer and Bathmann 1997; Bathmann et al. 1997). In this highly variable context, our data appear both consistent with and discrepant to those of other authors dealing with summer situations in the same area (see El-Sayed et al. 1983; Smith and Nelson 1985; Wilson et al. 1986; DeMaster et al. 1992; Innamorati et al. 1992a; Smith et al., 1996; Di Tullio and Smith 1996; Saggiomo et al. 1998). We therefore believe that a productive comparative effort should be postponed until an adequate coverage provides a more coherent picture for ecologically comparable subsystems.

As regards the size structure of phytoplankton communities, von Bröckel (1981) emphasized the role of nanophytoplankton responsible for about 90% of the production during the austral autumn in the northern part of the Antarctic Peninsula. Hewes et al. (1985) and Weber and El-Sayed (1987) have reported a general predominance of the nanophytoplankton fraction in the open waters of the Southern Ocean, which coincides with the data reported herein for the polynya. For the Southern Ocean, Ning et al. (1996) reported a predominance of microphytoplankton (mean >60%) in nutrient-rich areas and of picophytoplankton (>40%) in infertiles ones. Holm-Hansen and Mitchell (1991) showed that, in a coastal area of the Bransfield Strait, microphytoplankton was predominant during December, the nanophytoplankton fraction becoming predominant during the period from January throughout March. In two stations located in the outer part of the Ross Sea, Wilson et al. (1986) found that the fraction between 8 and 50 μm accounted for almost the whole population, the fractions >50 μm and between 3 and 8 μm were absent and picophytoplankton comprised between 0 and 30% of the community. Boyd et al. (1995) showed for the Bellingshausen Sea during the austral spring (seven stations) that the nanophytoplankton fraction (2–18 μm) was dominant in the water column at the ice and transitional stations, whereas in the ice-free offshore stations microphytoplankton prevailed, concomitantly with the highest integrated biomasses. For the Polar Front, the Antarctic Circumpolar Current, the Weddel Gyre Boundary and for the Marginal Ice Zone (13 stations), Detmer and Bathmann (1997) reported a pico- and nanophytoplankton contribution up to 90% in areas with low biomass concentrations, but less than 50% in others characterized by high biomass (>1.8 $\mu g l^{-1}$) and dominance of large sized cells.

As reported in Table 1, our observations showed that different size compositions of the phytoplankton communities characterized the three areas considered, Fig. 5 illustrating the percentage contribution, in terms of Chl a, of micro- (>10 μm), nano- (10–2 μm) and total picophytoplankton (<2 μm) fractions. In the polynya, the microphytoplankton fraction appears to strongly predominate; but the relative abundance of the micro- and nanophytoplankton fractions in this area (stations 1 to 8) is certainly biased by the presence of large quantities (up to 3 406 000 cell $l^{-1}$ at station 8) of the palmelloid form of *Phaeocystis* cfr. *antarctica* (Saggiomo et al., this Vol.), that is retained by a 10-μm membrane. The time series performed in this area showed low variations in total biomass and percentage contribution by size fractions. In this area, the biomass was evenly distributed along the water column, in consequence of the lack of stratification. The low phaeo/Chl a ratio found in the polynya suggests that *Phaeocystis* cfr. *antarctica*, particularly in its aggregated form, was not grazed and that it sedimented toward the bottom, as reported by DeMaster et al. (1992) and Di Tullio and Smith (1996).

In the MIZ, that showed at the beginning of the sampling the highest biomasses, the micro- and nanoplankton fractions were dominant. For this area, Saggiomo et al. (this Vol.) report a phytoplankton community dominated by diatoms (*Fragilariopsis curta* and *Pseudo-nistzchia subcurta*), which increase their relative concentration from stations 9 (47%) to 11 (97%); this trend is concomitant with the gradual disappearance of *Phaeocystis*. Time series performed at stations 9 and 11 (Fig. 7) showed total biomasses dropping to low levels and consisting almost entirely of nano- and picophytoplankton. Biomass levels up to 200 $mg\,m^{-2}$ were observed at the beginning of the time series. This quantity cannot, per se, be considered as particularly high for this ecological context, where the presence of very large amounts of krill (Azzali et al. 1998) suggests an active removal of cells. Krill grazing activity can, in fact, drastically reduce high biomasses of meso- and microphytoplankton within hours, shifting the size composition toward the nanoplankton fraction (Jacques and Panouse 1991). Also Di Tullio and Smith (1996) emphasize the role of krill in reducing phytoplankton biomasses, but referring to Marr (1962), exclude an important role of krill within the Ross Sea. On the contrary, we can assert that krill represents a basic factor in conditioning the Ross Sea pelagic food webs at least for the period under consideration. In fact, the association of krill to pack ice has been reported also for the Ross Sea and is confirmed by our direct and repeated observations of krill attached to overturned ice floes. The presence of remarkable masses of krill in the Ross Sea has been proved by the echo survey carried out during the same cruise by Azzali and coworkers in an area inclusive of the studied transect and characterized by a mean ice cover >8/10 (Azzali et al., this Vol.). The echo survey yielded a mean density of 82.6 tons of krill per nautical mile$^2$ (Azzali et al. 1998).

The distribution and the time variations just described for the polynya and MIZ phytoplankton assemblages must, in addition, be evaluated at the light of the data provided by Saggiomo et al. (1998), who showed that, for this same time series, the structure of the water column remained substantially constant. It is therefore reasonable to think that possible advective phenomena interested areas with similar hydrographic and ecological conditions.

In the third area, covered by pack ice, an almost opposite pattern was observed, as the bulk of the biomass consisted of picophytoplankton. In fact, the literature on the subject (Stretch et al. 1988; Smetacek et al. 1990) provides evidence of the selective role of grazing, mostly by krill, on the larger phytoplankton fraction, inhabiting both sympagic environment and the ice-covered water column. During the first half of the cruise, we frequently observed the presence of krill attached to ice floes overturned by the ship. The influence of grazing on the dimensional structure of the phytoplankton of pack-covered areas is further supported by the marked increase of the ratio phaeo/Chl a, higher in the northern part of the transect than in the polynya area, a fact we interpret as a sign of a sustained grazing activity. The biomass values for this area were four to eight times higher than those reported for Antarctic waters in winter (Brightman and Smith 1989), indicating an active growth under the ice cover or a release of sympagic flora from the pack ice, or both, being the microphytoplankton fraction predominant only at the surface, probably released from the pack ice. As regards the picophytoplankton fraction, the warning by Jacques and Panouse (1991) about the possibility that this fraction might be inclusive of fragments of bigger cells resulting from filtration procedures, as well as that raised by Gieskes and Elbrachter (1986), about the presence of substantial quantities of free floating chloroplasts in the nanophytoplankton fraction, are to be excluded on the base of the microscopic observations by Marino and Cabrini (1997). In addition Vanucci and Bruni (this Vol.) report the predominance, in this area, of a small (<2 µm), as yet unidentified cell.

At the resampled stations, particularly in the time series performed at station 15, substantial changes in the quantity and size structure of the biomass were not observed, with the picoplankton stations keeping its dominant role. The high photosynthetic efficiency of this fraction (Saggiomo et al., this Vol.) suggests an important role of microzooplankton in Antarctic trophic webs (von Bröckel 1981; Becquevort et al. 1992; Lancelot et al. 1993).

As regards the area covered by the echo-survey track, the observations obtained from the resampled stations indicated a clear spatial heterogeneity, along with a change in the dimensional structure of the community. The dimensional structure remained constant, with a predominance of picophytoplankton fractions, when the levels of biomass were low (<0.5 $\mu g\,l^{-1}$); at stations showing higher biomasses, with values exceeding 0.5 $\mu g\,l^{-1}$, a concomitant increase of the micro- and nanophytoplankton fractions was observed.

In the northernmost investigated area, located in the vicinity of Cape Adare MIZ, high surface levels of biomass, probably coinciding with an increased water column stability, were in agreement with the classically accepted model of functioning of Antarctic pelagic systems at the onset of summer. On the contrary, the high biomass concentrations observed by us during mid-spring were associated with well mixed or slightly stratified water columns.

## Concluding Remarks

Within the geographical area of the Ross Sea, three subsystems were identified during spring: the pack-covered waters dominated by picophytoplankton; the MIZ with a dominance of microphytoplankton; the polynya, with a prevalence of *Phaeocystis* cfr. *antarctica* in its aggregated form.

The observed variations of the dimensional structure of the spring phytoplankton communities, associated with the marginal-ice zone and the pack-covered water column, were to a large extent the consequence of the removal of the meso- and microphytoplankton by selective grazing. The observed prevalence of the picophytoplankton fraction in the water column under the pack may be interpreted as a product of selective removal of the larger cells attached to the ice surface by the krill *Euphasia superba,* an interpretation supported by the findings of Smetacek et al. (1990) in the Weddell Sea. The high phaeo/Chl a ratios we observed further support the hypothesis of the driving role of grazing in structuring the size composition of phytoplankton communities.

Due to the active removal of vegetal biomass by grazers, a common phenomenon in both the offshore and inshore environments, an efficient transfer of organic carbon through the different levels of the Antarctic food web can be hypothesized. This hypothesis seems supported by the findings of DeMaster et al. (1992), who demonstrated that the amount of carbon reaching the bottom is limited with respect to that of biogenic silica.

The oscillation in Chl a concentrations between high and low levels throughout the investigated area, with values comparable with those reported by El-Sayed et al. (1983), Wilson et al. (1986), DeMaster et al. (1992), Innamorati et al. (1992a,b), Smith et al. (1996) and Di Tullio and Smith (1996) for the austral summer in the Ross Sea, suggests the presence of recurrent blooms throughout spring and summer. Biomass concentrations confirm the key role of the MIZ in setting the conditions for the seasonal dynamics of the open waters system, even when it does not produce a stabilizing effect on the water column.

*Acknowledgements.* This work was carried out in the framework of the Italian National Programme for Antarctica (PNRA), whose support in the field and laboratory activities is gratefully acknowledged. We thank Massimo Azzali for kindly providing information from the of krill acoustic survey aggregation. The manuscript has benefited from the suggestions of two anonymous referees.

## References

Arrigo KR, McClain CR (1994) Spring phytoplankton production in the Western Ross Sea. Science 266: 261–263

Azzali M, Kalinowsky J, Cosimi G, Castagnani R, Saino N (1998) Estimation of biomass of krill Euphasia superba and birds and mammals census during the Xth Italian Expedition to Ross Sea, November-December 1994. In: Guglielmo L, Faranda F (eds) National Scientific Commission for Antarctica (ed), Oceanographic Campaign 1994–95, Data Rep Part II (in press)

Bathmann UV, Scharek R, Klaas C, Dubischar CD, Smetacek V (1997) Spring development of phytoplankton biomass and composition in major water masses of the Atlantic sector of the Southern Ocean. Deep Sea Res II 44(1–2): 51–67

Becquevort S, Mathot S, Lancelot C (1992) Interactions in the microbial community of the marginal ice zone of the northwestern Weddell Sea through size distribution analysis. Polar Biol 12: 211–218

Boyd PW, Robinson C, Savidge G, Williams PJL (1995) Water column and sea-ice primary production during austral spring in the Bellingshausen Sea. J Plankton Res 42(4–5): 1177–1200

Brightman RI, Smith WO Jr (1989) Photosynthesis-irradiance relationships of Antarctic phytoplankton during austral winter. Mar Ecol Prog Ser 53: 143–151

Comiso JC, McClain CR, Sullivan CW, Ryan JP, Leonard CL (1993) Coastal zone colour scanner pigment concentrations in the Southern Ocean and relationships to geophysical surface features. J Geophys Res 98: 2419–2451

DeMaster DJ, Dunbar RB, Gordon LI, Leventer AR, Morrison JM, Nelson DM, Nittrouer CA, Smith WO Jr (1992) Cycling and accumulation of biogenic silica and organic matter in high-latitude environments: the Ross Sea. Oceanography 5(3): 146–153

Detmer AE, Bathmann UV (1997) Distribution patterns of autotrophic pico- and nanoplankton and their relative contribution to algal biomass during spring in the Atlantic sector of Southern Ocean. Deep Sea Res II 44(1–2): 299–320

Di Tullio GR, Smith WO Jr (1996) Spatial patterns in phytoplankton biomass and pigment distributions in the Ross Sea. J Geophys Res 101(8): 18,467–18,477

El-Sayed SZ, Biggs DC, Holm-Hansen O (1983) Phytoplankton standing crop, primary productivity, and near-surface nitrogenous nutrient fields in the Ross Sea, Antarctica. Deep Sea Res 30(8A): 871–886

Gieskes WWC, Elbrachter M (1986) Abundance of nanoplankton-size chlorophyll-containing particles caused by diatom disruption in surface waters of the Southern Ocean (Antarctic Peninsula region). Neth J Sea Res 20: 291–303

Hewes CD, Holm-Hansen O, Sakshaug E (1985) Alternate carbon pathways at lower trophic levels in the Antarctic food web. In: Siegfried WR, Condy PR, Laws RM (eds) Antarctic nutrient cycles and food webs. Springer, Berlin Heidelberg New York, pp 277–288

Holm-Hansen O, Mitchell G (1991) Spatial and temporal distribution of phytoplankton and primary production in the western Bransfield Strait region. Deep Sea Res 38(8–9): 961–980

Holm-Hansen O, Lorenzen CJ, Holmes RW, Strickland JDH (1965) Fluorometric determination of chlorophyll. J Cons Perm Int Explor Mer 30(1): 3–15

Innamorati M, Lazzara L, Massi L, Mori G, Saggiomo V (1992a) Spatial and temporal distribution of phytoplankton size fractions in Antarctic waters: biomass and production, summer 1989–90. In: Ambrosetti W, Piccazzo M, Ruffoni Riva T (eds) Atti X Congr Ass Ital Oceanogr Limnol, Lang, Genova, pp 605–612

Innamorati M, Lazzara L, Massi L, Mori G, Nuccio C, Saggiomo V (1992b) Indagine sulla biomassa fitoplanctonica nel Mare di Ross in relazione ai fattori ambientali. In: Gallardo VA, Ferretti O, Moyano HI (eds) Oceanografia in Antartide. ENEA – PNRA-EULA, Concepcion, Chile, pp 235–252

Jacques G, Panouse M (1991) Biomass and composition of size fractionated phytoplankton in the Weddell-Scotia Confluence area. Polar Biol 11: 315–328

Jochem FJ, Mathot S, Quéguiner B (1995) Size-fractionated primary production in the open Southern Ocean in austral spring. Polar Biol 15: 381–392

Kiørboe T (1993) Turbulence, phytoplankton cell size and the structure of pelagic food webs. Adv Mar Biol 29: 1–72

Lancelot C, Mathot S, Veth C, de Baar H (1993) Factors controlling phytoplankton ice-edge blooms in the marginal ice-zone of northwest Weddell Sea during sea ice retreat 1988: field observations and mathematical modelling. Polar Biol 13: 377–387

Marino D, Cabrini M (1997) Distribution and succession of phytoplankton populations in the Ross Sea (Antarctica) during the austral spring 1994. In: Faranda F, Guglielmo L, Povero P (eds) National Programme for Antarctic Research, ROSSMIZE 1993–1995, Data Rep Part I, pp 307–341

Marr JWS (1962) The natural history and geography of the Antarctic krill (*Euphasia superba* Dana). Discovery reports 32. Institute of Oceanographic Sciences, Wormley, England, pp 33–464

Marra J (1996) Biogeochemical cycling in the Ross Sea: an introduction. J Geophys Res 101(8): 18,453–18,454

Nelson DM, Smith WO Jr, Gordon LI, Huber BA (1987) Spring distributions of density, nutrients, and phytoplankton biomass in the ice-edge zone of the Weddell-Scotia Sea. J Geophys Res 92: 7181–7190

Ning X, Liu Z, Genhai Z, Junxian S (1996) Size-fractionated biomass and productivity of phytoplankton and particulate organic carbon in the Southern Ocean. Polar Biol 16: 1–11

Saggiomo V, Massi L, Modigh M, Innamorati M (1992) Size-fractionated primary production in Terra Nova Bay (Ross Sea) during Austral Summer (1989–90). In: Gallardo VA, Ferretti O, Moyano HI (eds) Oceanografia in Antartide. ENEA – PNRA-EULA, Concepcion, Chile, pp 289–294

Saggiomo V, Carrada GC, Mangoni O, Ribera d'Alcalà M, Russo A (1998) Spatial and temporal variability of size-fractionated biomass and primary production in the Ross Sea (Antarctica) during austral spring and summer. J Mar Syst 17: 98–115

Savidge G, Harbour D, Gilpinand LC, Boyd PW (1995) Phytoplankton distributions and production in the Bellingshausen Sea, austral spring 1992. Deep Sea Res II 42(4–5): 1201–1224

Smetacek V, Scharek R, Nothig EM (1990) Seasonal and regional variation in the pelagial and its relationship to the life history cycle of krill. In: Kerry R, Hempel G (eds) Antarctic ecosystem: ecological change and conservation. Springer, Berlin Heidelberg New York, pp 103–114

Smith WO Jr, Nelson DM (1985) Phytoplankton bloom produced by a receding ice edge in the Ross Sea: spatial coherence with the density field. Science 227: 163–166

Smith WO, Gordon LI (1997) Hyperproductivity of the Ross Sea (Antarctica) polynya during austral spring. Geophys Res Lett 24: 233–236

Smith WO Jr, Nelson DM, Di Tullio GR, Leventer AR (1996) Temporal and spatial patterns in the Ross Sea: phytoplankton biomass, elemental composition, productivity and growth rates. J Geophys Res 101(8): 18,455–18,465

Stretch JJ, Hammer PP, Hammer WM, Michel WC, Cook J, Sullivan CW (1988) Foraging behaviour of Antarctic krill *Euphasia superba* on sea ice microalgae. Mar Ecol Prog Ser 44: 131–139

Sullivan CW, Arrigo KR, McClain CR, Comiso JC, Firestone J (1993) Distributions of phytoplankton blooms in the Southern Ocean. Science 262: 1832–1837

Von Bröckel K (1981) The importance of nanoplankton within the pelagic Antarctic ecosystem. Kiel Meeresforsch Sonderh 5: 61–67

Weber LH, El-Sayed SZ (1987) Contributions of net, nano- and picoplankton to the phytoplankton standing crop and primary productivity in the Southern Ocean. J Plankton Res 9(5): 973–994

Wilson DL, Smith WO Jr, Nelson DM (1986) Phytoplankton bloom dynamics of the western Ross Sea ice edge. I. Primary productivity and species-specific production. Deep Sea Res 33(10): 1375–1387

# CHAPTER 18

# Phytoplankton Biomass Related to Environmental Factors in the Ross Sea

M. Innamorati, G. Mori, L. Massi, L. Lazzara, and C. Nuccio[1]

**ABSTRACT**

Three years of investigation in the western Ross Sea reveal that despite the general abundance of phytoplankton biomass, the ratios between biomass and nutrient availability are low due to high nutrient concentrations. The N/P ratio is practically identical to that of Redfield; however, during blooms the ratio can be severely altered due to very low phosphate concentrations. Phytoplankton biomass during summer shows a bimodal cyclic pattern with two blooms in early and late summer. The spatial and temporal distribution of these blooms follows the latitudinal distribution of daily irradiance, which determines the advance of sea-ice melting from south to north forming an upper mixed layer rich in nutrients. This mixed layer is well illuminated and warmer than the deeper layer, thereby favouring the development of the early bloom. Afterwards, in the mixed water column, biomass decreases and reaches the summer minimum. A new stabilization of the water column is followed by a second bloom, less extensive than the first. In all 3 years, similar cycles have been recorded, giving corroborative evidence for the existence of a second bloom.

## Introduction

Studies on the ecology of phytoplankton in the Southern Ocean initially revealed the contrast between the large quantity of nutrients and the low concentration of phytoplankton (Holm-Hansen et al. 1977; El-Sayed et al. 1983; El-Sayed 1988). Jacques and Tréguer (1986) described the situation in the Southern Ocean as the Antarctic paradox, because of a high eutrophy which is capable of producing a quantity of biomass as small as that of most oligotrophic zones in the oceans. Geographic variety exists, moreover, between the coastal and open-sea areas (El-Sayed et al. 1983; Holm-Hansen and Mitchell 1991); in any case, phytoplankton blooms of remarkable intensity are reported in the Southern Ocean (El-Sayed 1984, 1988; Smith and Nelson 1985; Knox 1994).

As far as the Ross Sea is concerned, various studies have been carried out, such as those of Bunt and Lee (1970), Holm-Hansen et al. (1977), El-Sayed et al. (1983), and Wilson et al. (1986), also in areas close to Terra Nova Bay. The bay, however, has remained almost completely unknown with regards to various oceanographic aspects, among which the hydrography and ecology of phytoplankton.

Research by the Italian Antarctic National Research Program (PNRA) was initially aimed at supplying a picture of the marine ecosystems of the western Ross Sea. The results of the first expedition (R/V Polar Queen, 1987–88) indicated a more extensive biomass and a more complex seasonal cycle (Innamorati et al. 1990c) than had been expected from the data reported in the literature. The second expedition was therefore organized to include a neritic group, to conduct a series of daily surveys at several given stations in Terra Nova Bay (M/N Barken and R/V Malippo, 1989–90), and an oceanic group, to carry out surveys on a large spatial scale in the Ross Sea (R/V Cariboo, 1989–90). The third expedition (M/N Italica, 1994–95) reexamined the temporal series in the bay. All data obtained during the three expeditions have been published in Data Reports (Innamorati et al. 1990a, 1990b, 1991, 1994a; Nuccio et al. 1994; Lazzara et al. 1997), and several preliminary results have been comunicated at congresses and workshops (Innamorati et al. 1992a,b; Lazzara and

[1] Laboratorio di Ecologia, Dipartimento di Biologia Vegetale, Università di Firenze. Via P.A. Micheli, 1. 50121 Florence, Italy

Nuccio 1994; Lazzara et al. 1995, 1996; Nuccio et al. 1992; Saggiomo et al. 1992).

## Materials and Methods

The investigated area in 1987–88 (Fig. 1a), on board the R/V Polar Queen, covered a grid of 40 × 50 n mi with 40 stations spaced 5 n mi near the coast and 10 n mi offshore. Stations 16, 17, 49, 50 and 51 were all along 74°49′S up to 125 n mi from the coast.

During the 1989–90 expedition, the neritic group carried out its work at three stations (Fig. 1b): Terra Nova Bay (BTN; 74°41′42″S, 164°07′24″E), near the rocky shore, sampled at the surface daily; Mergellina (MRG), 250 m from the coast (74°41′33″S, 164°07′15″E; 50 m deep); Santa Maria Novella (SMN), at the centre of the bay (74°43′00″S, 164°16′00″E), where the bottom was 500 m deep. The oceanic group on board the R/V Cariboo, in the area from 60° to 75°S, sampled 27 stations (Fig. 1c), and several samples for

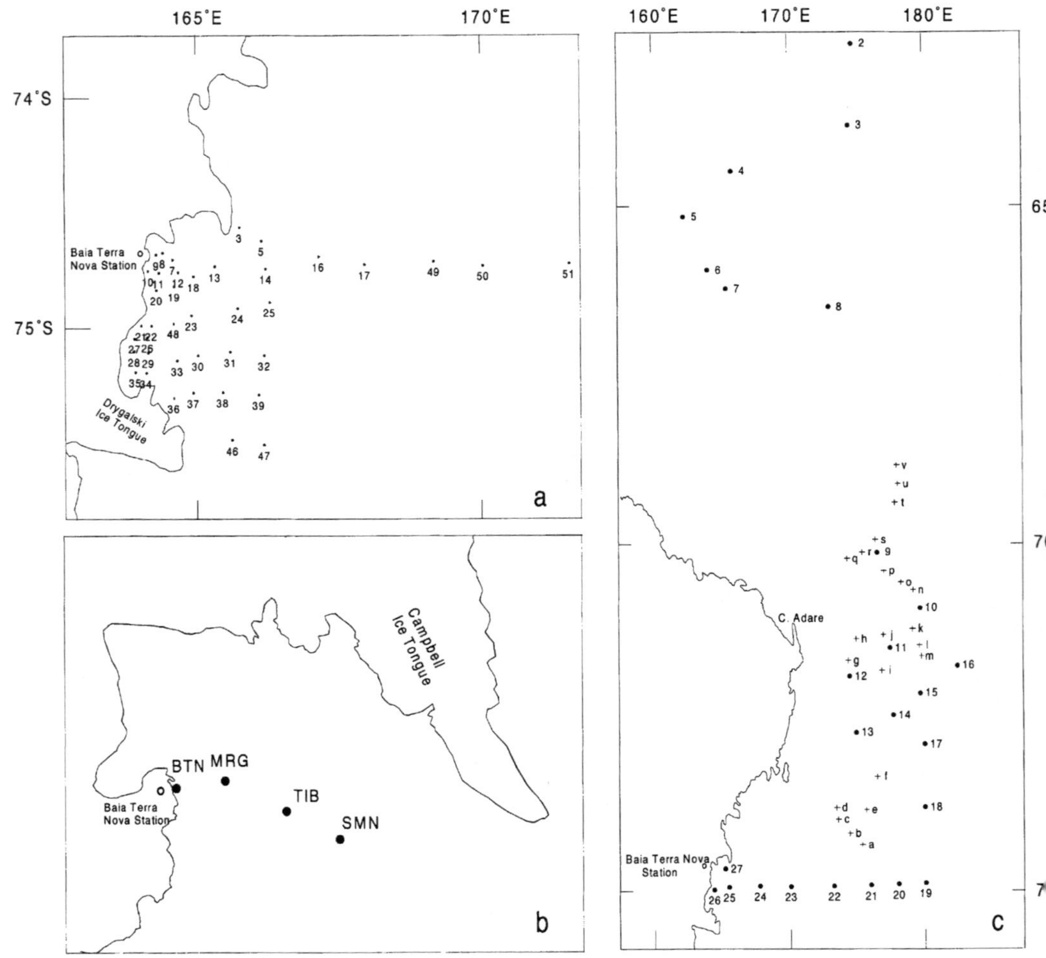


Fig. 1a–c. Maps of the stations. a 1987–88. b Coastal stations SMN, MRG, BTN (1989–90), TIB, BTN (1994–95). c 1989–90

chlorophaeopigments analysis were collected on the way back from Terra Nova Bay using a pumping system, located at 3 m from the sea surface (Fig. 1c; stations a–v).

In 1994–95, two stations were sampled: BTN, as in 1989–90, and Tiburtina (TIB; 74°42′05″S, 164°10′01″E), with a depth of 300 m (Fig. 1b).

Hydrographic data of the 1987–88 cruise are from Boldrin and Stocchino (1990), and those of the R/V Cariboo cruise are from Artegiani et al. (1992). In 1989–90 and 1994–95 CTD data were acquired by means of the KMS profiler Meerestechnik, equipped with a Dr. Haardt Back-Scat fluorometer.

Seawater was sampled using Niskin bottles for phytoplankton microscopic analysis (fixed with Lugol or formaldehyde to a final concentration of 1 and 4%, respectively), for the analysis of nitrites, nitrates, phosphates and silicates (Hansen and Grasshoff 1983) as well as for pigment determinations, after filtration on Whatman GF/F. Data on nutrients for 1987–88 and for the 1989–90 R/V Cariboo cruises are from Catalano and Benedetti (1990) and Catalano et al. (1991). Liposoluble pigments were determined, after extraction with 90% acetone, using a Kontron Uvikon 930 spectrophotometer. The chlorophyll a + phaeophytin a (Phaeo) = chlorophaeopigments (Chl) concentration was determined following Lorenzen (1967) modified by Riemann (1978). During the three expeditions, chlorophaeopigment analyses at station BTN were begun at the end of December-beginning of January, on the ship's arrival (Fig. 7). In 1989–90 and 1994–95, samples for the microscopic analysis of phytoplankton densities were taken from the beginning of December (Fig. 7).

Measurements of solar global irradiance were carried out by means of a CM5 solarimeter (Kipp and Zonen) continuously recorded and daily integrated (CC1 Kipp and Zonen) in 1987–88 and 1989–90, and in 1994–95 by means of a PAR quantameter equipped by a data logger (LI-192SA; LI-100 LI-COR), storing data every 10 min. The conversion from PAR ($\mu M\,s^{-1}\,m^{-2}$ of photons) to global solar irradiance ($W\,m^{-2}$) was obtained by dividing the former by 1.96 (Morel and Smith 1974; Jerlov 1976).

Surface and underwater downwelling, upwelling, scalar quantum PAR irradiances were measured by means of a photoprobe (Innamorati et al. 1994b) equipped with three cosine and one spherical meters (LI 192SA, LI 193SA LI-COR).

The depth of the euphotic layer, $z_e$, was calculated from $E_e/E_0 = 0.01$, where $E_e$ and $E_0$ are the downwelling quantum PAR irradiances at the euphotic depth and sea surface, respectively. For some stations, data for downwelling PAR were missing, so that, as the waters all belong to case I (Morel 1988), the euphotic zone depth was estimated from the orthogonal regression of log $z_e$ vs. log Chl (Morel 1988) calculated on the overall data set, as described in Lazzara et al. (this Vol.).

Images for pack-ice cover (NOAA-HRTP AVHRR) in the Ross Sea during 1994–95 were obtained from the ENEA-PNRA receiving station at Terra Nova Bay.

## Results

### Irradiance and Ice Covering

Increasing daily irradiance during summer at these high latitudes is the main factor determining the melting of pack ice and subsequently all other physical and biological events. For about 35 days before and after the summer solstice, daily irradiance was greater in the direction of the poles, and decreased with latitude in the direction of the tropics (Fig. 2). The melting pattern of sea ice reflected that for irradiance; it began from the south and proceeded towards the north, encountering a certain delay in the coastal zone compared to ice in the open sea. At the beginning of

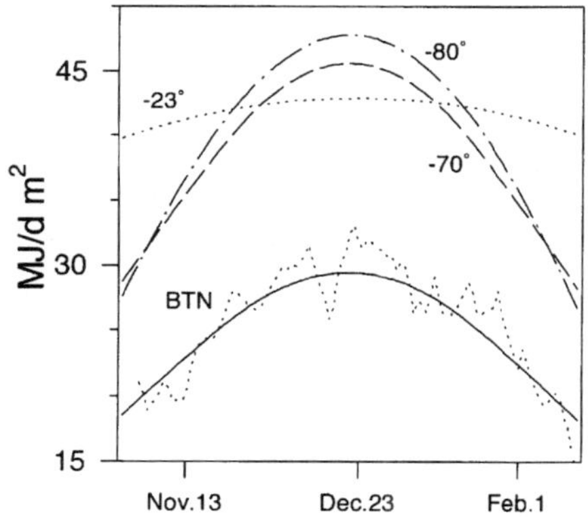


**Fig. 2.** Latitudinal variations of extraterrestrial daily irradiance (after Iqbal 1983) and mean of the global daily irradiance measured during the three expeditions at Terra Nova Bay with its interpolating function

December, in addition to the polynya of Terra Nova Bay, the central part of the Ross Sea was free of ice and, towards the beginning of January, this was also the case in coastal areas in the vicinity of the Terra Nova Bay Station (Fig. 3).

As a result of ice melting and warming of surface temperatures, salinity and density decreased, determining a stratification and stability of the water column that favoured the development of phytoplankton.

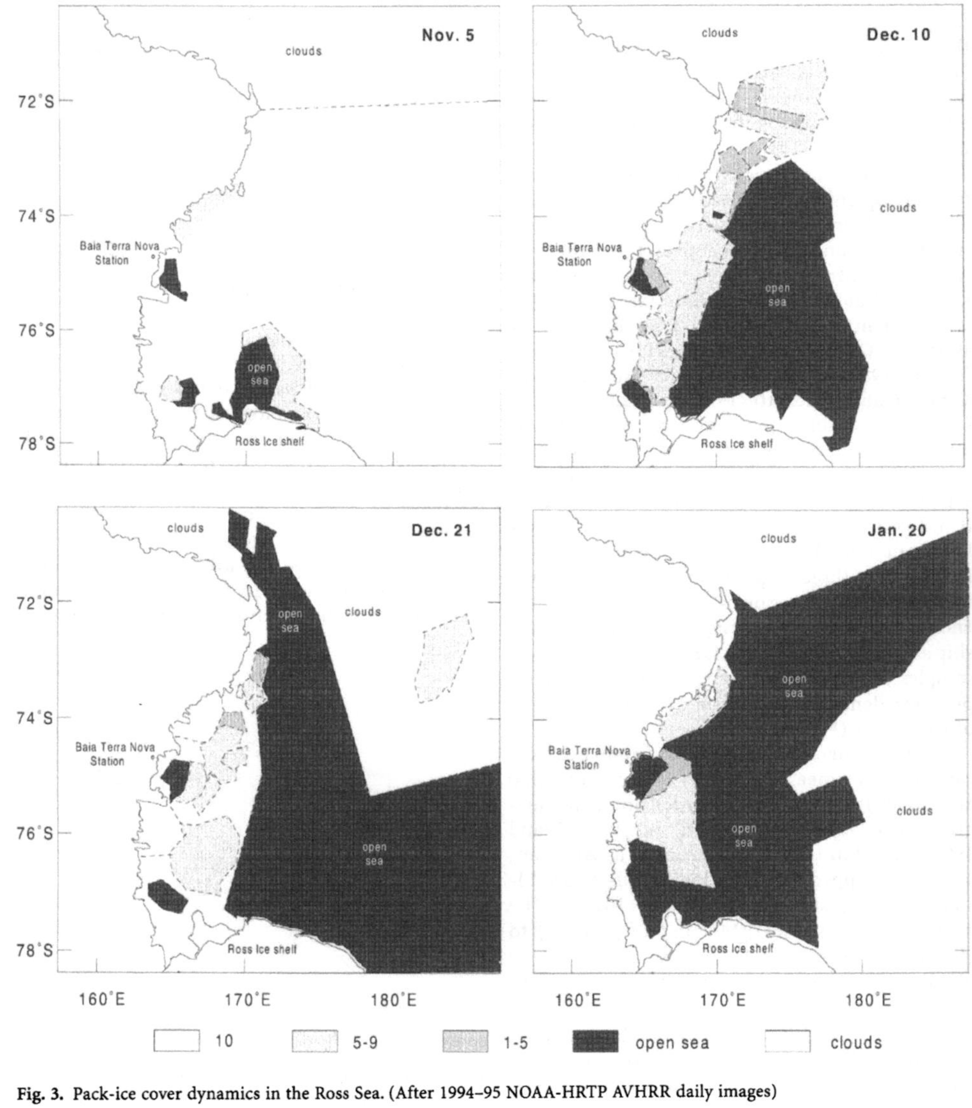


**Fig. 3.** Pack-ice cover dynamics in the Ross Sea. (After 1994–95 NOAA-HRTP AVHRR daily images)

## Vertical Structure of the Water

The characteristics of the water column of the stations in Terra Nova Bay (Fig. 1b) are very similar, and their temporal evolution was well represented by that of the SMN station (Fig. 4).

In the early days of January (Fig. 4a), the irradiance determined the melting of ice, formation of the pycnocline, heating of the water, which exceeded 2 °C at the surface, and formation of the thermocline. Up until the second half of January, stratification remained well defined in the waters that were ice-free. Starting from the second half of January (Fig. 4b), the new mixing and the decreased supply of fresh water tended to render the column homogeneous. In early February (Fig. 4c), due to the melting of floating ice, a new stratification phase occurred which was less developed than the preceding one. In fact, in the bay there had been a continuous accumulation and freeing again of large slabs of pack ice coming from the south, brought by winds and currents, and directed northward. The temperature decreased to –0.6 °C.

This same type of succession in the structure of the water column was noted also at other stations in the Ross Sea, beginning from the north in a southerly direction (Fig. 1c). Therefore, these stations were grouped and classified in a sequence according to the time elapsed since the liberation of the ice and, therefore, to the type of structure of the water column, independent of the chronological sequence with which they were detected. Other factors may also have affected the structure of the water column; nonetheless, in our opinion irradiance controlling ice melting was the determining factor.

The observed phases were the following:

1st phase: stations 5 to 8 (Fig. 5a); pack ice barely broken by the ship; this was a phase prior to the beginning of the sampling at SMN; it indicated the beginning of stratification.

2nd phase: stations 25 to 27 (Figs. 5b, 4a); the sea still had some ice present, complete stratification; the upper mixed layer (UML) was about 25 m thick; the surface temperature varied from 0 to 0.5 °C and, when stability continued, reached >1 °C.

3rd phase: stations 11 to 19 (Figs. 5c, 4b); the sea had been free of ice for about 2 weeks; the pycnocline was greatly reduced or non-existent, and was not influenced by the thermocline; a remixing of the surface layers with the underlying ones occurred.

4th phase: stations 20 to 23 (Figs. 5d, 4c); the sea had been free of ice for about 1 month; in some cases, the temperature of the surface layer exceeded 1.5 °C, forming a thermocline that produced only a weak pycnocline. Such stratified conditions were typically produced by heating of the surface layers (Fig. 5d), while at the stations of Terra Nova Bay, they originated essentially from the decrease in salinity (Fig. 4c).

The latitudinal succession of the phases was consistent with that of daily irradiance (Fig. 2). It began from the south, where it was greater and where the most advanced phases were found. It then proceeded in a northerly direction, where, due to less intense irradiance, more time was needed to melt the pack ice and where the initial phases were found. However, because of the faster advancement in the melting of ice towards the north and slower melting towards the west, as reported in Fig. 3 and also by Smith and Nelson (1985), stations 25, 26 and 27 were ice-free almost simultaneously with those of Terra Nova Bay, but later compared with those of equal latitudes off shore.

## Nutrients

During the first phase, nutrients (nitrates and phosphates) had a high concentration that was almost homogeneous throughout the water column (Fig. 5e). Subsequently, in the surface layer they were strongly consumed by the phytoplankton biomass, thus contributing to the formation of a nutricline (Figs. 4d,g, 5f,j). Later, the production of phytoplankton decreased, as a result of the lack of nutrients, causing an alteration in the ratios between nutrients in the water, compared to those necessary for phytoplankton assimilation. This alteration was caused primarily by phosphate concentration, which was so low that it could no longer be detected by the analytical method (Figs. 4d, 5f). Therefore phytoplankton biomass was reduced (Figs. 4h, 5j,k), also because of zooplankton grazing activity (Fonda Umani et al. 1992; Knox 1994). Thus, there was a decrease in the consumption of nutrients, whose concentrations were enhanced (Figs. 4e, 5g), as a result of a mixing of the water column probably caused by winds and waves, which rehomogenized the water column

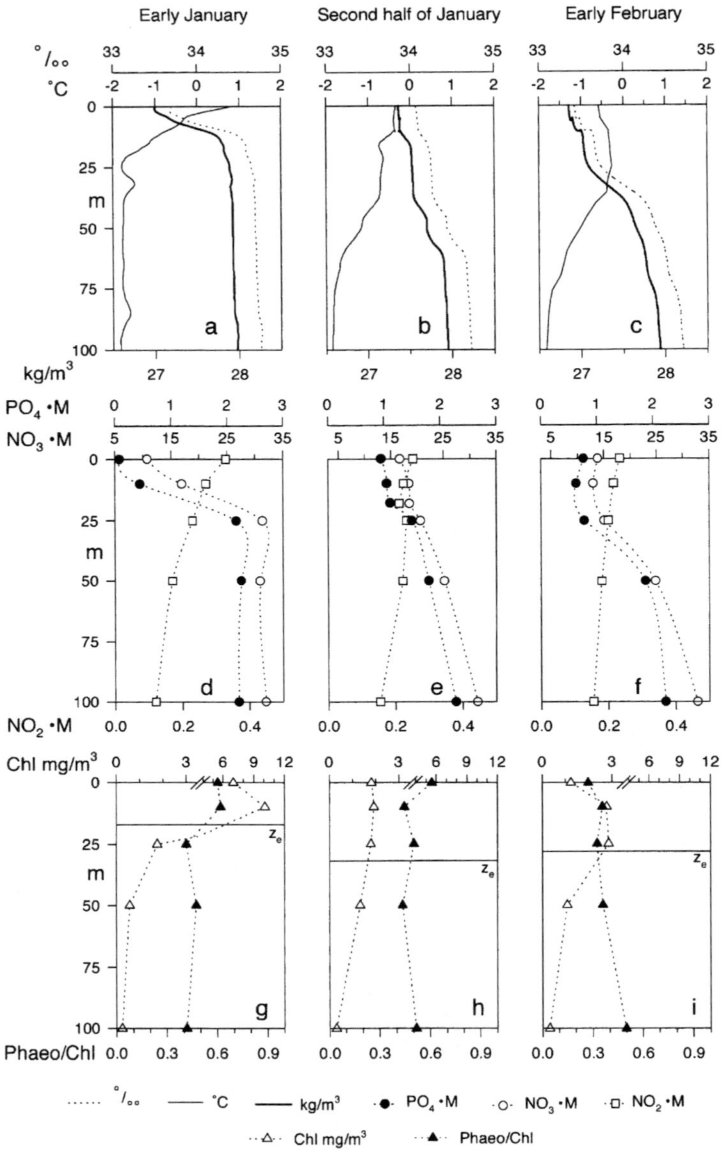


**Fig. 4a–i.** Representative types of vertical profiles of salinity (‰), temperature (°C) and density (kg $m^{-3}$) (**a, b, c**); phosphate (μM), nitrate (μM) and nitrite (μM) (**d, e, f**), Chl (mg $m^{-3}$) and Phaeo/Chl ratio (**g, h, i**) at SMN, 1989–90. $z_e$ represents the euphotic zone depth

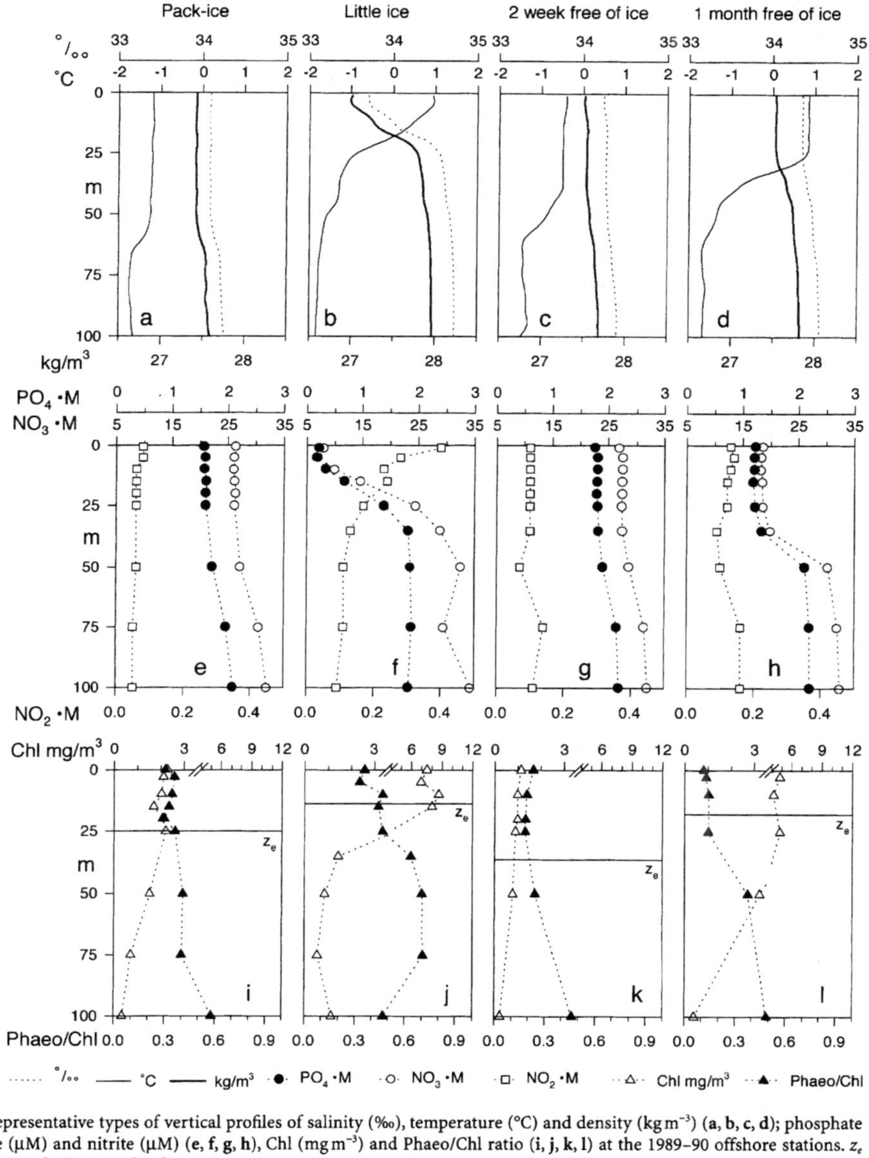

**Fig. 5a–i.** Representative types of vertical profiles of salinity (‰), temperature (°C) and density ($kg\,m^{-3}$) (**a, b, c, d**); phosphate (μM), nitrate (μM) and nitrite (μM) (**e, f, g, h**), Chl ($mg\,m^{-3}$) and Phaeo/Chl ratio (**i, j, k, l**) at the 1989–90 offshore stations. $z_e$ represents the euphotic zone depth

structure (Figs. 4b, 5c). In the end, a second increase in the phytoplankton biomass occurred (Figs. 4i, 5l), again accompanied by the consumption of nutrients and the development of a nutricline (Figs. 4f, 5h).

Similar dynamics were verifiable also for silicates, the concentrations of which were very high during the periods preceding (80 μM) and following (65 μM) the first bloom. Since diatoms were dominant in this period, a large consumption (8 μM) of silicates was recorded. The concentration decreased again, also during the second increase in phytoplankton biomass (down to 55 μM).

The nitrite distribution in the surface layer (Figs. 4d–f, 5d–f), behaved differently from other nutrients; its concentration increased as the phytoplankton biomass increased (Figs. 4g–i, 5i–l).

## Dynamics of the Biomass

The temporal evolution of environmental conditions, Chl, and Phaeo/Chl is shown in Fig. 6. In Fig. 7, the temporal evolution of phytoplankton biomass, assessed as Chl and cell densities, at Terra Nova Bay during the three summers is reported. From Figs. 4, 5 and 6 it is clear that the biomass maxima were in agreement with phases 2 and 4 of the water-column stratification, with a marked pycnocline and nutricline, and high absolute biomass values, up to 18 mg m$^{-3}$ of Chl and $10^7$ cell dm$^{-3}$.

The lack of Chl analyses at the beginning of December was sufficiently compensated for by the data on cell densities. The cycle in the development of biomass began with a minimum at the beginning of December (Fig. 7d,f), which subsequently increased to a maximum between December and January, and decreased once more to a second minimum between January and February. In all three expeditions, biomass showed this second minimum in close agreement with phase 3, during which there was also a diminishing of the stratification and nutrient concentrations (Figs. 4b,e,h, 5c,l, and 6 between 25 January and 1 February), which made it similar to phase 1. On the whole, the evolution of the biomass assumed the form of an undulatory cycle. However, after having performed one complete oscillation, it started a second. Indeed, Figs. 7; 4c,f,i, 5d,h,l and Fig. 6 between 1 February and 8 February show that there was a second increase in biomass. It is also evident that this second phase (Fig. 5d,h,l) was preceded by conditions (Fig. 5c,g,k) similar to those of phase 1 (Fig. 5a,e,i). In addition (Fig. 7e), the decrease shown by Phaeo compared to Chl in correspondence with the resumption of cell multiplication could be an indication that a young assemblage was in a developing phase.

This temporal pattern does not seem to be limited to Terra Nova Bay. A situation of scarce biomass and absence of nutricline and pycnocline were reported in the area between 74° and 67°S, in the early days of January 1990 (stations 11–19, Fig. 1c; Fig. 5c,g,k). Successively from January 13 to 16 (74°–73°S), 17 to 20 (72°–70°S), and from 21 to 22 (70°–67°S) enhancement of biomass up to an average of 2.5, 1.4 and 0.9 mg m$^{-3}$ of Chl (Fig. 8), respectively, was recorded, which characterised the zone as also being affected by the second end-of-season blooming. As in the first bloom, this second bloom occurred in different times and at different latitudes.

## Phytoplankton Assemblages

As far as the phytoplankton assemblages are concerned, those of the first bloom (Figs. 4a,d,g, 5b,f,j) were similar in all of the Ross Sea. They appeared at different times and were always characterized by diatoms dominated by the genus *Fragilariopsis* (Nuccio et al., this Vol.). The role of sympagic microalgae in this bloom is controversial and some authors (Krebs et al. 1987; Knox 1994) contend that blooms might be due to sympagic cells rather than free-living organism. However, as regards the interrelationship between sympagic and free-living organisms, the active colonization of the sea-ice environment by the water-column organisms should also be considered. Further, if it is feasible to think that the sympagic algae make a contribution, it is difficult to believe that they determine the entire pelagic bloom.

During the summer minimum, phytoplankton assemblages generally showed a greater species diversity. However, diatoms were still dominant close to the coast; offshore the assemblages become predominantly composed of phytoflagellates (Nuccio et al., this Vol.). Near the coast phytoplankton assemblages of the second bloom were dominated by *Fragilariopsis* cf. *curta* and some other diatoms. Offshore, diatoms of the genus *Nitzschia* dominated, together with *Phaeocystis* sp. (Nuccio et al., this Vol.).

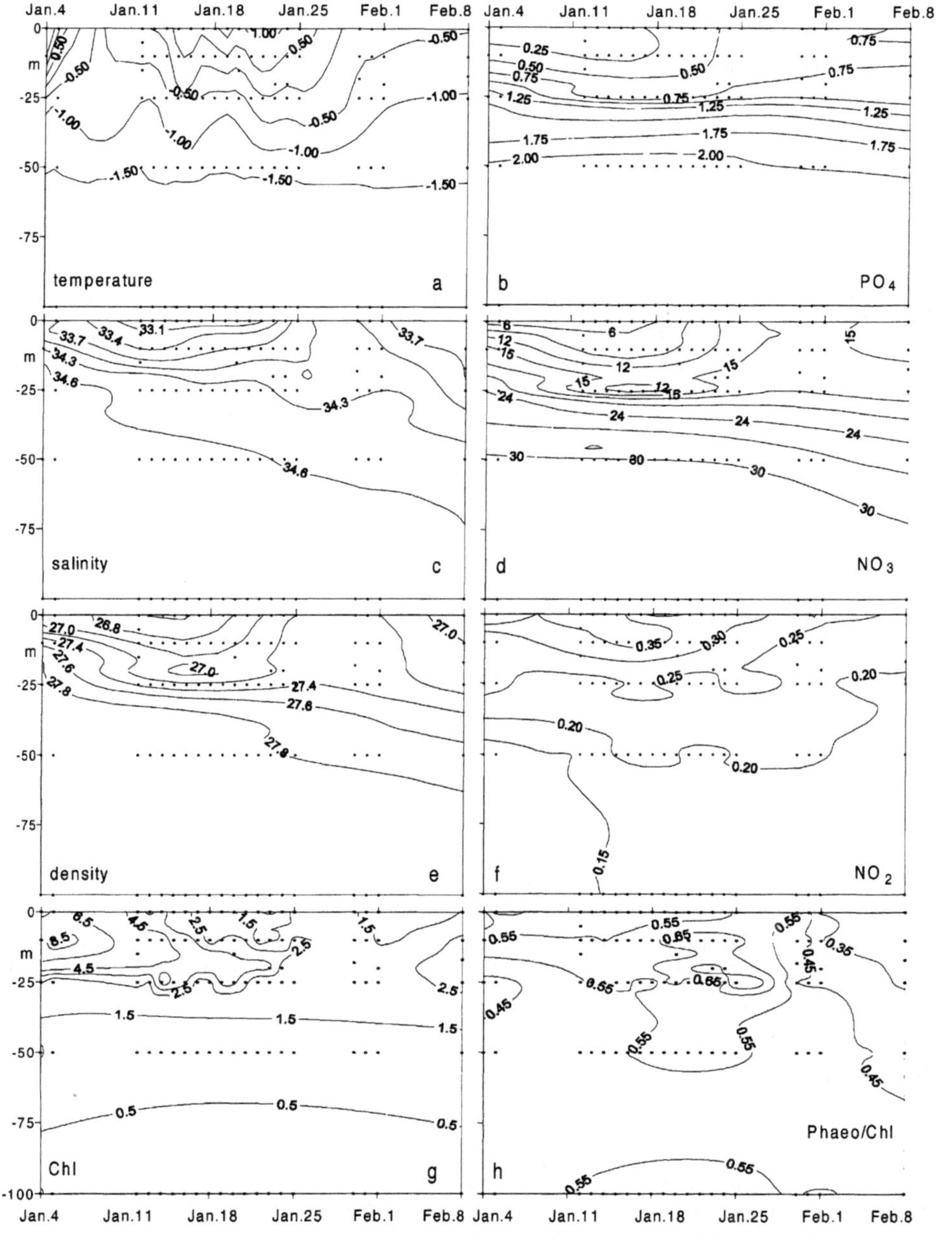


**Fig. 6a–h.** Isopleths of: **a** temperature (°C); **b** phosphate (μM); **c** salinity (‰); **d** nitrate (μM); **e** density ($kg\,m^{-3}$); **f** nitrite (μM); **g** Chl ($mg\,m^{-3}$) and **h** Phaeo/Chl ratio at SMN from January 4 to February 8, 1990

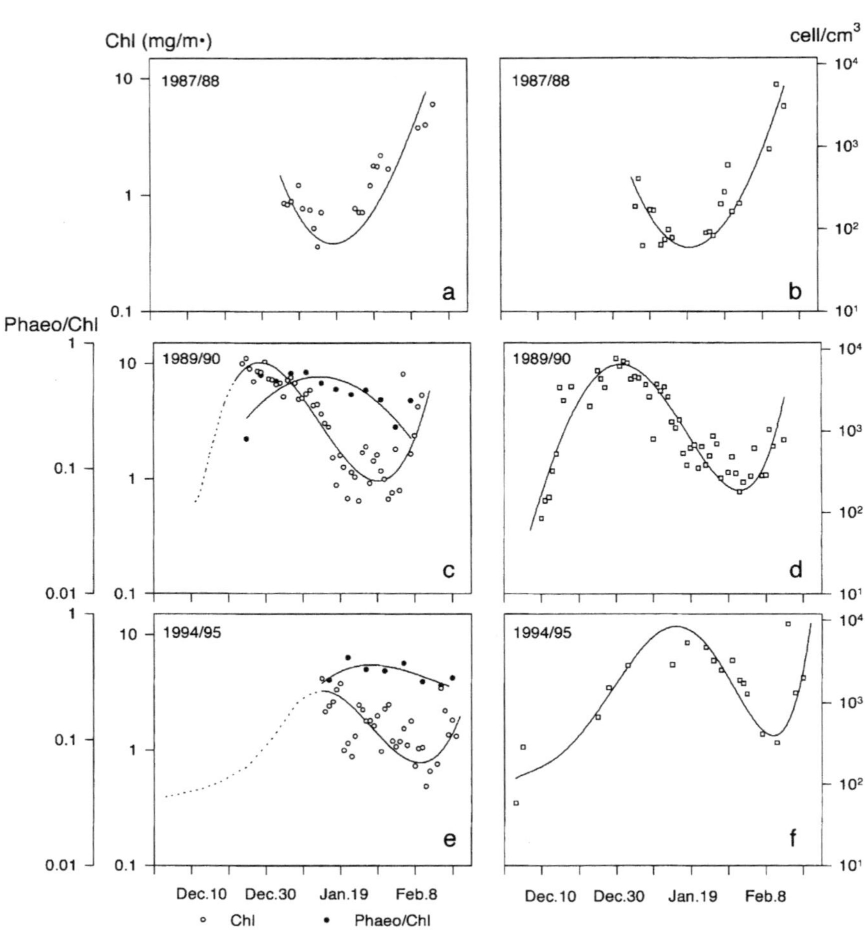


**Fig. 7a–f.** Chl (**a**, **c**, **e**) and cell density (**b**, **d**, **f**) temporal dynamics at Terra Nova Bay during the 3 years. Data are averages of daily samplings in the surface layer. *Continuous lines* are the interpolating function; *dotted lines* are the reconstruction of Chl dynamics based on cell density dynamics. *Filled symbols* in **c** and **e** represent the Phaeo/Chl temporal evolution ratio with its interpolating function. Data were averaged every 4 days

## Discussion

The results obtained during the three expeditions show that phytoplankton biomass was generally very high, and during the summer two blooms developed. In Terra Nova Bay, the first was between December and January and the second in February. Satellite images (Arrigo and McClain 1994a,b) and the bimodal annual pattern of phytoplankton depicted by Knox (1990) support these conclusions. The annual cycle of phytoplankton in the Ross Sea would therefore not appear as the classic unimodal pattern reported in the literature for polar regions.

The N/P ratio in the water was generally similar to the Redfield's theoretical level $N/P = 16$, and shows (Innamorati and Giovanardi 1992) the minimum variance we have registered. This balance between nutrients is altered during phytoplankton blooms when DIN and DIP concentrations decrease strongly; their ratio increases up to $N/P > 200$, probably as a consequence of a prefer-

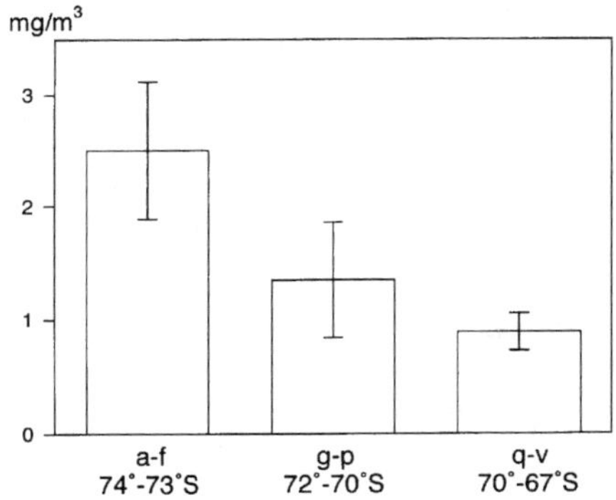


**Fig. 8.** Histograms representing the mean Chl concentrations of the three groups of stations (*a–f, g–p, q–v*) during the return journey to New Zealand at the end of January 1990. The corresponding range of latitudes is reported for each group of stations. The *error bars* indicate the ± one standard deviation range

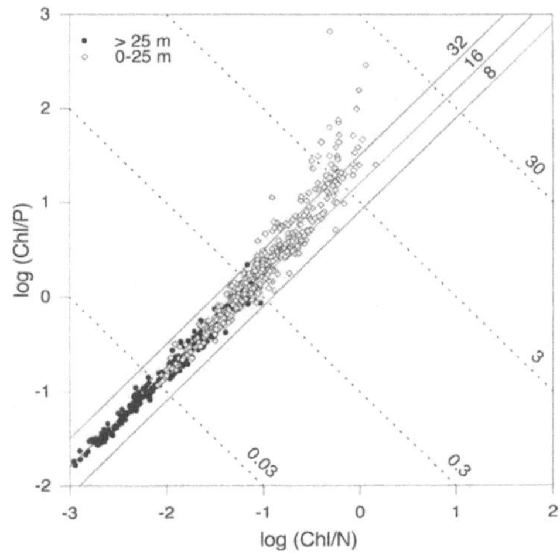


**Fig. 9.** Distribution of log(Chl/P) vs. log(Chl/N). P = phosphate; N = nitrate + nitrite The 45° straight lines represent the 32, 16 and 8 N/P ratios. The −45° straight lines represent the 0.03, 0.3, 3 and 30 Chl $\sqrt{NP}$ ratios

ential consumption of DIP (luxury consumption?), as noted in the upper layer of other seas we have investigated (Innamorati and Giovanardi 1992). At these very high ratios it is plausible to assume that phosphate must be recycled many times compared to nitrate (e.g. 200/16 = 12.5 times) in order to be assimilated in the N/P = 16 ratio.

The relation between the actual phytoplankton biomass, i.e. Chl, and the potential biomass, i.e. nutrients available in the water, can be represented by the nutrition ratios Chl/P and Chl/N. These ratios are generally lower than any we have found in other seas (Innamorati and Giovanardi 1992), but during blooms they increase up to Chl/P >100 and Chl/N >1 (Fig. 9); similar values are often found in eutrophic areas, such as the NW Adriatic Sea.

The low Chl/P and Chl/N ratios in non-bloom periods can be due to limiting factors other than major nutrient availability, such as low temperature (Eppley 1972; Knox 1994), scarcity of light due to vertical instability (Sverdrup 1953), scarcity of iron (Martin et al. 1990) or silicates (Sullivan et al. 1993). During bloom periods, on the contrary, phytoplankton biomass is very high, similar to production (Lazzara et al., this Vol.), suggesting that limiting factors may disappear. Our observations suggest that the main factor which determines a bloom is the daily irradiance that, as it increases, melts the ice and warms the water rich in nutrients, forming a less dense and eutrophic UML. Therefore the stability of the UML favours the availability of light to phytoplankton. Together with an increase in water temperatures, these conditions will favour high phytoplankton growth rates.

The spatial distribution of blooms in the Ross Sea follows the latitudinal pattern of increasing daily irradiance (Fig. 2), which determines the advancement of the melting of the sea ice from the south northward. Blooms in the Ross Sea do follow the melting of the sea ice as regards time and movement (Fig. 3). In the same way, the phases described in Figs. 4 and 5 occur in sequence, beginning and developing in different times at different latitudes, but always starting with the disappearance of the covering of the sea ice.

The major events influencing phytoplankton dynamics can be summarized as follows:

- the irradiance controlling melting of ice produces low-salinity surface waters and an UML is formed which allows phytoplankton growth; phytoplankton biomass reaches its maximum while nutrient concentrations (nitrate, phosphates and silicates) attain their minimum.
- the UML disappears as a consequence of meteorological events, and a well-mixed water condition emerges. In this period a minimum of phytoplankton biomass is recorded.

- a short period of stratification occurs, due to thermocline formation off-shore and low salinity layer in coastal waters, which sustains a second bloom.

In the peak phase (Figs. 4a,d,g, 5b,f,j) the euphotic zone can become less deep than that of the UML so that, in this layer, phenomena of photolimitation occur as the biomass itself severely reduces the PAR. At the same time, an increase in nitrite is observed in the UML. High concentrations of nitrite have been observed as a peculiarity of the Antarctic Ocean (Jacques and Tréguer 1986). In accordance with Harris (1986), this could be the consequence of the scarcity of light, which does not supply the phytoplankton with sufficient energy to reduce the nitrate to nitrite and this to ammonium, so that the cells accumulate nitrite and release it into the water. This mechanism could also be favoured by the concentrations of nitrate (7–8 $\mu$M), which are high compared to the extremely low levels of phosphate.

The summer minimum (Figs. 4b,e,h, 5e,g,k) occurs as a result of the sharp decrease in the concentration of nutrients, in particular of phosphate, intense feeding activity of the zooplankton (Fonda Umani et al. 1992; Knox 1994), and, above all, because of the disappearance in the stratification. Since the summer minimum occurred during all 3 years of our study, we believe that it is induced by meteorological conditions.

Offshore this phase is fairly long (mid-December–mid-January) and it is therefore probable that this was the condition which was most frequently sampled during the early Antarctic scientific expeditions, providing a picture of scarcity of biomass in the Ross Sea, to which the heterogeneity of the temporal and spatial distribution of the biomass may also have contributed.

The cycle proceeds with an ultimate stabilization of the water column (Figs. 4c,f,i, 5d,h,l) which may have different origins. Offshore, due to the persistance of good weather conditions in the periods of December and January, and in the absence of a marked haline and pycnic discontinuity, the thermocline is formed. In the coastal stations in February, there is, on the other hand, a decrease in salinity at the surface which is probably caused by the melting of floating sea ice transported by the currents northwards. The vertical stabilization is regularly followed by a second bloom, less extensive than the first, which occurs about half way through January at the offshore stations (Lazzara and Nuccio 1994) and towards the beginning of February in Terra Nova Bay.

The evidence for the presence of this second bloom is given by the fact that in all 3 years (Fig. 7), we have observed a minimum phase followed by an increase of phytoplankton biomass. The offshore sampling also showed a substantial increase in phytoplankton biomass; in areas characterized by average surface values of 0.3 mg $m^{-3}$ at the end of December (Innamorati et al. 1991), an average value of 2.5 mg $m^{-3}$ was recorded half-way through January (Fig. 8).

The fate of this second bloom is not known. It is probable that it will cease in the ascending phase because of a decrease in irradiance, and a fall in temperatures with the consequent reforming of ice. In any case, the evolution of environmental conditions and biomass define this second resurgence of cell multiplication as the development of a second bloom, the progress of which reaches fairly high quantitative levels of biomass. Finally, the temporal series in Terra Nova Bay and the spatial series in the Ross Sea both contribute to depict a model of temporal successions (as shown in Fig. 7d,f), which we believe can be extended to the entire Ross Sea.

## References

Artegiani A, Azzolini R, Paschini E, Creazzo S (1992) Physical oceanographic conditions in the Southern Pacific Ocean and in the Western Ross Sea. In: National Scientific Commission for Antarctica (ed) Oceanographic Campaign 1989–90, Data Rep II, Genova, pp 5–62

Arrigo KR, McClain CR (1994a) Spring phytoplankton production in the Western Ross Sea. Science 266: 261–263

Arrigo KR, McClain CR (1994b) Phytoplankton bloom in the Western Ross Sea as seen by the coastal zone color scanner (CZCS). EOS 75 (3): 201

Boldrin A, Stocchino C (1990) On the hydrological characteristics of Terra Nova Bay. In: National Scientific Commission for Antarctica (ed) Oceanographic Campaign 1987–88, Data Rep I, Genova, pp 161–238

Bunt JS, Lee CC (1970) Seasonal primary production in Antarctic sea ice at McMurdo Sound in 1967. J Mar Res 28: 304–320

Catalano G, Benedetti F (1990) Distribution of nutrients in the Terra Nova Bay and in the Ross Sea. In: National Scientific Commission for Antarctica (ed) Oceanographic Campaign 1987–88, Data Rep I, Genova, pp 61–83

Catalano G, Benedetti F, Goffart A, Iorio M (1991) Distribution of dissolved oxygen, pH, total alkalinity and nutrients in the Southern Oceans and Ross Sea (R/V Cariboo 1989–90 cruise). In: National Scientific Commission for Antarctica (ed) Oceanographic Campaign 1989–90, Data Rep I, Genova, pp 11–23

El-Sayed SZ (1984) Productivity of the Antarctic waters. A reappraisal. In: Holm-Hansen O, Bolis L, Gilles R (eds) Marine phytoplankton and productivity. Springer, Berlin Heidelberg New York, pp 19–34

El-Sayed SZ (1988) Productivity in Southern Ocean: a closer look. Comp Biochem Physiol 90: 489–498

El-Sayed SZ, Biggs DC, Holm-Hansen O (1983) Phytoplankton standing crop, primary productivity and near-surface nitrogenous nutrient fields in the Ross Sea, Antarctica. Deep Sea Res 30: 871–886

Eppley RV (1972) Temperature and phytoplankton growth in the sea. Fish Bull Nat Ocean Atmospheric Admin 70: 1063–1085

Fonda Umani S, Chemelli M, Monti M (1992) Microzooplankton populations temporal trend in Terra Nova Bay (austral summer 1989–90). In: National Scientific Commission for Antarctica (ed) Oceanographic Campaign 1989–90, Data Rep II, Genova, pp 153–178

Hansen HP, Grasshoff K (1983) Automated chemical analysis. In: Grasshoff K, Ehrardt M, Kremlin K (eds) Methods of seawater analysis, 2nd edn. V Ch Weinheim, 419 pp

Harris GP (1986) Phytoplankton ecology. Structure, function and fluctuation. Chapman and Hall, London

Holm-Hansen O, Mitchell BG (1991) Spatial and temporal distribution of phytoplankton and primary production in western Bransfield Strait. Deep Sea Res 38: 961–980

Holm-Hansen O, El-Sayed SZ, Franceschini G, Cuhel R (1977) Primary production and factors controlling phytoplankton growth in the Southern Ocean. In: Llano GA (ed) Adaptations within Antarctic ecosystem. Proc 3rd SCAR Symp Antarctic Biology. Smithsonian Institution Press, Washington, DC, pp 11–50

Innamorati M, Giovanardi F (1992) Interrelationships between phytoplankton biomass and nutrients in the eutrophicated areas of the North-Western Adriatic Sea. Sci Tot Environ (Suppl 1992): 235–250

Innamorati M, Mori G, Lazzara L, Nuccio C, Lici M, Vanucci S (1990a) Ecology of phytoplankton. In: National Scientific Commission for Antarctica (ed) Oceanographic Campaign 1987–88, Data Rep I, Genova, pp 161–238

Innamorati M, Mori G, Lazzara L, Nuccio C, Lici M, Vanucci S (1990b) Ecology of phytoplankton. In: National Scientific Commission for Antarctica (ed) Oceanographic Campaign 1987–88, Data Rep II, Genova, pp 9–63

Innamorati M, Mori G, Lazzara L, Vanucci S (1990c) Eutrofia ed oligotrofia nell'Oceano Antartico. Oebalia 16 (1): 153–162

Innamorati M, Lazzara L, Mori G, Nuccio C, Saggiomo V (1991) Phytoplankton ecology. In: National Scientific Commission for Antarctica (ed) Oceanographic Campaign 1989–90, Data Rep I, Genova, pp 141–252

Innamorati M, Lazzara L, Massi L, Mori G, Nuccio C, Saggiomo V (1992a) Indagine sulla biomassa fitoplanctonica nel mare di Ross in relazione ai fattori ambientali. In: Gallardo VA, Ferretti O, Moyano HI (eds) Oceanografia en Antarctica. ENEA-PNRA-EULA, Concepcion, Chile, pp 235–252

Innamorati M, Lazzara L, Mori G, Nuccio C, Saggiomo V (1992b) Spatial and temporal distributions of phytoplankton size fractions in Antarctic waters: biomass and production, summer 1989–90. Atti IX Congr Ass Ital Oceanogr Limnol, Lang, Genova, pp 605–612

Innamorati M, Lazzara L, Massi L, Mori G, Nuccio C, Saggiomo V, Cartei P, Franchini L, Gamba C (1994a) Phytoplankton ecology: particles, gilvin, fluorescence spectra, irradiance and phytoplankton species density in the Ross Sea. Oceanographic Campaign 1989–90. In: Faranda F, Guglielmo L (eds) National Scientific Commission for Antarctica Oceanographic Campaign 1989–90, Data Rep Part III, Genova, pp 101–178

Innamorati M, Massi L, Lazzara L, Cartei P (1994b) Biomassa fitoplanctonica e proprietà ottiche del mare. Biol Mar Medit 1(1): 83–88

Iqbal M (1983) An introduction to solar radiation. Academic Press, New York

Jacques G, Tréguer P (1986) Ecosystèmes pelagiques marins. Masson, Paris

Jerlov NG (1976) Marine optics. Elsevier oceanographic series. Elsevier, Amsterdam

Knox GA (1990) Primary production and consumption in McMurdo Sound, Antarctica. In: Kerry KR, Hempel G (eds) Antarctic ecosystems: ecological change and conservation. Springer, Berlin Heidelberg New York, pp 115–128

Knox GA (1994) The biology of Southern Ocean. Cambridge University Press, Cambridge

Krebs WN, Lipps JH, Burckle LH (1987) Ice diatoms floras, Arthur Harbor, Antarctica. Polar Biol 7: 163–171

Lazzara L, Nuccio C (1994) Relazioni tra particellato, biomassa e popolamenti fitoplanctonici nel Mare di Ross durante la fioritura estiva. Atti X Congr Ass Ital Oceanogr Limnol, Lang, Genova, pp 655–665

Lazzara L, Nuccio C, Massi L, Innamorati M (1995) Le microalghe simpagiche di Baia Terra Nova (Antartide) nell'estate 1994–95. G Bot Ital 129 (1): 425

Lazzara L, Innamorati M, Massi L, Mori G, Nuccio C (1996) Irradiance control of phytoplankton biomass and production in Terra Nova Bay (Western Ross Sea). EOS Trans AGU, 76 (3): 144

Lazzara L, Massi L, Nuccio C, Biondi N, Innamorati M (1997) Phytoplankton ecology: irradiance, particles, gilvin, pigments, absorption, fluorescence, production and species density in Terra Nova Bay, Ross Sea. In: Faranda F, Guglielmo L, Povero P (eds) Nat Progr Ant Res, National Scientific Commission for Antarctica (ed) Oceanographic Campaign 1994–95, Data Rep Part I, Genova, pp 229–279

Lorenzen CJ (1967) Determination of chlorophyll phaeopigments spectrophotometric equations. Limnol Oceanogr 12: 343–346

Martin JH, Fitzwater SE, Gordon RM (1990) Iron deficiency limits phytoplankton growth in Antarctic waters. Global Biogeochem Cycles 4: 5–12

Morel A (1988) Optical modeling of the upper ocean in relation to its biogenous matter content (case I waters). J Geophys Res 93 (C9): 10749–10768

Morel A, Smith RC (1974) Relation between total quanta and total energy for aquatic photosynthesis. Limnol Oceanogr 19: 591–600

Nuccio C, Innamorati M, Lazzara L, Mori G (1992) Popolamenti fitoplanctonici nella Baia di terra Nova, Mare di Ross. In: Gallardo VA, Ferretti O, Moyano HI (eds) Oceanografia en Antarctica. ENEA-PNRA-EULA, Concepcion, Chile, pp 235–252

Nuccio C, Innamorati M, Mori G, Lazzara L, Nizzi Grifi G (1994) Phytoplankton ecology: species density in the Ross Sea. In: National Scientific Commission for Antarctica (ed) Oceanographic Campaign 1987–88, Data Rep Part III, Genova, pp 5–48

Riemann B (1978) Carotenoid interference in the spectrophotometric determination of chlorophyll degradation products from natural populations. Limnol Oceanogr 23 (5): 1059–1066

Saggiomo V, Massi L, Modigh M, Innamorati M (1992) Size-fractionated primary production in Terra Nova Bay (Ross Sea) during the austral summer 1989–90. In: Gallardo VA, Ferretti O, Mojano HI (ed) Oceanografia en Antartica. ENEA-PNRA-EULA Concepcion, Chile, pp 289–294

Smith WO Jr, Nelson DM (1985) Phytoplankton biomass near a receding ice-edge in the Ross Sea. In: Siegfrid WR, Condy PR, Laws RM (eds) Antarctic nutrient cycles and food webs. Springer, Berlin Heidelberg New York, pp 70–77

Sullivan CW, Arrigo KR, McClain CR, Comiso JC, Firestone J (1993) Distribution of phytoplankton blooms in the Southern Ocean. Science 262: 1832–1837

Sverdrup HU (1953) On the condition for the vernal blooming of phytoplankton. J Cons Cons Perm Int Explor Mer 18: 287–295

Wilson DL, Smith WO Jr, Nelson DM (1986) Phytoplankton bloom dynamics of the Western Ross Sea ice-edge. I. Primary productivity and species-specific production. Deep Sea Res 33: 1375–1387

# Spatial and Temporal Distribution of Phytoplankton Assemblages in the Ross Sea

C. Nuccio, M. Innamorati, L. Lazzara, G. Mori, and L. Massi[1]

**ABSTRACT**

The temporal evolution and spatial distribution of phytoplankton assemblages were investigated, in coastal and open waters of the western Ross Sea, during three austral summers. Terra Nova Bay shows a first massive bloom (up to $10^7$ cell $l^{-1}$) of *Fragilariopsis* cf. *curta* between December and January in the receding ice-edge zone and another increase in February, also with the contribution of different species. Phytoplankton shows a patchy distribution, with areas of bloom dominated by diatoms (*Fragilariopsis*, *Nitzschia*) and *Phaeocystis* sp., and less rich zones, mainly dominated by dinoflagellates and other flagellates. The different phytoplankton assemblages show characteristics corresponding to different stages that alternate during the summer season in the different areas, but which are related to the temporal development of environmental conditions after the melting of the ice.

## Introduction

Studies of the phytoplankton distribution in the Ross Sea, particularly in the south-western region, showed as typical features an intense and extensive spring bloom of *Phaeocystis* (El-Sayed et al. 1983; Knox 1994) related to the polynya dynamics (Arrigo and Weiss 1996) and the following diatom-dominated summer bloom (Knox 1994; Garrison et al. 1996). Diatom blooms dominated by few taxa (*Fragilariopsis*, *Nitzschia*) in the receding ice-edge zone were reported during the summer in the western Ross Sea (Wilson et al. 1986; Carbonell-Moore 1994). Very little information about phytoplankton composition and the dynamics of blooms in the northern open waters has been obtained up until now, except for the reported observations from some inshore stations of Terra Nova Bay (Andreoli et al. 1995).

Our investigations on phytoplankton spatial and temporal distributions in large areas of the northwestern Ross Sea were carried out during three austral summers (1987–88, 1989–90, 1994–95). Previous results had shown the spatial variability between areas of high biomass, with few diatoms (*Fragilariopsis*, *Nitzschia*) or *Phaeocystis* blooms in different layers of the water column, and less rich zones, the different characteristics of some biomass indicators (chlorophyll, phaeopigments, particle size and number, cell densities) in different areas, and a new biomass increase in late summer (Innamorati et al. 1990c, 1992; Nuccio et al. 1992; Lazzara and Nuccio 1994).

The temporal evolution of the phytoplankton assemblages in Terra Nova Bay and their spatial distribution in the northern open waters of the Ross Sea are described for the first time in this chapter and are related to previous observations in order to depict the general features of their variations in these waters during summer.

## Materials and Methods

The first cruise (R/V Polar Queen; Dec. 1987–Jan. 1988) was conducted on a grid of 40 stations in Terra Nova Bay and along a transect as far as 231 km offshore (Fig. 1a). A second cruise (R/V Cariboo; Nov. 1989–Feb. 1990) covered the route from 45° to 75°S sampling 27 stations (Fig. 1c). In all the stations of the two cruises, samples were collected with a rosette sampler at various depths from the surface down to 200 m. Furthermore, during the Cariboo cruise, continuous measurements of in vivo fluorescence of chlorophyll a (Backscat fluorometer Dr. Haardt) were carried

[1] Laboratorio di Ecologia, Dipartimento di Biologia Vegetale, Università di Firenze. Via P.A. Micheli, 1. 50121 Florence, Italy

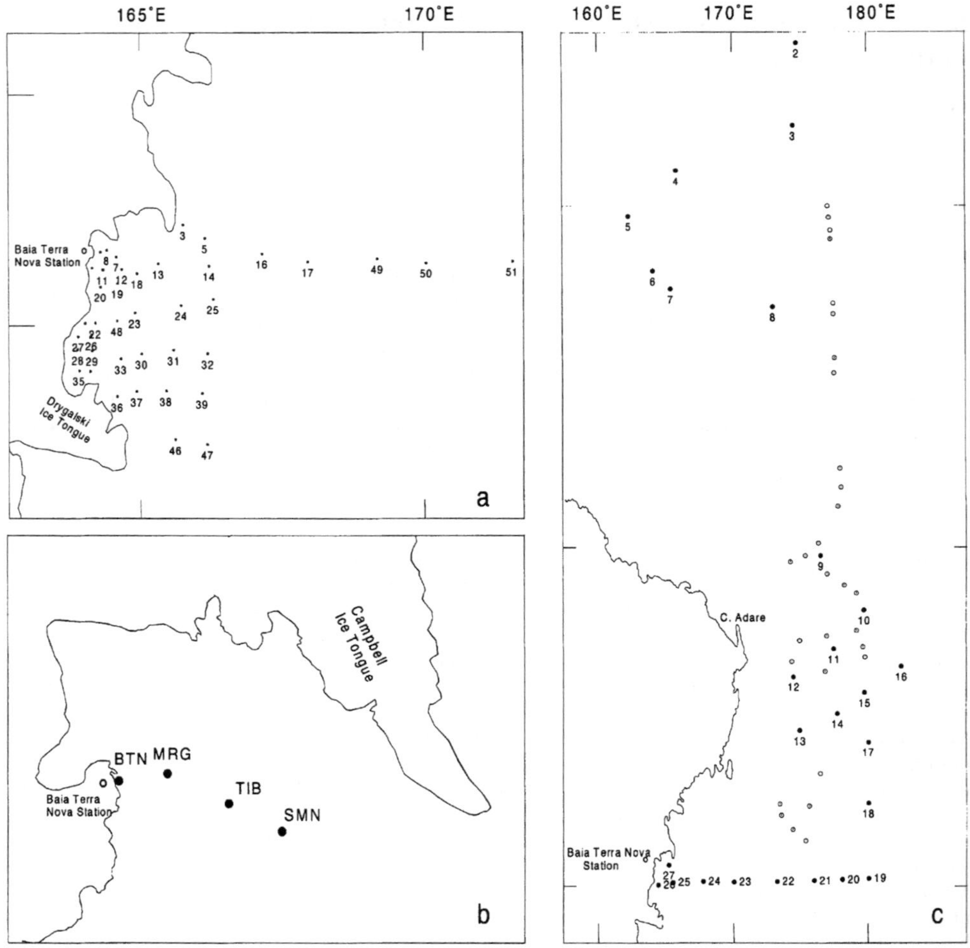


**Fig. 1a–c.** Maps of station locations. R/V Polar Queen cruise (Dec. 1987–Feb. 1988) in Terra Nova Bay (**a**); stations sampled from December to February 1989–90 and 1994–95: Baia Terra Nova (*BTN*), Mergellina (*Merg*), Santa Maria Novella (*SMN*), Tibertina (*TIB*) (**b**); R/V Cariboo cruise (Nov. 1989–Feb. 1990): stations (*closed circles*) sampled during the southwards route and samples (*open circles*) collected from surface layers during the way back (**c**)

out in the water pumped from a depth of 3 m and 79 samples were collected from this water going both southwards and northwards; only the samples from the way back (Fig. 1c) are discussed here. During this second expedition and in 1994–95, three (BTN, MRG, SMN) and two (BTN, TIB) stations (Fig. 1b), respectively, were sampled in Terra Nova Bay from December to February, so as to follow the temporal trend of the phytoplankton biomass and composition. Station BTN was sampled near the rocky shore only on the surface, MRG at 0, 10, 25 m and SMN and TIB at 0, 10, 25, 50, 100, 200 m from the R/V Malippo using Niskin bottles. All the data acquired are detailed in Innamorati et al. (1990a,b, 1991, 1994). Chlorophyll concentrations were determined, after filtration (Whatman GF/F) and extraction in acetone (final dilution to 90%), by means of spectrophotometric analysis (Kontron, Uvikon 930), following Lorenzen (1967) as modified by Riemann (1978) for separate determinations of chlorophyll a and phaeopigments. The sum of the two, named chlorophaeopigment (Chl), is used here. Surface and underwater downwelling, upwelling, and scalar quantum PAR irradiances were measured by means of a photoprobe equipped with three cosine and one spherical meters (LI-COR). Hydrographic data of the Polar Queen and Cariboo cruises are taken from Boldrin and Stocchino (1990) and Artegiani et al. (1992), respectively.

Samples (250 cm$^3$) for microscopic analysis were fixed with Lugol (1987–88 and 1989–90) or neutralized (sodium carbonate) formaldehyde (1994–95) to the final concentration of 1 and 4%, respectively. Seven hundred samples were analyzed by means of Nikon Diaphot and Zeiss IM35 inverted microscopes on sedimentation chambers (10, 25 or 50 cm$^3$). Taxonomical identification was performed following: Hendey (1937); Manguin (1960); Hasle (1964, 1965a,b); Balech (1976); Sournia et al. (1979); Priddle and Fryxell (1985); Medlin and Priddle (1990). *Phaeocystis* sp. occurred in the non-motile stage, but the single individuals were counted.

Table 1 provides a list of recorded taxa. Some considerations on taxonomic terminology used

**Table 1.** List of the recorded taxa

| | |
|---|---|
| Diatoms | Dinoflagellates |
| *Actinocyclus actinochilus* (Ehrenb.) Sim. | *Dinophysis* spp. |
| *Achnanthes* sp. | Gymnodiniaceae < 20 µm |
| *Amphiprora* cf. *kufferathii* Manguin | Gymnodiniaceae > 20 µm |
| *Asteromphalus heptactis* (Bréb.) Ralfs | Gymnodiniaceae > 100 µm |
| *Asteromphalus hookerii* Ehrenb. | *Gymnodinium* cf. *guttula* Balech |
| *Asteromphalus hyalinus* Karsten | *Gyrodinium lachryma* (Meun.) Kofoid & Swezy |
| *Asteromphalus parvulus* Karsten | *Gyrodinium* spp. > 20 µm |
| *Centric diatoms* spp. | Naked dinoflagellates < 20 µm |
| *Chaetoceros* cf. *"bulbosum complex"* sp. | Naked dinoflagellates > 20 µm |
| *Chaetoceros* cf. *criophilum* Castr. | *Oxytoxum criophilum* Balech |
| *Chaetoceros* cf. *dichaeta* Ehrenb. | *Prorocentrum* cf. *antarcticum* (Hada) Balech |
| *Chaetoceros* cf. *neglectum* Karsten | *Protoperidinium* cf. *adeliense* (Balech) Balech |
| *Chaetoceros* cf. *neogracile* Van Landingham | *Protoperidinium antarcticum* (Schim.) Balech |
| *Chaetoceros* spp. | *Protoperidinium applanatum* (Mangin) Balech |
| *Cocconeis imperatrix* A.S. | *Protoperidinium* cf. *archiovatum* (Balech) Balech |
| *Corethron criophilum* Castr. | *Protoperidinium* cf. *bellulum* (Balech) Balech |
| *Coscinodiscus* spp. | *Protoperidinium defectum* (Balech) Balech |
| *Cylindrotheca closterium* (Ehrenb.) Reim. & Lewin | *Protoperidinium* cf. *incertum* (Balech) Balech |
| *Dactyliosolen tenuijunctus* Manguin | *Protoperidinium* cf. *mediocre* (Balech) Balech |
| *Eucampia antarctica* (Castr.) Mangin | *Protoperidinium* cf. *rosaceum* (Balech) Balech |
| *Fragilariopsis* cf. *curta* (Van Heurck) Hustedt | *Protoperidinium* cf. *ps.antarcticum* (Balech) Balech |
| *Fragilariopsis* sp. 1 (8 µm) | *Protoperidinium* cf. *unipes* (Balech) Balech |
| *Fragilariopsis* spp. | *Protoperidinium* spp. |
| *Grammatophora* cf. *charcotii* M. Per. | Thecate dinoflagellates < 20 µm |
| *Licmophora* spp. | Thecate dinoflagellates > 20 µm |
| *Melosira* sp. | |
| *Naviculacea* sp. 20 µm | Prymnesiophyceae |
| *Nitzschia* sec. *Nitzschiella* sp. | *Phaeocystis* sp. (cf. *P. antarctica* Karsten) |
| *Nitzschia* cf. *subcurvata* Hasle | |
| *Nitzschia* sp. *(bilobatae* sec.) | Dictyochophyceae |
| *Nitzschia* spp. | *Dictyocha speculum* Ehrenb. |
| *Odontella weissflogii* (Jan.) Grunow | |
| *Pennate diatoms* > 20 µm | Other flagellates |
| *Proboscia alata* (Bright.) Sundström | *Cryptophycea* sp. 1 |
| *Pseudonitzschia* spp. | *Cryptophyceans* spp. |
| *Rhizosolenia chunii* Karsten | Phytoflagellate sp. 7 µm |
| *Rhizosolenia* cf. *hebetata* Bailey | Phytoflagellate sp. 10 µm |
| *Rhizosolenia truncata* Karsten | Phytoflagellates < 10 µm |
| *Thalassiosira* cf. *antarctica* Comber | Phytoflagellates < 20 µm |
| *Thalassiosira* sp. 43 µm | Undetermined 20 µm (*Halosphaera* rosettes?) |
| *Thalassiosira* spp. | |
| *Thalassiothrix antarctica* (Schim.) Karsten | |
| Undetermined diatoms spp. | |

need to be emphasized here. *Fragilariopsis* cf. *curta* is the form in valvar view that prevailed over all the others in the 1987–88 and 1989–90 samples and *Fragilariopsis* spp., the denomination used for the 1994–95 samples, includes both this form in valvar view and a great number of long chains in girdle view which ultimately could not be identified. *Nitzschia* sec. *Nitzschiella* sp. (from now on called *Nitzschia* sp. 1 for brevity) is a tiny form resembling *Cylindrotheca closterium*, but which we prefer not to assign yet to this latter species, that is quite recognizable, for its larger size, in the 1994–95 samples. *Fragilariopsis* sp. 1 is a very small form (length 7–9 μm) often in doublets, which we assume belongs to *Fragilariopsis cylindrus*. Detailed floristic lists are provided in Innamorati et al. (1994) and Nuccio et al. (1994).

The average ($\bar{x}$) cell densities (Table 2; Fig. 10) and Chl concentrations (Fig. 4) in the 0–25 m layer

**Table 2.** Biovolume and carbon content of selected diatoms and temporal variation at station SMN of their cell densities, biovolume and biomass in the layer 0–25 m

| | $\mu m^3$ | C pg cell$^{-1}$ | | | $\mu m^3$ | C pg cell$^{-1}$ |
|---|---|---|---|---|---|---|
| *Fragilariopsis* cf. *curta* | 465 | 39.80 | *Pennate diatom* 40 μm | | 1855 | 113.61 |
| *Nitzschia* sp. 1 | 68 | 9.27 | *Pennate diatom* 90 μm | | 3463 | 182.35 |
| *Pseudonitzschia* spp. | 524 | 43.58 | *Rhizosolenia truncata* | | 39891 | 1162.69 |
| *Eucampia antarctica* | 28000 | 889.09 | *Thalassiosira* 43 μm | | 25415 | 826.15 |
| cell $10^3$ l$^{-1}$ | Dec.31 | Jan.05 | Jan.12 | Jan.17 | Jan.23 | Jan.30 |
| *Fragilariopsis* cf. *curta* | 3500 | 3356 | 1888 | 804 | 564 | 710 |
| *Nitzschia* sp. 1 | 670 | 215 | 13 | 3 | 0 | 1 |
| *Pseudonitzschia* spp. | 278 | 676 | 724 | 26 | 1 | 9 |
| Total | 4448 | 4247 | 2625 | 833 | 565 | 720 |
| % of total cell density | 95 | 98 | 90 | 83 | 80 | 83 |
| *Eucampia antarctica* | 2 | 0 | 0 | 0 | 0 | 0 |
| *Pennate diatom* 40 μm | 4 | 2 | 7 | 0 | 5 | 9 |
| *Pennate diatom* 90 μm | 2 | 5 | 1 | 1 | 0 | 1 |
| *Rhizosolenia truncata* | 1 | 0 | 0 | 0 | 0 | 0 |
| *Thalassiosira* 43 μm | 0 | 0 | 0 | 0 | 1 | 1 |
| $\mu m^3$ $10^6$ l$^{-1}$ | Dec.31 | Jan.05 | Jan.12 | Jan.17 | Jan.23 | Jan.30 |
| *Fragilariopsis* cf. *curta* | 1627.5 | 1560.5 | 877.9 | 373.9 | 262.3 | 330.1 |
| *Nitzschia* sp. 1 | 45.6 | 14.6 | 0.88 | 0.2 | 0 | 0.068 |
| *Pseudonitzschia* spp. | 145.7 | 354.2 | 379.4 | 13.6 | 0.5 | 4.7 |
| Total | 1818.8 | 1929.3 | 1258.1 | 387.7 | 262.8 | 334.8 |
| *Eucampia antarctica* | 56 | 0 | 0 | 0 | 0 | 0 |
| *Pennate diatom* 40 μm | 7.4 | 3.7 | 12.9 | 0 | 9.3 | 16.7 |
| *Pennate diatom* 90 μm | 6.9 | 17.3 | 3.5 | 3.5 | 0 | 3.5 |
| *Rhizosolenia truncata* | 39.9 | 0 | 0 | 0 | 0 | 0 |
| *Thalassiosira* 43 μm | 0 | 0 | 0 | 0 | 25.4 | 25.4 |
| Total | 110.2 | 21 | 16.4 | 3.5 | 34.7 | 45.6 |
| Carbon μg l$^{-1}$ | Dec.31 | Jan.05 | Jan.12 | Jan.17 | Jan.23 | Jan.30 |
| *Fragilariopsis* cf. *curta* | 140.0 | 134.2 | 75.5 | 32.2 | 22.6 | 28.4 |
| *Nitzschia* sp. 1 | 6.0 | 1.9 | 0.12 | 0.03 | 0 | 0.009 |
| *Pseudonitzschia* spp. | 12.2 | 29.7 | 31.8 | 1.1 | 0.04 | 0.4 |
| Total | 158.2 | 165.8 | 107.4 | 333.3 | 22.6 | 28.8 |
| *Eucampia antarctica* | 1.8 | 0 | 0 | 0 | 0 | 0 |
| *Pennate diatom* 40 μm | 0.5 | 0.2 | 0.8 | 0 | 0.6 | 1.0 |
| *Pennate diatom* 90 μm | 0.4 | 0.9 | 0.2 | 0.2 | 0 | 0.2 |
| *Rhizosolenia truncata* | 1.2 | 0 | 0 | 0 | 0 | 0 |
| *Thalassiosira* 43 μm | 0 | 0 | 0 | 0 | 0.8 | 0.8 |
| Total | 3.9 | 11.0 | 1.0 | 0.2 | 14.0 | 2.0 |

were obtained as: $\bar{x} = \frac{1}{2z_n}\sum(x_a + x_b)(z_a - z_b)$, where $x_a$ and $x_b$ are the pairwise values in the adjacent sampling depth $z_a$ and $z_b$ at or above the depth $z_n$ considered. The percentages of phytoplankton classes reported for the 0–25- or 10–50-m layers (Figs. 2, 5, 6, 8) were calculated from the cell densities obtained with the same equation.

Cell volumes (Table 2) were computed from linear dimensions, equating shapes to standard geometric figures, or were taken from the literature (Hasle 1969). Biomass ($\mu g l^{-1}$ of carbon) was computed from the equation of Strathmann (1967).

## Results

The summer evolution of the phytoplankton assemblages (Figs. 2, 3), followed in the northwestern waters of Terra Nova Bay (station BTN, MRG, SMN; Fig. 1b), showed the main peak of the bloom between December and January. Biomass and cell densities followed similar trends (Figs. 2a, 3c,d, 4), up to $10\,mg\,m^{-3}$ of Chl and $9.8 \times 10^6$ cells $l^{-1}$, slowly decreasing with their maxima in deeper waters until the end of January and with a tendency to increase again in February.

The phytoplankton assemblages had the same composition down to the depth of 25 m (Fig. 2b), with a contribution of diatoms that was never lower than 80%. *Fragilariopsis* cf. *curta*, up to $7.6 \times 10^6$ cells $l^{-1}$, overwhelmed all other diatoms (on average, 83% of the class), and was followed by *Pseudonitzschia* spp. and *Nitzschia* sp. 1 that did not exceed an average of 10%. Only these pennate diatoms reached high densities; the centric species (*Proboscia alata*, *Rhizosolenia truncata*, *Thalassiosira* sp. 43 μm) showed only very scarce abundances. Dinoflagellates represented, on average, 5% of the total density, and increased in mid January up to $2.15 \times 10^5$ cells $l^{-1}$, mainly with naked forms. In the deeper layers (Fig. 2c,d) *Phaeocystis* sp. appeared, and dinoflagellates became relatively more important.

In the surface waters of st. BTN, where it was possible to take samples for a longer period of time, the early phase of the bloom (Fig. 3c,d) was evident from mid-December, when more species showed a growth phase (*Naviculacea* sp., *Fragilariopsis* sp. 1). *Fragilariopsis* cf. *curta* reached its highest contribution in the peak of the bloom. Afterwards, the bloom declined, mainly due to the

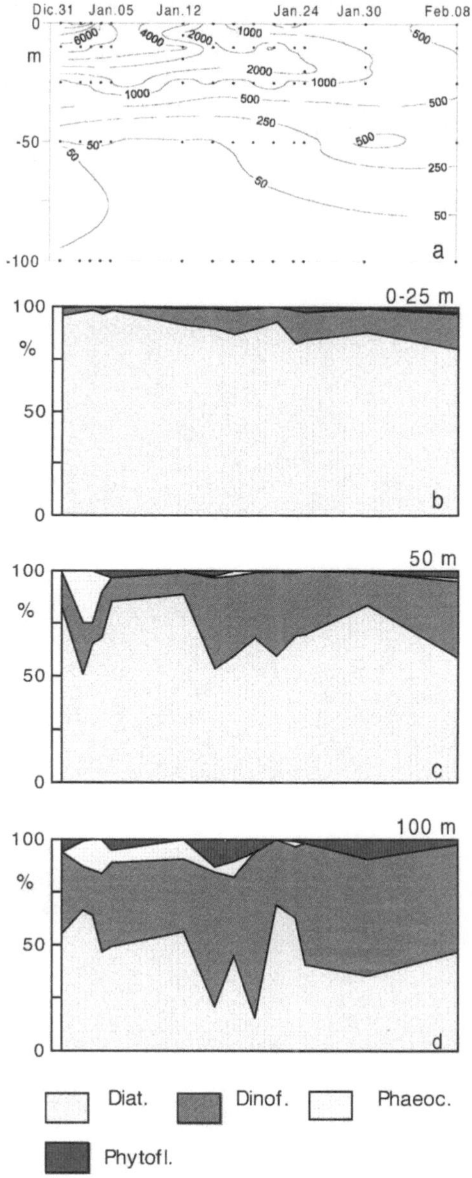


**Fig. 2a–d.** Vertical distribution of phytoplankton densities ($10^3$ cells $l^{-1}$) **(a)**, and percentage composition of phytoplankton classes: 0–25 m **(b)**; 50 m **(c)**; 100 m **(d)**, at st. SMN (Jan.–Feb. 1990)

decrease of the latter species. The increasing trend in February was again due to *Fragilariopsis* cf. *curta* and *Naviculacea* sp., and *Phaeocystis* was present in this period (Fig. 3a). Phytoflagellates showed their main increase at the beginning of the bloom, and a more or less constant decrease until February. Dinoflagellates showed little variability in their density, with a shift towards a greater number of thecate dinoflagellates (*Prorocentrum* cf. *antarcticum*, *Protoperidinium antarcticum*, *P. applanatum*, *P. defectum*). Not all the 1994–95 samples were analyzed but, although a temporal shift of maxima, the same trend (Fig. 3b,d) and a new increase in February, up to 9.0 × $10^6$ cells $l^{-1}$, due to *Fragilariopsis*, *Naviculacea* and *Pseudonitzschia*, appeared. It should be noted that the phytoflagellates (Fig. 3d) dominated during December, when they reached 70% of the total

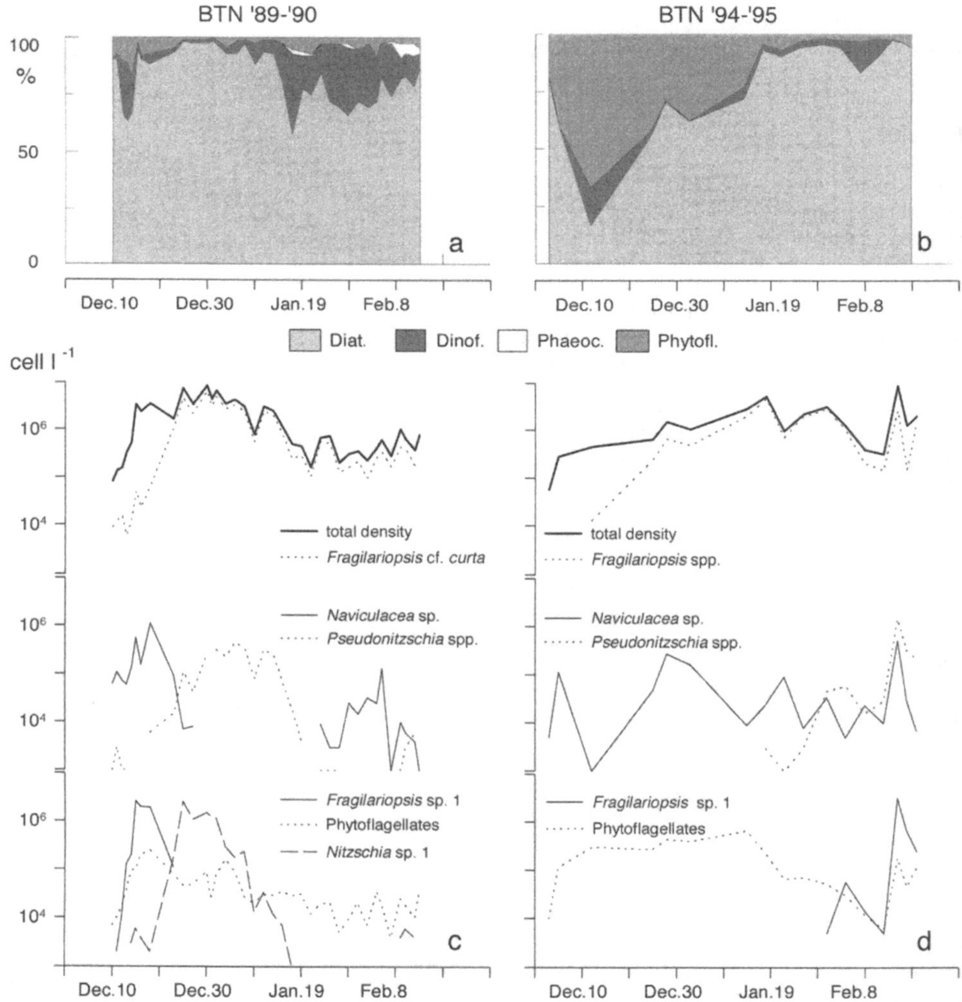


**Fig. 3a–d.** Surface percentage composition of phytoplankton classes and cell densities (cells $l^{-1}$) at station BTN during summer 1989–90 (**a, c**) and 1994–95 (**b, d**)

density, with forms <10 μm; their increase during the second half of February was due to a bloom of *Dictyocha speculum* with healthy cells forming chains. The greater variety of the February assemblages is reflected by the specific diversity ($\bar{H}$): more than 2 in this case, and not more than 1.8 bits cell$^{-1}$ during the first bloom.

The phytoplankton biomass confirmed this temporal evolution (Fig. 4), and an opposite trend between Chl concentrations and phaeopigment/Chl ratios appeared. The highest values of the ratio were in the declining phase of the first bloom, with ratios up to 0.7, while the lowest ones were evident in the early phases, at the beginning and at the end of the period. Furthermore, the dominant role of *Fragilariopsis* cf. *curta* and a few other diatoms in the bloom was evident not only from cell numbers (always the highest percentages of total phytoplankton densities) but also in terms of biomass (Table 2). During the temporal evolution at station SMN (Table 2), the diatoms of greater size were present with very low densities and their contribution, as cell volume or carbon concentration, mainly in the period of bloom, was

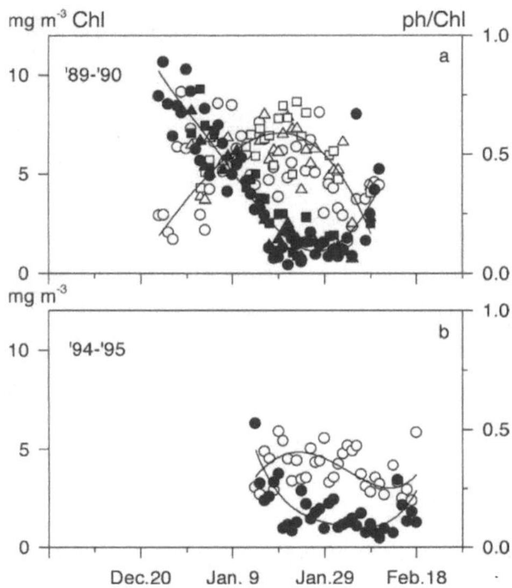


**Fig. 4a,b.** Temporal evolution of Chl concentrations (*closed symbols*) and phaeopigments/Chl (ph/Chl) ratios (*open symbols*) in the layer 0–25 m of station MRG (*square*) and station SMN (*triangle*), and surface of station BTN (*circles*) 1989–90 (a); surface samples of station BTN 1995 (b)

of at least one order of magnitude less than the one of the smallest dominant diatoms.

During the 1989–90 summer, at the same time as the temporal sampling in Terra Nova Bay, a much larger area of the Ross Sea (moving southwards from 20 Dec. to 7 Jan. and northwards in the first fortnight of January) was investigated. Between 70° and 75°S (Fig. 1c), low phytoplankton densities (Fig. 5a) generally less than $10^5$ cells l$^{-1}$ were found. The surface composition (Fig. 5b) was different from the other layers (Fig. 5c), because of the scarce presence of *Phaeocystis* sp. as far as the Ross Sea slope (stations 15–18), where there was evidence of the Circumpolar Deep Water upwelling (Jacobs et al. 1970; Artegiani and Paschini 1992) and a nutrient depletion (Catalano 1992). A slight increase, up to $3.0 \times 10^5$ cells l$^{-1}$, of *Phaeocystis* sp. and other flagellates (Fig. 5b) occurred here. The other characteristics were (Fig. 5b,c,d) a very low percentage of diatoms (*Fragilariopsis* cf. *curta*, *Nitzschia* sp. 1, unidentified pennates) and a general dominance of flagellates (naked dinoflagellates, unidentified phytoflagellates, cryptophyceans). Along 75°S quite a different situation (Fig. 6) was found. Moving westward, densities became very high, up to $4.0 \times 10^6$ cells l$^{-1}$ at station 23 and over $10^6$ cells l$^{-1}$ on the whole western side of the transect. The phytoplankton assemblages had the same composition down to 25 m (Fig. 6b), with the dominance of *Phaeocystis* sp. on the eastern side and of diatoms in the western one, where densities increased. Mainly *Nitzschia* sp. 1 dominated at station 22, and then decreased westward, while *Fragilariopsis* cf. *curta* reached the highest densities towards the coast. *Phaeocystis* sp. increased its relative importance in the deeper layers (Fig. 6c), just as did dinoflagellates, mainly Gymnodiniaceae and *Prorocentrum* cf. *antarcticum* (Fig. 6d).

Only samples from surface layers were collected in the same area on the way back, so that the spatial variation of surface phytoplankton assemblages (Fig. 7) during two consecutive periods, December and January, could be compared. Going southwards (Fig. 7a,c), from about 70°S to Terra Nova Bay (TNB), the surface phytoplankton densities showed the variations just mentioned in relation to the transects (Figs. 5a, 6a). At the beginning of the northward route (Fig. 7b,d), the bloom of *Fragilariopsis* in Terra Nova Bay was still evident. Afterwards, in the Ross Sea open waters, previously poor and dominated by flagellates, *Nitzschia* sp. 1 and *Phaeocystis* sp. produced the

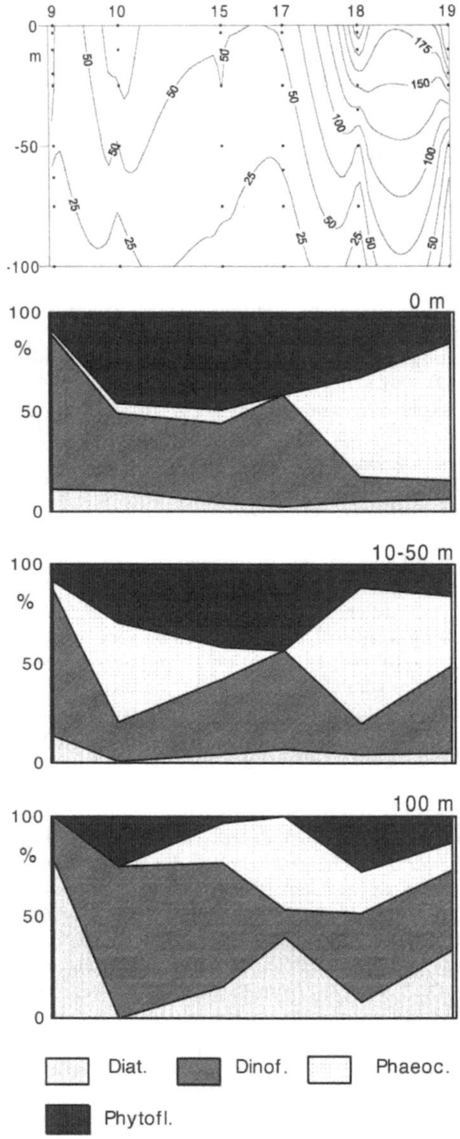


**Fig. 5a–d.** Vertical distribution of phytoplankton densities ($10^3$ cells $l^{-1}$) (**a**), and percentage composition of phytoplankton classes: 0 m (**b**), 10–50 m (**c**), 100 m (**d**), between 70° and 75°S (stations 9–19)

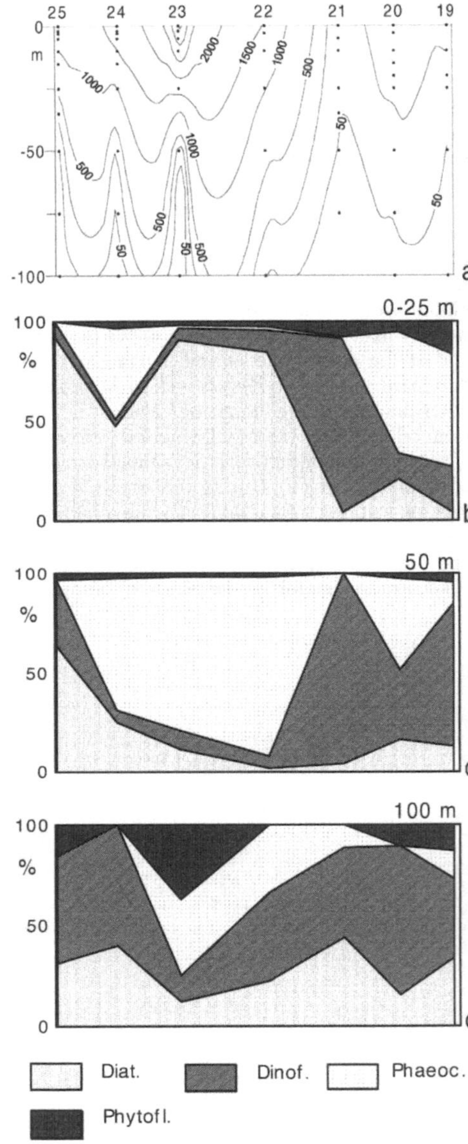


**Fig. 6a–d.** Vertical distribution of phytoplankton densities ($10^3$ cells $l^{-1}$) (**a**), and percentage composition of phytoplankton classes: 0–25 m (**b**), 50 m (**c**), 100 m (**d**), in the transect along 75°S (stations 19–25)

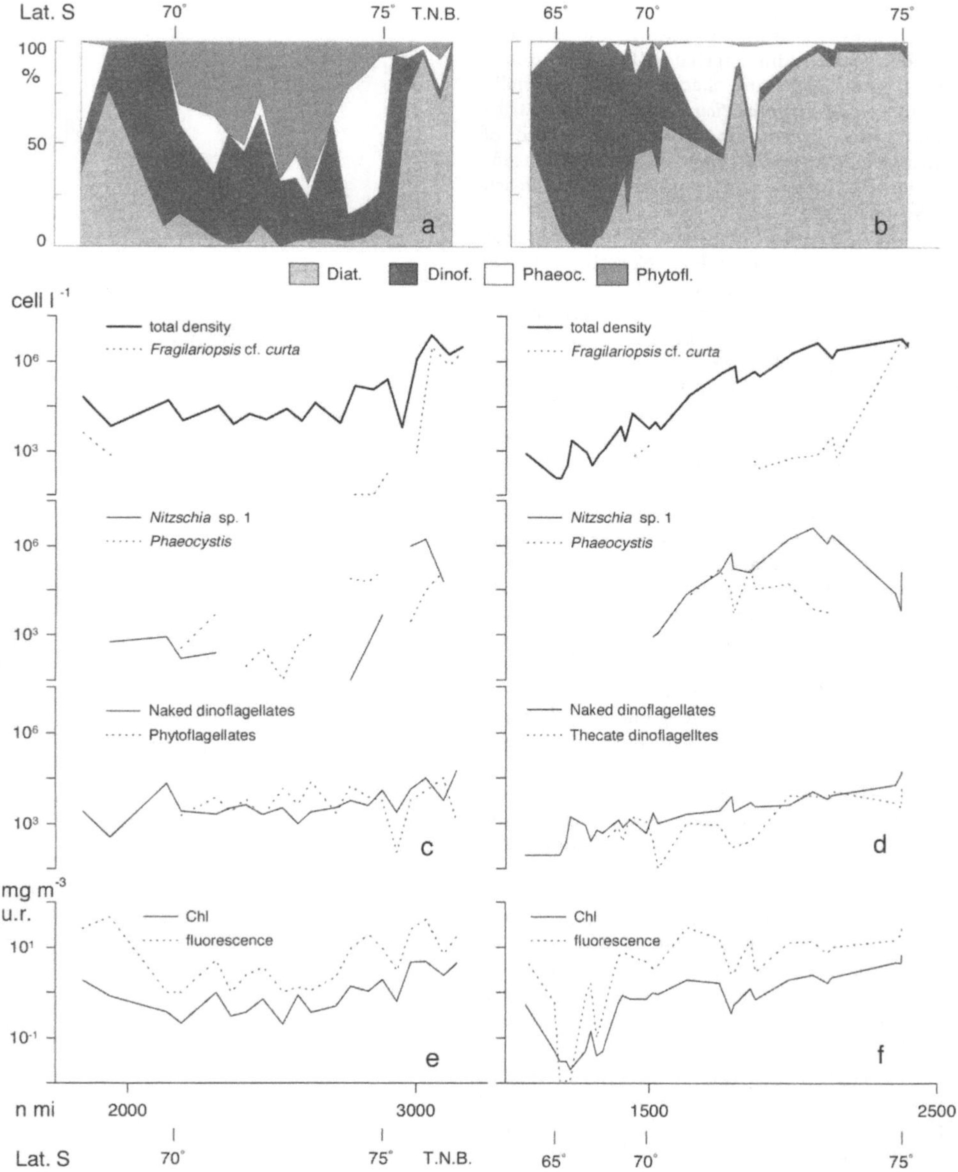


**Fig. 7a–f.** Surface percentage composition of phytoplankton classes, cell densities (cells $l^{-1}$), in vivo fluorescence continuous measurements and Chl concentrations from discrete samples along the southwards (**a, c, e**) and northwards (**b, d, f**) routes during the Cariboo cruise (Dec. 1989–Jan. 1990)

highest values (Fig. 7d), up to $2.6 \times 10^6$ cell $l^{-1}$. When densities decreased towards the northern oceanic regions, dinoflagellates (Fig. 7b,d) dominated with naked and thecate forms, especially *Prorocentrum* cf. *antarcticum*. This spatial distribution fitted in with the continuous record of surface in vivo fluorescence and the Chl concentrations of discrete samples (Fig. 7e,f) that reach almost 4 mg $m^{-3}$.

The phytoplankton spatial distribution was also investigated all over Terra Nova Bay (Fig. 1a) during January and February 1988. As a representative situation, Fig. 8 shows the transect towards the offshore waters (Fig. 1a) where the highest values of phytoplankton biomass, about 18 mg $m^{-3}$ of Chl, and density, $5.5 \times 10^6$ cells $l^{-1}$, were found. After the nearshore rich waters sampled at the beginning of January (stations 10, 11; Fig. 8a), a larger area (stations 12–16) had low densities of $<2.5 \times 10^5$ cells $l^{-1}$ at the end of January. The richest zone (stations 17–51) was observed during the first fortnight of February, with characteristics of a senescent assemblage for the high percentage of phaeopigments (up to 60% of Chl). In the 0–25 m layer, the phytoplankton assemblage (Fig. 8b) was dominated by *Phaeocystis* sp., up to $2.5 \times 10^6$ cells $l^{-1}$, and diatoms (*Nitzschia* sp. 1 and *Fragilariopsis* cf. *curta*) in the richest zone, dinoflagellates and other flagellates in the poorest one, *Fragilariopsis* cf. *curta* and *Nitzschia* sp. 1 in nearshore waters. From 25 to 50 m (Fig. 8c), *Phaeocystis* sp. increased its densities in this latter zone, while diatoms in the richest part and diatoms and flagellates in the other part of the transect were still dominant. *Phaeocystis* sp. generally increased from 25 to 50 m (Fig. 8d), as already assessed in the Ross Sea (El-Sayed et al. 1983; Smith et al. 1990; Nuccio et al. 1992), and showed a widespread distribution all over Terra Nova Bay (Fig. 9), in assemblages with less mature characteristics, as phaeopigments/Chl ratios and specific diversity, than the surface diatom ones (Nuccio et al. 1992).

As regards the taxonomic composition of phytoplankton, identification work is still in progress. Table 1 provides a list in which many adopted denominations (Gymnodiniaceae, naked or thecate dinoflagellates, phytoflagellates, etc.) consist of five to six different forms that have not yet been identified. Among the diatoms, only a few pennates (*Fragilariopsis*, *Nitzschia*, *Pseudonitzschia* and *Naviculacea*) dominated the blooms. Dinoflagellates were present mainly with nanoplanktonic naked forms, probably belonging

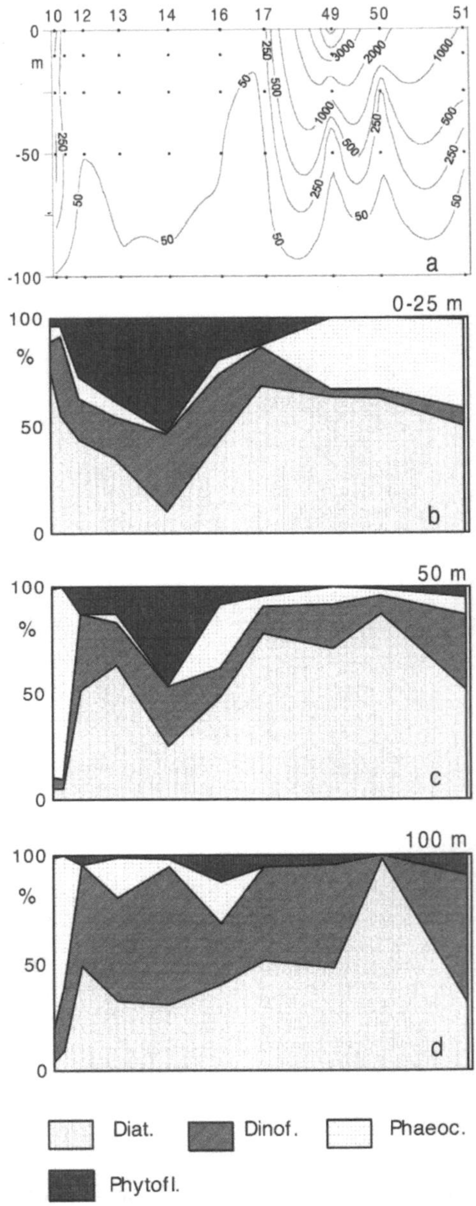


**Fig. 8a–d.** Vertical distribution of phytoplankton densities ($10^3$ cells $l^{-1}$) (**a**), and percentage composition of phytoplankton classes: 0–25 m (**b**), 50 m (**c**), 100 m (**d**), in the transect sampled during summer 1987–88 (stations 10–51)

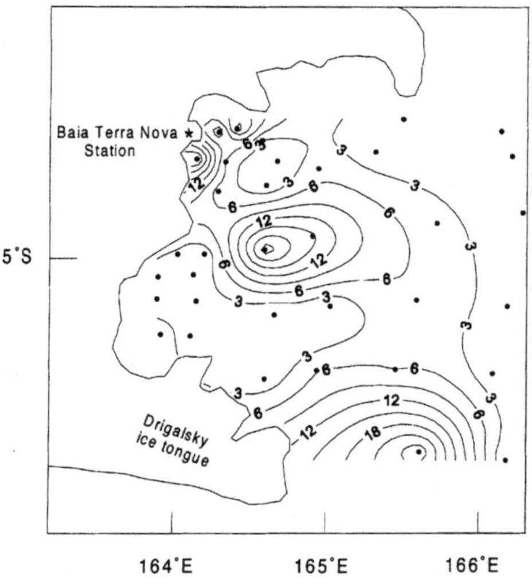


**Fig. 9.** Horizontal distribution of integrated (0–100 m) densities ($10^9$ cells $m^{-2}$) of *Phaeocystis* sp. in Terra Nova Bay

to the *Amphidinium*, *Gymnodinium*, *Gyrodinium* genera. Among thecate dinoflagellates, only *Protoperidinium* was present with many species (Table 1), and represented a consistent fraction of the heterotrophic biomass in Terra Nova Bay (Fonda Umani et al. 1998). The "phytoflagellates" listed in Table 1 probably consist of Prasinophyceans, with many small forms ≤5 μm with one or two flagella and some other tetraflagellates which had a remarkable presence at stations BTN (Fig. 3). In the open waters from 70° to 75°S (Fig. 7), forms with an asymmetric flagellar insertion, therefore assigned to Cryptophyceans, prevailed.

## Discussion

In recent years, it has been estimated, mainly from satellite images (Sullivan et al. 1993; Arrigo and McClain 1994), that the western Ross Sea is one of the most productive regions of the Southern Ocean, with intense blooms extending both in space and time, as recognized from our first findings (Innamorati et al. 1990c; 1992a). The bloom develops (Arrigo and McClain 1994) from the second half of December, over the large areas of the pack ice retreating from south to north and from open to coastal waters (Innamorati et al., this Vol.). During January the bloom declines, also if high pigment concentrations can cover large areas in Terra Nova Bay in late summer. Our data appear to be in agreement with this pattern.

Despite differences in cycle periods and amplitudes during the three summer seasons (probably related to the interannual variability of irradiance and time of ice melting), the phytoplankton cycle in the northwestern Ross Sea shows (Fig. 10) the main peak of the first bloom from the second fortnight of December to the beginning of January, when the daily irradiance reaches its maximum (Innamorati et al., this Vol.). The bloom of *Fragilariopsis* cf. *curta* develops over a large area (Fig. 6a) in the water column of the ice-edge zone (Fig. 11a,d) that is stabilized by the high irradiance and the lowered surface salinity due to the ice melting, with surface maxima in the upper mixed layer that causes a very shallow euphotic zone.

During the 1987–88 summer, only the declining phase of the *Fragilariopsis* bloom was detected (Fig. 10). The most common vertical structure of the phytoplankton assemblages shows (Figs. 8, 11g) the surface dominance of the senescent *Fragilariopsis* bloom and the underlying prevalence of *Phaeocystis* sp. in an early stage in a water column in which the thermal stratification replaced the haline one (Fig. 11l).

The subsequent decrease in the phytoplankton during January (Fig. 10) appears to be related to the change in environmental conditions: in the area of the bloom (Fig. 6a), nutrients were depleted [$NO_3$ 8.4 μM; $Si(OH)_4$ 17.8 μM; $PO_4$ 0.18 μM, Catalano 1992; $NO_3$ 4.2 μM; $PO_4$ 0.03 μM at SMN, Innamorati et al. 1991] and phosphorus in particular may have played a limiting role (N/P >>16; Innamorati et al., this Vol.). This consumption and the characteristics of these assemblages (Lazzara and Nuccio 1994): high phaeopigment/Chl and phaeopigment/particles ratios, average size of the particles and specific diversity, show that the bloom was in senescence.

The large area (Fig. 5) sampled about 2 weeks after ice melt (Hecq et al. 1992) is characterized (Fig. 11b,e) by a deep mixing layer, a deep euphotic zone and a flagellate assemblage in a growing phase (low phaeopigment/Chl ratios, small particles size, low specific diversity; Lazzara and Nuccio 1994) in which *Phaeocystis* sp. is increasing in the surface waters (Figs. 5, 6). The higher concentrations of pheophorbides (Hecq et al. 1992) in these waters and the numerous krill swarms observed along the Ross Sea slope during the Cariboo cruise

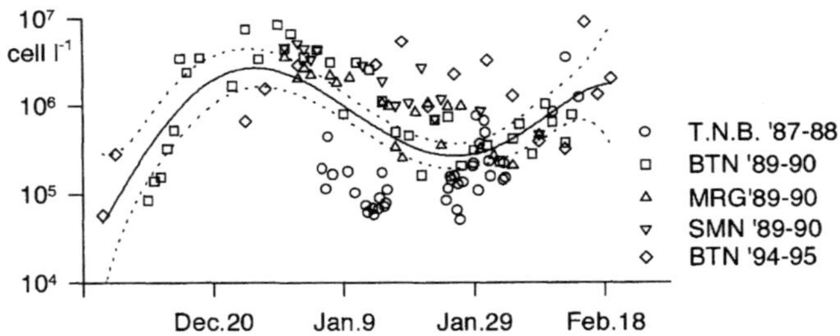


**Fig. 10.** Temporal evolution of surface (station BTN 1989–90 and 1994–95) and 0–25 m (Terra Nova Bay 1987–88; SMN and MRG 1989–90) phytoplankton densities (cells $l^{-1}$) in Terra Nova Bay during the three expeditions. Polinomial fitted curve 5th order ($P \leq 0.05$)

(Azzali et al. 1992) lead us to hypothesize, in agreement with other research from the Southern Ocean (Kopczynska 1992), that the phytoflagellate assemblages represent a stage following an initial diatom bloom that had occurred at the ice edge and had been grazed by zooplankton. In fact, some stations (stations 5, 6; Fig. 1c) sampled during the early days of December, in an area of just-broken pack ice, showed a much higher biomass, up to $6\,mg\,m^{-3}$ of Chl.

In Terra Nova Bay, flagellate assemblages with very low densities prevail towards the end of January (Figs. 8, 11h,m) after the diatom bloom and, at the same time, *Phaeocystis* sp. reaches its highest densities in the centre of the bay (Fig. 9), where the microzooplankton fraction, probably feeding on nanoplankton, reaches its maximum abundance (Fonda Umani et al. 1998). Afterwards, the late summer increase, in February 1988 and 1990 occurs as in 1995 (Fig. 10); therefore, it does not appear as an episodical trend.

Offshore, in the waters that had been free of ice longer (Fig. 11c,f), *Nitzschia* sp. 1 and *Phaeocystis* sp. were growing up to the bloom detected later in the same area in the first half of January (Fig. 7). Furthermore, the high phytoplankton biomass (stations 49–51; Fig. 8a) that developed in the surface mixed layer (Fig. 11i,n) of waters affected by ice melting and depleted nutrients ($NO_3$ 10.2 μM; $PO_4$ 0.29 μM; $SiO_2$ 26 μM, Catalano and Benedetti 1990) in February 1988, had a composition similar (*Nitzschia* sp. 1, *Phaeocystis* sp.) to the one blooming in January 1990 (Fig. 7b,d). It could therefore be its later senescent phase.

The waters in which the seasonal cycle was followed (Fig. 1b) appear to have a more stable phytoplankton composition (Figs. 2, 3) during time than the open waters (Figs. 5, 6, 8). It is possible that these waters are restricted by the cyclonic circulation in the Bay (Jacobs et al. 1970; Boldrin and Stocchino 1990) and affected mainly by surface thermal processes (Boldrin and Stocchino 1990) that support environmental conditions more stable during time than offshore.

The massive bloom of *Fragilariopsis* cf. *curta* confirms previous studies on the importance of this species in western Ross Sea ice-edge blooms (Wilson et al. 1986; Carbonell-Moore 1994) and as revealed from sediment traps (Leventer and Dunbar 1996). In February, more species (*Nitzschia* sp. 1, *Naviculacea* sp., *Fragilariopsis* sp. 1, *Pseudonitzschia* spp., *Phaeocystis* sp., *Dictyocha speculum*), together with *Fragilariopsis* increase. Also *Phaeocystis* shows an intense diffusion in the subeuphotic layers of stratified waters during the first period of the season (Figs. 2, 11c,g), but also up to the surface in deep mixed waters or later in the season (Figs. 3, 5, 8, 11b,i).

Summarizing the major findings, we can conclude that: the early summer bloom of *Fragilariopsis* cf. *curta* develops when conditions of irradiance and temperature are not limiting and the melting of pack ice and the warming of surface layers lead to water column stability that allows phytoplankton growth; this bloom extends over large areas of the retreating ice edge and its decline can be related to nutrient depletion, zooplankton grazing and to the loss of water stability mainly due to meteorological perturbations; after this bloom, flagellates dominate the phytoplankton assemblage in conditions of a mixed column water and a bloom of *Phaeocystis* spreads in Terra Nova Bay; another bloom, with phytoplankton composition different from the first one, develops when availability of new nutrients and conditions of water stability occur.

*Acknowledgements.* Thanks are due to Dr. G. Nizzi Grifi, Dr. L. Franchini, Dr. C. Gamba, Dr. L. Borgioli, who performed part of the microscopic analysis and to M. Colombarini, Dr. M. Nigro, Dr. P. Sparapani, who collected the phytoplankton samples before the arrival of our research group.

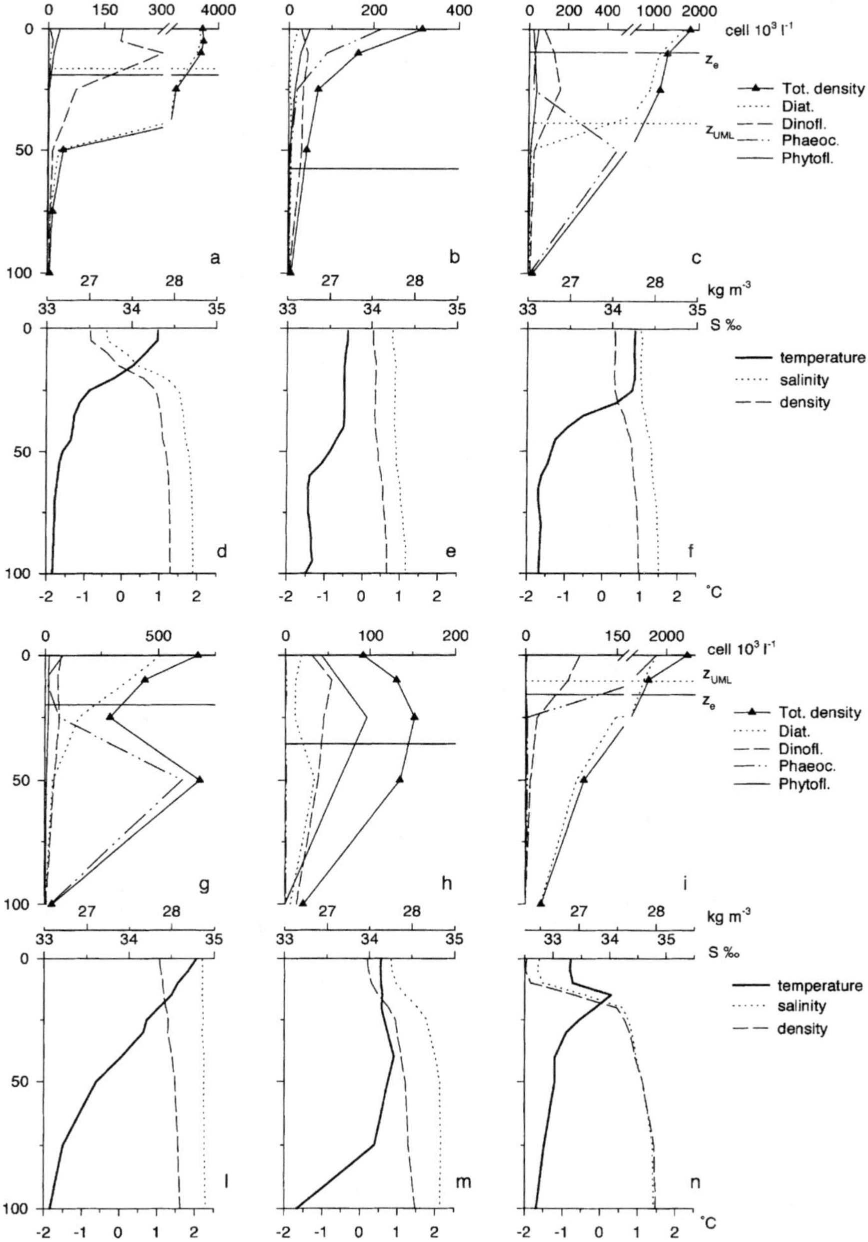


**Fig. 11a–n.** Representative profiles of phytoplankton classes and total densities (**a, b, c, g, h, i**) and corresponding profiles (**d, e, f, l, m, n**) of temperature, salinity and density. Station 26, Jan. 1990 (**a, d**); station 19, Dec. 1989 (**b, e**); station 22, Jan. 1990 (**c, f**); station 10, Jan. 1988 (**g, l**); station 14, Jan. 1988 (**h, m**); station 50, Feb. 1988 (**i, n**). The depths of upper mixed layer ($z_{UML}$) and euphotic zone ($z_e$) are indicated

## References

Andreoli C, Tolomio C, Moro I, Radice M, Moschin E, Bellato S (1995) Diatoms and dinoflagellates in Terra Nova Bay (Ross Sea-Antarctica) during austral summer 1990. Polar Biol 15: 465–475

Arrigo KR, McClain CR (1994) Spring phytoplankton production in the Western Ross Sea. Science 266: 261–263

Arrigo KR, Weiss AM (1996) Interannual variation in pigment concentration and primary production in the Western Ross Sea as seen by the Nimbus 7 Coastal Zone Color Scanner. EOS Trans AGU, 76 (3): 137

Artegiani A, Paschini E (1992) Condizioni oceanografiche nel settore neozelandese dell'Oceano Antartico durante l'estate australe 1989–90. In: Gallardo VA, Ferretti O, Moyano HI (eds) Oceanografía en Antártica. ENEA-PNRA-EULA, Concepción, Chile, pp 33–38

Artegiani A, Azzolini R, Paschini E, Creazzo S (1992) Physical oceanographic conditions in the Southern Pacific Ocean and in the Western Ross Sea. In: National Scientific Commission for Antarctica (ed) Oceanographic Campaign 1989/90, Data Rep Part II, Genova, pp 5–62

Azzali M, Kalinowski J, Castagnani R (1992) Biomass of krill in the Ross Sea (Pacific sector) in relation to BIOMASS FIBEX/SIBEX results. In: Gallardo VA, Ferretti O, Moyano HI (eds) Oceanografía en Antártica. ENEA-PNRA-EULA, Concepción, Chile, pp 331–338

Balech E (1976) Clave ilustrada de Dinoflagelados Antarticos. Publ Inst Antart Argent 11: 1–99

Boldrin A, Stocchino C (1990) On the hydrological characteristics of Terra Nova Bay. In: National Scientific Commission for Antarctica (ed) Oceanographic Campaign 1987–88, Data Rep Part I, Genova, pp 11–57

Carbonell-Moore MC (1994) Phytoplankton of an ice-edge bloom in the Ross Sea, with special reference to the elemental composition of antarctic diatoms. EOS Trans AGU, 75 (3): 201

Catalano G (1992) L'attività del gruppo di oceanografia chimica nell'ambito delle Campagne Oceanografiche Italiane in Antartide. In: Gallardo VA, Ferretti O, Moyano HI (eds) Oceanografía en Antártica. ENEA-PNRA-EULA, Concepción, Chile, pp 59–70

Catalano G, Benedetti F (1990) Distribution of nutrients in the Terra Nova Bay and in the Ross Sea. In: National Scientific Commission for Antarctica (ed) Oceanographic Campaign 1987/88, Data Rep Part I, Genova, pp 59–83

El-Sayed S, Biggs DC, Holm-Hansen O (1983) Phytoplankton standing crop, primary productivity, and near-surface nitrogenous nutrient fields in the Ross Sea, Antarctica. Deep Sea Res 30 8A: 871–886

Fonda Umani S, Monti M, Nuccio C (1998) Microzooplankton biomass distribution in Terra Nova Bay, Ross Sea (Antarctica). J Mar Syst 17: 289–303

Garrison DL, Mathot S, Gowing MM, Kunze H (1996) Phytoplankton and microzooplankton community structure in the Ross Sea polynya: November-December, 1994. EOS Trans AGU, 76 (3): 137

Hasle GR (1964) *Nitzschia* and *Fragilariopsis* species studied in the light and electron microscopes. I. Some marine species of *Nitzschiella* and Lanceolatae. Skr Nor Vidensk Akad Oslo Mat Naturvidents K1 NS 16: 1–48

Hasle GR (1965a) *Nitzschia* and *Fragilariopsis* species studied in the light and electron microscopes. II. The group *Pseudonitzschia*. Skr Nor Vidensk Akad Oslo Mat Naturvidents K1 NS 18: 1–45

Hasle GR (1965b) *Nitzschia* and *Fragilariopsis* species studied in the light and electron microscopes. III. The genus *Fragilariopsis*. Skr Nor Vidensk Akad Oslo Mat Naturvideusts Kl NS 21: 1–49

Hasle GR (1969) An analysis of the phytoplankton of the Pacific Southern Ocean: abundance, composition, and distribution during the Brattegg expedition, 1947–1948. Hvalradets Skr 52: 1–168

Hecq JH, Magazzù G, Goffart A, Catalano G, Vanucci S, Guglielmo L (1992) Distribution of planktonic components related to structure of water masses in the Ross Sea during the Vth ITALIANTARTIDE expedition. Atti IX Congr Ass Ital Oceanogr Limnol, Lang, Genova, pp 665–678

Hendey NI (1937) The plankton diatoms of the southern seas. Disc Rep 16: 151–364, VI–XIII

Innamorati M, Mori G, Lazzara L, Nuccio C, Lici M, Vanucci S (1990a) Ecology of Phytoplankton. In: National Scientific Commission for Antarctica (ed), Oceanographic Campaign 1987–88, Data Rep Part I, Genova, pp 161–238

Innamorati M, Mori G, Lazzara L, Nuccio C, Lici M, Catalano G, Benedetti F (1990b) Phytoplankton Ecology in the Ross Sea. In: National Scientific Commission for Antarctica (ed), Oceanographic Campaign 1987/88, Data Rep Part II, Genova, pp 9–63

Innamorati M, Mori G, Lazzara L, Vanucci S (1990c) Eutrofia ed oligotrofia nell'Oceano Antartico. Oebalia XVI-1: 153–162

Innamorati M, Lazzara L, Mori G, Nuccio C, Saggiomo V (1991) Phytoplankton ecology. In: National Scientific Commission for Antarctica (ed) Oceanographic Campaign 1989/90, Data Rep Part I, Genova, pp 141–252

Innamorati M, Lazzara L, Massi L, Mori G, Nuccio C, Saggiomo V (1992) Indagine sulla biomassa fitoplanctonica nel Mare di Ross in relazione ai fattori ambientali. In: Gallardo VA, Ferretti O, Moyano HI (eds) Oceanografía en Antártica. ENEA-PNRA-EULA, Concepción, Chile, pp 235–252

Innamorati M, Lazzara L, Massi L, Mori G, Nuccio C, Saggiomo V, Cartei P, Franchini L, Gamba C (1994) Phytoplankton ecology: particles, gilvin, fluorescence spectra, irradiance and phytoplankton specie density in the Ross Sea. Ocean Camp 1989–90. In: National Scientific Commission for Antarctica (ed) 1989–90, Data Rep Part III, Genova, pp 101–178

Jacobs SS, Amos AF, Bruchhausen PM (1970) Ross Sea oceanography and Antarctic bottom water formation. Deep Sea Res 17: 935–962

Knox GA (1994) The biology of the Southern Ocean. Cambridge University Press, Cambridge

Kopczynska EE (1992) Dominance of microflagellates over diatoms in the Antarctic areas of deep vertical mixing and krill concentration. J Plankton Res 14 (8): 1031–1054

Lazzara L, Nuccio C (1994) Relazioni tra particellato, biomassa e popolamenti fitoplanctonici nel Mare di Ross durante la fioritura estiva. Atti X Congr Ass Ital Oceanogr Liminol. Lang, Genova, pp 655–665

Leventer A, Dunbar RB (1996) Factors influencing the distribution of diatoms and other algae in the Ross Sea. J Geophysical Res 101 (C8): 18489–18500

Lorenzen CJ (1967) Determination of chlorophyll and phaeophytin spectrophotometric equations. Limnol Oceanogr 12: 343–346

Manguin E (1960) Les diatomées de la Terre Adélie. Campagne du Commandant Charcot 1949–1950. Ann Sci Nat (Bot) 12 (1): 223–363

Medlin LK, Priddle J (1990) Polar marine diatoms. British Antarctic Survey, Cambridge

Nuccio C, Innamorati M, Lazzara L, Mori G (1992) Popolamenti fitoplanctonici nella Baia di Terra Nova, Mare di Ross. In: Gallardo VA, Ferretti O, Moyano HI (eds) Oceanografí a en Antártica. ENEA-PNRA-EULA, Concepción, Chile, pp 253–262

Nuccio C, Innamorati M, Mori G, Lazzara L, Nizzi Grifi G (1994) Phytoplankton ecology: species density in the Ross Sea. Oceanographic Campaign 1987–88. In: National Scientific Commission for Antarctica (ed) Oceanographic Campaign 1987–88. Data Rep Part III, Genova, pp 5–48

Priddle J, Fryxell G (1985) Handbook of the common plankton diatoms of the Southern Ocean: centrales except the genus *Thalassiosira*. British Antarctic Survey, Cambridge

Riemann B (1978) Carotenoid interference in the spectrophotometric determination of chlorophyll degradation products from natural populations. Limnol Oceanogr 23 (5): 1059–1066

Smith WO Jr, Kelly HP, Rich JH (1990) Chlorophyll distribution and primary productivity in the Ross Sea, austral summer 1990. Antarct J US 25 (5): 179–182

Sournia A, Grall JR, Jacques G (1979) Diatomées et dinoflagellés planctoniques d'une coupe méridienne dans le sud de l'océan Indien (campagne Antiprod I du Marion-Dufresne, mars 1977). Bot Mar 22: 183–198

Strathmann RR (1967) Estimating the organic carbon content of phytoplankton from cell volume or plasma volume. Limnol Oceanogr 12: 411–418

Sullivan CW, Arrigo KR, McClain CR, Comiso JC, Firestone J (1993) Distributions of phytoplankton blooms in the Southern Ocean. Science 262: 1832–1837

Wilson DL, Smith OW Jr, Nelson DM (1986) Phytoplankton bloom dynamics of the Western Ross Sea ice edge. I. Primary productivity and species-specific production. Deep Sea Res 33 (10): 1375–1387

CHAPTER 20

# Ecological and Physiological Aspects of Primary Production in the Ross Sea

**V. Saggiomo[1], G.C. Carrada[2], O. Mangoni[2], D. Marino[1], and M. Ribera d'Alcalà[1]**

**ABSTRACT**

A cruise on board the R/V Italica was carried out in the Ross Sea during the austral spring 1994 to investigate the ecological mechanisms that characterize the marginal ice zone (MIZ) of the Ross Sea polynya. In this framework, size-fractionated primary production was studied to assess its spatial and temporal evolution in relation to irradiance, structure of the water column and processes of removal by consumers. Integrated total daily production levels were very variable, ranging from 131 to 2992 $mg\,C\,m^{-2}\,day^{-1}$. Highest values were measured in the polynya and in the MIZ, in situations where ice melting could not ensure water column stability.

Picoplankton represented the main primary productive component in the ice-covered areas, whereas micro- and nanophytoplankton dominated in the MIZ and in the open waters of the polynya. Total and fractionated production appeared more responsive to biological factors than to hydrographic ones. Selective grazing, mostly by krill, was responsible for the picoplankton prevalence in ice-covered areas, as a consequence of the active removal of the larger fractions attached to or released from the pack ice. The absence of such efficient grazing in the Terra Nova Bay MIZ and in the Ross Sea polynya accounts for the predominance, in the former, of a micro- and nanophytoplankton-based production and, in the latter, of a large concentration of *Phaeocystis* cf. *antarctica*. This species, in the colonial stage, and *Pseudo-nitzschia subcurvata*, were dominant in the polynya and MIZ, while *Fragilariopsis curta* represented the major part of the phytoplankton population of Terra Nova Bay. Photosynthetic characteristics of Ross Sea phytoplankton indicated a community adapted to low light regimes, as demonstrated by the poor correlation between photosynthetic capacity (P/B) and available irradiance.

## Introduction

A major part of the Southern Ocean is considered, on the basis of the research carried out in the past 30 years, a high nutrient-low chlorophyll (HNLC) area. A number of physical, chemical and biological variables have been considered as possible factors in controlling primary production processes: light (Smith and Sakshaug 1990; Nelson and Smith 1991), temperature (Tilzer and Dubinsky 1987), stability of the water column (Smith and Nelson 1985), nutrients (Sullivan et al. 1993), meteorological factors (Savidge et al. 1995; Boyd et al. 1995), grazing (Mitchell and Holm-Hansen 1991; Lancelot et al. 1993) and micronutrients (e.g. Fe). The role of iron, in particular, has been much discussed since the early 1990s and evidence is accumulating about its relevance for the ecology of the Southern Ocean (Martin et al. 1990; Jochem et al. 1995; De Baar et al. 1995; Di Tullio and Smith 1996) and in general for all HNLC areas (Behrenfeld et al. 1996). By contrast, coastal environments (Holm-Hansen and Mitchell, 1991), marginal ice edges (Smith and Nelson 1986; Sullivan et al. 1988) and frontal systems (Laubscher et al. 1993; Savidge et al. 1995; Bathmann et al. 1997) are considered areas of high primary production. Within this heterogeneous context, the Ross Sea is considered one of the most productive Antarctic systems, both on the basis of the high flux of biogenic material accumulating on its continental shelf (Dunbar et al. 1985; Ledford-Hoffman et al.

[1] Stazione Zoologica "A. Dohrn", Biological Oceanography Laboratory, Naples 80121, Italy
[2] Department of Zoology, University of Naples Federico II, Naples, Italy

1996; DeMaster et al. 1992) and for the high levels of primary production measured in situ (Wilson et al. 1986; Smith et al. 1996) or detected from satellite platforms (Sullivan et al. 1993; Comiso et al. 1993; Arrigo and McClain 1994) which showed chlorophyll concentrations four times higher than the average calculated for the entire Southern Ocean (Arrigo and McClain 1994; Nelson et al. 1996).

Information on primary production in the Ross Sea is mostly based on measurements carried out during the austral summer, among the highest for the Southern Ocean; however, this information is still insufficient to portray a reliable picture of the spatial patterns and temporal dynamics of this basic ecological process (El-Sayed et al. 1983; Smith and Nelson 1985; Wilson et al. 1986; DeMaster et al. 1992; Magazzù and Decembrini 1992; Saggiomo et al. 1992; Smith et al. 1996). Information on spring primary production is much more limited and essentially based on the results of two cruises carried out during the spring 1994 (Saggiomo et al. 1998; Smith and Gordon 1997) and on satellite imagery of a late spring situation (Arrigo and McClain 1994). The two cruises, conducted by an Italian and USA team, respectively, on the R/V Italica and R/V N. Palmer, represented the earliest entry of oceanographic vessels in the Ross Sea during the austral spring. An additional complication toward the achievement of a satisfactory understanding of primary production processes in the Ross Sea is represented by the markedly insufficient information concerning the dimensional structure of the phytoplankton communities, the seasonal dynamics of different size fractions, the photosynthetic characteristics of each size category, as well as the structure of the trophic web. For the Ross Sea, a model of a trophic web has been proposed in which meso- and microphytoplankton play a major role, while krill would play an almost irrelevant role, being limited to the northernmost part of the area in summer (Di Tullio and Smith 1996; Nelson et al. 1996).

The present chapter deals with the spatial and temporal evolution of the production processes of the different size fractions of the Ross Sea phytoplankton during the austral spring. Ecophysiological aspects of production in phytoplankton have also been investigated, in particular with regards to the relationship between P/B ratio and irradiance regimes.

## Materials and Methods

A cruise on board the R/V Italica was carried out in the Ross Sea during the austral spring 1994 from 14 November to 13 December. Twenty seven stations were sampled along a transect laying on 175°E, from 76.5°S to 72.0°S, and covering, in succession, the zones of drifting, broken and closed pack ice (Fig. 1). Several stations, as in the case of stations MP, 2, 4, 6, 9, 11, 12, 13, 14, 15, 17, 19 and 24 of the transect, were repeated at different time intervals in order to allow an evaluation of the temporal changes of size-fractionated primary production.

From 13 Dec. to 15 Dec., three hydrological stations were located along a transect from the Ross Sea polynya to Terra Nova Bay in the shore waters facing the Drygalski Ice Tongue (Fig. 1). Samples were collected by means of a SBE 32 Carousel sampler, equipped with 24 bottles (12-l), SBE 9/11 Plus CTD, SBE oxygen sensor, Chelsea Aquatrack III fluorometer, Sea Tech transmissometer and a Datasonics PSA 900 altimeter. At each station, profiles of down- and upwelling underwater PAR were obtained utilizing a photoprobe equipped with LI-COR quantum meters. For measurements of primary production, sampling depths were selected according to optical levels equivalent to 100, 40, 20, 10, 5, 1 and 0.13% of incident irradiance. Measurements of primary production were performed under in situ-simulated conditions. Each sample collected at the surface and the optical depths listed above was enclosed in light and dark polycarbonate (Nalgene) 450-ml bottles and incubated with 1 ml (20 μCi) of $NaH^{14}CO_3$. Incubations of 4–6 h were made on deck in incubators cooled by circulating surface seawater, at the different light levels obtained by means of neutral light screens (electroformed nickel screens, produced by Veco Int. Co. USA). After incubation, the separation into size fractions (total, micro-, nano- and picophytoplankton) was obtained by differential filtration. Subsamples of 150 ml were filtered through, 10- and 2-μm Nuclepore polycarbonate membranes and a Whatman GF/F filter.

PvsE relationship was studied in the deep fluorescence maximum (DFM) at four stations (13C, 14C, 28 and 30). Subsamples from DFM were confined in 100-ml Pyrex bottles, inoculated with 20 μCi of $NaH^{14}CO_3$ and then submitted for 2 h to 16 irradiance levels (from 100 to 0.13% incident

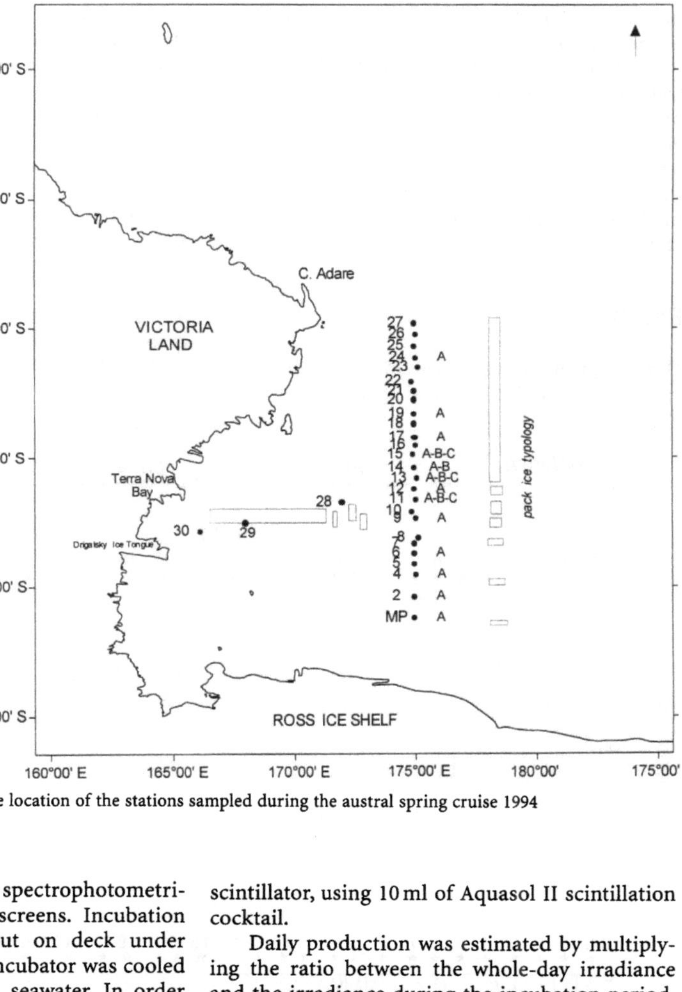


**Fig. 1.** The area investigated with the location of the stations sampled during the austral spring cruise 1994

irradiance) obtained utilizing spectrophotometrically checked neutral light screens. Incubation experiments were carried out on deck under natural light conditions; the incubator was cooled by circulating surface (–6 m) seawater. In order to check dark fixation from each sample, one subsample was incubated in darkness with the addition of four drops of a saturated solution of DCMU in seawater (Legendre et al. 1983). Filtration on 25-mm Whatman GF/F filters was carried out immediately after incubation under dim light conditions.

Radioactive content of each sample, as well as the samples for measurement of size-fractionated primary production, were measured, after acidification with 200 µl of HCL 0.1 N, within 24 h after filtration, in a Beckman LS 1801 liquid scintillator, using 10 ml of Aquasol II scintillation cocktail.

Daily production was estimated by multiplying the ratio between the whole-day irradiance and the irradiance during the incubation period. The integration of the values of primary production to the depth of, the 0.13% of surface radiation was obtained by the method of the trapezoidal rule. The PvsE data were fitted with the model of Platt et al. (1980):

$$P^B(E) = P^B_S[1 - \exp(-\alpha I/P^B_S)]\exp(-\beta I/P^B_S)$$

where $P^B_S$ is the potential light-saturated Chl-specific rate of photosynthesis in the absence of photoinhibition, in mg C (mg Chla)$^{-1}$ h$^{-1}$, $\alpha$ is the initial slope of PvsE curves, in mg C (mg Chla)$^{-1}$ h$^{-1}$

($\mu E\, m^{-2} s^{-1})^{-1}$ and $\beta$ is the index of photoinhibition (same units of $\alpha$).

Phytoplankton samples were fixed with neutralized formaldehyde to a final dilution of 3% and preserved in 500-ml dark glass bottles. Species abundance and composition were determined on 50-ml subsamples with a ZEISS IM 35 inverted microscope using the classical Utermöhl (1958) method. For pigment analysis see Carrada et al. (this Vol.).

## Results

During spring, the water column along the entire transect was well mixed and a very slight stabilization, induced by ice melting, could be observed in the northernmost part of the Ross Sea polynya (stations 9 to 11). The polynya was characterized by a homogeneous cover of thin, newly formed ice. For the entire sampling period, the area included between 75°S and 72°S (stations 12 to 27) was occupied by compact pack ice with a mean cover of 8–10/10.

Figure 2a illustrates density ($\sigma_\theta$) distribution and Fig. 2b spatial variations in primary production along with the depth of 1% of surface irradiance along the transect. Integrated total daily primary production values, their evolution in space and time, as well as the relative contribution of the different phytoplankton size fractions are reported in Fig. 3. The integrated daily values for

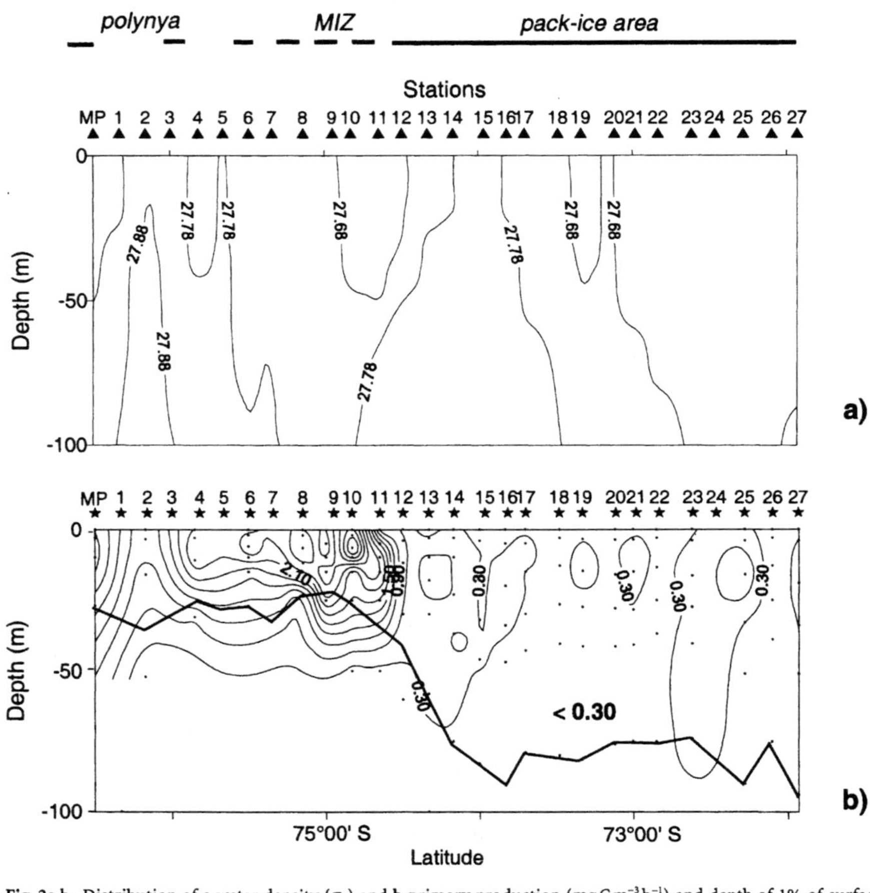


**Fig. 2a,b.** Distribution of **a** water density ($\sigma_\theta$) and **b** primary production ($mg\,C\,m^{-3} h^{-1}$) and depth of 1% of surface irradiance (*solid line*) along the S-N transect

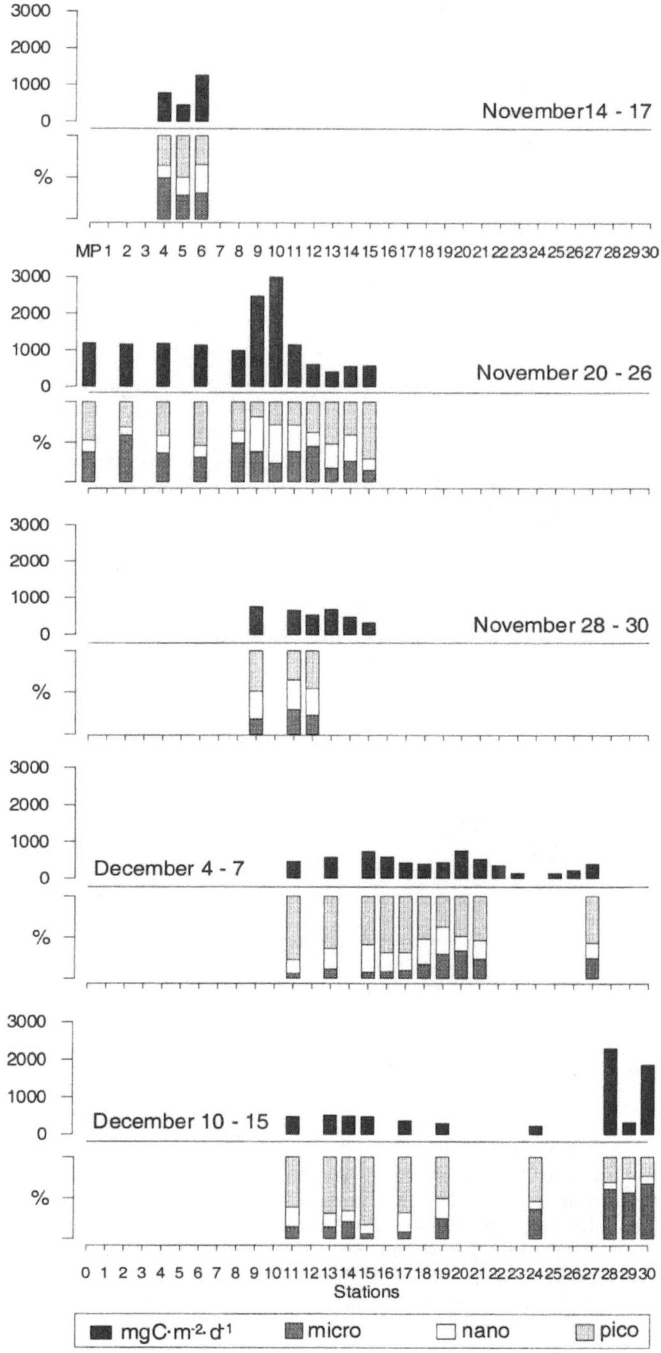


**Fig. 3.** Spatial and temporal variability of daily integrated primary production (mg C $m^{-2}$ $day^{-1}$) and percent contribution by size fractions for all stations

**Table 1.** Photosynthetic parameters and percent contribution by size fractions to total primary production at four selected stations

| | Station 13C | Station 14B | Station 28 | Station 30 |
|---|---|---|---|---|
| Depth | 30 m | 31 m | 20 m | 20 m |
| $\alpha$ | 0.091 | 0.090 | 0.031 | 0.050 |
| $\beta$ | 0.004 | 0.0039 | 0.0013 | 0.010 |
| $P^B_{max}$ | 2.076 | 2.20 | 1.099 | 1.61 |
| $E_m$ | 86 | 93 | 134 | 99 |
| $E_k$ | 23 | 24 | 35 | 23 |
| % micro- | 16.22 | 16.50 | 58.65 | 68.86 |
| % nano- | 14.60 | 19.50 | 3.53 | 0.17 |
| % pico- | 69.18 | 64.00 | 37.82 | 30.96 |
| Chl ($\mu g l$)$^+$ | 0.22 | 0.26 | 2.72 | 1.11 |

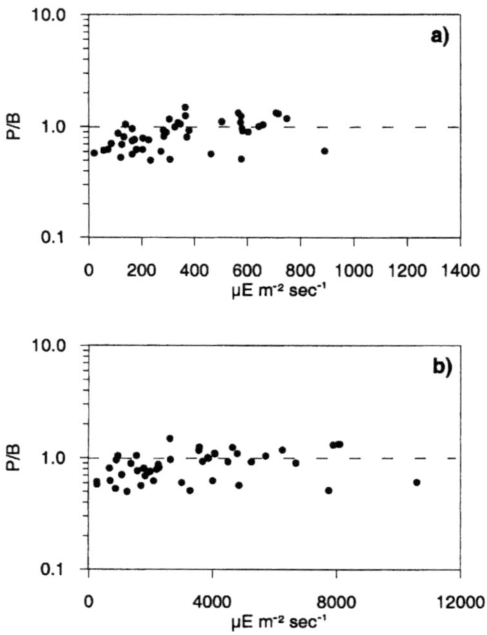


**Fig. 4a,b.** Pooled measurements of integrated P/B [mg C (mg Chl a)$^{-1}$ m$^{-2}$ day$^{-1}$] ratio against a surface irradiance and **b** integrated irradiance

all stations ranged from 131 mg C m$^{-2}$ day$^{-1}$ at station 25 to 2992 mg C m$^{-2}$ day$^{-1}$ at station 10. The three stations (28, 29, 30) sampled in mid-December along the transect from the central Ross Sea to Terra Nova Bay were located, respectively, along the western ice edge of the Ross Sea polynya, in the pack-ice-covered area and along the eastern margin of the Terra Nova Bay polynya. Integrated daily values of primary production for stations 28, 29 and 30 were 2302, 329 and 1863 mg C m$^{-2}$ day$^{-1}$, respectively. The photosynthetic parameters for samples from the deep fluorescence maximum of stations 13C, 14B, 28 and 30 are reported in Table 1.

The integrated P/B ratio (photosynthetic capacity) against surface and integrated irradiance is reported in Fig. 4a,b. Figure 5 reports the P/B ratio normalized for irradiance (efficiency) vs. irradiance of the total communities for each of the size fractions considered. In order to assess photosynthetic capacity in relation to light levels, efficiency (P/B/E) vs. photosynthetic capacity (P/B) was plotted for all samples analyzed, (in Fig. 6).

A total of 78 taxa – 43 diatoms, 25 dinoflagellates and 10 flagellates belonging to different taxonomic groups – were recorded in the examined samples. Cell concentrations varied between 156 cell ml$^{-1}$ (station 22) to 7746 cell ml$^{-1}$ (station 9) during the first sampling series. Lower cell numbers were recorded during the other sampling series, where cell concentrations ranged from 12 cell ml$^{-1}$ (station 15B) to 2194 cell ml$^{-1}$ (station 4A).

Phytoplankton assemblages recorded at most of the stations were dominated by few species, including the diatoms *Pseudo-nitzschia subcurvata* and *Fragilariopsis curta* and the prymnesiophycean *Phaeocystis* cf. *antarctica.* This last species was particularly abundant along the southern part of the main transect and at station 28. In contrast, a markedly different population was recorded at stations located in the middle part of the transect (stations 9 to 12), where diatoms gradually replaced *Phaeocystis* cf. *antarctica.* At

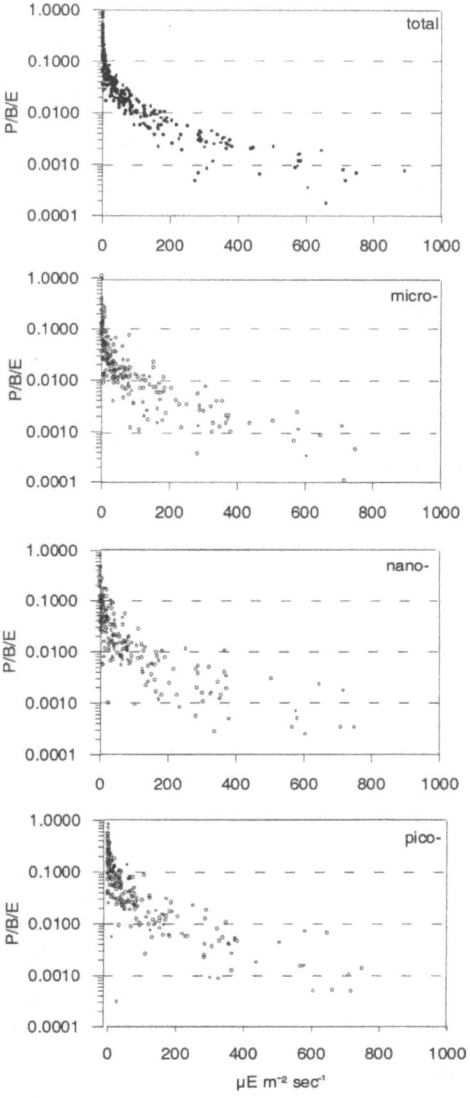

**Fig. 5.** Efficiency (P/B/E) [mg C (mg Chl a)$^{-1}$ h$^{-1}$ (μE m$^{-2}$ s$^{-1}$)$^{-1}$] against irradiance by total phytoplankton and size fractions

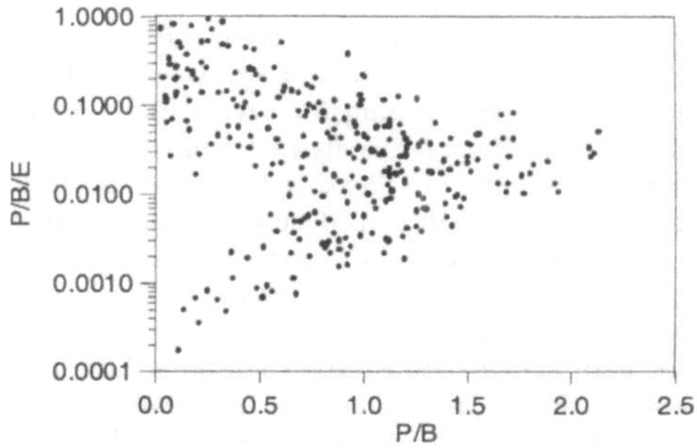

**Fig. 6.** Efficiency (P/B/E) [mg C (mg Chl a)$^{-1}$ h$^{-1}$ (μE m$^{-2}$ s$^{-1}$)$^{-1}$] plotted against photosynthetic capacity (P/B) [mg C (mg Chl a)$^{-1}$ h$^{-1}$] for all samples

these stations, diatoms represented more than 47% of the entire phytoplankton population with a maximum value of 97.9% at station 11. Finally, very low phytoplankton concentrations were measured at the stations located in the northern part of the transect where assemblages were dominated by diatoms and small flagellates.

## Discussion

The spatial variability of a number of biological, physical and chemical aspects of the Ross Sea ecosystem has already been pointed out (Di Tullio and Smith 1996; Nelson et al. 1996; Smith et al. 1996). Nelson et al. (1996) divided the Ross Sea into three different subsystems on the basis of seasonal cycles of ice cover and biological productivity: the southwestern part, characterized by intense diatom blooms and high accumulation of biogenic material in the upper sediment layers; the southeastern part, showing the highest spring daily production due to *Phaeocystis antarctica* blooms; the northern area with the same characteristics as the southwest area, but with a longer period of ice cover and relatively low production values. On the basis of biological and chemical environmental characteristics, El-Sayed et al. (1983) also suggested that the Ross Sea consists of three different parts: the consolidated pack-ice area, the loose pack ice and the open waters, each characterized by differences in euphotic depth, chlorophyll a, nitrate and ammonium concentrations and ratios Chl/phaeo, Chl/ATP and POC/PON.

During our cruise, three subsystems, corresponding to the pack-ice covered area, the marginal ice zone and the polynya, could be identified on the basis of phytoplankton biomass (Carrada et al., this Vol.), production levels and their dimensional structure and floristic composition. High values of primary production, always exceeding 1 g C $m^{-2}$ $day^{-1}$ and lasting for several days, were measured in the polynya area (stations MP to 7), when a bloom of *Phaeocystis* cf. *antarctica* occurred. It is worth considering that, during the same period, Smith and Gordon (1997) reported a sixfold higher primary production by the same species, the highest ever observed in the Ross Sea. The marginal ice zone (MIZ) (stations 8 to 11) showed the highest values of integrated daily primary production, reaching 2.9 g C $m^{-2}$ $day^{-1}$ at station 10. Stations 9 and 11 were revisited a few days later and production levels appeared drastically reduced, particularly at station 9. These values can be considered as particularly high given the presence of large amounts of krill (Azzali et al., this Vol.) inducing the presence of other species representative of higher trophic levels such, as birds and whales (see also Carrada et al., this Vol.). The presence of large amounts of krill, never observed at this latitude in the Ross Sea before the Italian cruise of spring 1994 (Azzali et al., this Vol.), plays an important role in controlling primary production (Wefer et al. 1989; Jacques and Panouse 1991; Lancelot et al. 1993) and in determining the size structure of the phytoplankton population by means of selective grazing. In fact, in the MIZ, from November 22 to 23, most of the primary production was due to the micro- and nanophytoplankton fractions (33% and 42%, respectively), while, by November 28 to 29, the production dropped to very low levels and was entirely sustained by the nano- and picophytoplankton fractions (34% and 42%, respectively). The same considerations could also be applied to the pack-ice covered area. Carrada et al. (this Vol.), on the basis of direct observations of krill attached to the lower surface of ice overturned by at the passage of the ship as well as on the high ratio phaeo/Chl a, suggest that sympagic algae attached or released from the ice were in large part grazed by krill. The selective grazing could, therefore, be the factor determining the size structure of phytoplankton populations as shown by Stretch et al. (1988) and Smetacek et al. (1990). In fact, production values were the lowest of the entire area, ranging from 0.1 (station 25) to 0.7 g C $m^{-2}$ $day^{-1}$ (station 13A) and were mainly sustained by the picophytoplankton fraction that showed integrated mean percent values of 56%. Only minor variations were recorded at the revisited stations. The primary production values recorded in the pack ice area were an order of magnitude higher than those reported for winter in open Antarctic waters (Brightman and Smith 1989), thus indicating active growth.

The situation observed along the E-W transect, including stations 28, 29 and 30, was representative of a spatiotemporal evolution of the Ross Sea polynya (station 28), of the pack-ice-covered area, characterized, at the time of sampling, by the absence of krill (station 29) and of the initial development of a bloom along the marginal ice zone of the Terra Nova Bay polynya (station 30). In fact, station 28, located at the western margin of the Ross Sea polynya, showed high primary production values, similar to those measured at the nearby station 10 and comparable to those reported by Smith et al. (1996) for the same area during summer. At both stations, the biomass was evenly distributed along the water column, with a deep chlorophyll maximum of 3.36 and 3.01 μg Chl a $l^{-1}$ located at 34 and 23 m depth, respectively (Carrada et al., this Vol.). In contrast to the situation described by Wilson et al. (1986) and Smith et al. (1996) for the same area of the Ross Sea in summer, a water column stabilization was not observed. In fact, at station 28 the difference in density ($\sigma_\theta$) was of 0.19 kg $m^{-3}$ in the upper 100 m and of only 0.03 kg $m^{-3}$ between 0 and 30 m. The same hydrographic structure characterized station 10. While the main objective of our activity was to describe the S-N evolution of the Ross Sea MIZ, the situation observed at station 28, sampled on December 14, was probably indicative of an E-W evolution of the Ross Sea polynya; here algae released from the pack ice were actively growing in a mixed water column. This contrasts with the model of bloom formation proposed by Smith and Nelson (1985) for the Ross Sea during summer. The absence of krill (Azzali et al., this Vol.) and the low phaeo/Chl a ratio (Carrada et al., this Vol.) were associated with the ongoing bloom. Surface samples showed a predominance of *Phaeocystis* cf. *antarctica* (718 cell $ml^{-1}$), and diatoms had concentrations of 290 cell $ml^{-1}$, of which 90% were *Pseudo-nitzschia subcurvata*. This same species was dominant at station 10 with 492 cell $ml^{-1}$ (total diatoms 508 cell $ml^{-1}$). The absence of a strong grazing activity was further supported by the data collected at station 29, representing conditions typical of an ice-covered site. Here, low levels of primary production along with a dominance of microphytoplankton, probably released from the ice, were observed, contrasting with the dynamics characterizing the stations located in the ice covered sector of the 175°E transect. Station 30, in the marginal ice zone of the Terra Nova Bay polynya, was more representative of coastal waters. In contrast to the situation at station 28, *Fragilariopsis curta* was dominant (869 cell $ml^{-1}$, total diatoms 1264 cell $ml^{-1}$), while *Phaeocystis* cf. *antarctica* was represented by only 104 cell $ml^{-1}$. Marino et al. (1995) also report *Fragilariopsis curta* as dominating in the diatom assemblages found in the oval faecal pellets (400-μm diameter) released by an unknown organism, abundant in Terra Nova Bay during summer, as well as in the gut of the ostracod *Alacia belgicae*. Innamorati et al. (1992) report, in the same area, a strong bloom starting in late spring, where small *Fragilariopsis* species were dominant (Nuccio et al. 1992). Phytoplankton composition analysis confirms that the polynya, the marginal ice zone and the pack-ice zone were characterized by very different phytoplankton assemblages, not only in quantitative terms but also in species composition. As regards spatial variations, the coastal areas appeared characterized by different species of diatoms, of which *Fragilariopsis curta* was dominant. This consideration is confirmed by the observations by Leventer and Dunbar (1996), who, analyzing samples from sediment traps positioned on the southwestern shelf of the Ross Sea, recorded two late-summer (February 1991 and 1992) blooms of *Fragilariopsis curta*.

Considering the physiological aspects of primary production, the photosynthetic capacity (integrated P/B ratio) in relation to surface and integrated irradiance levels seems to be independent from light level (Fig. 4a,b). The efficiency (P/B ratio normalized for irradiance) plotted against irradiance shows that Antarctic phytoplankton communities are adapted to low light levels.

Picophytoplankton showed the highest efficiency. In contrast with our observations on the efficiency of small size fractions, Jochem et al. (1995), on the base of information obtained during the austral spring in the Circumpolar Subantarctic Current and in the Weddell Gyre, consider the picoplankton fraction as the least efficient and explain the predominance of picophytoplankton in hydrographically unstable areas as the effect of its ability to efficiently utilize low light levels. They also maintain that this same fraction is also capable of growing in well-stabilized areas and therefore characterized by high irradiance regimes.

Our findings regarding the high efficiency of the picophytoplankton fraction seem supported by the high values of the P/E relationship within the fluorescence subsurface maximum at stations 13C and 14B, where the picophytoplankton fractions account for 69 and 64%, respectively (Table 1). The higher photosynthetic efficiency of picophytoplankton during the austral spring is confirmed by Savidge et al. (1995), who carried out specific measurements of P/E relationship for the different size fractions in the Bellingshausen Sea and by Ning et al. (1996), for the austral summer in the Atlantic and Indian sectors of the Southern Ocean. The photosynthetic capacity plotted against efficiency produces an interesting scattergram (Fig. 6). In fact, the photosynthetic capacity increases in direct relationship with increasing efficiency, up to values of 0.02 mg C (mg Chl a)$^{-1}$ μE $m^{-2} s^{-1} h^{-1}$, corresponding to about 20 μE $m^{-2} s^{-1}$, while it decreases at lower light levels.

Phytoplankton communities of the Ross Sea seem to have during spring an optimal interval of irradiance between 15–20 and 150–200 μE $m^{-2} s^{-1}$, showing low efficiences and photosynthetic capacities at high light levels and very high efficiencies and poor photosynthetic capacities at low light levels, phenomena which are indicative of photoinhibition and photolimitation, respectively. In this context, photosaturation occurs at an irradiance of about 20 μE $m^{-2} s^{-1}$ in spring; below this irradiance, the efficiency is very high, while the photosynthetic capacity decreases proportionally.

This behaviour of Antarctic phytoplankton with respect to light is confirmed by the photosynthetic parameters measured at the deep fluorescence maximum at four stations (Table 1). In fact, in three of the four stations considered, the photoadaptation index ($E_k$) ranged between 23 and 24 μE $m^{-2} s^{-1}$, actually very near to the value of 20 cited above; the irradiance at which photosynthesis is maximal ($E_m$) never exceeded 134 μE $m^{-2}$ $s^{-1}$. In addition, it must be noted that all samples showed a strong photoinhibition (β), particularly at station 30, which we interpret as the result of the past light history of the diatom dominated assemblage, recently released in ice-free waters and, thus, exposed to relatively much higher light levels. In contrast, at station 28 the observed lower photoinhibition and higher photoadaptation index are, in our opinion, due mostly to the strong dominance of the palmelloid form of *Phaeocystis* cf. *antarctica*.

A number of authors (Neori and Holm-Hansen 1982; Tilzer et al. 1986; Tilzer and Dubinsky 1987), mostly on the basis of in-vitro experiments, have emphasized the role of temperature on production processes in Antarctic phytoplankton. Our emphasis on light as the main factor influencing such processes is based on the fact that, during spring, there seems to be no significant correlation between photosynthetic capacity and efficiency with temperature or density, as shown by Fig. 7. On the other hand, the information on photosynthetic parameters reported above is in good agreement with that collected from four depths at 19 stations, during a summer cruise carried out in the same area in 1996, when the mean values of $E_k$ and Em were 24 (STD 7) and 121 (STD 29) $\mu E\,m^{-2}\,s^{-1}$, respectively (unpubl. data), and supported by the values reported in the literature both for algal cultures and for natural phytoplankton populations from different Antarctic seasons and areas (Jacques 1983; Sakshaug and Holm-Hansen 1986; Holm-Hansen and Mitchell 1991; Brightman and Smith 1989; Figueras et al. 1994; Savidge et al., 1995; Boyd et al. 1995; Dower et al. 1996; Lazzara et al., this Vol.).

## Concluding Remarks

During the austral spring, the Ross Sea was characterized by blooms occurring in unstable hydrographic conditions, a feature present throughout the investigated area, so that water column stability appears as an optimal but not necessary condition for phytoplankton growth. The primary production in the Ross Sea was mainly based on nano- and microphytoplankton, while picophytoplankton became predominant when strong selective grazing drastically reduced the other fractions and did not appear correlated with the levels of primary production and biomass, as already shown by Detmer and Bathmann (1997). Di already Tullio and Smith (1996) and Nelson et al. (1996) stressed that the role of krill was negligible in the Ross Sea, while, in our opinion, krill could play during spring, a major role, in controlling production processes in the entire offshore area north of the 75° parallel.

From our data and the observations made by other authors, the Ross Sea was characterized by both spring and summer blooms, while the lowest production values were confined to the second half of January (Smith et al. 1996). A strong bloom

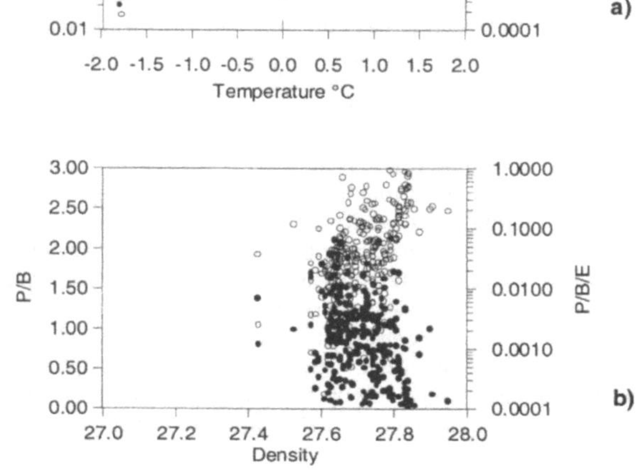


**Fig. 7a,b.** Efficiency (○) and photosynthetic capacity (●) against **a** temperature and **b** density

observed at Terra Nova Bay during the first half of February (Saggiomo, unpubl. data) was not likely to be a local phenomenon but, also on the basis of the observations of Leventer and Dunbar (1996), may represent a generalized feature for all the southwestern part of the Ross Sea. Our observations indicate an annual production for the Ross Sea even higher than the levels calculated by Nelson et al. (1996) and Arrigo and McClain (1994) and as much as fourfold higher than the average for the Southern Ocean. Our measurements of the photosynthetic capacity showed a strong adaptation to low light intensities both in spring and summer and in stable areas where the mixed layer exceeded the euphotic zone. This might explain the poor correlation of the P/B ratio in the euphotic zone with both incident and integrated light. The saturated light level equaled the average irradiance in the mixed layer, an observation which agrees with that of Figueras et al. (1994).

The spatial heterogeneity of the floristic composition of phytoplankton and its primary production levels in the Ross Sea could usefully be framed within by the hypothesis presented by Di Tullio and Smith (1996) implying the key role of micronutrients (i.e. Fe and Mn) in conditioning the dynamics of primary production in the pelagic system this area.

*Acknowledgements.* This work was carried out in the framework of the Italian National Programme for Antarctica (PNRA), whose support in the field and laboratory activities are gratefully acknowledged. We thank Massimo Azzali for kindly providing data from the acoustic survey of krill aggregations and Nello Russo for information on hydrographic parameters.

## References

Arrigo KR, McClain CR (1994) Spring phytoplankton production in the Western Ross Sea. Science 266: 261–263

Bathmann UV, Scharek R, Klaas C, Dubischar CD, Smetacek V (1997) Spring development of phytoplankton biomass and composition in major water masses of the Atlantic sector of the Southern Ocean. Deep Sea Res II 44(1–2): 51–67

Behrenfeld MJ, Bale AJ, Kolber ZS, Aiken J, Falkowski PG (1996) Confirmation of iron limitation of phytoplankton photosynthesis in the equatorial Pacific Ocean. Nature 383: 508–511

Boyd PW, Robinson C, Savidge G, Williams PJL (1995) Water column and sea-ice primary production during austral spring in the Bellingshausen Sea. J Plankton Res 42(4–5): 1177–1200

Brightman RI, Smith WO Jr (1989) Photosynthesis-irradiance relationships of Antarctic phytoplankton during austral winter. Mar Ecol Prog Ser 53: 143–151

Comiso JC, McClain CR, Sullivan CW, Ryan JP, Leonard CL (1993) Coastal zone colour scanner pigment concentrations in the Southern Ocean and relationships to geophysical surface features. J Geophys Res 98: 5419–2451

De Baar HJW, de Jong JTM, Bakker DCE, Loscher BM, Veth C, Bathmann U, Smetacek V (1995) Importance of iron for plankton blooms and carbon dioxide draw down in the Southern Ocean. Nature 373: 412–415

DeMaster DJ, Dunbar RB, Gordon LI, Leventer AR, Morrison JM, Nelson DM, Nittrouer CA, Smith WO Jr (1992) Cycling and accumulation of biogenic silica and organic matter in high-latitude environments: the Ross Sea. Oceanography 5(3): 146–153

Detmer AE, Bathmann UV (1997) Distribution patterns of autotrophic pico- and nanoplankton and their relative contribution to algal biomass during spring in the Atlantic sector of the Southern Ocean. Deep Sea Res II 44(1–2): 299–320

Di Tullio GR, Smith WO Jr (1996) Spatial patterns in phytoplankton biomass and pigment distributions in the Ross Sea. J Geophys Res 101(8): 18467–18477

Dower KM, Lucas MI, Phillips R, Dieckmann G, Robinson DH (1996) Phytoplankton biomass, P-I relationships and primary production in the Weddell Sea, Antarctica, during the austral autumn. Polar Biol 16: 41–52

Dunbar RB, Anderson JB, Domack EW, Jacobs SS (1985) Oceanographic influences on sedimentation along the Antarctic continental shelf. Antarct Res Ser 43: 291–313

El-Sayed SZ, Biggs DC, Holm-Hansen O (1983) Phytoplankton standing crop, primary productivity, and near-surface nitrogenous nutrient fields in the Ross Sea, Antarctica. Deep Sea Res 30(8A): 871–886

Figueiras FG, Pérez FF, Pazos Y, Rios AF (1994) Light and productivity of Antarctic phytoplankton during austral summer in an ice edge region in the Weddell-Scotia Sea. J Plankton Res 16: 233–253

Holm-Hansen O, Mitchell G (1991) Spatial and temporal distribution of phytoplankton and primary production in the western Bransfield Strait region. Deep Sea Res 38(8–9): 961–980

Innamorati M, Lazzara L, Massi L, Mori G, Saggiomo V (1992) Spatial and temporal distribution of phytoplankton size fractions in Antarctic waters: biomass and production, summer 1989–90. In: Ambrosetti W, Riccazzo M, Ruffoni Riva T (eds) Atti IX Congr Ass Ital Oceanogr Limnol, Lang, Genova, pp 605–612

Jacques G (1983) Some ecophysiological aspects of the Antarctic phytoplankton. Polar Biol 2: 23–33

Jacques G, Panouse M (1991) Biomass and composition of size fractionated phytoplankton in the Weddell-Scotia Confluence area. Polar Biol 11: 315–328

Jochem FJ, Mathot S, Quéguiner B (1995) Size-fractionated primary production in the open Southern Ocean in austral spring. Polar Biol 15: 381–392

Lancelot C, Mathot S, Veth C, de Baar H (1993) Factors controlling phytoplankton ice-edge blooms in the marginal ice-zone of northwest Weddell Sea during sea ice retreat 1988: field observations and mathematical modelling. Polar Biol 13: 377–387

Laubscher RK, Perissinotto R, McQuaid CD (1993) Phytoplankton production and biomass at frontal zones in the Atlantic sector of the Southern Ocean. Polar Biol 13: 471–481

Ledford-Hoffman PA, DeMaster DJ, Nittrouer CA (1996) Biogenic-silica accumulation in the Ross Sea and importance of Antarctic continental-shelf deposits in the marine silica budget. Geochim Cosmochim Acta 50(9): 2099–2110

Legendre L, Demers S, Yentsch CM, Yentsch CS (1983) The $^{14}C$ method: patterns of dark $CO_2$ fixation and DCMU correction to replace the dark bottle. Limnol Oceanogr 28: 996–1003

Leventer A, Dunbar RB (1996) Factors influencing the distribution of diatoms and other algae in the Ross Sea. J Geophys Res 101(8): 18489–18500

Magazzù G, Decembrini F (1992) I risultati di produzione primaria delle crociere oceanografiche 1987–88 e 1989–90 del Programma Nazionale di Ricerche in Antartide. In: Gallardo VA, Ferretti O, Moyano HI (eds) Oceanografia in Antartide. ENEA - PNRA - EULA, Concepcion, Chile, pp 273–284

Marino D, Montresor M, Mazzella L, Saggiomo V (1995) Diatom flora in faecal pellets from Terra Nova Bay (Antarctica). In: Marino D, Montresor M (eds) Proc 13th Int Diatom Symp, Bio Press Ltd, Bristol, pp 229–240

Martin JH, Fitzwater SE, Gordon RM (1990) Iron deficiency limits phytoplankton growth in Antarctic waters. Global Biogeochem Cycles 4: 5–12

Mitchell BG, Holm-Hansen O (1991) Observations and modelling of the Antarctic phytoplankton crop in relation to mixing depth. Deep Sea Res A 38: 981–1007

Nelson DM, Smith WO Jr (1991) Sverdrup revisited: critical depth, maximum chlorophyll levels, and the control of Southern Ocean productivity by the irradiance-mixing regime. Limnol Oceanogr 36: 1650–1661

Nelson DM, DeMaster DJ, Dunbar RB, Smith WO Jr (1996) Cycling of organic carbon and biogenic silica in the Southern Ocean: estimates of water-column and sedimentary fluxes on the Ross Sea continental shelf. J Geophys Res 101(8): 18519–18532

Neori A, Holm-Hansen O (1982) Effect of temperature on rate of photosynthesis in Antarctic phytoplankton. Polar Biol 1: 33–38

Ning X, Liu Z, Genhai Z, Junxian S (1996) Size-fractionated biomass and productivity of phytoplankton and particulate organic carbon in southern Ocean. Polar Biol 16: 1–11

Nuccio C, Innamorati M, Lazzara L, Mori G (1992) Popolamenti fitoplanctonici nella baia di Terra Nova, Mare di Ross. In: Gallardo VA, Ferretti O, Moyano HI (eds) Oceanografia in Antartide. ENEA - PNRA - EULA, Concepcion, Chile, pp 253–262

Platt T, Gallegos CL, Harrison WG (1980) Photoinhibition of photosynthesis in natural assemblages of marine phytoplankton. J Mar Res 38: 687–701

Saggiomo V, Massi L, Modigh M, Innamorati M (1992) Size-fractionated primary production in Terra Nova Bay (Ross Sea) during austral summer (1989–90) In: Gallardo VA, Ferretti O, Moyano HI (eds) Oceanografia in Antartide. ENEA - PNRA - EULA, Concepcion, Chile, pp 289–294

Saggiomo V, Carrada GC, Mangoni O, Ribera d'Alcalà M, Russo A (1998) Spatial and temporal variability of size-fractionated biomass and primary production in the Ross Sea (Antarctica) during austral spring and summer. J Mar Syst 17: 98–115

Sakshaug E, Holm-Hansen O (1986) Photoadaptation in Antarctic phytoplankton: variations in growth rate, chemical composition and P versus I curves. J Plankton Res 8: 459–473

Savidge G, Harbour D, Gilpinand LC, Boyd PW (1995) Phytoplankton distributions and production in the Bellingshausen Sea, austral spring 1992. Deep Sea Res II 42(4–5): 1201–1224

Smetacek V, Scharek R, Nöthig EM (1990) Seasonal and regional variation in the pelagial and its relationship to the life history cycle of krill. In: Kerry R, Hempel G (eds) Antarctic ecosystem: ecological change and conservation. Springer, Berlin Heidelberg New York, pp 103–114

Smith WO, Gordon LI (1997) Hyperproductivity of the Ross Sea (Antarctica) polynya during austral sping. Geophys Res Lett 24: 233–236

Smith WO Jr, Nelson DM (1985) Phytoplankton bloom produced by a receding ice edge in the Ross Sea: spatial coherence with the density field. Science 227: 163–166

Smith WO Jr, Nelson DM (1986) Importance of ice edge phytoplankton production in the southern ocean. BioScience 36: 251–257

Smith WO Jr, Sakshaug E (1990) Polar phytoplankton. In: Smith WO Jr (ed) Polar oceanography, part A: physical science, part B: chemistry, biology, geology. Academic Press, San Diego, pp 477–526

Smith WO Jr, Nelson DM, Di Tullio GR, Leventer AR (1996) Temporal and spatial patterns in the Ross Sea: phytoplankton biomass, elemental composition, productivity and growth rates. J Geophys Res 101(8): 18455–18465

Stretch JJ, Hammer PP, Hammer WM, Michel WC, Cook J, Sullivan CW (1988) Foraging behaviour of Antarctic krill *Euphasia superba* on sea ice microalgae. Mar Ecol Prog Ser 44: 131–139

Sullivan CW, McClain CR, Comiso JC, Smith WO Jr (1988) Phytoplankton standing crops within Antarctic ice edge assessed by satellite remote sensing. J Geophys Res 93: 487–489

Sullivan CW, Arrigo KR, McClain CR, Comiso JC, Firestone J (1993) Distributions of phytoplankton blooms in the Southern Ocean. Science 262: 1832–1837

Tilzer MM, Dubinsky Z (1987) Effects of temperature and day length on the mass balance of Antarctic phytoplankton. Polar Biol 7: 35–42

Tilzer MM, Elbrachter M, Gieskes WW, Beese B (1986) Light-temperature interactions in the control of photosynthesis in Antarctic phytoplankton. Polar Biol 5: 105–111

Utermöhl H (1958) Zur Vervollkommnung der quantitativen Phytoplankton-Methodik. Mitt Int Ver Theor Angew Limnol 9: 1–38

Wefer G, Fischer G, Fuetterer D, Gersonde R (1989) Seasonal particle flux in the Bransfield Strait, Antarctica. Deep Sea Res A 35: 891–898

Wilson DL, Smith WO Jr, Nelson DM (1986) Phytoplankton bloom dynamics of the western Ross Sea ice edge. I. Primary productivity and species-specific production. Deep Sea Res 33(10): 1375–1387

# Photosynthetic Parameters, Irradiance, Biooptical Properties and Production Estimates in the Western Ross Sea

L. Lazzara[1], V. Saggiomo[2], M. Innamorati[1], O. Mangoni[3], L. Massi[1], G. Mori[1], and C. Nuccio[1]

**ABSTRACT**
The primary production characteristics of phytoplankton coenoses were investigated during three summer periods (1987–88, 1989–90 and 1994–95), in Terra Nova Bay. The basic photosynthetic parameters of the P vs. E curves were analyzed, as well as the in vivo spectral absorption and fluorescence properties, which allow the maximum quantum yield for carbon uptake and for PSII fluorescence to be appreciated and compared, in order to evaluate the physiological conditions and the photoacclimation performance of this phytoplankton. Low light acclimation features (elevated values of $\alpha^B$, $P^B_{max}$ and $\phi_m$, with $E_k$ between 11 and 102 $\mu E\,m^{-2}\,s^{-1}$) are clearly present in these natural populations. High daily productions of the water column have been estimated through a biooptical model (0.1–3.1 $g\,C\,m^{-2}\,day^{-1}$), and a seasonal integrated value of 125 $g\,C\,m^{-2}$ has been appreciated at Terra Nova Bay, for the period from December through February.

## Introduction

The western Ross Sea is being recognized, in the past few years, as a site of elevated primary production and extensive phytoplankton blooms (Smith et al. 1990, 1996; Innamorati et al. 1992a,b), mainly after the use of satellite visible imagery (Sullivan et al. 1993; Arrigo and McClain 1994; Arrigo and Weiss 1996). Nevertheless, a lack of studies on primary production distribution, biooptics and photophysiology of phytoplankton coenoses in the Southern Ocean and especially in the Ross Sea, is still evident. The biooptical models of primary production used for remote sensing applications (Platt and Sathyendranath 1988; Morel and Berthon 1989; Morel 1991; Antoine and Morel 1996) are essentially based at present on photosynthetic parameters obtained on tropical or temperate phytoplankton and still few biooptical data (Mitchell and Holm-Hansen 1991) have been directly measured on antarctic coenoses.

Collection of data concerning distribution of phytoplankton biomass and production, together with photosynthetic parameters and biooptical properties of natural populations, was performed throughout three austral summers (Innamorati et al. 1992a; Lazzara et al. 1997) in the western Ross Sea. Objectives of this study are: the local validation of a biooptical model of primary production and the eventual identification of photosynthetic and optical parameters specific for the western Ross Sea phytoplankton; the spatial and temporal extension of primary production estimates in this region with the consequent contribute to the characterization of the western Ross Sea ecology.

## Methods

### Sampling

Investigations have been carried out in a large area of the Ross Sea, diversified in the course of three campaigns, as detailed in Innamorati et al. (this Vol.; Fig. 1). In 1988 the sampling area covered a grid extending 40 × 50 miles, centred around latitude 75 °S and 165 °E longitude, while in 1989–90 and 1995, respectively, three (BTN, SMN, MERG) and two (BTN, TIB) stations were sampled in Terra Nova Bay, in order to follow the temporal trend of

[1] Dipartimento di Biologia Vegetale, Università di Firenze, Via P.A. Micheli 1, 50121 Florence, Italy
[2] Stazione Zoologica A. Dohrn, Naples 80121, Italy
[3] Dipartimento di Zoologia , Università di Napoli Federico II, Naples, Italy

phytoplankton biomass and production and the related environmental factors. In 1989–90 a cruise was carried out over a much larger area of the western Ross Sea between the latitudes of 60° and 75°S and the longitudes of 165° and 177°E. Moreover, during the 1995–96 season, in Terra Nova Bay, four stations were sampled (stations 32, 33, 36, 38) for primary production measurements, as detailed by Saggiomo et al. (this Vol.).

## Pigments

The concentrations of liposoluble pigments were determined after filtration of 2 to 3 $dm^3$ of seawater on Whatmann GF/F filters, and extraction in acetone 100% (diluted to 90% for measurements), by means of both spectrophotometric analysis (Kontron, Uvikon 930) for chlorophaeopigments (chlorophyll a and phaeopigments, hereinafter named Chl) using the specific absorption coefficient of Jeffrey and Humphrey (1975), following Lorenzen (1967) modified by Riemann (1978) for separate determination of chlorophyll a and phaeopigments and by means of spectrofluorometric analysis (Perkin Elmer, LS-5b) for chlorophyll a and phaeophytin a determination, with purified chlorophyll a (Sigma Chemical) as standard (cf. Lazzara et al. 1997, for further details).

## Irradiance and Biooptical Properties

Measurements of surface global irradiance were performed by means of solarimeters (Kipp & Zonen, CM5) and of PAR quantum meters (LI-Cor, LI-192SB) continuously recorded (chart recorder) or stored every 10 min (data logger) during the three seasons (1987–88, 1989–90 and 1994–95) at Terra Nova Bay and in the Western Ross Sea in 1989–90. Out of the period of the campaigns, daily irradiance has been measured by solarimeters (Kipp & Zonen) at the meteorological stations ARGOS ID n. 7351 and n. 7353 of the Antarctic Metereological Observatory (PNRA). Intercomparison between sensors used in the diffent years, during days of minimum atmospheric attenuation (clear sky), gives less than 1.5% differences in daily integrated irradiances. A clear-sky irradiance curve has been interpolated with these maxima which gives, with respect to the theoretical values outside the atmosphere, a miminum atmospheric attenuation of 24%, whereas the real attenuation including clouds effect, over the period from December through February, averaged over the 3 years, was of 39% (see Fig. 5). In-water vertical profiles of downwelling, upwelling and scalar quantum PAR irradiance were measured by means of a photoprobe (Innamorati et al. 1994b) equipped with cosine and spherical quantum meters (LI-Cor), depth meter and reference surface sensor, for the three expeditions, in both Terra Nova Bay and in the western Ross Sea. Spectral downwelling and upwelling irradiances have been measured by means of a spectroradiometer (LI-Cor, 1800-UW) for several stations at different depths during the three seasons. The photosynthetically available part of global irradiance was considered to be 43% of total solar radiation (Mobley 1994), the ratio used to convert irradiance from energetic to quantic units was $Q/E = 2.77 \times 10^{18}$ (Morel and Smith 1974). The average loss for surface reflectance was considered to be 7%, as calculated by Morel (1991) for the summer solstice at 75°N, and an average ratio of the downwelling to scalar irradiance was found to be, from all our measurements, as $E_d/E_0 = 0.70$, and was used to obtain scalar PAR just below the surface [$PAR_0$ ($0^-$)].

## Spectral Absorption of Phytoplankton

Spectral absorption of phytoplankton ($a_{ph}$) was derived from that of total particulate matter ($a_p$) after methanol extraction (Kishino et al. 1986). Seawater samples up to 2 $dm^3$ were filtered on Whatman GF/F filters previously soaked in filtered seawater. Absorption of wet filters was immediately measured by means of a LI-COR spectroradiometer (LI-1800UW-12) equipped with an integrating sphere (LI-1800-12S). Detritus absorption ($a_d$) was measured on the same filters after methanol extraction of pigments for 24 h at 4 °C. Before measurements the bleached filters were soaked again for 1 h in filtered seawater to allow the extraction (even if partial) of the water soluble pigments (Bricaud et al. 1995). Finally phytoplankton absorption was calculated subtracting $a_d$ from $a_p$. The spectra are blank-subtracted and corrected for the pathlength amplification (β factor), following Bricaud and Stramski (1990).

## Spectral Fluorescence and Quantum Yield of Photosynthesis

In vivo fluorescence was measured directly on suspensions, for nearly the same amount of samples as for absorption, using a spectrofluorometer (Perkin-Elmer LS-5b) equipped with a red-sensitive photomultiplier tube and a quantum correction device (rhodamine B) which allows fluorescence excitation spectra to be corrected up to 630 nm. The measurements were performed with excitation / emission at 440/684 nm, after addition of the photosynthesis inhibitor DCMU (final concentration 20 μM) and 30 s of exposure to saturating irradiance, so that maximal fluorescence ($F_{max}$) was measured. Spectral values were measured on samples concentrated by gentle filtration in dim light on a 0.22-μm Nuclepore filter, to record the entire spectra with sufficient accuracy, and the spectra were scaled to the signal measured at fixed λ (440 ex/684 em). For each sample the signal of the filtered suspension was subtracted to avoid interference of Raman scattering (further details and discussion on the whole procedure are given in Lazzara et al. 1996). The spectral in vivo fluorescence yield of phytoplankton in relative values was calculated as the ratio of $F_{max}(\lambda)$ to $a_{ph}(\lambda)$, for the spectral domain 380–630 nm. An index of the photosynthetic conversion efficiency was also calculated, as FRI = $(F_{max} - F_o)/F_{max}$, being $F_o$ the fluorescence measured before the DCMU addition.

The maximum quantum yield of photosynthesis ($\phi_m$) was estimated from the ratio:

$$\phi_m = P_{max}/AQ, \tag{1}$$

where $P_{max}$ (mol C m$^{-3}$ day$^{-1}$) is the maximal production in the light limited portion of the PvsE curve, and AQ (Morel et al. 1987) is the total amount of absorbed quanta estimated from:

$$AQ = \int_{400}^{700} E(\lambda) a_{ph}(\lambda) d\lambda \tag{2}$$

and expressed in mol quanta m$^{-3}$ day$^{-1}$, with $\int E(\lambda)$ being equivalent to the irradiance at the onset of saturation ($E_k$).

## Primary Production

Primary production processes were investigated in the Ross Sea, in summer 1990, 1995 and 1996. A temporal series (from January 12 to February 8, 1990), with sampling intervals of about 4 days, was performed at stations MERG and SMN in Terra Nova Bay. The measurement of primary production rates were performed on samples incubated in situ at 0, 5, 10, 25 m, as well as at the level of the deep fluorescence maximum, for 4 h, around noon. A second set of measurements was carried out during the austral summer 1995 and 1996 (February 2 to 10). Stations TIB, 33, 36 and 38 were located in the same area of 1990 while station 32 was located in the southermost part of Terra Nova Bay. At each station, profiles of down- and upwelling underwater PAR were obtained using the above-mentioned photoprobe. Sampling depths were selected according to the optical levels (50, 20, 10, 5, 1 and 0.13% of incident irradiance) and a surface sample was also included. The measurement was performed by in situ-simulated conditions. Each sample collected from surface and the optical depths listed above was enclosed in light and dark polycarbonate (Nalgene) 450-ml bottles and incubated with 1 ml (20 μCi) of $NaH^{14}CO_3$ for 4–6 h on deck, in running surface seawater, at the different light levels obtained by means of neutral light screens (Electroformed Nickel screens, Veco Int. Co. USA). The radioactive content was measured, after acidification, within 24 h from filtration, in a Beckman LS 1801 liquid scintillator, using 10 ml of Aquasol II scintillation cocktail. In all stations sampled during February 1996, the P vs. E relationship was measured on four depths using a radial photosynthetron as in Babin et al. (1994). The photosynthetic parameters were calculated according to Platt et al. (1980).

## Depth-Integrated Biomass

The depth-integrated values of biomass for the euphotic zone were obtained, through trapezoidal integration, from chlorophyll profiles and underwater PAR measurements. From the underwater PAR profiles carried out during the four cruises, a good relationship was established (Fig. 1a; r = 0.91, N = 54) between $z_{eu}$, the depth of the euphotic zone (1% of surface PAR) and its chlorophaeopigment content $Chl_{ze}$, described by the regression $Z_{eu} = 34\ Chl_{ze}^{-0.502}$. Thus, when the depth of the euphotic zone had not been directly measured, it was estimated through the above-mentioned relationship and an iterative procedure. This started from an estimate of $z_{eu}$, based on surface values of Chl, which gives a first integrated value of Chl, and consequently a second estimate of $z_{eu}$ and so on, thus reaching convergence towards a final stable value of $z_{eu}$ and of its depth integrated value of chlorophaeopigments ($Chl_{tot}$).

The relationship between the mean concentration of chlorophaeopigments in a layer of one penetration depth ($Chl_{pd}$) and $Chl_{tot}$ is described by the regression: $Chl_{tot} = 52.4\, C_{pd}^{0.502}$ ($r = 0.91$, Fig. 1b). This relationship is slightly different from that described by Morel and Berthon (1989) for high latitude waters and uniform profiles. In fact, our samples are, in most cases, from stratified conditions (see Fig. 6, in Innamorati et al., this Vol.), where the biomass has a subsurface maximum in the euphotic layer and only a few samples are from well mixed waters, as in the previously quoted work. Therefore, the value of $Chl_{tot}$ has been estimated through the previous regression, when only surface samples had been collected.

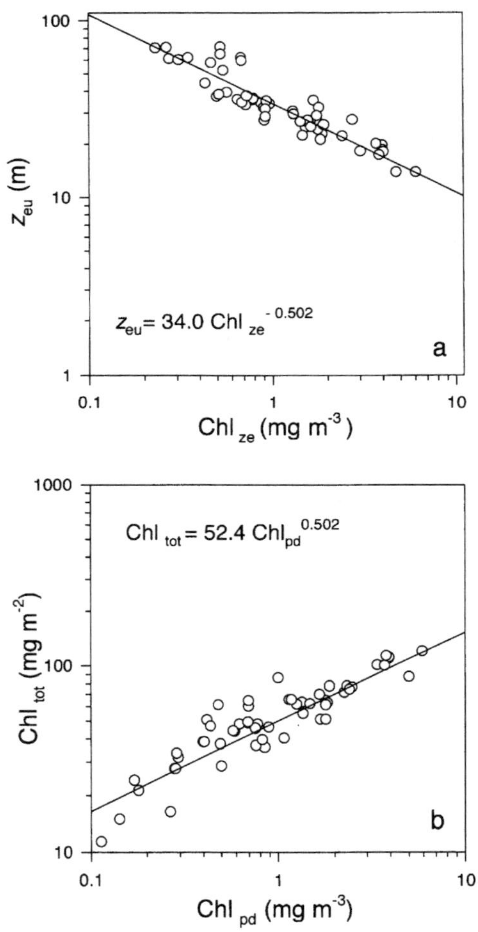


**Fig. 1. a** Relationship between depth of the euphotic zone ($z_{eu}$) and its chlorophaeopigments mean concentration ($Chl_{ze}$). **b** Relationship between $Chl_{tot}$ (integrated Chl content in the euphotic layer) and Chl concentration in the layer of one penetration depth ($Chl_{pd}$). The regression equations are given

To extend the temporal data set at Terra Nova Bay down to late spring, the phytoplankton cell density has been considered and the chlorophyll content of surface cells examined; it was found to undergo a relatively small change during the summer season, passing from 2 to 6 pg Chl/cell. More precisely, this ratio was remarkably stable around 2 pg/cell, from the end of December throughout January, whereas an increase was observed from the beginning of February. Considering that the phytoplankton community dominated by diatoms had not changed significantly during the first phase of the bloom, we could reasonably extend the value typical of the top of the bloom (2 pg Chl/cell) up to its beginning, on December 10 and calculate, from the cell density, the surface values of chlorophaeopigments ($Chl_{pd}$).

## Integrated Production Estimates

The depth-integrated values of primary production were obtained applying the biooptical model of Morel (1991; Antoine and Morel 1996), which makes use of the daily surface irradiance and of the chlorophyll content of the euphotic layer. The daily carbon production (PP, in $g\,C\,m^{-2}\,day^{-1}$) was computed from the following equation:

$$PP = (1/39)\rho\ Chl_{tot}\ PAR(0^+)\Psi^*, \qquad (3)$$

where 39 corresponds to the energy in kJ stored in 1 g of organic carbon, ρ is the average ratio of chlorophyll a / Chl, $Chl_{tot}$ is here the depth-integrated Chl content ($g\,Chl\,m^{-2}$) up to 0.1% of surface PAR, PAR ($0^+$) is the daily PAR irradiance impinging on the sea surface in energetic units ($MJ\,m^{-2}\,day^{-1}$), $\Psi^*$ is the column photosynthetic efficiency or cross-section for photosynthesis ($m^2\,g\,Chl^{-1}$).

The ratio of chlorophyll a / Chl generally showed no significant variation inside the euphotic zone and, when separately studied for each campaign, it was found to be sufficiently constant in Terra Nova Bay, for the 1987–88 campaign ($\rho = 0.88 \pm 0.05$) and for the 1989–90 campaign in the western Ross Sea ($\rho = 0.85 \pm 0.06$), with the only exception of much lower values ($0.45 < \rho <$

0.6) for three stations of each campaign. Therefore, these average values were used in Eq. (3) to compute daily primary production (PP). On the contrary, during the summer seasons of 1989–90 and 1994–95 in Terra Nova Bay, a regular variation was observed for ρ, progressing with the seasonal evolution of biomass from high values, during the bloom periods ($\rho = 0.90$) to minimum values ($0.4 < \rho < 0.5$) corresponding to periods of minimal biomass. For this reason, instead of an average ratio, as for 1987–88 and for the western Ross Sea, an interpolated time-dependent value of the ratio was applied in Terra Nova Bay during summer 1989/90 and 1994/95.

As far as concerns the photosynthetic cross-section, a reasonable mean value for the world ocean $\Psi'^{*} = 0.07\,m^2\,g\,Chl^{-1}$ (Morel 1978; Platt 1986) was used in the standard calculation (Figs. 7, 8) of the integrated primary production (PP) for this summer antarctic phytoplankton (see below for discussion).

## Results

The spectral absorption and fluorescence excitation properties of phytoplankton at Terra Nova Bay during 1989–90 and 1994–95 are well represented (Innamorati et al. 1994a; Lazzara et al. 1997, for a more detailed examination), by the sample shown in Fig. 2a (station TIB on February 4 1995) and used for production measurements. These spectra are typical of a diatom dominated phytoplankton community, as it was generally observed at Terra Nova Bay during summer in the euphotic layer (see Nuccio et al., this Vol.), and reveal the presence of chlorophyll a, chlorophyll c and fucoxanthin-like pigments. In Fig. 2b the fluorescence yield spectrum, with its nearly flat shape, accounts for a negligible presence of non-photosynthetic pigments in this phytoplankton, being so favourable to the simplifying assumption of coincidence between absorption and action spectrum for photosynthesis, adopted in the standard version of the biooptical model for integrated production (Antoine and Morel 1996).

The $P^B$ vs. irradiance relationship explored for a time series carried out in austral summer 1990 at Terra Nova Bay shows two different trends. The first (Fig. 3a) is related to the late phase of a spring bloom characterized by a well-stabilized water column, high daily irradiance, low nutrient concentrations and an intense grazing activity

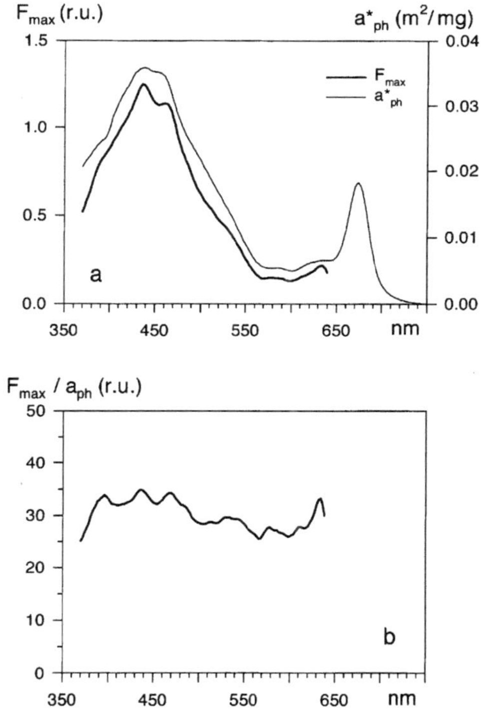


**Fig. 2.** **a** Absorption and fluorescence excitation spectra of phytoplankton at station TIB on February 4 1995. **b** Fluorescence yield spectrum ($F_{max}/a_{ph}$) of the above sample

(Marino et al. 1995). The second (Fig. 3b) coincides with the probable beginning of a late summer bloom; in this period the mixed layer was deeper and nutrient concentrations were higher before a strong storm occurred at the end of January. The lower $P^B_{max}$ values are related, in our opinion, to the observed differences in water column structure and to the reduced daily irradiance. The $P^B$ vs. E relationship measured in Terra Nova Bay during February 1996, represents the time evolution of photosynthetic parameters during a late summer bloom, when an integrated primary production up to $2.4\,g\,C\,m^{-2}\,day^{-1}$ was measured (Mangoni et al. 1999). The mean $P^B$ vs. E curves related to four depths for each station are reported in Fig. 3c,d,e, and f and show a quite similar trend. As for the offshore areas of Ross Sea, as well as other Antarctic areas in austral summer, phytoplankton commu-

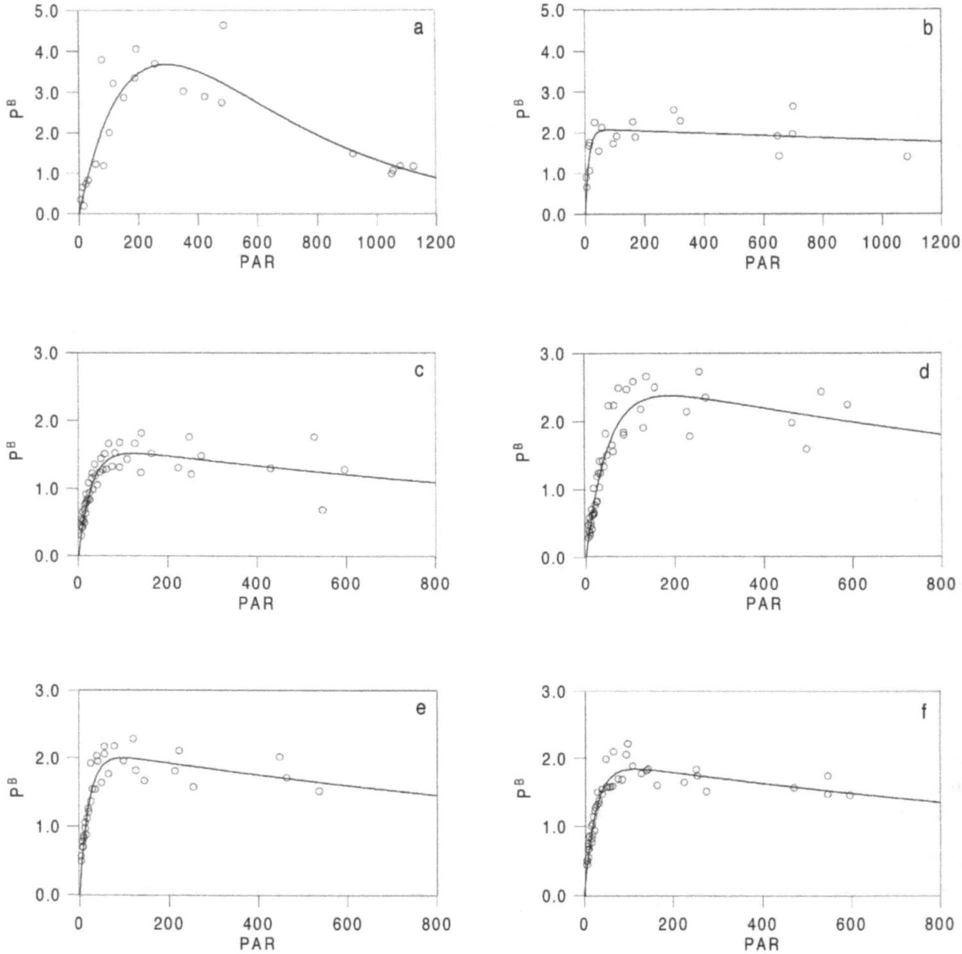


**Fig. 3a–f.** Time series of primary production normalized for chlorophyll *a* ($P^B$) *vs.* PAR during January (**a**) and February 1990 (**b**). $P^B$ *vs.* E curves measured at stations 32 (**c**), 33 (**d**), 36 (**e**) and 38 (**f**) during February 1996; the mean curves were obtained from experiments carried out at four depths for each station

nities were well adapted to low light level (Saggiomo et al., this Vol.; Holm-Hansen and Mitchell 1991; Figueiras et al. 1994). The photosynthetic efficiency, $\alpha^B$, is very high but lies in the same range of that reported for the Ross Sea and other regions in austral summer from the same authors. In contrast, $P^B_{max}$ of the 1990 time series and of February 1996 are from two to three times higher than that reported for Bransfield Strait region by Holm-Hansen and Mitchell (1991), but similar to that observed for the ice edge of Weddel Scotia Sea by Figueiras et al. (1994) and among the highest reported for the offshore Ross Sea by Saggiomo et al. (this Vol.).

The biooptical properties and the classical photosynthetic parameters describing the Pvs.E relationship for the phytoplankton coenoses of Terra Nova Bay are shown in Table 1. The maximum productivity ($P^B_{max}$) is always very high, up to 2.8 times of what would be expected at the

**Table 1.** Basic photosynthetic parameters from the $P^B$ *vs.* E relationship and biooptical properties of summer phytoplankton coenoses at Terra Nova Bay

| | $P^B_{max}$ gCgChl$^{-1}$d$^{-1}$ | $\alpha^B$ (gC/gChl)(E/m$^2$s)$^{-1}$ | $E_k$ µEm$^{-2}$s$^{-1}$ | $E_{max}$ µEm$^{-2}$s$^{-1}$ | $a^*_{avg}$ m$^2$/gChl | $a^*_{max}$ m$^2$/gChl | $\phi_m$ mol/mol |
|---|---|---|---|---|---|---|---|
| '90 January | 3.67 | 9.8 | 102 | 287 | 22.8 | 49.6 | 0.051 |
| '90 February | 2.06 | 46.7 | 11 | 69 | | | |
| Terra Nova Bay 1990 | 3.22 | 11.9 | 75 | 239 | | | |
| TNB '94 st 30 January | 1.60 | 13.3 | 33 | 102 | (22.8) | | (0.061) |
| TNB st.TIB 4.2.95 | 2.30 | 35.0 | 19 | 67 | 24.1 | 51.4 | 0.110 |
| TNB '96 st 32 February | 1.54 ± 0.2 | 16 ± 2.8 | 28 | 136 | (24.1) | | (0.059) |
| TNB '96 st 33 February | 2.38 ± 0.4 | 15.6 ± 3.1 | 45 | 180 | (24.1) | | (0.058) |
| TNB '96 st 36 February | 2.44 ± 0.7 | 31.6 ± 7.2 | 21 | 101 | (24.1) | | (0.096) |
| TNB '96 st 38 February | 1.84 ± 0.1 | 21.5 ± 1.9 | 24 | 114 | (24.1) | | (0.079) |
| Claustre et al. '97 Cryp | 1.40 | 7–9 | 50–53 | | 14.2 | | 0.091 |
| Claustre et al. '97 Diat | 3.80 | 12–14 | 76–86 | | 15.3 | | 0.112 |
| Model 0 °C[a] | 1.30 | 6.4 | 50 | | 16.0 | 35.5 | 0.070 |
| Model 20 °C[a] | 4.60 | 6.4 | 180 | | 16.0 | 35.5 | 0.070 |

[a] From Morel (1991) and Antoine and Morel (1996).

temperature of 0 °C, with a $Q_{10}$ = 1.88 (Eppley 1972), whereas the irradiance at the onset of saturation ($E_k$) is often very low (11–102 µE m$^{-2}$s$^{-1}$) and consequently the values of the photosynthetic efficiency ($\alpha^B$) are also generally high {10 – 47[gC(gChl)$^{-1}$ (Em$^{-2}$)$^{-1}$]}. The maximum quantum yield of photosynthesis ($\phi_m$) estimated from carbon production, spectral absorption and irradiance measurements, reaches rather high values for natural populations, in January ($\phi_m$ = 0.051) and close to the theoretical maximum in February ($\phi_m$ = 0.110). In 1994 and 1996 estimates of $\phi_m$ were made (values in parentheses) using the absorption values measured in the same months of 1990 and 1995, respectively and they lie in the same range (0.058–0.096 mol C/mol quanta). For a population of open water near the ice edge at McMurdo Sound, SooHoo et al. (1987) find quantum yields of photosynthesis ranging from 0.028 to 0.054 mol C (mol quanta)$^{-1}$, whereas Claustre et al. (1998) for two Antarctic coenoses dominated by diatoms and cryptophytes, respectively, report mean values for $\phi_m$, of 0.112 and 0.091 mol C (mol quanta)$^{-1}$.

The phytoplankton specific absorption coefficients from the whole data set, show spectral averages, $a^*_{avg}$ (400–700 nm), ranging from 17.7 to 26.9 m$^2$ g Chl$^{-1}$ and maxima, $a^*_{max}$, in the range 37.4 – 57.8 m$^2$ g Chl$^{-1}$, although just the absorption spectra contemporary with the $^{14}$C measurements are reported in Table 1. These absorption data are of a typical summer phytoplankton assemblage of the Western Ross Sea dominated by the diatom *Fragilariopsis*, as frequently observed in Terra Nova Bay (Nuccio et al., this Vol.). At McMurdo Sound SooHoo et al. (1987, see their Fig. 6 and Table II) with a population dominated by the prymnesiophyte *Phaeocystis pouchetii* find a mean specific absorption coefficient (irradiance-weighted and spectral average), °a, ranging from 15 to 29 m$^2$ g Chl$^{-1}$. These °a values are close to those measured at Terra Nova Bay and, under the realistic hypothesis of a similar spectral distribution of underwater irradiance, owing to the same Chl concentration and depth, the $a^*_{avg}$ should be similar as well. Claustre et al. (1998), on the basis of the pigments concentrations and of their specific "in vivo" absorption coefficients, reconstructed the absorption spectra of antarctic coenoses dominated by diatoms and cryptophytes and find that the irradiance-weighted coefficients °a are 9.7 and 7.1 m$^2$ g Chl$^{-1}$ respectively (Table 1) and the mean values of $a^*_{avg}$ are 15.3 and 14.2 m$^2$ g Chl$^{-1}$ for the two coenoses (H. Claustre, pers. comm.). Unfortunately, the low absorption coefficients so estimated are not completely comparable with those obtained by the filter-pad method. Mitchell and Holm-Hansen (1991) estimated the specific absorption of phytoplankton of Antarctic Peninsula at high Chl concentration (0.5 ≤ Chl ≤ 50 mg m$^{-3}$) using the particulate absorption normalized by the chlorophyll content. These spectra show high blue absorption with respect to the red, so that a detritus contribution, even if low, has to be supposed. In spite of this, the values of these specific particulate spectra are very low, with an average maximum value at 435 nm of 18 $m^2 g$ $Chl^{-1}$. The authors attribute this low specific absorption to a high package-effect "typical of larger cells with high cellular pigment concentrations expected for chronic low light adaptation in high nutrient waters". On the contrary, at Terra Nova Bay small cells (Nuccio et al., this Vol.) with a relatively low Chl content let think of a weak package effect for this phytoplankton. Also measurements of the spectral irradiance attenuation coefficient $k(\lambda)$, and of the chlorophyll-specific attenuation $k_c(\lambda)$, [where $k_c(\lambda) = (k(\lambda) - k_w(\lambda)]$ / Chl, with $k_w(\lambda)$ attenuation coefficient of pure water) were carried out at Terra Nova Bay and in the western Ross Sea during the three expeditions. These $k_c(\lambda)$ values, in accordance with the $k_c$ definition and its properties for green waters of the case I, are slightly higher than $a^*_{ph}$ [average (±s.d.) $k_c(440) = 88\, g\, Chl\, m^{-2}$ (±20, with n = 35)], and so confirm the reliability of this phytoplankton specific absorption, measured through the filter-pad method.

The standard model of Morel (1991) for integrated production in cold waters (0–1 °C) at high latitudes, implies lower values of $P^B_{max}$ and $E_k$, which in turn implies low values of the photosynthetic cross section ($\Psi^*$). An average value for $\Psi^*$ at 74 °S, between December and February, can be estimated as $\Psi^* = 0.04\, m^2 g\, Chl^{-1}$ for clear-sky conditions (Morel 1991; Antoine and Morel 1996) and has been used in a first application of the production model [Eq. (3)]. A comparison between the integrated primary production obtained from this procedure and the values resulting from $^{14}C$ production measurements gives the orthogonal regression: $PP(^{14}C) = 0.11 + 1.71\, PP(mod)$ ($r = 0.81$, $N = 17$, $p < 0.001$) which suggests a general underestimate (up to 50%) for the modelled values of integrated production. The simple regression of the integrated production from $^{14}C$ measurements with the product of $PAR(0^+)^* Chl_{tot}$, [see Eq. (3)] as shown in Fig. 4, after transformation of the production units in their energetic equivalent ($MJ\, m^{-2} day^{-1}$) allows estimating with our data, from the angular coefficient of the regression, an average value for the photosynthetic cross section of 0.068 ($m^2 g\, Chl^{-1}$), which is very close to the world ocean mean value (e.g. $0.07\, m^2 g\, Chl^{-1}$ in Morel 1991). Claustre et al. (1998), with an extensive data set of primary production of antarctic phytoplankton, also found for $\Psi^*$ the average value of $0.069\, m^2 g\, Chl^{-1}$, with maximal mean values of 0.114, when coenoses were dominated by diatoms and descending to 0.05, when dominated by flagellates or cryptophytes. As a consequence of the high values of $P^B_{max}$ and $\alpha^B$ which characterize this phytoplankton, also high values of the light utilization index $\Psi^*$ can be expected (Platt 1986) especially in light-limited conditions. The use of higher values of $\Psi^*$, like the average value of $0.07\, m^2 g\, Chl^{-1}$, is thus justified with this summer Antarctic phytoplankton, at least as a first approximation.

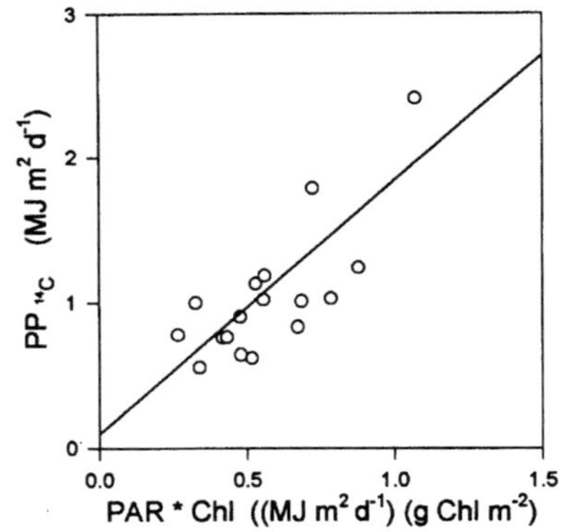


**Fig. 4.** Regression between integrated production from $^{14}C$ measurements (PP-$^{14}C$) and the product PAR ($0^+$) × $Chl_{tot}$, (PP-mod); the slope of the regression gives an estimate of the photosynthetic cross-section $\Psi^*$

However, a seasonal trend for $\Psi^*$, inverse to the irradiance availability with a minimum around the solstice, is predicted by the model of Morel (1991) and evident from the data of Claustre et al. (1998). It also appears from data at Terra Nova Bay, with an average value of $\Psi^* = 0.06$ (±0.01) $m^2 g\, Chl^{-1}$ in January and of 0.09 (±0.02) in February. Therefore, the integrated production (PP) has been calculated also with variable values of $\Psi^*$, obtained as a function of daily PAR (D. Antoine, pers. comm.) and enhanced so that their average was made equal to 0.068. The results in terms of modelled productions are quite similar, as averages, to the more rough estimates obtained with a constant value of $\Psi^*$, but their range of variation is slightly reduced (e.g. in 1989–90, instead of 0.35 – 3.14 it becomes 0.38 – 2.6 $g\, C\, m^{-2} day^{-1}$).

The temporal evolution of daily integrated irradiances at Terra Nova Bay during three summer seasons is shown in Fig. 5 and compared with the curve of maximum irradiance empirically determined (see methods) and corresponding to clear-sky conditions. Values range from a maximum of $36\,MJ\,m^{-2}\,day^{-1}$ (equivalent to $70\,E\,m^{-2}\,day^{-1}$ of PAR) near the solstice of December 1987, to minima below $5\,MJ\,m^{-2}\,day^{-1}$ in the second half of February. Generally high values have been measured with respect to clear-sky conditions (on average 80% for the three seasons), with fluctuations in the time scale of weeks, reflecting the average period for occurrence of storms or intense cloud coverages. In fact, relatively short periods (3–4 days) of maximal irradiance at the sea surface were frequently present during each of the three seasons and, exceptionally, as during the late spring of 1987–88, also periods of 10 to 14 days occurred. During the first 40 days of the season 87–88, the highest relative irradiance reached the sea surface (on average 94.4% of maximum), whereas in February 1988 it was the minimal (62%) compared to the same period in 1990 (78%) and 1995 (85%).

The column-integrated values of chlorophaeopigments at Terra Nova Bay are shown in their temporal evolution in Fig. 6; they range between 10 and $170\,mg\,Chl\,m^{-2}$. The top of the summer bloom was clearly reached at the end of December in 1989–90, whereas it seems difficult to place the seasonal maximum for the integrated biomass

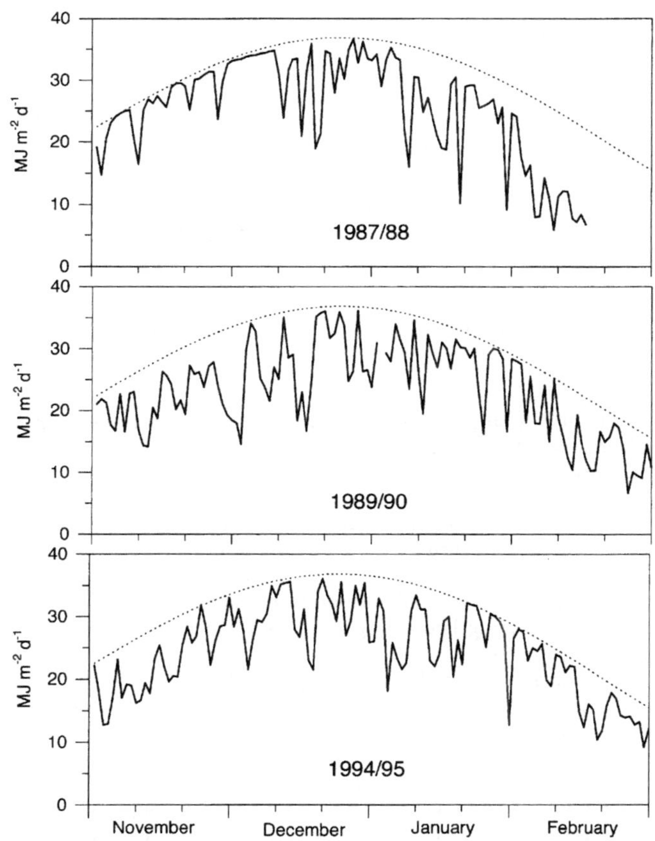


**Fig. 5.** Daily integrated irradiance at Terra Nova Bay during the years of investigations. The curve of maximum daily irradiance under clear-sky conditions (*dotted line*) is also reported

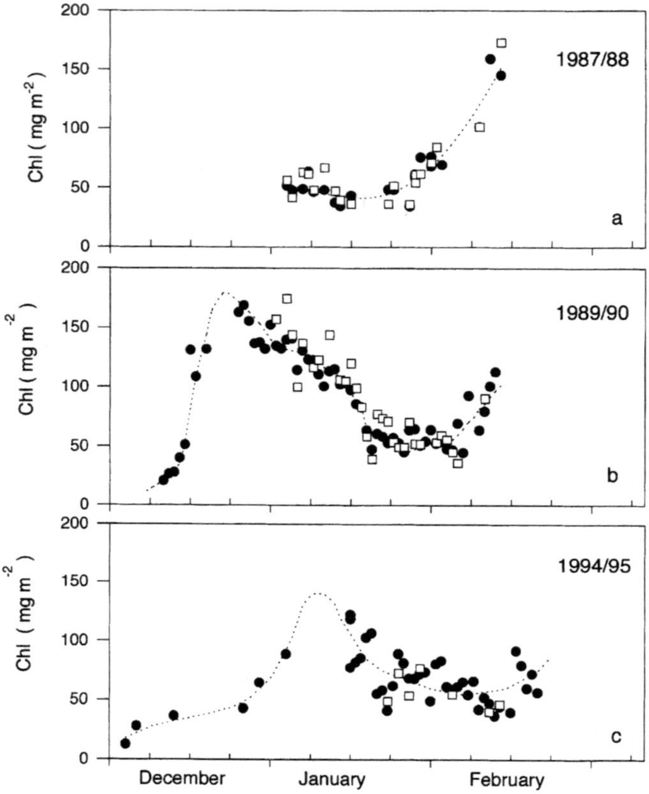


**Fig. 6a–c.** Temporal evolution of integrated chlorophaeopigments at Terra Nova Bay during the three summer seasons, as computed from $C_{pd}$ (*filled dots*) and from Chl integration up to 0.1% of surface PAR (*open squares*)

during 1987–88 and 1994–95, owing to the scarcity of sampling. On the other hand, it is more evident the position of the minimum biomass (less than 50 mg Chl $m^{-2}$) which occurs at mid January in 1987–88, at the end of January in 1989–90 and at the beginning of February in 1994–95. The presumably earlier season of 1987–88 can be related to the higher relative irradiance measured at the sea surface, during late spring 1987 (November and December, see Fig. 5).

The integrated production PP (g C $m^{-2}$ $day^{-1}$) calculated from the biooptical model, for the three seasons at Terra Nova Bay, is shown in Fig. 7. Values range from low to quite high daily productions (0.1 – 3.1 g C $m^{-2}$ $day^{-1}$) reaching the top with the bloom of December 1989. A similar seasonal pattern is evident for the three data series, as it was for biomass (Fig. 6) in spite of the lack of completeness in sampling the whole seasons. It has to be noted that the rise in biomass of late summer 1987–88 does not correspond to an increase in daily primary production (PP) owing to a double reason: the very low daily irradiance of that period, even less than 5 MJ $m^{-2}$ $day^{-1}$ in mid-February (Fig. 5), and the high proportion in phaeopigments which characterize the last three stations sampled in February 1988 (stations 49, 50, 51).

The integrated primary production values in the western Ross Sea during December 1989 (Fig. 8) exhibit more than a tenfold variation, from the lowest values of 0.21 g C $m^{-2}$ $day^{-1}$ (station 2) to the highest of 2.544 g C $m^{-2}$ $day^{-1}$ (station 22 along the 75°S transect). Three large areas show very high production (>1 g C $m^{-2}$ $day^{-1}$): the first at Terra Nova Bay (stations 25, 26 and 27), the second nearly at the same latitude (75°S) but at 173°E (about 220 n miles offshore, stations 20–22) and

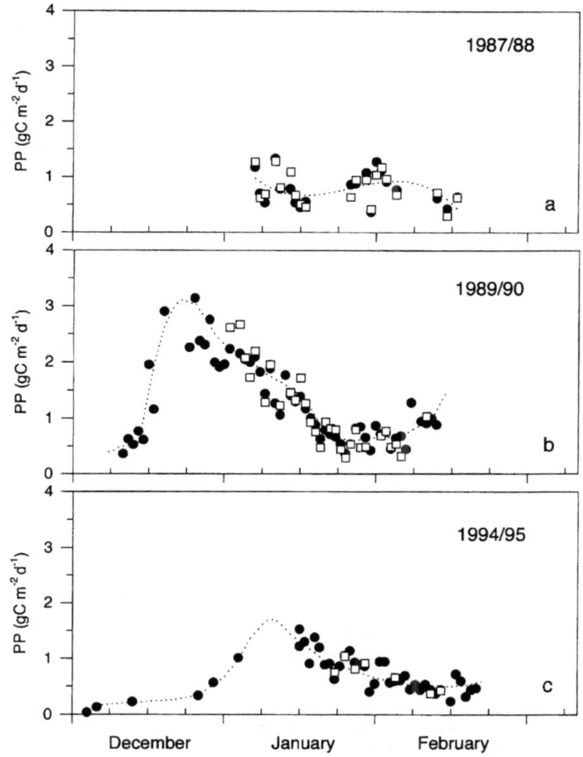


**Fig. 7a–c.** Temporal evolution of integrated production (PP) at Terra Nova Bay for the 3 years of investigations, as computed from $C_{pd}$ (*filled dots*) and from Chl integration up to 0.1% of surface PAR (*open squares*)

the third at the latitude of 65°S approaching the Balleny Islands in a wide region of seasonal ice melting (stations 5, 6). The lower values of daily production (<0.3 g C m$^{-2}$ day$^{-1}$) are observed to the north of the Antarctic convergence (62°S, station 2) and between 70° and 75°S in waters characterized by scarce biomass (less than 50 mg Chl m$^{-2}$), as was shown by Innamorati et al. (1992a).

## Discussion

The primary production of the three years can be compared through the average values of the daily production obtained during the whole period (Table 2), which ranges from 0.73 ± 0.21 g C m$^{-2}$ day$^{-1}$ in 1987–88 to 1.39 ± 0.79 g C m$^{-2}$ day$^{-1}$ in 1989–90. When limited to the period common to the three seasons (6 Jan through 10 Feb) the average productions do not differ so much, (0.79–0.98 g C m$^{-2}$ day$^{-1}$), despite the fact that the seasonal cycles appear somewhat shifted from one year to the other by approximatively 2 weeks.

These values of real and daily integrated productions appear high if compared to those even recently obtained in different antarctic waters, in the Weddell Sea (Dower et al. 1996) or in the Bellinghausen Sea (Savidge et al. 1995), but they are quite comparable to the direct measurements of primary production carried out in the whole Ross Sea. Wilson et al. (1986) obtained an average value in January and February 1983 of 0.962 g C m$^{-2}$ day$^{-1}$, Smith et al. (1990) for a transect at 72.5°S found an average value of 0.999 g C m$^{-2}$ day$^{-1}$ and of 1.37 at 76.5°S and more recently Smith et al. (1996) observed values ranging from 0.15 to 2.85 g C m$^{-2}$ day$^{-1}$, with the highest in January in the southernmost part of the Ross Sea.

An attempt to estimate the whole season production through a temporal integration (Table 2), for the months from December through February,

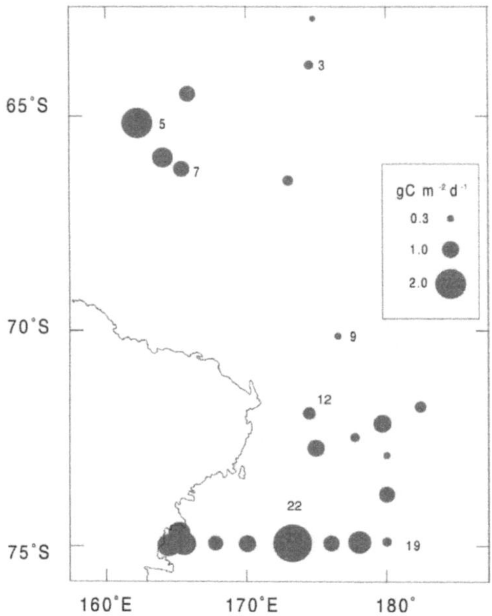


**Fig. 8.** Daily and column integrated primary production (PP) in the western Ross Sea during December 1989

**Table 2.** Time-integrated and average primary production (PP) at Terra Nova Bay for the 3 years of investigation. Data *in brackets* must be considered as underestimate of seasonal production

| | Whole period $gC\,m^{-2}$ day | Common period $gC\,m^{-2}$ day | Dec-Jan-Feb $gC\,m^{-2}$ |
|---|---|---|---|
| 1987–88 | $0.79 \pm 0.28$ | $0.79 \pm 0.28$ | (70.7) |
| 1989–90 | $1.39 \pm 0.79$ | $0.98 \pm 0.79$ | 125 |
| 1994–95 | $0.76 \pm 0.39$ | $0.76 \pm 0.39$ | (65.9) |

gives values of $125\,gC\,m^{-2}$ (9 weeks) in 1989–90, of 70.7 over 6 weeks in 1987–88 and of 65.9 in 1994–95, the last ones being probably biased by undersampling in the period of the bloom. A recent estimate of the Ross Sea production by CZCS imagery (Arrigo and McClain 1994) gives a value of $141\,gC\,m^{-2}$, between Dec. 10 and the Feb. 19 1979, in occasion of an intense and extensive bloom concerning the southern Ross Sea, which contrasts with the nearly ten fold lower values estimated for the entire Southern Ocean (Holm-Hansen et al. 1977; El-Sayed 1978) but also with values of 40 to $49\,gC\,m^{-2}year^{-1}$ appreciated in the Bellinghausen Sea (Boyd et al. 1995) or in the Weddell Sea (Smith and Nelson 1990). The high seasonal production of the Ross Sea seems now definitely assessed and has been recently confirmed by estimates of annual production from a several years study (Nelson et al. 1996), with values ranging from $216\,gC\,m^{-2}$ $year^{-1}$ in the southeastern part of the Ross Sea to $91\,gC\,m^{-2}year^{-1}$ in the northern part. The annual production of the southwestern part, where Terra Nova Bay is included, is $142\,gC\,m^{-2}year^{-1}$, but when limited to the period from December through February a value of approximatively 105 $gC\,m^{-2}$ can be appreciated, which falls in the range of seasonal production estimated at Terra Nova Bay (Table 2).

The seasonality of phytoplankton production at Terra Nova Bay is characterized by a main peak occurring, depending on the year, from the second half of December through the beginning of January and eventually a second late summer bloom in February. December is the period of the main phytoplanktonic bloom in the Ross Sea, as reported also by Arrigo and McClain (1994) and Nelson et al. (1996). During February, a second rise of production is less evident than for phytomass (Fig. 6, and also Innamorati et al. 1992b) but it must be pointed out that even in mid February daily productions higher than $1\,gC\,m^{-2}day^{-1}$ have been estimated. Similar, or higher, values of daily production during this month cannot be considered as exceptional in antarctic waters, having also been observed at Palmer station (Moline and Prezelin 1997) and in the Ross Sea (Nelson et al. 1996). In the latter case, the monthly average productions estimated do not show an evident second peak in February, probably owing to the interannual variability of this phenomenon.

As previously pointed out, the photosynthetic performance of this phytoplankton (see Table 1) reveals a general adaptation to low light regimes, as it is characterized by high values of the photosynthetic efficiency $\alpha^B$, by low values of the saturation irradiance $E_K$, high values of the maximum assimilation number $P^B_{max}$ and also by high values of the maximum quantum yield for photosynthesis, $\phi_m$. Similar values of these photosynthetic parameters have been measured in the Weddell Sea (Figueiras et al. 1994; Claustre et al. 1998); and they are, indeed, consistent with the higher value of the column photosynthetic cross-section $\Psi^*$ here observed (Fig. 4), which is suggested to be a general feature of Antarctic phytoplankton at Palmer Station (Claustre et al. 1998).

These features of phytoplankton production can be partially explained by a general low-light adaptation, but they could also be related to a low-temperature adaptation of the growth performance. In spite of low surface temperatures (−0.5 – 1.5 °C), the phytoplankton of Terra Nova Bay did reveal, from cell counts, quite high growth rates, at least during the initial phase of the bloom. Under the simplifying hypothesis of no losses due to grazing and sedimentation, a net growth rate of 0.8 doublings $day^{-1}$ is reached by surface phytoplankton during a period of 5 days in December 1989 and still of 0.2 doublings $day^{-1}$ at the beginning of the second bloom, in February 1990. Moreover, besides the high daily productions measured by $^{14}C$ assimilation (Mangoni et al. 1999) also high values of DCMU-induced variable fluorescence, with FRI higher than 0.70, were often observed at Terra Nova Bay, during the 1989–90 and 1994–95 expeditions.

Iron availability should not limit phytoplankton production in this area, as pointed out by Sullivan et al. (1993) owing to the important aeolian input of iron-rich dust from the coast to waters which are placed downstream of the dominant winds. Moreover, a 3-day incubation experiment on phytoplankton of Terra Nova Bay (Lazzara et al. 1997), with an iron-chelated addition (12 nM), led to generally high values of variable fluorescence and slightly lower (FRI = 0.44 ± 0.17) for the treated samples, than for the control (FRI = 0.61 ± 0.16).

The relationship between temperature and growth or productivity is obviously entailed by the temperature effects on the enzymatic activity of the Calvin cycle, but especially in the low-temperature domain, it is still debated whether and/or how much natural populations are limited in their growth by the depressed temperatures. Tilzer and Dubinsky (1987) pointed out that photosynthetic rates are less strongly affected by low temperatures than respiration rates are, so resulting in higher relative net productions for polar phytoplankton. Sullivan et al. (1993), using CZCS-satellite imagery, made the "counterintuitive conclusion that the highest phytoplankton crops occur in the coldest waters of the Southern Ocean". More recently, Beherenfeld and Falkowski (1997) obtained a statistical relationship between maximal productivity and sea surface temperature, which, used as base of a production model for satellite images, made it possible to appreciate an enhanced production for polar phytoplankton with respect to previous satellite estimates (Antoine et al. 1996; Longhurst et al. 1996).

The constant feature of a high production level for phytoplankton in the western Ross Sea is confirmed by the present study even at the seasonal time scale (3 months) and notwithstanding the presence of an interannual variability, which mainly concerns changes in the phase of the phytoplankton seasonal cycle (2 weeks, at least). Specific values of the photosynthetic parameters and of the biooptical properties seem to characterize the phytoplankton in the western Ross Sea, pointing out the need of further elaboration of specific regional algorithms to improve the estimate, through remote sensing, of phytoplankton biomass and production in this region.

*Acknowledgements.* We are grateful to P. Grigioni for providing data from the metereological stations AWS 7351 and 7753, to M. Nigro and to P. Sparapani for phytoplankton sampling at Terra Nova Bay during December 1994 and February 1995. H. Claustre and R. Casotti are acknowledged for comments and suggestions on the manuscript.

## References

Antoine D, Morel A (1996) Oceanic primary production. 1. Adaptation of a spectral light-photosynthesis model in view of application to satellite chlorophyll observations. Global Biogeochem Cycles 10: 43–55

Antoine D, André JM, Morel A (1996) Oceanic primary production. II. Estimation at global scale from satellite (Coastal Zone Color Scanner) chlorophyll. Global Biogeochem Cycles 10: 57–69

Arrigo KR, McClain CR (1994) Spring phytoplankton production in the Western Ross Sea. Science 266: 261–263

Arrigo KR, Weiss AM (1996) Interannual variation in pigment concentration and primary production in the Western Ross Sea as seen by the Nimbus 7 Coastal Zone Color Scanner. EOS Trans AGU, 76(3): 137

Babin M, Morel A, Gagnon R (1994) An incubator designed for extensive and sensitive measurements of phytoplankton photosynthetic parameters. Limnol Oceanogr 39: 694–702

Beherenfeld MJ, Falkowski PG (1997) Photosynthetic rates from satellite-based chlorophyll concentration. Limnol Oceanogr 42: 1–20

Boyd PW, Robinson C, Savidge G, Williams PJ (1995) Water column and sea-ice primary production during austral spring in the Bellingshausen Sea. Deep Sea Res 42: 1177–1200

Bricaud A, Stramski D (1990) Spectral absorption coefficients of living phytoplankton and nonalgal biogenous matter: a comparison between the Perù upwelling and the Sargasso Sea. Limnol Oceanogr 35: 562–582

Bricaud A, Babin M, Morel A, Claustre H (1995) Variability in the chlorophyll-specific absorption coefficients of natural phytoplankton: analysis and parametrization. J Geophys Res 100: 13321–13332

Claustre H, Moline MA, Prezelin B (1998) Sources of variability in the column photosynthetic cross section for Antarctic coastal waters. J Geophys Res (in press)

Dower KM, Lucas MI, Phillips R, Dieckmann G, Robinson DH (1996) Phytoplankton biomass, P-I relationships and primary production in the Weddell Sea, Antarctica, during the austral autumn. Polar Biol 16: 41–52

El-Sayed SZ (1978) Primary productivity and estimates of potential yields of the Southern Ocean. In: McWhinnie MA (ed) Polar research. Westview, Boulder, co, pp 141–160

Eppley RW (1972) Temperature and phytoplankton growth in the sea. Fish Bull 70: 1063–1085

Figueiras FG, Pérez FF, Pazos Y, Rios AF (1994) Light and productivity of Antarctic phytoplankton during austral summer in an ice edge region in the Weddell-Scotia Sea. J Plankton Res 16: 233–253

Holm-Hansen O, Mitchell G (1991) Spatial and temporal distribution of phytoplankton and primary production in the western Bransfield Strait region. Deep Sea Res 38: 961–980

Holm-Hansen O, El-Sayed SZ, Franceschini G, Cuhel R (1977) Primary production and factors controlling phytoplankton growth in the Southern Ocean. In: Llano GA (ed) Adaptations within Antarctic ecosystem. Smithsonian Institution Press, Washington, DC, pp 11–50

Innamorati M, Lazzara L, Massi L, Mori G, Nuccio C, Saggiomo V (1992a) Indagine sulla biomassa fitoplanctonica nel Mare di Ross in relazione ai fattori ambientali. In: Gallardo VA, Ferretti O, Moyano HI (eds) Oceanografia en Antartica. ENEA-PNRA-EULA, Concepcion, Chile, pp 235–252

Innamorati M, Lazzara L, Mori G, Nuccio C, Saggiomo V (1992b) Spatial and temporal distributions of phytplankton size fractions in antarctic waters: biomass and production, summer 1989/90. Atti IX Congr Ass Ital Oceanogr Limnol, pp 605–612

Innamorati M, Lazzara L, Massi L, Mori G, Nuccio C, Saggiomo V, Cartei P, Franchini L, Gamba C (1994a) Phytoplankton ecology: particles, gilvin, fluorescence spectra, irradiance and phytoplankton species density in the Ross Sea. Oceanographic Campaign 1989–90. In: Faranda F, Guglielmo L (eds) National Scientific Commission for Antarctica Oceanographic Campaign 1989–90, Data Rep Part III, Genova, pp 101–178

Innamorati M, Massi L, Lazzara L, Cartei P (1994b) Biomassa fitoplanctonica e proprietà ottiche del mare. Biol Mar Medit 1: 83–88

Jeffrey SW, Humphrey GF (1975) New spectrophotometric equations for determining chlorophylls a, b, c and c2 in algae, phytoplankton and higher plants. Biochem Physiol Plant 167: 191–194

Kishino M, Okami N, Takahashi M, Ichimura S (1986) Light utilization efficiency and quantum yield of phytoplankton in a thermally stratified sea. Limnol Oceanogr 31(3): 557–566

Lazzara L, Bricaud A, Claustre H (1996) Spectral absorption and fluorescence excitation properties of phytoplanktonic populations at a mesotrophic and an oligotrophic site in the tropical North Atlantic (EUMELI program). Deep Sea Res I, 43: 1215–1240

Lazzara L, Massi L, Nuccio C, Biondi N, Innamorati M (1997) Phytoplankton ecology: irradiance, particles, gilvin, pigments, absorption, fluorescence, production and species density in Terra Nova Bay, Ross Sea. In: Faranda F, Guglielmo L (eds) National Scientific Commission for Antarctica Oceanographic Campaign 1994–95, Data Rep Part I, Genova, pp 229–279

Longhurst A, Sathyendranath S, Platt T, Caverhill C (1996) An estimate of global primary production in the ocean from satellite radiometer data. J Plankton Res 17: 1245–1271

Lorenzen CJ (1967) Determination of chlorophyll and phaeo-pigments: spectrophotometric equations. Limnol Oceanogr 19: 591–600

Mangoni O, Carrada GC, Saggiomo V, Sgrosso S (1999) Relationship between environmental parameters and primary production during a late summer phytoplankton bloom in Terra Nova Bay (Ross Sea – Antartica). Atti XII Congr Ass Ital Oceanogr Limnol (in press)

Marino D, Montresor M, Mazzella L, Saggiomo V (1995) Diatom flora in faecal pellets from Terra Nova Bay (Antarctica). In: Marino D, Montresor M (eds) Proc 13th Int Diatom Symp Bio Press Ltd, Bristol, pp 229–240

Mitchell BG, Holm-Hansen O (1991) Bio-optical properties of Antarctic Peninsula waters: differentiation from temperate ocean models. Deep Sea Res 38: 1009–1028

Mobley CD (1994) Light and water. Radiative transfer in natural waters. Academic Press, San Diego, 592 pp

Moline MA, Prezelin B (1997) High-resolution time-series data for 1991/1992 primary production and related parameters at a Palmer LTER coastal site: implications for modeling carbon fixation in the Southern Ocean. Polar Biol 17: 39–53

Morel A (1978) Available, usable and stored radiant energy in relation to marine photosynthesis. Deep Sea Res I 25: 673–688

Morel A (1991) Light and marine photosynthesis: a spectral model with geochemical and climatological implications. Prog Oceanogr 26: 263–306

Morel A, Berthon JF (1989) Surface pigments, algal biomass profiles, and potential production of the euphotic layer: relationships reinvestigated in view of remote-sensing applications. Limnol Oceanogr 34: 1545–1562

Morel A, Smith RC (1974) Relation between total quanta and total energy for aquatic photosynthesis. Limnol Oceanogr 19: 591–600

Morel A, Lazzara L, Gostan J (1987) Growth rate and quantum yield time response for a diatom to changing irradiances (energy and color). Limnol Oceanogr 32: 1066–1084

Nelson DM, DeMaster DJ, Dunbar RB, Smith Jr WO (1996) Cycling of organic carbon and biogenic silica in the Southern Ocean: estimates of water-column and sedimentary fluxes on the Ross Sea continental shelf. J Geophys Res 101: 18519–18532

Platt T (1986) Primary production of the ocean water column as a function of surface light intensity: algorithms for remote sensing. Deep Sea Res 33: 149–163

Platt T, Sathyendranath S (1988) Oceanic primary production: estimates by remote sensing at local and regional scales. Science 241: 1613–1620

Platt T, Gallegos CL, Harrison WG (1980) Photoinhibition of photosynthesis in natural assemblages of marine phytoplankton. J Mar Res 38: 687–701

Riemann B (1978) Carotenoid interference in the spectrophotometric determination of chlorophyll degradation products from natural populations. Limnol Oceanogr 23: 1059–1066

Savidge G, Harbour D, Gilpin LC, Boyd PW (1995) Phytoplankton distributions and production in the Bellingshausen Sea, austral spring 1992. Deep Sea Res II 42: 1201–1224

Smith WO, Nelson DM (1990) Phytoplankton growth and new production in the Weddell Sea marginal ice zone in the austral spring and autumn. Limnol Oceanogr 35: 809–821

Smith WO, Kelly HP, Rich JH (1990) Chlorophyll distribution and primary productivity in the Ross Sea, austral summer 1990. Antarct J 25(2): 179–181

Smith WO, Nelson DM, DiTullio GR, Leventer AR (1996) Temporal and spatial patterns in the Ross Sea: phytoplankton biomass, elemental composition, productivity and growth rates. J Geophys Res 101: 18455–18465

SooHoo JB, Palmisano AC, Kottmeier ST, Lizotte MP, SooHoo SL, Sullivan CW (1987) Spectral light absorption and quantum yield of photosynthesis in sea ice microalgae and a bloom of *Phaeocystis pouchetii* from McMurdo Sound, Antarctica. Mar Ecol Prog Ser 39: 175–189

Sullivan CW, Arrigo KR, McClain CR, Comiso JC, Firestone J (1993) Distributions of phytoplankton blooms in the Southern Ocean. Science 262: 1832–1837

Tilzer MM, Dubinsky Z (1987) Effects of temperature and daylength on the mass balance of antarctic phytoplankton. Polar Biol 7: 35–42

Wilson DL, Smith WO, Nelson DM (1986) Phytoplankton bloom dynamics of the Western Ross Sea ice-edge. I. Primary productivity and species-specific production. Deep Sea Res 33: 1375–1387

# Distribution of the Main Microzooplankton Taxa in the Ross Sea (Antarctica): Austral Summer 1994

M. Monti and S. Fonda Umani[1]

**ABSTRACT**

The microzooplankton Antarctic community was studied during the Italian Expedition in the Ross Sea (November-December 1994). Water samples were collected, with sampling bottles, at surface, intermediate and bottom depths along an offshore transect (175°E between 72°00'S and 76°30'S) covering three distinct areas: a southern polynya, an intermediate marginal-ice zone and a northern ice-covered area. Microzooplankton ranged from a minimum of 1.92 ind $l^{-1}$ to a maximum of 265 ind $l^{-1}$. The community structure was very simple, constituted by typical Antarctic species. Heterotrophic dinoflagellates were abundant in the upper layers, reaching a maximum of 232 ind $l^{-1}$. They were mostly constituted by the genera *Protoperidinium* and *Gyrodinium*. Micrometazoans prevailed over ciliates at most of the stations and depths; Foraminiferida and Radiolaria increased in bottom layers. A clearly decreasing gradient in both abundance and species diversity was observed from the polynya to the northernmost stations still covered by ice, as well as from the surface to bottom.

## Introduction

Microzooplankton consists of heterotrophic organisms in the 20–200 µm size range and is considered one of the first links of the trophic system in consumer terms (Fenchel 1988). Indeed, microzooplankton can transfer organic matter produced by pico- and nano-plankton to larger-sized consumers. It was suggested by Landry et al. (1993) that their grazing can control some elements of the phytoplankton community, with grazing pressure able to remove >100% of primary production in open waters of the Marginal Ice Zone (Burkill et al. 1995).

The first Antarctic studies on microzooplankton date back to the Deutsche Sudpolar Expedition in 1901–1903 (Laackmann 1907; Schroder 1909; Popofsky 1909). Balech (1958, 1971, 1973) described the tintinnids collected in three cruises in subantarctic areas. Heinbokel and Coats (1986) reported the populations found in the Weddell Sea in November 1983 where maximum tintinnid abundances were always <17 ind $l^{-1}$. Littlepage (1968) in McMurdo Sound found tintinnid concentrations in excess of 100 ind $l^{-1}$. Boltovskoy et al. (1989) found, in the Weddell Sea, maxima of 859 ind $l^{-1}$. In the Bellingshausen Sea and the Bransfied Strait, Alder and Boltovskoy (1991) found maxima of 589 ind $l^{-1}$. The same authors (Alder and Boltovskoy 1993) found in the Weddel-Scotia Confluence area tintinnids and dinoflagellates concentrated at 50–90 m; foraminifers and radiolarians were most abundant in noticeably deeper waters peaking below 150 m. Stoecker et al. (1995) reported that heterotrophic nanoplankton and protozoan microzooplankton increased in biomass during a *Phaeocystis* bloom in McMurdo Sound (December 1990).

Heterotrophic (and mixotrophic) dinoflagellates seem to contribute significantly to the micro-heterotrophic biomass (Dodge and Priddle 1987; Bjornsen and Kuparinen 1991; Elbrächter and Zölfell 1993; Stoecker et al. 1995), sometimes accounting for more than 50% of the total heterotrophic biomass (Alder and Boltovskoy 1993), and more than 60% in the 0–150-m interval (Boltovskoy and Alder 1992). Their distribution is affected by high heterogeneity, with patchiness of relatively high abundance related to certain particular current patterns (Dodge and Priddle 1987).

[1] Laboratory of Marine Biology, A. Piccard 54, 34010, Santa Croce, Trieste, Italy

In McMurdo Sound, up to 97% of the microprotozoan biomass in the early austral spring was due to heterotrophic dinoflagellates (Lessard and Swift 1986). In the Weddell Sea, Nothing et al. (1991) have reported that heterotrophic dinoflagellates can be a significant biomass constituent of the protozooplankton during the austral summer. Garrison et al. (1999) found that ciliates and heterotrophic dinoflagellates were relatively more abundant at stations where small diatom and motile cells of *Phaeocystis* predominated, suggesting that these were suitable prey for grazers. Lessard et al. (1999), found that during an early spring cruise in November–December 1994 an abundant microprotozoan was a large athecate heterotrophic dinoflagellate *Gyrodinium lachryma*. The Antarctic populations of microzooplankton were therefore fairly well known, though data on the area explored by the Italian expedition were limited to one oceanographic cruise during January–February 1988 at Terra Nova Bay (Fonda Umani and Monti 1990) and two stations closer to the Italian Base (at Terra Nova Bay) sampled during January 1990 (Fonda Umani et al. 1992, 1998; Monti and Fonda Umani 1995).

The main purpose of the present research was to describe the community structure and assess microzooplankton density distribution along an offshore transect in the Ross Sea.

## Study Area

The study area, belonging to the Ross Sea, is classified as a coastal and continental shelf zone (CCSZ). Its hydrology is characterized during the summer by the presence of a stable pycnocline and steep temperature and salinity gradients. The surface layer, ascribable to Antarctic Surface Water (AASW), having a thickness not greater than 100 m, has temperature values that exceed 0 °C. Below the ASSW, a homogeneous water body belonging to the Ross Sea Shelf Water (RSSW) is present. Here, salinity increases slightly towards the bottom with an average temperature of −1.9 °C and salinity of 34.9 PSS-78 (Boldrin and Stocchino 1990).

In the coastal area, the surface layer is characterized by salinities >34.7 and temperatures ranging from 1.9 to 2.1 °C. This area is affected by the prevailing surface thermal process. Eastward, offshore surface waters are affected by dilution processes due to ice melting, typical of the Marginal Ice Zone (MIZ), with salinity values <34.6. Surface currents are constituted by two branches: the first incoming from NE, in the northernmost part of the Bay, and the second from E in the southernmost part (below 75°S). The first flux invades the central area and then turns N to constitute the coastal current. Generally the surface circulation is characterized by a clockwise cyclonic stable gyre (Stocchino and Manzella 1991).

During the 1994 cruise it was possible to identify three distinct areas along the sampling transect: the polynya was present southward of 75°00′S (stations from MP to 8), the MIZ corresponded to intermediate stations (9 to 11) and the pack ice-covered area northward (stations 12 to 27). The southern stations were characterized by instability and higher salinity due to wind mixing, the northern ones by pycnocline stability due to ice melting (Saggiomo et al. 1998).

## Materials and Methods

Samples were collected at 31 stations along an offshore transect (175°E between 72°00′S and 76°30′S), and, at two stations, located at Terra Nova Bay 74°58′S and 167°59′E (station 29) and 75°08′S and 166°07′E (station 30) (Fig. 1). Most of the stations along the offshore transect were repeated several times (Table 1).

Samples were collected at the surface, 20–50- and 150–200-m depth by means of a Sea Bird Electronics SBE 32 Carousel. At station 15B water samples were collected at 12 depths (0–570 m). Sampling depths at intermediate layers were chosen on the basis of physical structure of the water column (i.e. below the main pycnocline).

For each sample, 5 l (24 l at station 15B) of water were collected, immediately concentrated with 20-μm mesh, and reduced to 250 ml. Samples were preserved in a 4% formaldehyde solution buffered with sodium tetraborate. Subsamples (50 ml) were then examined in a settling chamber using a Leitz Labovert inverted microscope (magnitude ×200), according to the Utermöhl method (1958).

Loricate ciliates were classified according to Brandt (1906, 1907), Laackmann (1910) and Kofoid and Campbell (1929, 1939). Aloricate ciliates were grouped into separate taxa which were characterized on the basis of size, shape, visible ciliature and morphology (Hamburger and

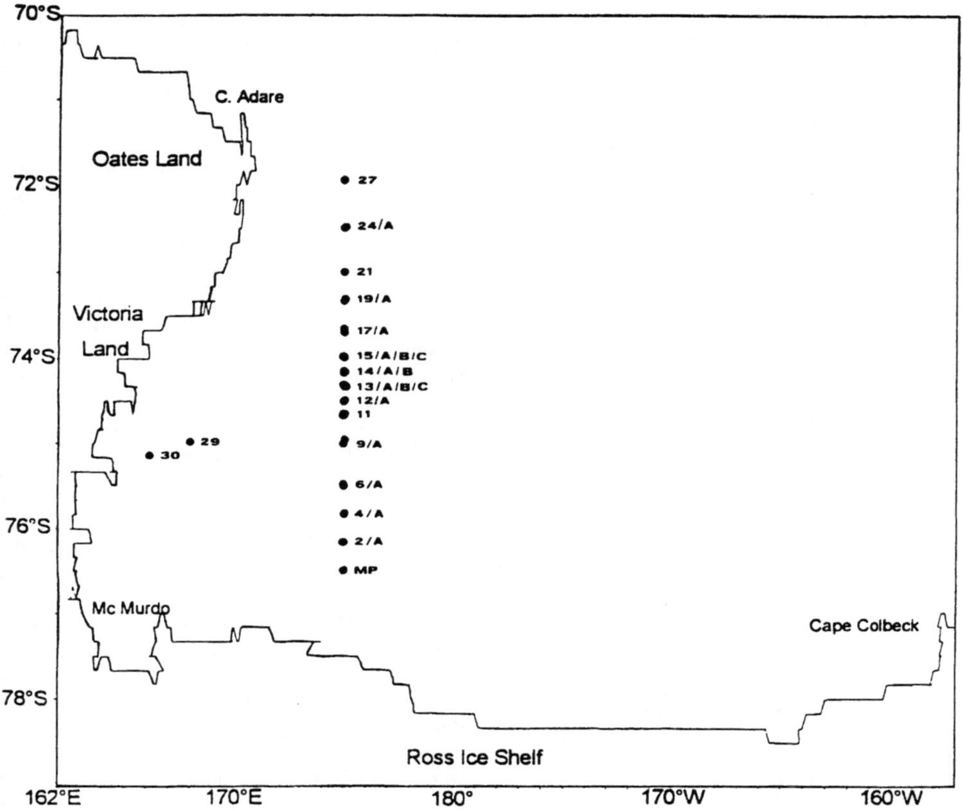


**Fig. 1.** Study area and sampling stations during the 1994 Italian cruise

Buddenbrock 1929; Corliss, 1979; Lynn and Montagnes 1988a,b). Dinoflagellates were classified according to Balech (1976). They included species belonging to the genus *Protoperidinium* and the species *Gyrodinium lachryma*, considered heterotrophic on the basis of literature descriptions (Lessard and Swift 1986; Dodge and Priddle 1987; Elbrächter and Zölfell 1993). Because of fixation procedures and method of collection, more delicate microzooplankters may have been lost (Gifford 1985; Stoecker et al. 1994). For this reason, oligotrich ciliates and athecate dinoflagellates may have been understimated in our counts.

At station 15B, biomass was estimated by measuring the linear dimension and equating shapes to standard geometric figures or combinations of these. Measures were taken for each organism. Conversion factors used to transform these volumes into organic carbon values were, dinoflagellates: pg C = $\mu m^3 \times 0.13$ (Edler 1979); tintinnids: pg C = $\mu m^3 \times 0.14$ (Putt and Stoecker 1989); all other groups: pg C = $\mu m^3 \times 0.08$ (Beers and Stewart 1970).

Cluster analysis based on the complete linkage method was computed by Matedit (Burba et al. 1992) on the matrix after calculating the correlation coefficient on species and the similarity ration on stations' depths.

## Results

Microzooplankton abundance ranged from 1.92 ind$l^{-1}$ (station 27, 200 m) to 265 ind$l^{-1}$ (station 30, 0 m). Maxima were recorded at the surface and minima at the bottom (Fig. 2a,b,c). The strong

**Table 1.** Station sampling dates and depths

| Station | Date | Depth (m) |
|---|---|---|
| MP | 14.11.94 | 0-30-150 |
| 2 | 15.11 | 0-20-200 |
| 2A | 20.11 | 0-23-260 |
| 4 | 16.11 | 0-25-200 |
| 4A | 21.11 | 0-31-200 |
| 6 | 17.11 | 0-25-150 |
| 6A | 20.11 | 0-27-200 |
| 9 | 22.11 | 0-25-200 |
| 9A | 28.11 | 0-27-200 |
| 11 | 23.11 | 0-25-200 |
| 12 | 23.11 | 0-30-200 |
| 12A | 29.11 | 0-26-200 |
| 13 | 25.11 | 0-18-200 |
| 13A | 29.11 | 0-20-200 |
| 13B | 4.12 | 0-42-210 |
| 13C | 13.12 | 0-43-200 |
| 14 | 25.11 | 0-37-230 |
| 14A | 30.11 | 0-24-200 |
| 14B | 12.12 | 0-31-200 |
| 15 | 26.11 | 0-32-200 |
| 15A | 30.11 | 0-30-200 |
| 15B | 4.12 | 0-4.5-16-26-38-75-115-150-200-250-300-570 |
| 15C | 12.12 | 0-37-200 |
| 17 | 3.12 | 0-43-200 |
| 17A | 11.12 | 0-50-200 |
| 19 | 5.12 | 0-41.5-200 |
| 19A | 11.12 | 0-30-200 |
| 21 | 6.12 | 0-41-200 |
| 24 | 7.12 | 0-40-200 |
| 24A | 10.12 | 0-38-72-128-200 |
| 27 | 8.12 | 0-51-200 |
| 29 | 15.12 | 0-37-200 |
| 30 | 15.12 | 0-34-150 |

decrease at the bottom was due to the reduction of all taxonomical groups other than foraminifers and radiolarians. On the other hand, the increase of these groups did not balance the general decrease. Abundance of microzooplankton was significantly related (101 data, $r^2 = 0.73$, $p > 0.01$) (Fig. 3) to total chlorophyll a distribution (Carrada et al. 1999).

We determined eight tintinnid taxa, for a total of four genera and five species (Table 2). Tintinnid abundance ranged from a minimum of $0\,ind\,l^{-1}$ to a maximum of $13.2\,ind\,l^{-1}$ (station 4, 0 m), mostly due to *Cymatocylis drygalskii*. Most of the tintinnids found had hyaline loricae and showed a high morphological variability, as described in previous studies (Boltovskoy et al. 1990; Wasik and Mikolajczyk 1994; Williams et al. 1994; Monti and Fonda Umani 1995).

Aloricate ciliates belonged mostly to the families Strombidiidae and Strobilidiidae and to the genera *Enchylis* and *Leegardiella*. Aloricate ciliate abundance ranged from $0\,ind\,l^{-1}$ to $64.8\,ind\,l^{-1}$ (station 19, 0 m; Fig. 2a,b,c); they were less abundant than in our previous findings in the same area (Fonda Umani and Monti 1990). Foraminifers, radiolarians and nauplii of copepods were also present. In Fig. 2, foraminifers and radiolarians are assembled together and indicated as "others". Radiolarians, (i.e. *Sticholonche zanclea*), were more abundant in the bottom layers and maximum values were always $<25\,ind\,l^{-1}$ with absolute maximum at station 12A (200 m) (Fig. 2a,b,c). Heterotrophic dinoflagellates were prevalent at the surface, and several times they were also prevalent at intermediate depths. They were mostly constituted by the genus *Protoperidinium*, with the species *P. pseudoantarcticum*, *P. defectum* and *Gyrodinium lachryma*. Abundant in the upper 50 m, they became scarce below 100-m depth, reaching maximum values of $232\,ind\,l^{-1}$ at the closest station to the coast (station 30, 0 m; Fig. 2a,b,c).

At the 12 depths sampled at station 15B we obtained a more defined vertical distribution; here, abundance of microzooplankton showed a maximum at 16 m with $30\,ind\,l^{-1}$ (Fig. 4), mainly due to aloricate ciliates. This fraction decreased below 75 m, never reaching $1\,ind\,l^{-1}$. Tintinnids were not abundant; maximum values were detected at 115 m ($1.68\,ind\,l^{-1}$) but this group was present throughout the water column. Below 75 m radiolarians increased, with a maximum at 200 m ($10.9\,ind\,l^{-1}$), mostly due to the species *S. zanclea*. Heterotrophic dinoflagellates were present mostly in the upper 100 m with a maximum at 38 m ($23.37\,ind\,l^{-1}$).

The vertical distribution of biomass showed a pattern different from that of abundance (Fig. 5). Values were higher in the upper 75 m, where a maximum of $732\,ng\,C\,l^{-1}$ was detected. The greatest amount of carbon content was due to dinoflagellates, in particular at 75 m *P. pseudoantarcticum* peaked up to $696\,ng\,C\,l^{-1}$. Below 75 m total values decreased constantly, except at 200 m where a relative increase ($50\,ng\,C\,l^{-1}$) was due to the presence of *S. zanclea*.

Data obtained sampling at the same stations once or more, after 1 or more days, generally indicated a decreasing pattern at surface and intermediate depths mostly due to a decrease in dinoflagellates, and a slight increase at the bottom

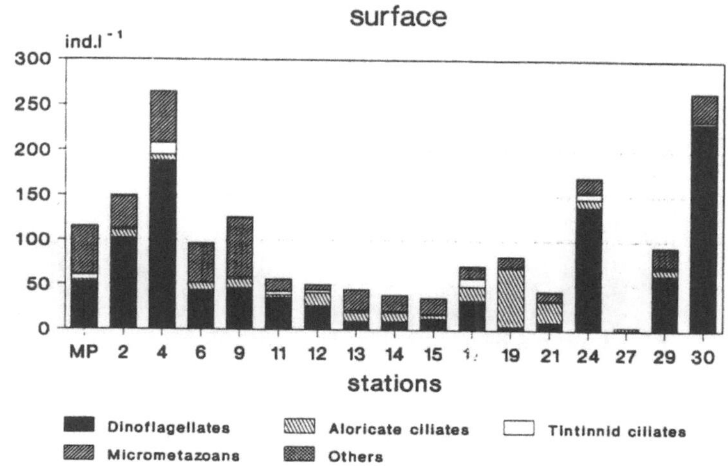


a

**Fig. 2a–c.** Abundance of dinoflagellates, aloricate ciliates, tintinnids, micrometazoans and "others" (ind $l^{-1}$) at surface (**a**), intermediate (**b**) and bottom (**c**) depths

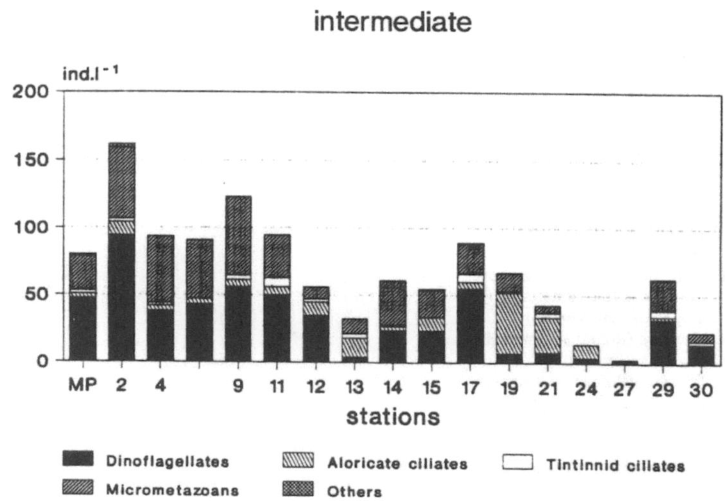


b

**Fig. 2a–c.** *Continued*

partially due to an increase in dinoflagellates (Fig. 6a,b,c). Cluster analysis was performed on a matrix species/station depths which included all depths at each station sampled for the first time. Analysis allowed us to identify four major groups of organisms (Fig. 7) and two major groups of station depths (Fig. 8). The first group included mainly surface and intermediate samples, the second mainly bottom samples and all depths at station 27. The first group was subdivided into three subgroups (Table 3) which were significantly different from one another on the basis of the taxonomical composition of their community. The first subgroup included all depths at the MP

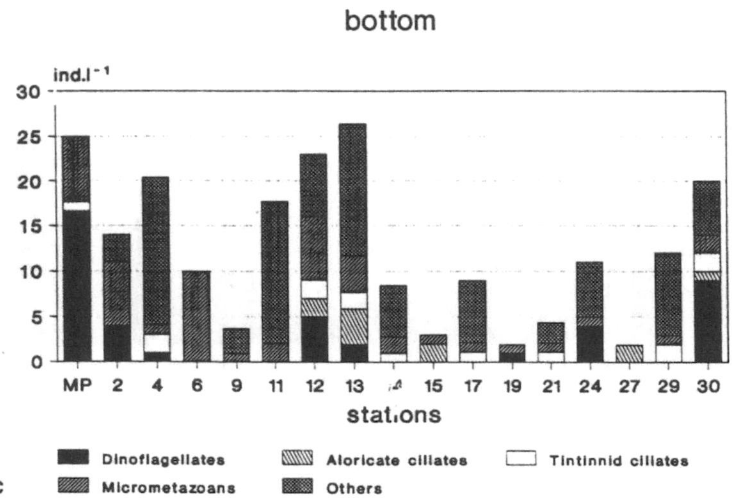


**Fig. 2a–c.** *Continued*

**Table 2.** Tintinnids sampled during the three Italian expeditions

| List of tintinnids | 1987–88 | 1989–90 | 1994–95 |
|---|---|---|---|
| *Codonellopsis gaussi* (Laackmann) | * | * | * |
| *C. glacialis* (Laackmann) | | * | * |
| *Codonellopsis* sp. (Laackmann) | | * | |
| *Laackmanniella naviculaefera* (Laackmann) | * | * | * |
| *L. prolongata* (Laackmann) | * | * | |
| *Laackmanniella* sp.(Laackmann) | * | | * |
| *Coxliella frigida* (Laackmann) | | * | |
| *Coxliella* sp. (Brandt) | | * | |
| *Cymatocylis conica* (Laackmann) | * | | |
| *C. convallaria* (Laackmann) | | | * |
| *C. drygalskii* (Laackmann) | * | * | * |
| *C. ecaudata* (Kofoid & Campbell) | | * | |
| *C. flava* (Laackmann) | * | | |
| *C. folliculus* (Kofoid & Campbell) | | * | |
| *C. glans* (Kofoid & Campbell) | | * | |
| *C. nobilis* (Laackmann) | | * | |
| *C. subconica* (Kofoid & Campbell) | | * | |
| *C. vanhoffeni* (Laackmann) | * | | * |
| *Cymatocylis* sp. (Laackmann) | * | * | * |
| *Ptychocylis* sp. (Brandt) | | * | |
| *Salpingella* sp. (Jorgensen) | * | * | * |

station, all surface and intermediate depths at the polynya, surface and intermediate depths in the marginal-ice zone and some intermediate samples from the ice-covered zone. It was characterized by taxa belonging to the third and fourth species group, in which the prevailing species were *P. pseudoantarcticum*, *P. defectum*, *G. lachryma*, *Cy. drygalskii*, nauplii of copepods and unidentified eggs. The second subgroup of stations included all surface and some intermediate samples in the ice-covered zone, and was characterized by the presence of taxa belonging to the second species group, namely *Enchylis* sp. and unidentified Strombidiidae. The third station group included

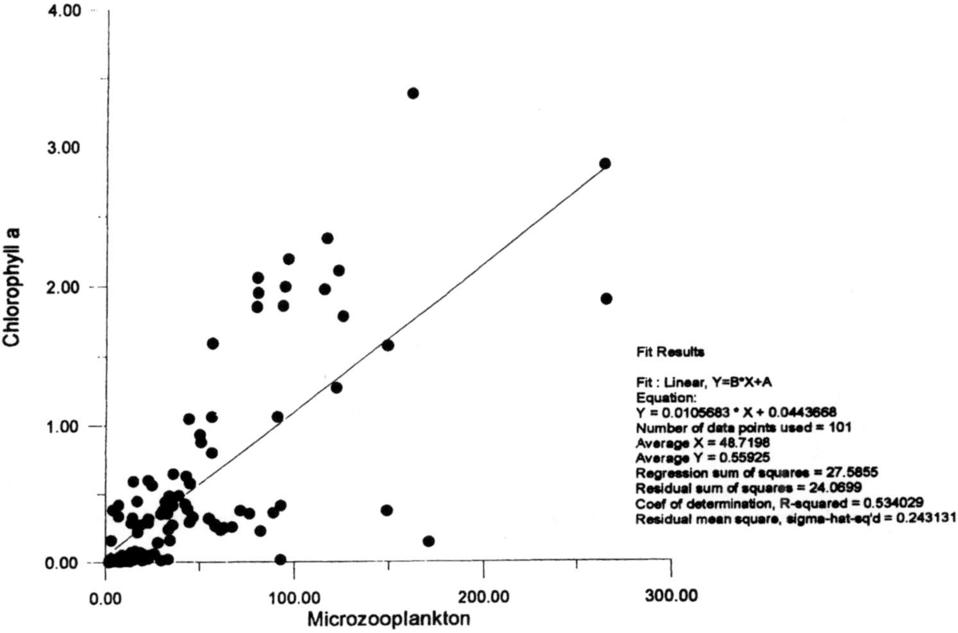


**Fig. 3.** Coefficient of determination for microzooplankton/total chlorophyll a

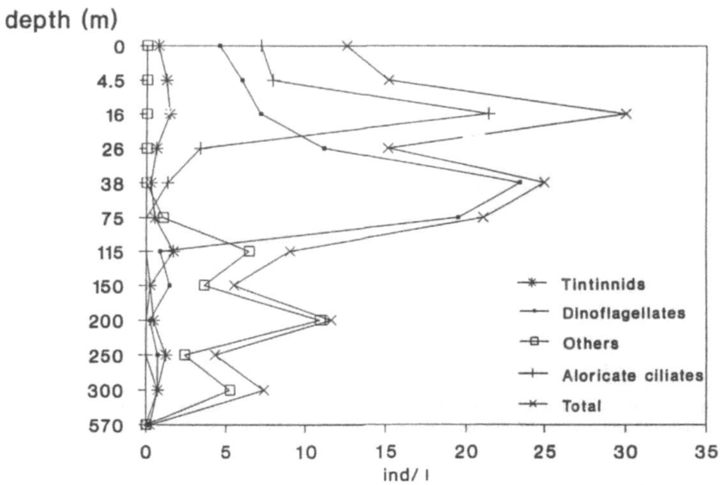


**Fig. 4.** Vertical distribution of tintinnids, dinoflagellates, "others", aloricate ciliate abundance and total abundance (ind $l^{-1}$) at station 15

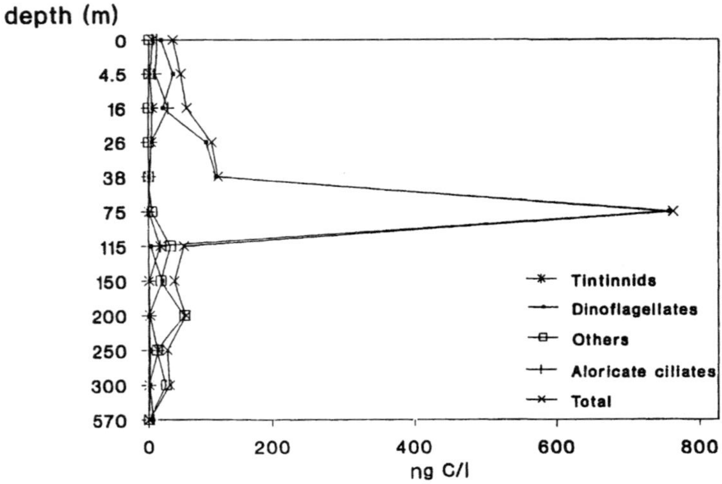


**Fig. 5.** Vertical distribution of tintinnids, dinoflagellates, "others", aloricate ciliate biomass and total biomass (ng C $l^{-1}$) at station 15

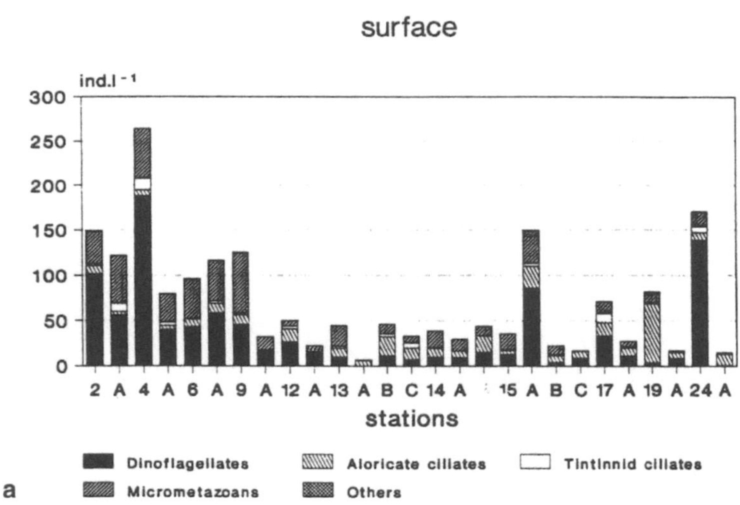


**Fig. 6a–c.** Abundance of dinoflagellates, aloricate ciliates, tintinnids, micrometazoans and "others" (ind $l^{-1}$) at surface (**a**), intermediate (**b**) and bottom (**c**) depths at the repeated stations (2, 4, 6, 9, 12, 13, 14, 15, 17, 19 and 24)

intermediate and one bottom sample in the ice-covered zone, and was characterized by low abundance of the second, third and fourth species groups. Bottom samples belonging to the second station group were mainly identified by the presence of the first species group such as *S. zanclea* and unidentified radiolarians, while all depths at station 27 as well as bottom samples at stations 15 and 19 were characterized by very scarce populations. A clearly decreasing gradient is shown in the table compiled on the basis of the results from cluster anlysis (Table 3), which shows the decreas-

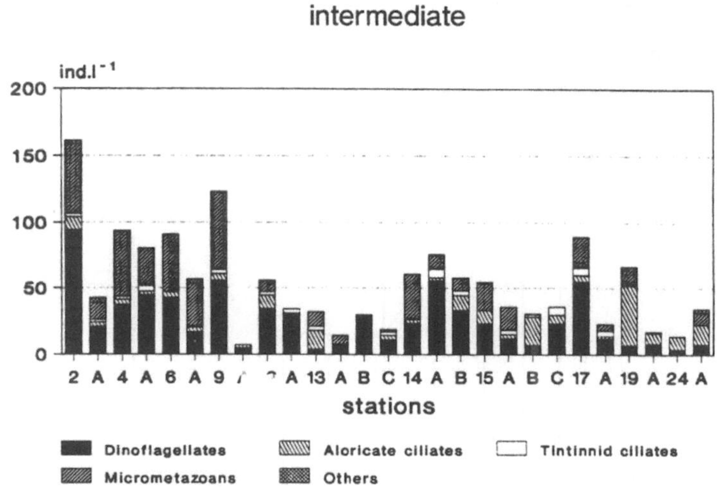


b

Fig. 6a–c. *Continued*

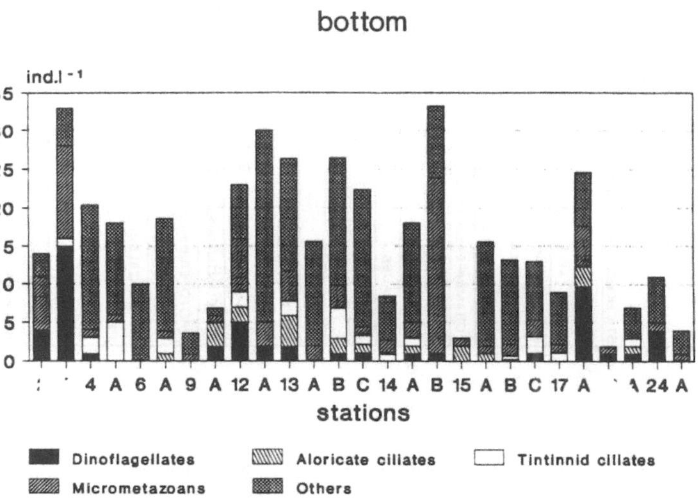


c

Fig. 6a–c. *Continued*

ing trend from the richest stations in the polynya to the most northern station 27. Except for bottom samples at all stations, where only radiolarians and *S. zanclea* seemed abundant, cluster analysis showed a kind of successional development from ice-covered stations to the most southern station passing through MIZ, increasing in number of species and abundance.

## Discussion and Conclusions

Microzooplankton are considered more linked to the microbial loop than to the classic marine food web, since they feed almost exclusively on the smallest fraction of producers such as bacteria and nanoflagellates (Landry and Hassett 1982; Burkill et al. 1993, 1995; Froneman and

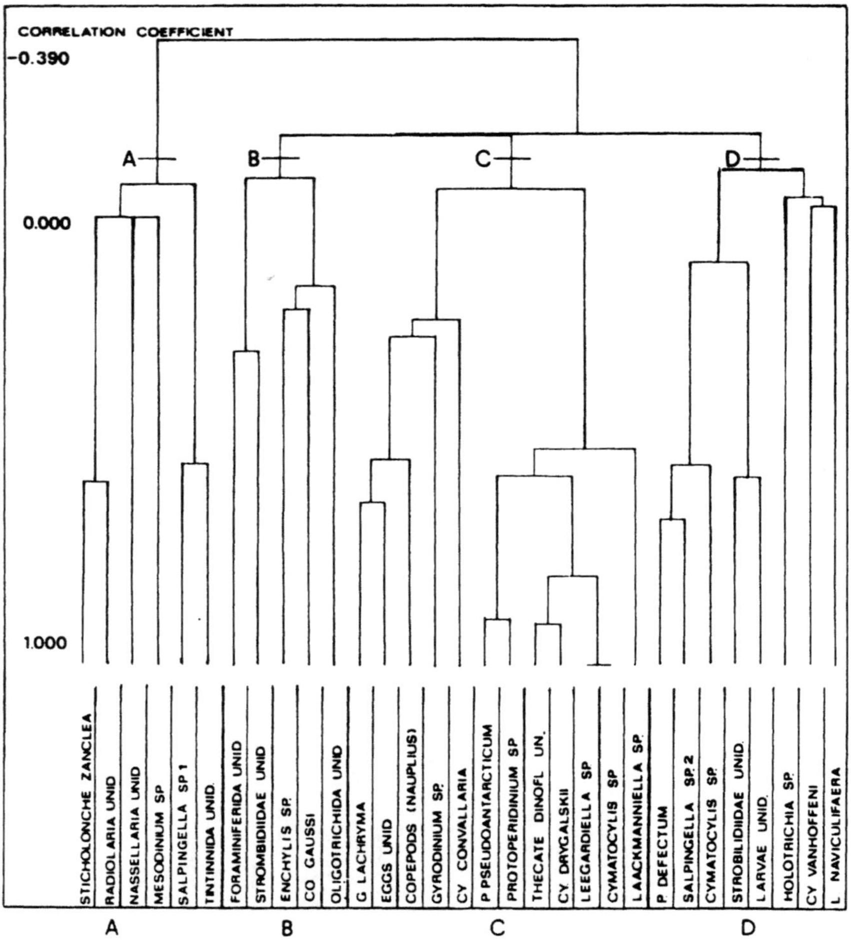


**Fig. 7.** Dendrogramm obtained by cluster analysis ordering the species groups

Perissinotto 1996). Several studies have indicated that the microbial loop becomes more important at the end of the austral spring when diatoms are substituted as the major producers and there is an increase in the heterotrophic fraction due to the abundant particulate organic matter produced earlier (Alder and Boltovoskoy, 1993; Garrison et al. 1998). Consequently, microzooplankton populations reach their maximum development during the period of microbial loop increase.

In this chapter we have only considered heterotrophic organism >20 μm because of our sampling and filtration methods. This obviously led to an understimation of the smallest organisms which are generally included in the microzooplankton community (Grey et al. 1997). Furthermore, we used buffered formaldehyde, which may have destroyed some of the ciliates and athecate dinoflagellates. Therefore, cell abundances from fixed samples might have been understimated. The degree of loss and shrinkage of preserved protists is taxon and fixative-specific, making it difficult to correct for whole assemblages (Stoecker et al. 1994).

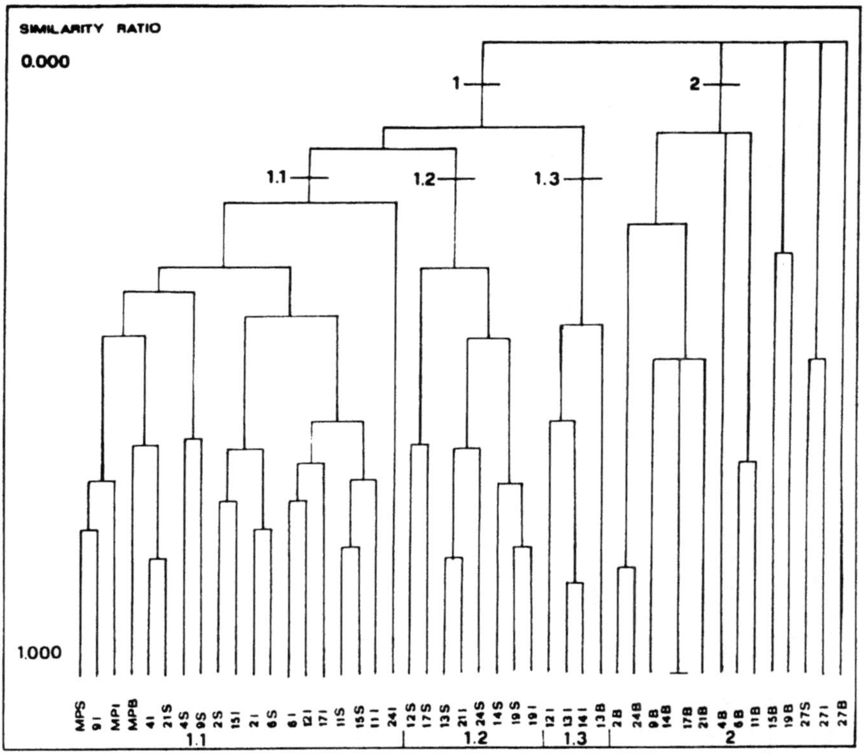


S = SURFACE
I = INTERMEDIATE
B = BOTTOM

**Fig. 8.** Dendrogramm obtained by cluster analysis ordering the station depth groups

Highest abundances were always due to high percentages of heterotrophic dinoflagellates in the microzooplankton, as was observed during the 1988 cruise in Terra Nova Bay (Fonda Umani et al. 1998). The sampling methods in the 1988 and 1994 cruises are comparable, except for dinoflagellates. During the 1988 cruise, dinoflagellates were analyzed from unfiltered water samples, according to the Utermhöl method (1958). On that occasion we found exceptionally high dinoflagellate abundances: up to 8000 ind $l^{-1}$ For the rest of the microzooplankton during the same cruise we found abundances as high as 2000 ind $l^{-1}$. These were among the highest values ever reported for Antarctic areas, besides the recent findings of Burkill et al. (1995).

During the 1994 cruise, microzooplankton abundance was 1 order of magnitude less than that of the previous cruise, also at stations 29 and 30, which were very close to those sampled in 1988. The 1988 cruise was at the end of the austral summer when the microbial loop is supposedly more important in an area such as Terra Nova Bay. The 1994 data, closer to those reported in the literature, were mostly obtained in an offshore area and in an earlier period than in 1988. Values found during the 1994 cruise, on the other hand, fall within the range observed by several authors in other areas (i.e. Littlepage 1968; Boltovsky et al. 1989; Alder and Boltovsky 1991, 1993; Garrison 1991; Leakey et al. 1994).

**Table 3.** Species/stations depths matrix following the result of cluster analysis. Refer to Fig. 1 for station locations

| Station | 24 | 14 | 19 | 19 | 12 | 13 | 14 | 13 | 2 | 24 | 9 | 14 | 17 | 21 | 4 | 6 | 11 | 15 | 19 | 27 | 27 | 27 |
|---|---|---|---|---|---|---|---|---|---|---|---|---|---|---|---|---|---|---|---|---|---|---|
| Depth | 0 | 0 | 0 | 41 | 200 | 18 | 37 | 200 | 200 | 200 | 200 | 230 | 200 | 200 | 200 | 150 | 200 | 200 | 200 | 0 | 51 | 200 |
| *Sticholonche zanclea* | | | | | 7.0 | | | 1.78 | 3.0 | 6.0 | 1.84 | 5.64 | 6.72 | 2.20 | 16.32 | 8.0 | 13.52 | | | | | |
| Radiolaria unid. | | | | | | | | 3.92 | | | | | | | | 2.0 | 2.80 | | | | | |
| Nassellaria unid. | | | | | | | | | | | | | | | | | | | | | | |
| *Mesodinium* sp. | | | | | | | | | | | | | | | | | | | | | | |
| *Salpingella* sp. 1 | | | | | | | | | | | | | | | 1.20 | | | | | | | |
| Tintinnida unid. | | | | | | | | | | | | | | | 1.20 | | | | | | | |
| Foraminiferida unid. | | | | | | | | | | | 0.92 | | | | | | | | | | | |
| Strombidiidae unid. | 1.80 | 1.0 | 2.16 | 1.0 | | | | | | | | | | | | | | | | 3.0 | | 1.40 |
| *Enchylis* sp. | 1.80 | | | | | | | | | | | | | | | | | | | | | |
| *Co. gaussi* | 2.16 | | | | 1.0 | | | | | | | | | | | | | | | | | |
| Oligotrichida unid. | 1.80 | | | | | 1.0 | | 1.96 | | | | | | | | | | 2.40 | | | | |
| *G. lachryma* | 12.96 | 3.0 | 2.16 | 1.0 | | | | | 1.0 | 1.0 | | | | | | | | | | 1.0 | | 0.52 |
| Eggs unid. | 16.20 | 16.0 | 1.80 | 8.0 | 3.0 | 9.0 | 8.46 | | 1.0 | | | | | | 1.20 | | | | | | | |
| Copepods nauplius | | | 1.80 | 2.0 | 4.0 | 2.0 | 25.38 | 3.92 | 5.0 | 1.0 | 0.92 | 1.88 | 1.12 | 1.10 | | | 2.80 | 1.20 | 1.0 | | | |
| *Gyrodinium* sp. | | | | | | | | | | | | | | | | | | | | | | |
| *Cy. convallaria* | | 1.0 | | | | | | | | | | | | | | | | | | | | |
| *P. pseudoantarcticum* | 52.92 | | | | | | | | 1.0 | 2.0 | | | | | 1.20 | | | | | | | |
| *Protoperidinium* sp. | 51.84 | 2.0 | 1.80 | 2.0 | 1.0 | 3.0 | 18.80 | 1.96 | 2.0 | 1.0 | | | | | | | | | | | 1.40 | |
| Thecate Dinofl. unid. | 19.44 | 4.0 | | 2.0 | 2.0 | 1.0 | 4.70 | | | | | | | | | | | | 1.0 | | | |
| *Cy. drygalskii* | 1.80 | | | | | | | | | | | | | | | | | | | | | |
| *Leegardiella* sp. | | | | | | | | | | | | | | | | | | | | | | |
| *Cymatocylis* sp. | | | | | | | | | | | | | | | | | | | | | | |
| *Laackmanniella* sp. | | | | | | | | | | | | | | | | | | | | | | |
| *P. defectum* | 1.80 | 1.0 | 1.80 | 2.0 | 2.0 | | | | | | | | | | | | | | | | | |
| *Salpingella* sp. 2 | 3.24 | | | 1.0 | 1.0 | 3.0 | 0.94 | 1.84 | | | | 0.94 | 1.12 | | | | | | | | | |
| Strobilidiidae unid. | 5.40 | 9.0 | 62.64 | 44.0 | 2.0 | 13.0 | 2.82 | 1.96 | | | | | | | | | | | | | | 0.96 |
| Larvae unid. | 1.8 | 2.0 | 1.8 | 4.0 | | | | | | | | | | | | | | | | | | |
| Holotrichia unid. | | | | | | | | | | | | | | | | | | | | | 0.52 | 0.96 |
| *Cy. vanhoffeni* | | | | | | | | | | | | | | 1.10 | | | | | | | | |
| *L. naviculifaera* | | | | | | | | | | | | | | | | | | | | | | |

During the 1994 cruise, highest values were observed at the surface and intermediate depths of the first stations (MP to 9) corresponding to a polynya, where Carrada et al. (1998) found the highest values of total chlorophyll a. Microzooplankton total abundance was significantly related to total chlorophyll a distribution, as reported by Gowing and Garrison (1991), Hansen (1991), Alder and Boltovoskoy (1993) and Monti and Fonda Umani (1995) in other Antarctic areas. Particularly in these southern stations, pico- and nanoplankton biomass accounted on average for more than 55% of total chlorophyll a (Saggiomo et al. 1998).

In the MIZ, microphytoplankton were dominant (>60% of biomass) and microzooplankton decreased also in the ice-covered northern area, where picoplankton biomass was >50% of total chlorophyll a. The only exception was station 24, where a significant increase occurred in nanoplankton biomass (Saggiomo et al. 1998). It appears that microzooplankton abundance is more closely related to nanoplankton than to picoplankton biomass, and is tightly coupled to nanoplankton distribution at surface and intermediate depths. In the MIZ we sampled during a spring diatom bloom (Marino and Cabrini 1999). In this case, the size of cells may have been too large to be eaten by most microzooplankters, even though some microzooplankton are known to eat relatively large-sized cells (Jacobson and Anderson 1986; Hansen 1992, Hansen and Nielsen 1997). In particular, some polar dinoflagellates can consume large diatoms (Stoecker et al. 1995).

The higher values in the polynya were related to a high percentage of *Phaeocystis* whose cell size is in the range of "palatable" food for microzooplankters, as observed by Garrison et al. (1999) and Stoecker et al. (1995). In the ice-covered zone, only a few small flagellates (Marino and Cabrini 1999) were available for microconsumers.

Qualitative and quantitative differences were considerable between upper and bottom layers. The deepest communities were characterized by low abundances and the presence of protozoans other than ciliates, as observed also by Alder and Boltovsky (1993), and in particular by radiolarians (e.g. *Sticholonche zanclea*). At surface and intermediate depths a clear increasing pattern of both species diversity and abundance was observed from the ice-covered zone to the polynya, showing a kind of successional development which did not seem to reach the final stage.

The tintinnid community was even simpler than our previous findings in the same area (Monti and Fonda Umani 1995). Particularly, the genus *Cymatocylis* was constituted by only three species instead of the eight found during the 1988 and 1990 cruises in Terra Nova Bay. Nevertheless, our data are in agreement with Heinbokel and Coats (1986), who found 18 species of tintinnids in the Weddell Sea, mostly belonging to the genera *Cymatocylis* and *Laackmanniella*, which are considered characteristic Antarctic genera (Pierce and Turner 1993).

Also, other ciliates were less diversified than in our previous cruises. During the 1988 cruise in Terra Nova Bay we found nine genera and close to the Italian base during the 1990 cruise seven genera were found, two of which, *Enchylis* and *Leegardiella*, were typical of offshore areas also during the 1988 cruise (Fonda Umani et al. 1992). Heterotrophic dinoflagellates were less abundant and less diversified than previously recorded (Fonda Umani et al. 1998), but in this case the different sampling method may have contributed to the difference. The two previous cruises were carried out in January-early February, which corresponded to the end of the first phytoplankton bloom and the beginning of the second (Innamorati et al. 1992; Saggiomo et al. 1998). Stoecker et al. (1995) observed that in the southwestern Ross Sea, athecate heterotrophic dinoflagellates increased 10–20-fold in number and 28-fold in biomass between late November and late January. Planktonic ciliates also increased in number and biomass in the same sampling period. Oligotrichous ciliates increased in average abundance from $<10^2$ cell $l^{-1}$ to $1-2 \times 10^3$ cell $l^{-1}$.

At the end of the productive season the microzooplankton community appeared better structured, being at the final stage of its successional development, while data obtained in the present study were related to a very early stage of the whole planktonic system in a more offshore area, where each year only one phytoplankton spring bloom was detected (Saggiomo et al. 1998). The slight decrease observed in the samples repeated after some days along the transect might support the hypothesis that in the more offshore area microzooplankton could never reach a high abundance and a relative high structural complexity as happens in the more inshore area of Terra Nova Bay, due to the lack of adequate food source or to a higher grazing pressure exerted by upper consumers (e.g. krill).

*Acknowledgements.* Thanks are due the crew of the R/V Italica for their professional help. This work was supported by the Italian National Antarctic Research Program.

## References


Alder VA, Boltovskoy D (1991) Microplanktonic distributional patterns west of Antarctic Peninsula, with special emphasis on the tintinnids. Polar Biol 11: 103–112

Alder VA, Boltovskoy D (1993) The ecology of larger microzooplankton in the Weddel-Scotia Confluence area: horizontal and vertical distribution patterns. J Mar Res 51: 323–344

Balech E (1958) Dinoflagelles et tintinnoides de la Terra Adelie. Vie Milieu 8: 382–408

Balech E (1971) Microplancton de la campana oceanografica Productividad III. Rev Mus Argent Cienc Nat Bernardino Rivadavia 3: 202

Balech E (1973) Segunda contribucion al conocimiento del microplancton del Mar de Bellinghausen. Contrib Inst Antart Argent 107: 1–63

Balech E (1976) Clave ilustrada de Dinoflagelados antarticos. Contrib Inst Antart Argent 11: 1–100

Beers JR, Stewart GL (1970) Numerical abundance and estimated biomass of microzooplankton. Bull Scripps Inst Oceanogr 17: 67–87

Bjornsen P, Kuparinen J (1991) Growth and herbivory by heterotrophic dinoflagellates in the Southern Ocean, studied by microcosm experiments. Mar Biol 106: 1–9

Boldrin A, Stocchino C (1990) On the hydrological characteristics of Terra Nova bay. In: National Scientific Commission for Antarctica (ed) Oceanographic Campaign 1987–88. Data Rep Part I, Genova, pp 11–57

Boltovskoy D, Alder VA (1992) Microzooplankton and tintinnid species-specific assemblage structures: patterns of distribution and year-to-year variations in the Weddell Sea (Antarctica). J Plankton Res 14: 1405–1423

Boltovskoy D, Alder VA, Spinelli F (1989) Summer Weddell Sea microplankton: assemblage structure, distribution and abundance, with special emphasis on the Tintinnina. Polar Biol 9: 447–456

Boltovskoy D, Dinofrio EO, Alder VA (1990) Intraspecific variability in Antarctic tintinnids: the *Cymatocylis affinis/convollaria* species group. J Plankton Res 12(2): 403–413

Brandt K (1906) Die Tintinnodeen der Plankton-Expedition. Tafelerklärungen nebst kurzer Diagnose der neuen Arten. Ergeb Atlant Ozean Planktonexped Humboldt-Stift, 3: 1–33

Brandt K (1907) Die Tintinnodeen der Plankton-Expedition. Systematischer Teil. Ergebn Atlant Ozean Planktonexped Humboldt-Stift, 3: 1–488

Burba N, Feoli E, Malaroda M, Zuccarello V (1992) Un sistema informativo per la vegetazione. Software per l'archiviazione della vegetazione italiana e per l'elaborazione di tabelle. Manuale di utilizzo dei programmi. Quaderni CETA 2: 78 pp

Burkill PH, Edwards ES, John AWG, Sleigh MA (1993) Microzooplankton and their herbivourous activity in the north-east Atlantic Ocean. Deep Sea Res II 42: 479–494

Burkill PH, Edwards ES, Sleigh MA (1995) Microzooplankton and their role controlling phytoplankton growth in the marginal ice zone of the Bellingshausen sea. Deep Sea Res II 42: 1277–1290

Carrada GC, Mangoni O, Russo GF, Saggiomo V, Sgrosso S (1999) Spatial and temporal distribution of phytoplankton pigments in the Ross Sea during the austral spring 1994. In: Faranda F, Guglielmo L, Povero P (eds) National Programme for Antarctic Research, Oceanographic Campaign 1994–95, Data Rep Part II, Genova (in press)

Corliss JO (1979) The ciliated protozoa. Pergamon, London, 455 pp

Dodge JD, Priddle J (1987) Species composition and ecology of dinoflagellates from the Southern Ocean near South Georgia. J Plankton Res 9(4): 685–697

Edler L (1979) Recommendations on methods for marine biological studies in the Baltic Sea. Phytoplankton and chlorophyll. Baltic Mar Biol 5: 1–38

Elbrächter M, Zölfell M (1993) Taxonomy and ecology of heterotrophic ice flagellates. Ber Polarforsch 121: 26–29

Fenchel T (1988) Marine plankton food chain. Annu Rev Ecol Syst 19: 19–38

Fonda Umani S, Monti M (1990) Microzooplankton populations in Terra Nova Bay (Ross Sea). In: National Scientific Commission for Antarctica (ed) Oceanographic Campaign 1987–88, Data Rep Part I, Genova, pp 241–254

Fonda Umani S, Chemelli M, Monti M (1992) Microzooplankton populations temporal trend in Terra Nova Bay (Austral Summer 1989/90). In: National Scientific Commission for Antarctica (ed) Oceanographic Campaign 1989–90, Data Rep Part II, Genova, pp 153–178

Fonda Umani S, Monti M, Nuccio C (1998) Microzooplankton biomass distribution in Terra Nova Bay, Ross Sea (Antarctica). J Mar Syst 17: 289–303

Froneman PW, Perissinotto R (1996) Structure and grazing of the microzooplankton communities of the subtropical convergence and warm-core eddy in the atlantic sector of the Southern Ocean. Mar Ecol Prog Ser 135: 237–245

Garrison DL (1991) An overview of the abundance and role of protozooplankton in Antarctic waters. J Mar Syst 2: 317–331

Garrison DL, Mathot S, Gowing MM, Kunze H, Lessard E (1999) Phytoplankton and microzooplankton community structure in the Ross Sea polynya: November –December 1994. Antarct JUS (in press)

Gifford DJ (1985) Laboratory culture of marine planktonic oligotrichs (Ciliophora, Oligotrichida). Mar Ecol Prog Ser 23: 257–267

Gowing MM, Garrison DL (1991) Austral winter distributions of large tintinnid and large sarcodinid protozooplankton in the ice-edge zone of the Weddel/Scotia Seas. J Mar Syst 2: 131–141

Grey J, Laybourn-Parry J, Leakey RJG, McMinn A (1997) Temporal patterns of protozooplankton abundance and their food in Ellis Fjord, Princess Elizabeth Land, Eastern Antarctica. Estuarine Coastal and Shelf Sci 45: 17–25

Hamburger C, Buddenbrock H (1929) Nordisches Ciliata mit Ausschluss der Titinnoidea. In: Nordisches Plankton, Zoologischer Teil, VII Protozoa. Verlag von Lipsius & Tischer, Kiel und Leipzig, pp 1–152

Hansen PJ (1991) Quantitative importance and trophic role of heterotrophic dinoflagellates in a coastal pelagic food web. Mar Ecol Prog Ser 77: 253–261

Hansen PJ (1992) Prey size selection, feeding rates and growth dynamics of heterotrophic dinoflagellates with special emphasis on *Gyrodinium spirale*. Mar Biol 114: 327–334

Hansen PJ, Nielsen TG (1997) Mixotrophic feeding of *Fragilidium subglobosum* (Dinophyceae) on three species of *Ceratium*: effects of prey concentration, prey species and light intensity. Mar Ecol Prog Ser 147: 187–196

Heinbokel JF, Coats DW (1986) Patterns of tintinnine abundance and reproduction near the edge of seasonal pack ice in the Weddell Sea. November 1983. Mar Ecol 33: 71–80

Innamorati M, Lazzara L, Massi L, Mori G, Saggiomo V (1992) Spatial and temporal distribution of phytoplankton size fractions in Antarctic waters: biomass and production, summer 1989-90. In: Ambrosetti W, Piccazzo M, Ruffoni Riva T (eds) Atti IX Congr Ass Ital Oceanogr Limnol, Lang, Genova, pp 605–612

Jacobson DM, Anderson DM (1986) Thecate heterotrophic dinoflagellates: feeding behavior and mechanism. J Phycol 22: 249–258

Kofoid CA, Campbell AS (1929) A conspectus of the marine and fresh-water Ciliata belonging to the suborder Tintinnoinea with descriptions of new species principally from the Agassiz Expedition to the eastern tropical Pacific, 1904-1905. Univ Calif Publ Zool 34: 1–403

Kofoid CA, Campbell AS (1939) Reports on the scientific results of the expedition to the Eastern Tropical Pacific. The Ciliata: the Tintinnoinea. Bull Mus Comp Zool Harv Univ 84: 1–473

Laackmann H (1907) Antarktische Tintinnen. Zool Antarct 31: 235–239

Laackmann H (1910) Die Tintinnodeen der Deutschen Südpolar-Expedition 1901–1903. Dtsch Südpolar-Exped 1901–1903, 11: 341–496

Landry MR, Hassett RP (1982) Estimating the grazing impact of marine microzooplankton. Mar Biol 67: 283–288

Landry MR, Monger BC, Selph KE (1993) Time-dependency of microzooplankton grazing and phytoplankton growth in subarctic Pacific. Prog Oceanogr 32: 205–222

Leakey RJG, Fenton N, Clarke A (1994) The annual cycle of planktonic ciliates in nearshore waters at Signy Island, Antarctica. J Plankton Res 16: 841–856

Lessard EJ, Swift E (1986) Dinoflagellates from the North Atlantic classified as phototrophic or heterotrophic by epifluorescence microscopy. J Plankton Res 8: 1209–1215

Lessard EJ, Foy MS, Garrison DL, Gowing MM (1999). Grazing on phytoplankton blooms in the Ross Sea polynya: November–December, 1994. Antarct JUS (in press)

Littlepage JL (1968) Plankton investigations McMurdo Sound. Antarct JUS 3: 162–163

Lynn DH, Montagnes DJS (1988a) Taxonomic descriptions of some conspicuous species of strobilidiine ciliates (Ciliophora: Choreotrichida) from the Isles of Shols, Gulf of Maine. J Mar Biol Assoc UK 68: 639–658

Lynn DH, Montagnes DJS (1988b) Taxonomic descriptions of some conspicuous species of Strombidiidae (Ciliophora: Oligotrichida) from the Isles of Shols, Gulf of Maine. J Mar Biol Assoc UK 68: 259–266

Marino D, Cabrini M (1999) Phytoplankton distribution in the Ross Sea (Antarctica) during austral spring and summer 1994. (In: Faranda F, Guglielmo L, Povero P (eds) National Programme for Antarctic Research, Oceanographic Campaign 1994–95, Data Rep Part II (in press)

Monti M, Fonda Umani S (1995) Tintinnids in Terra Nova Bay-Ross Sea during two austral summers (1987/88 and 1989/90). Acta Protozool 34: 193–201

Nothing E, von Bodungen B, Sui Q (1991) Phyto- and protozooplankton biomass during austral summer in surface waters of the Weddel Sea and vicinity. Polar Biol 11: 293–304

Pierce RW, Turner JT (1993) Global biogeography of marine tintinnids. Mar Ecol Prog Ser 94: 11–26

Popofsky A (1909) Die Radiolarien der Antarktis. Dtsch Sudpolar Exped 10: 306 pp

Putt M, Stoecker DK (1989) An experimentally determined carbon: volume ratio for marine oligotrichous ciliates from estaurine and coastal waters. Limnol Oceanogr 34(6): 1097–1103

Saggiomo V, Carrada GC, Mangoni O, Ribera d'Alcala M, Russo A (1998) Spatial and temporal variability of size-fractionated biomass and primary production in the Ross Sea (Antarctica) during austral spring and summer. J Mar Syst 17: 115–127

Schroder O (1909) *Sticholonche zanclea* und *Wagnerella borealis*. Deut Sudpolar Exp, 10: 317–322

Stocchino C, Manzella GMR (1991) Analisi delle correnti marine di Baia Terra Nova (Mare di Ross-Antartide). CNR Area di ricerca di Genova, pp 1–55

Stoecker DK, Gifford DJ, Putt M (1994) Preservation of marine planktonic ciliates: losses and cell shrinkage during fixation. Mar Ecol Prog Ser 110: 293–299

Stoecker DK, Putt M, Moisan T (1995) Nano- and microplankton dynamics during the spring *Phaeocystis* sp. bloom in McMurdo Sound, Antarctica. J Mar Biol Assoc UK 75: 815–832

Utermöhl H (1958) Zur Vervollkommnung der quantitativen Phytoplankton-Methodik. Mitt Int Ver Theor Angew Limnol 9: 38

Wasik A, Mikolajczyk E (1994) Annual cycle of tintinnids in Admiralty Bay with emphasis on seasonal variability in *Cymatocylis affinis/convollaria* lorica morphology. J Plankton Res 16(1): 1–8

Williams R, McCall H, Pierce RW, Turner JT (1994) Speciation of the tintinnid genus *Cymatocylis* by morphometric analysis of the loricae. Mar Ecol Prog Ser 107: 263–272

# Copepod Communities in the Pacific Sector of the Southern Ocean in Early Summer

T. Zunini Sertorio[1], P. Licandro[1], C. Ossola[2], and A. Artegiani[3]

**ABSTRACT**

Copepod composition, distribution, abundance and biomass were investigated in offshore waters of the South Pacific at the beginning of the Antarctic summer, during the Italian Antarctic expedition in 1989–90 with the RV Cariboo. The stations were scattered in a vast area extending from 62° to 73°S, within the longitudinal limits of 161°E and 177°W. Zooplankton samples were collected at different depth with an EZ-BIONESS-NET fitted with 500-μm mesh opening-closing nets. Depending on the hydrology and ice-cover conditions, four zones were considered in the study area from north to south, i.e. near the polar front, in the pack ice, with drifting ice and ice-free. In all areas except the pack-ice zone, the bulk of copepods occurred close to the surface. In the first 100 m, the assemblages mainly consisted of adult females of *M. gerlachei* and *C. acutus* already clustered at the surface, ready for spawning. On the pack-ice zone, however, the ascent of copepods to the surface had not yet begun or was less pronounced. The greater concentration of copepods was found in the upper 100 m of the southernmost area, i.e. the one that had been ice-free for longer. Below the 100-m depth copepod numbers and biomass abruptly diminished and copepod diversity notably increased. Recorded biomass of oceanic copepods was found to greatly vary in the upper 200 m, ranging from 0.02 to 2.3 g DW/m$^2$. It varied in the range 0.04–1.2 g DW/m$^2$ in the upper 1000 m. It was noticed that sampling methods may constitute a significant source of variability in determining zooplankton standing stocks. In the present study the highest concentrations of *M. gerlachei* were detected only analyzing the euphotic layer (0–200 m) divided in 20-m-thick sections, while they went undetected in the same euphotic layer divided in 100-m-thick sections.

## Introduction

Copepods were studied during the cruise in the SW sector of the Pacific by the Italian Antarctic expedition in 1989–90. Stations were located along the cruise track from New Zealand to the Italian base in Terra Nova Bay. This area is characterized by the presence of the Antarctic Circumpolar Current (ACC) which encircles, without interruption of emerged land, the Antarctic continent from west to east and which is driven by the westerlies that predominate at this latitude.

Several frontal zones are associated to the ACC. Starting from the north, the first frontal zone is the sub-Antarctic Front (SAF), i.e. the northernmost limit of the current, followed by the Polar Front (PF) which is divided into different subfronts. More to the south, there is the Southern ACC Front which, in this part of the Southern Ocean, approximately coincides with the southern boundary of the ACC (Orsi et al. 1995). Between New Zealand and the Antarctic Continent, the ACC lies between 56° and 66°S, while the PF is located around 63° (Orsi et al. 1995); however, these locations are subject to remarkable interannual fluctuations (Gordon 1988; Artegiani et al. 1997). The Polar Front (PF), where cold Antarctic water slips below the warmer sub-Antarctic one, is of particular biological interest as it has been found to be a marked transition zone for surface zooplankton (Deacon 1982 and references therein).

The southern limit of the ACC determines the division between two large-scale subsystems that are known to exist in the Southern Ocean. The

[1] Dipartimento per lo studio del Territorio e delle sue risorse, Università di Genova, Corso Rainusso 14, 16039 S. Margherita Ligure, Genoa, Italy

[2] Museo Nazionale dell'Antartide, Sezione di Genova Viale Benedetto XV, 5 16132 Genoa, Italy

[3] CNR – IRPEM Ancona Molo Mandracchio, 60125 Ancona, Italy

first, where waters are always ice-free, extends towards the north until the Subtropical Front, and the other, that is seasonally covered by ice, extends southwards in the area close to the continent. Here, on the continental shelf, a third ecosystem is recognized, where waters are permanently or nearly always covered by pack ice (Hempel 1985).

The copepod samples analyzed in the present study were all obtained in oceanic waters in the seasonal pack-ice zone, in areas still covered with pack ice, as well in areas free from ice. This study was undertaken to provide information on copepod composition, spatial distribution, abundance and biomass considering the hydrography and ice cover in this region.

In Antarctic waters, copepods have been investigated since the beginning of Antarctic research (see Park 1978 for a review). There are several papers that deal with the taxonomy, distribution and life cycles of copepods, but fewer studies that deal with biomass assessment. Open-ocean estimates of zooplankton biomass, including copepods, have been made by Hopkins (1971, 1985), Hopkins and Torres (1988), Boysen-Ennen et al. (1991), and Ward et al. (1995). Surface macroplankton biomass has been estimated by Pakhomov and McQuaid (1996). It is well known that copepods are important constituents of the zooplankton in all of the world oceans. In Antarctic waters, their biomass is considerable, and in some areas, it is probably more conspicuous than euphausiid biomass (Conover and Huntley 1991).

## Materials and Methods

The oceanographic survey was carried out on board the R/V Cariboo in early summer (December 1989–January 1990). The stations were in a sampling area extending from 60° to 75°S, within the longitudinal limits of 161°E and 177°W. However, the present study deals only with offshore stations located between 62° and 73°S, in which the same sampling and study methods were used. Due to sampling failures, not all of the hydrological stations shown in Fig. 1 have a corresponding biological station. The copepod stations are listed in Table 1. In all, 185 samples from 11 stations were examined.

Hydrological parameters (temperature and conductivity) were measured using a Neil Brown MKIII CTD (conductivity-temperature versus depth recorder), coupled with a 12-bottle (10-l) General Oceanics Rosette. For CTD calibration, field control and data processing procedures see Artegiani et al. (1992).

**Table 1.** Data on zooplankton stations

| Station | Date | Sampling start local GMT | Bottom depth (m) | Sampled water column (m) | Filtered volume ($m^3$) |
|---|---|---|---|---|---|
| 3 | 02-12-89 | 14.27 | 3220 | 0–200 | 390.1 |
| 5 | 05-12-89 | 10.38 | 3130 | 0–1000 | 1483.5 |
| 7[a] | 07-12-89 | 17.48 | 3505 | 0–200 | 460.6 |
| | 07-12-89 | 18.47 | 3530 | 0–1000 | 1088.8 |
| 8 | 14-12-89 | 08.06 | 3350 | 0–200 | 561.4 |
| | 14-12-89 | 09.32 | 3390 | 0–1000 | 1129.0 |
| 9 | 21-12-89 | 13.13 | 3285 | 0–1000 | 1287.9 |
| | 21-12-89 | 15.27 | 3290 | 0–200 | 633.6 |
| 10 | 23-12-89 | 15.02 | 1325 | 0–200 | 576.4 |
| | 23-12-89 | 16.30 | 1340 | 0–1000 | 1347.7 |
| 11 | 24-12-89 | 16.40 | 940 | 0–200 | 498.6 |
| | 24-12-89 | 18.00 | 845 | 0–800 | 1135.9 |
| 12 | 25-12-89 | 14.34 | 685 | 0–200 | 433.5 |
| | 25-12-89 | 15.46 | 575 | 0–500 | 1070.0 |
| 14 | 27-12-89 | 08.18 | 1575 | 0–200 | 615.6 |
| | 27-12-89 | 14.02 | 1545 | 0–1000 | 1518.7 |
| 15 | 28-12-89 | 08.21 | 2120 | 0–200 | 348.9 |
| | 28-12-89 | 09.33 | 2140 | 0–1000 | 1317.9 |
| 16 | 29-12-89 | 08.17 | 760 | 0–200 | 494.2 |
| | 29-12-89 | 09.31 | 750 | 0–700 | 1155.3 |

[a] In station 7[a] the 0–80 m layer is not subdivided.

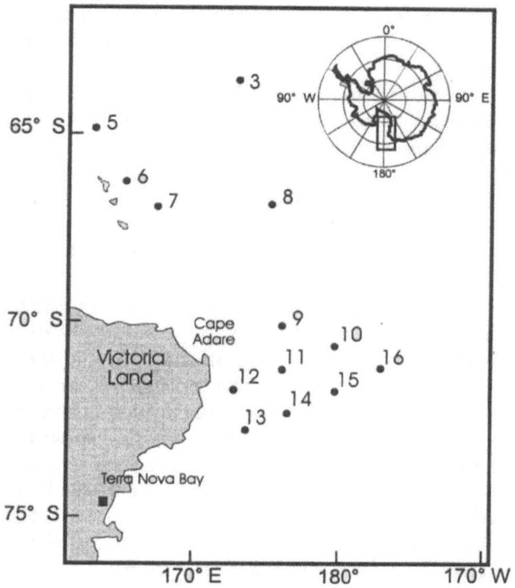


**Fig. 1.** Sampling area

Zooplankton samples were collected at different depths with an EZ-BIONESS-NET fitted with 500-μm mesh opening-closing nets. Samples were taken during the day between 08.00 and 20.00h. When possible, zooplankton was collected over two different water columns at each station: 0–200 m (euphotic layer) and 0–1000m. At station 3, however, only the 0–200m depth was sampled, and at station 5 only the 0–1000m. On three occasions (stations 11, 12, 16), due to the bathymetry, the net fished to depths lower than 1000m. Hereafter, stations sampled from the surface to ≥500m depth will be referred to as deep water columns.

In the 0–200m water columns, the net fished obliquely at nine different levels, the thickest one of 40m (200–160m) and all the others (160–140, 140–120, ... 20–0) of 20m. The 0–1000m water columns were also subdivided into nine strata, the deepest one of 200m (1000–800m) and all the remaining ones of 100m. However, in the water columns which did not reach 1000m (stations 11, 12, 16), this protocol was slightly modified. All samples were preserved in 4% formalin-buffered seawater.

Copepods were delivered to us by the Marine Sorting Center-Italia. Copepods were those present in aliquots drawn from entire zooplankton samples. The aliquots contained no less than 100 copepods, a number considered sufficient to obtain a representative population count. In the laboratory, all the specimens were counted and identified to the lowest possible taxonomic unit, separately taking into account adult males, adult females, and the different copepodite stages. Not all copepods have been identified to the species level. Copepodite stages CI–CIV of *Calanoides acutus* and *Calanus propinquus* have been pooled as adults and CV only were distinguished. The cyclopoids *Oithona* and *Oncaea* and the small calanoids *Ctenocalanus* were considered too small to be quantitatively sampled by the mesh size employed (Zmijewska 1987; Marin 1987; pers. experience). Their numerical importance in coastal and oceanic Antarctic waters is well known (e.g. Chojnacki and Weglenska 1984; Errhif et al. 1997), but unfortunately they have been excluded from the present quantitative analysis. We also have excluded the first copepodite stages of most calanoids but not those of *Euchaeta* which were large enough to be retained by 500-μm meshes (CII, mean = 1.8mm).

Copepods were measured to obtain their size distribution and a rough estimate of their dry weight. The dry weight of dominant forms was determined for each developmental stage - adult males and females, and the different copepodite stages - by rinsing in distilled water and drying a predetermined number of specimens to a constant weight (60°C). Dry weight values were then increased by 30% to account for loss of weight due to formalin preservation. Mean dry weight estimates and mean cephalothorax length of different copepodite stages were used to derive log-linear relationships. The regression derived for all copepods ($\log DW_{\mu g} = 3.197 \log L_{\mu m} - 8.791$) was used to attribute the biomass value to forms which, due to insufficient numbers, did not allow for direct dry weight assessment (see for details Zunini Sertorio et al. 1994). The total sample biomass was estimated by totaling the weight of each species as derived from direct measurement or from size-weight regressions. Abundance and biomass of copepods are expressed in terms of volume of water filtered, or alternatively, in terms of water surface.

Diversity index (Shannon-Weaver H′) and the evenness index (Pielou J′) were calculated over the 0–200m and 0–1000m water columns, with adults and copepodite CV of the species pooled together.

Stations were compared with respect to the distribution of species in the upper 100 m, where copepods were concentrated, and in the underlying 100–1000 m layer, where they were scarce but much more diversified. Q mode classification by hierarchical agglomerative clustering using group average linkage was performed with a PRIMER package (Clarke and Warwick 1994). The percentage similarity method was employed to discern the species responsible for Bray and Curtis similarity within groups, and dissimilarity between groups (Clarke 1993). We used 188 taxonomic descriptors, which represented all of the species sampled (adult males, females and the different development stages) and the upper (0–100 m) and lower (100–1000 m) layers at seven stations, since these were the only ones that reached the 1000 m depth. Densities (no. $m^{-2}$) were log-transformed, and the Bray and Curtis similarity measure was used to produce the association matrix.

## Results

### The Environment

Depending on the hydrologic features of the water column and the ice-cover conditions, four zones were considered in the study area:

1. The zone near the Polar Front (PF), free from ice, including station 3 located at 63°S.
2. The zone in the pack ice, near the Balleny Islands, including stations 5, 6, 7.
3. The oceanic zone with drifting ice, between 66° and 70°S, including stations 8, 9.
4. The ice-free zone, between 71° and 73°S, including two transects with a NE-SW direction: the north transect included stations 10, 11, 12, the south transect included stations 13, 14, 15, 16.

Vertical profiles of temperature and salinity from the surface to 1000 m (Fig. 2) showed discrepancies and similarities between stations, indicating that hydrological properties mainly differed in the upper 100 m of the water column. The thermal characteristics of the surface layer at station 3 closely resembled those of sub-Antarctic waters (T > 0 °C), whereas stations 8 and 9 were characterized by surface Antarctic waters (AASW) (T < −1.0, −1.5 °C) already subjected to summer warming. Intermediate waters (200–1200 m) at station 3 differed from stations 8 and 9 since they were slightly less salty, different from the Circumpolar Deep Water (CDW) found at stations 8 and 9 with higher salinity but lower oxygen content.

Stations 6 and 7, located in an area nearly continuously covered by pack ice (Schweitzer 1985), were characterized by a homogeneous upper layer to 100-m depth with extremely low temperatures (T° < −1.6 °C) whereas the intermediate layer (150–800 m) was dominated by CDW. At station 5, however, further away from the Balleny Islands, summer warming had already initiated.

Surface layers of the north and south transects were characterized by Antarctic surface water (AASW) that was clearly affected by summer warming (T > −1 °C). The northern transect formed by stations 10, 11 and 12 is a section across Scott Canyon, with station 12 on the Mawson Bank. In the subsurface layer, between the 40- and 100-m depths, there was a water mass with a well-marked temperature minimum (T < −1.6 °C) which corresponded to winter Antarctic surface water. The intermediate layer at stations 10 and 11 consisted of CDW with a temperature maximum > +1.3 °C. Antarctic Bottom Water characterized the deep layers at station 12.

The south transect (stations 13, 14, 15 and 16) was located close to the continental slope of the Ross Sea, crossing Scott Canyon, with station 13 on the Mawson Bank and Stn. 16 on the northern extension of Iselin Bank. The central part of the section was dominated by CDW: in fact, along the continental slope of the Ross Sea, CDW upwelled to the surface (Jacobs 1991) along the canyons that characterized the slope and invaded the Ross Sea following the different basins. These are ancient glacier flood plains that furrow this sea in a SW-NE direction.

Phytoplankton data of the Cariboo Cruise, borrowed from Innamorati et al. (1992), are shown in Fig. 2. Marked differences exist in Chl a concentration between the stations, the pack-ice zone being characterized by a much higher chlorophyll concentration.

### Copepods in the Euphotic Layer (0–200 m)

Table 2 shows the abundance (no. $m^{-2}$) and biomass values (g DW $m^{-2}$) of total copepods over the 0–200 m water columns. Data are integrated values of nine sampled layers. With the exception of station 12, the numbers of individuals and the dry weights are in the range $1.8–9.7 \times 10^3$ ind $m^{-2}$

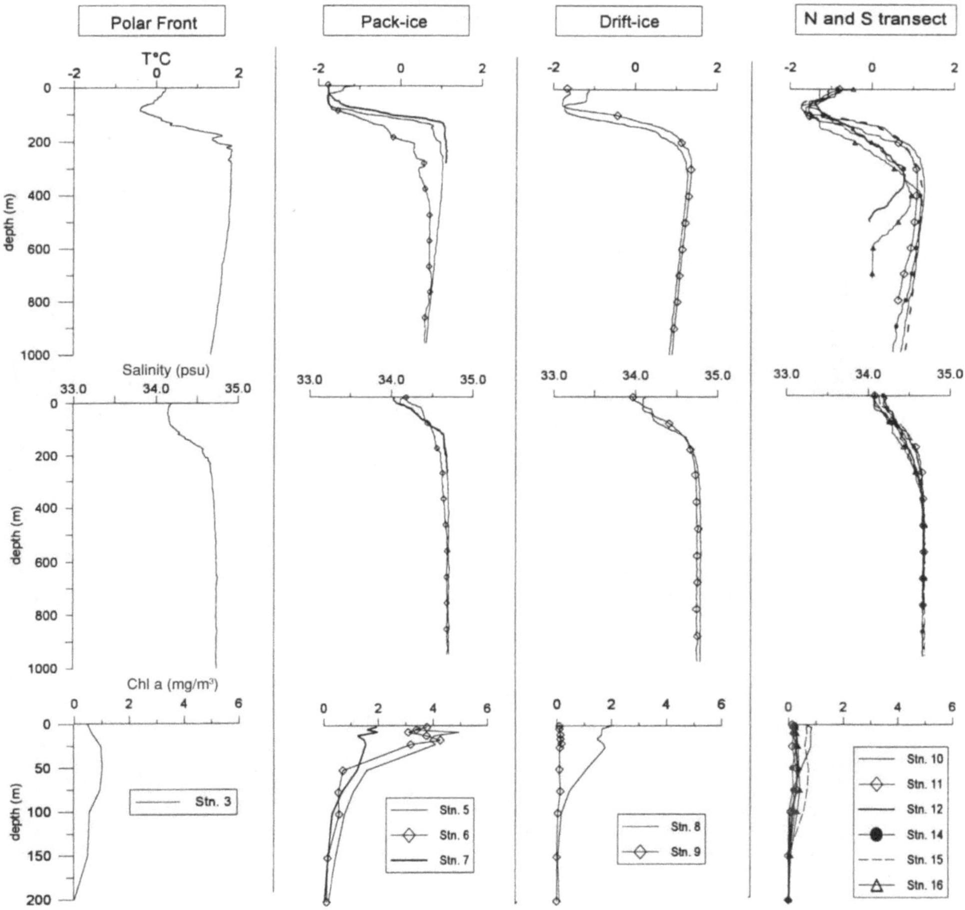


**Fig. 2.** Vertical profiles of temperature (T °C), salinity (psu) and chlorophyll a ($mg\,m^{-3}$) in different stations of the South Pacific, grouped by zones

and 0.3–2.3 g $m^{-2}$, respectively, showing high variability in the investigated area. At station 12, number and biomass were 1 order of magnitude lower than at the other stations ($0.2 \times 10^3$ ind $m^{-2}$ and 0.02 g $m^{-2}$). Due to this low biomass, this station has been included in the tables and figures but excluded from all further calculations.

At all stations except station 3, located near the Polar Front, at 63°S, *Metridia gerlachei, C. acutus, C. propinquus* and *Euchaeta* sp. CII–CIII together accounted for 96–98% of all copepod numbers. At all stations but station 3, the same species accounted for more than 85% of all copepod biomass (Fig. 3A, B). Station 3 revealed a more diverse composition in which also *R. gigas* and the Scolecithricidae family became numerically important with 13 and 7% of all copepod numbers, respectively.

Densities of main copepod components over 0–200 m water columns are given in Table 3. *M. gerlachei* was the main contributor to high numbers and the species with the most irregular distribution over the stations. Adult females were found in very dense concentrations at three

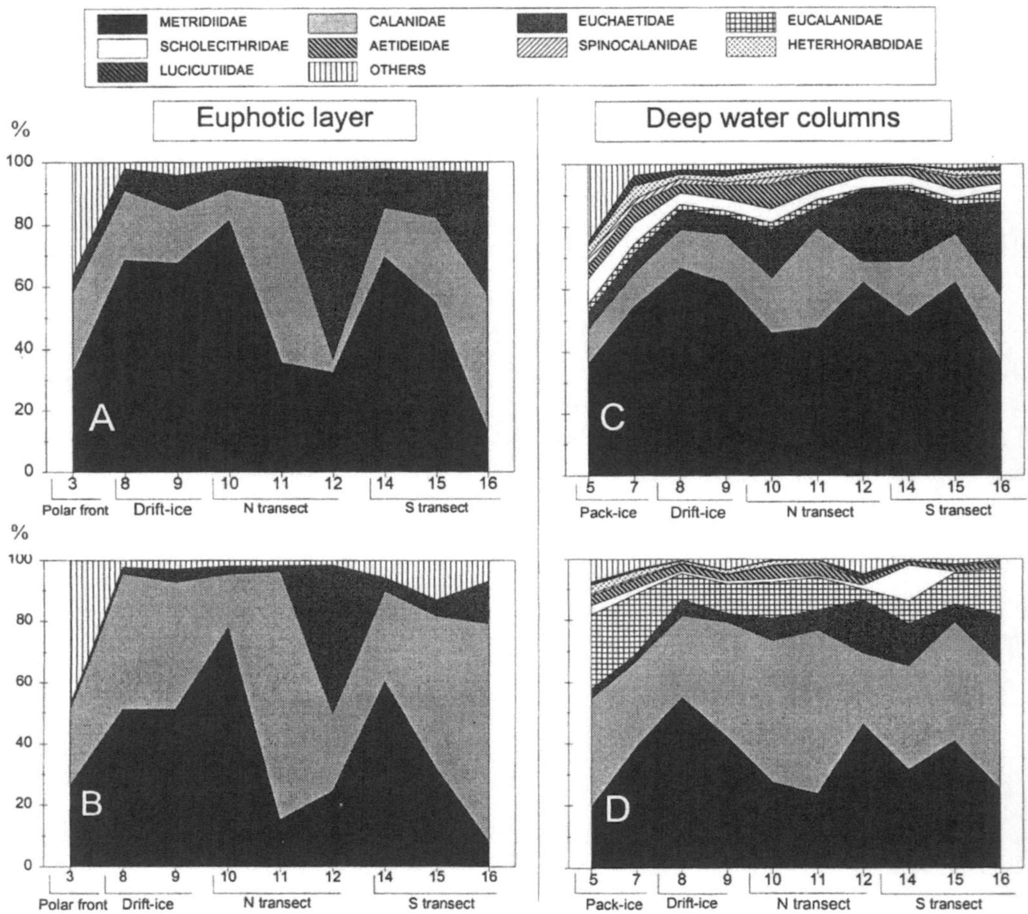


**Fig. 3A–D.** Copepod composition (%) of abundance (**A,C**) and biomass (**B,D**) for the euphotic layer (0–200 m) and deep water columns

stations of the transects (10, 14, 15). Marin (1987) also described high densities of *M. gerlachei* irregularly distributed in the Scotia Sea, on both sides of the Polar front. *C. acutus* was the second contributor in terms of copepod numbers. These two species were also the major biomass contributors, with *M. gerlachei* contributing up to 78% of total copepod dry weight (station 10) and *C. acutus* up to 80% when *M. gerlachei* was scarce (station 11).

The *M. gerlachei* population consisted mainly of adult females and CV copepodites; males were very rare. Deep water sampling revealed that at all the stations adult males were concentrated between 500–800 m. Copepodite CIV were scarce everywhere except at oceanic stations 8 and 9.

The population size of *C. acutus* was substantially greater than that of *C. propinquus*. Both species were present in the samples mainly as adult females and copepodite CV. Males of *C. propinquus* were rare and *C. acutus* ones were completely absent, thereby confirming the observations of several authors (Vervoort 1965; Zmijewska 1985; Marin 1988a) who found that *C. acutus* males are a very exceptional finding at

**Table 2.** Abundance (no. $m^{-2}$) and biomass (mg DW $m^{-2}$) of total copepods in offshore waters of the South Pacific

| Euphotic layer | | | | Deep water columns | | | |
|---|---|---|---|---|---|---|---|
| Station | Water column (m) | no. $m^{-2}$ | mg DW $m^{-2}$ | Station | Water column (m) | no. $m^{-2}$ | mg DW $m^{-2}$ |
| | | | | 5 | 0–1000 | 4070 | 455.5 |
| 3 | 0–200 | 4042 | 818.9 | 7 | 0–1000 | 2779 | 377.6 |
| 8 | 0–200 | 3735 | 676.1 | 8 | 0–1000 | 4912 | 707.6 |
| 9 | 0–200 | 2149 | 307.2 | 9 | 0–1000 | 4090 | 561.2 |
| 10 | 0–200 | 7288 | 1495.6 | 10 | 0–1000 | 3383 | 519.5 |
| 11 | 0–200 | 1945 | 500.0 | 11 | 0–800 | 2481 | 557.2 |
| 12 | 0–200 | 186 | 21.4 | 12 | 0–500 | 443 | 44.6 |
| 14 | 0–200 | 9705 | 2275.9 | 14 | 0–1000 | 4460 | 928.9 |
| 15 | 0–200 | 7545 | 1810.1 | 15 | 0–1000 | 6015 | 1193.4 |
| 16 | 0–200 | 1805 | 409.8 | 16 | 0–700 | 3167 | 688.1 |

**Table 3.** Abundance (no. $m^{-2}$) of the copepod species, Shannon-Weaver diversity index (H′), Pielou eveness index (J′) over 0–200-m water columns at different stations sampled in offshore waters of the South Pacific

| | Zone | Polar Front | Drift ice | | North transect | | | South transect | | |
|---|---|---|---|---|---|---|---|---|---|---|
| | Lat. S | 62°59′ | 66°16′ | 70°12′ | 71°12′ | 71°36′ | 72°10′ | 72°43′ | 72°19′ | 71°56′ |
| | Station | 3 | 8 | 9 | 10 | 11 | 12 | 14 | 15 | 16 |
| *Calanoides acutus* | ♀ | 92 | 395 | 224 | 439 | 877 | 0 | 842 | 947 | 445 |
| | CV | 158 | 292 | 119 | 202 | 121 | 0 | 255 | 446 | 193 |
| *Calanus propinquus* | ♀ | 1 | 76 | 6 | 31 | 4 | 2 | 260 | 476 | 50 |
| | ♂ | 0 | 4 | 0 | 0 | 0 | 0 | 0 | 17 | 4 |
| | CV | 31 | 24 | 7 | 17 | 8 | 3 | 85 | 98 | 64 |
| Calanidae | CIV | 740 | 40 | 12 | 38 | 6 | 2 | 7 | 29 | 22 |
| *Rhincalanus gigas* | ♀ | 302 | 11 | 3 | 9 | 6 | 0 | 127 | 113 | 20 |
| | ♂ | 34 | 0 | 0 | 0 | 0 | 0 | 0 | 0 | 0 |
| | CV | 200 | 0 | 1 | 1 | 1 | 0 | 0 | 0 | 3 |
| | CIV | 0 | 1 | 1 | 0 | 0 | 0 | 0 | 0 | 0 |
| | CIII | 0 | 0 | 0 | 0 | 0 | 0 | 1 | 0 | 0 |
| *Euchaeta* sp. | CV | 0 | 0 | 0 | 0 | 3 | 0 | 1 | 3 | 1 |
| | CIV | 0 | 0 | 2 | 36 | 5 | 0 | 69 | 121 | 19 |
| | CI-CIII | 222 | 258 | 233 | 452 | 198 | 112 | 1126 | 976 | 693 |
| *Metridia gerlachei* | ♀ | 1075 | 1392 | 590 | 5224 | 267 | 0 | 6199 | 2215 | 127 |
| | ♂ | 2 | 14 | 21 | 1 | 1 | 1 | 7 | 14 | 0 |
| | CV | 210 | 896 | 389 | 723 | 367 | 60 | 581 | 1882 | 113 |
| | CIV | 0 | 264 | 420 | 0 | 0 | 0 | 0 | 80 | 0 |
| Others | | 975 | 68 | 121 | 186 | 81 | 6 | 145 | 128 | 51 |
| Shannon-Weaver (H′) | | 2.5 | 1.1 | 1.3 | 0.6 | 1.7 | 0.5 | 1.0 | 1.5 | 1.7 |
| Pielou (J′) | | 0.66 | 0.35 | 0.32 | 0.17 | 0.55 | 0.31 | 0.29 | 0.38 | 0.58 |

shallow depths. A high proportion of males was reported only in deep samples (500–850 m) by Marin (1988a) in late winter. We found very few males in deeper waters: two at station 7 (300–600 m) and two at station 11 (200–300 m). Calanidae CIV (*C. acutus* and *C. propinquus*) was abundant only at station 3.

*Euchaeta* was found in the epipelagic layer almost exclusively as the first copepodite stages CII, CIII; it was much more abundant in the south transect stations. *R. gigas*, an endemic form in the Southern Ocean, was generally scarce in our samples except at station 3 and along the south transect.

All other species were poorly represented in the euphotic zone samples. Only one species was present at all stations (*Racovitzanus antarcticus*), three were rather common (*Scaphocalanus sub-

*brevicornis*, *Heterorhabdus* spp., *Haloptilus* sp.), while all the others were sporadic. These are all forms that, with the exception of *Haloptilus*, have their level of preference in the mesopelagic layer, as highlighted by deep-water sampling.

Vertical distribution data (Fig. 4) reveal that the bulk of the copepods was concentrated in the upper 100 m, with maximum numbers and biomass occurring at less than 80 m, at all stations but station 3. The four dominant species exhibited a different pattern of vertical distribution within the 0–200 m layer. Also in this case, station 3 was remarkably different since it showed two copepod density peaks, one of which was below 140 m.

## Copepods in Deeper Layers (0–1000 m or 0 ≥ 500 m)

Table 2 shows copepod numbers and biomass over the deep water column as integrated values obtained from nine layers. Standing stocks of copepods were in the range 2.4–6.0 × $10^3$ ind $m^{-2}$ and 0.4–1.2 g $m^{-2}$, with the exception of station 12,

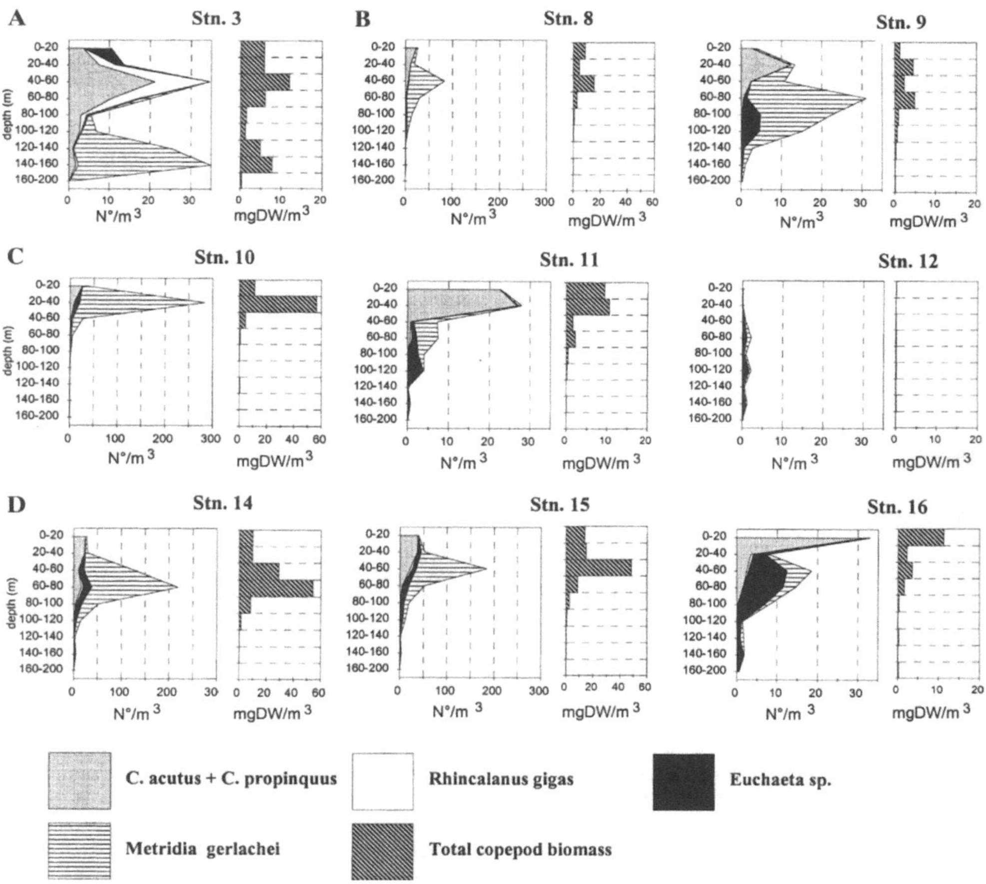


**Fig. 4A–D.** Vertical distribution of abundance of main copepods (no. $m^{-3}$) and total copepod biomass (mg DW $m^{-3}$) in the different layers of the 0–200-m water columns. **A** Polar Front. **B** Drift ice. **C** North transect. **D** South transect. Note the different scales

which was extremely poor ($0.4 \times 10^3 ind\,m^{-2}$ $-0.04\,g\,m^{-2}$), in line with what was noted for the epipelagic layer in the same area. Quantitative values for deep-water columns were of the same order of magnitude as those seen in the upper 200 m. In fact, in all areas except the pack-ice zone, the bulk of copepods was as close to surface as in summer. In the first 100 m, the assemblages mainly consisted of adult females of *M. gerlachei* and *C. acutus* already clustered at the surface, ready for spawning. Andrews (1966) and Vladimirskaya (1978) also observed that fertilized females were dominant in the surface population of *C. acutus* at the beginning of summer. On the pack-ice zone, however, the ascent of copepods to the surface had not yet initiated (station 7) or was less pronounced (station 5). Below the 100-m depth, copepod numbers and biomass abruptly diminished (Fig. 5), but copepod composition became richer in mesopelagic forms. Table 4 indicates only the gross taxonomic units for deep-water columns. The list of identified species is given at the end of the chapter, whereas a full report on copepod taxonomy will be published elsewhere. Figure 3C shows the composition by number of the copepod community over the deep-water columns and highlights the fact that the composition was very similar at all stations located south to 70°S that were either ice-free or with drifting ice, while copepod diversity was greater at stations 5 and 7 located further north in the zone covered with pack ice.

It is worth noting that the numerous mesopelagic species encountered in low densities below the 100-m depth display different types of vertical distributions, somewhat different in the pack ice zone as compared to other areas (Fig. 6).

## Cluster Analysis

Classification of all species in the 0–100 m and 100–1000 m layers of seven stations (5, 7, 8, 9, 10,

**Table 4.** Taxonomic composition (%) of the copepod assemblages over deep water columns at different stations sampled in oceanic waters of the South Pacific. The Shannon-Weaver diversity index (H′) and the Pielou eveness index (J′) values are also indicated

| Zone | Pack ice | | Drift pack | | North transect | | | South transect | | |
|---|---|---|---|---|---|---|---|---|---|---|
| Latitude S | 64°52′ | 66°32′ | 66°16′ | 70°12′ | 71°10′ | 71°39′ | 72°11′ | 72°43′ | 72°18′ | 71°57′ |
| Station | 5 | 7 | 8 | 9 | 10 | 11 | 12 | 14 | 15 | 16 |
| Water column (m) | 0–1000 | 0–1000 | 0–1000 | 0–1000 | 0–1000 | 0–800 | 0–500 | 0–1000 | 0–1000 | 0–700 |
| Calanidae | 11.0 | 10.7 | 12.5 | 15.4 | 17.8 | 32.0 | 6.6 | 17.7 | 15.6 | 20.7 |
| Eucalanidae | 2.8 | 3.2 | 1.8 | 1.6 | 2.0 | 2.8 | 0.4 | 1.7 | 1.8 | 3.4 |
| Paracalanidae | 0.0 | 0.0 | 0.0 | 0.1 | 0.0 | 0.0 | 0.0 | 0.0 | 0.0 | 0.0 |
| Clausocalanidae | 0.3 | 0.3 | 0.0 | 0.5 | 0.0 | 0.0 | 0.0 | 0.0 | 0.0 | 0.0 |
| Spinocalanidae | 3.4 | 1.9 | 1.7 | 0.9 | 1.9 | 0.2 | 0.0 | 0.2 | 1.1 | 0.0 |
| Aetideidae | 3.7 | 5.7 | 2.9 | 4.5 | 8.0 | 5.7 | 1.9 | 2.5 | 3.4 | 2.9 |
| Euchaetidae | 5.1 | 6.2 | 6.3 | 5.9 | 15.6 | 6.3 | 22.9 | 22.7 | 9.2 | 30.6 |
| Phaennidae | 0.3 | 0.6 | 0.0 | 0.0 | 0.0 | 0.0 | 0.0 | 0.0 | 0.0 | 0.0 |
| Scholecithricidae | 8.2 | 6.6 | 3.6 | 3.6 | 4.4 | 2.9 | 4.0 | 2.9 | 3.0 | 2.1 |
| Temoridae | 0.0 | 0.0 | 0.0 | 0.1 | 0.0 | 0.0 | 0.0 | 0.0 | 0.0 | 0.0 |
| Metridiidae | 36.4 | 54.6 | 66.7 | 62.2 | 46.0 | 47.7 | 62.2 | 51.2 | 62.3 | 36.8 |
| Centropagidae | 0.0 | 0.0 | 0.0 | 0.1 | 0.0 | 0.0 | 0.0 | 0.0 | 0.0 | 0.0 |
| Lucicutiidae | 1.5 | 3.3 | 1.6 | 2.0 | 0.8 | 0.5 | 0.8 | 0.1 | 0.7 | 1.3 |
| Heterhorabdidae | 3.3 | 4.3 | 1.4 | 1.8 | 2.5 | 1.3 | 0.5 | 0.6 | 1.4 | 1.3 |
| Augaptilidae | 1.0 | 0.0 | 0.0 | 0.2 | 0.1 | 0.0 | 0.0 | 0.0 | 0.1 | 0.1 |
| Arietellidae | 0.0 | 0.0 | 0.0 | 0.0 | 0.0 | 0.0 | 0.0 | 0.0 | 0.0 | 0.0 |
| Candacidae | 0.0 | 0.1 | 0.0 | 0.0 | 0.0 | 0.0 | 0.0 | 0.0 | 0.0 | 0.0 |
| Bathypontidae | 0.1 | 0.0 | 0.1 | 0.3 | 0.2 | 0.0 | 0.0 | 0.0 | 0.0 | 0.0 |
| Oithonidae | 0.1 | 0.4 | 0.4 | 0.0 | 0.0 | 0.0 | 0.0 | 0.0 | 0.0 | 0.0 |
| Oncaeidae | 8.8 | 0.1 | 0.0 | 0.0 | 0.0 | 0.0 | 0.0 | 0.0 | 0.1 | 0.0 |
| Unknown | 14.1 | 2.0 | 1.0 | 0.5 | 0.7 | 0.5 | 0.6 | 0.2 | 1.2 | 0.6 |
| Shannon–Weaver (H′) | 4.2 | 2.9 | 2.8 | 1.7 | 2.2 | 2.5 | 2.4 | 1.9 | 1.7 | 2.2 |
| Pielou (J′) | 0.68 | 0.61 | 0.58 | 0.39 | 0.44 | 0.53 | 0.51 | 0.44 | 0.36 | 0.56 |

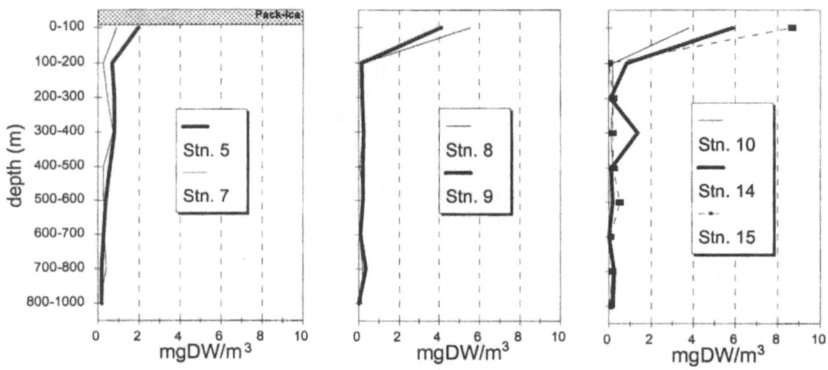


**Fig. 5.** Vertical distribution of total copepod biomass (mg DW $m^{-3}$) over deep-water columns

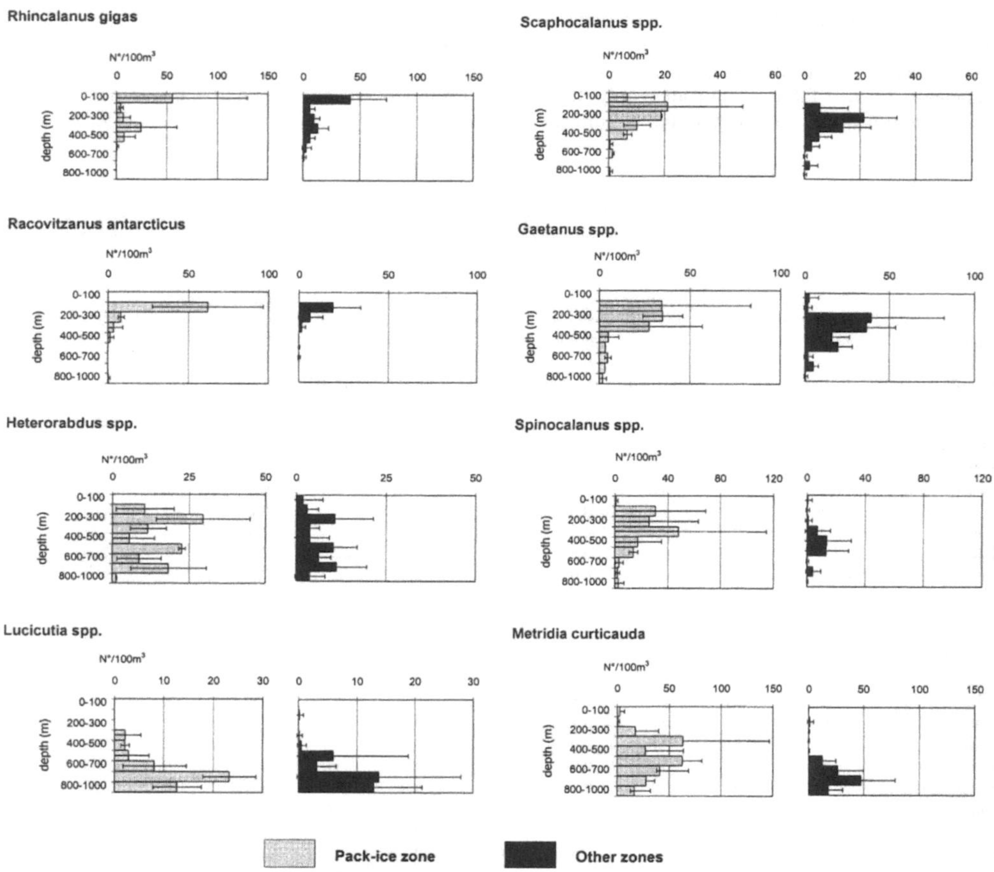


**Fig. 6.** Vertical distribution patterns within the top 1000 m of different mesopelagic copepods (no. 100 $m^{-3}$). The stations are arranged into two groups: the pack-ice zone (stations 5, 7) and the other zones (stations 8, 9, 10, 14, 15)

Fig. 7. Cluster analysis dendrogram of seven stations (depths: A 0–100 m; B 100–1000 m), using group average clustering from Bray-Curtis on log-transformed abundances of all copepods

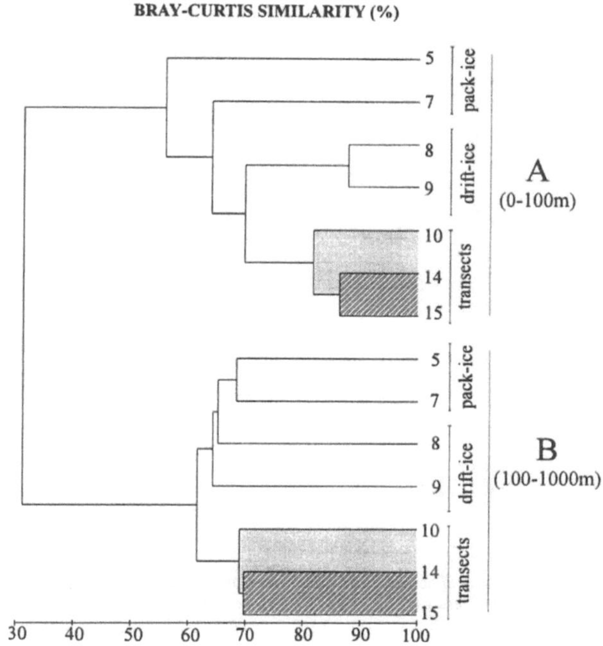


14, 15) located in three different zones revealed, as expected, that the upper and deeper layers were different. The first was richer in copepod numbers and the latter richer in copepod species (Fig. 7). Within the two major groups (Fig. 7A, B), the stations on the transects (10, 14, 15) were always grouped together and separated from those in the drifting ice zone (8, 9) and those in the pack ice (5, 7). The stations on the transects have an average similarity of 83% in the upper 100 m, and of 72% in the mesopelagic depths. Dissimilarity in the upper 100 m between the transect stations and the other stations more to the north was mostly due to juveniles of *M. gerlachei* (CIV), *Euchaeta* (CIII–CIV), *C. propinquus* (*CV*) and Calanidae (CIV), while in deep waters dissimilarity was attributed to juveniles of *Euchaeta* and *Gaetanus* spp. and to adults and juveniles of *Lucicutia* spp. and *Spinocalanus* spp.

The cluster highlights, for the first 100 m, differences between copepod assemblages that mirror the zones into which the area was subdivided for hydrological characteristics and ice cover. For the mesopelagic layer, cluster analysis shows that copepod assemblages at all stations were more similar than at the surface.

## Size Distribution

The size spectrum of copepods ranged between <1 to 9 mm. Size fraction distribution by number was remarkably different from distribution by weight. This is shown in Fig. 8, which shows copepod distributions for classes of length of 1 mm in the epipelagic zone (0–200 m) and in the underlying mesopelagic layer (200–1000 m). Figure 8 gives the average results of three water columns, located in three different zones (stations 5, 8, 10). In the epipelagic layer, the maximum biomass values were in the 3–4 mm size, corresponding to adults of the dominant species *M. gerlachei*, *C. acutus*, and of the common *C. propinquus*, while the maximum numbers were in the size 1–4 mm. At the upper end of the range with a size >8 mm, adult *R. gigas* were recorded, few in number but with a significant biomass.

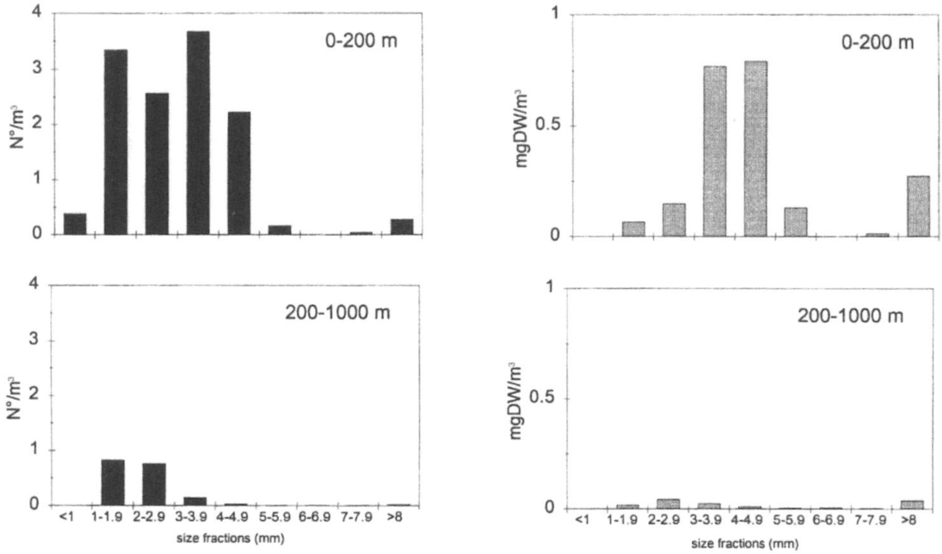


**Fig. 8.** Distribution of copepods abundance (no. $m^{-3}$) and biomass (mg DW $m^{-3}$) among size fractions of 1 mm in the 0–200 m and 200–1000 m layers. Average values from three stations (5, 8, 10)

In the mesopelagic layer, the maximum number of individuals was in the 1–2 mm range, while biomass, even though low, showed its maximum in the 2–3 mm range, with the sporadic presence in the >8-mm class (adult *Euchaeta*).

## Discussion

The study of copepods from stations scattered in a vast area extending from 62° to 73°S allows observations of some aspects of copepod distribution in oceanic waters of the South Pacific Ocean at the beginning of the Antarctic summer. At the time of sampling, in December, the ice was thawing and the area was partly covered with ice, and partly ice-free. It must be noted that in the Ross Sea thawing starts in the south, close to the continent, and then propagates northward (Hecq et al. 1992).

Based on the consideration that sea ice plays a major role in structuring Antarctic ecosystems (Eicken 1992) and that thawing, and especially the ice edge, greatly influences phytoplankton production (El-Sayed and Taguchi 1981; Wilson et al. 1986; Smith and Nelson 1986) and, as a consequence, zooplankton distribution and behaviour (Conover and Huntley 1991), the sampled area was divided into zones depending on their hydrological characteristics and ice cover. Differences among zones in the surface layer, that is Antarctic Superficial Waters (AASW), have already been pointed out by several authors in the study area for the chemical, phytoplanktonic and zooplanktonic characteristics of the waters (see Catalano 1992; Hecq and Guglielmo 1992; Innamorati et al. 1992; Fabiano et al. 1993).

With regard to copepods, noticeable differences were observed between zones in terms of species abundance and vertical distribution rather than copepod composition. At our stations, as expected, the copepod assemblage in the euphotic layer showed poor diversification (diversity index H′ = 0.5–1.7) as is generally observed in the Antarctic in summer where AASW is sampled, both in coastal and oceanic waters (Bradford 1971; Chojnacki and Weglenska 1984; Conover and Huntley 1991; Errhif et al. 1997). The only remarkable exception was station 3 close to the Polar Front (H′ = 2.5) where water with sub-Antarctic characteristics was observed. However, our sampled stations did show remarkable differences between zones with regard to species density and

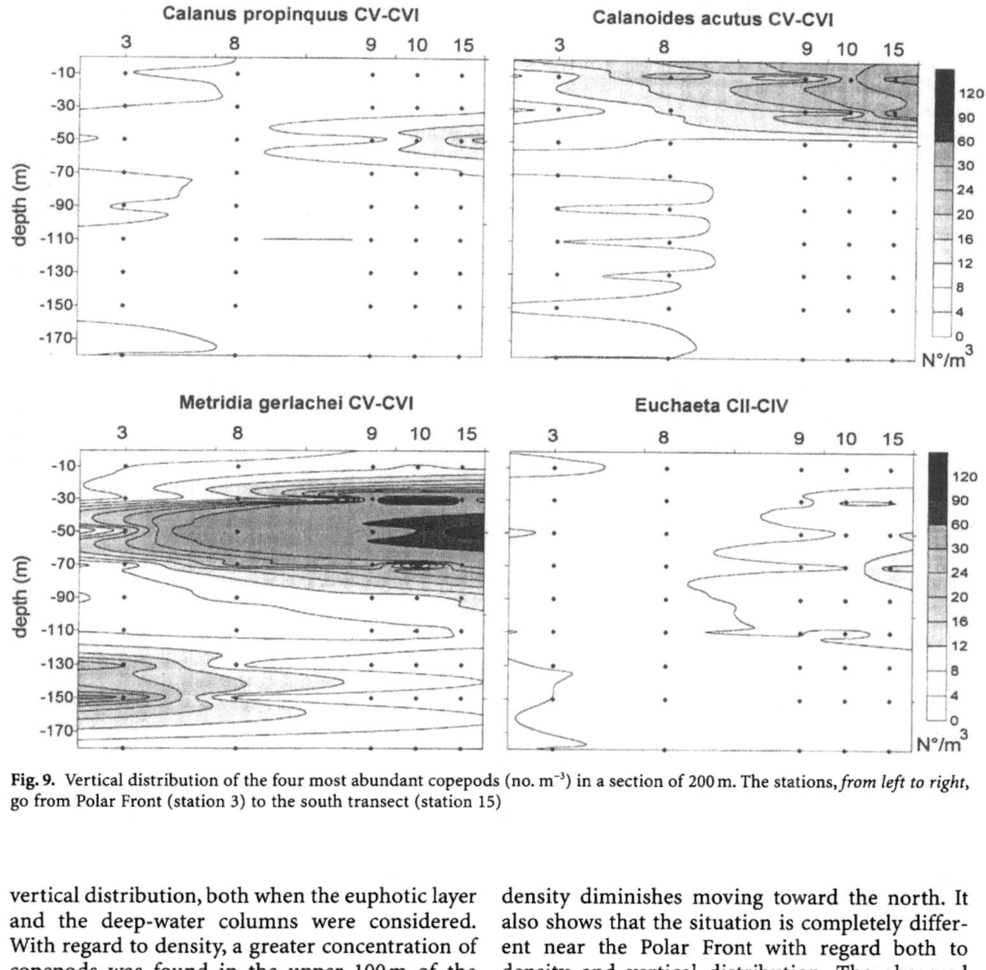


**Fig. 9.** Vertical distribution of the four most abundant copepods (no. $m^{-3}$) in a section of 200 m. The stations, *from left to right*, go from Polar Front (station 3) to the south transect (station 15)

vertical distribution, both when the euphotic layer and the deep-water columns were considered. With regard to density, a greater concentration of copepods was found in the upper 100 m of the southernmost area, i.e. the one that had been ice-free for longer. At all stations, epipelagic copepod assemblages were dominated by adult females and CV copepodites of *C. acutus*, *C. propinquus* and *M. gerlachei*. Very few copepodite stage CIV of Calanidae were observed throughout the sampled area, with the exception of station 3 where, on the contrary, they were particularly numerous. Figure 9 is a cross section along the N-S direction, from close to the Polar Front to the south transect, illustrating density and vertical distribution of the main species in the upper 200 m. It shows that *C. acutus*, *C. propinquus*, *Euchaeta* sp. and *M. gerlachei* all settle at different depths and that their density diminishes moving toward the north. It also shows that the situation is completely different near the Polar Front with regard both to density and vertical distribution. The observed distribution reflects the different behaviours of the dominant species, already highlighted in studies on their life cycles and feeding patterns (Ottestad 1932; Andrews 1966; Voronina et al. 1978; Hopkins 1987; Marin 1988a, b; Atkinson 1991; Schnack-Schiel et al. 1991; Schnack-Schiel and Hagen 1994; Drits et al. 1993; Hopkins et al. 1993a,b).

The difference between zones is even more evident when vertical distributions over deep-water columns are considered. The bulk of copepods resided near the surface in areas free from ice or with drifting ice, while copepods were more homogeneously distributed over the water column in the pack-ice zone, indicating that the ascent to the surface of seasonal migrants (*C. acutus, C. propinquus, R. gigas*) and of *M. gerlachei* had not yet started.

It is worth noting that low abundance was recorded at station 12, even though it was located in the transect zone. This further demonstrates the significant variability of Antarctic plankton, as reported by several authors (e.g. Hardy and Gunther 1935; Hopkins 1971) and may partly be attributable to the pronounced patchy distribution, so common in high latitudes.

Another interesting aspect that sampling allowed for, was data acquisition on oceanic copepod biomass. Recorded biomass of oceanic copepods was found to greatly vary at our stations in the upper 200 m, ranging from 0.02 to 2.3 g DW $m^{-2}$. Most probably our biomass data, obtained in December, do not represent seasonal maxima which seemingly occur later on in summer. Chojnacki and Weglenska (1984), for example, found that maximum copepod seasonal abundance in Cumberland Bay in Feb.–March was 1.5–9 times greater than that recorded in Dec.–Jan.

The biomass data obtained in this study were low, but fall within the range of data reported for oceanic Antarctic waters; these values are much lower than those of Antarctic coastal and fjord areas and of those recently estimated by Ward et al. (1995) for the north-west of South Georgia (Table 5). Sampling methods may constitute a significant source of variability in determining zooplankton standing stocks. The sampling of thin layers, with the BIONESS or LHPR, and their integration over a water column, may result in more significant biomass values. As Longhurst and Williams (1976) have pointed out, sampling in thin layers filtering a great quantity of water may detect zooplankton concentrations that would otherwise go undetected with a net obliquely sampling a thicker layer. This is what we have noticed, when

**Table 5.** Zooplankton and copepod biomass (dry weight) in different Antarctic zones, over different water columns

| Site | Period | Net meshes (μm) | Water column (m) | Zooplankton <20 mm ($g\,m^{-2}$) | Copepods ($g\,m^{-2}$) | Reference |
|---|---|---|---|---|---|---|
| **South Atlantic** | | | | | | |
| Croker Passage | Mar–Apr. 83 | 162 | 0–1000 | 2.90 | 2.69 | Hopkins (1985) |
| Weddell Sea, W | | | | | | Hopkins and Torres |
| Open sea | Mar. 86 | 162 | 0–1000 | 1.10 | 0.89 | (1988) |
| Pack zone | " | 162 | 0–1000 | 1.20 | 1.03 | |
| Weddell Sea, NW | | | | | | Boysen-Ennen |
| Oceanic zone | Feb. 83 | 320 | 0–300 | 2.30 | 1.36 | et al. (1991) |
| Shelf NE | " | 320 | 0–300 | 2.20 | 1.89 | |
| Shelf SW | " | 320 | 0–300 | 0.90 | 0.22 | |
| Scotia Sea | Jun–Aug. 88 | 162 | 0–1000 | 0.64–0.98 | | Hopkins et al. (1993) |
| Cumberland E Bay | | | | | | Ward (1989) |
| (S Georgia) | Sept. 83 | 330 | 0–220 | 2.56 | 2.21 | |
| | Jan. 87 | 330 | 0–220 | 11.99 | 8.61 | |
| South Georgia | | | | | | Ward et al. (1995) |
| Shelf | Jan. 90 | 200 | 0–200 | ~13.00 | | |
| Oceanic zone | | 200 | 0–800 | ~13.00 | | |
| **South Pacific** | | | | | | |
| McMurdo Sound (Ross Sea) | Feb. 83 | 162 | 0–800 | 1.5–3.4 | | Hopkins (1987) |
| Terra Nova Bay | | | | | | Zunini Sertorio |
| (Ross Sea) | Jan. 88 | 500 | 0–200 | | 2.27 | et al. (1994) |
| | " | 500 | 0–600 | | 5.59 | |
| Oceanic zone | | | | | | This study |
| Polar Front, 63°S | Dec. 89 | 500 | 0–200 | | 0.82 | |
| Pack ice, 64°–69°S | " | 500 | 0–1000 | | 0.42 | |
| Drift ice, 66°–70°S | " | 500 | 0–200 | | 0.49 | |
| " | " | 500 | 0–1000 | | 0.63 | |
| Open waters | | | | | | |
| 71°–73°S | " | 500 | 0–200 | | 1.30 | |
| " | " | 500 | 0–1000 | | 0.88 | |

analyzing the euphotic layer divided in 20-m-thick sections, and the same euphotic layer divided in 100-m-thick sections. We detected the highest *M. gerlachei* concentrations only with the former method.

## List of Species

Family Calanidae (all stations)
*Calanoides acutus* (Giesbrecht 1902); *Calanus propinquus* Brady 1883

Family Eucalanidae (all stations)
*Rhincalanus gigas* Brady 1883

Family Paracalanidae (station 9)
*Paracalanus* sp.

Family Clausocalanidae (stations 5,7,8,9,11)
*Clausocalanus* sp.; *Ctenocalanus* sp.; *Farrania frigida* (Wolfenden 1911)

Family Spinocalanidae (stations 5,7,8,9,10,11,14,15,16)
*Mimocalanus cultrifer* Farran 1908; *Mimocalanus inflatus* Davis 1949; *Spinocalanus abyssalis* Giesbrecht 1888; *Spinocalanus antarcticus* Wolfenden 1906; *Spinocalanus horridus* Wolfenden 1911; *Spinocalanus magnus* Wolfenden 1904; *Spinocalanus spinipes* Brodsky 1950; *Spinocalanus spinosus* Farran 1908; *Spinocalanus* sp.

Family Aetideidae (all stations)
*Aetideopsis antarctica* (Wolfenden 1908); *Aetideopsis minor* (Wolfenden 1911); *Aetideopsis* sp; *Aetideus australis* Vervoort 1957; *Chiridiella megadactyla* Bradford 1971; *Chiridiella* sp.; *Euchirella rostromagna* Wolfenden 1911; *Euchirella* sp.; *Gaetanus antarcticus* Wolfenden 1905; *Gaetanus brevispinus* (Sars 1900); *Gaetanus tenuispinus* (Sars 1900); *Gaetanus* sp., *Pseudochirella elongata* (Wolfenden 1905); *Pseudochirella hirsuta* (Wolfenden 1905)

Family Euchaetidae (all stations)
*Pareuchaeta antarctica* Giesbrecht 1902; *Euchaeta* sp.

Family Phaennidae (stations 5,7,8,9,15,16)
*Cephalophanes* sp; *Cornucalanus antarcticus* Brodsky & Zvereva 1950; *Onchocalanus magnus* (Wolfenden 1906);

Family Scolecithricidae (all stations)
*Amallophora* sp.; *Amallotrix* sp.; *Lophotrix simplex* Wolfenden 1911; *Mixtocalanus alter* (Farran 1929); *Mixtocalanus vervoorti* (Park 1980); *Racovitzanus antarcticus* Giesbrecht 1902; *Racovitzanus* sp.; *Scaphocalanus affinis* (Sars 1905); *Scaphocalanus antarcticus* Park 1982; *Scaphocalanus brevicornis* (Sars 1900); *Scaphocalanus echinatus* (Farran 1905); *Scaphocalanus farrani* Park 1982; *Scaphocalanus parantarcticus* Park 1982; *Scaphocalanus subbrevicornis* (Wolfenden 1911); *Scaphocalanus vervoorti* Park 1982; *Scaphocalanus* sp.; *Scolecithricella cenotelis* Park 1980; *Scolecithricella dentipes* Vervoort, 1951; *Scolecithricella emarginata* (Farran 1905); *Scolecithricella ovata* (Farran 1905); *Scolecithricella* sp.

Family Temoridae (station 9)
*Temora* sp.

Family Tharybidae (station 5)
*Undinella acuta* Vauper-Klein, 1970

Family Metridiidae (all stations)
*Metridia curticauda* Giesbrecht 1889; *Metridia gerlachei* Giesbrecht 1902; *Metridia* sp.; *Pleuromamma* sp.

Family Lucicutiidae (all stations)
*Lucicutia curta* Farran 1905; *Lucicutia macrocera* Sars 1920; *Lucicutia magna* Wolfenden 1903; *Lucicutia ovalis* (Giesbrecht 1889); *Lucicutia wolfendeni* Sewell 1932; *Lucicutia* sp.

Family Heterorhabdidae (all stations)
*Hemirhabdus* sp., *Heterorhabdus austrinus* Giesbrecht 1902; *Heterorhabdus farrani* Brady 1918; *Heterorhabdus pustulifer* Farran 1929; *Heterostylites longicornis* (Giesbrecht 1889)

Family Augaptilidae (stations 5,9,10,11,14,15,16)
*Euaugaptilus antarcticus* Wolfenden 1911; *Euaugaptilus nodifrons* (Sars 1905); *Haloptilus ocellatus* Wolfenden 1905; *Haloptilus oxycephalus* (Giesbrecht 1889); *Pachyptilus pacificus* Johnson 1936

Family Arietellidae (station 11)

Family Candaciidae (stations 5,7,9)
*Candacia falcifera* Farran 1929; *Candacia* sp.

Family Bathypontidae (stations 5,8,9,10,11)
*Temorites brevis* Sars 1900

Family Oithonidae (all stations)
*Oithona frigida* Giesbrecht 1902; *Oithona similis* Claus 1866

Family Onceidae (all stations)
*Oncaea curvata* Giesbrecht 1902; *Oncaea* sp.; *Lubbockia* sp.

*Acknowledgements.* We thank Prof. L. Guglielmo of the University of Messina, scientist responsible for the cruise, for the information he gave us regarding the sampling. We are grateful to Elena Markhaseva and Nina Vyskvartzeva, of the Zoological Institute of the Russian Academy of Sciences of St. Petersburg, for their cooperation in the identification of the copepods. We also thank Marina Giallain, of the DIP. TE. RIS. of the University of Genoa for her technical help in measuring and weighing the copepods.

## References

Andrews KJH (1966) The distribution and life history of Calanoides acutus (Giesbrecht). Discovery Rep 34: 117–162

Artegiani A, Azzolini R, Paschini E, Creazzo S (1992) Physical oceanographic conditions in the southern Pacific ocean in the Western Ross Sea. In: National Scientific Commission for Antarctica (ed) Oceanographic Campaign 1989–90, Data Rep, Part II, Genova, pp 5–62

Artegiani A, Russo A, Budillon G, Spezie G (1997) Upper ocean temperature observations along the P14 WOCE section in the Southern Ocean (XBT measurements). Proceedings of the International Conference on the Oceanography of the Ross Sea-Antarctica, Lerici 24–28 March 1997

Atkinson A (1991) Life cycles of Calanoides acutus, Calanus simillimus and Rhincalanus gigas (Copepoda: Calanoida) within the Scotia Sea. Mar Biol 109: 79–91

Boysen-Ennen E, Hagen W, Hubold G, Piatkowski U (1991) Zooplankton biomass in the ice-covered Weddell Sea, Antarctica. Mar Biol 111: 227–235

Bradford JM (1971) The fauna of the Ross Sea. 8. Pelagic Copepoda. Bull N Z Dep Sci Ind Res 206: 9–31
Catalano G (1992) L'attività del gruppo di Oceanografia chimica nell'ambito delle Campagne oceanografiche italiane in Antartide. In: Gallardo VA, Ferretti O, Moyano HI (eds) Oceanografia in Antartide. ENEA–PNRA, Rome, pp 59–70
Chojnacki J, Weglenska T (1984) Periodicity of composition, abundance, and vertical distribution of summer zooplankton (1977/1978) in Ezcurra Inlet, Admiralty Bay (King George Island, South Shetland). J Plankton Res 6: 997–1017
Clarke KR (1993) Non parametric multivariate analyses of changes in community structure. Aust J Ecol 18: 117–143
Clarke KR, Warwick RM (1994) Change in marine communities: an approach to statistical analysis and interpretation. Natural Environment Research Council, UK, Bourne Press, Bournemouth, UK, 144 pp
Conover RJ, Huntley M (1991) Copepods in ice-covered seas – distribution, adaptations to seasonally limited food, metabolism, growth patterns and life cycle strategies in polar seas. J Mar Syst 2: 1–41
Deacon GER (1982) Physical and biological zonation in the Southern Ocean. Deep Sea Res 29: 1–15
Drits AV, Pasternak AF, Kosobokova KN (1993) Feeding, metabolism and body composition of the Antarctic copepod Calanus propinquus Brady with special reference to its life cycle. Polar Biol 13: 13–21
Eicken H (1992) The role of sea ice in structuring the Antarctic ecosystems. Polar Biol 12: 3–13
El-Sayed SZ, Taguchi S (1981) Primary production and standing crop of phytoplankton along the ice edge in the Weddell Sea. Deep Sea Res 28: 1017–1032
Errhif A, Razouls C, Mayzaud P (1997) Composition and community structure of pelagic copepods in the Indian sector of the Antarctic Ocean during the end of the austral summer. Polar Biol 17: 418–430
Fabiano M, Povero P, Danovaro R (1993) Distribution and composition of particulate organic matter in the Ross Sea (Antarctica). Polar Biol 13: 525–533
Gordon AL (1988) Spatial and temporal variability within the Southern Ocean. In: Sahrhage D (ed) Antarctic Ocean and resources variability. Springer, Berlin Heidelberg New York, pp 41–56
Hardy AC, Gunther ER (1935) The plankton of the South Georgia whaling ground and adjacent waters, 1926–1927. Discovery Rep 11: 1–456
Hecq JH, Guglielmo L (1992) Structure and functioning of the Ross Sea pelagic ecosystem: an interdisciplinary approach. In: Gallardo VA, Ferretti O, Moyano HI (eds) Oceanografia in Antartide. ENEA-PNRA, Rome, pp 227–233
Hecq JH, Magazzù G, Goffart A, Catalano G, Vanucci S, Guglielmo L (1992) Distribution of planktonic components related to the structure of water masses in the Ross Sea during the Vth Italiantartide expedition. In: Albertelli G, Ambrosetti W, Piccazzo M, Ruffoni Riva T (eds) Atti 9° Congr Ass Ital Oceanogr Limnol, Lang, Genova, pp 665–678
Hempel G (1985) On the biology of polar seas, particulary the Southern Ocean. In: Gray S, Christiansen ME (eds) Marine biology of polar regions and effects of stress on marine organisms. Wiley, Chichester, pp 3–33
Hopkins TL (1971) Zooplankton standing crop in the Pacific sector of the Antarctic. Antarct Res Ser 17: 347–362
Hopkins TL (1985) The zooplankton community of Croker Passage, Antarctic Peninsula. Polar Biol 4: 161–170
Hopkins TL (1987) Midwater food web in McMurdo Sound, Ross Sea, Antarctica. Mar Biol 96: 93–106
Hopkins TL, Torres JJ (1988) The zooplankton community in the vicinity of the ice edge, western Weddell Sea, March 1986. Polar Biol 9: 79–87
Hopkins TL, Ainley DG, Torres JJ, Lancraft TM (1993a) Trophic structure in open waters of the marginal ice zone in the Scotia-Weddell confluence region during spring (1983). Polar Biol 13: 389–397
Hopkins TL, Lancraft TM, Torres JJ, Donnelly J (1993b) Community structure and trophic ecology of zooplankton in the Scotia Sea marginal ice zone in winter (1988). Deep Sea Res 40: 81–105
Huntley ME, Escritor F (1991) Dynamics of C. acutus (Copepoda: Calanoida) in Antarctic coastal waters. Deep Sea Res 38: 1145–1167
Innamorati M, Lazzara L, Massi L, Mori G, Nuccio C, Saggiomo V (1992) Indagine sulla biomassa fitoplanctonica nel Mare di Ross in relazione ai fattori ambientali. In: Gallardo VA, Ferretti O, Moyano HI (eds) Oceanografia in Antartide. ENEA-PNRA, Rome, pp 235–252
Jacobs SS (1991) On the nature and significance of the Antarctic slope front. Mar Chem 35: 9–24
Longhurst AR, Williams R (1976) Improved filtration systems for multiple-serial plankton samplers and their deployment. Deep Sea Res 23: 1067–1073
Marin V (1987) The oceanographic structure of the eastern Scotia Sea. IV. Distribution of copepod species in relation to hydrography in 1981. Deep Sea Res 34: 105–121
Marin V (1988a) Independent life cycles: an alternative to the asynchronism hypothesis for antarctic calanoid copepods. Hydrobiologia 167/168: 161–168
Marin V (1988b) Qualitative models of the life cycles of Calanoides acutus, Calanus propinquus, and Rhincalanus gigas. Polar Biol 8: 439–446
Orsi AH, Withworth T, Worth DN Jr (1995) On the meridional extent and fronts of the Antarctic circumpolar current. Deep Sea Res 42: 641–673
Ottestad P (1932) On the biology of some southern Copepoda. Hvalradets Skr 5: 1–61
Pakhomov EA, McQuaid CD (1996) Distribution of surface zooplankton and seabirds across the Southern Ocean. Polar Biol 16: 271–286
Park T (1978) Calanoid copepods belonging to the families Aetideidae and Euchaetidae from Antarctic and Subantarctic waters. Antarct Res Ser 27: 91–290
Schnack-Schiel SB, Hagen W (1994) Life cycle strategies and seasonal variations in distribution and population structure of four dominant copepod species in the eastern Weddell Sea, Antarctica. J Plankton Res 16: 1543–1566
Schnack-Schiel SB, Hagen W, Mizdalski E (1991) Seasonal comparison of Calanoides acutus and Calanus propinquus (Copepoda: Calanoida) in the southeastern Weddell Sea, Antarctica. Mar Ecol Prog Ser 70: 17–27
Schweitzer PN (1995) Monthly average polar sea ice concentration. US Geological Survey Digital Data Series DDS-27
Smith WO Jr, Nelson DM (1986) The importance of ice-edge phytoplankton productivity in the Southern Ocean. Bioscience 36: 351–357

Vervoort W (1951) Plankton copepods from the Atlantic sector of the Antarctic. K Ned Akad Wet Verh Afd Nat (Tweede Sectie) 47: 1–156

Vervoort W (1957) Copepods from Antarctic and sub-Antarctic plankton samples. Rep BANZ Antarct Res Exped 1929–1931 (Ser B) 3: 1–160

Vervoort W (1965) Notes on the biogeography and ecology of free-living, marine copepods. Monogr Biol 15: 381–400

Vladimirskaya YV (1978) Age composition of winter population of abundant copepod species in the southern part of the Scotia Sea. Okeanologyia 18: 202–204

Voronina NM, Vladimirskaya YV, Zmijewska MI (1978) On the seasonal variation in the age composition and the vertical distribution of the mass plankton species in the Southern Ocean. Okeanologyia 18: 512–517 (in Russian)

Ward P (1989) The distribution of zooplankton in an Antarctic fjord at South Georgia during summer and winter. Antarct Sci 1: 141–150

Ward P, Atkinson A, Murray AWA, Wood AQ, Williams R, Poulet SA (1995) The summer zooplankton community at South Georgia: biomass, vertical migration and grazing. Polar Biol 15: 195–208

Wilson DL, Smith WO Jr, Nelson DM (1986) Phytoplankton bloom dynamics of the western Ross Sea ice edge. I. Primary productivity and species-specific production. Deep Sea Res 33: 1375–1387

Wolfenden RN (1908) Crustacea VIII. Copepoda. In: National Antarctic Expedition 1901–1904, vol 4. Zoology. British Museum, London, pp 1–46, 7 pls

Zmijewska MI (1983) Copepoda (Calanoida) from the Prydz Bay (Antarctica, Indian Ocean Sector). Pol Polar Res 4: 33–47

Zmijewska MI (1985) Copepoda in the southern part of Drake Passage and in Bransfield Strait during early summer 1983–1984 (BIOMASS-SIEBEX, December-January). Pol Polar Res 6: 79–93

Zmijewska MI (1987) Horizontal and vertical distribution of Copepoda in the southern part of the Drake Passage and in the Bransfield Strait (BIOMASS-SIBEX 1983/1984. Pol Polar Res 8: 381–390

Zunini Sertorio T, Salemi Picone P, Bernat P, Cattini E, Ossola C (1990) Copepods collected in sixteen stations during the Italian Antarctic Expedition 1987–1988. In: National Scientific Commission for Antarctica (ed) Oceanographic Campaign 1987–88, Data Rep Part II, pp 67–125

Zunini Sertorio T, Licandro P, Ricci F, Giallain M (1992) A study of Ross sea copepods. In: National Scientific Commission for Antarctica (ed) Oceanographic Campaign 1989–90, Data Rep Part II, pp 217–246

Zunini Sertorio T, Ossola C, Licandro P (1994) Size, length-weight relationships and biomass of copepods in Antarctic waters (Terra Nova Bay, Ross Sea). In: Albertelli G, Cattaneo-Vietti R, Picazzo M (eds) Atti X Congr Ass Ital Oceanogr Limnol, Lang, Genova, pp 667–680

CHAPTER 24

# Planktonic Copepods in Terra Nova Bay (Ross Sea): Distribution and Relationship with Environmental Factors

**A. Carli[1], L. Pane[1], and C. Stocchino[2]**

**ABSTRACT**

This study presents the results obtained on copepods collected during the Italian oceanographic Antarctic expeditions in 1987–88 and 1989–90. The structure of copepod populations in Terra Nova Bay and its neighbouring Ross Sea zones was examined with particular attention to distribution and abundance of the large-sized species dominating the copepod zooplankton in this region. Analysis of the population structure has emphasized the active reproductive stage of *Calanoides acutus*, *Euchaeta antarctica* and *Metridia gerlachei*; particularly *M. gerlachei* larval stages outnumber adults in the late Antarctic summer. The research was carried out by considering the hydrologic conditions in the area during the Antarctic summer, showing significant differences in abundance among the various zones and zooplankton assemblages in some stations in relation to stable cyclonic gyres of water masses.

## Introduction

Terra Nova Bay is recognized as a marine area of primary scientific interest due to the presence of an extended polynya (Kurtz and Bromwich 1985), thereby offering the opportunity to study climatic relationships between sea ice and the atmosphere and the behaviour of different planktonic species with respect to ice. The summer plankton community in the Ross Sea ice-free area is mostly represented by copepods, particularly abundant between 50 and 100 m (Hecq et al. 1993; Knox 1994); such crustaceans represent the basic link in the marine food chain of the Southern Ocean pelagic areas characterized by scarce krill abundance (Voronina 1984, in Drits et al. 1993). Studies have been carried out on the taxonomy, distribution and biomass of copepods in the Ross Sea and coastal zones, such as McMurdo Sound (Bradford and Jones 1971; Bradford and Wells 1983; Foster 1989; Hopkins 1987); however, Terra Nova Bay and its neighbouring areas are poorly known (Carli et al. 1990, 1992a; Zunini Sertorio et al. 1990).

The present study examines the copepods in Terra Nova Bay and the Ross Sea during the Italian Antarctic expeditions in 1987–88 and 1989–90 and is part of a total zooplankton survey in this area (see Guglielmo et al. 1990, 1992). The aim of this research was to characterize the copepod population structure, with particular attention to vertical and horizontal distributions of the different species. Detailed information is given on large-sized species dominating Antarctic zooplankton biomass (Bathmann et al. 1993; Schnack-Schiel and Hagen 1994): *Metridia gerlachei*, *Calanoides acutus*, *Euchaeta antarctica* and their larval stages relative to environmental factors in Terra Nova Bay, its neighbouring areas and the Ross Sea during the Antarctic summer.

## Materials and Methods

Zooplankton samples were collected during the Italian Antarctic expeditions in 1987–88 and 1989–90, from 04.01.88 to 21.02.88 (R/V Polar Queen) and from 25.11.89 to 22.01.90 (R/V

[1] Università degli Studi di Genova, Istituto di Scienze Ambientali Marine
Viale Benedetto XV, 5, 16132 Genoa, Italy
[2] Istituto per l'Automazione Navale, CNR, Genoa
Via De Marini, 6, 16149 Genoa, Italy

Cariboo). A total of 60 stations were investigated: 33 stations during the first expedition, from Cape Washington (74°39′S) to Drygalski Ice Tongue (75°29′S) along longitude 167°59′E, and 27 stations during the second expedition, in the Pacific sector of the Southern Ocean from New Zealand to the Ross Sea. Samples were collected by means of an EZ-BIONESS NET, a multiple opening and closing net sampler fitted with ten nets with a square mouth opening of $0.25\,m^2$ (Sameoto et al. 1980). Two different mesh size nets were used: 250 µm and 500 µm (1987–88 expedition) and 500 µm (1989–90 expedition). The thickness of sample layers ranged from 10 to 100 m depending on the different depth of the water columns being examined; a sample for the whole water column was collected at each station. Samples with 500-µm mesh were taken in the coastal zone, while 250-µm mesh samples were collected in the Ross Sea. Detailed information about sampling strategies are reported in Guglielmo et al. (1990, 1992) and Crescenti et al. (this Vol.). Samples were kept in 4% formalin-buffered seawater. Copepods were sent to us by the Marine Sorting Center Institute of Marine Environmental Sciences, University of Genoa. The study of copepods was carried out by considering 23 stations during the first expedition and 4 stations, sampled twice (the second sample was indicated as bis) during the second expedition since all these stations were located in the same area (Fig. 1).

The laboratory study consisted in counting and identifying all specimens to species or genus level, by considering separately adults and larval stages (Carli et al. 1990, 1992a; Zunini Sertorio et al. 1990). Hydrological data were obtained using a Meerestechnik KMS multiparametric (CTD) probe capable of providing continuous vertical profiles with one degree of precision, resolution and a time constant ($P$ = 0.35% of the bottom scale, 3000 m; 0.01 m, 20 ms; T: ± 0.02 °C, 0.001 °C, 160 ms; C: ± 0.02 ms $cm^{-1}$; 0.001 ms $cm^{-1}$; 100 ms).

Copepod abundance was expressed in terms of volume of filtered water or alternatively in terms of water surface. Due to the unevenness of samplings at the different stations, and in order to standardize the largest number of available data, the 23 stations during the 1987–88 expedition were examined according to different combinations; particularly, the study of species composition was carried out by analyzing the 0–500 m water column (stations 27, 29, 18, 48, 32, 36), while the 0–150 m (stations 9, 10, 11a, 12, 18, 19, 20, 48, 27, 29, 35, 36, 3, 5, 16, 17, 32, 46) and the 150–500 m (stations 11, 12, 19, 22, 48, 27, 26, 29, 34, 18, 5, 16, 32, 39, 46) water columns were considered for horizontal distribution studies. Vertical distribution was investigated by referring to the levels of the most representative stations.

Data on the abundance of collected organisms at different depths (ind $10\,m^{-3}$) and integrated on the whole water column (ind $10\,m^{-2}$) were used for the statistical analysis; particularly, free-distribution tests (U-test of Mann-Whitney and variance analysis of Kruskal-Wallis) were carried out to evaluate differences among stations, due to the different localisations, during the two sampling periods. Cluster analysis-single linkage (SPSS Inc. Maplnfo package) was used to identify probable assembly of the most representative non-transformed data as abundance of *Calanus propinquus*, *Calanoides acutus*, *Metridia gerlachei* and *Euchaeta antarctica*, large-sized copepods, collected with both 250- and 500-µm mesh nets.

## Results

### Environmental Features of the Study Area

Before considering zooplankton results, it is useful to describe the main environmental features (sea bottom, temperature, salinity and sea current), probably affecting the distribution of these organisms in the study area. Terra Nova Bay lies between 74°35′S and 75°15′S and extends from the coastline to about 165°30′E (Fig. 1). Data on morphobathymetric features were obtained by hydrographic surveys during the 1986–87, 1987–88 and 1989–90 expeditions by the Italian Navy Hydrographic Institute of Genoa, Italy (Stocchino 1989, 1991). The sea bed is relatively deep, with depths of 500–600 m only a few miles offshore. In the northwestern area, there is a submarine plateau characterized, in its central part, by a very irregular pattern apparently engraved by a series of canyons. The main one of these penetrates the Gerlache Inlet in a northwest-southeast direction with depths of about 400–500 m. A poorly defined canyon extends south, from Inexpressible Island up to the Nansen Ice Sheet in the central zone; the above-mentioned submarine topography causes the upwelling of colder and denser water, which could support high productivity, and therefore zooplankton abundance in particular areas. As regards the hydrologic

**Fig. 1.** Study during the 1987–88 (+) and 1989–90 (°) cruises

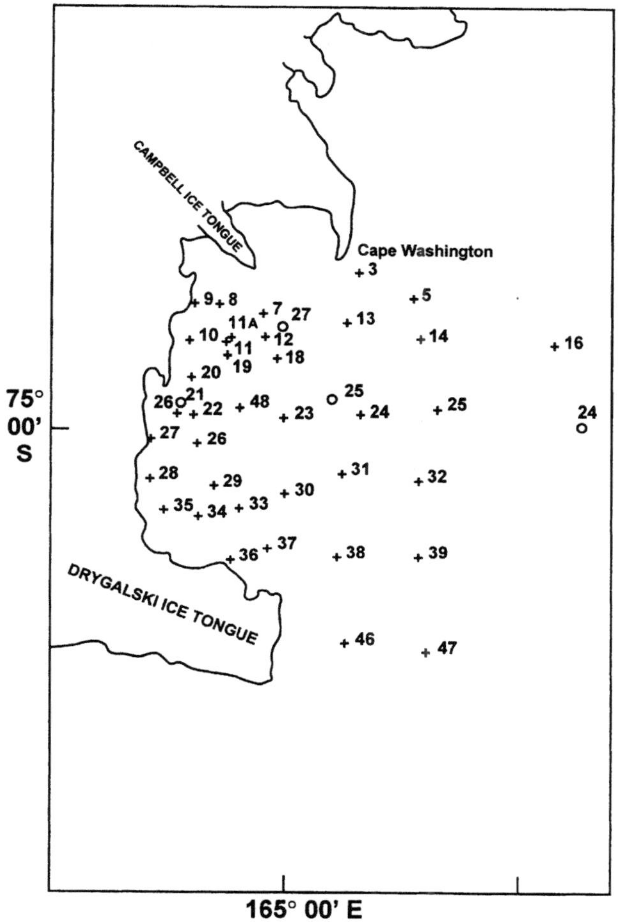


features, chemical and physical data were obtained during the 1986/1987 expedition (Stocchino and Lusetti 1990) and detailed analysis were carried out in the following year (Boldrin and Stocchino 1990). With respect to hydrologic features of Terra Nova Bay, recorded thermohaline values generally agreed with those observed in the Ross Sea during summer (Jacobs et al. 1970) and in McMurdo Sound (Barry 1988). Hydrologic data show that the study area can be divided into a coastal zone, with high temperature (up to 2.1 °C) and salinity (38.4 ppt) values at the surface and an open sea zone (east of the Cape Washington – Drygalski Ice Tongue linkage) characterized by low salinities (<34.6 ppt, up to 33.5 ppt) and temperatures (0.0 to 1.0 °C). Furthermore, the coastal zone shows that salinity and temperature values in the northern area are higher than those observed in the southern region (Fig. 2). The relatively high temperatures in the bay are probably due to several surface thermal processes such as, for example, the local heating and the combined action between wind and heat conduction by marine currents; low offshore salinity is probably due to melting ice and water mass circulation. An interesting cold water cell (−0.2 °C), probably due to wind-induced upwelling, was detected in the central part of the study area, in front of Cape Russell. A minimum

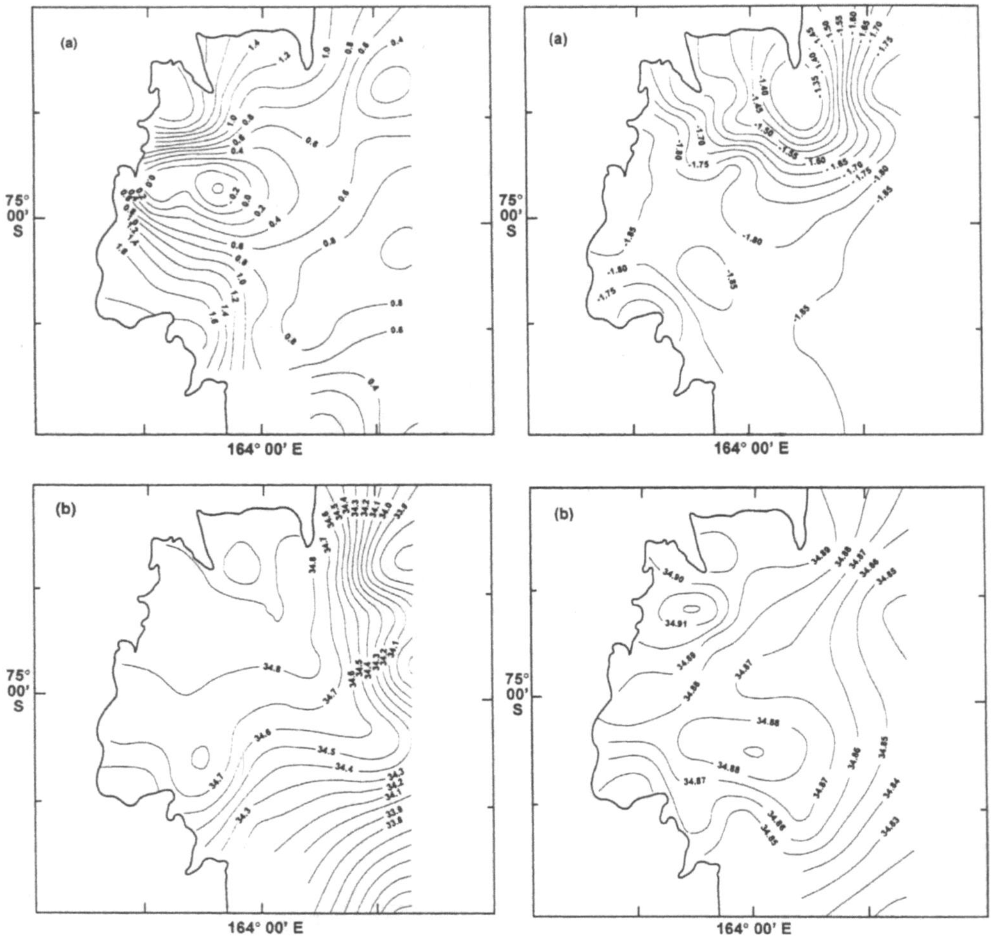


**Fig. 2a,b.** Distribution of temperature (**a**) and salinity (**b**) at the surface (*left*) and 100-m depths

temperature value (−2.0 °C) was observed below the surface layer, around stations 27, 29 and 48 and was identified as Ice Shelf Water, originating from ice melting in the Bay (Jacobs and Comino 1989). As regards the direction and speed of the current, the local circulation of water masses was very complex (Fig. 3). Owing to the peculiar location of Terra Nova Bay, water mass movements are not apparently affected by the Antarctic Coastal Current (ACC) which, on this side of the Ross Sea, turns, almost always, directly north and only partly east. Water movements are clearly affected, therefore, by the mixing processes resulting from the interaction of waters having different thermohaline characteristics: the circulation is then probably due to melting processes, to the presence of salinity gradients and to the upwelling of deep waters near the coast. Particularly interesting is the presence of narrow closed formations (2–3 km width), probably due to both the baroclinic structure, in accordance to temperature and salinity data, and the marked vertical shears in marine currents. These particular cyclonic gyres show an irregular distribution in the central study area;

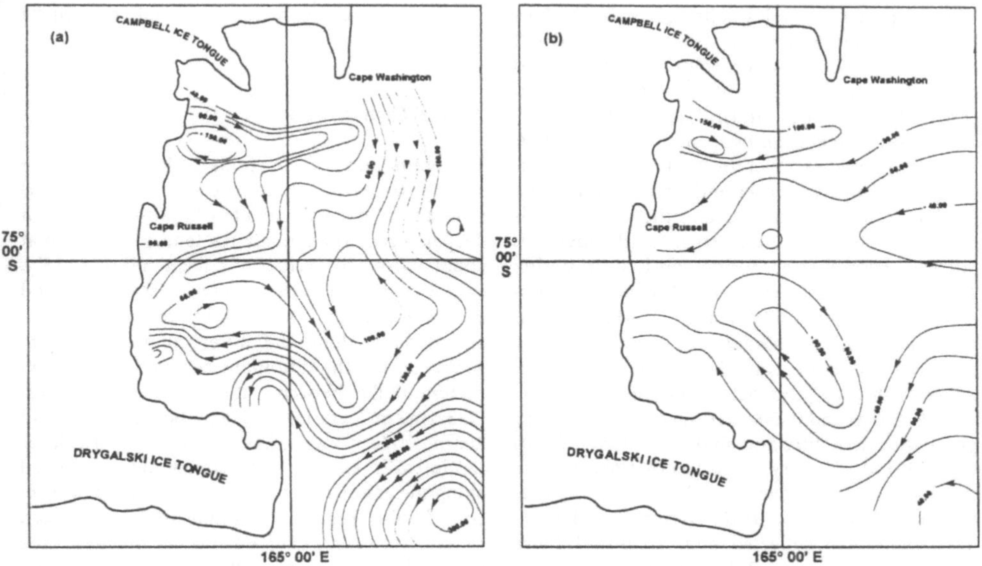


**Fig. 3a,b.** Dynamic topographies at the sea surface (**a**) and 100 m (**b**) depth, referring to the 500 dbar surface

some of them persist along the water column and could support zooplanktonic populations in restricted areas. The main one was observed in the northwestern zone (stations 9, 10, 11, 20) from the surface to 300 m; others were recorded in front of and along the northern edge of the Drygalski Ice Tongue, up to 150 m.

## Copepod Composition

The copepods recorded in the examined water layers belong to the Calanoida, with the following species: *Calanus propinquus* Brady 1883, *Calanoides acutus* Giesbrecht 1902, *Ctenocalanus* sp., *Microcalanus pygmaeus* Sars 1900, *Aetideopsis antarctica* Wolfenden 1908, *Euchaeta antarctica* Giesbrecht 1902, *Euchaeta* sp., *Stephus longipes* Giesbrecht 1902, *Racovitzanus antarcticus* Giesbrecht 1902, *Metridia gerlachei* Giesbrecht 1902, *Paralabidocera antarctica* Thompson 1898, and Cyclopoida, with *Oithona frigida* Giesbrecht 1902, *Oithona similis* Claus 1866, *Oncaea curvata* Giesbrecht 1902, *Oncaea* sp. Early data on copepods sampled during the 1987–88 and 1989–90 expeditions are already published (Carli et al. 1990, 1992a; Zunini Sertorio et al. 1990). The results of samplings (500-μm net) carried out in Terra Nova Bay (Polar Queen cruise) along the 0–500 m water column, at stations 27 and 29, show a total abundance of 16668.4 ind $m^{-2}$ and 17961.9 ind $m^{-2}$, respectively. *Metridia gerlachei* adults and larval stages CI-CV (51.3–54.1%), followed by *Calanoides acutus* (18.6–28.0%), *Calanus propinquus* (1.0–1.4%) and their copepodids CI-CV (2.4–3.3%), *Euchaeta antarctica* and juveniles CI-CV (12.2–15.2%), make up the bulk of the copepod population in Terra Nova Bay. Other copepods, such as *Ctenocalanus* sp. (2.4–3.4%), *Oithona similis* and *Oithona frigida* (2.6–3.8%) and *Oncaea curvata* (<0.1%) are scarce in the studied area. Samples collected at Terra Nova Bay during the Cariboo cruise in 1989–90 from the surface to 500-m water column (500-μm net, stations 24, 25, 26, 27) have a lower abundance than the samples from the previous expedition; in particular, the copepod abundance reaches only 5901.6 ind $m^{-2}$ at station 25. However, the composition of the copepod populations is similar to that previously observed, with *Metridia gerlachei* and larval stages dominating; during this expedition, the adult and juvenile *Racovitzanus antarcticus*, although poorly represented (2.3 ind $m^{-2}$) occurred at station 25.

With respect to samples collected at Terra Nova Bay using the 250-µm net (stations 18 and 48) along the 0–450 m water column, the copepod abundance and composition were different: copepods reach 36 379 ind $m^{-2}$ and 46 941 ind $m^{-2}$, respectively, *Ctenocalanus* sp. and *Oithona similis* were dominant (53.3% of the total collected organisms); *Metridia gerlachei* and their larval stages (23.2%), *Calanoides acutus* and their copepodids (11.4%), *Euchaeta antarctica* and juveniles (10.9%) were scarcer. Inside the bay, at station 12 (Zunini Sertorio et al. 1990), copepod abundance was higher (57 915 ind $m^{-2}$), *Microcalanus pygmaeus* and *Paralabidocera antarctica* also occurred at this site. Outside the bay, samples collected with a 250-µm net at stations 32 and 46, from 0–550 m and 0–500 m, respectively, showed a population structure and abundance (from 29 504 to 42 122 ind $m^{-2}$) very similar to that observed in the bay. The abundance of large-sized species (*Calanoides acutus, Metridia gerlachei* and *Euchaeta antarctica*) collected between 0 and 150 m (Fig. 4) ranged from 1381 (station 17) to 29 747 ind $m^{-2}$ (station 10), between 150 and 450 m from 595 (station 11) to 16 890 ind $m^{-2}$ (station 5). The study of the population structure of these three large-sized copepods in Terra Nova Bay with a 500-µm net showed a prevalence of adult *Metridia gerlachei* and *Calanoides acutus* compared with their larval stages (Fig. 5a). For *Euchaeta antarctica*, the situation was reversed, with juveniles reaching very high percentages (>96%). Samples collected inside and outside the bay with a 250-µm net showed a clear dominance of the three copepodid stages compared to adult specimens (Fig. 5b,c).

## Vertical Distribution

The abundance of copepods collected from the surface to 200 m in the 500-µm sampled stations ranges from a few ten to a few hundred ind $m^{-3}$. In deeper layers, the abundance tend to decrease at about 300 m, with values of a few ten ind $m^{-3}$ (Fig. 6a). *Metridia gerlachei, Calanoides acutus* and *Euchaeta antarctica* showed different distribution and density along the water column. *Metridia gerlachei* was the dominant copepod and occurs preferentially between the surface and 50 m (133.3 ind $m^{-3}$, station 27) and between 250 and 400 m, while it was scarce in the 550–600 m layer (2.7 ind $m^{-3}$, station 27); copepodids occurred along the whole water column, although they were considerably less abundant than adults (Fig. 6b,c). *Calanoides acutus* (Fig. 6d) occurs mostly in the surface layer (0–50 m), with abundances between 16.4 to 56.1 ind $m^{-3}$; larval stages (mostly CV) are scarcer than adults (max. 36 ind $m^{-3}$), but they showed the same vertical distributions (Fig. 6e). *Euchaeta antarctica* (Fig. 6f) was found in lower abundance (0.1–2.5 ind $m^{-3}$) below 250 m; on the contrary copepodids (Fig. 6g) occurred mostly between 100 and 200 m (max. 18 ind $m^{-3}$). The copepod abundance at Stations 12, 18, 48, sampled with 250-µm mesh nets, ranges from 100 to 500 ind $m^{-3}$ in the first 150 m; in the deeper layers abundance drastically decreased to a few ind $m^{-3}$ (Fig.

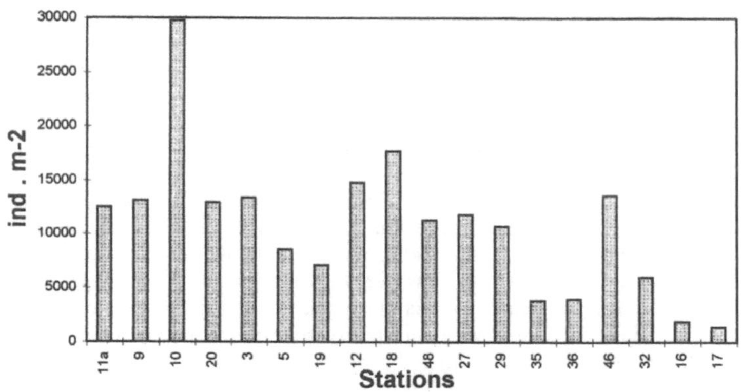


**Fig. 4.** Abundance of copepods (ind $m^{-2}$) from 0–150 m at different stations

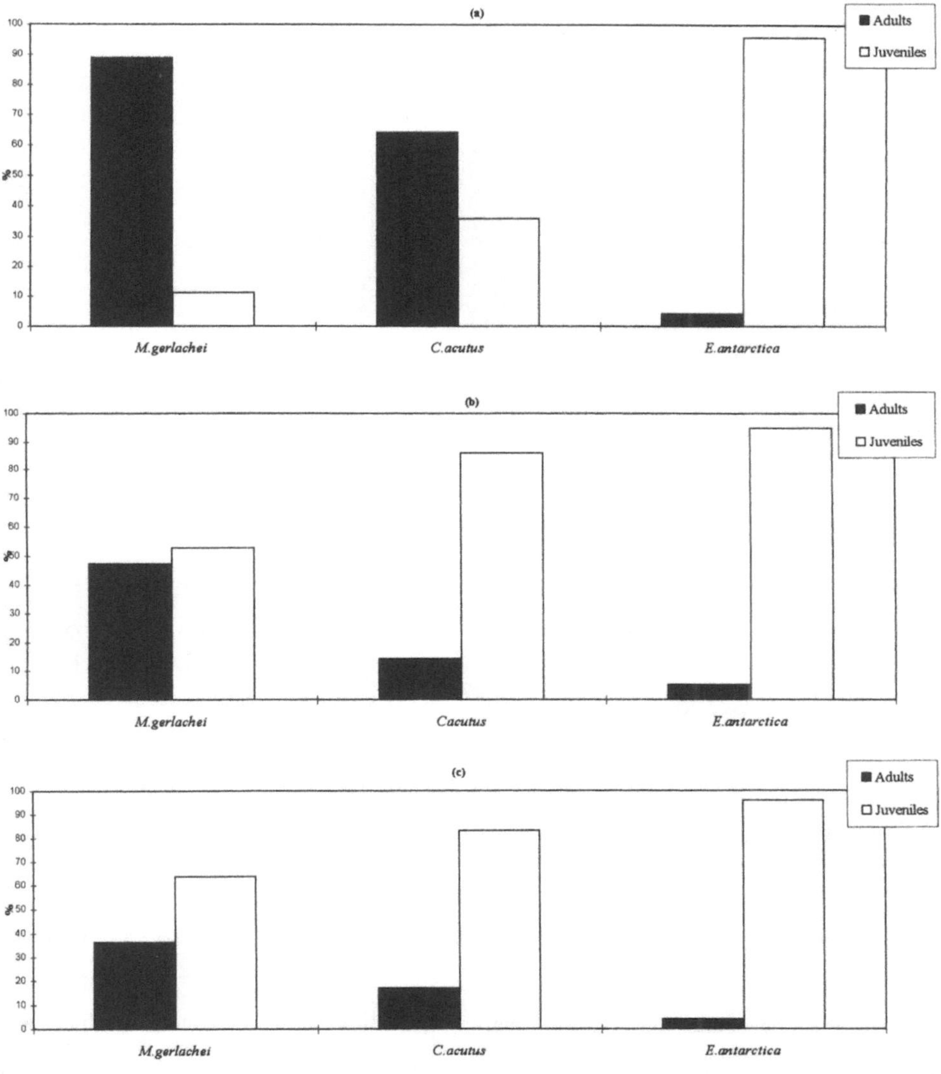


**Fig. 5a–c.** Percentage values of juveniles and adults of *M. gerlachei*, *C. acutus*, *E. antarctica* inside (**a** 500-μm net mesh, **b** 250-μm net mesh) and outside (**c** 250-μm net mesh) Terra Nova Bay

7a). *Metridia gerlachei* (Fig. 7b) mostly occurred in the 150–200 m layer (from 5 to 20 ind $m^{-3}$) and rapidly decreased below 200 m (<5 ind $m^{-3}$); copepodids are abundant between 100 and 150 m (over 75 ind $m^{-3}$) and decreased in the deeper layers up to a few ind $m^{-3}$ (Fig. 7c). *Calanoides acutus* (Fig. 7d) was more abundant between the surface and 150 m and was scarce in the deeper layers, and the larval stages (0–150 m, 160 ind $m^{-3}$) were almost absent below 150 m (Fig. 7e). *Euchaeta antarctica* (Fig. 7f) was scarce up to 150 m, and occurred with a few specimens between 250 and 400 m

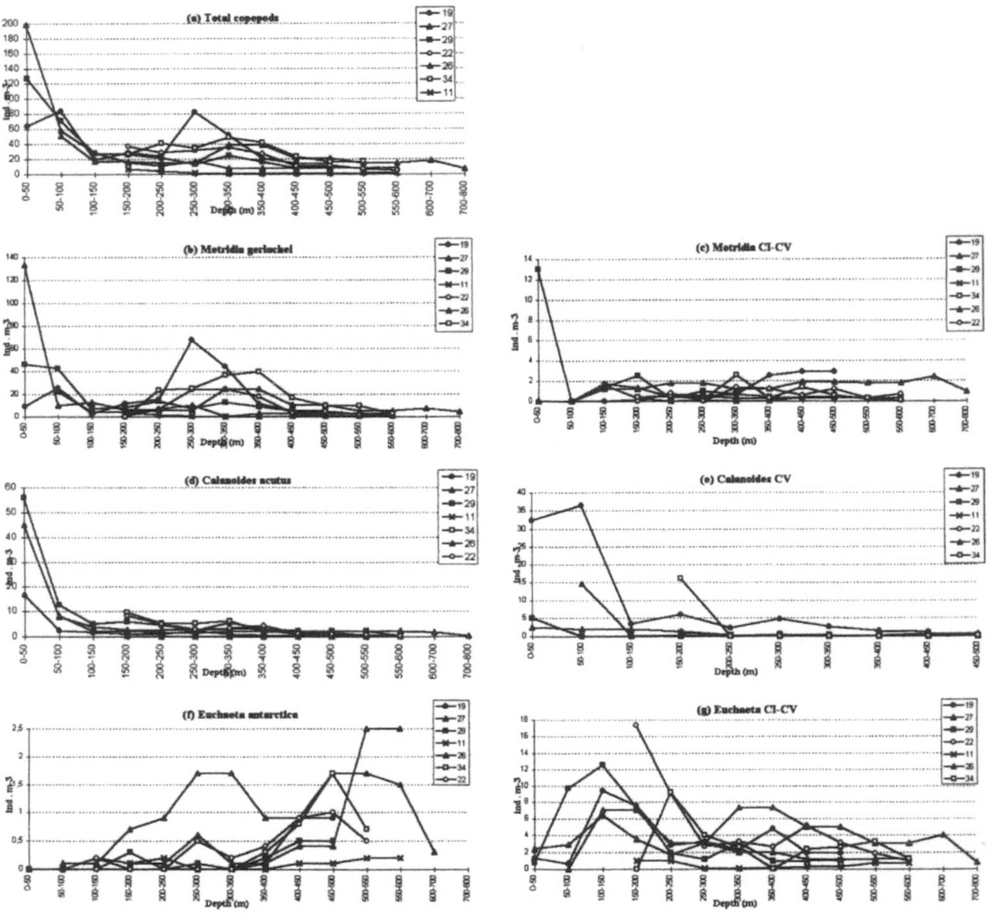


**Fig. 6a–g.** Total copepod abundance (ind $m^{-3}$) and main copepod vertical distribution (Terra Nova Bay, 500-μm net mesh)

(<1.2 ind $m^{-3}$). Copepodids (Fig. 7g) were more abundant than adults and mostly distributed between 50 and 150 m. The total population was dominated by small-sized species, mostly occurring in the surface layers. In particular, *Ctenocalanus* sp. was well represented between 0 and 150 m (270 ind $m^{-3}$), while it was almost absent below 150 m (Fig. 7h). Outside Terra Nova Bay, the total copepod abundance was lower than in the inner area (about 360 ind $m^{-3}$), and the maximum values for the different populations were recorded at the deeper layers (Fig. 8a). *Metridia gerlachei* (Fig. 8b) mostly occurred at about the 300-m level, while copepodids (Fig. 8c) were more distributed along the water column and more abundant in the surface layers. *Calanoides acutus* was infrequently recorded (<11 ind $m^{-3}$) over the whole water column (Fig. 8d), but larval stages were more abundant (100 ind $m^{-3}$) and located between the surface and 100 m (Fig. 8e). *Euchaeta antarctica* showed low abundance (<2.2 ind $m^{-3}$) and mostly occurred below 250–300 m (Fig. 8f); juvenile stages (Fig. 8g) occurred in the upper layers, between 150–200 m. *Ctenocalanus* sp. was well represented in the 100–150 m layer (about 260 ind $m^{-3}$) and was almost absent below 150 m (Fig. 8h).

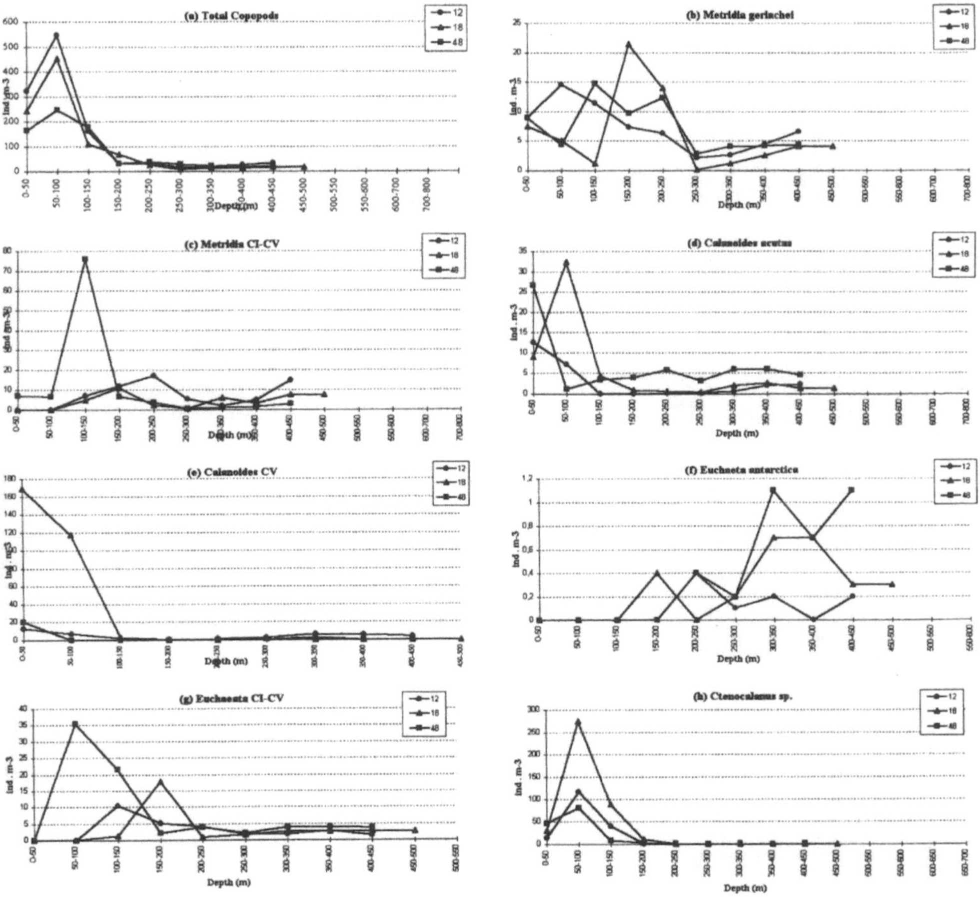


**Fig. 7a–h.** Total copepod abundance (ind $m^{-3}$) and main copepod vertical distribution (Terra Nova Bay, 250-μm net mesh)

## Conclusions

Of the 15 species sampled, typically representative of the copepod fauna in the Ross Sea (Bradford and Jones 1971), *Calanus propinquus*, *Calanoides acutus*, *Ctenocalanus* sp., *Euchaeta antarctica*, *Metridia gerlachei* and *Oithona similis* made up the bulk of the Terra Nova Bay planktonic copepods. These species have also been collected in other zones of the Southern Ocean (Conover and Huntley 1991; Knox 1994). In the coastal zone of Terra Nova Bay, copepods were mostly located in the 0–100 m layer. Cluster analysis emphasised groups at the same depth (Fig. 9), with maximum abundances in the first 50 m, where almost the entire phytoplankton biomass (Innamorati et al. 1990a,b), primary productivity (Magazzù and Decembrini 1990) and pheopigments (Lazzara and Nuccio 1992) occur and where high herbivorous zooplankton grazing has been reported (Hecq et al. 1993). Data on the Terra Nova Bay offshore zone, however, revealed the occurrence of species in the 50–100 m layer, according to data obtained in other pelagic zones of the Ross Sea and the Southern Ocean (Hecq et al. 1993; Knox 1994). Low copepod abundances recorded in Terra

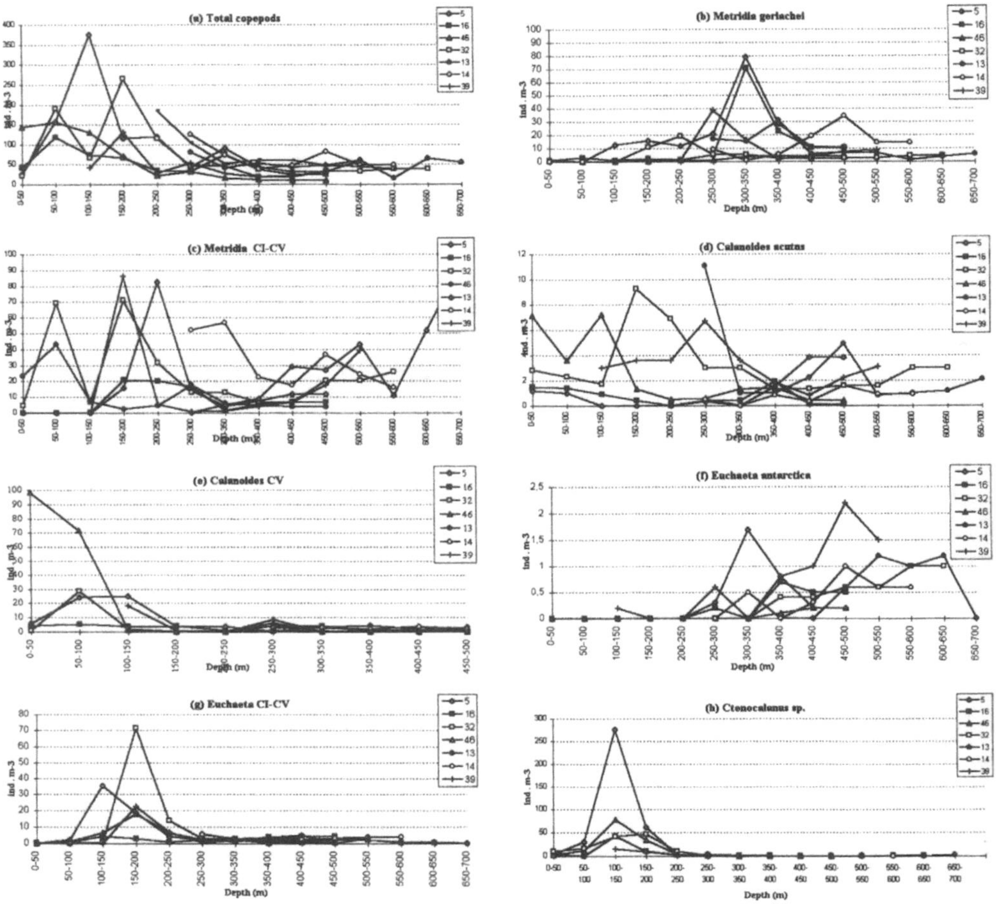


**Fig. 8a–h.** Total copepod abundance (ind $m^{-3}$) and main copepod vertical distribution (outside Terra Nova Bay, 250-μm net mesh)

Nova Bay during the Cariboo expedition were probably linked to the presence of a wide ice-covered zone, responsible for the limited growth of mesozooplankton populations (Hecq et al. 1993). On the basis of the results obtained with 250-μm net samples, the four large-sized calanoids represented the highest biomass, similar to previous studies in Terra Nova Bay (Zunini Sertorio et al. 1990) and in other Southern Ocean zones (Schnack-Schiel and Hagen 1994), while small-sized copepods made up the most abundant fraction of the total population. Such results point out the importance of the net mesh size employed, in obtaining a complete zooplankton population structure of these zones, as recently reported in other cruises (Hopkins 1985; Zwijewska 1985). In this connection, differences observed on density of large-sized copepods in the Terra Nova Bay stations could probably be related both to the different environmental conditions in the sampled areas (northern zone, stations 11a, 12, 9, 10; central zone, stations 19, 27, 18, 48; southern zone, stations 29, 35, 36), and to the different sampling strategies, such as the sampling direction of BIONESS, extremely important in the specimen collection relative to Terra Nova Bay dynamic features. In

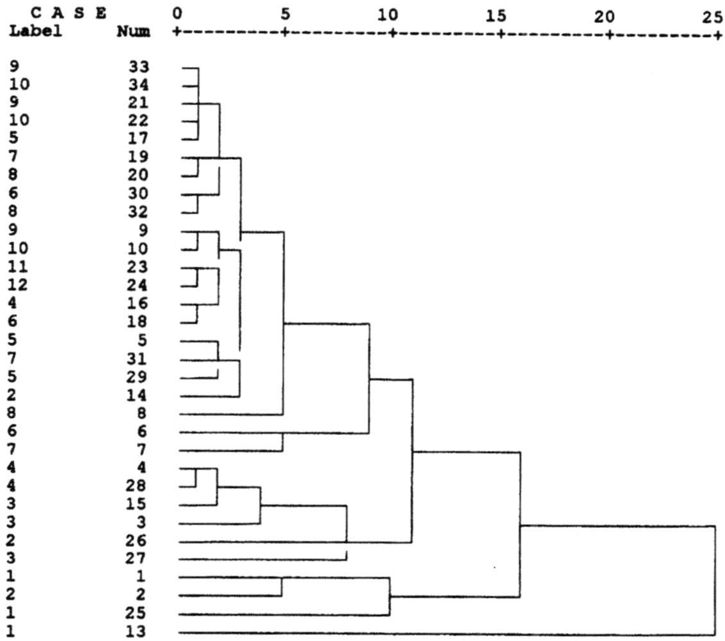


**Fig. 9.** Cluster analysis obtained by processing the depth/station matrix with rescaled distance cluster combine (*Label* number of layers from 0 to 600-m depth; *Num* Number of cases in each layer)

fact, stable cyclonic gyres, observed in the water mass circulation at Terra Nova Bay, could support patches of planktonic organisms as copepods or, for example, larval crustacean decapods (Carli and Pane 1990, 1992), ostracods (McKenzie et al. 1990; this Vol.) and microzooplankton (Fonda Umani and Monti 1990). In particular, copepod populations showed (from 0 to 150 m) decreasing abundances from north to centre and south (median: 13974, 11560 and 3940 ind $m^{-2}$, respectively), with statistically significant differences (Kruskal-Wallis, $p = 0.022$). Low abundances in the southern zone could be related both to low salinity, due to the ice melting of Drygalski Ice Tongue and Nansen Ice Sheet, and to sea bottoms being deeper than in the northern area. The active reproductive stages of *Calanoides acutus*, *Metridia gerlachei* and *Euchaeta antarctica* were confirmed by the presence of their larval stages. With respect to herbivorous *Calanoides acutus*, the copepodids (mainly CV) occurred mostly in the second sampling period, as observed in some Southern Ocean zones (Andrews 1966). Early larval stages of *Metridia gerlachei* increased in the second period (Carli et al. 1992b), when an increase of phytoplankton biomass was observed before the new formation of pack ice (Innamorati et al. 1992). There were statistically significant differences between the two sampling periods (Mann-Whitney U-test, $p < 0.019$) and a probable second period of hatching during late summer can be hypothesized. In fact, Huntley and Escritor (1992) refer to some hatching of *Metridia gerlachei* occurring between November and January.

*Acknowledgements.* Special thanks are due to Dr. Mirvana Feletti and Dr. Gian Luigi Mariottini of the Institute of Marine Environmental Sciences for their collaboration in the preparation of this manuscript, and to Dr. Antonio Marletta for the statistical elaboration of data. The authors would also like to thank the reviewers for their valuable comments.

## References

Andrews KJH (1966) The distribution and life history of *Calanoides acutus* (Giesbrecht). Discovery Rep 34: 117–162

Barry JP (1988) Hydrographic patterns in McMurdo Sound, Antarctica, and their relationship to local benthic communities. Polar Biol 8: 377–391

Bathmann UV, Makarov RR, Spiridonov VA, Rohardt G (1993) Winter distribution and overwintering strategies of the Antarctic copepod species *Calanoides acutus*, *Rhincalanus gigas* and *Calanus propinquus* (Crustacea, Calanoida) in the Weddell Sea. Polar Biol 13: 333–346

Boldrin A, Stocchino C (1990) On the hydrological characteristics of Terra Nova Bay (Ross Sea - Antarctica). In: National Scientific Commission for Antarctica (ed) Oceanographic Campaign 1987–88, Data Rep Part I, Genova, pp 11–57

Bradford JM, Jones NS (1971) The fauna of the Ross Sea. Part 8. N Z Oceanogr Inst Mem 59: 9–31

Bradford JM, Wells BJ (1983) New calanoid and harpacticoid copepods from beneath the Ross Ice Shelf, Antarctica. Polar Biol 2: 1–15

Carli A, Pane L (1990) Contribution to the study of larval crustacean decapods in the Terra Nova Bay (Ross Sea): *Notocrangon antarcticus* (Pfeffer 1887). In: National Scientific Commission for Antarctica (ed) Oceanographic Campaign 1987–88, Data Rep Part II, Genova, pp 161–167

Carli A, Pane L (1992) Larval crustacean decapods collected during the Italian oceanographic campaign in Antarctica 1989–90 Nat Sc Com Ant, Ocean Camp 1987–88, Data Rep II: 211–216

Carli A, Mariottini GL, Pane L (1990) Contribution to the study of copepods collected in Terra Nova Bay (Ross Sea). In: National Scientific Commission for Antarctica (ed) Oceanographic Campaign 1987–88, Data Rep Part II, Genova, pp 129–167

Carli A, Feletti M, Mariottini GL, Pane L (1992a) Contribution to the study of copepods collected during the Italian oceanographic campaign in Antarctica 1989–90. In: National Scientific Commission for Antarctica (ed) Oceanographic Campaign 1989–90, Data Rep Part II, Genova, pp 179–210

Carli A, Feletti M, Mariottini GL, Pane L (1992b) Distribuzione di *Metridia gerlachei* Giesbrecht, 1902 (Copepoda, Calanoida) nella Baia di Terranova (Mare di Ross). In: Ambrosetti W, Piccazzo M, Ruffoni Riva T (eds) Atti IX Congr Ass Ital Oceanogr Limnol, Lang, Genova, pp 623–633

Conover RJ, Huntley M (1991) Copepods in ice-covered seas – distribution, adaptations to seasonally limited food, metabolism, growth patterns and life cycle strategies in polar seas. J Mar Syst 2: 1–41

Drits AV, Pasternak AF, Kosobokova KN (1993) Feeding, metabolism and body composition of the Antarctic copepod *Calanus propinquus* Brady with special reference to its life cycle. Polar Biol 13: 13–21

Fonda Umani S, Monti M (1990) Microzooplanktonic populations in Terra Nova Bay (Ross Sea). In: National Scientific Commission for Antarctica (ed) Oceanographic Campaign 1987–88, Data Rep Part I, Genova, pp 241–254

Foster BA (1989) Time and depth comparisons of sub-ice zooplankton in McMurdo Sound, Antarctica. Polar Biol 9: 431–435

Guglielmo L, Costanzo G, Manganaro A, Zagami G (1990) Spatial and vertical distribution of zooplanktonic communities in the Terra Nova Bay (Ross Sea). In: National Scientific Commission for Antarctica (ed) Oceanographic Campaign 1987–88, Data Rep Part I, Genova, pp 257–398

Guglielmo L, Costanzo G, Zagami G, Manganaro A, Arena G (1992) Zooplankton ecology in the Southern Ocean. In: National Scientific Commission for Antarctica (ed) Oceanographic Campaign 1987–88, Data Rep Part II, Genova, pp 301–468

Hecq JH, Brasseur P, Goffart A, Lacroix G, Guglielmo L (1993) Modelling approach of the planktonic vertical structure in deep austral ocean. The example of the Ross Sea ecosystem. Progress in Belgian Oceanographic Research, Brussels, January 21–22, pp 235–250

Hecq JH, Brasseur P, Goffard A, Lacroix G, Guglielmo L (1993) Modelling approach of the planktonic vertical structure in deep austral ocean. The example of the Ross Sea ecosystem. In: Royal Academy of Belgium, Ministry of Public Health and Environment (eds) Progress in Belgian Oceanographic Research (Brussels, 21–22 January 1993), Brussels, pp 235–250

Hopkins TL (1985) Food web of an Antarctic midwater ecosystem. Mar Biol 89: 197–212

Hopkins TL (1987) Midwater food web in McMurdo Sound, Ross Sea, Antarctica. Mar Biol 96: 93–106

Huntley M, Escritor R (1992) Dynamics of *Calanoides acutus* (Copepoda: Calanoida) in Antarctic coastal waters. Deep Sea Res 38: 1145–1167

Innamorati M, Mori G, Lazzara L, Nuccio C, Lici M, Vanucci S (1990a) Ecology of phytoplankton. In: National Scientific Commission for Antarctica (ed) Oceanographic Campaign 1987–88, Data Rep Part I, Genova, pp 161–238

Innamorati M, Mori G, Lazzara L, Nuccio C, Lici M, Catalano G, Benedetti F (1990b) Phytoplankton ecology in the Ross Sea. In: National Scientific Commission for Antarctica (ed) Oceanographic Campaign 1987–88, Data Rep Part II, Genova, pp 9–63

Innamorati M, Lazzara L, Mori G, Nuccio C, Saggiomo EV (1992) Spatial and temporal distribution of phytoplankton size fractions in Antarctic waters: biomass and production, summer 1989/90. In: Albertelli G, Ambrosetti W, Picazzo M, Riva T (eds) Atti IX Congr Ass Ital Oceanol Limnol, Lang, Genova, pp 605–612

Jacobs SS, Comino JC (1989) Sea ice and oceanic processes on the Ross Sea continental Shelf. J Geophys Res 94: 195–211

Jacobs SS, Amos AF, Bruchhausen PM (1970) Ross Sea oceanography and Antarctic Bottom Water formation. Deep Sea Res 17: 935–962

Knox GA (1994) Zooplankton. In: Knox GA (ed) The biology of the Southern Ocean. Cambridge University Press, Cambridge, pp 67–80

Kurtz DD, Bromwich DH (1985) A recurring, atmospherically forced polynya in Terra Nova Bay. In: Jacobs SS (ed) Oceanology of the Antarctic Continental Shelf. American Geophysical Union, Washington, DC, pp 177–202

Lazzara L, Nuccio C (1992) Relazioni tra particellato, biomassa e popolamenti fitoplanctonici nel Mare di Ross durante la fioritura estiva. In: Albertelli G, Cattaneo-Vietti R, Piccazzo M (eds) Atti X Congr Ass Ital Oceanogr Limnol, Lang, Genova, pp 655–665

Magazzù G, Decembrini F (1990) Primary production and picoplankton C assimilation in the Ross Sea (Antarctica). In: National Scientific Commission for Antarctica (ed)

Oceanographic Campaign 1987–88, Data Rep Part I, Genova, pp 107–157
McKenzie KG, Benassi G, Naldi M, Ferrari I, Menozzi P (1990) Report on planktonic Ostracoda from the Ross Sea, Antarctica. In: National Scientific Commission for Antarctica (ed) Oceanographic Campaign 1987–88, Data Rep Part II, Genova, pp 171–229
Sameoto DD, Jaroszynski LO, Fraser WB (1980) BIONESS, a new design in multiple net zooplankton. Can J Fish Aquat Sci 37: 722–724
Schnack-Schiel SB, Hagen W (1994) Life cycle strategies and seasonal variations in distribution and population structure of four dominant calanoid copepod species in the eastern Weddell Sea, Antarctica. J Plankton Res 16: 1543–1566
Stocchino C (1989) Antartide, Mare di Ross – Baia Terra Nova. Carta batimetrica no 883, scala 1:100.000. Istituto Idrografico della Marina, Genova
Stocchino C (1991) Antartide, Mare di Ross – Baia Terra Nova. Carta batimetrica no 882, scala 1:50.000. Istituto Idrografico della Marina, Genova
Stocchino C, Lusetti C (1990) Prime osservazioni sulle caratteristiche idrologiche e dinamiche di Baia Terra Nova (Mare di Ross – Antartide). Istituto Idrografico della Marina, Genova, 55 pp
Zunini Sertorio T, Salemi Picone P, Bernat P, Cattini E, Ossola C (1990) Copepods collected in sixteen stations during the Italian Antarctic expedition 1987–1988. In: National Scientific Commission for Antarctica (ed) Oceanographic Campaign 1987–88, Data Rep Part II, Genova, pp 67–125
Zwijewska MI (1985) Copepoda in the southern part of Drake Passage and in Bransfield Strait during early summer 1983–1984 (BIOMASS-SIBEX, December–January). Pol Polar Res 6: 79–93

# Crustacean Decapod Larvae in Terra Nova Bay and in the Ross Sea (Cruises 1987–88 and 1989–90)

A. Carli and L. Pane[1]

**ABSTRACT**

Taxonomic studies, as well as density and distribution patterns of larval crustacean decapods collected during the Italian National Antarctic Expeditions in 1987–88 and 1989–90 were carried out. Specimens of *Acanthephyra pelagica* (Risso 1816) (Oplophoridae), *Chorismus antarcticus* (Pfeffer 1887) (Hippolythidae) and *Notocrangon antarcticus* (Pfeffer 1887) (Crangonidae) were classified. With respect to larval density and spatial distribution, it was found that *A. pelagica* occurs only at stations located between the continental slope and pelagic waters north of the Ross Sea, while almost all specimens of *N. antarcticus* and *C. antarcticus* were collected over the continental slope along the east-west 75°S transect and in Terra Nova Bay. The richest level was found between 20 and 100 m; in particular, the greatest density of *N. antarcticus* and *C. antarcticus* occurred between 20 and 50 m and that of *A. pelagica* between 50 and 100 m. Larval *N. antarcticus* were caught at stages I, II and III, *C. antarcticus* at stages II and III.

## Introduction

Crustacean decapods are quite scarce southwards of the Antarctic Convergence (Yaldwyn 1965, 1967); they also show low diversity (Clarke and Crame 1989) and belong almost exclusively to Natantia Caridea (Arntz and Gorny 1991; Gorny et al. 1992). The presence of Reptantia Crustacean Decapods has never been recorded (Yaldwyn 1965; Gutt et al. 1991), except for a specimen of *Paraloma spectabilis* Hansen (1908) in the Pacific Sector of the Southern Ocean around Scott Island (Birsthein and Vinogradov (1967). At present, *Chorismus antarcticus* (Caridea, Hippolythidae) and *Notocrangon antarcticus* (Caridea, Crangonidae) are the only two species found; they are both benthic, occur on the continental shelf and show a circumpolar distribution (Ward 1985; Gutt et al. 1991). *C.antarcticus* crosses the Antarctic Convergence, and is found as far as the Straits of Magellan (Garth et al. 1967; Zarenkov 1968). These two species are, therefore, found in several Antarctic zones: in the Weddell Sea (Vo 1988; Gutt et al. 1991; Gorny et al. 1992), and in the Ross Sea (Borradile 1910; Bage 1938; Ledoyer 1969); furthermore, they occur as "pabulum" for some nototheniid fish in Terra Nova Bay (Vacchi et al. 1994; La Mesa et al. 1997) and for Weddell seals in McMurdo Sound (Dearborn 1965, in Arntz and Gorny 1991). The biology of crustacean decapods, particularly their reproductive cycle (Maxwell 1977; Gorny et al. 1992; Yaldwyn 1966) and adaptation to low temperatures (Maxwell 1977; Clarke 1977, 1979; Clarke and Lakhani 1979; Maxwell and Ralph 1985) has been extensively studied during the last decades. As described by Calman (1907), Gurney (1937, 1942) and Makarov (1970), the reproductive cycle in both species goes through three stages. Adults of *Lebeus antarcticus* (Hippolythidae) occur in the Southern Ocean; they were described by Zarenkov (1968) and subsequently by Ward (1985), who found specimens in the stomach contents of Weddell seals. Recently (Tiefenbacher 1990; Arntz and Gorny 1991) specimens of the benthic *Eualus kinzeri* (Hippolytidae) were found in the Weddell Sea; however, no bibliographic data exist on the larvae of this species. As for the bathybenthic *Nematocarcinus lanceopes* Bate 1988 (Nematocarcinidae), which inhabit the Weddell Sea and the deep waters of the Indian sector of the Southern Ocean (Yaldwyn 1965), their development stages are not yet known (Arntz

[1] Università degli Studi di Genova, Istituto di Scienze Ambientali Marine
Viale Benedetto XV, 5, 16132 Genoa

and Gorny 1991; Gutt et al. 1991, 1994). Other pelagic Antarctic Crustacean Decapods, such as *Acanthephyra pelagica*, *Petalidium foliaceum*, *Pasiphea longispina* and other species of the genus *Pasiphea*, and *Hymenodora gracilis* (Yaldwyn 1965; Kirkwood 1984; Clarke and Holmes 1987; Arntz and Gorny 1991) are seldom found; the same can be said for their larvae, a description of which is still lacking (Gurney 1924; Williamson 1962).

Due to the lack of knowledge on the occurrence of larval Crustacean Decapods in the Southern Ocean and in the Ross Sea, and considering that they had never been collected in Terra Nova Bay, the aim of this work was to increase our knowledge on the density, distribution and biometry of the mesoplankton species collected during two Italian Antarctic Oceanographic expeditions.

## Study Area

The study area during the first expedition (1987–1988 austral summer, R/V Polar Queen) extended from Cape Washington to Drygalski Ice Tongue, and 150 miles off Terra Nova Bay (74°39′S, 75°29′S–167°59′E); in the second cruise (1989–1990 austral summer) it extended from New Zealand to the Ross Sea, Terra Nova Bay included (Fig. 1). The features of the study area are detailed by Crescenti et al. (this Vol.); based on hydrologic data relative to the stratification, the distribution of plankton organisms, markers of primary productivity, and on the relationship between temperature and salinity of surface and intermediate waters, the following zones were identified: sub-Antarctic, Antarctic Polar Front, Antarctic and Continental zone. The characteristics of these zones have been well described by Hecq et al. (1992) and Hecq and Guglielmo (1992).

As concerns Terra Nova Bay, in most of the hydrologic stations examined and common to the zooplankton samples, the water column seems to be vertically stratified and shows clear vertical temperature and salinity gradients (Figs. 2,3) reaching a maximum stability around 50 m (Boldrin and Stocchino 1990; Stocchino and Manzella 1991; Carli et al. (this Vol.).

## Materials and Methods

The zooplankton samples (n = 886) were collected (Fig. 1) using a Bedford Institute of Oceanography and Environmental Sensing System (EZ-NET BIONESS) equipped with ten nets (0.25-m$^2$ opening mouth, mesh size 250–500 µm). Two different mesh sizes were employed according to the different sampling strategies adopted during the two cruises; in 1987–88, both nets (250–500 µm) were utilized, in the second cruise only the 500-µm net was used. Samplings were carried out at different depths; further details on sampling strategies and station characteristics have been reported by Guglielmo et al. (1990, 1992) and Crescenti et al. (this Vol.). The sampler was lowered with the first net opened in order to sample the whole water column from the surface up to the maximum fixed depth; the results of such samplings are reported for each station (Table 1) and were considered only for the elaboration of the qualitative data. Zooplankton samples were fixed in 4% formalin buffered in seawater. Larval crustacean decapods were isolated by the Sorting Centre (Institute of Marine Environmental Sciences, University of Genoa). A total of 212 larvae and postlarvae (167 from the first expedition and 45 from the second) were classified to the species level according to Calman (1907), Lebour (1931), Gurney (1937, 1942, 1960), Williamson (1962), Herring (1967), Makarov (1973). The density (ind 10 m$^{-3}$, ind m$^{-2}$) was calculated for each station and depth and the developmental stages of classified species were also identified. The biometric evaluation of uninjured specimens was carried out by measuring the total length (TL: from the rostral apex to the base of the telson) and carapace length (CL: from the rear of the eye socket to the posterior dorsal edge of the carapace), according to Arntz and Gorny (1991) and Gorny et al. (1992), using a Leica M3C stereomicroscope with a micrometric grid. The regression line TL-CL was calculated and the data were statistically analysed using non-parametric tests ($r_s$ according to Spearman test) since the data were freely scattered (Bakus 1990).

## Results

### Crustacean Decapod Larvae

Crustacean decapod larvae and postlarvae were collected (Table 1) during the Polar Queen cruise in stations 8-10-11a-20-18-19-48-27-35-36, located in Terra Nova Bay, and stations 3-5-14-25-39-32-46-16-17 located on the Ross Sea continental shelf, and during the Cariboo cruise in stations 25-

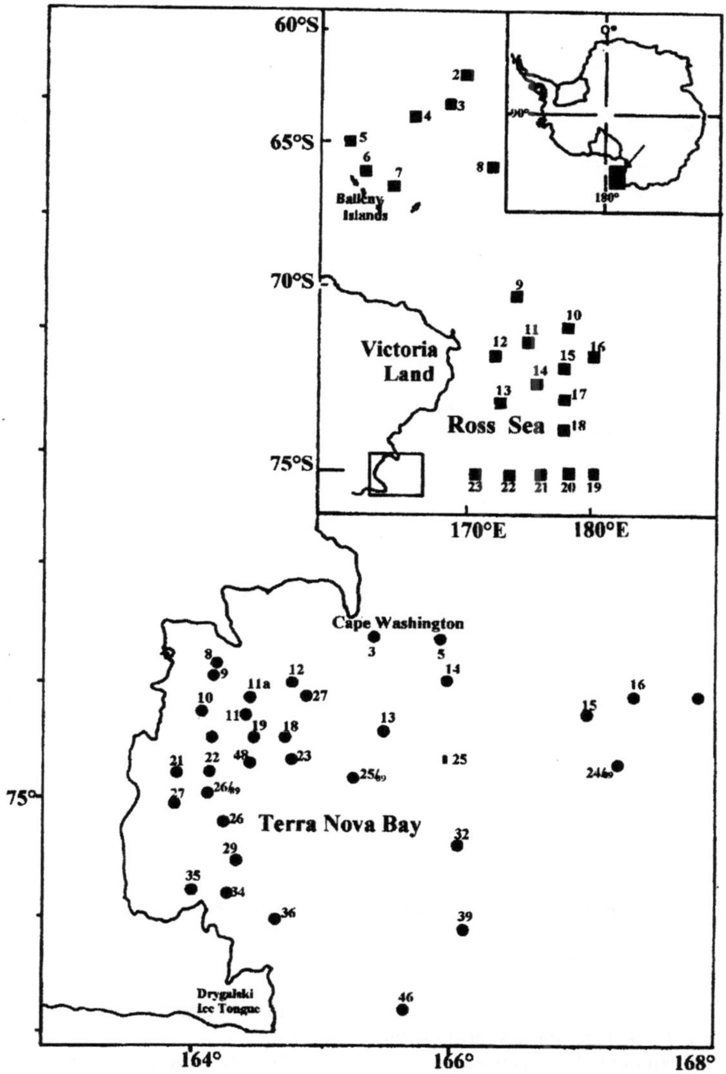


Fig. 1. Studied area and sampling stations (● = cruise 1987–88; ■ cruise 1989–90)

25bis-26bis-27-27bis in Terra Nova Bay, stations 22N-22P (N = night sample; P = p.m. sample) -22bis-23bis in the 75°S transect, and stations 10-10bis-11-11bis-15-16bis-17 in the pelagic zone (71°–73°S) west of Cape Adare-Victoria Land Shelf (some stations were sampled twice and the second sample was indicated as 'bis'). An adult *N. antarcticus*, around 20 mm TL, was also collected at station 13; decapod larvae were not found in samples collected south of the Antarctic Convergence and in the marginal-ice zone of Balleny Islands (stations 3, 4, 5, 6, 7, cruise 1989–90).

Taxonomic studies indicated that the larvae belonged to the benthic species *Notocrangon*

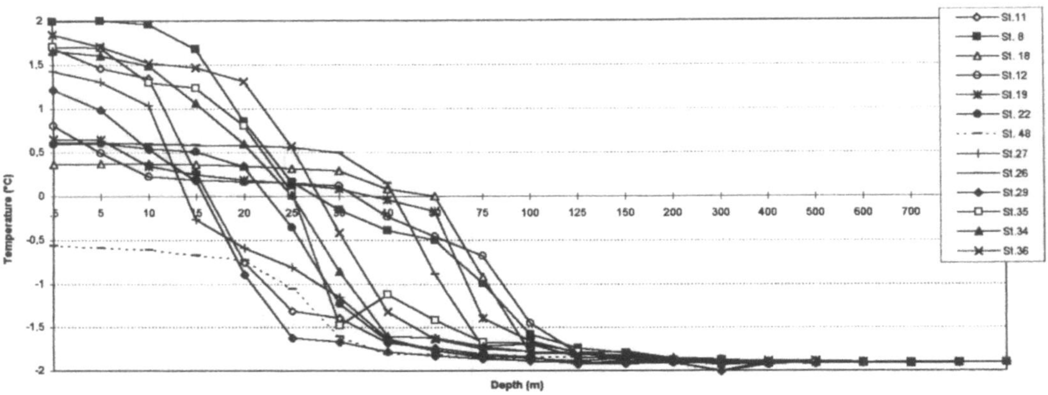


**Fig. 2.** Vertical profiles of temperature in the different stations of Terra Nova Bay. (Boldrin and Stocchino 1990)

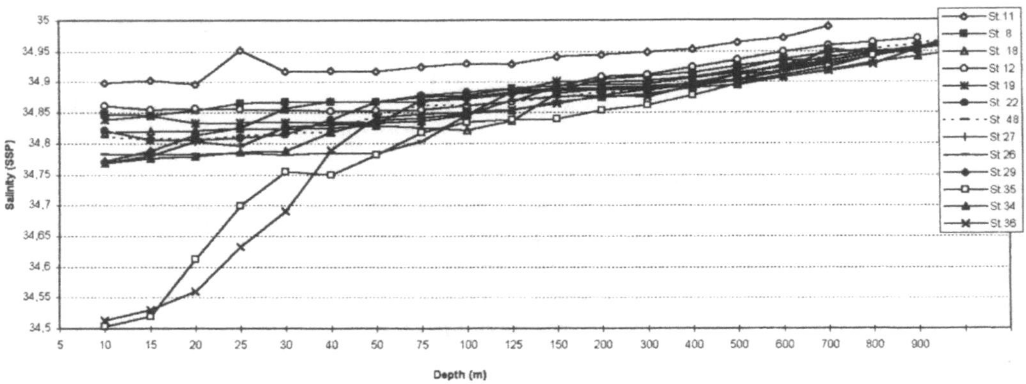


**Fig. 3.** Vertical profiles of salinity in the different stations of Terra Nova Bay. (Boldrin and Stocchino 1990)

*antarcticus* and *Chorismus antarcticus* and to the bathypelagic species *Acanthephyra pelagica*, which amounted respectively to 71.7% (n = 152), 12.3% (n = 26) and 16% (n = 34 ) of the total. *N.antarcticus* (n = 122) occurred in the first (2.5%), second (63.9%) and third (33.6%) larval stages (Carli and Pane 1990), while *C. antarcticus* (n = 21) in the second (52.0%), third (42.8%) larval stages and in the postlarval stage (4.8%). The *A. pelagica* larval stages were not clearly identified (probably second, third and juvenile), according to Gurney (1924).

## Abundance

*A. pelagica* occurred at only five stations located in the northern pelagic area of the Ross Sea; *N. antarcticus* and *C. antarcticus* were sampled in the Ross Sea continental shelf along the east-west 75°S transect and in Terra Nova Bay (Fig. 4a,b). Total larval density ranged from 0.2 ind $10\,m^{-3}$ to 23.6 ind $10\,m^{-3}$, but only for *N. antarcticus* ranged from 0.2 ind $10\,m^{-3}$ (0–100 m) to 11.8 ind $10\,m^{-3}$ (20–30 m); *C. antarcticus* ranged from 0.17 ind $10\,m^{-3}$ (30 m) to 2.15 ind $10\,m^{-3}$ (10–20 m), and *A. pelagica* from 0.2 ind $10\,m^{-3}$ (100–200 m) to 23.6 ind $10\,m^{-3}$ (40–60 m). The highest densities of *N. antarcticus*

**Table 1.** Abundance and density of larval crustacean decapods for each layer of the water column

| Station No. | Larvae depth (m) | Water filtered ($m^3$) | Larvae No. | Total larvae (ind $10\,m^{-3}$) | N. antarcticus (ind $10\,m^{-3}$) | C. antarcticus (ind $10\,m^{-3}$) | A. pelegica (ind $10\,m^{-3}$) |
|---|---|---|---|---|---|---|---|
| Cruise 87–88 | | | | | | | |
| 8 | 0–500[a] | | 1 | | | | |
| | 100–0 | | 4 | | | | |
| 10 | 30–0 | 8.2 | 1 | 1.22 | 0 | 1.22 | 0 |
| 11a | 20–10 | 9.3 | 3 | 3.22 | 1.07 | 2.15 | 0 |
| | 30–20 | 13.9 | 1 | 0.72 | 0.72 | 0 | 0 |
| | 40–30 | 15.3 | 2 | 1.31 | 1.31 | 0 | 0 |
| 20 | 60–40 | 17.3 | 1 | 0.6 | 0.6 | 0 | 0 |
| 19 | 0–500[a] | | | | | | |
| | 50–0 | 16.3 | 10 | 6.13 | 5.52 | 0.61 | 0 |
| 27 | 50–0 | 15.2 | 2 | 1.32 | 1.32 | 0 | 0 |
| 35 | 30–20 | 38.9 | 49 | 12.6 | 11.83 | 0.77 | 0 |
| | horiz. 30 | 23.7 | 7 | 2.95 | 2.53 | 0.42 | 0 |
| | horiz. 30 | 59.5 | 44 | 7.39 | 7.22 | 0.17 | 0 |
| | 40–30 | 37.2 | 7 | 1.88 | 1.61 | 0.27 | 0 |
| 36 | 30–20 | 13.9 | 2 | 1.44 | 0.72 | 0.72 | 0 |
| | 50–40 | 13.7 | 1 | 0.73 | 0.73 | 0 | 0 |
| 3 | 0–150[a] | | 1 | | | | |
| | 40–30 | 19.7 | 3 | 1.52 | 1.01 | 0.51 | 0 |
| | 50–40 | 26.0 | 2 | 0.77 | 0.77 | 0 | 0 |
| 13 | 0–700[a] | | ladult | | | | |
| 5 | 0–500[a] | | 2 | | | | |
| | 50–0 | 29.8 | 4 | 1.34 | 1.01 | 0.33 | 0 |
| 14 | 0–250[a] | | 1 | | | | |
| 25 | 0–500[a] | | 1 | | | | |
| | 500–0 | | 1 | | | | |
| 17 | 0–350[a] | | 1 | | | | |
| | 70–30 | 32.4 | 1 | 0.31 | 0.31 | 0 | 0 |
| 16 | 50–0 | 29.9 | 1 | 0.33 | 0.33 | 0 | 0 |
| 18 | 50–0 | 37.4 | 6 | 1.6 | 0.27 | 1.33 | 0 |
| | 100–50 | 37.8 | 1 | 0.26 | 0 | 0.26 | 0 |
| 48 | 50–0 | 29.0 | 2 | 0.69 | 0.69 | 0 | 0 |
| 46 | 50–0 | 33.7 | 1 | 0.3 | 0 | 0.3 | 0 |
| 39 | 0–550[a] | | 1 | | | | |
| 32 | 50–0 | 30.5 | 2 | 0.66 | 0.66 | 0 | 0 |
| Criuse 89–90 | | | | | | | |
| 10 | 40–20 | 49.6 | 2 | 0.4 | 0 | 0 | 0.4 |
| | 60–40 | 59.4 | 24 | 4.04 | 0 | 0 | 4.04 |
| | 80–60 | 69.4 | 8 | 1.15 | 0 | 0 | 1.15 |
| 10bis | 100–0 | 110.7 | 12 | 1.08 | 0 | 0 | 1.08 |
| 11 | 0–200[a] | 156.3 | 2 | | | | |
| | 100–80 | 64.4 | 6 | 0.9 | 0 | 0 | 0.9 |
| | 120–100 | 54.0 | 10 | 1.85 | 0 | 0 | 1.85 |
| 11bis | 100–50 | 54.8 | 2 | 0.36 | 0 | 0 | 0.36 |
| | 200–100 | 123.9 | 2 | 0.2 | 0 | 0 | 0.2 |
| 15 | 0–200[a] | 241.1 | 14 | | | | |
| | 60–40 | 15.9 | 38 | 23.64 | 0 | 0 | 23.64 |
| | 160–140 | 38.0 | 4 | 1.05 | 0 | 0 | 1.05 |
| 16bis | 250–200 | 94.0 | 4 | 0.43 | 0 | 0 | 0.43 |
| 17 | 120–100 | 37.8 | 2 | 0.53 | 0 | 0 | 0.53 |
| 22N | 150–100 | 87.9 | 2 | 0.23 | 0.23 | 0 | 0 |
| 22P | 100–50 | 75.4 | 2 | 0.26 | 0.26 | 0 | 0 |
| 22bis | 50–40 | 34.0 | 2 | 0.59 | 0 | 0.59 | 0 |
| 23bis | 0–100[a] | 189.7 | 4 | 0.21 | 0.21 | 0 | 0 |
| 25 | 0–800[a] | 238.3 | 2 | | | | |
| | 50–0 | 88.8 | 2 | 0.22 | 0.22 | 0 | 0 |
| 25bis | 20–0 | 49.5 | 2 | 0.4 | 0 | 0.4 | 0 |
| 26bis | 20–0 | 88.7 | 2 | 0.22 | 0.22 | 0 | 0 |
| 27 | 0–600[a] | 221.5 | 2 | | | | |
| 27bis | 0–100[a] | 92.5 | 2 | 0.22 | 0.22 | 0 | 0 |

[a] Sample carried out along the whole water column.

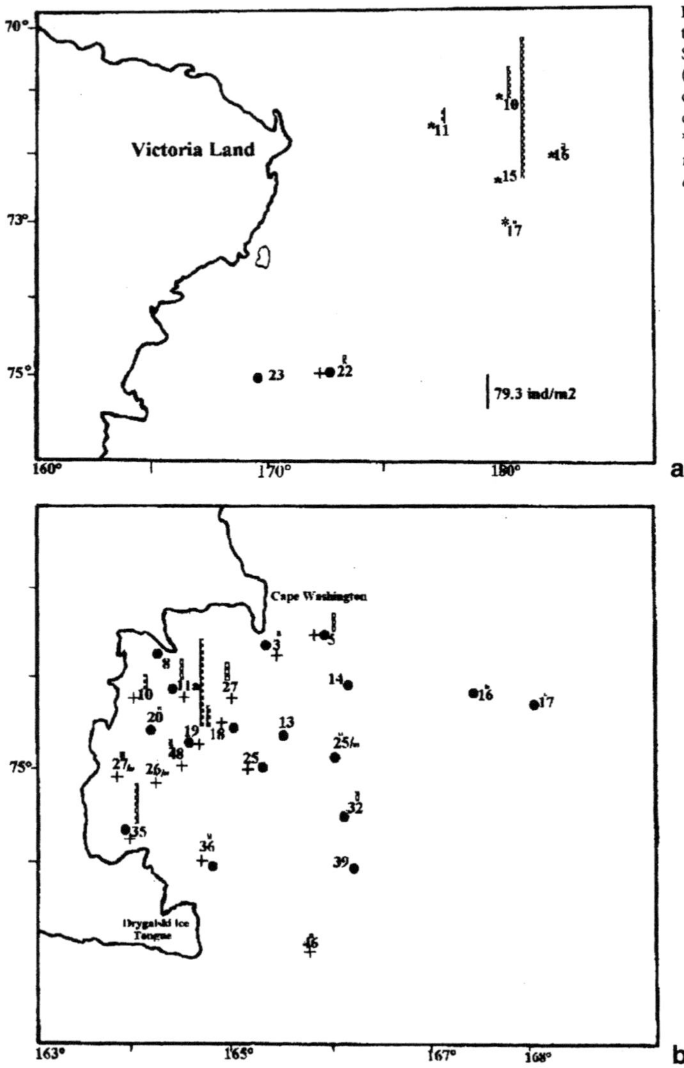


**Fig. 4a,b.** Stations where larval crustacean decapods occurred (**a** Ross Sea; **b** Terra Nova Bay) and density (ind $m^{-2}$) of larval crustacean decapods along the 0–200-m water column in the different stations. **Achanthephyra pelagica*, +*Chorismus antarcticus*, •*Notocrangon antarcticus*

and *C. antarcticus* were observed in the upper layers, particularly at stations 35 and 11a, respectively (Fig. 5).

## Vertical Distribution

Early larval stages of *A. pelagica* (Fig. 6) occurred mostly (70%) between 40 and 100 m. Late stages occurred more deeply, and up to 250 m (station 16bis); the highest density (23.6 ind 10 $m^{-3}$) was recorded at 40–60 m (station 15). Larval density (ind $m^{-2}$) between 200 m and the surface was highly variable; in the 71°–73°S sector of the Ross Sea (Fig. 4a) *A. pelagica* ranged from 49.4 ind $m^{-2}$ (station 15) to 1.1 ind $m^{-2}$ (station 17). *C. antarcticus* and *N. antarcticus* were present at low densities,

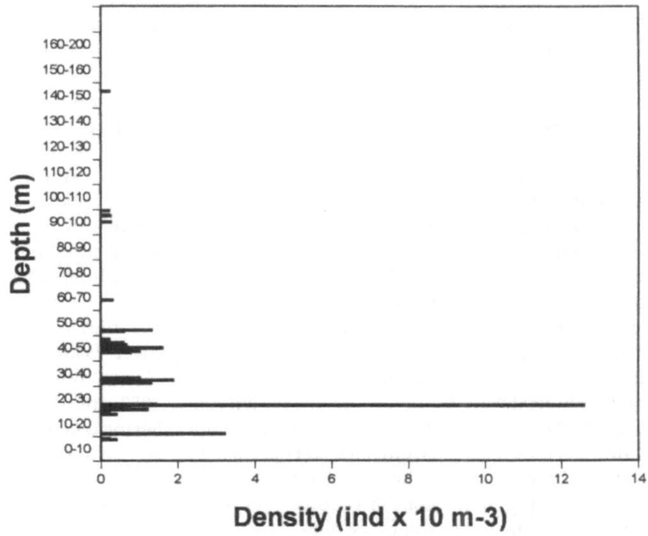


**Fig. 5.** Vertical distribution (ind $10\,m^{-3}$) of *N. antarcticus* and *C. antarcticus* in Terra Nova Bay and in the Ross Sea. Both species and all stations were combined

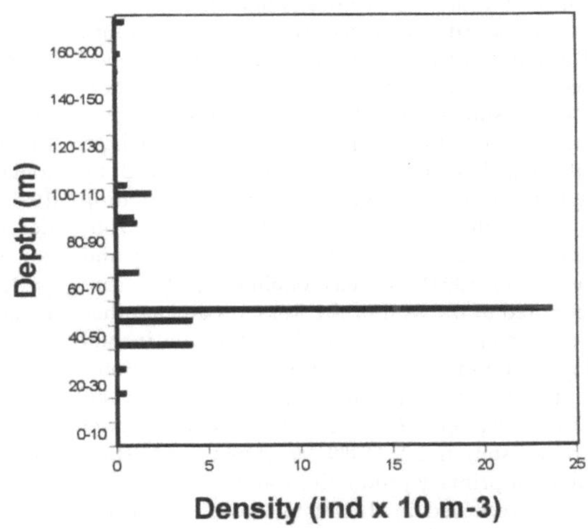


**Fig. 6.** Vertical distribution (ind $10\,m^{-3}$) of *A. pelagica* in the Ross Sea. All stations were combined

ranging from 1.2 ind $m^{-2}$ (station 17) to 6.7 ind $m^{-2}$ (station 5) along the 75°S transect; higher densities occurred inside the bay, where the maximum value of 30.7 ind $m^{-2}$ was recorded (station 19; Fig. 4b).

## Biometry

*N. antarcticus* (n = 92) and *C. antarcticus* (n = 21) had average lengths of 12.5 ± 1.1 mm TL, 2.9 ± 0.4 mm CL, and 12.2 ± 1.5 TL, 2.6 ± 0.3 mm CL, respectively. The minimum TL for *N. antarcticus* (first larval stage) was 8.6 mm, with 2.6 mm CL; the maximum TL (initial postlarval stage) was 15.2 mm with 3.7 mm CL; *C. antarcticus* ranged from 9.4 mm TL and 2.1 mm CL (second larval stage) to 14.5 mm TL and 3.0 mm CL (postlarval stage). The average length of *A. pelagica* was 11.31 mm TL and 2.98 mm CL; early larval stages measured 9.88 mm

TL and 2.59 mm CL, and late stages 15.56 mm TL and 4.23 mm CL.

## Discussion

The occurrence and distribution of larval crustacean decapods in the Ross Sea and Terra Nova Bay depend on several factors, such as the abiotic and trophic conditions, as well as adult biology. As regards environmental conditions, to the north of the Antarctic front there is a thermally homogeneous zone, with temperatures around 2.5 °C over 200 m depths, where the phytoplankton population is scarce and the low phaeophorbid concentrations indicate an area of low grazing activity; in this zone larvae of crustacean decapods were not found, as far as in the marginal ice zone of Balleny Islands, characterized by low water temperatures, high phytopigment concentrations, probably attributed to the liberation of ice algae by melting ice, where the mesozooplankton was low in density and the herbivorous zooplankton was not yet developed along the whole water column (Hecq et al. 1992; Innamorati et al. 1990). In open ice-free waters the total chloropigment concentration was similar to that recorded in the marginal ice zone; nevertheless, at a depth of 40–70 m, high chlorophyll a concentrations were replaced by high phaeoforbid a concentrations, corresponding to the highest copepod and herbivorous zooplankton densities (Hecq et al. 1993; Hecq and Guglielmo 1992). A thermohaline coastal front delimited to the west of the Ross Sea and another frontal zone was associated to the continental slope around station 17 (1989–90 expedition); another thermohaline front separating coastal and offshore western Ross Sea waters was observed around station 22 (1989–90 expedition). High values of primary productivity and phaeophorbid a were recorded in this front; these parameters were also found to be high along the east-west transect (75°S) as far as Terra Nova Bay (Hecq et al. 1992). In the Ross Sea, frontal and ice-free zones showed the highest densities of larval Crustacean Decapods; the bathypelagic species *A. pelagica* occurred mostly in the north-western area characterized by great depths, at the border of the continental shelf, in agreement with data on the Weddell Sea (Boysen-Ennen 1987), where larval stages were found only in oceanic waters. Larval stages of the benthic species *N. antarcticus* and *C. antarcticus* occurred in the Ross Sea in the frontal area (station 22) separating neritic and pelagic waters; according to Boysen-Ennen (1987) and Piatkowski (1987) these species were collected only within the continental shelf boundaries. The greatest amounts of *N. antarcticus* and *C. antarcticus* were collected within Terra Nova Bay, precisely in the northern section, where shallow depths and stable cyclonic gyres could favour patches of these organisms and of other zooplankters, and were found to be more persistent in the water column (Stocchino and Manzella 1991; Carli et al., this Vol.). The vertical larval distribution showed that *N. antarcticus* and *C. antarcticus* make up the bulk (over 95%) of sampled larvae between 20 and 50 m; the considerable abundance (Carli et al. 1990, 1992, and this Vol.) and the high biomass (Zunini Sertorio et al. 1994) of copepods in surface layers, supported by great phytoplankton density (Innamorati et al. 1990; Hecq et al. 1992) and high particulate organic matter (Fabiano et al. 1993) seem to support development of these carnivore larval Crustacean Decapods, hatching as schizopod stage which has an abbreviated planktonic phase. Recent studies carried out in the Ross Sea and Terra Nova Bay show high chlorophyll a concentrations during spring, a gradual decrease in summer (Smith and Nelson 1985; Innamorati et al. 1990) and a second phytoplankton bloom in February (Innamorati et al. 1990). Zooplankton communities are better developed in ice-free areas than in ice-covered zones (Guglielmo et al. 1990; Hecq et al. 1993); particularly ostracods (McKenzie et al. 1990; Benassi et al. 1992) and copepods show high density (Carli et al. 1990, 1992). Arntz and Gorny (1991) hypothesize that seasonal migration of female *C. antarcticus* and *N. antarcticus* from deeper to shallower areas (Weddell Sea) in spring may provide improved feeding conditions for the larvae.

Total size (TL) of the various developmental stages of *N. antarcticus* and *C. antarcticus* was intermediate to that indicated Gurney (1937) and Makarov (1973). Cephalothorax sizes (CL) at the second stages are similar to those recorded by Gorny et al. (1992) in aquarium-hatched organisms. The trend of linear regression between TL and CL was extremely significant (Figs. 7,8,9) both for *N. antarcticus* ($r_s = 0.72$), *C. antarcticus* ($r_s = 0.98$) and *A. pelagica* ($r_s = 0.79$). The quantitative data presented here are significantly higher than those reported in the literature (Gurney 1937; Makarov 1970; Boysen-Ennen 1987). These data

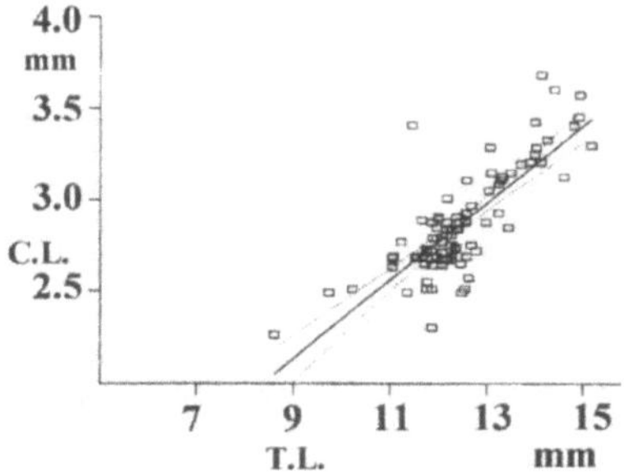


**Fig. 7.** *Notocrangon antarcticus*: total size (*T.L. mm*) vs. cephalothorax size (*C.L. mm*) ——linear regression; ……95% confidence interval

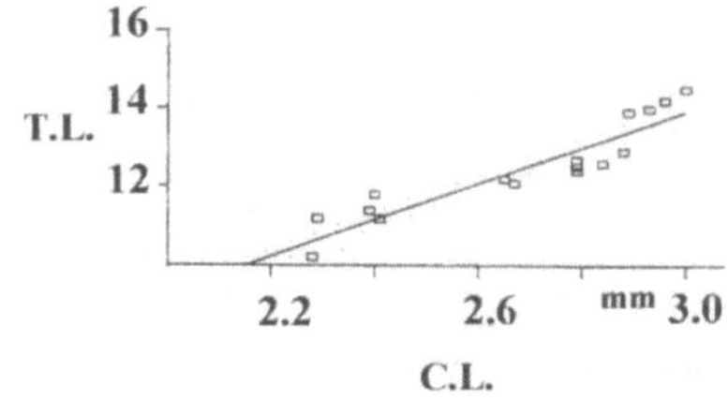


**Fig. 8.** *Chorismus antarcticus*: total size (*T.L. mm*) vs. cephalothorax size (*C.L. mm*) ——linear regression; ……95% confidence interval

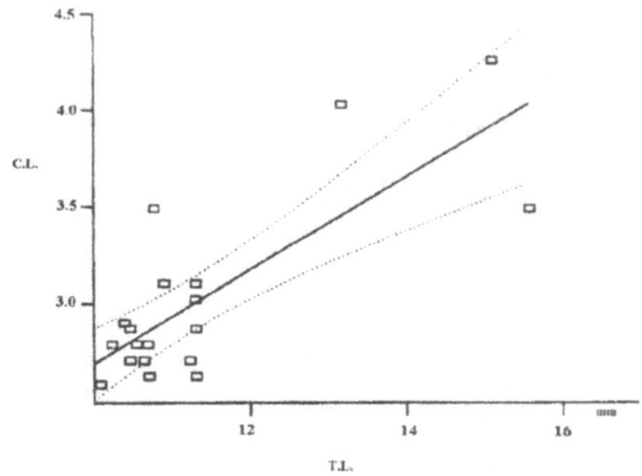


**Fig. 9.** *Acanthephyra pelagica*: total size (*T.L. mm*) vs. cephalothorax size (*C.L. mm*) ——linear regression; ……95% confidence interval

are also significant because the larvae of these organisms were rarely collected in the Weddell Sea, in spite of the fact that the density of adults turned out to be higher; this has been clearly demonstrated by recent photographic reports (Arnz and Gorny 1991).

This research has emphasized that both sampling strategy (Guglielmo et al. 1990, 1992) and sampling systems adopted during the oceanographic expeditions seem to be suitable for sampling these organisms. For instance, the exploration in thin layers from the surface to a 100-m level probably allowed a more effective detection of zooplankton patches especially in zones affected by stable cyclonic gyres; furthermore, the adopted EZ-NET BIONESS sampler, capable of collecting plankton samples in succession even over long distances, enables a more effective exploration of the area than traditional plankton nets. It is worth noting that in the 52 zooplankton samples collected at various depths in a coastal station in Terra Nova Bay during the 1994–95 expedition (WP2 net, 200-μm mesh), no crustacean decapod larvae were found (Pane, unpubl. data). Nevertheless, *N. antarcticus* and *C.antarcticus* adults form a significant portion in the stomach contents of *Trematomus loennbergii*, and even if less so in the stomachs of *Trematomus hansoni*, both of which nototheniid fish are found in Terra Nova Bay (La Mesa et al. 1997), thus emphasizing their importance within the marine food chain in this particular area.

## References

Arntz WE, Gorny M (1991) Shrimp (Decapoda, Natantia) occurrence and distribution in the eastern Weddell Sea, Antarctica. Polar Biol 11: 169–177

Bage F (1938) Crustacea Decapoda. Australasian Antarctic Expedition 1911–1914. Sci Rep Ser 6: 1–13

Bakus GJ (1990) Quantitative ecology and marine biology. Balkema, Rotterdam

Benassi G, Ferrari I, Gentile G, Menozzi P, McKenzie KG (1992) Planktonic Ostracoda in the Southern Ocean and in the Ross Sea: 1989–90 campaign. In: National Scientific Commission for Antarctica (ed) Oceanographic Campaign 1989–90, Data Rep Part II, Genova, pp 247–300

Birsthein YA, Vinogradov LG (1967) Occurrence of *Paralomis spectabilis* Hansen (Crustacea, Anomura) in the Antarctic. Biological Reports of the Soviet Antarctic Expedition (1955–1958). Acad Sci USSR Zool Inst 3: 390–396

Boldrin A, Stocchino C (1990) On the hydrological characteristic of Terra Nova Bay (Ross Sea-Antarctica). In: National Scientific Commission for Antarctica (ed) Oceanographic Campaign 1987–88, Data Rep Part I, Genova, pp 11–57

Borradile LA (1910) Crustacea I. Decapoda. British Antarctic "Terra Nova" Expedition, 1910. Nat Hist Rep, Zool III 2: 75–110

Boysen-Ennen E (1987) Zur Verbreitung des Meso- und Makrozooplanktons im Oberflächenwasser der Weddellsee (Antarktis). Ber Polarforsch 35: 126

Calman WT (1907) Crustacea I. Decapoda. National Antarctic Expedition 1901–1904. Nat Hist Rep, Zool II: 1–7

Carli A, Pane L (1990) Contribution to the study of larval crustacean decapods in the Terra Nova Bay (Ross Sea): *Notocrangon antarcticus* (Pfeffer, 1887). In: National Scientific Commission for Antarctica (ed) Oceanographic Campaign 1987–88, Data Rep Part II, Genova, pp 161–167

Carli A, Mariottini GL, Pane L (1990) Contribution to the study of Copepods collected in Terranova Bay (Ross Sea). In: National Scientific Commission for Antarctica (ed) Oceanographic Campaign 1987–88, Data Rep Part II, Genova, pp 129–159

Carli A, Feletti M, Mariottini GL, Pane L (1992) Contribution to the study of copepods collected during the Italian Oceanographic Campaign in Antarctica 1989–90. In: National Scientific Commission for Antarctica (ed) Oceanographic Campaign 1989–90, Data Rep Part II, Genova, pp 179–210

Clarke A (1977) Seasonal variations in the total lipid content of *Chorismus antarcticus* (Pfeffer) (Crustacea: Decapoda) at South Georgia. J Exp Mar Biol Ecol 27: 93–106

Clarke A (1979) On living in cold water: K-strategies in Antarctic benthos. Mar Biol 55: 111–119

Clarke A, Crame A (1989) The origin of the Southern Ocean marine fauna. Biota Geol Soc Spec Publ 47: 253–268

Clarke A, Holmes LJ (1987) Notes on the biology and distribution of *Pasiphaea* species from the Southern Ocean. Br Antarct Surv Bull 74: 17–30

Clarke A, Lakhani KH (1979) Measures of biomass, moulting behaviour and the pattern of early growth in *Chorismus antarcticus* (Pfeffer). Br Antarct Surv Bull 47: 61–88

Fabiano M, Povero P, Danovaro R (1993) Distribution and composition of particulate organic matter in the Ross Sea (Antarctica). Polar Biol 13: 525–533

Garth JS, Haig J, Yaldwyn JC (1967) The decapod Crustacea of the Royal Society expedition to Southern Chile, 1958–59. Trans R Soc N Z Zool 8, 16: 169–186

Gorny M, Arntz WE, Clarke A, Gore DJ (1992) Reproductive biology of caridean decapods from the Weddell Sea. Polar Biol 12: 111–120

Guglielmo L, Costanzo G, Manganaro A, Zagami G (1990) Spatial and vertical distribution of zooplanktonic communities in the Terra Nova Bay (Ross Sea). In: National Scientific Commission for Antarctica (ed) Oceanographic Campaign 1987–88, Data Rep Part I, Genova, pp 257–398

Guglielmo L, Costanzo G, Zagami G, Manganaro A, Arena G (1992) Zooplankton ecology in the Southern Ocean. In: National Scientific Commission for Antarctica (ed) Oceanographic Campaign 1989–90, Data Rep Part II, Genova, pp 301–468

Gurney R (1924) Crustacea. IX. Decapoda larvae. British Antarctic (Terra Nova) Expedition, 1910. Nat Hist Rep Zool 8: 37–202

Gurney R (1937) Larvae of decapod Crustacea IV. Hippolytidae. Discovery Rep 14: 351–404

Gurney R (1942) Larvae of decapod Crustacea. Ray Society, London, 306 pp

Gurney R (1960) Bibliography of the larvae of decapod Crustacea and larvae of decapod Crustacea. In: Cramer J, Swann HK (eds) Historiæ naturalis classica, vol 8. Wheldon and Wesley, Weinheim, 240 pp

Gutt J, Gorny M, Arntz W (1991) Spatial distribution of Antarctic shrimps (Crustacea: Decapoda) by underwater photography. Antarct Sci 3: 363–369

Gutt J, Ekau W, Gorny M (1994) New results on the fish and shrimp fauna of the Weddell Sea and Lazarev Sea (Antarctic). Proc NIPR Symp Polar Biol 7: 91–102

Hecq JH, Guglielmo L (1992) Structure and functioning of the Ross Sea pelagic ecosystem: an interdisciplinary approach. In: Gallardo VA, Ferretti O, Moyano HI (eds) Oceanografia en Antártica. ENEA-PNRA-EULA, Concepcion, Chile, Rome, pp 227–233

Hecq JH, Magazzù G, Goffart A, Catalano G, Vanucci S, Guglielmo L (1992) Distribution of planktonic components related to structure of water masses in the Ross Sea during the Vth Italiantartide Expedition. In: Ambrosetti W, Piccazzo M, Ruffoni Riva T (eds) Atti IX Congr Ass Ital Oceanogr Limnol, Lang, Genova, pp 665–678

Hecq JH, Brasseur P, Goffart A, Lacroix G, Guglielmo L (1993) Modelling approach of the planktonic vertical structure in deep austral ocean. In: Progress in Belgian oceanographic research (Brussels, January 21–22, 1993). Royal Academy of Belgium, National Committee of Oceanology, Brussds, pp 235–250

Herring PJ (1967) Observations on the early larvae of three species of *Acanthephyra* (Crustacea, Decapoda, Caridea). Deep Sea Res 14: 325–329

Innamorati M, Lazzara L, Mori G, Nuccio C, Saggiomo EV (1990) Spatial and temporal distributions of phytoplankton size fractions in Antarctic waters: biomass and production, summer 1989/90. In: Ambrosetti W, Piccazzo M, Ruffoni Riva T (eds) Atti IX Congr Ass Ital Oceanogr Limnol Lang, Genova, pp 605–612

Kirkwood JM (1984) A guide to the Decapoda of the Southern Ocean. ANARE Res Notes 11: 47

La Mesa M, Vacchi M, Castelli A, Diviacco G (1997) Feeding ecology of two nototheniid fishes, *Trematomus hansoni* and *Trematomus loennbergii*, from Terra Nova Bay, Ross Sea. Polar Biol 17: 62–68

Lebour MV (1931) The larvae of the Plymouth Caridea. I. The larvae of Crangonidae. II. The larvae of the Hyppolytidae. Proc Zool Soc Lond 56: 1–9

Ledoyer M (1969) Sur divers crustacés antarctiques (leptostracés, cumacés, mysidacés et caridés) recuillis en terre Adélie en 1961–63 et 1964–65. Crustaceana 17: 88–96

Makarov RR (1970) Biology of the Antarctic shrimp *Notocrangon antarcticus* (Decapoda, Crangonidae). Zool Zh 49: 28–37

Makarov RR (1973) Larval development of *Notocrangon antarcticus* (Decapoda, Crangonidae) Zool Zh 52: 1149–1155

Maxwell JGH (1977) The breeding biology of *Chorismus antarcticus* (Pfeffer) and *Notocrangon antarcticus* (Pfeffer) (Crustacea, Decapoda) and its bearing on the problems of the impoverished Antarctic decapod fauna. In: Llano GA (ed) Adaptations within Antarctic ecosystems. Proc 3rd SCAR Symp Antarct Biol, Gulf Publ, Houston, pp 335–342

Maxwell JGH, Ralph R (1985) Non-cold-adapted metabolism in the decapod *Chorismus antarcticus* and other sub-Antarctic marine crustaceans. In: Siegfried WR, Condy PR, Laws RM (eds) Antarctic nutrient cycles and food webs. Springer, Berlin Heidelberg New York, pp 397–406

McKenzie KG, Benassi G, Naldi M, Ferrari I, Menozzi P (1990) Report on planktonic Ostracoda from the Ross Sea Antarctica. In: National Scientific Commission for Antarctica (ed) Oceanographic Campaign 1987–88, Data Rep Part II, Genova, pp 171–229

Piatkowski U (1987) Zoogeographische Untersuchungen und Gemeinschaftsanalysen am antarktischen Makroplankton. Ber Polarforsch 34: 150

Smith WO, Nelson DM (1985) Phytoplankton bloom produced by a receding ice in the Ross Sea: spatial coherence with the density field. Science 227: 163–166

Stocchino C, Manzella GM (1991) Analisi delle correnti marine di Baia Terra Nova (Mare di Ross-Antartide). C.N.R. Area di Ricerca di Genova. Istituto Idrografico della Marina, Genova, 56 pp

Tiefenbacher L (1990) *Eualus kinzeri*, a new hippolytid from the Weddell Sea (Antarctica). Spixiana 13: 117–120

Vacchi M, La Mesa M, Castelli A (1994) Diet of two coastal nototheniid fish from Terra Nova Bay, Ross Sea. Antarct Sci 61: 61–65

Vo J (1988) Zoogeographie und Gemeinschaftsanalyse des Makrozoobenthos des Weddellmeeres (Antarktis) Ber Polarforsch 45: 1–45

Ward P (1985) New records of *Lebbeus antarcticus* (Hale) (Crustacea: Decapoda) from the Antarctic Peninsula. Br Antarct Surv Bull 69: 57–63

Williamson DI (1962) Crustacea Decapoda: Larvae III. Caridea, families Oplophoridae, Nematocarcinidae and Pasiphaeidae. Cons Int Explor Mer Zooplankton Sheet 92: 1–5

Yaldwyn JC (1965) Antarctic and subantarctic decapod Crustacea. In: van Mieghem J, van Oye P (eds) Biogeography and ecology in Antarctica. Junk, The Hague, pp 324–332

Yaldwyn JC (1966) Protandrous hermaphroditism in decapod prawns of the families Hippolytidae and Campylonotidae. Nature 209: 1366

Yaldwyn JC (1967) The decapod crustacea of the Royal Society Expedition to Southern Chile, 1958–59 Trans R Soc N Z Zool 8: 169–186

Zarenkov NA (1968) Crustacean Decapoda collected by the Soviet Antarctic Expedition in the Antarctic and Antiboreal region. In: Andriyashev AP, Ushakov PW (eds) Biological reports of the Soviet Antarctic expedition (1955–1958). Academy of Sciences of the USSR, Moscow, pp 153–201

Zunini Sertorio T, Ossola C, Licandro P (1994) Size, length-weight relationships and biomass of copepods in Antarctic waters (Terra Nova Bay, Ross Sea). In: Albertelli F, Cattaneo-Vietti R, Piccazzo M (eds) Atti X Congr Ass Ital Oceanogr Limnol, Lang, Genova, pp 667–680

# Occurrence of Mysidacea (Crustacea) in the Pacific Sector of Antarctica and Distribution of *Antarctomysis ohlini* in Terra Nova Bay (Western Ross Sea)

**N. Crescenti, L. Guglielmo, G. Costanzo, G. Zagami, and A. Granata**[1]

**ABSTRACT**

During the summer Italian Antarctic Expeditions of 1987–88 (Polar Queen cruise) and 1989–90 (Cariboo cruise), mysids were collected in Terra Nova Bay and the western Ross Sea (268 and 572 samples, respectively), by a multinet BIONESS. Four species (*Boreomysis brucei, Caesaromysis hispida, Dactylabblyops hogdsoni* and *Euchaetomera zurstasseni*) were caught in the Pacific sector of the Southern Ocean and western Ross Sea while *Antarctomysis ohlini* was the only species found in Terra Nova Bay. It reached maximum densities in the southern part of the bay in deep straits and trenches. The absence of *Antarctomysis maxima* in both BIONESS samples and in the gut contents of coastal benthic fishes suggests that *A. maxima* inhabited the outer shelf of Terra Nova Bay, while *A. ohlini* was confined in the deeper trenches of the bay. The generation time from egg to egg for *A. ohlini* in Terra Nova Bay was assumed to be 5 years. The growth rate of juveniles was estimated to be about 1 mm per month and the newly moulted young C0 caught at Terra Nova Bay in January were about 3 months old. Gut content analysis confirms that *A. ohlini* was a generalist during its life cycle. Diatoms were the most important group in the juvenile diet composition. The ice-algae *Fragilariopsis curta* dominate the diatom assemblages in all samples. Coprophagy seems to be the principal feeding in juveniles (age 0 and age 1+), whereas diet diversity increased with increasing size in the species.

## Introduction

Important contributions to the knowledge of the taxonomy and large-scale distribution of Mysidacea were made by Holt and Tattersall (1906), Zimmer (1914), Tattersall (1955), Birstein and Tchindonova (1962).

Holt and Tattersall (1906) were the first to describe an immature male of Mysidacea collected by the Discovery in the Ross Sea, attributing it erroneously to a new species of the genus *Mysis* (*M. maxima*). Subsequently, Coutière (1906), after examining adult specimens of Mysidacea collected during the Charcot expedition, created the new genus *Antarctomysis*. Two years later, Hansen (1908) discovered the species *A. ohlini* in material deriving from the Belgica expedition, including in it also the Discovery specimen.

The first information about the presence and distribution of *Antarctomysis ohlini* and *A. maxima* in the Ross Sea, due entirely to Tattersall (1923) and Tattersall (1955, 1961, 1965), was scanty and fragmentary. Having found an ovigerous female of 71 mm off Cape Bird, Ross Island (77°15′S–166°00′E), Tattersall (1923) noted that the distribution of *A. ohlini* extended to the southern part of the Ross Sea and that the species increased in size with latitude. Examining the Discovery material (1955) and other specimens deriving from the Ross Sea (1961, 1965), Tattersall provided a further contribution to the taxonomy and distribution of these two species. This author often mentioned that *A. ohlini* and *A. maxima* were found in the same locality and many times in the same samplings; he concluded that *A. ohlini* has the same distribution as *A. maxima* in the Ross Sea and that both species are mesoplanctonic with demersal habits, living on or in proximity to the bottom.

More recently, Ward (1984, 1985) provided detailed information about the biology of *A. ohlini* and *A. maxima* from South Georgia, while Siegel

[1] Dipartimento di Biologia Animale ed Ecologia Marina – Università di Messina. Salita Sperone 31, 98166 Messina, Italy

and Mühlenhardt-Siegel (1988) reported on the biology and the spatial and vertical distribution of the Mysidacea of the Antarctic Peninsula.

There are very few data concerning the role that these two species of *Antarctomysis* play in the Antarctic food web. From examination of the stomach contents, Mauchline (1980) and Ward (1984, 1985) classified the two species as omnivorous feeders, with clear phytophagous behaviour by the juveniles, coinciding with the increase of primary production. In the Ross Sea, *A. ohlini* was defined as a generalist species with a diet very similar to that of the amphipod *Orchomene plebs* (Hopkins 1987). From the composition of the diet, it is classified among the species that ingest 40 to 80% of Metazoa together with dense aggregates of *Coscinodiscus* and *Nitzschia*. Unlike *A. ohlini*, *A. maxima* constitutes an important dietary component of many Antarctic fishes (Permitin 1970; Permitin and Tarverdieva 1972; Kock 1981; Targett 1981). Tattersall (1961, 1965) reported a few specimens of *A. ohlini* and *A. maxima* in the stomach of a Weddell seal (locality G) and that the two species constituted food for penguins, seals and whales.

The present work was aimed at providing further knowledge of the distribution of the Mysidacea in the Pacific sector of the Antarctic Ocean and of the spatial distribution and trophic role of *A. ohlini* at Terra Nova Bay (Western Ross Sea).

## Study Area

According to Patterson and Whitworth (1990), the ocean circulation in the Antarctic sector of the Pacific is dominated by the continuous eastward-flowing Antarctic Circumpolar Current (ACC). This current causes the horizontal distribution of water properties in the Southern Ocean. Thus, various zones can be characterized on the basis of distinctive vertical stratification or temperature-salinity relationships of surface and intermediate waters. These zones are, from north to south, the Subantarctic Zone, the Polar Frontal Zone, the Antarctic Zone and the Continental zone.

The continental margin of the Ross Sea can be divided into two very well-defined parts: the shallowest eastern area, where the average depth is between 300 and 400 m at the shelf edge (De Witt 1970) and the deepest western area characterized by reliefs and depressions also down to 1000 m. The shelf break is located 400–700 m deep, with an average depth of more than 500 m.

A strong cyclonic gyre characterizes the summer surface layer circulation over the continental shelf, which means that the surface current flows westward along the ice shelf and then north along the coast of Victoria Land (Smith and Schnack-Schiel 1990). According to Patterson and Whitworth (1990), the Ross Sea shelf was characterized by these water masses: (1) Antarctic Surface Water (AASW, 0–50 m) with low salinity (<34.50‰) and, during the summer, relatively high temperatures at the surface (−1.5/2.0 °C); (2) Shallow Ice Shelf Water (SISW, 50–200 m) with salinity <34.60‰ and temperature <−1.9 °C; (3) Modified Circumpolar Deep Water (200–300 m) with salinity >34.50‰ and temperature >−1.6 °C; (4) Deep Ice Shelf Water (DISW, 300–600 m) with salinity >34.60‰ and temperature <−1.9 °C, and High Salinity Shelf Water (HSSW, 600 m to the bottom) with salinity >34.75‰ and temperature ~−1.9 °C, both characteristic of the SW area of the Ross Sea Shelf. During the Italian Antarctic expedition in 1990–91 (Spezie et al. 1993), the off-shore waters were characterized by a maximum temperature located at 300 m and salinity values between 34.40 and 34.80 psu.

Terra Nova Bay is an 80-km by 30-km-wide inlet in the southwest portion of the Ross Sea (Fig. 1a). It is delimited in the north by the volcano Mt. Melbourne and the narrow peninsula of Cape Washington. To the south it is bounded by the Drygalski Ice Tongue, arising from David Glacier, which extends into the Ross Sea for about 60 km (Fig. 1b).

Recent studies in the Ross Sea reported high concentrations of chlorophyll a during austral spring (Smith and Gordon 1997), with a progressive diminution to oligotrophic levels in summer (Smith and Nelson 1985a; Innamorati et al. 1990). Abundant phytoplanktonic blooms develop every year as a consequence of the melting ice edge (Smith and Nelson 1985b; Wilson et al. 1986; Hecq et al. 1993). Specific composition, abundance, vertical distribution, size and biomass of copepods collected at Terra Nova Bay are reported in Carli et al. (1990) and Zunini Sertorio et al. (1990). Tintinnid distribution was studied by Monti and Fonda Umani (1995).

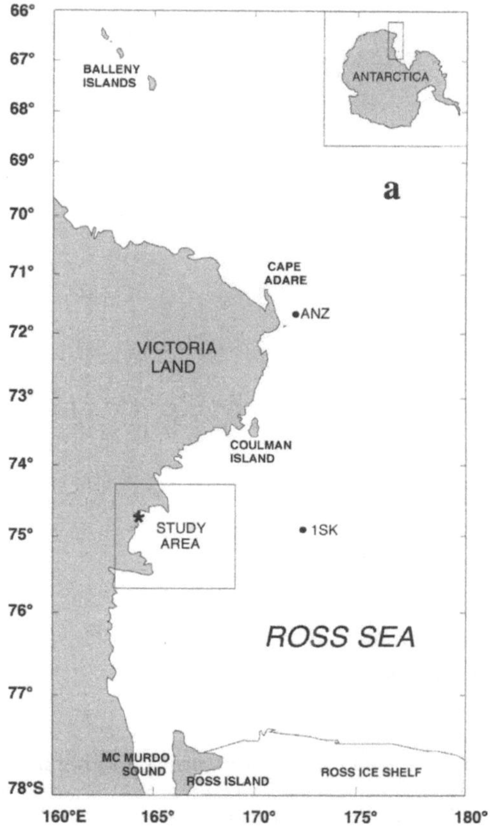


**Fig. 1a,b.** Sampling area and station locations during the 1987–88 Italian Antarctic expedition (Polar Queen cruise). **a** Stations sampled in the western Ross Sea. **b** Stations sampled in Terra Nova Bay. *Dotted line* indicates coast formed by permanent pack ice

## Sampling Procedure

### Polar Queen Cruise 1987–88

During the period between January 4 and February 21, 1988, zooplankton samples were taken at 33 stations in Terra Nova Bay (Ross Sea, Fig. 1b). The study area extended from Cape Washington (74°39′S) to the Drygalski Ice Tongue (75°29′S), along longitude 167°59′E. The sampling sites were located mainly in the continental shelf of the Ross Sea, while station 1SK was located ca. 150 miles off -shore from the coast and station ANZ near Cape Adare (Fig. 1a).

Zooplankton samples were taken with an EZ-BIONESS NET (Sameoto et al. 1980). This was a multiple opening and closing net sampler with ten nets and with a square mouth opening of $0.25\,m^2$. The mesh size of the nets used were 250 and 500 μm, depending on the stations. The BIONESS continuously measured temperature, salinity and depth during tows by means of an applied mycrosystems digital CTD. Flow through the nets was monitored with external and internal TSK flowmeters. The BIONESS was firstly deployed at low speed to the maximum depth programmed and then towed at speed of $1.5–2\,m\,s^{-1}$ while it was slowly lowered along an oblique path to the desired depth. The nets were opened and closed on command from the ship at depth intervals of 10–40 m from 100 and 300 m profiles and between 50- and 100-m intervals below 300 m to the maximum depth sampled. An integrated sample for the whole water column was also collected at each station. Sampling effort (such as stations sampled, bottom depth, trawling depth, volume

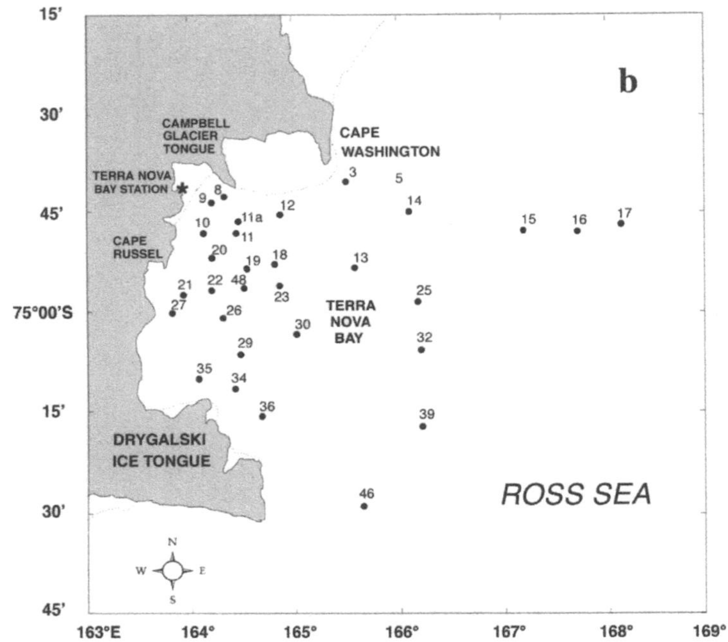


**Fig. 1a,b.** *Continued*

filtered, *A. ohlini* abundance 100 m$^{-3}$) is shown in Table 1.

Total effort was 33 hauls (268 samples) filtering 12 452 m$^3$ of water. Detailed information is reported in Guglielmo et al. (1990). Samples were preserved in a 4% buffered formalin and seawater solution.

### Cariboo Cruise 1989–90

During the period between November 25, 1989 and January 22, 1990 zooplankton and micronekton samples were taken at 27 stations located in the Pacific sector of the Southern Ocean, from New Zealand to the Ross Sea (Fig. 2). Station 1 was located in the Subantarctic Zone, stations 2 and 3 across the Polar Frontal Zone, stations 4 to 7 in the permanent ice zones (Balleny Islands) and stations 8 to 25 in the western Ross Sea.

Samples were collected by the same BIONESS multinet utilized in the Polar Queen cruise, equipped with a multiparametric probe (KMSII) and a Back-Scat fluorometer. The flow through the nets was monitored by external and internal acoustic Doppler flowmeters (ME-SM 11 H). Total effort was 57 hauls (572 samples) filtering 47 438 m$^3$ of water. Some stations were sampled twice in different depth intervals. Only 500-μm mesh nets were used. Detailed information is available in Guglielmo et al. (1992). Sampling effort (such as stations sampled, bottom depth, trawling depth, volume filtered, Mysid species 100 m$^{-3}$) is shown in Table 2.

## Sample Analysis

Mysids were sorted and identified in the laboratory from 268 samples resulting from 33 hauls (Polar Queen cruise) and from 572 samples resulting from 57 hauls (Cariboo cruise). According to the developmental/maturity stages described by Mauchline (1980) adult, immature and juvenile individuals were identified. Age group determination was based on length-frequency distribution. Only two gravid females were found and embryos were counted and measured under a

**Table 1.** Positive stations sampled, bottoms depth (m), trawling depth (m), volume filtered (m³), *Antarctomysis ohlini* abundance (N) for juveniles, immatures and adults during the 1987–88 Italian Antarctic expedition

| Station | Date (1988) | Bottom depth | Trawling depth (m) | VF ($m^3$) | *Antarctomysis ohlini* (N) | | |
|---|---|---|---|---|---|---|---|
| | | | | | Juveniles | Immatures | Adults |
| 8 | Jan. 05 | 570 | 0–500 | 48.3 | 1 | 0 | 0 |
| 9 | Jan. 06 | 560 | 300–260 | 13.4 | 0 | 0 | 1 |
| 11A | Jan. 07 | 530 | 100–80 | 13.6 | 0 | 0 | 1 |
| | | | 70–60 | 22.6 | 0 | 1 | 0 |
| | | | 40–30 | 28.3 | 1 | 0 | 0 |
| 19 | Jan. 10 | 660 | 500–400 | 51.8 | 3 | 0 | 0 |
| | | | 400–350 | 21.2 | 0 | 1 | 0 |
| 22 | Jan. 12 | 645 | 0–600 | 150.4 | 13 | 0 | 1 |
| | | | 600–550 | 17.8 | 3 | 0 | 0 |
| | | | 550–500 | 23.2 | 5 | 0 | 0 |
| | | | 500–450 | 33.9 | 3 | 0 | 0 |
| 26 | Jan. 13 | 930 | 800–700 | 17.6 | 0 | 2 | 0 |
| | | | 700–600 | 20.6 | 2 | 0 | 0 |
| | | | 600–500 | 32.5 | 7 | 0 | 0 |
| | | | 500–400 | 35.8 | 0 | 0 | 1 |
| | | | 400–300 | 38.2 | 0 | 0 | 1 |
| 27 | Jan. 13 | 815 | 600–500 | 20.2 | 5 | 1 | 0 |
| | | | 500–400 | 19.7 | 0 | 0 | 1 |
| 29 | Jan. 14 | 1010 | 0–500 | 36.4 | 1 | 0 | 0 |
| 34 | Jan. 15 | 1100 | 0–600 | 84.9 | 19 | 2 | 0 |
| | | | 600–550 | 31.3 | 49 | 2 | 1 |
| | | | 550–500 | 38.6 | 14 | 1 | 0 |
| | | | 500–450 | 37.1 | 0 | 1 | 0 |
| 13 | Jan. 25 | 810 | 550–500 | 44.2 | 1 | 1 | 0 |
| 5 | Jan. 26 | 700 | 0–500 | 96.5 | 1 | 0 | 0 |
| | | | 350–300 | 64.2 | 3 | 1 | 0 |
| | | | 50–0 | 29.8 | 1 | 0 | 0 |
| 14 | Jan. 26 | 875 | 600–500 | 50.0 | 0 | 0 | 1 |
| 25 | Jan. 26 | 1000 | 500–0 | 155.4 | 1 | 0 | 1 |
| 17 | Jan. 27 | 550 | 0–350 | 66.4 | 0 | 0 | 1 |
| 16 | Jan. 27 | 650 | 0–500 | 72.1 | 2 | 0 | 0 |
| | | | 500–400 | 75.4 | 0 | 1 | 0 |
| 18 | Jan. 28 | 655 | 0–500 | 58.6 | 1 | 0 | 0 |
| | | | 500–400 | 66.9 | 8 | 1 | 0 |
| | | | 400–350 | 51.5 | 6 | 0 | 0 |
| | | | 350–300 | 65.3 | 4 | 1 | 1 |
| | | | 150–100 | 44.3 | 1 | 0 | 0 |
| 23 | Jan. 28 | 1000 | 0–600 | 83.8 | 4 | 1 | 0 |
| | | | 600–560 | 36.6 | 8 | 0 | 0 |
| | | | 560–520 | 41.3 | 13 | 0 | 0 |
| | | | 520–480 | 35.3 | 5 | 0 | 0 |
| | | | 480–440 | 77.8 | 5 | 2 | 2 |
| | | | 440–400 | 75.1 | 9 | 0 | 0 |
| | | | 400–360 | 50.3 | 0 | 0 | 1 |
| 48 | Jan. 29 | 700 | 0–450 | 86.1 | 1 | 0 | 0 |
| | | | 350–300 | 44.3 | 1 | 0 | 0 |
| 30 | Jan. 29 | 1000 | 800–0 | 440.0 | 30 | 0 | 2 |
| 46 | Feb. 02 | 810 | 0–500 | 94.1 | 0 | 0 | 1 |
| | | | 500–400 | 188.1 | 0 | 1 | 0 |
| 39 | Feb. 02 | 840 | 550–500 | 36.6 | 1 | 0 | 0 |
| | | | 450–400 | 65.9 | 1 | 0 | 1 |
| | | | 400–350 | 75.9 | 0 | 0 | 1 |
| 32 | Feb. 02 | 895 | 0–650 | 88.1 | 0 | 0 | 1 |
| | | | 650–550 | 86.1 | 1 | 0 | 0 |
| | | | 550–450 | 100.3 | 7 | 0 | 0 |
| | | | 450–350 | 100.3 | 0 | 1 | 0 |
| 15 | Feb. 02 | 750 | 600–550 | 31.3 | 1 | 0 | 0 |
| | | | 550–500 | 58.3 | 3 | 0 | 0 |
| | | | 500–450 | 31.9 | 2 | 0 | 0 |
| | | | 400–350 | 35.1 | 0 | 0 | 1 |

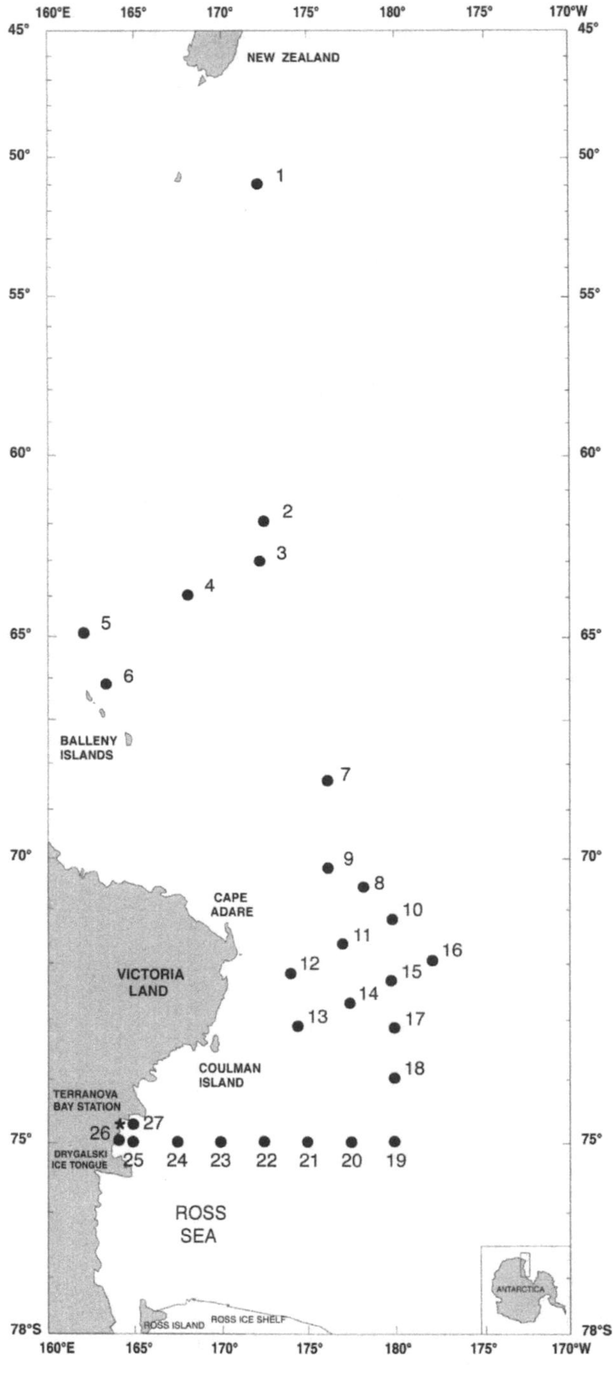


**Fig. 2.** Sampling area and station locations during the 1989–90 Italian Antarctic expedition (Cariboo cruise)

**Table 2.** Positive station sampled, bottom depth (m), trawling depth (m), volume filtered ($m^3$), and mysid species caught during the 1989–90 Italian Antarctic expedition

| *Station* (no.) | Date | Bottom depth (m) | Trawling depth (m) | VF ($m^3$) | Species | N |
|---|---|---|---|---|---|---|
| 2 | Nov. 29, 1989 | 4230 | 800–780 | 11.6 | Damaged | 1 |
| 4 | Dec. 03, 1989 | 3125 | 800–700 | 170.7 | Damaged | 1 |
| | | | 700–600 | 176.1 | Damaged | 1 |
| | | | 400–300 | 192.5 | Damaged | 3 |
| | | | 300–250 | 24.4 | Damaged | 3 |
| 5 | Dec. 05, 1989 | 3130 | 200–100 | 138.5 | *Caesaromysis hispida* | 1 |
| | | | 100–0 | 84.9 | *Caesaromysis hispida* | 1 |
| 6 | Dec. 06, 1989 | 2700 | 800–700 | 61.2 | *Dactylambliops hogdsoni* | 1 |
| | | | 700–600 | 63.3 | *Dactylambliops hogdsoni* | 1 |
| | | | 400–300 | 16.4 | *Dactylambliops hogdsoni* | 1 |
| 7 bis | Dec. 17, 1989 | 3530 | 700–600 | 108.7 | *Dactylambliops hogdsoni* | 2 |
| 9 | Dec. 21, 1989 | 3285 | 1000–800 | 180.7 | Damaged | 1 |
| 10 bis | Dec. 23, 1989 | 1340 | 700–600 | 148.5 | Damaged | 2 |
| 11 bis | Dec. 24, 1989 | 845 | 300–200 | 134.5 | *Euchaetomera zurstrasseni* | 1 |
| 14 bis | Dec. 27, 1989 | 1545 | 400–300 | 187.4 | *Euchaetomera zurstrasseni* | 1 |
| 15 bis | Dec. 28, 1989 | 2140 | 800–700 | 116.5 | Damaged | 4 |
| | | | 500–400 | 179.5 | Damaged | 3 |
| | | | 300–200 | 131.3 | *Euchaetomera zurstrasseni* | 4 |
| 16 bis | Dec. 29, 1989 | 750 | 700–600 | 179.0 | *Boreomysis sibogae* | 6 |
| | | | 600–500 | 199.1 | *Boreomysis sibogae* | 14 |
| | | | 500–400 | 173.1 | *Boreomysis sibogae* | 1 |
| 24 | Jan. 06, 1990 | 520 | 350–300 | 123.6 | *Antarctomysis ohlinii* | 2 |
| 25 | Jan. 07, 1990 | 885 | 800–700 | 238.3 | *Antarctomysis ohlinii* | 1 |
| | | | 600–500 | 146.5 | *Antarctomysis ohlinii* | 1 |
| | | | 400–300 | 162.7 | *Antarctomysis ohlinii* | 5 |
| 26 | Jan. 08, 1990 | 635 | 400–350 | 174.3 | *Antarctomysis ohlinii* | 2 |
| | | | 350–300 | 107.2 | *Antarctomysis ohlinii* | 9 |
| | | | 300–260 | 91.0 | *Antarctomysis ohlinii* | 2 |
| 27 | Jan. 12, 1990 | 735 | 400–350 | 96.6 | *Antarctomysis ohlinii* | 1 |

stereomicrosope after removal of brood pouches judged to be full (Crescenti et al. 1994). The total length (TL) of each stage (from the tip of the rostrum to the posterior end of the telson) was measured to the nearest mm.

All collected samples were taken into account to describe the size-frequency distribution. Abundance values represent a weighed mean of layers pooled from 100 and 500 m (±100 m) standardized as numbers per filtered volume (N 100 $m^{-3}$) calculated from total specimens counted and total volume filtered at each station.

A total of 119 specimens for preliminary diet analysis were used, divided into juveniles (n = 89), immatures (n = 12) and adults (n = 18). As reported by Hopkins (1987) for large animals with high potential for net feeding, only data from hindguts or intestines were used. The stomach contents were identified to lowest possible taxa, usually genera, but also to species and stage in many cases, and counted. The diet was presented in terms of incidents of food types in digestive tractis (Hopkins 1985).

## Results

The mysids which were found in the collected samples are listed below following the nomenclature of Müller (1993):

*Boreomysis brucei* Tattersall 1913
*Caesaromysis hispida* Ortmann 1893
*Dactylamblyops hogdsoni* Holt and Tattersall 1906
*Euchaetomera zurstrasseni* (Illig 1906)
*Antarctomysis ohlini* Hansen 1908

### Species Occurrence

Mean abundances (N 100 $m^{-3}$) and frequency of occurrence in all positive hauls are given in Tables 1 (Polar Queen cruise 1987–88) and 2 (Cariboo cruise 1989–90).

*Boreomysis brucei*
This species was found only at station 16 bis (Ross Sea, off Cape Adare), with 21 specimens, at depths between 400 and 700 m. Of these specimens, four were adult spent females with lengths between 26.9 and 29.5 mm, two immature females of 19 and 20 mm, a male of 30 mm, while the rest were juveniles. Specimens of this species have previously been reported in the Weddell Sea during the Scottish National Antarctic expedition (Tattersall 1913). The records of the Discovery (Tattersall 1955) report it as a deep-water form confined to the Antarctic and to the coldest waters of the South Atlantic. Siegel and Mühlenhardt-Siegel (1988) recovered only two mature females of 20 and 23 mm in samplings of a mean depth of ca. 90 m above the bottom of the Gerlache Strait (Palmer Archipelago) in winter.

*Caesaromysis hispida*
This species, found only at station 5 (65°S, north of the Balleny Islands), included only two small immature males of 2.0 and 2.3 mm found in a depth range from the surface to 200 m. The specific morphological characteristics, such as the carapace and abdominal somites armed with many spines, the eyes separated into two parts (anterior and postero-lateral areas), the lanceolate antennal scale, without spinous processes and bearing seven setolae on the distal margin, the exopodite of the first thoracic limb reduced to a small protuberance, along with the other characteristics of the species, leave no doubt that we are dealing with *Caesaromysis hispida*. However, this species has not previously been found so far south. In fact, despite a wide geographical distribution, there are only two reports for the South Pacific, both from the waters off Val Paraiso. The first was by Colosi (1916), who recovered a small male specimen and created a new genus and species (*Caesaromysis liguriae*); later, Tattersall (1955) maintained that it belonged to *Caesaromysis hispida*. The second report was by Birstein and Tchindonova (1958), who found a female specimen of 7 mm, with oostegites not yet developed, during the expedition carried out with the M/V OB at locality 423 (39°44′S–109°16′W) at a depth of 3228 m and in a depth range from the surface to 1200 m.

*Dactylamblyops hogdsoni*
Three specimens were recovered at station 6 near the Balleny Islands (66°S) in the 300–800-m layer and two specimens at station 7 bis at the entrance to the Ross Sea from the 600- to 700-m depth. These were juvenile specimens measuring between 3.5 and 4.0 mm. This species has a typical circumpolar distribution south of 50°. Zimmer (1914) reported it in the Southern Indian Ocean at seven localities SW of Heart Island (69°25′S–62°10′W) and Illig (1930) near the Bouvet Islands in a zone between 55° and 62°S at depths between 1500 and 2000 m. In the Discovery Report it is mentioned around South Georgia and along the frozen margin of the South Pacific and Indian Oceans (Tattersall 1955). Birstein and Tchindonova (1958) reported two female specimens of 13 and 13.5 mm and a male of 12 mm during the expedition carried out with the M/V OB at station 57 (64°03′S–161°59′E) in a depth range from the surface to 3000 m. Wittmann (1990) defined the species as circumpolar (53°–66°S). The depth at which we found it (300–700 m) confirms the preceding records which define *D. hogdsoni* as a mesopelagic species (Tattersall 1955; Siegel and Mühlenhardt-Siegel 1988).

*Euchaetomera zurstrasseni*
One juvenile specimen (2.8 mm) was recovered at station 11 in a depth range of 200 to 300 m, two juveniles (2.8 and 3.5 mm) at station 14 in a depth range of 300 to 400 m, an adult male of 7.7 mm, a spent female of 7.8 mm and two juveniles (2.6 and 2.8 mm) at locality 15 in a depth range of 200 to 300 m. This species is widely distributed in the Southern Ocean (Tattersall 1955) where it has been reported off Bouvet Island, to the NW of Graham Land, to the east of the Weddell Sea, in the Scotia Sea, in the Bellinghausen Sea, at the frozen margin of Enderby Land, as well as in South Georgia and in other neighbouring zones. Its recovery in the Ross Sea between 71° and 73°S extends its known limits of distribution in southern seas (Tattersall 1955; Mauchline and Murano 1977; Siegel and Mühlenhardt-Siegel 1988; Müller 1993).

*Antarctomysis ohlini*

## Distribution

*A. ohlini* was the only species found in the shelf waters of Terra Nova Bay. A total of 267 juveniles, 23 immatures and 23 adults were collected during the summer Polar Queen cruise (Table 1). In the Cariboo cruise *A. ohlini* was caught only in the inner stations (25, 26 and 27) of the bay (Table 2). Highest abundances were observed in the southern areas of Terra Nova Bay while it is lacking or rare in the northern part (Fig. 3). Maximum abundances were found at station 34 ($\bar{X}$ = 23.1 ind 100 m$^{-3}$; bottom depth = 1100 m). Off Terra Nova Bay *A. ohlini* occurred in very low numbers and was not sampled in deeper midwater samples beyond the continental slope of the Western Ross Sea (Table 2).

*Antarctomysis ohlini* was a mesopelagic form with a well defined vertical distribution pattern. Most records were from depth between 250 and 600 m while the population core occurred between 500 and 600 m (Fig. 4). No difference in the vertical distribution seems to occur among the age classes. A peak of immature specimens (age 3+) was observed between 250 and 300 m. The very rough bottom in the investigated area, characterized by deep trenches and sills, prevented adequate sampling strategies. For this reason BIONESS hauls were made more than 50 m above the bottom.

## Growth

Due to the low total abundance of *A. ohlini* (n = 229) and particularly of large specimens more than 12 mm, it was impossible to fit a statistically significant distribution mixture to the observed length frequency. A tentative age class distribution is presented in Fig. 5. Data for Terra Nova Bay were drawn from the summer months of 1987–88 and 1989–90. In January hauls, recently liberated juveniles show a distinct peak at 9.7 mm (±0.63 SD) representing age class 0. The 1-year-old juveniles reach a mean length of 22.4 mm (±1.69 SD), whereas at the end of the following growth period the immatures (age 2+) reach a mean length of 36.2 mm (±1.95 SD). At the end of the 3rd year, the mean was 47.2 mm (±2.61 SD), while the first

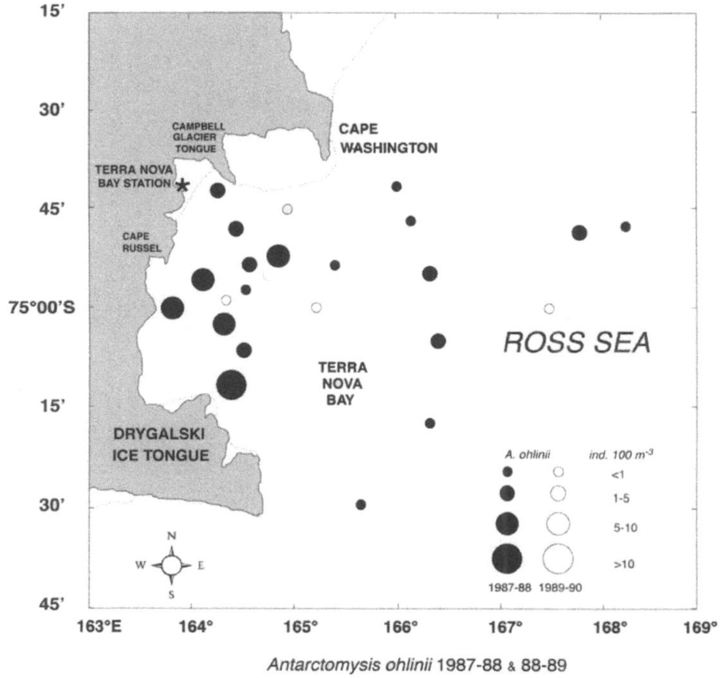


**Fig. 3.** Horizontal distribution and abundance (N 100 m$^{-3}$) of *Antarctomysis holini* collected in the 100–500 m (±100 m) layer in Terra Nova Bay during the 1987–88 and 1989–90 Italian Antarctic expeditions

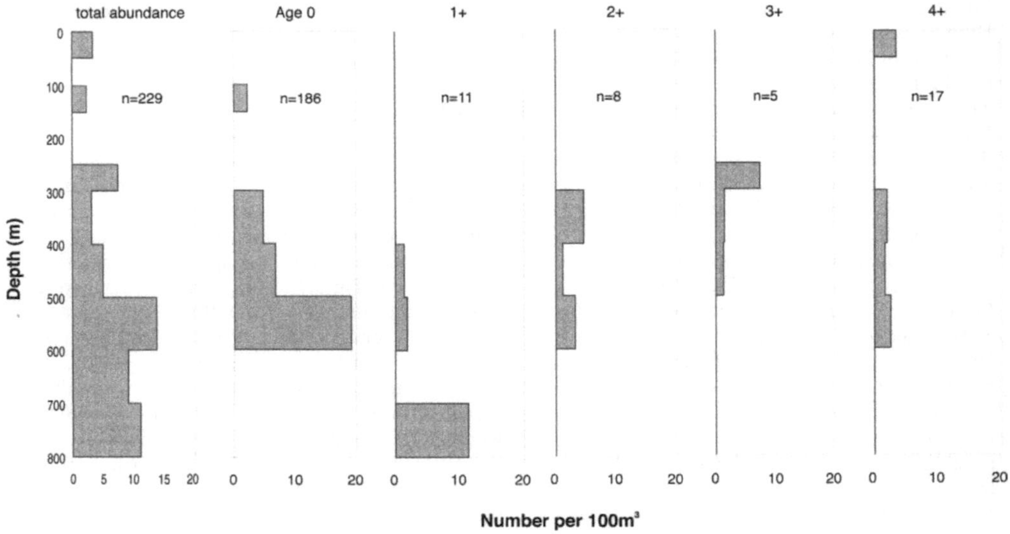


**Fig. 4.** Vertical age class profiles of *Antarctomysis ohlini* abundance (N $100\,m^{-3}$) in Terra Nova Bay during the 1987–88 Italian Antarctic expedition

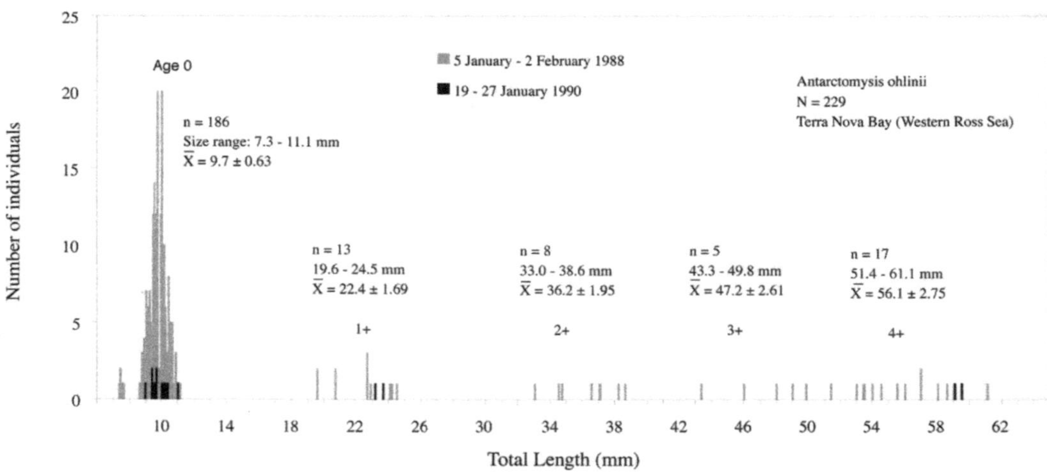


**Fig. 5.** Length frequency distribution of *Antarctomysis ohlini* in Terra Nova Bay during the 1987–88 and 1989–90 Italian Antarctic expeditions. *0 to 4+* indicate the tentative age classes

specimens mature around 50 mm. During the 4th year, a mean length of about 56.1 mm (±2.75 SD) was reached and the majority were spent females while only two gravid females were found. Stage II and stage III larvae were removed from those two females (Table 3), counted and measured (Crescenti et al. 1994). Stage II embryos had a mean length of 3.2 mm (±0.2 SD) with a number around 65 embryos in the marsupium. Stage III embryos had a mean length of 7.2 mm (±0.2 SD) with a number around 60 embryos.

**Table 3.** Fecundity of *Antarctomysis ohlini*

| Gravid females | Stage II | Stage III |
|---|---|---|
| No. investigated | 1 | 1 |
| Total length (mm) | 58.6 | 58.0 |
| Size (mm) of embryos in marsupium | 3.2 ± 0.2 | 7.2 ± 0.2 |
| No. of embryos in marsupium | 65 | 60 |
| No. of embryos measured | 55 | 50 |

In January, tentative growth rate estimates for Terra Nova Bay by size classes indicate mean differences varying from 0.91 to 1.15 mm month$^{-1}$ for classes I–III and around 0.74 mm for class IV. These values agree with the data reported by Ward (1985) for Moraine Fjord of South Georgia. This author assumed a growth rate of about 1.00–1.50 mm month$^{-1}$ for classes I–III and approximately 0.50 mm month$^{-1}$ for class IV.

Thus, according to Ward (1985) and Siegel and Mühlenhardt-Siegel (1988) the generation time from egg to egg for *A. ohlini* from Terra Nova Bay was assumed to be 5 years.

## Diet Composition

Preliminary gut content analysis (Table 4) confirms previous data (Mauchline 1980; Hopkins 1985, 1987; Ward 1985) and indicates that *A. ohlini* was an omnivorous feeder during its life cycle. Diet

**Table 4.** Frequency of occurrence (no. of guts) in the identifiable gut contents of Terra Nova Bay *Antarctomysis ohlini*

| No. of guts analyzed (n)<br>Size range (mm) | Juveniles<br>89<br>9.0–24.5 | Immatures<br>12<br>33.0–49.8 | Adults<br>18<br>51.4–61.1 |
|---|---|---|---|
| Diatoms | | | |
| *Nitzschia* sp. | 3 | | |
| *Coscinodiscus* sp. | 22 | | |
| *Fragilariopsis curta* | 34 | | |
| *Eucampia antarctica* | 25 | | |
| Dinoflagellats | | | |
| *Protoperidinium* sp. | 6 | | |
| Tintinnidi | | | |
| *Cymatocylis flava* | 31 | 8 | 6 |
| Radiolarians | 52 | | |
| Copepods | | | |
| *Calanoides acutus* (F) | | 1 | 4 |
| *Calanoides acutus* (M) | | | 1 |
| *Metridia gerlachei* (F) | | 4 | 3 |
| *Metridia gerlachei* (M) | | | 4 |
| *Metridia gerlachei* (copepodids) | 3 | 2 | 5 |
| *Oithona frigida* (F) | | 1 | |
| *Paraeuchaeta antarctica* (F) | | | 2 |
| *Other copepods* | 3 | 1 | 1 |
| Ostracods | | | |
| *Alacia belgicae* | 6 | 5 | 5 |
| Euphausiids larvae | | | 2 |
| Polychaets | | | |
| *Pelagobia longicirrata* | 36 | 11 | 13 |
| *Alciopidae* n.c. | | | 2 |
| Pteropods | | | |
| *Limacina helicina* | | 2 | 7 |
| Chaetognats | | | 4 |
| Mysids | | 1 | 1 |

diversity increased moderately with increasing size. Diatoms (mainly *Fragilariopsis curta*, *Eucampia antarctica* and *Coscinodiscus* sp.) and radiolarians (unidentified species diameter 10 µm) were the principal food source of juveniles together with tintinnids (*Cymatocylis flava*), juvenile ostracods (*Alacia belgicae*) and pelagic polychaetes (*Pelagobia longicirrata*). Copepods were insignificant.

The diet composition of immature specimens was different from that of juveniles. Diatoms and radiolarians were insignificant, whereas pteropods (*Limacina helicina*) and a few species of copepods (*Calanoides acutus* and *Oithona frigida*) were important dietary constituents. Diet diversity increased in adults. More copepod species constituted the gut contents (*Calanoides acutus*, *Metridia gerlachei*, *Paraeuchaeta antarctica*); also polychaetes Alciopidae, euphausiid larvae, chaetognaths and mysids were ingested.

## Discussion

Although the data were scarce, the Mysidacea collected in the present study can be classified by their habitat preference. All species were found south of the Antarctic Convergence and seem to have a well-defined distribution gradient.

From north to south (1989–90 Cariboo cruise), *Caesaromysis hispida* was found at a station located in the marginal ice zone (65°S). Only two immature males were caught between the surface and the 200-m depth. The present record extends the previous known southerly distribution limit.

*Dactylamblyops hogdsoni* was recorded at two stations located between 66° and 69°S in the permanent pack-ice zone. Also Tattersall (1955) found *D. hogdsoni* along the marginal ice zone of both the South Pacific and South Indian Oceans. The five juvenile specimens caught from 300 to 800 m confirms previous depth records (Tattersall, 1955; Siegel and Mühlenhardt-Siegel 1988) which assume that *D. hogdsoni* is a mesopelagic species.

*Boreomysis brucei* and *Euchaetomera zurstrasseni* were found off Cape Adare (71°–73°S) in the seasonal pack ice zone. This area corresponded to the continental slope and was characterized by particular hydrographic features, highest productivity and krill biomass (Hecq et al. 1992, 1993). *B. brucei* was caught at station 16 between 400 and 700 m, thus confirming a mesopelagic preference. Also *E. zurstrasseni* was found in the mesopelagic zone (200–400 m). The mixture of adult males, spent females and juveniles in the same samples indicates that *B. brucei* and *E. zurstrasseni* reproduces in this area, where higher productivity allows for a variety of food resources for young and adults.

*Antarctomysis ohlini* was the only species collected in Terra Nova Bay during both the 1987–88 and 1989–90 expeditions. It reached maximum densities in the southern part of Terra Nova Bay in deep straits and trenches. Our results confirm previous data (Siegel and Mühlenhardt-Siegel 1988) that the species does not appear to occur in deeper oceanic midwater samples beyond the continental slope of the continent. *A. ohlini* seems to be confined to Terra Nova Bay with the last record found at 168°E. Also Ward (1985) found *A. ohlini* confined to inner fjord zones of South Georgia.

Previous investigations in the Ross Sea provide little information both on the biology and ecology of this species. An increase in adult size with increasing latitude noted by Tattersall (1923) in the Ross Sea was not confirmed from our samples collected between 74°15′S and 75°15′S. Tattersall (1961, 1965) records *A. ohlini* from deeper waters of the Ross Sea during both the 1958–59 and 1959–60 seasons. He found one immature female (22 mm) caught with a triangular dredge (bottom 695 m), two immature females (23 mm) caught with a vertical plankton net from 800 m to the surface, one spent female (54 mm) caught with a trawl, two adult males (54–55 mm) caught with a fish trap and three specimens in the gut content of a Weddell seal. During the HMNZS Endeavour cruise, Tattersall (1965) found one mature male (59 mm) caught by Knudsen bottles from 574 to 543 m and one juvenile from 264 to 256 m. During the austral winter 1960, Tattersall (1965) caught *A. ohlini* in McMurdo Sound with a fish trap twice on the 560-m bottom depth.

From these results Tattersall (1961, 1965) stated that *A. ohlini* and *A. maxima* sometimes occurred in the same net haul and hypothesized that both species preferred an hyperbenthic habitat. In this study *A. ohlini* was not sampled together *A. maxima* in BIONESS samples. *A. maxima* did not even occur in the gut contents of some benthic fishes caught in the inner part of Terra Nova Bay (Vacchi et al. 1994; Vacchi and La Mesa 1995; La Mesa et al., 1997). This supports Ward's hypothesis (1985) that the two species may have some overlap in their general distribution but may be segregated to some extent.

In Terra Nova Bay, *A. ohlini* confirms its mesopelagic preference with a defined vertical distribution pattern. The population core occurred between 500 and 600 m while no depth differences seem to occur among age classes. It is probable that the distribution of the two species in Terra Nova Bay was similar to that described by Siegel and Mühlenhardt-Siegel (1988) in the generalized picture along a transect from the Drake Passage across the western Bransfield Strait of the Antarctic Peninsula. The absence of *A. maxima* both in BIONESS samples and in the gut content of coastal benthic fishes supports the hypothesis that *A. maxima* inhabits the outer shelf area of Terra Nova Bay, while *A. ohlini* was found in the deeper trenches of the bay.

The life cycle of *A. ohlini* from Terra Nova Bay seems to be similar to that described both by Ward (1985) from Moraine Fjord of South Georgia and Siegel and Mühlenhardt-Siegel (1988) from the Antarctic Peninsula. Assuming both that the length of stage III larvae in the marsupium were approximatively 7.2 mm ± 0.2 (Crescenti et al. 1994) and the growth rate of juveniles was estimated to be about 1 mm month$^{-1}$, the newly moulted young C0 caught at Terra Nova Bay in January (mean length 9.7 ± 0.63) were about 3 months old and the larvae were released during October.

The generation time from egg to egg for *A. ohlini* from Terra Nova Bay agree with Ward's (1985) data and is assumed to be 5 years. Thus, also in this case, *A. ohlini* confirms certain ecological adaptations in higher antarctic environments noted by Ward (1985) and Siegel and Mühlenhardt-Siegel (1988) (e.g. lower fecundity, more prolonged incubation time, larger number of embryos in marsupium, larger size of embryos).

Gut content analysis confirms that *A. ohlini* was a generalist during its life cycle. Ward (1985) stated that the release of young coincides with increased primary production. During austral spring (October–November) Terra Nova Bay was entirely covered by sea ice. Through mid-November and early December 1994 (Italian ROSSMIZE Project) highest values of both chlorophyll a (up to 3.8 µg C m$^2$ day$^{-1}$) and primary production (up to 2.8 mg C m$^2$ day$^{-1}$) were found at the marginal ice zones and in pack ice-covered waters in the vicinity of and inside Terra Nova Bay (Carrada et al. 1997; Saggiomo et al. 1997). During the same sampling period increased chlorophyll a concentrations and primary production (3–10 µg l$^{-1}$ and 3.5–6.0 g C m$^2$ day$^{-1}$, respectively) were measured in the Ross Sea polynya (Smith and Gordon 1997).

These data suggest that both marginal ice zones and polynya areas support the first stages of *A. ohlini*'s life. In fact, the presence of permanent polynyas at Terra Nova Bay in October could provide favourable food conditions for the development of early stages, after young surface release. During this period, nanophytoplankton predominated the marginal ice zone (up to 80%) and picophytoplankton reached maximum concentrations (60–80%) under the pack ice at about the 100-m depth (Carrada et al. 1997; Vanucci and Bruni 1997). Thus, early planktonic stages of *A. ohlini* concentrate in the surface mixed layer where feeding conditions (e.g. shape and size range of particles about 2–5 µm, Vanucci and Bruni, 1997) seem to be near optimum.

At this time, postlarvae may be collected in well-stratified Antarctic Surface Water (AASW) and the vertical distribution appears to be related to the forming thermocline. The AASW was shifted towards the edges of Cape Washington and Drygalski Ice Tongue, flowing along the coast and accumulation of larvae in the frontal systems and edges can be explained by this drift (Guglielmo et al. 1998). It is probable that moulted free-swimming juveniles migrated downward during the austral summer; in January the entire *A. ohlini* population occurred between 250 and 800 m mainly in the deeper southern part of the shelf in cold High Salinity Shelf Water (HSSW) (Spezie et al. 1993). From our BIONESS sampling times (07.00–24.00 h), it seems that *A. ohlini* did not migrate into surface waters during the daytime but remain instead in the deeper layers.

As noted by Hopkins (1985, 1987), diatoms were the most important group in the diet composition of juveniles. Of the diatoms found in the juvenile gut contents, *Fragilariopsis curta*, *Eucampia antarctica* and *Coscinodiscus* sp. were abundant. *F. curta* dominated the diatom assemblages in all samples collected in the ice-covered waters of the Ross Sea and Terra Nova Bay during the austral spring 1994 (Marino and Cabrini 1997). The same species also largely dominated the diatom assemblages (from 74 to 97% of the total cell number) of unknown zooplankton oval faecal pellets collected in the coastal area of Terra Nova Bay (Marino et al. 1995). *Eucampia antarctica* and *Coscinodiscus centralis* were found in the larger krill faecal strings collected in deeper traps of the Scotia and Weddel Seas (Gonzalez 1992). As observed in oval faecal pellets (Marino et al. 1995) all *F. curta* cells found in the gut content of *A. ohlini* had intact cytoplasm together with *Coscinodiscus* sp., while only fragments were found of *Nitzschia* sp. and *Eucampia antarctica*.

Our data support the hypothesis of a coprophagy behaviour of juvenile *A. ohlini*. It is possible that unknown zooplankton oval faecal pellets were recycled in the deeper layers. Also krill faeces were collected in large amounts in sediment traps suspended at depth below 500 m (Wefer 1989). Coprofagy would result in repackaging of the shredded mineral skeletons. Dense aggregates or "packets" of intermixed *Nitzischia* spp., *Coscinodiscus* diatoms, peridinians and tintinnids were found by Hopkins (1987) in the gut contents of *A. ohlini* in McMurdo Sound (Ross Sea).

Oval faecal pellets were more abundant in surface (0–50 m) waters of Terra Nova Bay (Marino et al. 1995) and in subsurface waters (50–150 m) of the Scotia and Weddell Seas (Gonzalez 1992). The dominance inside oval faecal pellets of sea-ice/planktonic species suggest that a planktonic crustacean, either an animal able to scrape food off the underside of ice or a filter feeder, could be the producer organism (Marino et al. 1995). Sinking of the oval faecal pellets from upper to deeper layers may be possible due to the presence of a double-layered peritrophic membrane which make the pellets resistent to mechanical stress, allowing them to reach the deep layers undamaged. Other suggestions indicate that zooplankton oval faecal pellets must be largely produced below the pycnocline (Longhurst and Harrison 1989) and in the entire upper 1000 m (Gonzalez 1992). In this sense, high numbers of *Fragilariopsis curta* were found in the gut of the planktonic ostracod *Alacia belgicae* in Terra Nova Bay (Marino et al. 1995).

*A. ohlini* can be considered an important macroplankton-micronekton species in the Terra Nova Bay midwater food web. Although the inner population occupied a similar deep range (250–600 m) in summer, no trophic overlapping seems to occur among the different age classes. *A. ohlini* may be considered an opportunist omnivorous feeder during its life cycle. Coprophagy seems to be the principal feeding mode in juveniles (age 0 and age 1+) while diet diversity increases with increasing size in this species. The zooplankton species ingested by all size classes (e.g. *Cymatocylis flava*, *Metridia gerlachei*, *Alacia belgicae* and *Pelagobia longicirrata*) were common in the midwater layers off Terra Nova Bay (Carli et al. 1990; Guglielmo et al. 1990; Benassi et al. 1992; Monti and Fonda Umani 1995).

In the future, it will be important to clarify the role of *A. ohlini* in the carbon flux of the Antarctic midwater food web, where these processes seem to be very significant.

*Acknowledgements.* Many thanks are due to officers and crew of the research vessels Polar Queen and Cariboo and technicians involved in the BIONESS sampling procedures for their excellent cooperation in field work. We are grateful to Prof. Alberto Castelli from the University of Sassari, Prof. Serena Fonda Umani from the University of Trieste and Dr. Donato Marino from the Stazione Zoologica di Napoli for their help in identifying polychaets, tintinnids and phytoplankton, respectively, in the gut contents. We are also grateful to anonymous referees for constructive comments during the preparation of the manuscript. Logistical and financial support for this chapter was given from the Italian National Project for Antarctic Research (PNRA).

## References

Benassi G, Naldi M, McKenzie KG (1992) Preliminary research on species distribution of planktonic ostracods (Halocyprididae) in the Ross Sea, Antarctica. J Crustacean Biol 12: 68–78

Birstein JA, Tchindonova JG (1958) Glubocovodniie Mysidii severo zapadnoi ciasti Tihogo Okeana (The deep sea mysids of the northwest Pacific Ocean). Tr Inst Okeanol 27: 258–355

Birstein JA, Tchindonova JG (1962) Mysidacea collected by the Soviet Antarctic expedition on the M/V Ob. Biol Rep Soviet Antarct Exped (1955–1958) 1: 58–68

Carli A, Mariottini GL, Pane L (1990) Contribution to the study of copepods collected in Terra Nova Bay (Ross Sea). In: National Scientific Commission for Antarctica (ed) Oceanographic Campaign 1987–88, Data Rep Part I, Genova, pp 129–167

Carrada GC, Mangoni O, Russo GF, Saggiomo V, Sgrosso S (1997) Spatial and temporal distribution of phytoplankton pigments in the Ross Sea during the austral spring 1994. In: Faranda F, Guglielmo L, Povero P (eds) National Programme for Antarctic Research, ROSSMIZE 1994–95, Data Rep I Part I, pp 197–228

Colosi G (1916) *Caesaromysides liguriae*, n.gen. n.sp.; nota preliminare. Monit Zool Ital 27: 136–139

Coutière MH (1906) Crustacés schizopodes et décapodes. J Charcot Exped Antarct Fr (1903–1905): 1–10

Crescenti N, Costanzo G, Guglielmo L (1994) Developmental stages of *Antarctomysis ohlini* hansen, 1908 (Mysidacea) in Terra Nova Bay, Ross Sea, Antarctica. J Crustacean Biol 14: 383–395

De Witt HH (1970) The character of the midwater fish fauna of the Ross Sea, Antarctica. In: Holdgate MW (ed) Antarctic ecology, vol 1. Academic Press, London, pp 305–314

Gonzalez HE (1992) The distribution and abundance of krill faecal material and oval pellets in the Scotia and Weddell

Seas (Antarctica) and their role in particle flux. Polar Biol 12: 81–91

Guglielmo L, Costanzo G, Manganaro A, Zagami G (1990) Spatial and vertical distribution of zooplanktonic communities in Terra Nova Bay (Ross Sea). In: National Scientific Commission for Antarctica (ed) Oceanographic Campaign 1987–88, Data Rep Part I, Genova, pp 257–398

Guglielmo L, Costanzo G, Zagami G, Manganaro A, Arena G (1992) Zooplankton ecology in the Southern Ocean. In: National Scientific Commission for Antarctica (ed) Oceanographic Campaign 1989–90, Data Rep Part II, Genova, pp 301–467

Guglielmo L, Granata A, Greco S (1998) Distribution and abundance of postlarval and juvenile *Pleurogramma antarcticum* (Pisces, Nototheniidae) off Terra Nova Bay (Ross Sea, Antarctica). Polar Biol 19: 37–51

Hansen HJ (1908) Schizopoda and Cumacea. In: Resultats du voyage du S.Y. Belgica en 1897–1898–1899 sous . . . A de Gerlache de Gomery. Anvers, JE Buschmann, Rapp Sci, Zool, 20 pp, 3pls

Hecq JH, Magazzu' G, Goffart A, Catalano G, Vanucci S, Guglielmo G (1992) Distribution of planktonic components related to structure of water masses in the Ross Sea during the Vth ITALIANTARTIDE Expedition. In: Ambrosetti W, Piccazzo M, Ruffoni Riva T (eds) Atti IX Congr Ass Ital Oceanogr Limnol, Lang, Genova, pp 665–678

Hecq JH, Brasseur P, Goffart A, Lacroix G, Guglielmo L (1993) Modelling approach of the planktonic vercical structure in deep Austral Ocean. The example of the Ross Sea ecosystem. Proceedings of the Annual Workshop on Belgian Oceanographic Research. Royal Academy of Belgium, Brussels, pp 235–250

Holt EW, Tattersall WM (1906) Preliminary notice of the Schizopoda collected by HMS Discovery in the Antarctic region. Ann Mag Nat Hist (7) 17: 1–11

Hopkins TL (1985) Food web of an Antarctic midwater ecosystem. Mar Biol 89: 197–212

Hopkins TL (1987) Midwater food web in McMurdo Sound, Ross Sea, Antarctica. Mar Biol 96: 93–106

Illig G (1930) Die Schizopoden der deutschen Tiefsee-Expedition. Rep Valdivia Exped 22: 397–625

Innamorati M, Mori G, Lazzara L, Vanucci S (1990) Eutrofia e oligotrofia nell'Oceano Antartico. Oebalia 1: 153–162

Kock KH (1981) Fischereibiologische Untersuchungen an drei antarktischen Fischarten: *Champsocephalus gunnari* Lönnberg, 1905, *Chaenocephalus aceratus* (Lönnberg, 1906) und *Pseudochaenichthys georgianus* Norman, 1937 (Notothenoidei, Channichthyidae). Mitt Inst Seefisch Bundesforschungsanst Fisch Hamburg 32: 226 pp

La Mesa M, Vacchi M, Castelli A, Diviacco G (1997) Feeding ecology of two nototheniid fishes, *Trematomus hansoni* and *Trematomus loennbergii*, from Terra Nova Bay, Ross Sea. Polar Biol 17: 62–68

Longhurst AL, Harrison WG (1989) The biological pump: profiles of plankton production and consumption in the upper ocean. Prog Oceanogr 22: 47–123

Marino D, Cabrini M (1997) Distribution and succession of phytoplankton populations in the Ross Sea (Antarctica) during the Austral spring 1994. In: Faranda F, Guglielmo L, Povero P (eds) National Programme for Antarctic Research, ROSSMIZE 1993–95, Data Rep Part I, Genova, pp 307–341

Marino D, Montresor M, Mazzella L, Saggiomo V (1995) Diatom flora in oval faecal pellets from Terra Nova Bay (Antarctica). In: Marino D, Montresor M (eds) Proc 13th In Diatom Symp, Bio Press Ltd, Bristol, pp 229–240

Mauchline J (1980) The biology of mysids and euphausiids. Adv Mar Biol 18: 1–681

Mauchline J, Murano M (1977) World list of the Mysidacea, Crustacea. J Tokyo Univ Fish 64 (1): 39–88

Monti M, Fonda Umani S (1995) Tintinnids in Terra Nova Bay-Ross Sea during two austral summers (1987/88 and 1989/90). Acta Protozool 34: 193–201

Müller HG (1993) World catalogue and bibliography of the recent Mysidacea. Wissenschaftler Verlag, Laboratory for Tropical Ecosystems Research & Information Service, Wetzlar, 491 pp

Patterson SL, Whitworth T (1990) Physical oceanography. In: Glasby GP (ed) Antarctic Sector of the Pacific. Elsevier, Amsterdam, pp 55–93

Permitin YE (1970) The consumption of krill by Antarctic fish. In: Holdgate ME (ed) Antarctic ecology, vol 1. Academic Press, London, pp 177–182

Permitin YE, Tarverdieva MI (1972) The food of some Antarctic fish in the South Georgia area. J Ichthyol (USSR) 12: 104–114

Saggiomo V, Carrada GC, Mangoni O (1997) Spatial and temporal distribution of total and size fractionated primary production in the Ross Sea during the austral spring 1994. In: Faranda F, Guglielmo L, Povero P (eds) National Programme for Antarctic Research, ROSSMIZE 1993–95, Data Rep Part I, Genova, pp 291–305

Sameoto DD, Jaroszynski LO, Fraser WB (1980) BIONESS a new design in multiple net zooplankton samplers. Can J Fish Aquatic Sci 37: 772–774

Siegel V, Mühlenhardt-Siegel U (1988) On the occurrence and biology of some Antarctic Mysidacea (Crustacea). Polar Biol 8: 181–190

Smith SL, Schnack-Schiel SB (1990) Polar zooplankton. In: Smith WO (ed) Polar oceanography, part B: Chemistry, biology and geology. Academic Press, London, pp 527–598

Smith WO, Gordon LI (1997) Hyperproductivity of the Ross Sea (Antarctica) polynia during austral spring. Geophys Res Lett 24: 233–236

Smith WO, Nelson DM (1985a) Phytoplankton bloom produced by a receding ice edge in the Ross Sea: spatial coherence with the density field. Science 227: 163–166

Smith WO, Nelson DM (1985b) Phytoplankton biomass near a receding ice-edge in the Ross Sea. In: Siegfried WR, Condy PR, Laws RM (eds) Antarctic nutrient cycles and food webs. Springer, Berlin Heidelberg New York, pp 70–77

Spezie G, Tucci S, Burdillon G, Corradi N, Piccazzo M, Fanucci F, Firpo M, Mirabile L, Ferrari M (1993) Water masses characteristics, sediment dispersal and seismic stratigraphy in the continental margin of the Ross Sea (Antarctica). Ann Ist Univ Nav Napoli 60: 32

Targett TE (1981) Trophic ecology and structure of coastal Antarctic fish communities. Mar Ecol Prog Ser 4: 243–263

Tattersall OS (1955) Mysidacea. Discovery Rep 28: 1–190

Tattersall OS (1961) Report on some Mysidacea from the deeper waters of the Ross Sea. Proc Zool Soc Lond 137: 553–571

Tattersall OS (1965) The fauna of the Ross Sea. IV. Mysidacea. NZ Dept Sci Ind Res Bull 167: 35 pp

Tattersall WM (1913) The Schizopoda, Stomatopoda, and non-Arctic Isopoda of the Scottish National Antarctic Expedition. Trans R Soc Edinb 49: 867–894

Tattersall WM (1923) Crustacea. VII. Mysidacea. In: British Antarctic (Terra Nova expedition, 1910, Natural history reports. London. Br Mus Nat Hist Zool 3: 273–304

Vacchi M, La Mesa M (1995) The diet of Antarctic fish *Trematomus newnesi* Boulenger, 1902 (Nototheniidae) from Terra Nova Bay, Ross Sea. Antarct Sci 7: 37–38

Vacchi M, La Mesa M, Castelli A (1994) Diet of two coastal nototheniid fish from Terra Nova Bay, Ross Sea. Antarct Sci 6: 61–65

Vanucci S, Bruni V (1997) Nanoplankton and picophytoplankton in the Ross Sea (Austral Spring 1994). In: Faranda F, Guglielmo L, Povero P (eds) National Programme for Antarctic Research, ROSSMIZE 1993-95, Data Rep Part I, Genova, pp 343–359

Ward P (1984) Aspects of the biology of *Antarctomysis maxima* (Crustacea: Mysidacea). Polar Biol 3 (2): 85–92

Ward P (1985) On the biology of *Antarctomysis ohlini* (Crustacea: Mysidacea) of South Georgia. Bull Br Antarct Surv 67: 13–23

Wefer G (1989) Particle flux in the ocean: present and past. In: Berger WH, Smetaceck VS, Wefer G (eds) Dahlem Konferenzen. Wiley, Chichester, pp 139–154

Wilson DL, Smith WO, Nelson DM (1986) Phytoplankton bloom dynamics of the western Ross Sea ice edge. I. Primary productivity and species-specific production. Deep Sea Res 33: 1375–1387

Wittmann KJ (1990) Mysidacea. In: Sieg J, Wägele JW (eds) Fauna der Antarktis. Parey, Berlin, pp 130–133

Zimmer C (1914) Die Schizopoden der deutschen Südpolar-Expedition 1901–1903. In: von Drygalski (ed) Deutsche Südpolar-Expedition 1901–1903, pp 377–445

Zunini Sertorio T, Salemi Picone P, Bernat P, Cattini E, Ossola C (1990) Copepods collectes in 16 stations during the Italian Antarctic expedition 1987–1988. In: National Scientific Commission for Antarctica (ed) Oceanographic Campaign 1987–88, Data Rep Part I, pp 67–125

CHAPTER 27

# Planktonic Ostracoda in the Ross Sea: Their Distribution and Associated Environmental Factors

K.G. McKenzie[1], I. Ferrari[2], and G. Benassi[2]

**ABSTRACT**

The Italian Progetto Antartide conducted cruises to Terra Nova Bay and the Ross Sea in 1987–1988 and 1989–1990; planktonic halocyprid ostracodes were among the groups sampled as part of the zooplankton programs. Four species: *Alacia belgicae* (Müller, 1906), *A. hettacra* (Müller, 1906), *Metaconchoecia isocheira* (Müller, 1906) and *M. skogsbergi* (Iles, 1953) were identified in 1987–1988 samples; *Boroecia antipoda* (Müller, 1906), *Conchoecissa symmetrica* (Müller, 1906) and *Proceroecia rivoltella* Benassi et al. 1994 were added in 1989–1990. The latter two species comprised single records only, well south of their known latitudinal ranges, but the other species represent the characteristic planktonic ostracode faunule to 1000 m in the western Ross Sea. *A. belgicae* is regarded as a facultative Antarctic species at midwater depths. We discuss the spatial distribution of species particularly with regard to their highest densities at the stations, sampling limitations, vertical migration, possible hybridization in *A. belgicae*; and deal briefly with important associated environmental factors such as the Antarctic Convergence, the water masses of the Ross Sea, the clockwise summer gyre in Terra Nova Bay and associated turbulence zones, the role of the Antarctic Circumpolar Current, melting ice; then draw some conclusions re the planktonic Ostracoda for any continuing program of investigations in the Ross Sea under the aegis of the Italian Progetto Antartide.

## Introduction

This chapter summarizes the results of research on planktonic Ostracoda collected during the two Italian Oceanographic expeditions to Antarctica in 1987–88 and 1989–90. During the earlier expedition, 311 zooplankton samples were collected at 33 stations, 31 of these in Terra Nova Bay and 2 farther offshore in the Ross Sea. During the second expedition, zooplankton was sampled from a region south of New Zealand in the Southern Ocean to the Ross Sea, viz. from 51° to 75°S: 572 samples were collected at 48 stations; only 4 of these stations were located in the sampling area of the 1987–88 expedition (Terra Nova Bay), but a greater number of them (35) were located between 71° and 75°S in the area of the Ross Sea (Fig. 1). Zooplankton sampling was carried out under the supervision of Prof. L. Guglielmo (University of Messina).

Species determinations were based on Müller (1906a,b), Poulsen (1973), Martens (1979) and Angel (1981). A bibliographic survey has shown that the most useful publications dealing with Antarctic planktonic Ostracoda are those of Müller (1906a,b, 1908), Brady (1907), Skogsberg (1920), Barney (1921), Poulsen (1969, 1973), Martens (1979), Deevey (1978, 1982, 1983), Angel (1981) and Kock (1992).

## Materials and Methods

Between 4 January and 21 February 1988, zooplankton was sampled at 27 stations in Terra Nova Bay, Ross Sea. The samples were taken simultaneously by EZ-NET BIONESS, a multiple opening and closing net sampler with ten nets, having a square mouth opening of 0.25 m$^2$. Two mesh size nets were used at different stations: 250 µm and

[1]Charles Stuart University – Riverina, School of Science and Technology, Wagga Wagga 2678, Australia
[2]Università degli Studi di Parma, Dipartimento di Scienze Ambientali, 43100 Parma, Italy

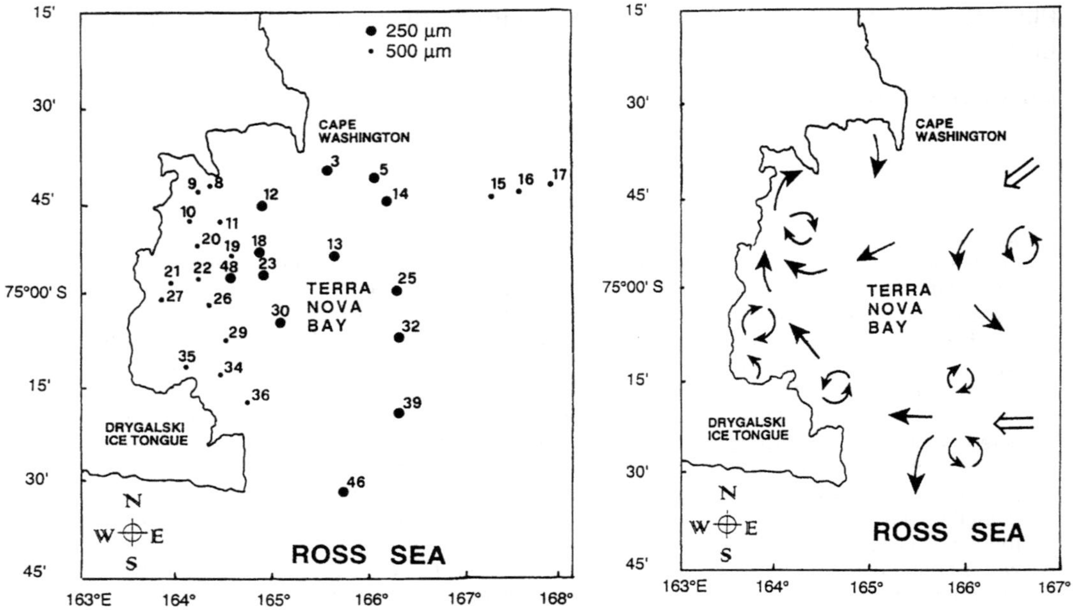


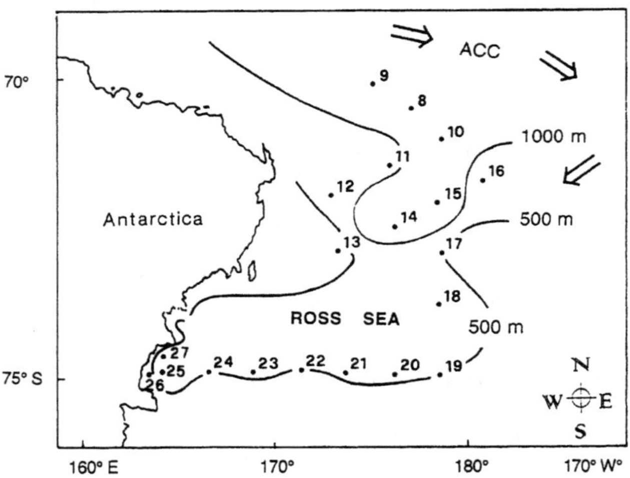


**Fig. 1,1–3.** Localities for stations occupied during the Italian expeditions to the Ross Sea in 1987–1988 (1, 2) and 1989–1990 (3), incorporating the 500-m and 1000-m isobaths (3), the summer clockwise gyre (2, *arrows*); the ACC (3, *broad arrows*); and net mesh sizes used (1). (In part, after Stocchino and Manzella 1991, Fig. 3c)

500 µm. Nets were opened and closed by shipboard command at depth intervals of 10- to 40 m for 100- and 300-m profiles and at 50- to 100-m intervals below 300 m to the maximum depth sampled (800 m). An integrated sample for the whole water column was also collected at each station.

All 500-µm samples were collected relatively near the coast, while the 250-µm samples were taken further offshore in the Ross Sea (Fig. 1). The sampling program used 500-µm nets from 5–15 January 1988 and 250-µm nets from 24 January to 2 February 1988. Stations 1SK (74°50′S, 172°40′E) and ANZ (72°18′S, 172°E), outside the sampling area proper, were sampled on 4 February and 21 February 1988, respectively, using 250-μm mesh nets. The samples were preserved in 4% buffered formalin.

The EZ-NET BIONESS also continuously measured temperature, salinity and depth during sampling tows via an Applied Microsystems digital CTD. In addition, flow through the nets was measured internally and externally by TSK flowmeters.

In 1989–1990, zooplankton were sampled at 27 stations from New Zealand to the Ross Sea between 25 November 1989 and 22 January 1990. Stations 8–27 were located in the western Ross Sea. Some stations were occupied twice. The maximum sampling depth was 1000 m. Only four-stations (24–27) fell within the sampling area of the 1987–1988 campaign (Fig. 1). On this cruise, the EZ-NET BIONESS multinet apparatus was equipped with a multiparametric probe (KMSII) and a Back-Scat fluorometer; flow through the nets was monitored by external and internal acoustic doppler flowmeters. Only 500-μm mesh nets were used. As in 1987–1988, temperature, salinity and depth were also continuously recorded. Again, the samples were preserved in 4% buffered formalin.

Sorting of the Ostracoda from both cruises was carried out at the Marine Sorting Center Italia (Istituto Scienze Ambientali Marine, University of Genova), and the specimens were remitted for study to the Istituto di Ecologia, University of Parma. There were 263 samples with 54 224 ostracodes from the 1987–1988 Polar Queen cruise and 273 samples with 12 124 (adjusted from the original mostly 1/2 aliquot counts) ostracodes from the 1989–1990 Cariboo cruise. The lesser number of specimens in 1989–1990 was mainly due to the exclusive use of 500-μm nets on that cruise. All Ostracoda, juveniles as well as adults, were identified to species and counted for each sample, except in cases where the preservation was too poor to enable determination.

The study materials are retained among the collections of the Dipartimento di Scienze Ambientali (former Istituto di Ecologia), University of Parma, except for two paratypes of a new species, *Proceroecia rivoltella* Benassi et al. 1994, which are held in the collections of the Australian Museum, Sydney, Australia.

## Results

### Ostracoda Species

The first results of research on Ostracoda sampled during the 1987–1988 and 1989–1990 Italian Progetto Antartide cruises have been published in expedition data reports (McKenzie et al. 1990; Benassi et al. 1992); and by Benassi et al. (1994a). Five species were identified in the sampling area between 71° and 75°S. They are: *Alacia belgicae* (Müller 1906b), *Alacia hettacra* (Müller 1906a), *Metaconchoecia skogsbergi* (Iles, 1953), *Metaconchoecia isocheira* (Müller 1906a) and *Boroecia antipoda* (Müller 1906a,b). The first four species were found during both expeditions – three of them were reported earlier from the Ross Sea by Brady (1907) and Barney (1921) – but *B. antipoda* occurred only in the 1989–90 samples.

*Alacia belgicae* is a relatively large species (males 2.3–2.75 mm, females 2.4–2.85 mm) with a regular shape and well-developed shoulder vault; height distinctly greater than half the length. It was the dominant species collected in Terra Nova Bay during the 1987–88 expedition, when it represented 82.7% of total Ostracoda abundance and reached very high densities with a maximum of 1975 total ind 100 $m^{-3}$ in a sample at a nearshore station. In 1989–90 it occurred in all the sampling stations located between 71° and 75°S with the highest densities (368 total ind 100 $m^{-3}$) in station 27 near the most southern latitude. It is a typical Antarctic endemic species, since there are no confirmed records beyond the Antarctic Convergence except for individuals carried north in Antarctic Bottom Water (Deevey 1982; Hartmann 1986). Analysis of carapace length data of both adult females and males of this species confirms Müller's (1906b) note that it appears to occur in two distinct size ranges (Fig. 3). The life cycle of *A. belgicae* has been described by Kock (1992).

*Alacia hettacra* is a smaller species (length about 1.8–2.0 mm in both sexes), regularly shaped with a well-developed shoulder vault and elongate posterodorsal RV hinge socket accommodating an elongate LV hinge tooth; height about half the length. *Alacia hettacra* had a maximum density of 10 total ind 100 $m^{-3}$ during the 1987–88 expedition. In 1989–90, it occurred in significantly higher densities offshore at latitudes between 62° and 74°S; its maximum density (120 total ind 100 $m^{-3}$) was at station 9bis (70°10′S) off Cape Adare. As with the previous species, the life cycle of *A. hettacra* has been described by Kock (1992).

*Metaconchoecia isocheira* is a small species (males about 0.9 mm, females 0.85–1.05 mm), broadly rounded ventrally-posteriorly with a mostly straight dorsum; height distinctly greater than half the length; shell smooth. *Metaconchoecia isocheira* is a common species in the Ross Sea. In 1987–88, it was 16.7% of the total Ostracoda abundance, attaining a maximum density of 37.4 total ind 100 $m^{-3}$. In the 1989–90 expedition it was found frequently from 64° to 75°S; its maximum density (29.9 total ind 100 $m^{-3}$) was in a station at 75°S. Its abundance probably is not truly reflected in the samples because, being a small-sized form, some early juvenile instars are likely to have been extruded through the 500-μm mesh net. It is regarded as an upper mesopelagic species at austral latitudes in all oceans.

*Metaconchoecia skogsbergi* is larger than *M. isocheira* (length 1.25–1.40 mm) and more elongate, but still has the characteristic rounded ventral-posterior margin of the genus; height less than half the length in males, about half the length in females. *Metaconchoecia skogsbergi* was reported only occasionally in the Ross Sea during 1987–88; in 1989–90 it occurred regularly from 51° to 72°S, its highest density being 33.2 total ind 100 $m^{-3}$ in a station at 71°S. Like the previous species, and for the same reason, its abundance is unlikely to be truly reflected in the samples.

*Boroecia antipoda* is a large species (males 2.8–3.0 mm, females 3.1–3.3 mm), elongate, with a well-developed shoulder vault creating a medial inflexure along the dorsum; posterior margin broadly rounded and having three to four posterodorsal denticles; greatest height posteromedial and about half the length. *Boroecia antipoda* was fairly abundant in a large number of samples collected during the 1989–1990 expedition from 63° to 73°S: its highest density was attained at 500–600 m (34.9 total ind 100 $m^{-3}$) in station 15bis at 72°20′S. It is a mesopelagic species reported from all oceans at austral latitudes. The ontogeny is given in Kock (1992).

## Surficial Distribution

The surficial (0–50 m) distribution of planktonic ostracodes in the Ross Sea can be abstracted from the primary data provided in expedition reports, noting that this depth zone was not sampled at all the stations (Guglielmo et al. 1990; McKenzie et al. 1990; Benassi et al. 1992). Further, in 1987–1988 500-μm nets were used in Terra Nova Bay whereas 250-μm nets replaced them offshore (Fig. 1). In 1989–1990, the Ross Sea stations were located either north of Terra Nova Bay (stations 9–18) or along an E-W profile at about 75°S. In this profile, stations 24–27 sampled approximately the same area as was traversed during the 1987–1988 campaign.

In sum, the near-shore 0–50 m was sampled in Terra Nova Bay at ten stations (22 samples) in 1987–1988 and at two stations (2 samples) in 1989–1990; in the area offshore and east of Terra Nova Bay it was sampled at eight stations (11 samples) in 1987–1988; in the area north of Terra Nova Bay it was sampled at six stations (7 samples) in 1989–1990; and along the E-W profile at 75°S it was sampled at three stations (3 samples); there is also the data from station 1SK in 1987–1988. The results indicate very low abundances of ostracodes in the surficial 50 m in any of these areas, irrespective of whether 500-μm or 250-μm nets were used (Table 1): the maximum density (43.6 total ind 100 $m^{-3}$) was recorded in 1989–1990 from 0–20 m at station 16.

## Other Distributions

### Offshore and East of Terra Nova Bay (250-μm Nets 1987–1988)

The 1987–1988 campaign sampled three N-S profiles and one E-W profile using 250-μm nets (Fig. 1). In the most inshore N-S profile, we lack adequate data for station 30. The two stations at about the shelf break (stations 12, 18) indicate a gradual increase with depth in the densities of *A. belgicae* and *M. isocheira*, the two most abundant species, maximizing between 300 and 450 m. Although station 23, sited just off the shelf edge, was sampled at different intervals (40 m instead of 50 m) it shows a similar trend with density values of around 100 total ind 100 $m^{-3}$ for *A. belgicae* from 300–600 m.

The next off-shore N-S profile comprised a shallow station near Cape Washington and two widely separated areas south of this (stations 3, 13, 46; Fig. 1). Station 3 (187 m) had low total ostracode densities and was impoverished in adults; station 13 showed increases in total Ostracoda and adults densities from 350–550 m with *A. belgicae*

**Table 1.** Total numbers of ostracodes collected from 0–50 m in the Ross Sea

| | A | Total ostracodes | B | Total ostracodes | C | Total ostracodes | D | Total ostracodes |
|---|---|---|---|---|---|---|---|---|
| 1987–1988 | | | | | | | | |
| Station | 9(1) | Nil | 3(4) | 1 | 10(1) | 2 | 19 bis(1) | 2 |
| | 10(1) | 1 | 5(1) | 7 | 11 bis(1) | 2 | 22(2) | 6 |
| | 11a(5) | Nil | 16(1) | 4 | 12(1) | 2 | | |
| | 19(1) | 4 | 17(1) | 1 | 14(1) | 2 | | |
| | 20(2) | Nil | 18(1) | Nil | 15(1) | 2 | | |
| | 27(1) | 3 | 32(1) | Nil | 16(2) | 8 | | |
| | 29(1) | Nil | 46(1) | 1 | | | | |
| | 32(1) | Nil | 48(1) | Nil | | | | |
| | 35(5) | Nil | | | | | | |
| | 36(4) | Nil | | | | | | |
| 1989–1990 | | | | | | | | |
| Station | 25(1) | 6 | | | | | | |
| | 27 bis(1) | 10 | | | | | | |
| Total | (24) | 24 | (11) | 14 | (7) | 18 | (3) | 8 |

A = Terra Nova Bay (both cruises); B = stations offshore and eastwards of Terra Nova Bay (1987–1988); C = stations in the northwest Ross Sea, offshore from Coulman Island and Cape Adare (1989–1990); D = stations on the 75°S E-W profile (1989–1990). The number of samples at each station are given in parentheses after the station number. Total ostracodes include juveniles. Original counts for 1989–1990 stations were based on 1/2 aliquots. (Benassi et al. 1992)

dominant, then a tailing off; station 46 similarly displayed increases in both total Ostracoda and adult densities between 250–500 m, but with *M. isocheira* adults more numerous between 250–300 m and *A. belgicae* adults more numerous from 350–500 m.

The most easterly N-S profile was also the most closely sampled (stations 5, 14, 25, 32, 39). Total Ostracoda densities increased markedly from 200 m down, with *A. belgicae* again dominating the assemblages in deeper water to the maximal sampling depth. Of interest here is the patchy occurrence in low densities of *A. hettacra.*

The 1987–1988 E-W profile extended yet further offshore from Terra Nova Bay (stations 15, 16, 17) – see McKenzie et al. (1990, Fig. 1). Total ostracode densities began to increase at around 150 m depth; *A. belgicae* dominated below 250–300 m, *M. isocheira* was occasionally more numerous than *A. belgicae* between 150–250 m and well represented throughout the water column below 250 m and *A. hettacra* occurred patchily at station 16.

## Terra Nova Bay (500-μm Nets, 1987–1988 and 1989–1990)

Some 15 stations were sampled in Terra Nova Bay during 1987–1988 and 4 more in 1989–1990. For 1987–1988, total ostracode densities generally began to increase at 150–200 m, with *A. belgicae* the predominant species everywhere, often in huge abundances recalling the description of Barney (1921). Two stations (35, 36) do not match this pattern. They were sited in the south of the Bay towards the Drygalski Ice Tongue and were impoverished in Ostracoda throughout the sampled water column. On the other hand, station 34, located between them (Fig. 1) had high total ostracode densities from 200 m to the maximal sampling depth (600 m). Note that these three stations in the Bay were off the shelf break.

In 1989–1990 (stations 24–27), *A. belgicae* again predominated (over 350 total ind 100 $m^{-3}$ between 220–300 m), *A. hettacra* was much less common and *M. isocheira* reached almost 30 total ind 100 $m^{-3}$ in deep midwater (400–450 m) at station 24.

Highest densities per sample for both main species in Terra Nova Bay, correlated with depth, are given in Table 2.

## Northwest and Easterly Offshore Areas (250-μm Nets, 1987–1988; 500-μm Nets, 1989–1990)

Stations 1SK and ANZ (1987–1988) displayed a reduction in densities of *A. belgicae*, with *M. isocheira* consistently the more numerous species in the samples, and total ostracode densities

**Table 2.** Maximum densities (total ind $100\,m^{-3}$) for *A. belgicae* and *M. isocheira* in Terra Nova Bay (500-μm nets)

| | | *A. belgicae* | *M. isocheira* | Depth zone (m) | Bottom (m) |
|---|---|---|---|---|---|
| 1987–1988 | | | | | |
| Station | 8 | 1485.71 | 25.40 | 200–260 | 570 |
| | 9 | 1974.68 | 25.32 | 120–150 | 560 |
| | 10 | 800.00 | – | 220–245 | 325 |
| | 11 | – | 5.52 | 150–200 | 640 |
| | 11 | 428.57 | – | 395–400 | 640 |
| | 19 | 1173.75 | 7.72 | 400–500 | 660 |
| | 21 | 1259.74 | 25.97 | 300–350 | 556 |
| | 22 | – | 28.09 | 300–350 | 645 |
| | 22 | 1179.94 | – | 450–500 | 645 |
| | 26 | 1358.97 | 23.50 | 395 | 930 |
| | 27 | – | 13.85 | 200–250 | 815 |
| | 27 | 1347.37 | – | 300–400 | 815 |
| | 29 | 852.46 | – | 350–400 | 1010 |
| | 34 | 1735.45 | – | 350–400 | 1100 |
| | 34 | – | 37.36 | 400–450 | 1100 |
| 1989–1990 | | | | | |
| Station | 24 | 288.44 | – | 220–260 | 520 |
| | 24 | – | 29.90 | 400–450 | 520 |
| | 25 | – | 8.33 | 100–200 | 885 |
| | 25 | 185.62 | – | 300–400 | 885 |
| | 26 | – | 13.16 | 220–260 | 635 |
| | 26 | 367.03 | – | 260–300 | 635 |
| | 27 | 368.05 | 8.09 | 250–300 | 735 |

increasing from about 150 m down to maximal sampling depth.

The samples taken in the northwest part of the Ross Sea during the 1989–1990 campaign, offshore from Coulman Island to Cape Adare (between about 70° and 74°S) had the same four species that were identified in the earlier cruise, plus *Boroecia antipoda*; and single records of *Conchoecissa symmetrica* (Müller 1906a) at station 9 and of *Proceroecia rivoltella* Benassi et al. 1994 at station 10. We will consider these single records later (cf. Discussion below). The easterly offshore samples were collected along an E-W profile, mostly following the shelf break at about 75°S and between 180°–170°E. The maximum sampling depth for 1989–1990 was 1000 m. This proved adequate except for stations 9, 10, 14 and 15, where the bottom was deeper – 3285, 1325, 1575 and 2120 m, respectively. The results featured a regular occurrence of *B. antipoda* in the northwest stations and the abundance of *A. hettacra* throughout both areas.

In general, the highest densities for *A. belgicae* were between 100 and 500 m, for *A. hettacra* between 0 and 300 m and for *M. isocheira* between 200 and 500 m. In the northwest stations, high-density values were recorded for *A. hettacra* from 100–300 m and for *B. antipoda* from 300–600 m, while *M. skogsbergi* occurred most commonly from 500–600 m (Tables 3, 5). Summarizing, "it appears that at higher latitudes and inshore *Alacia belgicae* dominates over *A. hettacra* at most depths, and *M. isocheira* is the other common species; whereas offshore and at lower latitudes. . . . *A. hettacra* dominates the surface waters while *M. skogsbergi* and *B. antipoda* take over at middle depths . . ." (Benassi et al. 1994a; see also Benassi et al. (1994b; Figs. 1,2).

## Discussion

### Sampling Limitations

Angel (1977a) drew attention to the sampling limitations in any oceanographic program. For example, in the sampling apparatus employed, both net mouth size and net mesh size are important vectors. In the Italian Antarctic campaigns of 1987–1988 and 1989–1990, the EZ-NET BIONESS multinet apparatus was set up for nets having a

**Table 3.** Highest densities for planktonic Ostracoda at stations 9–27 (total ind $100\,m^{-3}$) in the northwest and offshore (75°S) areas of the western Ross Sea (1989–1990). The original counts (Benassi et al. 1992) were mostly based on 1/2 aliquots

| Species | Station | Latitude | Bottom (m) | Depth (m) | Highest density (total ind/100 $m^3$) |
|---|---|---|---|---|---|
| **Northwest** | | | | | |
| *Alacia belgicae* | 16 | 72°S | 760 | 100–120 | 210.84 |
| *Alacia hettacra* | 9 bis | 70°S | 3290 | 120–140 | 120.09 |
| *Metaconchoecia isocheira* | 17 | 73°S | 535 | 180–220 | 14.93 |
| *Metaconchoecia skogsbergi* | 9 | 70°S | 3285 | 500–600 | 21.77 |
| *Boroecia antipoda* | 15 bis | 72°20′S | 2140 | 500–600 | 34.91 |
| **Easterly offshore** | | | | | |
| *Alacia belgicae* | 27 | 74°45′S | 735 | 250–300 | 368.05 |
| *Alacia hettacra* | 23 | 75°S | 345 | 160–180 | 23.18 |
| *Metaconchoecia isocheira* | 23 | 75°S | 345 | 260–300 | 58.13 |
| *Metaconchoecia skogsbergi* | 19 | 75°S | 460 | 0–400 | 1.90 |
| *Boroecia antipoda* | – | – | – | – | – |

0.5 m × 0.5 m square mouth opening but, as noted earlier (cf. Material and Methods above), two net mesh sizes, 250 and 500 μm, were used in the earlier campaign, whereas 500-μm nets were standard on the second cruise. This variation goes some way to explaining why the highest density of *Alacia belgicae* was 972 total ind $100\,m^{-3}$ (station 32, 250-μm nets) in 1987–1988 versus 288 total ind $100\,m^{-3}$ (station 24, 500-μm nets) in the same general area in 1989–1990, since it could be assumed that many more individuals would be able to elude a coarser than a finer net during a sampling tow. That a different net mesh was not the only explanation for such a variation will become evident during our further discussion below.

Other factors referred to by Angel (1977a) such as rate of tow, period of tow and volume of water filtered may be regarded as linked to biomass, since once the multinet system reaches its predetermined sampling depths and the net open command is initiated the tow normally continues until a satisfactory biomass capture is achieved. There would not be much purpose in continuing a tow when nets are clogged with biomass. A remaining consideration is whether or not the net mouth size which was used was appropriate for the zooplankton program. In this connection, it is important to recall that earlier expeditions to Terra Nova Bay and the western Ross Sea (Brady 1907; Barney 1921) captured precisely the same planktonic ostracode species as those taken during the Italian campaigns; although they used larger nets, non-automated sampling techniques and took longer to collect samples. Clearly, the 0.25-$m^2$ net mouth size was an effective sampler of all the Ross Sea planktonic ostracode species at depths to 1000 m.

Interspecies size and shape variations also influence whether or not a particular group is properly sampled. Angel (1977a, Fig. 2) demonstrated that the near spherical planktonic ostracode *Halocypria globosa* Claus, 1874 was retained on the sampling net at smaller sizes in greater numbers than the subcylindrical species *Porroecia spinirostris* (Claus 1874). In Terra Nova Bay and the Ross Sea, his data imply that some juvenile stages of the small species *M. isocheira* (0.85–1.05 mm) and *M. skogsbergi* (1.25–1.40 mm) which are subellipsoidal when mature and nearspherical as juveniles would tend to be retained on the nets whereas similar-sized and even some larger juveniles of the subcylindrical species *A. belgicae* (2.3–2.85 mm), *A. hettacra* (1.8–1.9 mm) and *B. antipoda* (2.8–3.3 mm) would slip through the mesh. The net result (no pun intended) is that the Ross Sea planktonic ostracode community of adults and juveniles is likely to be more correctly represented in our samples than if the smaller species had the same shape as the larger ones.

## Vertical Migration

In considering whether or not planktonic ostracodes use vertical migration and what they might use it for (Angel 1984, 1985, 1993; McKenzie et al. 1979), it seems reasonable to base the study on *Alacia belgicae* and on data from the region surveyed during the first, 1987–1988, Italian cruise plus those stations located in the same area during 1989–1990. There are sound reasons to support this choice: (1) taking 90 specimens in each highest density sample as a minimum, this still leaves 28 stations for which we have enough data (14 using 500-µm nets, 14 with 250-µm nets); (2) we have already shown that a positive correlation exists between ostracode density and depth, a prime variable for vertical migration (McKenzie et al. 1990, Fig. 14); (3) *A. belgicae* represents 82.7% of the entire 1987–1988 ostracode faunule, i.e. about 42 500 specimens, increasing to over 44 000 when those from the relevant 1989–1990 stations (24–27) are added; (4) using only the highest density samples still provides adequate numbers, i.e. 15 299 specimens, about 23% of total ostracodes, and nearly 35% of all *A. belgicae*; (5) both adults and juveniles occur in sufficient numbers to be analyzed separately, namely 1337 adults (805 females, 532 males) and 13 962 juveniles, and for the analyses of both categories to have some utility. Table 4 presents our results – the raw data have been modified in part for convenience of display in Fig. 2 as explained in Table 4. Note that depth in the western Ross Sea also serves to delineate the main water masses (Table 6), as we discuss later.

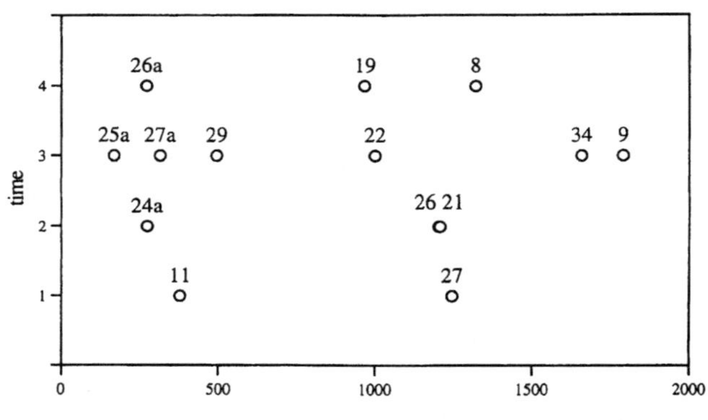


1

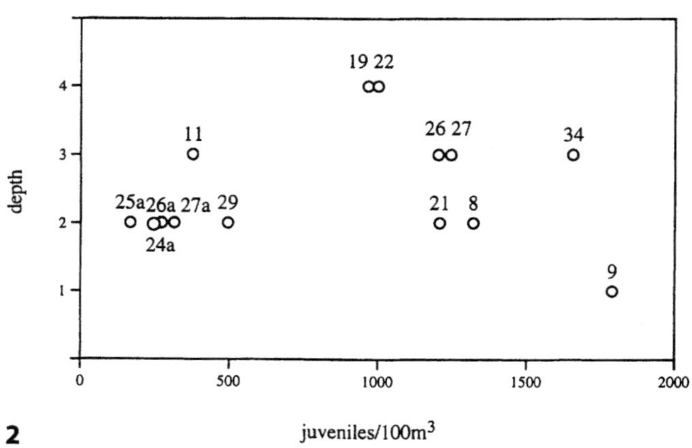


2

**Fig. 2.1–8.** Scattergrams for juveniles and adults of the highest density samples (>90 individuals) of *A. belgicae* per time of day and depth, at stations in Terra Nova Bay and the Ross Sea (500- and 250-µm nets). Stations 24a–27a are from the 1989–1990 campaign. Values of all numbers representing the time and depth variables are given in Table 4. Station numbers are plotted in the scattergrams

**Fig. 2.1–8.** *Continued*

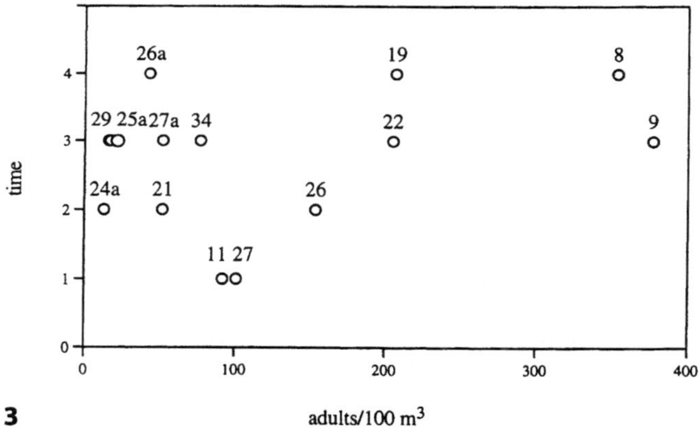


**3**

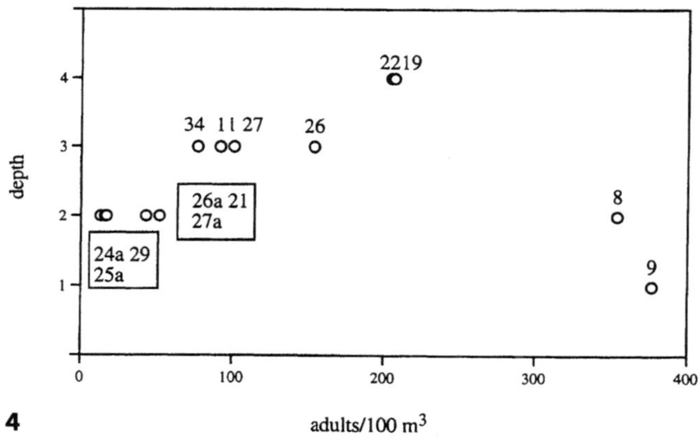


**4**

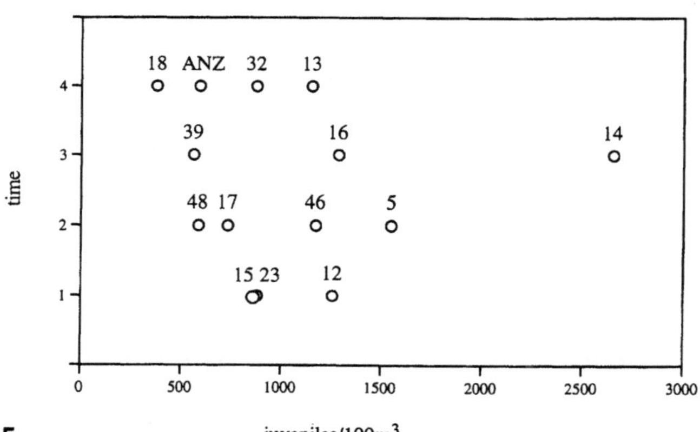


**5**

Fig. 2.1-8. *Continued*

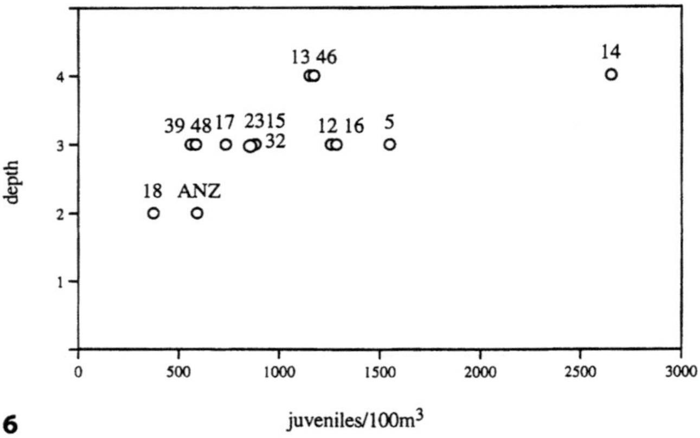


6

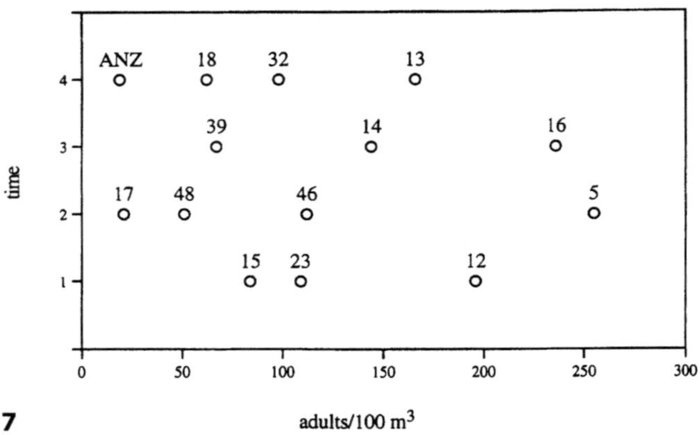


7

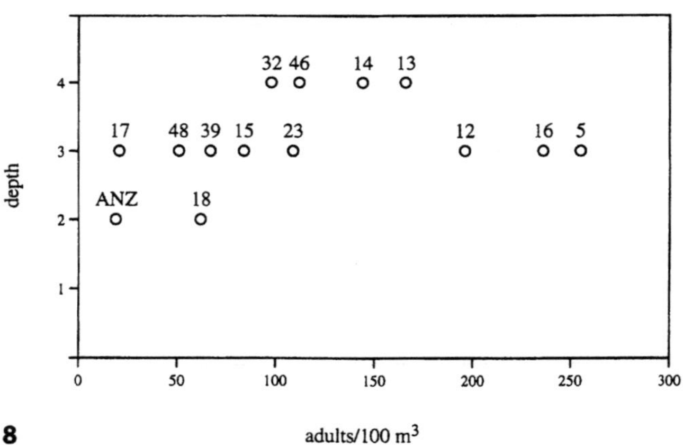


8

**Table 4.** Stations from Terra Nova Bay and near-lying Ross Sea in terms of total juveniles, adults and adults/100 m³ per highest density sample of *A. belgicae*, sample depth and time of day (McKenzie et al. 1990; Benassi et al. 1992). Net mesh sizes also indicated. Stations with <90 total individuals per highest density sample excluded. Original counts at 1989–1990 stations were based on 1/2 aliquots. (Benassi et al. 1992)

| | Station no. | Mesh (μm) | A Time | B Depth | C Total juvs | D Juvs/ 100 m³ | E Total adults | F Adults/ 100 m³ |
|---|---|---|---|---|---|---|---|---|
| 1987–1988 | | | | | | | | |
| | 8 | 500 | 4 | 2 | 416 | 1321 | 52 | 165 |
| | 11 | 500 | 1 | 3 | 82 | 378 | 11 | 51 |
| | 9 | 500 | 3 | 1 | 283 | 1791 | 29 | 183 |
| | 19 | 500 | 4 | 4 | 501 | 967 | 107 | 207 |
| | 21 | 500 | 2 | 2 | 93 | 1208 | 4 | 52 |
| | 22 | 500 | 3 | 4 | 339 | 1000 | 61 | 180 |
| | 26 | 500 | 2 | 3 | 564 | 1205 | 72 | 154 |
| | 27 | 500 | 1 | 3 | 592 | 1246 | 48 | 101 |
| | 29 | 500 | 3 | 2 | 156 | 495 | 5 | 16 |
| | 34 | 500 | 3 | 3 | 627 | 1659 | 29 | 77 |
| | 12 | 250 | 1 | 3 | 763 | 1259 | 119 | 196 |
| | 13 | 250 | 4 | 4 | 522 | 1152 | 75 | 166 |
| | 5 | 250 | 2 | 3 | 1074 | 1550 | 177 | 255 |
| | 14 | 250 | 3 | 4 | 1326 | 2652 | 72 | 144 |
| | 17 | 250 | 2 | 3 | 308 | 732 | 9 | 21 |
| | 16 | 250 | 3 | 3 | 970 | 1286 | 178 | 236 |
| | 18 | 250 | 4 | 2 | 553 | 374 | 92 | 62 |
| | 23 | 250 | 1 | 3 | 532 | 879 | 66 | 109 |
| | 46 | 250 | 2 | 4 | 1101 | 585 | 95 | 51 |
| | 48 | 250 | 2 | 3 | 462 | 1173 | 44 | 112 |
| | 39 | 250 | 3 | 3 | 369 | 560 | 44 | 67 |
| | 32 | 250 | 4 | 3 | 877 | 874 | 98 | 98 |
| | 15 | 250 | 1 | 3 | 324 | 880 | 31 | 84 |
| | ANZ | 250 | 4 | 2 | 427 | 591 | 14 | 19 |
| 1989–1990 | | | | | | | | |
| | 24a | 500 | 2 | 2 | 230 | 274 | 12 | 14 |
| | 25a | 500 | 3 | 3 | 274 | 168 | 28 | 17 |
| | 26a | 500 | 4 | 2 | 330 | 271 | 38 | 31 |
| | 27a | 500 | 3 | 2 | 312 | 315 | 52 | 53 |

The numbers in C,E are real numbers; A and B represent the following sets:

| A time of day (h) | B depth sampled (m) |
|---|---|
| 1 = 21.00–03.00 | 1 = 0–150 |
| 2 = 03.00–09.00 | 2 = 150–300 |
| 3 = 09.00–15.00 | 3 = 300–450 |
| 4 = 15.00–21.00 | 4 = 450 ≥ 600 |

An immediate conclusion from Table 4 is that only one high density sample occurs above 150 m. This is at station 9 (120–150 m), where 312 individuals (adults plus juveniles) were collected. We have already demonstrated the paucity of Ostracoda from the surficial 50 m in the western Ross Sea; Table 4 extends this to depths of over 100 m. Moguilevsky and Angel (1975) state that at high planktonic latitudes halocyprids are almost completely absent from surficial waters; our data reinforce their observation.

Figure 2.1 shows that time of day had no significant influence on the distribution of high density samples of juveniles since results are spread across all sampling times. The data for adults in Terra Nova Bay are similar (Fig. 2.3), those for adults and juveniles offshore and east of Terra Nova Bay (Figs. 2.5, 2.7) indicate even more spread with respect to time of day.

Regarding the depth variable, in Terra Nova Bay itself (Fig. 2.2), large numbers of juveniles favoured depths of 150–450 m. On the other hand,

Fig. 2.6 indicates some clustering of stations offshore, where 250 μm nets were used, around depths of 300–600 m for large numbers of juveniles; but only one sample (station 14, 500–600 m) represents a depth over 500 m. Possibly, the "real" preferred depth range indicated by Fig. 2.6 for high densities of *A. belgicae* juveniles offshore and east of Terra Nova Bay was nearer 400–500 m.

Considering the adults 100 $m^{-3}$ distributions, the results for depth show that most density peaks of Terra Nova Bay adults occurred between 150–450 m (Fig. 2.4), while offshore most density peaks of adults were found at depths of 300–600 m (Fig. 2.8). We have already suggested that the "real" preferred depth range offshore is 400–500 m. The broad spread across times and depths of the highest densities of *A. belgicae* is clearly featured in Table 4.

Considering now the ideas of Longhurst (1976) and Angel (1985) on vertical migration, Ross Sea halocyprids appear well placed in the water column to feed on sinking detritus, e.g. euphausiid faecal pellets and copepod debris, for optimal bioenergetic advantage at depths of around 150–300 m in Terra Nova Bay and 400–500 m off-shore. They occupy these midwater depths with chaetognaths, polychaetes and gammarideid amphipods; deeper water with mysidaceans (Table 6). The general dispersal of *A. belgicae* highest densities throughout the day suggests that they at least may be actively avoiding visual predation. We comment on the alternative explanations of patchiness (a general oceanic phenomenon) and horizontal dispersion/transport in the next section. If Ross Sea planktonic ostracodes are using vertical migration to optimize breeding opportunities, as seems possible given the spread of adult sizes of *A. belgicae* (cf. Hybridization below, Fig. 3) it has a poor correlation to time of day in the Antarctic summer (Table 4).

## Hybridization

Angel (1977b, Fig. 2) discussed and figured length-size distribution of large numbers of *Discoconchoecia elegans* (Sars 1865) at different depths; and noted that three latitudinally bound ranges in the mean lengths of adults of both sexes of this species had been recorded earlier by Poulsen (1973). Angel (1977b,c) interpreted his own data as being the result of hybridization between the largest and smallest of the three forms, producing a broad spread of both adult and juvenile sizes.

Müller (1906b) noted that *Alacia belgicae* appeared to occur in two size ranges. Our relevant 1987–1988 data are represented in Fig. 3, clearly based on large numbers of adult animals of both sexes; we do not have the data for juveniles. As with *D. elegans*, however, our results demonstrate a broad spread of adult sizes and additionally there are two apparent peaks in both sexes. Because *A. belgicae* is an Antarctic endemic species, a melange of latitudinal size variations cannot account for Fig. 3. We suggest instead that our results indicate hybridization between seasonal populations which have interacted either through vertical migration or via the mixing of demes as an effect of water turbulence. Another possibility is that *A. belgicae* may reproduce twice a year, in which case the brood produced when productivity is higher may grow better (comment by referee on our original submission). We note also that occasionally in 1987–1988 there were two high-density samples at one station separated by a full sampling interval (around 50 m) of lesser density, e.g. at station 22 (500-μm nets) and station 23 (250-μm nets). Our stations were too dispersed to discount the possibility that either obliquely horizontal movement/transport or patchiness might more readily account for such patterns than hybridization. It is one of several aspects of the biology of *A. belgicae* that would repay a more intensive investigation.

## Environmental Factors

The general environment of the Ross Sea has been well dealt with by other authors (Carli et al., this Vol.; Crescenti et al., this Vol.). Here, we make brief further reference to some factors which have particular importance in terms of the ostracode distributions.

## The Antarctic Convergence

The Antarctic Convergence in this part of the Southern Ocean oscillates between 60° and 65°S, and acts as an effective trap, preventing surficial-midwater horizontal escape/migration northwards of most Antarctic halocyprids. Occasionally, other species are found south of the Convergence, due to a number of temporary local factors such as wind and current. During 1989–1990, *Conchoecissa symmetrica* and *Proceroecia rivoltella* occurred at stations 9 and 10 (70°–71°S), well south of their normal latitudinal ranges. Previ-

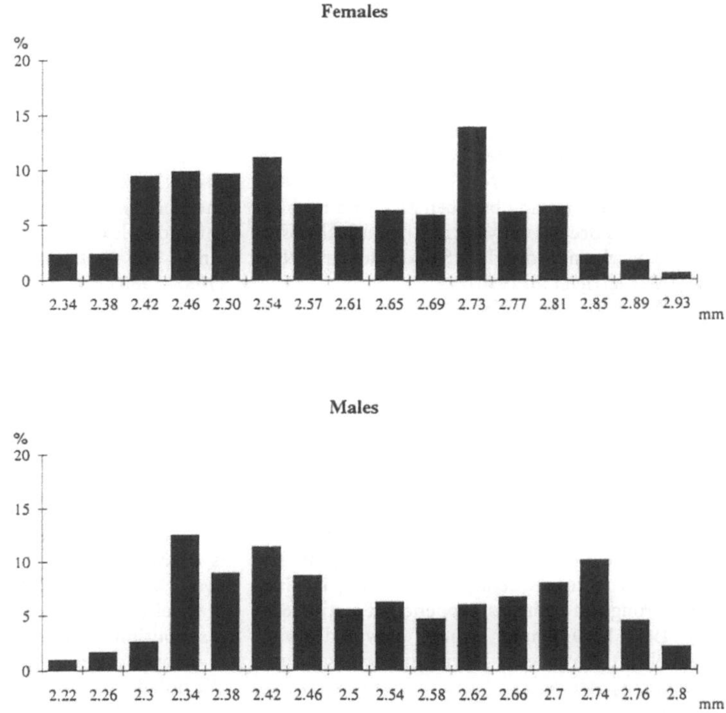


**Fig. 3.** Size distributions of adult males and females of *A. belgicae* (all stations) in Terra Nova Bay and the near-lying Ross Sea for the 1987–1988 Italian campaign. (After McKenzie et al. 1990)

ously, *C. symmetrica* had not been recorded below 64°S (Deevey 1982).

There is only one escape route for Antarctic and other high latitude austral taxa into northern regions and this is with cold Antarctic Bottom Water as it plunges below the warmer surface waters of the Antarctic Circumpolar Current (see next section). Without exception, all the Ross Sea Antarctic planktonic halocyprids and *C. symmetrica* are found in the deep waters of the Indian and Atlantic Oceans, in most cases to well north of the Equator (Deevey 1982, 1983; Hartmann 1986). We have hypothesized earlier (Benassi et al. 1994a) that this also happens with *P. rivoltella*.

## The Antarctic Circumpolar Current (ACC)

The deep-reaching counterclockwise ACC (thermal maximum at around 300 m) makes its presence felt at about the location of station 17 of the 1989–1990 Cariboo cruise (Patterson and Whitworth 1990). Figure 1.3 shows where a reentrant of the 1000-m isobath closely approaches the 500-m isobath in this area. As a result, a steep front develops at the interface between cold Antarctic Bottom Water (ABW) running off the Ross Sea shelf and the relatively warmer ACC water (Hecq and Goffart 1992). The natural egress for ABW – not as highly stratified off-shore as it in Terra Nova Bay (see next section) – is to travel downwards north by northeast along the deep ocean bottom, which here lies at about 3000–4000 m (Benassi et al. 1994a, Table 2).

Planktonic ostracode distributions bring out this relationship quite clearly, because *Boroecia antipoda* disappears from the assemblages south of station 17 (Benassi et al. 1994a, Table 3). Further, this is where *Alacia belgicae* makes its move northwards, via ABW and the deep ocean bottom, because it disappears north of station 10 at 71°10′S

(Benassi et al. 1994a, Table 3). *Alacia hettacra* and *Metaconchoecia isocheira* probably do the same, but they also maintain populations at midwater depths further north in austral latitudes (Angel 1981; Deevey 1983). *Metaconchoecia skogsbergi*, very rare at 74°–75°S, is a characteristic shallow to deep mesopelagic species from about 72°S northwards (Benassi et al. 1994a, Table 3; Gooday 1981); probably some mixing between ACC and Ross Sea shelf waters occasionally carries it south almost to the ice barrier at around 75°S and into Terra Nova Bay (McKenzie et al. 1990).

The ACC also assists the occasional move southwards of more northerly established species, such as *C. symmetrica* and *P. rivoltella* (see previous section) once they drift into its path.

## Pack and Melting Ice, Water Masses, the Summer Gyre in Terra Nova Bay

Pack ice sometimes caused problems with the orderly progress of the zooplankton program, even to forcing a sampling tow to be rapidly terminated with consequent poor efficiency in filtration through the nets, as happened at station 6 on the 1989–1990 cruise (notes provided by L. Guglielmo). However, the principal ice-associated effects are positive. Ice melting stimulates primary production leading to high concentrations of chlorophyll a in spring, under optimal light conditions (Hecq and Goffart 1992). Sea-ice algae melted into the water column also stimulate phytoplankton development. Indeed, abundant seasonal blooms in phytoplankton derive from the melting ice and, as most of the Ross Sea becomes ice-free, zooplankton feed on the phytoplankton and algae, and in their turn rapidly increase in abundance (Hecq and Guglielmo 1992; Hecq et al. 1993; Crescenti et al., this Vol.). The residue of the zooplankton bloom in the surficial layers (<100 m) filters downwards to assuage the bioenergetic needs of the midwater and deeper zooplankton, including planktonic halocyprid ostracodes.

Boldrin and Stocchino (1990) defined the Ross Sea water masses in terms of values of temperature, salinity, density, meteorological conditions, etc. They noted that at most 1987–1988 stations the water column was distinctly stratified, producing high temperature and salinity gradients – this stratification was particularly marked in the upper 100 m (Boldrin and Stocchino 1990, Fig. 5), which is also the zone of greatest primary production (Hecq and Goffart 1992).

Halocyprids occur mainly below this nutrient-enriched and highly stratified zone (Table 6) but still respond to turbulence created by the dynamic stratification below the surficial layers and to the bottom at each station, and between stations (Stocchino and Manzella 1991). This could be expected to lead to some patchiness in distributions both horizontally and vertically, but neither the quality nor the continuity of our data allow this complex problem to be fully resolved (cf. Angel and Fasham 1983). The unusual situation (1987–1988 cruise) noted earlier, where stations 35 and 36 were impoverished in Ostracoda throughout the water column whereas station 34 which lay between them had high ostracode densities, is probably best explained in terms of lowered salinities near the melting Drygalski Ice Tongue and nearby turbulence. Station 46, moderately rich in ostracodes at midwater depths appears to have been sited near another turbulence zone (Stocchino and Manzella 1991, Figs. 1.1, 1.2).

Summer surface circulation in Terra Nova Bay is characterized by a clockwise offshore cyclonic gyre (Boldrin and Stocchino 1990; Fig. 1.2). Its main effect vis à vis halocyprids is to regulate the distribution of *A. belgicae* and *M. isocheira* (99.4% of all ostracodes in Terra Nova Bay in 1987–1988). We noted earlier that these taxa are depth correlated (see also McKenzie et al. 1990, Fig. 14).

The water masses of Terra Nova Bay are shown in Table 6 derived from Crescenti et al. (this Vol.). We have already demonstrated that the nutrient-nich AASW (0–50 m) carried virtually no ostracodes (Table 1) and that their densities only began to pick up at depths around 150–200 m, i.e. towards the base of the SISW. As at lower latitudes, planktonic ostracodes are most numerous in the midwater of Terra Nova Bay, i.e. the MCDW (200–300 m) and DISW (300–600 m) (Fig. 2). Table 2 suggests that the highest densities of the two main species often occur in similar water masses/depths. Where the respective depths differ, this is less likely to indicate differential vertical migration than some local turbulence. Adult *A. belgicae* have about 15.6× the volume of adult *M. isocheira*, so when the surrounding water mass is disturbed some differences in their respective highest density depths could be anticipated.

In the area off-shore from Coulman Island and Cape Adare (1989–1990 Cariboo cruise), Fig. 1.3 shows that most stations were located between the 500-m and 1000-m isobaths while the other stations – 8, 9, 10, 14, 15 – occurred in the reentrant zone deeper than 1000 m. Ostracodes densities were less than in Terra Nova Bay. Table 5 gives the adjusted raw data. *Alacia hettacra*, *Boroecia antipoda* and *Microconchoecia skogsbergi* were the reentrant taxa occupying the ACC water mass. Table 5 makes it clear that the deeper-occurring *B. antipoda* and *M. skogsbergi* would be blocked at the steep front between the ACC and Ross Sea Shelf water. *A. hettacra*, a shallower-adapted species, was carried south beyond the interface zone and occurred all the way to around 75°S off the melting ice front.

Where the bottom occurred at 500–1000 m (Fig. 1.3) the halocyprids were *A. belgicae*, *A. hettacra* and *M. isocheira*. Greatest densities seem to have been associated with the equivalent of MCDW (Tables 5, 6). The stations 12, 13, 17, 18 which lay in the region of greatest disturbance at the ACC/Ross Sea Shelf water interface were all ostracode poor (<30 individuals in the highest density samples).

The offshore stations along the 75°S traverse (1989–1990) in front of the melting ice barrier showed highest densities of *A. belgicae* between 160–300 m (lower SISW-MCDW), i.e. somewhat higher in the water column than for 1987–1988. This is probably a seasonal effect linked to melting ice and the early summer enrichment of nutrient in surficial/subsurficial waters (Hecq and Goffart 1992).

**Table 5.** The species content, adults and juveniles of highest density samples, at stations in the Ross Sea offshore from Coulman Island and Cape Adare (1989–1990), with relevant depths. Minimum highest density 30 individuals

| Station (m) | Juveniles | Adults | Depth |
|---|---|---|---|
| 8 | 38AH | 12AH | 600–500 |
| 8 bis | 92AH | 6AH | 200–100 |
| 9 | 22AH, 14MS | 6AH, 30MS | 600–500 |
| 9 bis | 88AH | 22AH, 8BA | 140–120 |
| 10 | 40AH | 10AH | 100–80 |
| 10 bis | 2BA, 28MS | 8BA, 26MS | 600–500 |
| 11 | 44AH | 10AH | 120–100 |
| 11 bis | 6BA, 10MS | 2BA, 12MS, 2MI | 500–400 |
| Interface between ACC and Ross Sea Shelf water (see Fig. 1.3) | | | |
| 14 | 66AB | 2AB | 160–140 |
| 14 bis | 50AB, 154AH | 4AB, 28AH | 300–200 |
| 15 | 12AB, 10AH | 4AH | 120–100 |
| 15 bis | 22AH | 12AH | 400–300 |
| 16 | 132AB, 24AH | 8AB, 22AH, 4MI | 120–100 |
| 16 bis | 140AB, 16AH | 16AB, 28AH, 4MI | 200–100 |
| 19 | 94AB, 2AH | 4AB, 18MI | 200–160 |
| 22 | 118AB | 12AB, 4MI | 300–250 |
| 22N | 100AB | 24AB, 2AH | 250–200 |
| 22P | 236AB, 2AH | 20AB, 2AH, 16MI | 300–250 |
| 23 | 60AB | 12AB, 14AH, 26MI | 180–160 |
| 24 | 230AB, 2MI | 12AB, 2AH, 10MI | 260–220 |
| 25 | 274AB | 28AB, 2AH, 4MI | 400–300 |
| 26 | 330AB | 38AB, 2AH, 16MI | 260–220 |
| 27 | 312AB | 52AB, 8MI | 300–250 |

AH = *Alacia hettacra*; MS = *Metaconchoecia skogsbergi*; BA = *Boroecia antipoda*; MI = *Metaconchoecia isocheira*; AB = *Alacia belgicae*. The numbers have been adjusted from the original counts which were based mostly on 1/2 aliquot samples (Benassi et al. 1992). Sampling at station 22 was carried out at three different times in January 1990.

## Conclusions

McKenzie et al. (1990) made several recommendations for future zooplankton programs in the Ross Sea under the aegis of the Italian Progetto Antartide. Exclusive use of 500-μm nets in the subsequent Cariboo cruise certainly helps explain the lower numbers of planktonic ostracodes taken in 1989–1990; there were also some sampling problems due to less favourable seasonal factors (notes provided by L. Guglielmo). Further, although several 1000-m samples were collected in 1989–1990, they did not access the deeper ostracode fauna which probably includes at least *Metaconchoecia arcuata* (Deevey, 1978, 1982). The inclusion of the northwestern Ross Sea in the zoo-

**Table 6.** The vertical distribution of major taxa in zooplankton of the Ross Sea and the corresponding water masses with their relevant depths (incl Benassi and Ferrari, unpubl data)

| Water masses | Depths | Zooplankton groups |
|---|---|---|
| AASW (0–50 m) | | |
| | 50–150 m | Copepoda |
| SISW (50–200 m) | 150–500 m | |
| MCDW (200–300 m) | | Chaetognatha |
| | | Polychaeta |
| DISW (300–600 m) | | Ostracoda |
| | | Amphipoda (gammaridea) |
| HSSW (600–800 m) | 600–800 m | Mysidacea |
| | | Ostracoda |

AASW = Antarctic Surface Water; SISW = Shallow Ice Shelf Water; MCDW = Modified Circumpolar Deep Water; DISW = Deep Ice Shelf Water; HSSW = High Salinity Shelf Water (Crescenti et al., this volume). Most halocyprid Ostracoda and their highest densities occur in MCDW and DISW; there are also some concentrations offshore at greater depths (HSSW), where the mysid *Antarctomysis ohlini* can prey on *A. belgicae*.

plankton program for 1989–1990 clarified, as we had foreshadowed, the distributional relationships between *Alacia belgicae*, *Metaconchoecia isocheira* and other species (textual discussion, Tables 2, 5). The reduced catch in 1989–1990 precluded study of the ontogeny of either *A. belgicae* (cf. Kock 1992) or *M. isocheira*; nor did we attempt any detailed analysis of the two forms of *A. belgicae* (Fig. 3), although we have suggested a plausible explanation (cf. Hybridization). Our data indicated that available light (= time of day) is not a critical factor in the diurnal biological rhythms of Antarctic planktonic ostracodes during summer (cf. Vertical Migration). The use of multivariate statistical techniques remains a possibility; here we develop the relationships of Ross Sea halocyprids to water masses and depth – the statistical basis for this was established earlier (McKenzie et al. 1990, Fig. 14). As a facultative midwater Antarctic taxon, *Alacia belgicae* remains a potential monitor for future changes in the western Ross Sea ecosystem, whether these are due to climatic factors or to pollution should Antarctic resources come to be improperly exploited.

*Acknowledgements.* We thank Dr. M.V. Angel (Southampton Oceanography Centre), Prof. Dr. G. Hartmann (University of Hamburg), Prof. L.S. Kornicker (Smithsonian Institution) for reprints and advice: Dr. J. Lowry (Australian Museum) for use of facilities: Prof. L. Guglielmo (University of Messina) for providing data and references; Prof. F.M. Faranda (University of Genova) for supervision of ostracode sorting; Dr. V. Rossi, Dr. S. Sei and Dr. M. Vezzosi for advice and computational help; the ENEA Progetto Antartide program for funding participation by one of us (K.G.M.).

## References

Angel MV (1977a) Windows into a sea of confusion: sampling limitations to the measurement of ecological parameters in oceanic mid-water environments. In: Anderson NR, Zahuranec BJ (eds) Oceanic sound-scattering prediction. Plenum Press, New York, pp 217–248

Angel MV (1977b) Some speculation on the significance of carapace length in planktonic halocyprid ostracods. In: Loeffler H, Danielopol DL (eds) 6th Int Symp on ostracodes, Saalfelden. Junk, The Hague, pp 45–54

Angel MV (1977c) Studies on Atlantic halocyprid ostracods: vertical distributions of the species in the top 1000 m in the vicinity of 44°N, 13°W. J Mar Biol Assoc UK 57: 239–252

Angel MV (1981) Ostracoda. In: Boltovskoy D (ed) Atlas del zooplankton del Atlanctico sudoccidental. Publicacion INIDEP, Buenos Aires, pp 543–585

Angel MV (1984) The diel migrations and distributions within a mesopelagic comunity in the north east Atlantic. III. Planktonic ostracods, a stable component in the community. Prog Oceanogr 13: 319–351

Angel MV (1985) Vertical migrations in the oceanic realm: possible causes and probable effects. Contrib Mar Sci Texas 27 Suppl: 45–70

Angel MV (1993) Marine planktonic ostracods. In: Kermack DM, Barnes RSK, Crothers JH (eds) Synopses of the British fauna (New Series), vol 48. Field Studies Council, Shrewsbury, 239 pp

Angel MV, Fasham MJR (1983) Eddies and biological processes. In: Robinson AR (ed) Eddies in marine science. Springer, Berlin Heidelberg New York, pp 492–524

Barney RW (1921) Crustacea. V. Ostracoda. British Antarctic (Terra Nova) expedition, 1910. Nat Hist Rep Zool 3(7): 175–190

Benassi G, Ferrari I, Gentile G, Menozzi P, McKenzie KG (1992) Planktonic Ostracoda in the Southern Ocean and in the Ross Sea: 1989–90 Campaign. In: National Scientific Commission for Antarctica (ed) Oceanographic Campaign 1989–90, Data Rep Part II, Genova, pp 247–300

Benassi G, Ferrari I, Menozzi P, McKenzie KG (1994a) Planktic ostracodes from the Antarctic and Subantarctic collected by the 1989–1990 Italian Antarctic Expedition. Rec Aust Mus 46(1): 25–37

Benassi G, Ferrari I, Menozzi P, Rossi V, McKenzie KG (1994b) Distribuzione degli ostracodi planctonici in aree Antartiche e Subantartiche dell'Oceano Pacifico. Atti 10° Cong Assoc Ital Oceanol Limnol, Alassio 4–6 Nov 1992. CNR, Genova, pp 681–690

Boldrin A, Stocchino C (1990) On the hydrological characteristics of Terra Nova Bay. In: National Scientific Commission for Antarctica (ed) Oceanographic Campaign 1987–1988, Data Rep Part I, Genova, pp 11–57

Brady GS (1907) Crustacea. V. Ostracoda. Reports of the National Antarctic Expedition S.S. Discovery, 1901–1904. Nat Hist 3: 1–9

Claus C (1874) Die Gattungen und Arten der Halocypriden. Verh Zool Bot Ges Wien 24: 175–178

Deevey GB (1978) A taxonomic and distributional study of the planktonic Ostracoda collected on three cruises of the Eltanin in the South Pacific and the Antarctic Region of the South Pacific. Biol Antarctic Seas VIII. Antarct Res Ser 28(3): 43–70

Deevey GB (1982) A faunistic study of the planktonic ostracods (Mydocopa, Halocyprididae) collected on eleven cruises of the Eltanin between New Zealand, Australia, the Ross Sea and the South Indian Ocean. Biol Antarctic Seas X. Antarct Res Ser 32(5): 131–167

Deevey GB (1983) Planktonic ostracods (Myodocopa, Halocyprididae) from six Eltanin cruises in South Pacific and Antarctic waters. J Crustacean Biol 3(3): 409–416

Gooday AJ (1981) The *Conchoecia skogsbergi* species complex (Ostracoda, Halocyprididae) in the Atlantic Ocean. Bull Br Mus Nat Hist Zool 40(4): 137–209

Guglielmo L, Costanzo G, Manganaro A, Zagami G (1990) Spatial and vertical distribution of zooplanktonic communities in the Terra Nova Bay (Ross Sea). In: National Scientific Commission for Antarctica (ed) Oceanographic Campaign 1987–1988, Data Rep Part I, Genova, pp 257–398

Hartmann G (1986) Ostracoden aus der Tiefsee des Indischen Ozeans und der Iberischen See sowie von ostatlantischen sublitoralen Plateaus und Kuppen. Senckenb Marit 17(4/6): 303–304

Hecq J-H, Goffart A (1992) Results of phytopigments analysis by high performance liquid chromatography (HPLC) during the Vth Italiantartide expedition. In: National Scientific Commission for Antarctica (ed) Oceanographic Campaign 1989–90, Data Rep Part II, Genova, pp 123–152

Hecq J-H, Guglielmo L (1992) Structure and functioning of the Ross Sea pelagic ecosystem: an interdisciplinary approach. In: Gallardo VA, Ferretti O, Moyano HI (eds) Oceanografia in Antartide. ENEA-PNRA-EULA, Concepiòn, Chile, pp 227–233

Hecq J-H, Brasseur P, Goffart A, Lacroix G, Guglielmo L (1993) Modelling approach of the planktonic vertical structure in deep austral ocean: the example of the Ross Sea ecosystem. In: Progress in Belgian oceanographic research. IRMA, Brussels, pp 235–250

Kock R (1992) Ostracoden im Epipelagical vor der Antarktischen Halbinsel - ein Beitrag zur Systematik sowie zur Verbreitung und Populationsstruktur unter Berücksichtigung der Saisonalität. Ber Polarforsch 106: 1–209

Longhurst AR (1976) Vertical migration. In: Cushing DH, Walsh JJ (eds) The ecology of the seas. Blackwell, Oxford, pp 116–137

Martens JM (1979) Die pelagischen Ostracoden der Expedition Marchile I (Südost-Pazifik), II: Systematik und Vorkommen (Crustacea: Ostracoda: Myodocopida). Mitt Hamb Zool Mus Inst 76: 303–366

McKenzie KG, Milne PW, Bryen C, Freeman K (1979) Distribution of marine planktic and benthic Ostracoda in northern Australian waters and some correlated environmental factors. In: Kristic N (ed) Proc VIIth Int Symp Ostracodes, Beograd, Vol I, pp 233–236

McKenzie KG, Benassi G, Naldi M, Ferrari I, Menozzi P (1990) Report on planktic Ostracoda from the Ross Sea, Antarctica. In: National Scientific Commission for Antarctica (ed) Oceanographic Campaign 1987–88, Data Rep Part II, Genova, pp 171–229

Mogulievsky A, Angel MV (1975) Halocyprid ostracods in Atlantic neuston. Mar Biol 32: 295–302

Müller GW (1906a) Ostracoda. Wiss Ergeb Dtsch Tiefsee-Exped Valdivia 1898–1899 8 (Jena): 29–154

Müller GW (1906b) Ostracoden: résultats du voyage du S.Y. Belgica en 1897–1898–1899. Rapp Sci Zool JE Buschmann Anvers: 1–10

Müller GW (1908) Die Ostracoden der deutschen Südpolar-Expedition 1901–1903. X. Wiss Ergebn Deutsch Südpol-Exped Band 10, Zool Berl II: 58–181

Patterson SL, Whitworth T (1990) Physical oceanography. In: Glasby GP (ed) Antarctic sector of the Pacific. Elsevier, Amsterdam, pp 55–93

Poulsen EM (1969) Ostracoda-Myodocopa. III. A. Halocypriformes - Thaumatocypridae and Halocypridae. Dana Rep 75: 1–100

Poulsen EM (1973) Ostracoda-Myodocopa. III. B. Halocypriformes - Halocypridae Conchoeciinae. Dana Rep 84: 1–224

Skogsberg T (1920) Studies on marine ostracodes. I. Cypridinids, halocyprids and polycopids. Zool Bidr Upps Suppl 1: 1–784

Stocchino C, Manzella G (1991) Analisi delle correnti marine a Baia Terra Nova in rapporto alla circolazione oceanica e a quella nel Mare di Ross. CNR, Area Ricerca Genova, pp 1–56

CHAPTER 28

# Spatial Distribution and Feeding Habits of Larval and Juvenile *Pleuragramma Antarcticum* in the Western Ross Sea (Antarctica)

**A. Granata[1], L. Guglielmo[1], S. Greco[2], M. Vacchi[3], O. Sidoti[1], G. Zagami[1], and M. La Mesa[3]**


## ABSTRACT

During four spring-summer Italian Antarctic expeditions to Terra Nova Bay and the Western Ross Sea in 1988, 1989–90, 1994–95 and 1996, zooplankton and ichtyoplankton were collected by a multinet BIONESS (250 µm and 500 µm) and PHN (500 µm), respectively. Nototheniids were predominant in the whole study area, representing 97.3% of all larvae. *Pleuragramma antarcticum* was the dominant species sampled in the spring-summer cruises of 1987–88 (98.6%), 1989–90 (95.9%) and 1995–96 (99.8%), while it predominated in association with *Trematomus lepidorhinus* in 1994–95. The early life history of *Pleuragramma antarcticum* was described from 891 samples. Maximum abundances of *P. antarcticum* postlarvae were confined to Terra Nova Bay. The mean relative abundance in the whole area in summer 1988, for positive 0- to 150-m hauls, was 434 (±720 SD) specimens 100 $m^{-3}$ and 158 (±212 SD) specimens per 100 $m^3$ in January 1990 for positive 0–140-m hauls. Off Terra Nova Bay, postlarval abundance did not exceed 5–86 individuals 100 $m^{-3}$. Juvenile *Pleuragramma antarcticum* showed a more oceanic occurrence than postlarvae of the same species. Low concentrations between 0.1 and 12.5 ind 100 $m^{-3}$ (mean 1.66 ± 2.88 SD) were recorded in offshore waters between 170° and 175°E and in the southern part of Terra Nova Bay. Postlarvae (n = 4385) ranged between 7 and 23 mm SL (mean 13.9 ± 1.62 SD) and juveniles (n = 388) were between 28 and 53 mm SL (mean 40.0 ± 4.78 SD). A growth rate of 0.08 mm $day^{-1}$ was confirmed for the first year. The daily increments of *Pleuragramma antarcticum* postlarvae estimated from sequential sampling in 1987–88 from Terra Nova Bay and in 1995–96 from the southern western Ross Sea were about 0.21–0.25 mm. The co-occurrence of different postlarval size classes during the investigations confirm that two hatching periods occurred in Terra Nova Bay between early and late December. The presence of early stages in January 1988, 1990 and 1996 (7–9 mm SL) confirms that *Pleuragramma antarcticum* spawn in Terra Nova Bay. Highest concentrations of postlarvae occurred at the edge near Cape Washington and in the inshore waters of Drygalski Ice Tongue. These areas seem to represent a nursery ground of early stages, where the presence of permanent polynyas could provide favourable food conditions for the development of the first stages of life. More than 62% of the postlarvae were collected in well-stratified Antarctic Surface Waters (AASW) with temperature values that exceed 0 °C and vertical distribution appears to be related to the summertime thermocline. From these sites, early stages were drifted by a westward-flowing current of the limb of the Antarctic coastal current and the southern limb of the Ross Sea Gyre. These large gyres favoured postlarval retention mostly in northern and southern areas of Terra Nova Bay, preventing the drift north of Cape Washington, south of Drygalski Ice Tongue and in offshore waters beyond 170°E. Thus, the youngest planktonic stages of *Pleuragramma antarcticum* concentrate in the surface mixed layer of Terra Nova Bay where feeding conditions are favourable. In the size classes 8–17 mm the diet was dominated by calanoid eggs (35.5%), *Limacina* (32.1%) and tintinnids (17.6%). The cyclopoid copepod *Oncaea curvata* contributed to 7.1% of the diet.



[1]Dipartimento di Biologia Animale ed Ecologia Marina, Università di Messina. Salita Sperone 31, 98166 Messina, Italy

[2]Istituto Talassografico CNR, Spianata S. Raineri 86, 98123 Messina, Italy

[3]ICRAM Istituto Centrale per la Ricerca Scientifica e Tecnologica applicata al mare, Via L. Respighi 5, 00197 Rome, Italy

## Introduction

*Pleuragramma antarcticum* is the dominant pelagic fish in the high-Antarctic zone, where it plays a fundamental role in the food web of the Antarctic marine ecosystem (Hureau 1994). In both the Weddell and Ross Seas, it is an important food resource for top predators, such as seals and whales, as well as other notothenioid fishes (Emison 1968; Eastman 1985a,b; Plötz 1986). In the western Ross Sea, in summer, the pelagic fish fauna is dominated by this species (96.7% in biomass, De Witt 1970) and for this reason *P. antarcticum* is considered the most important element in the food web of the fish community (Takahashi and Nemoto 1984). *Pleuragramma antarcticum* is the most abundant species of the Weddell Sea ichthyoplankton (Hempel et al. 1983; Ekau et al. 1987; Piatkowski 1987; Boysen-Ennen and Piatkowski 1988; Hubold 1990, 1991; Piatkowski et al. 1990; White and Piatkowski 1993) and represents about 98% of the total ichthyoplankton living on the Weddell Sea continental shelf (Keller 1983).

The Ross Sea is the southernmost of the Antartic Seas, having as northern limit Cape Adare (170°S) and Cape Colbeck (158°W) and southern end the Ross Ice Shelf (RIS) at around 78.5°S (Fig. 1b). Compared to other Antarctic areas where the continental shelf is narrow or lacking, the Ross Sea has a wide and deep shelf (about 2.4 million km$^2$), the topographic characteristics of which have been described by Kennett (1968). The mean depth of the shelf is about 450 m with greater depths close to the coastal areas, up to 1000 m in the Drygalski Basin. According to Patterson and Whitworth (1990) the Ross Sea Shelf was characterized by these water masses: (1) Antarctic Surface Water (AASW, 0–50 m) with low salinity (<34.50 psu) and, during the summer, relatively high temperatures at the surface (–1.5/2.0 °C); (2) Shallow Ice Shelf Water (SISW, 50–200 m) with salinity <34.60 and temperature <–1.9 °C; (3) Modified Circumpolar Deep Water (MCDW, 200–300 m) with salinity >34.50 and temperature >–1.6 °C; (4) Deep Ice Shelf Water (DISW, 300–600 m) with salinity >34.60 and temperature <–1.9 °C, and High Salinity Shelf Water (HSSW, 600 m to the bottom) with salinity >34.75 and temperature ~–1.9 °C. The Ross Sea is entirely covered by sea ice for at least 9 months of the year.

Terra Nova Bay is an 80-km by 30-km wide inlet in the southwest part of the Ross Sea (Fig. 1a). It is delimited in the north by the volcano Mt. Melbourne and narrow peninsula of Cape Washington. To the south it is bounded by the Drygalski Ice Tongue, arising from David Glacier, which extends into the Ross Sea for about 60 km. During the zooplankton sampling, hydrological and current measurements were carried out in Terra Nova Bay (Boldrin and Stocchino 1990; Stocchino and Lusetti 1990; Innamorati et al. 1990; Stocchino and Manzella 1991) and Western Ross Sea (Spezie et al. 1993; Budillon et al., this Vol.; Picco et al., this Vol.; Russo, this Vol.). The coastal area was characterized by the presence of warmer (+0.9/1.5 °C) and more saline surface water (S > 34.7 psu, Boldrin and Stocchino 1990; Innamorati et al. 1990). Subsequent summer Italian expeditions (Stocchino and Manzella 1991; Spezie et al. 1993) confirmed the thermoaline values measured during the 1987–88 expedition and agree both with the general distribution of Ross Sea water masses under summer conditions (Jacobs et al. 1970) and with the values reported for McMurdo Sound (Berry 1988). According to Patterson and Withworth (1990), highest salinities seem to occur in coastal regions of persistent wintertime leads and polynyas which prolong the exposure of the sea to the cold Antarctic atmosphere. Terra Nova Bay was characterized by polynyas persisting also during wintertime (Bromwich and Kurtz 1984) formed and preserved by persistent katabatic winds which maintain the area free of ice (Bromwich and Kurtz 1982).

The presence of a very thin Warm Surface Water (+1.5 °C) along the Terra Nova Bay coastal area was confirmed also by Spezie et al. (1993) during austral summer 1990–91, probably due both to seasonal radiation and to the effect of katabatic winds producing an eastward advection of the newly formed ice, and the presence of the Drygalski Ice Tongue preventing the northward drift of the ice (Stocchino and Manzella 1991). The lowest temperature values (–0.3 °C) were found in the central area of the bay, probably due both to local eddies and to upwelling processes determined by katabatic winds. To the east of the adjoining Cape Washington and Drygalski Ice Tongue a surface salinity of <34.6 was recorded. The area is affected by dilution processes due to ice melting, with the added effect of a clockwise cyclonic gyre characterizing the summer surface circulation.

Along a coast-seaward hydrological transect from Terra Nova Bay Station (74°48′S–164°00′E) to

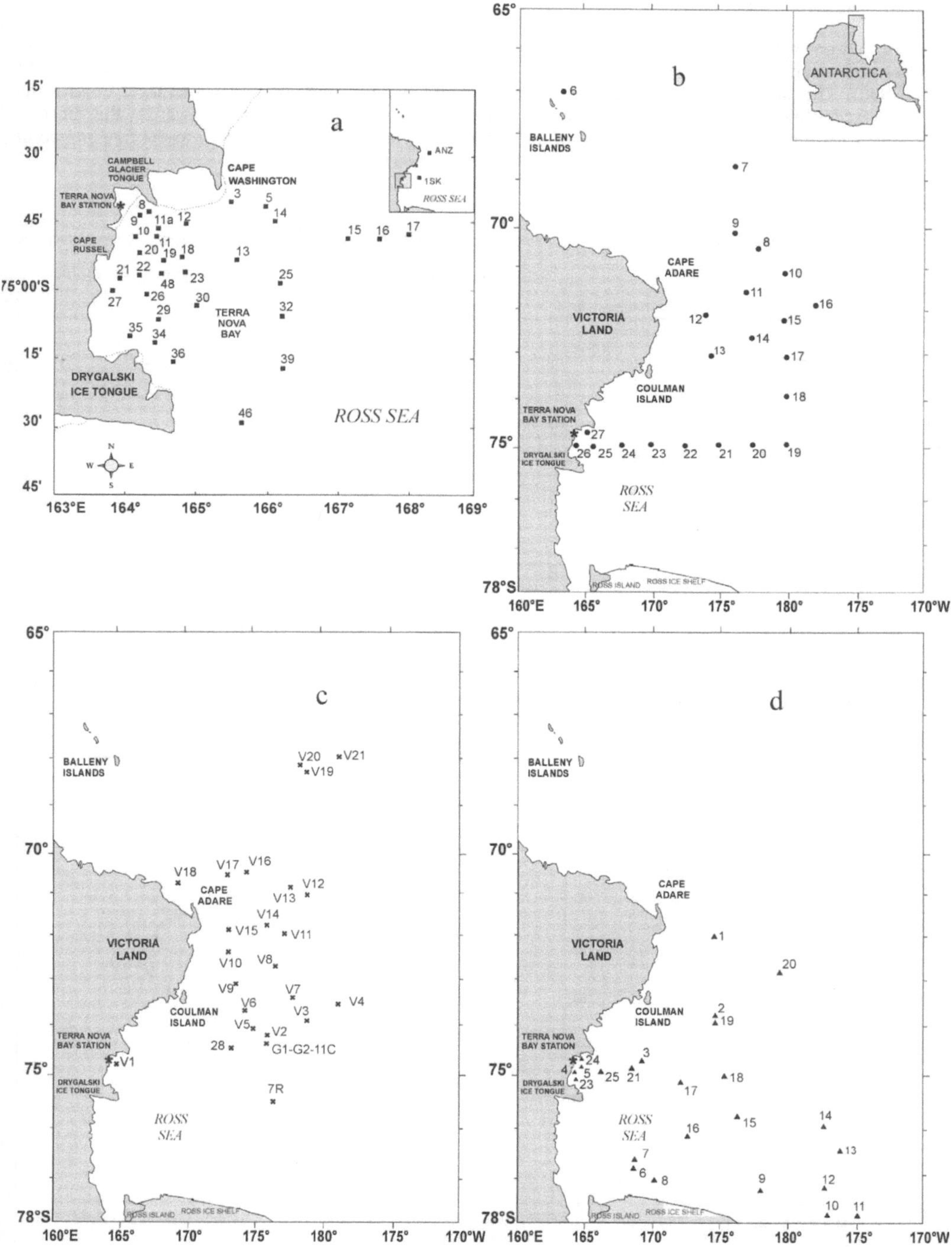


**Fig. 1a–d.** Sampling stations during cruises. **a** Polar Queen in January–February 1988. **b** Cariboo in November 1989-January 1990. **c** ROSSMIZE in November–December 1994. **d** Italica in January–February 1996, in the western Ross Sea and Terra Nova Bay. *Dotted line* indicates coast formed by permanent pack ice

ca. 125 nm offshore (172°00′E), a strong vertical stratification was observed (Innamorati et al. 1990). The surface layer ascribed to Antarctic Surface Water (AASW) had temperature values that exceed 0 °C. Temperature profiles of the water column during BIONESS sampling in 1987–88 (Fig. 2) confirm the presence of a definite summer thermocline, varying from 20 to 70 m. The AASW floats on a single homogeneous layer in wich the salinity increases slightly towards the bottom with temperatures very close to the freezing point (−1.92 °C).

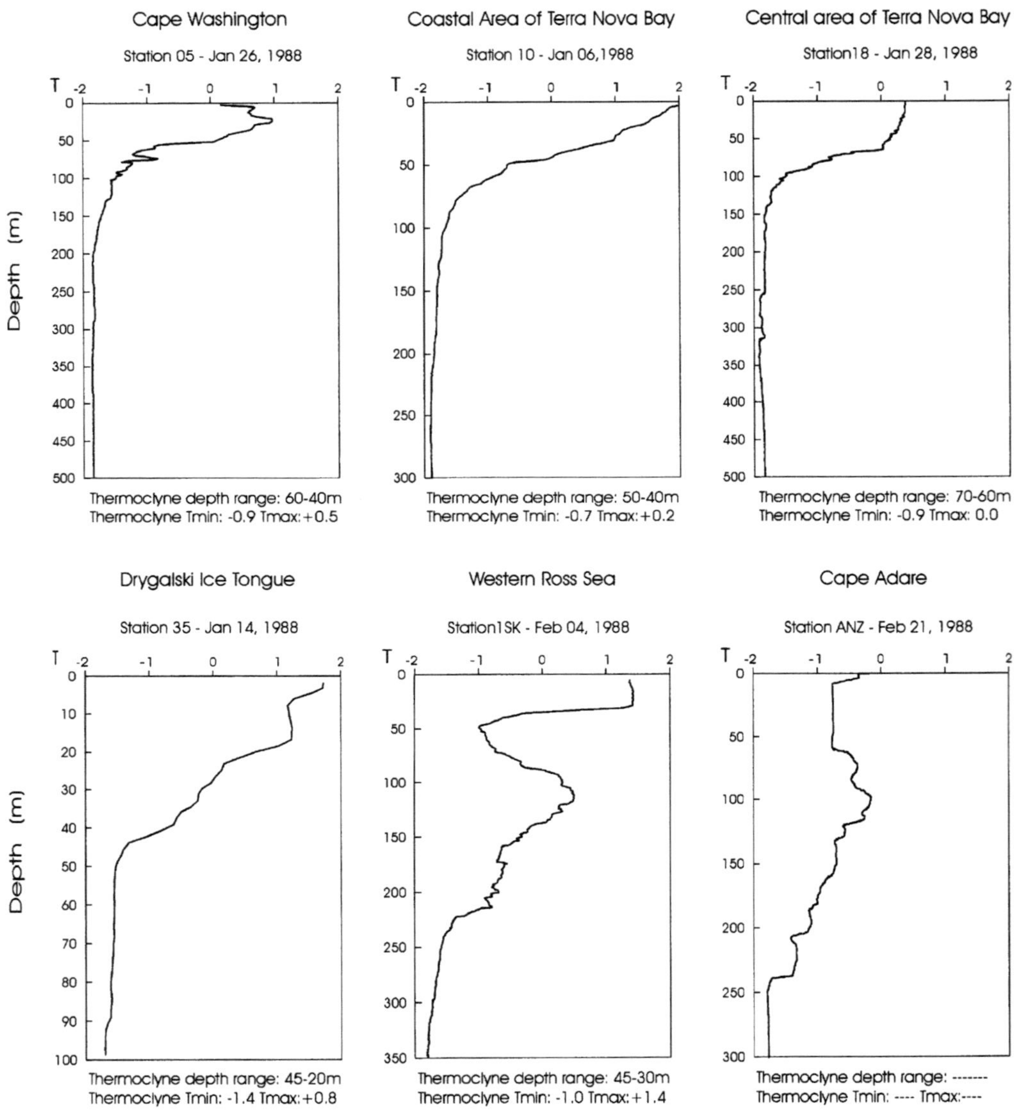


**Fig. 2.** Temperature profiles during BIONESS sampling in January–February 1988

Current measurements were carried out in Terra Nova Bay (Stocchino and Manzella 1991), in the Western Sector of the Ross Sea (see Russo, this Vol.) and in the Ross Continental Shelf (see Picco et al., this Vol.) by both CTD and ADCP current meters. The circulation at Terra Nova Bay was characterized by large-scale vortices. At the surface, the circulation was driven by local forcing (mainly katabatic winds); advective movements were detected at depth. At the surface, highest velocities were measured in the southern area (up to 8 cm $s^{-1}$) and offshore (up to 50 cm $s^{-1}$). A general schematic representation of the Terra Nova Bay surface circulation is given in Guglielmo et al. (1998), according to data and description by Stocchino and Manzella (1991). The investigated area was affected by two components of the flow coming from the northeast in the northern part, and from the east in the southern one. The first component is divided into two branches, the first flows to the central area, while the second turns east at the 75°S parallel. The second component partly flows out of the bay and partly reaches the northern border of the Drygalski Ice Tongue. This one branch then follows the coast flowing to the north and joins the first component, giving rise to a coastal current.

During the late spring-summer ROSSMIZE cruise (1994–95) in the western Ross Sea (see Russo, this Vol.) the polynya was expanding and a marginal ice zone was encountered on the Pennell Bank. Below it, an intense Warm Core (WMCO) was present. WMCO occupied also the northern Mawson Bank. Also Lower Circumpolar Deep Water (LCDW) was noted intruding on the continental shelf, and mixing with surface water in correspondence to the slope from which WMCO originates. Below the LCDW, a bottom boundary layer of Low Salinity Bottom Water (LSBW) has been detected; it could originate from Low Salinity Shelf Water (LSSW) outflowing from the northern end of the Joides Basin. On the two depressions crossed by the transect, both belonging to the Joides Basin, High Salinity Shelf Water (HSSW) forms a dome; in the southern depression, HSSW occupies almost the whole water column, while in the northern one it is present as a bottom layer, with LSSW upon it. During summer 1995–96 (see Picco et al., this Vol.) the Circumpolar Deep Water (CDW) entered the Ross Sea from the deep basin and circulated cyclonically along the Ross Sea shelf break. The Ice Shelf Water (ISW), outflowing from beneath the Ross Ice Shelf, had two preferential directions, westward and northward.

Despite the key role of *P. antarcticum* in Antarctic shelf ecosystem (Hubold 1984), few studies have been conducted on its life cycle and distribution in the Ross Sea (Regan 1916; De Witt 1970). More recently, spatial distribution and abundance of the larval and juvenile stages of *P. antarcticum* collected both in Terra Nova Bay and western Ross Sea, in summer, were reported in Guglielmo et al. (1998) and Vacchi et al. (1999), respectively. Other data are available from the McMurdo Sound (De Witt and Tyler 1960), Whales Bay (Ross Sea, Biggs 1982), Weddell Sea (Keller 1983; Hubold 1984, 1985, 1990), Bransfield Strait and off the Antarctic Peninsula (Kellermann and Kock 1984, 1988; Ascencio and Mujica 1986; Kellermann 1986a,b, 1989; Slosarczyk 1986; White and North 1987; Kellermann and Schadwinkel 1991; Loeb 1991; Morales-Nin et al. 1995). Feeding patterns of postlarval and juvenile *Pleuragramma antarcticum* were studied in the Weddell Sea (Hubold 1985; Hubold and Ekau 1990; Hubold and Hagen 1997), Antarctic Peninsula (Kellermann 1987), Bransfield Strait (Moreno et al. 1986) and western Ross Sea (De Witt and Hopkins 1977).

The purpose of this chapter was to both improve and extend our knowledge of the spatial distribution, abundance, growth and food composition of *P. antarcticum* in the Western Ross Sea, based on larval specimens obtained during late spring-summer Italian Antarctic cruises in 1987–88, 1989–90, 1994–95 and 1995–96.

## Materials and Methods

### Sampling Procedure

Postlarval and juvenile *Pleuragramma antarcticum* were collected in the zooplankton and micronekton samples by both an electronic multinet BIONESS (Sameoto et al. 1980) and a monowire midwater trawl Plankton Hamburg Net (PHN). The BIONESS was a multiple opening and closing net sampler with ten nets with a square mouth opening of 0.25 $m^2$ each. The mesh sizes of the nets used were 250 μm and 500 μm in both different cruises and stations. These nets were fine enough to collect all zooplankters over 300 μm in smallest dimension as well as micronekton up to 10 cm in length (Sameoto and Lewis 1980a,b). For this reason, both mesh sizes were used to gather *P. antarcticum* data. The BIONESS was towed at a speed of 1.5–2 $m s^{-1}$ as it was slowly lowered along an oblique path to the desired depth. Fishing upward, the nets opened and closed on command at 10- to 20-m intervals in the hauls programmed in the upper 200 m according to different water masses and between 30- and 100-m intervals below this zone to the maximum depth sampled. Due to time constraints on the ship, replicate tows were not made. To help compensate for the lack of replicate tows and to increase the precision of the samples, the tows were made as long as possible, ranging in length from ~1000 to ~6000 m. The rationale for this procedure was discussed by Wiebe (1971, 1972). Contamination of the nets by surface zooplankton was negligible, since the nets were all tightly closed before and after fishing. Towing the BIONESS with closed nets in the zooplankton maximum at about 20–30 m depth we found that contamination was insignificant.

The mouth area of the PHN was abouth 5 $m^2$ and mesh size of 500 μm. The volume of water filtered was measured by means of a TSK flowmeter placed on the mouth of the net. For each station, the time and depth of the tow were recorded by a Time Depth Recorder. The towing speed ranged between 1.5 and 2.5 knots and the time of fishing was from 45 and 60 min.

Ichthyoplankton was collected during the course of the Italian Antarctic expeditions 1987–88, 1989–90, 1994–95 and 1995–96.

## Polar Queen Cruise 1987–88

During a period between 4th January and 21st February 1988, zooplankton samples were taken at 33 stations in Terra Nova Bay (Ross Sea, Fig. 1a). The study area extended from Cape Washington (79°39′S) to Drygalsky Ice Tongue (75°29′S), along longitude 167°59′E. The sampling stations were located mainly on the continental shelf of the Ross Sea, while station 1SK was located ca. 150 miles offshore from the coast and at station ANZ near Cape Adare. The BIONESS continously measured temperature, salinity and depth during tows by a means of an applied microsystems digital CTD. Flow through the nets was monitored with external and internal TSK flowmeters. Differences in numbers of collected samples were due always to technical reasons, whereas the lack of surface samples in some stations (e.g. 22, 34, 13, 23, 39 and 15) was chosen at 50-m intervals in deep hauls. Detailed information on sampling strategies are available in Guglielmo et al. (1990). Sampling effort (stations sampled, bottom depth, positive hauls depth range, volume filtered, *P. antarcticum* abundance 100 $m^{-3}$) are reported in Guglielmo et al. (1998). Total effort was 33 hauls (268 samples) filtering 12 452 $m^3$ of water. Low volumes of filtered water in some stations and upper layers were affected by local microalgal blooms.

## Cariboo Cruise 1989–90

During the period between 25th November 1989 and 29th January 1990, zooplankton and micronekton samples were taken at 27 stations located in the Pacific Sector of the Southern Ocean, from New Zealand to the Ross Sea (Fig. 1b). Stations 6 and 7 were located in the permanent ice zones (Balleny Islands), station 8 to station 24 in the Western Ross Sea and stations 25, 26 and 27 in Terra Nova Bay. Samples were collected by the same BIONESS utilized in the Polar Queen cruise, equipped with a multiparametric probe (KMSII) and a Back-Scat fluorometer. The flow through the nets was monitored by external and internal acoustic doppler flowmeters (ME-SM11H). Total effort was 57 hauls (572 samples) filtering 47 438 $m^3$ of water. Some stations were sampled twice in different depth intervals. Only 500-μm mesh nets were used. Detailed information on sampling strategies are available in Guglielmo et al. (1992). Sampling effort (stations sampled, bottom depth, positive hauls depth range, trawling depth, volume filtered, *Pleuragramma antarcticum* abundance 100 $m^{-3}$) is shown in Table 1.

## ROSSMIZE Cruise 1994–95

The Italian Antarctic Oceanographic cruise ROSSMIZE (Ross Ice Marginal Ice Zone Ecology) on board the R/V Italica was carried out off Victoria Land in the northern sector of the Western Ross Sea from 2 November 1994 to 3 January 1995. During the cruise, ichthyoplankton samples were obtained by PHN from 21 November to 27 December 1994. The sampling operations provided several specimens of larval and juvenile fish collected at 26 stations between 67°50′S and 75°20′S latitude and 164°40′E and 179°20′W lon-

**Table 1.** Stations sampled, bottom depth (m), haul depth range (m), positive trawling depth (m), volume filtered ($m^3$), *P. antarcticum* abundance (N 100 $m^{-3}$) for age 0 and age 1+ during the 1989–90 Italian Antarctic expedition

| Station | Date | Bottom depth (m) | Hauls depth range (m) | Positive trawling depth (m) | VF ($m^3$) | *Pleuragramma antarcticum* (N 100 $m^{-3}$) Age 0 | Age 1+ |
|---|---|---|---|---|---|---|---|
| 13 | Dec 26, 1989 | 315 | 280–0 | 160–140 | 85.6 | 0.0 | 2.3 |
| 22 | Jan 4, 1990 | 545 | 500–0 | 0–500 | 189.9 | 0.0 | 2.1 |
| | | | | 0–100 | 137.0 | 0.0 | 2.2 |
| | | | | 50–0 | 98.7 | 0.0 | 1.0 |
| | | | | 40–30 | 42.6 | 0.0 | 70.4 |
| | | | | 30–20 | 43.0 | 0.0 | 7.0 |
| | | | | 20–0 | 92.1 | 0.0 | 2.2 |
| 23 | Jan 5, 1990 | 345 | 300–0 | 40–30 | 44.6 | 4.5 | 0.0 |
| | | | | 20–0 | 115.5 | 6.9 | 0.0 |
| 24 | Jan 6, 1990 | 520 | 450–0 | 0–100 | 220.5 | 1.4 | 0.0 |
| | | | | 60–50 | 44.6 | 2.2 | 0.0 |
| | | | | 50–40 | 40.8 | 12.3 | 0.0 |
| | | | | 40–30 | 35.3 | 8.5 | 0.0 |
| | | | | 30–20 | 32.1 | 53.0 | 0.0 |
| | | | | 20–0 | 78.6 | 1.3 | 0.0 |
| 25 | Jan 7, 1990 | 885 | 800–0 | 0–800 | 238.3 | 35.2 | 0.0 |
| | | | | 0–200 | 160.5 | 280.4 | 0.0 |
| | | | | 500–400 | 159.6 | 1.3 | 0.0 |
| | | | | 400–300 | 162.7 | 1.2 | 0.0 |
| | | | | 200–160 | 108.3 | 2.8 | 0.0 |
| | | | | 200–100 | 152.9 | 1.3 | 0.0 |
| | | | | 100–80 | 47.1 | 2.1 | 0.0 |
| | | | | 100–50 | 76.8 | 15.6 | 0.0 |
| | | | | 50–0 | 88.8 | 1099.1 | 0.0 |
| | | | | 40–20 | 45.9 | 45.8 | 0.0 |
| | | | | 20–0 | 49.5 | 367.7 | 0.0 |
| 26 | Jan 8, 1990 | 635 | 450–0 | 0–100 | 92.5 | 5.4 | 0.0 |
| | | | | 70–60 | 37.3 | 8.0 | 0.0 |
| | | | | 60–50 | 35.7 | 5.6 | 0.0 |
| | | | | 40–30 | 28.6 | 10.5 | 0.0 |
| | | | | 30–20 | 32.2 | 2944.1 | 0.0 |
| | | | | 20–0 | 88.7 | 1192.8 | 1.1 |
| 27 | Jan 12, 1990 | 735 | 600–0 | 0–600 | 221.5 | 1.4 | 0.0 |
| | | | | 0–100 | 92.5 | 2.2 | 0.0 |
| | | | | 200–150 | 92.6 | 6.5 | 0.0 |
| | | | | 150–100 | 74.4 | 5.4 | 0.0 |
| | | | | 90–80 | 29.3 | 0.0 | 3.4 |
| | | | | 60–50 | 35.7 | 2.8 | 0.0 |
| | | | | 40–30 | 28.6 | 3.5 | 0.0 |
| | | | | 30–20 | 32.2 | 6.2 | 0.0 |
| | | | | 20–0 | 88.7 | 3.4 | 0.0 |

gitude. The study area showing all the sampling stations is reported in Fig. 1c. Fishing depth usually ranged from the surface to 120 m, except for station 11C, where the net was hauled at 380 m depth. Detailed information on sampling strategies are available in Vacchi et al. (1999). Total effort was 26 hauls filtering 412 544 $m^3$ of water. Sampling effort (stations sampled, bottom depth, positive hauls depth range, trawling depth, volume filtered, *P. antarcticum* abundance 100 $m^{-3}$) is shown in Table 2.

## Italica Cruise 1995–96

During the period between 12 January and 5 February 1996, zooplankton and micronekton were taken by PHN at 25 stations located in the Western Ross Sea. The study area extended from Cape Adare (72°S) to the Ross Ice Shelf (77°–78°S) along longitude 175°W (Fig. 1d). Only four stations (4, 5, 23, 24) were sampled in Terra Nova Bay. The maximum depth sampled was 800 m (station 25). Total effort was 25 hauls filtering 620 089 $m^3$ of water. Sampling effort (stations sampled, bottom depth, positive trawling depth, volume filtered, *Pleuragramma antarcticum* abundance $100\,m^{-3}$) is shown in Table 3.

## Laboratory Procedure

All specimens were preserved in 4% buffered seawater formaldehyde solution. After the samples were sorted into major taxa in the laboratory, larval and juvenile stages of fish were identified to species following the descriptions of Hureau (1982), North and White (1982), Efremenko (1983), Kellermann (1990), and North and Kellermann (1990), and counted and measured for standard length (SL) to the nearest millimiter.

Large catches of *P. antarcticum* were subsampled, and 50 to 100 individuals were measured. Shrinkage due to formaldehyde preservation (ca. 10%) was not considered. Fishes were subdivided into size classes. Larvae (with yolk sac) were not collected, so all specimens are postlarvae or juveniles. The age groups were defined according to Hubold (1985), Kellermann (1986b) and Guglielmo et al. (1998). Postlarvae of *P. antarcticum* between 8 and 21 mm SL were attributed to age 0; old postlarvae and juveniles between 30 and 53 mm SL to age 1+, and juveniles between 70–80 mm SL to age 2+.

All collected samples were taken into account to describe both growth and size-frequency temporal distribution. Abundance values for postlarvae represent a weighted mean of pooled layers between 0–150 m standardized as numbers per filtered volume ($N\,100\,m^{-3}$) calculated from total specimens counted and total volume filtered at each station.

The stomach contents of the following specimens caught in summer 1988 at Terra Nova Bay were investigated:

**Table 2.** Stations sampled, bottom depth (m), haul depth range (m), positive trawling depth (m), volume filtered ($m^3$), *P. antarcticum* abundance ($N\,100\,m^{-3}$) for age 0 and age 1+ during the 1994–95 Italian Antarctic expedition

| Station | Date | Bottom depth (m) | Positive trawling depth (m) | VF ($m^3$) | *Pleuragramma antarcticum* ($N\,100\,m^{-3}$) Age 0 | Age 1+ |
|---|---|---|---|---|---|---|
| 28 | Dec 15, 1994 | 522 | 30–0 | 18471 | 0.000 | 0.931 |
| V2 | Dec 18, 1994 | 437 | 30–0 | 12504 | 0.000 | 0.008 |
| V6 | Dec 20, 1994 | 318 | 20–0 | 16348 | 0.000 | 0.245 |
| V15 | Dec 24, 1994 | 1650 | 80–0 | 22599 | 0.000 | 0.004 |
| V18 | Dec 26, 1994 | 2285 | 60–0 | 16203 | 0.000 | 0.012 |

**Table 3.** Stations sampled, bottom depth (m), haul depth range (m), positive trawling depth (m), volume filtered ($m^3$), *P. antarcticum* abundance ($N\,100\,m^{-3}$) for age 0, age 1+ and age 2+ during the 1995–96 Italian Antarctic expedition

| Station | Date | Bottom depth (m) | Positive trawling depth (m) | VF ($m^3$) | *Pleuragramma antarcticum* ($N\,100\,m^{-3}$) Age 0 | Age 1+ | Age 2+ |
|---|---|---|---|---|---|---|---|
| 1 | Jan 12, 1996 | 1485 | 760–0 | 57 702 | 0.000 | 0.002 | 0.000 |
| 2 | Jan 13, 1996 | 602 | 400–0 | 21 511 | 0.000 | 0.005 | 0.000 |
| 3 | Jan 13, 1996 | 322 | 250–0 | 23 200 | 44.875 | 0.034 | 0.000 |
| 4 | Jan 14, 1996 | 322 | 260–0 | 23 835 | 0.344 | 0.000 | 0.000 |
| 6 | Jan 19, 1996 | 839 | 550–0 | 26 760 | 0.187 | 0.000 | 0.000 |
| 10 | Jan 21, 1996 | 636 | 490–0 | 40 918 | 0.139 | 0.000 | 0.000 |
| 12 | Jan 22, 1996 | 656 | 350–0 | 21 093 | 0.000 | 0.000 | 0.005 |
| 15 | Jan 24, 1996 | 478 | 420–0 | 21 397 | 0.000 | 0.023 | 0.000 |
| 16 | Jan 24, 1996 | 659 | 600–0 | 22 501 | 0.151 | 0.027 | 0.004 |
| 18 | Jan 25, 1996 | 292 | 280–0 | 20 407 | 0.000 | 0.118 | 0.000 |
| 19 | Jan 26, 1996 | 613 | 600–0 | 23 178 | 0.043 | 0.000 | 0.000 |
| 21 | Jan 28, 1996 | 603 | 570–0 | 22 128 | 85.561 | 0.077 | 0.000 |
| 23 | Feb 02, 1996 | 900 | 700–0 | 24 750 | 542.234 | 0.000 | 0.000 |
| 24 | Feb 05, 1996 | 300 | 250–0 | 23 254 | 0.353 | 0.000 | 0.000 |
| 25 | Feb 05, 1996 | 886 | 800–0 | 21 997 | 865.573 | 0.000 | 0.000 |

- 91 postlarvae of 8–11 mm SL (mean: 9.98 ± 0.91 SD)
- 99 postlarvae of 12–14 mm SL (mean: 12.99 ± 0.79 SD)
- 90 postlarvae of 15–17 mm SL (mean: 15.94 ± 0.81 SD)
- 26 juveniles of 36–41 mm SL (mean: 39.62 ± 1.47 SD)
- 21 juveniles of 42–44 mm SL (mean: 43.24 ± 0.77 SD)
- 24 juveniles of 45–50 mm SL (mean: 46.46 ± 1.44 SD)

All specimens were caught in the top 150 m of the water column. Stomachs were dissected and the contents carefully washed into a petri dish. Food items were identified under a dissecting microscope to the lowest taxon possible (i.e. the species level for most of the copepods), measured to nearest 0.01 mm and counted. Total number of prey items (n) in each fish length class was used to calculate percentage of the total number of food items ingested (F), the frequency of occurrence in stomachs containing food expressed as percentage (FO) and percentage of the relative frequency of occurrence for each prey category (RFO). Moreover, the electivity index by Ivlev (Jacobs 1974) was used to analyze food preference:

$$D = \frac{r - p}{r + p - 2rp},$$

where r is the fraction of a food taxon in the stomach and p the fraction of the same food taxon in the plankton. Environmental copepod abundance data were obtained by the BIONESS from the same *Pleuragramma* samples: both stations 3 and 32 were used to analyze food selection of postlarvae (Zunini Sertorio et al. 1990) and juveniles (Carli et al. 1990), respectively.

## Results

### Ichtyoplankton Abundance and Species Composition

Early life stages of fish collected in all investigated cruises are reported in Table 4. A total of 34 437 fish larvae belonging to 5 families (14 species) were collected in Terra Nova Bay during the Antarctic summer 1987–88. The nototheniids were numerically dominant, representing 98.7% of all larvae. Amongst these, *Pleuragramma antarcticum* was the dominant species sampled both in number (33 964 individuals) and frequency of occurrence in the samples (98.6%). Among channichtyid species the most common icefish captured was *Chionodraco myersi*, composing 1.19% of total ichtyoplankton, while rare or single specimens of other species were collected. As shown in Table 5, in Antarctic summer 1987–88 *Pleuragramma antarcticum* was collected in 138 of the 299 samples, while appearing at all stations except station 1SK. Abundances varied between 0 and 27 075 specimens 100 $m^{-3}$ (mean of positive 0- to 150-m hauls: 437 ± 720 SD). Postlarvae were most abundant and frequent in the samples (99.8% of total catch), while low concentrations of juveniles were found (0.02%).

In late spring-summer 1989–90, a total of 4050 fish larvae of 19 species (8 families) were caught in the western Ross Sea, 97.8% of which in Terra Nova Bay. Nothothenioids were the most commonly occurring group comprising 96.5% of the individuals, and among these postlarvae and juvenile *P. antarcticum* were the dominant components comprising 95.9% of the total catch. The species was very abundant in the inner part of Terra Nova Bay (stations 25, 26 and 27), while only juveniles were collected at stations 13 and 22 in the western Ross Sea. *P. antarcticum* was collected in 41 of the 493 samples. Abundances varied between 0 and 2944 specimens 100 $m^{-3}$ (mean of positive 0- to 140-m hauls: 158 ± 212 SD). The most common icefish caught was *Chionobathyscus dewitti* (0.37%) followed by *Chionodraco rastrospinosus* and *C. hamatus*. Macruridae, Myctophidae, Bathygalidae and Paralepidae were well represented.

In spite of considerable sampling effort, in late spring-summer 1994–95 only 849 specimens of juvenile fishes of 21 species were collected off Victoria Land in the western Ross Sea. Nototheniids were the most diverse group being represented by six species, followed by bathydraconids and channichthyids, both with five species. Artedidraconids, even though they were not recognized at the species level, occurred with three different morphological types. The two last groups found in our samples were paralepidids and macrourids represented the former by *Notolepis annulata*, *N. coatsi*, *Notolepis* sp., and the latter by *Macrourus whitsoni*, respectively. The nototheniids were the most abundant group of fish as absolute abundance, representing 94.3% of all the individuals caught. *Trematomus lepidorhinus* was the most abundant species (67.1%), even though it was sampled in high numbers at only one station, followed by the Antarctic silverfish *P. antarcticum* (26.0%). No postlarvae of *P. antarcticum* were

**Table 4.** Larval fish composition and abundance in the samples collected in Terra Nova Bay and western Ross Sea during the four Italian Antarctic expeditions. Total number of larvae (N) and frequency of occurrence in the samples (F)

| Cruise | | 1987–88 | | 1989–90 | | 1994–95 | | 1995–96 | |
|---|---|---|---|---|---|---|---|---|---|
| Family (S spp.) | Species | N | F (%) | N | F (%) | N | F (%) | N | F (%) |
| Nototheniidae (15) | *Pleuragramma antarcticum* | 33964* | 98.63 | 3885* | 95.91 | 220 | 25.97 | 354380* | 99.79 |
| | *?Trematomus centronotus* | – | – | – | – | – | – | 1 | <0.01 |
| | *?Trematomus eulepidotus* | – | – | – | – | – | – | 41 | 0.01 |
| | *Trematomus lepidorhinus* | 31 | 0.09 | – | – | 570 | 67.30 | 51 | 0.01 |
| | *Trematomus loennbergi* | – | – | – | – | 3 | 0.35 | – | – |
| | *Trematomus newnesi* | 10 | 0.03 | 1 | 0.02 | 1 | 0.12 | 22 | 0.01 |
| | *Trematomus pennellii* | 1 | <0.01 | – | – | – | – | – | – |
| | *Trematomus scotti* | 5 | 0.01 | 6 | 0.15 | 3 | 0.35 | – | – |
| | *Trematomus* sp. | – | – | – | – | – | – | 1 | <0.01 |
| | *Pagothenia* sp. | – | – | – | – | – | – | 1 | <0.01 |
| | *Notothenia kempi.* | – | – | – | – | 2 | 0.24 | – | – |
| | *Notothenia* sp. | – | – | 1 | 0.02 | – | – | – | – |
| | *Nototheniops nudifrons* | – | – | 9 | 0.22 | – | – | – | – |
| | *Nototheniops* sp. | – | – | – | | – | – | 1 | <0.01 |
| | Nototheniidae | – | – | 1 | 0.02 | – | – | 115 | 0.03 |
| Channichthyidae (11) | *Chionodraco hamatus* | – | – | 2 | 0.05 | – | – | 39 | 0.01 |
| | *Chionodraco rastrospinosus* | – | – | 4 | 0.10 | – | – | 274 | 0.08 |
| | *Chionodraco myersi* | 410 | 1.19 | – | – | 14 | 1.65 | – | – |
| | *Pagetopsis macropterus* | 1 | <0.01 | – | – | 2 | 0.24 | – | – |
| | *Pagetopsis maculatus* | 3 | 0.01 | – | – | – | – | 7 | <0.01 |
| | *Dacodraco hunteri* | – | – | – | – | – | – | 58 | 0.02 |
| | *Chaenodraco wilsoni* | 1 | <0.01 | – | – | 1 | 0.12 | – | – |
| | *Cryodraco antarcticus* | – | – | – | – | 2 | 0.24 | 1 | <0.01 |
| | *Chionobathyscus dewitti* | – | – | 15 | 0.37 | 1 | 0.12 | 38 | 0.01 |
| | *Pseudochaenichthys georgianus* | – | – | – | – | – | – | 1 | <0.01 |
| | Channichthyidae | – | – | – | – | – | – | 3 | <0.01 |
| Bathydraconidae (8) | *Racovitzia glacialis* | 1 | <0.01 | 5 | 0.12 | – | – | 15 | <0.01 |
| | *Prionodraco evansii* | – | – | – | – | 2 | 0.24 | – | – |
| | *Akarotaxis nudiceps* | 1 | <0.01 | – | – | 1 | 0.12 | 1 | <0.01 |
| | *Gerlachea australis* | – | – | – | – | 3 | 0.35 | – | – |
| | *Gymnodraco acuticeps* | – | – | – | – | 4 | 0.47 | 16 | <0.01 |
| | *Parachaenichthys charcoti* | – | – | – | – | | | 1 | <0.01 |
| | *Cygnodraco mawsoni* | – | – | – | – | 3 | 0.35 | – | – |
| | *Psilodraco breviceps* | 7 | 0.02 | – | – | | | 1 | <0.01 |
| Artedidraconidae (5) | *Artedidraco mirus* | – | – | – | – | – | – | 10 | <0.01 |
| | *Artedidraco* sp. | 1 | <0.01 | 5 | 0.12 | – | – | 5 | <0.01 |
| | Species 1 | – | – | – | – | 4 | 0.47 | – | – |
| | Species 2 | – | – | – | – | 1 | 0.12 | – | – |
| | Species 3 | – | – | – | – | 3 | 0.35 | – | – |
| Macrouridae (1) | *Macrourus whitsoni* | – | – | 21 | 0.52 | 2 | 0.24 | – | – |
| Muraenolepididae (1) | *?Muraenolepis microps* | – | – | – | – | – | – | 4 | <0.01 |
| Myctophidae (3) | *Electrona antarctica* | 1 | <0.01 | 3 | 0.07 | – | – | 5 | <0.01 |
| | *Electrona carlsbergi* | – | – | 3 | 0.07 | – | – | – | – |
| | *Electrona* sp. | | | 16 | 0.40 | – | – | 1 | <0.01 |
| Bathylagidae (1) | *Bathylagus antarcticus* | – | – | 8 | 0.20 | – | – | 2 | 0.00 |
| Paralepididae (3) | *Notolepis annulata* | – | – | 5 | 0.12 | – | – | 14 | 0.00 |
| | *Notolepis coatsi* | – | – | 4 | 0.10 | 5 | 0.59 | 9 | 0.00 |
| | *Notolepis* sp. | – | – | 56 | 1.39 | – | – | – | – |
| Total specimens | | 34437 | | 4038 | | 847 | | 355118 | |
| Total species | | 14 | | 19 | | 21 | | 30 | |
| Number of hauls | | | 33 | | 27 | | 26 | | 24 |
| Number of samples | | | 299 | | 493 | | 26 | | 24 |
| Positive samples for *P. antarcticum* | | | 138** | | 41 | | 5 | | 5 |
| * Comprising damaged individuals n.: | | | 124 | | 15 | | | | 29 |

** Comprising 8 samples with damaged individuals

**Table 5.** Mean (and range) standard lengths (mm) and standard deviation (SD±) of the age groups of *Pleuragramma antarcticum* calculated from the length frequencies (N = number of fishes measured)

| Month | Age 0 | | | | | Age 1+ | | | | |
|---|---|---|---|---|---|---|---|---|---|---|
| | Days | N | Mean SL (mm) | SD± | Range (mm) | Days | N | Mean SL (mm) | SD± | Range (mm) |
| December, 1989 | – | – | – | – | – | 26 | 2 | 47.00 | – | 47 |
| December, 1994 | – | – | – | – | – | 15–20 | 212 | 38.04 | 4.07 | 28–49 |
| January, 1988 | 5–29 | 1582 | 12.68 | 1.93 | 8–20 | 10–30 | 42 | 42.55 | 3.67 | 36–48 |
| January, 1990 | 5–12 | 441 | 12.12 | 0.84 | 9–15 | 4–27 | 45 | 40.98 | 4.18 | 34–46 |
| January, 1996 | 13–28 | 184 | 13.10 | 2.68 | 7–19 | 12–28 | 62 | 42.10 | 5.31 | 31–51 |
| February, 1988 | 2 | 159 | 15.11 | 2.02 | 10–20 | 2 | 26 | 44.27 | 3.54 | 37–50 |
| February, 1996 | 2–5 | 2019 | 14.14 | 1.35 | 10–20 | – | – | – | – | – |

caught in this cruise, while very low abundances of juveniles were found in five stations of the northermost part of the western Ross Sea (mean of positive hauls: 0.24 ind 100 $m^{-3}$ ± 0.40 SD).

In Antarctic summer 1996, a very high number of postlarvae and juvenile fish (355 117 specimens of 30 species) were collected in the western Ross Sea, from Cape Adare to the Ross Ice Shelf. Postlarvae *Pleuragramma antarcticum* predominated overwhelmingly (99.8%) only at four stations located in Terra Nova Bay (stations 4, 21, 23 end 25). Abundances varied between 0 and 866 specimens 100 $m^{-3}$, while the species was rare or absent in the western Ross Sea. Low number of juvenile *P. antarcticum* were found in offshore waters.

Probably the differences in both abundance and species composition during the four cruises were partly due to different fishing methods.

## Length

During the Antarctic summer 1988, in Terra Nova Bay, the size of postlarvae ranged between 8 and 20 mm SL, based on a subsample of 1741 *Pleuragramma antarcticum*. Only 67 specimens were between 36 and 53 mm SL (mean 43.4 ± 3.19 SD). The length-frequency distribution in shown in Fig. 3a. The length of the majority of postlarvae ranged between 10 and 15 mm (mean 12.9 ± 2.06 SD) and a bimodal distribution with maxima at 14–15 mm and 45 mm was found. Length differences seem to be correlated to the geographical sites of larval fish capture (Guglielmo et al. 1998).

In late spring-summer 1989–90, a bimodal SL distribution was found. The postlarvae ranged between 9 and 15 mm SL (mean 12.1 ± 0.84 SD) based on a subsample of 441 specimens, while only 47 specimens between 34 and 50 mm (mean 41.2 ± 4.26 SD) were collected (Fig. 3b). Only one specimen was 57 mm SL (station 22, 0–500 m).

In late spring 1994, only juveniles were caught in the western Ross Sea by PHN net. The length-frequency distribution based on the total of specimens collected (n = 212) showed a unimodal diagram. Juveniles ranged between 28 and 49 mm SL (mean 38.0 ± 4.07 SD) (Fig. 3c).

In the Antarctic summer of 1996, the postlarvae ranged between 7 and 23 mm SL based on a subsample of 2203 specimens (mean 14.1 ± 1.53 SD, Fig. 3d). Few individuals were between 7 and 10 mm SL (1.9%), and only eight specimens between 20 and 23 mm SL. A total of 62 juveniles between 28 and 51 mm were found (mean 42.1 ± 5.31 SD), while only two specimens of 73 and 75 mm SL, respectively, were collected. In the same cruise, 25 adults between 100 and 190 mm SL were caught near the Ross Ice Shelf.

The sizes of *Pleuragramma antarcticum* collected by various gear both in the western Ross Sea and Terra Nova Bay ranged between 7 and 190 mm SL. The length-frequency distribution of *P. antarcticum* subsampled in the four cruises from the whole investigated area is shown in Fig. 3e. Postlarvae (n = 4385) ranged between 7 and 23 mm SL (mean 13.9 ± 1.62 SD) and juveniles (n = 388) were between 28 and 53 mm in length (mean 40.0 ± 4.78 SD).

## Age and Growth

Two size groups can be recognized from observed size-frequency distributions recorded by the

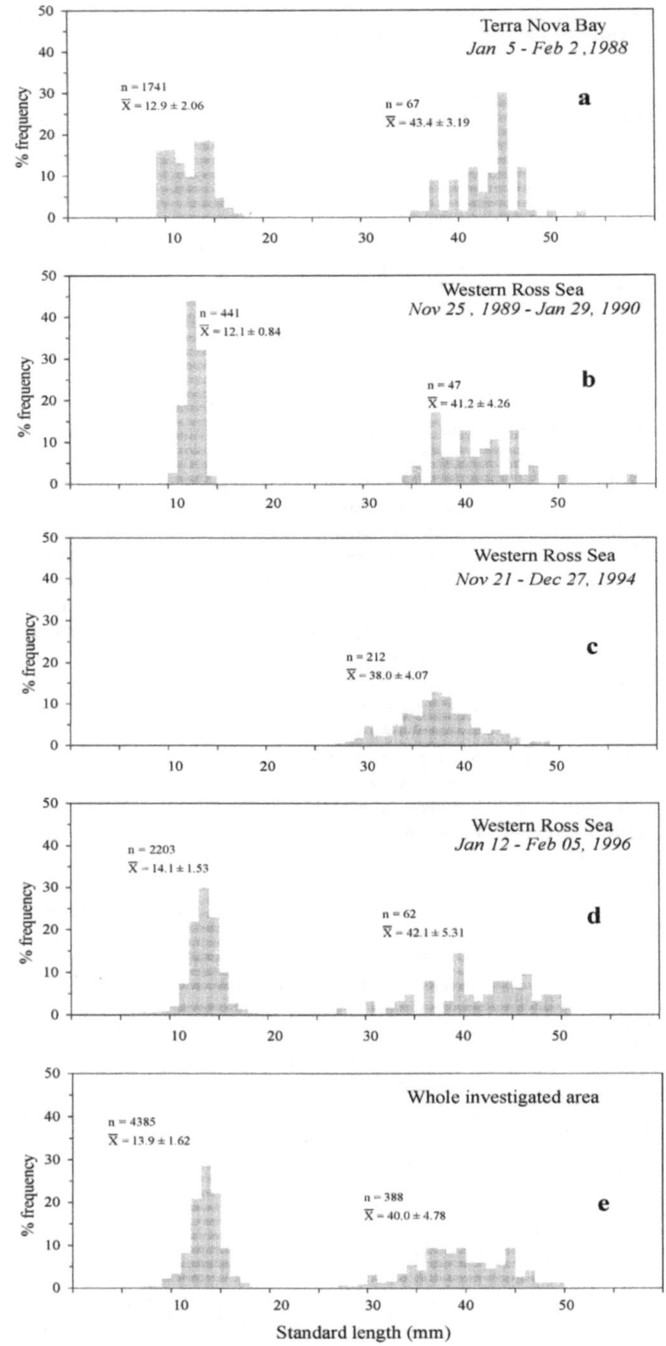


**Fig. 3a–e.** Length-frequencies distribution of postlarval and juvenile *Pleuragramma antarcticum* in Terra Nova Bay and western Ross Sea during the cruises. Number of fishes measured (*n*), mean standard length ($\bar{X}$)and standard deviation (±)

investigations. Postlarvae with a peak between 7 and 23 mm SL and juveniles with a peak between 28 and 53 mm conform to age classes 0 and 1+, respectively (Hubold 1984). The two specimens of 73 and 75 mm SL, respectively, collected in the western Ross Sea during the austral summer 1996 can be attributed to juveniles of age 2+.

Average growth rates over a period of 1 year may be estimated by using the differences between mean lengths of cooccurring postlarvae and juveniles. For the two age groups 0 and 1+ present in summer 1988, a difference of 30.5 mm (Fig. 3a) suggests an average growth rate of 0.083 mm day$^{-1}$ for the 1st year. A growth rate of 0.079 mm day$^{-1}$ results from a difference of 29.1 mm in late spring-summer 1990 (Fig. 3b). In summer 1996, the difference was 28.0 mm, giving an average growth rate of 0.076 mm day$^{-1}$ (Fig. 3d).

Daily growth rates of postlarval *Pleuragramma antarcticum* during the summer 1987–88 in Terra Nova Bay, as calculated from the modal length distribution during 29 days, were about 0.21 mm (Guglielmo et al. 1998). Figure 4 shows the single daily length frequency distribution of *P. antarcticum* during the summer 1995–96 in the southern Western Ross Sea. The first histogram has a maximum at 11 mm. Sixteen days later (January 28 1996) the former 11-mm length group was observed in the class size of 15 mm. Based on modal values, the mean daily increment of *P. antarcticum* postlarvae, over a 16-day time period, was about 0.25 mm. Considering that new larvae may hatch at a length of 6 mm (Regan 1916), postlarvae caught during the period 13–28 January 1996 were about 3–5 weeks old and had hatched in mid- to late December.

Postlarvae caugth in January 1990 (Fig. 5) confirm previous results (Guglielmo et al. 1998) that two hatching periods of *P. antarcticum* occurred in Terra Nova Bay between early and late December.

Mean (and range) standard lengths of postlarvae (age 0) and juveniles (age 1+) from the respective months are given in Table 5. The range of postlarvae was 8–20 mm in the period 5–29 January 1988 (mean: 12.7 ± 1.93 SD), 9–15 mm in the period 5–12 January 1990 (mean: 12.1 ± 0.84) and 7–19 mm in the period 13–28 January 1996 (mean: 13.1 ± 2.68). From 2 to 5 February 1996 the range of postlarvae were 10–20 mm (mean: 14.1 ± 1.35). The range of juveniles were 28–49 mm (mean: 38.0 ± 4.07) in the period 15–20 December 1994, 36–48 mm (mean: 42.5 ± 3.67) in the period 10–30 January 1998, 34–46 mm (mean: 41.0 ± 4.18) in the period 4–27 January 1990 and 31–51 mm from 12 to 28 January 1996 (mean: 42.1 ± 5.31). A range of 37–50 mm (mean: 44.3 ± 3.54) was found in juveniles captured 2 February 1996.

## Horizontal Distribution

Highest abundances of *P. antarcticum* postlarvae occurred in Terra Nova Bay. In summer 1988 most specimens were caught at stations 3 and 5 near Cape Washington (mean: 2108 ind 100 m$^{-3}$ ± 179 SD), at station 16 off Cape Washington (1346 ind 100 m$^{-3}$) and at station 35 near the Drygalski Ice Tongue (1403 ind 100 m$^{-3}$). Lower densities occurred at five stations (mean: 296 ind 100 m$^{-3}$ ± 280 SD) located between Cape Washington and Campbell Glacier Tongue (station 12), farther offshore Terra Nova Bay (station 17), in the central part of Terra Nova Bay (station 48) and in the southermost part of the Drygalski Ice Tongue (stations 36 and 46). Most of the lower densities were recorded in the northern basin of Terra Nova Bay between Campbell Glacier Tongue and Cape Russell (mean: 31 ind 100 m$^{-3}$ ± 23 SD).

Considerably fewer postlarvae were caught during 1989–90 and 1995–96, while no specimens were found during the 1994–95 cruise. Maximum densities were recorded in the inner stations near Drygalski Ice Tongue (station 26, 20–30 m, 2944 ind 100 m$^{-3}$ January 1990; station 23, 700–0 m, 542 ind 100 m$^{-3}$ February 1996) and off Terra Nova Bay (station 25, 50–0 m, 1099 ind 100 m$^{-3}$ January 1990; station 25, 800–0 m, 865 ind 100 m$^{-3}$ February 1996). Postlarvae were absent or rarely caught in waters north of Cape Washington, south of Drygalski Ice Tongue and in offshore waters beyond 170°E.

The maximum catches of age 0 individuals were made in waters ranging in depths from 187–1040 m, while the lowest concentrations seemed to coincide with the central basin shelf with a depth range between 257 and 700 m (mean 532 m). Figure 6 shows the standardized abundance of *P. antarcticum* postlarvae in 1987–88 and 1989–90 from BIONESS samples.

The older postlarvae and juveniles were much less abundant than postlarvae, but were frequently encountered in the whole investigated area. They were absent or rare in the central basin shelf area and in the northern portion of Terra Nova Bay. Low concentrations were encountered in the

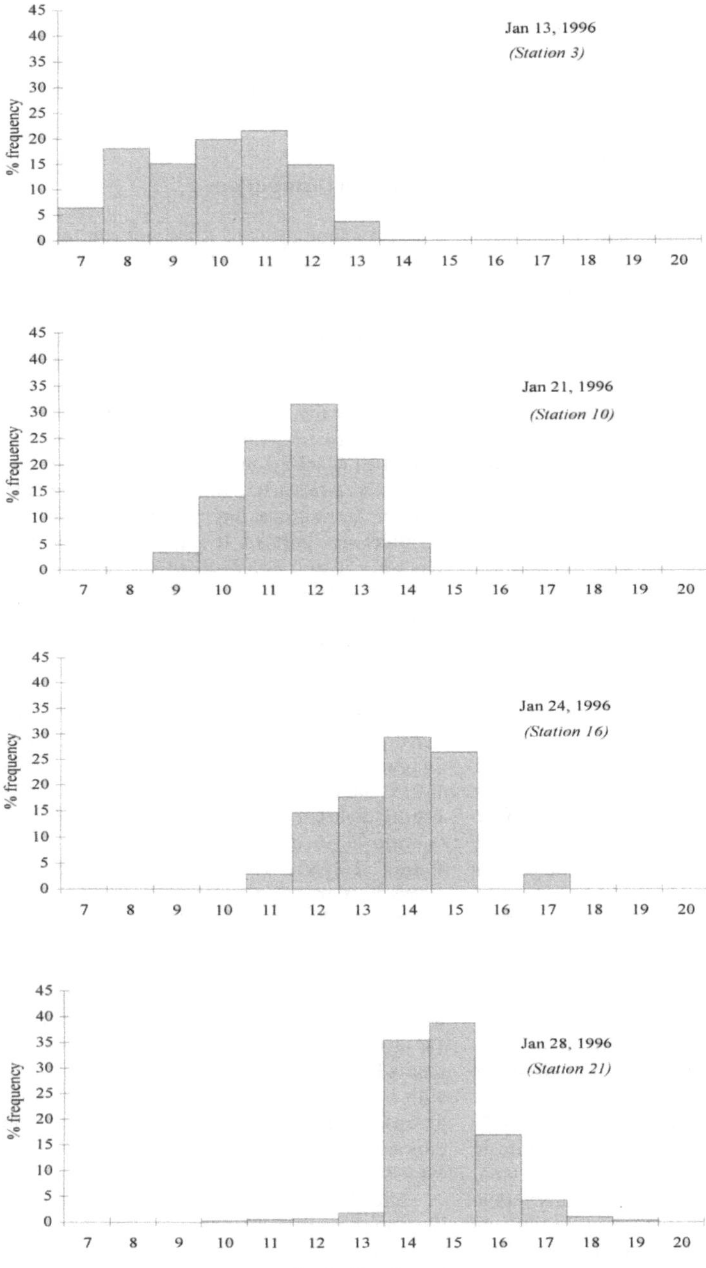


**Fig. 4.** Temporal length-frequencies distribution of postlarval *Pleuragramma antarcticum* during the summer 1996 in the western Ross Sea

**Fig. 5.** Temporal length-frequencies distribution of postlarval *Pleuragramma antarcticum* during the summer 1990 in Terra Nova Bay

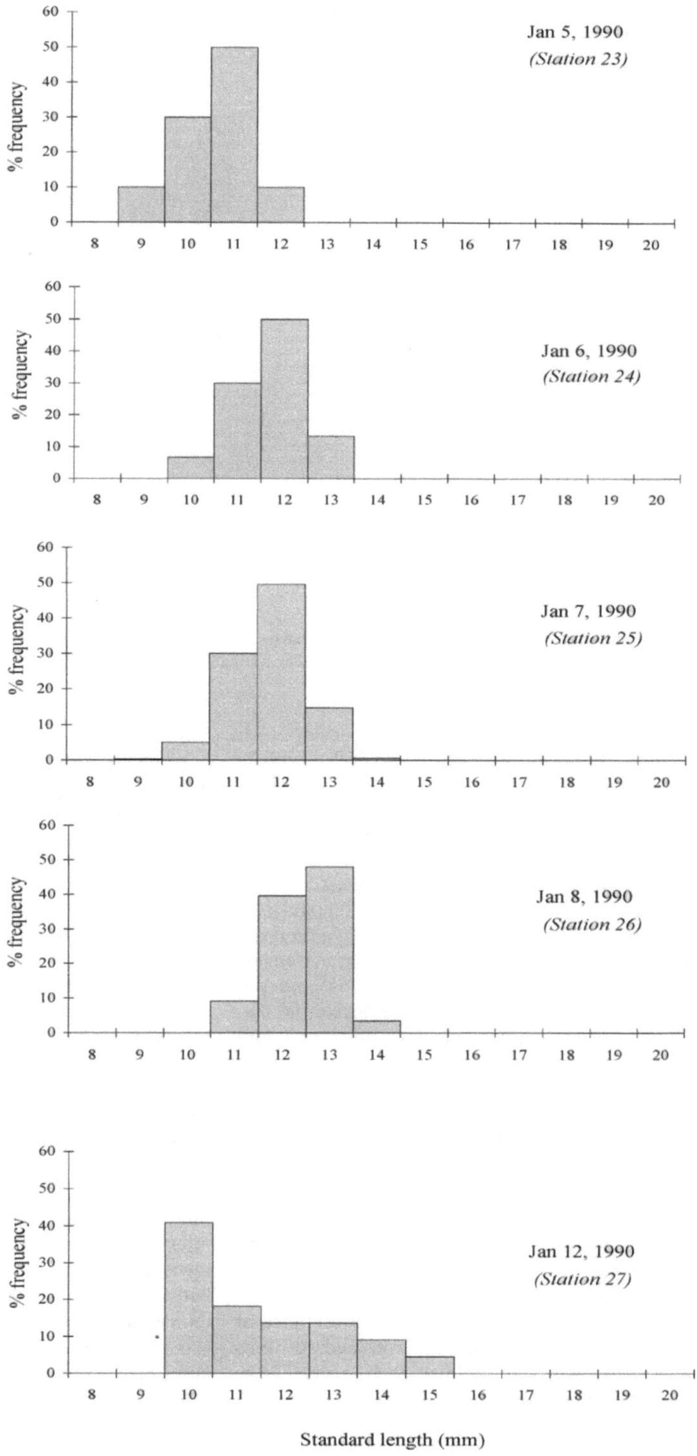

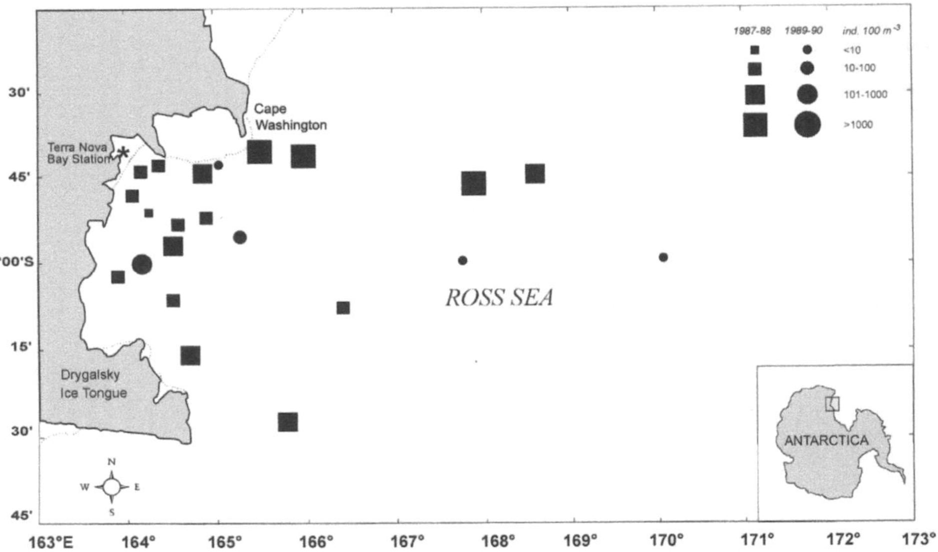


**Fig. 6.** Standardized horizontal distribution and abundance (N $100\,m^{-3}$) of postlarval *Pleuragramma antarcticum* collected by the BIONESS in the 0- to 150-m layer in Terra Nova Bay during the 1987–88 and 1989–90 cruises

1987–88 expedition between 0 and 150 m both in the southern part of Terra Nova Bay at station 32 (20 ind $100\,m^{-3}$) and near the Drygalski Ice Tongue at station 36 (2 ind $100\,m^{-3}$). This area seemed to coincide with the deeper trenches in the shelf. In offshore waters *P. antarcticum* juveniles were recorded in lower densities between 0.1 and 12.5 ind $100\,m^{-3}$ (mean $1.66 \pm 2.88$ SD) and their occurrence was restricted to stations along a transect north-south between 170° and 175°E (depth range: 751 m.). Figure 7 shows the horizontal distribution in abundances of older postlarvae and juveniles in the whole investigated area with respect to total water depth sampled. Very few individuals of age 2+ were found, while adults were caught only near the Ross Ice Shelf.

## Vertical Distribution

Only BIONESS hauls carried out both during the Polar Queen 1987–88 and Cariboo 1989–90 cruises provide information on the vertical distribution of the early stages. Since the main abundance of *P. antarcticum* postlarvae was confined to Terra Nova Bay, two stations for each cruise were selected from the investigated area.

*Pleuragramma antarcticum* postlarvae were found throughout the water column sampled but were most abundant in the upper 150 m (Fig. 8). Within the 0- to 150-m layer, the maximum abundance occurred at 0–50 m (1126 ind $100\,m^{-3}$ $\pm 1897$ SD, 1987–88; 407 ind $100\,m^{-3} \pm 545$ SD, 1989–90). *P. antarcticum* showed a well-defined vertical distribution pattern. Abundances decreased from the surface to 200-m depth, while very moderate densities were found below 200 m.

A marked water stratification was found in the whole investigated area. A seasonal thermocline can be traced by the 0.5 °C isotherm at 40–60 m in the northern part, by the −1 °C isotherm at 60–70 m in the central part and by the 0 °C isotherm at 20–40 m in the southern part of Terra Nova Bay. The abundance peaks of postlarvae were nearly always close to the thermocline, while juveniles were often distributed at greater depths. Juveniles were distributed within the top 200 m, but most catches of single specimens were made between 150- and 300-m depth. Increased abundance was recorded both at the southern station 32 (Polar Queen cruise, 1987–88) where the density amounted to 38.5 specimens $100\,m^{-3}$ in the 50–100 m layer and in the offshore station 22 during

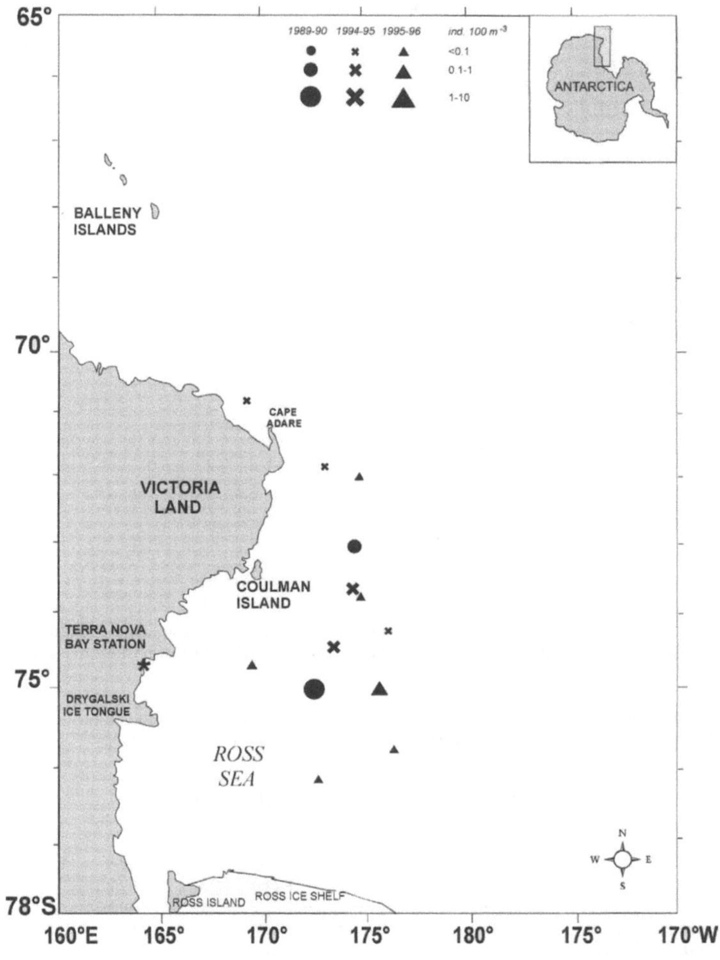


**Fig. 7.** Horizontal distribution and abundance (N $100\,m^{-3}$) of juvenile *Pleuragramma antarcticum* in the western Ross Sea during the 1989–90, 1994–95 and 1995–96 cruises. Abundances are combined values in respect to the total of sampled strata

the Cariboo 1989–90 cruise (70.4 ind $100\,m^{-3}$ in the 30–40 m layer).

Analysis of the size frequency distribution of the age 0 group of fish caught during the 1989/90 cruise confirms the trend in the depth distribution by size found in Terra Nova Bay during summer 1987/88 (Guglielmo et al. 1990). It was evident that early stages occupied the entire water column sampled, while older postlarvae inhabited less deep layers. The mean size increased with depth from the surface to 150 m and then seemed to decrease from 150 to 600 m depth with the largest postlarvae found in the 100 to 150 m layer (mean: 15.17 mm ± 2.04 SD).

## Feeding Habits

The diets of three size classes of postlarval *Pleuragramma antarcticum* were examined (Table 6). Of the 280 fishes investigated, 260 specimens (92.8%) contained food. The diet consisted of six distin-

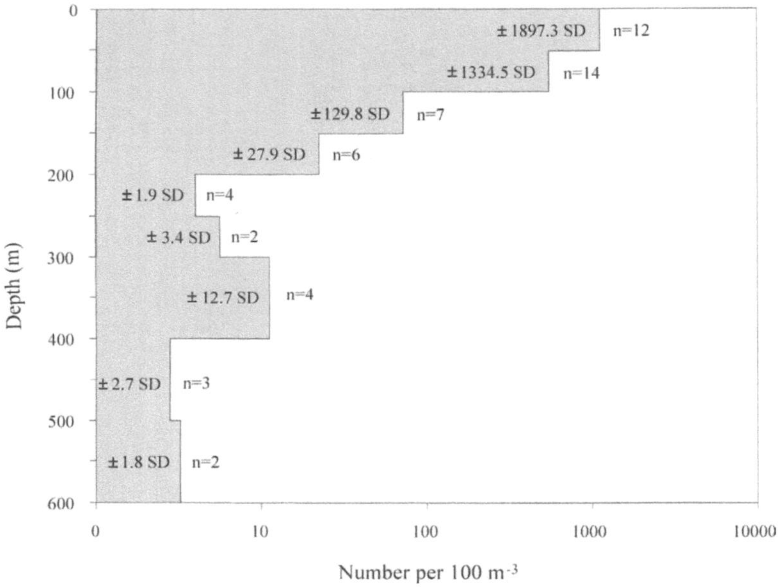


**Fig. 8.** Vertical profile of postlarval *Pleuragramma antarcticum* mean abundance (N $100\,m^{-3}$) for 50 (100)-m depth strata in Terra Nova Bay during the 1987–88 Italian Antarctic expedition. Number of positive hauls (*n*) and standard deviation (±*SD*)

**Table 6.** Food composition of three size classes of postlarval *Pleuragramma antarcticum* at Terra Nova Bay in summer 1998. Number of items (n) and percentage of total number of items (F)

| Prey items | *Pleuragramma antarcticum* size classes (SL mm) | | | | | |
|---|---|---|---|---|---|---|
| | 8–11 | | 12–14 | | 15–17 | |
| | n | F (%) | n | F (%) | n | F (%) |
| *Oncaea curvata* | 5 | 1.9 | 14 | 7.3 | 33 | 11.9 |
| *Metridia gerlachei* C II | | | 4 | 2.1 | 6 | 2.2 |
| Calanoid eggs | 75 | 28.5 | 103 | 53.9 | 82 | 29.5 |
| Copepod nauplii | 7 | 2.7 | 12 | 6.3 | 27 | 9.7 |
| Tintinnids | 50 | 19.0 | 33 | 17.3 | 46 | 16.5 |
| *Limacina* sp. | 126 | 47.9 | 25 | 13.1 | 84 | 30.2 |
| Σ prey items | 263 | | 191 | | 278 | |
| Stomachs examined | 91 | | 99 | | 90 | |
| Stomachs empty | 3 | | 8 | | 9 | |

guishable items, including two species of copepods, calanoid eggs, copepod nauplii, tintinnids and *Limacina* sp. The diet was dominated by calanoid eggs (35.5%), *Limacina* sp. (32.1%) and tintinnids (17.6%). The cyclopoid copepod *Oncaea curvata* and copepod nauplii contributed 7.1 and 6.3%, respectively. *Metridia gerlachei* (copepodite stage II) occurred only occasionally (1.4%). Prey sizes were between 0.13 and 0.90 mm (Table 7).

The stomach contents of three size classes of juvenile *P. antarcticum* are summarized in Table 8. Of the 71 fishes investigated, 65 specimens (91.5%) contained food. The principal food consisted of copepods with 98.2% of items. *Oncaea curvata*

(85.1%) was the most abundant and the most frequently consumed prey; second in abundance were copepodites III–IV of the calanoid *Metridia gerlachei* (11.4%). *Oithona similis* (1.7%) and euphausiid larval stages (1.7%), most probably of *Euphausia crystallorophias*, were ingested occasionally.

Prey sizes in *P. antarcticum* guts were clearly related to fish lenghs. Postlarval fishes from 8–17 mm fed primarily on prey between 0.13 and 0.51 mm length (Table 7, calanoid eggs, tintinnids and *Limacina*), while juvenile *P. antarcticum* from 36 to 50 mm SL switched to larger prey between 0.34 and 1.73 mm (Table 9, *Oncaea curvata* and *Metridia gerlachei* C III–IV). In this size range the prey length mode at 0.6 mm was due to the large number of cyclopoid copepod *Oncaea curvata* in the guts (85.1%). The much larger *Metridia gerlachei* C III–IV (1.41–1.73 mm length) were of minor importance numerically but had a

**Table 7.** Food composition of postlarval *Pleuragramma antarcticum* at Terra Nova Bay in summer 1988. Prey items' size range (mm), number of items (n), percentage of total number of items (F), percent frequency of occurrence (FO) and percent of relative frequency of occurrence (RFO)

| Prey items | Size (mm) | n | F % | FO (%) | RFO (%) |
|---|---|---|---|---|---|
| *Oncaea curvata* | 0.20–0.49 | 52 | 7.1 | 11.9 | 10.1 |
| *Metridia gerlachei* C II | 0.60–0.90 | 10 | 1.4 | 3.5 | 2.9 |
| Calanoid eggs | 0.13–0.18 | 260 | 35.5 | 48.5 | 40.9 |
| Copepod nauplii | 0.20–0.80 | 46 | 6.3 | 11.5 | 9.7 |
| Tintinnids | 0.27–0.51 | 129 | 17.6 | 22.3 | 18.8 |
| *Limacina* sp. | 0.30–0.48 | 235 | 32.1 | 20.8 | 17.5 |

**Table 8.** Food composition of three size classes of juvenile *Pleuragramma antarcticum* at Terra Nova Bay in summer 1998. Number of items (n) and percentage of total number of items (F)

| Prey items | *Pleuragramma antarcticum* size classes (SL mm) | | | | | |
|---|---|---|---|---|---|---|
| | 36–41 | | 42–44 | | 45–50 | |
| | n | F (%) | n | F (%) | n | F (%) |
| *Oncaea curvata* | 261 | 85.6 | 381 | 91.8 | 84 | 63.2 |
| *Oithona similis* | 6 | 2.0 | 7 | 1.7 | 2 | 1.5 |
| *Metridia gerlachei* C III-IV | 34 | 11.1 | 23 | 5.5 | 40 | 30.1 |
| *E. crystallorophias* calyptopes | 1 | 0.3 | 1 | 0.2 | 1 | 0.7 |
| *E. crystallorophias* furciliae | 3 | 1.0 | 3 | 0.7 | 6 | 4.5 |
| Σprey items | 305 | | 415 | | 133 | |
| Stomachs examined | 26 | | 21 | | 24 | |
| Stomachs empty | 3 | | 0 | | 3 | |

**Table 9.** Food composition of juvenile *Pleuragramma antarcticum* at Terra Nova Bay in summer 1988. Prey items' size range (mm), number of items (n), percentage of total number of items (F), percent frequency of occurrence (FO) and percent of relative frequency of occurrence (RFO)

| Prey items | Size (mm) | n | F % | FO (%) | RFO (%) |
|---|---|---|---|---|---|
| *Oncaea curvata* | 0.34–0.80 | 726 | 85.1 | 36.9 | 30.4 |
| *Oithona similis* | 0.35–0.69 | 15 | 1.7 | 10.8 | 8.8 |
| *Metridia gerlachei* C III–IV | 1.41–1.73 | 97 | 11.4 | 58.5 | 48.1 |
| *E. crystallorophias* calyptopes | 1.07–1.91 | 3 | 0.3 | 4.6 | 3.8 |
| *E. crystallorophias* furciliae | 2.81–2.92 | 12 | 1.4 | 10.8 | 8.9 |

frequency of occurrence of 58.5% and were frequently found in guts of *P. antarcticum* from 45 to 50 mm SL (30.1%). In the same size class larval stages of *Euphausia crystallorophias* (mainly furciliae 2.81–2.92 mm length), started to become an important prey (FO 10.8%).

Station 3 near Cape Washington in Terra Nova Bay was used to analyze food selection of postlarval *P. antarcticum* (Table 10). Calanoid eggs were not considered here because they were not covered by the mesh size of the BIONESS nets due to their small diameter. Positive selection was revealed for *Oncaea curvata*, while preference was weak for *Metridia gerlachei* CII. Though the mollusc *Limacina* sp. was an important component of the diet, it did not occur in the diets in the same proportion as that found in the natural zooplankton.

Food selection by juvenile *P. antarcticum* was analyzed at station 32 on the southern part of Terra Nova Bay (Table 10). Positive selection was very strong for the cyclopoid *Oncaea curvata* and low preference was shown for the calanoid *Metridia gerlachei* CIII–IV. The small cyclopoid *Oithona similis* showed high indices of negative selection.

## Discussion

The ichthyoplankton community of the Ross Sea was described for the first time by Regan (1916). He reported a total of six species during late spring and summer in the Ross Sea and Terra Nova Bay, most of which were recently revised (North and Kellermann 1989; Loeb et al. 1993). Our data strongly increase the number of species representing the larval fish assemblages of the western Ross Sea during the Antarctic late spring-summer. A total of 20 species were collected in Terra Nova Bay during the four Italian Antarctic cruises and 39 species in the western Ross Sea, from Cape Adare to the Ross Ice Shelf, of which only 16 species were in common with Terra Nova Bay samples. White and North (1987) reported a total of 29 species, including several oceanic species, in samples collected around the South Shetland Islands. Hubold (1990) reports a total of about 23 ichthyoplankton species, for the period October to February, in the Weddell Sea and Kellermann (1989) found a total of 26 fish larvae species during spring-summer in the Antarctic Peninsula.

Nototheniids were predominant in the whole study area, representing 97.3% of all larvae. *Pleuragramma antarcticum* was the dominant species sampled in spring-summer cruises 1987–88 (98.6%), 1989–90 (95.9%) and 1995–96 (99.8%), while it predominated in association with *Trematomus lepidorhinus* in 1994–95. *P. antarcticum* was also dominant in other sites of the Ross and Weddell Seas (Regan 1916; De Witt 1970; Hubold 1984, 1985, 1990, 1991; White and Piatkowski 1993; Guglielmo et al. 1998) and off the Antarctic Peninsula (Kellermann 1986a).

Maximum abundances of *P. antarcticum* postlarvae were confined to Terra Nova Bay. The mean relative abundance in the whole area in summer 1988, for positive 0- to 150-m hauls, was 434 (±720 SD) specimens 100 m$^{-3}$ (Guglielmo et al. 1998) and 158 (±212 SD) specimens 100 m$^{-3}$ in January 1990, for positive 0–140-m hauls. Off Terra Nova Bay, postlarval abundance did not exceed 5–86 ind 100 m$^{-3}$. High concentrations were found near Cape Washington in summer 1988 (1126 ind

**Table 10.** Food selection of postlarval and juvenile *Pleuragramma antarcticum* at stations 3 and 32, respectively, in Terra Nova Bay (summer 1988). Numbers (*n = 74; **n = 24) and fractions (r) of food items in the stomachs; abundance of food items per 100 m$^3$ (m) and fractions of items (p) in the plancton (*150–0 m, **200–0 m). Index of food selection (D) according to Jacobs (1974)

| Food taxa | n | r | m | p | D |
|---|---|---|---|---|---|
| Station 3* Postlarvae | | | | | |
| *Oncaea curvata* | 12 | 0.16 | 286 | 0.02 | 0.82 |
| *Metridia gerlachei* C II | 4 | 0.06 | 792 | 0.05 | 0.09 |
| *Limacina* sp. | 60 | 0.79 | 16474 | 0.94 | −0.61 |
| Station 32** Juveniles | | | | | |
| *Oncaea curvata* | 9 | 0.3 | 64 | 0.008 | 1.00 |
| *Oithona similis* | 4 | 0.13 | 3976 | 0.48 | −0.71 |
| *Metridia gerlachei* C III–IV | 17 | 0.57 | 4223 | 0.51 | 0.12 |

100 m$^{-3}$; 50–0 m) and in the inner stations close to Drygalski Ice Tongue in summers 1988, 1990 and 1996. In the southern and eastern Weddell Sea, postlarval abundance in January and February amounted up to 30–370 ind 100 m$^{-3}$ during several years with means of 15–34 100 m$^{-3}$ (Keller 1983; Hubold 1984). As reported by Kellermann (1986a), off the Antarctic Peninsula the abundance of postlarvae was 3.3–14.3 100 m$^{-3}$ and means were 0.5–1.6 100 m$^{-3}$ in February 1976 and 1982, and 7.7 100 m$^{-3}$ in March 1981.

In this study, juvenile *Pleuragramma antarcticum* showed a more oceanic occurrence than postlarvae of the same species. Juveniles were widely dispersed in the whole investigated area. Low concentrations between 0.1 and 12.5 ind 100 m$^{-3}$ (mean: 1.66 ± 2.88 SD) were recorded in offshore waters between 170° and 175°E and in the southern part of Terra Nova Bay. This area seemed to coincide with deeper trenches in the shelf.

The observed length-frequency distributions of *P. antarcticum* subsampled in the four cruises from the whole investigated area are in good agreement with data from the literature. A bimodal length distribution was found in Terra Nova Bay in late spring-summers 1987–88, 1989–90 and 1995–96. Postlarvae (n = 4385) ranged between 7 and 23 mm SL (mean 13.9 ± 1.62 SD) and juveniles (n = 388) were between 28 and 53 mm SL (mean 40.0 ± 4.78 SD). According to Regan (1916) and De Witt and Tyler (1960), the observed length ranges in the western Ross Sea were 10–20 mm (mean: 14.1 ± 1.35 SD) for postlarvae and 37–50 mm (mean: 44.3 ± 3.54 SD) for juveniles in early February. These data agree with other Antarctic regions (Keller 1983; Hubold 1984, 1985; Kellermann 1986a,b).

A growth rate of 0.08 mm day$^{-1}$ was confirmed for the 1st year as reported from Terra Nova Bay by Guglielmo et al. (1998) and from the Antarctic Peninsula by Kellermann (1986b). The daily increments of *Pleuragramma antarcticum* postlarvae estimated from sequential sampling in 1987–88 from Terra Nova Bay (Guglielmo et al. 1998) and in 1995–96 from the southern western Ross Sea were about 0.21–0.25 mm. These values agree with the larval growth rates reported from the Southern Weddell Sea (0.20–0.24 mm) by Keller (1983) and Hubold (1985), respectively.

The co-occurrence of different postlarval size classes during the investigations confirm the pevious hypothesis (Guglielmo et al. 1998) that two hatching periods occurred in Terra Nova Bay between early and late December. Regan (1916) stated that hatching of *P. antarcticum* in the Ross Sea occurred in early December and newly hatched larvae may be 6 mm. De Witt and Tyler (1960) presumed there was a longer hatching season, from early October to late December.

The presence of early stages in January 1988, 1990 and 1996 (7–9 mm SL) confirms previous data (Regan 1916; Guglielmo et al. 1998) that *Pleuragramma antarcticum* spawn in Terra Nova Bay. It is generally accepted that spawning occurs in the vicinity of continental ice shelves and spawning success is related to the presence and extent of a coastal polynya (Hubold 1984). In the southwestern Ross Sea the coast is almost enterily fringed by continental ice shelves. Permanent sea ice persists in some areas (e.g. Tethys Bay), while in winter seasonal sea ice covers the entire area except for zones where katabatic winds or upwelling permit development of polynyas. Terra Nova Bay was characterized by continental ice shelves and polynyas persisting also during winter-time (Bromwich and Kurtz 1984). Highest concentrations of postlarvae occurred at the edge near Cape Washington and in the inshore waters of Drygalski Ice Tongue. These areas seem to represent a nursery ground of early stages, where the presence of permanent polynyas could provide favourable food conditions for the development of the first stages of life.

*Pleuragramma antarcticum* is well adapted to a pelagic life (Eastman and De Vries 1982) and also has floating eggs. Eggs and larvae (with yolk sac) have not been found during the investigations. Due to the difficulty of sampling for the severe ice conditions, very little is known about spawning time, development time of the eggs and hydrography under winter sea ice. The eggs are probably spawned in winter (Hubold 1984) and spawning success is subject to year to year variations in relation to sea-ice conditions (Kellermann 1986a).

The first hatching period of *Pleuragramma antarcticum* occurred in Terra Nova Bay in early December (Regan 1916; Guglielmo et al. 1998), probably in the vicinity of the coastal polynya when the bay was completly covered by sea ice. During the 1987–88 cruise, a wide open polynya was present at Terra Nova Bay. As a typical characteristic of the Ross Sea, the ice-free surface propagates from south to north, with an early increase of the ice-free water exposed to sunlight and also an increase to the length of the ice edge (Hecq et al. 1992, 1993). Hatching may coincide with the stabilization of the water column (Hubold 1984).

After hatching, the larvae rise to the surface layer where feeding conditions are favourable. Thus, the youngest planktonic stages of *P. antarcticum* concentrate in the surface mixed layer of the northern and southern areas of Terra Nova Bay where chlorophyll a attains maximum values at 25–50 m near Cape Washington ice edge and at 0–10 m in inshore waters of Drygalski Ice Tongue (1.2–1.6 mg $m^{-3}$ and 2.0–2.4 mg $m^{-3}$, respectively, Innamorati et al. 1990).

More than 62% of the postlarvae were collected in well-stratified Antarctic Surface Water (AASW) with temperature values that exceed 0 °C and where vertical distribution appears to be related to the summertime thermocline. The AASW was situated towards the edge of Cape Washington and the Drygalski Ice Tongue, flowing along the coast. The accumulation of postlarvae in the frontal systems and eddies can be explained by this drift. From these sites, early stages were drifted by a westward-flowing current of the limb of the Antarctic coastal current and the southern limb of the Ross Sea Gyre. These currents become part of the clockwise gyre in Terra Nova Bay. The two interacting components give rise to high amplitude gyres that are favoured by the coastline. The spatial and temporal variability was very high. Periodicity at 1 and 5 days was associated, respectively, with diurnal tides and atmospheric forcing. These large gyres favoured postlarval retention mostly in northern and southern areas of Terra Nova Bay, preventing the northward drift to Cape Washington, south of Drygalski Ice Tongue and in offshore waters beyond 170°E.

According to Hubold (1984), our data demonstrate that the larger percentage of age 0 postlarvae remain during the first summer near the surface, where the presence of coastal Warm Surface Water (WSW), with temperatures exceeding 0 °C, and a well-established stratification provide a favourable environment for plankton development. White and Piatwowski (1993) stated that, in addition to occurring under favourable feeding conditions, the larval stages would gain a significant growth advantage by occupying the warmer strata.

Thus, the youngest planktonic stages of *Pleuragramma antarcticum* concentrate in the surface mixed layer of Terra Nova Bay where feeding conditions are favourable. In the size classes 8–17 mm the diet was dominated by calanoid eggs (35.5%), *Limacina* (32.1%) and tintinnids (17.6%). The cyclopoid copepod *Oncaea curvata* contributed to 7.1% of the diet. Prey sizes were clearly related to larval fish lengths and were between 0.13 and 0.51 mm. It is known that the early larvae of many nothothenoids feed on prey 0.05–0.75 mm (Balbotin et al. 1986; Kellermann 1986b, 1987; North and Ward 1989; North 1990). Though *Limacina helicina* was the typical and very abundant organism of the summer surface mesozooplankton in the shelf of Terra Nova Bay (83 687 ind 100 $m^{-3}$, Guglielmo et al. 1990), it did not occur in the diet of postlarval *P. antarcticum* in the same proportion as that found in the natural zooplankton. In January, cyclopoid copepods (mainly *Oithona similis* and *Oncaea curvata*) constituted approximately 22% of the surface zooplankton (2500 ind 100 $m^{-3}$) near Cape Washington (Zunini Sertorio et al. 1990), while very low numbers were recorded near Drygalski Ice Tongue (>100 ind 100 $m^{-3}$, Carli et al. 1990). In the Weddell Sea, youngest postlarvae of *P. antarcticum* are known to feed on cyclopoid copepods and small *Limacina* sp. (Hubold 1984, 1985; Kellermann 1986b) whereas in the Bransfield Strait and adjacent waters their diet was mainly *Oncaea* spp., calanoid copepod eggs, tintinnids and the eggs of *Euphausia superba* (Kellermann 1987).

Old postlarvae and juveniles (age 1+) of 28–53 mm SL were much less abundant than postlarvae and were widely dispersed in the whole investigated area. Low concentrations (0.1–12.5 ind 100 $m^{-3}$) were recorded in offshore waters between 170° and 175°E and in the southern part of Terra Nova Bay, near Drygalski Ice Tongue. Only two specimens of age 2+ (73 and 75 mm SL) were caught in summer 1996 in offshore waters at stations 16 and 12, respectively. Hubold (1985) stated that *P. antarcticum* specimens larger than 60 mm SL appear to occupy a different habitat than older postlarvae and juveniles 1+. Many juvenile specimens have been caught in association with krill swarms in the East Wind Drift (Slosarczyk 1982). Our data confirm that also juvenile *P. antarcticum* (age 1+) in the western Ross Sea occupied more pelagic waters than postlarvae, with preference of deeper layers of the southern area of Terra Nova Bay in cold HSSW.

Samplings under first-year sea ice (140-mm thickness) during November 1997, near the coast of Terra Nova Bay (Guglielmo, pers. comm.), underlined the presence of several 26–30-mm SL juveniles. According to Regan (1916) and Guglielmo et al. (1998), the length range of postlarvae by the end of first summer was about 15–25 mm;

then *P. antarcticum* young of the year have grown about 0.02–0.03 mm $day^{-1}$, from March to November of the 2nd year. By visual observations it was evident that these juvenile *P. antarcticum* feed upon the cryopelagic fauna. This confirms the results of Hubold and Hagen (1997) that juvenile *P. antarcticum* inhabit the surface layer during the winter where the low zooplankton abundance available is sufficient to support the population. The cyclopoid copepod *Oncaea curvata* was the dominant prey of juvenile *P. antarcticum* in Terra Nova Bay and is also a member of sea-ice copepods (Schnack-Schiel et al. 1998). After the ice melting, juveniles probably migrate somewhat offshore, which becomes their preferred habitat for the 2nd year.

Few adults (100–190 mm SL) were collected only in the vicinity of the Ross Ice Shelf coinciding with high concentrations of Antarctic krill (see Azzali and Kalinowski, this Vol.). Preliminary dietary analysis confirm the results of De Witt and Hopkins (1977) that in the Ross Sea *Pleuragramma* eat copepods and adult euphausiids, not fish, with copepods dominant by number. In order of abundance we found *Metridia gerlachei*, *Calanoides acutus*, *Calanus propinquus* and *Paraeuchaeta antarctica*. Thus, during its life cycle, *P. antarcticum* is principally a zooplankton feeder (De Witt 1970, 1971; De Witt and Hopkins 1977; Takahashi and Nemoto 1984) with a very important ecological role in the food chain of the western Ross Sea biological community.

*Acknowledgements.* Logistical and financial support for this chapter was given by the Italian National Project for Antarctic Research (PNRA). Many thanks go to the officers and crew of the research vessels Polar Queen, Cariboo and Italica and to the technicians involved in the BIONESS sampling procedures for their excellent cooperation in field work. We are grateful to G. Arena and V. Bonanzinga for their help in data and graphic elaborations and R. William Jr. for his patience during preparation of the manuscript.

## References

Ascencio V, Mujica A (1986) Ichthyoplankton of the Bransfield Strait (SIBEX-Phase II, Chile). Ser Cient Inst Nac Antart Chil 35: 81–86

Balbontin F, Garreton M, Neuling J (1986) Composicion del alimento y tamano de las presas en larvas de peces del estrecho Bransfield (SIBEX-Fase II-Chile) Ser Cient Inst Nac Antart Chil 35: 125–144

Berry JP (1988) Hydrographic patterns in McMurdo Sound Antarctica, and their relationship to local benthic communities. Polar Biol 8: 377–391

Biggs DC (1982) Zooplankton excretion and NH4+ cycling in near-surface waters of the Southern Ocean. I. Ross Sea, austral summer 1977–78. Polar Biol 1: 55–67

Boldrin A, Stocchino C (1990) On the hydrological characteristics of Terra Nova Bay. In: National Scientific Commission for Antarctica (ed) Oceanographic Campaign 1987–88, Data Rep Part I, Genova, pp 11–57

Boysen-Ennen E, Piatkowski U (1988) Meso- and macrozooplankton communities in the Weddell Sea, Antartica. Polar Biol 9: 17–35

Bromwich DH, Kurtz DD (1982) Experiences of Sott's northern party: evidence for a relationship between winter katabatic winds and the Terra Nova Bay polynya. Polar Rec 131: 137–146

Bromwich DH, Kurtz DD (1984) Katabatic wind forcing of the Terra Nova Bay polynya. J Geophys Res 88: 9717–9722

Carli A, Mariottini GL, Pane L (1990) Contribution to the study of copepods collected in Terra Nova Bay (Ross Sea). In: National Scientific Commission for Antarctica (ed) Oceanographic Campaign 1987–88, Data Rep Part II, Genova, pp 129–167

De Witt HH (1970) The character of the midwater fish fauna of the Ross Sea, Antarctica. In: Holdgate MW (ed ) Antarctic ecology, vol 1 Academic Press, New York, pp 305–314

De Witt HH (1971) Coastal and deepwater benthic fish of the Antarctic. In: Bushell VC (ed) Antarctic man folio series, folio 15. American Geographical Society, New York, pp 1–10

De Witt HH, Hopkins TL (1977) Aspects of the diet of the Antarctic silverfish *Pleuragramma antarcticum* Boulenger. In: Llano GA (ed) Adaptation within Antarctic ecosystems. Gulf, Houston, pp 557–567

De Witt HH, Tyler JC (1960) Fishes of the Stanford Antarctic Biological Research Program, 1958–1959. Stanford Ichthyol Bull 7: 162–199

Eastman JT (1985a) The evolution of neutrally buoyant notothenioid fishes: their specializations and potential interactions in the Antarctic marine food web. In: Siegfried WR, Condy PR, Laws RM (eds) Antarctic nutrient cycles and food webs. Proc 4th SCAR Symp Antarct Biol. Springer, Berlin Heidelberg New York, pp 430–436

Eastman JT (1985b) *Pleuragramma antarcticum* (Pisces, Nototheniidae) as food for other fishes in McMurdo Sound, Antarctica. Polar Biol 4: 155–160

Eastman JT, De Vries AL (1982) Buoyancy studies of notothenioid fishes in McMurdo Sound, Antarctica. Copeia 2: 385–393

Efremenko VN (1983) Illustrated guide to fish larvae of the Southern Ocean. Cybium 7: 1–74

Emison WB (1968) Feeding preferences of the Adélie penguin at Cape Crozier, Ross Island. Antarct Res Ser 12: 191–212

Ekau W, Hubold G, Wöhrmann A (1987) Fish and fishlarvae. Ber Polarforsch 39: 210–218

Guglielmo L, Costanzo G, Manganaro A, Zagami G (1990) Spatial and vertical distribution of zooplanktonic communities in Terra Nova Bay (Ross Sea). In: National Scientific Commission for Antarctica (ed) Oceanographic Campaign 1987–88, Data Rep Part I, Genova, pp 257–398

Guglielmo L, Costanzo G, Zagami G, Manganaro A, Arena G (1992) Zooplankton ecology in the Southern Ocean. In: National Scientific Commission for Antarctica (ed)

Oceanographic Campaign 1989–90, Data Rep Part II, Genova, pp 301–468

Guglielmo L, Granata A, Greco S (1998) Distribution and abundance of postlarval and juvenile *Pleuragramma antarcticum* (Pisces, Nototheniidae) off Terra Nova Bay (Ross Sea Antarctica). Polar Biol 19: 37–51

Hecq J-H, Magazzù G, Goffart A, Catalano G, Vanucci S, Guglielmo L (1992) Distribution of planktonic components related to structure of water masses in the Roos Sea during the Vth ITALIANTARTIDE expedition. In: Ambrosetti W, Piccazzo M, Ruffoni Riva T (eds) Atti IX Congr Ass Ital Oceanogr Limnol, Lang, Genova pp 665–678

Hecq J-H, Brasseur P, Goffart A, Lacroix G, Guglielmo L (1993) Modelling approach of the plancktonic vertical structure in deep Austral Ocean. The example of the Ross Sea ecosystem. Proceedings of the annual workshop on Belgian oceanographic research. Royal Academy of Belgium Brussels, pp 235–250

Hempel I, Hubold G, Kaczmaruk B, Keller R, Weigmann-Haass R (1983) Distribution of some groups of zooplankton in the inner Weddell Sea in summer 1979/80. Ber Polarforsch 9: 1–35

Hubold G (1984) Spatial distribution of *Pleuragramma antarcticum* (Pisces: Nototheniidae) near the Filchner and Larsen Ice Shelves (Weddell Sea, Antarctica). Polar Biol 3: 231–236

Hubold G (1985) The early life history of the high Antarctic Silverfish, *Pleuragramma antarcticum*. In: Siegfried WR, Condy PR, Laws RM (eds) Antarctic nutrient cycles and food webs. Springer, Berlin Heidelberg New York, pp 445–451

Hubold G (1990) Seasonal patterns of ichthyoplankton distribution and abundance in the southern Weddell Sea. In: Kerry KR, Hempel G (eds) Antarctic ecosystems. Ecological change and conservation. Springer, Berlin Heidelberg New York, pp 149–158

Hubold G (1991) Ecology of notothenioid fish in the Weddell Sea. In: Di Prisco G, Maresca B, Tota B (eds) Biology of Antarctic. Springer, Berlin Heidelberg New York, pp 3–21

Hubold G, Ekau W (1990) Feeding patterns of post-larval and juvenile notothenioids in the Southern Weddell Sea (Antarctica). Polar Biol 10: 255–260

Hubold G, Hagen W (1997) Seasonality of feeding and lipid content in juvenile *Pleuragramma antarcticum* (Pisces: Nototheniidae) from the southern Weddell Sea. In: Battaglia B, Valencia J, Walton WH (eds) Antarctic communities. Species, structure and survival. Cambridge University Press, New York, pp 277–283

Hureau JC (1982) Methods for studying early life history stages of Antactic fishes. Cybium 6: 3–11

Hureau JC (1994) The significance of fish in the marine Antarctic ecosystems. Polar Biol 14: 307–313

Innamorati M, Mori G, Lazzara L, Nuccio C, Lici M, Catalano G, Benedetti F (1990) Phytoplankton ecology in the Ross Sea. In: National Scientific Commission for Antarctica (ed) Oceanographic Campaign 1987–88, Data Rep Part II, Genova, pp 9–63

Jacobs J (1974) Quantitative measurement of food selection (a modification of the forage ratio and Ivlev's Electivity Index). Oecologia (Berl) 14: 413–417

Jacobs SS, Amos AF, Bruchhausen PM (1970) Ross Sea oceanography and Antarctic bottom water fomation. Deep Sea Res 17: 935–962

Keller R (1983) Contributions to the early life history of *Pleuragramma antarcticum* Boul. 1902 (Pisces, Notothenioidei) in the Weddell Sea. Meeresforschung 30: 10–24

Kellermann A (1986a) Geographical distribution and abundance of postlarval and juvenile *Pleuragramma antarcticum* (Pisces, Notothenioidei) off the Antarctic Peninsula. Polar Biol 6: 111–119

Kellermann A (1986b) On the biology of early life stages of notothenioid fishes (Pisces) off the Antarctic Peninsula. Ber Polarforsch 31: 149

Kellermann A (1987) Food and feeding of postlarval and juvenile *Pleuragramma antarcticum* (Pisces, Notothenioidei) in the seasonal pack ice zone of the Antarctic Peninsula. Polar Biol 7: 307–315

Kellermann A (1989) The larval fish community in the zone of seasonal ice cover and its seasonal and interannual variability. Arch Fischerei wiss 39: 89–109

Kellermann A (1990) Catalogue of early life stages of Antarctic notothenioid fishes. BIOMASS Sci Ser 10: 44–136

Kellermann A, Kock KH (1984) Postlarval and juvenile notothenioids (Pisces, Perciformes) in the Southern Scotia Sea and Northern Weddell Sea during FIBEX 1981. Meeresforschung 30: 82–93

Kellermann A, Kock KH (1988) Patterns of spatial and temporal distribution and their variation in early life stages of Antarctic fish in the Antarctic Peninsula region. In: Sahrhage D (ed) Antarctic ocean and resource variability. Springer, Berlin Heidelberg New York, pp 147–159

Kellermann A, Schadwinkel S (1991) Winter aspects of the ichthyoplankton community in Antarctic Peninsula waters. Polar Biol 11: 117–127

Kennett JP (1968) Fauna of the Ross Sea. VI. Ecology and distribution of Foraminifera. Bull N Z Dep Sci Ind Res 186: 48 pp

Loeb VJ (1991) Distribution and abundance of larval fishes collected in the western Bransfield Strait region, 1986–87. Deep Sea Res 38: 1251–1260

Loeb VJ, Kellermann AK, Koubbi P, North AW, White MG (1993) Antarctic fish assemblages: a review. Bull Mar Sci 53: 416–449

Morales-Nin B, Palomera I, Schadwinkel S (1995) Larval fish distribution and abundance in the Antarctic Peninsula region and adjacent water. Polar Biol 15: 143–154

Moreno C, Rueda T, Asencio G (1986) The trophic niche of *Pleuragramma antarcticum* in the Bransfield Strait, Antarctica. Quantitative comparison with other areas of the Southern Ocean. Ser Cient Inst Nac Antart Chil 35: 101–117

North AW (1990) Ecological studies of Antarctic fish with emphasis on early development of inshore stages of South Georgia. PhD Thesis, CNAA, UK; 319 pp

North AW, Ward P (1989) Initial feeding by Antarctic fish larvae during winter at South Georgia. Cybium 13: 357–364

North AW, White MG (1982) Key to fish postlarvae from the Scotia Sea, Antarctica. Cybium 6: 13–32

North AW, Kellermann A (1989) Key to the early stages of Antarctic fish. BIOMASS Sci Ser 10: 1–44

North AW, Kellermann A (1990) Key to the early stages of Antarctic fish. Ber Polarforsch 67: 1–44

Patterson SL, Whitworth T (1990) Physical oceanography. In: Glasby GP (ed) Antarctic sector of the Pacific. Elsevier, Amsterdam, pp 55–93

Piatkowski U (1987) Zoogeographical investigations and community analysis on Antarctic macroplankton. Ber Polarforsch 34: 73–81

Piatkowski U, White MG, Dimmler W (1990) Micronekton of the Weddel Sea: distribution and abundance. Ber Polarforsch 68: 73–81

Plötz J (1986) Summer diet of Weddell seals (*Leptonychotes weddelli*) in the Eastern and Southern Weddell Sea, Antarctica. Polar Biol 6: 97–102

Regan CT (1916) Larval and postlarval fishes. I. Antarctic and sub-Antarctic fishes. Nat Hist Rep Br Antarct "Terra Nova" Exp 1910 Zool 1: 125–156

Sameoto DD, Lewis MK (1980a) Zooplankton and micronekton associated with scattering layers on the Nova Scotia shelf and slope. Can Tech Rep Fish Aquat Sci 875: 44 pp

Sameoto DD, Lewis MK (1980b) Zooplankton and micronekton associated with acoustic scattering layers on the Nova Scotia shelf and slope during June, 1978. Can Tech Rep Fish Aquat Sci 936: 32 pp

Sameoto DD, Saroszynsky LO, Fraser WB (1980) BIONESS, a new design in multiple net zooplankton sampler. J Fish Res Board Can 37: 722–724

Schnack-Schiel SB, Thomas D, Dahms HU, Haas C, Mizdalski E (1998) Copepods in Antarctic sea ice. Antarct Res Ser 73: 173–182

Slosarczyk W (1982) Preliminary estimation of distribution and abundance of juvenile Nototheniodei within krill concentrations in the east Antarctic region. Mimeo, BIOMASS Working Party on Antarctic Fish Ecology, Hamburg

Slosarczyk W (1986) Attempts at a quantitative estimate by trawl sampling of distribution of postlarval and juvenile notothenioids (Pisces, Perciformes) in relation to environmental conditions in the Antarctic Peninsula region during SIBEX 1983-84 Mem Natl Inst Polar Res Special Issue 40: 299–315

Smith SL, Schnack-Schiel SB (1990) Polar zooplankton. In: Smith WO (ed) Polar oceanography, part B: Chemistry, biology, and geology. Academic Press, London, pp 527–598

Spezie G, Tucci S, Budillon G, Corradi N, Piccazzo M, Fanucci F, Firpo M, Mirabile L, Ferrari M (1993) Water masses characteristics, sediment dispersal and seismic stratigraphy in the continental margin of the Ross Sea (Antarctica). Ann Ist Univ Nav Napoli 60: 32 pp

Stocchino C, Lusetti C (1990) Prime osservazioni sulle caratteristiche idrologiche e dinamiche di Baia Terra Nova (Mare di Ross - Antartide). Istituto Idrografico della Marina, Genova, pp 1–55

Stocchino C, Manzella GMR (1991) Analisi delle correnti marine di Baia Terra Nova (Mare di Ross - Antartide). Consiglio Nazionale Ricerche, Roma, pp 1–56

Takahashi M, Nemoto T (1984) The food of some Antarctic fish in the Western Ross Sea in summer 1979. Polar Biol 3: 237–239

White MG, North AW (1987) Postlarval nototheniods and midwater fish collected during the SIBEX cruise by British Antarctic Survey, 1985. In: Kullander SO, Fernholm B (eds) Proc 5th Congr Eur Ichthyol, Swedish Museum of Natural History, Stockholm, 1985, pp 405–411

White MG, Piatkowski U (1993) Abundance, horizontal and vertical distribution of fish in eastern Weddell Sea micronekton. Polar Biol 3: 41–53

Wiebe PH (1971) A computer model study of zooplankton patchiness and its effects on sampling error. Limnol Oceanogr 16: 28–29

Wiebe PH (1972) A field investigation of the relationship between length of two, size of net and sampling error. J Cons Int Explor Mer 34: 268–275

Vacchi M, La Mesa M, Greco S (1999) Summer distribution and abundance of larval and juvenile fishes in the Western Ross Sea. Antarct Sci 11: 54–60

Zunini Sertorio T, Salemi Picone P, Bernat P, Cattini E, Ossola C (1990) Copepods collected in sixteen stations during the Italian Antarctic Expedition 1987–1988. In: National Scientific Commission for Antarctica (ed) Oceanographic Campaign 1987–1988, Data Rep Part II, Genova, pp 67–125

# CHAPTER 29

# A Modelling Approach to the Ross Sea Plankton Ecosystem

J.H. Hecq[1], L. Guglielmo[2], A. Goffart[1], G. Catalano[3], and H. Goosse[1]

**ABSTRACT**

A conceptual and numerical model adapted to the local plankton ecosystem (ECOHYDROMVG) has been developed on the basis of the information acquired during various oceanographic cruises in the Ross Sea. The construction of the biological/physical coupled 1-D model of the upper water column ecosystem is presented. The model takes into account the control of the plankton ecosystem by ice cover, ice edge dynamics, atmospheric forcing and surface circulation, nutrients being non-limiting factors. Ice-edge thickness, retreat and melting determine the initial conditions for the spring bloom (ice algae concentration, vertical structure of the water column) and allow the biological processes of primary production in the water column to start. The combined effects of wind and solar irradiance determine the depth of the upper mixed layer and quantitatively control photosynthetic levels.

In this approach, biological variables are limited to diatoms, *Phaeocystis* and smaller cells, depending on local conditions. Grazing activity forcing by *Euphausia superba*, copepods, *Limacina helicina* and amphipods control the phytoplankton variables.

The model is used to determine a standard state of the Ross Sea plankton ecosystem and numerical results accurately simulate the vertical stabilization of the water column within the Ross Sea marginal ice zones during the ice-melting period, and the biological structure of the upper layers. It simulates primary production and the selection of the type of phyto- and zooplankton communities. It is applied to local situations controlled by different wind conditions, ice cover and initial content of ice algae, and accurately reproduces field observations.

## Introduction

The productivity and biodiversity of the Southern Ocean ecosystem is to a large extent determined by the specific environmental features of the Seasonal Ice Zone. As shown by satellite images, highest plankton productivity is associated to that area (Comiso 1991) in the marginal ice zone where ice algae are released by ice melting and where the stabilization of the water column enhances productivity (Smith and Nelson 1985). The maximum phytoplankton bloom and ice-algae releasing occurs during the ice-melting period when radiation and photoperiod are highest. This melting process which releases a variable quantity of ice algae induces especially a stabilizing surface buoyancy flux, a subsequent decrease of the mixed layer depth and an increase of plankton production (Sullivan et al. 1988; Ackley and Sullivan 1994).

The food web in the Southern Ocean is unique among ocean ecosystems in that it is characterized by the dependence largely, in many areas, of a single key species, the Antarctic krill (*Euphausia superba*), which can strongly limit diatom standing stocks. Like many species of the food web, krill depend on sea ice during some or all of their life history and on hydrodynamical processes on a larger scale (Priddle et al. 1988).

The ring shape around the polar continent, isolation by circumpolar fronts, zonal transport by strong circumpolar currents coupled with the seasonal changes in the light regime and sea-ice cover impose a relative homogeneity in species composition and the concept of a Southern Ocean Seasonal Ice Zone Ecosystem has been suggested

[1] University of Liège, Belgium
[2] University of Messina, Italy
[3] Thalassographic Institute of Trieste, Italy

and can be applied to the Ross Sea. However, limitations and spatial and temporal variability occur in relation to local characteristics of depth, climate, hydrodynamics, wind and ice coverage.

To illustrate the concept of a Ross Sea Seasonal Ice Zone ecosystem and to test the influence of those physical constraints on the variability of the plankton ecosystem, a mechanistic modeleing approach was undertaken. To balance between excessively holistic and realistic approaches (Platt et al. 1981), a conceptual scheme of the typical Ross Sea plankton ecosystem is developed after presentation of the principal physical and ecological characteristics and discussion of the sources of variability. A coupled biological/physical/ ice numerical model of the water column is developed. This model ECOHYDRO-MVG parameterizes explicitly the water column vertical structure and mixing (Hecq et al. 1993) and takes into account the importance of ice melting, atmospheric constraints and the presence of ice algae (Goosse and Hecq 1994). A first approach was voluntarily limited to the first trophic levels of the food chain during the spring and summer. A standard run is developed and an application to various local situations is discussed in relation with data acquired during the cruises.

## The Conceptual Model of the Ross Sea Plankton Ecosystem

### The Physical Variability in the Ross Sea

In the Ross Sea, temperatures never exceed values higher than 3 to 4 °C in summer and approach −2 °C at the ice shelves. The only input of fresh water essential for the vertical stabilization of the upper layers of the water column is due to ice retreat.

The principal physical cause for the variability of the Ross Sea ecosystem is the existence of the Southern Ross Sea polynya generated by strong winds which advect the ice cover (Zwally et al. 1983) and by sensible heat coming from the ocean (Jacobs and Comiso 1989). During the winter and at the beginning of the spring, the ice-free surface propagates from south to north, as shown by weekly remote sensing pictures (Hecq et al. 1993, Fig. 1). As a consequence, in that area, the ice retreat and melting, and the exposure of the water column to sunlight appear earlier than in other areas. The characteristic extension of the marginal ice edge around the polynya seems to be responsible for the higher productivity, an increase in the productive season and of the length of the ice edge. Another smaller polynya is also observed in Terra Nova Bay, just north of the Drygalski Ice Tongue, and is due to offshore winds which remove ice as fast as it is formed (Gordon 1988). The mode of formation of the polynya imposes a spatial heterogeneity of ice cover and thus the initial timing of the seasonal plankton cycle varies on the same scale. Moreover, the ice-edge retreat is asymmetrical and the progression northwards is faster than eastwards and westwards. The vertical structure of the water column associated to the ice-edge melting is maintained for a longer time along the zonal borders of the polynya (for example Victoria Land) than in the north (central Ross Sea).

Moreover, the progression to the north of the ice edge of the polynya is not a continuous process due to differences in ice thickness between the south and north. At the end of winter, the polynya opens slowly in the south due to ice melting and opening of leads. Following successive wind events, the polynya continues to spread due to the northward transport of ice floes. These ice pieces accumulate around 75°N where the ice pack is thicker, and particularly in the western part, where they form a barrier between the southern Ross Sea and Terra Nova polynyas. The second phase of opening is very fast and does not last longer than the period of ice melting. The third phase occurs later, north of 75°S, and progresses for a longer time. At the end of the third phase (end of December), the polynya encounters the circumpolar ice edge and becomes contiguous with the whole Southern Ocean. This discontinuity in the progression of the polynya ice front leads to a strong variability in the light, stability of the water column, in the algae released in the water column, and influences the pattern of the global ecosystem.

Another source of variability is due to local wind conditions. Strong catabatic winds can increase the depth of the mixed layer and impose a decrease of phytoplankton biomass. Two extreme cases of variability are observed. In the permanent ice-free zone, north to the Antarctic divergence, because of strong winds and low vertical stability, the mixed layer is deep and plankton productivity poor. In the eastern Ross Sea, below the non-melting ice (permanent ice pack), the mixed layer is also deep due to

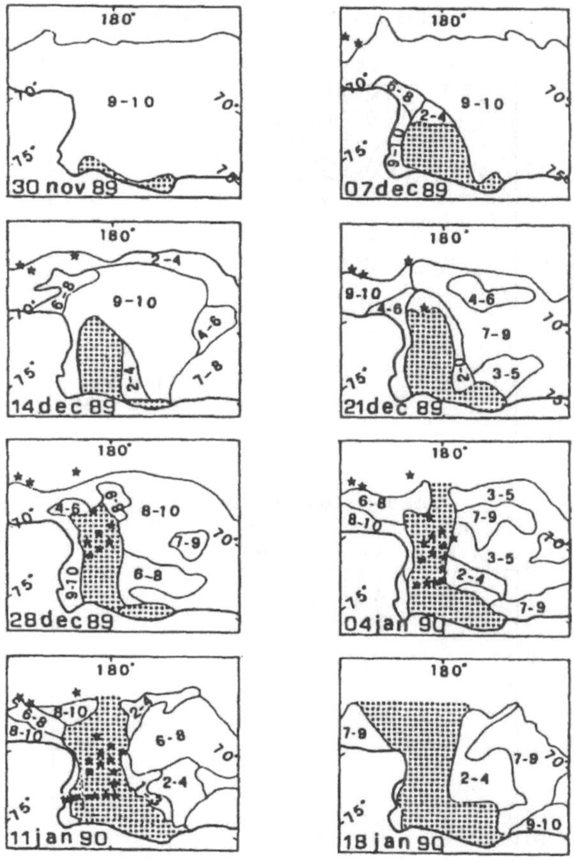


**Fig. 1.** Temporal evolution of ice cover in the southern part of the studied area during the cruise period. *Numbers* represent ice cover of the sea surface in tenths. Data are taken from satellite images published weekly by NAVY-NOAA Joint ICE Center. *Stars* represent the location of the hydrological stations until the date of the image. The *dotted area* is the ice-free area

the weak stratification and the high momentum transfer caused by drift ice (Niebauer and Smith 1989).

## Phytoplankton Variability in the Ross Sea

High phytoplankton concentrations and massive blooms from 10 to 40 mg $m^{-3}$ of chlorophyll a (Chl a) have been observed in the Ross Sea polynya, directly (Smith and Nelson 1985; Nelson and Smith 1986; Hecq et al. 1992, 1993; Smith and Gordon 1997; Goffart et al. 1999, Submtd) or by remote sensing (Comiso 1991; Arrigo and McClain 1994) in November when other austral regions are again ice-covered. These high Chl a concentrations are among the highest observed in the Austral Ocean and could justify a depletion in nutrients. However, nutrient concentrations are high and never limit primary production (Catalano et al. 1997). In winter, nitrate, phosphate and silicate concentrations can exceed 30 mmol $m^{-3}$, 2.5 mmol $m^{-3}$ and 100 mmol $m^{-3}$, respectively. These nutrient concentrations are high enough to support very high productivity in the Ross Sea where the extreme southern latitude (75–78°S) and climatic anticyclonic forcing (low cloud coverage) impose an intense and permanent solar irradiance during spring and summer.

An example of the vertical distribution of phytoplankton in the water column and its variability was illustrated during the ROSSMIZE cruise in November 1994 (Fig. 2). From 20 to 26 November 1994, a transect of vertical stations between the

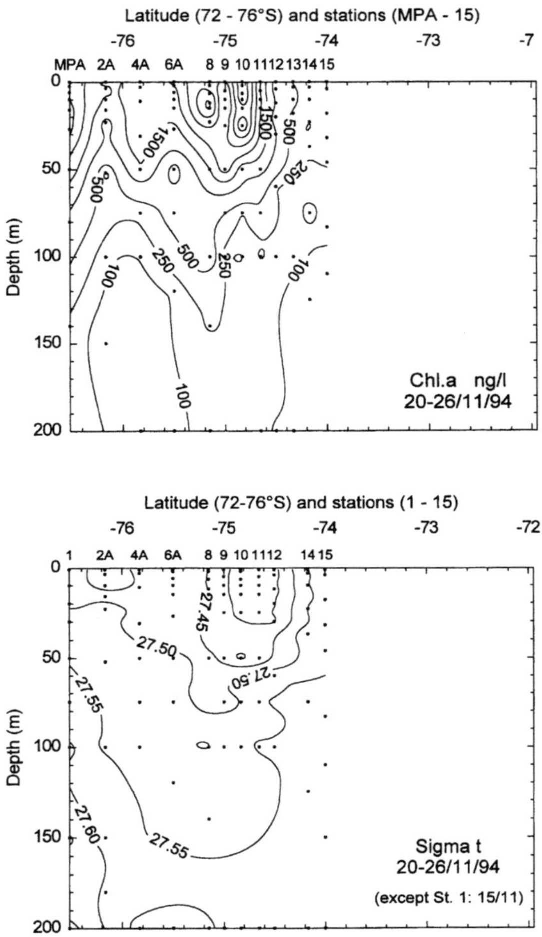


**Fig. 2.** Vertical distribution of Chl a and $\sigma_t$ in the southern Ross Sea polynya along the 175°E transect, from 76°30′S to 73°58′S (20–26 November 1994) ($\sigma_t$ computed from T and S data of the ROSSMIZE cruise; Russo et al., 1997)

Ross ice shelf (76°30′S) and Cape Adare (73°58′S) was sampled along 175°E (stations 10 miles apart, 10 sampling depths per station). Spatial and temporal successions of phytopigments, products of degradation by grazing (phaephorbids) detected by HPLC and ammonium excretion detected by an autoanalyzer were determined in the water column. Significant values of Chl a (>1 mg m$^{-3}$) are generally observed in the surface layer (0 to 40 m), where the stratification reached a maximum in the layer of $\sigma_t$ lower than 27.45. Pigment composition indicated that phytoplankton was dominated by prymnesiophytes (probably *Phaeocystis* sp.). Maximum Chl a values (>4 mg m$^{-3}$) were restricted to stations 9–12 associated with the northern edge of the polynya and with the northern slope of the Pennel Bank. In these particular areas, where ice melting reached a maximum, $\sigma_t$ values are lower than 27.4, and vertical stratification increases. Pigment composition revealed the dominance of diatoms in these areas. In the ice-pack area, (stations 12 to 15, Joides Basin) very low Chl a concentrations were observed (<0.5 mg m$^{-3}$) in the water column whereas high pigment content was measured in the ice. In this area, ice melting probably had not yet started.

The same transect was repeated 15 days later from 74°6′S to 72°25′S, some significant stations being resampled after a few days to investigate temporal variations (Fig. 3). Between stations 9 to 11, Chl a concentrations were one order of magnitude lower than 2 weeks, even if the stratification (distribution of $\sigma_t$) pattern was the same. The reduction was probably due to grazing activity as suggested by high phaeophorbid levels and ammonium concentrations, especially in stations 9 and 10. Pigment analysis in gut contents emphasized strong grazing activity in this area. In addition to an increase in phaeophorbids (indicators of faecal pellet production by zooplankton), an increase of ammonium was observed at the beginning (example station 11). In stations 17 to 27, very low Chl a concentrations were observed (<30 mg $m^{-2}$) in the highly mixed water column (mixed layer > 100 m), where rapid sedimentation of phytoplankton cells did not permit the initiation of a bloom. In this area, a deepening of the mixed layer may have been due to strong wind conditions.

The diversity of phytoplankton populations was identified on the basis of pigment assemblages determined by HPLC (Goffart et al. 1999, submitted). *Phaeocystis* dominated in the major part of the southern Ross Sea. High diatom concentrations dominated in coastal areas and in the mar-

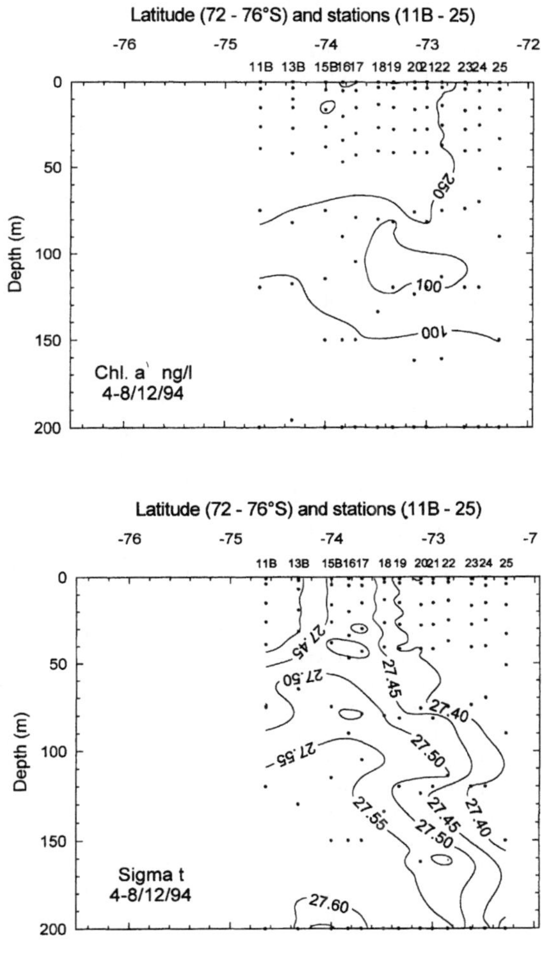


**Fig. 3.** Vertical distribution of Chl a and $\sigma_t$ in the southern Ross Sea polynya along the 175°E transect, from 74°40′S to 72°25′S (04–08 December 1994) ($\sigma_t$ computed from T and S data of the ROSSMIZE cruise; Russo et al., 1997)

ginal ice zone even if the concentrations are virtual (and replaced by phaeophorbids) in the case of high grazing activity (controlled overall by krill), especially in the central part of the Ross Sea at the northern limits of the polynya. Microphytoplankton consisted of large-sized diatoms that dominated in the periantarctic marginal ice zone. Nano- and pico-phytoplankton were more abundant in the periantarctic ice-free zone.

On the other hand, the results confirm the idea of the temporal succession of various nanophytoplankton groups like cryptophytes and prasinophytes and the appearance of picoplankton cells later in the season.

## Zooplankton Specificity of the Ross Sea

The principal characteristics of the Ross Sea zooplankton are the complete absence of benthic larvae, the low diversity (poor in species, rich in individuals), the existence of diurnal and seasonal migrations (Mackintosh 1973) and the reproduction of species, which coincides with the seasonal pulse of phytoplankton production occurring over a restricted period (Knox 1994). Other important characteristics are the large size of many of the zooplankters (typified by *Euphausia superba* but also other groups such as copepods) and their slow growth and long life cycles (J. Priddle, pers. comm.). Herbivorous zooplankton mostly occurs in the upper 50 m of the water column (≈the mixed layer) and consists largely of protozoans, and larval and juvenile stages of copepods for microzooplankton. Herbivorous copepods (*Calanus propinquus* and *C. acutus*) represent the mesozooplankton, and krill and the mollusc *Limacina helicina* dominate the macrozooplankton. The maximum grazing pressure on large-sized diatoms is due to these herbivorous macrozooplankton.

Omnivorous forms are presents below 50 m and are dominated essentially by the copepod *Metridia gerlachei*, amphipods and ostracods, respectively in coastal and offshore areas. Copepods (*Euchaeta antarctica*), molluscs (*Clione* sp.), chaetognaths and larvae of *Pleuragramma antarcticum* dominate carnivorous zooplankton. Below 500 m the omnivorous mysid *Antarctomysis ohlini* is the only dominant species.

Euphausiids are considered to comprise at least 50% of the standing crop of zooplankton (Knox 1983) with *Euphausia crystallorophias* in the coastal zone and *Euphausia superba* in offshore areas. Salps are essentially restricted north to the Antarctic Divergence. In the northern and central parts of the Ross Sea, the krill populations (*Euphausia superba*) were as abundant as those of the Atlantic and Indian sectors. In the central and southern parts of the Ross Sea, in the polynya completely surrounded by pack ice, numerous swarms were detected near the regions of plankton blooms and near the ice edge but were absent south to 75°S (Azzali and Kalinowski 1992). It could be suggested that the populations of *E. superba* were limited to areas deeper than 400 m probably in relation to ontogenetic migrations. Krill grazing remains to be experimentally determined. Of particular relevance to current hypothesis is the association of krill and sea ice (Smetacek et al. 1990). Evidence for winter habitat segregation between adult and larval stages exists (Quetin et al. 1996). In the fall, adults presumably migrated inshore, leaving the open waters beyond the continental slope void of postlarval krill. This pattern appears to change with the onset of spring. By October, adults can be found in the water column both in the open ocean and under the ice. In late winter and spring, it is observed that krill move toward the marginal ice zone and take advantage of both the algae in the ice and released by melting ice to supply energy for growth and egg production (Huntley et al. 1994). In late spring, a progression in krill school size and depth was seen moving from the ice to open waters. Few schools were detected in the water column under the ice; small schools occurred in the marginal ice zone and large schools were detected at greater depths in open waters. Generally, krill aggregations and other herbivorous plankton occurred mainly in the mixed layer in the depth ranging from 10 to 60–70 m. The omnivorous ostracods and amphipods were restricted below the pycnocline.

## The Biotic Assemblages in the Ross Sea

Variations in ecosystem patterns characterized by specific biota assemblages were identified on the basis of plankton composition in various areas of the Ross Sea. The *Phaeocystis* assemblage in the southern Ross Sea along the permanent ice shelf is dominated by these algae. Zooplankton is scarce and no predation by higher trophic levels occurs. The full trophic chain assemblage in the deep Ross

Sea includes the northern part of the polynya where reduced ice melting determines a vertical stability favourable to phytoplankton productivity. All groups are present (diatoms-*Phaeocystis*, copepods-krill) including birds and whales. The diatom-krill assemblage in the northern Ross Sea along the periantarctic marginal ice edge exists where depths exceed 500 m in the south and 1000 m in the north. The coastal diatom assemblage encountered in the Terra Nova Bay polynya, where *Limacina helicina* is the dominant herbivorous plankter. In this coastal area a mesozooplankton herbivorous food chain (diatoms, copepods, ostracods) exists but the plankton ecosystem is dominated overall by the macrozooplankton food chain (diatoms-molluscs, amphipods); larval stages of *Pleuragramma* represent the carnivorous level. Other assemblages exist such as the oligotrophic subantarctic assemblage, but these are poorly documented.

These different biotic assemblages correspond to local particularities and seasonal successions of different phases of the whole Southern Ocean seasonal ice-zone ecosystem controlled by specific physical conditions. They are not to be taken as different ecosystems but, more realistically, as different states of a typical ecosystem locally controlled by specific constraints or progressing from pack ice to ice-free waters. The marginal ice zone retreat and melting are not simultaneous in the whole Ross Sea, the central and southern part opening earlier than the lateral parts. Some areas (generally shallower than 500 m) seem to be inaccessible to large krill swarms and grazing pressure is modified. The early opening of the polynya and the early spring production period in the polynya could be an attractive force for krill and hence attract predators such as penguins, seals and whales. For unknown reasons, ice-algae content (as initial condition for the spring bloom) seems to be more important in the western part of the Ross Sea.

This typical ecosystem is summarized as a conceptual model (Fig. 4). The principal variables, processes and forcing functions identified in the Ross Sea seasonal ice shelf are included. The atmospheric forcing, sun energy, cloud cover, air temperature and wind stress will act on the ice-melting process and the liberation of low salinity water; water turbulence, mixed layer depth and stratification will also indirectly affect the release of ice algae and primary productivity. The ice algae growth is explicitly parametrized.

Four dominant phytoplankton variables are diatoms, *Phaeocystis*, nano- (small diatoms and other) and pico-phytoplankton. The zooplankton variables are copepods, krill, *Limacina*, amphipods and ostracods, and are explicitly included in the coupled model.

## The Numerical Model of the Ross Sea Plankton Ecosystem

In previous 1D coupled biological/physical models used to study the water structure influence on plankton ecosystems in the Southern Ocean (Veth 1991; Veth et al. 1992; Sakshaug et al. 1991; Svanson 1991), the effect of dishomogenous vertical profiles in turbulence induced by low wind stress, the role of low but significant deeper turbulence due to mesoscale processes and the importance of the sea ice-melting process for ecosystems needed a more realistic approach.

The present model (ECOHYDRO MVG) includes explicitly the effect of vertical profiles of turbulence induced by thermal exchange and wind stress. As proposed by Niebauer and Smith (1989) in the Arctic, the unidimentional model of the upper ocean water column is applied with a k-l turbulence closure. The melting dynamics are taken into account by adding a thermodynamic model of sea ice including a parameterization of the leads' surface (Goosse and Hecq 1994).

The hydrodynamical model assumes horizontal homogeneity and the momentum equations are restricted to:

$$\frac{\partial u}{\partial t} = fv + \frac{\partial}{\partial z}\left(\tilde{v}\frac{\partial u}{\partial z}\right) \tag{1}$$

$$\frac{\partial v}{\partial t} = -fu + \frac{\partial}{\partial z}\left(\tilde{v}\frac{\partial v}{\partial z}\right), \tag{2}$$

where u,v are the horizontal components of velocity, t the time, f the Coriolis parameter, z the coordinate along the vertical axis and $\tilde{v}$ the eddy viscosity.

For temperature T and salinity S in water, equations are:

$$\frac{\partial T}{\partial t} = \frac{\partial}{\partial z}\left(\tilde{\lambda}_T\frac{\partial T}{\partial z}\right) \tag{3}$$

$$\frac{\partial S}{\partial t} = \frac{\partial}{\partial z}\left(\tilde{\lambda}_s\frac{\partial S}{\partial z}\right). \qquad \tilde{\lambda}_T = \tilde{\lambda}_S = \tilde{\lambda} \tag{4}$$

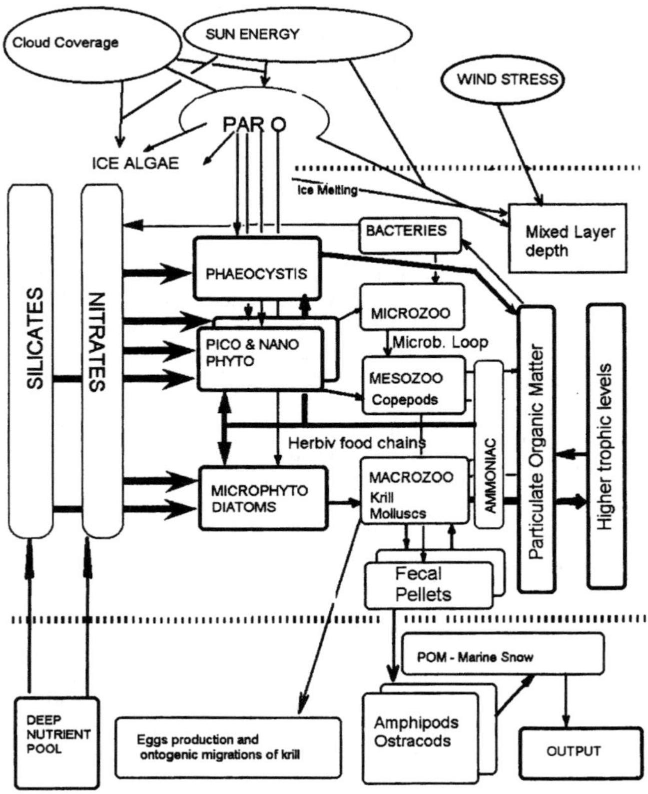


**Fig. 4.** Conceptual model (ECOHYDRO MVG) of the Ross Sea plankton ecosystem

$\tilde{\lambda}_T$ and $\tilde{\lambda}_S$ are, respectively, the eddy viscosity for temperature and salinity; these are supposed to be equal. The density ρ is computed by a simplified version of the International Equation of State of Sea Water (1980) (Pond and Pickard 1983). To compute the eddy diffusivities and viscosities and then close the system, an equation for turbulent kinetic energy, k is added:

$$\frac{\partial k}{\partial t} = \tilde{\nu} M^2 - \tilde{\lambda}.N^2 - \varepsilon + \frac{\partial}{\partial z}\left(\tilde{\nu}\frac{\partial k}{\partial z}\right), \qquad (5)$$

where N and M are the Prandl and Brunt-Väisälä frequency and ε is the energy dissipation rate (Nihoul 1984).

The sea-ice model is represented by an horizontally homogenous slab divided into two layer of ice (Semtner 1976; Shine and Henderson-Sellers 1985). In the three layers, the temperature is calculated by means of a one-dimensional heat equation. Temperature and heat fluxes are assumed to be continuous at the ice-snow interface.

To compute the top surface temperature, a balance between the fluxes is assumed.

$$(1 - I_0).F_{sw} - \varepsilon_{su}.F_{lw} + F_{LE} + F_H^+ F_{cs} - \varepsilon_{su}.\sigma T_s^4 = 0. \qquad (6)$$

$F_{sw}$ is the net shortwave flux. $I_0$ is the part of solar radiation which is not absorbed in the top 10 cm and does not contribute to surface melting (0% for snow, 17% for ice: Maykut and Untersteiner 1971). $\varepsilon_{su}$ is the emissivity of the top surface. $F_{lw}$, the incoming longwave radiation, is parameterized according to Parkinson and Washington (1979). $F_{LE}$ and $F_H$ are, respectively, sensible heat and latent heat flux. The values for the drag coefficient are $1.75 \times 10^{-3}$ for ice or snow (Parkinson and Washington 1979), $1.2 \times 10^{-3}$ for leads and $1.0 \times 10^{-3}$ for the ice-free ocean (Andreas and Murphy

1986). $F_{cs}$ is the conductive flux through ice or snow and $T_S$ is the surface temperature. If the balance requires a temperature greater than the melting point (273.15 K for snow, 273.05 K for ice), the temperature is held at this point and the excess energy is used for melting snow or ice.

At the ice bottom, every difference between the conduction flux in the ice and the flux from the water ($F_{wi}$) is compensated by accretion or melting of ice. The temperature at the bottom of the sea ice is supposed to be equal to the freezing point of seawater at the first ocean grid point. The freezing temperature T is determined as a function of salinity ($S_0$) according to Doronin and Kheisin (1977):

$$T = -(3 + 52.75\, S_0 + 0.045\, S_0^2 + 0.0004\, S_0^3)10^{-3}. \qquad (7)$$

The leads have an important role, particularly in the Ross Sea where sea surface is never completely ice-covered (Andreas and Ackley 1982). The solar energy received by the leads is much greater than on ice or snow because of the smaller albedo of water. An important part of the melting (lateral or basal) is induced by this energy support. The lead coverage depend on the heat budget of the lead:

$$F_{lead} = F_{sw}.(1 - I_w) + \varepsilon_{su}.F_{kw} - \varepsilon_{su}.\sigma.T_L^4 + F_H + F_{LE} + F_{wL} + F_{sal}. \qquad (8)$$

The notations and parameterizations are the same as those used in Eq. (6). $F_{wL}$ is the heat flux from the water. $I_w$ is the fraction of light which goes through the lead and is absorbed in the underlying water. $F_{sal}$ is the heat flux released by the lead because of the variation of its temperature due to changes in its salinity.

$$F_{sal} = -\rho.c_w.H_L.\frac{dT_L}{dt}, \qquad (9)$$

where $C_w$ is the specific heat of seawater.

In the period of interest, the budget is always positive and lateral melting is given by the changes in the fraction of leads $\left(\frac{\partial A}{\partial t}\right)$:

$$\frac{\partial A}{\partial t} = \frac{A.F_{lead}}{Q_i.h_i + Q_s.h_s}, \qquad (10)$$

where $h_i$ and $h_s$ are the thickness of ice and snow. $Q_i$ and $Q_s$ are the volumetric heat of fusion of ice ($110 \times 10^6\,J\,m^{-3}$) and snow ($302 \times 10^6\,J\,m^{-3}$). Details for the parametrization of leads and the determination of boundary conditions have been discussed in Goosse and Hecq (1994).

The biological processes are influenced by physical ones: advection and diffusion. The corresponding variables are thus governed by equations similar to those used for physical ones (T,S).

$$\frac{\partial B}{\partial t} = -\frac{\partial}{\partial z}(w.B) + \frac{\partial}{\partial z}\left(\tilde{\lambda}.\frac{\partial B}{\partial z}\right) + f_{biol}, \qquad (11)$$

where B is the mass of the biological variable and w a sedimentation or migration speed. Eddy diffusion is supposed to be equal to turbulent diffusion of heat and salt. $f_{biol}$ represents the local modifications due to biological processes.

$$f_{biol} = B.(P^B - R^B - M^B - G^B), \qquad (12)$$

where $P^B$, $R^B$, $M^B$, $G^B$ are the production rate and the metabolic, mortality and predation loss rates of the variable B. According to Platt et al. (1980), primary production P ($mg\,C\,m^{-3}h^{-1}$) can be described by:

$$P^B = P_s^B(1 - \exp(-\alpha \cdot PAR_z/P_s^B)) \cdot \exp(-\beta \cdot PAR_z/P_s^B) \qquad (13)$$

where

$$P_s^B \equiv P_{max}^B\left(\frac{\alpha+\beta}{\alpha}\right)\left(\frac{\alpha+\beta}{\beta}\right)^{(\alpha/\beta)}, \qquad (14)$$

where B is the phytoplankton biomass ($mg\,C\,m^{-3}$), $PAR_z$ is the photon flux density of photosynthetically active radiation at depth z ($\mu mol\,m^{-2}s^{-1}$). $P_{max}^B$ is the maximum production rate by biomass unit at the considered temperature and varies between 0.015 and 0.025 $h^{-1}$. For different groups of algae (Palmisano et al. 1986; Cota et al. 1991), the photosynthetic efficiency ($\alpha$) and the coefficient of photoinhibition ($\beta$) vary respectively between 0.1 and 0.02 and between 0.001 and 0.004 mg C mg Chl $a^{-1}h^{-1}$ ($\mu mol\,m^{-2}s^{-1})^{-1}$. (Jacques 1983; Mitchell and Holm-Hansen 1991; Sakshaug et al. 1991). These equation parameters seem to correctly represent primary production in the Ross Sea area and the adopted values seem to be reasonable for the Ross Sea in spring and summer (Wilson et al. 1986).

To compute $PAR_z$, an exponential decay is used. The attenuation coefficient is, according to Mitchell and Holm-Hansen (1991):

$$PAR_z = PAR_0.\exp(-KPAR.z) \qquad (15)$$

$$KPAR = 0.078 + 0.0136([\text{Chl a}] + [\text{phaeo}]), \qquad (16)$$

where $PAR_{0^-}$ is the active radiation above the sea surface and [Chl a] and [phaeo] are concentrations in chlorophyll a and phaeopigments. We suppose a carbon/chlorophyll ratio of 50. Metabolic losses (respiration + excretion) are considered to be proportional to the biomass and respiration rate taken at $3 \times 10^{-3}\,h^{-1}$ (Mitchell and Holm-Hansen 1991). In the period of interest, mortality is considered limited by biomass. Excepted for *Phaeocystis*, which are not consumed by zooplankton, the term for grazing is parametrized by:

$$\text{Grazing} = \text{Zp}.\text{Gr}_{\text{max}}.\frac{P}{kg + P}, \tag{17}$$

where Zp is zooplankton biomass ($mg\,C\,m^{-3}$). $Gr_{max}$ is the maximum grazing rate taken between 0.06 $day^{-1}$ (for the krill) and 0.2 $day^{-1}$ (for copepods) and kg is a constant taken at $200\,mg\,C\,m^{-3}$ (Conover and Huntley 1991). A part of the phytoplankton consumed ($\eta \approx 20\%$,) is not assimilated and represents faecal pellet production. The term of metabolic losses (respiration + excretion) and predation are supposed to be proportional to zooplankton biomass with rates of $1 \times 10^{-3}\,h^{-1}$ (Slagstad 1981; Conover and Huntley 1991).

Algal cells can live and reproduce in the ice, where biomass as high as $400\,mg\,Chl\,a\,m^{-3}$ can be present (Garrison and Buck 1991; Cota et al. 1991). A simplified equation for ice algae growth has been included in the model. When ice melts, ice algae are released in the water column and can act as seeds for primary production. Just-released algae may aggregate and sink rapidly to the bottom. It is difficult to determine the value of this flux and the sinking speed has been taken between 0.9 and $0.4\,m\,day^{-1}$.

The general equation applied to the zooplankton variables takes into account the part of the phytoplankton consumed ($\eta \approx 20\%$) but not assimilated which represents faecal pellet production. In the present state of the model, the terms of metabolic losses (respiration + excretion) and predation are supposed to be proportional to zooplankton biomass and loss rates of $1 \times 10^{-3}\,h^{-1}$ were considered (Conover and Huntley 1991).

## Results of the Model and Discussion

To simulate the theoretical standard state of the Ross Sea plankton ecosystem, the ECOHYDRO-MVG model has been tested with initial conditions and physical forcing constraints corresponding to mean values for the whole region (standard run). Initial thickness of ice and snow are 70 and 20 cm, respectively (Goosse and Hecq 1994), and initial percentage of leads is 15% (Jacobs and Comiso 1989). The initial profile of salinity and temperature is homogenous in the first 100 m; salinity equals 34, and temperature is at the melting point (−1.89 °C). Under this layer, salinity and temperature increase linearly until 150 m whereupon their values are fixed at 34.2 and −1.75 °C, respectively. Initial concentrations of phytoplankton and zooplankton in the water are $0.50\,mg\,C\,m^{-3}$ and $10\,mg\,C\,m^{-3}$. Initial concentrations of ice algae are $0.5\,g\,m^{-2}$. During winter, the macro and meso-zooplankton remain at greater depths with lower metabolic rates (Conover and Huntley 1991); upward migration seems to begin with an increase in sun radiation. In the present state of the model, zooplankton grazing is active during ice melting but is reduced when the sea is completely ice free. For other zooplankton groups, grazing is light-limited.

Parameters needed to compute atmospheric forcing are listed in Table 1. Variation of wind speed can be important to the mixed layer depth (Mellor and Kantha 1989). Wind speed is considered to oscillate sinusoidally around an average value with a period of 4 days and an amplitude of $1.5\,m\,s^{-1}$.

As shown in Fig. 5, at the end of winter, before 20 December, no ice melting occurs. The mixed layer is deep and values for turbulent kinetic energy are high over a large depth. The stabilizing buoyancy flux induced by melting implies a decrease in the thickness of the mixed layer. The results of the model indicate that ice melting begins in the second week of December and that the sea is ice-free at the end of the first week of January, which is a reasonable value for the Ross Sea (Jacobs and Comiso 1989).

The retreat is nearly linear; but, as the melt and flux increase with time, halocline strength increases. When the ice has completely disappeared, the absence of a buoyancy flux allows the wind to deepen the mixed layer. Note that wind strength increases with time in this season (Table 1).

The temperature of the upper layer remains near the melting point when ice is present and increases with time when it has retreated (Fig. 5). When atmospheric conditions become more rigorous, temperature decreases until freezing. The

**Table 1.** Atmospheric parameters

| | Sept. | Oct. | Nov. | Dec. | Jan. | Feb. | Mar. |
|---|---|---|---|---|---|---|---|
| Wind speed ($m\,s^{-1}$)[a] | 6 | 6 | 5.5 | 4.5 | 5 | 5.5 | 6 |
| Air temperature (°C)[a] | −20 | −15 | −8 | −3 | −2 | −7 | −14 |
| Relative humidity[b] | 0.65 | 0.65 | 0.65 | 0.65 | 0.65 | 0.65 | 0.65 |
| Cloud cover[c] | 0.54 | 0.55 | 0.55 | 0.55 | 0.60 | 0.64 | 0.70 |

[a] Savage and Stearns (1985), Creazzo (1992).
[b] Andreas and Ackley (1982).
[c] Comiso et al. (1993), Parkinson and Washington (1979).

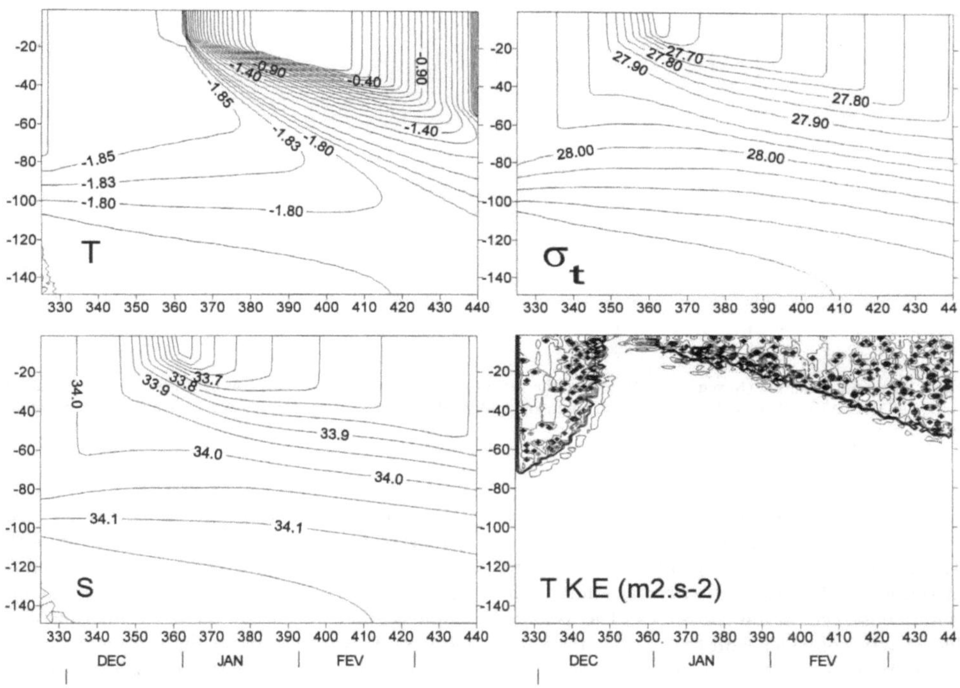


**Fig. 5.** Temporal evolution (Julian days, November to March) of the vertical distribution of the variables in the water column (standard run of ECOHYDRO MVG Model). *T* temperature (°C); *S* salinity (psu); $\sigma_t$ density; *TKE* turbulent kinetic energy $m^2\,s^{-2}$. Oscillations in curves (TKE) are due to wind oscillations

computed temperatures are a little too low. This is linked to the delay in the retreat of ice. However, water temperature rarely exceeds 1 °C in the Ross Sea (Jacobs and Comiso 1989).

The phytoplankton results of the standard run (Fig. 6) confirm the importance of stratification induced by melting for primary productivity. The beginning of ice algae growth corresponds to snow melting. When the ice retreats, the decrease of the mixed layer depth and the increase in light induce phytoplankton growth in the water column. When the ocean has just become ice-free, the phytoplankton is already at a high value (50 mg Chl a $m^{-2}$ – more than 1 mg Chl a $m^{-3}$) in the surface layer. As time proceeds, the available light decreases because of self-shading and the grazing

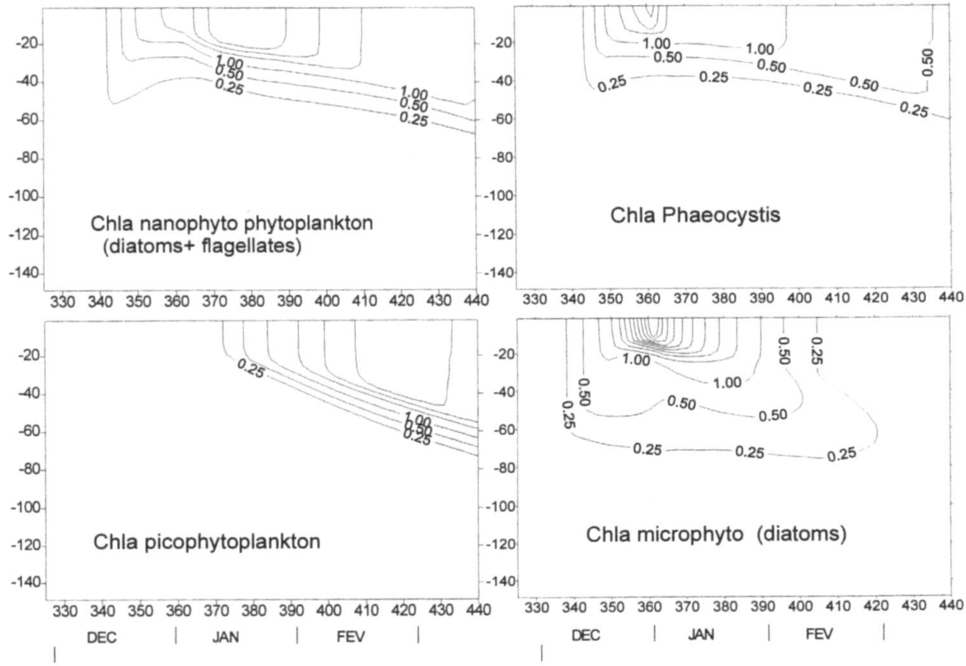


**Fig. 6.** Temporal evolution (November to March) of the vertical distribution of the variables in the water column (standard run of ECOHYDRO MVG model). Chl a of nanophytoplankton, *Phaeocystis*, picophytoplankton and microphytoplankton ($mg\,Chl\,a\,m^{-3}$)

pressure is more intense. In the water column, phytoplankton biomass reaches maximum values of $4\,mg\,Chl\,a\,m^{-3}$ (150 to $200\,mg\,Chl\,a\,m^{-2}$) at the end of December. This estimation corresponds to the values observed in the Ross Sea. The peak of *Phaeocystis* coincides with that of diatoms but is less important and occurs later in the season in relation with different production and grazing processes of these two groups. The maximum of nano and picoplankton occur later, respectively, in January and February. This latter result should be taken with care because of the limitations of the model at the end of summer. Zooplankton biomass grows when food is available and decreases when phytoplankton concentrations are too low.

The model has been tested in various typical areas or situations of the Ross Sea (Figs. 7, 8) and simulates explicitely the variability of the marginal ice zone and phytoplankton growth in function of specific conditions (Table 2).

1. In the southern part of the Ross Sea polynya (Fig. 7), ice thickness is only 20 cm and snow is negligible, with a consequent early stabilization of the water column; phytoplankton blooms earlier than in the standard run. However, biomass is two times lower than in the standard run even if initial algal concentrations are the same. This result suggests that the presence of ice algae and their role as seeding organisms is less important than the low phytoplankton growth rate due to low stratification.

2. In the northern part of the Ross Sea polynya (Fig. 7), where the wind is stronger, the mixed layer is deeper. An increase of $2\,m\,s^{-1}$ can be responsible for a decrease in phytoplankton biomass maximum of 20%. Local storm conditions, which often occur in the northern part of the Ross Sea polynya, can destroy the stratification induced by melting and nearly suppress the favourable conditions for phytoplankton growth, with consequently less apparent peaks. These

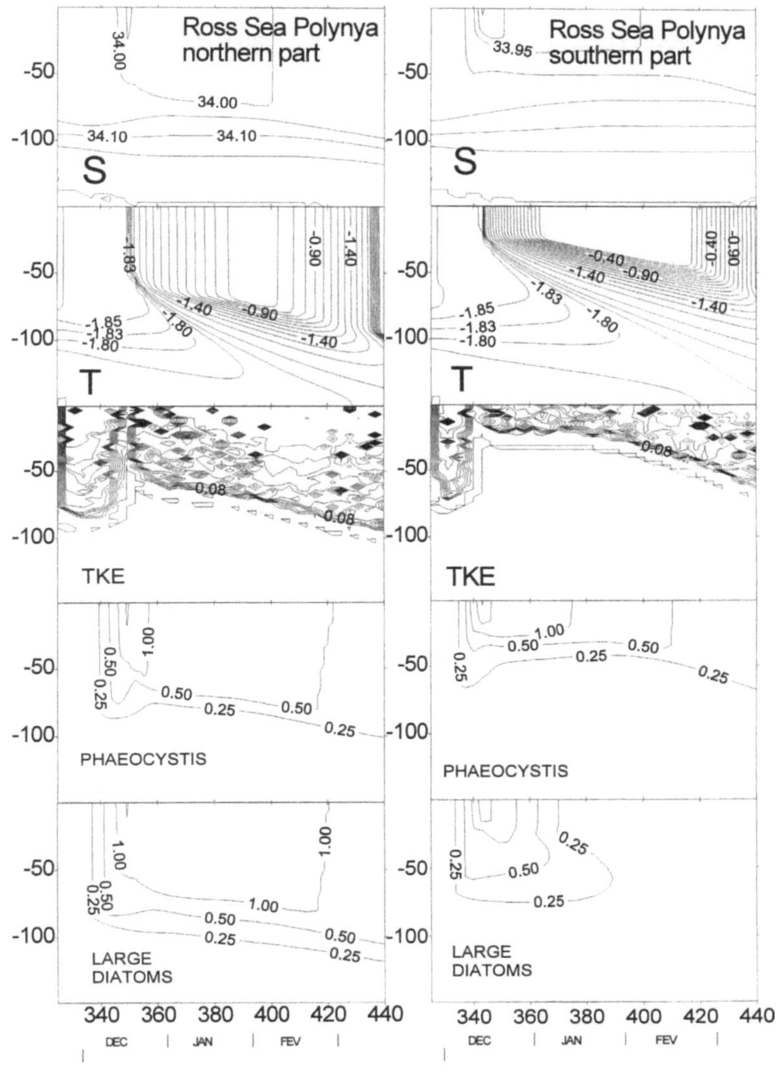


**Fig. 7.** Application of ECOHYDRO MVG model to the northern and southern parts of the Ross Sea polynya. For symbols see Figs. 5 and 6. Chl a units are mg Chl a $m^{-3}$

latter results simulate the situations observed during the ROSSMIZE cruise (Figs. 2 and 3).

3. In permanent ice-free areas (Fig. 8), the mixed layer is deep (>60 m) inducing a small photosynthetic activity. Even in the absence of grazing, net phytoplankton production is low. If grazing is included, phytoplankton changes would be nearly unobservable on this scale.

4. In permanent ice covered areas (Fig. 8), phytoplankton growth is delayed because of the ice, which diminishes available light; however, during ice melting, production increases and phy-

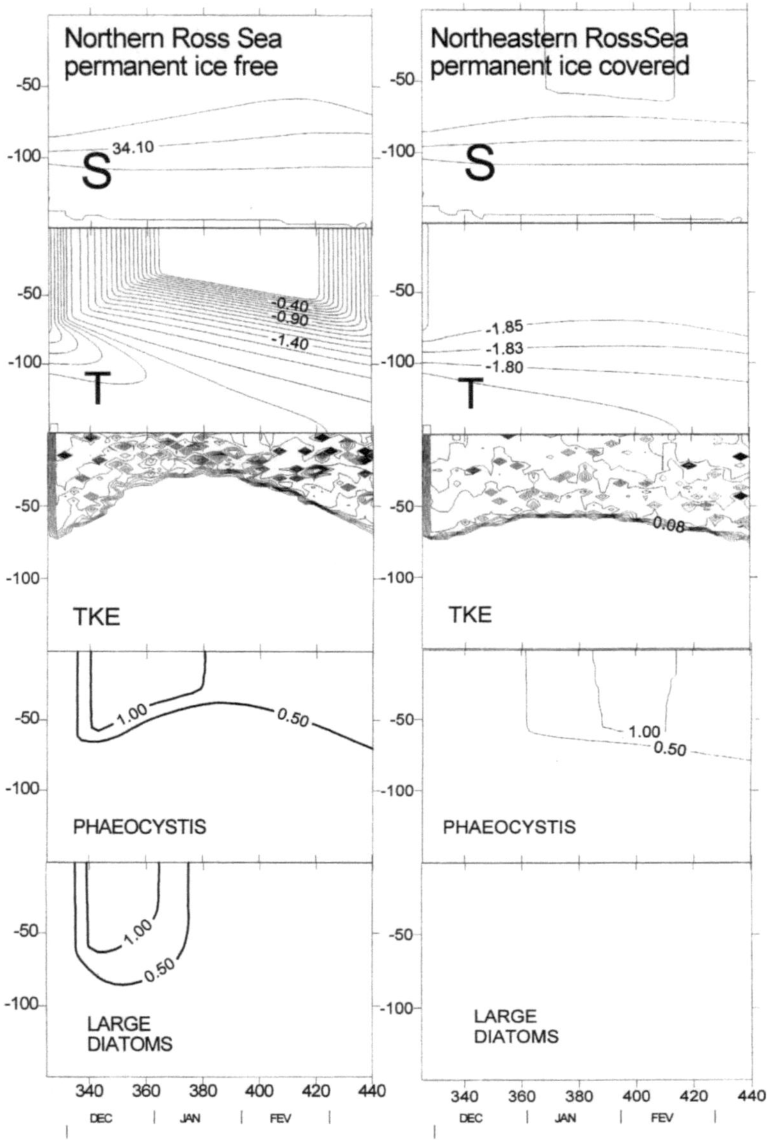


**Fig. 8.** Application of ECOHYDRO MVG model to the northern Ross Sea permanent ice-free zone and the northeastern permanent ice-covered zone. For symbols see Figs. 5 and 6. Chl a units are $mg Chl a m^{-3}$

**Table 2.** Initial and forcing values for standard run and local conditions of southern and northern part of the Ross Sea polynya and for hypothetical permanent ice-free and ice-covered area. *Hginit* and *hninit* are initial values of ice and snow thickness (m); *cleads* is the lead surface ratio (%) and *cdiv* is the ice divergence ratio (% per month). *Vwind* is the wind velocity ($m\,s^{-1}$)

| | Standard run | South polynya | North polynya | Ice-free | Ice permanent |
|---|---|---|---|---|---|
| *hginit* | 0.7 | 0.2 | 0.2 | 0 | 2 |
| *hninit* | 0.2 | 0.0 | 0.0 | 0 | 0.2 |
| *cleads* | 15 | 15 | 15 | 100 | 1 |
| *cdiv* | 15 | 25 | 25 | | 1 |
| *vwind* | 5.5 | 5.5 | 7.5 | 5.5 | 5.5 |

toplankton biomass reaches, in the absence of total ice melting, the values obtained where there is no ice.

Sensitivity of phytoplankton production to some physical parameters has been tested. The light available for phytoplankton will be modified if the ice melts sooner or if there is more cloud. Sedimentation can also play a role. But these parameters are less influential than the biological parameters previously mentioned. Biological results are very sensitive to the values of the parameters included in the formulation of photosynthetic production. Small local or geographical variations of $P^B$ and $\alpha$ can have strong effects on phytoplankton biomass and may explain some of the variability in the Ross Sea (Goosse and Hecq 1994). As a consequence, even small nutrient limitation, which induces a small reduction in photosynthesis, could reduce net growth. Note that a reasonable modification of the light attenuation coefficient has the same consequences as variation of $\alpha$, but quantitatively less important. It is thus very important in the future to select values of $P^B$ and $\alpha$ corresponding to the population and external conditions occurring in the studied area (Neori and Holm-Hansen 1982; Tilzer et al. 1986).

## Conclusions

As a conclusion, it appears that diverse ecological and hydrodynamic processes simultaneously control the variability of the Ross Sea Seasonal Ice Zone ecosystem. All trophic levels depend on sea ice at least during a part of their life histories. For all of these reasons, the ecosystem is essentially sensitive to environmental changes such as climate. An interdisciplinary modelling approach has been developed to understand how the biological and physical processes, interlinked at various scales, govern this ecosystem.

The major vertical interaction between physical and biological processes has been simulated. Bloom dynamics and water structure compare realistically with the few data available for such an environment. The introduction of an explicit ice model is an improvement. The impact of melting on the water column, which is essential for the understanding of bloom dynamics, is obtained from the model and not as an initial condition or artificial parameterization. Moreover, the evolution of biomass can be studied from the beginning of the bloom (under ice cover) until it decreases and not only after complete ice melting.

Ice acts mainly as a switch: if ice melting is present, production is intense and massive blooms are possible even with high losses (grazing). In the absence of ice melting, blooms are nearly impossible even with moderate losses. Phytoplankton biomass is especially influenced by modifications of biological parameters. Zooplankton is a potential powerful control factor of phytoplankton biomass but its stock and metabolism are scarcely known. Primary production is better known, but results are very sensitive to parameters included in its parameterization. Precise values are essential. Initial concentrations in the water and the presence of ice algae seems to be of lesser importance.

In a first step, the biological model has been simplified. It includes simple equations for phytoplankton and zooplankton variables. These are expressed in $mg\,C\,m^{-3}$. No variable is associated to nutrients because these are rarely limiting, even in blooms.

This model seems to be convenient to understand key interactions between ice, upper ocean and biological populations and their implications for melting, structuring of the water column and productivity. However, we have neglected 3-D phenomena such as ice dynamics or water advection, quantitatively important but less directly concerned in biological/physical vertical coupling. In the present step, complex relations between higher trophic levels have also been neglected.

The ECOHYDRO MVG model in its present state is applied only for the spring-summer period and is utilized for the simulation of the microscale physical processes of vertical stabilization of the water column at the level of the marginal-ice zone of the Ross Sea. It will need more developments for ice formation and biological processes inside the ice (photo-adaptation, UV, etc.).

*Acknowledgements.* This work is a part of the Belgian Research Program in the Antarctic funded by the Prime Minister's Services, Science Policy Office (SSTC Project Antar A4/DD/B21, Brussels, Belgium) in collaboration with the PNRA (Programma Nazionale di Ricerca in Antartide, Italy). J. H. Hecq is Research Assistant from FNRS (National Foundation for Scientific Research, Belgium). The authors thank Dr. T. Platt and Dr. J. Priddle for providing helpful criticisms of the manuscript.

## References

Ackley SF, Sullivan CW (1994) Physical controls on the development and characteristics of Antarctic sea ice biological communities - a review and synthesis. Deep Sea Res 41(10): 1583–1604

Andreas EL, Ackley SF (1982) On the differences in ablation seasons of Arctic and Antarctic sea ice. J Atmos Sci 39: 440–447

Andreas EL, Murphy B (1986) Bulk transfer coefficients for heat and momentum over leads and polynyas. J Phys Oceanogr 16: 1875–1883

Arrigo KR, McClain CR (1994) Spring phytoplankton production in the Western Ross Sea Science 266: 261–263

Azzali M, Kalinowski J (1992) Italian Antarctic acoustic survey of krill. In: Gallardo VA, Ferretti O, Moyano HI (eds) Oceanografia in Antartide. ENEA-PNRA-EULA, Concepcion, Chile, pp 321–330

Catalano G, Povero P, Fabiano M, Benedetti F, Goffart A (1997) Nutrient utilisation and particulate organic matter changes during summer in the upper mixed layer (Ross Sea, Antarctica). Deep Sea Res 44(1): 97–112

Comiso JC (1991) Satellite remote sensing of the polar oceans. J Mar Syst 2: 395–434

Comiso JC, McClain CR, Sullivan CW, Ryan JP, Leonard CL (1993) Coastal zone color scanner pigment concentration in the Southern Ocean and relationships to geophysical surface features. J Geophys Res 98(15): 2419–2451

Conover RJ, Huntley M (1991) Copepods in ice-covered seas. Distribution, adaptations to seasonally limited food, metabolism, growth patterns and life cycle strategies in polar seas. J Mar Syst 2(1/2): 1–41

Cota GF, Legendre L, Gosselin M, Ingram RG (1991) Ecology of bottom ice algae. I. Environmental controls and variability. J Mar Syst 2(3/4): 257–277

Creazzo S (1992) Meteorological observations. Oceanographic campaign 1989–1990, data report, part 2. Ministry of the University and Scientific and Technological Research Republica Italiana, Rome, pp 63–78

Doronin YP, Kheisin DE (1977) Sea ice. Amerind, Fairbanks, Alaska, 323 pp

Garrison DL, Buck KR (1991) Surface-layer sea ice assemblages in Antarctic pack ice during the austral spring: environmental conditions primary production and community structure. Mar Ecol Prog Ser 75: 161–172

Goffart A, Hecq JH, Catalano G (1999) Factors controlling the distribution of diatoms and *Phaeocystis* in the Ross Sea. Submitted to J Mar Syst.

Goosse H, Hecq JH (1994) Modelling the ice-ocean-plankton interactions in the Southern Ocean. J Mar Syst 5: 471–484

Gordon AL (1988) Spatial and temporal variability within the Southern Ocean. In: Sahrhage D (ed) Antarctic ocean and resources variability. Springer, Berlin Heidelberg New York, pp 41–56

Hecq JH, Magazzù G, Goffart A, Catalano G, Vanucci S, Guglielmo L (1992) Distribution of planktonic components related to vertical structure of water masses in the Ross Sea and the Pacific sector of the Southern Ocean. In: Albertelli G, Ambrosetti W, Picazzo M, Ruffoni-Riva T (eds) Atti IX Congr Ass Ital Oceanol Limnol, S Lang, Genova, pp 665–678

Hecq JH, Brasseur P, Goffart A, Lacroix G, Guglielmo L (1993) Modelling approach of the planktonic vertical structure in deep Austral Ocean. The example of the Ross Sea ecosystem. In: Progress in Belgian Oceanographic Research. Proc Annu Worksh on Belgian Oceanographic Research, Academy of Sciences, Brussels, pp 235–250

Huntley ME, Nordhausen W, Lopez MDG (1994) Elemental composition metabolic activity and growth of Antarctic krill *Euphausia superba* during winter. Mar Ecol Prog Ser 107: 23–40

Jacobs SS, Comiso JC (1989) Sea ice and oceanic processes on the Ross Sea continental Shelf. J Geophys Res 94: 18195–18211

Jacques G (1983) Some ecophysiological aspects of the Antarctic phytoplankton. Polar Biol 2: 27–33

Knox GA (1983) The living resources of the Southern Ocean: a Scientific overview. In: Vicuna FO (ed) Antarctic resources policy: scientific, legal and political issues. Cambridge University Press, Cambridge, pp 21–60

Knox GA (1994) The biology of the Southern Ocean. Cambridge University Press, Cambridge, 429 pp

Mackintosh NA (1973) Distribution of post-larval krill in the Antarctic. Discovery Rep 36: 95–156

Maykut GA, Untersteiner N (1971) Some results from a time-dependent thermodynamic model of sea ice. J Geophys Res 76: 1550–1575

Mellor GL, Kantha L (1989) An ice-ocean coupled model. J Geophys Res 94: 10937–10954

Mitchell BG, Holm-Hansen O (1991) Observation and modeling of the Antarctic phytoplankton crop in relation to mixing depth. Deep Sea Res 38(8/9): 981–1007

Nelson DM, Smith WO Jr (1986) Phytoplankton bloom dynamics of the western Ross Sea ice edge. II. Mesoscale cycling of nitrogen and silicon. Deep Sea Res 33(10): 1389–1412

Neori A, Holm-Hansen O (1982) Effect of temperature on rate of photosynthesis in Antarctic phytoplankton. Polar Biol 1: 33–38

Niebauer HJ, Smith WO Jr (1989) A numerical model of mesoscale physical-biological interactions in the Fram Strait Marginal Ice Zone. J Geophys Res 94(11): 16151–16175

Nihoul JCJ (1984) A three-dimensional general marine circulation model in a remote sensing perspective. Ann Geophys 2(4): 433–442

Palmisano AC, SooHoo JB, SooHoo SL, Kottmeier ST, Craft LL, Sullivan CW (1986) Photoadaptation in *Phaeocystis pouchetii* advected beneath annual sea ice in McMurdo Sound Antarctica. J Plankton Res 8(5): 891–906

Parkinson CL, Washington WM (1979) A large-scale numerical model of sea ice. J Geophys Res 84(1): 311–337

Platt T, Gallegos CL, Harrison WG (1980) Photoinhibition of photosynthesis in natural assemblages of marine phytoplankton. J Mar Res 38: 687–701

Platt T, Mann KH, Ulanowicz RE (eds) (1981) Mathematical models in biological oceanography. UNESCO Monogr Oceanogr Methodol 7: 156 pp

Pond S, Pickard GL (1983) Introductory dynamical oceanography, 2nd edn. Pergamon, Oxford, 329 pp

Priddle J, Croxall JP, Everson I, Heywood RB, Murphy EJ, Prince PA, Sear CB (1988) Large-scale fluctuations in distribution and abundance of krill: a discussion of possible causes. In: Sahrhage D (ed) Antarctic Ocean and resources variability. Springer Berlin Heidelberg New York, pp 169–182

Quetin LB, Ross RM, Frazer TK, Haberman KL (1996) Factors affecting distribution and abundance of zooplankton with an emphasis on Antarctic krill *Euphausia superba*. Antarct Res Ser 70: 357–371

Russo A, Gallarato A, Testa G, Corbo C, Pariante R (1997) Physical data collected during ROSSMIZE cruise (Ross Sea, November - December 1994). In: Faranda F, Guglielmo L, Povero P (eds.) National Programme for Antarctic Research. Rossmize 1993–1995. Data Report Part I: 25–110

Sakshaug E, Slagstad D, Holm-Hansen O (1991) Factor controlling the development of phytoplankton blooms in the Antarctic Ocean - a mathematical model. Mar Chem 35: 259–271

Savage ML, Stearns C (1985) Climate in the vicinity of Ross Island Antarctica. Antarct J US 20(1): 1–9

Semtner AJ Jr (1976) A model for the thermodynamic growth of sea ice in numerical investigations of climate. J Geophys Res 6: 379–389

Slagstad D (1981) Modeling and simulation of physiology and population dynamics of copepods. Effects of physical and biological parameters. Model Identification Control 2(3): 119–162

Shine KP, Henderson-Sellers A (1985) The sensitivity of a thermodynamic sea ice model to changes in surface albedo parameterization. J Geophys Res 90: 2243–2250

Smetacek V, Scharek R, Nöthig EM (1990) Seasonal and regional variation in the pelagial and its relationship to the life history cycle of krill. In: Kerry KR, Hempel G (eds) Antarcic ecosystems: ecological change and conservation. Springer, Berlin Heidelberg New York, pp 103–114

Smith WO, Gordon LI (1997) Hyperproductivity of the Ross Sea (Antarctica) polynya during austral spring. Geophys Res Lett 24(3): 223–236

Smith WO Jr, Nelson DM (1985) Phytoplankton bloom produced by a receding ice edge in the Ross Sea: spatial coherence with the density field. Science 227: 163–166

Sullivan CW, McClain CR, Comiso JC, Smith WO Jr (1988) Phytoplankton standing crops within an Antarctic ice edge assessed by satellite remote sensing. J Geophys Res 93: 12487–12498

Svanson A (1991) A simple primary production model for the NW Weddell Sea. Mar Chem 35: 347–354

Tilzer MM, Elbrächter M, Gieskes WW, Beese B (1986) Light-temperature interactions in the control of photosynthesis in Antarctic phytoplankton. Polar Biol 5: 105–111

Veth C (1991) The evolution of the upper water layer in the marginal ice zone austral spring 1988 Scottia-Weddell Sea. J Mar Syst 2(3/4): 451–464

Veth C, Lancelot C, Ober S (1992) On processes determining the vertical stability of surface waters in the marginal ice zone of the north western Weddell Sea and their relationship with phytoplankton bloom development. Polar Biol 12: 237–243

Wilson DM, Smith WO Jr, Nelson DM (1986) Phytoplankton bloom dynamics of the western Ross Sea ice edge - I Primary productivity and species-specific production. Deep Sea Res 33(10): 1375–1387

Zwally HT, Parkinson CL, Comiso JC (1983) Variability of Antarctic sea ice and changes in carbon dioxide. Science 220: 1005–1012

CHAPTER 30

# Characteristic Properties and Dynamic Aspects of Krill Swarms from the Ross Sea

**M. Azzali, J. Kalinowski, G. Lanciani, and G. Cosimi**[1]

**ABSTRACT**

In comparison with insect swarming, fish schooling and phytoplankton patches, studies on swarming behaviour of Antarctic krill, that take into account at least the ratio of random and non-random forces in terms of the density of organisms, do not exist. It is known that insects appear to disperse according to a reaction-diffusion model with a constant diffusion coefficient. The aim of this chapter is to verify if this mathematical approach, tested on insect swarming, can also be applied to krill swarms. Sixty-three swarms, registered without loss of information on the BioSonics hydroacoustic system during the Italian expeditions to the Ross Sea in the years 1989–1990 and 1994 and identified as *Euphausia superba* from net hauls and visual inspection of their echograms, were used to verify this hypothesis. The selected swarms had a mean vertical extent around 32 m and differed in all other parameters. In particular, mean swarm density ranged from 34 to 4914 krill $m^{-3}$. It was found that the mean vertical density distribution and the mean ratio of the diffusity and attractive forces of the krill swarms, whether considered by the density classes or all together, are consistent with insect model predictions. These results suggest that many dynamic aspects and physical characteristics of krill swarms can be similar to those of insect swarming. To proceed into krill swarming, however, other kinematic, biological and environmental data are required for a complete evaluation and an eventual improvement of the mathematical model.

## Introduction

Various insect species appear to disperse according to a reaction-diffusion model with a constant diffusion coefficient (Okubo 1980, 1986; Shigesada 1980). Kareiva (1983) gives the actual values for the diffusion coefficients which he measured from experiments on a variety of insect species. Can we use the same model to study krill swarming?

In insect swarming each insect is subject primarily to a random force, to which it opposes with a non-random force of attraction towards the centre of the swarm (Okubo 1980). Therefore the flight pattern of swarming individuals varies widely from irregular, zig-zag motion, that can be regarded as random flight, to a straight to- and-fro movement around the centre of swarm, that can be regarded as non-random flight (Downes 1969). Without this inward force of attraction, under the action of the random force, a swarm would diffuse very soon (Fick's law), like an assemblage of particles released at sea. Theories of oceanic diffusion (deduced by Fick's law) are expected to hold approximately for nutrients, phytoplankton, fish eggs and larvae in the early stages (Okubo 1980). In these cases, the random motions of the ambient water are the effective agents of the passive movement of the organisms (diffusion).

On the other hand, in a school of fish, the random force is negligible and predominant non-random forces have a behavioral nature (Horwood and Cushing 1977). Neighbouring fish have a tendency to equalize their velocities and orientation (arrayal force), their distance (mutual interaction forces) and their forward thrust (Van Olst and Hunter 1970; Radakov 1973; Okubo et al. 1977). Thus, for schooling fish, the movement appears to be dominated by deterministic factors: the motion of individual fish in a school remains synchronized and a school preserves a highly organized

[1] CNR/IRPeM., Molo Mandracchio, 60100 Ancona, Italy

structure under almost any hydrodynamic condition.

Although insect swarming, fish schooling and phytoplankton patches are extensively investigated, studies on swarming behaviour of pelagic zooplankton, and in particular of krill, that take into account at least the ratio of random and non-random forces in terms of the density of organisms, are nearly non-existent. An intuitive approach suggests that insect and zooplankton swarming process can be similar in many aspects and that mathematical models tested on insect swarming should also be applied in some zooplankton swarms. The aim of this chapter is to verify this hypothesis by using some parameters of the insect models that can be extracted from krill swarms by means of acoustic techniques. The acoustic data used in the chapter concerns only the internal structure of krill swarms such as the average density and the vertical distribution of krill density relative to the centre of mass. Migratory aspect of swarms, their geographical distribution, the biological characteristics (sex, length) of individuals in swarms are not analyzed here. Acoustic data, together with insect mathematical models, enable us to estimate some features of krill swarms such as the ratio of diffusivity and attractive forces and the vertical distribution around the dispersing centre. These results, although deduced from simple models that probably will require significant adjustment and refinement, may be used as a measure of diversity between krill (*Euphausia superba*) and other zooplankton (i.e. *Euphausia crystallorophias*) swarms.

## Swarming Models

Flows in water (and in air) are generally in turbulent state. Turbulence, a spatially and temporally random motion, is produced by forces that act on the water in horizontal direction (wind and currents at different depth) and in vertical direction (kinetic energy, freshwater from ice melting, light, heat, vertical current shear). Because the inputs of driving forces are asymmetric and act mainly on a water column, also the diffusive action of turbulence on plankton patterns is asymmetric and changes more rapidly in the vertical direction than in the horizontal (Evans 1977). In the model the direction of diffusivity is considered, although in the results mainly the vertical structures of swarms are discussed. Despite the diffusive action of water turbulence in vertical and horizontal directions as well as the apparently random swimming of individuals in swarms (Siegel and Kalinowski 1994), krill maintain their cohesion without appreciable dispersion over some days. To deal with this problem, a mechanism of reaction was introduced into the diffusion process which opposes the action of diffusion (reaction process).

### The Basics of the Diffusion Process

Turbulent diffusion is an irreversible and random phenomenon by which a system composed by a number of particles (physical or biological), initially concentrated in a small region (A), spreads out in time and occupies ever-increasing regions (B,C,...) around the initial one. The term irreversible means that the system, once evolved from a state A to a state B, will never return by itself to A. The term random means that, as a result of diffusion, the system, changing from state A to state B, tends to increase its disorder (i.e. forces of attraction among particles weaken as the distances between particles increase). When this phenomenon involves a large number of particles, resulting in some macroscopic and gross regular motion of the group, we think of it as a diffusion process. According to the classical theory of diffusion the amount of particles in the xi ($i = 1,2,3$) direction across a unit normal area in a unit time (i.e. the flux Ji) is proportional to the gradient of concentration of particles:

$$Ji = -Di(N)\frac{\partial N}{\partial xi}, \qquad (1)$$

where N is the density of particles (number/m$^3$), in general function of t and x, D is the diffusivity (m$^2$/t), in general function of N, t and x with $\frac{\partial D}{\partial N} > 0$. The minus sign indicates that diffusion occurs from high to low density.

The change in time of average density of particles is expressed by the classical Fick's equation of diffusion:

$$\frac{\partial N}{\partial t} = -\frac{\partial Ji}{\partial xi} = \frac{\partial}{\partial xi}\left(Di(N)\frac{\partial N}{\partial xi}\right). \qquad (2)$$

If D does not depend on xi and the homogenities of stimuli and environment produce a not-oriented response in organisms, Eq. (2) becomes:

$$\frac{\partial N}{\partial t} = D(N)\frac{\partial^2 N}{\partial x^2}. \quad (3)$$

This equation can be solved subject to the condition that initially No particles in a unit area are concentrated at x = 0, resulting in:

$$N(x,t) = \frac{No}{(2\pi)^{1/2}(2Dt)^{1/2}}\exp(-x^2/4Dt). \quad (4)$$

Clearly Eq. (4) is the relation for purely random dispersal Fickian diffusion. The variance ($\sigma^2 = 4Dt$) increases linearly with time; the standard deviation increases with square root of time. Organisms initially concentrated at a point x = 0 tend to spread indefinitely [i.e. the diffusion distance at time t is $(2Dt)^{1/2}$]

## Reaction-Diffusion Equations

Swarming is a phenomenon by which a number of organisms oppose the diffusion process, that tends to spread organisms from the centre, with a reaction process that acts to attract them to the centre. If a steady state is established, the effect of diffusion is negated and, as a result, organism concentration or antidiffusion is observed. Thus, the flux flow of organisms to a plane perpendicular to the xi axis must consist of two components, one random from high to low concentration, described by diffusion coefficient D, and the other non-random, in the opposite direction, towards the mass centre of the swarm. This non-random flow may be formulated as Ui N, where Ui denotes the component of mean speed of individuals passing through a plane perpendicular to the i direction and N is the organism's density around the plane. In general, both Ui and D can vary within a swarm and can depend on the density of organisms N. Thus, the total flux is given by:

$$Ji = UiN - Di(N)\frac{\partial N}{\partial xi}. \quad (5)$$

If the swarming is stationary, Ji is constant and the same through any plane. Furthermore, at the boundary of the swarm, N = 0 and accordingly the flux must vanish (Ji = constant = 0). Therefore, Eq. (5), within an almost stationary swarm becomes:

$$\frac{1}{N}\frac{\partial N}{\partial xi} = \frac{Ui}{Di(N)}, \quad (6)$$

where N generally depends on xi, while Ui and Di generally depend on xi and N. Equation (6) allows evaluation of the ratio of the mean speed of organisms and diffusivity from density distribution of organisms N (xi) within the swarm, measured acoustically along the xi axis. The diffusion equation [Eq. (2)] becomes the reaction diffusion Eq. (7), when the organisms oppose diffusion with a speed of flow Ui (reaction) from high to low densities:

$$\frac{\partial N}{\partial t} = -\frac{\partial}{\partial xi}(UiN) + \frac{\partial}{\partial xi}\left(Di(N)\frac{\partial N}{\partial xi}\right), \quad (7)$$

where the reaction Ui and the diffusivity Di in general depend on x, t and N. Therefore Eq. (7) holds: either the swarming is dynamic $\left(\frac{\partial N}{\partial t} \neq 0\right)$ or stationary $\left(\frac{\partial N}{\partial t} = 0\right)$.

## Models for Swarming by Reaction-Diffusion Equations

Reaction-diffusion equations form a reasonable base for studying insect and animal grouping (Okubo 1980, 1986; Shigesada 1980). The origin (x = 0) is taken at the peak density of the swarm that is assumed to be the centre of attraction for insects. As a simple model, the speed of attractive flow (Ui) is assumed constant along the direction i (Shigesada et al. 1979):

$$Ui = -Uosgn(xi), \quad (8)$$

where sign(xi) is a function equal to 1 if x > 0 and −1 if x < 0.

The diffusivity Di is assumed to depend on density N alone. A typical form for diffusivity Di(N) is (Okubo 1980):

$$Di(N) = Do\left[\frac{N(xi)}{No}\right]^m, \; m > 0, \quad (9)$$

where Do is the diffusivity for N(0) = No (peak of density taken at the origin of xi axis). As m > 0 the diffusivity increases with N = N(xi).

If swarm is in a steady state, the processes of diffusion and reaction balance each other, so that $\frac{\partial N}{\partial t} = 0$. Under this condition, substituting Eq. (8) and Eq. (9) into Eq. (7), we find:

$$0 = Uo\frac{\partial}{\partial xi}[\mathrm{sgn}(xi)N] + Do\frac{\partial}{\partial xi}\left[\left(\frac{N}{No}\right)^m \frac{\partial N}{\partial xi}\right]. \quad (10)$$

Integrating Eq. (10) once with respect to xi, we find the equation of flux as the swarming is quasistationary [Eqs. (5) and (6)]:

$$\frac{1}{N}\frac{\partial N}{\partial xi} = \frac{Uo}{Do}\mathrm{sgn}(xi)\left(\frac{N}{No}\right)^{-m} = \frac{Ui}{Di(N)}. \quad (11)$$

Equation (11) shows that the parameter $\frac{U}{D}$ is small near the origin, where the gradients are large, and large near the borders of a swarm, where the gradients are small. Therefore, the maximum inward force is exerted by insects at the periphery of a swarm. Integrating Eq. (11) once more with respect to xi and seeking a symmetric around the origin, we obtain:

$$N = No\left(1 - |xi|\frac{mUo}{Do}\right)^{1/m}, \quad |xi| \le L = \frac{Do}{mUo}$$
$$N = 0, \quad |xi| > L = \frac{Do}{mUo}. \quad (12)$$

Equation (12) shows that insect dispersal is finite in the xi direction. The maximum extent of a swarm in the xi direction L, obtained from the reaction-diffusion equation, approaches $\frac{Do}{mUo}$ as $|xi| \to \infty$. The spatial extent of the swarm L along the xi direction can be interpreted as the standard variation of the spatial distribution of insects $\frac{N(xi)}{No}$. Without the inward flow, UoN, the swarm would spread in the xi direction at the rate $(2Dt)^{1/2}$ [see Eq. (4)]. Depending on the density dependence of Di(N), the spatial distribution of insects:

$$\frac{N(xi)}{No} = \left(1 - |xi|\frac{mUo}{Do}\right)^{1/m} \quad (13)$$

can be leptokurtic ($m < 1$), that has higher amplitudes near its centre and towards its tail than normal distribution with the same mean and variance, or a plantykurtic ($m > 1$). The distribution of a special type of mosquito swarm observed by Okubo and Chiang (1974) is leptokurtic ($m = 1/2$).

When non-homogenities of stimuli and environment produce oriented response in animals, the extent of the swarm is not symmetric:

$$L(+) = \frac{Do}{m1Uo}, \quad \text{if } \mathrm{sgn}(xi) = 1$$
$$L(-) = \frac{Do}{m2Uo}, \quad \text{if } \mathrm{sgn}(xi) = -1. \quad (14)$$

The above model has been applied successfully to a variety of insect swarms (Okubo 1980) and can perhaps also be used for krill swarms. It assumes that individuals are attracted toward the region in the swarm of higher density [density-dependent model, Eq. (9)]. This implies that mutual communication between krill individuals plays a fundamental role in swarm formation and maintenance. Some possible mechanisms of communication are visual, mechanosensory and chemosensory contact (Siegel and Kalinowski 1994). On the contrary, in the density-independent models, animal individuals are independently attracted to a source of attraction such as female pheromones or baits.

## Materials and Methods

The 63 swarms presented in this chapter were selected from a set of 132 swarms on the basis of their vertical extent. These 132 swarms were recorded during the 1989–1990 surveys (78 swarms, Fig. 1) and the 1994 surveys (54 swarms, Fig. 2) in the Ross Sea to develop the DMVBS classifier, described by Azzali and Kalinowski (this Vol.). Most of these swarms were identified as *Euphausia superba* from testing trials, the others from visual evaluation of their echograms (Madureira et al. 1993; Kalinowski and Witek 1985). These 132 swarms had a vertical extent ranging from 8 to 82 m.

The vertical extent of the 63 swarms (about 48% of the total) taken for this analysis is included between 24 to 44 m (Tables 1, 2 and 3).

### Equipment

The swarm data were collected using a BioSonics dual beam, multiplexing system, calibrated at 38 and 120 kHz (Azzali and Kalinowski, this Vol.). Each frequency had a ping rate of $1.2\,s^{-1}$ (triggered alternatively every 0.6 s), and a pulse duration of 0.5 ms. The 38-kHz frequency was received with a 2.5-kHz bandpass and the 120-kHz frequency with a 5-kHz bandpass. The swarms were acoustically

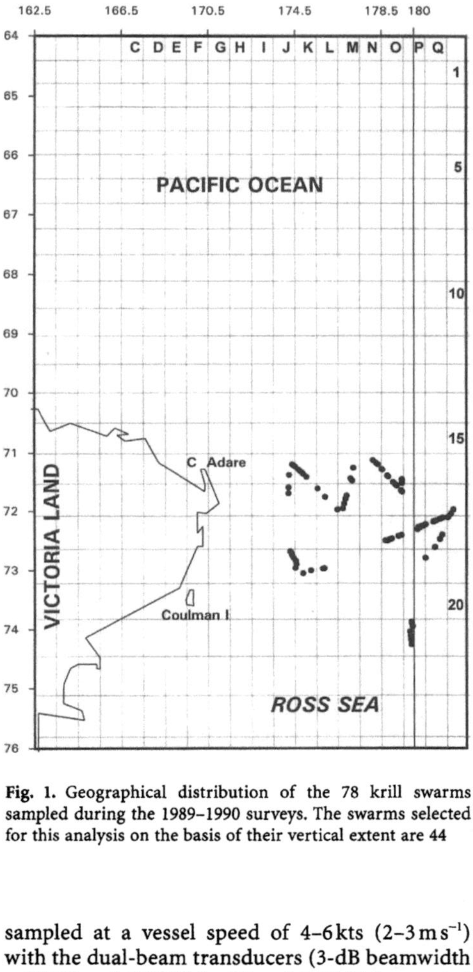


Fig. 1. Geographical distribution of the 78 krill swarms sampled during the 1989–1990 surveys. The swarms selected for this analysis on the basis of their vertical extent are 44

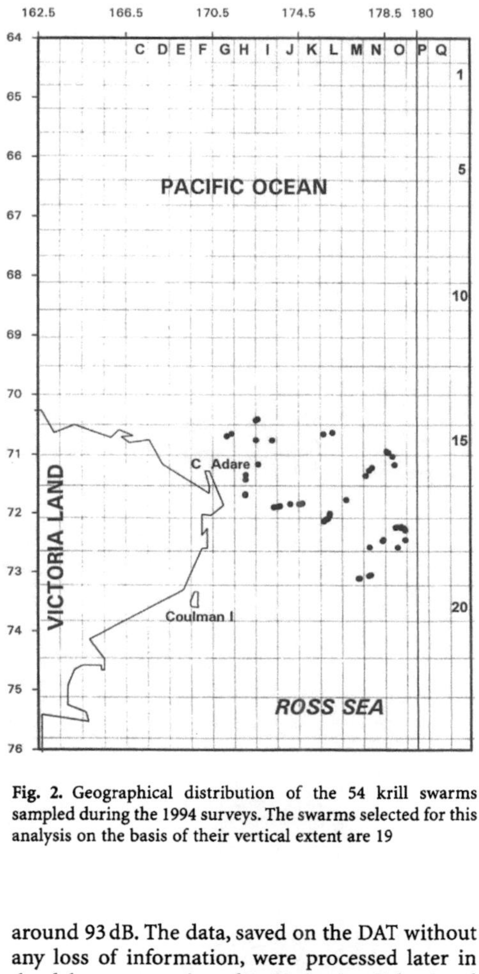


Fig. 2. Geographical distribution of the 54 krill swarms sampled during the 1994 surveys. The swarms selected for this analysis on the basis of their vertical extent are 19

sampled at a vessel speed of 4–6 kts (2–3 m s$^{-1}$) with the dual-beam transducers (3-dB beamwidth = 7°/18° and 10°/22° for 38 and 120 kHz) mounted on a 3-foot V-fin. Data used for this study were stored ping by ping at both frequencies on the BioSonics recording system, which consists of a BioSonics Recorder Interface (Model 171) coupled with a Digital Audio Tape (DAT). The BioSonics interface conditions the echosignal and multiplexes the dual-beam echosignals so that data of both channels (Narrow and Widebeams) at each frequency, and their associated timing and geographical information, are recorded on the two channels of DAT. For swarm analysis, only data from narrowbeam channels were used. The dynamic range of the tape-recording system was around 93 dB. The data, saved on the DAT without any loss of information, were processed later in the laboratory, using the BioSonics Echosignal Processing (a circuit board installed in a PC and operating in Microsoft Windows environment), implemented with a software developed by the acoustic group of IRPeM.

## Formation of the Image of a Swarm

When the vessel passed over a swarm, a "view" of it, at 120 and 38 kHz, was formed by the K successive echosignals V(t) that were returned to the echosounder from the swarm area (Fig. 3). The K echosignals were stored on the DAT with the appropriate timing, spatial and frequency context.

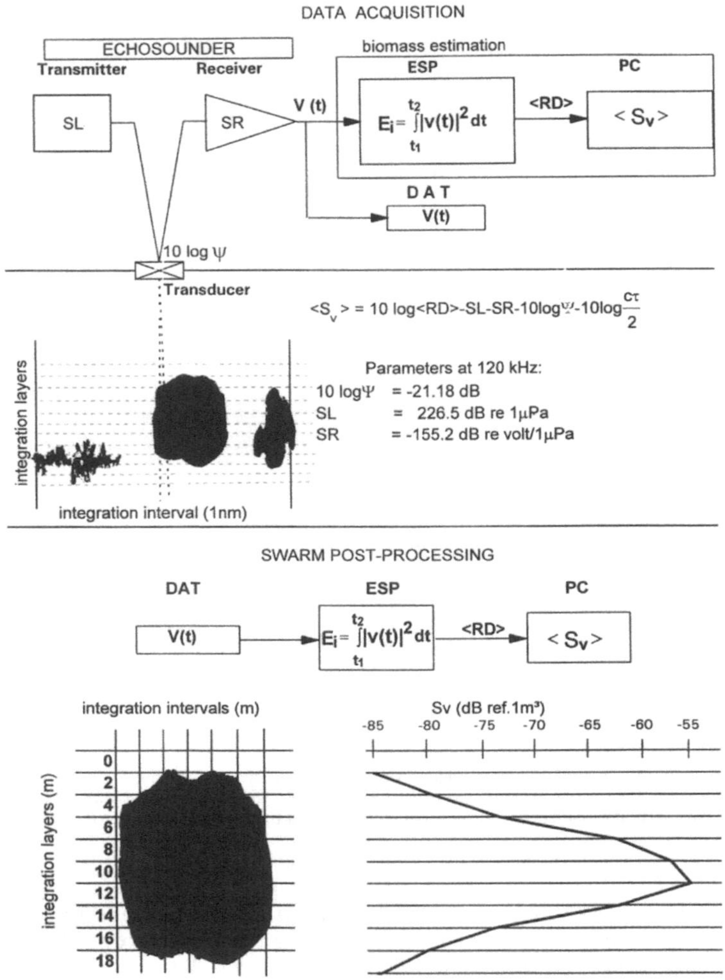


**Fig. 3.** Methodology of data acquisition and swarm postprocessing

After some processing on these data, using ESP software, the echo signal energy or relative density RD ($V^2$) was averaged over equally spaced incremental intervals of 2 m, taken along each water column k (k = 1,2, . . . K) of the swarm area. The $\langle RD \rangle$ values were converted in mean volume backscattering strength $\langle S_v \rangle$, at each frequency, for each column k and incremental interval, from the formula:

$$\langle S_v \rangle = 10 \log \langle RD \rangle - SL - SR - 10 \log \Psi - 10 \log c\tau/2,$$

where SL is the source level (dB), SR the sounder receiver (dB), $10 \log \Psi$ the beam pattern factor (dB), c the speed of sound in water ($m\,s^{-1}$) and $\tau$ the pulse width (s).

The depths (d1,d2) at which the DMVBS (i.e. $S_{v120} - S_{v38}$), returned by a 2-m interval of a water column k (k = 1,2 . . . K) within the swarm area, falling either above 14 dB or below 6 dB, determined the boundaries of the image of the swarm. These thresholds are the range of variation of the krill DMVBS classifier (i.e. mean DMVBS = 10.15 dB; SD = 1.6 dB), presented by Azzali and Kalinowski (this Vol.). The solid silhouette formed in this way was represented by a rectangular matrix of Z columns and X rows. A cell of the matrix had a resolution equal to 2 m on the vertical axis (z) and a two-ping interval on the horizontal axis (x). The $S_v$ value at 120 kHz, averaging over 2 m on the vertical axis and over two pings on the horizontal axis, was attributed to the cells within the solid silhouette, and zero value to the other cells of the matrix. The mean volume backscattering strength of each cell [i.e. $S_v$ (x,z)] was converted in absolute units N (krill/m$^3$) from the formula:

$$10 \log N(x,z) = \langle S_v(x,z)\rangle - \langle TS\rangle - \Delta R,$$

where ⟨TS⟩ is the mean target strength of the krill at 120 kHz.

Target strength of krill was calculated according to the CCAMLR recommendation (Anonymous 1991):

$$TS = -127.45 + 34.85 \log (\text{length in mm}).$$

The matrix N (x,z) is the image of the krill density distribution in the swarm area (Fig. 4).

## Data Analysis

The mean vertical and horizontal distribution of krill in the swarm area was obtained averaging the discretized N(x,z) values over each row (x) and over each column (z), respectively (Fig. 4). Vertical density distribution (and perhaps properties of the sea) usually were more irregular and changed more rapidly than horizontal density distribution. This chapter discusses only the mean vertical distribution of krill swarms. The 63 swarms selected for this analysis (Tables 1, 2 and 3) had a mean vertical extent around 32 m, ranging between 24 (swarm no. 10) and 44 m (swarm no. 2), and differed in all the other physical parameters: density ranged from 34 (swarm no. 1) to 4914 krill m$^{-3}$ (swarm no. 56), horizontal extent from 30 (swarm no. 59) to 748 m (swarm no. 28), and depth at the top from 4 (swarm no. 39) to 70 m (swarm no. 46). They were sampled in different years (1989–1990 and 1994), in different seasons (early summer and late spring), at different periods of day, and in different environmental conditions (seawater free and covered by ice). The 63 swarms were divided, on the basis of the mean swarm density calculated from the matrix N (z,x) (Fig. 4), in three classes:

1. Low density (<100 krill m$^{-3}$): 10 swarms. The mean density of these swarms was 80.74 krill

**Table 1.** Physical parameters of the ten swarms that form the class at low density (<100 krill m$^{-3}$)

| No. | Thickness (m) | Depth (m) | G. centre[a] (m) | Length (m) | M. density[b] (krill m$^{-3}$) | P. density[c] (krill m$^{-3}$) | P. depth[d] (m) |
|---|---|---|---|---|---|---|---|
| 1 | 30 | 6 | 21 | 304 | 34.07 | 71.29 | 26 |
| 2 | 44 | 14 | 36 | 316 | 99.85 | 285.10 | 48 |
| 3 | 38 | 6 | 25 | 157 | 79.89 | 187.07 | 28 |
| 4 | 26 | 16 | 29 | 76 | 87.30 | 145.88 | 28 |
| 5 | 38 | 8 | 27 | 140 | 99.46 | 191.87 | 26 |
| 6 | 28 | 8 | 22 | 206 | 69.34 | 143.22 | 26 |
| 7 | 34 | 8 | 25 | 558 | 96.16 | 206.54 | 36 |
| 8 | 28 | 12 | 26 | 109 | 75.47 | 129.72 | 36 |
| 9 | 32 | 6 | 22 | 318 | 87.10 | 192.72 | 20 |
| 10 | 24 | 10 | 22 | 109 | 78.80 | 141.58 | 22 |
| Mean | 32.20 | 9.40 | 25.50 | 229.30 | 80.74 | 169.50 | 29.60 |
| SD | 6.28 | 3.53 | 4.50 | 147.27 | 19.37 | 57.10 | 8.26 |
| Max | 44.00 | 14.00 | 36.00 | 558 | 99.85 | 285.10 | 48.00 |
| Min | 24.00 | 6.00 | 21.00 | 76 | 34.07 | 71.29 | 20.00 |

[a] G. centre = geometrical centre.
[b] M. density = swarm mean density.
[c] P. density = peak of the vertical density distribution.
[d] P. depth = depth of the peak density.

**Table 2.** Physical parameters of the 41 swarms that form the class at medium density (100 krill $m^{-3}$ <density of swarms <1000 krill $m^{-3}$)

| No. | Thickness (m) | Depth (m) | G. centre[a] (m) | Length (m) | M. density[b] (krill $m^{-3}$) | P. density[c] (krill $m^{-3}$) | P. depth[d] (m) |
|---|---|---|---|---|---|---|---|
| 11 | 42 | 24 | 45 | 284 | 309.74 | 796.16 | 48 |
| 12 | 38 | 26 | 45 | 155 | 283.79 | 1667.25 | 44 |
| 13 | 34 | 16 | 33 | 271 | 167.88 | 284.45 | 32 |
| 14 | 34 | 24 | 41 | 156 | 310.45 | 731.14 | 46 |
| 15 | 32 | 24 | 40 | 122 | 155.24 | 306.90 | 46 |
| 16 | 34 | 8 | 25 | 429 | 192.31 | 373.25 | 24 |
| 17 | 32 | 18 | 34 | 172 | 297.20 | 532.11 | 32 |
| 18 | 28 | 14 | 28 | 108 | 161.06 | 325.09 | 28 |
| 19 | 28 | 38 | 52 | 136 | 975.00 | 2322.74 | 62 |
| 20 | 26 | 28 | 41 | 58 | 218.27 | 674.53 | 42 |
| 21 | 26 | 26 | 39 | 187 | 619.44 | 1238.80 | 40 |
| 22 | 30 | 14 | 29 | 76 | 244.34 | 605.34 | 38 |
| 23 | 38 | 28 | 47 | 154 | 257.04 | 656.15 | 42 |
| 24 | 28 | 10 | 24 | 94 | 104.95 | 207.97 | 32 |
| 25 | 34 | 20 | 37 | 357 | 695.02 | 1309.18 | 38 |
| 26 | 32 | 32 | 48 | 265 | 633.87 | 1270.57 | 50 |
| 27 | 34 | 26 | 43 | 186 | 354.00 | 1000.00 | 42 |
| 28 | 28 | 24 | 38 | 748 | 304.79 | 497.74 | 40 |
| 29 | 38 | 20 | 39 | 347 | 223.87 | 803.53 | 48 |
| 30 | 28 | 38 | 52 | 265 | 533.33 | 1122.02 | 44 |
| 31 | 36 | 16 | 34 | 268 | 325.09 | 613.76 | 40 |
| 32 | 32 | 22 | 38 | 283 | 497.73 | 698.23 | 38 |
| 33 | 34 | 18 | 35 | 268 | 355.63 | 559.63 | 34 |
| 34 | 34 | 10 | 27 | 109 | 137.40 | 285.76 | 29 |
| 35 | 28 | 22 | 39 | 107 | 509.33 | 1185.77 | 36 |
| 36 | 28 | 14 | 27 | 188 | 159.22 | 262.42 | 26 |
| 37 | 32 | 18 | 34 | 59 | 305.49 | 805.38 | 38 |
| 38 | 32 | 12 | 28 | 253 | 178.65 | 320.63 | 28 |
| 39 | 32 | 4 | 20 | 110 | 470.98 | 1297.40 | 22 |
| 40 | 36 | 26 | 44 | 185 | 483.73 | 949.67 | 52 |
| 41 | 32 | 62 | 78 | 153 | 932.57 | 1857.58 | 72 |
| 42 | 28 | 16 | 30 | 367 | 888.47 | 2013.32 | 36 |
| 43 | 32 | 14 | 30 | 235 | 534.90 | 1073.07 | 42 |
| 44 | 28 | 8 | 21 | 300 | 451.55 | 955.14 | 26 |
| 45 | 28 | 16 | 30 | 170 | 590.69 | 1260.98 | 36 |
| 46 | 28 | 70 | 84 | 70 | 951.18 | 2185.15 | 82 |
| 47 | 32 | 46 | 62 | 62 | 624.02 | 2498.98 | 54 |
| 48 | 32 | 8 | 24 | 147 | 894.19 | 2744.94 | 30 |
| 49 | 36 | 8 | 26 | 70 | 198.45 | 387.54 | 22 |
| 50 | 36 | 24 | 42 | 76 | 455.06 | 934.40 | 32 |
| 51 | 40 | 12 | 32 | 157 | 787.32 | 1313.98 | 32 |
| Mean | 32.20 | 22.05 | 38.16 | 200.17 | 433.49 | 998.26 | 39.63 |
| SD | 6.06 | 13.51 | 13.43 | 129.57 | 252.53 | 652.54 | 12.43 |
| Max | 42.00 | 70.00 | 84.00 | 748 | 975.00 | 2744.94 | 82.00 |
| Min | 26.00 | 4.00 | 20.00 | 58 | 104.95 | 207.97 | 22.00 |

[a] G. centre = geometrical centre.
[b] M. density = swarm mean density.
[c] P. density = peak of the vertical density distribution.
[d] P. depth = depth of the peak density.

**Table 3.** Physical parameters of the 12 swarms that form the class at high density (>1000 krill $m^{-3}$). The image of swarm no. 59 is shown in Fig. 4

| No. | Thickness (m) | Depth (m) | G. centre[a] (m) | Length (m) | M. density[b] (krill $m^{-3}$) | P. density[c] (krill $m^{-3}$) | P. depth[d] (m) |
|---|---|---|---|---|---|---|---|
| 52 | 36 | 26 | 44 | 139 | 1454.02 | 4830.59 | 54 |
| 53 | 32 | 40 | 56 | 282 | 1722.81 | 3706.81 | 60 |
| 54 | 36 | 42 | 60 | 87 | 2583.45 | 8279.42 | 66 |
| 55 | 28 | 32 | 46 | 90 | 1236.01 | 2098.94 | 44 |
| 56 | 28 | 66 | 80 | 70 | 4914.30 | 10280.16 | 82 |
| 57 | 34 | 36 | 53 | 201 | 3001.04 | 5767.66 | 56 |
| 58 | 36 | 56 | 74 | 168 | 3808.71 | 6018.18 | 76 |
| 59 | 36 | 58 | 76 | 30 | 3287.03 | 5802.54 | 74 |
| 60 | 36 | 48 | 61 | 370 | 3287.03 | 4884.54 | 56 |
| 61 | 32 | 50 | 66 | 75 | 4718.93 | 8752.39 | 60 |
| 62 | 36 | 20 | 38 | 220 | 1232.55 | 2860.41 | 36 |
| 63 | 24 | 22 | 34 | 187 | 1793.72 | 1482.52 | 36 |
| Mean | 32.83 | 41.33 | 57.33 | 159.92 | 2753.30 | 5397.01 | 58.33 |
| SD | 4.13 | 14.78 | 15.02 | 99.07 | 1300.11 | 2699.74 | 14.82 |
| Max | 36.00 | 66.00 | 80.00 | 370 | 4914.30 | 10280.16 | 82.00 |
| Min | 24.00 | 20.00 | 34.00 | 30 | 1232.55 | 1482.52 | 36.00 |

[a] G. centre = geometrical centre.
[b] M. density = swarm mean density.
[c] P. density = peak of the vertical density distribution.
[d] P. depth = depth of the peak density.

$m^{-3}$ (SD 19.37 krill $m^{-3}$) and the mean vertical extent was 32.2 m (SD 6.28 m) (Table 1).

2. Medium density (100 krill $m^{-3}$ ≤medium density <1000 krill $m^{-3}$): 41 swarms. The mean density of these swarms was 433.49 krill $m^{-3}$ (SD 252.53 krill $m^{-3}$) and the mean vertical extent was 32.2 m (SD 6.06 m) (Table 2).
3. High density (>1000 krill $m^{-3}$): 12 swarms. The mean density of these swarms was 2753.30 krill $m^{-3}$ (SD 1300.11 krill $m^{-3}$) and the mean vertical extent was 32.8 m (SD 4.13 m) (Table 3).

The origin of the z axis (z = 0) was located at the peak (No) of the vertical density distribution of each swarm and the z axis was vertically oriented towards the surface. The peak density was taken also as reference density to normalize the density distribution [N(z)/No] of each swarm. The mean concentration distribution of a class ⟨N(z)/No⟩ was estimated from the normalized distribution of each swarm as a function of z, from the origin (z = 0) to the boundaries of each swarm, in a 2-m step. The ratio [U/Dz(N)] between the velocity of attraction and diffusion was estimated for each class from Eq. (6): $\left\langle \frac{U}{Dz(N)} \right\rangle = \frac{1}{\langle N(z) \rangle} \frac{\langle \Delta N(z) \rangle}{\Delta z}$, where Δz is the incremental interval on z axis (2 m).

The diffusion coefficient m of a class was obtained [Eq. (14)] from the ratio between the velocity of attraction and diffusion in the origin ⟨Uo/Do⟩, and the vertical extent of the class, referred to the origin of the z axis (⟨L(+)⟩,⟨L(−)⟩). The diffusion coefficient m and the ratio ⟨Uo/Do⟩, found experimentally for each class, were the parameters used to calculate the theoretical density distribution [Eq. (13)], the theoretical vertical extent [Eq. (12)], and the theoretical ratio U/Dz(N) [Eq. (11)] of insect swarms. The degree of fit of the equation representing insect density distribution and the krill density distribution of each class was estimated as follows:

$$I = \frac{\sum_{i=1}^{M} [N(zi) - \langle N(\Delta zi) \rangle]^2}{\sum_{i=1}^{M} [N(zi)]^2},$$

where N(zi) is the density distribution of insect swarms (theoretical), N(Δzi) is the discretized density distribution of a class of krill swarm (experimental), M is the number of 2-m intervals (Δz) included between the two boundaries of the class.

The index I has value 0 if and only if the fit is exact.

## Image of the swarm No. 59

| depth (m) | z | 5 | 10 | 15 | 20 | 25 | |
|---|---|---|---|---|---|---|---|
| 60 | 14 | | | | 376.404 | 1132.954 | 754.679 |
| 62 | 12 | | | 1085.349 | 5333.502 | 1621.830 | 2680.227 |
| 64 | 10 | | | 5710.807 | 8001.279 | 1904.956 | 5205.680 |
| 66 | 8 | | 930.479 | 8124.971 | 6843.744 | 1523.346 | 4355.635 |
| 68 | 6 | | 1946.771 | 8269.652 | 7799.295 | 1695.989 | 4927.927 |
| 70 | 4 | | 6065.997 | 8915.611 | 5964.380 | 1438.626 | 5596.154 |
| 72 | 2 | 585.154 | 9566.576 | 9245.747 | 6712.544 | 1059.678 | 5433.940 |
| 74 | 0 | 931.439 | 9573.299 | 11011.245 | 6587.828 | 908.892 | 5802.541 |
| 76 | -2 | 732.565 | 5804.720 | 9801.803 | 7897.474 | 1012.521 | 5049.816 |
| 78 | -4 | 467.485 | 5490.507 | 9822.151 | 6654.654 | 419.223 | 4570.804 |
| 80 | -6 | 1769.667 | 6910.536 | 8128.375 | 4038.478 | 4.181 | 4170.247 |
| 82 | -8 | 1795.880 | 6869.548 | 9027.874 | 2565.342 | 0.355 | 4051.800 |
| 84 | -10 | 694.064 | 4466.581 | 8999.480 | 531.228 | | 3672.838 |
| 86 | -12 | 262.776 | 4828.616 | 5151.571 | 35.605 | | 2569.642 |
| 88 | -14 | 560.783 | 5721.009 | 972.211 | | | 2418.001 |
| 90 | -16 | 83.601 | 237.967 | 10.078 | 4.898 | | 84.136 |
| 92 | -18 | 13.519 | 120.831 | 13.232 | 10.970 | | 39.638 |
| | | 717.903 | 4895.245 | 6518.135 | 4334.852 | 1060.213 | |

length (m)

(A)

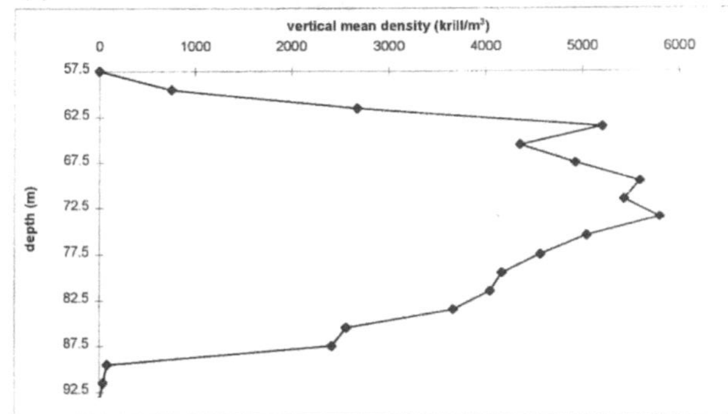


(B)

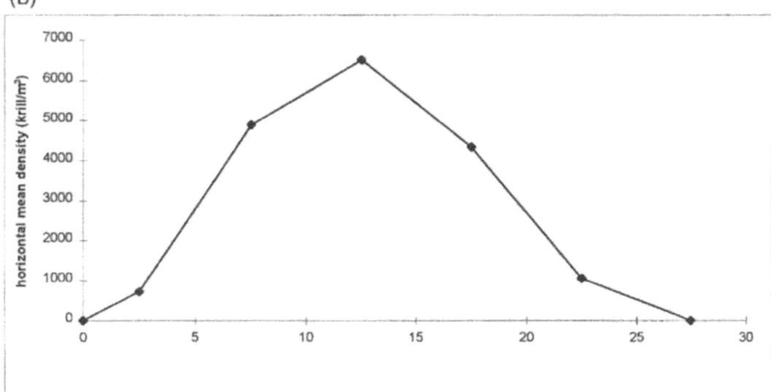


Peak of vertical density distribution: 5802.54krill/$m^3$.
Distance from the upper boundary: 14 m.
Distance from the deeper boundary: 18 m.

Peak of horizontal density distribution: 6518.14 krill/$m^3$.

Mean swarm density: 3808.71 krill/$m^3$.

**Fig. 4A,B.** Images of krill swarm no. 59. The boundaries of the image are determined by the DMVBS classifier (i.e. $6\,dB \le S_{v120} - S_{v38} \le 14\,dB$). Each cell that falls *within the solid silhouette* is filled with the value of the krill density estimated at 120 kHz, zero values are attributed to the other cells. The mean of all the cells within the solid silhouette is called mean swarm density. The vertical **(A)** [horizontal **(B)**] density distribution is obtained by averaging the density values of the cells of each row (column) that forms the image of the swarm

## Results

Figure 5 shows the vertical density distribution ⟨N(z)/No⟩ for swarms with densities lower than 100 krill $m^{-3}$. The peak of density No is 163.20 krill $m^{-3}$, SD 57.72 krill $m^{-3}$ (Table 4). It is situated 3 m deeper than the geometric centre. The krill density distribution has amplitudes towards the tails higher than the theoretical distribution. The ratio of the diffusivity and the attractive forces ⟨U/D⟩, calculated from the vertical concentration distribution [Eq. (6)], is presented in Table 4 and Fig. 6. The maximum inward forces are exerted by the krill at the periphery of the swarms, according to the theory [Eq. (11)]:

$\langle U/D \rangle = -1.83\ m^{-1}$ from the upper edge towards the centre of attraction.

$\langle U/D \rangle = 3.10\ m^{-1}$ from the deeper edge towards the centre of attraction.

Taking the mean value of ⟨U/D⟩ at the boundaries ($\approx 2.5\ m^{-1}$) and assuming a vertical component of U of $0.13 \cos 45° = 0.09\ m\,s^{-1}$ ($0.13\ m\,s^{-1}$ is the mean swimming speed of krill for migration, according to Hammer 1984), it is possible to obtain a tentative value of the diffusivity D of an individual near the edge: $0.036\ m^2\,s^{-1}$. This leads to an estimation of the rate at which a swarm would spread $[2(2D)^{1/2}]$, without a component of inward flow, of $0.54\ m\,s^{-1}$. The concentration distribution of krill swarms at low density is described by the parameters (Table 8):

$$\langle Uo/Do \rangle = 0.16\ m^{-1};\ \langle L \rangle = 24\ m;\ \langle m \rangle = 0.26.$$

The model based on these parameters seems to be fairly adequate for experimental data. The degree of fit is better than 8% (Fig. 5, Table 8): I = 0.078.

Figure 7 shows the vertical density distribution for krill swarms with density ranging from 100 to 1000 krill $m^{-3}$. The peak of density is 998.26 krill $m^{-3}$, SD 652.54 krill $m^{-3}$ (Table 5). The geometric and the attractive centres are almost in line, as in the insect model. In the upper part of the krill distribution the amplitudes are very close

**Table 4.** The class formed by the ten krill swarms at low density. The vertical density distribution $[N(z)/N_0]$ and the ratio (U/D) of the diffusivity and the attractive forces, estimated from the experimental data and calculated from the insect swarming model, are presented

| z (m) | <N(z)> (krill $m^{-3}$) | S.D. (krill $m^{-3}$) | $\langle N(z)/N_0 \rangle$ | $N(z)/N_0$ | z + 1 (m) | <U/D> ($m^{-1}$) | U/D ($m^{-1}$) |
|---|---|---|---|---|---|---|---|
| 26 | 0.92 | 2.92 | 0.006 | | 27 | −1.83 | |
| 24 | 2.34 | 7.41 | 0.014 | 8.46E−05 | 25 | −0.43 | −3.17 |
| 22 | 5.69 | 12.90 | 0.035 | 9.80E−04 | 23 | −0.42 | −1.18 |
| 20 | 10.97 | 18.88 | 0.067 | 4.28E−03 | 21 | −0.32 | −0.73 |
| 18 | 27.60 | 39.98 | 0.169 | 1.23E−02 | 19 | −0.43 | −0.53 |
| 16 | 35.70 | 41.03 | 0.219 | 2.82E−02 | 17 | −0.13 | −0.41 |
| 14 | 36.67 | 37.68 | 0.225 | 5.55E−02 | 15 | −0.01 | −0.34 |
| 12 | 42.38 | 40.81 | 0.260 | 9.85E−02 | 13 | −0.07 | −0.29 |
| 10 | 54.59 | 41.76 | 0.335 | 1.62E−01 | 11 | −0.13 | −0.25 |
| 8 | 65.74 | 45.85 | 0.403 | 2.52E−01 | 9 | −0.09 | −0.22 |
| 6 | 79.09 | 47.12 | 0.485 | 3.74E−01 | 7 | −0.09 | −0.20 |
| 4 | 92.88 | 34.77 | 0.569 | 5.34E−01 | 5 | −0.08 | −0.18 |
| 2 | 119.88 | 39.27 | 0.735 | 7.41E−01 | 3 | −0.13 | −0.16 |
| 0 | 163.20 | 57.72 | 1.000 | 1.00E+00 | 1 | −0.15 | −0.15 |
| −2 | 113.36 | 51.40 | 0.695 | 7.41E−01 | 0 | 0 | 0 |
| −4 | 82.51 | 55.64 | 0.506 | 5.34E−01 | −1 | 0.18 | 0.15 |
| −6 | 56.53 | 42.29 | 0.346 | 3.74E−01 | −3 | 0.16 | 0.16 |
| −8 | 50.18 | 46.47 | 0.307 | 2.52E−01 | −5 | 0.19 | 0.18 |
| −10 | 45.34 | 54.36 | 0.278 | 1.62E−01 | −7 | 0.06 | 0.20 |
| −12 | 45.07 | 69.00 | 0.276 | 9.85E−02 | −9 | 0.05 | 0.22 |
| −14 | 21.36 | 39.37 | 0.131 | 5.55E−02 | −11 | 0.00 | 0.25 |
| −16 | 11.65 | 26.81 | 0.071 | 2.82E−02 | −13 | 0.36 | 0.29 |
| −18 | 1.33 | 3.11 | 0.008 | 1.23E−02 | −15 | 0.29 | 0.34 |
| −20 | 0.23 | 0.72 | 0.001 | 4.28E−03 | −17 | 0.80 | 0.41 |
| −22 | 0.00 | 0.00 | 0.00 | 9.80E−04 | −19 | 0.71 | 0.53 |
| −24 | 0.00 | 0.00 | 0.00 | 8.46E−05 | −21 | 3.10 | 0.73 |
| −26 | 0.00 | 0.00 | 0.00 | 8.92E−09 | −23 | | 1.18 |
| | | | | | −25 | | 3.17 |

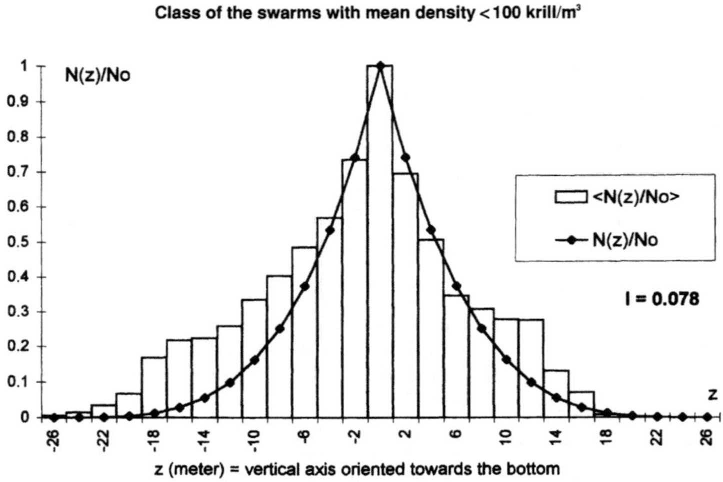


**Fig. 5.** Vertical density distribution of the class formed by the low-density swarms (<100 krill $m^{-3}$). The density distribution is normalized in respect to the density peak (163.2 krill $m^{-3}$) and compared with the density distribution of insects. The degree of fit is around 8%

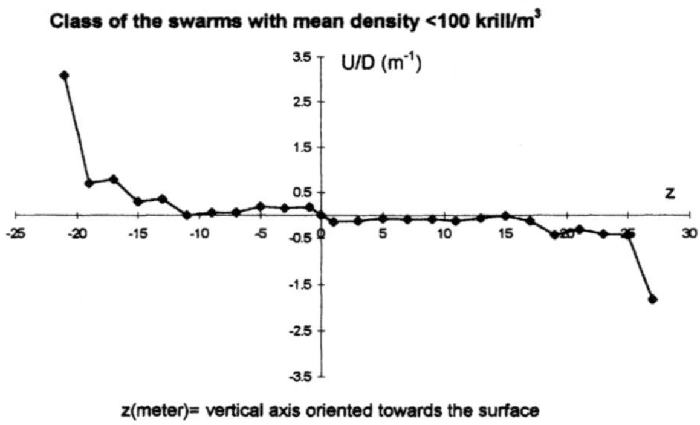


**Fig. 6.** The ratio of the diffusivity and the attractive forces evaluated for the class of krill swarms at low density. The maximum inward forces are exerted by the individuals at the edges of the swarms

to the theoretical, while in the deeper part the amplitudes are slightly emphasized with respect to the theoretical curve. The ratio U/D is shown in Fig. 8 and in Table 5. The inward force is felt strongly at the swarm edges:

$$\langle U/D \rangle = -5.79\ m^{-1} \text{ from the upper edge towards the centre of attraction}$$

$$\langle U/D \rangle = 2.76\ m^{-1} \text{ from the deeper edge towards the centre of attraction}$$

On average, the inward flows are about 70% stronger than in the low-density swarms ($\approx 4.3\ m^{-1}$). The same assumptions used for low-density swarms lead to a tentative value of the diffusivity of an individual near the edge D of 0.021 $m^2 s^{-1}$. Without a component of inward flow, the swarms at medium density would diffuse at the rate of 0.41 $m s^{-1}$. The parameters used in the theory for this class of density are (Table 8):

$$\langle Uo/Do \rangle = 0.15\ m^{-1};\ \langle L \rangle = 25\ m;\ \langle m \rangle = 0.27.$$

The comparison between the theory and the observation is good (Fig. 7, Table 8). The degree of fit is about 3%: I = 0.031.

**Table 5.** The class formed by the 41 krill swarms at medium density. The vertical density distribution [$N(z)/N_0$] and the ratio (U/D) of the diffusivity and the attractive forces, estimated from the experimental data and calculated from the insect swarming model, are presented

| z (m) | $\langle N(z)\rangle$ (krill m$^{-3}$) | SD (krill m$^{-3}$) | $\langle N(z)/N_0\rangle$ | $N(z)/N_0$ | z + 1 (m) | $\langle U/D\rangle$ (m$^{-1}$) | U/D (m$^{-1}$) |
|---|---|---|---|---|---|---|---|
| 26 | 0.47 | 2.13 | 0.0005 | | 27 | −5.79 | |
| 24 | 8.62 | 37.46 | 0.0086 | 4.92E−06 | 25 | −0.90 | |
| 22 | 21.41 | 74.43 | 0.0214 | 3.37E−04 | 23 | −0.43 | −1.91 |
| 20 | 36.38 | 93.92 | 0.0364 | 2.33E−03 | 21 | −0.26 | −0.95 |
| 18 | 50.79 | 88.86 | 0.0509 | 8.32E−03 | 19 | −0.17 | −0.63 |
| 16 | 93.94 | 147.30 | 0.0941 | 2.14E−02 | 17 | −0.30 | −0.47 |
| 14 | 128.63 | 183.54 | 0.1289 | 4.57E−02 | 15 | −0.16 | −0.38 |
| 12 | 187.78 | 235.64 | 0.1881 | 8.56E−02 | 13 | −0.19 | −0.31 |
| 10 | 258.10 | 315.55 | 0.2586 | 1.47E−01 | 11 | −0.16 | −0.27 |
| 8 | 361.60 | 322.89 | 0.3622 | 2.35E−01 | 9 | −0.17 | −0.23 |
| 6 | 496.75 | 420.97 | 0.4976 | 3.57E−01 | 7 | −0.16 | −0.21 |
| 4 | 623.08 | 487.35 | 0.6242 | 5.19E−01 | 5 | −0.11 | −0.19 |
| 2 | 784.05 | 501.18 | 0.7854 | 7.31E−01 | 3 | −0.11 | −0.17 |
| 0 | 998.26 | 652.54 | 1.0000 | 1.00E+00 | 1 | −0.12 | −0.16 |
| −2 | 693.25 | 471.28 | 0.6945 | 7.31E−01 | 0 | 0 | 0 |
| −4 | 504.93 | 406.50 | 0.5058 | 5.19E−01 | −1 | 0.18 | 0.16 |
| −6 | 305.80 | 340.11 | 0.3063 | 3.57E−01 | −3 | 0.16 | 0.17 |
| −8 | 232.63 | 310.99 | 0.2330 | 2.35E−01 | −5 | 0.25 | 0.19 |
| −10 | 134.08 | 199.93 | 0.1343 | 1.47E−01 | −7 | 0.14 | 0.21 |
| −12 | 95.05 | 216.21 | 0.0952 | 8.56E−02 | −9 | 0.27 | 0.23 |
| −14 | 48.22 | 124.03 | 0.0483 | 4.57E−02 | −11 | 0.17 | 0.27 |
| −16 | 33.35 | 124.35 | 0.0334 | 2.14E−02 | −13 | 0.33 | 0.31 |
| −18 | 15.62 | 60.72 | 0.0157 | 8.32E−03 | −15 | 0.18 | 0.38 |
| −20 | 6.48 | 35.51 | 0.0065 | 2.33E−03 | −17 | 0.36 | 0.47 |
| −22 | 9.76 | 61.54 | 0.0098 | 3.37E−04 | −19 | 0.41 | 0.63 |
| −24 | 2.71 | 17.13 | 0.0027 | 4.92E−06 | −21 | −0.20 | 0.95 |
| −26 | 0.33 | 1.86 | 0.0003 | | −23 | 0.57 | 1.91 |
| | | | | | −25 | 0.78 | |
| | | | | | −27 | 2.76 | |

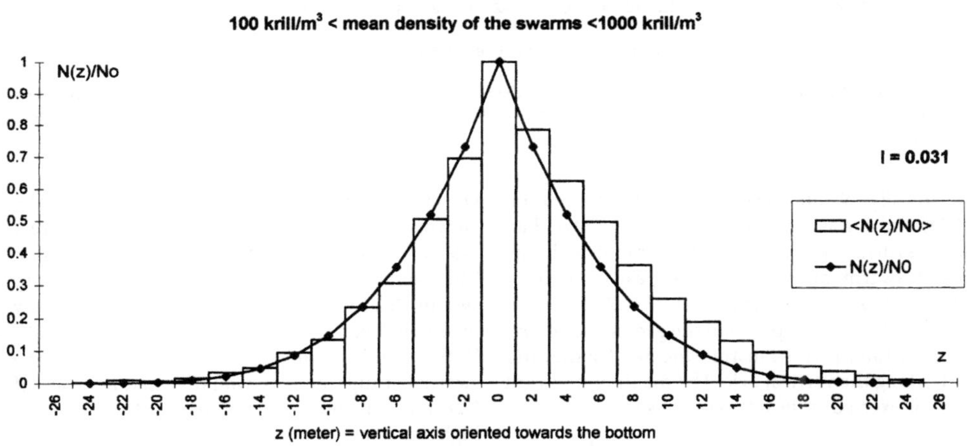


**Fig. 7.** Vertical density distribution of the class formed by the medium density swarms (100 krill m$^{-3}$ <density of swarms <1000 krill m$^{-3}$). The density distribution is normalized with respect to the density peak (998.26 krill m$^{-3}$) and compared favourably with the density distribution of insects. The degree of fit is around 3%

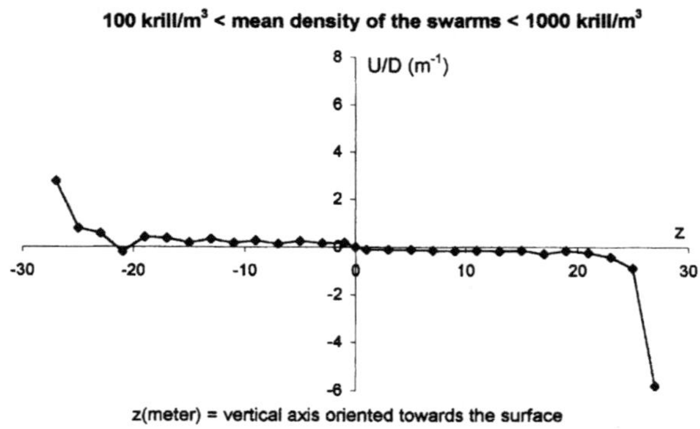


**Fig. 8.** The ratio of the diffusivity and the attractive forces evaluated for the class of krill swarms at medium density. The maximum inward forces are exerted by the individuals at the edges of the swarms

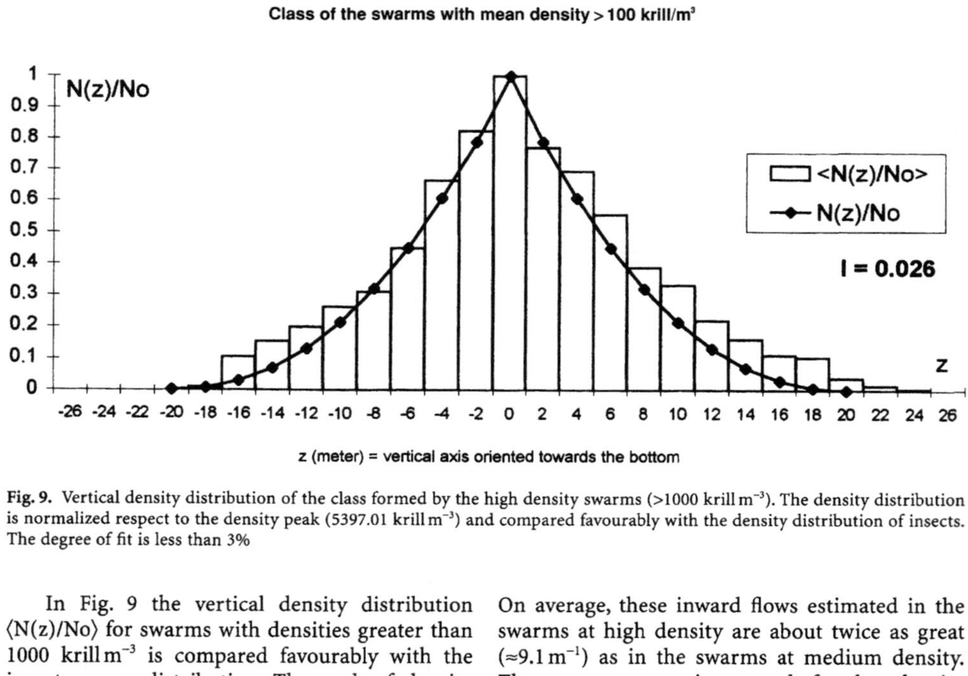


**Fig. 9.** Vertical density distribution of the class formed by the high density swarms (>1000 krill $m^{-3}$). The density distribution is normalized respect to the density peak (5397.01 krill $m^{-3}$) and compared favourably with the density distribution of insects. The degree of fit is less than 3%

In Fig. 9 the vertical density distribution ⟨N(z)/No⟩ for swarms with densities greater than 1000 krill $m^{-3}$ is compared favourably with the insect swarm distribution. The peak of density (Table 6) is 5397 krill $m^{-3}$, SD 2699.75 krill $m^{-3}$. It is situated 3 m deeper than the geometric centre. Figure 10 and Table 6 illustrate the trend of the ratio U/D along the vertical axis. The maximum inward force is exerted at the edges of the swarm:

$$\langle U/D \rangle = -6.34\ m^{-1} \text{ from the upper edge towards the centre of attraction.}$$

$$\langle U/D \rangle = 11.77\ m^{-1} \text{ from the deeper edge towards the centre of attraction.}$$

On average, these inward flows estimated in the swarms at high density are about twice as great (≈9.1 $m^{-1}$) as in the swarms at medium density. The same assumptions used for low-density swarms lead to a tentative value of D (diffusivity of an individual near the edge) of 0.010 $m^2 s^{-1}$. Without this component of inward flow, the swarms at high density would spread at a rate of 0.28 $m s^{-1}$. The observed parameters for swarms at high density are (Table 8):

$$\langle Uo/Do \rangle = 0.11\ m^{-1};\ \langle L \rangle = 20\ m;\ \langle m \rangle = 0.44.$$

The model of concentration distribution based on these parameters is very close to the experimental

**Table 6.** The class formed by the 12 krill swarms at high density. The vertical density distribution [$N(z)/N_0$] and the ratio (U/D) of the diffusivity and the attractive forces, estimated from the experimental data and calculated from the insect swarming model, are presented

| z (m) | <N(z)> (krill m$^{-3}$) | SD (krill m$^{-3}$) | <$N(z)/N_0$> | $N(z)/N_0$ | z + 1 (m) | <U/D> (m$^{-1}$) | U/D (m$^{-1}$) |
|---|---|---|---|---|---|---|---|
| 26 | 1.10 | 3.80 | 0.0002 | | 27 | −6.34 | |
| 24 | 26.11 | 90.45 | 0.0048 | | 25 | −0.92 | |
| 22 | 83.27 | 284.40 | 0.0154 | | 23 | −0.52 | |
| 20 | 206.41 | 559.72 | 0.0382 | 4.78E−05 | 21 | −0.43 | |
| 18 | 558.15 | 1002.60 | 0.1034 | 6.84E−03 | 19 | −0.46 | −1.81 |
| 16 | 600.96 | 1177.60 | 0.1114 | 2.89E−02 | 17 | −0.04 | −0.70 |
| 14 | 863.94 | 1215.82 | 0.1601 | 6.94E−02 | 15 | −0.18 | −0.43 |
| 12 | 1182.64 | 1114.65 | 0.2191 | 1.30E−01 | 13 | −0.16 | −0.31 |
| 10 | 1794.66 | 1714.16 | 0.3325 | 2.13E−01 | 11 | −0.21 | −0.25 |
| 8 | 2091.84 | 1652.68 | 0.3876 | 3.20E−01 | 9 | −0.08 | −0.20 |
| 6 | 2995.35 | 1854.75 | 0.5550 | 4.50E−01 | 7 | −0.18 | −0.17 |
| 4 | 3751.17 | 2183.95 | 0.6950 | 6.07E−01 | 5 | −0.11 | −0.15 |
| 2 | 4165.60 | 2409.28 | 0.7718 | 7.90E−01 | 3 | −0.05 | −0.13 |
| 0 | 5397.01 | 2699.75 | 1.0000 | 1.00E+00 | 1 | −0.13 | −0.12 |
| −2 | 4456.96 | 2385.62 | 0.8258 | 7.90E−01 | 0 | 0.00 | 0.00 |
| −4 | 3583.57 | 2449.72 | 0.6640 | 6.07E−01 | −1 | 0.10 | 0.12 |
| −6 | 2427.08 | 1780.23 | 0.4497 | 4.50E−01 | −3 | 0.11 | 0.13 |
| −8 | 1673.55 | 1678.70 | 0.3101 | 3.20E−01 | −5 | 0.19 | 0.15 |
| −10 | 1417.64 | 1745.29 | 0.2627 | 2.13E−01 | −7 | 0.18 | 0.17 |
| −12 | 1073.38 | 1526.47 | 0.1989 | 1.30E−01 | −9 | 0.08 | 0.20 |
| −14 | 834.42 | 1420.58 | 0.1546 | 6.94E−02 | −11 | 0.14 | 0.25 |
| −16 | 565.41 | 1121.93 | 0.1048 | 2.89E−02 | −13 | 0.13 | 0.31 |
| −18 | 46.82 | 94.04 | 0.0087 | 6.84E−03 | −15 | 0.19 | 0.43 |
| −20 | 0.77 | 2.05 | 0.0001 | 4.78E−05 | −17 | 0.85 | 0.70 |
| | | | | | −19 | 0.97 | 1.81 |
| | | | | | −21 | 11.77 | |

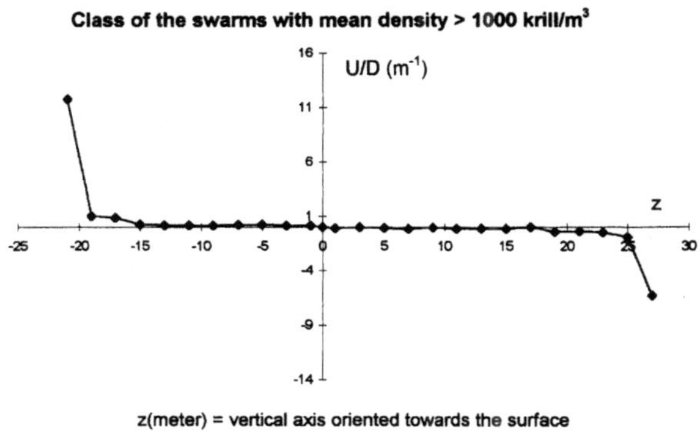


**Fig. 10.** The ratio of the diffusivity and the attractive forces evaluated for the class of krill swarms at high density. The maximum inward forces are exerted by the individuals at the edges of the swarms

data (Fig. 9, Table 8): the degree of fit is better than 3%: I = 0.026.

In Fig. 11 the vertical concentration distribution of all the 63 swarms taken for this analysis is shown. The geometric and attractive centres are in line, as in the class at medium density. The peak of density is 1702.32 krill m$^{-3}$; SD 2219.01 krill m$^{-3}$ (Table 7). Figure 12 and Table 7 illustrate the trend of the ratio U/D along the vertical axis. The inward flows at the edges of the swarms are rather similar:

$$\langle U/D \rangle = -5.54\ \mathrm{m^{-1}}\ \text{from the upper edge towards the centre of attraction}$$

$$\langle U/D \rangle = 7.50\ \mathrm{m^{-1}}\ \text{from the deeper edge towards the centre of attraction}$$

**Table 7.** The set of all the 63 swarms selected for this analysis. The vertical density distribution ($N(z)/N_0$) and the ratio (U/D) of the diffusivity and the attractive forces, estimated from the experimental data and calculated from the insect swarming model, are presented

| z (m) | <N(z)> (krill m$^{-3}$) | SD (krill m$^{-3}$) | <$N(z)/N_0$> | $N(z)/N_0$ | z + 1 (m) | <U/D> (m$^{-1}$) | U/D (m$^{-1}$) |
|---|---|---|---|---|---|---|---|
| 26 | 0.66 | 2.61 | 0.0004 | | 27 | −5.54 | |
| 24 | 10.96 | 49.24 | 0.0064 | 2.18E−04 | 25 | −0.89 | |
| 22 | 30.70 | 136.53 | 0.0180 | 2.13E−03 | 23 | −0.47 | −0.98 |
| 20 | 64.73 | 257.31 | 0.0380 | 8.08E−03 | 21 | −0.36 | −0.66 |
| 18 | 143.75 | 474.13 | 0.0844 | 2.08E−02 | 19 | −0.38 | −0.47 |
| 16 | 181.27 | 550.30 | 0.1065 | 4.33E−02 | 17 | −0.12 | −0.36 |
| 14 | 254.09 | 611.74 | 0.1493 | 7.88E−02 | 15 | −0.17 | −0.30 |
| 12 | 354.19 | 650.64 | 0.2081 | 1.31E−01 | 13 | −0.16 | −0.25 |
| 10 | 518.48 | 990.24 | 0.3046 | 2.03E−01 | 11 | −0.19 | −0.22 |
| 8 | 644.21 | 1031.77 | 0.3784 | 2.99E−01 | 9 | −0.11 | −0.19 |
| 6 | 906.38 | 1338.26 | 0.5324 | 4.22E−01 | 7 | −0.17 | −0.17 |
| 4 | 1134.75 | 1634.87 | 0.6666 | 5.78E−01 | 5 | −0.11 | −0.16 |
| 2 | 1322.73 | 1783.71 | 0.7770 | 7.69E−01 | 3 | −0.08 | −0.14 |
| 0 | 1702.32 | 2219.01 | 1.0000 | 1.00E+00 | 1 | −0.13 | −0.13 |
| −2 | 1318.10 | 1884.84 | 0.7743 | 7.69E−01 | 0 | 0.00 | 0.00 |
| −4 | 983.01 | 1612.05 | 0.5775 | 5.78E−01 | −1 | 0.13 | 0.13 |
| −6 | 670.29 | 1176.06 | 0.3938 | 4.22E−01 | −3 | 0.15 | 0.14 |
| −8 | 478.13 | 953.24 | 0.2809 | 2.99E−01 | −5 | 0.19 | 0.16 |
| −10 | 364.48 | 912.61 | 0.2141 | 2.03E−01 | −7 | 0.17 | 0.17 |
| −12 | 274.71 | 772.98 | 0.1614 | 1.31E−01 | −9 | 0.13 | 0.19 |
| −14 | 193.71 | 682.96 | 0.1138 | 7.88E−02 | −11 | 0.14 | 0.22 |
| −16 | 131.25 | 527.76 | 0.0771 | 4.33E−02 | −13 | 0.17 | 0.25 |
| −18 | 19.30 | 64.47 | 0.0113 | 2.08E−02 | −15 | 0.19 | 0.30 |
| −20 | 4.40 | 28.68 | 0.0026 | 8.08E−03 | −17 | 0.74 | 0.37 |
| −22 | 6.35 | 49.65 | 0.0037 | 2.13E−03 | −19 | 0.63 | 0.47 |
| −24 | 1.76 | 13.82 | 0.0010 | 2.18E−04 | −21 | −0.18 | 0.66 |
| −26 | 0.22 | 1.51 | 0.0001 | | −23 | 0.57 | 0.98 |
| | | | | | −25 | 0.78 | 3.08 |
| | | | | | −27 | 7.50 | |

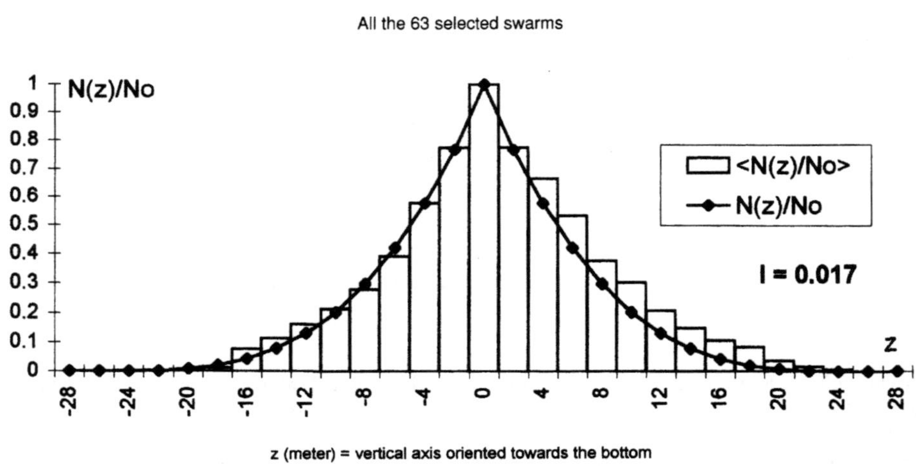


**Fig. 11.** Vertical density distribution of all the 63 swarms as a whole. The density distribution is normalized respect to the density peak (1702.32 krill m$^{-3}$) and compared favourably with the density distribution of insects. The degree of fit is less than 2%

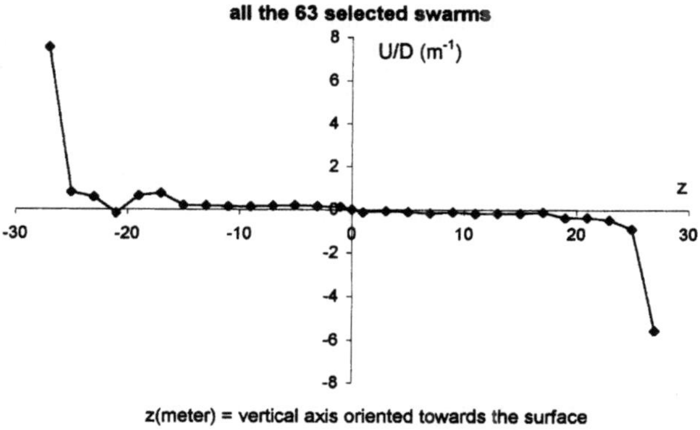


**Fig. 12.** The ratio of the diffusivity and the attractive forces evaluated for all the 63 swarms as a whole. The maximum inward forces are exerted by the individuals at the edges of the swarms

On average, the inward flow at the edges of all the swarms is ≈6.5 $m^{-1}$. Under the same assumptions used above, we obtain a tentative value of the diffusivity D of an individual near the edge of 0.014 $m^2 s^{-1}$. Without this component of inward flow, the swarms would spread at the rate of 0.34 $m s^{-1}$. The parameters extracted from all the swarms used for this analysis are (Table 8):

$$\langle Uo/Do \rangle = 0.13\,m^{-1};\ \langle L \rangle = 26\,m;\ \langle m \rangle = 0.30.$$

The model compared with all the experimental data seems to be adequate. The degree of fit is better than 2%: I = 0.017.

## Summary and Conclusions

In this chapter the question whether the models of insect swarming can be used to study krill swarming was posed. A simple model, that has been extensively applied to insect swarming, was chosen to compare the spatial density patterns of airborne insects and the vertical density distributions of 63 krill swarms. These swarms, recorded in the Ross Sea in the 1989–1990 and 1994 echo surveys, were characterized by a rather similar vertical extent (from 24 to 44 m), and differed in all other physical parameters. In particular, the swarm mean densities ranged from 34 to 4914 krill $m^{-3}$. The swarms were divided in three classes (low, medium and high density). The vertical density distribution of each class and of all the swarms as a whole was analyzed and compared favourably with the model. The degree of fit ranged from 8% (low-density class) to 2% (all the swarms). The coefficient of diffusion ⟨m⟩, estimated for each class and for all the swarms, ranged from 0.25 (low-density class) to 0.58 (high-density class). Therefore the vertical density distribution of krill swarms is leptokurtic ($m < 1$), as in many insect concentration distribution (Okubo 1980). A leptokurtic distribution may be the result of a different behaviour/size/sex of the individuals in a swarm, some dispersing more rapidly than others (James 1978). The ratio of the diffusivity and the attractive forces U/D was estimated in terms of krill density. In general, the inward force is felt more strongly at the swarm edges, according to the model [Eq. (11)], and this force increases with the mean density of swarms (≈2.5 $m^{-1}$ for low-density swarms, ≈4.3 $m^{-1}$ for medium-density swarms, ≈9.1 $m^{-1}$ for high-density swarms, and 6.5 $m^{-1}$ for all the swarms). All the parameters estimated from krill swarms are asymmetric with respect to the peak of density. The values calculated from the attraction centre towards the surface and towards the bottom are different, while in the model they are assumed equal.

The above results suggest that many kinematic aspects of krill swarming are very similar to those of insect swarming. Therefore mathematical models that have been favourably applied to insect swarming should also be applicable to krill swarming. To proceed into modelling krill swarms, other kinematic, biological, and environmental data, beside density, are needed. For example, information on krill speed and orientation in a swarm is necessary to estimate the parameter U, and will enable us to estimate the energy expended in swarming; the identification of krill size and sex distribution in a swarm will help us to understand why the vertical density distribu-

**Table 8.** Summary of the parameters (the ratio $<U_0/D_0>$, the diffusion coefficient $<m>$, the maximum vertical extents from the peak of density L) and of the degree of fit I, estimated for each class and for all the 63 swarms as a whole. The parameters are asymmetric. For each parameter is indicated the orientation [from the peak of density whether towards the surface (↑) or towards the bottom (↓)], and the mean value. The mean values of the parameters were used to calculate the vertical density distribution, the vertical extent, and the ratio U/D of the swarming insects

| Parameters | Experimental data | | | Mean data |
|---|---|---|---|---|
| Density < 100 krill m$^{-3}$ (10 swarms) | | | | |
| $U_0/D_0$ (m$^{-1}$) | ↑ 0.1500 | | ↓ 0.1800 | 0.1600 |
| L (m) | ↑ 26 | | ↓ 20 | 24 |
| m | ↑ 0.25 | | ↓ 0.27 | 0.26 |
| l | | 0.078 | | |
| Density from 100 to 1000 krill m$^{-3}$ (41 swarms) | | | | |
| $U_0/D_0$ (m$^{-1}$) | ↑ 0.1202 | | ↓ 0.1803 | 0.1502 |
| L (m) | ↑ 26 | | ↓ 26 | 25 |
| m | ↑ 0.32 | | ↓ 0.21 | 0.27 |
| l | | 0.031 | | |
| Density > 1000 krill m$^{-3}$ (12 swarms) | | | | |
| $U_0/D_0$ (m$^{-1}$) | ↑ 0.1288 | | ↓ 0.0954 | 0.1121 |
| L (m) | ↑ 26 | | ↓ 18 | 20 |
| m | ↑ 0.30 | | ↓ 0.58 | 0.44 |
| l | | 0.026 | | |
| All the selected swarms (63) | | | | |
| $U_0/D_0$ (m$^{-1}$) | ↑ 0.1306 | | ↓ 0.1274 | 0.1290 |
| L (m) | ↑ 26 | | ↓ 26 | 26 |
| m | ↑ 0.30 | | ↓ 0.30 | 0.30 |
| l | | 0.017 | | |

tion in a swarm is, in general, leptokurtic. In addition, information on vertical shear, water turbulence, sea ice, light and the presence of predators will help us to understand the observed asymmetry of the vertical structure of krill swarms and the differences/interactions between the vertical and horizontal structures of a swarm.

## References

Anonymous (1991) Report of the Tenth Meeting of the Scientific Committee (SC-CCAMLR - X). CCAMLR, Hobart, Australia, 427 pp

Downes JA (1969) The swarming and mating flight of Diptera. Annu Rev Entomol 14: 271–298

Evans GT (1977) Biological effects of vertical-horizontal interactions. In: Steele JH (ed) Spatial pattern in plankton communities. Proc NATO Conf on Marine biology, Erice, Italy, 13–21 November 1977. Plenum Press, New York, pp 157–181

Foot KG, Everson I, Watkins JL, Bone DG (1990) Target strengths of Antarctic krill (*Euphausia superba*) at 38 and 120 kHz. J Acoust Soc Am 87(1): 16–24

Hammer WM (1984) Aspects of schooling in *Euphausia superba*. J Crustacean Biol 4 (Special Issue no 1): 67–74

Horwood JW, Cushing DH (1977) Spatial distribution and ecology of pelagic fish. In: Steele JH (ed) Spatial pattern in plankton communities. Proc NATO Conf on Marine biology, Erice, Italy, 13–21 November 1977. Plenum Press, New York, pp 355–383

James IR (1978) Estimation of the mixing proportion in a mixture of two normal distributions, rapid measuraments. Biometrics 34: 265–275

Kalinowski J, Witek Z (1985) Scheme for classifying aggregations of Antarctic krill. BIOMASS handbook No 27. SCAR/SCOR/ABO/ACMRR, Group of Specialist on Living Resources of the Southern Occean ecology, 12 pp

Kareiva PM (1983) Local movement in herbivorous insects: applying a passive diffusion model to mark-recapture field experiments. Oecologia (Berl) 57: 322–327

Madureira LSP, Everson I, Murphy EJ (1993) Interpretation of acoustic data at two frequency to discriminate between

Antarctic krill (*Euphausia superba* Dana) and other scatterers. J Plankton Res 15(7): 767–802

Okubo A (1980) Diffusion and ecological problems: mathematical models. Springer, Berlin Heidelberg New York, p 254

Okubo A (1986) Dynamical aspects of animal grouping: swarms, schools, flocks and herds. Adv Biophys 22: 1–94

Okubo A, Chiang HC (1974) An analysis of the kinematics of swarming of *Anarete pritchardi* Kim (Diptera: Cecidomyiidae). Res Popul Ecol 16: 1–42

Okubo A, Sakamoto W, Inagaki T, Kuroki T (1977) Studies on the schooling behavior of fish. Note on the dynamics of fish schooling. Bull Jpn Soc Sci Fish 43: 1369–1377

Radakov DV (1973) Schooling in the ecology of fish. Wiley, New York

Shigesada N (1980) Spatial distribution of dispersing animals. J Math Biol 9: 85–96

Shigesada N, Kawasaki K, Teramoto E (1979) Spatial segregation of interacting species. J Theor Biol 79: 83–99

Siegel V, Kalinowski J (1994) Krill demography and small-scale processes: a review. In: El-Sayed SZ (ed) Southern Ocean ecology: the BIOMASS perspective. Cambridge University Press, Cambridge, pp 145–163

Van Olst JC, Hunter JR (1970) Some aspects of the organization of fish schools. J Fish Res Board Can 27: 1225–1238

CHAPTER 31

# Spatial and Temporal Distribution of Krill *Euphausia superba* Biomass in the Ross Sea (1989–1990 and 1994)

M. Azzali and J. Kalinowski[1]

**ABSTRACT**

This chapter aims to improve the knowledge of krill populations in the Ross Sea, an area that has received little attention in the past and which has been thought to be krill-deficient. In the early summer 1989–90, two surveys were conducted in ice-free waters with quasistationary environmental conditions. In the late spring 1994, two more surveys were conducted under conditions of complete to partial ice cover. Concurrent with the acoustic surveys, censuses were taken of the birds and mammals. This chapter gives a description of the instruments used and of data-acquisition and data-processing methods. The results indicate that, contrary to prevalent thinking, the Ross Sea is not krill-deficient. Very dense concentrations of *Euphausia superba* were found in the southern part of the Ross Sea, up to 76°S in latitude, an area presented in the literature as krill-absent. In these areas, density peaks of krill consumers (minke whales, crabeater seals and Adelie penguins) were found. The mean density of *E. superba* estimated in 1989–90 and 1994 is of the same order of magnitude as that estimated in the Indian and Atlantic sectors, respectively. On the basis of these results, the authors suggest that the large-scale distribution of krill around Antarctica, presented in the literature, should be supplemented by the part of the Ross Sea extending to the north from the latitude of 76°S.

## Introduction

The purpose of this chapter is to present new results on the spatial and temporal distribution of *Euphausia superba* in the Ross Sea and to estimate its size.

Krill populations of the Pacific sector, and in particular of the Ross Sea, have received very little attention in the past. British Discovery expeditions (1935, 1939 and 1950–51) gave the first information on krill in this area. On the basis of the results of these expeditions, Marr (1962) concluded that *Euphausia superba* is almost completely absent in the Ross Sea shelf, where it is replaced by another euphasiid, *Euphausia crystallorophias*. Marr's conclusion, confirmed later by Mackintosh (1973), Lubimova et al. (1984) and Shimadzu (1984), effectively diverted krill research towards areas in the Atlantic and Indian Ocean sectors. For example, the First International Biomass Experiment (BIOMASS/FIBEX 1981) included only these two areas (El-Sayed 1994), and assumed that the Pacific sector, especially the Ross Sea, was too poor or sparsely populated. Mackintosh (1973, quoted in Amos 1984) also observed that while the Ross Sea seems to be an ideal environment for krill the only major population in the region is over deep waters to the north of the pack ice in the East Wind Drift within the Ross Sea Gyre.

Despite these earlier studies, analyses of the stomach contents of minke whales in the Ross Sea (Ichii 1990) suggest that krill are indeed present in the southern part of the Ross Sea up to the Ross Ice Shelf. Moreover, the annual consumption of krill by minke whales in the Ross Sea was estimated to be about 3 million tons (Tamura et al. 1997).

Aiming to improve the knowledge of krill populations in the Ross Sea, four acoustic surveys of *Euphausia superba* were conducted during the Italian expeditions of 1989–1990 and 1994 to estimate their biomass and to describe their spatial and temporal distributions. The results of these surveys indicate that dense concentrations of krill are present in regions of the Ross Sea that before Italian expeditions were considered krill-deficient.

[1] CNR-IRPeM, Largo Fiera della Pesca, 2, Ancona, Italy

In this chapter, after a description of equipment and methodology used in the two expeditions, the abundance and spatial distribution of krill is presented and analyzed on a yearly, seasonal and daily scale. These results are then discussed and compared with those from other regions of the Antarctic.

## Materials and Methods

From 26 December 1989 to 25 January 1990, two acoustic surveys were conducted in ice-free waters, with quasistationary environmental conditions (Azzali et al. 1991, 1992; Azzali and Kalinowski 1992).

In 1994, two more surveys were conducted from 9 November to 28 December in conditions of complete to partial ice cover and relatively dynamic environmental conditions (i.e. receding ice edge, and krill predation from minke whales; Azzali et al. 1996).

### Acoustic Equipment

The surveys were conducted using a scientific echosounder (BioSonics 102), operating at 38 and 120 kHz in multiplexing mode, an echointegrator (BioSonics 221), a dual-beam processor (BioSonics 281), and an echosignal processor (BioSonics ESP), running on a PC computer. Each frequency had a ping rate of $1.2\,s^{-1}$. Pulses of 0.5 ms were alternatively transmitted every 0.6 s and received with 2.5- and 5.0-kHz band pass filter, respectively, at 38 and 120 kHz. The alternate outputs were directed to different chart recorders (BioSonics 111). Volume backscattering strengths ($S_v$) at both frequencies were integrated over 6 m bins (25 bins from 10 to 160 m), averaged each mile and recorded on the hard disc of the computer. Geographic position and vessel speed were determined from a Global Positioning System and logged each mile. The dual-beam echo signals at 120 and 38 kHz and their associated timing information were recorded ping by ping on different digital audio tapes (DAT), each connected to the echosounder through a tape-recorder interface (BioSonics 171). The dual-beam transducers (3 dB beam widths = 7°/18° and 10°/22° for 38 and 120 kHz, respectively) were mounted on a 3-feet V-fin (1989–1990 surveys) and in the hull of the vessel protected by a dome (1994 surveys). All the dual-beam transducers, whether hull-mounted or installed in the towed body, contained preamplifiers to optimize signal to noise ratios and demultiplexing relays that permitted the use of the same cable for both transducers. The acoustic equipment was fully calibrated in Italy, before and after each expedition, at a water temperature around 17 °C, under the same equipment settings as during the surveys. Calibration was carried out according to the method described in MacLennan and Simmonds (1992), using a 33- and a 36-mm tungsten carbide sphere for 120 and 38 kHz, respectively. The difference between the calibrated TS transducer gain measured on the towed body and the hull-mounted transducers was ~0.4 and ~1.6 dB for 38 and 120 kHz, respectively. It is recognized that: (1) calibrations should be performed under the same environmental conditions as during the survey (in the Ross Sea the temperature is typically −1 °C); (2) transducer efficiencies may decrease substantially with decreasing water temperature (Demer and Hewitt 1993, 1996); and (3) calibrating in waters warmer than survey conditions may lead to significant underestimation of krill abundance (Demer 1994; Brierley et al. 1997). Therefore, the abundance values presented in this chapter should be considered as minimum estimations of krill biomass. Krill abundance was calculated at 120 kHz. At this frequency the minimum $S_v$ was −92 (dB re $1\,m^3$) in ice-free water and −85 (dB re $1\,m^3$) in ice-covered water (Kalinowski and Azzali 1996).

### Net Sampling

In the 1989–1990 expedition krill aggregations detected by the echosounder were sampled using a $0.25\text{-}m^2$ EZ-NET-BIONESS plankton net (Guglielmo 1992) with 10 nets of 240- and 500-µm mesh and towed at speeds of 2.5 to 3 knots. Because the net was ineffective at sampling krill (<10 individuals per haul), the samples were combined (144 krill) to estimate the length distribution (mean = 40.8 mm; SD 3.6 mm). Total lengths (L) were measured from the anterior edge of the eye to the tip of the telson and rounded to the nearest mm. In the 1994 expedition, krill were sampled using either a $0.6\text{-}m^2$ EZ-NET-BIONESS towed at 2.5 to 3 kts or a $4\text{-}m^2$ Plankton Hamburg Net (PHN) with 1 net of 500-µm mesh towed at 4 kts. In general, the EZ-NET-BIONESS was used from 23 November (earliest possible date, due to ice) to 14 December, and the PHN was used

**Table 1.** Location of the net samples from 23 November–15 December 1994, showing the net used (B = EZ-NET-BIONESS; PHN = Plankton Hamburg Net), depth where samples were taken, number of individuals of *Euphausia superba* (*E.s.*) and *Euphausia crystallorophias* (*E.c.*), presence of other zooplankton (C = copepods; A = amphipods)

| Date | Beginning h:min | Depth m | Location | | Catch composition | | | | Net |
|---|---|---|---|---|---|---|---|---|---|
| | | | Lat. (S) | Long (E) | *E.s.* | *E.c.* | C | A | |
| 23/11 | 10:00 | 60–120 | 74°35′ | 175°00′ | 56000 | 0 | No | No | PHN |
| 28/11 | 23:30 | 40–60 | 75°00′ | 175°00′ | 101 | 0 | No | No | B |
| 29/11 | 14:30 | 20–40 | 74°25′ | 175°00′ | 23 | 0 | No | No | B |
| 04/12 | 10:00 | 40–80 | 74°40′ | 175°00′ | 0 | 5 | Yes | Yes | B |
| 11/12 | 18:00 | 20–40 | 73°38′ | 174°59′ | 12 | 3 | Yes | No | B |
| 12/12 | 18:00 | 20–150 | 74°12′ | 175°02′ | 5 | 6 | Yes | Yes | B |
| 13/12 | 15:00 | 10–160 | 74°17′ | 174°58′ | 0 | 0 | No | No | B |
| 13/12 | 20:30 | 20–40 | 74°38′ | 175°01′ | 12 | 2 | No | No | B |
| 14/12 | 19:50 | 20–60 | 74°43′ | 171°35′ | 0 | 0 | No | No | B |
| 14/12 | 23:30 | 20–60 | 74°45′ | 171°13′ | 2 | 79881 | No | No | PHN |

**Table 2.** Location of the net samples from 18–28 December 1994, showing the net used (B = EZ-NET-BIONESS; PHN = Plankton Hamburg Net), depth where samples were taken, number of individuals of *Euphausia superba* (*E.s.*) and *Euphausia crystallorophias* (*E.c.*), presence of other zooplankton (C = copepods; A = amphipods)

| Date | Beginning h:min | Depth m | Location | | Catch composition | | | | Net |
|---|---|---|---|---|---|---|---|---|---|
| | | | Lat. (S) | Long | *E.s.* | *E.c.* | C | A | |
| 18/12 | 15:10 | 30–40 | 74°36′ | 175°00′E | 12 | 0 | No | No | PHN |
| 18/12 | 23:50 | 100–120 | 74°05′ | 179°05′E | 1852 | 0 | No | No | PHN |
| 19/12 | 11:00 | 15–40 | 73°42′ | 179°45′E | 0 | 0 | No | No | PHN |
| 19/12 | 17:00 | 15–80 | 74°08′ | 176°45′E | 0 | 0 | No | No | PHN |
| 20/12 | 00:30 | 20–130 | 74°19′ | 173°28′E | 0 | 0 | No | No | PHN |
| 20/12 | 09:00 | 15–30 | 73°40′ | 173°27′E | 3571 | 0 | No | No | PHN |
| 20/12 | 17:20 | 20–40 | 73°14′ | 176°47′E | 38 | 0 | No | No | PHN |
| 21/12 | 16:50 | 30–40 | 72°46′ | 175°00′E | 0 | 3 | Yes | No | PHN |
| 21/12 | 23:40 | 20–40 | 73°07′ | 172°46′E | 2739 | 0 | No | No | PHN |
| 22/12 | 09:00 | 30–40 | 73°32′ | 172°26′E | 2212 | 0 | Yes | No | PHN |
| 22/12 | 15:30 | 20–40 | 72°11′ | 174°42′E | 2 | 0 | No | No | B |
| 23/12 | 00:10 | 20–40 | 71°55′ | 176°05′E | 0 | 0 | No | No | PHN |
| 23/12 | 16:00 | 20–40 | 71°03′ | 178°48′E | 0 | 0 | No | No | PHN |
| 24/12 | 00:00 | 20–40 | 71°03′ | 177°37′E | 0 | 0 | No | No | PHN |
| 24/12 | 09:00 | 20–40 | 71°43′ | 175°26′E | 0 | 0 | No | No | PHN |
| 24/12 | 17:00 | 20–80 | 71°57′ | 172°48′E | 1785 | 0 | No | No | PHN |
| 25/12 | 17:30 | 20–40 | 70°43′ | 173°36′E | 22 | 0 | No | No | PHN |
| 25/12 | 14:00 | 20–80 | 70°40′ | 174°40′E | 4 | 0 | No | No | PHN |
| 25/12 | 23:50 | 20–100 | 70°45′ | 171°00′E | 2 | 0 | No | No | PHN |
| 27/12 | 09:30 | 20–80 | 68°15′ | 179°00′W | 15 | 0 | Yes | Yes | PHN |
| 27/12 | 14:30 | 20–80 | 68°09′ | 178°42′W | 4 | 0 | No | No | PHN |
| 27/12 | 21:00 | 20–60 | 67°46′ | 179°15′W | 112 | 0 | No | No | PHN |

thereafter. Hauls lasted 30 min. The PHN was more efficient at sampling krill, allowing length distributions to be estimated for several stations (Table 1: 23 November to 15 December; Table 2: 17–28 December). The mean length of krill was calculated from a random subsample of 100 individuals if the catches were large, otherwise from all individuals. In the period from 9 to 23 November the mean length of krill (42.34 mm; SD 4.45 mm) obtained from all samples taken together (953 individuals) was used for biomass calculation. In the period from 23 November to 28 December, mean length of krill was taken from a successful station and used until the next successful station. Krill weights (g) were calculated from Morris et al. (1988):

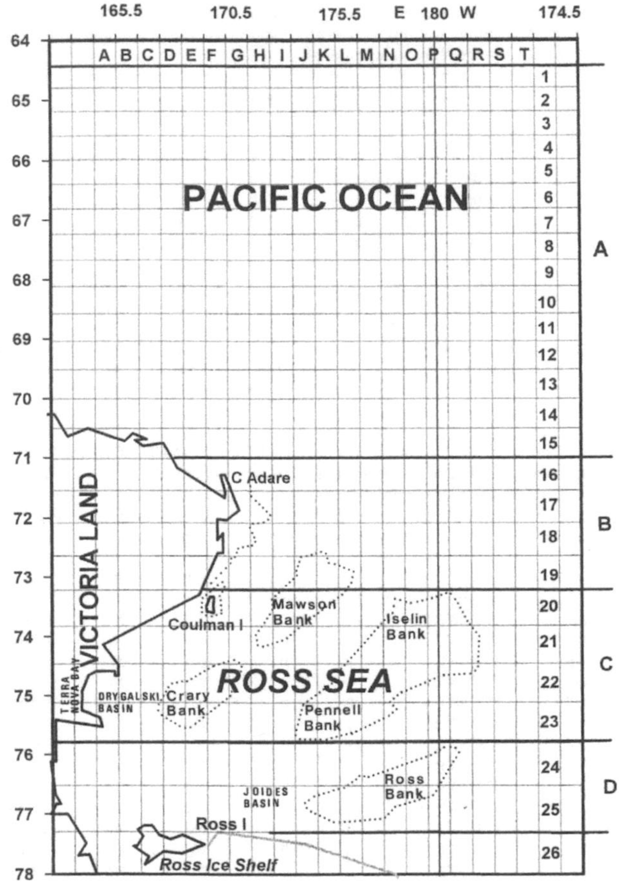


**Fig. 1.** Division of the investigated area in rectangles and geographical regions. The area of each rectangle is 600 nm². Geographical regions are: *A* oceanic water; *B* continental slope; *C* continental shelf; *D* continental shelf adjacent to the Ross Ice Shelf

$$w = 3.85\ 10^{-6} \times L^{3.20}.$$

In addition, the presence of *Euphausia crystallorophias* was noted and the mean lengths/weights were calculated for each sample (Tables 1, 2).

## The Geographical Area and the Acoustic Surveys

The area surveyed acoustically was approximately the western part of the Ross Sea (east from longitude 180°) and a strip of the Pacific Ocean extending from longitude 176°E to 176°W. According to Anonymous (1986b) and Lewis and Perkin (1985), the investigated area can be divided into four regions (Fig. 1):

1. (Fig. 1A) Oceanic water covering the very deep waters of the Pacific Ocean and extending up to about 71°S. The surveyed rectangles extend in latitude from 1 to 15.
2. (Fig. 1B) Continental slope where oceanic waters mix with shelf waters coming from the Ross Sea, extending between latitudes 71°S and 73°13′S. The surveyed area covers rectangles 16 to 19.
3. (Fig. 1C) Continental shelf extending between latitudes 73°S and 75°48′S, and including Mawson Bank, Iselin Bank, Pennell Bank, Crary Bank and "Drygalsky Basin" (Anonymous 1986b). The surveyed area extends from rectangle 20 to 23.

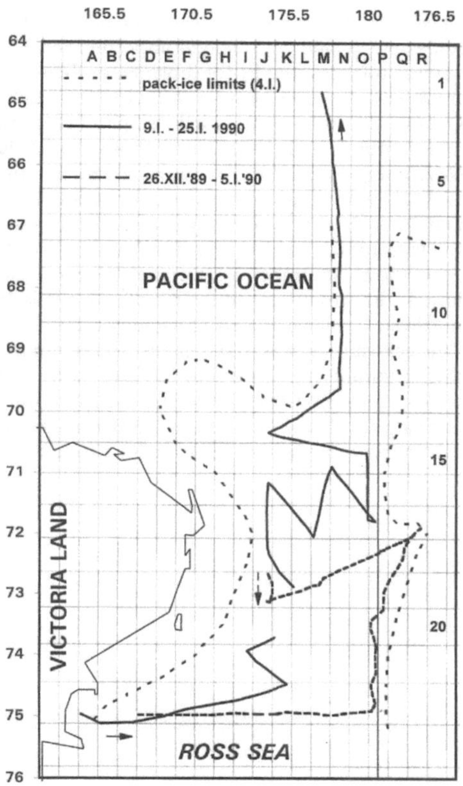


Fig. 2. Vessel's track during the 26 December 1989–25 January 1990 expedition

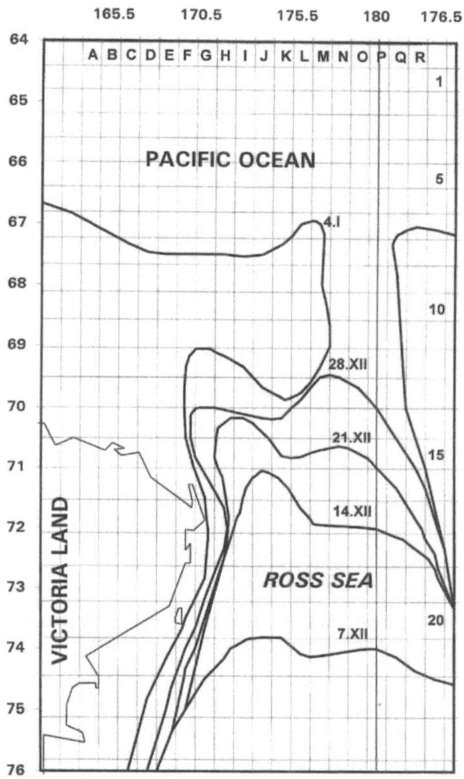


Fig. 3. Process of ice melting in the Ross Sea during the early austral summer of 1989–1990

4. (Fig. 1D) Continental Shelf adjacent to the Ross Ice Shelf Barrier (CSARIS), including Joides Basin, Ross Bank, and extending from latitude 75°48′ up to the Ice Shelf Barrier. The surveyed area covers rectangles 24 and 25.

The 1989–1990 surveys were conducted in early summer at speeds of 4.5 to 5 kts, in step with the retreat of sea ice (Figs. 2, 3).

The 26 December–5 January survey (584 nm) made use of the cruise track between oceanographic stations. Data were collected in the southern part of the continental slope (280 nm) and on the continental shelf (304 nm). Observations began in front of Cape Wheatstone (72°49′S, 174°30′E) and were conducted southwards to Terra Nova Bay (75°S, 167°30′E).

The 12–25 January survey (1109 nm) was carried out along the predetermined zig-zag design, that was strongly affected by storms. The eastern and western limits of the vessel's tracks were generally determined by pack ice. Data were collected on the continental shelf (239 nm), in the continental slope (368 nm), and in the Pacific Ocean (502 nm). An echo survey was carried out from Terra Nova Bay (TNB) northwards to Scott Island (64°26′S, 177°08′E).

The 1994 surveys were conducted at speeds of 5 to 9 kts, in late spring when the Ross Sea was still covered by ice (Figs. 4, 5, 6). The 9 November–15 December survey (2497 nm) was done between oceanographic stations; however, the track taken from one station to another was very irregular because of the ice. From the 9–12

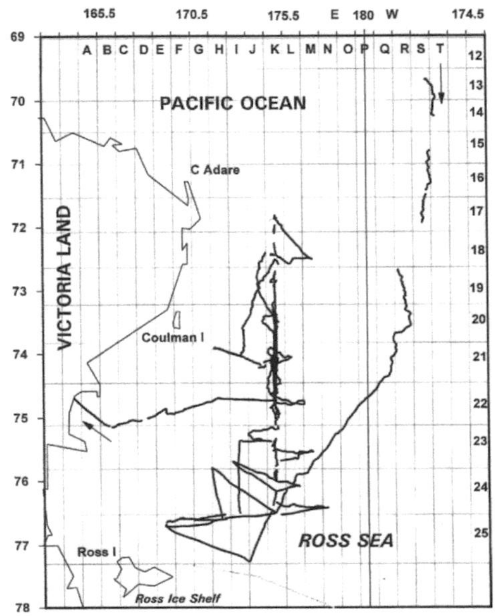


**Fig. 4.** Vessel's track during the 9 November–15 December 1994 expedition. The average ice sea cover was 80%

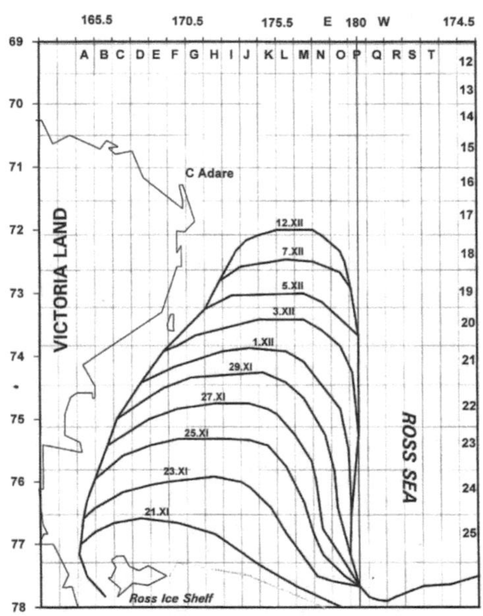


**Fig. 5.** Process of ice melting in the Ross Sea during the late austral spring of 1994

November, the echo survey was done between latitudes 69°40′ and 76°S behind an icebreaker. The sea was fully covered by ice (only 5% of the ship's track was free). Data were collected in the Pacific Ocean (71 nm), continental slope (114 nm), and continental shelf (239 nm). From the 13–20 November, measurements were done in the CSARIS region (607 nm), between latitudes 77°30′ and 75°30′S. The ice coverage was irregular and the vessel route had features of a "random walk". From the 21 November–9 December, the vessel followed the front ice zone, moving irregularly northwards approximately along longitude 175°E. The echosurvey was done many times around this longitude from latitudes 75° and 72°S. Data were collected on the continental shelf (709 nm), and slope (230 nm). From the 11–15 December, the echo survey was carried out from longitude 175°E

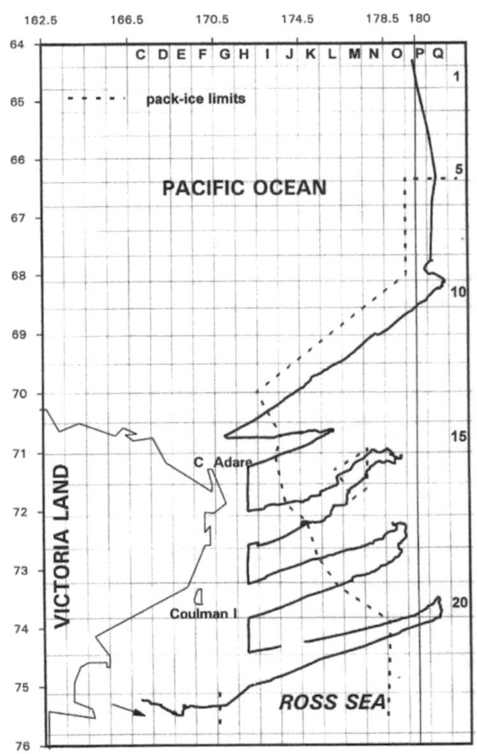


**Fig. 6.** Vessel's track during the 17–28 December 1994 expedition. The average ice sea cover was 37%

along latitude 74°30′S up to TNB, in ice-free waters. Acoustic observations were performed on the continental shelf (527 nm).

The 17–28 December survey (1883 nm) was carried out along previously planned transects (a loose zig-zag route). Data were collected on the continental shelf (577 nm), and slope (745 nm), and in the Pacific Ocean (561 nm). The survey began at the geographical position of 75°14′S, 166°56′E and was over at 64°18′S, 179°52′E. In the surveyed area, 37% of the ship's track was covered by ice. The eastern and western limits of the track were determined by pack ice.

Summarizing, during the two expeditions (1989–1990 and 1994) the Ross Sea was surveyed four times. Twice (26 Dec. 1989–5 Jan. 1990 and 9 Nov.–15 Dec. 1994) the echo survey began in the north and made use of the track between oceanographic stations. Twice (12–25 Jan. 1989 and 17–28 Dec. 1994) the echo survey began in the south and had a predetermined design (zig-zag). The 1989–1990 surveys were conducted in ice-free water and in quasistationary environmental conditions. In contrast, the 1994 surveys were conducted in waters covered by ice and in relatively dynamic conditions.

## Censuses of the Birds and Mammals

Concurrent with the two acoustic surveys of 1994, censuses were taken of the birds and mammals for 30 min of every hour, irrespective of ship speed (Azzali et al. 1996). Included in the record (Table 3) were animals observed from the flying bridge (14 m above sea level) which were located within a 400-m-wide strip to one side of the vessel.

## Data Analysis

Postprocessing of data was carried out ashore on the same computer used for data acquisition. Raw data on relative biomass, collected at 120 and 38 kHz in square voltages ($V^2$), were converted into mean volume backscattering strength ($S_v$) and expressed in dB, using a postprocessing BioSonics software (Anonymous 1990). Each of the 6-m bins containing scatters was filtered for noise by subtracting mean $S_v$ values from the adjacent segments which were void of signals (Kalinowski and Azzali 1996). Most of the noise was attributed to ice, and increased with percentage ice cover (from 0 to 40 dB for no ice to complete ice cover, respectively). Filtered data at 38 and 120 kHz were used for the successive analysis.

The discrimination of *Euphausia superba* from other zooplankton involved two stages of processing. In the first stage, the two-frequencies method (Mitson et al. 1996) was used for making a rough discrimination of krill from other scatters mile by mile. It is recognized that this method is still in its infancy. It has some limiting conditions, including a uniform composition of zooplankton in the chosen echo integration bins for $S_v$. This is sometimes a simplification of real conditions. However, visual analysis of echograms can overcome this limitation. Additionally, to insure that the same water volume is insonified, the pulse

**Table 3.** Mean density and location of peak density of birds and mammals, related to the krill densities detected in the same rectangles. Mean density of birds and mammals are related to an area around 440 nm$^2$, investigated from 9 November–28 December 1994

| Birds and mammals | Density | | Peak density location | | | | |
|---|---|---|---|---|---|---|---|
| | Mean ind nm$^{-2}$ | SD ind nm$^{-2}$ | Day | Rectangle | Ice cover % | Peak density ind nm$^{-2}$ | Krill density t nm$^{-2}$ |
| Antarctic petrel | 11.893 | 34.270 | 02/12 | K,18 | 82 | 282.994 | 7.35 |
| Adelie penguin | 9.590 | 24.233 | 09/11 | T,13 | 52 | 254.843 | 923.01 |
| Snow petrel | 2.406 | 4.812 | 09/11 | T,13 | 52 | 51.388 | 923.01 |
| Emperor penguin | 2.372 | 7.321 | 13/11 | E,24 | 89 | 72.115 | 1.67 |
| Crabeater seal | 0.756 | 3.025 | 29/11 | K,22 | 86 | 41.935 | 139.86 |
| Minke whale | 0.447 | 2.784 | 23/11 | L,22; | 83 | 37.535 | 323.12 |
| | | | | K,22 | | 33.239 | 2633.20 |
| South Polar skua | 0.206 | 0.859 | 14/11 | I,24 | 80 | 9.659 | 0.06 |
| Weddell seal | 0.103 | 0.481 | 10/11 | T,14 | 50 | 4.297 | 849.90 |

transmissions at both frequencies should be synchronized and the transducer beam widths should be equal. This is not the case in these surveys. However, water volume effects were reduced by in situ measurements of the difference in $S_v$ on selected krill aggregations, as described below, and averaging $S_v$ over many pings (generally all pings occurring over a distance of 1 nautical mile). During the second stage of processing, the echograms of the scatters attributed to krill in the first stage were scrutinized and the krill classification refined using personal experience and DAT data. In the two-frequency method the difference in mean volume backscattering strength (DMVBS) obtained by simultaneous integration at 120 and 38 kHz is assumed to be dependent on species and size only and independent of instrumentation.

It was calculated that for organisms up to 1 cm in length this value was about 20 dB (Kalinowski and Azzali 1992). On the basis of in situ experiments conducted on 132 aggregations identified as *Euphausia superba* either directly from net sampling or indirectly from analysis of echograms (Kalinowski and Witek 1985a,b; Madureira et al. 1993), it was found that the mean difference between $S_v$ of krill swarms at 120 and 38 kHz was 10.15 dB, with a standard deviation of 1.6 dB. The DMVBS range of variation was from 6 to 14 dB. Figure 7 illustrates the procedure and results of the experiment. Data recorded on DAT were used to find this DMVBS classifier. Scatters that exceeded $S_v$ values at 38 kHz from 6 to 14 dB were separated from other scatters and their echograms at both frequencies scrutinized. Visual inspection improved the two-frequency method mostly in places where other scatter entities became mixed with krill aggregations. Here the DVMBS values were often close to one limit of the expected range. The krill aggregations (compact and dark marks elongated vertically; Madureira et al. 1993; Kalinowski and Witek 1985b) were isolated on echo charts from other zooplankton (less dense and loosely bound marks) and/or other scatters. These various sources of noise were then excluded from the density estimation by processing those data recorded on DAT, relative to krill echograms. In addition, the scrutinizing method and DAT data were used to extract physical characteristics from krill aggregations (minimum depth, height, length, average density). In this chapter, only the characteristics of those aggregations classified as swarms (Siegel and Kalinowski 1994; Kalinowski and Witek 1985b) are reported. The mean volume backscattering strength $\langle S_v \rangle$ (dB re 1 m$^3$) of krill at 120 kHz, in each depth interval over one nautical mile, was converted in mean area density $\langle \rho_a \rangle$ (g m$^{-2}$) from the formula:

$$10 \log \langle \rho_a \rangle = \langle S_v \rangle - \langle TS \rangle + 10 \log \Delta R + 10 \log \langle w \rangle,$$

where $\Delta R$ is the width of integrator depth interval (m), $\langle w \rangle$ is the mean weight of individual krill (g) and $\langle TS \rangle$ is the mean target strength of krill (dB) at 120 kHz calculated from the relationship (Anonymous 1991): $TS = -127.45 + 34.85 \log L$.

The mean area density of krill for the entire integrated depth column was calculated from the $\langle S_v \rangle$ values of individual depth intervals (Johannenson and Mitson 1983).

## Abundance Estimation

The biomass of krill was estimated with the method called Elementary Statistical Sampling Rectangle (ESSR, Johannesson and Mitson 1983). The surveyed area was stratified through lines of latitude and longitude. The strata are the rectangles formed by the intersection of the longitudinal and latitude lines (Dalen and Nakken, unpubl. data; Simmonds and MacLennan, unpubl. data). The rectangles must be neither too large, because the density is assumed stationary within each stratum, nor too small, because the cruise track must pass through each rectangle at least once (MacLennan and Simmonds 1992). For krill in the Ross Sea it was found satisfactory to divide the surveyed area into rectangles of constant area (A = 600 nm$^2$), spaced 1° in longitude and with variable intervals in latitude. A rectangle was identified by a capital letter (A, B . . . T) and a number (1, 2 . . . 25), that indicated the intervals along lines of longitude and latitude, respectively (Fig. 1). There were about 200 rectangles in the surveyed area. If there are $n_k$ samples in the rectangle k (k = 1 to 200), then the mean abundance $B_k$ is estimated from the average density within the rectangle multiplied by the area A of the rectangle:

$$B_k = A\left(\sum_i \rho_{ik}\right)/n_k; i = 1, 2 \ldots n_k,$$

where $\rho_{ik}$ is biomass density over ith nautical mile in kth rectangle.

The calculation of standard deviation in each rectangle is equally straightforward.

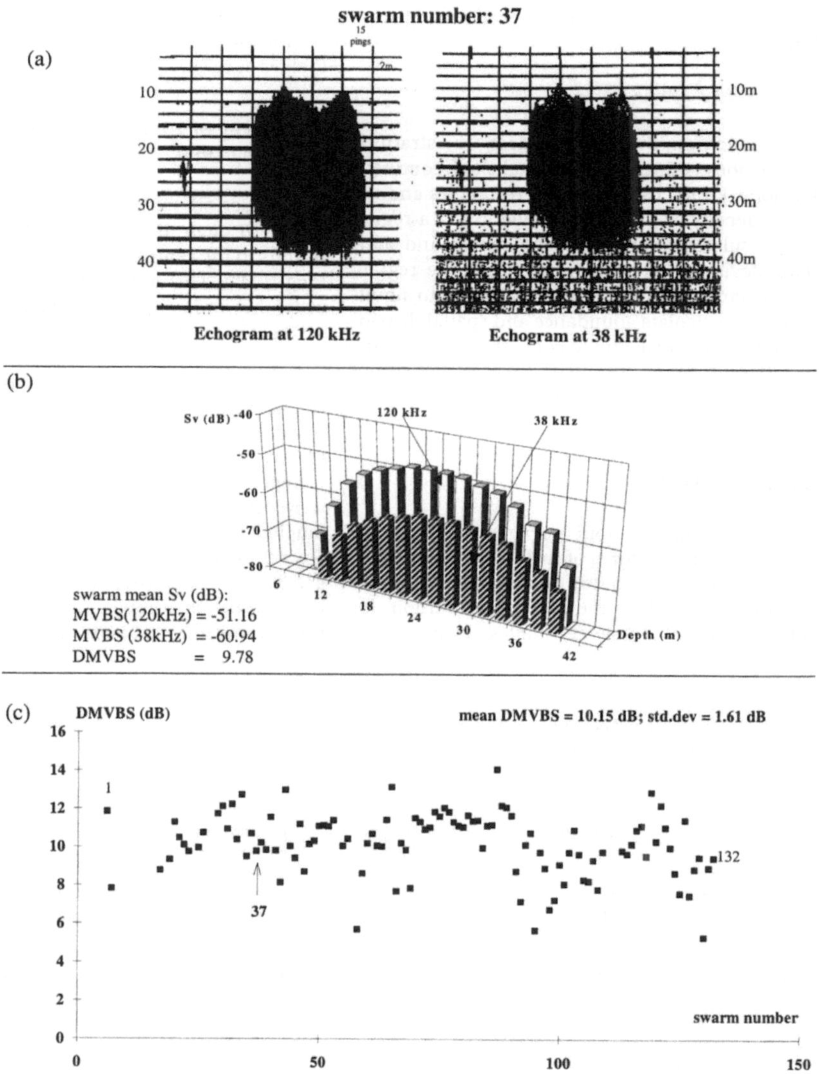


**Fig. 7a–c.** Procedure used to estimate the mean difference between $S_v$ values of krill swarms at 120 and 38 kHz (i.e. mean DMVBS). **a** The *solid silhouette* of a krill swarm is represented at each frequency by a rectangular matrix of $S_v$ (cells within the silhouette) and zeros (other cells). A cell has a resolution equal to 2 m on the vertical axis and to 15 ping intervals on the horizontal axis. **b** Each matrix is merged into one column, averaging the $S_v$ values of each row. A swarm MVBS at 120 and 38 kHz is calculated averaging the $S_v$ values of the respective column. **c** The DMVBS of a swarm is represented by a *point* on the graph. The mean DMVBS is calculated on 132 krill swarms

The total biomass, B, of krill in M rectangles is:

$$B_k = M \sum_k B_{kj};\ k = 1,2,\ldots .M.$$

In addition, the surveyed area was stratified into regions (CSARIS, continental shelf, continental slope, Pacific Ocean) by the rectangles and the mean density and standard deviation of a region was calculated combining mean density and standard deviation of each rectangle of the region. The rectangular grid method allowed us to simultaneously calculate abundance and spatial distribution (map) of krill and to compare directly maps of different surveys (acoustic surveys carried out in different periods, surveys on predators, net samplings).

## Survey Results

The results of the four surveys are reported sequentially in the form of krill density and standard deviation estimated within the rectangles (600 $nm^2$) crossed by the vessel. In addition, for combination of these rectangles, estimates of krill density, standard deviation, and biomass within the regions of the Ross Sea are provided (Tables 4, 5, 7).

### The Survey of 26 December 1989–5 January 1990

In the period from 6–25 December, the vessel sailed through the pack ice from the Balleny Islands to the central part of the Ross Sea (lat. 72°, long. 174°). This area was not surveyed acoustically; however, krill individuals were observed several times on the pieces of ice broken up by the vessel.

The southern part of the continental slope (rectangles 17–19; Fig. 8), where ice had recently receded, was the first area acoustically investigated (26–30 Dec). In places where depth rapidly changed and near ice edges, numerous swarms were detected. A peak in density was found in rectangle P,18 (Fig. 8): 111.37 $t\,nm^{-2}$ (SD 220.85 $t\,nm^2$). Over the continental slope, the average density was 41.58 $t\,nm^{-2}$ (SD 28.6 $t\,nm^{-2}$). The total biomass in the rectangles sampled acoustically (6000 $nm^2$) was estimated as 250 000 t (Table 4).

**Table 4.** Krill biomass and density estimated in the two consecutive echo surveys carried out in the early summer 1989–1990. The surveyed rectangles are shown

| Region | Rectangles | No. of integrated nautical miles | Density $t\,nm^{-2}$ | | Area $nm^2$ | Biomass mln t |
|---|---|---|---|---|---|---|
| | | | Mean | SD | | |
| **26 December 1989–5 January 1990** | | | | | | |
| Continental slope | R,17<br>N,O,P,Q,18<br>J,K,L,M,P,19 | 280 | 41.58 | 28.6 | 6000 | 0.250 |
| Continental shelf | P,20<br>P,21<br>D,E, . . . ,P,22 | 304 | 25.63 | 50.98 | 9000 | 0.230 |
| **9–25 January 1990** | | | | | | |
| Pacific Ocean | M:2,3,4,13,15<br>N:5,6, . . . ,13,15<br>K,L,14<br>O,15 | 502 | 34.95 | 36.02 | 10 800 | 0.377 |
| Continental slope | J,K, . . . ,O,16<br>J,L,M,O,P,17<br>J,K,18<br>K,19 | 368 | 63.5 | 50.56 | 7800 | 0.495 |
| Continental shelf | J,20<br>J,K,21<br>A,B, . . . ,K,22 | 239 | 8.26 | 11.83 | 8400 | 0.069 |

The continental shelf was the second area investigated (from 31 Dec.–5 Jan.; rectangles 20–22 in Fig. 8). Also in this area swarms were observed but with lower density and smaller dimensions than in the continental slope zone. The average density was 25.63 t nm$^{-2}$; SD 50.98 t nm$^{-2}$ (38.4% lower than on the continental slope of the Ross Sea). The total biomass in the rectangles that the vessel passed across (9000 nm$^2$) was estimated as 230 000 t (Table 4).

In the whole investigated area (continental slope and continental shelf) during the first echo survey the average density of krill was 32.0 t nm$^{-2}$ and the biomass in the rectangles crossed by transects (15 000 nm$^2$) was estimated as 480 000 t. The nautical miles sailed by the vessel were 280 over the continental slope and 304 in the continental shelf zone.

## The 12–25 January 1990 Survey

The first part (12–15 Jan.) of the second echo survey covered approximately the same area (continental shelf, 8400 nm$^2$) which had been sampled 10 days before. However, the average density (8.26 t nm$^{-2}$; SD 11.83 t nm$^{-2}$) and the number of swarms found were much lower than before. The total biomass was estimated as 69 000 t.

The continental slope region was investigated between the 16–19 January. Three weeks before, the northern part of this area was covered by pack ice, but visual observations had suggested occurrence of krill aggregations, as mentioned above. This was confirmed by the echosurvey: numerous and dense aggregations of krill were found in rectangles J,16 and J,17 in front of Cape Adare (Fig. 9). In rectangle J,17 maximum values for density

**Table 5.** Krill biomass and density estimated in the echosurvey carried out from 9 November–15 December 1994. The surveyed rectangles are shown

| Region | Period | Rectangles | No. of integrated nautica miles | Density t nm$^{-2}$ | | Area nm$^{-2}$ | Biomass mln t |
|---|---|---|---|---|---|---|---|
| | | | | Mean | SD | | |
| Pacific Ocean | 09/11 | T,13,14<br>S,15 | 71 | 656.46 | 400.74 | 1800 | 1.181 |
| Continental slope | 10/11 | S,16,17<br>R,19 | 114 | 70.42 | 9.83 | 1800 | 0.127 |
| Continental shelf | 11/11 | R,20<br>Q,21<br>O,P,22<br>M,N,23 | 239 | 139.87 | 216.83 | 3600 | 0.500 |
| CSARIS | 12–15/11 | H,I, . . . ,M,24<br>E,F, . . . ,J,25 | 348 | 10 | 9.29 | 7200 | 0.072 |
| | 16–20/11 | I,J,24<br>E,F, . . . ,J,25 | 259 | 4.49 | 5.80 | 8400 | 0.038 |
| Continental shelf | 20–22/11 | J,K,23<br>K,L,24 | 47 | 6.59 | 5.22 | 2400 | 0.0158 |
| | 23/11 | K,22 | 68 | 2633 | 8507 | 600 | 1.58 |
| | 23/11 | L,22 | 30 | 323.12 | 871.03 | 600 | 0.194 |
| | 24–26/11 | J,K,21 | 134 | 3.67 | 1.03 | 1200 | 0.0044 |
| Continental shelf | 28–30/11 | H, . . . ,K,21<br>K,22 | 157 | 62.82 | 53.11 | 3000 | 0.188 |
| Continental slope | 02–03/12 | J,K,18<br>J,19 | 84 | 32.71 | 23.97 | 1800 | 0.098 |
| Continental shelf | 01/12 | H,21<br>K,20 | 41 | 1103.81 | 927 | 600 | 0.662 |
| | 04–05/12 | H,I,K,21<br>K,22 | 232 | 35 | 17.97 | 3600 | 0.189 |
| Continental slope | 07–09/12 | K,17<br>K,L,M,18<br>K,19 | 146 | 40.59 | 22.49 | 3000 | 0.122 |
| Continental shelf | 06/12 | K,20,21,22 | 161 | 34 | 14 | 1800 | 0.061 |
| Continental shelf | 11–15/12 | K,20,21,22<br>A,B, . . . ,J,22 | 366 | 63.79 | 57.39 | 8400 | 0.536 |

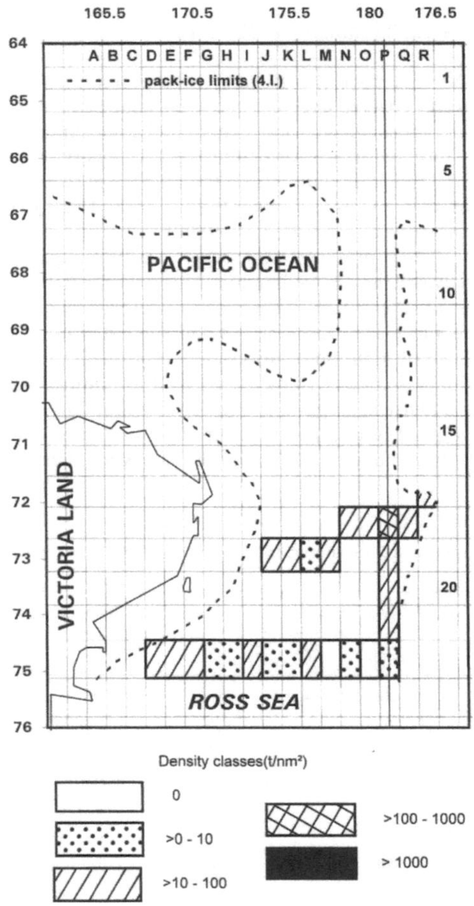


**Fig. 8.** Distribution of mean krill biomass density by rectangles during the survey carried out from 26 December 1989 to 5 January 1990

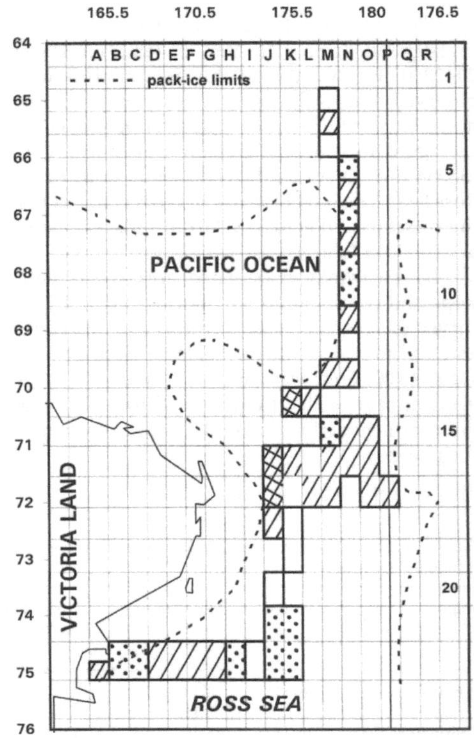


**Fig. 9.** Distribution of mean krill biomass density by rectangles during the survey carried out from 9 to 25 January 1990. Symbols as in Figure 8

were measured (204.89 $t\,nm^{-2}$; SD 358.65 $t\,nm^{-2}$). The average density in the continental slope zone was 63.54 $t\,nm^{-2}$ (SD 50.56 $t\,nm^{-2}$). The total biomass in the rectangles investigated (7800 $nm^2$) was estimated as 495 000 t.

The krill biomass estimated over the continental shelf and slope taken together (16 200 $nm^2$) was 564 000 t (Table 4). This value was close to the one found in the first survey in the same area (480 000 t).

The oceanic water zone (rectangles 1–15) was surveyed between the 20–25 January. From lat. 69° to 66°S, the survey was carried out along a channel in the pack ice. In this region, irregular forms began to appear, while the presence of swarms, believed to be krill, became more and more occasional. The mean density of krill measured in the oceanic water zone was 34.95 $t\,nm^{-2}$; SD 36.02 $t\,nm^{-2}$. The total biomass in the rectangles crossed by the vessel (10 800 $nm^2$) was estimated as 377 000 t.

In the whole investigated area (27 000 $nm^2$, including Pacific waters), during the second echo survey the average density was 34.85 $t\,nm^{-2}$ and the total biomass was 931 000 t.

## Comparison Between the Surveys of 1989 and 1990

In both surveys, the largest concentrations of krill biomass were found in the continental slope zone,

**Table 6.** Analysis of net samples taken from the continental shelf (C) in the period 23 November–14 December 1994. Some samples were taken from the same places to check the persistence in time of dense krill aggregations

| Date | Beginning h:min | Rectangle | *Euphausia superba* | | | *Euphausia crystallorophias* | | | Region |
|---|---|---|---|---|---|---|---|---|---|
| | | | Sample Number | Mean length mm | SD mm | Sample Number | Mean length mm | SD mm | |
| 23/11 | 10:00 | K,22 | 100 | 42.4 | 3.54 | – | – | – | C |
| 28/11 | 23:30 | K,22 | 100 | 42.5 | 4.41 | – | – | – | C |
| 29/11 | 14:30 | K,21 | 23 | 41.9 | 3.77 | – | – | – | C |
| 04/12 | 10:00 | K,22 | – | – | – | 5 | 28 | 0.93 | C |
| 11/12 | 18:00 | K,20 | 12 | 42.9 | 1.56 | 3 | 30 | – | C |
| 12/12 | 18:00 | K,21 | 5 | 43.2 | 4.21 | 6 | 29.5 | 0.46 | C |
| 13/12 | 20:30 | K,22 | 12 | 44.5 | 3.66 | 2 | 25.0 | – | C |
| 14/12 | 23:30 | G,22 | 2 | 45.0 | – | 100 | 25.1 | 3.66 | C |

between latitudes 71° and 73°S (rectangles 16–19), where oceanic waters mix with shelf waters creating conditions favourable for krill concentration formation (Parsons et al. 1977; Kalinowski and Witek 1985b; Makarov et al. 1988; Rakusa-Suszczewski 1988; Witek et al. 1988). The average density estimated on the continental shelf and slope was very close in the two surveys: 32.0 and 34.8 t $nm^{-2}$, respectively. However, the krill distribution pattern changed consistently between the two surveys carried out at an interval of a few weeks from one another. The biomass density on the continental shelf decreased from 26.5 to 8.3 t $nm^{-2}$ (–208%), while on the continental slope it increased from 41.58 to 63.5 t $nm^{-2}$ (+52.8%). In the period from the 14 December–4 January, pack ice retreated from lat. 72° to 67°S (Fig. 2). These results show that krill biomass distribution can change quickly, and suggest that ice retreat may be associated with these changes in distribution.

## Krill Swarms in the Surveys of 1989/1990

The krill swarms found on the continental shelf and slope during the 1989/90 surveys were 637. The average size of swarms was in horizontal extent 52.5 m (SD 66.57 m) by 11.0 m (SD 8.6 m), in vertical extent with a mean density of 167 g $m^{-3}$ or approximately 278.4 ind $m^{-3}$ (SD 160.1 g $m^{-3}$ or 266.9 ind $m^{-3}$). The average depth of swarms, measured from the top, was 30 m (SD 19.6 m). The Ross Sea was investigated in the absence of darkness; no significant diurnal vertical migration was observed of the aggregations (Azzali et al. 1992). The minimum depth of swarms measured at "night-time" (between 22:00 and 05:00 h) was 30.6 m (SD 23.4 m), very close to the one measured in "daytime" (between 05:00 and 22:00): 29.8 m; SD 16.4 m.

## The Survey of 9 November–15 December 1994

The results of this survey, carried out in conditions of marked variability of ice cover, are summarized in Table 5 (acoustic data) and Table 6 (net sampling data). The distribution of krill concentrations is illustrated in Fig. 10.

In the first stage of the investigation (9–12 November; Table 5) the vessel sailed behind an icebreaker from northwest (lat. 69°40′S; long. 176°W) to southeast (lat. 76°S; long. 175°45′E). The mean ice cover was 95%. Many times, krill individuals were observed on the pieces of ice broken up by the icebreaker. The survey began from the deep waters of the Pacific Ocean (rectangles T,13; T,14; S,15). In rectangles T,13 and T,14, where ice percentage cover was only 52%, the krill biomass density was very high: 923 t $nm^{-2}$ (SD 843 t $nm^{-2}$) and 850 t $nm^{-2}$, respectively. During the synoptic visual survey on birds and mammals around these rectangles, density peaks of Adelie penguins (253.7 ind $nm^{-2}$), Snow petrels (51.6 ind $nm^{-2}$) and Weddell seals (4.1 ind $nm^{-2}$) were found (Table 3). The estimated biomass on an area of 1800 $nm^2$ was 1 181 000 t. This area of the Pacific Ocean, abundant in krill and krill consumers such as Adelie penguins and snow petrels (Koch and Shimadzu 1994), is close to that described by Marr (1962) and Mackintosh (1973) as krill-abundant. In the eastern part of the continental slope zone (west of long. 180°), rectangles S,16 ; S,17 and R,19, with an average ice cover of 75%, were surveyed. The average density was 70.42 t $nm^{-2}$ (SD 9.83 t $nm^{-2}$) and the estimated biomass on an area of 1800 $nm^2$ was 127 000 t.

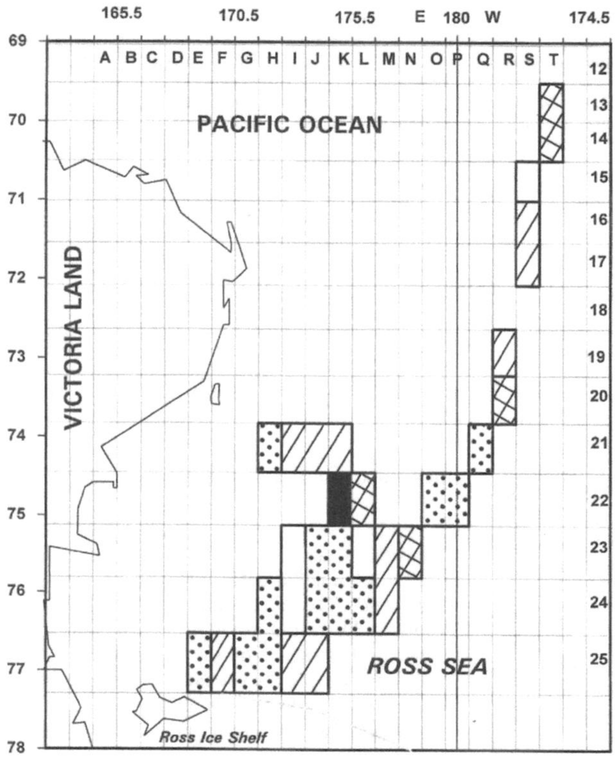


**Fig. 10.** Distribution of mean krill biomass density by rectangles during the survey carried out from 9 to 30 November 1994. The highest values of the means are shown in the rectangles sampled more times. Symbols as in Figure 8

In the continental shelf zone, dense aggregations were found in rectangles R,20 (543 t nm$^{-2}$; SD 1189 t nm$^{-2}$) and N,23 (112 t nm$^{-2}$; SD 588 t nm$^{-2}$), near Iselin and Pennel Banks, respectively. In rectangle M,23 density was 30.3 t nm$^{-2}$ (SD 52.43 t nm$^{-2}$). In the other rectangles, density was very low (less than 3 t nm$^{-2}$). The average krill density in the continental shelf zone was 139.87 t nm$^{-2}$ (SD 216.83 t nm$^{-2}$) and the krill biomass on an area of 3600 nm$^2$ was estimated as 500 000 t.

Between the 12 and 20 November, it was possible to take measurements only in the polynya adjacent to the Ross Ice Shelf (CSARIS, rectangles 24 and 25), because of the very difficult ice conditions. The polynya, an ice-free area located between the winter pack ice limits and the Ross Ice Barrier, was investigated twice. The first time, from the 12–15 November, 12 rectangles of the polynya zone were sampled. Average krill density was 10 t nm$^{-2}$ (SD 9.29 t nm$^{-2}$) and estimated biomass on an area of 7200 nm$^2$ was 72 000 t. During the survey on birds and mammals around rectangle E,25 (where krill was absent and ice coverage 89%) density peaks of Emperor penguin (72.4 ind nm$^{-}$) and South Polar skua (9.7 ind nm$^{-2}$) were found (Table 3). The second time, from the 16–20 November, 14 rectangles, including 9 previously investigated rectangles, were sampled. Average krill density was 4.49 t nm$^{-2}$ (SD 3.06 t nm$^{-2}$). Estimated biomass on an area of 8400 nm$^2$ was 38 000 t. In the CSARIS region, small scatters [$S_v(120) - S_v(38) > 14$ dB] were prevalent on scatters believed to be krill. It can be concluded that the CSARIS region, considered as krill-absent (Marr 1962; Deacon 1984; Guglielmo et al. 1990), is indeed poorly krill-populated (average density from 10 to 4.49 t nm$^{-2}$). However, the consistent presence of consumers, which do not depend on krill or not entirely on it (Polar skua, Emperor penguin), and of plankton smaller than krill suggest that other species may be abundant in this area.

From the 20 November–15 December, the dynamics of ice receding northwards (Fig. 5) strongly influenced the course of the vessel and probably krill density distribution, that in this period cannot be assumed stationary at least within the rectangles of the continental shelf zone.

In the period 20–26 November, 8 rectangles (K,24 and L,24, K,23 and J,23, K,22 and L,22; K,21 and J,21; Fig. 9) were sampled (number of integrated miles: 279). In the southern (rectangles 24 and 23) and northern (rectangles 21) parts of the continental shelf zone, mean krill density was very low, respectively 3.67 t $nm^{-2}$ (SD 1.03 t $nm^{-2}$) and 6.59 t $nm^{-2}$ (SD 5.22 t $nm^{-2}$). Estimated biomass in these two parts of the continental shelf (3600 $nm^2$) was 16000 t. In rectangles K,22 and L,22, on the Pennell Bank, a gigantic group of swarms was found. It was distributed in an area where the ice was melting, limited in the south by the receding ice edge and in the north by winter ice. Krill extended from the surface up to 110 m. In rectangle K,22, mean krill density on 68 integrated miles was 2635.50 t $nm^{-2}$ (SD 8507.04 t $nm^{-2}$); in rectangle L,22, mean krill density on 30 integrated miles was 323.12 t $nm^{-2}$ (SD 871.03 t $nm^{-2}$). The estimated biomass in the two rectangles (1200 $nm^2$) equaled 1773968 t; it was 98% of the krill biomass estimated in the 8 rectangles mentioned above (4800 $nm^2$). The bulk of the super swarm was found on 23 November in an ice-free area of some square nautical miles around the geographic position 74°36′S; 174°59′E (rectangle K,22; Fig. 10). Krill density in this small area, that was encircled by a herd of about 250 minke whales, was calculated as 51761 t $nm^{-2}$. However, density peaks of minke whales (37.4 ind $nm^{-2}$, Table 3) were found between the Pennell and the Iselin Banks (rectangle L,22). Krill distribution detected acoustically was corroborated by the results from net hauls, carried out close to the bulk of krill (lat. 74°35′; long. 175°), at a depth ranging between 120 and 60 m (Table 1). The net used was the PHN. The haul produced the largest catch of the expedition (around 56000 individuals). The mean total length of krill, measured in a random subsample of 100 individuals, was 42.4 mm; SD 3.54 mm (Table 6). No other species, other than *Euphausia superba*, was caught.

In the period from 28 to 30 November, krill biomass on an area of 3000 $nm^2$, that included rectangles J,21; K,21 and K,22 previously investigated, was estimated as 188000 t. Rectangle K,22 was sampled on the 29 November, while the ice edge was receding from this area. Mean krill density calculated on 49 nautical miles was 139.86 t $nm^{-2}$, about 20 times lower than the one found 6 days before. Around this rectangle we observed a peak in the density of crabeater seals (41.7 ind $nm^{-2}$), a consistent presence of Adelie penguins (101.1 ind $nm^{-2}$) and minke whales (19.9 ind $nm^{-2}$). Two net samplings (on 28 and 29 November) were taken in this location, using the EZ-NET-BIONESS. Both catches were monospecific. Krill mean length obtained from net hauls of 28 November (42.5 mm; SD 4.41 mm) and 29th November (41.6 mm; SD 4.04 mm) was similar to that of 23 November (Table 6). Figure 10 illustrates the distribution of krill concentrations detected from the 9–30 November.

In the period from 1–5 December, an area of 4200 $nm^2$ (273 integrated miles) was surveyed in the continental shelf region and an area of 1800 $nm^2$ (84 integrated miles) in the continental slope region (Table 5, Fig. 11). Krill biomass estimated in rectangles J,18; K,18; and J,19 within the continental slope was 98000 t (mean density: 32.71 t $nm^{-2}$; SD 23.97 t $nm^{-2}$). In rectangle K,18 (krill density 7.35 t $nm^{-2}$; SD 30.93 t $nm^{-2}$; ice coverage: 82%) a peak in the density of Antarctic petrels (281.77 ind $nm^{-2}$) was found (Table 3). Krill biomass estimated in the continental shelf region, that included rectangles J,21; K,21, and K,22 previously investigated, was 851000 t. It was concentrated (77.8%) in rectangle H,21 (mean density: 1103.8 t $nm^{-2}$; SD 927 t $nm^{-2}$), close to Mawson Bank. In the other rectangles, mean krill density was 35 t $nm^{-2}$ (SD 17.97 t $nm^{-2}$). In particular, the mean density of krill in rectangle K,22 sampled on the 5 December was 29.70 t $nm^{-2}$ (SD 64.02 t $nm^{-2}$), more than 120 times lower than that measured on the 23 November. This enormous diminution in krill density in a small temporal (12 days) and spatial (600 $nm^2$) scale may be attributed to the tremendous impact of predators (mostly minke whales) on krill biomass and to the effect of the retreating ice edge on krill spatial distribution.

In the period from 6–9 December, the exploration of the continental slope region proceeded in parallel to the retreating ice cover. Mean krill density calculated on 146 nautical miles was 40.59 t $nm^{-2}$ (SD 22.49 t $nm^{-2}$). Krill biomass estimated on an area of 3000 $nm^2$ (rectangles K,18; L,18; M,18; K,17; K,19) was 122000 t (Table 5, Fig. 11). In this period, 3 rectangles in the continental shelf region (K,20; K,21; K,22), already

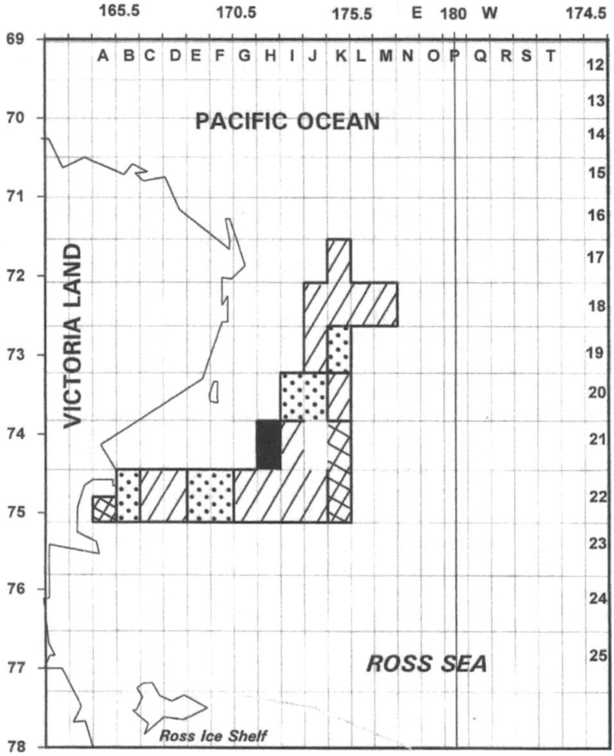


**Fig. 11.** Distribution of mean krill biomass density by rectangles during the survey carried out from 1 to 15 December 1994. The highest values of the means are shown in the rectangles sampled more times. Symbols as in Figure 8

investigated 5 days before, were surveyed. The results were close to the previous ones (mean krill density: 34 t nm$^{-2}$; SD 14 t nm$^{-2}$).

In the period 11–15 December, the echo survey was carried out from rectangles K,21 and K,22 (long. 175°E, close to Pennell Bank) to rectangle A,22 (long 164°E, near TNB), within a strip delimited by lat. 74°30′ and lat. 75°S (Fig. 11). Ice had receded from this area since the 1 December (Fig. 5). The sampled nautical miles were 366. Mean krill density was 63.79 t nm$^{-2}$ (SD 57.39 t nm$^{-2}$). Krill biomass, estimated on an area of 8400 nm$^2$, was 536 000 t. Dense aggregations were found in rectangles K,21 (136.2 t nm$^{-2}$; SD 222.9 t nm$^{-2}$), K,22 (184.94 t nm$^{-2}$; SD 159.79 t nm$^{-2}$) and A,22 (112.41 t nm$^{-2}$; SD 224.9 t nm$^{-2}$). In the other rectangles, the biomass density was either low or very low. During this transect, three successful hauls were carried out in the central part of the continental shelf (rectangles K,20 K,21; K,22), using the EZ-NET-BIONESS, and one haul in the western part of the continental shelf (rectangle G,22), using the PHN (Table 1). The catches in rectangles K,20, K,21 and K,22 provided similar data: a mixture of *Euphausia superba* and *E. crystallorophias* (Table 6). Mean length of the krill found in the three hauls was similar to those found in the same area on the 23, 28 and 29 November; but no other organisms except krill were caught in the hauls in November. The haul carried out in rectangle G,22 on the Crary Bank, produced the largest catch of *E. crystallorophias* of the expedition (79 881 individuals; mean length: 25.1 mm; mean weight: 0.125 g). The catch was almost monospecific (Table 6). Mean krill density on Crary Bank (rectangles G,22; F,22; E,22) was very low (less than 5 t nm$^{-2}$). In this zone and in the Drygalski Basin (rectangle B,22), scatters smaller than krill ($TS_{120} - TS_{38} > 14$ dB) were found on several occasions. Figure 11 illustrates the distribution of krill concentrations found in the period from the 1–15 December.

**Table 7.** Krill biomass and density estimated in the echo survey carried out from 17–28 December 1994. The surveyed rectangles are shown

| Regions | Rectangles | No. of integrated nautical miles | Density t nm$^{-2}$ | | Area nm$^2$ | Biomass mln t |
|---|---|---|---|---|---|---|
| | | | Mean | SD | | |
| Pacific Ocean | P,1,2,3,10,11<br>Q,4, . . . ,10<br>O,11; P,11<br>M,12; N,12 | 561 | 34.11 | 57.25 | 15000 | 0.511 |
| Continental slope | H,I,M,N,O,16<br>H,I, . . . ,M,17<br>H,I, . . . ,L,18<br>N,18; O,18 | 745 | 124.55 | 126.55 | 16200 | 2.018 |
| Continental shelf | I,J,K,L,20<br>P,20; Q,20<br>I,21; K, . . . ,P,21<br>H,I, . . . ,L,22<br>C,D, . . . ,G,23 | 577 | 66.11 | 74.91 | 14400 | 0.952 |

**Table 8.** Analysis of net samples taken from the continental shelf (C), continental slope (B), and Pacific Ocean (A) in the period 18–28 December 1994

| Date | Beginning h:min | Rectangle | *Euphausia superba* | | | *Euphausia crystallorophias* | | | Region |
|---|---|---|---|---|---|---|---|---|---|
| | | | Sample Number | Mean length mm | SD mm | Sample Number | Mean length mm | SD mm | |
| 18/12 | 15:10 | K,22 | 12 | 46.8 | 1.03 | – | – | – | C |
| 18/12 | 23:50 | O,21 | 100 | 41 | 3.52 | – | – | – | C |
| 20/12 | 09:00 | I,21 | 100 | 43.7 | 2.78 | – | – | – | C |
| 20/12 | 17:20 | M,20 | 38 | 41.1 | 6.15 | – | – | – | C |
| 21/12 | 16:50 | K,19 | – | – | – | 3 | 26.8 | – | B |
| 21/12 | 23:40 | H,19 | 100 | 44.7 | 3.08 | – | – | – | B |
| 22/12 | 09:00 | H,20 | 100 | 43.6 | 2.79 | – | – | – | B |
| 22/12 | 15:30 | K,18 | 2 | 43.8 | – | – | – | – | B |
| 24/12 | 17:00 | H,17 | 100 | 41.1 | 6.25 | – | – | – | B |
| 25/12 | 17:30 | I,15 | 22 | 41.6 | 5.50 | – | – | – | A |
| 25/12 | 14:00 | K,15 | 4 | 40 | – | – | – | – | A |
| 25/12 | 23:50 | G,15 | 2 | 41.2 | – | – | – | – | A |
| 27/12 | 09:30 | Q,10 | 15 | 34.9 | 5.89 | – | – | – | A |
| 27/12 | 14:30 | R,10 | 4 | 39.6 | – | – | – | – | A |
| 27/12 | 21:00 | Q,9 | 100 | 41.5 | 4.06 | – | – | – | A |

## The Survey of the 17–28 December 1994

In the period from the 17–28 December, the acoustic survey covered the area between lat. 75°30′ and 64°S, and long. 170°E and 179°W. Environmental conditions were less variable than in the previous survey. Hauls were carried out at preselected stations, while in the previous survey several hauls were carried out in the same places to check the persistence in time of dense krill aggregations. The results of this survey are summarized in Table 7 (acoustic data) and Table 8 (net sampling data). The distribution of krill concentrations is illustrated in Fig. 12.

The continental shelf was the first region sampled (17–20 December). Mean krill density, calculated on 577 nautical miles, was 66.11 t nm$^{-2}$ (SD 74.91 t nm$^{-2}$). Total biomass, estimated on 24 rectangles crossed by the vessel (14400 nm$^2$), was 952000 t. Numerous and dense krill swarms were found on the Iselin Bank, near the eastern ice border (rectangles O,21; N,21; L,21). The peak density was found in rectangle O,21 (321.49 t nm$^{-2}$; SD 383.32 t nm$^{-2}$). A total of seven net samples were carried out in the continental shelf area (Table 2). *Euphausia superba* was caught in four hauls, while *E. crystallorophias* was never caught (Table 8) The catches on the Iselin Bank (rectangle O,21; mean krill length 41.0 mm; SD 3.52 mm) and Mawson Bank (rectangle I,20; mean krill length: 43.7; SD 2.78) were rich (Table 2). Krill with mean lengths of 46.8 mm (SD 1.03 mm) were found in rectangle K,22. It was the maximum value detected in the expedition. Krill caught in rectangle M,20 had similar mean lengths to those in rectangle O,21 (41.1 mm; SD 6.15 mm). Around Mawson Bank (rectangle M,20), a consistent number of Antartic petrel (249.5 ind $nm^{-2}$) was observed.

In the period 21–24 December, the continental slope region was surveyed. Mean krill density, calculated on 745 nautical miles, was 124.55 t $nm^{-2}$ (SD 126.5 t $nm^{-2}$). Total krill biomass, estimated on 27 rectangles crossed by the vessel (16 200 $nm^2$) was 2 018 000 t. Dense concentrations of krill were found in the shallow water in front of Cape Adare (rectangles H,16; I,16; I,17; J,17; K,17) and Cape Hallet (J,18). A peak of krill density, found in rectangle H,16, was very high: 532.40 t $nm^{-2}$; SD 1855.51 t $nm^{-2}$. However, the observed krill predators (Adelie penguins, snow petrels and Antarctic petrels) were not abundant in this area. The haul stations over the continental slope were nine (Table 2). *Euphausia superba* was found in three hauls carried out in shallow waters (rectangles H,19; H,20; H,17), where catches were abundant (>1700 individuals). Krill mean lengths were similar in the three places (respectively, 44.7, 43.6, 41.1 mm), as shown in Table 8. Few individuals of *Euphausia crystallorophias* (3) and *E. superba* (2) were caught in the hauls carried out close to the slope, in rectangles K,19 and K18, respectively (Table 8). No zooplankton was caught in four hauls carried out in deep waters (rectangles L,17; N,16; M,16; K,16).

In the period 25–28 December, the southern waters of the Pacific (from lat. 70° to lat. 64°30′S) were surveyed. This area was still covered by pack ice, that compelled the captain to change the planned survey track. Mean krill density, calculated on 561 nautical miles, was 34.11 t $nm^{-2}$ (SD 57.25 t $nm^{-2}$). Total krill biomass, estimated on 25 rectangles, was 511 000 t. Krill distribution was irregular (Fig. 12). The largest concentrations were found in rectangles Q,6 (252.73 t $nm^{-2}$; SD 855.54 t $nm^{-2}$) and Q,5 (133.10 t $nm^{-2}$; SD 443.43 t $nm^{-2}$) between lat. 67° and 66°S, and long 179° and 178°W. Isolated swarms occurred northwest from Cape Adare (rectangles G,15; I,15; I,14).

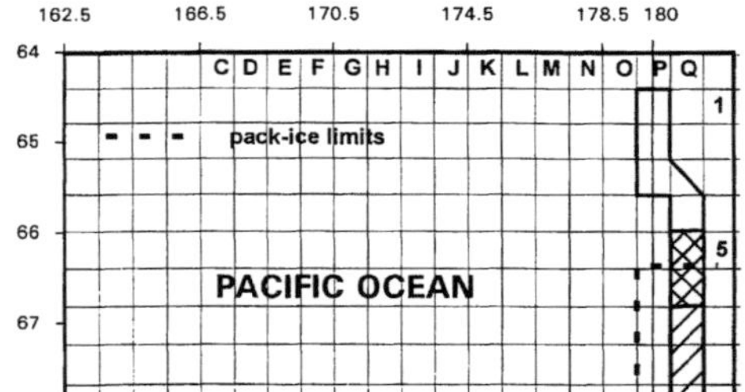


**Fig. 12.** Distribution of the mean krill biomass density by rectangles during the survey carried out from 17–28 December 1994. Symbols as in Figure 8

The six hauls, carried out under difficult conditions because of the ice, were successful, but produced very small catches of *Euphausia superba*. No other zooplankton was captured (Tables 2, 8). Mean length of krill caught in stations northwest from Cape Adare (rectangles K,15; I,15; G,15) had similar mean lengths (respectively, 40.0, 41.6, 41.2 mm). Krill sampled in rectangle Q,10 represented the smallest lengths found in the expedition (mean length: 34.9 mm; SD 5.89). Krill caught in rectangles R,10 and Q,9 had similar mean lengths (respectively 39.6 and 41.5 mm).

## Comparison Between the Surveys of 1994

The two consecutive estimates of krill biomass in the Ross Sea (continental shelf and continental slope) were similar: 3 120 000 t in the first survey and 2 970 000 t in the second. In contrast, the spatial patterns and geographic distribution of krill biomass in the first survey differed greatly from those of the second survey.

In the first survey, 96.2% of krill biomass (3 000 000 t) was found over the continental shelf, a region presented in the literature as krill-absent. The remaining 3.8% was recorded on the continental slope. In the central part of the continental shelf, around 2 500 000 t of krill were concentrated in small areas, enclosed by ice (rectangles K,22; L,22; and H,21, sampled between the 23 November–1 December). These enormous aggregations of krill, as well as the spots where they were detected, persisted for only a few days. In the western part of the continental shelf, along the transect towards TNB, around 500 000 t of krill were distributed on an area of 8400 $nm^2$, surveyed in the period 11–15 December. Mean krill density on the continental shelf was 245.0 $t\,nm^{-2}$. On the continental slope, around 120 000 t of krill were found on an area of 3000 $nm^2$, surveyed between 5 and 9 December. Mean krill density on the continental slope was 43.3 $t\,nm^{-2}$.

In the second survey, 32% of krill biomass (950 000 t) was found on the continental shelf and 68% (2 020 000 t) on the continental slope. On the continental shelf, krill biomass was distributed on an area of 14 400 $nm^2$, surveyed between the 17 and 20 December. Krill mean density was 65.9 $t\,nm^{-2}$. On the continental slope, krill biomass was distributed on an area of 16 200 $nm^2$, surveyed between 21 and 25 December. Mean krill density was 124.7 $t\,nm^{-2}$. Total krill biomass estimated in this second survey was distributed over a larger area (30 600 $nm^2$; average density = 97.1 $t\,nm^{-2}$) than in the first survey (13 200 $nm^2$; average density = 236.4 $t\,nm^{-2}$). The bulk of this biomass was found on the continental slope, whereas in the first survey it was on the continental shelf. This scattering and movement northwards of krill biomass, observed between November and December, occurred in parallel to ice melting and ice edge receding northwards. A similar but less dramatic change in the spatial distribution of krill in the Ross Sea was observed between the two consecutive surveys carried out between December 1989 and January 1990, as mentioned above. Also the spatial distribution of krill predator populations changed quickly, in parallel to ice-cover melting. Maximum densities of birds and mammals were found during the first survey. In particular, density peaks of minke whales, crabeater seals, and Adelie penguins were observed around the small ice-free areas, enclosed by ice, where high densities in krill aggregations were detected. In contrast, during the second survey high concentrations of birds or mammals were not observed, even in areas abundant in krill (in front of Cape Adare and Cape Hallet).

### Krill Swarms in the Surveys of 1994

Swarms found in the Ross Sea (continental shelf and continental slope) during the 1994 surveys and identified as *Euphausia superba* were 891. The average size of swarms in horizontal extent was 58.2 m (SD 67.0 m) and in vertical extent 15.6 m (SD 10.4 m). Mean density was 676.1 $g\,m^{-3}$ or 1121.10 $ind\,m^{-3}$ (SD 596.7 $g\,m^{-3}$ or 963.4 $ind\,m^{-3}$). Mean depth of swarms, measured from the top, was 36.9 m (SD 24.2 m). These swarms, found in late spring when light was very dim all day, had wider dimensions and a mean density four times higher than those found in the summer of 1989–1990. Moreover, these swarms, compared to the ones of 1989–1990, exhibited a diurnal vertical migration. The minimum depth of swarms measured between 22:00 and 05:00 h ("night-time") was 31.6 m (SD 17.4 m), while that measured between 05:00 and 22:00 h ("daytime") was 43.5 m (SD 27.3 m).

## Discussion

Inspection of data on spatial and temporal distribution of krill (1989–1990, and 1994 surveys) and krill predators (1994 surveys) suggest that krill movements through the areas of the Ross Sea (the so-called krill flux) and krill/predator interactions are controlled mainly by ice dynamics. Moreover, the data from the 1994 surveys suggest that the spatial distributions of *Euphausia superba* and *E. crystallorophias* may be influenced by seasonal sea ice, meltwater and ice-free water cycles. For example, in rectangle K,22, when it was enclosed by ice (23 December) and while ice was receding from it (28 and 29 December) only *Euphausia superba* was captured, but when this area was ice-free (13 January) a mixture of *Euphausia superba* and *E. crystallorophias* was caught.

The data on krill abundance over the continental shelf and slope of the Ross Sea come from two surveys carried out in early summer

1989–1990 and from two surveys carried out in late spring 1994. Krill biomass estimated in the 1989–1990 was around five times less than that estimated in 1994. Moreover, the 1989–1990 swarms differed from the ones in 1994 in size, density and behaviour. As the surveys occurred in different years and seasons and under different environmental conditions, there are insufficient data to separate the effects of year, season and environment on changes in krill biomass and behaviour. The data on krill abundance in the Pacific Ocean come from three surveys. The first was carried out in January 1990 in ice-free water (mean krill density 34.9 t $nm^{-2}$, calculated on an area of 10 800 $nm^2$). The second survey was carried out in November 1994, when waters were covered by ice (mean krill density 656.5 t $nm^{-2}$, calculated on an area of 1800 $nm^2$). The third survey was carried out in December 1994 while sea ice was melting (mean density 34.11 t $nm^{-2}$, calculated on an area of 15 000 $nm^2$). These results show that also in the Pacific Ocean high densities of krill can concentrate in small ice-free areas when water is covered by ice and this behaviour can affect acoustic estimation. The CSARIS region was sampled twice between the 12 and 20 November 1994. Estimated krill abundance was very low in both surveys (between 4.5 and 10 t $nm^{-2}$), but krill may not have been absent from this area.

It is difficult to compare these data from the Ross Sea with other regions of the Antarctic reported in the literature. This is due to problems in the value of krill target strength, higher until the end of the 1980s (formula in Anonymous 1986a) than those used in this chapter (formula in Anonymous 1991), and the calibration of acoustic instruments. We have recalculated the data from the literature by multiplying these by a factor of 5.62. This constant represents the average difference between target strength calculated by the two formulas mentioned above.

Regarding the Pacific sector of the Antarctic, the only data come from the acoustic survey carried out by the Japanese vessel Umitaka Maru (Anonymous 1986a) during the BIOMASS/FIBEX program. The mean biomass density, estimated along longitude 165°E, was 8 t $nm^{-2}$. This value is much lower than that obtained during the Italian expeditions in the strip between long. 176°E and 176°W.

In the Indian sector, Hampton (1985) carried out acoustic surveys between long. 15° and 30°E. He reports mean biomass density of 27.4 t $nm^{-2}$ and total krill biomass, estimated on an area of 180 373 $nm^2$, corresponding to 4 626 000 t. Higginbottom et al. (1988) surveyed the area between long. 58° and 70°E and between lat. 60° and 70°S in various summers and obtained the values of mean biomass density: 21.2 t $nm^{-2}$; 59.7 t $nm^{-2}$; 55.9 t $nm^{-2}$. Shirakihara et al. (1986) surveyed the area between long. 65° and 75 E, and lat. 62° and 70°S. The mean biomass density was 60 t $nm^{-2}$. These mean biomass densities in the Indian sector are of the same order of magnitude as those found in the Ross Sea in the summer of 1989/90.

The highest values of krill biomass density come from the Atlantic sector. Daly and Macaulay (1988) found that in the northwestern Weddell Sea, during the austral spring of 1983, krill biomass density ranged from 18.8 t $nm^{-2}$ to 1276.4 t $nm^{-2}$ under the ice and from 187.1 to 1871.2 t $nm^{-2}$ in open waters. Kalinowski (1982, 1984) in the eastern part of the Bransfield Strait found a krill biomass density of 485 t $nm^{-2}$ in March 1981. These data from the Atlantic sector, usually considered as the part of the Antarctic which is most abundant in krill, are of the same order of magnitude as those found in the Ross Sea in the spring of 1994.

In the literature, there is also information on superswarms found in the Atlantic and Indian sectors. One of the largest superswarms was found in February–March 1981 near Elephant Island (61°S, 55°W). The total biomass of this swarm was estimated as 2 100 000 t, distributed over an area of 131 $nm^2$ (Macaulay et al. 1984). Superswarms were found in the austral summers 1976–77 and 1978–79 in the area adjacent to South Georgia, near Clerk Rocks (Kalinowski and Witek 1985a). In the Indian sector, a superswarm was found near Prinz Bay, in March 1987 (Higginbottom and Hosie 1989). The physical dimension and biomass of these superswarms are similar to those found on the 23 November 1994 in the Ross Sea. Diurnal vertical migration of swarms, as often reported from the Atlantic and Indian sectors (Kalinowski 1978; Kalinowski and Witek 1980; Everson 1982, 1983; Siegel and Kalinowski 1994), were observed in the 1994 surveys, but not in the 1989–1990 surveys, when light was intense all day.

## Conclusions

On the basis of the results of the Italian expeditions of 1989–1990 and 1994 to the Ross Sea, the following conclusions have been drawn:

1. Contrary to what is generally believed, the Ross Sea is not krill-deficient. In particular, very high concentrations of *Euphausia superba* were found in the southern part of the continental shelf (November 1994), which is considered as krill-absent in the literature.
2. The biomass of *Euphausia superba* in the Ross Sea, estimated during late spring 1994, was around 3 million tons and the mean biomass density ranged from 250 t nm$^{-2}$ (first survey) to 100 t nm$^{-2}$ (second survey). These values are similar to those given in the literature for the Atlantic sector. The density of *Euphausia superba* estimated in the same area during the early summer of 1989–1990 was around 33 t nm$^{-2}$. This value is of the same order of magnitude as those found in the Indian sector.
3. During late spring, the biomass of *Euphausia superba* was concentrated over the continental shelf (between lat. 73° and 75°S), but at the beginning of summer the main part of krill biomass occurred on the continental slope (between lat. 71° and 73°).
4. When the sea was almost fully covered by ice or when the ice edge was receding northwards, populations of *Euphausia superba* were restricted in narrow ice-free spots, enclosed in the ice, where very dense aggregations were formed. Around these areas, density peaks of krill-consumer populations (minke whales, crabeater seals, Adelie penguins) were found. The temporal persistence of these particular areas was on a "weather scale" (<15 days).
5. Biomass of *Euphausia superba* was the main part of zooplankton biomass on the continen-

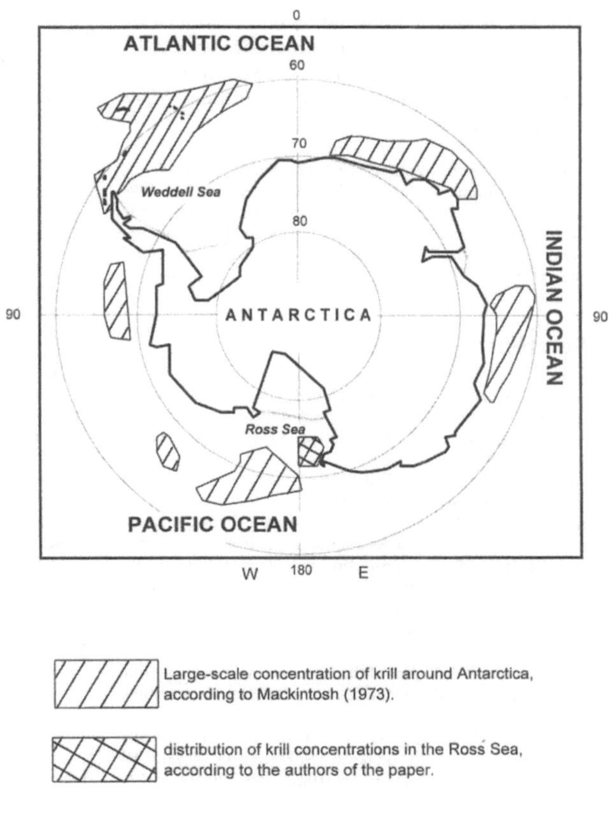


**Fig. 13.** Large-scale concentrations of krill, according to Mackintosh (1973), with the part of the Ross Sea supplemented by the authors of the present chapter

tal slope and central part of the continental shelf. Southwards (from lat.75 ° up to the Ross Sea Ice Barrier) and westwards (from long. 171°, along lat. 75°, up to Terra Nova Bay) this relation progressively changed in favour of smaller plankton and of *E. crystallorophias*.
6. Data on size, density and diurnal vertical migration of krill swarms found in the 1989–1990 and 1994 surveys are rather different. It is necessary to acquire more information in this field, including those on environmental parameters, during future research activities in this area.
7. Further research in the Ross Sea is required to assess krill biomass and its fluctuations over the years, and to describe changes in krill spatial patterns in relation to changes in ice cover, and to determine the interactions between these changes, krill predators population and other euphausiid species.

Finally, we wish to recommend that, on the basis of our results, information on the large-scale distribution of krill around Antarctica should be supplemented with those from the Ross Sea in the area extending to the north from latitude 75°, as shown in Fig. 13.

*Acknowledgements.* We thank Dr. Nicola Saino, who made the field observations on birds and mammals during the 1994 surveys and processed the data, and the scientists who made the measurements on krill samples. We also thank two anonymous referees for their detailed comments and helpful suggestions.

## References

Amos AF (1984) Distribution of krill (*Euphausia superba*) and the hydrography of the Southern ocean: large-scale processes. J Crustacean Biol 4 (Special Issue no 1): 306–329

Anonymous (1986a) Post-FIBEX Acoustic Workshop. BIOMASS Rep Ser 40: 126 pp

Anonymous (1986b) 4065 British Admiralty Chart. Southern Ocean, Ross Sea to New Zealand, 1:10000000 (at the Equator), Mercator projection. Published at Taunton 19th Dec 1986 under the superintendence of Rear-Admiral R.O. Morris, Hydrographer of the Navy

Anonymous (1990) Echo signal processor operator's manual. BioSonics, Seattle

Anonymous (1991) Report of the 10th Meeting of the Scientific Committee (SC-CAMLR-X). CCAMLR, Hobart, Australia, 427 pp

Azzali M, Kalinowski J (1992) Italian Antarctic acoustic survey of krill. In: Gallardo VA, Ferretti O, Moyano HI (eds). Oceanografia in Antartide. ENEA-PNRA-EULA, Concepcion, Chile, pp 321–330

Azzali M, Kalinowski J, Castagnani R (1991) How much krill is there? Results of the 5th Italian Expedition to the Ross Sea, 1989/90. BIOMASS Newsletter, BIOMASS/SCAR/SCOR/LABO/ACMRR, Texas A&M University, vol 13, No 1 pp 6–7

Azzali M, Kalinowski J, Castagnani R (1992) Biomass of krill in the Ross Sea (Pacific Sector) in relation to BIOMASS FIBEX/SIBEX results. In: Gallardo VA, Ferretti O, Moyano HI (eds) Oceanografia in Antartide, ENEA-PNRA-EULA, Concepcion, Chile, pp 331–338

Azzali M, Kalinowski J, Saino N (1996) Estimation of biomass of krill *Euphausia superba* and birds and mammals census during the Xth Italian Expedition to Ross Sea, November–December 1994. DOC WG-EMM-96/63, Agenda Item No 3 (CCAMLR, Hobart, Australia), 21 pp

Brierley AS, Goss C, Watkins JL, Woodroffe P (1997) Variations in echosounder calibration with temperature and some possible implications for acoustic surveys of krill biomass. Doc WG-EMM-97/52, July 1997, Agenda Item No 3 (CCAMLR, Hobart, Australia), 8 pp

Daly KL, Macaulay MC (1988) Abundance and distribution of krill in the ice-edge zone of the Weddell Sea, austral spring 1983. Deep Sea Res 35(1): 21–41

Deacon GER (1984) Water movements and Antarctic krill. J Crustacean Biol 4 (Special Issue no 1): 13–15

Demer DA (1994) Accuracy and precision of acoustic surveys of Antarctic krill. PhD Thesis, University of California, San Diego, 144 pp

Demer DA, Hewitt RP (1993) Calibration of an acoustic echo-integration system in a deep tank, with gain comparisons over standard sphere material, water temperature and time. SC-CAMLR-Select, Sci Pap 9: 127–144

Demer DA, Hewitt RP (1996) Uncertainity in echosounder calibration. Doc WG-EMM 97/40 (CCAMLR, Hobart, Autralia), 17 pp

El-Sayed SZ (1994) History, organization and accomplishments of the BIOMASS programme. In: El-Sayed SZ (ed) Southern Ocean ecology. The BIOMASS perspective. Cambridge University Press, Cambridge, pp 1–8

Everson I (1982) Diurnal variations in mean volume backscattering strength of an Antarctic krill (*Euphausia superba*) patch. J Plankton Res 4(1): 155–162

Everson I (1983) Variations in vertical distribution and density of krill swarms in the vicinity of South Georgia. Memoirs of National Institute of Polar Research Special Issue no 27. Proceedings of the BIOMASS Colloquium in 1982. National Institute of Polar Research, Tokyo, pp 84–92

Guglielmo L (1992) The research on zooplankton and micronekton in the National Antarctic Research Program. In: Gallardo VA, Ferretti O, Moyano HI (eds) Oceanografia in Antartide. ENEA-PNRA-EULA, Concepcion, Chile, pp 298–307

Guglielmo L, Costanzo G, Manganaro A, Zagami G (1990) Spatial and vertical distribution of zooplanktonic communities in the Terra Nova Bay (Ross Sea). In: National Scientific Commission for Antarctica (ed) Oceanographic Campaign 1987–88, Data Rep Part I, Genova, pp 257–398

Hampton I (1985) Abundance, distribution and behaviour of *Euphausia superba* in the Southern Ocean between

15° and 30°E during FIBEX. In: Siegfried WR, Condy PR, Laws RM (eds) Antarctic nutrient cycles and food webs. Springer, Berlin Heidelberg New York, pp 294–303

Higginbottom IR, Hosie GW (1989) Biomass and population structure of a large aggregation of krill near Prydz Bay, Antarctica. Mar Ecol Prog Ser 58: 197–203

Higginbottom IR, Kerry KR, Wayte SE (1988) Hydroacoustic surveys of the distribution and abundance of krill: Prydz Bay region – FIBEX, ADBEX II and SIBEX II, MV *Nella Dan*. Australian National Antarctic Research Expeditions: ANARE Res Notes 62: 1–46

Ichii T (1990) Distribution of Antarctic krill concentrations exploited by Japanese krill trawlers and Minke whales. Proceedings of the NIPR Symposium on Polar Biology, no 3. National Institute of Polar Research Tokyo, pp 36–56

Johannsson KA, Mitson RB (1983) Fisheries acoustics: a practical manual for biomass estimation. FAO Fish Tech Pap 240: 249

Kalinowski J (1978) Vertical migration of krill in the region of South Georgia, February–March 1976. Pol Arch Hydrobiol 25(3): 573–583

Kalinowski J (1982) Distribution and stock of krill in the Drake Passage and Bransfield Strait during BIOMASS/FIBEX expedition. Pol Polar Res 3: 243–251

Kalinowski J (1984) The biomass of krill in the eastern part of Bransfield Strait, March 1981. Pol Polar Res 5(1-2): 99–105

Kalinowski J, Azzali M (1992) Possibility of discrimination various groups of species using dual-frequency hydroacoustic system. In: Underwater acoustics. Collected papers. Proc 9th Symp on Hydroacoustics, Gdynia, 2–5 June, Gdynia, AMW (Poland), pp 219–229

Kalinowski J, Azzali M (1996) Influence of ice-cover for precision estimation of antarctic krill biomass. In: Underwater acoustics, Collected papers. Proc 13th Symp on Hydroacoustics, Gdynia. 7–10 May Gdynia, AMW (Poland), pp 123–134

Kalinowski J, Witek Z (1980) Diurnal, vertical distribution of krill aggregations in the Western Atlantic. Pol Polar Res 1: 127–146

Kalinowski J, Witek Z (1985a) Elementy biologii, formy grupowego wystepowania kryli. Studia i Materialy MIR B52, Gdynia, 162 pp (in Polish)

Kalinowski J, Witek Z (1985b) Scheme for classifying aggregations of Antarctic krill. BIOMASS hand-book no 27, SCAR/SCOR/IAOB/ACMRR, Group of Specialists on Living Resources of the Southern Ocean, 12 pp

Koch KH, Shimadzu Y (1994) Trophic: relatioships and trends in population size and reproductive parameters in Antarctic high-level predators. In: El-Sayed SZ (ed) Southern Ocean ecology. Cambridge University Press, Cambridge, pp 287–312

Lewis EL, Perkin RG (1985) The winter oceanography of McMurdo Sound. Oceanology of the Antarctic Continental Shelf. Antarct Res Ser 43: 145–165

Lubimova TG, Makarov RR, Maslennikov VV, Shevtsov W, Shust KV (1984) The ecological pecularities , stocks and the role of *E. superba* in the trophic structure of the Antarctic ecosystem. Selected Papers Presented to the Scientific Committee of CCAMLR, 1981–1984, Part II, Hobart: 391: 505

Macaulay MC, English TS, Mathisen OA (1984) Acoustic characterization of swarms of Antarctic krill (*Euphausia superba*) from Elephant Island and Bransfield Strait. J Crustacean Biol 4 (Special Issue no 1): 16–44

Mackintosh NA (1973) Distribution of postlarval krill in the Antarctic. Discovery Rep 36: 95–156

MacLennan DN, Simmonds EJ (1992) Fishery acoustics. Chapman and Hall, London, 325 pp

Madureira LSP, Everson I, Murphy EJ (1993) Interpretation of acoustic data at two frequencies to discriminate between Antarctic krill (*Euphausia superba* Dana) and other scatters. J Plankton Res 15(7): 787–802

Makarov RR, Maslennikov VV, Solyankin EV, Spiridonov VA, Yakovlev VN (1988) Variability in population density of antarctic krill in the Western Scotia Sea in relation to hydrological conditions. In: Sahrhage D (ed) Antarctic ocean and resources variability. Springer, Berlin Heidelberg New York, pp 231–236

Marr JWS (1962) The natural history and geography of the Antarctic krill (*Euphausia superba* Dana). Discovery Rep 32: 33–464

Mitson RB, Simard Y, Goss C (1996) Use of a two-frequency algorithm to determine size and abundance of plankton in three widely spaced locations. ICES J Mar Sci 53(2): 209–215

Morris DJ, Watkins J, Ricketts C, Buchholz F, Priddle J (1988) An assessment of length and weight measurements of Antarcic krill, *Euphausia superba*. Br Antarct Surv Bull 79: 27–50

Parsons TR, Takahashi M, Hargrave B (1977) Biological oceanographic processes. Pergamon, Oxford, 322 pp

Rakusa-Suszczewski S (1988) Differences in the hydrology, biomass and species distribution of plankton, fishes and birds in the Bransfield Strait and Drake Passage during FIBEX 1981 and SIBEX 1983/84. In: Sahrhage D (ed) Antarctic ocean and resources variability. Springer, Berlin Heidelberg New York, pp 214–218

Shimadzu Y (1984) A brief summary of Japanese fishing activity in relation to Antarctic krill, 1972/73–1982/83. Selected Papers Presented to the Scientific Committee of CCAMLR, 1982–1984, Part I, Hobart: 439–452

Shirakihara K, Nakayama K, Komaki Y (1986) Acoustic estimates of krill biomass in r.v. Kaiyo Maru SIBEX I survey area (Indian sector of the Southern Ocean). Mem Natl Inst Polar Res Special Issue 40: 140–152

Siegel V, Kalinowski J (1994). Krill demography and small-scale processes: a review. In: El-Sayed SZ (ed) Southern Ocean ecology. Cambridge University Press, Cambridge, pp 145–163

Tamura T, Ichii T, Fujise Y (1997) Consumption of krill by Minke whales in areas IV and V of the Antarctic. WG-EMM-97/17, July 1997, Agenda Item Nos 6&4: 9p

Witek Z, Kalinowski J, Grelowski A (1988) Formation of Antarctic krill concentrations in relation to hydrodynamic processes and social behaviour. In: Sahrhage D (ed) Antarctic ocean and resources variability. Springer, Berlin Heidelberg New York, pp 237–244

CHAPTER 32

# The Coastal Fish Fauna of Terra Nova Bay, Ross Sea, Antarctica

**M. Vacchi[1], M. La Mesa[1], and S. Greco[2]**

**ABSTRACT**

The coastal fish community of Terra Nova Bay (Ross Sea) was studied from samples collected off the Italian Antarctic station Terra Nova Bay down to about 700 m depth from 1987 to 1995. Species composition, abundance and vertical distribution are reported and the biological characteristics such as feeding habits, reproduction, age and growth of the common species have been studied. Twenty-six fish species were recorded, the most common being the nototheniid *Trematomus bernacchii* and the channichthyid *Chionodraco hamatus*. As for the vertical distribution of fish, species diversity was higher in shallow than in deep waters. The fish community comprises species which exhibit three different trophic niches: *T. bernacchii* and *Trematomus pennellii* are mostly benthos feeders, *Trematomus hansoni* and *Trematomus loennbergi* are mostly piscivorous and *Trematomus newnesi* feeds preferentially on plankton. Differences in reproductive strategies have been found during the sampling period, i.e. January and February, when *C. hamatus* and *T. hansoni* were in a spawning condition, whereas *T. bernacchii* was in a resting-early development stage. The two most common species, *T. bernacchii* and *C. hamatus*, show a high longevity, living more than 20 and 10 years, respectively. Moreover, in both species the females attain slightly older age and grow faster than males.

## Introduction

The first samples of the fish fauna of the Ross Sea were collected during the 1839–1843 British expedition led by James Clark Ross (Richardson 1844–1848). The next substantial contributions to the knowledge of the fish community of this Antarctic sector were due to the British Antarctic expedition Terra Nova at the beginning of this century (Regan 1914). After a gap of more than 40 years, ichthyological studies in the Ross Sea started again mostly by shore-based research conducted at McMurdo Sound by American and New Zealand scientists. These studies concerned systematics, zoogeography, biology and physiology (Wohlschlag 1960, 1964; Dearborn 1965; DeWitt 1971; DeVries and Eastman 1981; Eastman and DeVries 1982; Montgomery and Macdonald 1987; Macdonald et al. 1988).

Other ichthyological investigations were carried out during Antarctic cruises in the 1960s and 1970s. Among them, relevant data on the fish fauna of the Ross Sea were collected during the NZ oceanographic survey on HMNZS Endeavour (Reseck 1961), the USA Eltanin oceanographic cruise (DeWitt 1970), and the Japanese exploratory fish survey on board the Daini Bashu-Maru (Iwami and Abe 1981).

In recent years, Italian scientists have performed several research programmes on coastal fish of the Ross Sea as a contribution to the Italian Programme for Antarctic Research (PNRA). Most studies were carried out at Terra Nova Bay (western sector of the Ross Sea, Victoria Land Coast), where the Italian Antarctic station Terra Nova Bay was built in 1986–87. Ichthyological investigations at Terra Nova Bay included biochemical and physiological approaches (di Prisco et al. 1991; Tota et al. 1991), ecotoxicology studies (Focardi et al. 1992), as well as studies on phylogenetic and evolutionary relations (Morescalchi et al. 1992; Bargelloni et al. 1994). Ecological and

[1] ICRAM-Istituto Centrale per la Ricerca scientifica e tecnologica Applicata al Mare, Via Lorenzo Respighi 5, 00197, Rome, Italy
[2] CNR-Istituto Talassografico Sperimentale, Spianata San Raineri 86, 98122 Messina, Italy

faunistic studies of the coastal fish fauna of Terra Nova Bay were the main aims of the Nekton team, who started ichthyological surveys as of the third Italian Antarctic expedition in 1987–88 (Vacchi et al. 1992). The research was particularly concerned with improving the knowledge of biological and ecological characteristics of the most abundant species, such as feeding habits (Vacchi et al. 1994; Vacchi and La Mesa 1995; La Mesa et al. 1997), reproduction (Vacchi et al. 1996) and age and growth (Vacchi et al. 1992; La Mesa et al. 1996). The present chapter provides additional information on the coastal fish fauna of the 3rd, 6th and 10th Italian Antarctic expeditions.

## Materials and Methods

The research on coastal fishes was conducted near the Italian station Terra Nova Bay (74°41′42″S and 64°07′25″E), from the Gerlache Inlet to Cape Russell, in the north-western Ross Sea (Fig. 1) during three Italian Antarctic expeditions. The sampling area is characterized by shallow hard bottoms down to the 20 m depth, where two macroalgal species (*Iridaea cordata* and *Phyllophora antarctica*) and a few animal taxa (mainly polychaetes, molluscs and peracarid crustaceans) dominate the benthic association (Gambi et al. 1994). In the soft bottom, that appears at about the 20-m depth and extends down to the 600-m depth, the benthic community is represented by facies of *Adamussium colbecki* and by sponges, cnidarians, bryozoans and polychaetes (Di Geronimo et al. 1992).

Fish samples were collected by trammel and gill nets, both with 32-mm mesh size (side), and longlines and traps. Fishing activities were mainly carried out during the day for an average of 4–5 h. The samplings in the Polar Queen expedition were made from 13 January to 13 February 1988, from a 6-m-long aluminium boat, and consisted of 61 hauls made from 16- to 300-m depth (Vacchi et al. 1992).

During the Cariboo expedition, a total number of 90 fishing operations were conducted from 19 December 1990 to 9 February 1991 (Vacchi et al. 1991); the availability of a larger boat equipped with a hydraulic winch allowed us to extend the depth range sampled. The hauls were taken along four transects perpendicular to the coastline and extended offshore, ranging between 16 and 682-m depths (Fig. 1). On December 16, 1994, we collected

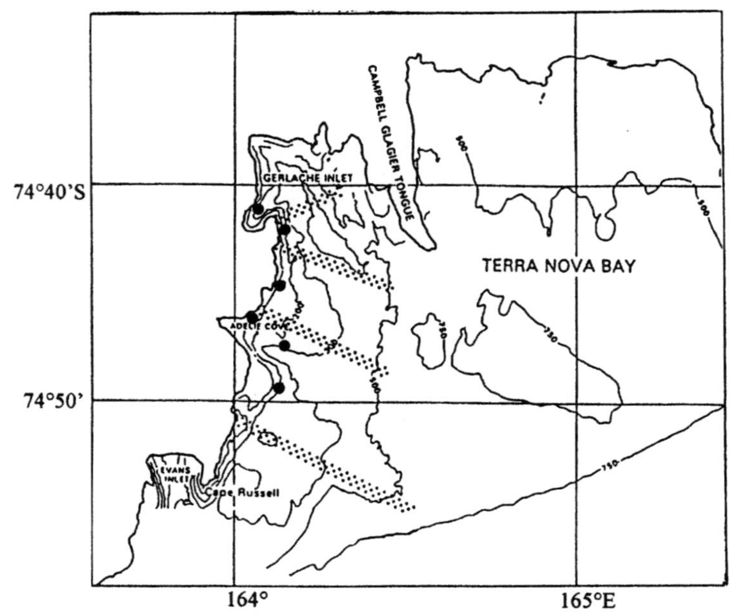


**Fig. 1.** The study area of Terra Nova Bay showing the sampling stations during the 1987/88 expedition (●) and 1990/91 expedition (grid stripes)

one sample of fish by gill net at the 150-m depth, during a short pause of the Italian oceanographic cruise ROSSMIZE near the Station.

All fishes were identified after Fischer and Hureau (1985) and Gon and Heemstra (1990). The specimens of the most common species were measured as total length (TL) to the nearest mm below. Total weight was determined to the nearest gram below. Each specimen was gutted and its gonads weighed and staged according to a five-point scale (Everson 1977; Anonymous 1989). With the aim to detect growth rates and spatial migrations of the most abundant species, 477 specimens of *Trematomus bernacchii*, caught mostly by hooks during the 1990–91 expedition, were tagged and released near the station.

The feeding habits of some abundant species, such as the nototheniids *T. bernacchii*, *T. hansoni*, *T. newnesi*, *T. loennbergii* and *T. pennellii* were determined by stomach content analysis (Vacchi et al. 1994; La Mesa et al. 1997). Some reproductive characteristics such as gonado-somatic index (GSI), fecundity and egg size were studied in *T. bernacchii*, *T. hansoni* and *Chionodraco hamatus* (Vacchi et al. 1996). Studies on age and growth, by means of otolith (sagittae) reading, were performed on *T. bernacchii* and *C. hamatus* (Vacchi et al. 1992; La Mesa et al. 1996).

## Results

### Species Composition and Vertical Distribution

The fishing activities of the Nekton team provided 19 species (11 genera) belonging to 4 families. In addition, one whole specimen of *Paraliparis* sp. was found in the stomach of the largest *Dissostichus mawsoni* specimen caught (Table 1). The following species collected from other colleagues are also reported in Table 1:

**Table 1.** Fish species found in Terra Nova Bay

| Family Species | No. of specimens | Size range (TL in mm) | Our samples | Other investigators |
|---|---|---|---|---|
| Nototheniidae | | | | |
| *Dissostichus mawsoni* Norman | 3 | 1200–1670 | + | |
| *Notothenia coriiceps* Richardson | 1 | –(*) | | + |
| *Pleuragramma antarcticum* Boulenger | 6 | –(*) | | + |
| *Trematomus bernacchii* Boulenger | 2107 | 141–345 | + | |
| *Trematomus eulepidotus* Regan | 23 | 155–320 | + | |
| *Trematomus hansoni* Boulenger | 169 | 210–420 | + | |
| *Trematomus lepidorhinus* Pappenheim | 1 | –(*) | | + |
| *Trematomus loennbergii* Regan | 94 | 105–330 | + | |
| *Trematomus newnesi* Boulenger | 198 | 210–249 | + | |
| *Trematomus nicolai* (Boulenger) | 5 | 185–263 | + | |
| *Trematomus pennellii* Regan | 195 | 165–255 | + | |
| *Trematomus scotti* (Boulenger) | 1 | 120 | + | |
| *Pagothenia borchgrevinki* (Boulenger) | ~20 | –(*) | | + |
| Bathydraconidae | | | | |
| *Cygnodraco mawsoni* Waite | 15 | 421–554 | + | |
| *Gymnodraco acuticeps* Boulenger | 14 | 366–423 | + | |
| *Prionodraco evansii* Regan | 1 | 138 | + | |
| Channichthyidae | | | | |
| *Chaenodraco wilsoni* Regan | 1 | 280 | + | |
| *Chionodraco hamatus* (Lonnberg) | 435 | 180–440 | + | |
| *Chionodraco myersi* DeWitt and Tyler | 6 | 297–336 | + | |
| *Cryodraco* sp. Dollo | 13 | 392–510 | + | |
| *Neopagetopsis ionah* Nybelin | 1 | –[a] | | + |
| *Pagetopsis macropterus* Boulenger | 16 | 230–304 | + | |
| Artedidraconidae | | | | |
| *Artedidraco loennbergi* Roule | 1 | –[a] | | + |
| *Histiodraco velifer* Regan | 1 | 169 | + | |
| *Pogonophryne scotti* Regan | 2 | 274–280 | + | |
| Liparididae | | | | |
| *Paraliparis* sp. | 1 | –[b] | + | |

[a] Not reported.
[b] Specimen damaged.

- One specimen of *Artedidraco loennbergi* found in the sediment of a bucket set at the 283-m depth, during the activities of benthic ecologists near the station in 1988.
- An adult *Notothenia coriiceps* caught in January 1991 by a small trammel net at the 80–90-m depth (Focardi, pers. comm.).
- One specimen of both *Neopagetopsis ionah* and *Trematomus lepidorhinus* collected by gill net during the 1995–96 expedition at the 200- and 150-m depths, respectively (Ungaro and Varotto 1997).
- Several specimens of *Pagothenia borchgrevinki* fished from holes or natural cracks in the fast ice by hand line during the 1994–95 and 1995–96 expeditions (Ungaro and Varotto 1997; di Prisco, pers. comm.).
- A few adult and subadult specimens of *Pleuragramma antarcticum* caught by trammel-net during the 1992–93 expedition (Nascetti, pers. comm.).

Nototheniidae were the most numerous family, with 12 species, followed by Channichthyidae, with 6 species. Nototheniidae were also dominant in number, with 85.4% of total catches and *Trematomus bernacchii* as the dominant species (Table 1). The three specimens of the giant nototheniid *Dissostichus mawsoni* (up to 60 kg in weight) were caught only during the first half of the sampling period of the 1990/1991 expedition (from late December to mid-January) in the 530–655-m depth range (Vacchi and Greco 1994). Among channichthyids, *C. hamatus* was the most common species, whereas the most abundant bathydraconid was *Cygnodraco mawsoni*. Artedidraconidae were represented by one specimen of *Histiodraco velifer* and two of *Pogonophryne scotti*. The bulk of catches consisted mainly of adult individuals, as shown in Table 1. *T. bernacchii*, *C. hamatus*, *T. hansoni* and *T. pennellii* were very common in the area, each species occurring in more than 25% of samples collected by nets. *T. newnesi*, *T. loennbergii*, *C. mawsoni*, *Pagetopsis macropterus*, *Cryodraco* sp., *T. eulepidotus*, *Gymnodraco acuticeps* and *T. nicolai* can be considered rather common species, showing a frequency of occurrence between 5 and 25% of the hauls. The other species, found in less of the 5% of the samples, may be considered occasional species (Table 2).

As for the vertical distribution, higher species diversity was found between 0- and 200-m depths. However, *T. bernacchii* and *T. hansoni* were

**Table 2.** Frequency of occurrence (%) of the fish species and mean number of specimens per haul at the different depth levels (standard deviation in brackets), sampled by trammel and gill nets

| Species | F (%) | 0–100 m 28 hauls | 101–200 m 27 hauls | 201–400 m 9 hauls | >400 m 9 hauls |
|---|---|---|---|---|---|
| Nototheniidae | | | | | |
| *Dissostichus mawsoni* | 1.4 | | | | * |
| *Trematomus bernacchii* | 86.3 | 35.1 (±29.9) | 18.3 (±23.6) | 2 (±2.2) | 0.2 (±0.4) |
| *Trematomus eulepidotus* | 9.6 | * | * | * | |
| *Trematomus hansoni* | 52.1 | 1.7 (±2.5) | 1.2 (±2.1) | 4.6 (±10.1) | 3.7 (±2.4) |
| *Trematomus loennbergii* | 13.7 | – | – | 1.1 (±3.0) | 9.3 (±6.6) |
| *Trematomus newnesi* | 20.5 | 4.9 (±18.7) | 0.3 (±0.5) | – | – |
| *Trematomus nicolai* | 5.5 | * | | | |
| *Trematomus pennellii* | 45.2 | 1.3 (±3.1) | 3.9 (±4.4) | 0.6 (±1.0) | – |
| *Trematomus scotti* | 1.4 | | | | * |
| Bathydraconidae | | | | | |
| *Cygnodraco mawsoni* | 13.7 | * | * | | |
| *Gymnodraco acuticeps* | 8.2 | * | * | | |
| *Prionodraco evansii* | 1.4 | | * | | |
| Channichthyidae | | | | | |
| *Chaenodraco wilsoni* | 1.4 | | * | | |
| *Chionodraco hamatus* | 69.9 | 5.4 (±7.1) | 7.8 (±11.5) | 2.1 (±2.0) | – |
| *Chionodraco myersi* | 4.1 | | | | * |
| *Cryodraco* sp. | 11.0 | * | * | | |
| *Pagetopsis macropterus* | 12.3 | * | * | * | |
| Artedidraconidae | | | | | |
| *Histiodraco velifer* | 1.4 | * | | | |
| *Pogonophryne scotti* | 2.7 | | * | | |

sampled over the whole depth range, *T. bernacchii* being the shallower (Table 2). On the other hand, *C. hamatus* and *T. pennellii* were collected in the first three depth levels (see Table 2), with a relative abundance peak in the 100–200-m depth range. Some species, such as *T. nicolai*, *T. newnesi*, *Cygnodraco mawsoni*, *Gymnodraco acuticeps* and *Cryodraco* sp. seem confined in shallow waters, whereas other species, such as *T. loennbergii*, *Chionodraco myersi* and *Dissostichus mawsoni*, were caught only in deep waters.

## Feeding

Thirty seven prey taxa, of which 22 were polychaetes, were identified in the diet of *T. bernacchii* (Vacchi et al. 1994). Besides polychaetes (*Barrukia cristata* and *Amphicteis cfr. midas*), bivalves (*Adamussium colbecki*) and euphausiids were the main food source. Nototheniids and amphipods seem to be secondary prey groups (Fig. 2). In *T. pennellii*, polychaetes (in particular *Amithas membranifera*, *Barrukia cristata* and *Aglaophamus ornatus*) were the main prey; amphipods were also well represented, whereas fish eggs constituted a secondary food. The diet similarity between *T. bernacchii* and *T. pennellii* was 23% by number and 26% by weight (Vacchi et al. 1994; Fig. 2).

Stomach content analysis of *T. hansoni* revealed that this species is mainly piscivorous. Fish prey were often represented by juveniles and provided a dietary coefficient about ten times higher than the secondary prey, i.e. the fish eggs

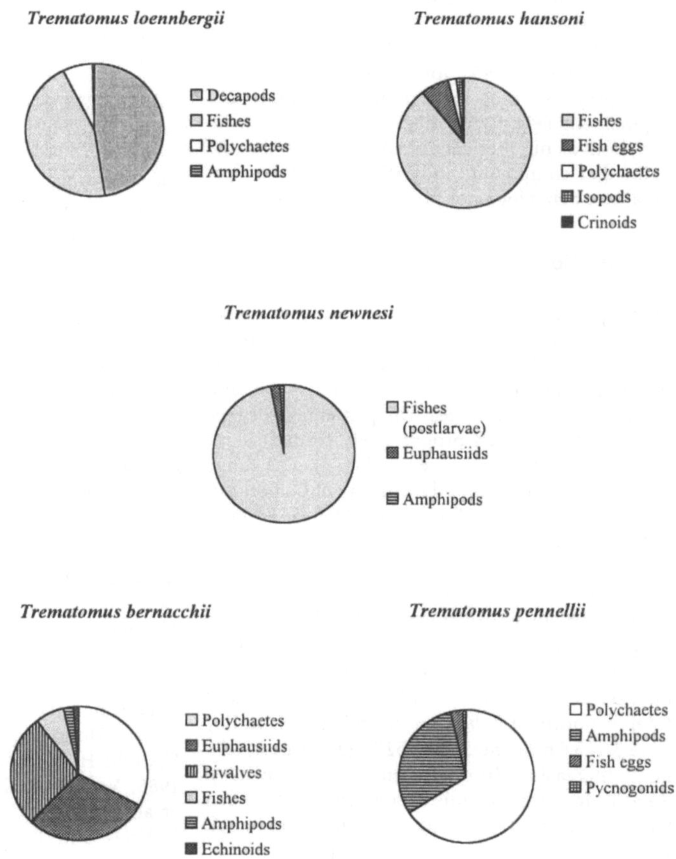


**Fig. 2.** Relative abundance (expressed by graphic representation of the dietary coefficient Q of the most common prey of the nototheniid species collected in this study

(Fig. 2). Polychaetes and decapods (*Chorismus antarcticus* and *Notocrangon antarcticus* together) were secondary food items.

In *T. loennbergii*, a total of 174 food items belonging to 28 taxa were identified, most of which were polychaetes (La Mesa et al. 1997). In this species, the decapods *C. antarcticus* and *N. antarcticus*, as well as fishes, represented the main food (Fig. 2).

The diet similarity between *T. hansoni* sampled below 300-m depth and *T. loennbergii* was 27.3% by number and 49.0% by weight, whereas between *T. hansoni* sampled above 300-m depth and *T. loennbergii*, it was 48.6% by number and 65.9% by weight. The difference in food composition observed in the diet of *T. hansoni* was probably due to the change of the benthic prey species in relation to depth of sample. This influenced the value of food similarity between the two nototheniids.

In the stomach content of *T. newnesi*, Vacchi and La Mesa (1995) found mostly zooplanktonic preys, such as fish larvae, amphipods and euphausiids (Fig. 2). Postlarval stages of *T. newnesi*, both as number caught and biomass, was the main diet component, indicating a marked cannibalistic behaviour not previously detected.

### Reproduction

Most specimens of *C. hamatus* caught in Terra Nova Bay during December–February were in spawning condition, with a gonado-somatic index (GSI) ranging from 7 to 14.4% in females and from 1.5 to 1.8% in males. The percentage of ripe specimens decreased progressively from December to February in both sexes (Fig. 3). The range of mean egg diameter and fecundity of *C. hamatus* was respectively 4–4.6 mm and 2360–3317 eggs (2.9–7.9 eggs $g^{-1}$ of total weight being the range of relative fecundity) (Vacchi et al. 1996).

Data on *T. hansoni* suggested that, in the same period, also this species spawns: GSI in females ranged from 9.9 to 11.8% and in males from 0.15 to 0.23%. A high percentage of females were gravid but no ripe males were found (Fig. 4). Egg size and total fecundity of *T. hansoni* from Terra Nova Bay were 3.1–3.6 mm and 5149–6527 eggs, respectively, both increasing from December to February. Relative fecundity of this species ranged between 9.8 and 16.5 eggs $g^{-1}$ of total weight (Vacchi et al. 1996).

*T. bernacchii* showed a decrease in percentage of spent females and an increase of specimens in an early developing stage of ovarian maturity from December to February. Almost all the males were in early gonad development stages and no ripe specimens were found during the whole sampling period (Fig. 5; Vacchi et al. 1996).

### Age and Growth

Age estimate was made by otolith section reading in *T. bernacchii* and *C. hamatus* (La Mesa et al. 1996; Vacchi et al. 1992). Of the 457 right otoliths of *T. bernacchii* examined, 211 otoliths of females ranging from 157 to 333 mm TL and 122 otoliths of males ranging from 154 to 256 mm TL were aged. The age range was 7–21 years for females and 8–16 years for males. The Von Bertalanffy growth curve (Fig. 6) was fitted to the length-at-age data for females and males and is summarized in the following equations:

$$Lt = 273.5[1 - e^{-0.109(t+2.10)}] \text{ (males, n = 122)}$$

$$Lt = 422.2[1 - e^{-0.055(t+1.92)}] \text{ (females, n = 211)}$$

The yearly length increment was very low, about 7 mm for males and 11 mm for females. The values of the Pauly Growth Performance Index (P) were 1.46 for males and 1.82 for females.

Our *C. hamatus* samples comprised mostly adults, with females ranging from 320 to 430 mm TL and males from 300 to 400 mm TL. Of 118 specimens of *C. hamatus* analyzed, 30 females and 70 males were aged, providing an estimated age range of 5–10 years. For this species, the Von Bertalanffy growth function was not fitted to the experimental data because, dealing only with relatively old individuals, such a fitting would probably have smoothed the extrapolation curve.

## Discussion

The faunistic composition and species diversity in Terra Nova Bay do not differ very much from the other coastal fish communities inhabiting other sites of east Antarctica, such as Lutzow-Holm Bay (Naito and Iwami 1982), the Vestfold Hills region and Mawson Sea (Williams 1988; Vacchi et al. 1996), Adélie Land (Hureau et al. 1965) and McMurdo Sound (DeWitt 1971; DeVries and Eastman 1981). Some differences concerning the presence/absence of a few species

**Fig. 3.** Seasonal pattern of maturity stages frequency (above) and gonado-somatic index (below) of *Chionodraco hamatus*

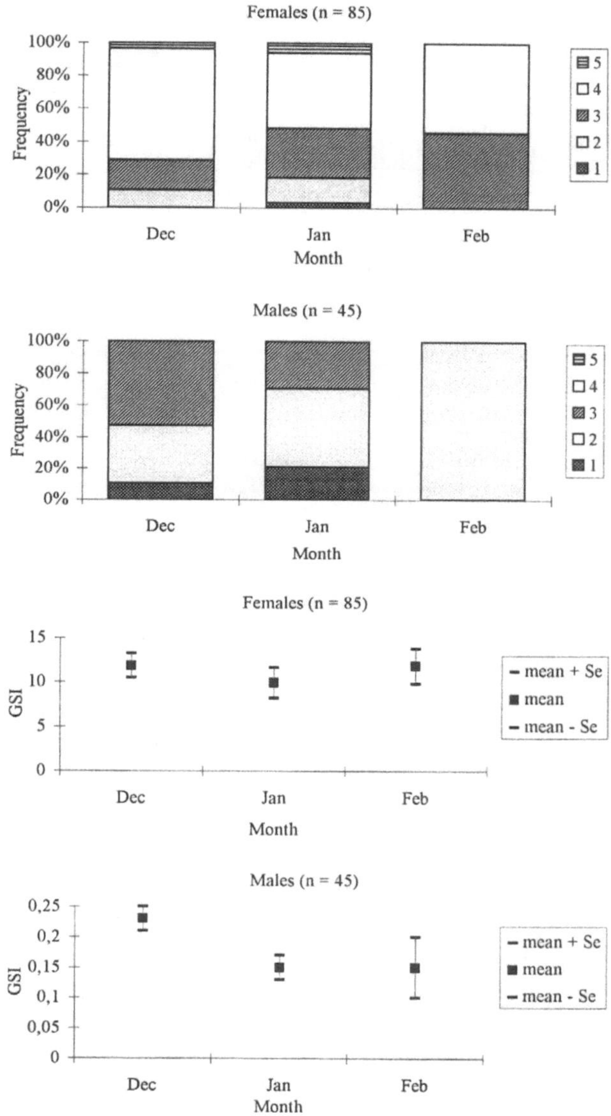


in the different areas can be explained more in terms of different selection of fishing gear used rather than a real paucity of some species. For instance, the lack of icefish in McMurdo Sound is most likely related to the particular fishing technique adopted at this site (lines and traps from drilled holes in the pack ice). Indeed, as we can see from our catches, channichthyids do not seem vulnerable to lines and traps. This is confirmed at Davis station, where 99% of the specimens of the icefish *C. hamatus* were caught by gill net (Williams 1988).

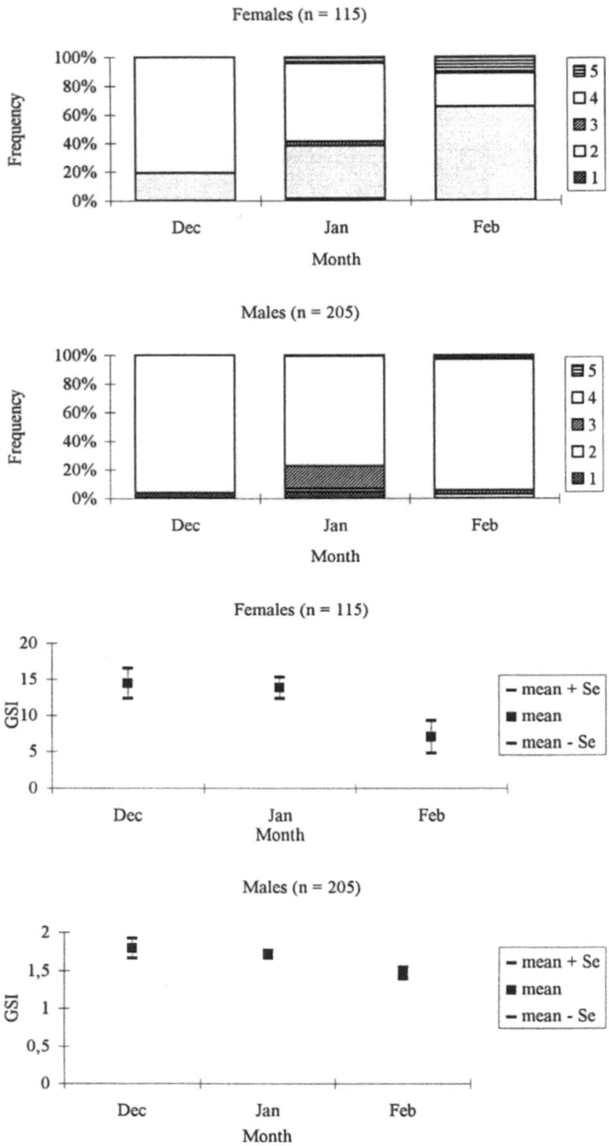


**Fig. 4.** Seasonal pattern of maturity stages frequency (above) and gonado-somatic index (below) of *Trematomus hansoni*

The capture at Terra Nova Bay of only one adult individual of *Notothenia coriiceps* can be considered as the southernmost finding of this species (Miller 1993). The paucity of this species distinguishes Terra Nova Bay from other Antarctic coastal areas, such as, for instance, Adélie Land, where *N. coriiceps* is very abundant in shallow waters (Hureau et al. 1965).

The analysis of biometric characteristics of *Cryodraco* specimens stressed some difficulties

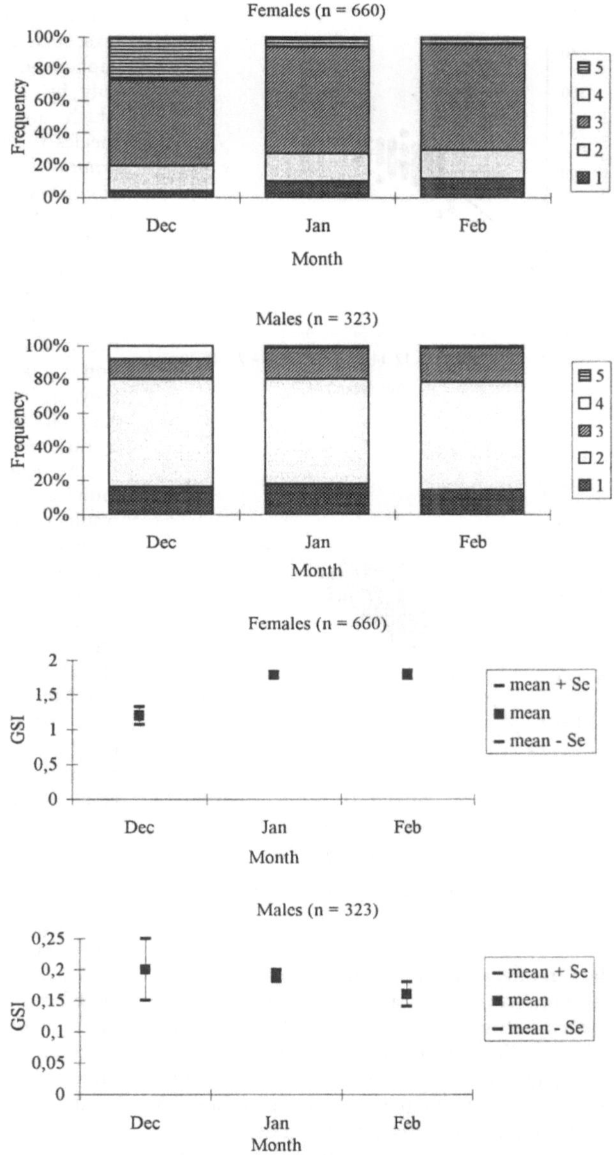


**Fig. 5.** Seasonal pattern of maturity stages frequency (above) and gonado-somatic index (below) of *Trematomus bernacchii*

in obtaining a clear taxonomic diagnosis of the species belonging to this genus. The specimens caught in Terra Nova Bay showed some, but not all, of the distinctive characters of *Cryodraco atkinsoni* described by Regan in 1914 (La Mesa and Vacchi 1997).

Considering the species richness in relation to depth, our results partially disagree with literature

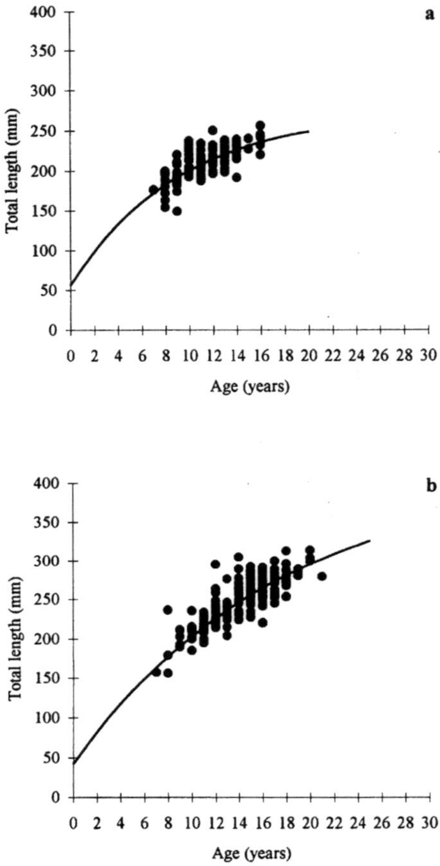


**Fig. 6.** Von Bertalanffy growth curve for males (**a**) and females (**b**) of *T. bernacchii*

data from east Antarctica, that indicate the greatest number of fish species occurring in deep waters, between 250 and 600 m (DeWitt 1971). However, the higher species diversity in shallow waters found at Terra Nova Bay (Table 2) may be explained mostly by the more intensive sampling made in the waters less than 200 m depth.

As for the feeding pattern of the most common fish of Terra Nova Bay, the comparison of the trophic niche was carried out between species showing the same areal distribution. The low diet similarity observed between *T. bernacchii* and *T. pennellii* indicates a partition of food resources. Some organisms are preyed by only one species, pycnogonids and gastropods (*T. pennellii*), echinoids (*T. bernacchii*) being an example. Other taxa, such as molluscs and amphipods, fishes and euphausiids, are consumed in significant amounts only by *T. bernacchii*. As regards polychaetes, which are the most common prey of both species, the burrow-dwelling species *Aglaophamus ornatus*, *Ophelina gymnopige* and *Scoloplos marginatus* were mainly preyed on by *T. pennellii*. Finally, the diet richness of *T. pennellii* and *T. bernacchii* was outlined, as well as the low trophic overlap between the two species. The same situation was found when comparing *T. hansoni* and *T. loennbergii*. Our data on the diet of these two species suggest that the interspecific competition is partially avoided either by taking different preys or by taking different amounts of the same prey. Moreover, the same prey was often consumed by different size categories of the two species.

As regard the pelagic fish species commonly found at Terra Nova Bay, namely *T. newnesi*, our results on feeding habit of adult fish confirm the zooplanktonic feeding behaviour of this species, reported in previous studies on juveniles (Casaux et al. 1990; Daniels 1982; Targett 1981). In the stomach contents of our specimens we found a high number of postlarval stages of fish (638 individuals), of which the majority were *T. newnesi*. This high level of cannibalistic behaviour has not been detected previously, although cannibalism has been reported in some larval stages of *T. newnesi* caught in the Bransfield Strait (Balbontin et al. 1986).

Our estimates of age and growth of *T. bernacchii* at Terra Nova Bay do not agree with those of Wohlschlag (1962) at McMurdo Sound and Hureau (1970) at Adélie Land, who determined both higher growth rates and lower estimates of longevity. Besides the natural intraspecific variability due to different environmental conditions at different sampling sites, the different ageing methodologies used (scales for McMurdo specimens and whole otoliths for Adélie Land specimens), could account for the discrepancy in the results. Conversely, the growth rates and the maximum age obtained by Ekau (1988) on *T. bernacchii* specimens from the Weddell Sea are generally in agreement with our data. Ekau found a maximum age of 21 years and a K value of 0.088 from the Von Bertanlaffy growth formula, which fall between our values for males and females. As concerns the age and growth of *C. hamatus*, not previously studied, our data show that both sexes grow very slowly in the sampled length range, but females are, on average, larger than males of the same age and increase their size at a greater rate.

Finally, our results on the reproduction patterns of *T. bernacchii*, *T. hansoni* and *C. hamatus* are in good agreement with the existing information on the spawning period of the High Antarctic Zone species, which show a high proportion of spring and especially summer spawning species (Kock 1992). Moreover, the low fecundity (both relative and total) along with the relatively large yolky eggs that are probably spawned annually as a single batch, seem to be common to the three species, and, on a wider scale, they represent some of the reproductive strategies most frequently adopted by the Antarctic fish fauna (Kock and Kellermann 1992).

*Acknowledgements*. This study was financially supported by the Italian National Programme for Antarctic Research (PNRA).

## References


Anonymous (1989) Report of the Working Group on Fish Stock Assessment. In: Report of the 8th Meeting of the Scientific Committee of the Commission for the Conservation of the Antarctic Marine Living Resources, 6–10 November 1989, CCAMLR, Hobart, Australia, Annex 6, pp 183–295

Balbontin F, Garreton M, Neuling J (1986) Composicion del alimento y tamaño de las presas en larvas de peces del estrecho Bransfield (SIBEX – Fase II, Chile). Ser Cient Inst Antarct Chil 35: 125–144

Bargelloni L, Ritchie PA, Patarnello T, Battaglia B, Lambert DM, Meye A (1994) Molecular evolution at subzero temperatures: mitochondrial and nuclear phylogenies of fishes from Antarctica (suborder Notothenioidae) and the evolution of antifreeze glycopeptides. Mol Biol Evol 11 (6): 854–863

Casaux RJ, Mazzotta AS, Barrera-Oro ER (1990) Seasonal aspects of the biology and diet of nearshore nototheniid fish at Potter Cove, South Shetland Islands, Antarctica. Polar Biol 11: 63–72

Daniels RA (1982) Feeding ecology of some fishes of the Antarctic Peninsula. Fish Bull 80: 575–588

Dearborn JH (1965) Reproduction in the nototheniid fish *Trematomus bernacchii* Boulenger at McMurdo Sound, Antarctica. Copeia 3: 302–308

DeVries AL, Eastman JT (1981) Physiology and ecology of notothenioid fishes of the Ross Sea. J R Soc N Z 11: 329–340

DeWitt HH (1970) The character of the midwater fish fauna of the Ross Sea, Antarctica. In: MW Holdgate (ed) Antarctic ecology, vol 1. Academic Press, London, pp 305–314

DeWitt HH (1971) Coastal and deep-water benthic fishes of the Antarctic. Antarct Map Folio Ser Am Geophys Un 15: 1–10

Di Geronimo I, Cattaneo-Vietti R, Gambi MC, Casazza G, Cormaci M, Scammacca B (1992) Prime osservazioni sulle comunità bentoniche costiere di Baia Terra Nova (Mare di Ross, Antartide): bionomia e distribuzione. In: Albertelli G, Ambrosetti W, Piccazzo M, Ruffoni Riva T (eds) Atti IX Congr Ass Ital Oceanogr Limnol, Lang, Genova, pp 635–646

di Prisco G, D'Avino R, Caruso C, Tamburrini M, Camardella L, Rutigliano B, Carratore V, Romano M (1991) The biochemistry of oxigen transport in red-blooded Antarctic fish. In: di Prisco G, Maresca B, Tota B (eds) Biology of Antarctic fish. Springer, Berlin Heidelberg New York, pp 26–281

Eastman JT, DeVries AL (1982) Buoyancy studies of notothenioid fishes in McMurdo Sound, Antarctica. Copeia 2: 385–393

Ekau W (1988) Ecomorphology of nototheniid fish from the Weddell Sea, Antarctica. Ber Polarforsch 51: 1–144

Everson I (1977) The living resources of the Southern Ocean. FAO, GLO/SO/77/Rome, vol 1, pp 1–156

Fischer W, Hureau JC (1985) FAO species identification sheets for fishery purposes. Southern Ocean (fishing areas 48, 58 and 88) (CCAMLR Convention Area). FAO Rome, vol 2, pp 233–470

Focardi S, Lari L, Marsili L (1992) PCB congeners, DDTs and hexachlorobenzene in Antarctic fish from Terra Nova Bay (Ross Sea). Antarct Sci 4 (2): 151–154

Gambi MC, Lorenti M, Russo GF, Scipione MB (1994) Benthic associations of the shallow hard bottoms off Terra Nova Bay, Ross Sea: zonation, biomass and population structure. Antarct Sci 6: 449–462

Gon O, Heemstra PC (1990) Fishes of the Southern Ocean. JLB Institute of Ichthyology, Grahamstown, 462 pp

Hureau JC (1970) Biologie comparée de quelques poissons antarctiques (Nototheniidae). Bull Inst Oceanogr Monaco 68: 1–244

Hureau JC, Arnaud P, Dousset C (1965) A propos de deux nouvelles collections de poissons recoltées en Terre Adélie (Antarctique) en 1964 et 1965. Bull Mus Natl Hist Nat 37: 941–950

Kock KH (1992) Antarctic fish and fisheries. Cambridge University Press, Cambridge

Kock KH, Kellermann A (1992) Review. Reproduction in Antarctic notothenioid fish. Antarct Sci 3: 125–150

Iwami T, Abe T (1981) The collection of fishes trawled in the Ross Sea. Antarct Rec 71: 130–141

La Mesa M, Vacchi M (1997) Morphometry of *Cryodraco* specimens of Terra Nova Bay. Cybium 21 (4): 363–368

La Mesa M, Arneri E, Giannetti G, Greco S, Vacchi M (1996) Age and growth of the nototheniid fish *Trematomus bernacchii* Boulenger from Terra Nova Bay, Antartica. Polar Biol 16: 139–145

La Mesa M, Vacchi M, Castelli A, Diviacco G (1997) Feeding ecology of two nototheniid fishes, *Trematomus hansoni* and *Trematomus loennbergii*, from Terra Nova Bay, Ross Sea. Polar Biol 17: 62–68

Macdonald JA, Montgomery JC, Wells RMG (1988) The physiology of McMurdo Sound fishes: current New Zealand research. Comp Biochem Physiol 90B: 567–578

Montgomery JC, Macdonald JA (1987) Sensory tuning of lateral line receptors in Antarctic fish to the movements of planktonic prey. Science 235: 195–196

Morescalchi A, Pisano E, Stanyon R, Morescalchi MA (1992) Cytotaxonomy of Antarctic teleosts of the *Pagothenia/ Trematomus* complex (Nototheniidae, Perciformes). Polar Biol 12 (6-7): 553-558

Naito Y, Iwami T (1982) Fish fauna in the northeastern parts of Lützow-Holm Bay with some notes on the stomach contents. Mem Natl Inst Polar Res Spec Issue 23: 64-72

Regan CT (1914) Fishes. British Antarctic (Terra Nova) Expedition, 1910. Nat Hist Rep Zool 1 (1): 1-54

Reseck J (1961) A note on fishes from the Ross Sea, Antarctica. N Z J Sci 4: 107-115

Richardson J (1844-1848) Ichthyology. In: Richardson J, Gray JS (eds) The zoology of the voyage of HMS Erebus and Terror under the command of Captain Sir James Clark Ross during 1839-43. 2 (2) 1844: pp 1-16; 1845: pp 17-52; 1846: pp 53-74; 1848: vii, 75-139. Janson & Sons, London, 2 (2): 1-139, 62 plates

Targett TE (1981) Trophic ecology and structure of coastal Antarctic fish communities. Mar Ecol Prog Ser 4: 243-263

Tota B, Agnisola C, Schioppa R, Acierno P, Harrison P, Zummo G (1991) Structural and mechanical characteristics of the heart of the icefish *Chionodraco hamatus* (Lonnberg). In: di Prisco G, Maresca B, Tota B (eds) Biology of Antarctic fish. Springer, Berlin Heidelberg New York, pp 204-219

Ungaro A, Varotto V (1997) Ecological and genetic data concerning the differentiation of population. Biodiversity and speciation. In: Tamburrini M, D'Avino R (eds) Newsletter of the Italian Biological Research in Antarctica, austral summer 1995-1996, Centro Audiovisivi e Stampa Università di Camerino, Camerino, Italy, vol 1, pp 27-29

Vacchi M, Greco S (1994) Capture of the giant Nototheniid fish *Dissostichus mawsoni* in Terra Nova Bay (Antarctica): Notes on the fishing equipment and the specimens caught. Cybium 18 (2): 199-203

Vacchi M, La Mesa M (1995) The diet of Antarctic fish *Trematomus newnesi* Boulenger, 1902 (Nototheniidae) from Terra Nova Bay, Ross Sea. Antarct Sci 7 (1): 37-38

Vacchi M, La Mesa M, Tarulli E (1992) Investigation by fixed gears on ichthyofauna of Terra Nova Bay (Ross Sea, Antarctica). In: Albertelli G, Ambrosetti W, Picazzo M, Ruffoni Riva T (eds) Atti IX Congr Ass Ital Oceanogr Limnol, Lang, Genova, pp 659-664

Vacchi M, Greco S, La Mesa M (1991) Ichthyological survey by fixed gears in Terra Nova Bay (Antarctica). Fish list and first results. Mem Biol Mar Oceanogr 1: 197-202

Vacchi M, Romanelli M, La Mesa M (1992) Age structure of *Chionodraco hamatus* (Teleostei, Channichthyidae) samples caught in Terra Nova Bay, East Antarctica. Polar Biol 12: 735-738

Vacchi M, La Mesa M, Castelli A (1994) Diet of two coastal nototheniid fish from Terra Nova Bay, Ross Sea. Antarct Sci 6: 61-65

Vacchi M, Williams R, La Mesa M (1996) Reproduction in three species of fish from the Ross Sea and Mawson Sea. Antarct Sci 8 (2): 185-192

Williams R (1988) The inshore fishes of the Vestfold Hills region, Antarctica. Hydrobiologia 165: 161-167

Wohlschlag DE (1960) Metabolism of an Antarctic fish and the phenomenon of cold adaptation. Ecology 41 (2): 287-292

Wohlschlag DE (1962) Antarctic fish growth and metabolic differences related to sex. Ecology 43: 589-597

Wohlschlag DE (1964) Respiratory metabolism and ecological characteristics of some fishes in McMurdo Sound, Antarctica. Antarct Res Ser 1: 33-62

CHAPTER 33

# ROSSMIZE Expedition: Distribution and Biomass of Birds and Mammals in the Western Ross Sea

N. Saino[1] and L. Guglielmo[2]

**ABSTRACT**

We studied the at-sea distribution of birds and mammals in the Ross Sea during the Italian ROSSMIZE expedition, in mid- and late spring 1994. During three different censusing periods (8–13 November, 14 November–15 December, 17–24 December 1994), we made 423 strip-transect counts, totalling an area of 1380 km$^2$. We counted 9499 individuals belonging to 5 species of birds and 545 individuals belonging to 4 species of mammals, and detected clear spatial patterns in density distribution. Correlation analysis of densities between species revealed that distribution patterns were related to pack-ice cover, feeding ecology and other factors. Biomass estimates per unit area were found to be similar to those previously observed in the Ross Sea for most of the concerned species. Exceptional was the minke whale, which was found to have biomass concentrations as much as five times larger than those previously reported. Moreover, the distribution of the main biomass of whales was markedly different from that previously reported. The data reported here will serve as a basis for a better understanding of the circa-annual evolution of the Ross Sea ecosystem and the pattern of distribution of top consumers in relation to distribution and abundance of krill.

## Introduction

To achieve a proper understanding of the structure and functioning of marine food webs in polar regions and to predict the effect of natural or anthropogenic perturbations on these ecosystems, accurate estimates of the distribution and biomass of top-level consumers are needed. Many land-based and at-sea studies on birds and, to a smaller extent, mammals have been completed (e.g. Croxall 1984; Laws 1977; van Franeker 1996). Nonetheless, studies of upper-trophic-level predators, such as birds and mammals, are rare in pelagic waters (Ainley et al. 1991). In the Ross Sea, in particular, extensive data are available for the at-sea distribution and biomass of birds and mammals only during the very late spring and summer (Ainley et al. 1984; Ainley 1985). The lower trophic levels in the Ross Sea, as well as the chemical and physical properties in the water column, undergo rapid changes during spring, as the polynyas widen and the sea surface becomes free from ice (Smith and Nelson 1985; Hecq et al. 1992; Smith and Gordon 1997). These changes are likely to have a profound effect on the community of apex predators. However, as far as we are aware, no published data exist on the distribution of birds and mammals during early spring.

The Italian ROSSMIZE (Ross Sea Marginal Ice Zone Ecology) expedition was a multidisciplinary diachronic investigation of biotic and abiotic components of marine ecosystems marginal to pack ice. The aims of the study of top predators during the ROSSMIZE project were: (1) to investigate abiotic factors that affect the distribution and biomass of birds and mammals, and their seasonal variation; (2) to study the temporal change in the top predator community in the marginal ice zones; (3) to investigate the relationships between krill (*Euphausia superba*) distribution and abundance, as determined by acoustic methods, and the distribution of krill predators. In the present chapter we mainly address the first two issues.

## Methods

We censused birds and mammals from the R/V Italica during the ROSSMIZE project, from 8

[1] Dipartimento di Biologia, Sez. Zoologia Scienze Naturali, Università di Milano, Via Celoria 26, I-20133 Milan, Italy
[2] Dipartimento di Biologia Animale ed Ecologia Marina, Università di Messina, Salita Sperone 31, I-98166 S. Agata, Messina, Italy

November until 24 December 1994. The first period corresponded to a rapid southwards cruise to reach the polynya north of the Ross Ice Shelf (8–13 November; Fig. 1); the second period lasted from 14 November until 15 December and was mainly designed to study the marginal ice zone during its northwards shift, and the third period lasted from 17 December to 24 December, and was specifically designed for bird and mammal censusing and krill acoustic survey (see also Fig. 1).

Birds and mammals were counted during continuous 30-min sessions followed by periods of no censusing lasting from 30 min to 1.5 days. Observations were made from the upper bridge, approximately 14 m above the sea surface, only when the ship was travelling. The distance travelled by the ship during an observation session, determined by satellite navigation, was considered to be a censusing strip transect (see also below). Overall, censused strip transects totalled 1379.5 km$^2$. All birds and mammals spotted by a single observer in the fore 90° sector on one side of the ship, within a distance of 400 m, were tallied. In a minority of the observation sessions, owing to poor visibility, censusing range had to be reduced to 300 or 200 m. Observations were always carried out using a 10 × 40 binocular or a 30× telescope. In all, we carried out 423 30-min observation sessions. Ice conditions were recorded every 5 min visually as the percentage of sea surface covered by ice in the censusing strip. We used the mean value of the six estimates in each transect. Censusing methods were similar to those adopted by Ainley et al. (1984) and Ainley (1985; Ainley et al. 1991), except for the following: (1) strip width was 400 m instead of 300 m; (2) we had no restriction on ship cruising speed (Ainley et al. did not use data collected at <5 kts); and (3) in the analysis of count data we decided to consider the maximum number of Antarctic petrels (*Thalassoica antarctica*) and snow petrels (*Pagodroma nivea*) because focal observations of some individuals revealed that this would have overestimated the number of birds encountered also owing to the low speed of the ship. Snow and Antarctic petrels apparently were attracted to the ship and some individuals that could be identified by field marks followed the ship for up to 1.5 h. Hence, some individuals might have been counted more than once, within a transect and during consecutive transects. To prevent double counting of the same individuals during the same transect, we considered the maximum number of birds that could be simultaneously sighted along the transect. In this respect our counts are clearly conservative with respect to the number of birds actually encountered in the transect strip. However, it should also be considered that, given the marked attractive effect of the ship on these species, the number of individuals encountered in the surveyed strip might have provided a large overestimate of the true density of these species. Hence, while the total number of petrels observed would certainly have provided an overestimated density, considering the maximum number of birds sighted could mitigate the effect of attraction to the ship on density estimates. To count penguins we carefully observed all the open sea areas with binoculars.

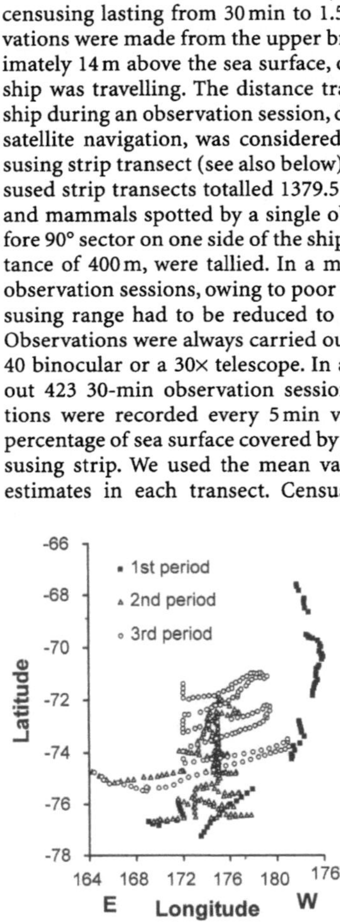


**Fig. 1.** Tracks of the transects (the position of the ship at the beginning of each transect is shown) during the three censusing periods. *1st period* 8–13 November; *2nd period* 14 November–15 December; *3rd period* 17–24 December

The density of birds and mammals in each transect was expressed as the number recorded divided by the area surveyed, which was calculated as the product of the distance travelled by the ship and the strip width.

Density data were converted into biomasses per unit area by assuming as mean body mass of individuals of each species those reported by Laws (1977) for mammals, and Cooper and Woehler (1984) for birds.

### Statistical Analysis

Autocorrelation analysis was used to explore the data with respect to independence of density estimates in successive transects. Simple and partial Kendall's τ non-parametric correlation coefficient was used to analyze the pairwise correlations between latitude, longitude, ice cover and species densities. Polynomial regression analysis was used to investigate the relationship between time of day and observed densities. Density of crabeater seals was corrected for diurnal change in sighting probability due to the hauling-out cycle by analysis of residuals. Cluster analysis by UPGMA agglomeration technique was used to summarize the pairwise correlations between species densities.

## Results

Autocorrelation analysis extended to a maximum interval of ten transects showed poor correlations between density estimates in successive transect counts. The only exception were Antarctic petrels, for which an autocorrelation coefficient of 0.40 emerged between each transect and the following one; for the other species the corresponding value was smaller than 0.30. Overall, autocorrelation coefficients rapidly decreased in value as the lag between transects increased. These results suggest that for most of the species, each transect could be treated as an independent observation. For simplicity we will assume here that this was the case also for the Antarctic petrel.

### Population Densities in Relation to Latitude, Longitude and Ice Cover

The species that we consider in this study are: emperor penguins (*Aptenodytes forsteri*, 808 individuals counted), Adélie penguins (*Pygoscelis adeliae*, 3531 ind), Antarctic petrels (*Thalassoica antarctica*, 4234 ind), snow petrels (*Pagodroma nivea*, 843 ind), South Polar skuas (*Catharacta maccormicki*, 83 ind), crabeater seals (*Lobodon carcinophagous*, 273 ind), Weddell seals (*Leptonychotes weddellii*, 29 ind), minke whales (*Balaenoptera acutorostrata*, 174 ind) and killer whales (*Orcinus orca*, 69 ind). Some other species were encountered during the census, including the southern giant petrel (*Macronectes giganteus*, 4 ind counted), the Wilson's storm petrel (*Oceanites oceanicus*, 9 ind), and the Ross seal (*Ommatophoca rossii*, 2 ind). However, the number of individuals counted was too small to allow statistical analyses and these species have been therefore excluded from the present chapter. Overall, we counted 5499 birds belonging to five species and 545 mammals belonging to four species.

For some of these species we observed spatial variation in density (Table 1). In the first period, we found strong, statistically significant relationships between latitude and density of Emperor penguins and South Polar skuas, which became more abundant as the ship approached the Ross Ice Shelf. Snow petrels and crabeater seals showed the opposite pattern. Density of emperor penguins and South Polar skuas also showed a strong negative correlation with longitude, which reflected the distance from Victoria Land. Both snow petrels and crabeater seals were more frequently observed far from the continent.

In the second period, crabeater seals, as well as Antarctic petrels, were more abundant in the northern part of the area. The latter species also showed an increase far off Victoria Land. The

**Table 1.** Kendall's τ correlation coefficients between ice cover or density of birds and mammals, and latitude or longitude respectively, in the three censusing periods. Sample sizes (= number of transects) are reported at the food of the table

| | 8–13 Nov. | | 14 Nov.–15 Dec. | | 17–24 Dec. | |
|---|---|---|---|---|---|---|
| | Lat. | Long. | Lat. | Long. | Lat. | Long. |
| Ice cover | −0.06 | −0.05 | 0.17*** | 0.04 | 0.04 | 0.26*** |
| Emperor penguin | −0.33*** | −0.32*** | −0.09 | −0.17*** | −0.11 | 0.07 |
| Adélie penguin | −0.04 | −0.07 | 0.09 | −0.07 | −0.14* | −0.03 |
| Antarctic petrel | 0.13 | 0.02 | 0.45*** | 0.13* | 0.16** | 0.20*** |
| Snow petrel | 0.25** | 0.26 | 0.03 | 0.13** | 0.08 | 0.18*** |
| South Polar skua | −0.32*** | −0.35*** | −0.09 | −0.11* | 0.03 | −0.07 |
| Crabeater seal | 0.31*** | 0.40 | 0.15** | 0.02 | 0.10 | 0.33*** |
| Weddell seal | 0.03 | 0.09 | 0.02 | −0.05 | 0.05 | 0.06 |
| Minke whale | −0.02 | 0.02 | 0.00 | 0.08 | −0.07 | −0.04 |
| | n = 71 | | n = 214 | | n = 138 | |

* $P < 0.05$; ** $P < 0.01$; *** $P < 0.001$.

density of emperor penguins and South Polar skuas was significantly lower, while that of snow petrels was significantly greater in eastern compared to western latitudes.

During the third census, Adélie penguins were more abundant in the southern part of the area. Conversely, Antarctic petrels were more abundant in the northern Ross Sea. Density of Antarctic and snow petrels and crabeater seals was strongly and positively correlated with the distance from Victoria Land.

Pack-ice distribution clearly affected the distribution of some of the species under investigation, at least during the second and third censusing periods. The density of crabeater seals was strongly and positively correlated with the proportion of sea surface covered by pack ice during both the second and third periods ($\tau = 0.19$ and $\tau = 0.32$, respectively; $P < 0.001$ in both cases). The densities of emperor and Adélie penguins were also larger in areas with extensive ice cover during the third period ($\tau = 0.25$ and $\tau = 0.34$, respectively; $P < 0.001$ in both cases).

When the density data were pooled across censusing periods and analyzed in relation to ice cover, very clear patterns of variation emerged both for birds and mammals (Fig. 2). Penguins and skuas showed a strong positive association with pack ice, whereas a negative correlation was found for Antarctic petrels ($P$ values always smaller than 0.005). No statistically significant correlation was found for the snow petrel, whose densities appeared to be less affected by ice cover; this aerial species was even frequently observed in areas completely covered by ice, presumably owing to its habit of foraging in the open sea areas temporarily left by the cruising ship. Crabeater and Weddell seals were also at significantly larger densities as pack-ice cover increased. No statistically significant correlation of pack-ice cover was found with

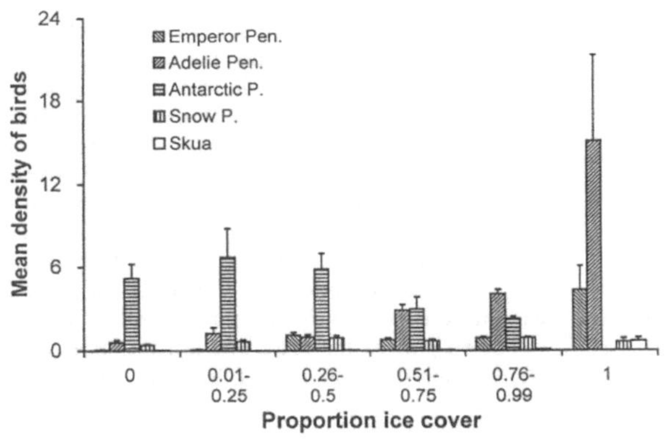


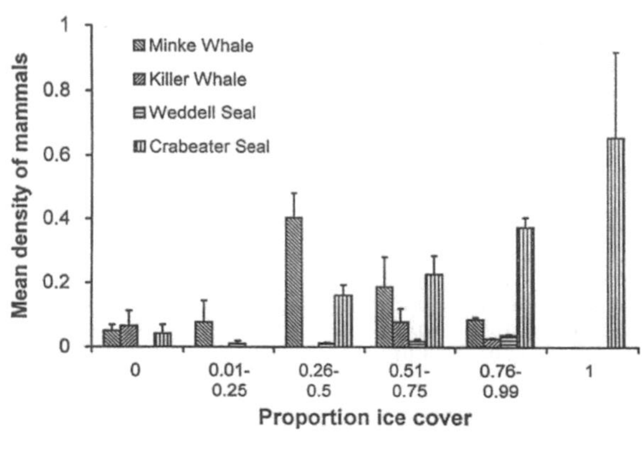


**Fig. 2.** Density of birds and mammals (ind km$^{-2}$) as a function of the proportion of pack-ice cover in the censusing strip. The first class of ice cover (*0*) indicates open seawater while the last (*1*) indicates no open seawater present

either minke or killer whale densities. Indeed, both of these species reached their maximum densities in areas with intermediate values. The result for minke whales, however, was strongly influenced by the observation of very dense schools during six transects on November 23 when we censused a total of 87 whales in an area of 19.7 $km^2$. Including whales that were observed outside the censusing sessions and strips, more than 400 individuals were counted during that single day.

The pattern of correlation between species density and geographic position or ice cover, respectively, might have been confounded if ice cover was strongly correlated with latitude or longitude. To disentangle the effects of ice cover and geography on species densities, we analyzed the correlations between ice cover and geographic position of the transect. Ice cover was found to be significantly larger in the northern part of the area censused during the second period ($\tau = 0.17$, $P < 0.001$) and in the western part of that censused during the third period ($\tau = 0.26$, $P < 0.001$), whereas no significant correlations were found in the other cases (unsigned $\tau$ values always non-significant and smaller than 0.07). We therefore ran partial correlation analyses of species density in relation to latitude (second period) and longitude (third period) while controlling for the effect of ice cover. The results from these analyses confirmed the significant correlations emerging from simple correlation analysis (Table 1). Similarly, partial correlation analyses of species density in relation to ice cover while controlling for geographic position confirmed the significant relationships reported above. In addition, in the third period a significant decline of Adelie penguin westwards while controlling for ice cover ($\tau_{par} = -0.13$, $P < 0.05$) and a positive correlation of Weddell seal density with ice cover while controlling for longitude ($\tau_{par} = 0.14$, $P < 0.5$) were observed.

## Relationships Among Densities of Species

The association between species was summarized by cluster analysis in which Kendall's $\tau$ correlation coefficient was used as an estimate of similarity in distribution between pairs of species. Overall, we found only weak correlations between species and no clear and generally consistent patterns of association (Fig. 3). Indeed, during the first period, only a few statistically significant pairwise correlations emerged (not shown). A subset of three species, including crabeater seal, minke whale, and Adélie penguin, constituted a well-defined guild whose densities were significantly correlated. This is obviously consistent with information available on their trophic ecology, since all of them rely on krill as the main food source. The number of Weddell seals might have been too low to infer general patterns of association with other species and compute reliable estimates of density for this species.

In the second period the strongest correlations arose between the Adélie penguins and crabeater seals and between snow petrels and minke whales. In the third period, the distributions of the two petrels were strongly associated, as well as those of the Adélie penguin and the crabeater seal.

When density data were pooled across censusing periods, significant positive correlations were found between species. Density of Adélie penguins was correlated with that of Antarctic petrels and crabeater seals ($\tau$ values of 0.22 and 0.18, respectively; $P < 0.005$ in both cases). Moreover, density of snow petrels increased with that of Antarctic petrels ($\tau = 0.15$, $P < 0.01$), crabeater seals ($\tau = 0.10$, $P < 0.01$), and minke whales ($\tau = 0.18$, $P < 0.001$). Density of crabeater seals was positively correlated with that of minke whales ($\tau = 0.10$, $P < 0.05$).

These results clearly suggest that the at-sea distributions of the main krill predators in spring are associated and presumably partly controlled by the distribution of their prey.

## Estimates of Biomass per Unit Area

To correct for potential biases in these estimates due to variation in detectability during the day, we first checked for the relationship between observed densities and time of the day for each species separately. Since a non-linear relationship was expected (see Erickson et al. 1971; Ainley 1985), we regressed density on first- and second-order polynomial terms of time of day. Only for the crabeater seal was the polynomial regression model statistically significant; the best-fit regression line reached a maximum approximately at midday. The equation of the line fitted to these data by polynomial regression analysis was: density

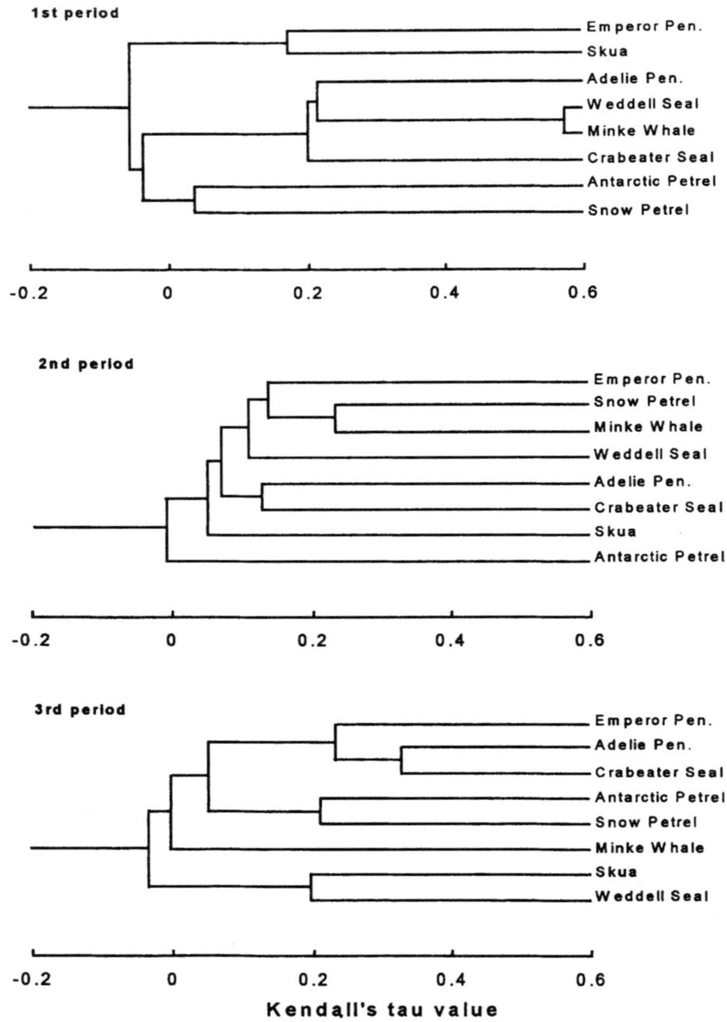


**Fig. 3.** Results of a cluster analysis, by UPGMA agglomeration procedure, showing the relationships between species densities during the three censusing periods. Values of Kendall's $\tau$ correlation coefficient were used as estimates of similarity in density between pairs of species

(ind km$^{-2}$) = −0.169 + 0.11 h$^{-1}$ − 0.0044 h$^{-2}$ ($F_{2,240}$ = 10.41, $P < 0.0001$). This function has an absolute maximum at 12:30 h. Since there appeared to be no reason to assume that this variation was due to the ship cruising in areas with particularly high densities of seals, we interpreted this pattern as due to circadian rhythms in hauling-out behaviour (see also Erickson et al. 1971; Ainley 1985). The absolute maximum of the fitted line (=0.53 ind km$^{-2}$) was used to correct the density data at other times of the day. To correct for this effect we calculated the difference between mean density observed when detectability was largest and the density value fitted to any particular time of the day by the polynomial regression analysis. The value thus obtained was then added to the density values recorded in transects cruised at that particular hour. Uncorrected density estimates in the three periods and biomass estimates, corrected as described above for crabeater seals, are reported in Table 2. The large biomass value estimated for the second period for mammals was mainly due to large schools of minke whales that were sighted on 23 November in an area where a very large krill swarm was detected by acoustic methods.

Bird populations were more dense between 173°E and 177°E, north of 74°S (Fig. 4). However, the largest biomass concentrations were observed south of 74°S because there Emperor penguins were more abundant. Mammals were particularly concentrated, in terms of both individuals and biomass, between 172°E and 177°E, and south of 74°S. The overall distribution of upper trophic level biomass (not shown) was determined by mammal biomass (Fig. 4), since birds represented only 3.3% of the biomass of mammals.

## Discussion

### Potential Sources of Bias in Density Data

As previously mentioned, several biases potentially affected our estimates of at-sea density of both birds and mammals. Abundance of penguins could have been underestimated because some of the swimming or diving individuals might have been missed in our counts even when open sea areas were carefully scanned using binoculars. Petrels are known to be attracted to, follow and circle around ships. Since we considered only the maximum number of individuals that could be sighted simultaneously during each transect, our counting procedure obviously provided a conservative estimate of the number of individuals that actually passed the ship within the surveyed area. However, the number of birds counted may also have been inflated by the attractive effect exerted by the ship. Although no direct test of the relative importance of these two effects on the final density estimates is possible, we believe that the procedure we adopted partly obliterated the two contrasting

**Table 2.** Top: mean (SE of the mean) density, expressed as individuals km$^{-2}$ of birds and mammals during the three censusing periods. Bottom: mean (SE of the mean) biomass, expressed as (g m$^{-2}$) times $10^3$

| | 8–13 Nov. | 14 Nov.–15 Dec. | 16–24 Dec. | Total |
|---|---|---|---|---|
| Density | | | | |
| Emperor penguin | 1.41(0.51) | 0.78(0.10) | 0.17(0.08) | 0.69(0.51) |
| Adélie penguin | 2.55(1.12) | 4.16(0.54) | 0.76(0.14) | 2.78(0.34) |
| Antarctic petrel | 0.01(0.01) | 2.49(0.64) | 6.73(1.05) | 3.46(0.48) |
| Snow petrel | 0.48(0.22) | 0.95(0.11) | 0.44(0.05) | 0.70(0.07) |
| South Polar skua | 0.09(0.04) | 0.08(0.02) | 0.02(0.01) | 0.06(0.01) |
| Crabeater seal[a] | 0.45(0.08) | 0.59(0.08) | 0.34(0.03) | 0.49(0.04) |
| Weddell seal | 0.00(0.00) | 0.03(0.01) | 0.01(0.01) | 0.02(0.01) |
| Minke whale | 0.02(0.01) | 0.22(0.08) | 0.04(0.01) | 0.13(0.04) |
| Killer whale | 0.02(0.01) | 0.04(0.02) | 0.07(0.04) | 0.05(0.02) |
| Biomass | | | | |
| Emperor penguin | 45.8(16.6) | 25.5(3.1) | 5.6(2.6) | 22.4(3.4) |
| Adélie penguin | 11.2(4.9) | 18.3(2.4) | 3.4(0.6) | 12.3(1.5) |
| Antarctic petrel | 0.01(0.01) | 1.9(0.5) | 5.2(0.8) | 2.7(0.4) |
| Snow petrel | 0.1(0.1) | 0.3(0.03) | 0.1(0.01) | 0.2(0.02) |
| South Polar skua | 0.1(0.04) | 0.1(0.02) | 0.02(0.01) | 0.1(0.01) |
| Crabeater seal[a] | 67.6(12.3) | 88.8(11.4) | 51.2(4.9) | 73.0(6.4) |
| Weddell seal | 0.8(0.8) | 6.9(1.7) | 1.7(1.1) | 4.2(0.9) |
| Minke whale | 142.9(87.6) | 1553.0(540.3) | 246.9(97.5) | 889.9(277.1) |
| Killer whale | 61.2(43.4) | 178.1(80.3) | 270.2(161.4) | 188.5(66.9) |
| **Birds** | 57.2(17.4) | 46.1(4.0) | 14.3(3.0) | 37.6(3.8) |
| **Mammals** | 272.6(99.4) | 1826.2(546.9) | 570.0(175.8) | 1155.6(285.3) |

[a] After correction of density values for circadian rhythms in hauling-out behaviour (see Results).

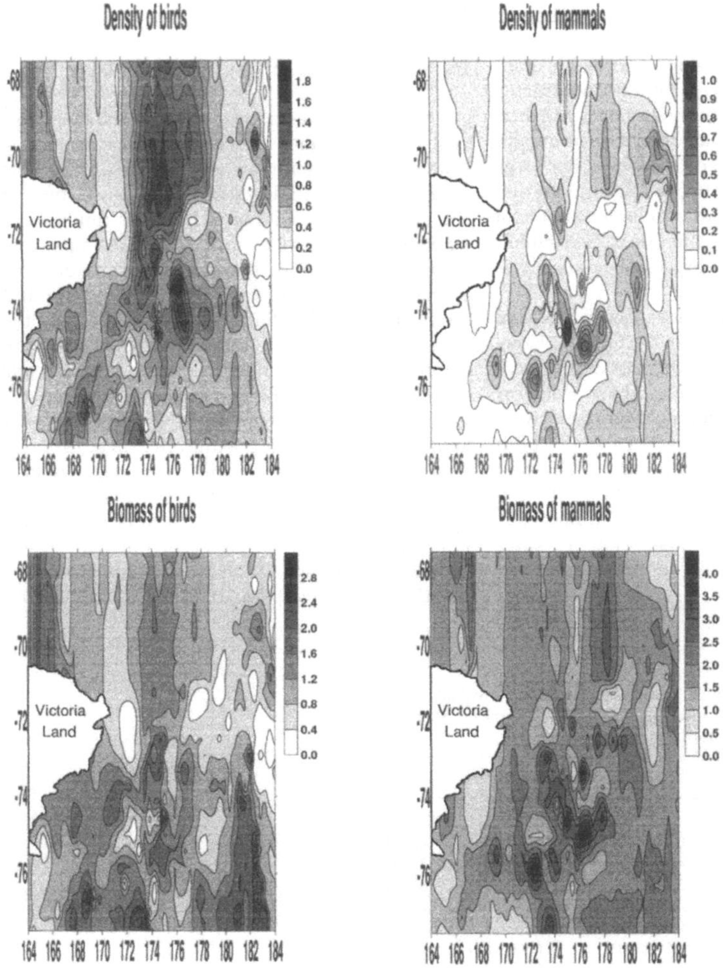


**Fig. 4.** Distribution maps of density and biomass of birds and mammals during the three censusing periods pooled. Maps are obtained by computer-assisted Kriging contouring. For clarity of representation, density data (ind km$^{-2}$) and biomass data per unit area (g m$^{-2} \times 10^3$) were transformed into $\log_{10}$ (density + 1) or $\log_{10}$ (biomass per unit area + 1), respectively

sources of bias, and thus provided reliable estimates of actual petrel density. We corrected for the circadian hauling-out cycle of crabeater seals by considering that individuals sighted when detectability was largest also represented the best estimate of the number of seals actually present in a particular area. Minke whale counts were not corrected in a similar fashion; hence, densities could have been underestimated. These considerations clearly suggest that the density and biomass estimates we obtained should be considered as minima for most of the species and, notably, for the species with the largest per capita biomass. On the other hand, minke whales were found to have a very aggregated dispersion, and random effects might have strongly affected the estimates of their density and, therefore, the biomass estimates of apex consumers.

## Density and Biomass in the Present and Previous Studies

The biomass of birds in the present study was mainly concentrated in the southwestern Ross Sea, with the largest values observed between 173°E and 177°E north of 74°S. This discrepancy is obviously explained by the fact that Emperor penguins, that contributed 60% of total avian biomass, were at their highest density in the southern polynya close to the Ross Ice Shelf. The maximum concentration of mammals, both in terms of density and biomass, was found south of 74°S, between 172°E and 177°E. Whales contributed a very large fraction of the total mammal biomass per unit area estimates.

The distribution of the biomass of apex predators we observed differs markedly from that reported later in the season by Ainley (1985), who recorded two maxima, one over the western part of the Ross Sea and the other over the continental slope. This difference could be clearly caused by the whale distribution. We found very large numbers of whales as far south as 74°S, on November 23, indicating that, at least in that particular year, whales penetrated the Ross Sea earlier than previously described (Ainley 1985). Results from acoustic surveys of krill and direct sampling (Azzali et al. pers. comm.; Guglielmo et al., unpubl. data) indicated very large swarms of krill where a large proportion of the whales were encountered. Minke whales were apparently associated with the marginal ice zone (see also Laws 1977; Ainley and DeMaster 1990) over the northern edge of the Ross Sea polynya, as indicated by their maximum concentration in areas with intermediate pack-ice cover.

The present study is the first in which density of top predators has been studied during spring in the Ross Sea and, therefore, provides an opportunity to analyze changes in seasonal patterns. Compared to the biomass estimates by Ainley (1985), those from the present study are similar for Emperor and Adélie penguins, but they are markedly lower for Antarctic and snow petrels. The differences (approximately 50% of Ainley's estimate for Antarctic petrels and 75% for snow petrels) seem too large to be due solely to counting methods. Lower densities in the present study are most likely caused by confinement of the study area to the western Ross Sea.

Overall estimates of bird biomass are closely similar to the lower limit of the range of estimates reported by Ainley (1985) and larger than those reported by Laws (1977). Estimates of seal biomass are consistent with those in Ainley (1985), a major discrepancy between the two studies being the different proportions of crabeater and Weddell seals. This is possibly a result of Ainley's study including more counts close to the continental coastline, where Weddell seals are known to be more abundant owing to their link with the fast ice where they breed. Our estimates for killer whale biomass per unit area was also similar to that obtained later in the season by Ainley (1985). Our biomass estimates, therefore, are largely similar to those previously reported for most of the concerned species during summer but indicate that total top predator biomass during spring might have been much larger than expected because of the precocious penetration of minke whales in the southernmost part of the Ross Sea.

Pack ice clearly affected the distribution of most of the species and this occurred in the predicted direction. Penguins and seals were more abundant where pack ice was extensive, while the opposite was true for Antarctic petrels. More detailed analysis of species density in relation to pack-ice cover while statistically controlling for the effect of latitude or longitude confirmed the patterns observed by simple correlation analyses, thus suggesting that the correlations we found between ice cover and density of the species were not spurious. Similarly, the geographic pattern of variation of species densities we observed was not confounded by ice cover, since it was confirmed by partial correlation analyses in which we statistically controlled for the effect of this potentially confounding variable.

The weak patterns of association between species were not strictly consistent in different periods. Indeed, even when statistically significant correlations were found, correlation coefficient values were low, suggesting that other concomitant variables differentially influenced species distributions.

In conclusion, this study provided a description of the at-sea distribution and biomass of birds and mammals in the western Ross Sea. Overall, the estimate of the concentration of biomass is consistent with those previously reported with the only notable exception of minke whales. On the contrary, the distribution of the biomass was markedly different in this study owing to the contribution of minke whales, whose biomass was found to be five times larger during the present study than in a previous one. Distribution of pack ice apparently had a marked effect on species' distribution. Data on krill distribution and abundance from acoustic surveys and direct sampling during the ROSSMIZE project will soon be available and will allow the statistical modelling of top predator biomass distribution as a function of oceanographic and geographic variables, pack-ice cover and distribution of prey.

*Acknowledgements.* We are grateful to David Ainley, who kindly commented on a draft of this chapter.

## References

Ainley GD (1985) Biomass of birds and mammals in the Ross Sea. In: Siegfried WR, Condy PR, Laws RM (eds) Antarctic nutrient cycles and food webs. Springer, Berlin Heidelberg New York, pp 498–515

Ainley GD, DeMaster DP (1990) The upper trophic levels in polar marine ecosystems. In: Smith WO, Jr (ed) Polar oceanography, part B: Chemistry, biology, and geology. Academic Press, New York, pp 599–625

Ainley GD, O'Connor EF, Boekelheide RJ (1984) The marine ecology of birds in the Ross Sea, Antarctica. Ornithol Monogr no 32. American Ornithological Union, Washington, DC

Ainley GD, Fraser WR, Smith WO, Hopkins TL, Torres JJ (1991) The structure of upper level pelagic food webs in the Antarctic: effect of phytoplankton distribution. J Mar Syst 2: 111–122

Cooper J, Woehler EJ (1984) Consumption of Antarctic krill (*Euphausia superba*) by seabirds during summer in the Prydz Bay region, Antarctica. In: Laws RM (ed) Antarctic ecology, vol 2. Academic Press, Orlando, pp 247–260

Croxall JP (1984) Seabirds. In: Laws RM (ed) Antarctic ecology, vol 2. Academic Press, Orlando, Fl, pp 533–616

Erickson AW, Siniff DB, Cline DR, Hofman RJ (1971) Distributional ecology of antarctic seal. In: Deacon G (ed) Symposium on Antarctic ice and water masses. SCAR, Cambridge, pp 55–76

Hecq JH, Magazzù G, Goffart A, Catalano G, Vanucci S, Guglielmo L (1992) Distribution of planktonic components related to structure of water masses in the Ross Sea during the V ITALIANTARTIDE expedition. In: Ambrosetti W, Piccazzo M, Ruffoni Riva T (eds) Atti IX Congr Ass Ital Oceanogr Limnol, Lang, Genova, pp 665–678

Laws RM (1977) The significance of vertebrates in the Antarctic marine ecosystem. In: Llano GA (ed) Adaptations within Antarctic ecosystems. Proc 3rd SCAR Symp Antarctic Biology. Smithsonian Institution, Washington, DC, pp 411–438

SCAR Birds Biology Subcommittee (1995) Ad hoc working group on seabirds at sea methodology. Glasgow, 26–27 March 1995

Smith WO, Gordon LI (1997) Hyperproductivity of the Ross Sea (Antarctica) polynya during austral spring. Geophys Res Lett 24: 233–236

Smith WO, Nelson DM (1985) Phytoplankton bloom produced by a receding ice edge in the Ross Sea: spatial coherence with the density field. Science 227: 163–166

van Franeker JA (1996) Pelagic distribution and numbers of the Antarctic petrel *Thalassoica antarctica* in the Weddell Sea during spring. Polar Biol 16: 565–572

CHAPTER 34

# Morphogroup Analysis of Recent Agglutinated Foraminifers off Terra Nova Bay, Antarctica (Expedition 1987–1988)

**D. Violanti**[1]

**ABSTRACT**
Statistical analyses of benthic agglutinated foraminifers from surface sediments off Terra Nova Bay (Ross Sea, Antarctica), collected during the 1987–1988 expedition (PNRA), are presented here. Percentage variations of epifaunal and infaunal morphogroups and a Q-mode analysis indicate that almost all the foraminiferal assemblages are dominated by epifaunal to interstitial small *Portatrochammina antartica*. Infaunal milioline-shaped *Miliammina earlandi* is dominant only in a nearshore area. Redepositional processes seem only in part reflected by the distribution of agglutinated tests. Only the extremely high percentages of small trochamminids, probably resuspended in the water column, and of the stout miliamminids may be related to winnowed sediments. Infaunal cylindrical (*Nodulina, Reophax*) and milioline-shaped (*Miliammina*) morphotypes are frequent on the inner shelf and slopes of the banks and are inferred to be mobile forms, well adapted to unstable bottoms. The flattened tapered *Pseudobolivina antarctica* seems positively correlated with increasing depth, and suggests oxygen-depleted muds and a sluggish circulation in the deep Drygalski Basin. All the dominant morphotypes (trochamminids, hormosinids, *Miliammina earlandi* and *Pseudobolivina antarctica*) are inferred to be opportunists, able to inhabit different microhabitats and exploit various feeding resources.

## Introduction

Foraminifers are marine protozoans with agglutinated or calcareous tests, that are very sensitive to ecological factors. Benthic foraminifers are extensively studied to interpret both ancient and modern environments. Numerous studies have described foraminiferal assemblages from Antarctic surface sediments (McKnight 1962; Anderson 1975; Ward et al. 1987; Mackensen et al. 1990) and have linked their distribution to environmental parameters such as the Carbonate Compensation Depth (CCD) (Kennett 1968; Milam and Anderson 1981), bottom-water masses (Ishman and Domack 1994), substrate type, oxygen content, and productivity (Osterman and Kellogg 1979; Asioli 1995). The Antarctic marine environment is characterized by high seasonality, high accumulation rates of silica and organic carbon (Nelson 1992). The Antarctic area is a key area influencing global physicochemical and biological cycles.

The distribution of agglutinated foraminifers with siliceous or organic cement is independent of the CCD and can be of great importance to improve our knowledge of the Antarctic ecosystem. The results here reported included morphogroup analyses of total and living agglutinated taxa from surface sediments of Terra Nova Bay and the Ross Sea, collected during the 1987/1988 Cruise of the Polar Queen (National Program for Antarctic Research, PNRA) (Fig. 1, Plate 1).

Shelf water masses are stratified during the austral summer: (1) Antarctic Surface Water (AASW), flows till a depth of 100 m (T = –1.9/–2.1 °C, salinity 34.1‰) (Boldrin and Stocchino 1990); (2) the dense, deeper Ross Sea Shelf Water (RSSW) (T = –1.9 °C, salinity 43.8‰), is generated by the freezing of seawater and trapped in shelf depressions. The oceanic Circumpolar Deep Water (CDW), with relatively high temperatures (T = 1.17 °C), high silica and low oxygen content, frequently intrudes the shelf (Jacobs 1989). Dissolved oxygen ranges in the bottom waters from 6.5 to 6.8 $ml\,l^{-1}$ (Innamorati et al. 1990). Nutrient concentrations are high over the whole area (Catalano and Benedetti 1990).

[1] Dipartimento di Scienze della Terra, Università di Torino, via Valperga Caluso, 35, 10125 Torino, Italy

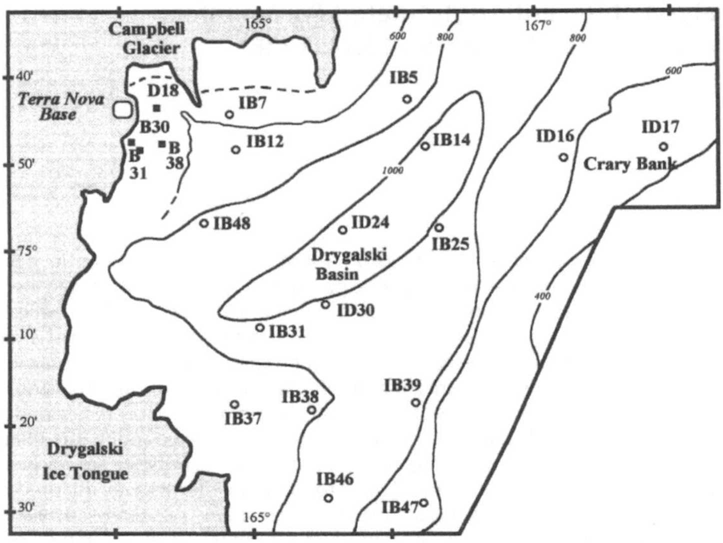


**Fig. 1.** Position of the studied samples in Terra Nova Bay and Ross Sea. *Circles* IB, ID samples; *squares* B, D samples

Based on the occurrence of calcareous and agglutinated taxa, Violanti (1996) indentified three foraminiferal assemblages: assemblage I, above the CCD, dominated by calcareous taxa is restricted to a nearshore area, assemblage II and assemblage III are both below the CCD, and are respectively characterized by dominantly or exclusively agglutinated tests.

Morphogroup analysis is an increasingly useful method for ecological and paleoecological research, as it is independent of changes in habitat of individual species and of taxonomical determinations (Severin 1983). Different test forms, even if not so straightforwardly related to microhabitat preferences and feeding strategies as previously suggested, can provide basic information about environmental parameters.

## Materials and Methods

Surface sediments were collected with a 65-l Van Ween grab and a Picard or a naturalist dredge (Table 1). On board, about 1 l sediment from each sample was preserved with buffered formaldehyde. In the laboratory, each sample was subsampled, stained with Rose Bengal to separate living (stained) and dead (unstained) specimens (Walton 1952), and gently washed through a 63-µm sieve. Countings of >63-µm foraminifers

**Plate 1**

1. *Portatrochammina antarctica* (Parr). Sample ID24, spiral view
2. *Deuterammina multiloculata* (Hoeglund). Sample IB25, spiral view of a loosely coiled specimen
3. *Portatrochammina antarctica* (Parr). Sample ID24, umbilical view. Specimen with a broken last chamber, but with partly preserved umbilical flap
4. *Pseudotrochammina bullata* (Hoeglund). Sample ID30, lateral view
5. *Ammodiscus incertus* (d'Orbigny). Sample IB38, side view
6. *Hemisphaerammina bradyi* Loeblich & Tappan. Sample IB7; specimen attached to a terrigenous clast
7. *Saccammina sphaerica* Brady. Sample IB14, apertural view
8. *Pseudobolivina antarctica* Wiesner. Sample IB25, side view
9. *Miliammina earlandi* Loeblich & Tappan. Sample IB5, front view
10. *Nodulina dentaliniformis* (Brady). Sample IB14, side view
11. *Labrospira jeffreysii* (Williamson). Sample B31, side view
12. *Recurvoides contortus* Earland. Sample IB14, lateral view

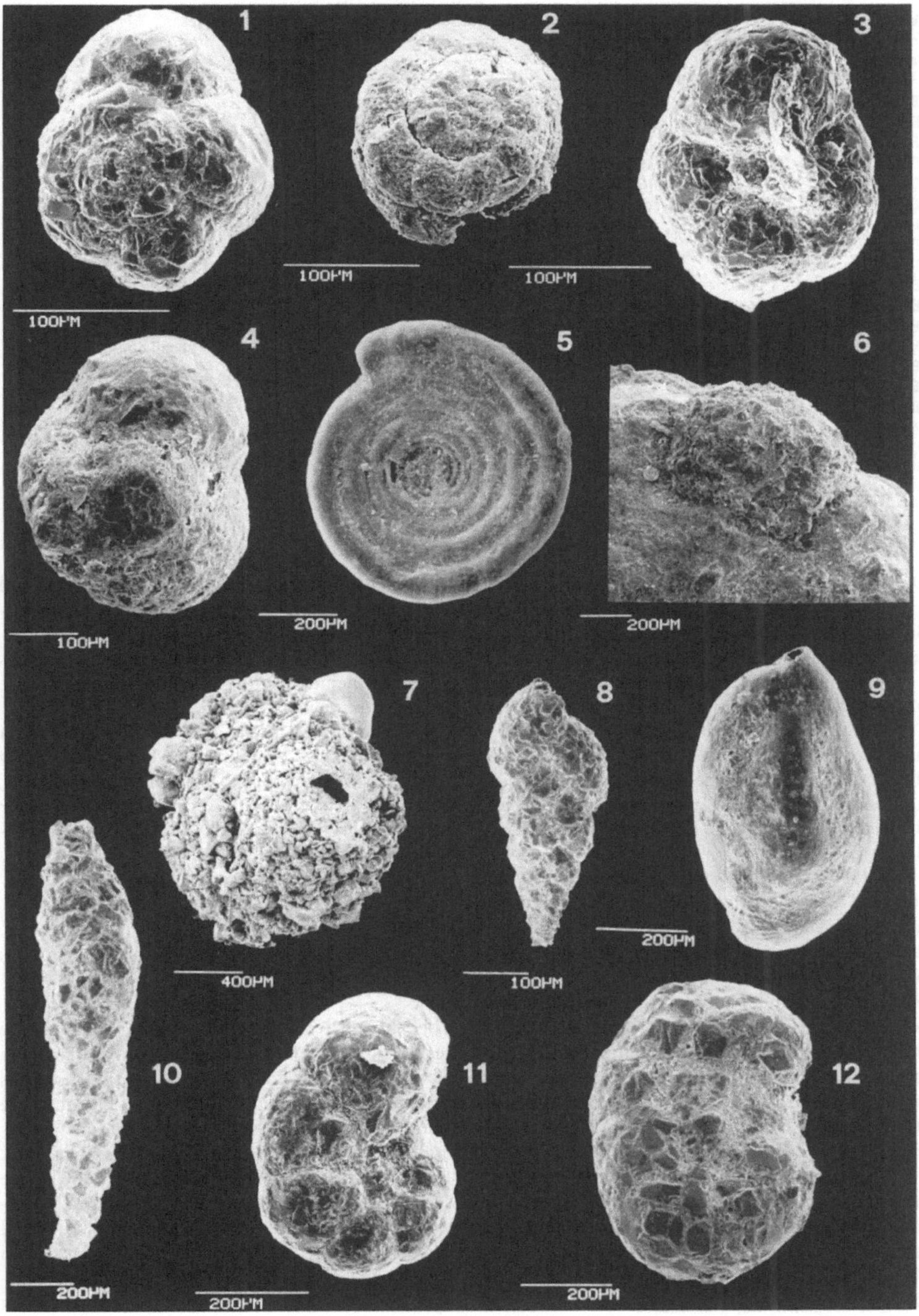
1
2
3
100µM
100µM
100µM
4
5
6
100µM
200µM
200µM
7
8
9
400µM
100µM
200µM
10
11
12
200µM
200µM
200µM

**Plate 1**

**Table 1.** Geographical coordinates, depth, sampling device and sediment type of the studied samples

| Sample no. | Latitude S | Longitude E | Depth (m) | Instrument | Sediment |
|---|---|---|---|---|---|
| IB 5 | 74°42′00 | 166°04′50 | 710 | Van Veen grab | Mud |
| IB 7 | 74°44′50 | 164°48′00 | 578 | " | Muddy sand |
| IB 12 | 74°47′90 | 164°49′80 | 820 | " | Muddy sand with cobbles |
| IB 14 | 74°47′90 | 166°08′50 | 1010 | " | Muddy sand with cobbles |
| ID 16 | 74°48′20 | 167°17′30 | 681 | Naturalist dredge | Mud |
| ID 17 | 74°48′00 | 167°17′30 | 525 | " " | Mud |
| ID 24 | 74°58′00 | 165°38′00 | 1100 | " " | Mud |
| IB 25 | 74°57′20 | 166°17′20 | 1000 | Van Veen grab | Gravelly clay |
| ID 30 | 75°09′00 | 165°00′30 | 850 | Picard dredge | Sandy-muddy gravel |
| IB 31 | 75°06′40 | 165°29′10 | 1000 | Van Veen grab | Sandy mud with cobbles |
| IB 37 | 75°16′70 | 164°51′80 | 770 | " " | Muddy sand |
| IB 38 | 75°17′80 | 165°26′60 | 716 | " " | Muddy sand |
| IB 39 | 75°17′60 | 166°08′80 | 850 | " " | Sandy mud |
| IB 46 | 75°29′00 | 165°31′80 | 849 | " " | Sandy mud with gravel |
| IB 47 | 75°30′00 | 166°13′20 | 680 | " " | Sandy mud with gravel |
| IB 48 | 74°57′20 | 164°37′20 | 953 | " " | Muddy sand with gravel |
| D 18 | 74°43′40 | 164°19′40 | 552 | Picard dredge | Muddy sand with gravel |
| B 30 | 74°46′90 | 164°03′10 | 163 | Van Veen grab | Sand |
| B 31 | 74°47′30 | 164°04′60 | 283 | " " | Sand |
| B 38 | 74°47′60 | 164°12′40 | 278 | " " | Sand with cobbles |

were carried out on fractions consisting of from 112 to 388 total tests (Table 2). According to Streeter (1973), also counts as low as 100 can be reliable for quantitative studies.

Specimens numbers were normalized to 1 ml of wet sediment (Jennings and Helgadottir 1994). The Total Foraminiferal Concentration (TFC = number of total tests $l^{-1}$ ml of wet sediment), the Living Foraminiferal Concentration (LFC = number of living tests $l^{-1}$ ml of wet sediment) and the living/total tests ratio (L/T) were used as abundance measures.

Total and living species percentages were summed in morphological groups, following Jones and Charnock (1985), Nagy et al. (1995) and Kaminski et al. (1995). Similar test morphologies of modern benthic taxa have been related to epifaunal (rounded trochospiral, plano-convex trochospiral, biconvex trochospiral and milioline) or shallow to deep infaunal microhabitats (spherical, rounded planispiral, flattened ovoid, flattened tapered, tapered and cylindrical) (Corliss and Chen 1985; Corliss 1991; Kaminski et al. 1995).

The sampling methodology prevented observation of taxa vertical distribution within sediments and damaged most of the loosely cemented tests of tubular *Botellina*, *Halyphysema* spp. etc. Tubular morphotypes were then omitted from the statistical analyses.

A Q-mode cluster analysis was performed with the program package Statistica, using unweighted pair-group average and Euclidean distance clustering techniques. Living percentages were not included in the cluster analysis, because of their low values.

Taxonomical notes and distributional data about samples here analyzed have been previously reported (Violanti 1996). For the present study, countings of seven samples yielding a very poor microfauna were carried out and added to the previously published data.

## Results

### Grain-Size Distribution

The >63-μm residue highest values (59–71%) are reached in either the samples from inner Terra Nova Bay (B30-D18) or offshore, at depths greater than 950 m (IB48) (Table 2). In these samples, sand and cobbles are abundant, deriving by active morainic reworking (Brambati et al. 1989). Sand percentages from 20 to 50% are shown by samples from 500–900 m depth, collected from the inner shelf slopes off Campbell Glacier (IB5, IB7, IB12) and off the Drygalski Ice Tongue (ID30-IB38). Sand fraction percentages less than 20% characterize the samples from

**Table 2.** For each sample, volume and weight of stained sediment, percentages of total >63-µm residue, fraction of counted subsample, number of total and living benthic tests, TFC = total foraminiferal concentration, LFC = living foraminiferal concentration and living/total ratio are given

| Sample no. | Wet v. (ml) | Weight (g) | % Fract. >63 µm | Counted fract. | % Count. fract. | Tot. B | Liv. B | TFC | LFC | L/T (%) |
|---|---|---|---|---|---|---|---|---|---|---|
| IB 5 | 50 | 84.84 | 41.56 | 3/16 | 18.75 | 309 | 60 | 32.96 | 6.40 | 19.41 |
| IB 7 | 60 | 85.14 | 32.46 | 5/16 | 31.25 | 120 | 15 | 6.40 | 0.80 | 12.50 |
| IB 12 | 60 | 84.13 | 35.15 | 1/8 | 12.50 | 124 | 24 | 16.53 | 3.20 | 19.35 |
| IB 14 | 60 | 83.23 | 19.76 | 1/32 | 3.12 | 306 | 65 | 163.20 | 34.66 | 21.24 |
| ID 16 | 60 | 73.24 | 1.20 | 1/8 | 12.50 | 388 | 92 | 51.73 | 12.26 | 23.71 |
| ID 17 | 60 | 74.73 | 1.27 | 3/16 | 18.75 | 300 | 74 | 26.66 | 6.57 | 24.66 |
| ID 24 | 50 | 70.50 | 11.34 | 1/32 | 3.12 | 364 | 62 | 232.96 | 39.68 | 17.03 |
| IB 25 | 50 | 68.76 | 3.56 | 1/16 | 6.25 | 290 | 36 | 92.80 | 11.52 | 12.41 |
| ID 30 | 60 | 77.16 | 25.15 | 9/128 | 7.03 | 318 | 92 | 75.37 | 21.80 | 28.93 |
| IB 31 | 60 | 82.03 | 16.92 | 1/32 | 3.12 | 304 | 46 | 162.13 | 24.53 | 15.13 |
| IB 37 | 60 | 83.60 | 25.15 | 1/8 | 12.50 | 299 | 80 | 39.86 | 10.66 | 26.75 |
| IB 38 | 60 | 83.91 | 30.61 | 1/16 | 6.25 | 315 | 75 | 84.00 | 20.00 | 23.80 |
| IB 39 | 60 | 78.94 | 16.03 | 3/32 | 9.37 | 302 | 61 | 53.68 | 10.84 | 20.19 |
| IB 46 | 60 | 82.70 | 5.94 | 1/16 | 6.25 | 315 | 66 | 84.00 | 17.60 | 20.95 |
| IB 47 | 60 | 81.09 | 9.16 | 1/16 | 6.25 | 309 | 107 | 82.40 | 28.50 | 34.62 |
| IB 48 | 50 | 94.95 | 64.56 | 3/8 | 37.50 | 121 | 37 | 6.45 | 1.97 | 30.57 |
| D 18 | 50 | 90.36 | 60.29 | 3/16 | 18.75 | 132 | 41 | 14.08 | 4.37 | 31.06 |
| B 30 | 50 | 86.69 | 59.62 | 3/16 | 18.75 | 143 | 48 | 15.25 | 5.12 | 33.56 |
| B 31 | 60 | 99.42 | 59.30 | 1/4 | 25.00 | 112 | 16 | 7.46 | 1.06 | 14.28 |
| B 38 | 50 | 97.05 | 71.23 | 1/2 | 50.00 | 113 | 15 | 4.76 | 0.60 | 12.60 |

Drygalski Basin (IB14-IB31) or from the deep slopes of Crary Bank (IB39-IB47). The smallest percentage of sand (1.2%) characterizes samples ID16 and ID17 from the Crary Bank slopes, at 500–700 m depths, where terrigenous supply is nearly absent.

## Foraminiferal Abundance

TFC values range between 4.76 and 232.96 (Table 2). The lowest values (<20) occur in the samples with >63-µm fraction percentages greater than 30%, from inner Terra Nova Bay (B30-B38, D18) and from the northern inner shelf (IB7, IB12, IB48). TFC values ranging from 25 to 92 occur in most samples, with a wide range of geographical settings and >63-µm percentages. The greatest TFC values are related to rather low percentages of the >63-µm fraction and are found in the deepest samples from Drygalski Basin (IB14, ID24, IB31), from 1000–1100 m depth.

LFC values range between 0.80 and 39.68. The distribution of LFC shows a good correlation with the TFC, only displaying a narrow range of values. The LFC in the nearshore sandy samples (B38-IB7) are markedly different from those in the deeper samples.

The L/T ratio shows a different pattern in comparison with the grain-size values. The lowest L/T values (<18%) are observed either in very deep samples (ID24-IB31) or in shallow samples with coarse sediments (IB7, B38). In the same way, L/T values >26% occur in a wide range of >63-µm percentages and geographical setting.

## Morphogroup Analysis

Agglutinated taxa with similar test shapes have been combined in nine morphotypes (Tables 3, 4, 5), based on morphogroups of Jones and Charnock (1985) and Nagy et al. (1995). Quantitative data are available on diskette upon request to the author.

Plano-convex trochospiral forms (group 1) are inferred to be mainly epifaunal (Kitazato 1984) and to some extent infaunal (Boltovskoy 1966), herbivores and omnivores (Jones and Charnock 1985). Group 1 is represented by the moderately compressed *Portatrochammina antarctica* and the small *Portatrochammina* sp. The similar-shaped *Portatrochammina wiesneri* and *P.* cf. *quadricamerata* and the rather more conical *P. malovensis* occur in most samples in low percentages, and are

**Table 3.** Agglutinated foraminiferal morphotypes distinguished in Terra Nova Bay samples, their dominant taxa and inferred microhabitats

| Test form | Morphotype | Dominant taxa | Microhabitat |
|---|---|---|---|
| Plano-convex trochospiral | 1 | *Deuterammina multiloculata*<br>*Portatrochammina antarctica*<br>*Portatrochammina* sp. | Epifaunal (interstitial) |
| Globular trochoid | 2 | *Adercotryma glomerata*<br>*Ammoglobigerina globigeriniformis* | Epifaunal |
| Disc-shaped coiled | 3 | *Ammodiscus incertus*<br>*Glomospira gordialis* | Epifaunal |
| Hemispherical, irregular, attached | 4 | *Hemisphaerammina bradyi*<br>*Tholosina bulla*<br>*Thurammina cariosa*<br>*Thurammina papillata*<br>*Tolypammina vagans* | Epifaunal |
| Spherical | 5 | *Astrammina rara*<br>*Crithionina pisum*<br>*Psammosphaera fusca*<br>*Psammosphaera parva*<br>*Saccammina sphaerica* | Infaunal |
| Flattened tapered<br>+<br>Tapered | 6 | *Pseudobolivina antarctica*<br>*Textularia earlandi*<br>*Verneuilina minuta* | Infaunal |
| Milioline shaped | 7 | *Miliammina earlandi*<br>*Miliammina lata*<br>*Miliammina obliqua* | Infaunal |
| Uniserial, cylindrical | 8 | *Hormosinella distans*<br>*Hormosinella ovicula gracilis*<br>*Lagenammina atlantica*<br>*Lagenammina difflugiformis*<br>*Nodulina dentaliniformis*<br>*Pseudonodosinella nodulosa*<br>*Reophax pilulifer*<br>*Reophax spiculifer*<br>*Reophax scorpiurus*<br>*Reophax subfusiformis* | Infaunal |
| Biumbilicate | 9 | *Labrospira jeffreysii*<br>*Haplophragmoides canariensis*<br>*Recurvoides contortus* | Infaunal - epifaunal |

summed to *P. antarctica*. Watch-shaped *Deuterammina multiloculata*, inclusive of scarce *D. glabra*, *D. grisea*, *D. rotaliformis* and *Lepidodeuterammina ochracea*, contributes to the plano-convex group in a lesser amount. Morphotype 1 strongly dominates the total (Fig. 2) and living assemblages (Tables 4, 5), with the exception of IB5, where it is outnumbered by milioline-shaped, cylindrical and biumbilicate morphotypes. It attains its maximum values (52–63%) at the shallow stations with coarse substrate (B30 to B38), displaying very low TFC and LFC and variable L/T ratios.

Morphotype 2 comprises globular trochoid forms, inferred to be epifaunal herbivores, detritivores and omnivors (Jones and Charnock 1985). They are here represented by globular trochamminids, such as the free-living *Pseudotrochammina bullata*, inclusive of *P. conica*, and by scarce *Adercotryma glomerata*. Group 2 is rare and reaches percentages >5% only at stations IB5, IB7 and D18, collected from 500–700 m depth on the northern inner shelf (Tables 4, 5). These samples have coarse grain size, very low values of TFC and LFC, and different L/T ratios. Living specimens never comprise more than 2%.

Morphotype 3 includes the disc-shaped, mostly planispiral-coiled *Ammodiscus incertus* and *Glomospira gordialis*, suggested to be epifaunal herbivores and deposit feeders. Group 3 is very rare and attains percentages >1% only in IB25, ID30 and IB47, from the deep, eastern sides of

**Table 4.** Percentages of morphotypes distinguished in the total agglutinated assemblages of Terra Nova Bay samples

| Group | IB5 | IB7 | IB12 | IB14 | ID16 | ID17 | ID24 | IB25 | ID30 | IB31 |
|---|---|---|---|---|---|---|---|---|---|---|
| 1 | 11.60 | 46.66 | 36.27 | 50.30 | 39.83 | 28.90 | 45.20 | 40.90 | 42.50 | 40.30 |
| 2 | 5.80 | 11.66 | 3.22 | 0.00 | 0.90 | 0.60 | 1.30 | 3.70 | 0.90 | 3.20 |
| 3 | 0.00 | 0.83 | 0.00 | 0.90 | 0.70 | 0.90 | 0.70 | 1.20 | 1.50 | 0.90 |
| 4 | 0.30 | 0.83 | 3.22 | 0.90 | 0.80 | 1.00 | 0.70 | 0.30 | 5.20 | 1.80 |
| 5 | 4.40 | 0.83 | 1.61 | 1.20 | 0.60 | 1.00 | 0.90 | 0.90 | 2.40 | 0.00 |
| 6 | 7.70 | 9.99 | 6.45 | 22.40 | 12.80 | 9.90 | 24.70 | 12.70 | 18.80 | 15.00 |
| 7 | 25.20 | 14.16 | 27.41 | 10.00 | 15.30 | 21.00 | 7.80 | 19.10 | 5.90 | 10.50 |
| 8 | 18.00 | 8.32 | 15.30 | 5.60 | 18.40 | 22.30 | 11.20 | 9.30 | 10.80 | 11.60 |
| 9 | 16.40 | 6.66 | 2.41 | 3.80 | 6.10 | 7.60 | 5.70 | 3.00 | 6.20 | 3.90 |

| Group | IB37 | IB38 | IB39 | IB46 | IB47 | IB48 | D18 | B30 | B31 | B38 |
|---|---|---|---|---|---|---|---|---|---|---|
| 1 | 28.70 | 50.70 | 41.60 | 31.60 | 28.30 | 28.92 | 37.86 | 52.43 | 58.02 | 63.85 |
| 2 | 2.30 | 2.50 | 0.90 | 2.20 | 0.90 | 3.30 | 6.05 | 2.09 | 3.56 | 0.00 |
| 3 | 0.00 | 0.90 | 0.00 | 0.90 | 1.20 | 0.00 | 0.00 | 0.00 | 0.00 | 0.00 |
| 4 | 2.60 | 2.10 | 0.30 | 0.60 | 0.90 | 13.21 | 18.17 | 15.36 | 0.00 | 1.68 |
| 5 | 0.60 | 2.10 | 1.50 | 1.50 | 0.90 | 4.95 | 3.77 | 6.27 | 1.78 | 5.04 |
| 6 | 8.00 | 13.20 | 24.40 | 22.80 | 25.20 | 0.00 | 1.51 | 6.99 | 0.00 | 9.24 |
| 7 | 26.40 | 14.80 | 11.50 | 18.30 | 14.40 | 17.35 | 12.11 | 2.09 | 11.60 | 12.58 |
| 8 | 18.70 | 8.00 | 10.20 | 14.10 | 16.70 | 14.04 | 18.17 | 3.48 | 4.46 | 0.00 |
| 9 | 8.30 | 3.10 | 7.50 | 4.60 | 6.10 | 9.91 | 1.50 | 6.28 | 17.85 | 0.00 |

**Table 5.** Percentages of morphotypes distinguished in the living agglutinated assemblages of Terra Nova Bay samples

| Group | IB5 | IB7 | IB12 | IB14 | ID16 | ID17 | ID24 | IB25 | ID30 | IB31 |
|---|---|---|---|---|---|---|---|---|---|---|
| 1 | 2.90 | 11.65 | 8.05 | 11.00 | 11.10 | 8.30 | 8.40 | 6.40 | 12.50 | 8.50 |
| 2 | 1.90 | 0.00 | 0.00 | 0.00 | 0.80 | 0.00 | 0.20 | 0.00 | 0.30 | 0.30 |
| 3 | 0.00 | 0.00 | 0.00 | 0.00 | 0.80 | 0.30 | 0.20 | 0.30 | 0.60 | 0.00 |
| 4 | 0.00 | 0.00 | 0.00 | 0.00 | 0.50 | 0.00 | 0.50 | 0.30 | 0.90 | 0.30 |
| 5 | 0.00 | 0.00 | 0.00 | 0.00 | 0.00 | 0.00 | 0.00 | 0.30 | 0.30 | 0.00 |
| 6 | 0.00 | 0.00 | 3.22 | 6.40 | 3.60 | 2.30 | 3.50 | 2.00 | 8.40 | 3.20 |
| 7 | 3.50 | 0.80 | 0.00 | 2.20 | 0.80 | 4.00 | 0.20 | 2.00 | 0.60 | 0.30 |
| 8 | 5.50 | 0.00 | 6.45 | 0.00 | 4.10 | 5.30 | 2.90 | 0.60 | 3.60 | 1.50 |
| 9 | 5.40 | 0.00 | 0.00 | 0.30 | 1.20 | 2.00 | 0.50 | 0.00 | 1.20 | 0.00 |

| Group | IB37 | IB38 | IB39 | IB46 | IB47 | IB48 | D18 | B30 | B31 | B38 |
|---|---|---|---|---|---|---|---|---|---|---|
| 1 | 9.30 | 12.00 | 8.20 | 11.00 | 11.00 | 5.78 | 8.32 | 18.16 | 12.49 | 8.40 |
| 2 | 1.00 | 1.90 | 0.00 | 0.00 | 0.30 | 0.00 | 0.00 | 0.69 | 0.00 | 0.00 |
| 3 | 0.00 | 0.30 | 0.00 | 0.30 | 0.60 | 0.00 | 0.00 | 0.00 | 0.00 | 0.00 |
| 4 | 1.00 | 0.60 | 0.00 | 0.00 | 0.30 | 9.08 | 8.32 | 9.08 | 0.00 | 0.00 |
| 5 | 0.00 | 0.30 | 0.00 | 0.90 | 0.00 | 2.47 | 0.00 | 0.00 | 0.00 | 0.00 |
| 6 | 4.00 | 6.90 | 7.50 | 4.40 | 10.30 | 0.00 | 0.00 | 0.00 | 0.00 | 0.00 |
| 7 | 4.00 | 0.60 | 2.30 | 0.60 | 2.50 | 0.82 | 5.30 | 0.00 | 0.00 | 0.84 |
| 8 | 4.00 | 0.30 | 1.20 | 2.40 | 5.90 | 4.12 | 9.09 | 2.79 | 0.00 | 0.00 |
| 9 | 2.60 | 0.00 | 0.00 | 0.60 | 2.50 | 4.13 | 0.00 | 0.00 | 0.00 | 0.00 |

Drygalski Basin (Tables 4, 5). Living specimens never exceed 1%.

Morphotype 4 comprises the hemispherical and irregular shapes of permanently attached *Hemisphaerammina bradyi*, *Thurammina* spp. and *Tolypammina vagans*, also reported as epifaunal, passive herbivores or deposit feeders. Total and living specimens are abundant only in a few samples with scarce microfaunas (IB48, D18 and B30) (Tables 4, 5).

Morphotype 5 comprises spherical forms as *Astrammina rara*, *Crithionina pisum*, *Psammo-*

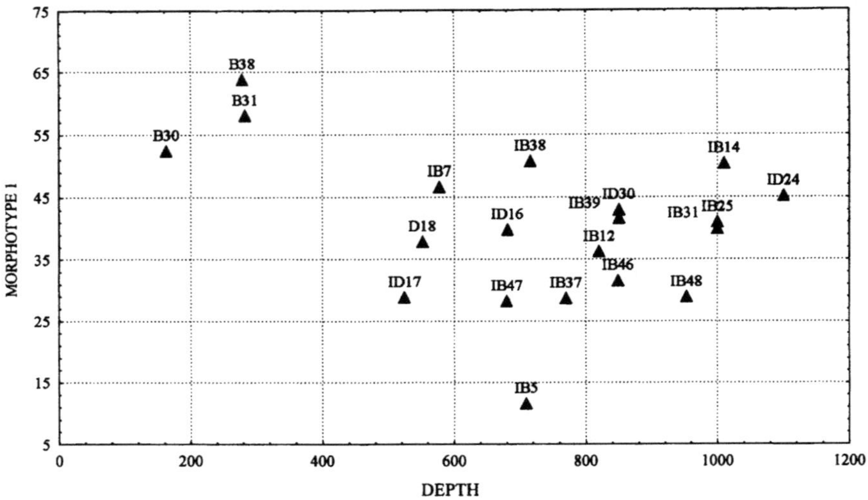


Fig. 2. Percentage distribution of morphotype 1 (plano-convex trochospiral forms) versus water depth

*sphaera* spp. and *Saccammina sphaerica*, that inhabit the top 1–4 cm of bottom sediments, exploiting different sources of nutrition (DeLaca 1986). Morphotype 5 attains values >3% only in the total assemblages at stations IB5, IB48, D18 and B30 (Tables 4, 5). Living specimens are usually absent, common only at station IB48.

Morphotype 6 includes the flattened tapered forms, mainly referred to as infaunal dwellers, although evidence was given for epifaunal habit of some species (Haward and Haynes 1976). Group 6 is represented by the small-sized *Pseudobolivina antarctica* and *Textularia earlandi*. Rare tapered *Verneuilina minuta* were included in group 6. Group 6 displays very low values in shallow samples, but it is abundant at the deep stations IB14, ID24, from the Drygalski Basin, and IB39 and IB47, from the slopes of Crary Bank (Fig. 3; Tables 4, 5). *Pseudobolivina antarctica* and *Textularia earlandi* contribute in a different way to the total group: the cosmopolitan *T. earlandi* is common (4–9%) only at stations IB7 and B38. On the contrary, *P. antarctica*, typical of Antarctic (Pflum 1966), and sub-Antarctic waters (Echols 1971) is abundant in the deepest samples from 1000–1100 m depth. Living specimens are absent in coarse samples and frequent between 700 and 1000 m.

Morphotype 7 comprises the milioline-shaped, thick tests of *Miliammina earlandi*, *M. lata* and *M. oblonga*. *Miliammina* species are reported to be infaunal (Frankel 1975; Haynes 1981). Group 7 shows low values both in the shallowest samples (B30-B38) and in the deepest ones from Drygalski Basin (IB39-ID24) (Fig. 4; Tables 4, 5). The highest values are reached in the samples from 600–800-m depth, on the inner shelf slopes (IB5, IB12) or off the Drygalski Ice Tongue (IB37). Also ID17, collected from siliceous sediments on Crary Bank slopes, displays high values of group 7. Living specimens seldom attain >1% percentages and mostly occur at depth from 500–800 m.

Morphotype 8 includes uniserial elongate, cylindrical forms, as *Lagenammina atlantica*, *L. difflugiformis*, *Nodulina dentaliniformis*, *Reophax pilulifer* and *R. spiculifer*. Hormosinids are reported as infaunal detrital, bacterial scavengers (Jones and Charnock 1985). Kaminski et al. (1988) interpreted a deep infaunal habitat for these mobile forms, able to inhabit the deepest favourable layers, but not necessarily restricted to them. Group 8 shows very low values in the shallowest samples. Low to moderate values characterize the deepest samples, whereas higher percentages (10%, IB7 to 18%, IB37) (Tables 4, 5) are measured in most samples from 700–900 m depth (Fig. 5), displaying intermediate values of sand fraction, TFC and LFC. The maximum value (22%) is recorded in the total assemblage of station ID17. Living specimens show a distribution rather similar to that of total tests.

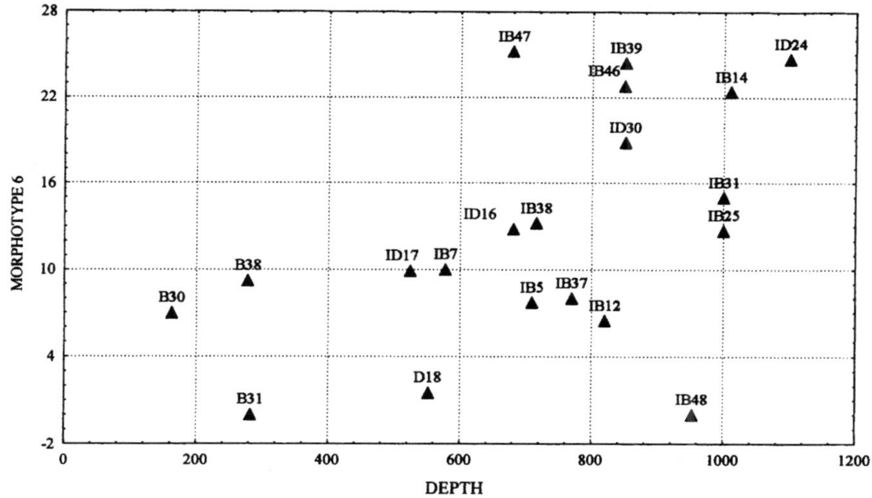


**Fig. 3.** Percentage distribution of morphotype 6 (flattened tapered forms) versus water depth

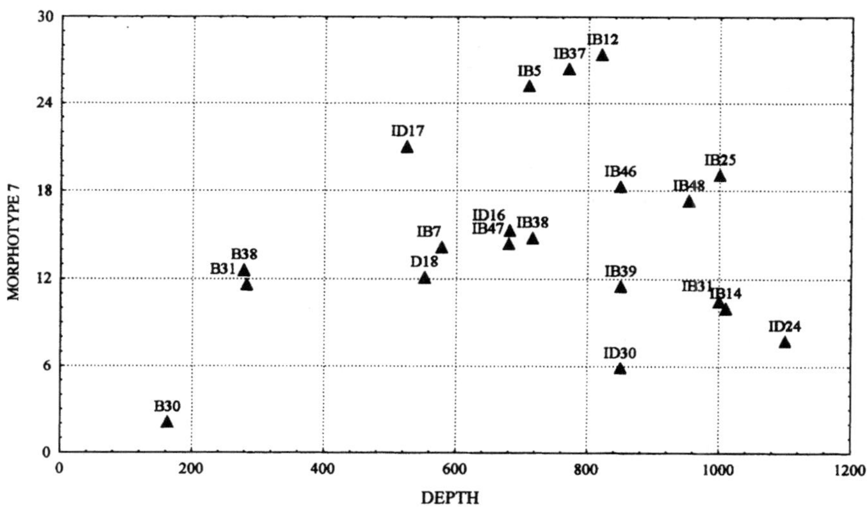


**Fig. 4.** Percentage distribution of morphotype 7 (milioline-shaped forms) versus water depth

Morphotype 9 includes the biumbilicate forms, represented by the rounded planispiral *Labrospira jeffreysii* and *Haplophragmoides canariensis*, and by the more asymmetrical *Recurvoides contortus*. Data about the ecological preferences of these taxa suggest an epifaunal to infaunal habitat (Murray 1991). Group 9 shows very different values in the coarse samples, as D18 (1.5%) or B31 (17.8%) and IB5 (16.4%) (Tables 4, 5). In the other samples, percentages range from 2 to 10%, without evident correlation with depth, grain-size or foraminiferal abundance measures. Living forms are rare, common only at stations IB5 and IB48.

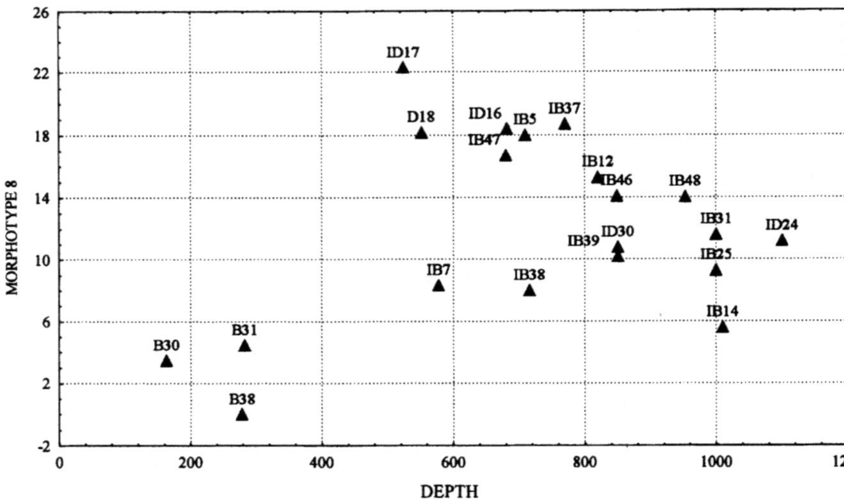


Fig. 5. Percentage distribution of morphotype 8 (cylindrical forms) versus water depth

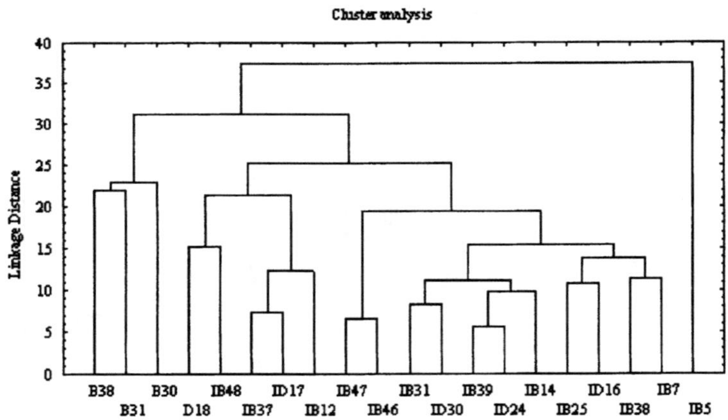


Fig. 6. Dendrogram from Q-mode cluster analysis. Three main clusters and five subclusters are distinguished

## Cluster Analysis

Clustering based on the average linkage method was performed on total foraminiferal data, in an attempt to recognize different agglutinated assemblages and factors influencing their distribution.

In the Q-mode dendrogram three main clusters and five subclusters can be recognized, from left to right of Fig. 6:

Cluster A includes the three shallowest samples, from inner Terra Nova Bay, characterized by very high sand fraction percentages, low TFC and LFC, maximum values of total and living morphotype 1 (plano-convex trochamminids), and variable amounts of the other morphotypes.

Cluster B includes five samples collected from 500–900 m depth, with group 1 ranging from about 28 to 37%, low TFC and LFC. Two subclusters can be distinguished. Samples D18 and IB48 have high percentages of sand, L/T ratio and high group 4 (attached forms) percentages.

Samples IB37 to IB12 differ by the lesser amount of >63-µm fraction, the medium L/T ratio and the higher percentages of groups 7 (milioline-shaped) and 8 (cylindrical).

Cluster C is the largest of the clusters and includes samples with frequent to abundant morphotype 6 (flattened, tapered), following in abundance the always dominant group 1. Three subclusters can be distinguished. Samples IB46, IB47 display low grain size and medium values of TFC and LFC, group 1 with percentages about 28–31%, very abundant group 6 (22–25%) and rather high percentages of groups 7 and 8 (14–18%). Living specimens belong mostly to groups 1, 6 and 8. Samples IB31-IB14 differ by the medium to high TFC and LFC, the higher values of group 1 (40–50%) and the lower percentages of groups 7 and 8. Living specimens still represent groups 1, 6 and 8. Samples IB25-IB7 display similar values of group 1 (40–50%) and percentages of groups 6, 7 and 8 very close to each other, ranging from about 8 to 19%. Living specimens mostly represent group 1 in IB7, and are more diversified in the other samples.

Sample IB5 has a low similarity with all other assemblages (high linkage distance): it is the only sample with low values of morphotype 1 (about 11%). Morphotype 7 (milioline-shaped) is dominant, followed by abundant groups 8 and 9. On the contrary, living specimens belong mostly to groups 8 and 9. This sample shows rather high percentage sand, low TFC and LFC and a medium L/T ratio.

## Discussion

Much of the study area is affected by strong currents and redepositional processes and only the Drygalski Basin bottom and Crary Bank are reported as almost free of terrigenous deposition (Brambati et al. 1989). The >63-µm grain-size percentages, the abundance measures (TFC, LFC and L/T) and the morphogroup analyses can be used to detect the influence of bottom morphologies and depositional processes on the distribution of foraminiferal assemblages.

The assemblage of sample IB5, collected on the inner shelf slopes, is dominated by infaunal group 7 (*Miliammina* spp.) and cylindrical group 8. It is clearly separated from all the other samples by cluster analysis and can be referred to the *Miliammina arenacea* (= *M. earlandi*) association, described from Antarctic muds and sands at 0–2100 m depth (Murray 1991). The morphotype distribution and the Q-mode cluster analysis show that the faunal composition of the other samples is quite similar, correlatable with the *Portatrochammina antarctica* association (Murray 1991), widespread in Antarctic muds and fine sands from 9–1275 m depth. The differences between samples mainly concern percentage variations of morphotypes. Clusters are only in part correlatable with assemblages II and III previously described: samples of assemblage II, distinguished based on the occurrence of living calcareous specimens, cannot be separated on the basis of agglutinated taxa. The correspondence is instead evident for the Drygalski Basin samples of cluster C and of subassemblage III C (Violanti 1996), even if the large agglutinated taxa such as *Astrorhiza arenaria*, *Botellina labyrinthica*, *Rhabdammina linearis*, etc., common to frequent in these deep stations, were omitted from the statistical analysis.

Problems arise in the evaluation of winnowing influencing the foraminiferal faunas. Whereas transport appears to be one of the primary depositional factors in the area, similar patterns of TFC and LFC values and of samples specific composition suggest a rather low amount of allochtonous tests in many assemblages. Transport processes would affect agglutinated taxa to a different extent and are probably closely related to the species life habit, test form and mechanical strength.

Epifaunal to interstitial morphotype 1 (plano-convex trochamminids) and, secondly, epifaunal groups 2 (globular trochamminids), 3 (disc-shaped ammodiscids) and 4 (attached forms) are strongly dominant (up to 70%, B30) in the coarse samples collected from reworked morainic sediments of inner Terra Nova Bay (cluster A), displaying very low TFC and LFC. Transport from shallower depths could be responsible for these values, that can be related to high sedimentation rates and to faunal dilution in the terrigenous debris. Trochamminid tests are very fragile, but are small enough to be suspended in the water column by bottom currents or biological activity. It seems possible that winnowing may affect epifaunal forms more than infaunal ones.

Low TFC and LFC, lower group 1 percentages and frequent infaunal morphotype 7 (milioline-shaped *Miliammina earlandi*) and morphotype 8 (cylindrical hormosinids) are found in the samples located along the eastern side of the

Drygalski Basin, where high terrigenous supply and slumpings from the Drygalski Ice Tongue and from the Campbell Glacier have been recorded. Epifaunal attached forms, globular trochamminids and the shallow infaunal spherical forms (group 5) attain relatively high occurrences in these samples, included in cluster B. The two subclusters (Fig. 6) are distinguished based on the dominance of epifaunal taxa (IB48, D18) or infaunal ones (IB12, ID17, IB37). Morphotype 7 shows a low proportion of living tests: the rather thick tests of *M. earlandi* could be more resistant than the other agglutinated forms and the dominance of dead individuals may be related to transport processes. Morphotypes 7 and 8 often show a negative correlation of their maximum values. Both groups are reported as infaunal forms and an interspecific competition between them can be inferred.

Increasing percentages of infaunal group 6 (flattened, tapered), medium to very high TFC and LFC and low L/T ratios characterize most samples of cluster C. The three subclusters group samples with different proportions of epifaunal and infaunal taxa: infaunal taxa, dominated by *Pseudobolivina antarctica*, reach their maximum in two samples (IB46, IB47) from the depression between the inner slope and Crary Bank, probably influenced by the high salinity and sluggish circulation of bottom waters (Jacobs 1989). Epifaunal forms are given mainly by group 1 and are somewhat more abundant than the infaunal ones in the rather homogeneous subcluster of the deepest samples (IB14-IB31). These stations, collected from the low-energy environment of Drygalski Basin and the deep slopes of Crary Bank, show little sign of reworking and are almost free from terrigenous transport. Assemblages appear less diversified, owing to the scarcity of epifaunal morphotypes 2, 3 and 4 and of the infaunal group 5. Very abundant group 1 (40–50%) and frequent infaunal groups 6, 7 and 8 are the factors linking other samples with less uniform foraminiferal faunas, collected from the inner shelf slopes, and also comprise cluster C.

All the dominant to frequent morphogroups, both epifaunal to interstitial or infaunal, are considered to be opportunistic species (r-strategists), able to exploit different microhabitats and trophic resources, most likely detritus and bacteria, in response to seasonal environmental changes. Many trochamminids (*Lepidodeuterammina ochracea*, *T. inflata*, *T.* ex gr. *squamosa*) are cosmopolitan, feeding on diatoms and their dissolved organic products (Frankel 1974) and are also able to rapidly colonize experimental substrates (Schafer et al. 1996). Flattened tapered *P. antarctica*, abundant in fine sediments as similar calcareous opportunists as *Brizalina*, *Bolivina*, could tolerate low-oxygen, organic-rich sediments. *M. earlandi*, frequently reported as *M. arenacea*, seems endemic to Antarctic waters (Schroeder-Adams 1990). It was the dominant taxon in recolonizing the mud sediment of Deception Island (South Shetland Islands) caldera, after volcanic eruptions (Finger and Lipps 1981). The similar species *M. fusca*, common in worldwide brackish waters, lives in the subsurface sediment, feeding on microrganisms and organic debris (Frankel 1975). Also *Reophax* and hormosinids are considered to have a competitive advantage in low-energy, oxygen-depleted sediments (Haynes 1981; Kaminski et al. 1995) as the similar cylindrical shapes of calcareous foraminifers, inferred to be anaerobic forms (Bernhard 1986; Kaiho 1991).

## Conclusions

Recent agglutinated foraminiferal assemblages off Terra Nova Bay are dominated by small plano-convex trochospiral trochamminids, mostly represented by *Portatrochammina antarctica* and can be referred to as a *P. antarctica* association (Murray 1991). Differences between samples concern the relative abundances of epifaunal/interstitial and infaunal morphotypes and can be related to bottom morphology, water salinity and density, oxygen content and sedimentary pattern.

The active redepositional processes of the studied area seem only in part reflected by the distribution of agglutinated tests, most likely broken during sediment flows. Only the extremely high percentages of small trochamminids, probably resuspended in the water column, and the stout miliamminids may be considered proxies of winnowed sediments.

Infaunal cylindrical (*Nodulina*, *Reophax*) and milioline-shaped (*Miliammina*) morphotypes are frequent on the inner shelf and slopes of the banks. These taxa are most probably mobile forms, well adapted to unstable, frequently winnowed sediments. The flattened tapered *Pseudobolivina antarctica* seems positively correlated with increasing depth, and suggests oxygen-depleted muds and a sluggish circulation in the bottom of Drygalski Basin.

All the dominant morphotypes (trochamminids, hormosinids, *Miliammina earlandi* and *Pseudobolivina antarctica* are inferred to be opportunists, able to inhabit different microhabitats and exploit different trophic resources (detritus, bacteria, dissolved organic matter).

*Acknowledgements.* The author is grateful to I. Di Geronimo for kindly providing samples from the 1987/1988 PNRA expedition; to D. Mana, B. Loi, A. Rosso and R. Sanfilippo for helpful collaboration; and to D. Alesi and E. Morani (ENEA), for literature searches. The author thanks M.A. Kaminski and R. Coccioni for critically reviewing the manuscript. The research was supported by grants from the ENEA, PNRA, Biological Oceanography, Respons. Prof. I. Di Geronimo.

## References

Anderson JB (1975) Factors controlling $CaCO_3$ dissolution in the Weddell Sea from foraminiferal distribution patterns. Mar Geol 19: 315–332

Asioli A (1995) Living (stained) benthic Foraminifera distribution in the western Ross Sea (Antarctica). Palaeopel 5: 201–214

Bernhard JM (1986) Characteristic assemblages and morphologies of benthic foraminifers from anoxic, organic rich deposits: Jurassic through Holocene. J Foraminiferal Res 16(3): 207–215

Boldrin A, Stocchino C (1990) On the hydrological characteristics of Terra Nova Bay. In: National Scientific Commission for Antarctica (ed) Oceanographic Campaign 1987/1988 Data Rep Part I, Genova, pp 11–57

Boltovskoy E (1966) Depth at which foraminifera can survive in sediments. Contr Cushman Found Foraminiferal Res 17: 43–45

Brambati A, Fanzutti GP, Finocchiaro F, Simeoni U (1989) Sediments and sedimentological processes in the Ross Sea continental shelf (Antarctica): results and preliminary conclusions. Boll Oceanogr Teor Appl 7(1–2): 159–188

Catalano G, Benedetti F (1990) Distribution of nutrients in the Terra Nova Bay and in the Ross Sea. In: National Scientific Commission for Antarctica (ed) Oceanographic Campaign 1987/1988 Data Rep Part I, Genova, pp 61–83

Corliss BH (1985) Microhabitats of benthic foraminifera within deep-sea sediments. Nature 314(6010): 435–438

Corliss BH (1991) Morphology and microhabitat preferences of benthic foraminifera from the northwest Atlantic Ocean. Mar Micropaleontol 17: 195–236

Corliss BH, Chen C (1985) Morphotype patterns of Norwegian Sea deep-sea benthic foraminifera and ecological implications. Geology 16: 716–719

DeLaca TE (1986) The morphology and ecology of *Astrammina rara*. J Foraminiferal Res 16(3): 216–223

Echols RJ (1971) Distribution of foraminifera in sediments of the Scotia Sea area, Antarctic waters. Antarct Res Ser 15: 93–168

Finger KL, Lipps JH (1981) Foraminiferal decimation and repopulation in an active volcanic caldera, Deception Island, Antarctica. Micropaleontology 27(2): 11–39

Frankel L (1974) Observations and speculations on the habitat and habits of *Trochammina ochracea* (Williamson) in subsurface sediments. J Paleontol 48(1): 143–148

Frankel L (1975) Pseudopodia in surface and subsurface dwelling *Miliammina fusca* (Brady). J Foraminiferal Res 5(3): 211–217

Haward NJB, Haynes JR (1976) *Chlamys opercularis* (Linnaeus) as a mobile substrate for foraminifera. J Foraminiferal Res 6(1): 30–38

Haynes JR (1981) Foraminifera. Macmillan, New York

Innamorati M, Mori G, Lazzara L, Nuccio C, Lici M, Catalano G, Benedetti F (1990) Phytoplankton ecology in the Ross Sea. In: National Scientific Commission for Antarctica (ed) Oceanographic Campaign 1987/1988 Data Rep Part II, Genova, pp 9–63

Ishman SE, Domack EW (1994) Oceanographic controls on benthic foraminifers from the Bellingshausen margin of the Antarctic Peninsula. Mar Micropaleontol 24: 119–155

Jacobs SS (1989) Marine controls on modern sedimentation on the Antarctic continental shelf. Mar Geol 85(2/4): 121–153

Jennings AE, Helgadottir G (1994) Foraminiferal assemblages from the fjords and shelf of Eastern Greenland. J Foraminiferal Res 24(2): 123–144

Jones RW, Charnock MA (1985) "Morphogroups" of agglutinating foraminifera, their life positions and feeding habits and potential applicability in (paleo)ecological studies. Rev Paleobiol 4(2): 311–320

Kaiho K (1991) Global changes of Paleogene aerobic/anaerobic benthic foraminifera and deep-sea circulation. Palaeogeogr Palaeoclimatol Palaeoecol 83: 65–85

Kaminski MA, Grassle JF, Whitlatch RB (1988) Life history and recolonization among agglutinated foraminifera in the Panama Basin. Abh Geolog Bundesanstalt Wien 41: 229–244

Kaminski MA, Boersma A, Tyszka J, Holbourn AEL (1995) Response of deep-water agglutinated foraminifera to dysoxic conditions in the California Borderland basins. Grybowski Found Spec Publ 3: 131–140

Kennett JP (1968) The fauna of the Ross Sea, part 6: Ecology and distribution of foraminifera. N Z Dep Sci Ind Res Mem 46: 1–48

Kitazato H (1984) Microhabitats of benthic foraminifera and their application to fossil assemblages. Benthos 83: 339–344

Mackensen A, Grobe H, Kuhn G, Futterer DK (1990) Benthic foraminiferal assemblages from the eastern Weddell Sea between 68° and 73°S: distribution, ecology and fossilization potential. Mar Micropaleontol 16: 241–283

McKnight WM (1962) The distribution of foraminifera off parts of the Antarctic coast. Am Paleontol Bull 44(201): 65–158

Milam RW, Anderson JB (1981) Distribution and ecology of recent benthonic foraminifera of the Adelie-George V continental shelf and slope, Antarctica. Mar Micropaleontol 6: 297–325

Murray JW (1991) Ecology and paleoecology of benthic foraminifera. Longman, London

Nagy J, Gradstein FM, Kaminski MA, Holbourn AEL (1995) Foraminiferal morphogroups, paleoenvironments and new taxa from Jurassic to Cretaceous strata of Thakkhola, Nepal. Grybowski Found Spec Publ 3:181–209

Nelson D (1992) Biogeochemical cycles of organic and siliceous matter in the Southern Ocean. In: Gallardo VA, Ferretti O, Movano HI (eds) Oceanografia in Antartide. ENEA-PNRA-EULA, Conception, Chile, pp 44–57

Osterman LE, Kellogg TB( 1979) Recent benthic Foraminiferal distributions from the Ross Sea, Antarctica; relation to ecologic and oceanographic conditions. J Foraminiferal Res 9: 250–269

Pflum CE (1966) The distribution of foraminifera in the Eastern Ross Sea, Amundsen Sea and Bellingshausen Sea, Antarctica. Am Paleontol Bull 50(226): 151–209

Schafer CT, Cole FE, Frobel D, Rice N, Buzas MA (1996) An *in situ* experiment on temperature sensitivity of nearshore temperate foraminifera. J Foraminiferal Res 26(1): 53–63

Schroeder-Adams CJ (1990) High latitude agglutinated Foraminifera: Prydz Bay (Antarctica) VS. Lancaster Sound (Canadian Arctic). NATO ASI Ser C 327: 315–343

Severin KP (1983) Test morphology of benthic foraminifera as a discriminator of biofaces. Mar Micropaleontol 8: 65–76

Streeter S (1973) Bottom water and benthonic foraminifera in the North Atlantic glacial-interglacial contrasts. Q Res 3: 131–141

Violanti D (1996) Taxonomy and distribution of recent benthic foraminifers from Terra Nova Bay (Ross Sea, Antarctica), Oceanographic Campaign 1987/1988. Palaeontol Ital 83: 25–71

Walton WR (1952) Techniques for the recognition of living foraminifera. Cushman Found Foraminiferal Res Contrib 3(2): 56–60

Ward BL, Barrett PJ, Vella P (1987) Distribution and ecology of benthic foraminifera in McMurdo Sound, Antarctica. Palaeogeogr Palaeoclimatol Palaeoecol 58: 139–153

CHAPTER 35

# The Macrophytobenthos of Terra Nova Bay

**M. Cormaci, G. Furnari, and B. Scammacca**[1]

**ABSTRACT**

The results of a study on the benthic macroalgal flora and vegetation of Terra Nova Bay (Ross Sea) are presented for the years 1987–88 and 1989–90. The flora is rather poor (nine Rhodophyceae, four Fucophyceae, four Chlorophyceae), but richer than that known from adjacent areas. The vegetation is on the whole characterized by a prevalence of shade-adapted species among which *Iridaea cordata*, *Phyllophora antarctica* and *Clathromorphum lemoineanum* dominated in the upper infralittoral, lower infralittoral and circalittoral zones, respectively. Moreover, we report on the results of a study, carried out in the years 1993–94, on the biomass of *Iridaea cordata* and *Phyllophora antarctica*, which represent the most important primary producers among the macroalgae of the Ross Sea.

## Introduction

The first studies dealing with the phytobenthos of the Ross Sea (discovered in the year 1841) were those by Barton (1902), Foslie (1905, 1907) and Gepp and Gepp (1905, 1907). A further contribution was made by Levring (1945), who studied the algae collected during the Norwegian expedition in 1924. More important contributions, based on both samplings and observations made by SCUBA, were made by Zaneveld (1966a,b, 1968a,b) along the coast of Ross Island, Victoria Land and the large western Ross Sea Islands (where a marine algal zonation was described as well), and by Zaneveld and Sanford (1980), who recorded three crustose corallinaceous species collected in three different stations in the Ross Sea. A more recent study is that by Miller and Pearse (1991), in which quantitative data on the distribution and biomass of some seaweeds from McMurdo Sound were reported.

Our preceding studies on the macrophytobenthos of Terra Nova Bay include those on the specific composition (Cormaci et al. 1992a), the extension (Cormaci et al. 1993), the structure and the vertical distribution of the phytobenthic communities along the coast from the Italian Base to Cape Russell (Cormaci et al. 1992b), as well as the biomass of *Iridaea cordata* (Cormaci et al. 1996) and *Phyllophora antarctica* (Cormaci 1996; Cormaci et al. 1998). Here we report the results from the cruises carried out in 1987–88 and 1989–90 in this area.

### Materials and Methods

Floristic and vegetational samplings were made during the Antarctic expeditions 1987–88 and 1989–1990 along a tract of coast of ca. 30 km, from the Italian base to Cape Russell. Sampling for the biomass study was made from the end of December 1993 to mid-February 1994 in two stations named Wharf IB (near the Italian base) and Faraglione (located 3 km to the south).

Almost all collections were made by SCUBA, but some of them were made by dredge and/or grab while algal communities were mapped with a wire-guided video camera Pluto 21 (Cormaci et al. 1993).

Both floristic and vegetational samplings were made along depth transects perpendicular to the coast. For the biomass study, thalli of both *I. cordata* and *P. antarctica* were collected on the top and on the vertical surfaces of boulders, at the 4- to 8-m depths, where the species forms homogeneous populations.

[1] Dipartimento di Botanica della Università di Catania, via A. Longo 19 95125 Catania, Italy

**Table 1.** Ranges of weight classes established for both *Iridaea cordata* and *Phyllophora antarctica*

| Class number | Weight range in g | |
|---|---|---|
| | *I. cordata* | *P. antarctica* |
| 1 | up to 0.25 | up to 0.03 |
| 2 | >0.25 to 0.5 | >0.03 to 0.06 |
| 3 | >0.5 to 1 | >0.06 to 0.125 |
| 4 | >1 to 2 | >0.125 to 0.25 |
| 5 | >2 to 4 | >0.25 to 0.5 |
| 6 | >4 to 8 | >0.5 to 1 |
| 7 | >8 to 16 | >1 to 2 |
| 8 | >16 to 32 | >2 |
| 9 | >32 to 64 | |
| 10 | >64 | |

In order to gather information on biomass variations during the short austral summer, sampling was repeated every 10–15 days. Unfortunately, at Faraglione, because of bad climatic conditions, sampling was conducted only in mid-January. The wet weight of each thallus was measured. Ten weight classes were established for *I. cordata* and eight for *P. antarctica* (Table 1). Moreover, the blade area of each thallus of *I. cordata* was measured.

## Results

### Flora

The flora of Terra Nova Bay consists of nine Rhodophyceae, four Fucophyceae and four Chlorophyceae (Cormaci et al. 1992a). Among the species found (Table 2), *Iridaea cordata*, *Phyllophora antarctica*, *Plocamium cartilagineum* and *Monostroma hariotii* are widely distributed in the Ross Sea as well as in the whole Antarctic Continent; others, such as *Geminocarpus geminatus* and *Urospora penicilliformis*, widely distributed throughout Antarctica, are quite rare in the Ross Sea: still others, such as *Ballia sertularioides* and *Petroderma maculiforme*, were recorded from Antarctica for the first time; finally, *Clathromorphum lemoineanum* (described by Mendoza and Cabioch 1985 from Terra del Fuego, the Falkland Islands, the South Orkneys and the Antarctic Peninsula), *Gainia mollis* (described by Moe 1985 on specimens from Antarctic Peninsula, South Shetlands and the Vestfold Hills; East Antarctica) and *Lola irregularis*, described by Zaneveld (1966c) from Cape Bird (Ross Island, Ross Sea), were recorded for the first time after their description.

### Vegetation

As mentioned above, the study of the vegetation was based on collections made along depth transects perpendicular to the coast. This allowed us to distinguish four different zones which, according to the terminology of Pérès and Picard (1964; already used by Arnaud 1974 in his study of the vegetation of Terre Adélie), were named supralittoral, midlittoral, infralittoral and circalittoral zones (Cormaci et al. 1992b). They correspond to the epilittoral, littoral, sublittoral and elittoral regions, respectively, distinguished by Zaneveld (1966a) in his description of the marine algal zonation of Victoria Land.

In the supralittoral zone, the vegetation showed more terrestrial than marine characteristics. In fact, seawater moistness was very scarce due to the predominance of offshore winds, while the presence of freshwater from ice melting was common. The only algae found in this zone were blue-green algae and diatoms.

In the midlittoral zone, the vegetation was prevalently characterized by blue-green algae. Only two macroalgal species were present: *Prasiola crispa* and *Urospora penicilliformis*. The last species formed dense populations prevalently inside crevices. It should be noted that *Prasiola crispa*, which in wave-swept stations is very common in the supralittoral zone (Arnaud 1974; Womersley 1984; Zaneveld 1966a) occurred only in the midlittoral zone in Terra Nova Bay because of the rather sheltered coast subject to katabatic winds.

In the infralittoral zone, two subzones were distinguished: the upper infralittoral (from the lowest tide level to ca. 12 m depth) and the lower infralittoral (12 to ca. 25 m depth). Two different communities lived in the upper infralittoral. The first, present on surfaces exposed to ice-scouring action, consisted of several species of diatoms that nearly covered such surfaces. The second, occurring in furrows between boulders, was characterized by *Iridaea cordata* (also with high cover values) and by some frequent small species such as *Monostroma hariotii* and *Geminocarpus geminatus* (epiphytic on *I. cordata*). Less frequent species were *Phyllophora antarctica* (sometimes with high cover values), *Ballia sertularioides* and *Plocamium cartilagineum*.

**Table 2.** Macroalgal flora of the Ross Sea. In Roman type synonyms or misapplied names. Abbreviations refer to papers in which taxa were reported: B = Barton (1902); C = Cormaci et al. (1992a); F = Foslie (1905); G = Gepp and Gepp (1907); L = Levring (1945); M = Miller and Pearse (1991); Za = Zaneveld (1966a); Zc = Zaneveld (1966c); ZS = Zaneveld and Sanford (1980)

| | | | | | | | | |
|---|---|---|---|---|---|---|---|---|
| *Ballia callitricha* (C. Agardh) Kützing | B | | | | Za | | | |
| *Ballia sertularioides* (Suhr) Papenfuss | | | | | | | | C |
| *Clathromorphum lemoineanum* Mendoza et Cabioch | | | | | | | | C |
| *Desmarestia aculeata* (Linnaeus) Lamouroux[a] | B | | | | | | | |
| *Desmarestia chordalis* Hooker et Harvey | | | | | | | | |
| Desmarestia rossii Hooker et Harvey | B | | | | | | | |
| *Desmarestia confervoides* (Bory) Ramírez et Peters | | | | | | | | |
| D. viridis Lamouroux[b] | B | | | | | | | |
| D. willii Reinsch[c] | | | | | Za | | | |
| *Desmarestia menziesii* J. Agardh | | | | | Za | | | |
| Desmarestia harveyana A. et E. S. Gepp | | G | | | | | | |
| *Erythrotrichia carnea* (Dillwyn) J. Agardh | | | | | | | | C |
| *Gainia mollis* Moe | | | | | | | | C |
| *Geminocarpus geminatus* (Hooker fil. et Harvey) Skottsberg | | | | | | | | C |
| Ectocarpus geminatus Hooker fil. et Harvey | | G | | | | | | |
| *Gracilaria* sp. | | G | | | | | | |
| *Gymnogongrus antarcticus* Skottsberg | | | | | | | | C |
| *Hildenbrandia lecannellieri* Hariot | | | | | Za | | | |
| *Himantothallus grandifolius* (A. et E. S. Gepp) Zinova | | | | | | | | |
| Lessonia grandifolia A. et E. S. Gepp | | G | | | | | | |
| Phyllogigas grandifolius (A. et E. S. Gepp) Skottsberg | | | | | Za | | | |
| *Iridaea cordata* (Turner) Bory | | | | | | | M | C |
| Iridaea micans Bory | | G | | | | | | |
| Iridaea obovata Kützing | | | | | Za | | | |
| *Leptophytum coulmanicum* (Foslie) Adey | | | | | | ZS | M | |
| Lithothamnion coulmanicum Foslie | | | F | | | | | |
| *Lithoderma antarcticum* Skottsberg cf. | | | | | | | | C |
| *Lola irregularis* Zaneveld | | | | | Zc | | | C |
| *Monostroma hariotii* Gain | | | | | Za | | M | C |
| *Palmaria decipiens* (Reinsch) Ricker | | | | | | | | |
| Gracilaria dumontioides A. et E. S. Gepp | | G | | | | | | |
| Gracilaria simplex A. et E. S. Gepp | | G | | | | | | |
| Leptosomia simplex (A. et E. S. Gepp) Kylin | | | | | Za | | | |
| *Petroderma maculiforme* (Wollny) Kuckuck | | | | | | | | C |
| *Phycodrys antarctica* (Skottsberg) Skottsberg | | | | | Za | | | C |
| *Phycodrys quercifolia* (Bory) Skottsberg | | | | | | | | |
| Delesseria quercifolia Bory | | G | | | | | | |
| *Phyllophora antarctica* A. et E. S. Gepp | | G | | | Za | | M | C |
| *Phymatolithon foecundum* (Kjellman) Düwel et Wegeberg | | | | | | | | |
| Leptophytum foecundum (Kjellman) Adey | | | | | | ZS | | |
| *Phymatolithon tenue* (Rosenvinge) Düwel et Wegeberg | | | | | | | | |
| Leptophytum laeve (Strömfelt) Adey | | | | | | ZS | | |
| *Plocamium cartilagineum* (Linnaeus) Dixon | | | | | | | | C |
| P. coccineum (Hudson) Lyngbye | B | G | | | Za | | | |
| *Prasiola crispa* (Lightfoot) Meneghini | | | | | | | | C |
| ssp. *antarctica* (Kützing) Knebel f. *antarctica* | | | | | Za | | | |
| *Scytothamnus fasciculatus* (Hooker fil. et Harvey) Cotton | | | | L | | | | |
| *Spongoclonium orthocladum* A. et E. S. Gepp[d] | | G | | | | | | |
| *Syringoderma* sp. | | | | | | | | C |
| *Ulothrix flacca* (Dillwyn) Thuret | | | | | Za | | | |
| *Urospora penicilliformis* (Roth) Areschoug | | | | | | | | C |
| *Zonaria* sp. | | G | | | | | | |
| Foliose red alga (Callophyllis cf.) | | | | | | | M | |
| Crustose brown alga | | | | | | | M | |
| Uniseriate, unbranched filamentous green alga | | | | | | | M | |

[a] This is probably a misidentification since *D. aculeata* is distributed only in the Northern Hemisphere (Fletcher 1987; Silva et al. 1996).

[b] This is a name misapplied to *D. confervoides*. In fact, according to both Ricker (1987) and Ramírez and Peters (1992), *D. viridis*, for the moment, is a species restricted to the Northern Hemisphere.

[c] The conspecificity of *D. confervoides* and *D. willii* was proposed by Ramírez and Peters (1992: 2438).

[d] According to Moe and Silva (1979) and Ricker (1987), this species does not belong to the genus *Spongoclonium* Sonder but more probably to *Antarcticothamnion* Moe *et* Silva.

The lower infralittoral was physiognomically characterized by a dense canopy of *Phyllophora antarctica*. Among the other species *Ballia sertularioides*, *Plocamium cartilagineum* and *Gainia mollis* (the last in the understorey) were the most frequent.

In the upper metres of the circalittoral zone (25–30-m depth) the cover of erect species (*Phyllophora antarctica* and *Plocamium cartilagineum*) decreased considerably, while the calcareous crustose *Clathromorphum lemoineanum* dominated. In deeper waters, erect species disappeared entirely and *C. lemoineanum* remained the only macroalga showing high cover values to 70-m depth.

### Biomass of *Iridaea cordata* and *Phyllophora antarctica*

On the whole, 935 thalli of *I. cordata* and 3841 thalli of *P. antarctica* were measured. In all samples, *Ballia sertularioides* and *Geminocarpus geminatus* were present, with no relevant biomass values.

The plant density of *I. cordata* depended mainly on the number of thalli belonging to the lowest weight classes (Cormaci et al. 1996). At the end of December the estimated density was 952 plants $m^{-2}$. It reached highest values in mid-January (1228 plants $m^{-2}$) then declined to minimum values of 644 plants $m^{-2}$ in February.

Conversely, biomass depended mainly on the occurrence of thalli of the highest weight classes. In fact, the highest biomass value (3439 $g\,m^{-2}$) estimated at the end of December was mainly due to the occurrence of some heavy old thalli of 9–10 weight classes (24 thalli $m^{-2}$, i.e. 2.5% of the total number of thalli) and of a high number of thalli of six to eight weight classes. The lowest value (2225 $g\,m^{-2}$) was estimated at the end of January when such heavy old thalli were lacking.

The total blade area of thalli of each sampling was also influenced by the presence/absence of large thalli of the heaviest classes. In fact, the highest values estimated at the end of December (10 $m^2\,m^{-2}$) and in mid-February (10.7 $m^2\,m^{-2}$) were due to the presence of very large thalli, even though scarce in number; while the lowest value (8.7 $m^2\,m^{-2}$) was estimated at the end of January when such thalli were absent.

The plant density of *P. antarctica* depended, like that of *I. cordata*, mainly on thalli of the lowest weight classes (Cormaci 1996; Cormaci et al. 1998). From the end of December to mid Juanuary it showed a slight increase (from 8832 to 9172 thalli $m^{-2}$), then a slight decrease at the end of January (8192 thalli $m^{-2}$), while in February a significant decline (4384 thalli $m^{-2}$) was registered.

Conversely, biomass depended mainly on the occurrence of thalli of the highest weight classes. In fact, at the end of January, when the highest biomass value was recorded (1548 $g\,m^{-2}$), 62% of it came from thalli belonging to weight classes 6 to 8, which numerically represented only 8% of the total number of thalli. The lowest value (899 $g\,m^{-2}$) was instead estimated at the end of December when, although a high number of thalli of weight classes 3 to 5 and a very high number of thalli of weight classes 1–2 were counted, no thalli belonging to weight class 8 were present.

## Discussion

### Flora

The macrophytobenthos of the Antarctic Continent consists of about 120 species (Clayton 1994). Such a poor presence (compared to the southern coasts of Australia where there are over than a thousand species; Ricker 1987), is due to the particular environmental conditions of polar regions. Also in Arctic regions, there is a small number of species (109 species on the eastern coasts of Greenland and 120 species on Novaya Zemlya; Lüning 1990). The presence of endemic species is, however, much higher in Antarctic regions (about a third of the floral contingent) than in Arctic ones (just 5%). This is due, above all, to the lack of coast links with cold-temperate regions, present, on the contrary, in the Arctic (Lüning 1990).

Although the Antarctic environmental conditions where benthos grows can be considered quite uniform from several points of view (Picken 1985), the richness of the flora varies considerably, depending on the different regions. In west Antarctica, for example, which extends to a latitude of about 65°S (Graham Land, Antarctic Peninsula), the flora is much richer than in east Antarctica, where only 25–30% of the Antarctic flora has been found. A much poorer flora (only 5–10%) is found in the latitudes which are higher than 70°S (such as in Terra Nova Bay, Ross Sea) to the minimum of seven taxa (three of which have been identified only at the class level) recorded by Miller and Pearse (1991) from McMurdo Sound (Ross Sea). However, there is nothing surprising if, in floras so poor in species, the identification of taxa is often uncertain and incomplete. Because of the clear difficulties in sampling and the short sampling periods when the study cruises can be performed, the only available specimens are often very young or sterile thalli.

According to the literature, the Ross Sea algal flora (Table 2) consists of 37 taxa (equal to about 30% of the Antarctic flora). Many of these are rather rare and/or with a distribution rather limited in this sea. The following are worthy of mention: *Desmarestia chordalis* (as *D. rossii*) (reported only by Barton 1902); *Phycodrys quercifolia* (as *Delesseria quercifolia*), and *Spongoclonium orthocladum* (reported only by Gepp and Gepp 1907); *Hildenbrandia lecannellieri*, and *Ulothrix flacca* (reported only by Zaneveld 1966a); *Scytothamnus fasciculatus* (reported only by Levring 1945).

The floristic richness recorded at Terra Nova Bay (17 species) comprises approximately half the flora of the Ross Sea. Therefore, Terra Nova Bay can be considered the richest area in species both in this area and in the stations located to the North. This result seems to contrast Zaneveld's (1966a) quotation: "... there is a constant decrease in the number of algal species proceeding southward in the Ross Sea ...", which may simply be due to the very careful sampling made in this area.

Among the species found, the following are worthy of consideration:

*Clathromorphum lemoineanum* Mendoza et Cabioch (Corallinales, Corallinaceae)

This species, widely distributed in west Antarctica (Mendoza and Cabioch 1985), in east Antarctica is present only at Terra Nova Bay, where it is quite common in the circalittoral zone down to the 60–70 m depth, with high cover values as well.

In addition to *C. lemoineanum*, three other Corallinaceae: *Phymatolithon foecundum*, *P. tenue*, and "*Leptophytum*"[1] *coulmanicum* are reported from the Ross Sea (Table 2). Although they show almost the same ecology, their distribution areas do not seem to overlap. In fact, *P. foecundum*, widely distributed in the high latitudes of the Northern Hemisphere, is found in the Southern Hemisphere from Cape Hallett to the Macquarie Islands with two stations in the Ross Sea: Moubray Bay and Cape Hallett, where none of the other species has been recorded (Zaneveld and Sanford 1980); *P. tenue*, is also quite common in high and medium latitudes of the Northern Hemisphere, while in the Southern Hemisphere it is recorded only from Franklin I. (Ross Sea), where no other calcareous species have been found (Zaneveld and Sanford 1980); finally, "*Leptophytum*" *coulmanicum*, which is present in Coulman Island and the Macquarie Islands (Zaneveld and Sanford 1980), in the Ross Sea is recorded only from McMurdo Sound, where it is found in different stations among which Cape Armitage (77°51′S, 166°40′E), representing the southernmost station for a calcareous species (Miller and Pearse 1991).

*Gymnogongrus antarcticus* Skottsberg (Gigartinales, Phyllophoraceae)

A few specimens of *G. antarcticus* showing cystocarps embedded in the thallus were collected at Terra Nova Bay at 6 m depth (Cormaci et al. 1992a). To date, the tetrasporophyte of this species has never been reported. However, according to Ricker (1987, p 200), studies of *G. antarcticus* have revealed encrusting stages from the germination of carpospores which, although they had not yet produced any reproductive structures, could presumably represent the sporophyte phase of this species. Silva and DeCew (1992), on the basis of their studies in culture, have proposed the new genus *Ahnfeltiopsis* to arrange those Phyllophoraceae "... with erect unisexual gametophytes and crustose tetrasporophytes ..." while they have referred to *Gymnogongrus* those species which, as the type species [*G. griffithsiae* (Turner) Martius], are tetrasporoblastic. As a consequence, *G. antarcticus* should be transferred to *Ahnfeltiopsis*. However, we do not propose here the new combination of this species under the genus *Ahnfeltiopsis*, because we agree with the conclusions reached by Fredericq and Ramírez (1996) who, studying *rbc*L sequence analysis of 38 taxa of Phyllophoraceae, found that the species of *Gymnogongrus* with tetrasporoblasts did not group together in a single clade but were scattered among several clades. Moreover, they stated that the species with tetrasporoblasts could be geneti-

[1] According to Düwel and Wegeberg (1996), the genus *Leptophytum* Adey is a nomenclatural synonym of *Phymatolithon* Foslie. Nevertheless, the new combination of this species under the genus *Phymatolithon* is not here proposed, because the holotype of *Lithothamnion coulmanicum* (basyonym of *Leptophytum coulmanicum*) has not yet been studied.

cally more closely related to species with cystocarps and encrusting tetrasporophytes than to other species with tetrasporoblasts and concluded that " . . . presence or absence of a tetrasporoblast cannot form the basis for recognition of different genera . . . " and that " . . . more detailed morphological studies are needed before *Ahnfeltiopsis/Gymnogongrus* can be formally segregated into several genera . . . ". Such a conclusion was also shared by Masuda et al. (1996) in their study on life history of *G. griffithsiae* from Ireland.

*Iridaea cordata* (Turner) Bory (Gigartinales, Gigartinaceae)
This species is common throughout the infralittoral zone of Terra Nova Bay, but more abundant in the upper infralittoral.
The distribution of this species has been controversal for a long time. This was principally due to the poor characterization of this genus and to the misidentification of the species because of its considerable polymorphism. For example, according to Chapman (1979), *I. cordata* should occur in the Northern Hemisphere while *I. micans* Bory (that, however, proved to be a synonym of *I. cordata*) in the Southern Hemisphere. Conversely, according to Scagel et al. (1989), *I. cordata* should occur in the Southern Hemisphere, but *I. spendens* (Setchell et Gardner) Papenfuss [now *Mazzaella splendens* (Setchell et Gardner) Fredericq] in the Northern Hemisphere.

Real differences concerning the structure of cystocarps and the origin of tetrasporangial sori between *Iridaea* spp. from Pacific coast of North America and *I. cordata* from South America, have been revealed by Leister and Hommersand (1991), while a revised description of the genus *Iridaea* as well as of its related genera within the Gigartinaceae, based on reproductive, developmental and morphological criteria, was more recently given by Hommersand et al. (1993). Finally, according to phylogenetic and biogeographic studies of the Gigartinaceae based on sequence analysis of *rbc*L, carried out by Hommersand et al. (1994), the genus *Iridaea* is actually restricted to the Southern Hemisphere, with *I. cordata* occurring at Cape Horn, Antarctica, Falkland Is., Scotia Sea and Is. West Wind Drift.

*Phyllophora antarctica* A. et E.S. Gepp (Gigartinales, Phyllophoraceae)
At Terra Nova Bay, the species is common throughout the infralittoral, but more abundant in the lower infralittoral zone.

As far as the southern oceans are concerned, Papenfuss (1964) listed four species of *Phyllophora*, all present in the Antarctic and/or sub-Antarctic regions: *P. antarctica*, *P. ahnfeltioides* Skottsberg, *P. appendiculata* Skottsberg and *P. abyssalis* Skottsberg. According to recent observations by Fredericq and Ramírez (1996) on specimens from King George I. (South Shetland Is.), the last species, considered by Skottsberg (1953) and by Ricker (1987) probably not specifically distinct from *P. antarctica*, could be referred to as *Hymenocladiopsis crustigena* Moe; while *P. appendiculata* was considered as a synonym of *Gymnogongrus turquetii* Hariot by Fredericq and Ramírez (1996).

Therefore, in the southern oceans, only the following two species of *Phyllophora* occur: *P. antarctica* and *P. ahnfeltioides*. These two species were placed in this genus because of cystocarps borne on pinnules. However, according to Fredericq and Ramírez (1996), the taxonomic importance of this character at a generic level is questionable, since from their study *P. antarctica* and *P. ahnfeltioides* are divided into a clade cluster widely divergent from the Northern Hemisphere *Phyllophora* clade. Moreover, they hypothesized for these two Antarctic species an arrangement into a new genus, since they form spermatangia in pairs by surface cortical cells rather than in pits. As a consequence, the genus *Phyllophora* would now be restricted to the Northern Hemisphere.

Although Kylin and Skottsberg (1919) reported only blade-like sterile thalli and female gametophytes bearing cystocarps of *P. antarctica*, both Ricker (1987) and Miller and Pearse (1991) hypothesized that it could show a heteromorphic life history. In particular, Miller and Pearse thought that the non-calcareous crusts living in the "*Leptophytum*" *coulmanicum* zone (20-m depth and below) were tetrasporophytes of *P. antarctica* rather than bladeless gametophytic basal crusts, and that there was a spatial separation between free-living life history phases in that species.

In our opinion, however, considering that of the above species only female gametophytes bearing cystocarps are known, while fertile male plants are very rare [the only specimen was collected at Cape Evans by Miller and Pearse (1991) and those, collected in the same locality by De Laca in (1976) are described by Fredericq and Ramírez (1996)], a complete apogamic life history, like that described in *Gigartina stellata* (Stackhouse) Batters in culture by Chen et al. (1974) and in *Mastocarpus papillatus* (C. Agardh) Kützing by Zupan and West (1988) [as *Gigartina papillata*

(C. Agardh) J. Agardh] and by Polanshek and West (1988), is presumable. If confirmed, this should represent another adaptative strategy of this species to the Antarctic environment since, as stated by Miller and Pearse (1991), apogamy "... bypasses fertilization, sexual recombination (and the establishment of a free-living diploid phase) and multiplies the numbers of individuals with a successful genotype".

*Plocamium cartilagineum* (Linnaeus) Dixon (Plocamiales[2], Plocamiaceae)
This species, frequent in Terra Nova Bay in the infralittoral zone with tetrasporangial thalli, is widely distributed throughout the world (Ricker 1987), including the sub-Antarctic (Kylin and Skottsberg 1919) and Antarctic (Lamb and Zimmermann 1977) zones. Although the species shows a considerable variation in both habit and dimensions, according to Womersley (1994), it can be distinguished for its consistent branching pattern, sessile cystocarps and morphology of the stichidia.

Nevertheless, in a study dealing with some red algae, among which some species of *Plocamium* (*P. cartilagineum* from USA, Chile, Ireland and Antarctica included), Fredericq et al. (1996) observed that "... Relationship among the small number of *Plocamium* species sampled are fully resolved in the minimal *rbc*L tree but are not well supported. Although no particular species distribution pattern is evident from this limited data set, the use of *Plocamium cartilagineum* (L.) Dixon for taxa from different parts of the world is not supported".

In a recent study on the upper survival temperatures (USTs) of 15 Antarctic red algae, Bischoff-Bäsmann and Wiencke (1996) classified *P. cartilagineum* as a stenothermal species growing only between 0 and <5 °C and exhibiting extremely low USTs (7 °C). These values are definitely lower than those observed in the same species coming from other seas: USTs 23 °C (Lüning and Freshwater 1988). According to Bischoff-Bäsmann and Wiencke (1996), *P. cartilagineum* from Antarctica should be considered either as an ecotype or belonging to a closely related species. This last hypothesis should agree with the results of the studies carried out by Fredericq et al. (1996).

[2] This order has been proposed by Saunders and Kraft (1994).

## Vegetation

It is well known that the zonation and geographical distribution of macroalgae is determined principally by light and temperature (Wiencke 1996), as well as by other factors such as substratum type (hard or soft) and hydrodynamics. In the Antarctic region, the vegetation is also strongly influenced by the extreme variations in irradiance, the seasonal changes in day length, the low water temperature with an annual range usually ranging from −1.8 and 2 °C and by the presence of ice (Kirst and Wiencke 1995). This last factor influences the environment in the following ways: it causes a major erosion of the substratum that does not allow a stable macroalgal settlement (Neushul 1965; Picken 1985); it forms a stream (freshwater inflows) due to the meltwater from the land creating a bare zone in the upper infralittoral zone (Chung et al. 1994); it forms a screen absorbing most of the sunlight. In the Ross Sea, environmental conditions are particularly severe. In fact, the sea is generally covered with ice 10 months out of 12, therefore provoking a strong decrease in irradiance even in shallow waters. Zaneveld (1966a) measured an incident light reduction to 1.39% of the surface value in the months of November and December, at a depth of 20 m under a 2-m-thick ice cover.

These very harsh environmental conditions are responsible for the low specific diversity of the Ross Sea flora and for the presence of a poorly diversified and structured vegetation. The Ross Sea vegetation is particularly simple in structure, since there is no presence of large brown algae (only at Cape Hallet, on the northern board, can *Himantothallus grandifolius* be found).

Despite a greater richness in flora compared to the other areas of the Ross Sea, the vegetation shows a structure and a vertical zonation pattern similar to those described by Zaneveld (1966a) in various stations of Victoria Land and the Ross Island and by Miller and Pearse (1991) at McMurdo Sound.

It should be noted that *I. cordata* and *P. antarctica*, characterizing the upper and the lower infralittoral zones, respectively, are shade-adapted species that live as understorey species in lower latitude areas, whereas in Terra Nova Bay, as well as in the Ross Sea south of Cape Hallet (Zaneveld 1966a;

Miller and Pearse 1991), they live as overstorey species. This is probably due to the long winter darkness of this high latitude, as well as to the fading of summer light, due to the long persistence of ice cover. The presence of a dense canopy of the high shade-adapted species *I. cordata* in the upper infralitoral zone may be related to the different ultrastructure of the chloroplasts of this species, in plants from ice-free water or ice-covered water (Foltran et al. 1996). In fact, the species is capable of avoiding the risks of photodamage due to an increase in light intensity, "... by reducing its photosynthetic membranes and the inserted pigments (i.e. chlorophyll a and carotenoids), and disbanding, without demolition, the PSII peripheral antenna components" (Foltran et al. 1996).

## Biomass

On the basis of the above data which, together with those by Miller and Pearse (1991), are the only available biomass data on macroalgae from the Ross Sea, the following considerations can be made. The high number of thalli of *I. cordata* of medium-high weight classes recorded at the beginning of summer shows that this species continues to grow, although at a low rate, during the long Antarctic winter. This capability, formerly reported by Zaneveld (1968a,b) for this species (as *I. obovata*), was recently demonstrated in laboratory tests carried out by Weykam et al. (1997).

The very high number of thalli of *P. antarctica* of weight classes 1–2 recorded at the end of December shows that the production of such thalli started a short time before; while the high number of thalli of weight classes 3 to 5 shows that this species, as also observed by Zaneveld (1968a,b), continues to grow during the Antarctic winter.

Finally, taking into account that the above two species form populations with similar characteristics also along the coasts of Victoria Land, from Cape Hallet to McMurdo Sound (Zaneveld 1966a; Miller and Pearse 1991), and that biomass of the other macroalgae present is not relevant, *I. cordata* and *P. antarctica* represent the most important primary producers among the macroalgae of that Sea.

*Acknowledgements.* The authors are grateful to both Dr. G. Casazza (who took part in the expedition 1989–90) and Dr. M. Catra (who took part in the expedition 1993–94). This work was supported by ENEA (Rome), under the Italian Antarctica Project.

## References

Arnaud PM (1974) Contribution à la bionomie marine benthique des régions antarctiques et subantarctiques. Téthys 6: 465–656

Barton ES (1902) Algae. In: Report on the collections of Natural History made in the Antarctic regions during the voyage of the "Southern Cross". London, British Museum Naturral History, p 320

Bischoff-Bäsmann B, Wiencke C (1996) Temperature requirements for growth and survival of Antarctic Rhodophyta. J Phycol 32: 525–535

Chapman VJ (1979) The marine algae of New Zealand. III, 4: Gigartinales. Cramer, Vaduz

Chen LC-M, Edelstein T, McLachlan J (1974) The life history of *Gigartina stellata* (Stackh.) Batt. (Rhodophyceae, Gigartinales) in culture. Phycologia 13: 287–294

Chung H, Oh YS, Lee IK, Kim D-Y (1994) Macroalgal vegetation of Maxwell Bay in King George Island, Antarctica. Korean J Phycol 9: 47–58

Clayton MN (1994) Evolution of the antarctic marine benthic algal flora (review). J Phycol 30: 897–904

Cormaci M (1996) The macrophytobenthos of Terra Nova Bay (Ross Sea, Antarctica): results of three Italian scientific expeditions. G Bot Ital 129: 1235–1243

Cormaci M, Furnari G, Scammacca B (1992a) The benthic algal flora of Terra Nova Bay (Ross Sea, Antarctica). Bot Mar 35: 541–552

Cormaci M, Furnari G, Scammacca B, Casazza G (1992b) Il fitobenthos di Baia Terra Nova (Mare di Ross, Antartide). In: Gallardo VA, Ferretti O, Moyano HI (eds) Oceanografia in Antartide. ENEA-PNRA-EULA, Concepción, Chile, pp 395–408

Cormaci M, Furnari G, Scammacca B (1993) Carta della vegetazione marina di Baia Terra Nova (Mare di Ross, Antartide). Biol Mar Suppl Notiz SIBM 1: 313–314

Cormaci M, Furnari G, Scammacca B, Alongi G (1996) Summer biomass of *Iridaea cordata* (Gigartinaceae, Rhodophyta) from Antarctica. In: Lindstrom SC, Chapman DJ (eds) Proceedings of 15th Int Seaweed Symp, Chile, Kluwer, Dordrecht, pp 267–272

Cormaci M, Furnari G, Scammacca B, Alongi G, Catra M (1998) Summer biomass of a population of *Pyllophora antarctica* (Phyllophoraceae, Rhodophyta) from Antarctica. Hydrobiologia 362: 85–91

Düwel L, Wegeberg S (1996) The typification and status of *Leptophytum* (Corallinaceae, Rhodophyta). Phycologia 35: 470–483

Fletcher RL (1987) Seaweeds of the British Isles, vol 3: Fucophyceae (Phaeophyceae), part 1. British Museum (Natural History), London, X + 359 pp

Foltran A, Maranzana G, Rascio N, Scarabel L, Talarico L, Andreoli C (1996) *Iridaea cordata* from Antarctica: an ultrastructural, cytochemical and pigment study. Bot Mar 39: 533–542

Foslie M (1905) Den botaniske samling. Det Kgl Norske Vidensk Selsk Skr Mus Aarsberet: 15–18

Foslie M (1907) Antarctic and subantarctic Corallinaceae. Botanik 4: 1–16

Fredericq S, Ramírez ME (1996) Systematic studies of the Antarctic species of the Phyllophoraceae (Gigartinales, Rhodophyta) based on *rbc*L sequence analysis. In: Lindstrom SC, Chapman DJ (eds) Proceedings of 15th Int Seaweed Symp, Chile 1995. Kluwer, Dordrecht, pp 137–143

Fredericq S, Hommersand HM, Freshwater DW (1996) The molecular systematics of some agar- and carrageenan-containing marine red algae based on *rbc*L sequence analysis. In: Lindstrom SC, Chapman DJ (eds) Proceedings of 15th Int Seaweed Symp, Chile 1995. Kluwer, Dordrecht, pp 125–135

Gepp A, Gepp ES (1905) Antarctic Algae. J Bot 43: 105–109

Gepp A, Gepp ES (1907) Marine Algae. I. Phaeophyceae and Florideae. Nat Hist 3: 1–15

Hommersand MH, Guiry MD, Fredericq S, Leister GL (1993) New perspectives in the taxonomy of the Gigartinaceae (Gigartinales, Rhodophyta). In: Chapman ARO, Brown MT, Lahaye M (eds) Proceedings of 14th Int Seaweed Symp, France 1992. Kluwer, Dordrecht, pp 105–120

Hommersand MH, Fredericq S, Freshwater DW (1994) Phylogenetic systematic and biogeography of the Gigartinaceae (Gigartinales, Rhodophyta) based on sequence analysis of *rbc*L. Bot Mar 37: 193–203

Kirst GO, Wiencke C (1995) Ecophysiology of polar algae (Review). J Phycol 31: 181–199

Kylin H, Skottsberg C (1919) Zur Kenntnis der subantarktischen und antarktischen Meeresalgen. II. Rhodophyceen. Wiss Ergeb Schwed Südpolar-Exped 1901–1903 4(5): 1–88

Lamb IM, Zimmermann MH (1977) Benthic marine algae of the Antarctic Peninsula. Antarct Res Ser Ashington 23: 129–229

Leister GL, Hommersand MH (1991) "*Iridaea*" from Pacific North America compared to *Iridaea cordata* from South America. J Phycol 27 Suppl:43 (Abstr 240)

Levring T (1945) Marine algae from some Antarctic and sub-Antarctic islands. Lunds Univ Arsskr N F Avd 2 47: 1–36

Lüning K (1990) Seaweeds: their environment, biogeography, and ecophysiology. Wiley, New York

Lüning K, Freshwater W (1988) Temperature tolerance of northeast Pacific marine algae. J Phycol 24: 310–315

Masuda M, Kogame K, Guiry MD (1996) Life history of *Gymnogongrus griffithsiae* (Phyllophoraceae, Gigartinales) from Ireland: implications for life history interpretation in the Rhodophyta. Phycologia 35: 421–434

Mendoza ML, Cabioch J (1985) Critique et comparison morphogénétique des genres *Clathromorphum* et *Antarcticophyllum* (Rhodophyta, Corallinaceae). Conséquences biogéographiques et systématiques. Cah Biol Mar 26: 251–266

Miller KA, Pearse JS (1991) Ecological studies of seaweeds in McMurdo Sound, Antarctica. Am Zool 31: 35–48

Moe RL (1985) *Gainia* and Gainiaceae, a new genus and family of crustose marine Rhodophyceae from Antarctica. Phycologia 24: 419–428

Moe RL, Silva PC (1979) Morphological and taxonomic studies on Antarctic Ceramiaceae (Rhodophyceae). I. *Antarcticothamnion polysporum* gen. et sp. nov. Br Phycol J 14: 385–405

Neushul M (1965) Diving observations of subtidal Antarctic marine vegetation. Bot Mar 8: 234–243

Papenfuss GF (1964) Catalogue and bibliography of Antarctic and sub-Antarctic benthic marine algae. Antarct Res Ser 1: 1–76

Pérès JM, Picard J (1964) Nouveau manuel de bionomie benthique de la Méditerranée. Recl Trav St Mar Endoume 31: 5–137

Picken GB (1985) Marine habitats – benthos. In: Bonner WN, Walton DWH (eds) Key environments Antarctica. Pergamon, Oxford

Polanshek AR, West JA (1988) Culture and hybridization studies on *Gigartina papillata* (Rhodophyta). J Phycol 13: 141–149

Ramírez ME, Peters AF (1992) The South American species of *Desmarestia* (Phaeophyceae). Can J Bot 70: 2430–2445

Ricker RW (1987) Taxonomy and biogeography of Macquarie Island seaweeds. British Museum (Natural History), London

Saunders GW, Kraft GT (1994) Small-subunit rRNA gene sequences from representatives of selected families of the Gigartinales and Rhodymeniales (Rhodophyta). I. Evidence for the Plocamiales ord. nov. Can J Bot 72: 1250–1263

Scagel RF, Gabrielson PW, Garbary DJ, Golden L, Hawkes W, Lindstrom SC, Oliveira JC, Widdowson TB (1989) A synopsis of the benthic marine algae of British Columbia, Southeast Alaska, Washington and Oregon. Phycological contribution 3. Department of Botany, University of British Columbia, Canada

Silva PC, DeCew TC (1992) *Ahnfeltiopsis*, a new genus in the Phyllophoraceae (Gigartinales, Rhodophyceae). Phycologia 31: 576–580

Silva PC, Basson PW, Moe RL (1996) Catalogue of the benthic marine algae of the Indian Ocean. Univ Calif Publ Bot 79: xiv + 1–1259

Skottsberg C (1953) On two collections of Antarctic marine algae. Ark Bot 2: 531–566

Weykam G, Thomas DN, Wiencke C (1997) Growth and photosynthesis of the Antarctic red algae *Palmaria decipiens* (Palmariales) and *Iridaea cordata* (Gigartinales) during and following extended periods of darkness. Phycologia 36: 395–405

Wiencke C (1996) Recent advances in the investigation of Antarctic macroalgae (Review article). Polar Biol 16: 231–240

Womersley HBS (1984) The marine benthic flora of southern Australia. Part I. Hand-book of the flora and fauna of South Australia. Government Printer, Adelaide

Womersley HBS (1994) The marine benthic flora of southern Australia. Rhodophyta, part IIIA: Bangiophyceae and Florideophyceae (Acrochaetiales, Nemaliales, Gelidiales, Hildenbrandiales and Gigartinales sensu latu). Australian Biological Resources Study, Canberra

Zaneveld JS (1966a) The occurrence of benthic marine algae under shore fast ice in the western Ross Sea, Antarctica. In: Young EG, McLachlan JL (eds) Proceedings of the Fifth International Seaweed Symposium, Canada 1965. Pergamon, Oxford, pp 217–231

Zaneveld JS (1966b) Vertical zonation of Antarctic and subantarctic marine algae. Antarct J US 1: 211–213

Zaneveld JS (1966c) *Lola irregularis* (Chlorophyta, Cladophoraceae): a new species from the Ross Sea. J Phycol 2: 45–47

Zaneveld JS (1968a) Sub-ice observations of Ross Sea benthic marine algae. Antarct J US 3: 127–128

Zaneveld JS (1968b) The horizontal and vertical distribution of the benthic marine algae along a transect from McMurdo Station, Ross Island, to the Balleny Islands. In Antarct Map Folio Ser Am Geogr Soc NY 10: 10–12

Zaneveld JS (1969) Factors controlling the delimitation of littoral benthic marine algal zonation. Am Zool 9: 367–391

Zaneveld JS, Sanford RB (1980) Crustose corallinaceous algae (Rhodophyta) of the New Zealand and United States scientific expedition to the Ross Sea, Balleny Islands, and Macquarie Ridge, 1965. Blumea 26: 205–231

Zupan RJ, West JA (1988) Geographic variation in the life history of *Mastocarpus papillatus* (Rhodophyta). J Phycol 24: 223–229

CHAPTER 36

# Spatial and Vertical Distribution of Benthic Littoral Communities in Terra Nova Bay

R. Cattaneo-Vietti[1], M. Chiantore[1], M.C. Gambi[2], G. Albertelli[1], M. Cormaci[3], and I. Di Geronimo[4]

**ABSTRACT**

The littoral benthic communities of the Ross Sea, except for McMurdo Sound, are poorly known. An analysis of the composition and distribution of the communities in Terra Nova Bay has been conducted down to the 150-m depth between Tethys Bay and Adelie Cove.

These communities show a clear zonation according to depth and on the basis of some guide species. On hard bottoms, the upper zone down to the 2–3 m depths, due to the presence of drifting pack ice, shows a very poor community, mainly composed by cyanobacteria and diatoms and the amphipod *Paramoera walkeri*. From 2–4 m to 70 m, the algal communities constitute well-defined belts. These algal belts represent shelter and a food source for diversified and abundant vagile fauna, among which the echinoid *Sterechinus neumayeri* and the asteroid *Odontaster validus*. From 70–120 m, a complex community of sponges and anthozoans characterizes the rocks, reaching high values in biomass and species richness.

Soft bottoms generally start at the 20–30 m depth and are constituted by coarse sands and gravel where *Alcyonium* sp. is often found. Herein the communities are characterized by bivalves (*Laternula elliptica, Yoldia eightsi, Adamussium colbecki*) and polychaetes (*Tharyx cincinnatus, Aglaophamus ornatus, Spiophanes tcherniai, Leitoscoloplos mawsoni*). Much deeper, on muddy sands at about the 150-m depth, the polychaete *Laonice weddellia* characterizes the community, and brachiopods can be found on small gravel.

## Introduction

Antarctic benthic communities have been described as one of the most stable marine ecosystems in the world and the distribution of the most characteristic species is considered essentially circumpolar (Hedgpeth 1971; Knox and Lowry 1977; Richardson and Hedgpeth 1977; Voss 1988). Our present knowledge on the structure and composition of the littoral communities is so limited that the existence of comparable communities around the continent is still doubtful (White 1984). Recent data are available mainly from the Antarctic Peninsula (Klöser et al. 1994; Barnes 1995a,b). Arntz et al. (1994) made a review regarding knowledge of benthic communities, with particular reference to the Weddell Sea.

Studies on the benthic communities of the Ross Sea started with the works of Zaneveld (1966a,b), Bullivant (1967a,b) and Dearborn (1967). Further investigations were carried out by several authors (Zaneveld 1968; Dayton et al. 1970, 1974; Lowry 1976; Dayton and Oliver 1977a,b; Oliver and Slattery 1985; Battershill 1989; Dayton 1990; Miller and Pearse 1991), always around McMurdo Sound. Arnaud (1974) has detailed the littoral communities of Adelie Land westwards of the Ross Sea.

This bulk of information is too geographically limited to give a clear picture of the composition and distribution of the littoral benthic communities in the Ross Sea. Also the scarcity of ice-free zones along the coast is limiting in comparative studies. The aim of this chapter is to present a first general account on the zonation and the composition of the main benthic communities present both on hard and soft bottoms at Terra Nova Bay, between Tethys Bay and Adelie Cove, on the upper

[1] Istituto di Scienze Ambientali Marine, Università di Genova, Corso Rainusso 14, 16038 Santa Margherita Ligure (Genoa), Italy
[2] Lab. Ecologia del Benthos – Stazione Zoologica di Napoli, Punta San Pietro Ischia-Porto, Naples, Italy
[3] Dipartimento di Botanica, Università di Catania, Via A. Longo 19, 95125 Catania, Italy
[4] Istituto Policattedra di Oceanologia e Paleoecologia, Università di Catania, Corso Italia 55, 95129 Catania, Italy

continental shelf, up to the 150-m depth, an area recently proposed as a possible Antarctic Specially Protected Area (ASPA).

## Material and Methods

In the framework of the Italian Antarctic Research Programme (PNRA), studies on benthos have been conducted at Terra Nova Bay during several austral summers (1987–88, 1989–90, 1993–94; Fig. 1). Sampling on shallow hard bottoms (up to 30–35 m depth) was made by SCUBA diving (Cormaci et al. 1992a–c, 1996; Gambi et al. 1994). In deeper soft and hard bottoms, quantitative and qualitative samples were collected using a Van Veen grab (65 l) or naturalist's dredges from 16 to 200 m (Di Geronimo et al. 1992; Cattaneo-Vietti and Gambi 1995). General descriptions or detailed illustrations of some particular communities were made also using two types of ROVs: Pluto 21 (Di Geronimo and Rosso 1991) and Roby 2 (Bono et al. 1994).

## Distribution and Zonation

The shoreline at Terra Nova Bay is a granite cliff which can reach a depth of 70–80 m. Small pebble beaches are present in this area, at Road Bay, close to the Italian base, and at Adelie Cove (Simeoni et al. 1989), where an Adelie penguin (*Pygoscelis adeliae*) rookery occurs.

Generally the littoral sediments (Table 1) have their upper limit around 20 m depth and include poorly sorted coarse sands, interspersed with boulders of various sizes, and gravel. Deeper, between 50 and 150 m, the sand is finer and its percentage increases with depth (Brambati et al. 1989). Sparse rocks and shoals are common down to 100–120 m. Locally, muddy sands may become dominant, due to dynamic conditions, as occurs in Adelie Cove, a sheltered area where fine sediments rich in organic matter, due to the rookery inputs, are found.

### The *Littoral Fringe*

(= Infralittoral Zone, according to Genoa System Zonation, in Lüning 1990)

At Terra Nova Bay, this belt is generally azoic due to the severe physical conditions, but ephemeral communities develop inside small pools present along the rocky coast, mainly fed by fusion waters from the land. During summer, these micro-habitats are characterized by cyanobacteria and pennate diatoms (Fig. 2), whose growth is sus-

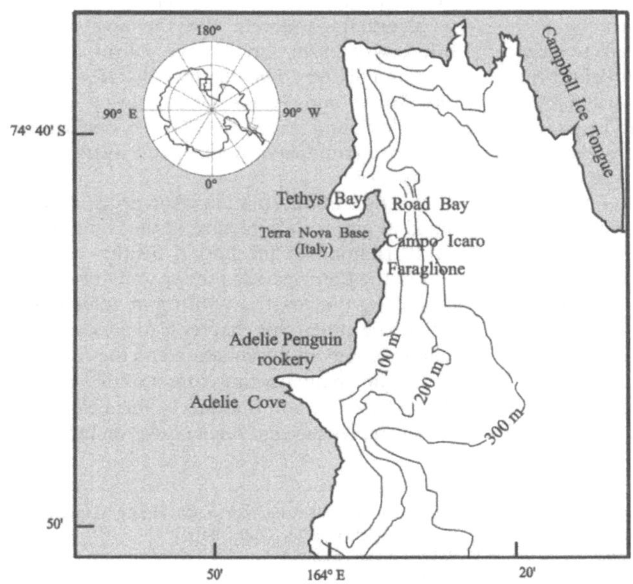


**Fig. 1.** Map of Terra Nova Bay showing study sites

**Table 1.** Granulometric features of the sediments (%) along the coast of Terra Nova Bay

| Locality | Lat S | Long E | Depth (m) | Chl a ($\mu g g^{-1}$) | Very coarse sand and gravel $\phi > 1$ | Coarse-medium sand $1 > \phi > 0.212$ | Fine-very fine sand $0.063 < \phi < 0.212$ | Silt and clay $\phi < 0.063$ |
|---|---|---|---|---|---|---|---|---|
| Road Bay | 74°41.98′ | 164°07.72′ | 55 | 2.24 | 73.0 | 21.3 | 5.6 | 0.1 |
| Faraglione | 74°42.54′ | 164°07.35′ | 43 | – | 3.5 | 40.4 | 55.7 | 0.4 |
| Faraglione | 74°43.03′ | 164°07.45′ | 50 | 5.53 | 8.5 | 37.2 | 54.2 | 0.1 |
| Faraglione | 74°43.00′ | 164°07.50′ | 50 | 9.85 | 14.1 | 41.5 | 44.3 | 0.1 |
| Faraglione | 74°42.24′ | 164°07.79′ | 52 | 3.80 | 14.7 | 56.3 | 28.8 | 0.2 |
| Faraglione | 74°43.07′ | 164°07.37′ | 50 | 6.02 | 0.9 | 30.1 | 68.7 | 0.3 |
| Faraglione | 74°42.58′ | 164°09.14′ | 223 | 9.66 | 1.7 | 22.1 | 74.3 | 1.9 |
| Campo Icaro | 74°42.70′ | 164°08.20′ | 132 | – | 1.2 | 21.0 | 76.9 | 0.9 |
| Campo Icaro | 74°42.63′ | 164°08.24′ | 155 | – | 2.3 | 24.5 | 72.5 | 0.7 |
| Adelie Cove | 74°46.32′ | 163°57.90′ | 40 | 14.13 | 5.7 | 53.7 | 40.1 | 0.5 |
| Adelie Cove | 74°46.42′ | 163°59.06′ | 37 | 20.07 | 0.6 | 34.6 | 63.7 | 1.1 |
| Adelie Cove | 74°46.33′ | 163°58.03′ | 67 | 22.83 | 0.5 | 48.2 | 50.2 | 1.1 |
| Adelie Cove | 74°47.05′ | 164°02.67′ | 146 | 7.81 | 0.3 | 14.8 | 83.5 | 1.4 |
| Adelie Cove | 74°46.17′ | 164°10.78′ | 219 | 3.58 | 0.0 | 36.0 | 63.1 | 0.9 |

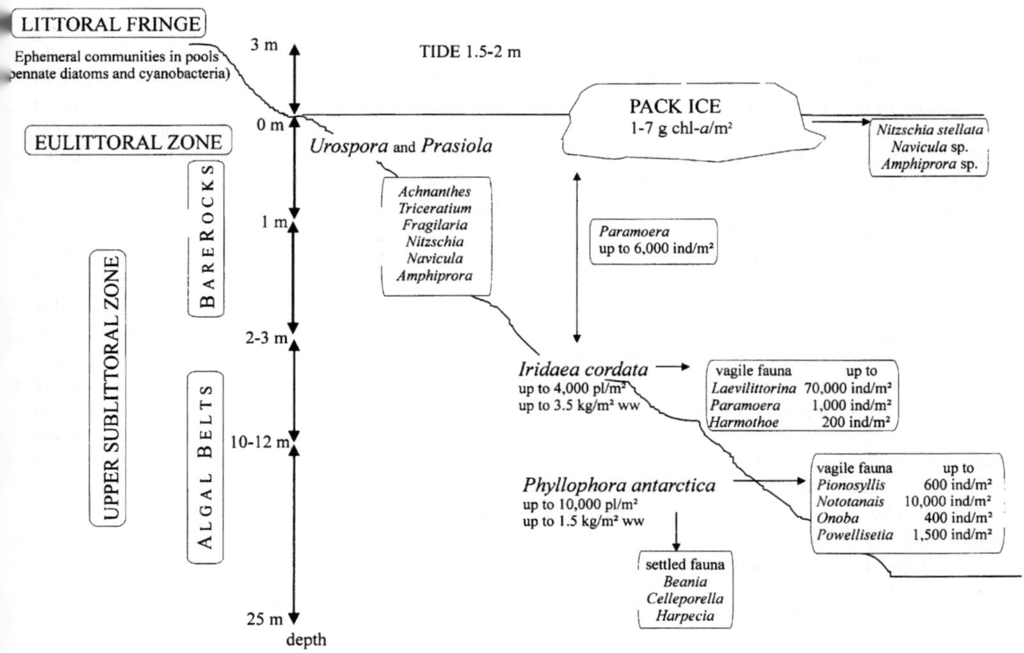


**Fig. 2.** Scheme of the main upper hard-bottom communities living at Terra Nova Bay, up to 25 m depth. Biomass and density values of the key species are reported

tained by the guano of penguins (*Pygoscelis adeliae*) and skuas (*Catharacta maccormicki*) washed straight into the pools.

### The *Eulittoral Zone*

(= Midlittoral Zone, according to Genoa System Zonation, in Lüning 1990)

The tidal range at Terra Nova Bay is 1.5–2 m (Stocchino and Lusetti 1988). In the eulittoral or intertidal zone, ice is the major factor influencing benthic colonization and only cyanobacteria and diatoms, such as *Achnanthes* spp., *Triceratium* sp., *Fragilaria* spp., *Nitzschia* spp., *Navicula* spp. (Gambi and Mazzella 1992) were found (Fig. 2). In the most sheltered areas, avoiding the abrasive action of ice, the green algae *Prasiola crispa* and *Urospora penicilliformis* can be present, the latter showing high coverage values (Cormaci et al. 1992a,b) at the end of summer.

### The *Upper Sublittoral Zone*

(= Infralittoral Zone, according to Genoa System Zonation, in Lüning 1990)

In this zone, in the first 2–3-m depths, the benthic community remains very poor, again due to the action of fast ice, ice foot and drifting pack ice. It is mainly colonized by epilithic diatoms and by the amphipod *Paramoera walkeri*, which can be considered the key species of this zone. In fact, the species reaches its highest denisty values (6000 ind $m^{-2}$) and biomass in this zone (Gambi et al. 1994). This species is also common beneath the pack ice, grazing on the dense sympagic diatom community which seems to be very differentiated. In mid-January, the bottom ice is dominated by the long chain *Amphiprora* cf. *kufferathii*, while the interior ice microalgae essentially belong to *Nitzschia stellata*. The surface community is exclusively dominated by *Navicula* sp. These communities can reach high biomass values (about 1–7 g Chl a $m^{-3}$) (Lazzara et al. 1995).

Below this bare substrate (Fig. 2), the benthic macroflora is represented by two dominant rhodophytes, *Iridaea cordata* and *Phyllophora antarctica*. The bottom up to the 10–12-m depth is colonized mainly by *Iridaea cordata*, sometimes together with the green alga *Monostroma hariotii*, while *Geminocarpus geminatus* is frequently found as an epiphyte.

*I. cordata* reaches its highest density (4336 plants $m^{-2}$) in mid-January. The presence of large thalli at the end of December and high biomass values (about 3.5 kg $m^{-2}$ wet wt) at the beginning of summer suggest that *I. cordata* grows, although at a low rate, also during the Antarctic winter (Cormaci et al. 1996). Its associated vagile fauna is qualitatively poor but quantitatively rich, mainly represented by the polychaete *Harmothoe spinosa* (200 ind $m^{-2}$), the gastropod *Laevilitorina antarctica* (up to 70 000 ind $m^{-2}$), the amphipod *Paramoera walkeri* (up to 1000 ind $m^{-2}$) and isopods belonging to the genus *Munna*.

Deeper, from 10–12 to 25 m, *Phyllophora antarctica* replaces the former species and constitutes dense and expanded beds. *P. antarctica* reaches its highest density (9760 plants $m^{-2}$) in mid-January, and its highest biomass value (about 1.5 kg $m^{-2}$ wet wt) at the end of January. *P. antarctica* is often fully colonized by sessile organisms, mainly represented by hydroids on which the nudibranch *Cuthona georgiana* feeds (Cattaneo-Vietti 1991), and serpulids and bryozoans. Among these latter, *Beania livingstonei*, *Celleporella antarctica* and *Harpecia spinosissima* are frequent and abundant (Rosso 1991). The associated vagile fauna, despite the highest species richness recorded, shows lower density and biomass values in comparison to the upper *Iridaea* belt, due to the dominance of small-sized taxa, such as the polychaete *Pionosyllis* cf. *comosa* (600 ind $m^{-2}$), the tanaid *Nototanais dimorphus* (10 000 ind $m^{-2}$), the gastropods *Onoba gelida* (400 ind $m^{-2}$) and *Powellisetia deserta* (1500 ind $m^{-2}$) and the bivalve *Philobrya sublaevis*, according to Gambi et al. (1994). Another rhodophyte, *Plocamium cartilagineum*, is frequently found in this zone (Cormaci et al. 1992a,b) where the asteroid *Odontaster validus* and the echinoid *Sterechinus neumayeri* are very common. Large sessile animals, such as the soft coral *Alcyonium* sp. and the anemone *Urticinopsis antarcticus*, are only occasionally observed.

### The *Lower Sublittoral Zone*

(= Circalittoral Zone, according to Genoa System Zonation, in Lüning 1990)

#### Hard Bottoms

The lower sublittoral zone (about 25–150 m depth) is characterized by a great heterogeneity of substrates favouring the occurrence of both hard- and

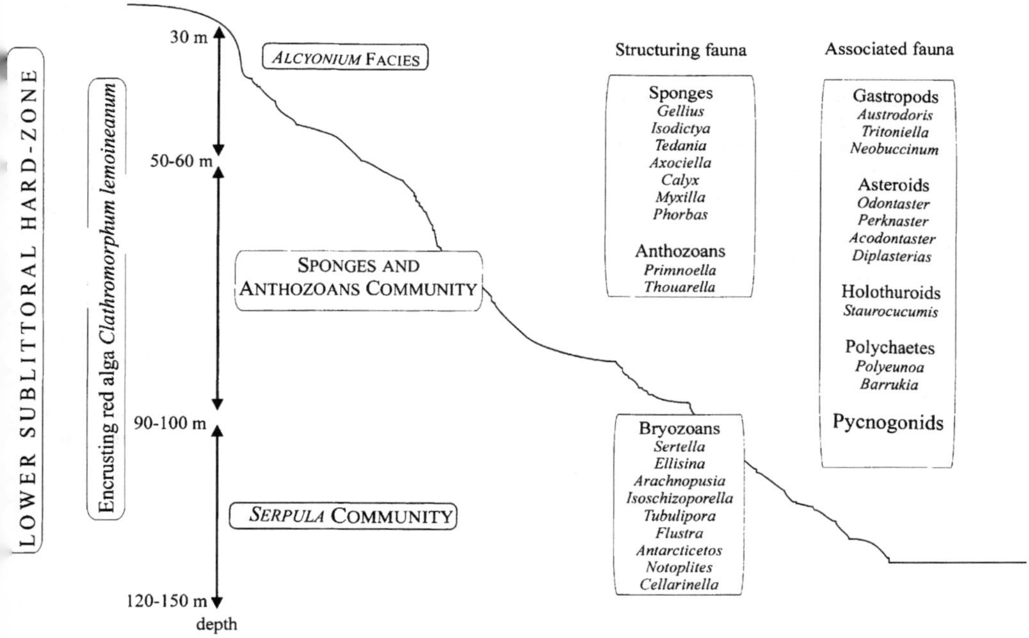


**Fig. 3.** Scheme of the main lower sublittoral hard-bottom communities living at Terra Nova Bay, from 30 to 120–150 m depth

soft-bottom communities. On rocky substrates, the upright algal colonization is replaced by the calcareous crustose coralline alga, *Clathromorphum lemoineanum* (Fig. 3). Zoobenthos is dominant with diversified filter-feeder communities, characterized in the upper zone by byssed *Adamussium* and, below, by sponges, soft corals (*Alcyonium* sp.) gorgonians (*Thouarella* and *Primnoella* species) and holoturoids. Among the sponges, which can reach very high biomass values (over 2–3 kg m$^{-2}$ wet wt), the commonest ones are *Tedania charcoti, Axociella nidificata, Calyx arcuarius, Isodictya erinacea, I. cactoides, I. conulosa, Gellius rudis, Gellius* sp., *Myxilla elongata* and *Phorbas glaberrima.* The hexactinellids (*Rossella racovitzae, R. nuda* and *Scolymastia joubini*), so common at McMurdo Sound, are uncommon in this bathymetric belt. The white holothuroid, *Staurocucumis liouvillei,* shows a strong acrophilic habitus and is frequently found on top of gorgonians and sponges.

Other invertebrates, such as serpulid and terebellid polychaetes, sipunculids, isopods, amphipods, bryozoans, bivalves, ophiuroids, holothuroids (e.g. *Psolidium* sp.) and ascidians (e.g. *Cnemidocarpa verrucosa*), are important components of this community, which develops mainly around the 100–120 m depths.

Numerous predators, such as the polychaete *Barrukia cristata*, the asteroids *Odontaster validus, Perknaster fuscus antarcticus, Acodontaster conspicuus, Diplasterias brucei,* the pycnogonids and the nudibranchs *Austrodoris kerguelenensis* and *Tritoniella belli* (see Cattaneo-Vietti 1991), are frequent, while the polynoid polychaete *Polyeunoa leavis* is often hosted by *Thouarella* and *Primnoella* gorgonians.

Below the 120–130 m depths, the hard-bottoms become very sparse and are mainly colonized by the polychaete *Serpula narconensis* and several bryozoans: the encrusting *Ellisina antarctica* and *Arachnopusia decipiens*, the erected *Reteporella* sp., *Isoschizoporella similis, Antarcticetos bubeccata, Notoplites* spp. and *Cellarinella* spp. and the rooted *Flustra angusta* and *F. vulgaris* (Rosso 1992), suggesting the great heterogeneity of the microenvironment.

## Soft Bottoms

Generally, at Terra Nova Bay, soft sediments have their upper limit at around the 20-m depth (Fig. 4). In the upper 20–40 m, the sediments are characterized by gravel and coarse sand on which, occasionally, some thalli of *I. cordata* and *P. antarctica* can be found, floating on the bottom. Here, the infaunal community is characterized by the large bivalve *Laternula elliptica*, the smaller *Limopsis* sp. and by the large-sized carnivore nephtyid polychaete *Aglaophamus ornatus* which can reach high abundance values (300 ind $m^{-2}$) (Gambi et al. 1997). The echinoid *Sterechinus neumayeri*, a grazer, and the carnivorous starfish *Odontaster validus* are very common.

Between 30–70 m, the substrate becomes finer and shows dense populations of benthic diatoms with biomass values around 3–8 µg Chl a $g^{-1}$. In sedimentation areas, such as Adelie Cove, strongly influenced by a penguin rookery, values can reach 20 µg Chl a $g^{-1}$.

The soft bottoms between 20–30 and 60–70 m depths are sometimes completely covered by the large scallop *Adamussium colbecki*, which locally constitutes a key species, reaching high biomass (100–120 g $m^{-2}$ dry wt soft tissues) and density (60–80 ind $m^{-2}$) values (Cattaneo-Vietti et al. 1997). The number of empty shells covering the bottom is always high, mainly below the 70-m depth. A typical microcommunity, composed by foraminiferans (Mullineaux and De Laca 1984; Berkman 1994), bryozoans (*Ellisina antarctica*, *Micropora brevissima*, *Arachnopusia decipiens* and *Aimulosia antarctica*) and by the spirorbid *Paralaeospira levinseni*, lives on the lower valve of *Adamussium*. In this community, volcano sponges (*Rossella* spp. and *Scolymastia joubini*), up to 1.2 m tall, live together with large predators such as the gastropod *Neobuccinum eatoni* and the nemertine *Parborlasia corrugatus*.

Generally, the very fine sediments below 120 m depth are characterized by tube-building

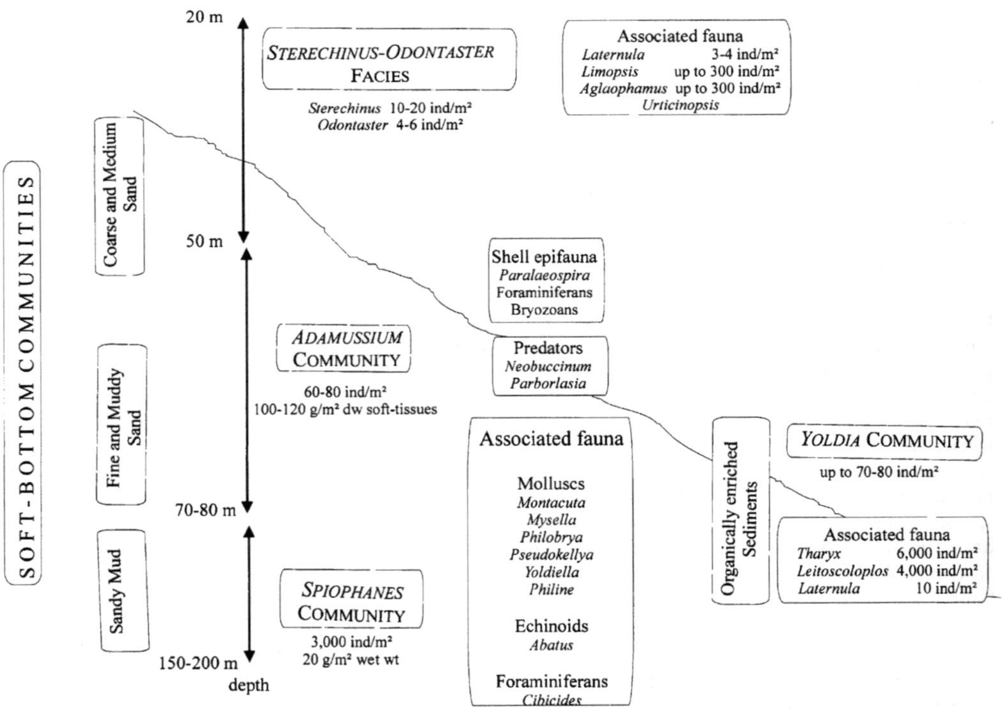


**Fig. 4.** Scheme of the soft-bottom communities living at Terra Nova Bay, from 20 to 150 m depth. Biomass and density values of the key species are reported

polychaetes, mainly *Spiophanes tcherniai*, which can reach abundance values of 3000 ind $m^{-2}$ and exceed 20 g $m^{-2}$ wet wt (Gambi et al. 1997). In this community, the opisthobranch *Philine alata* feeds on foraminiferans, mainly *Cibicides lobatulus* (see Violanti 1990) and, probably, on *Spiophanes*. Here, numerous small bivalves, such as *"Montacuta" nimrodiana*, *Yoldiella antarctica*, *Philobrya sublaevis*, *Pseudokellya cardiformis* and *Mysella gibbosa* (see Albertelli et al. 1994; Russo and Gambi 1994), are frequent. Upon the sediments rich in organic matter, the spatangoid *Abatus* sp. is also common, reaching 2–4 ind $m^{-2}$. In this belt, large amounts of hexactinellid and tetractinellid sponge spicules can be found free in the sediments or densely packed in spicule mats as described in other geographic areas by Dearbon (1967), Koltun (1968), Arnaud (1974) and Barthel (1992). Siliceous spicules can represent up to 50% of the total dry weight of sediment, reaching density values up to 500 000 spicules $g^{-1}$. A rich and peculiar community, characterized by crustaceans, polychaetes and the bivalve *Limatula hodgsoni* can live in this mat. These mats underline the important role of sponges, in addition to that played by diatoms, to determine the sediment texture and silica content of sediments. Much deeper, at about 150 m, the most abundant polychaete is *Laonice weddellia* (up to 800 ind $m^{-2}$) (Gambi et al. 1997), while on small pebbles, brachiopods can be occasionally found.

Muddy sediments can be found also in shallower bottoms, showing a higher organic content, as at Adelie Cove. In such areas, *Adamussium colbecki*, generally common at these depths, is less abundant, while the deposit feeder *Yoldia eightsi* reaches density values of 70–80 ind $m^{-2}$ together with the large bivalve *Laternula elliptica*. Among polychaetes, the limivore and surface deposit feeders *Leitoscoloplos mawsoni* and *Tharyx cincinnatus* can reach abundance values up to 4000–6000 ind $m^{-2}$, respectively, exceeding 60 g $m^{-2}$ wet wt (Gambi et al. 1997).

Finally, it is worth mentioning that notothenioid fishes, especially the *Trematomus* group, characterize the benthic community of this zone (Vacchi et al. 1991). *T. bernacchi* and *T. pennelli* (see Vacchi et al. 1994), *T. hansoni* and *T. loennbergii* (see La Mesa et al. 1997) exert an important role in the benthic food web, as consumers of many invertebrate species, mainly bivalves, crustaceans and polychaetes, as demonstrated by gut content analysis (Vacchi et al. 1994).

## Conclusions

In the Ross Sea, littoral benthic communities, on both hard and soft bottoms, show generally high specific richness and density (Dell 1972; Oliver and Slattery 1985; Battershill 1989) which generally increase with depth as they are influenced in shallow waters by the physical impact of ice and iceberg scour (Dearbon 1968; Dell 1972; White 1984; Clarke 1990; Arnaud 1992; Arntz et al. 1994).

In summer at Terra Nova Bay, the percent coverage of the hard-bottoms is always very close to 100%, and the role of ice in structuring and disturbing benthic communities is limited to very shallow waters (2–10 m depth). Anchor ice formation, so remarkable at McMurdo (Dayton et al. 1969, 1970; Dayton 1989), has never been observed.

Algal communities are quite homogeneous on a large spatial scale and well structured in zones, characterized by *Iridaea cordata* and *Phyllophora antarctica* in the upper sublittoral and by *Clathromorphum lemoineanum* in the lower sublittoral. The first two zones, particularly those characterized by *Phyllophora antarctica*, represent shelter and a food source for numerous invertebrates, showing relatively low values of species richness and frequently high values of abundance and biomass. The higher species richness measured in the *Phyllophora* belt is probably related to the less selective environment in which this alga develops best, the higher level of shelter that it exerts, and the higher number of microenvironments that it provides for benthic animals (Gambi and Mazzella 1992).

The distribution and composition of the algal communities at Terra Nova Bay show affinities and significant differences in comparison to other areas described in the Ross Sea. At Terra Nova Bay, the large brown algae *Himantothallus* and *Desmarestia*, present at Cape Adare (Zaneveld 1966b), are lacking and the algal belts are shifted towards the surface, particularly the sciaphilic *Phyllophora antarctica*, which in Adelie Land generally grows on deeper bottoms or underneath large brown algae (Arnaud 1974). At McMurdo Sound (Miller and Pearse 1991), the abundance of these algae is less remarkable and their bathymetric range is narrower; the encrusting alga *Clathromorphum lemoineanum* seems to be substituted by *Leptophyllum*.

All this suggests a phytogeographical gradient from north to south in the Ross Sea, from Cape

Adare to McMurdo Sound, with emerging algal belts ascending northwards, probably due to the shortening of the photoperiod and the longer persistence of the pack ice causing a decrease of light intensity at depth.

Below 30–40 m, soft and hard bottoms coexist, being affected by both physical and biological factors: continuous input of pebbles, stones and terrestrial detritus falling down from floating ice, irradiance, topography of the bottom, local water movements and grazing (Gambi et al. 1994), presence of spicule mats and organic matter flux (Albertelli et al. 1996, 1998). All this contributes to create a mosaic of environmental conditions, which induces a high degree of heterogeneity, well described and characterized by different communities (Table 2), and by some guide species, such as *Adamussium colbecki*, several sponges and *Spiophanes*. At the 70–120 m depths, the sponge and anthozoan community is certainly the most complex community described in the Ross Sea, with high values of biodiversity and biomass and strong biotic interactions (predation, epi- and endosymbiosis, competition).

Zoobenthic communities in Terra Nova Bay also show similarities and differences in compari-

**Table 2.** Key coastal benthic species at Terra Nova Bay (up to 200–300 m depth)

| | Adult size | Substrate | Depth range (m) | Average abundance | Feeding strategy | Community |
|---|---|---|---|---|---|---|
| Macroalgae: | | | | | | |
| Rhodophyta | | | | | | |
| *Iridaea cordata* | 20–25 cm | Rock and pebbles | 3–16 | Up to 4000 thalli $m^{-2}$ | Autotrophic | Upper Sublittoral Zone |
| *Phyllophora antarctica* | 10–15 cm | Rock and pebbles | 5–35 | Up to 10 000 thalli $m^{-2}$ | Autotrophic | Upper Sublittoral Zone |
| *Clathromorphum lemoineanum* | – | Rock | 10–100 | Encrusting | Autotrophic | Lower Sublittoral Zone |
| Porifera | | | | | | |
| *Gellius rudis* | Massive (20 cm) | Rock | 50–140 | Very abundant | Filter feeder | Hard-bottom filter feeders |
| *Tedania charcoti* | Globular lobate (40 cm) | Rock | 80–200 | Very abundant | Filter feeder | Hard-bottom filter feeders |
| *Axociella nidificata* | Variable in shape (15 cm) | Rock | 70–300 | Frequent | Filter feeder | Hard-bottom filter feeders |
| *Calyx arcuarius* | Fan-like (40 cm) | Rock | 100–400 | Abundant | Filter feeder | Hard-bottom filter feeders |
| *Isodictya erinacea* | Erect cylindrical hispid (40 cm) | Rock | 80–200 | Very abundant | Filter feeder | Hard-bottom filter feeders |
| *Inflatella belli* | Spherical (15 cm) | Rock | 70–300 | Abundant | Filter feeder | Hard-bottom filter feeders |
| Cnidaria | | | | | | |
| *Thouarella* sp. | 400–500 mm | Rock | 80–120 | Common (2–3 ind $m^{-2}$) | Filter feeder | Hard-bottom filter feeders |
| *Primnoella* sp. | 700–1000 mm | Pebbles | 40–80 | Very abundant (4–6 ind $m^{-2}$) | Filter feeder | Hard-bottom filter feeders |
| Nemertina | | | | | | |
| *Parborlasia corrugatus* | 1000–2000 mm | Sand | 40–100 | Occasional | Carnivorous | *Adamussium* community |
| Polychaeta | | | | | | |
| *Spiophanes tcherniai* | 20–30 mm | Muddy sand | 100–150 | Up to 2500 ind $m^{-2}$ | Deposit feeder | Polychaete deposit-feeders |
| *Leitoscoloplos mawsoni* | 20–30 mm | Muddy sand | 50–150 | Up to 4000 ind $m^{-2}$ | Deposit feeder | Polychaete deposit-feeders |
| *Aglaophamus ornatus* | 50–100 mm | Coarse sand to mud | 20–150 | Up to 300 ind $m^{-2}$ | Infaunal carn. | Lower Sublittoral Zone |
| *Serpula narconensis* | 80–120 mm | Sparse rocks-muddy sand | 120–150 | Common | Filter feeder | *Serpula* community |

**Table 2.** *Continued*

| | Adult size | Substrate | Depth range (m) | Average abundance | Feeding strategy | Community |
|---|---|---|---|---|---|---|
| Mollusca | | | | | | |
| *Neobuccinum eatoni* | 90 mm | Sand | 20–50 | Occasional | Carnivorous | *Adamussium* community |
| *Laevilitorina antarctica* | 2–3 mm | Macroalgae | 3–12 | Up to 70 000 ind $m^{-2}$ | Mesograzer | Upper Sublittoral Zone |
| *Adamussium colbecki* | 60–80 mm | Mainly fairly fine | 10–80 | 40–60 ind $m^{-2}$ | Filter feeder | *Adamussium* community |
| *Laternula elliptica* | 8–12 cm | Muddy sand | 10–80 | <20 ind $m^{-2}$ | Filter feeder | Coarse sands |
| *Yoldia eightsi* | 20–30 mm | Muddy sand | 10–80 | Up to 70–80 ind $m^{-2}$ | Deposit feeder | Muddy fine sands |
| Bryozoa | | | | | | |
| *Reteporella* sp. | 100 mm | Sparse rocks | 50–150 | Frequent | Filter feeder | *Serpula* community |
| Crustacea | | | | | | |
| *Paramoera walkeri* | 5 mm | Rock and ice | 0–5 | Up to 6000 $m^{-2}$ | Micrograzer | Upper Sublittoral Zone |
| Echinodermata | | | | | | |
| *Sterechinus neumayeri* | 50 mm | Rock and sand | 5–100 | Very abundant (10–20 ind $m^{-2}$) | Grazer | Ubiquitous |
| *Abatus* sp. | 50–60 mm | Muddy sand | 40–120 | Frequent (2–4 ind $m^{-2}$) | Detritus feeder | *Spiophanes* community |
| *Odontaster validus* | 70–90 mm | Rock and sand | 5–50 | Frequent (4–8 ind $m^{-2}$) | Carnivorous | Ubiquitous |
| *Perknaster fuscus* | 300–500 mm | Sand | 80–120 | Occasional | Spongivorous | |
| *Staurocucumis liouvillei* | 40–60 mm | Acrophilic | 70–120 | Very abundant | Filter feeder | Hard-bottom filter feeders |

son with other Antarctic areas. For example, the tubiculous *Spiophanes* and the large bivalves *Laternula* and *Adamussium* constitute similar communities in other localities (White and Robins 1972; Arnaud 1974; Stockton 1984; Oliver and Slattery 1985; Berkman 1990; Ahn 1993), but at Terra Nova Bay this latter locally reaches very high density values (Cattaneo-Vietti et al. 1997), playing a relevant role in the organic carbon cycle (Chiantore et al. 1998).

The sponge community living at Terra Nova Bay, even if quite similar to the McMurdo Sound Mixed and Glass Sponge Communities described by Bullivant (1967b) between the 70- and 180-m depths and to that starting below the 30–35-m depths of zone III described by Dayton et al. (1974) at McMurdo, seems to have a different structure. The main differences are observed in the abundance of single species of the hexactinellids, which are rarer at these depths at Terra Nova Bay.

The qualitative description of some of the above communities needs further improvement, since not all the benthic taxocoenoses have been taxonomically analyzed at the species level. This fact, coupled with the high degree of spatial heterogeneity and patchiness of the various communities, makes it difficult to provide a thematic map of the benthic communities at Terra Nova Bay. An improvement will result from a more intensive use of SCUBA diving techniques and ROVs in the future.

The establishment of a permanent base at Terra Nova Bay also makes it important to protect this area and to provide a control site to monitor existing impacts. Long-termtransects have already been established (Gambi et al. 1994) to monitor natural changes in coastal benthic communities and to evaluate the role of permanent penguin colonies on adjacent marine environments. In conclusion, this area, because of its vicinity to the summer base, should be considered an Antarctic Specially Protected Area (ASPA) to minimize environmental impacts.

*Acknowledgements.* This research was funded by the Italian Programma Nazionale di Ricerca in Antartide (PNRA) within the research project Ecology and Biogeochemistry of the Southern Ocean.

## References

Ahn I-Y (1993) Enhanced particle flux through the biodeposition by the Antarctic suspension-feeding bivalve *Laternula elliptica* in Marian Cove, King George Island. J Exp Mar Biol Ecol 171: 75–90

Albertelli G, Arnaud PM, Cattaneo-Vietti R (1994) Ecological aspects of the macrobenthos of the Ross Sea: the coastal and deep-sea bivalve molluscs of Terra Nova Bay. In: National Scientific Commission for Antarctica (ed) Oceanographic Campaign 1987–88, 1989–90, Data Rep Part III, Genova, pp 49–66

Albertelli G, Cattaneo-Vietti R, Chiantore M (1996) The influence of the organic matter flux on an *Adamussium colbecki* bed at Terra Nova Bay (Ross Sea, Antarctica). In: Albertelli G, Demaio A, Piccazzo M (eds) Atti XI Congr Ass Ital Oceanogr Limnol, Lang, Genova, pp 299–310

Albertelli G, Cattaneo-Vietti R, Chiantore M, Pusceddu A, Fabiano M (1998) Food availability to an *Adamussium* bed during the austral summer 1993/94 (Terra Nova Bay, Ross Sea). J Mar Syst 17: 425–434

Arnaud PM (1974) Contribution à la bionomie marine benthique des régions antarctiques et subantarctiques. Téthys 6(3): 465–656

Arnaud PM (1992) The state of the art in Antarctic benthic research. In: Gallardo VA, Ferretti O, Moyano HI (eds) Oceanografia in Antartide. ENEA-PNRA-EULA, Concepción, Chile, pp 341–346

Arntz WE, Brey T, Gallardo VA (1994) Antarctic zoobenthos. Oceanogr Mar Biol Annu Rev 32: 241–304

Barnes DKA (1995a) Sublittoral epifaunal communities at Signy Island, Antarctica. I. The ice-foot zone. Mar Biol 121: 555–563

Barnes DKA (1995b) Sublittoral epifaunal communities at Signy Island, Antarctica. II. Below the ice-foot zone. Mar Biol 121: 565–572

Barthel D (1992) Do hexactinellids structure Antarctic sponge associations? Ophelia 36(2): 111–118

Battershill CN (1989) Distribution and abundance of benthic marine species at Cape Armitage, Ross Island, Antarctica. Initial results. N Z Antarct Rec 9(2): 35–52

Berkman PA (1990) The population biology of the Antarctic scallop, *Adamussium colbecki* (Smith 1902) at New Harbor, Ross Sea. In: Kerry KR, Hempel K (eds) Antarctic ecosystems, ecological change and conservation. Springer, Berlin Heidelberg New York, pp 281–288

Berkman PA (1994) Epizoic zonation on growing scallop shells in McMurdo Sound, Antarctica. J Exp Mar Biol Ecol 179: 49–67

Bono R, Bruzzone G, Caccia M, Grassia F, Spirandelli E, Veruggio G (1994) Roby goes to Antarctica. Oceans 94-OSATES: Ocean engineering for today's technology and tomorrow's preservation, Brest, pp 621–625

Brambati A, Fanzutti GP, Finocchiaro F, Simeoni U (1989) Sediments and sedimentological processes in the Ross Sea continental shelf (Antarctica): results and preliminary conclusions. Boll Ocean Teor Appl 7: 159–188

Bullivant JS (1967a) New Zealand Oceanographic Institute Ross Sea Investigations 1958–60: general account and station list. N Z Oceanogr Inst Mem 32: 9–29

Bullivant JS (1967b) Ecology of the Ross Sea benthos. N Z Oceanogr Inst Mem 32: 49–75

Cattaneo-Vietti R (1991) Nudibranch molluscs from the Ross Sea, Antarctica. J Molluscan Stud 57: 223–228

Cattaneo-Vietti R, Gambi MC (1995) Ecological studies on benthos at Terra Nova Bay (Ross Sea, Antarctica): an overview of the Italian research programme (1987–92). Ambiente Antartide CNR 6: 18–23

Cattaneo-Vietti R, Chiantore M, Albertelli G (1997) The population structure and ecology of the Antarctic scallop *Adamussium colbecki* at Terra Nova Bay (Ross Sea, Antarctica). Sci Mar 61(2): 15–24

Chiantore M, Cattaneo-Vietti R, Albertelli G, Misic C, Fabiano M (1998) Role of filtering and biodeposition by *Adamussium colbecki* in circulation of organic matter in Terra Nova Bay (Ross Sea, Antarctica). J Mar Syst 17: 411–424

Clarke A (1990) Temperature and evolution. Southern Ocean cooling and the Antarctic marine fauna. In: Kerry KR, Hempel G (eds) Antarctic ecosystems. Ecological change and conservation. Springer, Berlin Heidelberg New York, pp 9–22

Cormaci M, Furnari G, Scammacca B, Casazza G (1992a) Il fitobenthos di Baia Terra Nova (Mare di Ross, Antartide): osservazioni sulla flora e sulla zonazione dei popolamenti. In: Gallardo VA, Ferretti O, Moyano HI (eds) Oceanografia in Antartide. ENEA-PNRA-EULA, Concepción, Chile, pp 395–408

Cormaci M, Furnari G, Scammacca B (1992b) The benthic algal flora of Terra Nova Bay (Ross Sea, Antarctica). Bot Mar 35(6): 541–552

Cormaci M, Furnari G, Scammacca B (1992c) Carta della vegetazione marina di Baia Terra Nova (Mare di Ross, Antartide). Biol Mar 1: 313–314

Cormaci M, Furnari G, Scammacca B, Alongi G (1996) Summer biomass of a population of *Iridaea cordata* (Gigartinaceae, Rhodophyta) from Antarctica. Hydrobiologia 326/327: 267–272

Dayton PK (1989) Interdecadal variation in an Antarctic sponge and its predators from oceanographic climate shifts. Science 245: 1484–1486

Dayton PK (1990) Polar benthos. In: Smith WO (ed) Polar oceanography, part B: Chemistry, biology and geology. Academic Press, London, pp 631–685

Dayton PK, Oliver JS (1977a) Antarctic soft-bottom benthos in oligotrophic and eutrophic environments. Science 197: 55–58

Dayton PK, Oliver JS (1977b) Long-term experimental benthic studies in McMurdo Sound. Antarct J US 13: 136–137

Dayton PK, Robillard GA, De Vries AL (1969) Anchor ice formation in McMurdo Sound, Antarctica, and its biological effects. Science 163: 273–274

Dayton PK, Robillard GA, Paine RT (1970) Benthic zonation as a result of anchor ice at Mc Murdo Sound, Antarctica. In: Holdgate MW (ed) Antarctic ecology I. Academic Press, London, pp 244–258

Dayton PK, Robillard GA, Paine RT, Dayton LB (1974) Biological accommodation in the benthic community of McMurdo Sound, Antarctica. Ecol Monogr 44: 105–128

Dearborn JH (1967) Stanford University Invertebrates Studies in the Ross Sea (1958–61): general account and station list. N Z Dep Sci Ind Res Bull 176: 31–48

Dearborn JH (1968) Benthic invertebrates. Aust Nat Hist 4: 134–139

Dell RK (1972) Antarctic benthos. Adv Mar Biol 10: 1–216

Di Geronimo I, Rosso A (1991) First Italian Oceanographic expedition in the Ross Sea, Antarctica. Benthos: a preliminary report. In: National Scientific Commission for Antarctics (ed) Oceanographic Campaign 1987–88, 1989–90, Data Rep Part I, Genova, pp 407–421

Di Geronimo I, Cattaneo-Vietti R, Gambi C, Casazza G, Cormaci M, Scammacca B (1992) Prime osservazioni sulle comunità bentoniche costiere di Baia Terra Nova (Mare di Ross, Antartide): bionomia e distribuzione. In: Albertelli G, Ambrosetti W, Piccazzo M, Ruffoni Riva T (eds) Atti IX Congr Ass Ital Oceanogr Limnol, Lang, Genova, pp 635–646

Gaino E, Bavestrello G, Cattaneo-Vietti R, Sarà M (1994) Scanning electron microscope evidence for diatom uptake by two Antarctic sponges. Polar Biol 14: 55–58

Gambi MC, Mazzella L (1992) Quantitative and functional studies on coastal benthic communities of Terra Nova Bay (Ross Sea, Antarctica): hard bottoms. In: Gallardo VA, Ferretti O, Moyano HI (eds) Oceanografia in Antartide. ENEA-PNRA-EULA, Concepción, Chile, pp 409–415

Gambi MC, Lorenti M, Russo GF, Scipione MB (1994) Benthic associations of the shallow hard bottoms off Terra Nova Bay, Ross Sea: zonation, biomass and population structure. Antarct Sci 6(4): 449–462

Gambi MC, Castelli A, Guizzardi M (1997) Polychaete populations of the shallow soft bottoms off Terra Nova Bay (Ross Sea, Antarctica): distribution, diversity and biomass. Polar Biol 17: 199–210

Hedgpeth JW (1971) Perspectives of benthic ecology in Antarctica. In: Quam Lo (ed) Research in the Antarctic. American Association for the Advancement of Science, Washington, DC, pp 93–136

Klöser H, Mercuri G, Laturnus F, Quartino ML, Wiencke C (1994) On the competitive balance of macroalgae at Potter Cove (King George Island, South Shetland). Polar Biol 14: 11–16

Knox GA, Lowry JK (1977) A comparison between the benthos of the Southern Ocean and the North Polar Ocean with special reference to the amphipods and the polychaeta. In: Dunbar MJ (ed) Polar oceans. Arctic Institute of North America, Calgary, pp 423–462

Koltun VM (1968) Spicules of sponges as an element of the bottom sediments of the Antarctic. Symposium on Antarctic Oceanography (SCAR), Santiago de Chile. Scott Polar Research Institute, Cambridge, pp 121–123

La Mesa M, Vacchi M, Castelli A, Diviacco G (1997) Feeding habits of two nototheniid fishes, *Trematomus hansoni* and *Trematomus loennbergii*, from Terra Nova Bay, Ross Sea. Polar Biol 17: 62–68

Lazzara L, Nuccio C, Massi L, Innamorati M (1995) Le microalghe simpagiche di Bia Terra Nova (Antartide), nell'estate 1994/95. G Bot Ital 129: 425

Lowry JK (1976) Studies on the macrobenthos of the Southern Ocean. PhD Thesis, University of Canterbury, Christchurch

Lüning K (ed) (1990) Seaweeds: their environment, biogeography and ecophysiology. Wiley-Interscience, New York

Miller KA, Pearse JS (1991) Ecological studies of seaweeds in McMurdo Sound, Antarctica. Am Zool 31: 35–48

Mullineaux LS, De Laca TE (1984) Distribution of Antarctic benthic foraminifers settling on the pecten *Adamussium colbecki*. Mar Biol 3: 185–189

Oliver JS, Slattery PN (1985) Effects of crustacean predators on species composition and population structure of soft-bodied infauna from McMurdo Sound, Antarctica. Ophelia 24(3): 155–175

Richardson MD, Hedgpeth JW (1977) Antarctic soft-bottom macrobenthic community adaptation to a cold stable highly productive, glacially affected environment. In: Llano S (ed) Adaptations within Antarctic ecosystems. Smithsonian Institution, Washington, DC, pp 181–196

Rosso A (1991) Infralittoral Bryozoa associated to macroalgae from the First Italian Antarctic Oceanographic Expedition (Terra Nova Bay, Ross Sea). Bol Soc Biol Concepcion 62: 179–186

Rosso A (1992) Bryozoa from Terra Nova Bay (Ross Sea, Antarctica). In: Gallardo VA, Ferretti O, Moyano HI (eds) Oceanografia in Antartide. ENEA-PNRA-EULA, Concepción, Chile, pp 359–369

Russo GF, Gambi MC (1994) First quantitative data on coastal soft bottoms populations off Terra Nova Bay (Ross Sea, Antarctica): bivalve molluscs. In: National Scientific Commission for Antarctica (ed) Oceanographic Campaign 1987–88, 1989–90, Data Rep Part III, Genova, pp 187–190

Simeoni U, Baroni C, Meccheri M, Taviani M, Zanon G (1989) Coastal studies in Northern Victoria Land (Antarctica): holocene beaches of Inexpressible Island, Tethys Bay and Edmonson Point. Boll Ocean Teor Appl 7(1–2): 5–16

Stocchino C, Lusetti C (1988) Le costanti armoniche di marea di Baia Terra Nova (Mare di Ross, Antartide). Pubbl Ist Idrografico della Marina FC 1128: 1–23

Stockton WL (1984) The biology and ecology of the epifaunal scallop *Adamussium colbecki* on the west side of McMurdo Sound, Antarctica. Mar Biol 78: 171–178

Vacchi M, Greco S, La Mesa M (1991) Ichthyological survey by fixed gears in Terra Nova Bay (Antarctica). Fish list and first results. Mem Biol Mar Oceanogr 19: 197–202

Vacchi M, La Mesa M, Castelli A (1994) Diet of two coastal nototheniid fish from Terra Nova Bay, Ross Sea. Antarct Sci 6(1): 61–65

Violanti D (1990) Distribution of living benthic foraminifera from Terra Nova Bay (Antarctica) – preliminary data. In: National Scientific Commission for Antarctica (ed) Oceanographic Campaign 1987–88, Data Rep Part II, Genova, pp 233–237

Voss J (1988) Zoogeographie und Gemeinschaftsanalyse des Makrozoobenthos des Weddellmeeres (Antarktis). Ber Polarforsch 45: 1–145

White MG (1984) Marine benthos. In: Laws RM (ed) Antarctic ecology II. Academic Press, London, pp 421–461

White MG, Robins MW (1972) Biomass estimates from Borge Bay, South Orkney Islands. Bull Br Antarct Surv 31: 45–50

Zaneveld JS (1966a) The occurrence of benthic marine algae under shore fast-ice in the Western Ross Sea, Antarctica. In: Proc 5th Int Seaweed Symp, Halifax, 25–28 Aug 1965. Pergamon Press, London, pp 217–231

Zaneveld JS (1966b) Vertical zonation of Antarctic and sub-Antarctic benthic marine algae. Antarct J US 1: 211–213

Zaneveld JS (1968) Benthic marine Algae, from Ross Island to Balleny Islands. Antarct Map Folio Ser Am Geogr Soc 10: 10–12

CHAPTER 37

# Shallow-Water Bryozoans and Serpuloideans from the Ross Sea (Terra Nova Bay, Antarctica)

A. Rosso and R. Sanfilippo[1]

**ABSTRACT**

Bryozoans and serpuloideans were studied during the Italian Antarctic Expeditions (1987–1994) in the Terra Nova Bay area (Ross Sea, Antarctica). Only shallow-water samples (upper shelf, down to 200 m) were considered; these gave a preliminary zonation pattern. Both bryozoan and serpuloidean epibiosis was lacking in the Mesolittoral and Upper Infralittoral biotopes. Some deeper assemblages were distinguished: an algal one in the Lower Infralittoral Zone, well characterized by encrusting, r-strategist bryozoan species, most of which appear to be restricted to shallow waters. A second assemblage of the Circalittoral-Pseudobathyal hard bottoms (60–140 m) was very rich and diversified, both in species and morphological adaptations. A third assemblage of the Circalittoral-Pseudobathyal soft bottoms was extremely poor but quite diversified, owing to the heterogeneity of the bottom grain size. A fourth assemblage consists in the epibiosis of *Adamussium colbecki* was rich but poorly diversified, dominated by encrusting species and by the spirorbid *Paralaeospira levinseni*.

## Introduction

Both spatial and vertical distribution of benthic littoral assemblages in the Ross Sea (McMurdo Sound area) have been mainly investigated by American scientists. Recently, within the framework of the Italian Antarctic Program, the benthos of Terra Nova Bay has also been studied by Italian scientists (Di Geronimo and Rosso 1990; Di Geronimo et al. 1992; Gambi et al. 1994; Cattaneo-Vietti et al., this Vol, for a review).

At present, the investigated areas are few and cover a small spatial scale. Little is known about the composition and structure of single phyla in such assemblages. Only a few papers deal with bryozoans and serpuloideans.

Despite the recent efforts in bryozoan systematics (Moyano 1965, 1967, 1969, 1969–70, 1970, 1972a,b, 1974, 1975, 1985, 1986, 1989, 1996a; Moyano and Gordon 1980; Hayward and Taylor 1984; Hayward and Thorpe 1987, 1988a–d, 1989a,b, 1990; Hayward and Ryland 1990, 1991; Hayward 1991; 1992; 1993; 1995, for a review; Rosso 1992a, 1994; Hayward and Winston 1994) and biogeography (Moyano, 1965–66, 1996b), sometimes even the bathymetric distribution of a single species is difficult to infer. Only a few papers have been devoted to bryozoan assemblages, focusing on bryozoan epibiosis on macroinvertebrates (Moyano 1972b, 1981; Barnes 1994; Barnes and Clarke 1995; Rosso and Sanfilippo 1996) on growth rates (Winston 1983; Barnes 1995a) and on the composition of prevailing shallow-water assemblages (Moyano 1978, 1984; Rosso 1990, 1991, 1992b; Winston and Hayward 1994; Barnes 1995b,c). As one of the main components of the Antarctic benthos, bryozoans have been involved in schemes of benthic zonation (Bullivant 1967; Dayton et al. 1974; Arnaud 1974).

Serpuloids are less known in the area and have been recorded by few authors. Studies on these organisms are mainly systematic, with a few biogeographical notes (Hartman 1966, 1967; Zibrowius 1968; Harris 1969; Knight-Jones and Walker 1972; Knight-Jones et al. 1973; Knight-Jones 1978; Hartman-Schroder and Rosenfeldt 1988; Knight-Jones and Knight-Jones 1994; Cantone 1995). Cantone and Sanfilippo (1992) gave data on the bathymetric distribution of several species.

[1]Dipartimento di Scienze Geologiche, Sezione di Oceanologia e Paleoecologia, Universita di Catania, Corso Italia 55, 95129 Catania, Italy

**Table 1.** Station list

| Expeditions | | | |
|---|---|---|---|
| 1987–88 | 1989–90 | | 1993–94 |
| 880804 | C1 bis | F4 bis | ALB 2 |
| 120288 | C3 | F4 tris | ALB 4 |
| 880606 | C4 | G2 | ALB 5 |
| 881006 | D 1D | G3 | ADA 2 |
| 120288 | D 3D | H 2D | ADAB 1 |
| 880810 | D 4D | H 3D bis | FAR POR 2 |
| 880312 | D 5D | MM 31 | FAR POR 4 |
| 880412 | D 78 | MM63bis | FAR |
| 880612 | D 150 | MM64 | TOR 5 |
| 880012 | E 135 | MM75 | Palam Nascet |
| 881015 | E1 bis | MM79 | |
| 880715 | E1 D | MM91 | |
| 880022 | E2 Da | MM106 | |
| 880122 | E2 Db | MM135 | |
| 880322 | F 1D | MM178 | |
| 881025 | F 2 bis | Z 90 | |
| 880133 | F3 | Z 100 | |
| 880535 | F3 bis | | |

The aim of this chapter is to outline a distribution scheme of bryozoan and serpuloidean assemblages on rocky and soft bottoms in the Ross Sea upper shelf (down to 200 m) from the material collected during the Italian expeditions in 1987–1994.

## Materials and Methods

Samples were collected during several oceanographic expeditions performed under the aegis of the PNRA (National Program for Antarctic Research) in Terra Nova Bay. Stations fall within a coastal belt, from the Campbell Glacial Tongue south to Cape Russell (Table 1).

Samples are from both rocky and soft bottoms. In the Terra Nova Bay area, rocky bottoms are widespread from the coastline to 70–80 m, becoming sparse below. Soft bottoms are rare and pebbly in shallow waters, becoming more and more widespread below 20 m. There, they consist of gravels or coarse and fine sands; muddy sands may also occur in deeper waters, but also at shallow depths (Brambati et al. 1989). Ice-rafted cobbles, pebbles and also large blocks are common.

A first group of samples (1987/88 expedition) was collected by scuba diving from 4 to 35 m (Di Geronimo and Rosso 1990; Cormaci et al. 1990). All of these come from rocky bottoms in the Infralittoral Zone. Sorting was mainly performed by algologists of the Istituto e Orto Botanico (Catania University).

Other samples were collected by means of grabs and dredges during the 1989–90 (Balduzzi and Cattaneo-Vietti 1991) and 1993/94 expeditions. The present study concerns only the shallowest stations, ranging from 40 to ca. 200 m. Samples come both from rocky (locally, with muddy coverings) and soft bottoms, with scattered blocks. The latter mainly consists of sandy sediments with cobbles and pebbles, and rarely of muddy sands or gravels. A few of these are notably rich in the scallop *Adamussium colbechi*, whose valves are colonized mainly by bryozoans and serpuloideans (Rosso and Sanfilippo 1996). Samples were made available by the Marine Sorting Centre Italia (Genoa University).

## Results

Bryozoans and serpuloideans were not sampled at all stations but the 62 studied samples (Table 1) yielded a large number of specimens of both groups (Figs. 1, 2).

Bryozoans were mainly represented by living colonies or fragments, whereas dead specimens were rarely found. As a whole, 73 living species were identified (13 Cyclostomatida and 60 Cheilostomatida, 38 of which belonged to the Infraorder Ascophorina *sensu* Gordon, 1989). Thanatocoenoses comprised only a few of these (15) and no additional species.

In the living assemblages, despite a high total number of species (Table 2), the specific richness per sample was quite low, ranging from 0 (in two samples) to 18, with a mean value of 5.3. Only a few species were frequent, i.e. *Celleporella antarctica* (29 samples), *Ellisina antarctica* (25 samples) and *Harpecia spinosissima* (23 samples) and, subordinately, *Beania erecta* var. *livingstonei* and *Arachnopusia decipiens* (both in 15 samples), *Tubulipora* sp. (14 samples), *Micropora brevissima*

**Fig. 1.** **A** *Reteporella frigida* (Waters), reteporiform. **B** *Fasciculipora ramosa* d'Orbigny, vinculariform. **C** *Idmidronea magna* Androsova, vinculariform. **D** *Smittina directa* (Waters), vinculariform. **E** *Larvapora* cf. *mawsoni* (Livingstone), adeonoform. **F** *Neofungella claviformis* (Waters), stout vinculariform. **G** *Swanomia membranacea* (Thornely), cellarinelliform. **H** *Cellarinella nutti* Rogick, cellarinelliform. **I** *Cellarinella marguerita*e Rogick, cellarinelliform. **L** *Antarcticaetos bubeccata* (Rogick), cellariiform. **M** *Adelascopora jeqolqa* Moyano, cellariiform. **N** *Vermiliopsis notialis* Monro

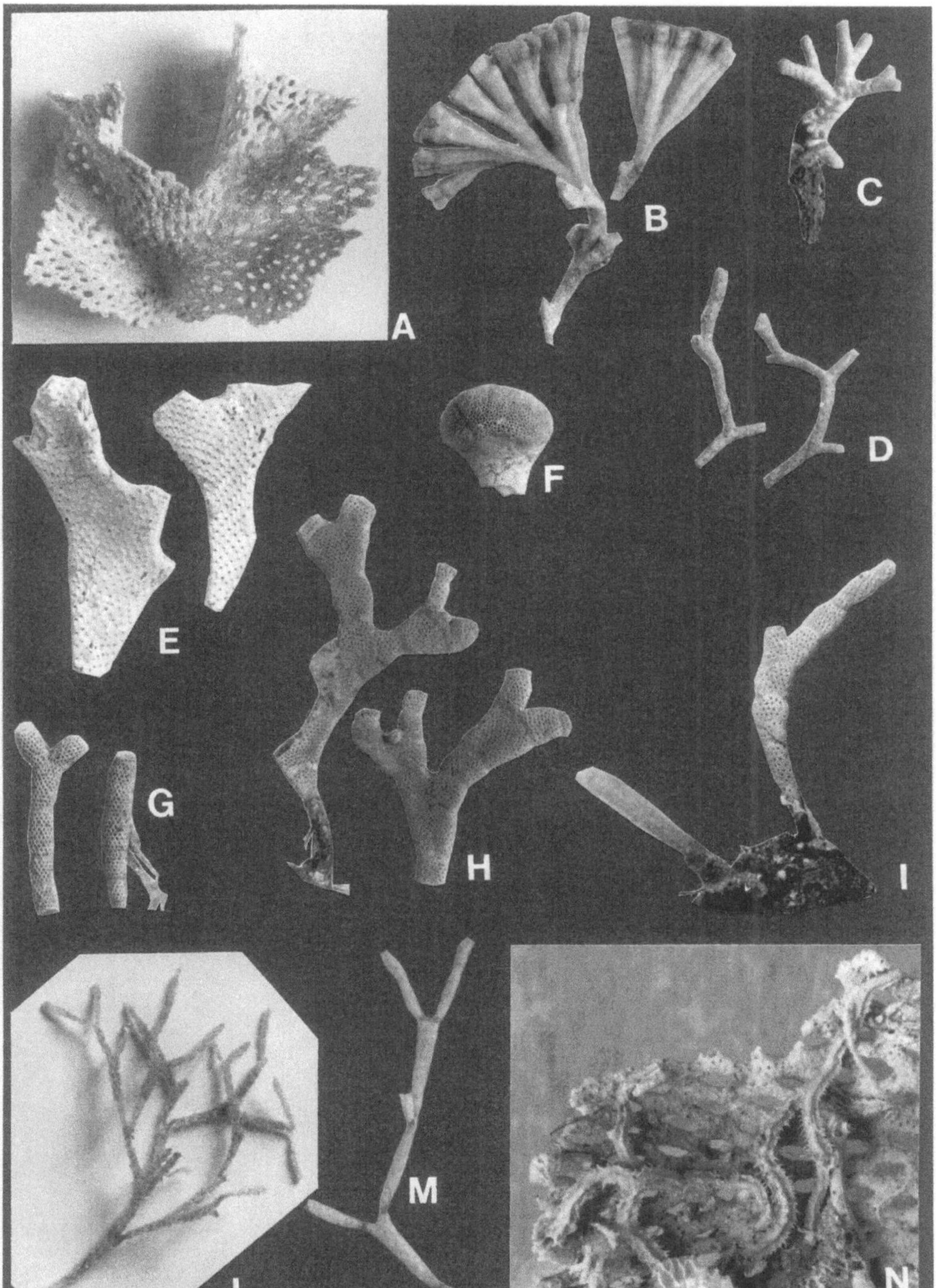


**Fig. 1 A–N.**

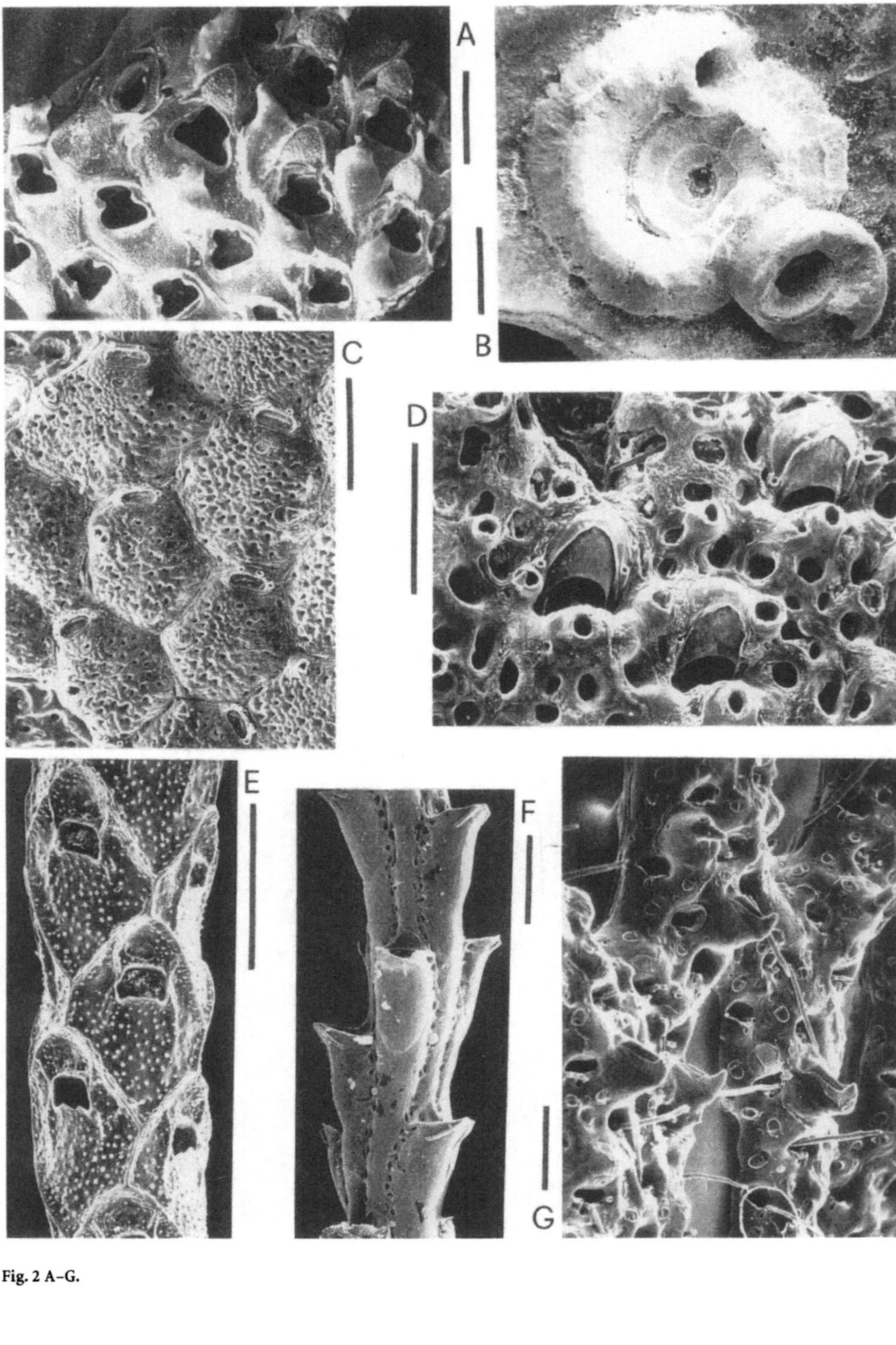


**Fig. 2 A–G.**

**Table 2.** Systematic list of bryozoans and serpuloideans found in the shallow water assemblages of the Terra Nova Bay area (Ross Sea). Numbers indicate the frequence of single taxa within each recognized assemblage. In brackets, data concerning thanatocoenoses

| Zone | Lower Infralittoral | Circalittoral and Upper Pseudobathyal | | | |
|---|---|---|---|---|---|
| Depth | 4–35 m | 63–143 m | 43–183 m | 40–100 m | 26–150 |
| Bottom type | Rocky | Rocky | Soft | *A. colbecki* | Unknown |
| No. of analyzed samples | 20 | 20 | 12 | 4 | 6 |
| Bryozoa | | | | | |
| *Crisia* sp. | 1 | | | | |
| *Stomatopora* sp. | | 1 | | | |
| ? *Hastingsia* sp. | | 3 | | | |
| *Entalophora* sp. | | | | | |
| ? *Tubulipora tubigera* Busk | 1 | | 1 | 1 | |
| *Tubulipora* sp. | 14 | | | | |
| *Idmidronea magna* Androsova | | 1 | | | |
| *Idmidronea* sp. A | 1 | 2 | 2 + (1) | | |
| *Defrancia sarsii* Borg | | 1 | | | |
| *Fasciculipora ramosa* d'Orbigny | | 2 | | | |
| *Neofungella claviformis* (Waters) | | 2 | 1 | | |
| *Hornera* cf. *lasarevi* Androsova | | 1 | 1 | | |
| *Lichenopora tubicen* Borg | 9 | | .(1) | | |
| *Lichenopora* sp. | | | | .(2) | |
| *Harpecia spinosissima* (Calvet) | 20 | | 2 + (1) | | 1 |
| *Isosecuriflustra tenuis* (Kluge) | | 1 | | | |
| *Austroflustra vulgaris* (Kluge) | | | | | 1 |
| *Ellisina antarctica* Hastings | 16 | 3 | 2 + (1) | 3 + (3) | 1 |
| *Amphiblestrum inermis* (Kluge) | | 5 | 1 | 2 + (1) | |
| *Amphiblestrum rossi* Hayward & Thorpe | | 2 | | | |
| ? *Chaperiopsis cervicornis* (Busk) | | 1 | | | |
| *Chaperiopsis quadrispinosa* (Kluge) | | 2 | | | |
| *Exallozoon simplicissimum* (Kluge) | | 8 | | 1 + (1) | 2 |
| *Camptoplites bicornis* (Busk) | 1 | 3 | | | |
| *Camptoplites rectilinearis* Hastings | | 1 | | | |
| *Camptoplites tricornis* (Waters) | | 2 | | | 1 |
| *Cornucopina polymorpha* (Kluge) | | | | | 1 |
| *Himantozoum antarcticum* (Calvet) | | 2 | 2 | | 1 |
| *Caberea darwinii* Busk | | 5 | | | 1 |
| *Notoplites drygalsky* (Kluge) | | 4 | 2 | | 1 |
| *Notoplites tenuis* (Kluge) | 1 | 7 | | | 2 |
| *Notoplites* sp. | 1 | | | | |
| *Beania erecta* var. *livingstonei* Hastings | 11 | 2 | 1 | | |
| *Micropora brevissima* Waters | 5 | 5 | | 2 + (1) | |
| *Cellaria aurorae* Livingstone | | 1 | | | |
| *Swanomia membranacea* (Thornely) | | | 1 + (1) | | |
| *Larvapora mawsoni* (Livingstone) | | 1 | | | |
| *Figularia discors* Hayward & Taylor | | 1 | | | |
| *Arachnopusia decipiens* Hayward & Thorpe | | 9 + (1) | 3 | 2 + (1) | 1 |
| *Exochella avicularis* Hayward | | 1 | | | |
| *Exochella hymenae* (Rogick) | | 3 + (1) | | 2 + (2) | |
| *Antarcticaetos bubeccata* (Rogick) | | 2 | | | |
| *Cellarinella margueritae* Rogick | | 1 | | | |
| *Cellarinella njegovanae* Rogick | | 2 | | | |
| *Cellarinella nutti* Rogick | | | 2 | | |
| *Hippothoa* sp. A | | | 1 | | |
| *Hippothoa* sp. | 1 | | | | |

**Fig. 2. A** *Amphiblestrum inermis* (Kluge). **B** *Metalaeospira pixelli* Harris. **C** *Microporella stenoporta* Hayward and Taylor. **D** *Arachnopusia decipiens* Hayward and Thorpe. **E** *Cellaria aurorae* Livingstone. **F** *Antarcticaetos bubeccata* Rogick. **G** *Reteporella frigida* (Waters). *Bar* 500 μm

**Table 2.** *Continued*

| Zone | Lower Infralittoral | Circalittoral and Upper Pseudobathyal | | | |
|---|---|---|---|---|---|
| Depth | 4–35 m | 63–143 m | 43–183 m | 40–100 m | 26–150 |
| Bottom type | Rocky | Rocky | Soft | *A. colbecki* | Unknown |
| No. of analyzed samples | 20 | 20 | 12 | 4 | 6 |
| *Celleporella antarctica* Moyano & Gordon | 20 | | 4 + (1) | 2 + (4) | 3 |
| *Dakariella dowbroni* Rogick | | 1 + (1) | | | |
| *Toretocheilum absidatum* Rogick | | 1 | | | |
| *Smittina abditavicularis* Rogick | | 4 | | | |
| *Smittina alticollarita* Rogick | | .(1) | 1 | | |
| *Smittina antarctica* (Waters) | | 1 | | | |
| *Smittina directa* (Waters) | | 1 | | | |
| *Smittina rogickae* Hayward & Taylor | | 3 | | | |
| *Smittoidea albula* Hayward & Taylor | | 3 | | | |
| *Smittoidea ornatipectoralis* Rogick | | 2 | | | |
| *Aimulosia antarctica* (Powell) | 4 | 1 | 2 | 3 + (1) | |
| *Trypticocirrus phylactelloides* (Calvet) | | 1 | | | |
| *Pemmatoporella marginata* (Calvet) | | 2 | | | |
| *Rhamphosmittina bassleri* (Rogick) | | 1 | | | |
| *Hippomonavella* cf. *pellucida* (Calvet) | | 1 | | | |
| *Microporella stenoporta* Hayward & Taylor | 1 | 4 + (2) | 1 | 3 + (2) | |
| *Fenestrulina antarctica* Hayward & Thorpe | | 2 | | | |
| *Fenestrulina exigua* (Waters) | | 4 | | | |
| *Fenestrulina proxima* (Waters) | 3 | | | | |
| *Fenestrulina* sp. | | 1 | | | |
| *Adelascopora jeqolqa* Moyano | | | 1 | | |
| *Osthimosia milleporoides* (Calvet) | 1 | 1 | | | 1 |
| *Reteporella frigida* (Waters) | | 10 + (2) | 2 | | |
| *Reteporella hippocrepis* (Waters) | | 2 | | | |
| *Reteporella* sp. | 1 | | | | |
| *Orthoporidra brachyrhyncha* Moyano | | | 1 | | |
| *Orthoporidra sternorhyncha* Moyano | | 2 + (1) | 1 | | |
| Serpuloidea | | | | | |
| *Serpula narconensis* Baird | 1 + (1) | 11 + (5) | 3 + (2) | | |
| *Vermiliopsis notialis* Monro | | 2 + (1) | | | |
| *Metalaeospira* cf. *armiger* Vine | .(1) | 1 + (3) | | | |
| *Metalaeospira pixelli* Harris | | 2 | | | |
| *Paralaeospira lebruni* (Caullery & Mesnil) | | .(2) | | | |
| *Paralaeospira levinseni* (Caullery & Mesnil) | | | | 3 + (4) | |
| *Paralaeospira* sp. A | .(1) | .(3) | | .(1) | |
| *Protolaeospira* sp. A | | .(2) | | | |
| *Romanchella* sp. A | | 1 | | | |
| *Helicosiphon biscoeensis* Gravier | .(1) | | | | |
| *Spirorbidae* sp. | | .(4) | | | |

and *Reteporella frigida* (12 samples), *Exallozoon simplicissimum* (11 samples), *Notoplites tenuis* and *Aimulosia antarctica* (10 samples). Most species were found in 2 to 9 samples, and 30 were recorded only once. Each species was generally represented by one or few colonies/fragments per sample, except for some markedly rich samples mainly from shallow stations.

Serpuloideans were mainly represented by empty tubes, whereas living specimens were rare. Overall, two serpulid and four spirorbid living species were identified. Thanatocoenoses yelded five additional spirorbids. The total number of species per sample was low, ranging from 0 to 4, with a mean value of 0.6. No species were found in about 1/3 of the stations, mainly shallower than 60 m. Deeper, the number of both species and living specimens was higher. *Serpula narconensis* was the most frequent species (15 samples). *Paralaeospira levinseni* occurred in three samples, and *Vermiliopsis notialis* and *Metalaeospira pixelli* in two samples. Finally, *Metalaeospira* cf. *armiger* and *Romanchella* sp. A were found only once.

## Discussion

The studied samples cover only the upper horizon of the continental shelf which, in the Ross Sea, extends down to ca. 500 m. This shallow horizon comprises both rocky and soft bottoms which are locally covered by drop stones and sandy-muddy sediments discharged by icebergs. This bottom feature, coupled with a weak sediment sorting due to essentially constant hydrologic conditions, causes an apparent, general uniformity in benthic assemblages (see Uschakov 1963; Pérès 1982).

The presently studied bryozoans and serpuloideans allow a preliminary identification of distint assemblages, some of which appear well characterized. In agreement with Di Geronimo and Rosso (1990), zonation follows the Genoa system (Arnaud 1974; Pérès 1982), partially amended to include the Pseudobathyal Zone of Andriashev (1978) which extends below the Circalittoral Zone, from ca. 70 m (i.e. from the deepest algal records – the crustose corallinacean *Clathromorphum lemoineanum*) in the area.

### Mesolittoral and Upper Infralittoral Zones

Bryozoans and serpuloideans were almost absent from these zones, down to ca. 4 m, because of the ice action preventing their settlement. Similar results were described by Barnes (1995a) at Signy Island where rocky shores are bare and bryozoans sporadically occur only below 2 m.

### Lower Infralittoral Zone

The studied area mainly consists of hard bottoms heavily colonized by macroalgae, often encrusted by sessile epibionts, among which bryozoans are very abundant. Several samples (20, ranging from 4 to 35 m) fall within this zone. Assemblages were largely dominated by bryozoans, while serpuloideans only occasionally occurred. Epiphytes mainly encrusted *Phyllophora antarctica* and, to a lesser extent, *Iridea cordata* and *Plocamium cartilagineum* (Rosso 1991). Bryozoan richness was relatively high with a total of 19 species. Values per sample ranged from two to ten species, with a mean value of 5.6.

This assemblage was well characterized by *Harpecia spinosissima* and *Celleporella antarctica*, which were very abundant in all samples. Other species were also common and frequent such as *Ellisina antarctica*, *Beania erecta* v. *livingstonei*, *Tubulipora* sp., *Lichenopora tubicen* and *Micropora brevissima*, followed by *Aimulosia antarctica* and *Fenestrulina proxima*. All other species appeared only occasionally. Some species (*Tubulipora* sp., *Lichenopora tubicen* and *Fenestrulina proxima*) were found only in this assemblage.

*Harpecia spinosissima* and *Celleporella antarctica* preferentially settled on algae and seemed to be restricted to shallow waters not exceeding 50–80 m. These two species represent a characteristic feature of the macroalgal bryozoan epibiosis also in other areas of the Ross Sea and Antarctic Peninsula (Winston and Hayward 1994), although the latter also colonizes infralittoral macroinvertebrates (Barnes and Clarke 1995). *H. spinosissima*, limited to algal substrata, may be considered as indicative of this assemblage.

Membraniporiforms largely dominated at species (12/19) and especially at specimen level. Five other morphotypes were present, although sporadically, except for the erect, flexible cellariiforms, with three species.

The assemblage, as a whole, was characterized by the presence of the bryozoans *H. spinosissima* and *C. antarctica* and, generally, by r-strategist species. They developed a great number of small colonies (often settled near the assumed parental ones), rapidly reaching the mature stage, thus ensuring a rapid exploitation of ephemeral algal substrata (Rosso 1991).

### Circalittoral and Upper Pseudobathyal Zones

The circalittoral zone comprises rocky and soft bottoms, locally colonized by dense populations of *Adamussium colbecki*. Both bryozoan and serpuloidean assemblages appear to live preferentially on the rocky substrata or on soft bottoms interspersed by pebbles and cobbles. Living and dead valves of *A. colbecki* seem to represent a suitable substratum.

#### Rocky Bottoms

As much as 20 stations were sampled on rocky bottoms, where sandy or muddy coverings may locally occur at depths ranging from about 60 to 140 m. Specific richness was higher than in the infralittoral bottoms.

Bryozoans were represented by a total of 55 species. Richness per sample ranged from 0 to 18 species, with a mean value of 6.7. Despite this, the number of specimens was low and species were often represented by a single colony.

Serpuloideans were better represented, compared with the infralittoral bottom assemblages, comprising five species. The number of species per sample was low, ranging from 0 to 2, with a mean value of 0.8. Like bryozoans, specimens were rare, except for *Serpula narconensis*.

This assemblage appeared well characterized by the frequency of the bryozoans *Reteporella frigida*, *Arachnopusia decipies* and *Exallozoon simplicissimum*, together with *Notoplites tenuis* and the serpuloidean *Serpula narconensis*. All of these species are easily recognizable due to their large size and peculiar growth morphologies. The dark red-coloured colonies of *E. simplicissimum* are also worthy of note. This last species seems to prefer circalittoral rocky bottoms, whereas the others also occur on the nearby soft bottoms and on deeper biotopes, thus being remarkably ubiquitous. Other less frequent bryozoans, such as *Amphiblestrum inermis*, *Caberea darwini*, *Notoplites drygalsky*, *Micropora brevissima* and *Microporella stenoporta*, also seem to have the same wide ecological distribution. Several species, probably restricted to hard bottoms, were only found in this group of samples, although sporadically. They mainly consisted of encrusting membraniporiform species, among which *Smittina abditavicularis*, *Fenestrulina exigua*, *S. rogickae* and *Smittoidea albula* were the most frequent. Some erect rigid species with a cemented base (*Defrancia sarsi*, *Fasciculipora ramosa*, *Smittina directa* and *Reteporella hippocrepis*) or erect rooted species (*Antarcticaetos bubeccata*, *Isosecuriflustra tenuis*, *Cellarinella njegovanae* and *C. margueritae*) also belonged to the same group. Among serpuloideans, other than *S. narconensis*, all of the recorded species were restricted to these hard substrata.

Among bryozoans, 8 morphotypes were distinguished. As in the previous group, membraniporiforms (29 species) strongly prevailed, followed by erect growth types. The latter ones mainly comprised rigid habits, with vinculariforms (8 species), adeonelliforms, reteporiforms (both with 2 species) and cellarinelliforms (3 species). Erect flexible species belonged both to the flustriforms (5 species) and cellariiforms (4 species). The latter three morphotypes were rooted and, depending on the rhyzoidal morphology, can directly colonize rocky bottoms and sediment pockets. The structure of the zoarial association suggests a spatial heterogenity in bottom type and texture, as already stressed by Rosso (1990, 1992b) for deeper assemblages.

### Soft Bottoms

This group comprised 12 stations sampled on gravel sands, sands and muddy sands, ranging from about 40 to 180 m. As a whole, 21 bryozoan and only 1 serpuloidean species were found. Bryozoans showed a species richness lower than in the previous groups, ranging from 0 to 9 with a mean value of 2.8. Species were generally represented by single specimens, as also noted for the only serpuloidean species sampled.

Among the bryozoans, the most frequent species (only three samples) was the ubiquitarian *A. decipiens* while all others were sporadically present in one to two samples. Among these, the rooted *Cellarinella nutti* and *Adelascopora jeqolqa* are worthy of note, as they seem to colonize soft bottoms. In contrast, large cemented species, such as *Neofungella claviformis*, *Hornera* cf. *lasarevi* or *Reteporella frigida*, seem to be linked to cobbles and pebbles interspersed in the soft bottom. This heterogeneity also accounts for the highly diversified growth form association (seven morphotypes), despite the strong reduction in species number, compared with hard-bottom assemblages. The unique recording of the serpuloidean species, *S. narconensis*, was infrequent and markedly rare. The low species richness and markedly low abundance of colonies per sample appear to be the main features of this assemblage.

### *Adamussium colbecki* Bottoms

This group comprises only four samples, from 40 to 100 m. Epibiotic assemblages were notably homogeneous although poorly diversified, as only 11 bryozoan and 1 serpuloidean species were recovered (Rosso and Sanfilippo 1996). Bryozoan species richness ranged from 5 to 7, with a mean value of 5.7. *Ellisina antarctica*, *Aimulosia antarctica* and *Microporella stenoporta* were the most frequent species with three records while *Amphiblestrum inermis*, *Micropora brevissima*, *Arachnopusia decipiens*, *Celleporella antarctica* and *Exochella hymanae* were present in two samples. Among these, only *E. antarctica*, followed by *M. brevissima* and *C. antarctica* (often settled on the inner surfaces of empty *Adamussium* valves), were abundant. None of the recorded species was peculiar to this assemblage, as they also occurred in other infralittoral or circalittoral biotopes. It is worthy of note that all of these species settle on both rigid and flexible substrata but are absent from soft, fine-grained bottoms. All the bryozoan species have a membraniporiform habit well adapted to hard, clean, very light and mobile substrata.

The only recorded serpuloidean species, the spirorbid *Paralaeospira levinseni*, was frequent (three samples) and abundant. This species was exclusively found on the valves of *A. colbecki*.

This assemblage was characterized by a low species richness, by the presence of *P. levinseni* and, among the bryozoans, by the exclusive presence of species with a membraniporiform habit.

## Conclusions

The distribution patterns seem to be strongly affected by physicochemical and edaphic factors. Among these, substratum morphology and exposure heavily affect shallow-water rocky biotopes (Barnes 1995c) while depth, bathymetry and edaphic factors produce larger variations in biota composition and, in particular, on bryozoan assemblages, as already noted by Barnes (1995b). Biotic factors, such as the local presence of other organisms, also largely influence benthic assemblages.

Zonation of the investigated groups was also linked to the presence of ice, both along the coastline, where scouring ice action prevents the colonisation at shallower depths, as well as offshore and in deeper and waters due to the input of fine and coarse (blocks) sediments discharged by ice melting. These inputs interact with the currents of the bay, creating a large-scale mosaic of environmental situations which, in turn, cause a patchiness in the benthos distribution. It is therefore difficult to establish a clear-cut bathymetric zonation. The proposed assemblage scheme represents a first attempt to delineate such a zonation pattern which must be integrated by more material in the investigated area and around the Antarctic shelf.

Finally, a comparison with other general, already established, zonation patterns (Bullivant 1967; Dayton et al. 1974; Arntz et al. 1994) seems quite hazardous due to the scantiness and extreme generality and heterogeneity of the available literature data.

*Acknowledgements.* The authors are grateful to Prof. Di Geronimo (Catania University) for usefull discussions and to Mr Dafnis (Stazione Zoologica Anton Dohrn, Naples) and Mr Falco (Cosenza University) for scanning micrographs. The research was supported by a grant from ENEA, PNRA, Biologic Oceanography, to I. Di Geronimo.

## References

Andriashev AP (1978) Some additions of the vertical zonation of marine bottom fauna, Marine Benthic Ecosystems II: 351-360

Arnaud PM (1974) Contribution à la bionomie marine benthique des régions antarctiques et subantarctiques. Téthys 6(3): 465-656

Arntz WE, Brey V, Gallardo A (1994) Antarctic zoobenthos. Oceanogr Mar Biol Annu Rev 32: 241-304

Balduzzi A, Cattaneo-Vietti R (1991) Littoral benthic communities of Terra Nova Bay: sampling data report. In: National Scientific Commission for Antarctica (ed) Oceanographic Campaign 1989-90, Data Rep Part I, Genova, pp 337-342

Barnes DKA (1994) Communities of epibiota on two erect species of Antarctic Bryozoa. J Mar Biol Assoc UK 74: 863-872

Barnes DKA (1995a) Seasonal and annual growth in erect species of Antarctic bryozoans. J Exp Mar Biol Ecol 188: 181-198

Barnes DKA (1995b) Sublittoral epifaunal communities at Signy Island, Antarctica. I. The ice-foot zone. Mar Biol 121: 555-563

Barnes DKA (1995c) Sublittoral epifaunal communities at Signy Island, Antarctica. II. Below the ice-foot zone. Mar Biol 121: 565-572

Barnes DKA, Clarke A (1995) Epibiotic communities on sublittoral macroinvertebrates at Signy Island, Antarctica. J Mar Biol Assoc UK 75: 689-703

Brambati A, Fanzutti GP, Finocchiaro F, Simeoni U (1989) Sediments and sedimentological processes in the Ross Sea continental shelf (Antarctica): results and preliminary conclusions. Boll Oceanol Teor Appl 7 (1-2): 159-188

Bullivant JS (1967) Ecology of the Ross Sea benthos. NZ Dep Sci Ind Res Bull 176: 50-77

Cantone G (1995) Polychaetes Sedentaria of Terra Nova Bay (Ross Sea, Antarctica). Capitellidae to Serpulidae. Polar Biol 15: 295-302

Cantone G, Sanfilippo R (1992) Polychaeta from Terra Nova Bay (Ross Sea, Antarctica). In: Gallardo VA, Ferretti O, Moyano HI (eds) Oceanografia in Antartide. ENEA-PARA-EULA, Conception, Chile, pp 372-381

Cormaci M, Furnari G, Scammacca B (1990) Data report on seaweeds from Terra Nova Bay (Ross Sea) Antarctic. In: National Scientific Commission for Antarctica (ed) Oceanographic Campaign 1987-88, Data Rep Part I, Genova, pp 445-453

Dayton PK, Robilliard GA, Paine RT, Daiton LB (1974) Biological accommodation in the benthic community of McMurdo Sound, Antarctica. Ecol Monogr 44: 105-128

Di Geronimo I, Rosso A (1990) First Italian Oceanographic Expedition in the Ross Sea, Antarctica. Benthos: a preliminary report. In: National Scientific Commission for

Antarctica (ed) Oceanographic Campaign 1987–88, Data Rep Part I, Genova, pp 407–421
Di Geronimo I, Cattaneo-Vietti R, Gambi MC, Casazza G, Cormaci M, Scammacca B (1992) Prime osservazioni sulle comunità bentoniche costiere della Baia Terra Nova (Mare di Ross, Antartide): bionomia e distribuzione. In: Albertelli G, Ambrosetti W, Piccazzo M, Ruffoni Riva T (eds) Atti IX Congr Ass Ital Oceanogr Limnol, Lang, Genova, pp 635–646
Gambi MC, Lorenti M, Russo G, Scipione MB (1994) Benthic associations of the shallow hard bottoms off Terra Nova Bay, Ross Sea: zonation, biomass and population structure. Antarct Sci 6(4): 449–462
Gordon DP (1989) The marine fauna of New Zealand: Bryozoa: Gymnolaemata (Cheilostomida, Ascophorina) from the western South Island Continental Shelf and Slope. NZ Oceanogr Inst Mem 97: 1–157
Harris T (1969) *Spirorbis* species (Polychaeta: Serpulidae) from the South Atlantic. Discovery Rep 35: 135–178
Hartman O (1966) Polychaeta Myzostomidae and Sedentaria of Antarctica. Antarct Res Ser 7: 158 pp
Hartman O (1967) Polychaetous Annélides collected by the USNS Eltanin and Staten Island cruises, chiefly from Antarctic seas. Allan Hancock Monogr Mar Biol 2: 387 pp
Hartmann-Schroeder G, Rosenfeldt P (1988) Die Polychaeten der Polar Stern – Reise ANT III/2 in die Antarktis 1984, part 2: Cirratulidae bis Serpulidae. Mitt Hamb Zool Mus Inst 83: 67–106
Hayward PJ (1991) Systematic studies on some Antarctic and sub-Antarctic Ascophora (Bryozoa: Cheilostomata). Zool J Linnean Soc 101: 299–335
Hayward PJ (1992) Some Antarctic and sub-Antarctic species of Celleporidae (Bryozoa, Cheilostomata). J Zool Lond 226: 283–310
Hayward PJ (1993) New species of cheilostomate Bryozoa from Antarctica and the Subantarctic southwest Atlantic. J Nat Hist 27: 1409–1430
Hayward PJ (1995) Antarctic Cheilostomatous Bryozoa. Oxford University Press, Oxford, 355 pp
Hayward PJ, Ryland JS (1990) Some Antarctic and sub-Antarctic species of Microporellidae (Bryozoa; Cheilostomata). J Nat Hist 24: 1263–1287
Hayward PJ, Ryland JS (1991) New and little known bryozoa from Antarctica and the southwest Atlantic. Bull Mus Natl Hist Nat 13A (3–4): 241–261
Hayward PJ, Taylor PD (1984) Fossil and recent Cheilostomata (Bryozoa) from the Ross Sea, Antarctica. J Nat Hist 18: 71–94
Hayward PJ, Thorpe JP (1987) The systematic position of *Smittia inclusa* Waters, an endemic Antarctic Bryozoan. J Nat Hist 21: 1469–1476
Hayward PJ, Thorpe JP (1988a) Species of *Chaperiopsis* (Bryozoa, Cheilostomata) collected by Discovery investigations. J Nat Hist 22: 45–69
Hayward PJ, Thorpe JP (1988b) New genera of Antarctic cheilostome Bryozoa. Cah Biol Mar 29: 277–296
Hayward PJ, Thorpe JP (1988c) A new family of cheilostome Bryozoa endemic to Antarctica. Zool J Linnean Soc 93: 1–18
Hayward PJ, Thorpe JP (1988d) Species of *Arachnopusia* (Bryozoa: Cheilostomata) collected by Discovery investigations. J Nat Hist 22: 773–799
Hayward PJ, Thorpe JP (1989a) Membraniporoidea, Microporoidea and Cellarioidea (Bryozoa, Cheilostomata) collected by Discovery investigations. J Nat Hist 23: 913–959
Hayward PJ, Thorpe JP (1989b) Systematic notes on some Antarctic Ascophora (Bryozoa, Cheilostomata). Zool Scripta 18 (3): 365–374
Hayward PJ, Thorpe JP (1990) Some Antarctic and sub-Antarctic species of Smittinidae (Bryozoa: Cheilostomata). J Zool Lond 222: 137–175
Hayward PJ, Winston JE (1994) New species of cheilostomate Bryozoa collected by the US Antarctic Research Program. J Nat Hist 28: 237–246
Knight-Jones P (1978) New Spirorbidae (Polychaeta: Sedentaria) from the East Pacific, Atlantic, Indian and Southern Oceans. Zool J Linncan Soc 64: 202–240
Knight-Jones P, Knight-Jones EW (1994) Spirorbidae (Polychaeta) from Signy Island, South Orkneys, including three new species. Ophelia 40(2): 75–94
Knight-Jones P, Walker JM (1972) Spirorbinae (Serpulidae: polychaeta) on limpets from the South Orkney Islands. Br Antarct Surv Bull 31: 33–40
Knight-Jones EW, Knight-Jones P, Bregazzi P (1973) *Helicosiphon biscoeensis* Gravier and its relationship with other Spirorbinae. Zool J Linnean Soc 52: 9–21
Moyano G (1965) Bryozoa collectados durante la Expedicion Antárctica Chilena 1964–1965. Primera parte: Familia Sclerodomidae. Inst Antártico Chileno 5: 29 pp
Moyano G (1965–66) Distribución y profundidades de las especies exclusivamente antárticas de Bryozoa cheilostomata recolectadas por la decimonovena expedicion antártica chilena 1964–65. Bol Soc Biol Concepciön 15: 113–123
Moyano G (1967) Bryozoa collectados por la Expedicion Antártica Chilena 1964–65. II. Familia Corymboporiidae Smitt, 1866 (Bryozoa, Cyclostomata). Inst Antártico Chileno p 17
Moyano G (1969) Bryozoa collectados por la Expediciön Antártica Chilena 1964–65. III. Familia Cellariidae Hincks, 1880. Bol Soc Biol Concepción 41: 41–77
Moyano G (1969–70) Bryozoa collectados por la Expedición Antártica Chilena 1964–65. IV. Familia Arachnopusiidae Jullien, 1888. Bol Soc Biol Concepción 42: 257–285
Moyano G (1970) Una familia, tres géneros y una especie nuevos para la Antártica (Bryozoa Cheilostomata). Cah Biol Mar 11: 153–166
Moyano G (1972a) Familia Flustridae: ensayo de redistribucion de sus especies a nivel génerico. Bol Soc Biol Concepción 44: 73–101
Moyano G (1972b) Briozoos epizoos de *Laetmonice producta* Grube (Annelida, Polychaeta). Ser Cient Inst Antártico Chileno 27: 12 pp
Moyano G (1974) Descripción de *Cellarinella latilaminata* n. sp. y clave para las especies de Cellarinellidae (Bryozoa, Cheilostomata). Bol Soc Biol Concepción 47: 181–186
Moyano G (1975) *Romancheina asymmetrica* sp. nov., nueva especie de la Antártica y clave para las especies Antárticas y magallánicas de Exochellidae (Bryozoa, Cheilostomata). Bol Soc Biol Concepción 49: 61–69
Moyano G (1978) Bryozoa de Bahías Antárticas: algunos aspectos ecológicos. Ser Cient Inst Antárt Chil 24: 35–60
Moyano G (1981) Bryozoa epizoos de *Psolus charcoti* Vaney, 1907 (Holothuroidea, Psolidae). Ser Cient Inst Antártico Chil 27: 5–11

Moyano G (1984) On small Bryozoan collection from near Ross Island, Antarctica. Ser Cient Inst Nac Antárt Chil 31: 75–83

Moyano G (1985) Bryozoa Lekythoporidae: discusion general y nuevas especies de los géneros *Catadysis* y *Orthoporidra* de Chile Austral y de la Antártida. Gayana Zool 49 (3–4): 103–149

Moyano G (1986) Estructura y sistemática del Briozoo Antártico Flustriforme *Kymella polaris* (Waters, 1904). Bol Soc Biol Concepcion 57: 21–35

Moyano G (1989) Briozoos microporelidos celariformes y flustriformes de la Antártica. Bol Soc Biol Concepcion 60: 161–172

Moyano G (1996a) On a new species of Flustridae from Antarctica (Bryozoa, Cheilostomata). Gayana Zool 60 (2): 73–78

Moyano G (1996b) Holocene bryozoan links between Australia, New Zealand, southern South America, and Antarctica – a preliminary evaluation. In: Gordon DP, Smith AM, Grant-Mackie JA (eds) Bryozoans in space and time. Natl Inst Water and Atmospheric Research, Wellington, pp 207–219

Moyano G, Gordon DP (1980) New species of Hippothoidae (Bryozoa) from Chile, Antarctica and New Zealand. J R Soc N Z 10 (1): 75–95

Pérès JM (1982) A comprehensive, integrated treatise on life in oceans and coastal waters. Marine ecology, V. Ocean management, part I. Wiley & Sons, Pitman Press, Bath, Avon, 642 pp

Rosso A (1990) Bryozoan community of IB3 station (Ross Sea, Antarctica). In: National Scientific Commission for Antarctica (ed) Oceanographic Campaign 1987–88, Data Rep Part I, Genova, pp 423–438

Rosso A (1991) Infralittoral bryozoa associated to macroalgae from the First Italian Antarctic Oceanographic Expedition (Terra Nova Bay, Ross Sea). Bol Soc Biol Concepcion 62: 179–186

Rosso A (1992a) *Melicerita digeronimoi* sp. nov.: a new Antarctic bryozoan. Bol Soc Biol Concepcion 63: 185–192

Rosso A (1992b) Bryozoa from Terra Nova Bay (Ross Sea, Antarctica). In: Gallardo VA, Ferretti O, Moyano HI (eds) Oceanografia in Antartide. ENEA-PNRA-EULA, Conception, Chile, pp 359–369

Rosso A (1994) Bryozoa of the First Italian Antarctic Oceanographic Expedition (Terra Nova Bay, Ross Sea). I. Flustridae Smitt, 1867. J Nat Hist 28: 695–713

Rosso A, Sanfilippo R (1996) Bryozoans and Serpuloideans encrusting *Adamussium colbecki* of Terra Nova Bay (Ross Sea, Antarctica. Int. Workshop "Ross Sea Ecology", Taormina, 14–16 May 1996

Uschakov PV (1963) Quelques particularités de la bionomie benthique de l'Antarctique de l'Est. Cah Biol Mar 4: 81–89

Winston JE (1983) Patterns of growth, reproduction and mortality in bryozoans from the Ross Sea, Antarctica. Bull Mar Sci 33 (3): 688–702

Winston JE, Hayward PJ (1994) Bryozoa of the U.S. Antarctic Research Program: preliminary report. In: Hayward PJ, Ryland JS, Taylor PD (eds) Biology and palaeobiology of bryozoans. Olsen and Olsen, Fredensborg, pp 205–210

Zibrowius H (1968) Remarques sur les polychaetes serpulidae antarctiques *Vermiliopsis notialis* Monro, 1930 et *Vermiliopsis glacialis* Monro, 1930. Bull Inst R Sci Nat Belg 44; 38: 1–9

CHAPTER 38

# Spatio-Temporal Variability in the Structure of Benthic Populations in a Physically Controlled System off Terra Nova Bay: The Shallow Hard Bottoms

M.C. Gambi, M.C. Buia, L. Mazzella, M. Lorenti, and M.B. Scipione[1]

## ABSTRACT

Along the coast off Terra Nova Bay (Ross Sea), the shallow hard bottoms are subjected to high levels of physical disturbance due to sea-ice scouring, and are characterized by a dense covering of two red macroalgae, *Iridaea cordata* and *Phyllophora antarctica*. These algae show a typical distribution, with *I. cordata* occurring at shallower depths (2–15 m) and *P. antarctica* progressively replacing the former species in deeper waters (5–35 m). Several vagile invertebrate species often occur often in relatively high numbers associated to these algae. The aim of this chapter is to give an overview of the spatial variability in the structure and distribution of the benthic communities of this peculiar biotope of the bay and to provide a preliminary evaluation of the interannual variability in some structural parameters in relation to the environment. Most of the samples were collected by SCUBA diving (scraping fixed surfaces of 40 × 40 cm) in seven different sites located around the Italian base during the three summer Antarctic expeditions in 1989/90, 1993/94 and 1994/95.

More than 83 000 specimens of macrobenthic organisms were collected, belonging to 62 taxa; the most diversified and abundant groups were polychaetes (19 taxa, 4.2% of the total), molluscs (11 taxa, 56.6%), amphipods (14 taxa, 10%), tanaids (2 taxa, 21.6%) and isopods (7 taxa, 7.5%). The most abundant species, that can be considered the core of the community, were *Harmothoe* spp. (mainly belonging to two forms of *H. spinosa*), *Pionosyllis comosa*, *Kefersteinia fauveli* and *Ophryotrocha claparedii* (polychaetes), *Laevilitorina antarctica*, *Powellisetia deserta*, *Onoba gelida* and *O. turqueti*, and *Philobrya sublaevis* (molluscs), *Nototanais dimorphus* (crustacean tanaids), *Munna antarctica*, *Paramunna rostrata* and *Austrosignum* sp. (isopods), *Oradarea acuminata*, *O. rossii*, *Paramoera walkeri* and *Prostebbingia serrata* (amphipods), and *Sterechinus neumayeri* and *Odontaster validus* (echinoderms). A great homogeneity was recorded in the various sites examined with regards to species composition. The main differences among stations were due to the relative abundance of common species and to the occurrence of quite rare ones. Correlations (Spearman rank correlation) among the community structural variables and depth showed that both the *Iridaea* and the total algal biomass decreased significantly ($P < 0.05$) depending on the depth, whereas both the number of animal taxa and the diversity (H′) increased with depth ($P < 0.01$). Animal density increased with increasing *Phyllophora* as well as total algal biomass, while diversity H′ decreased according to the same variables. Distribution at the community and species levels revealed a well-defined zonation pattern with depth, covering and architecture of the two dominant macroalgae. Species richness and diversity were higher in the *Phyllophora*-associated community, where habitat complexity and shelter, due to both algal architecture and reduced environmental stress, were higher. The highest faunal abundance (over 82 000 ind $m^{-2}$) was observed in association with the *Iridaea* covering, where the harsher environmental conditions selected only few taxa. Inter-annual differences in algal biomass and animal density were recorded in the few stations where we could compare three years of observations. The samples at site Faraglione showed the strongest fluctuations with a remarkable reduction in animal diversity and density (almost one order of magnitude) between the summers of 1990 and 1994, followed by relatively high values again in 1995. These interannual differences, more pronounced at the shallower

[1] Stazione Zoologica A. Dohrn di Napoli; Laboratorio di Ecologia del Benthos, I-80077 Ischia, Naples, Italy

stations (2–4 m depth) than at the deeper ones, could be related to variability in sea-ice development and persistence. As regards sea-ice conditions, in summer 1994 there was a delay of about 40 days in the melting of the seasonal fast- and pack ice in Terra Nova Bay. Notwithstanding these numerical changes, the basic structure of the community was maintained over time, probably due to the high motility potential of the species and to their reproductive strategies.

## Introduction

The shallow hard-bottom biotopes of the Antarctic coastal zone are poorly known because these environments are often not easily accessible to investigation and direct observation and sampling. These nearshore zones are subject, more than other Antarctic areas, to high levels of physical disturbance, due mainly to sea-ice scouring (seasonal pack ice, fast and anchor ice) (Dayton et al. 1970; Dayton 1989), and to a higher seasonal variability in light intensity, salinity and temperature. The study of these biotopes, physically controlled and more sensitive to local climatic fluctuations, is of particular importance to understand the influence of the unique polar abiotic parameters on distribution (zonation and patchiness) and structure at the community level, as well as on the ecophysiology and life history of individual species. In addition, the benthic assemblages of these biotopes are often characterized by conspicuous covering of both micro- (diatoms) and macroalgae, and therefore may contribute largely to the primary production of the coastal system. In Antarctica, benthic algae represent an important source of energy for associated grazers (Dhargalkar et al. 1988) and, as plant detritus, for deeper faunal assemblages (Reichardt 1987), being advected and deposited by ice drift and water movement. The vagile fauna associated with these macroalgae may be an important trophic link between primary producers and higher consumers, such as demersal predatory fish. For these reasons the seaweed-dominated populations of the hard bottoms are included among the key communities within the SCAR international programme EASIZ (SCAR 1994).

Along the coast off Terra Nova Bay (Ross Sea), the shallow hard bottoms are quite accessible to SCUBA diving during the austral summer (Taviani and Amato 1989). They are characterized by a dense covering of two red macroalgae, *Iridaea cordata* and *Phyllophora antarctica*, that show a typical distribution, with *I. cordata* colonizing shallower depths (2–15 m) and *P. antarctica* progressively replacing the former species in deeper waters (5–35 m) (Cormaci et al. 1992a,b; Gambi et al. 1994). Associated to these algae are several animal vagile species, belonging mainly to the carnivorous polychaetes, mesograzer molluscs, crustacean peracarids and echinoderms (Gambi and Mazzella 1991), that often occur with relatively high densities and biomass (Gambi et al. 1994).

The aim of this chapter is to give an overview of the structure and spatial distribution of the benthic communities of this peculiar biotope along the bay and to evaluate the interannual variability in some structural parameters that can be related to the local environmental conditions, including sea-ice dynamics and algal patchiness.

## Study Sites, Materials and Methods

Seven different sites were sampled in the shallow hard bottoms along the Bay during three summer Antarctic expeditions in 1989–90, 1993–94 and 1994–95 (Fig. 1). The sites are distributed from Adelie Cove, about 12 miles south of the Italian base, to Tethys Bay (Fig. 1), north to the base, and along a depth range from 0.5 m to 16 m. All samples were collected along rocky cliffs or on big erratic stones (e.g. sites 1 and 4 in Fig. 1) by SCUBA diving, except sample H1D (site 2) that was collected with a rectangular dredge (Charcot-type) from a sandy bottom covered with free macroalgal thalli, and the sample at Adelie Cove (site 1) where a big stone covered with algae was sampled by chance with a Van Veen grab (65-l volume). In this latter sample it was possible to collect separately the thalli of *Iridaea*, those of *Phyllophora* and the underlying sediment, in order to study the microdistribution of the associated fauna.

At site 3 (Fig. 1), hereafter named “Faraglione”, a transect was performed from 0.5 m to 16 m both in 1989–90 and in 1993–94 to study the depth distribution (Gambi et al. 1994). At this site, during the 1994–95 survey, samples were taken only at two depths, 3 and 16 m, and at increasing algal covering (30, 50 and 70% covering) and standing crop, in order to study the role

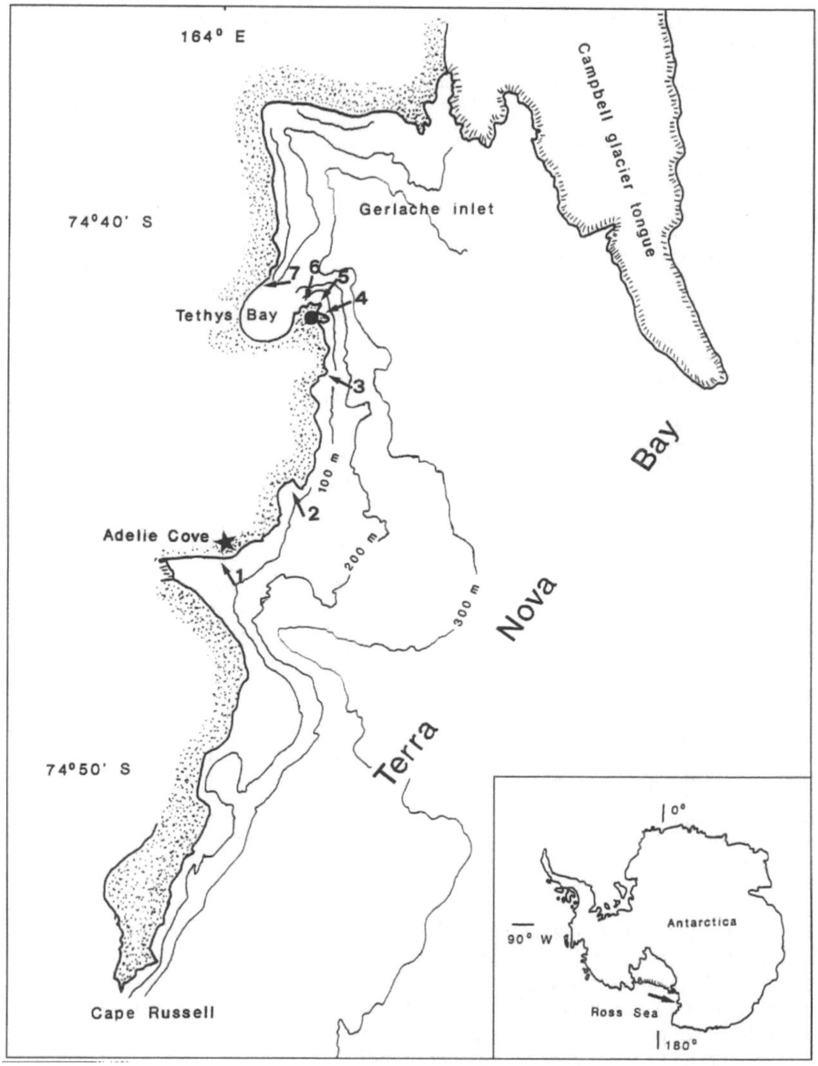


**Fig. 1.** Map of Terra Nova Bay with the location (*arrows*) of the sites where hard-bottom samples were collected: *1* Adelie Cove (Ad90); *2* H1D (dredge sample); *3* Faraglione (numbered samples); *4* base dock (Bd samples); *5* Tethys Bay (TB90); *6* Tethys Bay under pack 1990 (Tbp90); *7* Tethys Bay under pack 1995 (Tbp95). The *black circle* indicates the location of the Italian base; the *star on land* indicates the location of the Adelie penguin rookery at Adelie Cove

of algal biomass patchiness in structuring the associated community (Buia et al. in press). Finally, two samples were collected under the fast ice within Tethys Bay (Tbp90; Tbp95, Table 1). All samples, excepth H1D, were collected by scraping fixed surfaces (40 × 40 cm) and collecting both macroalgae and associated fauna. All the biological material was fixed in 4% neutralized formalin. For the macroalgae, biomass was measured as wet weight in formalin. Vagile animals were separated

**Table 1.** Actual values of some structural parameters of the benthic communities in the studied samples. All values (except H1D) refer to $0.16\,m^2$

| Sample | Location | Depth (m) | *Iridaea* biomass (gww) | *Phyll.* biomass (gww) | Total algal biomass (gww) | No. of animal ind | No. of animal taxa | H' |
|---|---|---|---|---|---|---|---|---|
| 1a90 | Faraglione | 0.5 | – | – | – | 1017 | 2 | 0.02 |
| 1a94 | Faraglione | 0.5 | – | – | – | 240 | 2 | 0.02 |
| 2a90 | Faraglione | 2 | 295.3 | 24 | 319 | 13126 | 21 | 0.65 |
| 2b90 | Faraglione | 2 | – | – | – | 533 | 3 | 0.62 |
| 2c90 | Faraglione | 2 | 378.1 | – | 378.1 | 1679 | 13 | 1.29 |
| 2a94 | Faraglione | 2 | – | – | – | 583 | 3 | 0.14 |
| 2bis94 | Faraglione | 4 | 400.4 | – | 400.4 | 2876 | 9 | 0.53 |
| 2a95 | Faraglione | 3 | 101.5 | 14 | 116 | 506 | 18 | 0.86 |
| 2b95 | Faraglione | 3 | 192.5 | – | 192.5 | 1199 | 15 | 0.46 |
| 2c95 | Faraglione | 3 | 291 | – | 291 | 4704 | 19 | 0.46 |
| 3a90 | Faraglione | 6 | 9.5 | 16 | 25.5 | 2429 | 21 | 2.19 |
| 3c90 | Faraglione | 6 | 172.8 | 18.1 | 190.9 | 1890 | 13 | 1.41 |
| 3a94 | Faraglione | 6 | 132 | – | 132 | 2049 | 11 | 0.66 |
| 4a90 | Faraglione | 12 | – | 75 | 75 | 2370 | 26 | 2.02 |
| 4c90 | Faraglione | 12 | – | 122.8 | 122.8 | 3776 | 23 | 1.69 |
| 4a94 | Faraglione | 12 | – | 14.5 | 14.5 | 668 | 13 | 1.18 |
| 4bis94 | Faraglione | 12 | – | 12.3 | 12.3 | 618 | 13 | 1.70 |
| 5a90 | Faraglione | 16 | 2.8 | 24 | 26.8 | 579 | 22 | 1.38 |
| 5c90 | Faraglione | 16 | – | 129.3 | 129.3 | 2315 | 29 | 2.16 |
| 5a94 | Faraglione | 16 | – | 5.3 | 5.3 | 804 | 17 | 1.31 |
| 5a95 | Faraglione | 16 | – | 96.2 | 96.2 | 3162 | 28 | 1.82 |
| 5b95 | Faraglione | 16 | – | 102.4 | 102.4 | 940 | 21 | 1.78 |
| 5bis95 | Faraglione | 16 | – | 113.6 | 113.6 | 357 | 17 | 1.70 |
| 5c95 | Faraglione | 16 | – | 116.7 | 116.7 | 1832 | 21 | 1.62 |
| Ac90 | Adelie Cove | 10 | 195.8 | 159 | 354.8 | 13325 | 20 | 1.15 |
| H1D* | near Adelie | 16 | 89.3 | 3 | 92.3 | 6236 | 19 | – |
| Bd90 | Base dock | 5 | – | 110 | 110 | 6776 | 15 | 1.79 |
| Bd94 | Base dock | 5 | 234.1 | 82.7 | 316.8 | 5009 | 21 | 1.25 |
| Bd95 | Base dock | 5.5 | 70.5 | 151.9 | 222.4 | 1741 | 17 | 1.43 |
| TB90 | Tethys Bay | 6 | – | 95.9 | 95.9 | 1191 | 13 | 1.86 |
| TBp90 | Tethys pack | 11 | – | – | – | 863 | 15 | 0.83 |
| TBp95 | Tethys pack | 6.8 | – | 16.8 | 16.8 | 124 | 15 | 2.08 |

* dredge sample.

into main taxa (phyla or orders), identified at the highest taxonomic level possible (species or genus) and counted. Faunal species richness (number of taxa) and diversity (Shannon-Wiener index, $H' = -\Sigma i\ pi \ln pi$), as well as plant biomass (wet weight) and animal density (number of individuals) were calculated in each of the studied samples. Spearman rank pairwise correlation was calculated for each of the considered variables in relation also to depth. Multivariate analysis of faunal data (MDS technique) was performed on a matrix based on the number of individuals of the main taxa identified, transformed in double square root. Since the taxonomic analysis was not complete in all samples considered (e.g. isopods, amphipods and polychaetes), the MDS was performed on a reduced set of taxa (55). The analysis was performed at least five times and the ordination model selected was the one with the lowest stress value. On the MDS results, a cluster analysis (dissimilarity, based on the Euclidean distance) was performed.

## Results

### Species Composition, Community Structure and Distribution at Different Spatial Scales

Thirty-one different samples were analyzed at all sites and years considered (Table 1). Macroalgae consisted of two main species of Rhodophyta, *Iridaea cordata* and *Phyllophora antarctica*. The main structural (e.g. biomass) and functional parameters (photosynthetic features) of the two algal species, as well as of their associated animal communties are summarized in both Tables 1 and 2.

In association with the dominant macroalgae, more than 83 000 specimens of macrobenthic organisms, belonging to 62 taxa (identified at various taxonomic levels, from phylum to species), were collected (Table 3). Among these, the most diversified taxonomic groups were polychaetes (19 taxa), amphipod crustaceans (14 taxa), and molluscs (11 taxa; 8 gastropods and 3 bivalves). From a quantitative point of view, the dominant groups were molluscs (56.6% of the individuals, mainly gastropods) and tanaid crustaceans (21.6%), followed by amphipods (10%), isopods (7.5%) and polychaetes (4.2%). In order to facilitate analysis of the spatial and temporal distribution of individual species and the reading of Table 3, faunal abundance data of various samples and years collected at Faraglione, have been grouped according to depth range. Twenty-five species were found only in one or two samples, with a few individuals. Among the remaining taxa, the most abundant, that occurred almost in all years and samples studied (Table 3), were *Harmothoe* spp. (mainly belonging to two forms of *H. spinosa*), *Pionosyllis comosa*, *Kefersteinia fauveli* and *Ophryotrocha claparedii* (polychaetes); *Laevilitorina antarctica*, *Powellisetia deserta*, *Onoba gelida* and *O. turqueti*, and *Philobrya sublaevis* (molluscs); *Nototanais dimorphus* (tanaids); *Munna antarctica*, *Paramunna rostrata* and *Austrosignum* sp. (isopods); *Oradarea acuminata*, *O. rossii*, *Paramoera walkeri* and *Prostebbingia serrata* (amphipods). The echinoderms *Sterechinus neumayeri* and

**Table 2.** Main structural and functional features of algal facies and associated animal communities. Data on photosynthetic features are from L. Mazzella (unpubl. data)

| | *Iridaea cordata* | *Phyllophora antarctica* |
|---|---|---|
| Algal structural features | | |
| Morphology | Simple, laminar | Complex, branched |
| Depth distribution (m) | Shallow (2–20 m) | Deeper (5–35 m) |
| Max Biomass (g ww $m^{-2}$) | 2500 | 990 |
| Mean Biomass (SD) | 1190.6 (767.5) | 446.6 (329.3) |
| Mean total algal biomass | Shallow samples (2–6 m) | Deep samples (6–16 m) |
| (g ww $m^{-2}$) (SD) | 1403 (703) | 575.6 (532.5) |
| Photosynthetic features | | |
| Photoinhibition ($\mu E\,m^{-2}\,s^{-1}$) | 300 | 150 |
| Ik ($\mu E\,m^{-2}\,s^{-1}$) | 80 | 100 |
| Max photosynthesis rate ($\mu mol\,O_2\,g^{-1}\,min^{-1}$) | 0.17 | 0.15 |
| Q10 | 1.25 | 2.20 |
| Mean production ($mg\,C\,m^2\,h^{-1}$) | 120.7 | 40.19 |
| Animal populations | | |
| Max no. of taxa | 21 | 29 |
| Mean no. of taxa (SD) | 16.3 (5.5) | 19.8 (5.6) |
| Max diversity (H') | Low (1.29) | Higher (2.16) |
| Max abundance (no.ind $m^{-2}$) | Up to 80 000 | Up to 20 000 |
| Max biomass (g ww $m^{-2}$) | Up to 96.4 | Up to 20.4 |
| Main associated taxa | *Harmothoe* spp.<br>*Munna* spp.<br>*Paramoera walkeri*<br>*Laevilitorina antarctica*<br>*Sterechinus neumayeri* | *Harmothoe* spp.<br>*Kefersteinia fauveli*<br>*Pionosyllis comosa*<br>*Nototanais dimorphus*<br>*Oradarea* spp.<br>*Onoba* spp.<br>*Powellisetia deserta*<br>*Philobrya sublaevis*<br>*Sterechinus neumayeri*<br>*Odontaster validus* |
| Degree of patchiness | High | Lower |
| Interannual variability | Higher | Lower |

**Table 3.** List and abundance of the animal taxa found in the studied stations and years. The data at Faraglione (Fa) for the years 1990 and 1994 have been grouped according to depth range: shallow (0.5–6 m) and deep (12–16 m)

| Locality/year | Fa90 | Fa90 | Fa94 | Fa94 | Fa95 | Fa95 | Bd90 | Bd94 | Bd95 | Ac90 | H1D90 | TB90 | TBp90 | TBp95 |
|---|---|---|---|---|---|---|---|---|---|---|---|---|---|---|
| Depth (m) | 0.5–6 | 12–16 | 0.5–6 | 12–16 | 3 | 16 | 5 | 5 | 5.5 | 10 | 16 | 6 | 11 | 6.8 |
| Alcyonaria (colonies) | | | | | | | | | | | | | | 5 |
| Oligochaeta | | | | | | | | | | 2 | | | | |
| Polychaeta | | | | | | | | 79 | | | | | | |
| *Harmothoe spinosa* form A Kinberg | 119 | 111 | | | | | | | | | | | | |
| *Harmothoe spinosa* form B Kinberg | 157 | 184 | | | | | | | | | | | | |
| *Harmothoe* spp. | | | 30 | 108 | 40 | 168 | 93 | 49 | 16 | 55 | 136 | 16 | 21 | 1 |
| *Harmothoe* cf. *monroi* (Uschakov) | | 1 | | | | | | | | | | | | |
| *Harmothoe* sp. 1 | 3 | | | | | | | | | | | | | |
| *Eunoe anderssoni* (Bergstrom) | 8 | 6 | | | | | | | | | | | | |
| *Aglaophamus tryssophyllus* Hartman | | | | | | | | | | 3 | 31 | | 1 | |
| *Pionosyllis* cf. *comosa* Gravier | 43 | 290 | 7 | 1 | 15 | 161 | 456 | | 132 | 135 | 2 | 170 | 4 | 29 |
| *Autolytus* sp. | | 2 | | | | | | | | | | | | |
| *Kefersteinia fauveli* Averincev | 3 | 16 | | | | 60 | | 7 | 5 | | | | | 3 |
| *Ophryotrocha claparedii* (Studer) | 3 | | 24 | | | 3 | | | | 5 | 86 | | 1 | |
| Cirratulidae gen. sp. | | | 1 | | | 1 | | | | | | | | |
| *Capitella capitata* (complex) Fabricius | 2 | 1 | | | | | | | | | | | | |
| Capitellidae gen. sp. | | | 2 | | 3 | 2 | | | | | | | | |
| Spionidae gen. sp. | | | | | | | | | | 4 | | | | |
| Maldanidae gen. sp. | | | | | | | | | | 12 | | | | |
| Terebellidae gen. sp. | | 8 | | | | 11 | 5 | | 1 | 2 | | 1 | | |
| *Helicosiphon biscoensis* Gravier | 1 | 38 | | | | 1 | | | | | | | | |
| Serpulidae gen. sp. | | 6 | | | | 3 | | | | | | | | |
| Spirorbidae gen. sp. | | 62 | | | | | | | | | | | | |
| Mollusca Gastropoda | | | | | | | | | | | | | | |
| *Laevilitorina antarctica* (Smith) | 13 283 | 1874 | 2868 | 1099 | 5724 | 1796 | 747 | 3274 | 1004 | 5726 | 3744 | 445 | 3 | 1 |
| *Onoba turqueti* (Lamy) | 42 | 140 | | 63 | 4 | 114 | 306 | 2 | 16 | 9 | 638 | 17 | 1 | |
| *Onoba gelida* (Smith) | 66 | 140 | | 23 | 1 | 222 | 85 | 23 | 18 | 34 | 13 | 3 | 1 | |
| *Powellisetia deserta* (Smith) | 341 | 598 | 1 | 147 | 20 | 837 | 715 | 80 | 273 | 26 | 23 | 141 | 3 | |
| *Margarites refulgens* (Smith) | | 42 | | 6 | 3 | 60 | 114 | | 11 | 24 | | 37 | 2 | |
| *Falsimargarita iris* (Smith) | | 1 | | | | | | | | | | | | |
| *Neobuccinum eatoni* (Smith) | | | | | | | | | | | | | 1 | |
| *Cuthona georgiana* (Pfeiffer) | | 1 | | | | | | | | | | | | |
| Mollusca Bivalvia | | | | | | | | | | | | | | |
| *Philobrya sublaevis* (Pelseneer) | 63 | 152 | | | 1 | 16 | 237 | 2 | 2 | 6 | | 36 | | |
| *Adamussium colbecki* (Smith) (juv.) | | | | | 3 | | | | | | | | | |
| Bivalvia gen. sp. | | | 6 | 5 | | | | | | 2 | 2 | | | |
| Brachiopoda | | | | | | | | | | | | | | 1 |
| Crustacea Tanaidacea | | | | | | | | | | | | | | |
| *Nototanais dimorphus* (Beddard) | 2066 | 3058 | 17 | 305 | 249 | 2186 | 2962 | 468 | 148 | 5432 | 95 | 135 | 655 | 32 |
| *Nototanais antarcticus* Hodgson | | | 10 | 17 | | | | 1 | | | 225 | | | |
| Crustacea Isopoda | | | 10 | 167 | | | 994 | 24 | 15 | 1809 | 310 | 155 | 143 | |
| *Munna antarctica* Pfeffer | 730 | 194 | | | | | | | | | | | | |
| *Munna* sp. (juv.) | 159 | 75 | | | 47 | 43 | | 90 | | | | | | 16 |
| *Paramunna rostrata* (Hodgson) | 55 | 39 | | | | | | | | | | | | |

**Table 3.** *Continued*

| Locality/year<br>Depth (m) | Fa90<br>0.5–6 | Fa90<br>12–16 | Fa94<br>0.5–6 | Fa94<br>12–16 | Fa95<br>3 | Fa95<br>16 | Bd90<br>5 | Bd94<br>5 | Bd95<br>5.5 | Ac90<br>10 | H1D90<br>16 | TB90<br>6 | TBp90<br>11 | TBp95<br>6.8 |
|---|---|---|---|---|---|---|---|---|---|---|---|---|---|---|
| Paramunnidae gen. sp. | | | | | 24 | 254 | | 288 | 25 | | | | | 5 |
| *Austrosignum* sp. | 494 | 63 | | | | | | | | | | | | |
| *Austrofilius furcatus* Hodgson | 5 | 13 | | | 1 | 6 | | | | | | | | 2 |
| *Gnathia* sp. | | | | | | | | | | | | | | 3 |
| Crustacea Amphipoda | 48 | | | | | 1 | 55 | | | 34 | 204 | 32 | 25 | |
| *Haplocheiria plumosa* Stebbing | | 16 | | | | 14 | | | | | | | | |
| *Oradarea acuminata* Thurston | 86 | 137 | | | 187 | 188 | | 137 | 54 | | | | | 12 |
| *Oradarea rossii* Thurston | 32 | 31 | 5 | 37 | 7 | 18 | | 15 | 1 | | | | | |
| *Paramoera walkeri* (Stebbing) | 2129 | 11 | 2785 | 95 | 48 | 1 | | 316 | | | 559 | | | |
| *Prostebbingia serrata* Schellenberg | 61 | 346 | | 1 | 8 | 20 | | 9 | 3 | | | | | |
| *Orchomenella pinguides* (Walker) | 3 | 2 | | | | 1 | | | | | | | | |
| *Thaumathelson herdmani* Walker | | | | | | | | | | | | | | 1 |
| Stenothoidae gen. sp. | | | | | 1 | 2 | | | | | | | | |
| Stenothoidae gen. sp. 1 | | | | | 1 | | | | | | | | | |
| Stenothoidae gen. sp. 2 | | | | | | 1 | | | | | | | | |
| Stenothoidae gen. sp. 3 | | | | | | 1 | | | | | | | | |
| Hyalidae gen. sp. | | | | | 1 | | | | | | | | | |
| Lysianassidae gen. sp. | | | 1 | 1 | | | | | | | | | | |
| *Heterophoxus* sp. | | | | 1 | | | | | | | | | | |
| Pycnogonida | | | | | | | 2 | | | | | | 2 | |
| Echinodermata | | | | | | | | | | | | | | |
| *Sterechinus neumayeri* (Meissner) | 3 | 4 | | 8 | 1 | 13 | | | | | 2 | | | |
| *Odontaster validus* (Koehler) | | 3 | | 4 | 2 | 10 | | | | | 1 | 1 | | 2 |
| *Diplasterias brucei* (Koehler) | | 1 | | | | | | | | | | | | |
| Ophiuroidea gen. sp. | | | | | | | 1 | | | | 1 | | | |
| Pisces | | | | | | | | | | | | | | |
| *Trematomus* sp. | | | | | 1 | | | | | | | | | |

*Odontaster validus*, generally very abundant in these environments, as observed by SCUBA diving (Taviani and Amato 1989), or by ROV imaging (Cattaneo-Vietti et al., this Vol.), were quite scarce in our samples due to the reduced sampling surface considered, with respect to the size of these organisms (Gambi and D'Agostino 1994). These latter species, as well as those cited above, can be considered as typical components of the algal communities, showing different distribution patterns and densities, according to depth, algal type and patchiness in algal biomass.

In general, the fauna of the shallower samples (0.5–6 m), dominated by the *I. cordata* facies, was characterized by relatively fewer species (max. = 21, mean = 16.3 ± 5.5) (Table 1), mainly *Laevilitorina antarctica*, *Munna* spp., and *Paramoera walkeri* (Table 2). However, in this algal facies some of the highest densities (up to 82 000 ind $m^{-2}$) and biomass (up to 96 g $m^{-2}$ wet weight) were recorded, due mainly to the above mentioned taxa and also to *Harmothoe* spp. and *Nototanais dimorphus* (Gambi and D'Agostino 1994; Gambi et al. 1994). At the deeper stations (10–16 m), dominated by *P. antarctica*, species richness was higher (max. = 29, mean = 19.8 ± 5.6) (Table 1), and characterized also by other common taxa such as *Pionosyllis comosa*, *Kefersteinia fauveli*, *Powellisetia deserta*, *Onoba turqueti*, *O. gelida*, *Philobrya sublaevis*, *Prostebbingia serrata*. However, both the densities and the biomass were lower than in the *Iridaea* facies (Gambi et al. 1994). The only exception was sample Ac90 at Adelie Cove, where the high density recorded (up to 83 000 ind $m^{-2}$) was probably related to the influence of a nearby Adelie penguin rookery that enhanced the organic matter input to the bottom. In this area, in fact, very high densities were also observed for the bivalve and polychaete populations of the soft bottoms (Russo and Gambi 1994; Gambi et al. 1997). On the other hand, the lowest densities were recorded in the two samples collected under the pack ice, and at some stations in the Faraglione area sampled in summer 1994 (Table 1). Correlations among the community structural variables and depth (Spearman rank order correlation, Table 4) showed that both the *Iridaea* and the total algal biomass decreased significantly with depth ($P < 0.05$), whereas, both the number of animal taxa and the diversity (H′) increased with depth. Animal abundance increased with increasing biomass of *Phyllophora* as well as total algal biomass; this was probably the reason why diversity H′ decreased accordingly. The number of taxa was also positively correlated to both number of individuals and diversity (Table 3).

Despite these depth-related differences, a great homogeneity was recorded at the various sites examined with regards to species composition. Local differences at the various sites and samples were due to the relative dominance of some common taxa or to the presence of quite rare ones. The distribution pattern of these samples is summarized in the MDS analysis (Fig. 2) and in the resulting cluster analysis (Fig. 3). The MDS ordination grouped the majority of the shallower stations in the positive part of the F1 axis and negative of F2, whereas the majority of the deeper stations were concentrated in the negative part of F1 and positive of F2 (Fig. 2). The cluster analysis revealed the presence of at least seven main larger clusters (reported also in Fig. 2): cluster A grouped the shallowest stations (0.5–6 m depth) at Faraglione where macroalgal covering was null or very low (see Table 1); cluster B included three shallow samples characterized by the high biomass of *I. cordata* and high density of *Laevilitorina antarctica*; cluster C included shallow and intermediate samples and some deeper samples of 1994, where algal biomass (mainly *I. cordata*) and animal density were again relatively low; cluster D represented the two samples collected under the fast ice at Tethys Bay; clusters E and F, which were separated at a very low dissimilarity level, grouped most of the deep samples (12–16 m depth) at Faraglione sampled in 1990 and 1995, and cluster G the two samples collected in 1990 at Adelie Cove and at the base dock that were particularly rich and diversified (Table 3). Sample 3a90 was isolated from all the others.

The distribution pattern at the small spatial scale was studied at Adelie Cove (Ac90) where it was possible to analyze separately the thalli of *Iridaea* and *Phyllophora*, and the underlying algal turf and sediment. The results showed once again that faunal composition and density were related to algal type. The three subsamples showed clear differences in animal composition and abundance (Fig. 4). The *Iridaea* thalli hosted almost exclusively the gastropod *Laevilitorina antarctica*, mainly represented by adult specimens (La Magna 1995), while isopods and tanaids were also

**Table 4.** Coefficients (and probability, *P*) of the Spearman rank correlation analysis among the studied structural community variables and depth. Significant coefficients are indicated in bold character

| | *Iridaea* | Phyll. | Total | No. ind | No. taxa | H′ |
|---|---|---|---|---|---|---|
| Depth | **−0.63** | 0.15 | **−0.60** | −0.01 | **0.60** | **0.67** |
| P | 0.01** | 0.50 n.s. | 0.00** | 0.92 n.s. | 0.00** | 0.00** |
| *Iridaea* | | 0.29 | **0.90** | 0.50 | −0.34 | **−0.58** |
| P | | 0.43 n.s. | 0.00** | 0.07 n.s. | 0.23 n.s. | 0.03* |
| *Phyll.* | | | **0.65** | **0.44** | 0.37 | 0.14 |
| P | | | 0.00** | 0.04* | 0.08 n.s. | 0.51 n.s. |
| Total | | | | **0.56** | −0.08 | **−0.58** |
| P | | | | 0.03* | 0.67 n.s. | 0.00** |
| No. ind. | | | | | **0.44** | 0.11 |
| P | | | | | 0.01** | 0.54 n.s. |
| No. taxa | | | | | | **0.56** |
| P | | | | | | 0.00** |

n.s. = not significant.
* significant $P < 0.05$.
** highly significant $P < 0.01$.

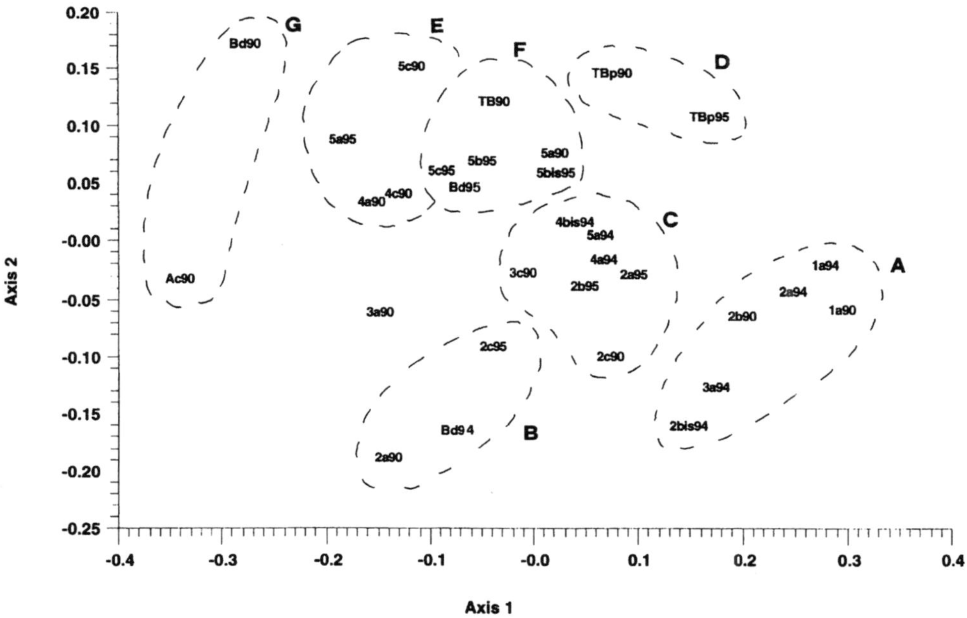


**Fig. 2.** MDS ordination model of studied samples according to the first two axes. Stress = 0.11 (for headings of samples see Table 1). The groups identified with the cluster analysis are circled and indicated with the same capital letters as in Fig. 3

abundant on the *Phyllophora* thalli. Polychaetes and juveniles gastropods occurred mainly on the underlying sediment accumulated at the base of the thalli (La Magna 1995). These microdistribution patterns reproduced the previously described depth related pattern quite well.

### Interannual Variability

Interannual differences in algal biomass and animal density were recorded in some samples at Faraglione as well as at the Base dock (Bd samples) where we could compare 3 years of observations. The stations at Faraglione showed the strongest fluctuations with a remarkable reduction in animal diversity and density (almost one order of magnitude) between the summers of 1990 and 1994, followed by high values in 1995 (see Table 1). These interannual differences seemed more pronounced at the shallower stations (2–6 m depth), characterized by the dominance of *I. cordata* where the most abundant species were *Laevilitorina antarctica*, *Nototanais dimorphus* and *Paramoera walkeri* as compared to deeper stations. Temporal variability at the base dock stations was less pronounced. These differences may have been due to both fluctuations in sea ice development and persistence, but also to patchiness of algal covering and biomass. As regards sea ice conditions, in summer 1994 a delay of about 40 days in the melting of the seasonal fast ice and pack ice was observed at Terra Nova Bay (Fig. 5). This delay could have affected algal development and consequently the colonization and density of the associated fauna.

## Discussion

The benthic associations of the shallow hard bottoms off Terra Nova Bay were characterized by relatively few species, but high densities, and the diversity showed consistently low values. In other Antarctic and sub-Antarctic zones, the presence of large perennial brown algae, absent in the Ross Sea (Cormaci et al. 1992a), allows for the development of much more complex and diversified communities (Arnaud 1974; Klöser et al. 1996).

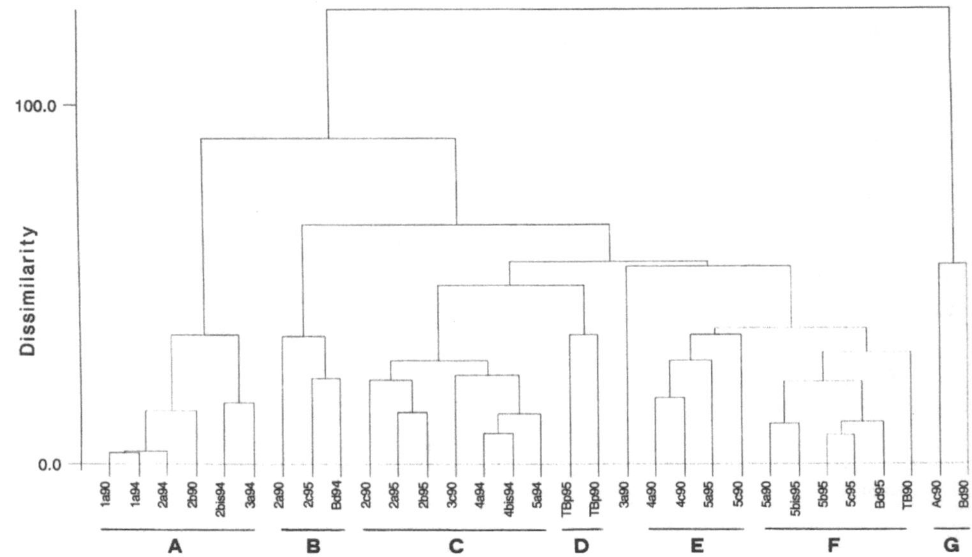


**Fig. 3.** Cluster analysis of the MDS data (based on dissimilarity) of the studied samples (for headings of samples see Table 1). The main clusters have been *underlined* and lettered

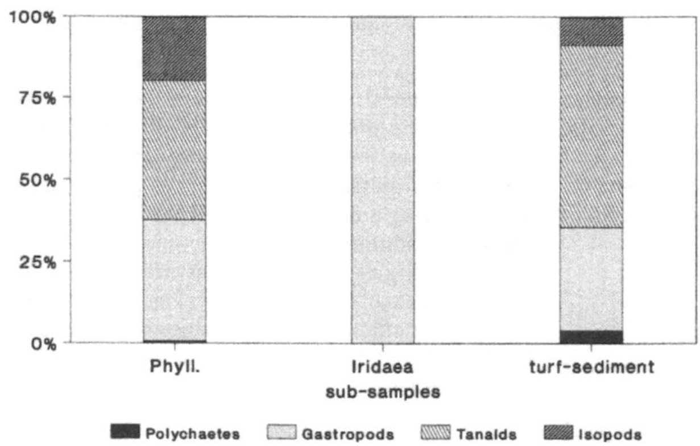


**Fig. 4.** Relative dominance (% number of individuals) of the main faunal groups in the sub-samples of Adelie Cove (Ad90): *Phyll Phyllophora antarctica*; *Iridaea Iridaea cordata*; *turf-sediment* underlying algal turf and sediment

The clear depth zonation showed by the two dominant macroalgal species seemed related not only to physical disturbance due to sea ice scouring, but also to eco-physiological adaptations to the light regime. Experiments on the photosynthetic features of the two dominant macroalgae (L. Mazzella, unpubl. data) revealed, in fact, that *P. antarctica* was better adapted to live in deeper waters with lower light intensity than *I. cordata.* Both species had maximum photosynthetic rates between 0.15 and 0.20 μmol $O_2 g^{-1}$ (at T = 0°C) with saturation occurring at low irradiance levels ($100\,\mu E\,m^{-2}\,s^{-1}$) (Table 2, from L. Mazzella, unpubl. data).

The diversity of the vagile fauna increased with increasing depth depending on algal type and standing crop. The highest number of species and diversity was observed in association with the *Phyllophora* belt, probably in relation to the calmer environment which offers shelter and a

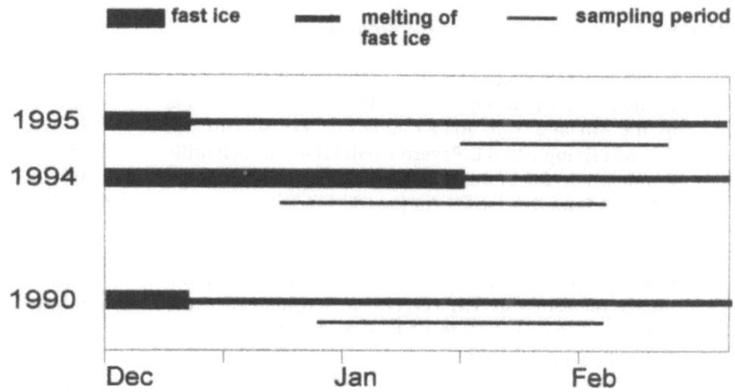


**Fig. 5.** Temporal evolution of fast-ice melting recorded during SCUBA diving sampling at Faraglione (Terra Nova Bay) in the austral summers 1990, 1994 and 1995

higher number of microhabitats for small, motile animals. *Phyllophora*, with branched, slender and fringed thalli, has a more complex morphology and architecture than the simple-bladed *Iridaea*. A similar pattern was observed at Vestfold Hills in the zoobenthos associated with the macrophytes, where the maximum diversity and biomass was recorded in the *Phyllophora* facies (Dhargalkar et al. 1988). On the other hand, the maximum density observed in the *Iridaea* facies was only partly related to the greater availability of food for the few taxa that occur in these stressful environmental conditions. The analysis of trophic guilds revealed, in fact, that most of the vagile species were mesoherbivores or detritivores feeding on living and dead algal tissues or on plant epiphytes (Gambi and Mazzella 1991). Some *in situ* experiments on plant detritus (Costantini et al. 1996), and laboratory experiments on the grazing of macroalgae by *Laevilitorina antarctica* and *Paramoera walkeri* (M.C. Buia et al. unpubl. data), seem to confirm this trophic behaviour.

Finally, it is interesting to point out that the basic zonation pattern observed in relation to depth was also observed at the small scale (within the same station as at Adelie Cove) suggesting a microniche differentiation among the various species according to algal type to reduce interspecific competition. With regards to temporal variability, it is interesting to point out that, besides the strong numerical fluctuations recorded in the different years, the basic structure of the community was maintained over time. Previous studies on benthic communities at long temporal scales are quite scarce in the Ross Sea area (Dayton 1989; Battershill 1990); such studies show dramatic changes in community structure, mainly due to the disturbing action of marine ice. Ice disturbance at shallow depths in Antarctic coastal zones exterts a strong selective pressure on organisms, and its effects on benthic community structure have been well documented (see Dayton 1990 for a review). These studies, however, considered mainly a community dominated by sessile filter feeders (sponges, cnidarians), with relatively long life spans. In our case study, most of the species were quite motile (e.g. *Paramoera walkeri*, *Harmothoe* spp.), with a relatively high migration potential along depth (e.g. *Nototanais dimorphus*: see Gambi et al. 1994), and probably have a much shorter life span (Pearse et al. 1991). Previous autoecological analysis of these species revealed the occurrence of different and contrasting reproductive strategies, including direct development (within egg capsules) (e.g. *Laevilitorina antarctica*, *Powellisetia deserta*) and brooding of embryos (e.g. *Philobrya sublaevis*) (Gambi et al. 1994). This latter feature, in particular, may also allow the restoration of dense populations, starting from relatively few mature specimens, and can be viewed as a prerequisite to colonize these fluctuating and disturbed environments.

*Acknowledgements.* Thanks are due to Dr. R. La Magna for collaboration in data collection, and to Mr M. Catra for help in SCUBA diving sampling during the 1993/94 expedition. This study was funded by the PNRA and conducted in the framework of the project Ecology and Biogeochemistry of the Southern Ocean.

## References

Arnaud PM (1974) Contribution à la bionomie marine benthique des régions antarctiques et subantarctiques. Téthys 6(3): 465–656

Battershill CN (1990) Temporal changes in Antarctic marine benthic community structure. Antarct Rec N Z 10(1): 23–27

Buia MC, Mazzella L, Gambi MC, Catra M Benthic populations of the shallow hard bottoms off Terra Nova Bay: results of the 9th and 10th Italian Antarctic expeditions. In: Faranda F, Guglielmo L, Povero P (eds) National Scientific Commission for Antarctica Oceanographic Campaign 1994–95, Data Rep Part II (in press)

Cormaci M, Furnari G, Scammacca B, Casazza G (1992a) Il fitobenthos di Baia Terra Nova (Mare di Ross, Antartide): osservazioni sulla flora e sulla zonazione dei popolamenti. In: Gallardo VA, Ferretti O, Moyano HI (eds) Oceanografia in Antartide. ENEA-PNRA-EULA, Concepcion, Chile, pp 395–408

Cormaci M, Furnari G, Scammacca B (1992b) The benthic algal flora of Terra Nova Bay (Ross Sea, Antarctica). Bot Mar 35: 541–552

Costantini ML, Fazi S, Rossi L (1996) Size distribution of the amphipod *Paramoera walkeri* (Stebbing) along a depth gradient in Antarctica. Hydrobiologia 337: 107–112

Dayton PK (1989) Interdecadal variation in an Antarctic sponge and its predators from oceanographic climate shifts. Science 245: 1484–1486

Dayton PK (1990) Polar Benthos. In: Smith WO (ed) Polar oceanography, part B: Chemistry, biology and geology. Academic Press, London, pp 631–685

Dayton PK, Robilliard GA, Paine RT (1970) Benthic zonation as a result of anchor ice at McMurdo Sound, Antarctica. In: Holdgate MW (ed) Antarctic ecology I. Academic Press, London, pp 244–258

Dhargalkar VK, Burton HR, Kirkwood JM (1988) Animal associations with the dominant species of shallow water macrophytes along the coastline of Vestfold Hills, Antarctica. Hydrobiologia 165: 141–150

Gambi MC, D'Agostino T (1994) Biomass estimates of some zoobenthic species associated to seaweeds in the shallow hard bottoms off Terra Nova Bay (Ross Sea, Antarctica). In: Albertelli G, Cattaneo-Vietti R, Piccazzo M (eds) Atti X Congr Ass Ital Oceanogr Limnol, Lang, Genova, pp 691-698

Gambi MC, Mazzella L (1991) Second Italian oceanographic expedition in the Ross Sea, Antarctica: preliminary report on the quantitative and functional studies of benthic communities of Terra Nova Bay. In: National Scientific Commission for Antarctica (ed) Oceanographic Campaign 1989/90, Data Rep Part I, Genova, pp 349–367

Gambi MC, Lorenti M, Russo GF, Scipione MB (1994) Benthic associations of the shallow hard bottoms off Terra Nova Bay (Ross Sea, Antarctica): zonation, biomass and population structure. Antarct Sci 6(4): 449–462

Gambi MC, Castelli A, Guizzardi M (1997) Polychaete populations of the shallow soft bottoms off Terra Nova Bay (Ross Sea, Antarctica): distribution, diversity and biomass. Polar Biol 17: 199–210

Klöser H, Quartino ML, Wiencke C (1996) Distribution of macroalgae and macroalgal communities in gradients of physical conditions in Potter Cove, King George Island, Antarctica. Hydrobiologia 333: 1–17

La Magna R (1995) Benthic associations of the coastal hard bottoms off Terra Nova Bay (Ross Sea, Antarctica): spatial and temporal analysis and autoecology of some plant and animal species. PhD Thesis, University "Federico II", Naples 160 pp (in Italian)

Pearse JS, McClintock JB, Bosch I (1991) Reproduction of Antarctic benthic marine invertebrates: tempos, modes and timing. Am Zool 31: 65–80

Reichardt WT (1987) Burial of Antarctic macroalgal debris in bioturbated deep-sea sediments. Deep Sea Res 34(10): 1761–1770

Russo GF, Gambi MC (1994) First quantitative data on coastal soft bottom populations off Terra Nova Bay (Ross Sea, Antarctica): Bivalve molluscs. In: National Scientific Commission for Antarctica (ed) Oceanographic Campaign 1989–90. Data Rep Part II, Genova, pp 187–190

SCAR (1994) Coastal and shelf ecology of the Antarctic sea-ice zone (CS-EASIZ), a SCAR programme of marine research for the coastal and shelf ecosystem of Antarctica. SCAR rep 10: 1–20

Taviani M, Amato E (1989) Diving in Antarctica: third Italian expedition in Terra Nova Bay. Boll Oceanogr Teor Appl 7: 43–53

CHAPTER 39

# The Role of Sponges in the Terra Nova Bay Ecosystem

**R. Cattaneo-Vietti[1], G. Bavestrello[1], C. Cerrano[1], E. Gaino[2], L. Mazzella[3], M. Pansini[1], and M. Sarà[1]**

**ABSTRACT**

One hundred and twenty five demosponge species are known from the Ross Sea, 49 of which have been recorded in Terra Nova Bay within the framework of the PNRA ecological project. The most common species are *Tedania charcoti, Axociella nidificata, Calyx arcuarius, Isodictya erinacea, I. cactoides, I. conulosa, Gellius rudis, Gellius* spp., *Myxilla elongata* and *Phorbas glaberrima*. Two of the 49 species we found are new for Antarctica: *Esperiopsis informis* and *Isodictya conulosa*. Most of the sponges were collected at 70–120 m depth, where a great substrate heterogeneity allows the coexistence of different biocoenoses.

The success of sponges in habitats characterized by fluctuating food supply during the year is difficult to understand. In fact, in winter, oligotrophic conditions in the water column could represent a metabolic constraint for a filter feeder. A possible explanation could be the direct uptake of diatoms that manage to live for a long time within the sponge tissues. Sponges, therefore, by transferring energy from the water column to the benthos, play a key role in the Antarctic environment. An unusual physical phenomenon has been detected, by studying the sponge/diatom association, in the spicules of the hexactinellid *Rossella racovitzae*: spicules conduct light as natural optical fibres.

Sponges, and particularly hexactinellids, greatly affect sediment quality by forming mats, up to 1.5-m thick, because their siliceous spicules take a very long time to dissolve. At Terra Nova Bay the occurrence of spicules in the sediments shows three different arrangements: (1) free spicules in the sediments, (2) sphaeroid balls (aegagropila) of densely packed spicules, (3) spicule mats.

## Introduction

Sponges represent a major component of the Antarctic zoobenthos, remarkably contributing to the species richness of this continent, where 350 species of demosponges are known (Sarà et al. 1992). Some of these are characterizing elements of several assemblages (Dayton et al. 1974; Barthel et al. 1990) where they can reach biomass values of about 2–4 kg $m^{-2}$ wet wt (Beliaev and Ushakov 1957), comparable with the highest values found for sponges in tropical seas (Wilkinson 1987).

In the Ross Sea, sponge communities, between 70 and 180 m depth, have been described by Bullivant (1967) as the McMurdo Sound Mixed and the Glass Sponge assemblages. At McMurdo, Dayton et al. (1974) described a littoral belt, at 30–60 m depth, in which sponges are the keystone species. In addition, sponge spicules constitute conspicuous mats (Koltun 1968; White et al. 1985; Barthel 1992), thereby affecting both sediment texture and soft-bottom community structure.

The aim of this chapter is to focus on the results obtained within the Italian Programme for Antarctic Research (PNRA) about the Porifera living at Terra Nova Bay (Ross Sea).

## Taxonomic and Ecological Aspects

Forty-nine species, belonging to 7 orders and 15 families, have been recorded at Terra Nova Bay in waters 300–350 m depth (Table 1), out of the 125 demosponge species known from the Ross Sea

[1] Dipartimento per lo Studio del Territorio e delle sue Risorse dell' Università di Genova, Viale Benedetto XV 5, I-16132 Genova, Italy

[2] Dipartimento di Biologia Animale ed Ecologia dell' Università di Perugia, Via Elce di Sotto, I-06123 Perugia, Italy

[3] Lab. Ecologia del Benthos – Stazione Zoologica Anton Dohrn, Ischia, Naples, Italy

**Table 1.** List of demosponge species from Terra Nova Bay so far identified

| | |
|---|---|
| 1. | *Plakina monolopha* Schulze 1880 |
| 2. | *Plakina trilopha* Schulze 1880 |
| 3. | *Cinachyra barbata* Sollas 1886 |
| 4. | *Tetilla leptoderma* Sollas 1886 |
| 5. | *Pseudosuberites antarcticus* (Carter 1876) |
| 6. | *Pseudosuberites nudus* Koltun 1964 |
| 7. | *Sphaerotylus borealis antarcticus* Kirkpatrick 1908 |
| 8. | *Suberites montiniger* Carter 1880 |
| 9. | *Suberites caminatus* Ridley & Dendy 1887 |
| 10. | *Latrunculia biformis* Kirkpatrick 1908 |
| 11. | *Stylocordyla borealis* Loven 1868 |
| 12. | *Eurypon miniaceum* Thiele 1905 |
| 13. | *Homaxinella balfourensis* (Ridley & Dendy 1887) |
| 14. | *Homaxinella flagelliformis* (Ridley & Dendy 1886) |
| 15. | *Asbestopluma belgicae* Topsent 1902 |
| 16. | *Mycale acerata* Kirkpatrick 1907 |
| 17. | *Mycale fibrosa* Boury-Esnault & Van Beveren 1982 |
| 18. | *Mycale tridens* Hentschel 1914 |
| 19. | *Esperiopsis informis* Stephens 1915 |
| 20. | *Isodictya cactoides* (Kirkpatrick 1908) |
| 21. | *Isodictya conulosa* (Ridley & Dendy 1886) |
| 22. | *Isodictya erinacea* (Topsent 1916) |
| 23. | *Inflatella belli* (Kirkpatrick 1908) |
| 24. | *Ectyodoryx antarctica* (Hentschel 1914) |
| 25. | *Ectyodoryx nobilis* (Ridley & Dendy 1886) |
| 26. | *Ectyodoryx ramilobosa* (Topsent 1916) |
| 27. | *Ectyomyxilla* sp. |
| 28. | *Iophon radiatus* Topsent 1902 |
| 29. | *Iophon unicornis* Topsent 1907 |
| 30. | *Iophon* sp. |
| 31. | *Lyssodendoryx flabellata* Burton 1929 |
| 32. | *Myxilla asigmata* (Topsent 1902) |
| 33. | *Myxilla elongata* Topsent 1917 |
| 34. | *Myxodoryx hanitschi* (Kirkpatrick 1907) |
| 35. | *Tedania charcoti* Topsent 1907 |
| 36. | *Tedania tantula* (Kirkpatrick 1907) |
| 37. | *Phorbas glaberrima* (Topsent 1917) |
| 38. | *Axociella nidificata* (Kirkpatrick 1907) |
| 39. | *Artemisina tubulosa* Koltun 1964 |
| 40. | *Clathria toxipraedita* Topsent 1913 |
| 41. | *Calyx arcuarius* (Topsent 1913) |
| 42. | *Haliclona dancoi* (Topsent 1902) |
| 43. | *Haliclona penicillata* (Topsent 1908) |
| 44. | *Microxina benedeni* (Topsent 1901) |
| 45. | *Gellius pilosus* Kirkpatrick 1907 |
| 46. | *Gellius rudis* Topsent 1901 |
| 47. | *Gellius* sp. 1 |
| 48. | *Gellius* sp. 2 |
| 49. | *Dendrilla membranosa* Pallas 1776 |

(Sarà et al. 1992). A species belonging to the genus *Iophon* is probably new for science and will be described in a separate chapter. Among the most common demosponges, *Tedania charcoti*, *Gellius rudis* and *Axociella nidificata*, are present in more than half of the total samples, *Calyx arcuarius*, *Isodictya erinacea* and *Gellius* spp., in more than 40%, *Isodictya cactoides*, *I. conulosa*, *Myxilla elongata* and *Phorbas glaberrima*, in more than 30%.

A large part of these species was collected, using a triangular dredge, between 80 and 120 m depth, inside a wide belt characterized by a great substrate heterogeneity allowing the coexistence of different communities.

Only three species (*Inflatella belli*, *Calyx arcuarius* and *Axociella nidificata*) have been recorded deeper, below 300 m depth. On sparse rocky outcrops (Fig. 1), the community is dominated by a diversified filter-feeder assemblage characterized by sponges, anthozoans and holothuroids (Di Geronimo et al. 1992; Cattaneo-Vietti et al., this Vol.). This community develops within the Encrusting Algal Zone, characterized by the coralline alga *Clathromorphum lemoineanum* and shows a high richness in species and biomass, which can be only partially compared to that described for McMurdo Sound, between 30 and 60 m depth (III zone of Dayton et al. 1974). According to our data, the Terra Nova Bay community seems to differ from that of McMurdo in composition and structure because it is dominated by demosponges, whereas hexactinellids and calcareous sponges are represented by only a few species, with a low density. Among the volcano species (Dayton et al. 1974), *Rossella racovitzae* is uncommon and *R. nuda* and *Scolymastra joubinii* are occasionally found (Fig. 1).

## Biogeographic Aspects

Appropriate oceanographic conditions are required to allow a long-range dispersal of sponge species, because these sessile organisms have also short living larval stages. The main biogeographic characteristics of the Antarctic sponge fauna can be defined by the following characteristics (Burton 1932; Koltun 1970): (1) a high level of species-endemism coupled with negligible genus endemism; (2) a circumpolar distribution for many species; (3) a closer faunistic relationship with southern South America and Falkland Islands, than with South Africa, Australia and New Zealand. More recently, Sarà et al. (1992) highlighted typical Antarctic demosponge fauna indicated as the Antarctic Faunistic Complex (AFC), which is homogeneous all along the continent. The distinction made by several authors (Ekman

**Fig. 1.** Underwater photograph showing the sponge assemblage at Terra Nova Bay (95 m depth). Scale bar = 1 m

1935; Andriashev 1965; Kussakin 1967; Knox 1970) into western and eastern biogeographic subunits seems to be artificial. The ordination pattern from the correspondence analysis of sponge species from Antarctica and neighbouring areas (Sarà et al. 1992) shows that Magellan, Falklands and South Georgia tend to be close to the continental sector west of the Antarctic Peninsula. This analysis confirms the existence of faunal exchanges between Antarctica and South America via the Scotia Arc, proposed by Knox and Lowry (1977) as one of the mechanisms for the origin and dispersal of the Antarctic marine fauna.

The biogeographic analysis of the demosponges from Terra Nova Bay concerns 45 species, because four taxa were identified only to the generic level. Most of these (36 species) were already known from the Western Ross Sea, while only 3 species (*Pseudosuberites antarcticus*, *Suberites caminatus* and *Isodictya erinacea*) were reported from the Eastern Ross Sea sector. Of the remaining 6 species, *Homaxinella flagelliformis* was known from Kerguelen and Magellan area, and has been recently recorded from Terra Nova Bay (Pansini et al. 1994) and from the Weddel Sea (Gutt and Koltun 1995). *Mycale tridens* and *Artemisina tubulosa* were already known from other Antarctic areas (Sarà at al. 1992) and *Mycale fibrosa* from the western side of the Antarctic Peninsula and from Kerguelen Islands. *Esperiopsis informis* was known from South Africa and *Isodictya conulosa* from South Africa, South America and Australia. Both are new for Antarctica.

The available data support the circumpolar distribution of many Antarctic sponges. In addition, the new record for Antarctica of two species known from South Africa is remarkable, because the biogeographic affinity of the two areas is thought to be low.

## Relationship Between Sponges and Autotrophic Organisms

It is difficult to explain the success of sponges in Antarctic habitats which are characterized by a quantitatively fluctuating food supply during the year. Oligotrophic conditions in the water column in winter (Matsuda et al. 1987) could represent, in fact, a metabolic constraint for filter-feeding organisms. Antarctic sponges take up and store planktonic and benthic diatoms by means of the so-called pinacoderm entrapping activity (Gaino et al. 1994). This strategy, if confirmed, would imply that these organisms can rely on an alternative energetic source during oligotrophic months, considering also the large amount of polysaccharids produced by diatoms (Edgar 1983). In addition, this entrapping (Fig. 2) supports the existence of a functional adaptation of Antarctic sponges which, in summer months, rely on a food uptake mechanism that bypasses the aquiferous system, preventing a possible clogging of the choanocyte chambers caused by large amounts of organic matter that settle on their surface (Fabiano et al. 1997).

Relationships between sponges and autotrophic organisms (cyanobacteria, zooxanthellae

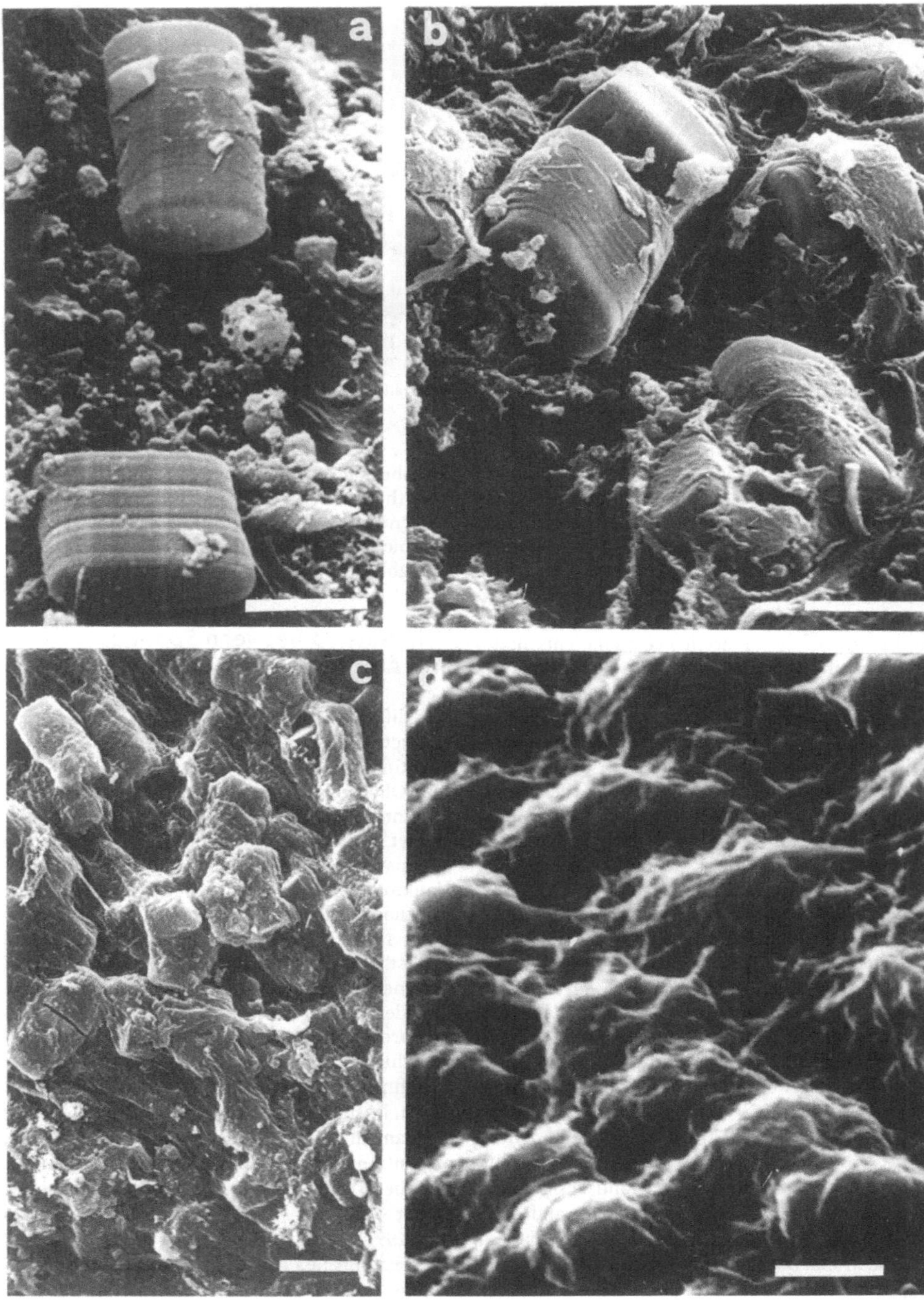


**Fig. 2a–d.**

and zoochlorellae) are well known in tropical (Wilkinson 1980) and temperate (Sarà 1971; Sarà and Vacelet 1973; Gaino et al. 1976; Arillo et al. 1993) waters where they represent a food source, in addition to organic particles coming from the water column (Wilkinson 1979, 1983). However, few data are available on possible feeding relationships between sponges and diatoms (Sarà 1966; Cox and Larkum 1983).

A few species of pennate and centric diatoms (Fig. 3), typical of Antarctic waters (Krebs 1983) and well adapted at very low irradiance levels (Palmisano et al. 1985), were found in *Suberites montiniger*, *Tedania charcoti*, *Phorbas glaberrima* and *Isodictya conulosa*. Benthic diatoms were represented by *Achnanthes* sp. (Fig. 3a,c) and *Pseudogomphonema* sp., whereas *Fragilariopsis curta* (Fig. 2) and *Fragilariopsis* sp. can be considered sympagic species, which represent the typical components of the sea-ice diatom flora (Krebs et al. 1987). Centric diatoms belong to the genera *Porosira* (Fig. 3a,b), *Coscinodiscus* and *Rhizosolenia*.

*Fragilariopsis* species were frequently found associated with all the examined sponges and reached a remarkable concentration in *Phorbas glaberrima* (Fig. 2) and *Tedania charcoti*. The assemblage living in *Suberites montiniger* (Fig. 3a–c) was characterized by *Achnanthes* sp. and *Porosira* sp., whereas *Pseudogomphonema* sp. was rarer and observed only in *Isodictya conulosa*. *Navicula* and *Nitzschia* were occasionally observed.

The occurrence of diatoms in sponge tissue was confirmed by chlorophyll values recorded from the sponge body. However, remarkable variations were measured among the considered species: the sole calcareous sponge examined (*Leuconia* sp.) (Table 2) showed low values of chlorophyll whereas the highest concentrations were found in the demosponges *Isodictya erinacea*, *I. conulosa*, *Inflatella belli*, *Phorbas glaberrima* and *Myxodoryx hanitschi*. The comparison of three specimens of *Myxodorix hanitschi* showed significant differences in Chl a content with values ranging from 5.5 to 16.1 $\mu g g^{-1}$ WW. In addition, chlorophyll concentrations were not uniform within the sponge body. The outer ectosome was richer in Chl a than the inner choanosome, characterized by a greater amount of Chl c in *Phorbas glaberrima*. By contrast, in *Iophon* sp. and particularly in *Inflatella belli*, the highest concentrations of Chl a and Chl c were found in the choanosome.

The amount of Chl a was about 2 to 4 times higher than Chl c; in *Phorbas glaberrima* the ratio Chl a/Chl c decreased from 4.7 to 1.3 from the ectosome to the choanosome. The very high values of Chl c, double those of Chl a (Table 2) recorded in *Suberites montiniger*, could be related to presence of *Achnanthes* sp., the dominant associated diatom in this sponge.

By studying these sponge-diatom associations, an unusual physical phenomenon was detected: the long spicules of the hexactinellid *Rossella racovitzae* were able to channel light as natural optical fibres (Cattaneo-Vietti et al. 1996). Presumably, this light-conducting system plays a metabolic role, affecting the occurrence, growth and distribution of associated diatoms inside the sponge body. Indeed, SEM observations showed numerous diatoms adherent to the spicules, thereby making such a physical phenomenon essential for diatom survival in rather deep waters at very low levels of light intensity.

## Spicule Mats

Particular attention has been paid to elucidate the role of sponge spicules in the composition of the littoral sediments of Terra Nova Bay. Even though siliceous sponges are known to be involved in the formation of important silica deposits (Bullivant 1967; Dearborn 1967; Koltun 1968; Dayton 1979; Barthel 1992; Frignani et al. 1992), their role has not been deeply investigated. The sponge siliceous spicules, especially those of the hexactinellids, take a long time to dissolve, can form mats, up to 1.5 m thick, contributing to a significant enrichment of biogenic silica (BSi) in the sediment and causing consequently a loss of available silica in the water column.

At Cape Armitage (Ross Island) these mats are mainly composed of *Cinachyra antarctica* spicules (Battershill 1989), while the thick spicule mats described from deeper waters by Dayton et al. (1974) are built up by an accumulation of spicules

**Fig. 2a–d.** SEM view of settled diatoms (*Fragilariopsis curta*) on the ectosome of *Phorbas glaberrima* (**a**), and different steps of the their uptake (**b–c**) and final incorporation (**d**). *Bars* 100 μm

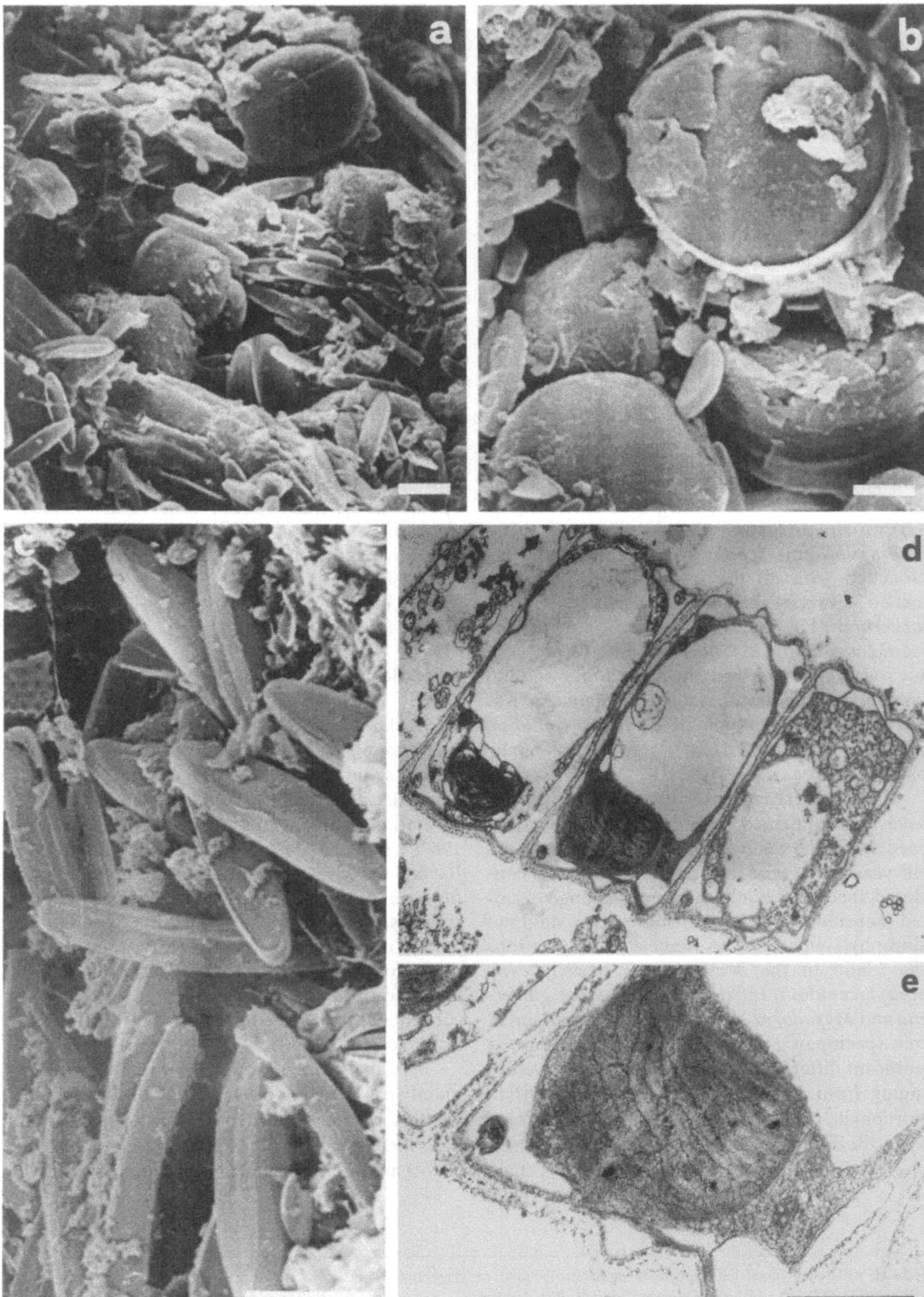


Fig. 3a–e.

**Table 2.** Chlorophyll values in Terra Nova Bay sponges collected from 70 to 100 m depth

| Species | No. of specimens | Chl a (μg/g) | Chl c (μg/g) |
|---|---|---|---|
| Calcispongiae | | | |
| *Leuconia* sp. | 3 | 0.67 ± 0.1 | 0.27 ± 0.1 |
| Demospongiae | | | |
| *Microxina benedeni* | 3 | 2.00 ± 0.3 | 1.53 ± 0.8 |
| *Artemisina tubulosa* | 3 | 2.67 ± 0.5 | 0.99 ± 0.1 |
| *Haliclona dancoi* | 3 | 2.86 ± 0.3 | 1.09 ± 0.3 |
| *Pseudosuberites nudus* | 3 | 2.95 ± 0.5 | 2.62 ± 0.6 |
| *Axociella nidificata* | 3 | 3.41 ± 0.2 | 0.99 ± 0.1 |
| *Gellius pilosus* | 6 | 3.43 ± 0.3 | 0.80 ± 0.1 |
| *Gellius rudis* | 3 | 5.28 ± 1.6 | 0.85 ± 0.1 |
| *Gellius* sp. 1 | 3 | 2.41 ± 1.0 | 0.90 ± 0.3 |
| *Dendrilla membranosa* | 3 | 5.68 ± 1.4 | 2.77 ± 1.2 |
| *Suberites montiniger* | 6 | 6.01 ± 0.9 | 11.12 ± 2.2 |
| *Tedania charcoti* | 6 | 6.11 ± 1.6 | 0.30 ± 0.7 |
| *Myxilla elongata* | 9 | 6.11 ± 2.0 | 1.67 ± 0.4 |
| *Isodictya erinacea* | 3 | 7.42 ± 0.9 | 3.52 ± 0.5 |
| *Myxodoryx hanitschi* | 9 | 9.80 ± 2.2 | 4.34 ± 0.8 |
| *Isodictya conulosa* | 3 | 10.73 ± 2.9 | 3.59 ± 0.8 |
| *Iophon* sp. (ectosome) | 3 | 1.56 ± 0.4 | 0.88 ± 0.4 |
| *Iophon* sp. (choanosome) | 3 | 3.68 ± 1.2 | 1.13 ± 0.5 |
| *Inflatella belli* (ectosome) | 3 | 0.69 ± 0.3 | 0.44 ± 0.1 |
| *Inflatella belli* (choanosome) | 3 | 9.00 ± 2.4 | 5.96 ± 1.7 |
| *Phorbas glaberrima* (ectosome) | 6 | 9.79 ± 3.2 | 2.10 ± 0.6 |
| *Phorbas glaberrima* (choanosome) | 6 | 5.44 ± 0.6 | 4.35 ± 1.3 |

coming from many sponge species. At Terra Nova Bay, spicule mats are formed by hexactinellid spicules particularly from *Rossella* sp. (Fig. 4a) while in the sediments from Cape Confusion, tetractinellid spicules, especially from *Cinachyra* sp., prevail (Fig. 4b).

The spicules found in Terra Nova Bay sediments show various aggregation types: (1) loose spicules, often broken, mixed with the sediment; (2) densely packed spicules forming sphaeroid balls (aegagropila) very variable in diameter, 10–90 mm in diameter (Fig. 4c), (3) stratified spicule mats, often formed by large amounts of aegagropila (Fig. 4d) with a silica ponderal value ranging between 50 and 80% of the total sediment weight. Aegagropila and mats were found along the coast from Tethys Bay to Adelie Cove, within a bathymetric range from 90 to 120 m depth, while free spicules were found practically in all the considered stations with densities varying from less than 5000 up to 500000 spicule $g^{-1}$ DW of sediment (Fig. 5). In addition, spicule mats were also detected in isolated sampling areas along the Campbell glacier tongue and in front of the Cape Confusion glacier at depths between 200 and 250 m. The amount of settled spicules gradually increases from the coast towards the open sea, showing a significant relationship with the filter-feeder assemblages of the hard bottom belt dominated by sponges between 80 and 120 m depth.

SEM analyses confirm the difficulty to dissolve opal at low temperatures. No signs of etching were observed on recent spicules (Fig. 4b), as well as on subfossil ones, collected in a deposit of the Hell Gate Glacier tongue.

**Fig. 3a–e.** SEM and TEM micrographs of diatoms living inside Antarctic sponges. **a** SEM general view of assemblages of *Achnanthes* sp. (**c**) and *Porosira* sp. (**b**) in *Suberites montiniger*. **d, e** TEM micrographs of incorporated diatoms. Note the sequence of the frustules (**d**) and the detail of intact photosynthetic apparatus (**e**). *Bars* **a** 20 μm; **b–e** 10 μm

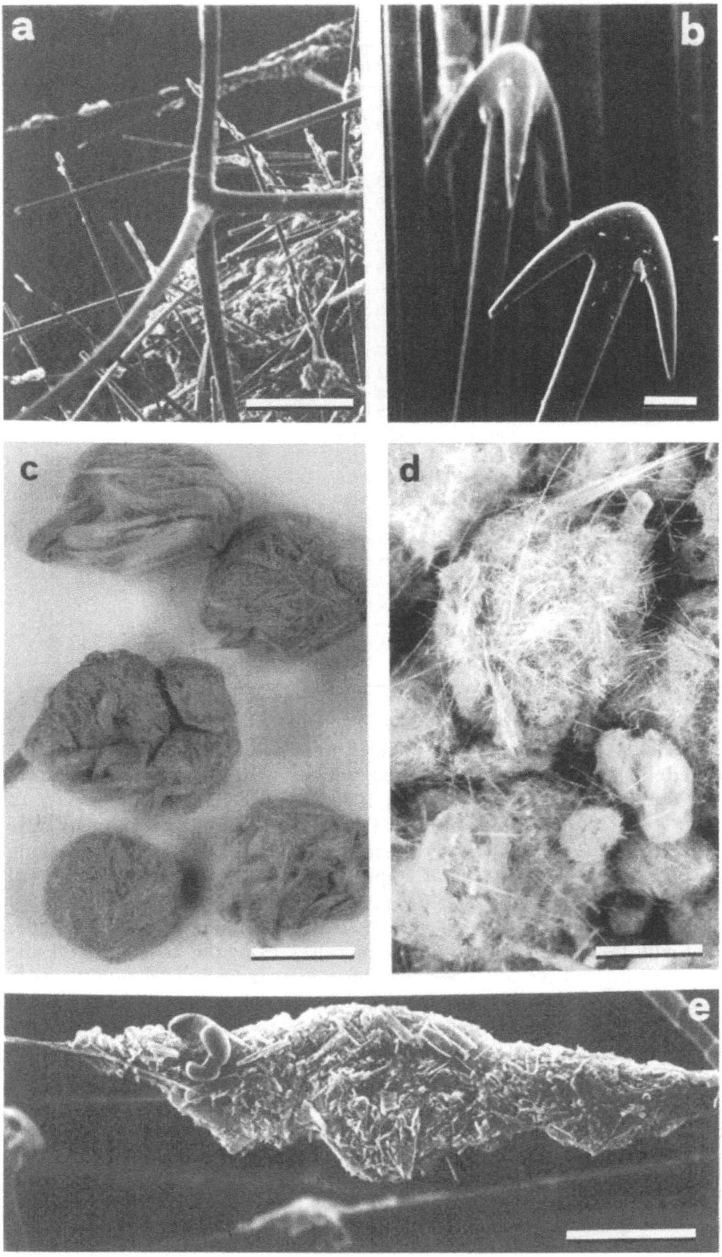


**Fig. 4a–e.** Spicule mats at Terra Nova Bay. SEM micrographs of a spicule mat constituted by hexactinellid (**a**) and tetractinellid (**b**) spicules. Free aegagropila in the sediments (**c**) and a stratified spicule mat constituted by innumerable aegagropila (**d**). Note diatoms growing on a burrowed spicule (**e**). *Bars* **a,e** 1 mm; **b** 100 µm; **c–d** 1 cm

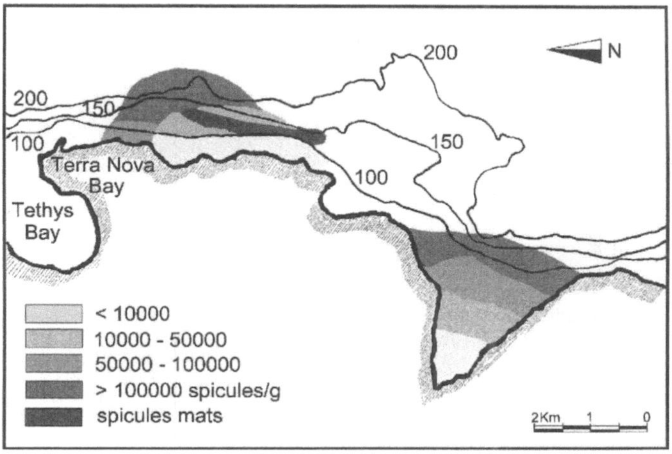


**Fig. 5.** Sponge spicule distribution in the sediments along the coast from Tethys Bay to Adelie Cove. The amount of spicules gradually increases from the coast towards the hard-bottom belt characterized by the sponge community, between 80 and 120 m depth

In order to verify the presence of free spicules also in off-shore sediments, additional observations were carried out on samples coming from the center of the northern part of the Joides Basin (site B: 74°S–175°E) and the northern flank of the Mawson Bank (site C: 72°30′S–175°E), where oceanographic moorings where deployed during the 1994 ROSSMIZE cruise (Labrozzi et al. 1998). Site B (586 m depth) can be considered a decantation area with very muddy sediments, while site C (456 m depth) was characterized by coarse sediment in an area of strong currents. The amount of free spicules found in site B ranged around 200000 spicule $g^{-1}$ DW of sediment, but was far lower in C: 4800 spicule $g^{-1}$ DW. These data agree with the remarkable BSi differences found in the sediment of the same sites by Labrozzi et al. (1998).

A possible ecological role of the spicules in the sediments may be advanced. The Chl a (Fig. 6), biomass and species richness values are higher in areas rich in spicules than in sites characterized by spicule-free sediments. A possible explanation may be bound to the sponge spicule capacity of creating a 3-D substratum on which larger diatom population can develop (Table 2), supporting therefore a richer associated fauna as observed by Dearbon (1967) at McMurdo. This process could be enhanced by the opal spicule capacity of conveying light as natural optical fibres (Cattaneo-Vietti et al. 1996) allowing diatoms to live on the spicules (Fig. 4e).

## Conclusions

In Antarctica, sponges can be considered as a key taxon, because of their high biomass and capacity to structure, by the spicules, soft-bottom sediments. In Antarctic assemblages, composed of long-living species, varied biological interactions prevail over physical factors. This is particularly true for the peculiar trophic strategy adopted by many sponges, able to take up and preserve dense concentrations of diatoms within their bodies. This strategy allows them to overcome severe food fluctuations typical of the Antarctic environment. The nature of these associations needs further investigation. It could be reasonable to suppose a symbiotic relationship, likewise observed for other autotrophic organisms associated to sponges in tropical and temperate oligotrophic habitats (Wilkinson 1979), even though a true endosymbiosis causes remarkable changes in the organization of diatom frustules (Reimer and Lee 1988).

The conspicuous sponge spicule mats found at Terra Nova Bay stress the important role of sponges in determining sediment texture with strong ecological, geological and sedimentological

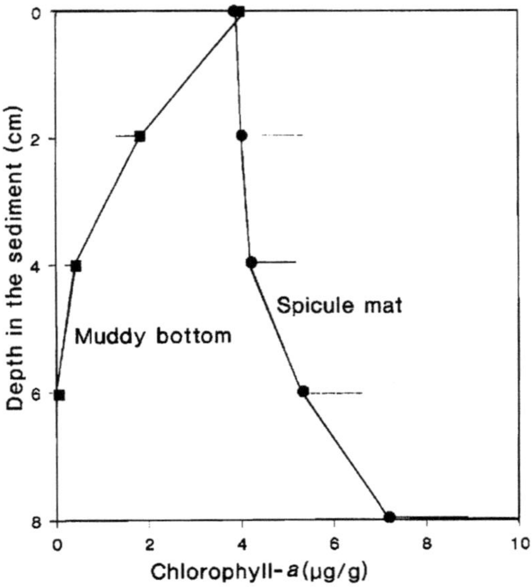


**Fig. 6.** Different trends of Chl a values at 120 m depth in muddy sediments with low amounts of spicules and in spicule mats (standard deviations are represented by *horizontal lines*)

implications (White et al. 1985). In fact, spicules create a more rich and structured 3-D substratum which favours the development of specific assemblages. The distribution of spicule mats is linked to some historical, biological and ecological factors. Most of these are constituted by hexactinellid spicules, a group not so frequent now in the bay to justify these large deposits. As a consequence, both spicule distribution and mat formation could reflect historical events (mass mortality of some species, iceberg scouring), in which currents have played a complex role. In fact, seafloor currents, according to their intensity, are able to disperse, aggregate and accumulate spicules along the coast, as suggested by the different shape of mat structures and by the patchy distribution of spicules.

The fact that opal sponge spicules may act as optical fibres, conveying light inside the sponge body, suggests that this physical phenomenon could be strictly linked to sponge evolution. In fact, autotrophic symbiosis in ancestral sponges was established probably during the late Precambrian (Wilkinson 1983), when a great expansion of siliceous sponges occurred. We can speculate that symbiosis, the winning strategy of this taxon, could be supported by this peculiar property of spicules.

*Acknowledgements.* This research was supported by the Italian Programma Nazionale di Ricerca in Antartide (PNRA) within the framework of the research project Ecology and Biogeochemistry of the Southern Ocean. The authors thank M.C. Gambi (Stazione Zoologica di Napoli), M. Ravaioli (CNR, Bologna) and N. Corradi (Università di Genova) for their help in sampling operations. Laura Paganelli and Paola Allegretti helped in data analysis.

## References

Andriashev AP (1965) A general review of the Antarctic fish fauna. In: van Oye P, van Meighem J (eds) Biogeography and ecology of Antarctica. Junk, The Hague, pp 491–550

Arillo A, Bavestrello G, Burlando B, Sarà M (1993) Metabolic integration between symbiotic cyanobacteria and sponges: a possible mechanism. Mar Biol 117: 159–162

Barthel D (1992) Do hexactinellids structure Antarctic sponge association? Ophelia 36: 11–118

Barthel D, Tendal O, Panzer K (1990) Ecology and taxonomy of sponges in the Eastern Weddell Sea shelf and slope communities. Ber Polarforsch 68: 120–130

Battershill CN (1989) Distribution and abundance of benthic marine species at Cape Armitage, Ross Island, Antarctica. Initial results. N Z Antarct Rec 9: 35–52

Beliaev GM, Ushakov PV (1957) Certain regularities in the quantitative distribution of the bottom fauna in Antarctic waters. Am Inst Biol Sci 112: 116–119

Bullivant JS (1967) Ecology of the Ross Sea benthos. N Z Oceanogr Inst Mem 32: 49–75

Burton M (1932) Sponges. Discovery Rep 6: 237–392

Cattaneo-Vietti R, Bavestrello G, Cerrano C, Sarà M, Bennati U, Giovine M, Gaino E (1996) Optical fibres in an Antarctic sponge. Nature 383: 397–398

Cox G, Larkum AWD (1983) A diatom apparently living in symbiosis with a sponge. Bull Mar Sci 33: 943–945

Dayton PK (1979) Observations on growth, dispersal and population dynamic of some sponges in McMurdo Sound, Antarctica. Coll Int CNRS 291: 271–282

Dayton PK, Robilliard GA, Paine RT, Dayton LB (1974) Biological accomodation in the benthic community of McMurdo Sound, Antarctica. Ecol Monogr 4 + 4: 105–128

Dearbon JH (1967) Food and reproduction of *Glyptonotus antarcticus* (Crustacea: Isopoda) at McMurdo Sound, Antarctica. Trans R Soc N Z 8: 163–168

Di Geronimo I, Cattaneo-Vietti R, Gambi MC, Casazza G, Cormaci M, Scamacca B (1992) Prime osservazioni sulle comunità bentoniche costiere di Baia Terra Nova (Mare di Ross, Antartide): Bionomia e distribuzione. In: Albertelli G, Ambrosetti W, Piccazzo M, Ruffoni Riva T (eds) Atti IX Congr Ass Ital Oceanogr Limnol, Lang, Genova, pp 635–646

Edgar LA (1983) Mucilage secretion of moving diatoms. Protoplasma 118: 44–48

Ekman S (1935) Tiergeographie des Meeres. Akademische Verlagsgesellschaft, Leipzig, 542 pp

Fabiano M, Chiantore M, Povero P, Cattaneo-Vietti R, Pusceddu A, Misic C, Albertelli G (1997) Short-term variations in particulate matter flux in Terra Nova Bay, Ross Sea, Antarctica. Antarct Sci 9: 143–149

Frignani M, Langone M, Ragaglia L, Ravaioli M (1992) Organic matter and biogenic silica in bottom sediments of the Western Ross Sea (Antarctica). In: Albertelli G, Ambrosetti W, Piccazzo M, Ruffoni Riva T (eds) Atti IX Congr Ass Ital Oceanogr Limnol, Lang, Genova, pp 573–582

Gaino E, Pansini M, Pronzato R (1976) Osservazioni sull'associazione tra una Cianoficea crococcale e la Demospongia *Chondrilla nucula*. Arch Oceanol Limnol 18: 545–552

Gaino E, Bavestrello G, Cattaneo-Vietti R, Sarà M (1994) Scanning electron microscopy evidence for diatom uptake by Antarctic sponges. Polar Biol 14: 1–4

Gutt J, Koltun VM (1995) Sponges of the Lazarev and Weddell Sea, Antarctica: explanations for their patchy occurrence. Antarct Sci 7: 227–234

Knox GA (1970) Antarctic marine ecosystems. In: Holdgate MW (ed) Antarctic ecology, vol 1. Academic Press, London, pp 69–96

Knox GA, Lowry JK (1977) A comparison between the benthos of the Southern Ocean and the North Polar Ocean with special reference to the Amphipods and the Polychaeta. In: Dunbar MJ (ed) Polar oceans. Arctic Institute of North America, Calgary, pp 423–462

Koltun VM (1968) Spicules of sponges as an element of the bottom sediments of the Antarctic. Symposium on Antarctic Ocenography (SCAR), Santiago de Chile. Scott Polar Research Institute, Cambridge, pp 121–123

Koltun VM (1970) Sponges of the Arctic and Antarctic: a faunistic review. In: Fry WG (ed) Biology of the Porifera. Academic Press London, pp 285–297

Krebs WN (1983) Ecology of neritic marine diatoms, Arthur Harbour, Antarctica. Micropaleontology 29: 267–297

Krebs WN, Lipps JK, Burchle LH (1987) Ice diatom floras, Arthur Harbour, Antarctica. Polar Biol 7: 163–171

Kussakin OG (1967) Isopoda and Tanaidacea from the coastal zones of the Antarctic and Subantarctic. Biol Rep Soviet Antarct Exped (1955–58): 220–239

Labrozzi L, Langone L, Frignani M, Ravaiolo M (1998) Burial rates for biogenic silica, organic C and N at three sites of the Ross Sea (Antarctica). In: Piccazzo M (ed) Atti XII Congr Ass Ital Oceanogr Limnol, Lang, Genova, pp 121–130

Matsuda O, Ishikawa S, Kawakuchi K (1987) Seasonal variations of downward flux of particulate organic matter under Antarctic fast ice. Proc NIPR Symp Polar Biol 1: 23–34

Palmisano AC, SooHoo JB, White DC, Smith GA, Stanton GR, Burckle RH (1985) Shade-adapted benthic diatoms beneath Antarctic sea ice. J Phycol 21: 664–667

Pansini M, Calcinai B, Cattaneo-Vietti R, Sarà M (1994) Demosponges from Terra Nova Bay (Ross Sea, Antarctica): 1987/88 and 1989/89 P.N.R.A. Expeditions. In: Faranda FM, Guglielmo L (eds) National Scientific Commission for Antarctica, Oceanographic Campaigns 1987–88 and 1989–90, Data Rep III, Genova, pp 97–100

Reimer CW, Lee JJ (1988) New species of endosymbiotic diatoms (Bacillariophyceae) inhabiting larger Foraminifera in the Gulf of Elat (Red Sea), Israel. Proc Natl Acad Nat Sci Philad 140: 339–351

Sarà M (1966) Associazioni fra Poriferi e alghe in acque superficiali del litorale marino. Ric Sci Roma 36: 277–282

Sarà M (1971) Ultrastructural aspects of the symbiosis between two species of the genus *Aphanocapsa* (Cyanophyceae) and *Ircinia* (Demospongiae). Mar Biol 11: 214–221

Sarà M, Vacelet J (1973) Ecologie des démosponges. In: Grassé PP (ed) Traité de zoologie: anatomie, systematique, biologie: spongiaires. Masson, Paris, pp 462–576

Sarà M, Balduzzi A, Barbieri M, Bavestrello G, Burlando B (1992) Biogeographic traits and checklist of Antarctic demosponges. Polar Biol 12: 559–585

White DC, Smith GA, Stanton GR (1985) Biomass, community structure and metabolic activity of the microbiota in the benthic marine sediments, and sponge spicule mats. Antarct J 19: 125–126

Wilkinson CR (1979) Nutrient translocation from symbiontic cyanobacteria to coral reef sponges. In: Levi C, Boury-Esnault N (eds) Biologie des spongiares. CNRS, Paris, pp 373–380

Wilkinson CR (1980) Cyanobacteria symbiontic in marine sponges. In: Schwemmler W, Schenk HEA (eds) Endocytobiology, endosymbiosis and cell biology, vol 1. De Gruyter, New York, pp 553–563

Wilkinson CR (1983) Net primary productivity in coral reef sponges. Science 219: 410–412

Wilkinson CR (1987) Interocean differences in size and nutrition of coral reef sponge populations. Science 236: 1654–1657

# Benthic Polychaetes off Terra Nova Bay and Ross Sea: Species Composition, Biogeography, and Ecological Role

G. Cantone[1], A. Castelli[2], and M.C. Gambi[3]

**ABSTRACT**

A synthesis on species composition and biogeography of benthic polychaetes off Terra Nova Bay is presented, together with a discussion on the main pattern of species distribution, and structure of both soft- and hard-bottom assemblages. A total of 132 species of polychaetes (29 families) were found at Terra Nova Bay, 46 of which (including some new taxa) were recorded for the first time in the Ross Sea. Considering previous historical and more recent systematic and ecological studies on polychaetes, a total of 235 species (33 families) are to date known for the whole Ross Sea. From a biogeographical point of view, a literature comparison on 108 taxa revealed that 31% of the species found at Terra Nova Bay are composed of truly Antarctic species (7% of which are endemic to the Ross Sea), 39% by Antarctic-sub-Antarctic, 12% by austral hemisphere, and 10% by cosmopolitan species, while 8% showed a disjunct distribution, occurring also in Arctic and other boreal areas. From a quantitative view point, only a few species were numerically dominant: *Tharyx cincinnatus, Spiophanes tcherniai, Laonice weddellia, Leitoscoloplos mawsoni, Aglaophamus trissophyllus* and *Maldane sarsi antarctica* in the soft bottoms; *Harmothoe spinosa, Kefersteinia fauveli* and *Pionosyllis comosa* on the shallow hard bottoms. Both poorly and well-structured polychaete assemblages occurred along the bay, mainly related to the degree of physical disturbance and bottom-sediment complexity. As regards the ecological role of polychaetes in the benthic system of the bay, their classification into feeding guilds revealed the occurrence of various feeding types with different distribution according to biotope and depth: carnivores (mainly polynoids and nephtyids) dominated hard bottoms, while deposit feeders dominated soft bottoms; among these, limivores (mainly maldanids and capitellids) showed an increase with depth. These latter forms have a strong potential for bioturbation of superficial sediments. Finally, gut content analysis of some nototheniid fishes (*Trematomus* spp.) revealed that polychaetes, both epifaunal, such as *Barrukia cristata*, and tube-dwelling, such as ampharetids, are largely consumed by these bottom predators. Burrow-dwelling species are more rarely preyed upon, but, when present, they represent useful indicators of the feeding habits of fish.

## Introduction

Polychaetes represent an important component, in term of number of species and individuals, of the Southern Ocean benthic communities in both deep and nearshore biotopes (Lowry 1976; Knox 1977). These organisms contribute largely to the overall biodiversity of Antarctic ecosystems (Arntz et al. 1994, 1997) and show a general wide bathymetric distribution (Hartman 1967) and an extensive range of geographical distribution (Knox and Lowry 1977). Hartman (1964, 1966) in her classical works on Antarctic polychaetes listed a total of 457 species for the Antarctica, that in a following survey (Hartman 1967) became 527. Knox (1977) listed 557 species, but in following studies (Lowry 1975, 1976; Knox and Lowry 1977) stated that the total number of polychaete species in the Antarctic continent was strongly underestimated and may be higher than 800. Finally, White (1984) in his classical review of Antarctic benthos, reported a total of 650 species of polychaetes.

[1] Dipartimento di Biologia Animale, Università di Catania, via Androne 81, Catania, Italy
[2] Dipartimento di Zoologia e Antropologia, Università di Sassari, Sassari, Italy
[3] Stazione Zoologica A. Dohrn, Laboratorio di Ecologia del Benthos, Ischia, Naples Italy

Many benthic studies in coastal areas provide information on the distribution and ecology of polychaete populations, including some quantitative analyses (see Sicinski and Janowska 1993; Gambi et al. 1997 for review and references). However, despite their high biodiversity and quantitative importance, Antarctic polychaetes are still relatively poorly studied as regards their taxonomy and biogeography, as well as their ecological role.

After the international geophysical year 1957, except for a few studies and reviews at the family level (Lowry 1975, 1976; Blake 1984), most of the available data on polychaetes of the Ross Sea come mainly from more general benthic surveys (e.g. Bullivant and Dearborn 1967; Oliver and Slattery 1985). After the opening of the Italian base at Terra Nova Bay in 1985, knowledge on polychaetes of this zone increased as regards to taxonomy and biogeography (Cantone and San Filippo 1992; Cantone 1994, 1995; Cantone and Di Pietro 1998), and distribution and ecology (Castelli 1992; Gambi et al. 1994, 1997). Recently, Knox and Cameron (1998) have provided a general manual of identification of polychaetes in the Ross Sea, and a first check-list of the species recorded in this basin. These AA, in revising the previous (historical) systematic and ecological works and adding their own original data, listed a total of 174 benthic species (excluding Myzostomidae) and 10 pelagic species of polychaetes.

The aim of this chapter is to give a first check-list of benthic polychaetes recorded at Terra Nova Bay, as well as in the whole Ross Sea, together with a biogeographic analysis based on the literature data. Patterns of species distribution in the bay and structure of polychaete assemblages in soft and hard bottoms are also discussed. The ecological importance of polychaetes in the benthal system is reviewed, as regards composition and distribution of feeding types, and their potential effects on bioturbation of superficial sediments, and role of these invertebrates in the food web, as a component of the diet of demersal nototheniid fishes (Vacchi et al. 1994; La Mesa et al. 1997).

## Study Area, Materials and Methods

Data originated from two benthic surveys conducted off Terra Nova Bay during the Italian PNRA expeditions in 1987/88 and 1989/90 (Cattaneo-Vietti and Gambi 1995). Polychaetes were obtained in 1987–1988 mainly in qualitative benthic samples (Di Geronimo and Rosso 1990; Cantone 1995): 46 Van Veen-grab and 21 Charcot-dredge samples were collected on soft bottoms both in the inner and outer zone of the bay, from 163 to 1100 m depth, (Cantone 1995); 5 supplementary samples were collected on hard bottoms by SCUBA diving. During the 1989/90 expedition, 14 quantitative Van Veen-grab samples were collected in soft bottoms along two coastal transects, from 23 to 274 m depth, located near the Italian base: the first transect in front of the Faraglione area (2.5 km south of the base), the second in front of Adelie Cove (about 20 km south of the base) (Bvia et al. in press). On hard bottoms, samples were collected by SCUBA diving, along a depth transect from 2 to 16 m at Faraglione (Gambi et al. 1994), and in a few other nearby areas (Gambi et al., this Vol.). All the collected soft-bottom material was sieved through a 1-mm mesh size, fixed in 4% formol and preserved in 70% alcohol.

All taxa classified at the species level were assigned to a biogeographic group, according to literature data on species distribution (Hartman 1964, 1966, 1978; Mackie 1987; Cantone 1994, 1995; Giangrande and Gambi 1997). Six biogeographic groups have been defined: species occurring in the continental Antarctica (AN) and species endemic of the Ross Sea (RS), that represent truly Antarctic taxa; species recorded both in Antarctic and sub-Antarctic areas (AS) (Antarctic Peninsula, sub-Antarctic Islands, Magellan region of southern America); species occurring also in the austral hemisphere coasts (AU) (South America, South Africa, Oceania); cosmopolitan species (CO), and finally taxa with a disjunct distribution (DI), occurring also in the Arctic and other boreal areas.

Species were also assigned to feeding and ecological guilds to evaluate their main position in the food web and their potential role in sediment bioturbation. Feeding guilds were defined according to Fauchald and Jumars (1979) and Gambi et al. (1997), identifying six main categories: carnivores (C), including scavengers, surface-deposit feeders, that have been distinguished as discretely motile (DSdf) and sessile (SSdf) forms, subsurface-deposit feeders or limivores (L), filter feeders (Ff) and omnivores (Om). Qualitative dominance (% no. of species) and quantitative dominance (% no. of individuals) of each feeding guild was calculated for each sample; samples were then grouped according to type of substrate (hard or soft bottoms) and depth range.

The role of polychaetes in the diet of some demersal nototheniid fishes has been studied in the framework of several investigations, performed at Terra Nova Bay between 1988 and 1991, aimed at investigating the feeding habits and trophic ecology of these bottom predators (Vacchi et al. 1994; La Mesa et al. 1997). The polychaetes found in the stomachs of the fish species studied have been divided into several ecological groups based on their potential accessibility as prey for fishes. Three major ecological groups have been defined, preyed upon with a different feeding mode and with an increasing difficulty for fishes: epibenthic, tube-dwelling, and burrow-dwelling species. The first group comprises typically epifaunal species belonging to Polynoidae, and small infaunal species living near the bottom surface, mainly belonging to Spionidae. The second group comprises sedentary or sessile tubicolous families (e.g. Ampharetidae) which generally are partially or completely buried within the sediment. The third group comprises burrow-dwelling, infaunal species, mainly belonging to Nephtyidae, Glyceridae and Orbiniidae, living in galleries deep in the sediment, which are difficult prey for the nototheniid fishes (Vacchi et al. 1994; La Mesa et al. 1997).

## Results

### Taxonomy and Biogeography

As a whole, a total of 132 species belonging to 29 families were recorded in all the analyzed samples at Terra Nova Bay. This list also includes some newly described species or subspecies, such as *Aricidea bispinata* Cantone, *Levinsenia duodecimbranchiata* Cantone, *Myriochele terranovensis* Cantone & Di Pietro, *Praxillella praetermissa antarctica* Cantone, *Perkinsiana borsibrunoi* Giangrande & Gambi, and *Perkinsiana milae* Giangrande & Gambi (Table 1). Twenty taxa, marked by capital letters, have been identified only at the genus level (Table 1). The two forms of *Harmothoe spinosa* (form A and B) have been considered as a single taxon.

Forty-six of the species found (including some new taxa) were recorded for the first time for the Ross Sea (marked by an asterisk in Table 1). The most abundant and diversified families were the Terebellidae (13 species), Maldanidae (12), Polynoidae (11), Spionidae (10), Ampharetidae (9) and Syllidae (7).

The species found by other AA in the Ross Sea are listed in Table 2 (Cantone 1994; Knox and Cameron 1998). Among these, *Perkinsiana antarctica* Kinberg, considered a circumantarctic species, was not reported, since the genus *Perkinsiana* in Antarctica has been recently reviewed. The Ross Sea specimens belong to three different species: *P. milae*, *P. borsibrunoi*, and *P. littoralis*, one of these (*P. milae*) endemic of the Ross Sea (Giangrande and Gambi 1997). Finally, the species *Leitoscoloplos kerguelensis* (McIntosh) and subspecies *L. kerguelensis minutus* (Hartman), have not been reported in the list, as a recent revision of the genus (Mackie 1987) has led to the distinction of several morphologically distinct species which are often geographically disjunct. The Ross Sea representatives of the genus *Leitoscoloplos* belong to the endemic species *L. mawsoni*. The above taxonomic considerations as well as a few others are reported in the notes at the end of both Table 1 and Table 2.

Comparison of our data at Terra Nova Bay with previous findings (Cantone 1994; Knox and Cameron 1998), has yielded a total of 235 species of polychaetes known for the entire Ross Sea (Tables 1, 2). For comparison and assessment of total polychaete biodiversity, only the taxa classified at the species level, and those likely to be new species have been considered.

Biogeographic analysis of the species found at Terra Nova Bay was based on previous literature information and conducted on 108 of the taxa found (Table 1). The analysis showed that 31% of the taxa found are composed by truly Antarctic species, 7% of which are endemic of the Ross Sea (AN and RS); 39% by Antarctic-sub Antarctic, 12% by austral hemisphere, and 10% by cosmopolitan species, while 8% showed a disjunct distribution (Fig. 1). The species regarded as endemic of the Ross Sea, in addition to four of the six newly described species or subspecies (except *Perkinsiana borsibrunoi*), were also represented by *Leitoscoloplos mawsoni*, *Spio obtusa*, *Phyllodoce adarensis*, and *Myriochele wilsoni*.

### Distribution and Ecological Role

The most abundant and widely distributed species in the analyzed samples were: *Leitoscoloplos*

**Table 1.** List of the polychaete species collected at Terra Nova Bay during Italian PNRA expeditions; the biogeographic category is indicated (below table) The asterisk (*) indicates the species found only in our samples and not previously reported for the Ross Sea

Orbiniidae
*Leitoscoloplos mawsoni* (Benham) RS[1]
*Scoloplos marginatus* (Ehlers) AU

Spionidae
*Laonice antarcticae* Hartman DI
*Laonice weddellia* (Hartman) AU
*Laonice* sp. A
**Prionospio* cf *patagonica* Augener AU
*Scolelepis eltaninae* Blake AS
*Scolelepis* sp. A
*Scolelepis* sp. B
*Scolelepis* sp. C
*Spio obtusa* Ehlers RS
*Spiophanes tcherniai* (Fauvel) AS

Chaetopteridae
**Spiochaetopterus typicus* Sars CO

Paraonidae
*Aedicira* sp. A
**Aricidea bispinata* Cantone RS
**Aricidea neosuecica nipponica* Imajima DI
**Levinsenia duodecimbranchiata* Cantone RS
*Levinsenia gracilis* (Tauber) CO[2]

Cossuridae
**Cossura* sp. A

Cirratulidae
*Chaetozone andersenensis* (Augener) AN
*Tharyx cincinnatus* (Ehlers) AS
*Tharyx epitoka* Monro DI
*Tharyx fusiformis* Monro AN

Capitellidae
**Capitella* cf. *capitata* (complex) (Fabricius) CO
*Notomastus latericeus* Sars CO

Maldanidae
**Axiothella antarctica* Monro AS
**Asychis amphiglypta* (Ehlers) AS
*Isocirrus yungi* (Gravier) AS
*Maldane sarsi antarctica* Arwidsson AS
**Maldanella antarctica* McIntosh AS
**Maldanella grossa* (Baird) AS
*Nicomache lumbricalis* (Fabricius) DI
*Notoproctus oculatus antarcticus* Arwidsson AN
*Praxillella kerguelensis* (McIntosh) AS
*Praxillella praetermissa antarctica* Cantone RS
*Rhodine loveni* Malmgren CO
*Rhodine* sp. A

Opheliidae
*Kesun abyssorum* Monro AN
*Ophelina breviata* (Ehlers) AN
*Ophelina gymnopyge* (Ehlers) AS
*Tachytrypane jeffreysi* McIntosh CO
*Travisia kerguelensis* McIntosh AS

Scalibregmatidae
**Oligoscalibregma hartmanae* (Hartman) AN
*Scalibregma inflatum* Rathke CO
*Sclerocheilus antarcticus* Ashworth AN

Phyllodocidae
*Phyllodoce adarensis* Benham RS[3]
*Phyllodoce longipes* Kinberg AU[3]
*Phyllodoce patagonica* Kinberg AU[3]
*Eteone* sp. A

Glyceridae
*Glycera capitata* (complex) Oersted CO
**Glycera kerguelensis* McIntosh AS

Sphaerodoridae
*Sphaerodorum parvum* Ehlers AS

Hesionidae
**Gyptis incompta* Ehlers AN
**Kefersteinia fauveli* Averincev AN

Syllidae
**Autolytus simplex* Ehlers AS
*Autolytus* sp. A
*Exogone* sp. A
*Sphaerosyllis* cf. *tetralix* (Eliason) CO
*Pionosyllis comosa* Gravier AS
*Pionosyllis maxima* Monro AS
*Trypanosyllis gigantea* (McIntosh) AU

Nephtyidae
*Aglaophamus trissophyllus* (Grube) AN[4]

Polynoidae
*Antinoella antarctica* (Bergstrom) AS[5]
**Antinoella setobarba* (Monro) AS
*Barrukia cristata* (Willey) AS
*Eucranta mollis* (McIntosh) AS
**Eunoe anderssoni* (Bergstrom) AS
*Eunoe* sp. A
*Harmothoe magellanica* (McIntosh) AS
**Harmothoe monroi* (Ushakov) AN
*Harmothoe spinosa* form A Kinberg AU
*Harmothoe spinosa* form B Kinberg AU
*Polyeunoa laevis* McIntosh AS

Amphinomidae
*Paramphinome australis* Monro AS

Lumbrineridae
**Lumbrineris antarctica* Monro AS
**Lumbrineris* cf. *cingulata* Ehlers AS
**Lumbrineris magalhaensis* Kinberg AN
*Lumbrineris tetraura* Schmarda CO
*Lumbrineris* sp. A

Dorvilleidae
**Pettiboneia* sp. A

**Table 1.** *Continued*

*Pettiboneia* sp. B
*Meiodorvillea* sp. A
*Ophryotrocha claparedii* Studer AS

Oweniidae
*Myriochele heeri* Malmgren DI
*Myriochele longicollaris* Hartmann-Schroeder and Rosenfeldt AN
*Myriochele joinvillensis* Hartmann-Schroeder and Rosenfeldt AN
*Myriochele scotiae* Hartman AN
*Myriochele wilsoni* (Blake) RS
*Myriochele terranovensis Cantone G. and Di Pietro RS

Flabelligeridae
*Flabelligera bicolor* (Schmarda) AU
*Flabelligera gourdoni* Gravier AS
*Flabelligera mundata* Gravier AN
*Ilyphagus coronatus* Monro AN
*Ilyphagus wyvillei* (McIntosh) AN
*Pherusa kerguelarum* (Grube) AS

Fauveliopsidae
*Fauveliopsis brevis* Hartman AU

Ampharetidae
*Ampharete kerguelensis* McIntosh AS
*Amphicteis gunneri antarctica* Hessle AS
*Amphicteis* cf. *midas* (Gosse)
*Amythas membranifera* Benham AN
*Anobothrella antarctica* (Monro) AN
*Anobothrella patagonica* (Kinberg) AU[6]
*Anobothrella* sp. A
*Grubianella* sp. A
*Neosabellides elongatus* (Ehlers) AN

Trichobranchidae
*Octobranchus* sp. A
*Terebellides stroemi kerguelensis* McIntosh AU

Terebellidae
*Amphitrite cirrata* Muller DI
*Axionice spinifera* (Ehlers) AS
*Axionice godfroyi* (Gravier) AN
*Hauchiella tribullata* (McIntosh) DI
*Lanicides bilobata* (Grube) AN
*Lysilla loveni* (Malmgren) CO
*Pista corrientis* McIntosh DI
*Pista* sp. A[7]
*Polycirrus kerguelensis* (McIntosh) AS
*Proclea graffii* (Langerhans) DI
*Terebella ehlersi* Gravier AS
*Thelepus* sp. A[8]
*Thelepides venustus* Levenstein AN

Sabellidae
*Jasmineira caeca* Ehlers AS
*Perkinsiana borsibrunoi* Giangrande and Gambi AN[9]
*Perkinsiana littoralis* (Hartman) AS[9]
*Perkinsiana milae* Giangrande and Gambi RS[9]
*Oriopsis limbata* (Ehlers) AS
*Oriopsis magna* Banse AS
*Euchone pallida* Ehlers AU

Serpulidae
*Helicosiphon biscoensis* Gravier AS
*Serpula narconensis* Baird AN

Spirorbidae
*Metalaeospira pixelli* (Harris) AS
*Paralaeospira* sp. A

Biogeographical legend:
RS: Ross Sea
AN: Antarctic Continent
AS: Antarctic and sub-Antarctic
AU: Austral Hemisphere
CO: Cosmopolitan
DI: Disjunct Distribution

[1] Reported as *Haploscoloplos kerguelensis* (McIntosh), and *H. kerguelensis minutus* Hartman by other AA (see Knox and Cameron 1998).
[2] Reported as *Paraonis gracilis* in Knox and Cameron (1998).
[3] Reported as *Anaitides adaresis*, *A. longipes* and *A. patagonica* in Knox and Cameron (1998).
[4] This species in our previous works has been reported as *Aglaophamus ornatus* Hartman, but Knox (in prep.) (cited in Knox and Cameron 1998) indicates that *A. ornatus* is a synonimous of *A. trissophyllus* (Grube).
[5] As *Austrolaenilla antarctica* in Knox and Cameron (1998).
[6] As *Anobothrus patagonicus* in Knox and Cameron (1998).
[7] As *Pista cristata* (Muller) in our previous works.
[8] As *Thelepus cincinnatus* (Fabricius) in our previous works and in Knox and Cameron (1998).
[9] The three species are confused and reported in Knox and Cameron (1998) as *Potamilla antarctica* Kinberg (see Giangrande and Gambi 1997).

**Table 2.** List of the polychaete species previously found in other areas of the Ross Sea by various authors, most of them reported in Knox and Cameron (1998)

Spionidae
*Nerinopsis hystricosa* Ehlers
*Laonice cirrata* (Sars)
*Spio filicornis* (Muller)
*Spiophanes bombyx* (Claparède)
*Spiophanes kroyeri* Grube
*Spiophanes* sp.

Chaetopteridae
*Phyllochaetopterus socialis* (Claparède)

Cirratulidae
*Chaetozone* sp.
*Cirratulus cirratus* (Muller)
*Tharyx* sp.

Apistobranchidae
*Apistobranchus* sp.

Paraonidae
*Paraonis belgicae* Fauvel[1]

Maldanidae
*Axiothella quadrimaculata* Augener
*Axiothella* sp.
*Clymenella antarctica* Knox and Cameron
*Euclymene watsoni* Gravier
*Euclymene grossa* (Baird)
*Maldane* sp.
*Praxillella* sp.

Oweniidae
*Owenia fusiformis* Delle Chiaje

Scalibregmatidae
*Hyboscolex longiseta* Schmarda
*Oncoscolex dicranochaetus* Schmarda
*Oligobregma collare* (Levenstein)
*Oligobregma notiale* Blake
*Pseudoscalibregma* sp.

Phyllodocidae
*Eteone aurantiaca* Schmarda
*Eulalia* sp.
*Steggoa hunteri* (Benham)
*Steggoa magalhaensis* (Kinberg)
*Austrophyllum charcoti* (Gravier)
*Anaitides bowersi* (Benham)
*Anaitides madeirensis* (Langerhans)
*Anaitides* sp.

Hesionidae
*Ophiodromus comatus* (Ehlers)
*Syllidia inermis* (Ehlers)

Syllidae
*Autolytus charcoti* Gravier
*Autolytus maclearanus* McIntosh
*Autolytus longstaffi* Ehlers
*Eurisyllis ehlersi* Benham
*Eusyllis kerguelensis* McIntosh
*Pionosyllis stylifera* Ehlers
*Syllides articulosus* Ehlers
*Syllides* sp.
*Typosyllis brachychaeta* (Schmarda)[2]
*Typosyllis armillaris* (Muller)[2]
*Typosyllis brachycola* (Ehlers)
*Typosyllis hyalina* (Grube)
*Typosyllis pennelli* Knox and Cameron
*Typosyllis prolixa* (Ehlers)
*Syllis amica* Quatrefages
*Exogone heterosetosa* McIntosh
*Exogone minuscula* Hartman

Nereididae
*Neanthes kerguelensis* (McIntosh)
*Nicon ehlersi* Hartman
*Platynereis australis* (Schmarda)

Nephtyidae
*Micronephthys* sp.

Aphroditidae
*Aphrodita rossi* Knox and Cameron
*Laetmonice producta* Grube

Polynoidae
*Antinoella hastulifera* (Bergstrom)[3]
*Antinoella* sp.[3]
*Barrukia curviseta* (Monro)
*Eulagisca corrientis* McIntosh
*Eulagisca gigantea* Monro
*Eunoe opalina* McIntosh
*Gorekia crassicirris* (Willey)
*Harmothoe crosetensis* (McIntosh)
*Harmothoe ernesti* Augener
*Harmothoe exanthema bergstromi* Monro
*Harmothoe hartmanae* Uschakov
*Hermadion ferox* Baird
*Hermadion magalhaensis* Kinberg
*Eunoe abyssorum* McIntosh
*Polynoe antarctica* Kinberg

Euphrosinidae
*Euphrosine armadilloides* Ehlers
*Euphrosine monroi* Kudenov
*Euphrosinella cirratoformis* (Averincev)
*Euphrosinopsis antipoda* Kudenov

Eunicidae
?*Eunice pennata* (Muller)

Onuphidae
*Nothria anoculata* Orensanz

Lumbrineridae
*Augeneria tentaculata* Monro
*Lumbrineris kerguelensis* (McIntosh)
*Paraninoe antarctica* (Monro)

Dorvilleidae[4]
*Ophryotrocha notialis* (Ehlers)

**Table 2.** *Continued*

| | |
|---|---|
| Ampharetidae | *Streblosoma bairdi antarctica* Monro |
| *Amage sculpta* Ehlers | *Thelepides koehleri* Gravier |
| *Melinnoides nelsoni* Benham | |
| *Samitha* (?) *speculatrix* Ehlers | Trichobranchidae |
| *Grubianella antarctica* McIntosh | *Octobranchus phyllicomus* Hartman |
| *Phyllocomus crocea* Grube | *Trichobranchus glacialis antarcticus* Hessle |
| | |
| Terebellidae | Sabellidae |
| *Amphitrite kerguelensis* McIntosh | *Myxicola sulcata* Ehlers |
| *Proclea* sp. | *Myxicola* sp.[5] |
| *Lanice flabellum* (Baird) | *Sabella oatesiana* Benham |
| *Lanice* sp. | *Euchone analis* (Kroyer) |
| *Leaena antarctica* McIntosh | *Fabricia* sp. |
| *Leaena wandelensis* Gravier | *Potamethus scotiae* (Pixell) |
| *Leaena collaris* Hessle | |
| *Nicolea chilensis* (Schmarda) | Serpulidae |
| *Pista mirabilis* McIntosh | *Chitinopomoides wilsoni* Benham |
| *Pista godfroyi* (Gravier) | |
| *Terebella* (?) *bilineata* Baird | Spirorbidae |
| *Polycirrus antarcticus* (Willey) | *Leodora perrieri* (Caullery and Mesnil) |
| *Lysilla loveni mcintoshi* Gravier | *Paralaeospira antarctica* Pixell |
| *Nicolea venustula* (Montagu) | *Spirorbis* spp. |

[1] Reported as *Aedicira belgicae* (Fauvel) in Knox and Cameron (1998).
[2] This species, reported as *Syllis brachychaeta* by Benham (1927), is considered a synonimous of *Typosyllis armillaris* by Knox and Cameron (1998).
[3] Reported as *Austrolaenilla hastulifera* and *Austrolaenilla* sp. by Knox and Cameron (1998).
[4] As Iphitimidae in Knox and Cameron (1998).
[5] Probably belonging to *M. sulcata* (Knox and Cameron 1998).

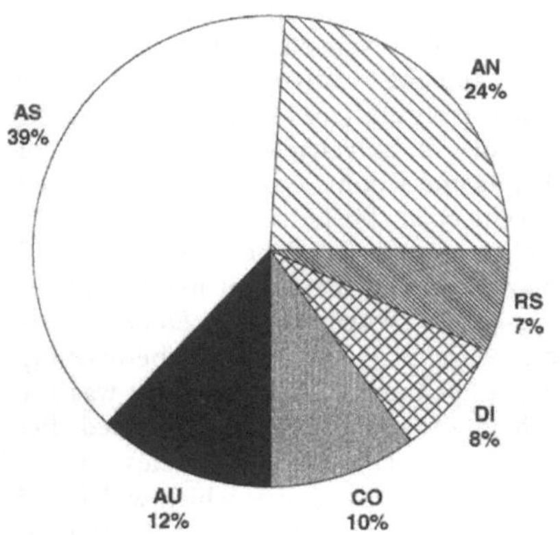


**Fig. 1.** Dominance (no. of species) of the biogeographic categories of Terra Nova Bay polychaetes. *AN* Continental Antarctica; *RS* species endemic to the Ross Sea; *AS* Antarctic-sub-Antarctic; *AU* austral hemisphere; *CO* cosmopolitan; *DI* disjunct distribution

*mawsoni, Laonice weddellia, Spiophanes tcherniai, Tharyx cincinnatus, Barrukia cristata* (abundant in moderately deep soft bottoms), *Maldane sarsi antarctica* (abundant in deeper soft bottoms), *Aglaophamus trissophyllus*, and *Harmothoe spinosa* (both forms A and B) (abundant mainly on shallow soft and hard bottoms). *A. trissophyllus*, in particular, was the species with the widest depth distribution, occurring in soft bottoms from 16 up to 1100 m. Many species (e.g. *Lumbrineris magalhaensis; Eucranta mollis*) were distributed mainly from 200 m to the limit of the shelf (600–800 m).

In the coastal soft bottoms, polychaetes represented, in terms of abundance, 76% of the macrobenthic fauna (Gambi and Castelli 1994). However, only the few above-listed species were numerically important and accounted for about 75% of the total abundance. The richest and best-structured assemblages occurred in fine sands below the 120 m depth, where the bottom is a mosaic of soft sediments and rocks, and habitat complexity is higher. Lowest diversity values were observed at shallow depths (>50 m) characterized by gravel and coarse sands (Gambi et al. 1997). Furthermore, the highest abundance and biomass of polychaetes occurred in some stations located inside of Adelie Cove, a small, sheltered bay, where the high density of some deposit feeder and burrower species (*Tharyx cincinnatus* and *Leitoscoloplos mawsoni*) is probably related to sediment organic enrichment due to the Adelie penguin rookery. In this area, the hard-bottom community associated with the macroalgae also showed extremely high densities (Gambi et al., this Vol.). The species richness and diversity pattern of polychaete assemblages in the coastal soft bottoms seems clearly related to the degree of physical disturbance and habitat complexity, mainly bottom-sediment heterogeneity.

On the hard bottoms, polychaetes accounted only for 4.2% of the abundance of the fauna sampled in association with the dominant macroalgae (Gambi et al. 1994), and their density and diversity were much lower than in the soft bottoms, with only 19 species being recorded. However, in these biotopes polychaetes showed relatively high biomass (Gambi and D'Agostino 1994) and represented the few carnivores found in the community (Gambi et al. 1994). The most abundant species recorded were *Harmothoe spinosa*, *Kefersteinia fauveli* and *Pionosyllis comosa*. The former species was abundant along the whole depth transect (2–16 m), and was represented mainly by small, probably juvenile specimens (Gambi, unpubl. data), indicating that these algal-dominated biotopes represent a favourable recruitment area for the species. The latter two species were more abundant at deeper stations (12–16 m) in association mainly with the red macroalga *Phyllophora antarctica* (Gambi et al. 1994).

The ecological role of polychaetes in the benthic system of the Bay was also analyzed as regards their feeding behaviour and its potential effects on sediment bioturbation. Polychaetes at Terra Nova Bay showed a relatively wide range of feeding guilds and life habits. Almost all the guilds identified have been found along the depth range examined. The relative qualitative (no. of species) and quantitative (no. of individuals) dominance of the various guilds according to substrate and depth are shown in Fig. 2A and B. The hard-bottom assemblages showed the dominance of carnivores (mainly polynoids and nephtyids), while in soft bottoms even though this guild was numerically reduced (Fig. 2B), it was still represented by several species (Fig. 2A). The mobile forms of surface deposit feeders (DSdf) showed a constant pattern in terms of number of species at all depths; the highest numerical dominance was observed in the 20–200-m zone and progressively decreased with depth, especially below 750 m. Sessile forms of surface deposit feeders (SSdf) showed an irregular trend with highest qualitative dominance in the 200–500 m zone (Fig. 2A). Limivores (L) progressively increased with depth, showing the highest dominances below 1000 m. Both filter feeders (Ff) and omnivores (Om) showed very low dominances, as number of species and individuals, at all depths (Fig. 2A,B). As regards their potential for sediment bioturbation, the feeding-guild of the limivores included some head-down feeding forms (e.g. maldanids and capitellids), defined as conveyor-belt forms (sensu Rhoads 1974), that largely contributed to sediment mixing by displacement of deep layers to the surface, favouring the flux of materials at the water-sediment interface.

The role of polychaetes in the benthic food web has been evaluated in relation to the feeding ecology of some species of demersal nototheniid fishes (Vacchi et al. 1994; La Mesa et al. 1997). Polychaetes were the most important prey for *Trematomus centronotus* Regan, and *T. bernacchii* Boulenger, and an important source of food for two other nototheniid species, *T. hansoni* Boulenger (mainly feeding on fishes) and *T. loennbergi* Regan (mainly feeding on crustacean decapods). About 40 different polychaete taxa have been identified in the stomach contents of these fishes (Vacchi et al. 1994; La Mesa et al. 1997). The relative abundance of polychaetes in the diet of each fish species, and the relative distribution of the main ecological groups (see Methods), showed the important role of polychaetes as indicators of feeding habits and trophic ecology of the studied fishes (Vacchi et al. 1994; La Mesa et al. 1997). Stomach content analysis of the four nototheniid species showed the occurrence of mainly epifaunal species (e.g. *Barrukia cristata*, *Eunoe anderssoni* and *Polyeunoa laevis*) or tube-dwelling forms (e.g. ampharetids). *T. centronotus* was the species which showed the more diversified diet and higher percentage of polychaetes in the stomach; it was the only species which widely fed on burrow-dwelling forms (mainly *Aglaophamus trissophyllus*, *Scoloplos marginatus*, *Ophelina gymnopyge*, *Leitoscoloplos mawsoni* and *Tharyx cincinnatus*). This indicates that this fish was scavenging relatively deep sediment layers when feeding.

## Discussion and Conclusions

The species richness recorded at Terra Nova Bay (132 species) was relatively high, expecially

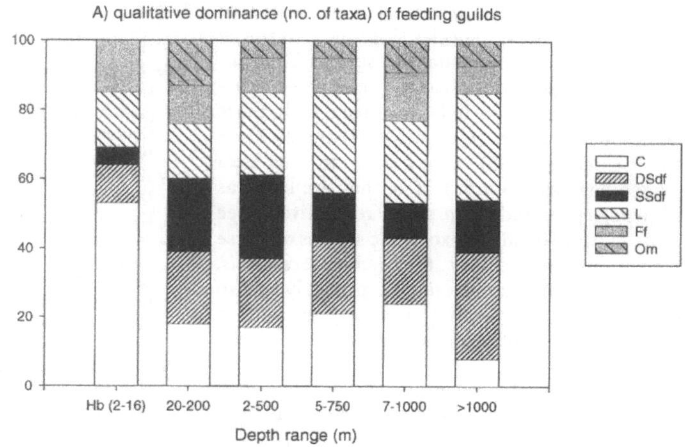


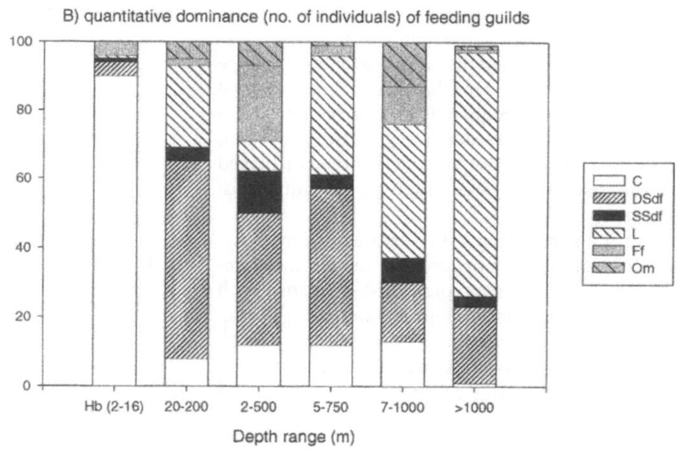


**Fig. 2A,B. A** Qualitative dominance (no. of taxa), and **B** quantitative dominance (no. of individuals) of the polychaete feeding guilds identigoed at Terra Nova Bay at different depth range (m). *C* Carnivores; *DSdf* discretely motile surface deposit feeders; *SSdf* sessile surface deposit feeders; *L* limivores or subsurface-deposit feeders; *Ff* filter feeders; *Om* omnivores

when compared to that known for the whole Ross Sea (174 species, 30 families) (Cantone 1994; Knox and Cameron 1998). Our data, combined with previous information in the literature, indicate the occurrence of 235 species of polychaetes, belonging to 33 families, for the whole Ross Sea. This is a relatively large number compared to other geographic areas of similar size, such as the Weddell Sea with 153 species and 30 families (Hartman 1978).

From a biogeographical point of view, it is interesting to point out the conspicuous occurrence at Terra Nova Bay of species belonging to the Antarctic province (AN + RS + AS), that accounted for 70%. This percentage is higher compared to that recorded for other Antarctic polychaetes (Knox and Lowry 1977; White 1984), and is comparable to that observed in the Antarctica for other zoological groups (Sarà et al. 1992; Arntz et al. 1997). Knox and Cameron (1998) reported 15 species as occurring only in the Ross Sea. Based on our findings, another 7 species can be added to this list for a total of 22 species endemic for the Ross Sea.

The occurrence of some cosmopolitan species and of others with a disjunt distribution, points out the need to increase the study of the Antarctic polychaete fauna to clarify the taxonomic status of these taxa, as well as that of many still unidentified species. As an example, the cosmopolitan species

*Kefersteinia cirrata* (Keferstein) is attributed to the antarctic endemic species *K. fauveli* (Averincev 1972). As regards our data, the species *Thelepus cincinnatus* and *Pista cristata* considered "cosmopolitan" and reported previously with these names in our studies (Cantone 1994) and by Knox and Cameron (1998), may belong to new taxa (Hutchings, pers. comm.). Here they are listed as *Thelepus* sp. A and *Pista* sp. A, respectively (see Table 1). More detailed taxonomic studies on these species, as well as on the genera *Scolelepis*, *Lumbrineris* and *Pettiboneia*, will probably lead to the description of some new species with a restricted distribution, possibly endemic to the Ross Sea.

From an ecological point of view, our data confirm for some species the wide depth and habitat distribution, as observed also by Knox and Cameron (1998). However, the data are still fragmentary for most taxa to confirm the high degree of euribathy shown generally by Antarctic polychaetes, and that for some families (e.g. polynoids) has been interpreted as a phylogenetic feature (Brey et al. 1996). On the other hand, several species showed a higher degree of habitat selection, especially those occurring in the shallower, more coastal areas. But this pattern may be biased by the relatively scarce number of samples of this type.

In the soft bottoms off Terra Nova Bay, polychaetes play an important role as sediment reworkers. In particular, conveyor-belt forms, with the highest potential of bioturbation, were particularly abundant below the 500-m depth (Fig. 2). This observation is consistent with data on soft bottom macrobenthos collected in some selected sites off the Ross Sea shelf, and studied in the framework of a different research programme (Gambi and Bussotti 1999). Finally, polychaetes exert an important role in the benthic food web as preferential prey of common demersal fishes.

## References


Arntz W, Brey T, Gallardo VA (1994) Antarctic zoobenthos. Oceanogr Mar Biol Annu Rev 32: 241–304

Arntz WE, Gutt J, Klages M (1997) Antarctic marine biodiversity: an overview. In: Battaglia B, Valencia J, Walton DWH (eds) Antarctic communities: species, structure and survival. Cambridge University Press, Cambridge, pp 3–14

Averincev VG (1972) Sezonnye izmeneya v sublitoralnoy faune mnogoshchetinkovych chervey (Polychaeta) Morya Davisa. Issledovaniya Fauny Morey 28(36): 4–70

Benham WB (1927) Polychaeta. British Antarctic (Terra Nova) Expedition, 1910. Nat Hist Rep Zool 7(3): 47–182

Blake J (1984) Polychaeta Oweniidae from Antarctic Sea collected by the United States Antarctic Research Programme. In: Hutchings PA (ed) Proc 1st Int Polychaete Conf Sydney (Australia), July 1983. Linnean Society N.S. Wales, Sidney, pp 112–117

Brey T, Dahm C, Gorny M, Klages M, Stiller M, Arntz W (1996) Do Antarctic benthic invertebrates show an extended level of eurybathy? Antarct Sci 8(1): 3–6

Buia MC, Mazzella L, Gambi MC, Catra M (in press) Benthic populations of the shallow hard bottoms off Terra Nova Bay: results of the 9th and 10th Italian Antarctic expeditions. In: Faranda F, Gugliclmo L, Povero P (eds) National Scientific Commission for Antarctica Oceanographic Campaign 1994–95, Data Rep Part II (in press)

Bullivant JS, Dearborn JH (eds) (1967) Fauna of the Ross Sea. N Z Dep Sci Ind Res Bull 176: 1–76

Cantone G (1994) Polychaete fauna of the Ross Sea (Antarctica). Animalia 21(1–3): 19–26

Cantone G (1995) Polychaeta Sedentaria of Terra Nova Bay (Ross Sea, Antarctica). Capitellidae to Serpulidae. Polar Biol 15: 295–302

Cantone G, Di Pietro N (1998) A new species of *Myriochele* (Polychaeta, Oweniidae) from Antarctica, with considerations on the Antarctic oweniids. Polar Biol 19: 421–423

Cantone G, Sanfilippo R (1992) Polychaeta from Terra Nova Bay (Ross Sea, Antarctica). In: Gallardo VA, Ferretti O, Moyano HI (eds) Oceanografia in Antartide. ENEA-PNRA-EULA, Concepcion, Chile, pp 371–381

Castelli A (1992) Primi dati sulla distribuzione batimetrica dei Policheti erranti della baia di Terra Nova (Mare di Ross, Antartide). Oebalia 17 Suppl 2: 331–333

Cattaneo-Vietti R, Gambi MC (1995) Ecological studies on benthos at Terra Nova Bay (Ross Sea): an overview of the Italian Antarctic Research Programme (1987–92). Ambiente Antartide 6: 18–23

Di Geronimo I, Rosso A (1990) First Italian Oceanographic Expedition in the Ross Sea (Antarctica). Benthos: preliminary report. In: National Scientific Commission for Antarctica (ed) Oceanographic Campaign 1987/88. Data Rep Part I, Genova, pp 407–421

Fauchald K, Jumars PA (1979) The diet of worms: a study of polychaete feeding guilds. Oceanogr Mar Biol Annu Rev 17: 193–284

Gambi MC, Bussotti S (1999) Composition, abundance and stratification of soft bottom macrobenthos in selected areas of the Ross Sea shelf (Antarctica). Polar Biol 21: 347–354

Gambi MC, Castelli A (1994) First quantitative data on coastal soft bottoms populations off Terra Nova Bay (Ross Sea, Antarctica): polychaetes. In: Faranda F, Guglielmo L (eds) National Scientific Commission for Antarctica (ed) Oceanographic Campaign 1989–90, Data Rep Part II, Genova, pp 179–186

Gambi MC, D'Agostino T (1994) Biomass estimates of some zoobenthic species associated to seaweeds in the shallow hard bottoms off Terra Nova Bay (Ross Sea, Antarctica). In: Albertelli G, Cattaneo-Vietti R, Piccazzo M (eds) Atti X Congr Ass Ital Oceanogr Limnol, Lang, Genova, pp 691–698

Gambi MC, Lorenti M, Russo GF, Scipione MB (1994) Benthic associations of the shallow hard bottoms off Terra Nova Bay, Ross Sea: zonation, biomass and population structure. Antarct Sci 6(4): 449–462

Gambi MC, Castelli A, Guizzardi M (1997) Polychaete populations of the shallow soft bottoms off Terra Nova Bay (Ross Sea, Antarctica): distribution, diversity and biomass. Polar Biol 17: 199–210

Giangrande A, Gambi MC (1997) The genus *Perkinsiana* (Polychaeta Sabellidae) in the Antarctica with the description of the new species *P. milae* and *P. borsibrunoi*. Zool Scr 26(3): 267–278

Hartman O (1964) Polychaeta Errantia of Antarctica. Antarct Res Ser 3: 1–131

Hartman O (1966) Polychaeta Myzostomidae and Sedentaria of Antarctica. Antarct Res Ser 7: 1–158

Hartman O (1967) Benthic zonation in Antarctica, as displayed by marine annelids (Polychaeta) based on published and new records, from intertidal to hadal depths. J.A.R.E. Scientific reports, special issue no.1. Proc Symp on Pacific-Antarctic Sciences. Dep Polar Res, National Science Museum, Ueno Park, Tokyo, pp 192–204

Hartman O (1978) Polychaeta from the Weddell Sea Quadrant, Antarctica. Antarct Res Ser 26: 125–223

Knox GA (1977) The antarctic Polychaete fauna: its characteristics, distribution patterns, and evolution. In: Llano GA (ed) Adaptation within Antarctic ecosystem. Proc 3rd SCAR Symp on Antarctic Biology, Washington, Smithsonian Institution; Washington, DC, pp 1111–1127

Knox GA, Cameron DB (1998) The marine fauna of the Ross Sea: Polychaeta. National Institute of Water and Atmospheric Research (NIWA); NIWA Biodiversity Memoir 108 (formerly New Zealand Oceanographic Institute Memoirs), 125 pp

Knox GA, Lowry JK, (1977) A comparison between the benthos of the Southern Ocean and the North Polar Ocean with special reference to the Amphipoda and the Polychaeta. In: Dunbar MJ (ed) Polar Oceans. Arctic Institute of North America, Calgary, pp 423–462

La Mesa M, Vacchi M, Castelli A, Diviacco G (1997) Feeding ecology of two nototheniid fishes, *Trematomus hansoni* and *Trematomus loennbergii*, from Terra Nova Bay, Ross Sea. Polar Biol 17: 62–68

Lowry JK (1975) Soft bottom macrobenthic community of Arthur Harbour, Antarctica. Antarct Res Ser 23: 1–19

Lowry JK (1976) Studies on the macrobenthos of the Southern Ocean. PhD Thesis, University of Canterbury, Christchurch, New Zealand, 176 pp

Mackie A (1987) A review of species currently assigned to the genus *Leitoscoloplos* Day, 1977 (Polychaeta: Orbiniidae), with descriptions of species newly referred to *Scoloplos* Bainville, 1828. Sarsia 72: 1–28

Oliver JS, Slattery PN (1985) Effects of crustacean predators on species composition and population structure of soft-bodied infauna from McMurdo Sound, Antarctica. Ophelia 24: 155–175

Rhoads DC (1974) Organism-sediment relations on the muddy sea floor. Mar Biol Annu Rev 12: 263–300

Sarà M, Balduzzi A, Barbieri M, Bavestrello G, Burlando B (1992) Biogeographic traits and check list of Antarctic Demospongiae. Polar Biol 12: 559–585

Sicinski J, Janowska E (1993) Polychaetes of the shallow sublittoral of Admiralty Bay, King George Island, South Shetland Islands. Antarct Sci 5: 161–167

Vacchi M, La Mesa M, Castelli A (1994) Diet of two nototheniid fish from Terra Nova Bay, Ross Sea. Antarct Sci 6: 61–64

White MG (1984) Marine benthos. In: Laws RM (ed) Antarctic ecology II. Academic Press, London, pp 421–461

# The Population Structure and Ecology of the Antarctic Scallop *Adamussium colbecki* in Terra Nova Bay

M. Chiantore, R. Cattaneo-Vietti, P. Povero, and G. Albertelli[1]

**ABSTRACT**

In Terra Nova Bay, the scallop *Adamussium colbecki* (Mollusca Bivalvia) constitutes large beds up to 70–80 m in depth, reaching high values of density (50–60 ind $m^{-2}$) and biomass (120 g $m^{-2}$ DW soft tissues). Its population structure and biometrics have been studied during the summer of 1989/90, 1993/94, 1994/95 and 1995/96. The population of *A. colbecki* was studied in terms of abundance values and of size frequency distribution (modal length class: 70–75 mm). The growth rate of this species was assessed by X-ray methods and averaged 8 mm $year^{-1}$. Biometrical measures were performed in order to assess the influence of the summer increase in food supply on the life cycle of the scallop, showing a strong coupling between food supply and gonad development. On the basis of the gonadosomatic index values (GSI) and of the occurrence of planktonic larvae some speculations are made on the reproductive behaviour of this scallop.

In situ video recordings allowed the analysis of the swimming behaviour of *Adamussium*, providing evidence of the role of such behaviour as an escape reaction to predators.

Laboratory experiments were carried out in order to assess filtering and biodeposition rates, averaging, at +1 °C temperature, 1 l $h^{-1}$ $g^{-1}$ (DW soft tissues) and 5–6 mg DW $gDW^{-1}$ $day^{-1}$, respectively. This allowed estimation of the $C_{org}$ flux, through biodeposition by *A. colbecki*, which is about 21 mg C $m^{-2}$ $day^{-1}$ under *in situ* conditions. On the basis of organic matter flux and laboratory estimates of *Adamussium* feeding activity, it was possible to assess the important role of the scallop in coupling the energy flux from the water column to the sea bed, processing about 14% of total carbon flux from the water column to the sediments, with an assimilation efficiency of 36%.

## Introduction

*Adamussium colbecki* (Mollusca, Bivalvia) is an endemic Antarctic scallop and is considered one of the commonest bivalves in the Antarctic benthos (Nicol 1966). At present, it is considered to have a circumpolar distribution and a depth range across the continental shelf from 0 to nearly 1500 m (Dell 1972), although large assemblages were found and studied up to 60–80 m depth only in McMurdo Sound (Stockton 1984; Berkman 1990), at Terra Nova Bay (Amato 1990; Berkman and Nigro 1992; Di Geronimo et al. 1992; Nigro 1993; Albertelli et al. 1994; Pranovi et al. 1994), Syowa Station (Nakajima et al. 1982) and Stonington Island (Ralph and Maxwell 1977).

Because of its unusually large size and its very high density and biomass values, this species has received a great deal of interest as a potentially exploitable benthic marine living resource (Fischer and Hureau 1985), as object of a "mussel watch monitoring project" (Berkman and Nigro 1992) and as a littoral key species in the framework of the SCAR-EASIZ Programme (SCAR 1994).

At Terra Nova Bay, since 1990, much work has been done on its population structure (Nigro 1993; Pranovi et al. 1994; Cattaneo-Vietti et al. 1997), growth estimates (Pranovi et al. 1994; Cattaneo-Vietti et al. 1997), concentration of heavy metals (Mauri et al. 1990; Viarengo et al. 1994; Nigro et al., 1997) and other contaminants and, more recently, about its behaviour and role in the littoral ecosystem of Terra Nova Bay (Ansell et al. 1998; Chiantore et al. 1998).

Its very high values of density and biomass are sustained by the summer phytoplankton bloom,

[1] Istituto di Scienze Ambientali Marine, Università di Genova, Corso Rainusso 14, 16038 Santa Margherita Ligure, Genoa, Italy

which provides a potentially high input of biogenic material both to the pelagic and benthic domains, reaching the sediments mostly as faecal pellets (Fabiano et al. 1997). This downward flux provides a regular seasonal signal, inducing significant metabolic responses in the benthic organisms (Karl et al. 1991; Cattaneo-Vietti et al. 1997). At the same time, suspension-feeding bivalve communities can play a significant role in coupling pelagic and benthic organic material flow as well as controlling materials recycling, enhancing energy flow from the water column to the benthic level and affecting the quality and quantity of particulate organic matter through biodeposition.

The aim of the present chapter is to summarize the knowledge regarding the ecology and the population structure of *A. colbecki*, recently acquired in the framework of the Italian Antarctic Research Programme (PNRA), considering the large abundance and role of this scallop in the shallow waters of Terra Nova Bay.

## Study Site and Sampling

The study of the population structure of *Adamussium colbecki* has been made using samples collected in Road Bay, close to the Italian base during the austral summers of 1989–90, 1993–94, 1994–95 and 1995–96, on a 40–50 m deep bottom, where the substrate is fairly fine (fraction above 212 µm representing 94.2%). Samples were collected by Charcot-Picard or naturalist dredges.

The distribution and abundance of this scallop in Terra Nova Bay was assessed in the same years from Van Veen grab samples (surface: 60 × 35 cm) and through visual observations that were carried out using a remotely operated vehicle (ROV).

Grab samples were immediately fixed in 10% buffered formalin, while dredge samples were transported into tanks in the aquarium of the Italian base. Some specimens, after acclimation, were used to conduct measurements of filtering and biodeposition rates (Chiantore et al. 1998).

During the austral summer of 1993/94, a study was carried out in the same area at 40 m depth in order to evaluate hydrological data and to assess the evolution of the biochemical composition of particulate organic matter and of sinking organic matter reaching the sediments, directly affecting benthic communities (Albertelli et al. 1996, 1998; Fabiano et al. 1997).

## Results

### Distribution, Abundance and Size Structure

In Terra Nova Bay, dense assemblages of *A. colbecki* are limited to a narrow belt from Tethys Bay to Adelie Cove, between 40 and 80 m depth (Fig. 1), showing locally high densities. *Adamussium* specimens can also be found from 15 m depth, but its density shows a decreasing trend with decreasing depth. Around 70–80 m depth, the bottom is completely covered by empty shells. Some areas, characterized by fine sediments and particularly enriched in organic matter, such as inside Adelie Cove, show lower abundance values of this scallop, which prefers coarse and medium grain sediments.

Where dense beds occur, its population density may be appreciated by ROV videos (Fig. 2), showing overlapping individuals. The population is mainly dominated by large adult individuals, which reach an average density of 60 ind $m^{-2}$ and a total biomass around 120 g $m^{-2}$ (dry weight soft tissues) (Cattaneo-Vietti et al. 1997). Juveniles have a scattered distribution and are mainly found attached through byssus threads to adult valves, where a microcommunity mainly composed of benthic diatoms on the upper valve and foraminiferans, bryozoans and the spirorbid *Paralaeospira levinsenii* on the inferior valve, can also be found (Rosso 1992; Berkman 1994).

Analysis of size-frequency distributions during several austral summers (Fig. 3) shows high persistency of large-sized individuals (modal length class: 70–75 mm) while some differences can be found in abundance of smaller size classes. Nevertheless, their presence was observed throughout the study period, suggesting continuous recruitment to the population due to consistent annual supply of primary production to the benthos. In 1989–90, an almost complete absence of the smallest-size classes was observed, while some years later (1993–94 and 1994–95) younger individuals were well represented and the previous gap shifted to size classes between 40 and 65 mm in length. In summer 1995–96, a deficiency of all young classes was again observed.

Finally, in all considered years a persistent deficiency of size classes 40–55 mm was observed,

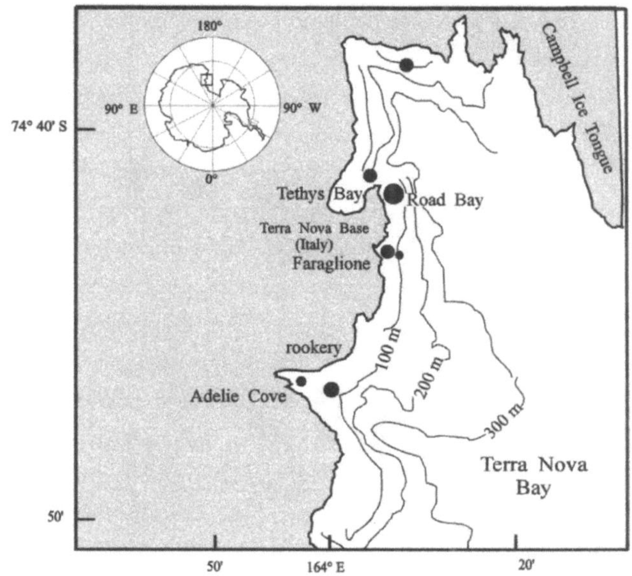


**Fig. 1.** Distribution of *Adamussium colbecki* along Terra Nova Bay. *Different-sized/circles* are proportional to average density of the scallop. •5, ●10, ●15, ●20, ● > 25 ind m$^{-2}$

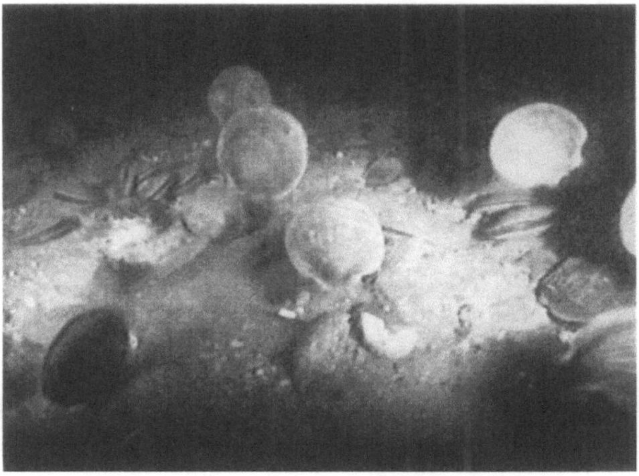

**Fig. 2.** ROV picture showing very dense aggregation of scallops at 40 m depth, at Road Bay. (Courtesy G. Verruggio, IAN-CNR, Genoa)

in accordance with observations by Nakajima et al. (1982) and Berkman (1990).

## Growth and Biometrics

X-ray photographs have allowed the measurement of the surface of each growth ring, to calculate their average diameter, and, consequently, to assess the growth increment from each growth ring to the next, that is the year growth increment, for individuals of different size. The average growth increment is 8.1 ± 1.0 mm year$^{-1}$ till the fifth ring, while this value falls to 5.1 ± 1.5 mm year$^{-1}$ in larger individuals (Cattaneo-Vietti et al. 1997).

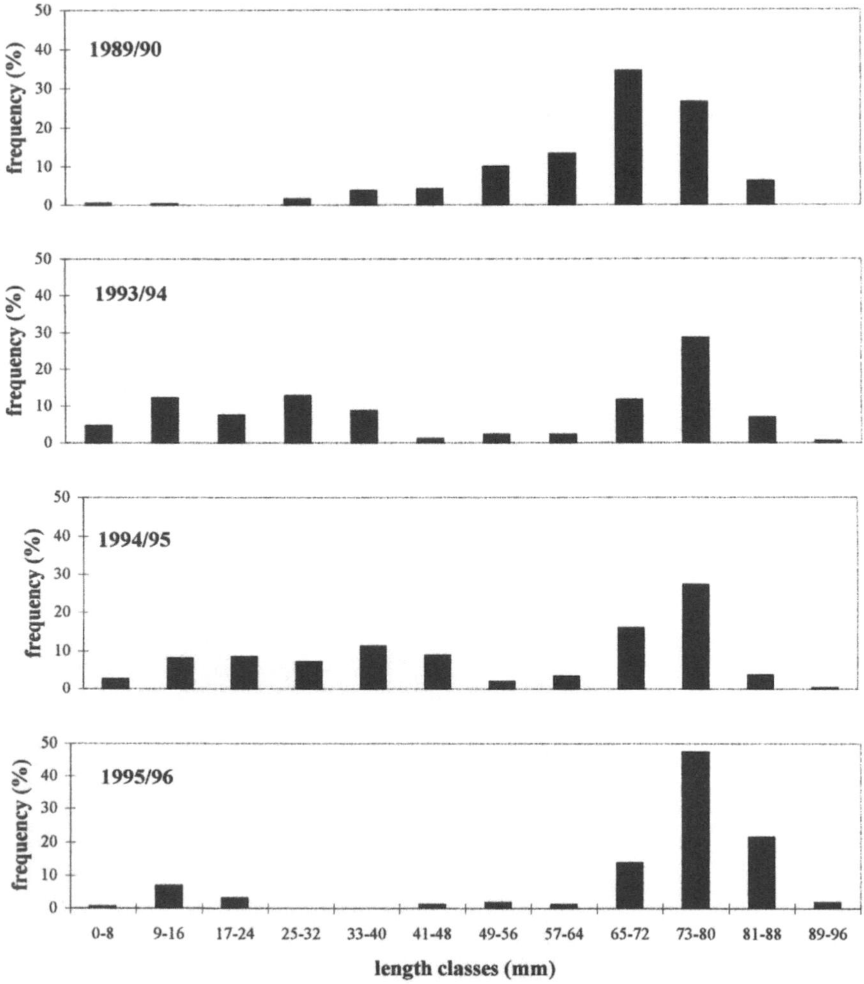


**Fig. 3.** Size frequency distribution of *Adamussium colbecki* at Road Bay, close to the Italian base, during the different study years

Correlation analyses between main biometrical features were calculated over the complete size range of collected individuals, showing changes in shell shape and in per cent contribution of different organs to overall biomass with age. In particular, the ratio between shell length and height (aspect-ratio) turned out to be particularly interesting, showing an inversion during growth (Cattaneo-Vietti et al. 1997). Until 45 mm length, this ratio is approximately below 1, and organisms are fundamentally oval-shaped, while over 50 mm length the shape is subcircular and length/height ratio is >1. This change in shape seems to occur when byssally attached juveniles are released from the adults. Increase of gonad and muscle weight and decrease of shell weight relative to total dry weight, which occur at this length, suggest the start of the adult phase and consequent sexual maturation (Cattaneo-Vietti et al. 1997).

The relation between muscle dry weight and shell length (Cattaneo-Vietti et al. 1997), becoming scattered in specimens larger than 50 mm, highlights the role of the muscle as an energy reserve available during gamete maturation (Barber and Blake 1981; Bricelj et al. 1987).

## Behaviour

Adult scallops are free-living, while immature individuals (up to 40–45 mm shell length) are found attached by byssus threads to the upper shell valve of the larger ones.

Although *A. colbecki* is known to swim (Stockton 1984; Berkman 1990; Cattaneo-Vietti et al. 1997), details of its swimming performance have been only recently described, making use of recordings of its swimming activity in situ at Terra Nova Bay (Ansell et al. 1998), using a remotely-operated vehicle (ROV). Swimming is performed through contraction of the "quick" adductor muscle, during the adduction cycle, forcing water out of the shell cavity through jet orifices that are formed in the mantle on either sides of the hinge so that the shell is propelled with the ventral margin leading.

The significance of swimming in *A. colbecki* must clearly be similar to its role in other pectinids, especially in escape from predators (Peterson et al. 1982). Predators of *A. colbecki* are known to include the gastropod *Neobuccinum eatoni*, the nemertean *Paraborlasia corrugatus*, starfishes and fishes such as *Trematomus bernacchii* (Naito and Iwami 1982; Vacchi et al. 1994). Other reasons for swimming performances could be avoidance of unfavourable conditions and local dispersal (Ralph and Maxwell 1977; Stockton 1984). Despite this evidence, diving observations (Berkman, pers. comm.) suggest that scallop movement may be relatively uncommon throughout its life: this fact could be due to different local conditions (currents, organic enrichment) that can probably affect scallop behaviour and physiology.

Juveniles are passively carried by the larger adults, remaining firmly attached during the swimming bout. Clearly, young scallops benefit from the potential of the adult to perform faster and more extensive swims, saving energy. The most vulnerable stage in the postsettlement life of such species could be that immediately following detachment of the juveniles from their adult hosts, also because of their shell shape (aspect-ratio < 1, umbonal angle and size of jet orifices), which does not allow high swimming effectiveness (Ansell et al. 1998). Consequently, predation could selectively be more relevant on just detached medium sized individuals (Cattaneo-Vietti et al. 1997).

## Reproductive Biology

Little is known about the spawning behaviour in *Adamussium colbecki*. Small size eggs (20–45 μm) found in January (Berkman et al. 1991) in McMurdo Sound seem to indicate that spawning may take place during the austral spring. In Terra Nova Bay, the GSI increase in females at the end of January (Cattaneo-Vietti et al. 1997) suggests that gonad maturation takes place in late summer, at the end of the primary production period (Fig. 4a,b,c). This fact could be also supported by evidence of different levels of gonadal Cd in females compared to males observed in the austral summer (Mauri et al. 1990), which is probably related to gametogenesis. Consequently there seems to be a shift in the reproductive cycle from McMurdo Sound to Terra Nova Bay, probably linked to differences in irradiance or food availability, as observed for other Antarctic invertebrate populations (Pearse 1965).

The small eggs and the low ratio (0.36) between first (PI) and second (PII) prodissoconchs suggest the occurrence of a planktotrophic larva in *Adamussium* (Berkman et al. 1991), although there are no data concerning the length of the larval stage. In Terra Nova Bay, in the austral summer of 1995–96, larvae collected under the pack-ice (Fig. 5a,b) showed PI = 125 μm and PII = 250 μm. This suggests these larvae still had to spend a long time prior to settling, as their PII length was much shorter than that measured on juveniles (PI: 124 μm; PII: 346 μm) by Berkman et al. (1991).

## Role of *Adamussium colbecki* in the Transport of Organic Matter to the Sediments

In Terra Nova Bay, food supply to the benthic community shows a clear pattern, according to main physical and biological events taking place along the water column: sea-ice breaking-up and melting and occurrence of primary and secondary production, as detected both by the analysis of suspended and sinking particulate matter and by fluorescence depth profiles (Albertelli et al. 1996,

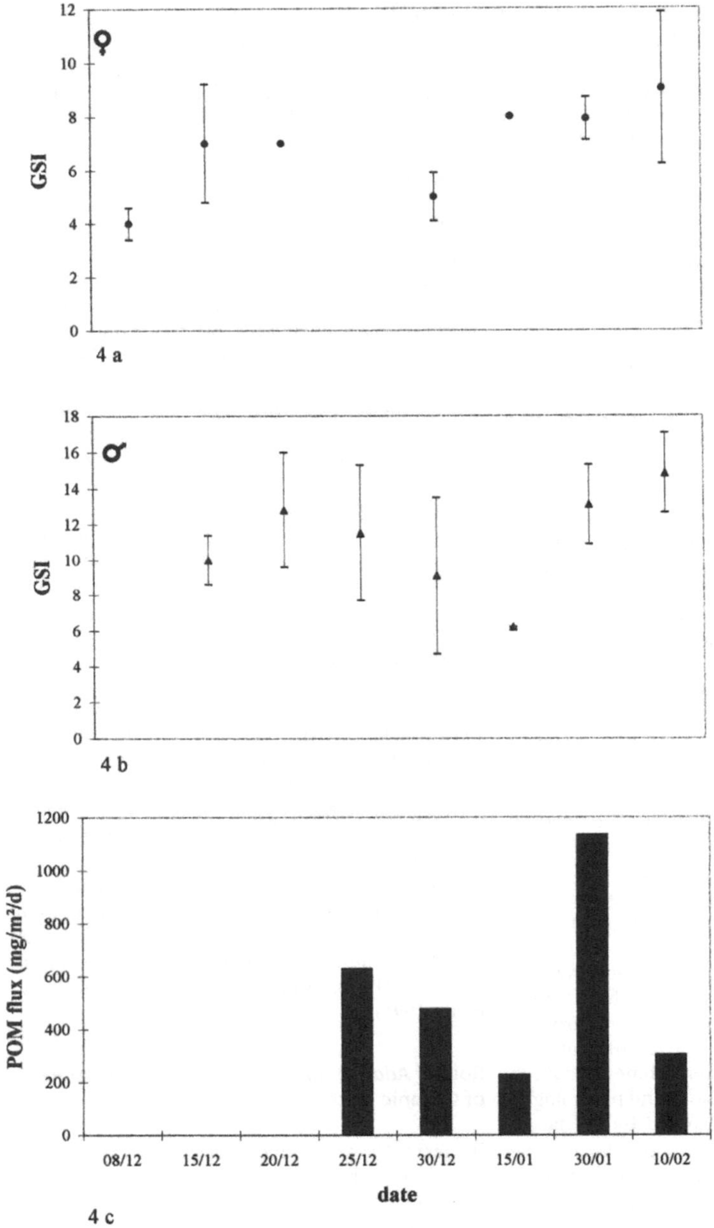


**Fig. 4a–c.** Gonadosomatic index (GSI) values: **a** in females; **b** in males; **c** changes in particulate organic matter (POM) flux during the summer season

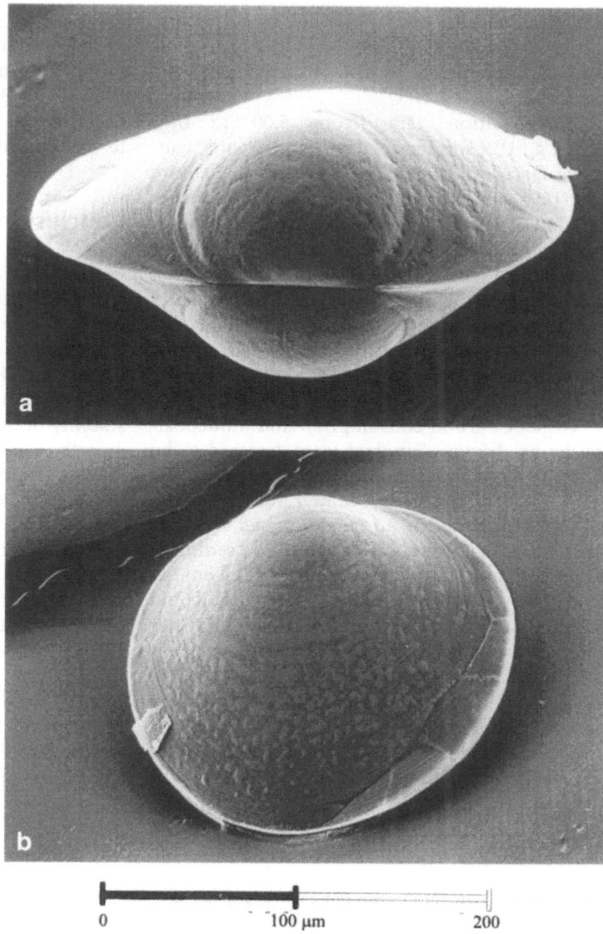


**Fig. 5a,b.** SEM micrographs of *Adamussium* larvae. **a** the sharp boundary between PI and PII is evidenced. **b** Dorsal view. *Bar* 100 μm

1998; Fabiano et al. 1997). A relevant role is played by zooplankton grazing, as indicated by high phaeopigment flux and abundance of faecal pellets in trapped material.

During the summer (about 60 days), the total energy transfer from the water column to the sediment, through the sinking particles, can be considered to be about 1000 kJ $m^{-2}$ (Albertelli et al. 1998). Such seasonal food supply plays a primary role in sustaining the biomass of the *Adamussium* population, although adult individuals can also take advantage of organic matter associated with resuspended sediments through shell clapping (Davis and Marshall 1961; Berkman et al. 1991).

To evaluate the amount of sinking material directly utilised by scallops and the material rejected as faeces, filtering and biodeposition rates were estimated, as well as biochemical composition of faecal material relative to seston. Filtering rate of *A. colbecki* ($1.12 \pm 0.30\,l\,h^{-1}\,g^{-1}$ DW soft tissues; Chiantore et al. 1998) is relatively low, if compared to that of other filibranchs, probably due to low temperatures and to seston features. In fact, low temperatures slow basal metabolic activities (i.e. growth rate; Clarke 1983), and high quality (i.e. C:N = 5.82) and quantity (POM flux up to 1100 mg m$^{-2}$ day$^{-1}$; Fig. 4c) of seston allow lower pumped volume to sustain such a metabolic demand.

Biodeposition rate (from 5.00 to 6.72 mg DW g$^{-1}$ DW soft tissues day$^{-1}$) and organic content of faeces of *Adamussium colbecki* (Chiantore et al. 1998) are consistent with those of another common Antarctic suspension feeding bivalve, *Laternula elliptica* (Ahn 1993), under quite similar experimental conditions. Biochemical analyses of faeces suggest the presence of still high quality organic material in *A. colbecki* faeces, probably due to very high concentration of suspended organic matter, leading to incomplete exploitation of the food source.

Organic carbon consumption and biodeposition by an *Adamussium* bed were assessed to be 33 mg m$^{-2}$ day$^{-1}$ and 21 mg C m$^{-2}$ day$^{-1}$ respectively: about 36% of all the organic matter flux (measured as POC) sinking to the *Adamussium* bed is assimilated while 64% of this is driven to the sediment as faecal matter, which still maintains good energy values (i.e. C:N ratio) and, consequently, could constitute an important food source for other benthic fauna, especially deposit feeders.

These data are in agreement with Knox (1990), who estimated that epifaunal suspension feeders in depths above 200 m, in the McMurdo Sound ecosystem, consume as much as 10% of the phytoplankton as it sediments to the bottom, as well as a proportion of the sea-ice microalgae released to the water column. Nevertheless, the major proportion of the phytoplankton net production, up to 80%, is not consumed and is either decomposed in the water column or sedimented to the bottom.

## Conclusions

In Terra Nova Bay, *Adamussium colbecki* is a key species of littoral soft bottoms, from 20 to 80 m depth. The presence of adults, belonging to different overlapping size classes, is characteristic of stable and K-selected populations (MacArthur 1960; White 1984), and is linked to the coupling of stability of the physical environment and the slow growth rate, which leads to the evolution of an efficient, stable community structure with a low turnover rate (White 1984) and large standing stocks, maximizing share of food input under the low-temperature regime of the Antarctic (Brey and Clarke 1993).

*Adamussium* biometrics at Terra Nova Bay are in agreement with growth estimates by Stockton (1984), Berkman (1990) and Ralph and Maxwell (1977; Table 1). Small differences are probably related to different levels in food availability, as observed in other scallops from temperate waters (Bayne and Newell 1983; McDonald and Thompson 1985).

**Table 1.** Main features of *Adamussium colbecki* populations

| Site | Depth (m) | Average density ind m$^{-2}$ | Average biomass g wet wt m$^{-2}$ | Maximum length (mm) | Modal length class (mm) | Age | Growth parameters | | | Growth rate (mm year$^{-1}$) | Reference |
|---|---|---|---|---|---|---|---|---|---|---|---|
| | | | | | | | K | $H_{max}$ | $t_0$ | | |
| Stonington Island (Antarctic Peninsula) | 10 | | | | | >6 | 0.24 | 90 | 0.150 | | Ralph and Maxwell (1977) |
| Syowa Station (Enderny Land) | 1–15 | Up to 112 | 2000 | 89 | 20–30<br>70–80 | | | | | | Nakajima et al. (1982) |
| McMurdo Explorers Cove (Ross Sea) | 4–6<br>30 | 85<br>20 | 1600<br>340 | 80–90 | 75–80 | >13 | 0.12 | 105 | - | | Stockton (1984) |
| McMurdo New Harbor (Ross Sea) | 3<br>30 | 45<br>20 | 1600<br>400 | 120 | 75 | >20 | 0.09 | 128 | −0.765 | | Berkman (1990) |
| Terra Nova Bay (Ross Sea) | 15<br>30 | 2<br>59 | | 95 | 75 | | | | | | Nigro (1993) |
| Terra Nova Bay (Ross Sea) | 0–30 | | | 81 | 70 | >20 | 0.19 | 93 | 0.621 | | Pranovi et al. (1994) |
| Terra Nova Bay (Ross Sea) | 40–100 | Up to 90 | Up to 1500 | 94 | 70–75 | | | | | 5.1–8.1 | Present study |

Changes in growth pattern have been detected (Nakajima et al. 1982; Berkman 1994; Cattaneo-Vietti et al. 1997), suggesting detachment from adults taking place at 30–50 mm in length and reproductive maturity being achieved (Nakajima et al. 1982; Berkman 1990; Cattaneo-Vietti et al. 1997).

Supposed discrepancies in spawning timing between McMurdo and Terra Nova Bay suggest that the reproductive cycle is controlled by both endogenous and exogenous factors such as light, temperature and food availability, as in other scallops (Bricelj et al. 1987), mussels (Newell et al. 1982) and cockles (Newell and Bayne 1980). In particular, in bivalves, the summer food supply influences the species physiology in terms of ingestion rate, digestion and assimilation efficiencies and differential utilization of biochemical compounds (Ansell 1974; Newell and Bayne 1980; Kreeger 1993): a high-quality food supply could induce rapid growth (Thompson and Nichols 1988) and support energy storage and gonad maturation (Barber and Blake 1981), while metabolic demands may also be satisfied by resuspended detritic material (Davis and Marshall 1961; Vernet 1977; Shumway et al. 1987).

Reproduction in late summer/winter, occurrence of a planktotrophic larva (Sastry 1979) and fluctuations in the water column food supply (Mileikovsky 1971) could strongly influence the recruitment success and explain the interannual variability in occurrence of spats (Stockton 1984; Berkman 1990; Berkman et al. 1991; Nigro et al. 1997). However, evidence for other Antarctic invertebrates (Pearse et al. 1991) suggest alternative food supplies to planktotrophic larvae (Olson et al. 1987), such as dissolved organic matter (DOM: Shilling and Bosch 1994) and bacterioplankton (Rivkin et al. 1986), thus reducing starvation problems.

It is now clear that the prolonged winter season can influence both spawning and larval survival, preventing adults from reaching sexual maturation and larvae from finding enough nourishment. Moreover, the concomitant occurrence of changes in growth pattern, sexual maturation and probably selective feeding by fishes could partly explain the differences in population structure in terms of unsuccessful recruitment or high mortality rate of juveniles.

*Acknowledgements.* This work was supported by Progetto Nazionale Ricerche in Antartide (PNRA) funded by the Italian Government. We are grateful to Dr. Carlo Ossola for zooplankton sampling and putting at our disposal *Adamussium* larvae.

## References

Ahn I-Y (1993) Enhanced particle flux through the biodeposition by the Antarctic suspension-feeding bivalve *Laternula elliptica* in Marian Cove, King George Island. J Exp Mar Biol Ecol 171: 75–90

Albertelli G, Arnaud PM, Cattaneo-Vietti R (1994) Ecological aspects of the macrobenthos of the Ross Sea: the coastal and deep-sea bivalve molluscs of Terra Nova Bay. In: Faranda F, Guglielmo L (eds) National Scientific Commission for Antarctica (ed) Oceanographic Campaign 1987–88, Ocean Camp 1989–90, Data Rep Part III, Genova, pp 49–66

Albertelli G, Cattaneo-Vietti R, Chiantore M (1996) The influence of the organic matter flux on an *Adamussium colbecki* bed at Terra Nova Bay (Ross Sea, Antarctica). In: Albertelli G, DeMaio A, Piccazzo M (eds) Atti XI Congr Ass Ital Oceanogr Limnol, Lang, Genova, pp 299–310

Albertelli G, Cattaneo-Vietti R, Chiantore M, Pusceddu A, Fabiano M (1998) Food availability to an *Adamussium* bed during the austral summer 1993/94 (Terra Nova Bay, Ross Sea). J Mar Syst 17: 425–434

Amato E (1990) Environmental impact assessment at sea. In: National Scientific Commission for Antarctica (ed) Oceanographic Campaign 1987–88, Data Rep Part I, Genova; pp 479–503

Ansell AD (1974) Seasonal changes in biochemical composition of the bivalve *Chlamys septemradiata* from the Clyde Sea Area. Mar Biol 25: 85–99

Ansell AD, Cattaneo-Vietti R, Chiantore M (1998) Swimming in the Antarctic scallop *Adamussium colbecki*: analysis of in situ video recordings. Antarct Sci 10(4): 369–375

Barber BJ, Blake NJ (1981) Energy storage and utilization in relation to gametogenesis in *Argopecten irradians concentricus* (Say). J Exp Mar Biol Ecol 52: 121–134

Bayne BL, Newell RC (1983) Physiological energetics of marine molluscs. In: Wilbur KM, Saleuddin AS (eds) The Mollusca 4 (1). Academic Press, New York, pp 407–515

Berkman PA (1990) The population biology of the Antarctic scallop, *Adamussium colbecki* (Smith, 1902) at New Harbor, Ross Sea. In: Kerry KR, Hempel K (eds) Antarctic ecosystems. Ecological change and conservation. Springer, Berlin Heidelberg New York, pp 281–288

Berkman PA (1994) Epizoic zonation on growing scallop shells in McMurdo Sound, Antarctica. J Exp Mar Biol Ecol 179: 49–67

Berkman PA, Nigro M (1992) Trace metal concentrations in scallops around Antarctica: extending the mussel watch programme to the Southern Ocean. Mar Pollut Bull 24: 322–323

Berkman PA, Waller TR, Alexander P (1991) Unprotected larval development in the Antarctic scallop *Adamussium colbecki* (Mollusca: Bivalvia: Pectinidae). Antarct Sci 3: 151–157

Brey T, Clarke A (1993) Population dynamics of marine benthic invertebrates in Antarctic and subantarctic environments: are there unique adaptations? Antarct Sci 5(3): 253–266

Bricelj VM, Epp J, Malouf RE (1987) Intraspecific variation in reproductive and somatic growth cycles of bay scallops *Argopecten irradians*. Mar Ecol Prog Ser 3: 123–137

Cattaneo-Vietti R, Chiantore M, Albertelli G (1997) The population structure and ecology of the Antarctic scallop, *Adamussium colbecki* at Terra Nova Bay (Ross Sea, Antarctica). Sci Mar 61(2): 15–24

Chiantore M, Cattaneo-Vietti R, Albertelli G, Misic C, Fabiano M (1998) Role of filtering and biodeposition by *Adamussium colbecki* in circulation of organic matter in Terra Nova Bay (Ross Sea, Antarctica). J Mar Syst 17: 411–424

Clarke A (1983) Life in cold water: the physiological ecology of polar marine ectotherms. Oceanogr Mar Biol Annu Rev 21: 341–453

Davis RL, Marshall N (1961) The feeding of the bay scallop, *Aequipecten irradians*. Proc Natl Shellfish Assoc 52: 25–29

Dell RK (1972) Antarctic benthos. Adv Mar Biol 10: 1–212

Di Geronimo IS, Cattaneo-Vietti R, Gambi MC, Casazza G, Cormaci M, Scammacca B (1992) Prime osservazioni sulle comunità bentoniche di Baia Terra Nova (Mare di Ross, Antartide): bionomia e distribuzione. In: Albertelli G, Ambrosetti W, Piccazzo M, Ruffoni Riva T (eds) Atti IX Congr Ass Ital Ocean Limnol, Lang, Genova, pp 635–646

Fabiano M, Chiantore M, Cattaneo-Vietti R, Albertelli G, Pusceddu A, Povero P (1997) Short-term variations in particulate matter flux in Terra Nova Bay, Ross Sea, Antarctica. Antarct Sci 9(2): 143–149

Fischer W, Hureau JC (1985) FAO species identification sheets for fisheries purposes. Food and Agriculture Organization of the United Nations, Rome, pp 95–105

Karl DM, Tilbrook BD, Tien G (1991) Seasonal coupling of organic matter production and particle flux in the western Bransfield Strait, Antarctica. Deep Sea Res 38(8/9): 1097–1126

Knox GA (1990) Primary production and consumption in McMurdo Sound, Antarctica. In: Kerry KR, Hempel HG (eds) Ecological change and conservation. Springer, Berlin Heidelberg New York, pp 115–128

Kreeger DA (1993) Seasonal patterns in utilization of dietary protein by the mussel *Mytilus trossulus*. Mar Ecol Prog Ser 95: 215–232

MacArthur R (1960) On the relative abundance of species. Am Nat 94: 25–34

Mauri M, Orlando E, Nigro M, Regoli F (1990) Heavy metals in the Antarctic scallop *Adamussium colbecki*. Mar Ecol Prog Ser 67: 27–33

McDonald BA, Thompson RJ (1985) Influence of temperature and food availability on the ecological energetics of the giant scallop *Placopecten magellanicus*. I Growth rates of shell and somatic tissue. Mar Ecol Prog Ser 25: 279–294

Mileikovsky SA (1971) Types of larval development in marine bottom invertebrates, their distribution and ecological significance. Mar Biol 10: 193–213

Morton B (1980) Swimming in *Amusium pleuronectes* (Bivalvia: Pectinidae). J Zool 190: 375–404

Naito Y, Iwami T (1982) Fish fauna in the northern parts of Lüzov-Holm Bay with some notes on the stomach contents. Mem Nat Inst Polar Res 23: 64–72

Nakajima Y, Watanabe K, Naito Y (1982) Diving observations of the marine benthos at Syowa Station, Antarctica. Natl Inst Polar Res: 44–54

Newell RIE, Bayne BL (1980) Seasonal changes in the physiology, reproductive condition and carbohydrate content of the cockle *Cardium* (=*Cerastoderma*) *edule* (Bivalvia: Cardiidae). Mar Biol 56: 11–19

Newell RIE, Hilbish TJ, Koehn RK, Newell CJ (1982) Temporal variation in the reproductive cycle of *Mytilus edulis* L. (Bivalvia, Mytilidae) from localities on the east of the United States. Biol Bull 162: 299–310

Nicol D (1966) Descriptive ecology and geographic distribution of some Antarctic pelecypods. Bull Am Paleontol 51(231): 1–102

Nigro M (1993) Nearshore population characteristics of the circumpolar Antarctic scallop *Adamussium colbecki* (Smith, 1902) at Terra Nova Bay (Ross Sea). Antarct Sci 5(4): 377–378

Nigro M, Regoli F, Rocchi R, Orlando E (1997) Heavy metals in Antarctic molluscs. In: Battaglia B, Valencia J, Walton DWH (eds) Antarctic communities: species, structure and survival. Cambridge University Press, Cambridge, pp 409–411

Olson RR, Bosch I, Pearse JS (1987) The hypothesis of Antarctic larval starvation examined for the asteroid *Odontaster validus*. Limnol Oceanogr 32(3): 686–690

Pearse JS (1965) Reproductive periodicities in several contrasting populations of *Odontaster validus* Koehler, a common Antarctic asteroid. Antarct Res Ser 5: 39–85

Pearse JS, McClintock JB, Bosch I (1991) Reproduction of Antarctic benthic marine invertebrates: tempos, modes and timing. Am Zool 31: 65–80

Peterson CH, Ambrose WG, Hunt JH (1982) A field test of the swimming response of the bay scallop (*Argopecten irradians*) to changing biological factors. Bull Mar Sci 32: 939–944

Pranovi F, Marcato S, Zanellato R (1994) Analisi biometriche e biologia di popolazione del mollusco antartico *Adamussium colbecki* a Baia Terra Nova, Mare di Ross. Atti dell'Istituto Veneto di Scienze, Lettere ed Arti 152 (1993–1994): 123–136

Ralph R, Maxwell JGH (1977) Growth of two Antarctic lamellibranchs: *Adamussium colbecki* and *Laternula elliptica*. Mar Biol 42: 171–175

Rivkin RB, Bosch I, Pearse JS, Lessard EJ (1986) Bacterivory: a novel feeding mode for asteroid larvae. Science 233: 1311–1314

Rosso A (1992) Bryozoa from Terra Nova Bay (Ross Sea, Antarctica). In: Gallardo VA, Ferrettio O, Moyano HI (eds) Oceanografia in Antartide. ENEA-PNRA-EULA, Concepcion, Chile, pp 359–369

Sastry AN (1979) *Pelecypoda* (excluding *Ostreidae*). In: Giese AC, Pearse JS (eds) Reproduction of marine invertebrates. Academic Press, New York, pp 113–292

SCAR (1994) Coastal and shelf ecology of the Antarctic sea-ice zone. Science plan and implementation plan. SCAR Rep 10: 1–20

Shilling FM, Bosch I (1994) Pre-feeding embryos of Antarctic and temperate echinoderms use dissolved organic material for growth and metabolic needs. Mar Ecol Prog Ser 109: 173–181

Shumway S (ed) (1991) Scallops: biology, ecology and aquaculture. Developments in aquaculture and fisheries science, vol 21. Elsevier, Amsterdam

Shumway SE, Elvin R, Schick DF (1987) Food resources related to habitat in the scallop *Placopecten magellanicus* (Gmelin, 1971): a qualitative study. J Shell Res 6: 89–95

Stockton WL (1984) The biology and ecology of the epifaunal scallop *Adamussium colbecki* on the west side of McMurdo Sound, Antarctica. Mar Biol 78: 171–178

Thompson JK, Nichols FH (1988) Food availability controls seasonal cycle of growth in *Macoma balthica* (L.) in San Francisco Bay, California. J Exp Mar Biol Ecol 116: 43–61

Vacchi M, La Mesa M, Castelli A (1994) Diet of two coastal nototheniid fish from Terra Nova Bay, Ross Sea. Antarct Sci 6(1): 61–65

Vernet M (1977) Alimentacion de la vieira tehuelche (*Chlamys tehuelchus*). Comision Nacional de Estudios Geo-Heliofisicos, Centro Nacional Patagonico, Puerto Madryn, Argentina, unpublished report, 26 pp

Viarengo A, Canesi L, Pertica M, Mancinelli G, Accomando R, Ponzano E, Orunesu M (1994) Cellular and molecular mechanisms of heavy metal cation homeostasis in the cold-adapted Antarctic scallop *Adamussium colbecki* (Smith 1902). Proc of the 2nd meeting on Antarctic Biology, Padova, Dipartimento di Biologia dell'Università, 26–28 Feb, 1992, pp 257–266

White MG (1984) Marine benthos. In: Laws RM (ed) Antarctic ecology II. Academic Press, London, pp 421–461

# Bioconcentration of Polychlorinated Biphenyls in the Pelagic Food Chain of the Ross Sea

S. Corsolini and S. Focardi[1]

**ABSTRACT**

The phenomenon of bioconcentration of polychlorinated biphenyls in the pelagic food chain of the Ross Sea, Antarctica is described. The mean concentration of the sum of 50 congeners of PCB was 1 $ng\,g^{-1}$ in phytoplankton, 4.23 $ng\,g^{-1}$ in zooplankton, 9.39 $ng\,g^{-1}$ (body burden) in silverfish and 124.76 $ng\,g^{-1}$ in penguin liver. The values of bioconcentration factor (BCF) show that the largest increment in PCB concentration was from water to phytoplankton, and from fish to birds. This confirms the transfer of PCBs from ice pack to biota and their bioconcentration in the food web.

## Introduction

The presence of persistent contaminants such as polychlorobiphenyls (PCBs) in terrestrial and marine ecosystems is due to their dispersal by atmospheric and hydrologic movements (Tanabe and Tatsukawa 1986; Iwata et al. 1993). These compounds may be transported to great distances from their sources by air currents (Knap 1990). Thus, they can be found in remote areas such as the Arctic and the Antarctic (Risebrough et al. 1976; Norstrom et al. 1988; Wolff 1990; Focardi et al. 1995). Atmospheric and marine contamination is different in the two hemispheres, with the Northern Hemisphere more polluted than the Southern. This is due to the fact that the industrialized countries of Europe and North America were the greatest producers and users of PCBs from the second world war until the 1970s. The concentrations of PCBs in the surface layers of the oceans are quite homogeneous in the middle latitudes of the Northern Hemisphere, with a maximum in the tropical region (Table 1). Lower values have been recorded in the Southern Hemisphere with a minimum in the Antarctic region (Tanabe and Tatsukawa 1983). It has also been found that organic matter of euphasiids contain high levels of PCBs. In very productive waters, PCBs therefore sediment in the deeper layers (Elder and Fowler 1977).

As far as PCB composition is concerned, congeners with two or three chlorines are reported to prevail at low latitudes and congeners with five or six chlorines at higher latitudes (Tanabe et al. 1983). In fact, PCBs with few chlorines are more water-soluble and those with many chlorines have more affinity for the organic component of suspended material.

Pollutants are transported by air mass from industrialized areas to remote areas where they fall out by wet and dry deposition. In Antarctica they are incorporated in the snow and ice. The ice pack can be described as a reservoir of persistent contaminants which it releases during the summer melt. The contaminants are transferred from the ice to the seawater and then to organisms. They bioconcentrate in the food chain, and become a potential risk for marine life. In the 1994–95 Antarctic summer we collected samples in the Ross Sea to evaluate the transfer of polychlorinated biphenyls from ice to the marine food web. The Antarctic trophic webs typically consist of few levels, for example: phytoplankton, zooplankton (copepods, salps, euphasiids, fish larvae, chaetognaths), fish (myctophids, *Pleuragramma antarcticum*), birds (Adelie penguins *Pygoscelis adeliae*, Emperor penguins *Aptenodytes forsteri*, other species) and marine mammals (baleen whales, crabeater seals *Lobodon carcinophagus*, Weddell seal *Leptonychotes weddelli*, leopard seals

[1] Dipartimento di Biologia Ambientale, Università di Siena via delle Cerchia, 3-53100 Siena, Italy

*Hydrurga leptonyx* and killer whales *Orcinus orca*) (Hempel 1985). Krill and *P. antarcticum* are the key species as they are principal food items for several species of birds and marine mammals. *P. antarcticum* feed on adult *Euphausia crystallorophias*; birds, seals and whales feed on zooplankton, krill and *P. antarcticum* (Hubold 1985). Previous studies have demonstrated that in polluted coastal areas PCBs affect reproductive and immune function in higher trophic predators such as marine mammals (Kannan et al. 1993; Borrell et al. 1996). Polychlorinated biphenyls including non-, mono- and di-*ortho* congeners (Fig. 1) were analyzed in the food web of the Ross Sea to evaluate their bioconcentration in the food chain. 2,3,7,8-TCDD equivalents (TEQs) were assessed based on coplanar congener concentrations and toxicity equivalence factors (TEFs).

**Table 1.** Concentration of polychlorobiphenyls (range in $ng\,l^{-1}$) in seawater samples from different latitudes[a]

| Latitude[b] | Concentration ($ng\,l^{-1}$) |
|---|---|
| 40°59′N | 0.23–0.49 |
| 38°19′N | 0.33–0.50 |
| 29°59′N | 0.35–0.83 |
| 16°39′N | 0.35–0.64 |
| 15°04′S | 0.12–0.21 |
| 69°S 39°E | 0.035–0.069 |

[a] After Tanabe and Tatsukawa (1983).
[b] Longitude from 104°49′E to 145°45′E.

## Materials and Methods

Sampling was carried out in the Ross Sea during the austral summer of 1994–95. The sampling area lies between 72° and 75°S and 170° and 178°E (Fig. 2). Sampling was conducted on board the Italian vessel Italica as part of the tenth expedition promoted by the Italian National Research Program in Antarctica. All samples were collected in November and December. The samples collected were phytoplankton (mainly diatoms), zooplankton (including copepods, anphipods and krill), silverfish *Pleuragramma antarcticum*, Adelie penguin *Pygoscelis adeliae*. Specimens of *E. crystallorophias* and *E. superba* were analyzed separately. Muscle and liver samples of *P. antarcticum* were obtained. Liver samples of Adelie penguins were collected on a previous expedition (1990–91). Table 2 shows some details of the samples collected.

Samples were digested with 1N KOH-ethanol solution and the extract was then transferred to hexane (Tanabe et al. 1987). The concentrated hexane layer was cleaned on a silica gel column and eluted with n-hexane. This eluate was concentrated in a rotary evaporator and an aliquot of 3 ml was reserved for GC-ECD and GC-MS to determine 50 PCB isomers. Alkali-alcohol digestion results in the decomposition of nona- and deca-chlorobiphenyls, therefore they can not be determined quantitatively. Hexane extracts were cleaned with 10% fuming sulfuric acid and rinsed in distilled water. The remaining 3 ml was run

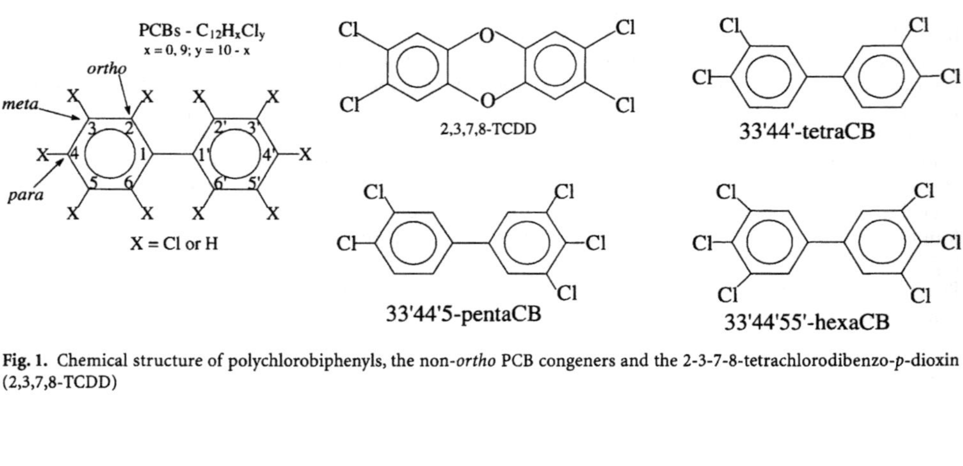


**Fig. 1.** Chemical structure of polychlorobiphenyls, the non-*ortho* PCB congeners and the 2-3-7-8-tetrachlorodibenzo-*p*-dioxin (2,3,7,8-TCDD)

through a column packed with activated carbon to separate the non-*ortho* coplanar PCB congeners 3,3′,4,4′ (PCB77), 3,3′,4,4′,5 (PCB126) and 3,3′,4,4′,5,5′ (PCB169). Non-*ortho* coplanar PCBs were eluted with benzene and ethylacetate mixture. Analysis of total PCBs was performed with a gas chromatograph equipped with $_{63}$Ni electron capture detector. A SPB-5 bonded phase capillary column was used. A mixture of specific isomers was used for calibration, recovery evaluation and confirmation.

Coplanar PCBs were quantified using the same gas chromatographic conditions as for GC/ECD. Recovery rates for non-*ortho* substituted PCBs were: PCB-77: (n = 6), mean 98%, SD 10%; PCB-126: (n = 6), mean 93%, SD 18%; PCB-169: (n = 6), mean 87%, SD 18%.

The results obtained for the planar PCB congeners were used to calculate the toxic potential as TEQs in fish and penguin. The 2,3,7,8-TCDD toxic equivalents (TEQs) of coplanar congeners were estimated by the Toxic Equivalency Factors (TEFs) approach proposed by Ahlborg et al. (1994) for PCB congeners 77, 126, 169, 105, 118, 156 170 and 180 and by Safe (1990) for PCBs 153, 137, 138, 128, 194 and 60.

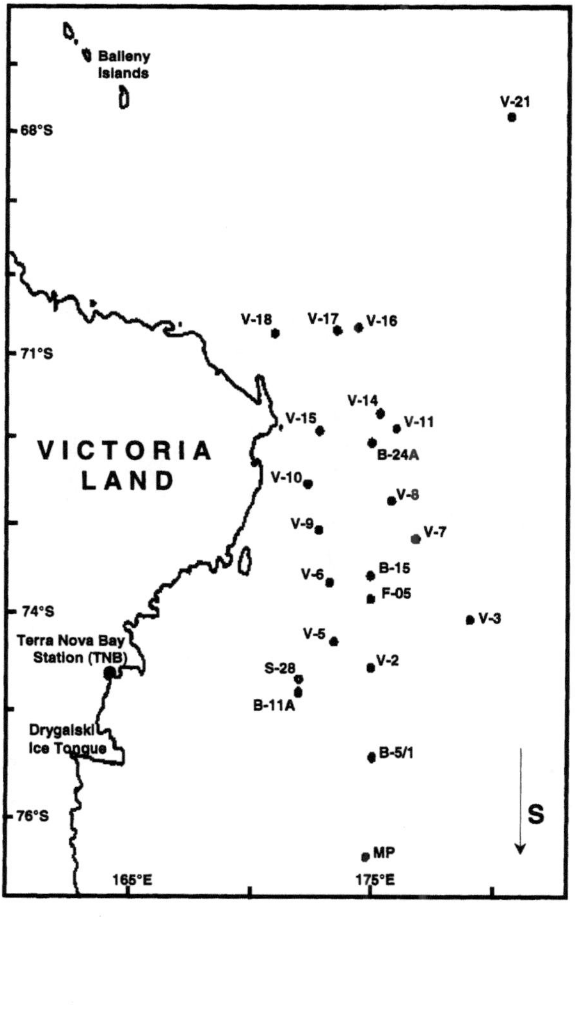


**Fig. 2.** The sampling area

**Table 2.** Date and site of sampling

| Sample | Date | Site[a] | Description |
|---|---|---|---|
| F1 | 18/12/94 | V2 | Diatoms |
| F2 | 19/12/94 | V3 | |
| F3 | 20/12/94 | V5 | |
| F5 | 22/12/94 | V11 | |
| F6 | 23/12/94 | V14 | |
| F8 | 24/12/94 | V17 | |
| Z2 | 20/12/94 | V5 | Amphipods, copepods |
| Z3 | 20/12/94 | V7 | |
| Z4 | 21/12/94 | V8 | |
| Z5 | 22/12/94 | V9 | |
| Z10 | 25/12/94 | V16 | |
| Z18 | 27/12/94 | V21 | |
| Ec1 | 15/12/94 | S28 | *Euphausia crystallorophias* |
| Ec2 | 20/12/94 | V6 | |
| Ec3 | 22/12/94 | V10 | |

| Sample | Date | Site[a] | Description |
|---|---|---|---|
| Es1 | 23/11/94 | F05 | *Euphausia superba* |
| Es2 | 23/11/94 | F05 | |
| Es3 | 19/11/94 | V3 | |
| Es4 | 20/12/94 | V6 | |
| Es5 | 20/12/94 | V7 | |
| Es7 | 22/12/94 | V10 | |
| Es8 | 24/12/94 | V15 | |
| Es9 | 27/12/94 | V21 | |
| Pa1 | 16/11/94 | B5/1 | *Pleuragramma antarcticum* |
| Pa2 | 20/11/94 | MP | |
| Pa3 | 26/11/94 | B15 | |
| Pa4 | 26/11/94 | B15 | |
| Pa5 | 28/11/94 | B11A | |
| Pa6 | 09/12/94 | B24A | |
| Pad | 1990 | TNB | Pygoscelis adeliae Weight = 4125 g; sex unknown |

[a] See Fig. 2.

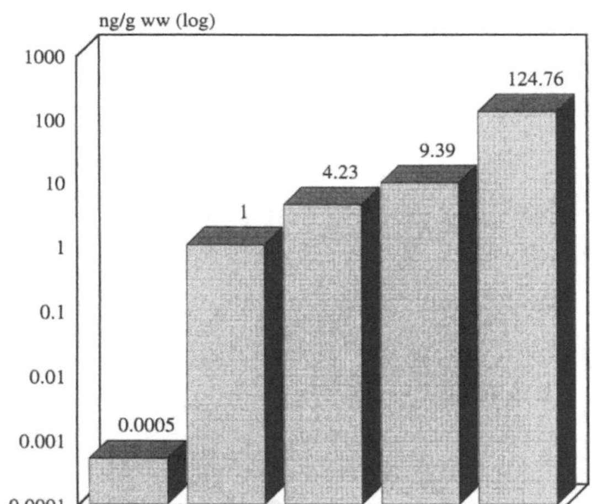


| | SW | P | Z | Pa | Padl |
|---|---|---|---|---|---|
| PCB | 0.0005 | 1 | 4.23 | 9.39 | 124.76 |
| BCF | | 2000 | 8460 | 18780 | 249520 |

SW = seawater
P = phytoplankton
Z = zooplankton
Pa = *P. antarcticum* (body burden)
Padl = *P. adeliae* (liver)

**Fig. 3.** Mean concentration (ng $g^{-1}$ wet weight) of the sum of polychlorobiphenyls in Antarctic organisms and relative bioconcentration factors (BCF)

## Results and Discussion

The results are expressed in $ng\,g^{-1}$ wet weight. The mean concentration of the sum of the 50 PCB congeners (Fig. 3) was $1\,ng\,g^{-1}$ in phytoplankton, and increased fourfold in zooplankton ($4.23\,ng\,g^{-1}$). It was $9.39\,ng\,g^{-1}$ (body burden) in whole fish and $124.76\,ng\,g^{-1}$ in penguin liver. A concentration of $0.5\,pg\,l^{-1}$ has been reported in seawater (Fuoco et al. 1994). These results clearly show the expected biomagnification at increasing levels in the food web. Our results in *Euphausia* sp. were of the same order of magnitude as those described by Risebrough et al. (1976). Strömberg et al. (1990) found $60\,ng\,g^{-1}$ wet weight in the fat of Adelie penguin and $0.17\,ng\,g^{-1}$ wet weight in *Trematomus bernacchii*. The largest increment in PCB concentration was from water to phytoplankton and from fish to birds. The progressive amplification of the quantity of contaminants in the organisms is due to bioconcentration and biomagnification and was evident when the bioconcentration factor (BCF) in the organisms was calculated. The BCF is defined as the ratio between the concentration of a pollutant in the organism and that in the water. This increasing ratio confirms the transfer of PCBs from seawater to phytoplankton and their bioconcentration from one level of the food web to the next (Fig. 3).

PCB concentrations in *E. superba* were examined in relation to the date of sampling (Fig. 4) to see whether there was any relationship. Figure 4a shows the samples and the dates of sampling. There was no evidence of an increase in PCB concentrations with melting of ice. Figure 4b shows the results divided into three main periods. Similar levels can be observed in November and mid-December and a slight increase in late December. We should examine two facts in considering these results. The transfer from seawater to organisms could be slow and not detectable in a short period and the ice melting occurs at different times in different sites.

We also looked for a relationship between the accumulation of PCBs and the size of plankton. In order to do this, we divided the organisms into five classes (Fig. 4c): (1) amphipods and copepods, (2) *E. crystallorophias*, (3) *E. superba* from 35 to 39 mm, (4) *E. superba* from 40 to 44 mm, (5) *E. superba* from 45 to 49 mm. There was a decrease in levels of contamination (Fig. 4c) from the smallest to the biggest organisms, albeit in a short range of concentrations (4.23 to $2.39\,ng\,g^{-1}$). The surface of small components of plankton can adsorb particulate organic material containing PCBs deriving from ice melting and this could explain the higher concentration depending on the surface/volume ratio. This hypothesis is confirmed by the fact that the samples were collected during the melt while contaminants were being released into the seawater (Fuoco et al. 1991).

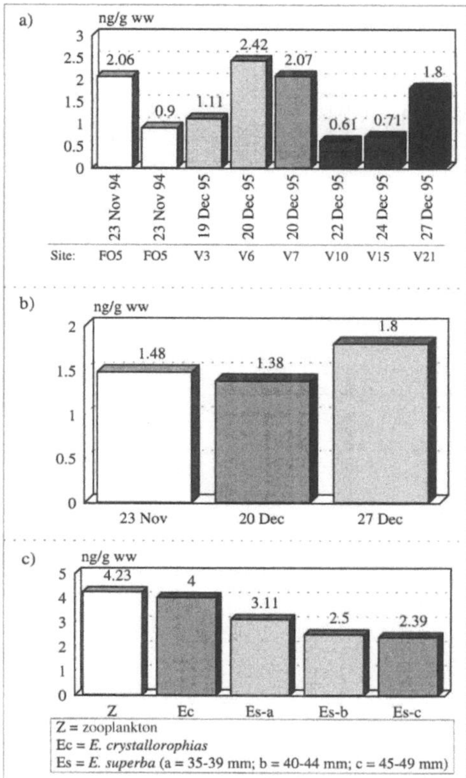


**Fig. 4a–c.** $\Sigma$PCBs mean concentration ($ng\,g^{-1}$ wet weight) in *E. superba* in relation to the date of sampling (**a**, **b**) and in relation to the size of their body (**c**)

As far as PCB composition is concerned, the percentages of the 50 isomers with respect to total PCBs are shown in Fig. 5 and 6. In seawater the percentage of each congener was quite low, below 10%, whereas at the first level of the food

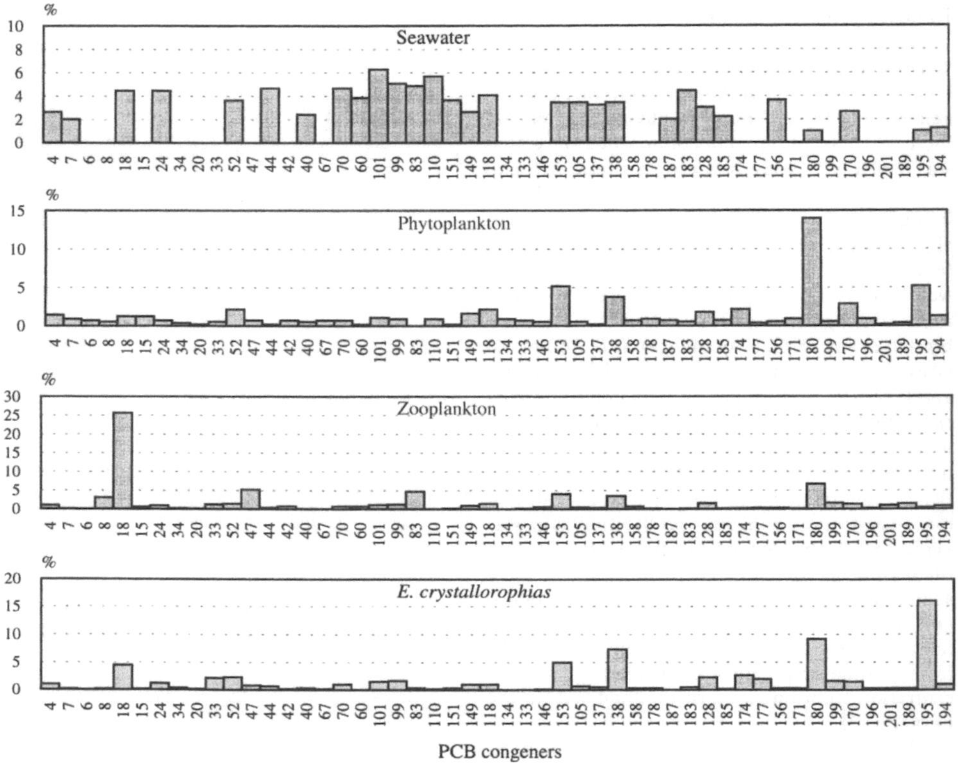


**Fig. 5.** Fingerprints of seawater (Fuoco et al. 1994), phytoplankton, zooplankton and *E. crystallorophias*

chain (phytoplankton) there were already some differences. In general, PCBs with few chlorines decreased in concentration and those with many increased, especially PCBs 153, 138, 180 and 195. This trend is similar at subsequent levels of the food web. In zooplankton and krill the same pattern was observed and this trend is more evident in fish and birds. Some differences were observed between muscle and liver of *P. antarcticum*; some congeners were more abundant in one tissue than in another, depending on lipid composition. The pattern in the liver of fish and birds was slightly different: PCB 180, 195 and 128 were higher in fish and PCB 138 and 153 predominated in birds. The percentage of PCB 118 was higher than that of PCB 149 in Adelie penguins while in fish it was the opposite. Our results agree with those of De Boer and Wester (1991) and confirm the inability of penguins to metabolize PCB 118.

We evaluated the BCF of a few congeners in order to better comprehend their behaviour in organisms (Fig. 7a). An increasing trend was evident from lower to higher levels of the food chain. It was especially clear for the more persistent PCBs 138 and 153. PCB 138, 153 and 180 had the highest BCF in all organisms. This confirms the increasing accumulation of these pollutants from seawater to organisms and in the higher levels of the food web.

The PCB isomer patterns observed at different levels of the food web (Fig. 7b) revealed that PCBs with two to five chlorines decreased in percentage from seawater to the higher levels of the food chain while highly chlorinated PCBs increased. This trend can be explained by the fact that

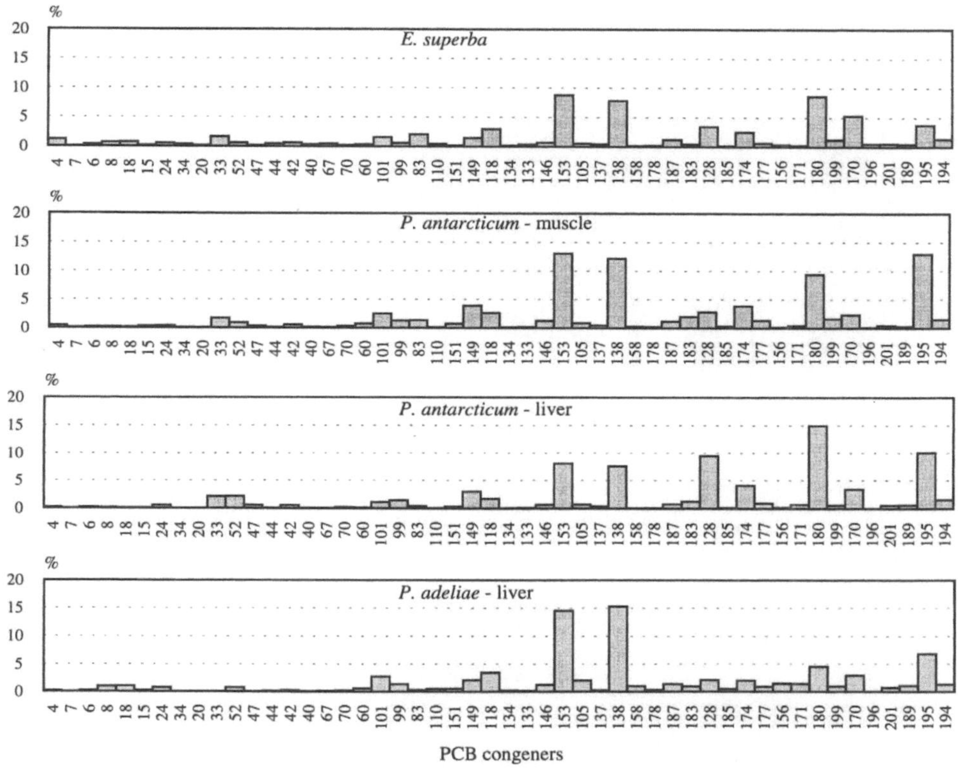


**Fig. 6.** Fingerprints of *E. superba*, muscle and liver of *P. antarcticum* and *P. adeliae* (liver)

the enzyme-detoxifying systems of organisms metabolize the congeners with few chlorines. The finding of PCBs with few chlorines in gill-breathing organisms such as zooplankton and Euphausiacea may be because they take up xenobiotics following respiration. Equilibrium with the environment is achieved easily and quickly since lower trophic organisms possess very weak drug-metabolizing enzyme systems. The largest contribution to total PCBs was by hexa- and hepta-chlorobiphenyls, which accounted for about 70% of the total, in krill, silverfish (*P. antarcticum*) and penguins. The high percentage of hepta-chlorobiphenyls in phytoplankton was due to PCB 180 (36% of PCBs).

The concentrations of non-*ortho* coplanar PCB and di- and mono-*ortho* congeners are reported in Table 3; these results were used to calculate TEQs. The toxicity of these three classes of congeners is di-*ortho* < mono-*ortho* < non-*ortho* PCBs. The three coplanar congeners were analyzed in krill (PCB 77 0.03 $pg\,g^{-1}$, PCB 126 0.15 $pg\,g^{-1}$, PCB 169 not detected). In muscle of silverfish we found 6.55 $pg\,g^{-1}$ of PCB 126 and in liver 2.11 $pg\,g^{-1}$ of PCB 77, 6.63 $pg\,g^{-1}$ of PCB 126 and 0.24 of PCB 169. Focardi et al. (1995) found 280 $pg\,g^{-1}$ wet wt of PCB 77 in *C. hamatus* caught in Terra Nova Bay near the Italian base. The lower concentrations found in the present study are probably because the sampling site is far from any human settlement. In Adelie penguins (liver) we found 250, 80 and 20 $pg\,g^{-1}$ of PCBs 77, 126 and 169, respectively. In general, there was an increasing trend of the sum of di-, mono- and non-*ortho* PCBs from the lower to the higher levels of the food web (Fig. 8a). The order of contribution to total planar PCB concentrations was di- > mono- > non-*ortho* PCB with di-*ortho* PCBs accounting for about 90%.

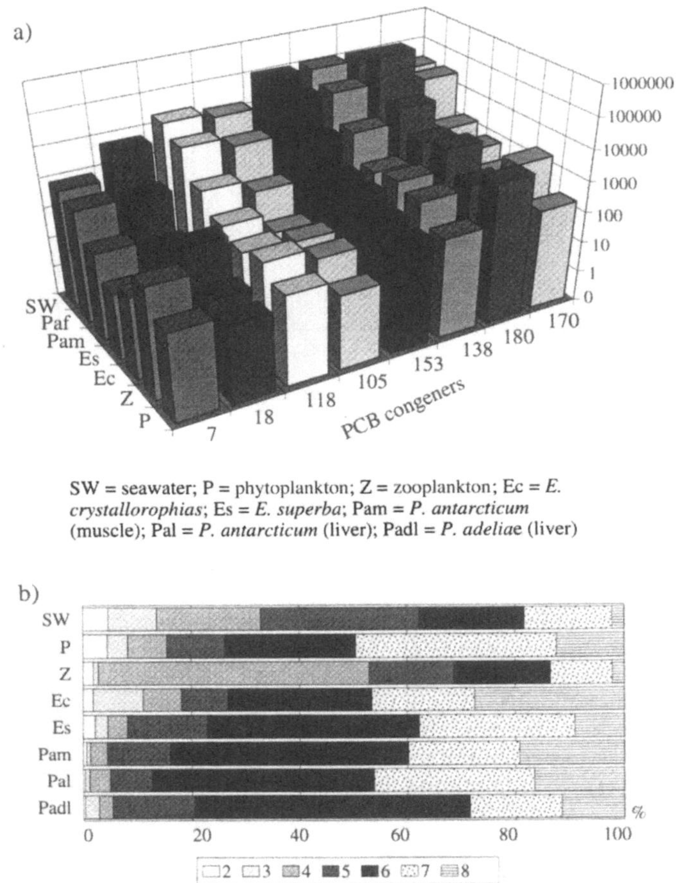


**Fig. 7a,b.** Bioconcentration factors of some PCB congeners in the Antarctic organisms (**a**) and PCBs composition (%) in the organisms (**b**)

Total TEQ values (Table 3) calculated for di-, mono- and non-*ortho* substituted PCBs were 1.23 $pg\,g^{-1}$ and 3.61 $pg\,g^{-1}$ in the muscle and liver, respectively, of *P. antarcticum* and 11.98 $pg\,g^{-1}$ in the liver of *P. adeliae*. The main contributor to the total TEQs was mono-*ortho* PCBs in the liver of *P. antarcticum* and non-*ortho* PCBs in muscle of *P. antarcticum* and in the liver of *P. adeliae* (Fig. 8b).

These levels are low compared to those found in species from other areas of the planet, especially the mid-latitudes of the Northern Hemisphere (Daelemans et al. 1993; Corsolini et al. 1995). They probably do not endanger these Antarctic organisms, even if these animals are not used to chemical stress and may have a very low sensitivity threshold. In silverfish the main contribution (>50% in muscle) to total TEQ was due to mono-*ortho* PCBs, which are considered to be the most toxic, after non-*ortho* congeners.

## Conclusions

These results confirm the presence of persistent pollutants, such as polychlorobiphenyls, in

**Table 3.** Mean concentration and TEQ values of di-, mono- and non-*ortho* PCBs (pg g$^{-1}$ wet weight; nd = not detected; nc = not calculated)

| PCB | TEF | *P. antarcticum* (muscle) | | *P. antarcticum* (liver) | | *P. adeliae* (liver) | |
|---|---|---|---|---|---|---|---|
| | | Conc. | TEQ | Conc. | TEQ | Conc. | TEQ |
| Di-*ortho* | | | | | | | |
| 153 | 0.00002[a] | 2300.00 | 0.05 | 25500.00 | 0.51 | 15060.00 | 0.30 |
| 137 | 0.00002[a] | 100.00 | <0.01 | 800.00 | 0.02 | 230.00 | <0.01 |
| 138 | 0.00002[a] | 2350.00 | 0.05 | 21900.00 | 0.44 | 15880.00 | 0.32 |
| 128 | 0.00002[a] | 700.00 | 0.01 | 6600.00 | 0.13 | 2130.00 | 0.04 |
| 180 | 0.00001[b] | 1700.00 | 0.02 | 18000.00 | 0.18 | 4670.00 | 0.05 |
| 170 | 0.00010[b] | 350.00 | 0.04 | 4700.00 | 0.47 | 3010.00 | 0.30 |
| 194 | 0.00002[a] | 230.00 | <0.01 | 200.00 | <0.01 | 350.00 | 0.01 |
| Total | | 7730.00 | 0.17 | 77700.00 | 1.75 | 41330.00 | 1.02 |
| Mono-*ortho* | | | | | | | |
| 60 | 0.001[a] | 300.00 | 0.30 | 400.00 | 0.40 | 600.00 | 0.60 |
| 118 | 0.0001[b] | 600.00 | 0.06 | 3900.00 | 0.39 | 3560.00 | 0.36 |
| 105 | 0.0001[b] | 160.00 | 0.02 | 2000.00 | 0.20 | 2080.00 | 0.21 |
| 156 | 0.001[b] | 20.00 | 0.02 | 200.00 | 0.20 | 1460.00 | 1.46 |
| Total | | 1080.00 | 0.40 | 6500.00 | 1.19 | 7700.00 | 2.63 |
| Non-*ortho* | | | | | | | |
| 77 | 0.0005[b] | nd | nc | 2.11 | <0.01 | 250.00 | 0.13 |
| 126 | 0.10[b] | 6.55 | 0.66 | 6.63 | 0.66 | 80.00 | 8.00 |
| 169 | 0.01[b] | nd | nc | 0.24 | <0.01 | 20.00 | 0.20 |
| Total | | 6.55 | 0.66 | 8.98 | 0.67 | 350.00 | 8.33 |
| Total | | 8816.55 | 1.23 | 84208.98 | 3.61 | 49380.00 | 11.98 |

[a] Safe (1990).
[b] Ahlborg et al. (1994).

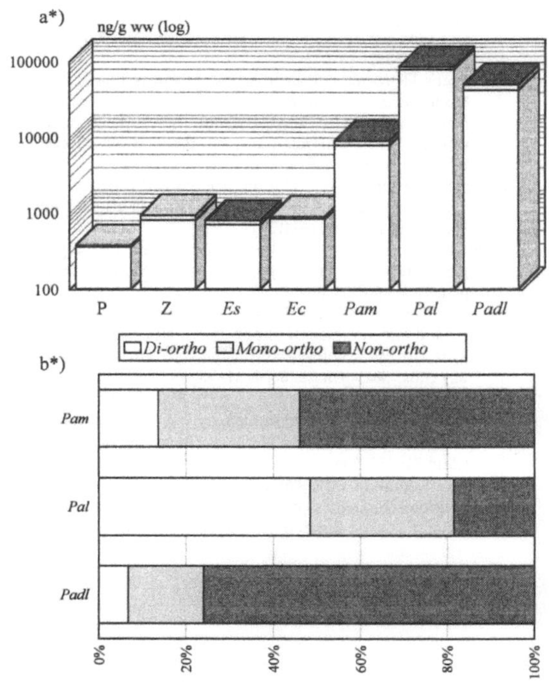


* P = phytoplankton; Z = zooplankton; Ec = *E. crystallorophias*; Es = *E. superba*; Pam = *P. antarcticum* (muscle); Pal = *P. antarcticum* (liver); Padl = *P. adeliae* (liver)

**Fig. 8a,b.** Mean concentration of di-, mono- and non-*ortho* PCB (%; **a**) and their contribution to the total TEQs (**b**)

Antarctica, though at low concentrations compared to other areas of the world. The transfer of PCBs from ice to seawater and then to organisms is in line with their concentrations in the samples and the values of the bioconcentration factors, which increased from seawater to the higher levels of the food chain. Highly toxic coplanar PCBs were also found in the fauna of this pristine environment, albeit in low concentrations.

*Acknowledgements.* This research was funded by the Italian Antarctic Research Programme (PNRA). We are grateful to Dr. Silvia Aurigi for her help with the chemical analysis.

## References

Ahlborg UG, Becking GC, Birnbaum LS, Brouwer A, Derks HJGM, Feeley M, Golor G, Hanberg A, Larsen JC, Liem AKD, Safe SH, Schlatter C, Wærn F, Younes M, Yrjänheikki E (1994) Toxic equivalency factors for dioxin-like PCBs. Chemosphere 28: 1049–1067

Borrell A, Aguilar A, Corsolini S, Focardi S (1996) Evaluation of toxicity and sex-related variation of PCB levels in Mediterranean striped dolphin affected by an epizootic. Chemosphere 32(12): 2359–2369

Corsolini S, Focardi S, Kannan K, Tanabe S, Borrell A, Tatsukawa R (1995) Congener profile and toxicity assessment of polychlorinated biphenyls in dolphins, sharks and tuna fish from Italian coastal waters. Mar Environ Res 40: 33–53

Daelemans FF, Mehlum F, Lydersen C, Schepens PJC (1993) Mono-*ortho* and non-*ortho* substituted PCBs in arctic ringed seal (*Phoca hispida*) from the Svalbard area: analysis and determination of their toxic threat. Chemosphere 27: 429–437

De Boer J, Wester P (1991) Chlorobiphenyls and organochlorine pesticides in various sub-Antarctic organisms. Mar Pollut Bull 22(9): 441–447

Elder DL, Fowler SW (1977) Polychlorinated biphenyls: penetration into the deep ocean by zooplancton fecal pellet transport. Science: 459

Focardi S, Corsolini S, Bargagli R (1995) Isomer-specific analysis and toxic evaluation of polychlorinated biphenyls in Antarctic fish, seabirds and Weddell seal from Terra Nova Bay (Ross Sea). Antarct Sci 7: 31–35

Fuoco R, Colombini MP, Abete C (1991) Evaluation of pack melting effect on polychlorobiphenyl content in sea water samples from Terra Nova Bay-Ross Sea (Antarctica). Ann Chim 81: 383–394

Fuoco R, Colombini MP, Abete C (1994) Determination of polychlorobiphenyls in environmental samples from Antarctica. Int J Environ Anal Chem 55: 15–25

Hempel G (1985) Antarctic marine food webs. In: Siegfried WR, Condy PR, Laws RM (eds) Antarctic cycles and food webs. Springer, Berlin Heidelberg New York, pp 266–270

Hubold G (1985) The early life-history of the high-Antarctic silverfish *Pleuragramma antarcticum*. In: Siegfried WR, Condy PR, Laws RM (eds) Antarctic nutrient cycles and food webs. Springer, Berlin Heidelberg New York, pp 445–451

Iwata H, Tanabe S, Sakai N, Tatsukawa R (1993) Distribution of persistent organochlorines in the oceanic air and surface seawater and the role of the ocean on their global transport and fate. Environ Sci Technol 27: 1080–1098

Kannan K, Tanabe S, Borrell A, Aguilar A, Focardi S, Tatsukawa R (1993) Isomer-specific analysis and toxic evaluation of polychlorinated biphenyls in striped dolphins affected by an epizootic in the western Mediterranean Sea. Arch Environ Contam. Toxicol 25: 227–233

Knap AH (ed) (1990) The long-range atmospheric transport of natural and contaminant substances. Kluwer, Dordrecht, pp 259–301

Norstrom RJ, Simon M, Muir DCG, Schweinsburg RE (1988) Organochlorine contaminants in Arctic marine food chains: identification, geographical distribution, and temporal trends in polar bears. Environ Sci Technol 22: 1063–1071

Risebrough RW, Walker W, Schmidt TT, DeLappe BW, Connors CW (1976) Transfer of chlorinated biphenyls to Antarctica. Nature 264: 738–739

Safe S (1990) Polychorinated biphenyls (PCBs), dibenzo-*p*-dioxins (PCDDs), dibenzofurans (PCDFs) and related compounds: environmental and mechanistic considerations which support the development of Toxic Equivalency Factors (TEFs). CRC Crit Rev Toxicol 21: 51–88

Strömberg JO, Anderson LG, Björk G, Bonner WN, Clark AC, Dick AL, Ernst W, Limbert DWS, Peel DA, Priddle J, Smith RIL, Walton DWH (1990) State of the marine environment in Antarctica. UNEP Regional Seas Rep Stud 129

Tanabe S, Tatsukawa R (1983) Chlorinated hydrocarbons in the Southern Ocean. In: Nemoto T, Matsuda T (eds) Proceedings of the BIOMASS Colloquium in 1982. National Institute of Polar Research, Tokyo, pp 64–76

Tanabe S, Tatsukawa R (1986) Distribution, behaviour and load of PCBs in the oceans. In: Waid JS (ed) PCBs and the environment, vol 1. CRC Press, Boca Raton, pp 143–161

Tanabe S, Hidaka H, Tatsukawa R (1983) PCBs and chlorinated hydrocarbon pesticides in Antarctic atmosphere and hydrosphere. Chemosphere 12: 277–288

Tanabe S, Kannan N, Wakimoto T, Tatsukawa R (1987) Method for the determination of three toxic non-*ortho*chlorine substituted coplanar PCBs in environmental samples at part-per-trillion levels. Int J Environ Anal Chem 29: 199–213

Wolff EW (1990) Signals of atmospheric pollution in polar snow and ice. Antarct Sci 2: 189–205

# Concentration Changes in Cadmium and Lead in Antarctic Coastal Seawater (Ross Sea) During the Austral Summer and Their Relationship with the Evolution of Biological Activity

G. Scarponi[1,3], G. Capodaglio[1], C. Barbante[1,2], G. Toscano[1], M. Cecchini[1], A. Gambaro[1,2], and P. Cescon[1,2]

**ABSTRACT**

Studies were made of the seasonal evolution of cadmium and lead distribution in the water column of Wood Bay (Ross Sea, Antarctica) during the 1993–94 austral summer. Anodic stripping voltammetry was used to carry out on-site measurements, immediately after the collection and filtration of samples. Before the phytoplankton bloom, the concentrations of both metals were uniform along the water column, with mean values of 0.64 (SD 0.07) and 0.031 (SD 0.006) nmol $l^{-1}$ for cadmium and lead, respectively. A subsequent depletion in metal concentration was observed in shallow waters. By the end of the season, the surface concentration of cadmium had decreased to about 0.1 nmol $l^{-1}$. As regards lead, vertical distribution was less affected by season and the mean surface concentration decreased to about 0.01 nmol $l^{-1}$ over the same period. Results are evaluated in relation to the evolution of biological activity in the area, and are examined and compared with those obtained during a similar study carried out in the Terra Nova Bay area (Gerlache Inlet) 3 years previously.

## Introduction

The main processes controlling trace metal distribution in the sea are the input from natural sources, e.g. aeolian, fluvial and hydrothermal inputs, and removal via sedimentary deposition and burial after various internal recycling as a result of biogeochemical processes (Whitfield and Turner 1987). Removal is controlled by the uptake from particulate material produced principally by phytoplankton in the photic layer and regeneration into the deep layers where the particles are largely redissolved (Bruland 1983; Whitfield and Turner 1987; Chester 1990). However, human activities can significantly affect the spatial and vertical distributions of some trace metals in the sea (Bruland 1983).

According to their particular biogeochemical behaviour, trace heavy metals in seawater can be subdivided into accumulated, recycled and scavenged metals (Bruland 1983). Accumulated metals show higher concentrations in seawater with respect to their crustal abundance and a uniform vertical profile. Recycled metals tend to be involved in the biological particle cycle and their vertical profile shows surface depletion and deep regeneration (typical nutrient profile). Scavenged metals interact strongly with particulate matter and are removed via sedimentary deposition; by consequence they show very low concentrations relative to their crustal abundance and exhibit different profiles according to their primary input; in particular a surface maximum is displayed when they enter the sea through enriched materials (possibly anthropogenic), e.g. via aeolian dust or from various coastal sources.

Polar seas, and Antarctic seas in particular, are subject to meteorological, climatic and general environmental conditions which differentiate them markedly from all other marine ecosystems. Changes in the above conditions during the year are so great and rapid, especially in the summer period, that they give rise to sudden changes in the distribution of trace components, which are unusual for oceanic areas; these are of extreme interest to general oceanographic scientific research and particularly stimulating to chemico-environmental investigation.

In spite of this, little research has been conducted on the distribution and seasonal evolution

[1] Dipartimento di Scienze Ambientali, Università Ca' Foscari di Venezia, I – 30123 Venice, Italy
[2] Centro di Studio sulla Chimica e le Tecnologie per l'Ambiente, CNR, c/o Università Ca' Foscari di Venezia, I, 30123 Venice, Italy
[3] Present address: Istituto di Scienze del Mane, Università di Ancona, I-60131 Ancona, Italy

of chemical species both at the surface and along the water column, particularly as regards heavy metal traces in Antarctic seas (Bruland 1980; Mart et al. 1982; Orren and Monteiro 1985; Capodaglio et al. 1989, 1991a,b, 1994; Westerlund and Öhman 1991; Saager et al. 1992; Flegal et al. 1993; Nolting and de Baar 1994; Scarponi et al. 1997). Possible reasons for this include the difficulty of access and prolonged stays in these areas, together with analytical difficulties such as risk of sample contamination during collection, treatment, storage and analysis, and the need for high sensitivity of the technique due to the extremely low concentrations involved coupled with capability to carry out measurements in the very complex seawater matrix.

During the 1990–91 austral summer, we carried out a first study on the summer seasonal evolution of Cd and Pb in the water column of Gerlache Inlet (Terra Nova Bay, Ross Sea) (Scarponi et al. 1997). Results showed that during the summer there was a considerable net decrease in metal concentration, higher for Cd than for Pb, in the first 50–100 m. This behaviour was associated mainly with the interaction of the metals with biological systems (Cd) and with particulate matter (Pb) during algal blooms, though water circulation in the Bay could have played an important role (Scarponi et al. 1997).

To study the summer evolution of metal profiles in Antarctic coastal areas further, the experiment was repeated during the 1993–94 expedition in Wood Bay, which is more distant from the Italian station than Terra Nova Bay. Water samplings were carried out approximately every 2 weeks from November 7, 1993 to February 10, 1994. Immediately after collection, samples were analyzed for Cd and Pb by two of the authors (C.B and G.T.) using the clean chemistry laboratory of the Italian station at Terra Nova Bay.

Cadmium, which is not known to be essential biologically, exhibits a nutrient-type distribution (recycled element); it is involved in shallow regeneration cycles similar to those of phosphate and nitrate and its distribution is influenced by processes of production and remineralization of biogenic particles (Bruland and Franks 1983; Frew and Hunter 1992; Bruland et al. 1994). Lead is a typical scavenged element substantially influenced by local inputs (Bruland 1983).

This chapter presents the results obtained directly on site concerning the seasonal changes in Cd and Pb in the seawater of Wood Bay, and compares them with previous observations carried out in Terra Nova Bay (Gerlache Inlet) (Scarponi et al. 1997). It discusses the relationship of the observed trends with the summer evolution of biological activity at the two sites.

## Experimental

### Sampling

During the 1993–94 austral summer, seawater samples were collected along the water column in Wood Bay (see Fig. 1, sites 1, 2 and 3; sea floor about 380 m in sites 1 and 2, and 280 m in site 3) on Nov. 7, 19 and 28, and Dec. 15 (site 1), Jan. 9 and 31 (site 2), and Feb. 10 (site 3). Site 1 (74°22.120′S, 165°24.058′E) was covered by pack ice throughout the expedition period. On Jan. 9 and 31 it was decided to move to site 2 (74°23.513′S, 165°25.961′E; about 2 km from site 1), which was free of ice. On Feb. 10 site 2 was inaccessible as it was occupied by floating ice, and samples were collected at site 3 (74°27.766′S, 165°31.984′E; 11.2 km from site 1).

Figure 1 also shows the location in Terra Nova Bay where the investigation was carried out in 1990–91, i.e. station B (74°40′07″S, 164°07′15″E) (Scarponi et al. 1997).

Subsurface water (about 0.5 m below the seawater level) was collected by a submersible Teflon pump and temporarily stored in a 50-l polyethylene tank until filtration (see below). Deeper samples were collected in 20- or 30-l Go-Flo samplers (General Oceanics, USA), coated with Teflon internally and with pressurization capability. A non-metallic (Kevlar) hydrowire and a Teflon-covered ballast attached to the wire at least 10 m below the sampler, were used. In the presence of pack ice, samples were collected through a hole drilled manually with a metallic ice corer (Duncan, UK, Model BTC).

Samples were filtered through membrane filters of 0.45-μm pore size immediately after sampling (within 2–3 h), by pressurizing (pressure lower than 0.5 bar) the polyethylene tank or the Go-Flo bottle with pure nitrogen (chromatographic grade) and pushing the sample through a Teflon filtration apparatus (Sartorius, holder for in-line pressure filtration, SM 16540; cellulose nitrate membrane filter, SM 11306, 142-mm diameter). After discarding a first aliquot of about 3–5 l of water, filtered samples were collected directly in Teflon containers, acidified with ultra-

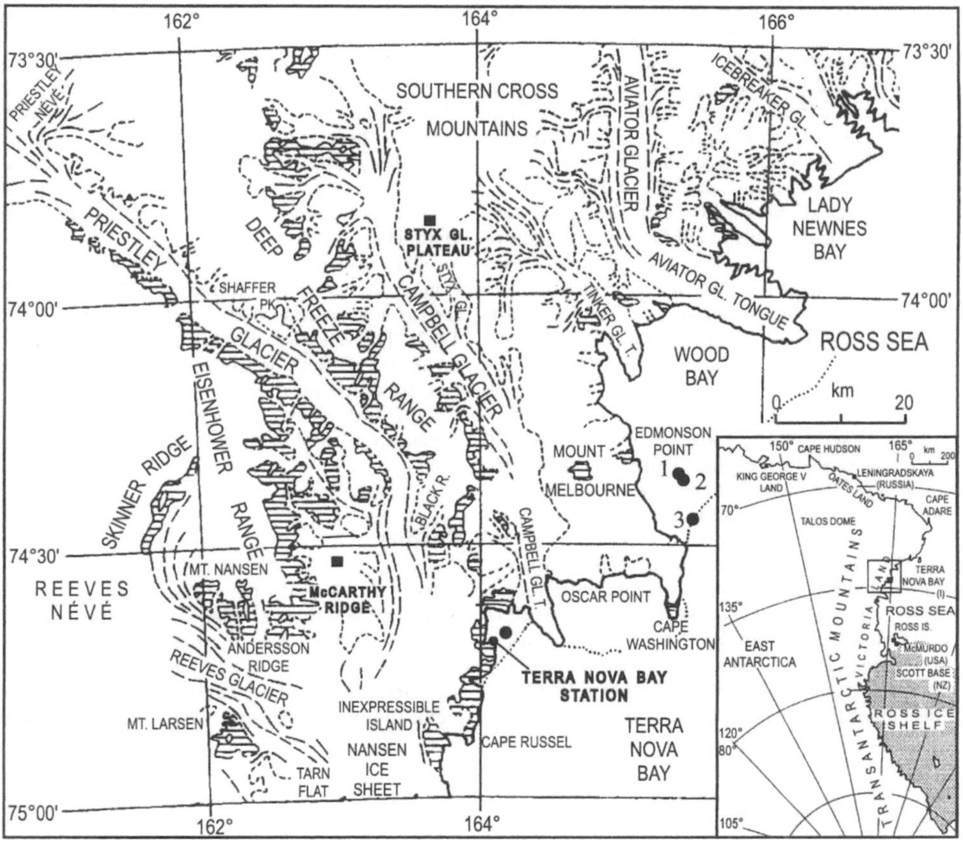


**Fig. 1.** Sampling positions for seawater collection in Wood Bay (sites 1, 2 and 3, 1993–94 expedition) and Terra Nova Bay (site B, Gerlache Inlet, 1990–91 expedition)

pure HCl and stored at room temperature until analysis (carried out on site within 1–4 days). Aliquots of samples were also collected in 2-l FEP bottles and stored frozen (−20 °C) not acidified for speciation analysis to be performed later in Italy (not considered here).

Polyethylene tanks, Go-Flo bottles, filtration apparatus, filters and storage bottles were acid-cleaned and conditioned to minimize sample contamination from the material used; the detailed cleaning procedures have been reported previously (Scarponi et al. 1996). Handling and treatment of the samples were carried out in the clean chemistry laboratory (Class 100) available at the Terra Nova Bay station in Antarctica.

Chlorophyll concentration was measured by a multiparametric probe (ME, Meerestechnik), equipped with a back scattering fluorometer (Dr. Haardt, Backscat).

## Voltammetric Analysis

Metal concentrations were determined by differential pulse anodic stripping voltammetry (DPASV) carried out on a thin mercury film electrode (TMFE). An EG&G 384B polarographic analyzer was used. The electrochemical cell (EG&G, mod. Rotel 2), has already been described (Scarponi et al. 1996; Mart et al. 1980). The cell is equipped with a rotating glassy carbon disc electrode (RGCDE), on which the TMFE is electrochemically plated, an Ag/AgCl,KCl(sat.) reference electrode and a platinum auxiliary electrode. The mercury film was prepared daily immediately prior to each analysis using the procedure previously described (Capodaglio et al. 1989, 1991a, 1994; Scarponi et al. 1996, 1997).

Total dissolved metal determinations were carried out on samples acidified with ultrapure HCl for at least 24 h. The acid treatment was carried out directly in the Teflon vessel used as the cup of the electrochemical cell adding 100 µl of HCl (32%, NIST, Gaithersburg, MD, USA) to approximately 50 ml of the sample (pH about 2). A more precise measurement of the volume was carried out subsequently, at the end of the determination (see below). Tests carried out to verify the efficiency of the acid-treatment procedure confirmed the complete release of the metals from the organic ligands (Capodaglio et al. 1989, 1991a).

Voltammetric measurements were performed applying the following experimental conditions: deposition potential −0.950 V, deposition time 15–30 min (depending on the actual concentrations), electrode rotation rate 4000 rpm, equilibration period 30 s, mode DPASV, scan rate 10 mV $s^{-1}$, pulse amplitude 50 mV, pulse frequency 5 $s^{-1}$, final potential −0.180 V. After the voltammetric scan the electrode was allowed to remain at a potential of −0.200 V for 5 min to ensure complete removal of the amalgamated metals. Quantification was obtained by the multiple standard additions method; three spikes of metal standard solutions were always made. At the end of the measurement cycle the sample volume was detected with greater precision (±0.5 ml) using a graduated cylinder.

The reagent blank was lower than the detection limit when HCl ultrapure NIST was used for the acid treatment of samples (Scarponi et al. 1997). The detection limit is strongly dependent on mercury film conditions and on the deposition time; the mean values were 6 and 5 pmol $l^{-1}$ for Cd and Pb respectively (obtained from 3× signal/noise) when a 20-min deposition time was used.

## Results

Tables 1 and 2 report the results obtained for total dissolved cadmium and lead concentrations as a function of the depth for the samples collected during the 1993–94 expedition (Wood Bay). Measurement replicates were carried out only at occasional intervals because very little time was

**Table 1.** Total dissolved cadmium concentration in seawater samples collected in Wood Bay (Ross Sea) during the 1993–94 Italian expedition. In parentheses original replicates, when available. Measurements carried out in Antarctica immediately after sampling

| Depth m | Cadmium concentration, nmol $l^{-1}$ | | | | | | |
|---|---|---|---|---|---|---|---|
| | Nov. 7 | Nov. 19 | Nov. 28 | Dec. 15 | Jan. 9 | Jan. 31 | Feb. 10 |
| 0.5 | 0.76 (0.80, 0.74, 0.75, 0.73) | 0.62 | 0.66 | 0.62 (0.64, 0.60) | 0.26 (0.26, 0.27) | 0.091 | [b] |
| 10 | 0.73 (0.75, 0.71, 0.72) | 0.61 (0.64, 0.59, 0.61) | 0.70 | 0.47 | 0.12 | 0.056 | 0.15 |
| 25 | 0.68 (0.67, 0.68) | 0.55 | [b] | 0.52 | 0.16 (0.18, 0.14, 0.17) | 0.099 | 0.16 |
| 50 | 0.71 (0.71, 0.76, 0.66) | 0.61 | 0.71 (0.70, 0.72) | 0.62 (0.59, 0.65) | 0.18 (0.19, 0.18) | 0.25 | 0.23 |
| 100 | 1.33[a] | 0.54 (0.56, 0.51) | 0.61 | 0.56 (0.56, 0.57) | 0.55 | 0.65 (0.76, 0.54) | 0.58 |
| 200 | [b] | 0.55 | 0.54 | [b] | [b] | 0.51 | 0.66[d] |
| 350 | 0.70 (0.72, 0.70, 0.67) | 0.63 | 0.66 | 0.65 (0.57, 0.68, 0.69) | 0.21[c] | 0.63 | [b] |

[a] Contaminated sample.
[b] Sample not collected.
[c] Depth 310 m.
[d] Depth 250 m.

**Table 2.** Total dissolved lead concentration in seawater samples collected in Wood Bay (Ross Sea) during the 1993–94 Italian expedition. In parentheses original replicates, when available. Measurements carried out in Antarctica immediately after sampling

| Depth m | Lead concentration, nmol $l^{-1}$ | | | | | | |
|---|---|---|---|---|---|---|---|
| | Nov. 7 | Nov. 19 | Nov. 28 | Dec. 15 | Jan. 9 | Jan. 31 | Feb. 10 |
| 0.5 | 0.032 (0.038, 0.029, 0.027, 0.034) | 0.032 | 0.030 | 0.022 (0.024, 0.019) | 0.014 | 0.022 | [b] |
| 10 | 0.030 (0.038, 0.025, 0.027) | 0.030 (0.030, 0.031, 0.029) | 0.036 | 0.018 | 0.017 | 0.0096 | 0.013 |
| 25 | 0.026 (0.029, 0.024) | 0.031 | [b] | 0.023 | 0.0076 (0.0086, 0.0068, 0.0075) | 0.010 | 0.018 |
| 50 | 0.132[a] (0.117, 0.159, 0.122) | 0.022 | 0.024 (0.021, 0.027) | 0.026 (0.025, 0.027) | 0.025 (0.027, 0.022) | 0.0093 | 0.013 |
| 100 | 0.252[a] | 0.026 (0.030, 0.022) | 0.029 | 0.017 (0.015, 0.018) | 0.020 | 0.017 (0.019, 0.014) | 0.025 |
| 200 | [b] | 0.037 | 0.030 | [b] | [b] | 0.270[a] | 0.027[d] |
| 350 | 0.045 (0.053, 0.033, 0.050) | 0.087[a] | 0.034 | 0.035 (0.036, 0.034, 0.034) | 0.037[c] | 0.018 | [b] |

[a] Contaminated sample.
[b] Sample not collected.
[c] Depth 310 m.
[d] Depth 250 m.

available for laboratory work during the expedition. However, an idea of the repeatability of the measurements can be obtained from the original results of replicates reported in Tables 1 and 2. RSD of 10–20% is typical.

## Cadmium

Figure 2 shows the distribution of total dissolved cadmium in the water column of Wood Bay during summer 1993–94. For comparison it also reports the profiles observed in Terra Nova Bay (Gerlache Inlet) during the analogous investigation carried out in 1990–91 (Scarponi et al. 1997).

At the very beginning of the summer season, until about end-November (and presumably during the winter), the vertical profile of Cd was uniform, with a mean concentration of 0.64 (SD 0.07) nmol$l^{-1}$. This value did not appear significantly different from that reported previously for Terra Nova Bay, i.e. 0.71 (SD 0.10) nmol$l^{-1}$ (Scarponi et al. 1997). The relatively high concentration level and the homogeneity were possibly accounted for by the upwelling or vertical diffusion of enriched deep water to the surface and by the absence of an efficient scavenging process or uptake by organisms. Values are in agreement with typical concentrations observed in deep waters of Antarctic (Westerlund and Öhman 1991; Nolting and de Baar 1994) and sub Antarctic regions (Frew and Hunter 1992), and similar uniform profiles were observed in upwelling areas of the Weddell Sea (Westerlund and Öhman 1991; Nolting and de Baar 1994).

Later in the season (see Table 1), at the same time as the development of biological activity and ice melting (see metal and chlorophyll profiles of December 15 and subsequent), a marked depletion of the cadmium concentration was observed. The depletion, which extended gradually to a depth of 50 m during the bloom and the breaking/melting of pack ice, led to a minimum concentration of 0.056 nmol$l^{-1}$ at a depth of 10 m (see profile of January 31). The last profile (February 10) showed an increase of Cd concentration in the 50-m upper layer, that could denote the beginning of a reverse process which was expected to lead to the (winter) uniform profile. However, the increase was not large enough and not supported by a sufficient number of data to emphasize a clear trend.

Other researchers reported the presence of the depleting phenomenon in Antarctic (Weddell Sea) (Westerlund and Öhman 1991; Nolting and de Baar 1994) and Subantarctic (Frew and Hunter 1992) regions, even if no temporal trend for the same sites was reported.

Comparison of the results of the present study, referred to Wood Bay and the 1993–94 season, with those previously obtained in the Gerlache Inlet in 1990–91 (see Fig. 2), shows remarkably consistent summer Cd evolution in the two

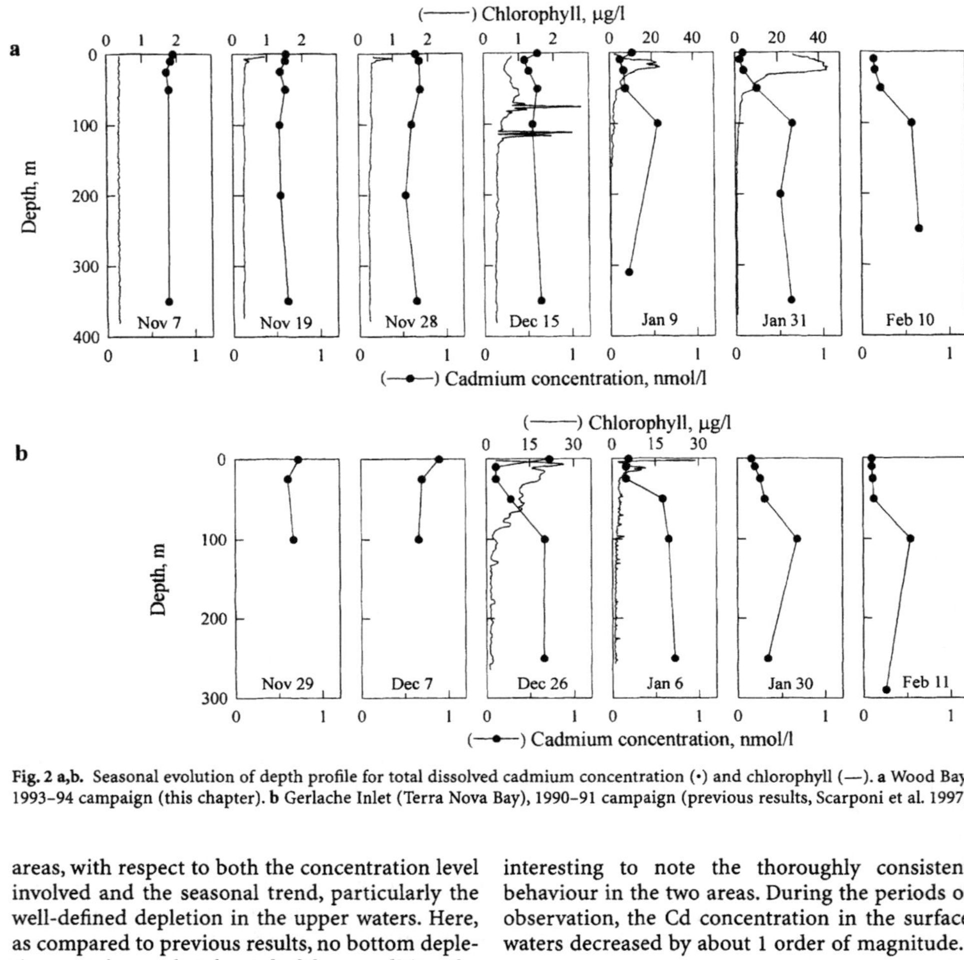


**Fig. 2 a,b.** Seasonal evolution of depth profile for total dissolved cadmium concentration (•) and chlorophyll (—). **a** Wood Bay, 1993–94 campaign (this chapter). **b** Gerlache Inlet (Terra Nova Bay), 1990–91 campaign (previous results, Scarponi et al. 1997)

areas, with respect to both the concentration level involved and the seasonal trend, particularly the well-defined depletion in the upper waters. Here, as compared to previous results, no bottom depletion was observed at the end of the expedition, the mean value under 100 m is 0.62 $nmol\,l^{-1}$, practically constant as of November. This discrepancy could be explained in terms of the presence of a local bottom depleting phenomenon in the Gerlache Inlet which was not present in Wood Bay (difference possibly related to different hydrodynamic regimes for the two areas); however, the effect of analytical bias in the previous measurements cannot be excluded.

For a better comparison of the two areas, the changes in Cd concentration in the surface layers (upper 10 m) are shown in Fig. 3, from November to February (summers 1993–94 and 1990–91, for Wood Bay and Gerlache Inlet, respectively). It is interesting to note the thoroughly consistent behaviour in the two areas. During the periods of observation, the Cd concentration in the surface waters decreased by about 1 order of magnitude.

## Lead

Table 2 and Fig. 4 show that the total dissolved lead concentrations in the water column of Wood Bay varied from 0.0076 to 0.045 $nmol\,l^{-1}$. These values were comparable with literature data reported for the Weddell Sea (Mart et al. 1982) (0.015–0.062 $nmol\,l^{-1}$) and for the Weddell and Scotia Seas (Flegal et al. 1993) (0.010–0.103 $nmol\,l^{-1}$). Considering our previous investigations carried out in the Ross Sea, particularly in the area of Terra Nova Bay, during several expeditions (Scarponi et al. 1997), the present data compares well with values

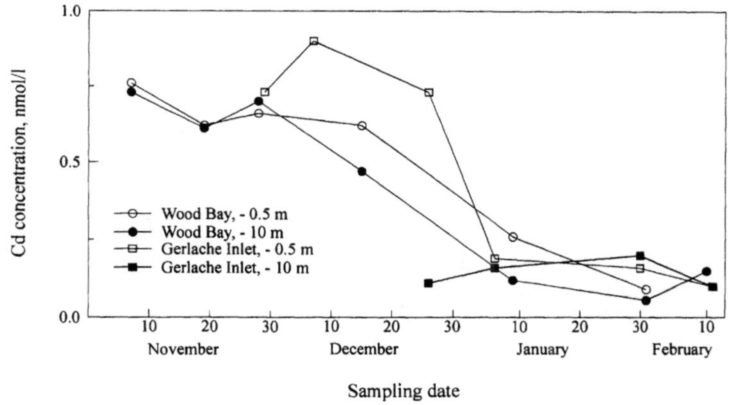


**Fig. 3.** Summer variation of total dissolved Cd concentration in the upper 10-m layer of Wood Bay (1993–94 season) and Gerlache Inlet (Scarponi et al. 1997) (1990–91 season)

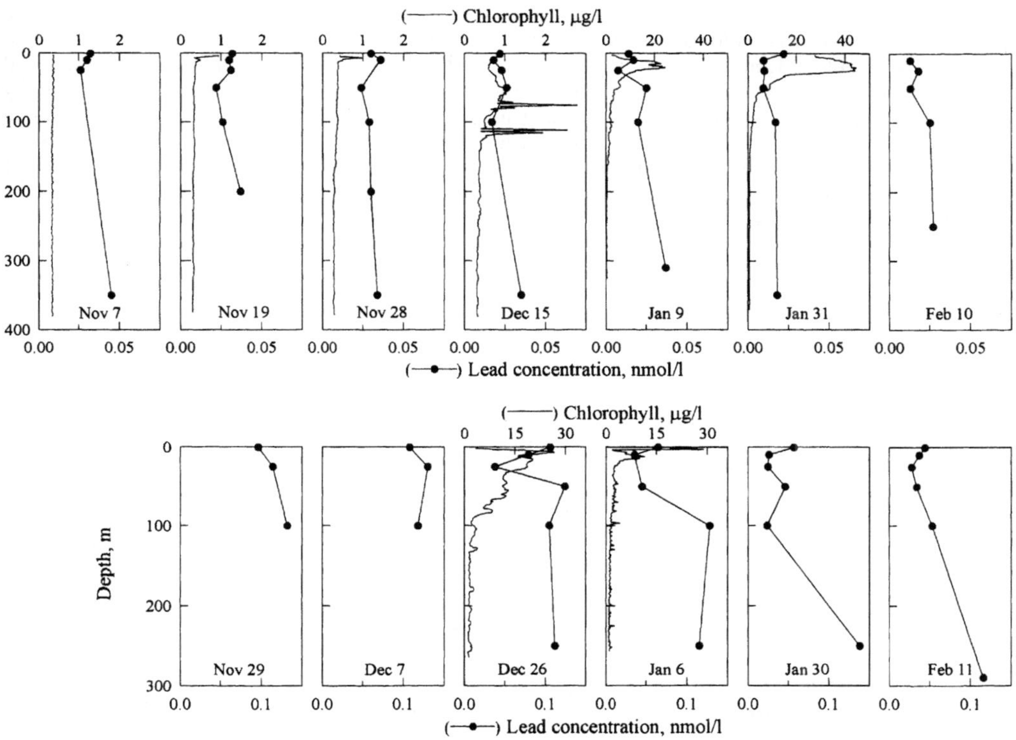


**Fig. 4 a,b.** Seasonal evolution of depth profile for total dissolved lead concentration (•) and chlorophyll (—). **a** Wood Bay, 1993–94 campaign (this chapter). **b** Gerlache Inlet (Terra Nova Bay), 1990–91 campaign (previous results, Scarponi et al. 1997)

referred to the open bay (in $nmol\,l^{-1}$, expedition 1987–88: 0.025–0.038; 1988–89: 0.018–0.093; 1989–90: 0.015–0.044) while they are two to three times lower than those obtained in the coastal area of Gerlache Inlet (1987–88: 0.061–0.114; 1990–91: 0.024–0.132 $nmol\,l^{-1}$, these last shown in Fig. 4). The increased lead concentration in the inner area of Gerlache Inlet with respect to the open area of Terra Nova Bay was already recognized and discussed by us in terms of geochemical factors (Capodaglio et al. 1989). In this respect it is worth noting that recent, preliminary data on sediments collected in the same areas (Giordano et al. 1998) show that the sediments of the Gerlache Inlet are in fact enriched in lead with respect to those of Wood Bay, thus confirming our previous hypothesis.

The comparison between the two areas is better highlighted if one considers the change in the lead concentration at the surface, in the 10-m upper layer of water, during the summer (Fig. 5). It can be noted that in Gerlache Inlet (summer 1990–91) lead concentration decreased from about 0.10 to about 0.04 $nmol\,l^{-1}$, while in Wood Bay (summer 1993–94) the decrease was from about 0.030 to about 0.015 $nmol\,l^{-1}$. It is interesting to note that, in spite of the considerable difference in the Pb concentration level in the two areas (factor ~3) the net decreasing trend in December and the corresponding decreased fraction (50–60%) were common to the two sites, testifying the common process leading to summer concentration depletion.

The seasonal behaviour of lead in the water column of Wood Bay, even in the presence of a relatively high analytical variability, due to the extremely low concentration involved, can be clearly observed in Fig. 4, from which a full comparison with Gerlache Inlet can also be obtained.

The vertical profile of lead concentration in Wood Bay appeared practically uniform at the beginning of the season until November 28, with a mean concentration of 0.031 (SD 0.06) $nmol\,l^{-1}$. A significant depletion of the concentration was already observed in the subsurface layers on December 15 (values ~0.02 $nmol\,l^{-1}$) when the biological activity increased rapidly (see chlorophyll profile in Fig. 4). During the summer, the depletion process continued in these layers (in parallel with the increasing biological activity as observed in chlorophyll profiles) to reach a concentration of about half of the initial value and, in contrast to what happened for Cd, the depletion extended deeper and deeper during the season, practically down to the bottom. At the end of the season, a tendency to a new increase of Pb concentration in the water column was observed (see the latter profile, February 10 in Fig. 4) beginning from the deepest layers.

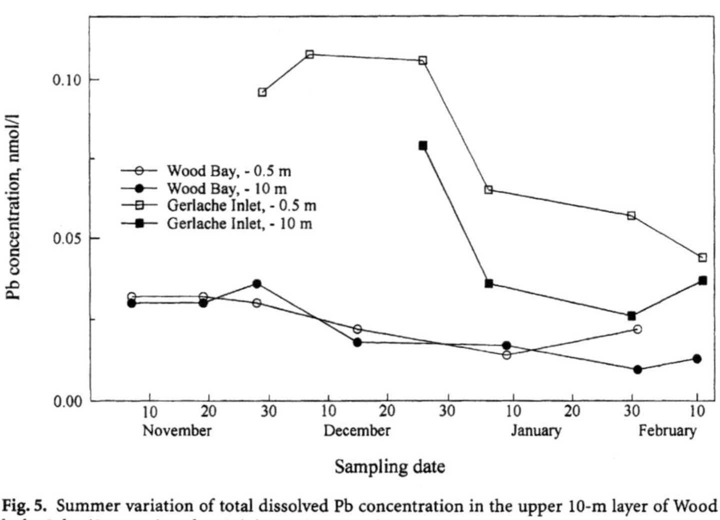


**Fig. 5.** Summer variation of total dissolved Pb concentration in the upper 10-m layer of Wood Bay (1993–94 season) and Gerlache Inlet (Scarponi et al. 1997) (1990–91 season)

Again referring to the whole water column, the comparison with previous data obtained in the Gerlache Inlet (1990–91 season, see Fig. 4) showed a substantially similar general behaviour of the summer Pb evolution in the two areas, even if the concentration levels were different and in Gerlache Inlet the depletion did not seem to reach the bottom. This latter observation, however, was in agreement with the peculiar characteristics of the Gerlache Inlet and with our interpretation of the high Pb levels in this area in terms of geochemical factors (see above).

## Discussion

Interpretation of data presented here on the summer evolution of Cd and Pb in the water column of the Ross Sea in the Victoria Land coastal area (Terra Nova Bay and Wood Bay) should account for the different geochemical characteristics of the two metals, the different interaction processes with phytoplankton and particles, and the seasonal evolution of biological processes during the investigation.

Cadmium is a typical recycled metal and lead a typical scavenged metal (Bruland 1983). At the beginning of the season, both biological activity and particle density are low; in this situation, both metals exhibit uniform vertical distributions at relatively high concentration levels consistent with coastal regions (see difference in Pb concentration between Gerlache Inlet and Wood Bay), with upwelling hydrodynamic regimes, reduced biological uptake and scavenging. Subsequently, a general decrease in metal content is observed both for Cd (shallow waters) and for Pb (shallow and deep waters). The amount of this depletion, the seasonal change and the water layer most affected are markedly dependent on the metal considered.

Cadmium, as a nutrient-type element, is subject to uptake from biological systems and is subsequently discarded with detritus and fecal pellets. Consequently, a rapid depletion of the metal is observed in the euphotic layer with the development of the phytoplankton bloom. The concentration of the upper 50-m layer continues to decrease during the summer season and the phenomenon does not affect deeper strata. Note that the highest chlorophyll concentrations are located at a depth of 10–30 m (Fig. 2). This layer therefore presents the highest Cd depletion; at the end of summer we expected a reverse trend as a consequence of the reduced biological activity and upwelling regime typical of Antarctic areas. In this respect, it is interesting to note that particulate Cd concentration (measured in the same Wood Bay samples) increased during the summer by a factor of ~10 (from ~0.01 to ~0.10 $\mu g\, g^{-1}$) in the upper 50-m layer (Frache et al. 1997).

Lead, a scavenged metal, tends to adsorb on suspended matter and its interaction with quickly flowing particles determines its removal through sedimentation processes (Chester 1990). In the coastal area of the Ross Sea the largest particles in the suspended matter, even if formed in the same zone of maximum biological production, are spread along the whole water column and reach maximum concentrations during the senescent phase of a bloom (Innamorati et al. 1991; Lazzara and Nuccio 1994). This explains the slowly decreasing trend of Pb concentration observed in both studied sites, which became more consistent and extended to almost all the vertical profile in February, when larger particles became prevalent.

A depletion of lead concentration in Antarctic seawater (Weddell and Scotia Seas) has also been reported and interpreted in terms of particulate scavenging processes related to intense primary production (Flegal et al. 1993).

Finally, it is known that the effect of the presence of local or remote anthropogenic sources of Pb (atmospheric input or immission from coastal areas) results in a surface enrichment of the metal (Bruland, 1983). From the present study the absence of a marked surface increase of Pb concentration with respect to the deep baseline values indicates the absence both in Terra Nova Bay and Wood Bay of significant Pb pollution coming from remote or local areas.

*Acknowledgements.* This work was supported by the Italian National Research Programme in Antarctica (PNRA). Thanks are due to the researchers of the project on Environmental Contamination for their accurate work in the field during the 1993–94 expedition and to all the members of the Italian technical staff for their assistance at sea.

## References

Bruland KW (1980) Oceanographic distributions of cadmium, zinc, nickel and copper in the North Pacific. Earth Planet Sci Lett 47: 176–198

Bruland KW (1983) Trace elements in seawater. In: Riley JP, Chester R (eds) Chemical Oceanography, vol 8. Academic Press, London, pp 157–220

Bruland KW, Franks RP (1983) Mn, Ni, Cu, Zn and Cd in the western North Atlantic. In: Wong CS, Boyle E, Bruland KW, Burton JD, Goldberg ED (eds) Trace metals in seawater. Plenum Press, New York, pp 395–414

Bruland KW, Orians KJ, Cowen JP (1994) Reactive trace metals in the stratified central North Pacific. Geochim Cosmochim Acta 58: 3171–3182

Capodaglio G, Toscano G, Scarponi G, Cescon P (1989) Lead speciation in the surface waters of the Ross Sea (Antarctica). Ann Chim 79: 543–559 [Correction (1990) Ann Chim 80: 393]

Capodaglio G, Scarponi G, Toscano G, Cescon P (1991a) Cadmium complexation in surface seawater of Terra Nova Bay (Antarctica). Ann Chim 81: 279–296

Capodaglio G, Scarponi G, Cescon P (1991b) Lead speciation in the Antarctic Ocean. Anal Proc 28: 76–77

Capodaglio G, Toscano G, Scarponi G, Cescon P (1994) Copper complexation in the surface seawater of Terra Nova Bay (Antarctica). Int J Environ Anal Chem 55: 129–148

Chester R (1990) Marine geochemistry. Unwin Hyman, London

Flegal AR, Maring H, Niemeyer S (1993) Anthropogenic lead in Antarctic seawater. Nature 365: 242–244

Frache R, Baffi F, Ianni C, Soggia F (1997) Dissolved and particulate metals in the pack ice melting process in the Ross Sea (Antarctica). Ann Chim 87: 367–374

Frew RD, Hunter KA (1992) Influence of Southern Ocean waters on the cadmium-phosphate properties of the global ocean. Nature 360: 144–146

Giordano R, Lombardi G, Ciaralli L, Beccaloni E, Sepe A, Ciprotti M, Costantini S (1998) Major and trace elements in Antarctic sediments from the Terra Nova Bay. Italian expedition 1993–94. Int J Environ Anal Chem (submitted)

Innamorati M, Lazzara L, Massi L, Mori G, Nuccio C, Saggiomo EV (1991) Indagine sulla biomassa fitoplanctonica nel Mare di Ross in relazione ai fattori ambientali. In: Gallardo VA, Ferretti O, Moyano HI (eds) Oceanografia in Antartide. ENEA-PNRA-EULA, Conception, Chile, pp 235–252

Lazzara L, Nuccio C (1994) Relazioni tra particellato, biomassa e popolamenti fitoplanctonici nel Mare di Ross durante la fioritura estiva. In: Albertelli G, Cattaneo-Vietti R, Piccazzo M (eds) Atti X Congr Ass Ital Oceanogr Limnol, Lang, Genova, pp 655–680

Mart L, Nurnberg HW, Valenta P (1980) Prevention of contamination and other accuracy risks in voltammetric trace metal analysis of natural waters. III. Voltammetric ultratrace analysis with a multicell system designed for clean bench working. Fresenius Z Anal Chem 300: 350–362

Mart L, Rutzel H, Klahre P, Sipos L, Platzek U, Valenta P, H. Nurnberg W (1982) Comparative studies on the distribution of heavy metals in the oceans and coastal waters. Sci Total Environ 26: 1–17

Nolting RF, de Baar HJW (1994) Behaviour of nickel, copper, zinc and cadmium in the upper 300 m of the transect in the Southern Ocean (57°–62°S, 49°W). Mar Chem 45: 225–242

Orren MJ, Monteiro PMS (1985) Trace element geochemistry in the Southern Ocean. In: Siegfried WR, Condy PR, Laws RM (eds) Antarctic nutrient cycles and food webs. Springer, Berlin Heidelberg New York, pp 30–37

Saager PM, de Baar HJW, Howland RJ (1992) Cd, Zn, Ni and Cu in the Indian Ocean. Deep Sea Res 39: 9–35

Scarponi G, Capodaglio G, Barbante C, Cescon P (1996) The anodic stripping voltammetric titration procedure for study of trace metal complexation in seawater. In: Caroli S (ed) Element speciation in bioinorganic chemistry. Wiley, New York, pp 363–418

Scarponi G, Capodaglio G, Turetta C, Barbante C, Cecchini M, Toscano G, Cescon P (1997) Evolution of cadmium and lead contents in Antarctic coastal seawater during the austral summer. Int J Environ Anal Chem 66: 23–49

Westerlund S, Öhman P (1991) Cadmium, copper, cobalt, nickel, lead and zinc in the water column of the Weddell Sea, Antarctica. Geochim Cosmochim Acta 55: 2127–2146

Whitfield M, Turner DR (1987) The role of particles in regulating the composition of seawater. In: Stumm W (ed) Aquatic surface chemistry; chemical processes at the particle-water interface. Wiley, New York, pp 457–493

# Subject Index

# Genera and Species Index

MIX
Papier aus verantwortungsvollen Quellen
Paper from responsible sources
FSC® C105338

If you have any concerns about our products,
you can contact us on
**ProductSafety@springernature.com**

In case Publisher is established outside the EU,
the EU authorized representative is:
**Springer Nature Customer Service Center GmbH**
**Europaplatz 3, 69115 Heidelberg, Germany**

Printed by Libri Plureos GmbH
in Hamburg, Germany